AF746464

UNIVERSITY OF NORTHUMBRIA
AT NEWCASTLE LIBRARY
WITHDRAWN

UNIVERSITY OF NORTHUMBRIA
AT NEWCASTLE LIBRARY
WITHDRAWN

NEWCASTLE UPON TYNE
POLYTECHNIC
LIBRARY

REFERENCE ONLY

ERRATA

VOLUME 10

Page	Line	For:	Read:
887	5	mol wt 167.04	mol wt 168.04
908	1st eq.	13 HF	15 HF
913	Table 6 footnote [f]	3 52 MPa	3.52 MPa
918	Sulindac structure	see text	delete dotted circle
925	23	(82)	(83)
932	ref. 60	entire ref.	D. A. Sutton, S. E. Drewes, and U. Welz, *Biochem. J.* **130,** 589 (1972).
934	ref. 174	*Anqu.*	*Angew.*

KIRK-OTHMER

ENCYCLOPEDIA OF CHEMICAL TECHNOLOGY

Third Edition

VOLUME 10

Ferroelectrics
to
Fluorine Compounds, Organic

EDITORIAL BOARD

HERMAN F. MARK
Polytechnic Institute of New York

DONALD F. OTHMER
Polytechnic Institute of New York

CHARLES G. OVERBERGER
University of Michigan

GLENN T. SEABORG
University of California, Berkeley

EXECUTIVE EDITOR

MARTIN GRAYSON

ASSOCIATE EDITOR

DAVID ECKROTH

KIRK-OTHMER

ENCYCLOPEDIA OF CHEMICAL TECHNOLOGY

THIRD EDITION

VOLUME 10

FERROELECTRICS
TO
FLUORINE COMPOUNDS, ORGANIC

A WILEY-INTERSCIENCE PUBLICATION
John Wiley & Sons
NEW YORK • CHICHESTER • BRISBANE • TORONTO

66(03)

12 300 923 01

Ref

Copyright © 1980 by John Wiley & Sons, Inc.

All rights reserved. Published simultaneously in Canada.

Reproduction or translation of any part of this work beyond that permitted by Sections 107 or 108 of the 1976 United States Copyright Act without the permission of the copyright owner is unlawful. Requests for permission or further information should be addressed to the Permissions Department, John Wiley & Sons, Inc.

Library of Congress Cataloging in Publication Data:

Main entry under title:
Encyclopedia of chemical technology.

At head of title: Kirk-Othmer.
"A Wiley-Interscience publication."
Includes bibliographies.
1. Chemistry, Technical—Dictionaries. I. Kirk, Raymond Eller, 1890–1957. II. Othmer, Donald Frederick, 1904—
III. Grayson, Martin. IV. Eckroth, David. V. Title: Kirk-Othmer encyclopedia of chemical technology.

TP9.E685 1978 660′.03 77-15820
ISBN 0-471-02063-X

Printed in the United States of America

CONTENTS

EDITORIAL STAFF FOR VOLUME 10

Executive Editor: **Martin Grayson**
Associate Editor: **David Eckroth**
Production Supervisor: **Michalina Bickford**
Editors: **Galen J. Bushey** **Caroline I. Eastman** **Anna Klingsberg** **Leonard Spiro**

CONTRIBUTORS TO VOLUME 10

Terry E. Acree, *Cornell University, Geneva, New York,* Flavor characterization
T. A. Antkowiak, *Firestone Tire and Rubber Co., Akron, Ohio,* Fluoroalkoxyphosphazenes under Fluorine compounds, organic
Frank F. Aplan, *The Pennsylvania State University, University Park, Pennsylvania,* Flotation
Gary Astrologes, *Halocarbon Corporation, Hackensack, New Jersey,* Fluorinated acetic acids; Fluoroethanols under Fluorine compounds, organic
John M. Avento, *M&T Chemicals, Inc., Rahway, New Jersey,* Antimony and other inorganic compounds under Flame retardants
R. B. Badachhape, *Rice University, Houston, Texas,* Direct fluorination under Fluorine compounds, organic
A. M. Benson, *Shell Development Company, Houston, Texas,* Fluid mechanics
John G. Blumberg, *PQ Corporation, Lafayette Hill, Pennsylvania,* Fillers
Max M. Boudakian, *Olin Chemicals, Rochester, New York,* Fluorinated aromatic compounds under Fluorine compounds, organic
John A. Caughlan, *Mallinckrodt, Inc., St. Louis, Missouri,* Fine chemicals

L. E. Cross, *The Pennsylvania State University, University Park, Pennsylvania,* Ferroelectrics

E. Lea Crump. *Gulf Oil Chemical Company, Orange, Texas,* Film and sheeting materials

Richard D. Danielson, *Minnesota Mining & Manufacturing Co., St. Paul, Minnesota,* Fluoro ethers and amines under Fluorine compounds, organic

George L. Drake, Jr., *United States Department of Agriculture, New Orleans, Louisiana,* Flame retardants for textiles

Richard E. Eibeck, *Allied Chemical Corporation, Buffalo, New York,* Sulfur under Fluorine compounds, inorganic; Perfluoroalkylsulfur fluorides under Fluorine compounds, organic

James S. Falcone, Jr., *PQ Corporation, Lafayette Hill, Pennsylvania,* Fillers

F. Fischetti, Jr., *Fritzsche Dodge and Olcott, Inc., New York, New York,* Flavors and spices

M. P. Freeman, *Dorr-Oliver Incorporated, Stamford, Connecticut,* Filtration

William G. French, *3M Company, St. Paul, Minnesota,* Fiber optics

John F. Gall, *Philadelphia College of Textiles and Science, Philadelphia, Pennsylvania,* Aluminum; Calcium; Hydrogen under Fluorine compounds, inorganic

Warren R. Griffin, *Air Force Materials Laboratory, Dayton, Ohio,* Perfluoroalkylene triazines under Fluorine compounds, organic

Richard A. Guenthner, *3M Company, St. Paul, Minnesota,* Fluorinated higher carboxylic acids; Perfluoroalkane sulfonic acid under Fluorine compounds, organic

K. H. Häerdtl, *Philips Forschungslaboratorium Aachen GmbH, Aschen, Federal Republic of Germany,* Ferroelectrics

H. S. Halbedel, *The Harshaw Chemical Company, Cleveland, Ohio,* Ammonium; Boron-fluoroboric acid and other fluoroborates; Lithium; Magnesium; Potassium under Fluorine compounds, inorganic

F. Halverson, *American Cyanamid Company, Stamford, Connecticut,* Flocculating agents

R. H. Hauge, *Rice University, Houston, Texas,* Direct fluorination under Fluorine compounds, organic

E. O. Huffman, *Consultant, Florence, Alabama,* Fertilizers

R. J. Lagow, *University of Texas, Austin, Texas,* Direct fluorination under Fluorine compounds, organic

E. R. Larsen, *Dow Chemical U.S.A., Midland, Michigan,* Halogenated flame retardants under Flame retardants

Stuart M. Lee, *Ford Aerospace & Communications Corp., Palo Alto, California,* Film deposition techniques

Charles B. Lindahl, *Ozark-Mahoning Company, a subsidiary of the Pennwalt Corp., Tulsa, Oklahoma,* Introduction; Antimony; Arsenic; Barium; Germanium; Phosphorus; Tin; Titanium; Zinc under Fluorine compounds, inorganic

John W. Lyons, *National Bureau of Standards, Washington, D.C.,* Flame retardants, an overview

John B. MacChesney, *Bell Laboratories, Inc., Murray Hill, New Jersey,* Fiber optics

John L. Margrave, *Rice University, Houston, Texas,* Direct fluorination under Fluorine compounds, organic

Donald R. Martin, *University of Texas, Arlington, Texas,* Boron under Fluorine compounds, inorganic

G. Q. Martin, *Shell Development Company, Houston, Texas,* Fluid mechanics

John N. McGovern, *The University of Wisconsin, Madison, Wisconsin,* Fibers, vegetable

Whitney Mears, *Allied Chemical Corporation, Buffalo, New York,* Sulfur-sulfur fluorides under Fluorine compounds, inorganic

Dayal T. Meshri, *Ozark-Mahoning Company, a subsidiary of the Pennwalt Corp., Tulsa, Oklahoma,* Introduction; Antimony; Cobalt; Copper; Iron; Lead; Mercury; Nickel; Silver; Tin; Titanium; Zirconium under Fluorine compounds, inorganic

William J. Middleton, *E. I. du Pont de Nemours & Co., Inc., Wilmington, Delaware,* Hexafluoroacetone under Fluorine compounds, organic

Thomas E. Nappier, *The Harshaw Chemical Company, Cleveland, Ohio,* Ammonium; Boron-fluoroboric acid and fluoroborates; Lithium; Magnesium; Potassium under Fluorine compounds, inorganic

David I. Netting, *ARCO Chemical Co., Glenolden, Pennsylvania,* Fillers

H. P. Panzer, *American Cyanamid Company, Stamford, Connecticut,* Flocculating agents

A. David Pearson, *Bell Laboratories, Murray Hill, New Jersey,* Fiber optics

Timothy V. Peters, *Consultant, Princeton, New Jersey,* Fibers, elastomeric

†**Odgen R. Pierce,** *Dow Corning Corporation, Midland, Michigan,* Fluorine compounds, organic, introduction

Paul R. Resnick, *E. I. du Pont de Nemours & Co., Inc., Wilmington, Delware,* Perfluoroepoxides under Fluorine compounds, organic

William J. Roberts, *Consultant, Bernardsville, New Jersey,* Fibers, chemical

J. A. Rogers, Jr., *Fritzsche Dodge and Olcott, Inc., New York, New York,* Flavors and spices

Bruce E. Smart, *E. I. du Pont de Nemours & Co., Inc., Wilmington, Delaware,* Fluorinated aliphatic compounds under Fluorine compounds, organic

Leonard H. Smiley, *PQ Corporation, Lafayette Hill, Pennsylvania,* Fillers

Irvine J. Solomon, *IIT Research Institute, Chicago, Illinois,* Oxygen under Fluorine compounds, inorganic

J. S. Son, *Shell Development Company, Houston, Texas,* Fluid mechanics

C. V. Sternling, *Shell Development Company, Houston, Texas,* Fluid mechanics

Robert M. Talcott, *Dorr-Oliver Incorporated, Stamford, Connecticut,* Filtration

David P. Tate, *Firestone Tire and Rubber Co., Akron, Ohio,* Fluoroalkoxyphosphazenes under Fluorine compounds, organic

Irving Touval, *M&T Chemicals, Inc., Rahway, New Jersey,* Antimony and other inorganic compounds under Flame retardants

Samuel M. Tuthill, *Mallinckrodt, Inc., St. Louis, Missouri,* Fine chemicals

Kenneth Wachter, *Olin Corporation, New Haven, Connecticut,* Sodium under Fluorine compounds, inorganic

Edward D. Weil, *Stauffer Chemical Company, Dobbs Ferry, New York,* Phosphorus compounds under Flame retardants

Charles Willus, *Dorr-Oliver Incorporated, Stamford, Connecticut,* Filtration

Andrew J. Woytek, *Air Products and Chemicals, Inc., Tamaqua, Pennsylvania,* Fluorine; and Halogens; Molybdenum; Nitrogen; Rhenium; Tantalum; Tungsten under Fluorine, compounds, inorganic

Frederick A. Zenz, *Consultant, Garrison, New York,* Fluidization

† Deceased.

NOTE ON CHEMICAL ABSTRACTS SERVICE REGISTRY NUMBERS AND NOMENCLATURE

Chemical Abstracts Service (CAS) Registry Numbers are unique numerical identifiers assigned to substances recorded in the CAS Registry System. They appear in brackets in the *Chemical Abstracts* (CA) substance and formula indexes following the names of compounds. A single compound may have many synonyms in the chemical literature. A simple compound like phenethylamine can be named β-phenylethylamine or, as in *Chemical Abstracts,* benzeneethanamine. The usefulness of the Encyclopedia depends on accessibility through the most common correct name of a substance. Because of this diversity in nomenclature careful attention has been given the problem in order to assist the reader as much as possible, especially in locating the systematic CA index name by means of the Registry Number. For this purpose, the reader may refer to the CAS Registry Handbook-Number Section which lists in numerical order the Registry Number with the *Chemical Abstracts* index name and the molecular formula; eg, 458-8-8, Piperidine, 2-propyl-, (*S*)-, $C_8H_{17}N$; in the Encyclopedia this compound would be found under its common name, coniine [*458-88-8*]. The Registry Number is a valuable link for the reader in retrieving additional published information on substances and also as a point of access for such on-line data bases as Chemline, Medline and Toxline.

In all cases, the CAS Registry Numbers have been given for title compounds in articles and for all compounds in the index. All specific substances indexed in *Chemical Abstracts* since 1965 are included in the CAS Registry System as are a large number of substances derived from a variety of reference works. The CAS Registry System identifies a substance on the basis of an unambiguous computer-language description of its molecular structure including stereochemical detail. The Registry Number is a machine-checkable number (like a Social Security number) assigned in sequential order to each substance as it enters the registry system. The value of the number lies in the fact that it is a concise and unique means of substance identification, which is independent of, and therefore bridges, many systems of chemical nomenclature. For polymers, one Registry Number is used for the entire family; eg, polyoxyethylene (20) sorbitan monolaurate has the same number as all of its polyoxyethylene homologues.

Registry numbers for each substance will be provided in the third edition cumulative index and appear as well in the annual indexes (eg, Alkaloids shows the Registry Number of all alkaloids (title compounds) in a table in the article as well, but the intermediates have their Registry Numbers shown only in the index). Articles such as Analytical methods, Batteries and electric cells, Chemurgy, Distillation, Economic evaluation, and Fluid mechanics have no Registry Numbers in the text.

Cross-references are inserted in the index for many common names and for some systematic names. Trademark names appear in the index. Names that are incorrect, misleading or ambiguous are avoided. Formulas are given very frequently in the text to help in identifying compounds. The spelling and form used, even for industrial names, follow American chemical usage, but not always the usage of *Chemical Abstracts* (eg, *coniine* is used instead of *(S)-2-propylpiperidine, aniline* instead of *benzenamine,* and *acrylic acid* instead of *2-propenoic acid*).

There are variations in representation of rings in different disciplines. The dye industry does not designate aromaticity or double bonds in rings. All double bonds and aromaticity are shown in the *Encyclopedia* as a matter of course. For example, tetralin has an aromatic ring and a saturated ring and its structure appears in the

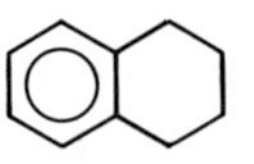

Encyclopedia with its common name, Registry Number enclosed in brackets, and parenthetical CA index name, ie, tetralin [*119-64-2*] (1,2,3,4-tetrahydronaphthalene). With names and structural formulas, and especially with CAS Registry Numbers the aim is to help the reader have a concise means of substance identification.

CONVERSION FACTORS, ABBREVIATIONS, AND UNIT SYMBOLS

SI Units (Adopted 1960)

A new system of measurement, the International System of Units (abbreviated SI), is being implemented throughout the world. This system is a modernized version of the MKSA (meter, kilogram, second, ampere) system, and its details are published and controlled by an international treaty organization (The International Bureau of Weights and Measures) (1).

SI units are divided into three classes:

BASE UNITS

length	meter† (m)
mass‡	kilogram (kg)
time	second (s)
electric current	ampere (A)
thermodynamic temperature§	kelvin (K)
amount of substance	mole (mol)
luminous intensity	candela (cd)

† The spellings "metre" and "litre" are preferred by ASTM; however "-er" are used in the Encyclopedia.

‡ "Weight" is the commonly used term for "mass".

§ Wide use is made of "Celsius temperature" (t) defined by

$$t = T - T_0$$

where T is the thermodynamic temperature, expressed in kelvins, and T_0 = 273.15 K by definition. A temperature interval may be expressed in degrees Celsius as well as in kelvins.

SUPPLEMENTARY UNITS

plane angle	radian (rad)
solid angle	steradian (sr)

DERIVED UNITS AND OTHER ACCEPTABLE UNITS

These units are formed by combining base units, supplementary units, and other derived units (2–4). Those derived units having special names and symbols are marked with an asterisk in the list below:

Quantity	*Unit*	*Symbol*	*Acceptable equivalent*
* absorbed dose	gray	Gy	J/kg
acceleration	meter per second squared	m/s^2	
* activity (of ionizing radiation source)	becquerel	Bq	1/s
area	square kilometer	km^2	
	square hectometer	hm^2	ha (hectare)
	square meter	m^2	
* capacitance	farad	F	C/V
concentration (of amount of substance)	mole per cubic meter	mol/m^3	
* conductance	siemens	S	A/V
current density	ampere per square meter	A/m^2	
density, mass density	kilogram per cubic meter	kg/m^3	g/L; mg/cm^3
dipole moment (quantity)	coulomb meter	C·m	
* electric charge, quantity of electricity	coulomb	C	A·s
electric charge density	coulomb per cubic meter	C/m^3	
electric field strength	volt per meter	V/m	
electric flux density	coulomb per square meter	C/m^2	
* electric potential, potential difference, electromotive force	volt	V	W/A
* electric resistance	ohm	Ω	V/A
* energy, work, quantity of heat	megajoule	MJ	
	kilojoule	kJ	
	joule	J	N·m
	electron volt†	eV†	
	kilowatt-hour†	kW·h†	

† This non-SI unit is recognized by the CIPM as having to be retained because of practical importance or use in specialized fields (1).

Quantity	*Unit*	*Symbol*	*Acceptable equivalent*
energy density	joule per cubic meter	J/m^3	
* force	kilonewton	kN	
	newton	N	$kg{\cdot}m/s^2$
* frequency	megahertz	MHz	
	hertz	Hz	1/s
heat capacity, entropy	joule per kelvin	J/K	
heat capacity (specific), specific entropy	joule per kilogram kelvin	J/(kg·K)	
heat transfer coefficient	watt per square meter kelvin	$W/(m^2{\cdot}K)$	
* illuminance	lux	lx	lm/m^2
* inductance	henry	H	Wb/A
linear density	kilogram per meter	kg/m	
luminance	candela per square meter	cd/m^2	
* luminous flux	lumen	lm	cd·sr
magnetic field strength	ampere per meter	A/m	
* magnetic flux	weber	Wb	V·s
* magnetic flux density	tesla	T	Wb/m^2
molar energy	joule per mole	J/mol	
molar entropy, molar heat capacity	joule per mole kelvin	J/(mol·K)	
moment of force, torque	newton meter	N·m	
momentum	kilogram meter per second	kg·m/s	
permeability	henry per meter	H/m	
permittivity	farad per meter	F/m	
* power, heat flow rate, radiant flux	kilowatt	kW	
	watt	W	J/s
power density, heat flux density, irradiance	watt per square meter	W/m^2	
* pressure, stress	megapascal	MPa	
	kilopascal	kPa	
	pascal	Pa	N/m^2
sound level	decibel	dB	
specific energy	joule per kilogram	J/kg	
specific volume	cubic meter per kilogram	m^3/kg	
surface tension	newton per meter	N/m	
thermal conductivity	watt per meter kelvin	W/(m·K)	
velocity	meter per second	m/s	
	kilometer per hour	km/h	
viscosity, dynamic	pascal second	Pa·s	
	millipascal second	mPa·s	
viscosity, kinematic	square meter per second	m^2/s	

Quantity	*Unit*	*Symbol*	*Acceptable equivalent*
	square millimeter per second	mm^2/s	
volume	cubic meter	m^3	
	cubic decimeter	dm^3	L(liter) (5)
	cubic centimeter	cm^3	mL
wave number	1 per meter	m^{-1}	
	1 per centimeter	cm^{-1}	

In addition, there are 16 prefixes used to indicate order of magnitude, as follows:

Multiplication factor	*Prefix*	*Symbol*	*Note*
10^{18}	exa	E	
10^{15}	peta	P	
10^{12}	tera	T	
10^{9}	giga	G	
10^{6}	mega	M	
10^{3}	kilo	k	
10^{2}	hecto	h[a]	[a] Although hecto, deka, deci, and centi are SI prefixes, their use should be avoided except for SI unit-multiples for area and volume and nontechnical use of centimeter, as for body and clothing measurement.
10	deka	da[a]	
10^{-1}	deci	d[a]	
10^{-2}	centi	c[a]	
10^{-3}	milli	m	
10^{-6}	micro	μ	
10^{-9}	nano	n	
10^{-12}	pico	p	
10^{-15}	femto	f	
10^{-18}	atto	a	

For a complete description of SI and its use the reader is referred to ASTM E 380 (4) and the article Units and Conversion Factors which will appear in a later volume of the *Encyclopedia*.

A representative list of conversion factors from non-SI to SI units is presented herewith. Factors are given to four significant figures. Exact relationships are followed by a dagger. A more complete list is given in ASTM E 380-76(4) and ANSI Z210.1-1976 (6).

Conversion Factors to SI Units

To convert from	*To*	*Multiply by*
acre	square meter (m^2)	4.047×10^3
angstrom	meter (m)	1.0×10^{-10} †
are	square meter (m^2)	1.0×10^2 †
astronomical unit	meter (m)	1.496×10^{11}
atmosphere	pascal (Pa)	1.013×10^5
bar	pascal (Pa)	1.0×10^5 †

† Exact.

To convert from	*To*	*Multiply by*
barrel (42 U.S. liquid gallons)	cubic meter (m^3)	0.1590
Bohr magneton, μ_β	J/T	9.274×10^{-24}
Btu (International Table)	joule (J)	1.055×10^{3}
Btu (mean)	joule (J)	1.056×10^{3}
Btu (thermochemical)	joule (J)	1.054×10^{3}
bushel	cubic meter (m^3)	3.524×10^{-2}
calorie (International Table)	joule (J)	4.187
calorie (mean)	joule (J)	4.190
calorie (thermochemical)	joule (J)	4.184†
centipoise	pascal second (Pa·s)	1.0×10^{-3} †
centistoke	square millimeter per second (mm^2/s)	1.0†
cfm (cubic foot per minute)	cubic meter per second (m^3/s)	4.72×10^{-4}
cubic inch	cubic meter (m^3)	1.639×10^{-5}
cubic foot	cubic meter (m^3)	2.832×10^{-2}
cubic yard	cubic meter (m^3)	0.7646
curie	becquerel (Bq)	3.70×10^{10}†
debye	coulomb·meter (C·m)	3.336×10^{-30}
degree (angle)	radian (rad)	1.745×10^{-2}
denier (international)	kilogram per meter (kg/m)	1.111×10^{-7}
	tex‡	0.1111
dram (apothecaries')	kilogram (kg)	3.888×10^{-3}
dram (avoirdupois)	kilogram (kg)	1.772×10^{-3}
dram (U.S. fluid)	cubic meter (m^3)	3.697×10^{-6}
dyne	newton (N)	1.0×10^{-5} †
dyne/cm	newton per meter (N/m)	1.0×10^{-3} †
electron volt	joule (J)	1.602×10^{-19}
erg	joule (J)	1.0×10^{-7} †
fathom	meter (m)	1.829
fluid ounce (U.S.)	cubic meter (m^3)	2.957×10^{-5}
foot	meter (m)	0.3048 †
footcandle	lux (lx)	10.76
furlong	meter (m)	2.012×10^{-2}
gal	meter per second squared (m/s^2)	1.0×10^{-2} †
gallon (U.S. dry)	cubic meter (m^3)	4.405×10^{-3}
gallon (U.S. liquid)	cubic meter (m^3)	3.785×10^{-3}
gallon per minute (gpm)	cubic meter per second (m^3/s)	6.308×10^{-5}
	cubic meter per hour (m^3/h)	0.2271
gauss	tesla (T)	1.0×10^{-4}
gilbert	ampere (A)	0.7958
gill (U.S.)	cubic meter (m^3)	1.183×10^{-4}
grad	radian	1.571×10^{-2}
grain	kilogram (kg)	6.480×10^{-5}
gram force per denier	newton per tex (N/tex)	8.826×10^{-2}
hectare	square meter (m^2)	1.0×10^{4} †
horsepower (550 ft·lbf/s)	watt (W)	7.457×10^{2}

† Exact.

‡ See footnote on p. xiv.

To convert from	*To*	*Multiply by*
horsepower (boiler)	watt (W)	9.810×10^3
horsepower (electric)	watt (W)	$7.46 \times 10^{2\dagger}$
hundredweight (long)	kilogram (kg)	50.80
hundredweight (short)	kilogram (kg)	45.36
inch	meter (m)	$2.54 \times 10^{-2\dagger}$
inch of mercury (32°F)	pascal (Pa)	3.386×10^3
inch of water (39.2°F)	pascal (Pa)	2.491×10^2
kilogram force	newton (N)	9.807
kilowatt hour	megajoule (MJ)	$3.6^\dagger$
kip	newton (N)	4.48×10^3
knot (international)	meter per second (m/s)	0.5144
lambert	candela per square meter (cd/m^2)	3.183×10^3
league (British nautical)	meter (m)	5.559×10^3
league (statute)	meter (m)	4.828×10^3
light year	meter (m)	9.461×10^{15}
liter (for fluids only)	cubic meter (m^3)	$1.0 \times 10^{-3\dagger}$
maxwell	weber (Wb)	$1.0 \times 10^{-8\dagger}$
micron	meter (m)	$1.0 \times 10^{-6\dagger}$
mil	meter (m)	$2.54 \times 10^{-5\dagger}$
mile (U.S. nautical)	meter (m)	$1.852 \times 10^{3\dagger}$
mile (statute)	meter (m)	1.609×10^3
mile per hour	meter per second (m/s)	0.4470
millibar	pascal (Pa)	1.0×10^2
millimeter of mercury (0°C)	pascal (Pa)	$1.333 \times 10^{2\dagger}$
minute (angular)	radian	2.909×10^{-4}
myriagram	kilogram (kg)	10
myriameter	kilometer (km)	10
oersted	ampere per meter (A/m)	79.58
ounce (avoirdupois)	kilogram (kg)	2.835×10^{-2}
ounce (troy)	kilogram (kg)	3.110×10^{-2}
ounce (U.S. fluid)	cubic meter (m^3)	2.957×10^{-5}
ounce-force	newton (N)	0.2780
peck (U.S.)	cubic meter (m^3)	8.810×10^{-3}
pennyweight	kilogram (kg)	1.555×10^{-3}
pint (U.S. dry)	cubic meter (m^3)	5.506×10^{-4}
pint (U.S. liquid)	cubic meter (m^3)	4.732×10^{-4}
poise (absolute viscosity)	pascal second (Pa·s)	$0.10^\dagger$
pound (avoirdupois)	kilogram (kg)	0.4536
pound (troy)	kilogram (kg)	0.3732
poundal	newton (N)	0.1383
pound-force	newton (N)	4.448
pound per square inch (psi)	pascal (Pa)	6.895×10^3
quart (U.S. dry)	cubic meter (m^3)	1.101×10^{-3}
quart (U.S. liquid)	cubic meter (m^3)	9.464×10^{-4}
quintal	kilogram (kg)	$1.0 \times 10^{2\dagger}$

†Exact.

To convert from	*To*	*Multiply by*
rad	gray (Gy)	$1.0 \times 10^{-2\dagger}$
rod	meter (m)	5.029
roentgen	coulomb per kilogram (C/kg)	2.58×10^{-4}
second (angle)	radian (rad)	4.848×10^{-6}
section	square meter (m^2)	2.590×10^{6}
slug	kilogram (kg)	14.59
spherical candle power	lumen (lm)	12.57
square inch	square meter (m^2)	6.452×10^{-4}
square foot	square meter (m^2)	9.290×10^{-2}
square mile	square meter (m^2)	2.590×10^{6}
square yard	square meter (m^2)	0.8361
stere	cubic meter (m^3)	$1.0^\dagger$
stokes (kinematic viscosity)	square meter per second (m^2/s)	$1.0 \times 10^{-4\dagger}$
tex	kilogram per meter (kg/m)	$1.0 \times 10^{-6\dagger}$
ton (long, 2240 pounds)	kilogram (kg)	1.016×10^{3}
ton (metric)	kilogram (kg)	$1.0 \times 10^{3\dagger}$
ton (short, 2000 pounds)	kilogram (kg)	9.072×10^{2}
torr	pascal (Pa)	1.333×10^{2}
unit pole	weber (Wb)	1.257×10^{-7}
yard	meter (m)	$0.9144^\dagger$

† Exact.

Abbreviations and Unit Symbols

Following is a list of commonly used abbreviations and unit symbols appropriate for use in the *Encyclopedia*. In general they agree with those listed in *American National Standard Abbreviations for Use on Drawings and in Text* (*ANSI Y1.1*) (6) and *American National Standard Letter Symbols for Units in Science and Technology* (*ANSI Y10*) (6). Also included is a list of acronyms for a number of private and government organizations as well as common industrial solvents, polymers, and other chemicals.

Rules for Writing Unit Symbols (4):

1. Unit symbols should be printed in upright letters (roman) regardless of the type style used in the surrounding text.

2. Unit symbols are unaltered in the plural.

3. Unit symbols are not followed by a period except when used as the end of a sentence.

4. Letter unit symbols are generally written in lower-case (eg, cd for candela) unless the unit name has been derived from a proper name, in which case the first letter of the symbol is capitalized (W,Pa). Prefix and unit symbols retain their prescribed form regardless of the surrounding typography.

5. In the complete expression for a quantity, a space should be left between the numerical value and the unit symbol. For example, write 2.37 lm, *not* 2.37lm, and 35 mm, *not* 35mm. When the quantity is used in an adjectival sense, a hyphen is often used, for example, 35-mm film. *Exception:* No space is left between the numerical value and the symbols for degree, minute, and second of plane angle, and degree Celsius.

6. No space is used between the prefix and unit symbols (eg, kg).

7. Symbols, not abbreviations, should be used for units. For example, use "A," not "amp," for ampere.

8. When multiplying unit symbols, use a raised dot:

N·m for newton meter

In the case of W·h, the dot may be omitted, thus:

Wh

An exception to this practice is made for computer printouts, automatic typewriter work, etc, where the raised dot is not possible, and a dot on the line may be used.

9. When dividing unit symbols use one of the following forms:

$$\mathrm{m/s} \text{ or } \mathrm{m{\cdot}s^{-1}} \text{ or } \frac{\mathrm{m}}{\mathrm{s}}$$

In no case should more than one slash be used in the same expression unless parentheses are inserted to avoid ambiguity. For example, write:

$$\mathrm{J/(mol{\cdot}K)} \textit{ or } \mathrm{J{\cdot}mol^{-1}\cdot K^{-1}} \textit{ or } \mathrm{(J/mol)/K}$$

but *not*

J/mol/K

10. Do not mix symbols and unit names in the same expression. Write:

joules per kilogram *or* J/kg *or* $\mathrm{J{\cdot}kg^{-1}}$

but *not*

joules/kilogram *nor* joules/kg *nor* $\mathrm{joules{\cdot}kg^{-1}}$

ABBREVIATIONS AND UNITS

A	ampere
A	anion (eg, H*A*)
a	atto (prefix for 10^{-18})
AATCC	American Association of Textile Chemists and Colorists
ABS	acrylonitrile–butadiene–styrene
abs	absolute
ac	alternating current, *n.*
a-c	alternating current, *adj.*
ac-	alicyclic
ACGIH	American Conference of Governmental Industrial Hygienists
ACS	American Chemical Society
AGA	American Gas Association
Ah	ampere hour
AIChE	American Institute of Chemical Engineers
AIP	American Institute of Physics
alc	alcohol(ic)
Alk	alkyl
alk	alkaline (not alkali)
amt	amount
amu	atomic mass unit
ANSI	American National Standards Institute
AP	atomic orbital
APHA	American Public Health Association
API	American Petroleum Institute
aq	aqueous
Ar	aryl
ar-	aromatic

as-	asymmetric(al)
ASH-RAE	American Society of Heating, Refrigerating, and Air Conditioning Engineers
ASM	American Society for Metals
ASME	American Society of Mechanical Engineers
ASTM	American Society for Testing and Materials
at no.	atomic number
at wt	atomic weight
av(g)	average
bbl	barrel
bcc	body-centered cubic
Bé	Baumé
bid	twice daily
Boc	*t*-butyloxycarbonyl
BOD	biochemical (biological) oxygen demand
bp	boiling point
Bq	becquerel
C	coulomb
°C	degree Celsius
C-	denoting attachment to carbon
c	centi (prefix for 10^{-2})
ca	circa (approximately)
cd	candela; current density; circular dichroism
CFR	Code of Federal Regulations
cgs	centimeter-gram-second
CI	Color Index
cis-	isomer in which substituted groups are on same side of double bond between C atoms
cl	carload
cm	centimeter
cmil	circular mil
cmpd	compound
COA	coenzyme A
COD	chemical oxygen demand
coml	commercial(ly)
cp	chemically pure
cph	close-packed hexagonal
CPSC	Consumer Product Safety Commission
D-	denoting configurational relationship
d	day
d	differential operator
d-	dextro-, dextrorotatory
da	deka (prefix for 10^1)
dB	decibel
dc	direct current, *n.*
d-c	direct current, *adj.*
dec	decompose
detd	determined
detn	determination
dia	diameter
dil	dilute
dl-; DL-	racemic
DMF	dimethylformamide
DMG	dimethyl glyoxime
DOE	Department of Energy
DOT	Department of Transportation
dp	dew point; degree of polymerization
dstl(d)	distill(ed)
dta	differential thermal analysis
(*E*)-	entgegen; opposed
ϵ	dielectric constant (unitless number)
e	electron
ECU	electrochemical unit
ed.	edited, edition, editor
ED	effective dose
EDTA	ethylenediaminetetraacetic acid
emf	electromotive force
emu	electromagnetic unit
eng	engineering
EPA	Environmental Protection Agency
epr	electron paramagnetic resonance
eq.	equation
esp	especially
esr	electron-spin resonance
est(d)	estimate(d)
estn	estimation
esu	electrostatic unit
exp	experiment, experimental
ext(d)	extract(ed)
F	farad (capacitance)
f	femto (prefix for 10^{-15})
FAO	Food and Agriculture Organization (United Nations)

fcc	face-centered cubic
FDA	Food and Drug Administration
FEA	Federal Energy Administration
fob	free on board
fp	freezing point
FPC	Federal Power Commission
frz	freezing
G	giga (prefix for 10^9)
G	gravitational constant
g	gram
(g)	gas, only as in $H_2O(g)$
g	gravitational acceleration
gem-	geminal
glc	gas-liquid chromatography
g-mol wt; gmw	gram-molecular weight
grd	ground
Gy	gray
H	henry
h	hour; hecto (prefix for 10^2)
ha	hectare
HB	Brinell hardness number
Hb	hemoglobin
HK	Knoop hardness number
HRC	Rockwell hardness (C scale)
HV	Vickers hardness number
hyd	hydrated, hydrous
hyg	hygroscopic
Hz	hertz
^{i}(eg, Pr^i)	iso (eg, isopropyl)
i-	inactive (eg, *i*-methionine)
IACS	International Annealed Copper Standard
ibp	initial boiling point
ICC	Interstate Commerce Commission
ICT	International Critical Table
ID	inside diameter; infective dose
IPS	iron pipe size
IPT	Institute of Petroleum Technologists
IPTS	International Practical Temperature Scale (NBS)
ir	infrared
IRLG	Interagency Regulatory Liaison Group
ISO	International Organization for Standardization
IUPAC	International Union of Pure and Applied Chemistry
IV	iodine value
J	joule
K	kelvin
k	kilo (prefix for 10^3)
kg	kilogram
L	denoting configurational relationship
L	liter (for fluids only)(5)
l-	*levo-*, levorotatory
(l)	liquid, only as in $NH_3(l)$
LC_{50}	conc lethal to 50% of the animals tested
LCAO	linear combination of atomic orbitals
lcl	less than carload lots
LD_{50}	dose lethal to 50% of the animals tested
LED	light emitting diode
liq	liquid
lm	lumen
ln	logarithm (natural)
LNG	liquefied natural gas
log	logarithm (common)
LPG	liquefied petroleum gas
ltl	less than truckload lots
lx	lux
M	mega (prefix for 10^6); metal (as in MA)
M	molar
m	meter; milli (prefix for 10^{-3})
m	molal
m-	meta
max	maximum
MCA	Chemical Manufacturers' Association
MEK	methyl ethyl ketone
meq	milliequivalent
mfd	manufactured
mfg	manufacturing
mfr	manufacturer
MIBC	methylisobutyl carbinol
MIBK	methyl isobutyl ketone
MIC	minimum inhibiting concentration
min	minute; minimum
mL	milliliter

MLD	minimum lethal dose
MO	molecular orbital
mo	month
mol	mole
mol wt	molecular weight
mom	momentum
mp	melting point
MR	molar refraction
ms	mass spectrum
mxt	mixture
μ	micro (prefix for 10^{-6})
N	newton (force)
N	normal (concentration)
N-	denoting attachment to nitrogen
n (as n_D^{20}	index of refraction (for 20°C and sodium light)
ⁿ (as Buⁿ), *n*-	normal (straight-chain structure)
n	nano (prefix for 10^{-9}
na	not available
NAS	National Academy of Sciences
NASA	National Aeronautics and Space Administration
nat	natural
NBS	National Bureau of Standards
neg	negative
NF	*National Formulary*
NIH	National Institutes of Health
NIOSH	National Institute of Occupational Safety and Health
nmr	nuclear magnetic resonance
NND	New and Nonofficial Drugs (AMA)
no.	number
NOI-(BN)	not otherwise indexed (by name)
NOS	not otherwise specified
nqr	nuclear quadrople resonance
NRC	Nuclear Regulatory Commission; National Research Council
NRI	New Ring Index
NSF	National Science Foundation
NTA	nitrilotriacetic acid
NTP	normal temperature and pressure (25°C and 101.3 kPa or 1 atm)
NTSB	National Transportation Safety Board
O-	denoting attachment to oxygen
o-	ortho
OD	outside diameter
OPEC	Organization of Petroleum Exporting Countries
OSHA	Occupational Safety and Health Administration
owf	on weight of fiber
Ω	ohm
P	peta (prefix for 10^{15})
p	pico (prefix for 10^{-12})
p-	para
p.	page
Pa	pascal (pressure)
pd	potential difference
pH	negative logarithm of the effective hydrogen ion concentration
pmr	proton magnetic resonance
POP	polyoxypropylene
pos	positive
pp.	pages
ppb	parts per billion
ppm	parts per million
PPO	poly(phenyl oxide)
ppt(d)	precipitate(d)
pptn	precipitation
Pr (no.)	foreign prototype (number)
pt	point; part
PVC	poly(vinyl chloride)
pwd	powder
qv	quod vide (which see)
R	univalent hydrocarbon radical
(*R*)-	rectus (clockwise configuration)
rad	radian; radius
rds	rate determining step
ref.	reference
rf	radio frequency, *n.*
r-f	radio frequency, *adj.*
rh	relative humidity
RI	Ring Index
RT	room temperature

^{s} (eg, Bu^s); *sec-*	secondary (eg, secondary butyl)
S	siemens
(*S*)-	sinister (counterclockwise configuration)
S-	denoting attachment to sulfur
s-	symmetric(al)
s	second
(s)	solid, only as in $H_2O(s)$
SAE	Society of Automotive Engineers
SAN	styrene-acrylonitrile
sat(d)	saturate(d)
satn	saturation
SCF	self-consistent field; standard cubic feet
Sch	Schultz number
SFs	Saybolt Furol seconds
SI	Le Système International d'Unités (International System of Units)
sl sol	slightly soluble
sol	soluble
soln	solution
soly	solubility
sp	specific; species
sp gr	specific gravity
sr	steradian
std	standard
STP	standard temperature and pressure (0°C and 101.3 kPa)
SUs	Saybolt Universal seconds
syn	synthetic
^{t} (eg, Bu^t), *t-*, *tert-*	tertiary (eg, tertiary butyl)
T	tera (prefix for 10^{12}); tesla (magnetic flux density)
t	metric ton (tonne) temperature
TAPPI	Technical Association of the Pulp and Paper Industry
tex	tex (linear density)
TGA	thermogravimetric analysis
THF	tetrahydrofuran
tlc	thin layer chromatography
TLV	threshold limit value
trans-	isomer in which substituted groups are on opposite sides of double bond between C atoms
TSCA	Toxic Substance Control Act
Twad	Twaddell
UL	Underwriters' Laboratory
USDA	United States Department of Agriculture
USP	*United States Pharmacopeia*
uv	ultraviolet
V	volt (emf)
var	variable
vic-	vicinal
vol	volume (not volatile)
vs	versus
v sol	very soluble
W	watt
Wb	Weber
Wh	watt hour
WHO	World Health Organization (United Nations)
wk	week
yr	year
(*Z*)-	zusammen; together

Non-SI (Unacceptable and Obsolete) Units		*Use*
Å	angstrom	nm
at	atmosphere, technical	Pa
atm	atmosphere, standard	Pa
b	barn	cm^2
bar†	bar	Pa
bbl		m^3

† Do not use bar (10^5Pa) or millibar (10^2Pa) because they are not SI units, and are accepted internationally only for a limited time in special fields because of existing usage.

Non-SI (Unacceptable and Obsolete) Units		*Use*
bhp	brake horsepower	W
Btu	British thermal unit	J
bu	bushel	m^3; L
cal	calorie	J
cfm	cubic foot per minute	m^3/s
Ci	curie	Bq
cSt	centistokes	mm^2/s
c/s	cycle per second	Hz
cu	cubic	exponential form
D	debye	C·m
den	denier	tex
dr	dram	kg
dyn	dyne	N
dyn/cm	dyne per centimeter	mN/m
erg	erg	J
eu	entropy unit	J/K
°F	degree Fahrenheit	°C; K
fc	footcandle	lx
fl	footlambert	lx
fl oz	fluid ounce	m^3; L
ft	foot	m
ft·lbf	foot pound-force	J
gf den	gram-force per denier	N/tex
G	gauss	T
Gal	gal	m/s^2
gal	gallon	m^3; L
Gb	gilbert	A
gpm	gallon per minute	(m^3/s); (m^3/h)
gr	grain	kg
hp	horsepower	W
ihp	indicated horsepower	W
in.	inch	m
in. Hg	inch of mercury	Pa
in. H_2O	inch of water	Pa
in.-lbf	inch pound-force	J
kcal	kilogram-calorie	J
kgf	kilogram-force	N
kilo	for kilogram	kg
L	lambert	lx
lb	pound	kg
lbf	pound-force	N
mho	mho	S
mi	mile	m
MM	million	M
mm Hg	millimeter of mercury	Pa
mμ	millimicron	nm
mph	miles per hour	km/h
μ	micron	μm
Oe	oersted	A/m
oz	ounce	kg
ozf	ounce-force	N
η	poise	Pa·s
P	poise	Pa·s
ph	phot	lx
psi	pounds-force per square inch	Pa
psia	pounds-force per square inch absolute	Pa
psig	pounds-force per square inch gauge	Pa

Non-SI (Unacceptable and Obsolete) Units		*Use*
qt	quart	m^3; L
°R	degree Rankine	K
rd	rad	Gy
sb	stilb	lx
SCF	standard cubic foot	m^3
sq	square	exponential form
thm	therm	J
yd	yard	m

BIBLIOGRAPHY

1. The International Bureau of Weights and Measures, BIPM, (Parc de Saint-Cloud, France) is described on page 22 of Ref. 4. This bureau operates under the exclusive supervision of the International Committee of Weights and Measures (CIPM).
2. *Metric Editorial Guide* (*ANMC-75-1*), American National Metric Council, 1625 Massachusetts Ave. N.W.; Washington, D.C. 20036, 1975.
3. *SI Units and Recommendations for the Use of Their Multiples and of Certain Other Units* (*ISO 1000-1973*), American National Standards Institute, 1430 Broadway, New York, N. Y. 10018, 1973.
4. Based on *ASTM E 380-79* (*Standard for Metric Practice*), American Society for Testing and Materials, 1916 Race Street, Philadelphia, Pa. 19103, 1979.
5. *Fed. Regist.*, Dec. 10, 1976 (41 FR 36414).
6. For ANSI address, see Ref. 3.

R. P. LUKENS
American Society for Testing and Materials

F

Continued

FERROELECTRICS

The term ferroelectric was first used by Valasek (1) to emphasize the analogy between the nonlinear hysteretic dielectric properties of Rochelle salt [*304-59-6*] (sodium potassium tartrate tetrahydrate, $NaKC_4H_4O_6.4H_2O$) and the magnetic behavior of ferromagnetic iron. More recently (2–3) the concept has been generalized and the term ferroic has been used to describe all materials that exhibit one or more phases showing a twin or domain structure in which the individual domain states may be reoriented by applied magnetic, electric, or elastic stress fields or combinations of such fields (see Magnetic materials).

The parameters, the state shifts, and the resulting hysteretic responses for the primary ferroic ferromagnets, ferroelectrics, and ferroelastics are shown in Figure 1. It is evident that in the ferroelectric the domain states differ in orientation of spontaneous electric polarization, and that the ferroelectric character is established when it is evident that the states can be transformed one to another by a suitably directed electric field. Spontaneous polarization (P_s), remanent polarization (P_r), and coercive field (E_c) are defined by analogy with corresponding magnetic quantities. Clearly, it is the reorientability of the domain state polarizations that distinguishes ferroelectrics as a subgroup from the larger class of pyroelectric crystals in the 10 polar-point symmetries (4).

For many years it appeared that ferroelectricity was a rare and unusual phenomenon, and Rochelle salt a strange accident of nature. However, with the discovery of the simple perovskite ferroelectric barium titanate [*12047-27-7*], $BaTiO_3$, in the early 1940s and the development of important practical applications, much more effort was focused on the problem, and by the late 1950s and early 1960s many other compounds in different structural families were shown to have ferroelectric phases.

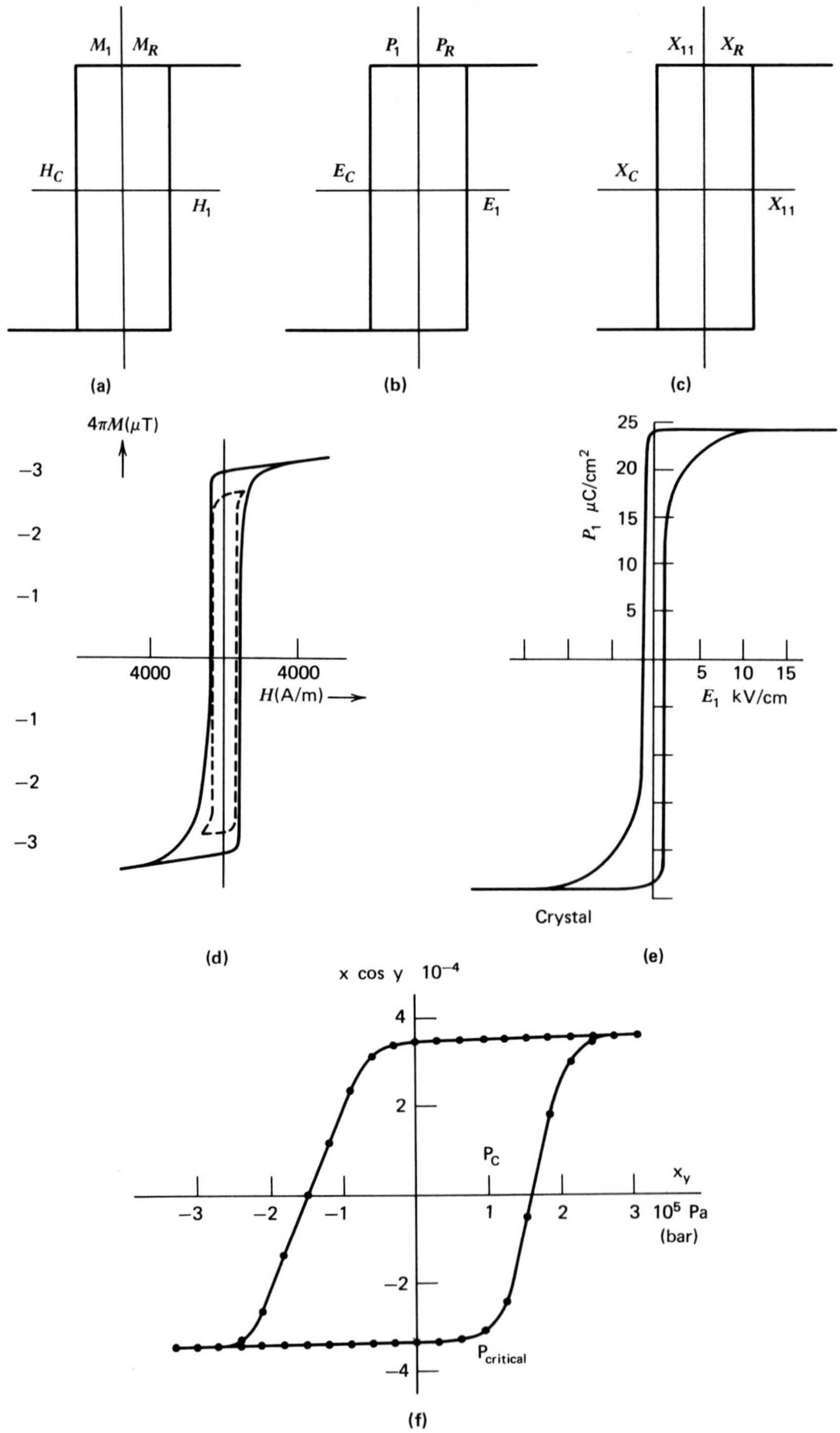

Figure 1. Primary ferroics. (**a**) ferromagnetic; (**b**), ferroelectric; (**c**), ferroelastic; (**d**), ceramic; $Co_{0.1}Fe(II)_{0.9}Fe_2O_4$, 25°C; (**e**), single crystal, $BaTiO_3$, 25°C; (**f**), single crystal, $Pb_3P_{1.6}V_{0.4}O_8$, 25°C. To convert tesla to gauss, divide by 10^{-4}; to convert A/m to oersted, divide by 79.58.

At present there are more than 38 different structural families of ferroelectrics. As an example, within the perovskite, in addition to barium titanate, there are more than 60 ferroelectric compounds, and crystalline solid solutions between more than 150 of these different end-member compositions have been explored.

An excellent compendium of experimental information on ferroelectric crystals is given by Landolt-Börnstein (5–6). The listing of materials is complete up to 1973, and the experimental data have been screened by authors of international repute.

Recent theoretical developments in the field of ferroelectricity have been admirably covered in a text that also reviews application of the single crystals to pyroelectric and electrooptic devices (7).

Oxide ferroelectrics and their applications in ceramic form for piezoelectric devices are also described in considerable detail in ref. 8.

Properties

In many (not all) ferroelectrics the domain state spontaneous polarization is a decreasing function of temperature, going to zero at a phase transition temperature T_c which is called the Curie point. Above T_c a number of ferroelectrics exhibit very high, temperature-dependent dielectric susceptibility (η) such that one component of the susceptibility tensor follows a Curie-Weiss law

$$\eta_{ii} = \frac{C}{T - \theta}$$

where C is called the Curie constant and θ the Curie-Weiss temperature. Again, the terminology is taken directly from the magnetic analogy. It must be noted, however, that for the dielectric case the Curie constant C is ca 5 to 7 orders of magnitude larger than in the ferromagnet. In practice this means that very high useful dielectric susceptibilities (permittivities) persist for a wide range of temperature above T_c in the paraelectric phase. In fact, it is this high intrinsic "softness" in the dielectric response that is the phenomenon most used in the practical application of ferroelectrics, and polycrystalline ceramic ferroelectric dielectrics with relative permittivities ranging up to 10,000 are very widely used in compact disk, tubular, and multilayer capacitors (see Dielectrics below).

Although the intrinsic coupling between changes of polarization and lattice deformation (strain) through the piezoelectric or electrostrictive effects (the a_{ij}, b_{ij}, or Q_{ijkl} constants) of the paraelectric or prototype structure are quite normal, the large spontaneous polarizations of the domains in many ferroelectrics lead to substantial spontaneous strains (ca 1% in $BaTiO_3$ at room temperature), and it is not surprising to find that a number of ferroelectric crystals are also full or partial ferroelastics such that all, or just some, of the electric domain states may be reoriented by uniform elastic stress. Perhaps of more practical importance is the fact that the very high polarizability of the ferroelectric gives large polarization changes at low fields, so that the field-related piezoelectric and electrostriction constants (d_{ij}, e_{ij}, and M_{ijkl}) are also very large. Thus a second major area of application uses this high electromechanical coupling for efficient transduction between electrical and mechanical signals, in sonics and ultrasonics (qv), NDE (non-destructive evaluation), acceleration and pressure sensing, fusing and spark ignition, and many other transducer applications.

Ferroelectric piezoelectrics are also used extensively for broad-band filter application, but not as yet for critical frequency control. A current challenge in ferroelectricity is to find materials with temperature-compensated elastic properties so that the strong electromechanical coupling coefficients may be used to fabricate precisely tuned bulk and surface acoustic wave filters and signal processing devices.

In ferroelectrics both P_s and P_r are strong functions of temperature, particularly at temperatures close to T_c, thus these crystals and poled ceramics have very high pyroelectric coefficients and may be applied in thermometry and in bolometric sensing devices (point detectors) of infrared radiation. A large research effort is at present focused on extending the application of pyroelectrics to long wavelength ir imaging using both vidicon and detector array schemes (see Infrared technology; Temperature measurement).

Since many ferroelectrics are high-band-gap insulating crystals, they can have good transparency in both the visible and near ir spectral regions. Qualitatively, it may be expected that the large dielectric polarizability will lead to an ability to modify the refractive index ellipsoid (indicatrix) under electric fields, and thus to high linear and quadratic electrooptic coefficients. There is much current research seeking to develop effective broad-band modulation and switching techniques for both bulk and guided optical waves by using ferroelectric electrooptic structures. Recently, polycrystalline ceramic ferroelectrics have been processed to very high densities and good optical transparency. These materials possess new parameter combinations for modulation and imaging devices.

In the single-domain state, many ferroelectric crystals also exhibit high optical nonlinearity and this, coupled to the large standing optical anisotropies (birefringences) that are often available, makes the ferroelectrics interesting candidates for phase-matched optical second harmonic generation (SHG).

Finally, a most important area of application utilizes the interaction between the dielectric polarization and the electrical transport processes in ferroelectrics. In single crystals the effects of the domain polarizations upon the drift and retrapping of photogenerated carriers give most interesting photoferroelectric effects. Of more immediate applicability, however, are the large effects of the dielectric changes at the ferroelectric phase transition upon the potential barriers at grain boundaries in suitably prepared semiconducting ceramic ferroelectrics (see Semiconductors). These barium titanate-based compositions show very strong positive temperature coefficients of resistivity (PTC effects) and are finding wide application in temperature and current control for domestic, industrial, and automotive applications (see Semiconducting PTC Materials).

Classifications

A ferroelectric crystal is a material that exhibits one or more ferroelectric phases in a realizable range of temperature and pressure. In the ferroelectric phase, the crystal is spontaneously electrically polarized and the polarization has more than one possible equilibrium orientation. To establish ferroelectricity it must be demonstrated that the polarization can be reoriented between orientation states by a realizable electric field.

Two facets of this rigorous definition can lead to practical difficulty: (*1*) it is often difficult to demonstrate that the persistent electrical polarizations in a crystal are an equilibrium property of the material, and not just long-lived induced electret states;

and (2) the requirement of reorientability leads to a conditional definition, which hinges on the interpretation of what should be a realizable field for any given insulator.

Since the conditions for dielectric breakdown are often difficult to establish and impossible to calculate, there must be a "fuzziness" at the edges of the subject for certain materials. This problem is severely compounded if the crystal is semiconducting or conducting, or if the structure is complex so that individual domains are difficult to identify.

There is some debate as to whether certain liquid crystal systems and polymers such as the poly(vinylidene difluorides) should be included in the class of ferroelectrics, and the problems are essentially in each case related to these facets of the definition.

Symmetry Classification. A comprehensive classification of all possible ferroelectrics has been given on symmetry grounds (9–10). It has been shown that the family of rotation and rotation/inversion operations required to get from each domain state to all other domain states, taken with the domain symmetry in the ferroelectric crystal, defines a prototype symmetry from which the spontaneous deformation parameters of the domain state can be unequivocally deduced. Conversely, for each possible prototypic point symmetry, the consequences of the appearance of spontaneous (switchable) polarization can be tabulated, so that all possible ferroelectric species accessible from a given prototype point symmetry are immediately evident.

An example of the derivation for a prototype of $4/mmm$ (D_{4h}) point symmetry is given in Figure 2, and the resultant ferroelectric species are tabulated in Table 1 using the notation of Shuvalov (10).

In many ferroelectrics it is intuitively obvious how to define P_s, since the prototype is unpolarized and thus all polarization is switchable. For ferroelectrics derived from polar prototypes, however, the definition of spontaneous polarization is not so simple and the concept of a prototype is more important. For ferroelastic crystals a similar tabulation has been given (11–12) and here the definition of the prototype is vital since spontaneous strain has no meaning without reference to the prototypic form.

It may be noted that similar symmetry-based compilations have now been given for all ferromagnetic species (13) and recently, in general, for all possible ferroic species (14–15).

Other Classification Schemes. ***Hard-Soft Ferroelectrics.*** In the earlier ferroelectric literature a classification based on the qualitative elastic properties is often suggested. Hard materials like the ceramic oxides with perovskite, pyrochlore, tungsten bronze, and bismuth oxide layer structures were contrasted with the softer organic and water-soluble compounds like Rochelle salt, potassium dihydrogen phosphate [*7778-77-0*] and other ferroelectric phosphates, arsenates, sulfates, nitrates, and nitrites.

Unfortunately, however, in these systems the dielectric properties are not commensurate with the mechanical properties, and the oxides are in general much softer dielectrically (Curie constant C ca 10^5 vs C ca 10^3 in water-soluble crystals). With the evolution of the subject and the development of many new materials with intermediate properties, this classification has now largely fallen into disuse.

Improper or Extrinsic Ferroelectrics. In most ferroelectric crystals, the parameter that characterizes the phase transition into the ferroelectric form (the primary order parameter for the transition) is the electric polarization, P. Elastic, piezoelectric,

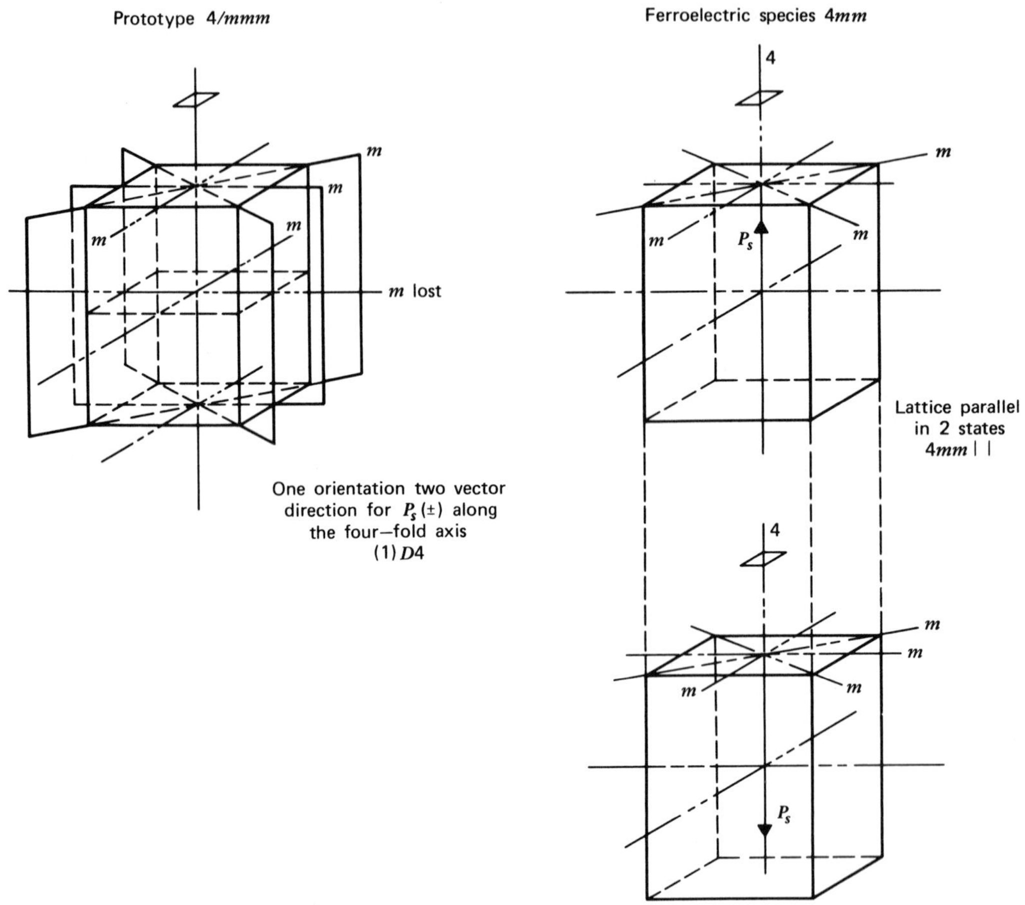

Figure 2. Pictorial representation of Shuvalov species $4/mmm(1)D4F4mm\|$.

optical, and thermal changes that occur may be regarded as essentially secondary consequences of the appearance of the nonzero P vector.

Recently, it has become apparent from studies of dielectrics in the gadolinium molybdate [*14985-14-9*] ($Gd_2(MoO_4)_3$) (16–18) and langbeinite [*67320-08-5*] ($Cd_2(NH_4)_2(SO_4)_3$) (19) families that ferroelectricity itself might arise as a secondary consequence in a phase change dominated by a completely different order parameter. If, as is the case in the examples quoted above, the order parameter is of different symmetry from the polarization, the dielectric properties in the prototypic phase might be quite normal and Curie-Weiss behavior entirely absent. Crystals in these families where the temperature-dependent P_s is not accompanied by high dielectric permittivity are efficient pyroelectric voltage generators and are under intensive study for pyroelectric imaging applications (20) (see Electrophotography).

Antiferroelectrics. In view of the analogy with magnetic materials, it is natural to inquire whether in dielectrics there are systems that exhibit antipolar ordering of elementary electric dipoles and thus could be considered analogous to the antiferromagnets. Unfortunately, however, in viewing almost any ionic crystal with a centric structure, sets of antipolar dipoles can be picked out and in this simple sense the classification is not meaningful.

Table 1. Shuvalov Species for Point Symmetry 4/*mmm*[a,b]

Syngony of initial phase	Kind of transition, kind of ferroelectric phase	Number of subtypes	Corresponding simple form of initial phase	Order of transition
tetragonal	$4/mmm(1)D4F4mm\|$		pinacoid	II
	$4/mmm(2)D2Fmm2\|$	2	tetragonal prism	II
	$4/mmm(4)A4Fm\|$		ditetragonal prism	I
	$4/mmm(4)A2Fm\|$	2	tetragonal bipyramid	I
	$4/mmm(8)A1F1\|$		ditetragonal bipyramid	I

[a] Ref. 10.

[b] In the Shuvalov notation the first group of symbols gives the prototype symmetry (4/*mmm*). Parentheses enclose the number of orientations; twice this number is the number of vector directions for P_s. The symbol *D* indicates the orientations are definite or *A*, arbitrary, with respect to the following axis or plane. The symbol *F* indicates a ferroic species, and the final block of symbols the point symmetry of the domain state. The symbol ‖ indicates that the lattices of domain states are parallel. If the symbol is absent they are mutually tilted. Simple forms are the corresponding solid figures whose faces are normal to the vector directions of P_s. The final column indicates whether the transition into the ferroic state can be continuous, II, or must be discontinuous (first-order), I.

There are, however, within a number of the ferroelectric structure families, crystals that exhibit phase changes into ferroelastic forms in which the volume of the primitive unit cell is an integral multiple of that in the prototype structure and appears to be well described as a superlattice of two interpenetrating but compensating ferroelectric-like displacement systems.

For several of these structures the similarity is further reinforced by the fact that the structure may be switched to a strongly polar ferroelectric form by a suitably oriented high electric field, and it may be demonstrated by x-ray analysis that the superlattice is now eliminated.

The term antiferroelectric is now reserved for such crystals. It is recognized that the dielectric anomalies at the phase change in these crystals are a characteristic of what would be a lower temperature ferroelectric form which has been frustrated by the slightly more stable antipolar arrangement occurring first.

The phenomenon of doubled, or split hysteresis, which occurs at the field-induced antiferroelectric–ferroelectric transitions driven by cyclic fields is an expected consequence of the phase change, but is in no way definitive without corroborating structural evidence. Similar looking double loops can occur in normal ferroelectrics, owing to defect dipoles pinning a stable domain structure, and at paraelectric–ferroelectric transitions when the behavior is markedly first order.

Domains and Switching

An internal structure of spontaneously electrically polarized domains is a characteristic feature of the ferroelectric phase, so that for most practical applications of ferroelectrics it is necessary to ascertain how this domain structure will modify the required properties and, often, how it may be controlled or eliminated to optimize performance.

The planes along which individual domains conjoin are termed domain walls, and the process of polarization reversal or reorientation under high fields is accomplished by the motion of existing walls, or by the creation and motion of new domain walls.

For each ferroelectric species, conditions for the permitted orientations of P_s in the domain structure are dictated by the prototype symmetry, which also determines whether the state will be fully, partially, or nonferroelastic. If contiguous domains are strain-distinct, the conjoining ferroelectric–ferroelastic wall has only a limited family of possible orientations which are rigorously prescribed by the conditions for dimensional compatibility between the two domains along the wall plane. Usually (but not always) the permitted wall is a common high-symmetry plane in the prototype, which was lost on the appearance of P_s. Conditions of compatibility have been generally formulated for all ferroelastic–ferroelectric walls (21–22). These ferroelastic domains in transparent ferroelectric crystals are readily observed in the polarizing microscope. Domains that are distinct in strain, a second-rank polar tensor property, must also be distinct in all other second-rank polar tensor properties, and are thus necessarily distinct in optical permeability. Examples for several different species are given in Figure 3.

The switching behavior of ferroelastic–ferroelectric walls depends markedly on the magnitude and the nature of the spontaneous domain strains, which can vary widely even between ferroelectrics in the same structure family. In cases where the strains are large, switching that is not carefully controlled can cause stress concentrations sufficient to crack the crystal, and fatigue effects are often evidenced on repeated switching. The problem is most acute in polycrystalline ceramic ferroelectrics, which have to be poled to a high saturation remanence for piezoelectric applications (see under Applications). In these systems both the elastic and electrical boundary conditions on the individual crystallites are complex functions of the crystallite size, size distribution, ferroelectric phase assemblage, minor phase additions, and poling conditions. Residual stresses left after poling do cause slow continuing changes in the domain assemblage and the continuous aging which is an undesirable feature in the properties of these ceramics.

For nonferroelastic domains, ie, antipolar domain structures originating from centric prototypes, all conjoining planes are elastically compatible and wall structures generally appear more tortuous. Generally, planes parallel to P_s appear to be preferred and would be expected to be so since the normal component of displacement across such planes is preserved. Figure 3**c** and **d** show typical antipolar structures in $BaTiO_3$ and in triglycine sulfate [*513-29-1*], $(NH_2CH_2COOH)_3.H_2SO_4$ (TGS).

Theoretical calculations (23–24) and experimental observation (25–26) confirm that the width of such simple ferroelectric walls is exceedingly narrow, with the polarization changing abruptly across only a few unit cells of the structure.

Very extensive studies have been made of the switching behavior of simple antipolar ferroelectric domains for $BaTiO_3$ (27), TGS (28), and other simple ferroelectrics (29). The kinetics can be described by measuring the time taken for reversal of the polarization under different levels of d-c driving field E and has been shown to follow an equation of the form

$$t_s \propto \exp(\alpha/E)$$

where t_s is the time required for switchings, α is a constant with the dimensions of a field (called the activation field), and E is the applied driving field.

The equation is valid for several materials over quite a wide range of fields, but at very high field levels the timing appears to change over to a power law of the form

$$t_s \propto E^{-n}$$

where n varies from 1 to 7 in different systems (30).

In $BaTiO_3$ the bulk switching relation has been effectively correlated to the individual domain wall motion which appears to follow a relation (31)

$$v = v_\infty \exp(-\delta/E)$$

where the constant δ appears to depend somewhat on the defect concentration in the crystal, v is the wall velocity under driving field E, and v_∞ a constant for the crystal.

It was noted quite early (32) that since the domain wall is very narrow, the energy gained by moving the wall just one lattice constant is necessarily very much smaller than the wall energy itself, and thus true continuous sidewise motion of the wall, as occurs for many magnetic domain walls, is most unlikely.

The model that does appear to explain the behavior is that the apparent sidewise motion of walls is caused by the nucleation and growth of step-like protrusions on existing 180° walls. An analysis of this model was given (33) to explain the exponential behavior in the low field region. The model was then extended (34) to account for the power law behavior, and a more recent general formulation covering both regimes has been published (35).

The narrow domain wall has most important consequences in the major practical application of ferroelectrics in capacitor dielectrics. Since critical size nuclei occur only very infrequently at low fields, reversible wall motion does not contribute in a major way to the dielectric response, and the capacitor engineer must manipulate the soft single-domain ferroelectric and the paraelectric permittivity to satisfy system requirements.

Similarly, the device engineer who wishes to exploit the apparent bistability of the dielectric hysteresis must be aware that the ferroelectric coercivity is due to the "knee" on an exponential curve, and that given time or frequent reverse pulsing, the ferroelectric may change polarization state for fields much below the apparent 60 Hz coercivity.

Ferroelectric Materials

Perovskites. The simple cubic perovskite structure, which is the high temperature form for many mixed oxides of the ABO_3 type, was one of the first simple structures to exhibit compounds with ferroelectric properties and is still probably the most important ferroelectric prototype.

The very simple cubic structure (point symmetry $m3m$) (Fig. 4) is made up of a regular array of corner-sharing oxygen octahedra with smaller highly charged cations like Ti, Sn, Zr, Nb, Ta, W, etc, occupying the central octahedral B site, and lower-charged larger cations like Na, K, Rb, Ca, Sr, Ba, Pb, etc, filling the interstices between octahedra in the larger 12-coordinated A sites.

The simple ABO_3 compounds, $BaTiO_3$, $PbTiO_3$ [*12060-00-3*], $PbZrO_3$ [*12060-01-4*], $NaNbO_3$ [*12034-09-2*], $KNbO_3$ [*12030-85-2*], $AgNbO_3$ [*12309-96-5*], WO_3 [*1314-35-8*], have interesting high temperature ferroelectric or antiferroelectric phases. The perovskite structure is, however, also tolerant of a very wide range of multiple cation substitution on both A and B sites, so that many more complex compounds such as $(K_{1/2}Bi_{1/2})TiO_3$, $Pb(Fe_{1/2}Ta_{1/2})O_3$, $Pb(Co_{1/4}Mn_{1/4}W_{1/2})O_3$ can be prepared.

Barium titanate was the first perovskite-type compound shown to be ferroelectric

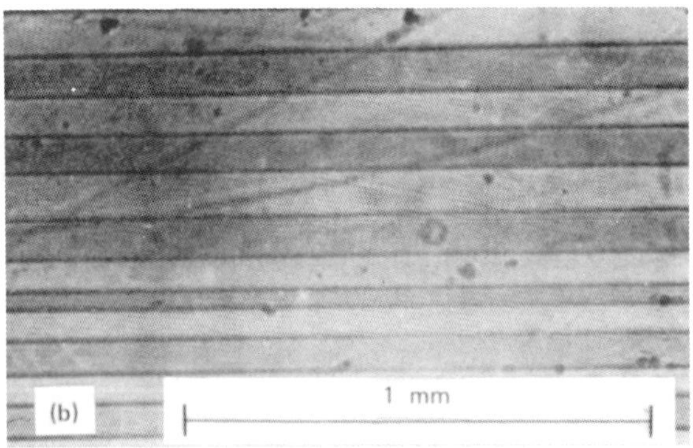

Figure 3. Examples of ferroelastic–ferroelectric and antiparallel domains. (**a**) Ferroelastic–ferroelectric 90 domains in $BaTiO_3$. Crystal dimensions are ca 0.7 × 0.7 × 0.05 mm. Polarized NaD light. Pins indicate orientations of tetragonal *c* axis. (**b**) Ferroelastic–ferroelectric domains in gadolinium molybdate $Gd_2(MoO_4)_3$. (×50) (**c**) Antiparallel domains in triglycine sulfate. Top and bottom surface of a plate cut perpendicular to the polar *b* axis, etched in water to reveal the structure. (**d**) Antiparallel domain in $BaTiO_3$ *c* domain crystal revealed by etching in HF.

(ca 1943) and is to date the most thoroughly investigated ferroelectric material (36). This is essentially a consequence of its relatively simple lattice structure (see Fig. 4), allowing, by means of simple models, an understanding of ferroelectric phenomena. In addition, $BaTiO_3$ proved to be ideal in applications because its electrical properties can be controlled within a wide range by means of mixed crystal formation and doping.

The characteristic feature of the $BaTiO_3$ unit cell (Fig. 4) is the TiO_6-octahedra, which because of their high polarizability, essentially determine the dielectric prop-

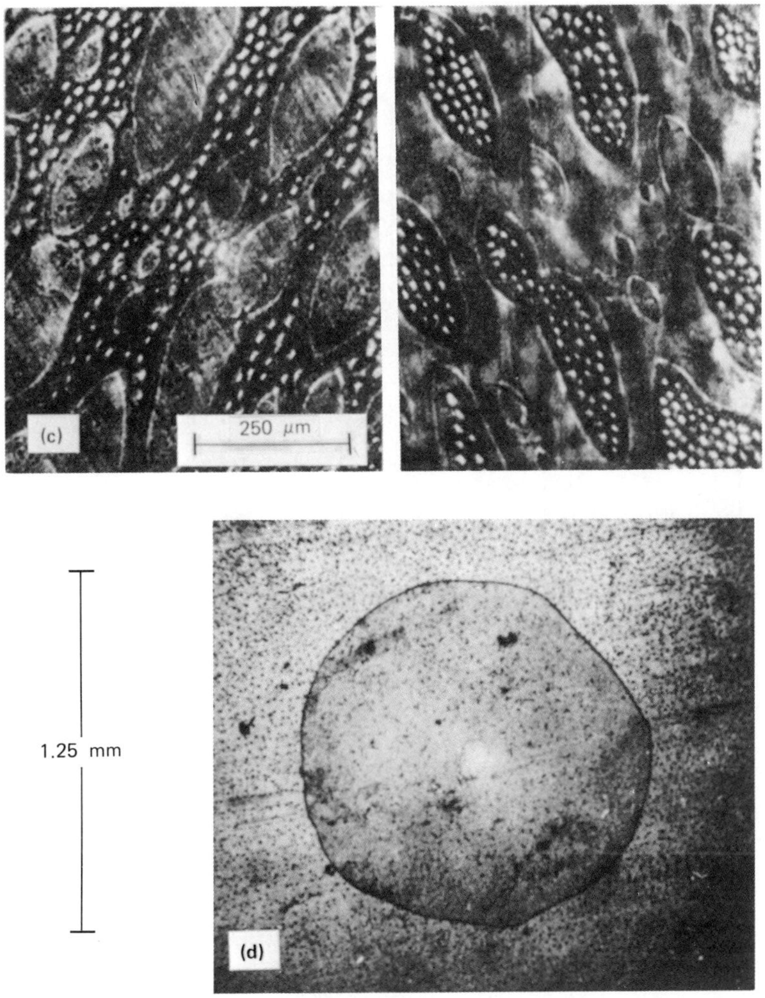

Figure 3. (*Continued*)

erties. The high polarizability is due to the fact that the small Ti^{4+} ions have relatively more space within the oxygen octahedra. The cubic unit-cell of Figure 4 with the Ti^{4+} ion in the center of the oxygen octahedron is, however, stable only above the Curie point T_c of about 130°C. Below T_c the Ti^{4+} ions occupy off-center positions. This transition to the off-center position at T_c results in a series of important physical consequences (Fig. 5). The crystal structure changes from cubic ($T > 130°C$) via tetragonal ($+5°C < T < +130°C$; $c/a = 1.01$) and orthorhombic ($-90°C < T < +5°C$) to rhombohedral ($T < -90°C$). At the same time a spontaneous polarization P_s (26 $\mu C/cm^2$ at room temperature) appears, the direction of which is in the tetragonal phase along one of the 6 edges, in the orthorhombic phase along one of the 12 surface diagonals, and in the rhombohedral phase along one of the 8 space diagonals of the ideal cubic unit cell. The direction of P_s can be switched by high electrical field (ca 1–2 kV/cm) between the different crystallographic allowed positions which are characteristic in each ferroelectric phase.

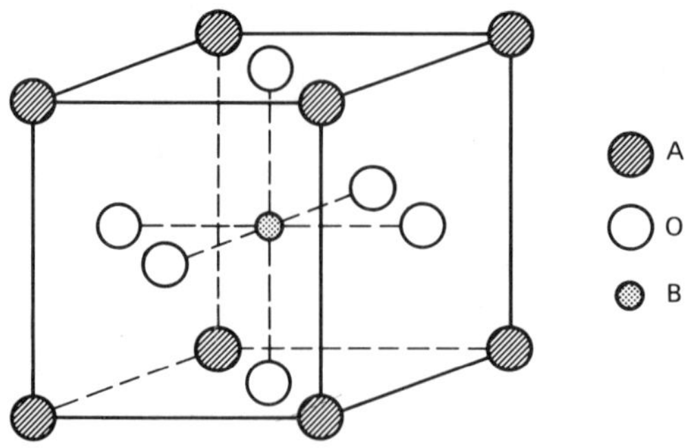

Figure 4. Cubic ($m3m$) prototype structure of perovskite-type ABO_3 compounds. For $BaTiO_3$, A = Ba^{2+}, B = Ti^{4+}, and O = O^{2-}.

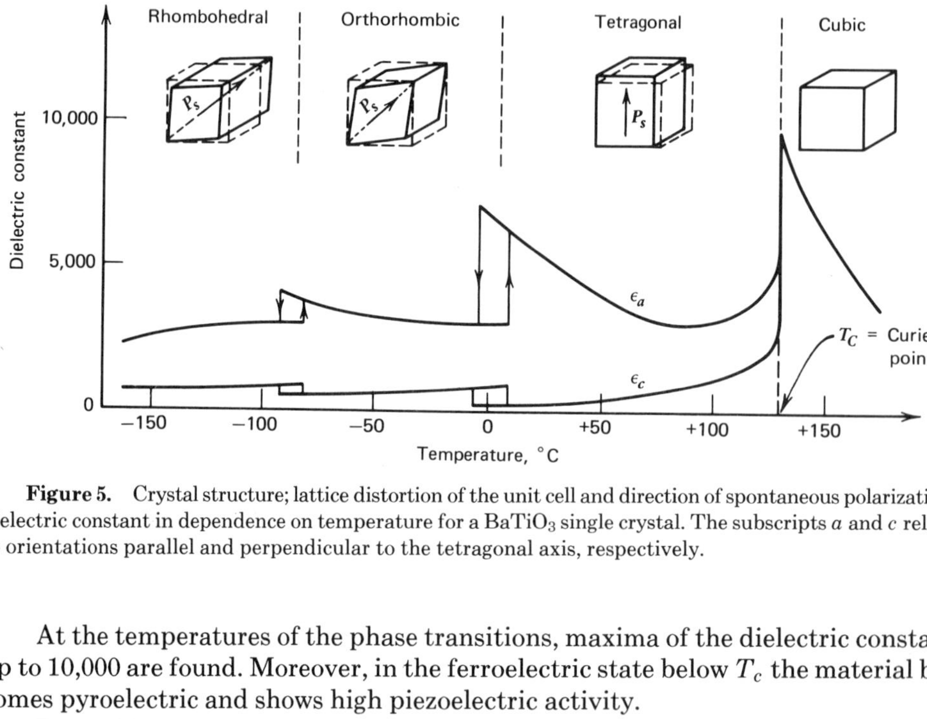

Figure 5. Crystal structure; lattice distortion of the unit cell and direction of spontaneous polarization; dielectric constant in dependence on temperature for a $BaTiO_3$ single crystal. The subscripts a and c relate to orientations parallel and perpendicular to the tetragonal axis, respectively.

At the temperatures of the phase transitions, maxima of the dielectric constant up to 10,000 are found. Moreover, in the ferroelectric state below T_c the material becomes pyroelectric and shows high piezoelectric activity.

Perovskite-type compounds and especially $BaTiO_3$ have the ability to form extensive solid solutions. By this means a wide variety of materials with continuously changing electrical properties can be produced in the polycrystalline ceramic state. By substituting Ba^{2+} ions with Pb^{2+} ions, T_c can be increased linearly up to 490°C for a 100% Pb substitution. In the same manner, T_c can be continuously decreased by the substitution of Ba^{2+} with Sr^{2+} or of Ti^{4+} with Zr^{4+} or Sn^{4+} (Fig. 6). Simultaneously with the change of T_c by formation of solid solutions, the low temperature phase transitions between the tetragonal–orthorhombic and orthorhombic–rhombohedral phases are shifted in a rather complex manner.

$PbZrO_3$–$PbTiO_3$-Based Materials. Since the middle of the 1950s, solid solutions of $PbZrO_3$–$PbTiO_3$ ceramics with perovskite structure have gained rising interest because of their superior piezoelectric properties (8,37–38). The phase diagram of the

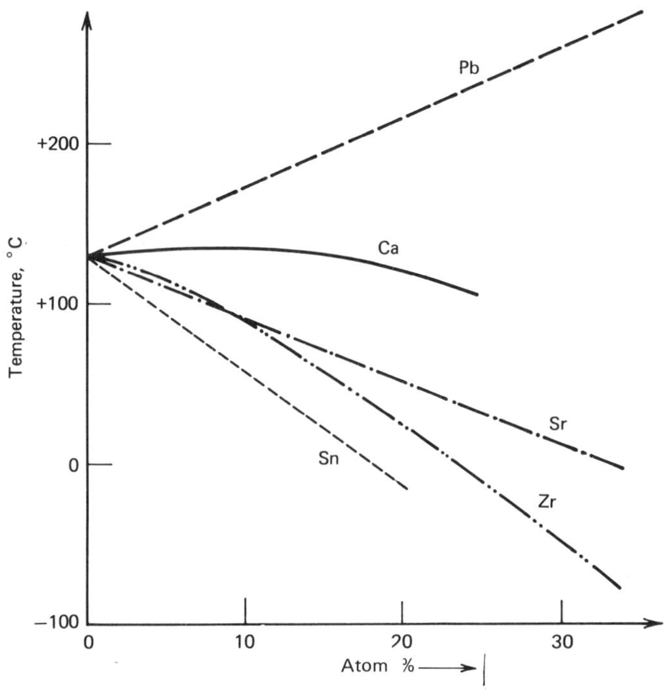

Figure 6. Shift of the Curie point of $BaTiO_3$ by substitution of Ba^{2+} ions by Pb^{2+}, Ca^{2+}, or Sr^{2+} ions, or of Ti^{4+} ions by Zr^{4+} or Sn^{4+} ions (8).

$Pb(Zr_xTi_{1-x})O_3$ system is shown in Figure 7. At high temperatures above T_c, the ideal cubic paraelectric structure is stable, where no ferroelectric phenomena such as spontaneous polarization appear. At room temperature the materials are ferroelectric and show, for Ti-rich compositions ($0 \leq x \leq 0.52$), a tetragonal distortion of the unit cell, whereas compositions with lower Ti content ($0.52 \leq x \leq 0.94$) have rhombohedrally distorted unit cells. Both phases are separated by a morphotropic phase boundary at $x = 0.48$. Compositions near the Zr side of the system ($0.94 \leq x \leq 1$) are antiferroelectric with orthorhombic structure. The direction of the spontaneous polarization P_s is along one of the edges of the unit cell for tetragonal distorted compositions and along one of the space diagonals for rhombohedral distorted compositions.

Today a third or even a fourth and fifth phase of a complex perovskite as, for example, $Pb(Mg_{1/3}Nb_{2/3})O_3$ is added in addition to $PbTiO_3$ and $PbZrO_3$ when forming the solid solution (39). In the same manner ca 20 different elements with similar ionic radii can be substituted in place of Mg or Nb, leading to a huge number of possible combinations and a multitude of compositions having, however, generally comparable properties. At low concentrations of a complex perovskite addition, the phase relationships of the quasi–binary composition are maintained. Increasing amounts of a complex perovskite reduces T_c and pseudocubic phases begin to appear.

Taken together with the fact that there is often a wide range of crystalline solid solubility between end-member compositions, and that the ferroelectric and antiferroelectric Curie temperatures and consequent properties appear to mutate continuously with fractional cation substitution, the perovskite system has the desired

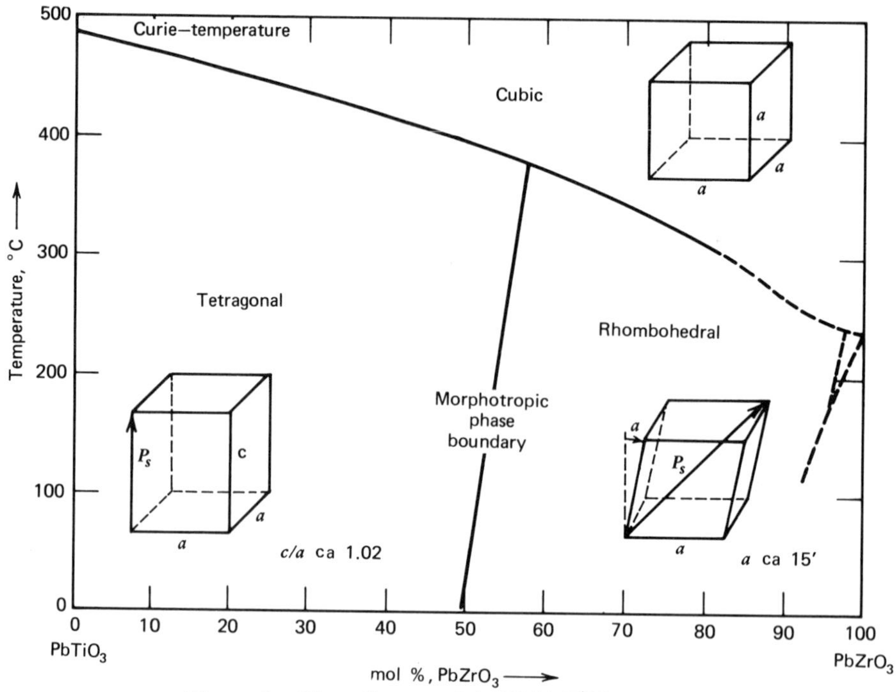

Figure 7. Phase diagram of the $Pb(Zr,Ti)O_3$ system.

versatility for the very extensive applications discussed in the following sections on ferroelectric ceramics.

Perovskite-Related Octahedral Structures. ***Lithium Niobate.*** $LiNbO_3$ [*12031-63-9*] and $LiTaO_3$ [*12031-66-2*] have a structure related to the perovskites, but of lower symmetry ($\bar{3}m$), due to the fact that the oxygen octahedra share edges (Fig. 8). In sequence along the trigonal c axis in the low temperature ferroelectric phase, the octahedra are occupied in the sequence Nb(Ta), vacancy, Li, Nb(Ta), vacancy, and Li (41).

Because of the very high Curie temperatures (ca 1200°C in $LiNbO_3$, 620°C for $LiTaO_3$), the prototypic character has not been well established in the niobate, but judging from the tantalate it does appear that the Nb ion is in the center of the oxygen octahedron, whereas the lithium ion is in dynamical equilibrium between displaced positions above and below the triangular ring of oxygens (40).

The ion displacements in the ferroelectric phase are exceedingly large, corresponding to the observed spontaneous polarizations of 50 $\mu C/cm^2$ in the tantalate and ca 70 $\mu C/cm^2$ in the niobate. Because of these large displacements, poling is usually carried out at elevated temperatures (42), and switching at room temperature was only recently demonstrated with very high pulsed fields. (43).

Lithium niobate is widely used in the single-domain form for both piezoelectric (44) and electrooptic devices (45). The tantalate has a wide application in simple pyroelectric point detectors (46) and, more recently, for acoustic wave devices (47).

Tungsten Bronze Structures. The tungsten bronze family has an oxygen octahedral framework structure that is much more open than the simpler perovskite structure (Fig. 9). The basic tetragonal unit cell (Fig. 9) contains 10 BO_6 octahedra corner-linked

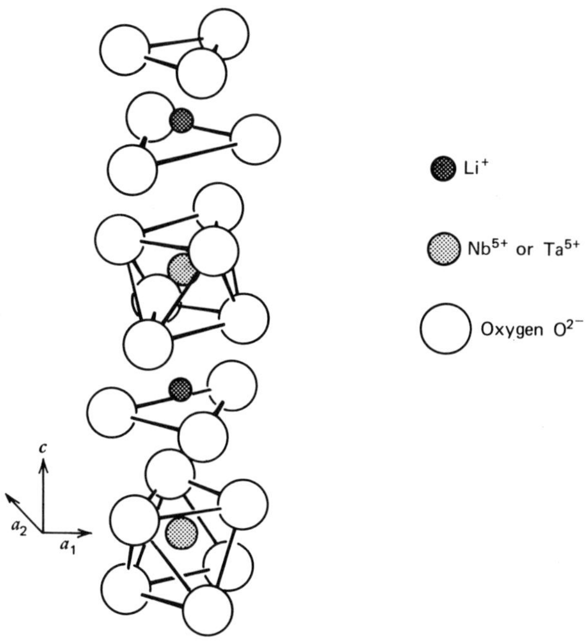

Figure 8. Structure of ferroelectric $LiNbO_3$ and $LiTaO_3$ (40).

to form square, triangular, and pentagonal tunnels that extend through the structure along the 4-fold symmetry (c) axis (48). The c axis linkage is regular, but in the a and b directions the linkage is markedly puckered (see also Copper alloys).

There are two types of A sites (A_1, A_2) that can be occupied by large monovalent, divalent, or trivalent ions, two different B sites (B_1, B_2) that are usually occupied by smaller, highly charged cations (Nb, Ta, W, Ti), and a small C site which may be vacant or occupied by Li (49–51).

Lead metaniobate [*12034-88-7*], $PbNb_2O_6$ (52), and metatantalate [*12065-68-8*], $PbTa_2O_6$ (53), appear to be anomalous as bronzes, and in both these useful ferroelectrics the bronze form is a metastable structure at room temperature (54).

In general, the bronzes appear to be solid solutions of at least two components, and a certain degree of disorder appears conducive to stabilizing the structure.

Among compounds of interest besides $PbNb_2O_6$ and $PbTa_2O_6$ are $Ba_xSr_{1-x}Nb_2O_6$ and $Ba_2NaNb_5O_{15}$ [*12323-03-4*]. The metaniobate is still widely used in poled ceramic form as a low Q piezoelectric transducer (55–56), the $Ba_{0.5}Sr_{0.5}Nb_2O_6$ [*37185-09-4*] composition in single-crystal form in pyroelectric point detectors (57–58), and the barium sodium niobate and various similar derivative compositions for their very high nonlinear optical properties (59–60).

Bismuth Oxide Layer Structures. The structural arrangements in these materials are typified by $PbBi_2Nb_2O_9$ [*12431-20-8*] (Fig. 10). The structure may be visualized simply as a stacking of regular perovskite-like $(PbNb_2O_7)^{2-}$ double sheets of corner-linked NbO_6 octahedra interleaved by $(Bi_2O_2)^{2+}$ layers along the tetragonal 4-fold axis. It appears possible to develop structures of this character with single (Bi_2WO_6) [*13595-86-3*], double ($PbBi_2Nb_2O_9$), triple ($Bi_4Ti_3O_{12}$) [*12010-77-4*], and up to five perovskite layers interleaving the bismuth oxide sheets (61). Extensive cation replacement appears possible within the perovskite layers, though the conditions are

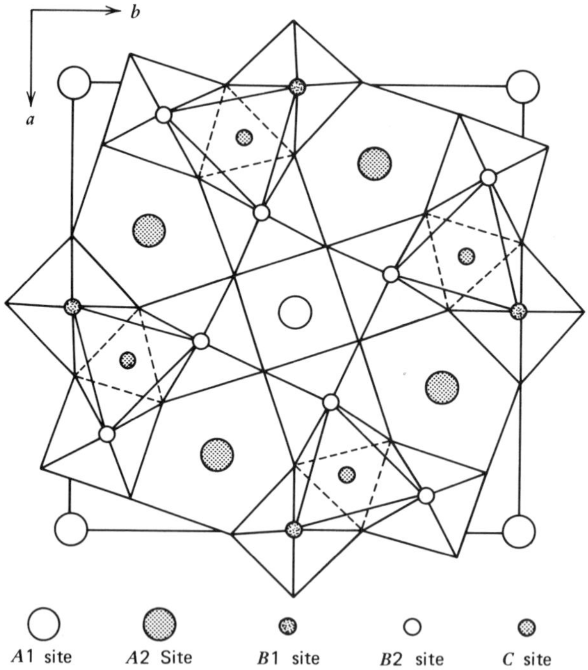

Figure 9. Projection of the tungsten bronze structure looking down the tetragonal *c* axis (48).

rather more restrictive than in pure perovskites (62), but bismuth cannot be replaced by any comparable trivalent ion in the $(Bi_2O_2)^{2+}$ sheets (63).

In almost all these sheet structures, the ferroelectric phases have spontaneous polarizations that lie in the plane of the perovskite sheets, and the ferroelectric symmetry is *mm*2 (64).

Bismuth titanate is unique in transforming directly at 675°C to a low-symmetry monoclinic ferroelectric phase (65) in which the domain states have a polarization vector tilted slightly (ca 4°) out of the plane of the sheet (66).

This tilted vector gives a very small monoclinic shear to the unit cell (less than 2 minutes of arc) but a major tilt to the optical indicatrix (ca 60°) (67). The very unusual optical situation in this crystal has been used for optical image storage (68–69) and to develop an optical page composer for holographic information storage (70) (see Holography).

Pyrochlore Structures. Ferroelectricity was first discovered in $Cd_2Nb_2O_7$ [*12187-14-3*] in 1952 (71). The structure is again based upon a corner-sharing arrangement of NbO_6 octahedra with the larger Cd^{2+} ions and the extra set of oxygens occupying the interstices. A projection of the structure onto the [110] plane is shown in Figure 11. This family, like the lithium niobate family, is very restricted in the cation substitutions that can be made, and similar properties have been found only in the $Cd_2Ta_2O_7$ [*12050-35-0*] and corresponding lead compounds $Pb_2Ta_2O_7$ [*12065-68-8*], $Pb_2Nb_2O_7$ [*12059-64-2*]. In all cases the Curie temperatures are low, below 200 K (72).

Very recently, very strong ferroelectric behavior ($T_c > 1300$°C) in $Sr_2Nb_2O_7$ [*12201-67-1*] has been observed (73), but the structure type is monoclinic and more complex than the cubic pyrochlore.

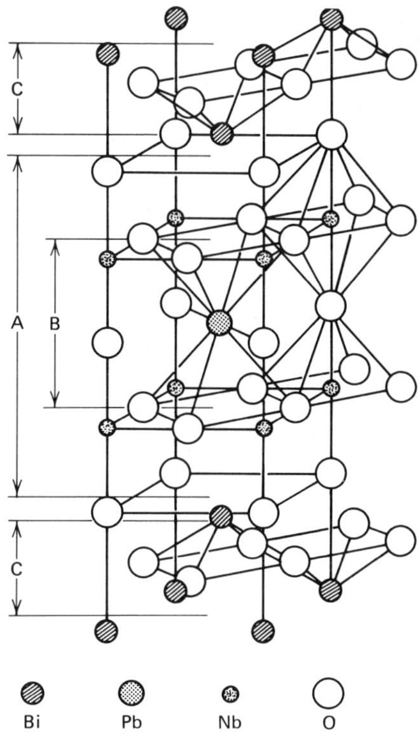

Figure 10. One-half of the tetragonal (4/*mmm*) unit cell of $PbBi_2Nb_2O_9$. A denotes the perovskite layer $(PbNb_2O_7)^{2-}$; B denotes the unit of a hypothetical $PbNbO_3$; C denotes the $(Bi_2O_2)^{2+}$ layers.

KDP Family. Potassium dihydrogen phosphate and the isomorphous rubidium and cesium phosphates RbH_2PO_4 [*13774-16-8*], CsH_2PO_4 [*18649-05-3*], and arsenates (RbH_2AsO_4 [*13464-57-8*], CsH_sAsO_4 [*16331-85-4*]) were the first major family of ferroelectrics to be discovered (74–75). The structure is relatively simple (Fig. 12), being based on an arrangement of PO_4 tetrahedra in layers so that the upper corners of the tetrahedra in one layer are hydrogen bonded to the lower corners of the tetrahedra in the next successive layer. The PO_4 tetrahedra alternate with potassium atoms at distances $c/2$ apart in the tetragonal unit cell (76).

It is of interest that the hydrogen bonds which play a major role in the ferroelectric ordering, lie in planes orthogonal to what becomes the polar axis of the ferroelectric orthorhombic phase. The symmetry above T_c is $\bar{4}2m$ and in the prototypic structure the protons are tunneling between off-center positions along the bond direction.

In the single-domain ferroelectric phase, the protons order along the bonds in such a manner that they are all near upper or lower oxygens of the tetrahedral phosphate groups, and it is this coupling that pushes the phosphorus and potassium from the high-symmetry positions giving rise to a switchable moment along the c axis.

The key role of the hydrogen bonds in the transition leads, as expected, to a strong dependence of the phase transition temperature upon deuteration (77) and DKDP, KD_2PO_4 [*13761-79-0*], is used in some electrooptic devices (78) because of its larger longitudinal electrooptic coefficients in the paraelectric phase.

In fact, all major applications of KDP family materials use properties in the

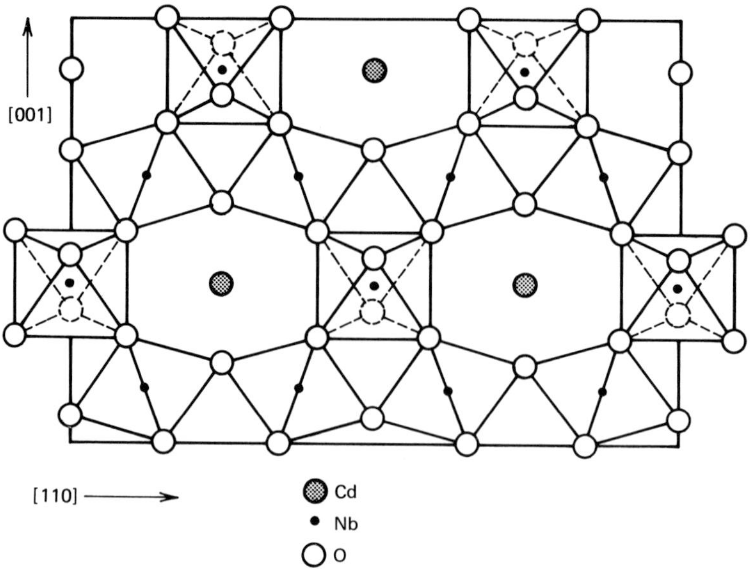

Figure 11. Schematic projection of the cubic ($m3m$) structure of the $Cd_2Nb_2O_7$ on [110] plane. The Cd ions that overlap the Nb ions at (1/8, 1/8) etc, and the seventh set of oxygens that are located above and below the Cd ions are not shown.

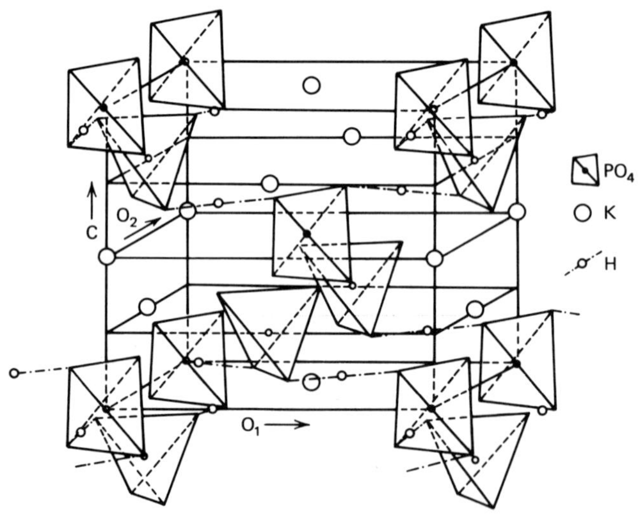

Figure 12. Unit cell of KH_2PO_4 in tetragonal $\bar{4}2m$ prototype structure (76).

paraelectric prototypic form. Frequently, the properties are far from ideal, but the tractable water solution growth of KDP has permitted large crystals of excellent quality, which are difficult if not impossible to duplicate in the more desirable refractory oxides.

Triglycine Sulfate. Strictly speaking, triglycine sulfate (79–80) (TGS) is not representative of a wide range of isomorphous structures, although the isomorphous selenate [920-25-2] (81) and fluoroberyllate [2396-72-7] (82) do have similar prototypic structures. There is, however, a much wider range of glycine-containing ferroelectrics (5,6) in which the glycinium molecule plays a major role and (TGS) is suitably illustrative.

A projection of the TGS structure along the *c* axis of the monoclinic unit cell is shown in Figure 13. The unit cell contains three inequivalent glycine molecules labeled I, II, and III, and all participate to some extent in the development of the ferroelectric moment which is confined to the *b* direction ($\pm P_b$). A major part of this moment does, however, appear associated with the planar glycinium ion which makes an angle of 12.5° with the plane at $b = 1/4$. Above T_c this plane (and the corresponding plane at $b = 3/4$) become mirror planes owing to random statistical distribution of the glycinium I ions, which order at the Curie temperature (79).

TGS has a near second-order phase change to the ferroelectric form at 49°C, and exhibits a spontaneous polarization P_s ca 4 $\mu C/cm^2$ at room temperature (80). The convenient location of T_c and the easy water solution growth of very perfect TGS crystals make this a popular single crystal material for pyroelectric detectors (81), and particularly for the demanding application in pyroelectric vidicon systems (82).

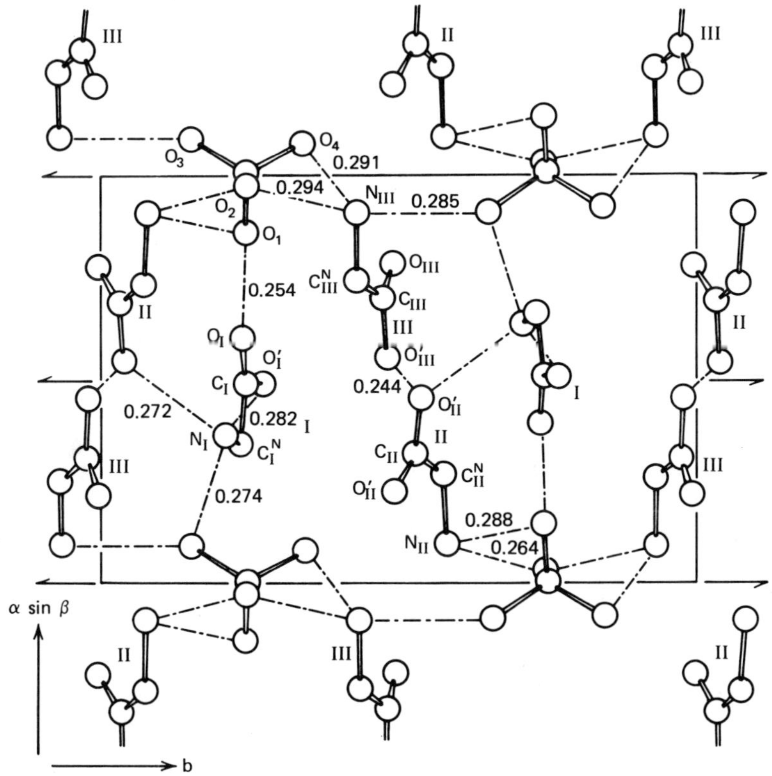

Figure 13. Projection of the structure of triglycine sulfate along the monoclinic *c* axis (81). Bond lengths are shown in nanometers.

Preparation of Ferroelectric Ceramics

Ferroelectric ceramics are composed of oxide compounds having generally the perovskite structure (Fig. 4), eg, $BaTiO_3$ or $Pb(Zr,Ti)O_3$. They are produced in successive steps as follows (see also Ceramics).

Powder Preparation. ***Conventional (Mixed Oxide) Methods.*** As raw materials, mostly oxides (TiO_2, ZrO_2, PbO) or carbonates ($BaCO_3$, $SrCO_3$, $CaCO_3$) are used which have to meet certain application-dependent specifications with respect to purity and grain size distribution. The final powder product with perovskite structure is formed by a solid state reaction (calcination) of these starting raw materials (8,83). A typical example is

$$BaCO_3 + TiO_2 \rightarrow BaTiO_3 + CO_2\uparrow$$

Usually the required product is a fully homogeneous compound having perovskite structure with small particle size (<1 μm). The solid-state reaction occurs by solid state diffusion with a relatively slow diffusion rate. Therefore, small diffusion distances or high calcination temperatures are necessary to complete the reaction. Since, however, high calcination temperatures will result in coarse-grained powders having unsuitable sintering properties, the diffusion distance must be minimized. This can be achieved by intimate mixing or milling procedures (dry or wet using water, alcohol etc) of the ingredients. Mixing and milling times are also limited, however, because long times lead to increasing impurity concentrations by mechanical abrasion of the milling equipment.

The powder preparation usually proceeds with due regard to the principal facts just mentioned, as follows: (*1*) weighing of the raw materials; (*2*) intimate mixing or milling (dry or wet) for 1 to 10 h using porcelain, SiO_2, or Al_2O_3 materials for equipment; (*3*) calcination 1 to 10 h, 1050 to 1200°C (800 to 900°C for Pb-containing materials); and (*4*) again intimate mixing or milling (dry or wet) for 1 to 10 h for a further homogenization and for crushing of agglomerates.

Wet Chemical Methods. Preparation by wet chemical methods affords a modern means for achieving high homogenization on the atomic scale during calcination of the different elements that form the perovskite compound. Moreover, these methods include the possibility of producing very pure materials.

Currently several methods for wet chemical preparation of ferroelectric ceramics are used (mostly on laboratory scale):

(*1*) Thermolysis of compounds consisting of the critical elements in an exactly defined stoichiometric ratio in a closely packed arrangement on the atomic scale. Examples are: preparation of $LaFeO_3$ [*12022-43-4*] from $La[Fe(CN)_6]$ (84) or of $BaTiO_3$ from $BaTiO(C_2O_4)_2.4H_2O$ (85).

(*2*) The homogeneous distribution of the different ions in a liquid solution that is nearly ideal, even on an atomic scale, can be transferred into the solid state by proper chemical processes: coprecipitation of the different ions from aqueous solutions (eg, preparation of $Pb(Zr,Ti)O_3$ (86); precipitation from metal alcoholates by hydrolysis (eg, $BaTiO_3$) (87) and $(Pb,La)(Zr,Ti)O_3$ (88) (see Alkoxides, metal); precipitation from aqueous solutions by spraying into a liquid in which there is no solubility (eg, preparation of $BaTiO_3$ by spraying an aqueous Ba–Ti-citrate solution into concentrated ethanol) (89); reaction of an aqueous citrate solution with a polyhydric alcohol (eg, ethylene glycol) to form an organic glass (polyester resin) (90).

The preparation methods according to (*1*) are restricted to the existence of a suitable compound and are not suited for the preparation of mixed crystals or doped materials. The last two methods according to (*2*) seem to be most universally applicable.

The formation of the perovskite compound again occurs by a calcination process that works at much lower temperatures if compared with the conventional mixed-oxide methods.

Shaping. Various amounts of binders (mostly organic compounds) are added to the powders for better rheological properties. "Green" ceramic bodies (density ca 50–60%) can then be easily shaped by a large variety of conventional ceramic processes such as pressing (disks), extrusion (tubes), rolling, and slip casting. Thin sheets (<50 μm) are made by pouring a controlled thickness of special slip onto a polished metal belt, then the slip is dried and stripped.

Sintering. The green ferroelectric ceramics are sintered at high temperatures for 1–10 h to achieve a high density, 95–100% of the theoretical. After sintering, the ceramic body is built up of closely packed grains with randomly distributed crystallographic orientation and with grain diameters in the 1–100 μm range. Proper sintering is most important for the quality of the final products and is generally carried out in electric furnaces for a better control of temperature and atmosphere. $BaTiO_3$-type materials are normally sintered at 1300–1450°C. $Pb(Zr,Ti)O_3$-bodies typically sinter in the 1150–1300°C range. These latter ceramics may lose PbO, which can be prevented by maintaining a partial pressure of this oxide in the ware, eg, by including pellets of $PbZrO_3$ in a refractory enclosure in which the ware is fired.

The whole sintering cycle (rising, holding, cooling time, temperature, atmosphere) determines in a rather complex way the microstructure of the finished ceramics and this, in turn, influences strongly the resulting ferroelectric properties.

For certain applications (eg, electrooptic and SAW (surface acoustic wave) devices) fully dense ceramics without residual pores are necessary. The usual way to achieve such full density is by hot pressing, which means sintering with the application of uniaxial or isostatic pressure (91–92). This can be done by using a high temperature mold of refractory oxides or metal and a source of heat (resistance wire or induction heating). Problems usually arise from the reaction of the powder to be compacted with the refractory oxides or metal of the mold. The temperature of operation is typically about 100°C lower than that required in the absence of pressure. The pressure range is typically about 10–50 MPa (100–500 atm).

Device Fabrication. The geometrical shape of the sintered ceramics corresponds in many cases to the shape of the final products (eg, disks). Frequently, however, where close dimensional tolerances are required, the final shape has to be achieved by cutting, sawing, and lapping with diamond tools that might contribute considerably to the production costs (see Abrasives).

Since the ferroelectric ceramics will be operated as electronic devices, they have to be furnished with electrically conductive electrodes. A series of different techniques for the application of the electrodes are used, all of them having their own special merits and drawbacks: (*1*) screening techniques such as forming a layer of metallic silver (applied in dispersed liquid or paste form and sintered at about 800°C to give a continuous conductive layer intimately bonded to the ceramic surface); (*2*) vacuum evaporation of gold or copper (see Film deposition techniques); (*3*) application of electroless nickel or gold plating (see Electroless plating). If necessary, leads are soldered or clamped to the electrodes (see also Electrical connectors).

After electroding, many devices have to be adjusted to their proper electrical specifications. This may be done (eg, in ceramic capacitors) by sandblasting part of the electrodes to adjust to the proper capacitance value or by reducing the geometrical shape of an electromechanical resonator to achieve the proper resonance frequency.

The last step in the device fabrication is encapsulation (eg, dipping a ceramic capacitor in an organic paste) to protect the device against moisture, dust, or mechanical damage and to prevent electrical breakdown (see Microencapsulation).

Economic Aspects

Table 2 shows the estimated world production in pieces per year and world turnover in dollars per year of ferroelectric ceramics for 1978. The trend is increasing production for all three types of applications.

The prices of the different devices are highly variable and range from the 1 cent region for a simple disk or miniature chip capacitor up near the 1 dollar region for encapsulated large-capacitance multilayer capacitors, PTC degaussing units, or large piezoceramic rings for ultrasound generation.

The overall price of a device depends on the cost of the different production steps: raw materials, powder preparation, preparation of ceramic specimens, and device fabrication. Because of the very different qualities of the finished products, it is impossible to give exact figures for the price fraction of the different production steps. It is, however, obvious that the larger the amount of ceramic material in a ferroelectric device, the higher the price fraction of the first production steps, and the lower the amount of ceramic the higher the price fraction of the last production steps. Therefore, only very rough order-of-magnitude estimates of the price fraction can be made for typical products:

Production stage	*Fraction of total cost, %*
raw materials	0.1–10; typically 1%
powder preparation	0.1–10; typically 2%
specimen preparation	5–50; typically 20%
device fabrication	40–90; typically 75%

Applications

Piezoelectric. Immediately after sintering, ferroelectric ceramics exhibit no piezoelectric effect. To exploit the latent piezoelectric properties of these ceramics, the originally isotropic spread of orientations of the spontaneous polarization within the

Table 2. Estimated World Production of Ferroelectric Ceramics for 1978

	Pieces/y	10^6 \$/y
piezoceramics		50
all ceramic capacitors	24×10^9	750
multilayer capacitors	1.5×10^9	250
PTC ceramics	150×10^6	50

various grains has to be aligned by a so-called poling process. Poling is accomplished by applying a large electrical field (higher than the coercive field E_C) at elevated temperatures (100–200°C) and cooling to room temperature with the field applied. By this means, piezoelectric response in a ceramic (inherently isotropic) material is achieved through the reorientability of P_s, which is characteristic of a ferroelectric.

The piezoelectric activity induced by the poling process is directly proportional to the degree of orientation of the grains which come up to the maximum attainable value (80–90% of the single-crystal value) for $Pb(Zr,Ti)O_3$ materials with rhombohedral structure. For $Pb(Zr,Ti)O_3$ materials with tetragonal structure, ca 50% orientation is characteristic. A large degree of orientation, F, high spontaneous polarization P_s combined with high remanent polarization P_r, $P_r = F{\cdot}P_s$, and high dielectric constant, are the basic requirements for maximum piezoelectric activity (93–94).

In this respect $Pb(Zr,Ti)O_3$ ceramics have much superior properties to the earlier $BaTiO_3$-based piezoceramic materials. However, high piezoelectric properties in the $Pb(Zr,Ti)O_3$ system (expressed by the electromechanical coupling factor k: the square of k can be used as a figure of merit) are found only for compositions near the morphotropic phase boundary (38,95) (Fig. 14). The reason for this is the high dielectric constant ϵ_{33}^T and the relatively large remanent polarization P_r which are attainable in this composition region.

Moreover, $Pb(Zr,Ti)O_3$ ceramics show the big advantage that their material parameters can be optimized for specific application by a controlled doping, and currently only such doped ceramics are produced. There exists a multitude of different doping ions that can be classified into three groups:

(*1*) Donor dopants, eg, La, Nb, Sb, or W, which are incorporated at a lattice site

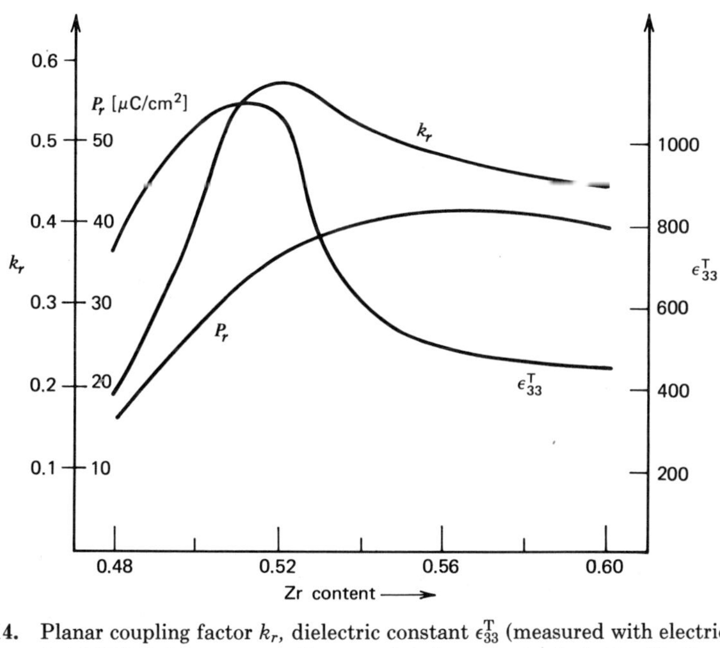

Figure 14. Planar coupling factor k_r, dielectric constant ϵ_{33}^T (measured with electrical field along poling direction and at 16 kHz below resonance frequency) and remanent polarization P_r of undoped ceramic $Pb(Zr_xTi_{1-x})O_3$ samples.

of lower valency (eg, La^{3+} at Pb^{2+} sites). These donor dopings lead to the so-called soft materials exhibiting increased coupling factors (up to 0.7) and dielectric constants (up to 3000) but also increased electrical and mechanical losses (94,96).

(*2*) The second group of doping ions is formed by the transition metals, eg, Fe, Mn, Ni, Co, which are incorporated at a lattice site of higher valency (eg, Fe^{3+} at Ti^{4+} sites), thus showing acceptor character. These acceptor dopants have an effect just opposite to the above donors. They reduce the dielectric constant, the coupling factor and also reduce the losses and improve the aging properties. Moreover, they increase the stability of the ceramics with respect to electrical or mechanical depolarization. Therefore, these acceptor-doped ceramics are called hard materials (94,96–100).

(*3*) On transgressing the solubility limit of the doping species, eg, at high concentrations of Mn, or if one dopes with ions that barely dissolve into the perovskite lattice, eg, Al, then a second phase is formed at the grain boundaries and this, by means of space-charge effects, also increases the stability of the ceramics (101–103).

Piezoelectric ceramics (piezoceramics) are used as piezoelectric devices in a multitude of compositions and geometrical shapes for a large variety of different applications (8,83,104–105).

As transducers, piezoceramics become increasingly useful at higher frequencies, and therefore, they are applied in the field of acoustics and especially in the field of ultrasonics (qv), transforming electrical into mechanical energy or vice versa. They are well adapted to receive or transmit elastic energy into liquids (sonar and medical applications, ultrasonic cleaning) or solids (ultrasonic welding, soldering, delay lines) that are comparable in density to the piezoelectric body, but they require mechanical transformer systems such as horns to match well into low-density media like air. For all transducer-type applications, a hard acceptor-doped (mostly Fe) $Pb(Zr,Ti)O_3$ is used which can be operated up to high power levels at elevated temperatures without depolarization.

Acceptor-doped hard solid solutions of $PbTiO_3$–$PbZrO_3$-complex perovskite (eg, $Pb(Mn_{1/3}Nb_{2/3})O_3$ are used mostly for ignition applications. Large stresses applied in the poling direction of piezoelectric slugs generate sparks in the voltage range up to 20 kV/cm for ignition of natural gas in space heaters, cooking stoves, or in cigarette lighters. Owing to the acceptor dopant, the stresses (impact and squeeze systems are used) can be applied thousands of times without serious degradation in the form of depolarization (106).

Piezoceramics also have major applications as resonators in devices determining or limiting the operating frequency of electrical networks. Contrary to transducer or ignition applications, resonators are operated at low driving levels. The problems in this area involve low mechanical losses (equivalent to high mechanical quality factors) and low temperature and time dependence (aging) of the resonance frequency. Optimum properties are obtained with Mn-doped $PbTiO_3$–$PbZrO_3$-complex perovskite (eg, $Pb(Zn_{1/3}Nb_{2/3})O_3$ ceramics of tetragonal crystal structure).

The ceramic resonators are frequently in the form of bandpass filters having high bandwidths of up to 10% of the central frequency and are possible because of the large coupling factors of the materials. Today they are widely used in large quantities in the intermediate frequency stage of AM and FM radio receivers (452 kHz and 10 MHz) (107).

The direct piezoelectric effect is exploited in phonograph pickup elements where a mechanical force is converted to an electrical charge. Highest conversion ratios are obtained with donor-doped soft materials showing high coupling factors and high dielectric constants. In the same type of materials, the inverse piezoelectric effect is utilized for "actuators" where an electrical field is converted into a change of the geometrical dimensions (8).

Dielectrics. Owing to their ferroelectric nature, $BaTiO_3$-based ceramics show high dielectric constants and therefore these materials are suited for and widely used in capacitor applications. The biggest advantage of the ceramic capacitor is compactness owing to the high capacitance per unit volume.

However, like single crystals (See Fig. 5), the dielectric constant of $BaTiO_3$ ceramics has a strong temperature dependence and a pronounced maximum at the Curie point (Fig. 15). By a partial substitution of Ba by Sr or Ca or of Ti by Zr or Sn (Fig. 6), the Curie point can be shifted to room temperature, resulting in materials with a maximum of the dielectric constant of 10,000 to 15,000 at this temperature (Fig. 15). Unfortunately, however, these simple materials suffer from high dielectric losses and show large temperature and field dependences, which worsen for high dielectric constant compositions. These unfavorable properties can be suppressed to a certain degree by: (*1*) optimum doping; (*2*) proper mixed crystal formation; and (*3*) exact control of the microstructure.

By doping with transition metal ions such as Fe or Mn, it is possible to stabilize the domain walls and to reduce the dielectric losses (108). Besides a shift of the Curie point to desired temperature regions, mixed crystal formation broadens the peak of the dielectric constant (Fig. 15), and thus reduces the strong temperature dependence (109–110).

As demonstrated in Figure 15, a change of the microstructure is an important means of influencing and controlling the electrical characteristics of a ferroelectric ceramic. Owing to the difference in microstructure, dielectric constants of about 1500

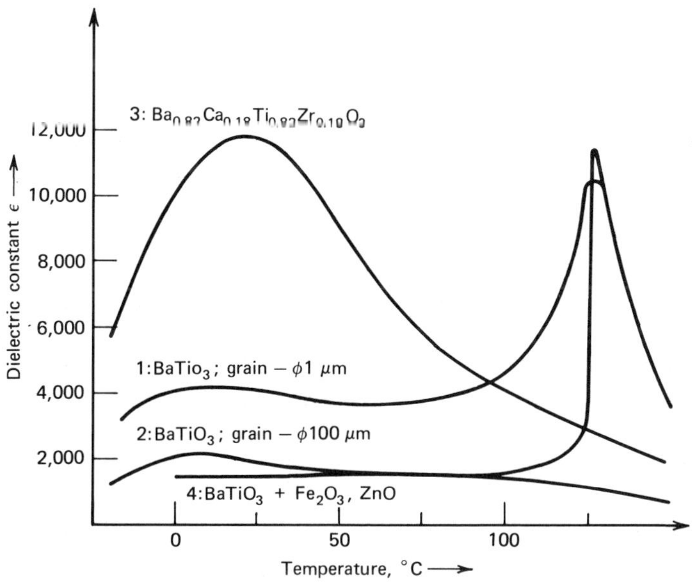

Figure 15. Temperature of dependence the dielectric constant of different $BaTiO_3$-based ceramics for capacitor applications. φ = mean grain diameter.

are found for coarse-grained (mean grain diameter 100 μm) and of about 3500 for fine-grained (mean grain diameter 1 μm) ceramic specimens. Moreover, the Curie peak of fine-grained specimens is broadened with respect to the coarse-grained sample (111).

It is quite likely that optimum improvements in the dielectric characteristics can be achieved by combining all three effects. By adding relatively large amounts of Fe_2O_3 and ZnO (112) to $BaTiO_3$ (resulting in extremely fine-grained doped ceramic materials and with second phases finely dispersed at the grain boundaries), the Curie peak can be fully suppressed (Fig. 15).

Extremely high dielectric constants can be obtained if the so-called intergranular layer capacitors are used. They consist of semiconducting grains divided by high-resistance barrier layers at each grain boundary. In that case the electric field concentrates in the barrier layers and the true thickness of the dielectric consists only of the sum of these high-resistance barrier layers, resulting in effective dielectric constants up to 100,000. There are, however, large losses and strong temperature and field dependences (113–114).

From these various materials ceramic capacitors are produced ranging from the pF to the μF region in a multitude of geometrical shapes and different applications (83). The most important are disk capacitors, tube capacitors, miniature chip capacitors, and multilayer capacitors.

Disk capacitors are generally produced by pressing the powder together with some binder in a disk-shaped die. Depending on operating voltage and capacitance value, the geometrical shape of the disk generally varies between 0.5 mm and 4 mm in thicknesses and between 4 and 10 mm in diameter. Since it is difficult to produce disk capacitors having a thickness below 0.5 mm, their capacitance values are relatively low, but they are suitable for large operating voltages ($\geq$ 250 V). Screen-printed silver normally serves as electrode material.

Tube capacitors are produced by extrusion. The geometrical dimensions range from 2 to 4 mm in diameter and from 8 to ca 40 mm in length. Normally, the electrodes are screen-printed silver.

The advantage of miniature chip capacitors is their low thickness (0.1–0.3 mm) resulting in a high capacitance per surface area at reduced operating voltages (typically 60 V). The thin ceramic sheet of this capacitor type is generally produced by rolling techniques from which rectangular chips are punched or cut out. Evaporated copper is often used as electrodes.

During recent years the multilayer capacitor has gained increasing importance (115). With this capacitor type, high capacitances per unit volume can be achieved by piling up several thin (20–50 μm) ceramic foils. These foils are produced by casting techniques. They are provided with screened metal electrodes in the "green" unsintered state, then laminated, pressed, and sintered together with the metal electrodes (Fig. 16). Because the dielectric materials require sintering temperatures of about 1300°C and atmospheres of oxygen or air to attain good dielectric properties, noble metal electrodes, mainly palladium, have to be used. The capacitance range of multilayer capacitors is 100 nF–1 μF.

Semiconducting—Positive Temperature Coefficient (PTC) Materials. By substituting some 0.1–0.4 mol % of Ba^{2+} ions by La^{3+} ions, or Ti^{4+} ions by Sb^{5+} or Nb^{5+} ions, $BaTiO_3$ can be transferred into an n-type semiconductor with a resistivity of about 10 Ω·cm at room temperature. However, as shown in Fig. 17, the resistivity rises very

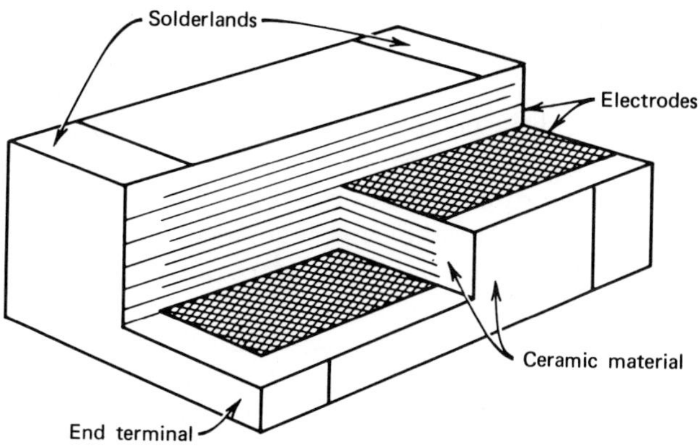

Figure 16. Schematic representation of a multilayer capacitor.

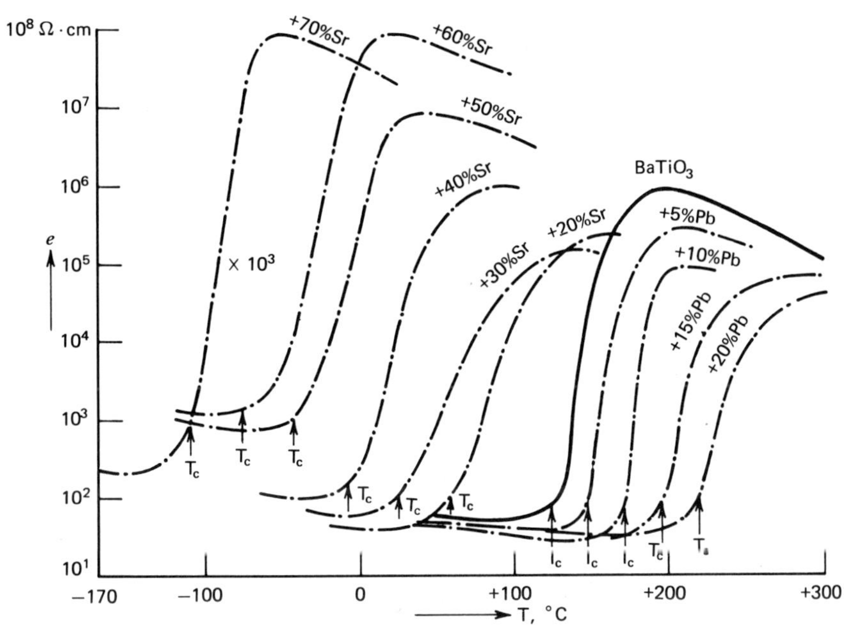

Figure 17. Temperature dependence of the resistivity of semiconducting $BaTiO_3$ and of mixed crystals of $(Ba,Sr)TiO_3$ and $(Ba,Pb)TiO_3$ after doping with 0.3 mol % Sb. The arrows mark the Curie point T_c of the respective compositions.

rapidly (up to 60%/°C) just at the Curie point T_c, thus forming a resistor material with a large positive temperature coefficient (PTC effect) (116). The range of the PTC effect can be continuously shifted to the desired temperature region by mixed crystal formation (see Figs. 6 and 17) (117).

This rather uncommon PTC behavior is caused by barrier layers at the grain boundaries. They originate from defects with acceptor character at the grain boundaries which trap conduction electrons and thus form a depletion layer with a high-resistance potential barrier (118–120). The height of the potential barrier is governed by the ferroelectric properties of $BaTiO_3$. Below T_c the spontaneous polarization P_s compensates partially or fully for the defects (depending on their concentration) and removes the depletion layer, so that the potential barrier is not fully effective, resulting in high conductivity. Above T_c, where no spontaneous polarization exists, the depletion layer and the potential barrier at the grain boundary become fully effective. The height of the potential barrier is dependent on the reciprocal of the dielectric constant which, according to the Curie-Weiss law, falls with increasing temperature. Consequently, the potential barrier increases with rising temperature and the resistivity, owing to the exponential dependence of the resistivity on the barrier height, increases by several orders of magnitude.

PTC ceramics are normally produced in disks or rectangular bars which are supplied with low-resistance electrodes with or without soldered beam leads. The devices are used in a wide variety of controlling or measuring applications such as current limiters, degaussing units in color television sets, self-controlling heating elements, for overload protection or for temperature measurement and control.

BIBLIOGRAPHY

"Ferroelectrics" in *ECT* 2nd ed., Vol. 9, pp. 1–25, by Edward C. Henry, General Electric Company.

1. J. Valasek, *Phys. Rev.* **17,** 475 (1921).
2. R. E. Newnham and L. E. Cross, *Mater. Res. Bull.* **9,** 927 (1974).
3. *Ibid.,* 1021 (1974).
4. J. F. Nye, *Physical Properties of Crystals,* Clarendon Press, Oxford, 1957.
5. Landolt-Börnstein, *Ferroelectric and Antiferroelectric Substances,* Vol. 3, New Series Group III, Springer Verlag, Berlin, 1969.
6. Landolt-Börnstein, *Ferroelectric and Antiferroelectric Substances,* Vol. 9, New Series Group III, Springer Verlag, Berlin, 1975
7. A. Glass and M. E. Lines, *Principles and Applications of Ferroelectrics and Related Materials,* Clarendon Press, Oxford, 1977.
8. B. Jaffe, W. R. Cooke, Jr., and H. Jaffe, *Piezoelectric Ceramics,* Academic Press, New York, 1971.
9. K. Aizu, *J. Phys. Soc. Jpn.* **20,** 959 (1965); *Phys. Rev.* **146,** 423 (1966).
10. L. A. Shuvalov, *J. Phys. Soc. Jpn.* **28S,** 38 (1970).
11. K. Aizu, *J. Phys. Soc. Jpn.* **27,** 387 (1969).
12. K. Aizu, *J. Phys. Soc. Jpn.* **28,** 706 (1970).
13. K. Aizu, *Phys. Rev.* **B2,** 754 (1970).
14. K. Aizu, *J. Phys. Soc. Jpn.* **32,** 1287 (1972).
15. K. Aizu, *J. Phys. Soc. Jpn.* **34,** 121 (1973).
16. H. L. Borchardt and P. E. Bierstedt, *J. Appl. Phys.* **38,** 2057 (1967).
17. L. E. Cross, A. Fouskova, and S. E. Cummins, *Phys. Rev. Lett.* **21,** 812 (1968).
18. V. Dvorak, *Phys. Status Solidi* **46B,** 763 (1971).
19. V. Dvorak, *Phys. Status Solidi* **52B,** 93 (1972).
20. B. Singer, "Theory and Performance Characteristics of Pyroelectric Imaging Tubes" in *Advances in Image Pickup and Display,* Vol. 13, Academic Press, London, 1977.
21. J. Fousek and V. Janovec, *J. Appl. Phys.* **40,** 135 (1969).
22. J. Fousek, *Czech. J. Phys.* **21,** 955 (1971).
23. V. A. Zhirnov, *Sov. Phys. JETP* **8,** 882 (1959).
24. W. Kinase and H. Takahashi, *J. Phys. Soc. Jpn.* **12,** 464 (1957).
25. T. Yatohashi, G. Honjo, and M. Tanaka, *J. Phys. Soc. Jpn.* **28,** 386 (1970).
26. M. Tanaka and G. Honjo, *J. Phys. Soc. Jpn.* **19,** 954 (1964).
27. W. J. Merz, *Phys. Rev.* **95,** 690 (1954).
28. E. Fatuzzo and W. J. Merz, *Phys. Rev.* **116,** 61 (1959).
29. E. Fatuzzo and W. J. Merz, *Ferroelectricity,* North Holland Press, Amsterdam, 1967, p. 197.
30. I. Hatta and A. Sawada, *Jpn. J. Appl. Phys.* **4,** 389 (1965).
31. R. C. Miller and A. J. Savage, *J. Appl. Phys.* **31,** 632 (1960).
32. R. Landauer, *J. Appl. Phys.* **28,** 227 (1957).

33. R. C. Miller and G. Weinreich, *Phys. Rev.* **117,** 1460 (1960).
34. H. L. Stadler and P. J. Zachmanidis, *J. Appl. Phys.* **34,** 3255 (1963).
35. M. Hayashi, *J. Phys. Soc. Jpn.* **33,** 616 (1972).
36. F. Jona and G. Shirane, *Ferroelectric Crystals,* Pergamon Press, New York, 1962, p. 108.
37. E. Sawaguchi, *J. Phys. Soc. Jpn.* **8,** 615 (1953).
38. B. Jaffe, R. S. Roth, and S. Marzullo, *J. Res. Nat. Bur. Stand.* **55,** 239 (1955).
39. H. Ouchi, K. Nagano, and S. Hayakawa, *J. Am. Ceram. Soc.* **48,** 630 (1965).
40. S. C. Abrahams and co-workers, *J. Phys. Chem. Solids* **34,** 521 (1973).
41. S. C. Abrahams, W. C. Hamilton, and A. Sequeira, *J. Phys. Chem. Solids* **28,** 1693 (1967).
42. K. Nassau, *J. Phys. Chem. Solids* **27,** 983 (1966).
43. I. Camlibel, *J. Appl. Phys.* **40,** 1690 (1969).
44. A. J. Slabodnik, Jr. *Proc. IEEE* **64,** 581 (1976).
45. I. P. Kaminow, *Ferroelectrics* **10,** 247 (1976).
46. S. T. Liu, *Ferroelectrics* **10,** 83 (1976).
47. F. W. Ainger, *Ferroelectrics* **10,** 75 (1976).
48. P. B. Jamieson, S. C. Abrahams, and J. L. Bernstein, *J. Chem. Phys.* **48,** 5048 (1968).
49. L. G. VanUitert and co-workers, *Mater. Res. Bull.* **3,** 47 (1968).
50. S. C. Abrahams, P. B. Jamieson, and J. L. Bernstein, *J. Chem. Phys.* **54,** 2355 (1971).
51. B. A. Scott and co-workers, *Mater. Res. Bull.* **5,** 47 (1970).
52. G. Goodman, *J. Am. Ceram. Soc.* **36,** 368 (1953).
53. G. A. Smolensky and A. I. Agranovskaya, *Dokl. Akad. Nauk SSSR* **97,** 237 (1954).
54. M. B. Francombe, *Acta Crystallogr.* **13,** 131 (1960).
55. U.S. Pat. 2,805,165 (Apr. 25, 1955), G. Goodman (to General Electric Company).
56. B. Lewis and L. A. Thomas, *Proc. Int. Conf. Solid State Electronics Brussels* **4,** 883 (1960).
57. A. M. Glass, *J. Appl. Phys.* **40,** 4699 (1969).
58. R. B. Maciolek and S. T. Liu, *J. Electron. Mater.* **2,** 191 (1973).
59. J. E. Geusic and co-workers, *Appl. Phys. Lett,* **11,** 269 (1968).
60. R. G. Smith, *J. Appl. Phys.* **39,** 4030 (1968).
61. R. E. Newnham, R. W. Wolfe, and J. F. Dorrian, *Mater. Res. Bull.* **6,** 1029 (1971)
62. R. E. Newnham and R. W. Wolfe, *J. Electrochem. Soc.* **116,** 832 (1969).
63. R. A. Armstrong and R. E. Newnham, *Mater. Res. Bull.* **7,** 1025 (1972).
64. L. E. Cross and R. Pohanka, *Mater. Res. Bull.* **6,** 939 (1971).
65. S. E. Cummins and L. E. Cross, *J. Appl. Phys.* **39,** 3268 (1968).
66. S. E. Cummins and T. E. Luke, *Ferroelectrics* **3,** 125 (1972).
67. S. E. Cummins, *J. Appl. Phys,* **37,** 2510 (1966).
68. S. E. Cummins and T. E. Luke, *IEEE Trans. Electron Devices* **ED18,** 761 (1971).
69. S. E. Cummins and B. H. Hill, *Proc. IEEE* **58,** 938 (1970).
70. S. E. Kenneman and co-workers, *Appl. Phys. Lett.* **17,** 173 (1970).
71. W. R. Cook, Jr. and H. Jaffe, *Phys. Rev.* **88,** 1426 (1952).
72. V. A. Isupov, *Izv. Akad. Nauk SSSR, Ser. Fiz.* **21,** 402 (1957).
73. S. Nanamatsu, M. Kimura, and T. Kawamura, *J. Phys. Soc. Jpn.* **38,** 817 (1975).
74. G. Busch and P. Scherrer, *Naturwissenschaften* **23,** 737 (1935).
75. G. Busch, *Helv. Phys. Acta* **11,** 269 (1938).
76. J. West, *Z. Krist.* **74,** 306 (1930).
77. W. Baulte, *Helv. Phys. Acta* **15,** 373 (1942).
78. G. Morie, *Ferroelectrics* **10,** 9 (1976).
79. B. T. Matthias, C. E. Miller, and J. P. Remeika, *Phys. Rev.* **104,** 849 (1956).
80. S. Triebwasser, *IBM J. Res. Dev.* **2,** 212 (1958).
81. S. Hoshino and co-workers, *Phys. Rev.* **107,** 1255 (1957).
82. R. Pepinsky, Y. Okaya, and F. Jona, *Bull. Am. Phys. Soc.* **2,** 220 (1957).
83. A. Bauer and co-workers, *Technologie und Anwendungen von Ferroelektrika,* Academische Verlagsgesellschaft Geest & Portig K.-G., Leipzig, 1976.
84. P. K. Gallagher, *Mater. Res. Bull.* **3,** 225 (1968).
85. W. S. Clabaugh, E. M. Swiggard, and R. Gilchrist, *J. Res. NBS, Res. Paper* **2677,** 289 (1956).
86. V. M. McNamara, *J. Can. Ceram. Soc.* **34,** 103 (1965).
87. K. S. Mazdiyasni, R. T. Dollot, and J. S. Smith II, *J. Am. Ceram. Soc.* **52,** 523 (1969).
88. G. H. Haertling, *J. Am. Ceram. Soc.* **54,** 303 (1971).
89. B. I. Mulder, *Am. Ceram. Soc. Bull.* **49,** 990 (1970).

90. U.S. Pat, 3,330,697 (Aug. 26, 1963), M. P. Pechini (to Sprague Electric Company).
91. G. Haertling, *Am. Ceram. Soc. Bull.* **43,** 875 (1964).
92. K. H. Härdtl, *Am. Ceram. Soc. Bull.* **54,** 201 (1975).
93. K. Carl and K. H. Härdtl, *Phys. Status Solidi A* **8,** 87 (1971).
94. K. H. Härdtl, *Ferroelectrics* **12,** 9 (1976).
95. D. Berlincourt, C. Cmolik, and H. Jaffe, *Proc. IRE* **48,** 220 (1960).
96. F. Kulcsar, *J. Am. Ceram. Soc.* **42,** 343 (1959).
97. H. Thomann, *Z. Angew. Phys.* **20,** 554 (1966).
98. T. B. Weston, A. H. Webster, and V. M. McNamara, *J. Am. Ceram. Soc.* **52,** 253 (1969).
99. N. Uchida and T. Ikeda, *Jpn. J. Appl. Phys.* **6,** 1292 (1967).
100. K. Carl and K. H. Härdtl, *Ferroelectrics* **17,** 473 (1978).
101. K. Okazaki and K. Sakata, *Electrotech. J. Jpn.* **7,** 13 (1962).
102. S. Takahashi and M. Takahashi, *Jpn. J. Appl. Phys.* **11,** 31 (1971).
103. H. Thomann, *Ferroelectrics* **4,** 141 (1972).
104. H. Jaffe, *Proceedings on Electron Devices, IEEE* **ED-16,** 557 (1969).
105. W. P. Mason, ed., *Handbook on Physical Acoustics,* Vol. 1, Part A, Academic Press, New York, 1964, Chapt. III.
106. Y. Kodama and M. Osada, *Sumitomo Electric. Techn. Rev.* **10,** 102 (1967).
107. H. H. Schüssler, *Proceedings on Sonics and Ultrasonics, IEEE* **SU-21,** 257 (1974).
108. H.-J. Hagemann, *J. Phys. C* **11,** 3333 (1978).
109. V. V. Kirilov and V. A. Isupov, *Ferroelectrics* **5,** 3 (1975).
110. A. J. Burggraaf and K. Keizer, *Mater. Res. Bull.* **10,** 521 (1975).
111. W. R. Buessem, L. E. Cross, and A. K. Goswami, *J. Am. Ceram. Soc.* **49,** 33 (1966).
112. H. Thomann and H. Heydrich, *Ferroelectrics* **7,** 357 (1974).
113. S. Waku, *Rev. Electr. Comm. Lab.* **5,** 689 (1967).
114. H. Brauer, *Z. Angew. Phys.* **29,** 282 (1970).
115. V. F. Capozzi, *Multilayer Ceramic Capacitor Materials and Manufacture,* Oxy Metal Industries Corp., Nutley, N.J., 1975.
116. Ger. Pat. 929,350 (1955), P. W. Haayman, R. W. Dam, and H. A. Klasens.
117. E. Andrich and K. H. Härdtl, *Philips Tech. Rev.* **26,** 119 (1965).
118. W. Heywang, *Solid State Electron.* **3,** 51 (1961).
119. G. H. Jonker, *Solid State Electron.* **7,** 895 (1964).
120. J. Daniels and co-workers, *Philips Res. Rep.* **31,** 487 (1976).

L. E. CROSS
The Pennsylvania State University

K. H. HÄRDTL
Philips Forschungslaboratorium Aachen GmbH

FERTILIZERS

Vegetation, like all living things, requires certain foods for its survival and growth. Fertilizers are materials added to the soil, and sometimes to foliage, to supply nutrients to sustain plants and promote their abundant and fruitful growth. The elements that constitute these plant foods are divided into three classes: (*1*) primary—nitrogen (N), phosphorus (usually expressed as P_2O_5), and potassium (expressed as K_2O); (*2*) secondary—calcium (Ca), magnesium (Mg), and sulfur (S); and (*3*) minor or so-called micronutrients—iron (Fe), manganese (Mn), copper (Cu), zinc (Zn), boron (B), and molybdenum (Mo). In addition to their role as nutrients, calcium and magnesium are important in adjusting the pH and tilth of the soil.

All the nutrient elements are present to some extent in soil or other natural materials in forms useful for plants but, with the frequent exception of micronutrients, such supply is generally not adequate for sustained and economic production of crops. Fertilizer is, therefore, one of the most important products of the chemical industry.

Although the beneficial effects of plant nutrient elements have long been known, the mechanisms of their function in plant physiology are not fully understood, and the plant physiology of crop nutrition is outside the scope of this article. Reference 1 contains details on functions of the micronutrients. The historical and other general aspects of fertilizers are not included here because they have been published frequently. Emphasis in this article is on the advances in and nature of fertilizer technology since the mid 1960s, placing in perspective questions of the food-population linkage vs fertilizer raw material resources, the energy requirements for fertilizer production, and the environmental restraints to production.

During the decade 1967–1977 fertilizer production technology continued to advance rapidly. Increasing recognition of population growth and the important role of fertilizer in food production were major contributors to this trend. Means for greatly increasing the use of fertilizers in developing countries where population growth rates are highest is receiving intensive study (2–5). The long-range outlook for fertilizer demand is clearly one of continued growth.

The Growing Demand for Fertilizer. Experts have used a variety of methods for estimating the future demand for fertilizer. The quantities forecast differ but there is agreement that the need for food and, therefore, the demand for fertilizer will continue to grow in the foreseeable future. Population growth will be the dominant cause for increased demand on a world basis but another important cause is the need to improve the food status of the world, especially in many developing nations. The food–fertilizer–population linkage is the important underlying basis supporting demand. For this reason a brief look at population growth will help to understand the expected large demand for fertilizer in the future.

United Nations data for world population growth project populations of 4.2 and 6.3 billion for the years 1980 and 2000, respectively (3), a 50% increase over that 20-yr period. Considering the need to improve the food status of populations, an estimate of double the 1975 food needs by the year 2000 may not be too high. Assuming a direct proportion between expected increased food production and fertilizer demand leads to a probable demand approaching 200×10^6 metric tons of fertilizer nutrients N, P_2O_5, and K_2O by the year 2000. Even greater demand has been projected by some (4).

TVA estimates for world consumption of N, P_2O_5, and K_2O for 1980 are tabulated below in 10^6 t.

	N	*P_2O_5*	*K_2O*	*Total*
low	53.1	28.3	23.5	104.9
high	60.8	34.1	27.8	122.7
average				114.0

The Fertilizer Institute estimates United States use of fertilizer in 1979–1980 to be at least 48×10^6 t (6).

It now appears that the TVA projections for 1980 will be met and come close to meeting projected world needs (7) at an estimated investment cost of some 25 billion dollars (U.S.) between 1974 and 1980. Projected demand for fertilizer in subsequent decades raises a serious question whether fertilizer production can keep pace unless population growth is curbed.

Recycled Nutrients. Potentially recyclable nutrients are largely in the form of the organic materials—night soil, animal manure, sewage sludge, composts, slaughter house wastes, and crop residues. There are some who believe that only naturally derived organic material should be used as fertilizer. Organic fertilizers do, indeed, have valuable properties as soil amendments. In particular, their humus content enhances the efficiency of mineral fertilizers, improves soil structure, increases water retention by soil, and decreases soil erosion.

Generally, recyclable organic fertilizers have very low nutrient content. The nutrient content of fresh manures of cattle, swine, and poultry ranges from 0.5–3% N, 0.25–3% P_2O_5, and 0.5–1.5% K_2O. Nutrient content of typical municipal compost is 1.5–2% N, 1–2.5% P_2O_5, and 0.1–3% K_2O. Some of these materials are good sources of micronutrients but some may also contain high levels of potentially dangerous heavy metals such as cadmium, lead, zinc, and chromium—especially at high soil application of ≥ 10 t/hm^2.

The amounts of nutrients potentially available for recycling worldwide are difficult to estimate but they certainly are a substantial fraction of the requirements for fertilizers. For the United States alone the sum of the N, P_2O_5, and K_2O in the excrement from humans, cattle, swine, and poultry amounts to approximately some 40% of the fertilizer nutrients used annually. Addition of crop residues brings the total to about one half of the fertilizer nutrients used. Nationally, and on a world basis, the actual amounts currently recycled in agriculture is only a very small fraction of that potential. Recycling has been practiced in some developing nations for centuries. FAO (8) estimates that at present organic waste constitutes only about 10% of all fertilizer used, and that this might potentially be raised to 25%.

There are serious problems in using all animal wastes in the United States, especially cattle manure. These wastes are most often concentrated in areas such as feed lots, creating two serious problems: the prohibitive cost of transporting such low-analysis fertilizer over long distances, and the pollution of runoff and ground water. Moreover, in areas of intensive agriculture the nitrogen requirements for crops may be ten times the amount available in the quantity of manure the soil can tolerate.

It appears that any increase in the contribution of recycled nutrients to the demand for fertilizer in the United States will be small over the next decade. For reasons of conservation and of improved environmental quality, every effort must be made to enhance the technology for recycling nutrients and to improve the economics of its practice. Both are essential for any substantial increase in use of recycled nutrients in agriculture.

Effectiveness of Fertilizers. Considering the magnitude of the projected demand for fertilizer, and the parallel demands on capital, raw materials, and energy, it is unfortunate that crops do not utilize applied fertilizer very efficiently. Nitrogen utilization by the first crop following application ranges from a low of about 30% for vegetables to a high of about 60% for grain crops. On average, no more than one half of applied fertilizer nitrogen is used by crops.

Most nitrogen fertilizer is water soluble and, in time, much of it can be leached from most soils, especially during nongrowing seasons. Moreover, because of nitrification and denitrification reactions in the soil, oxides of nitrogen and elemental nitrogen are volatilized from the soil. Usually very little of the fertilizer nitrogen applied in one season remains in the soil for use by later crops.

Water-soluble phosphates are quickly converted in the soil to less soluble forms. These materials are not leached from most soils but because of their lower solubility their availability to plants is lower also. Phosphate utilization is a function of the nutrient status of the soil, the crop, the weather, and other factors. Its uptake by the first crop following its application ranges from about 6 to 30%. However, most of the remaining phosphate stays in the soil and can be used by future crops but at a reduced rate. When the phosphate status of the soil is very high, as in some soils of the United States and western Europe, fair crop response can be obtained for periods of two or more years without annual dressings of phosphate. Even so, for most situations of intensive cropping, good yields require frequent or annual applications of phosphate fertilizer.

Potash fertilizers usually are water-soluble chloride, sulfate, or phosphate. These salts undergo cation exchange with soil minerals and usually are not seriously leached. The surface layers of potassium are readily extractable by plants but with time the potassium becomes increasingly imbedded in the crystal lattice of the clay mineral and becomes inaccessible for plants (9). Average soils in temperate regions contain as much as 30,000 kg/hm^2 of K in the 0–30-cm depth but 90–95% of this soil potassium is so firmly imbedded that it is of little value for agriculture on a practical time scale (9). As with phosphate, the potassium status of most soils is highly dependent on the recent agricultural use of the land.

In planning fertilizer use, farmers should be guided by the soil fertility status, a condition measurable by chemical analysis of the soil. The agricultural advisory service for a specific area recommends fertilizer practices for the crop to be planted. The effectiveness of increased use of fertilizer to increase crop production is demonstrated by much experience. Nearly one half of the increased crop production in the

United States since 1940 is credited to the increased use of fertilizer. The level of response to applied fertilizer is an inverse function of the nutrient status of the soil and also depends upon other agricultural practices. For example, in the United States experience for the period 1940–1955, 58% of increased crop production was attributable to fertilizer, and for the period 1955–1964 (a period following increased fertilizer use) the amount attributable to fertilizer was 36% (10).

Achieving the crop production potential depends not only on fertilizer but also on the crop variety, adequate soil moisture, control of insects and disease, and generally improved practices. Experience shows, however, that increased use of fertilizer is the main single factor in rapid increases in crop yield.

Another factor that affects the amount of fertilizer used is its financial return for the individual farmer. This is determined by a complex array of factors too extensive and complex for evaluation here but widespread experience in some of the developed nations has been that the net return per dollar of fertilizer investment ranges from about $1.50–10.00. TVA calculated the average in the United States for the period 1940–1974 to be $7.03 (10). A more detailed analysis of this aspect of the subject can be found in *The United Nations Fertilizer Manual* (3).

Structure and Control of the Industry

The organization of the fertilizer industry varies around the world. Some of the factors affecting the industry organization in a country are its size, national agricultural needs, the nature of the local agriculture, the availability of raw materials, and political considerations. As the United States is a major and growing factor in world fertilizer and food production, and information concerning the United States industry is accessible, this discussion is limited to the United States industry.

Structure. In 1977 the United States fertilizer industry furnished domestic consumers ca 46×10^6 metric tons (gross weight) of fertilizer. The industry is made up of some 90 producers of the basic materials, ammonia, phosphoric acid, and potash; and ca 12,000 mixers and retail distributors, of which 90% are bulk-blend, fluid-mix, and granulation plants (11). This is a dramatic change from the traditional fertilizer marketing pattern of 20 years ago when the role of basic producers was to supply intermediate products to mixing plants. These were then owned and operated by wholesale distributor-mixers who combined the basic materials into fertilizer mixtures that were distributed to retail outlets for sale to the farmer. Today, the principal marketing pattern consists of the basic producers distributing to large-scale ammoniation plants, smaller dry-bulk blenders, and liquid- and suspension-mix plants.

Two segments of the fertilizer industry at the local level pertain almost exclusively to the small business sector and account for a very high portion of the total fertilizer delivered and applied to farms. These are the bulk-blending and mixed-fluid fertilizer retail outlets. There are approximately 5400 bulk-blending plants and 2500 fluid-mixing plants throughout the United States. At the local level these firms make the price of fertilizer products highly competitive. Much of the fertilizers marketed through these outlets are either custom applied or are applied by the farmer, using equipment made available by the dealer.

Regionally, 59% of fertilizer plants are in the north central states, where 46% of all fertilizer materials was distributed in fiscal year 1976.

Private corporations own the largest percentage of the domestic fertilizer production capacity. The large corporations usually are engaged in other enterprises so that fertilizer sometimes represents a small but significant portion of total sales.

Farmer cooperatives play an increasingly important role in all phases in the fertilizer production and marketing scheme. From 1950–1970 the market share of cooperatives grew 15–32%. As of early 1976, cooperatives operated 18 anhydrous ammonia plants, representing ca 20% of United States capacity. They operated 8 wet-process phosphoric acid plants having ca 25% of United States capacity, 1 potash mine having 3%, urea plants of 28%, 8 ammonium nitrate plants having 17%, 11 ammonium phosphate plants of 28%, and 3 concentrated superphosphate plants having 30% of United States capacity.

During the early period of production of synthetic ammonia, production capacity was concentrated mainly within a few corporations. There were seven anhydrous ammonia producers in 1940 with a production capacity of 4.31×10^5 t, and the four largest firms had 96% of the total production capacity. Production capacity in 1951–1975 increased from less than 1.8×10^6 t to nearly 18×10^6 t, and the share of the top eight firms declined from 90 to 40%.

The phosphate segment of the fertilizer industry consists of the producers of phosphate rock, phosphoric acid, phosphatic fertilizers, and of sulfuric acid. Concentration of ownership of rock production capacity has fluctuated from over 100 companies in the late 1800s to only 11 in 1913. Currently, 22 firms operate 29 mines with a combined annual capacity of about 55×10^6 t of rock. The 10 largest firms have over 80% of the industry's total capacity.

There has been an increased tendency to vertical integration in the phosphate industry. At present, 11 firms produce both phosphoric acid and phosphate rock, 10 produce rock and concentrated superphosphate, and 13 produce rock and ammonium phosphate.

Potash production in the United States has declined since 1966 as imports from Canada have begun to satisfy United States demand. Rich deposits in Saskatchewan, Canada, were developed in the late 1960s and subsequently have become the principal source of United States potash. In 1971 Canadian mines supplied most of the United States potash imports and 70% of the potash used. Currently, imports meet over 75% of domestic demand. In 1976 15 potash firms operated in the United States and Canada with a capacity of about 10×10^6 t K_2O.

Price and Profit. During the early 1960s and through 1965, the average fertilizer industry net profit before interest charges and taxes was just over 10% of gross sales. Beginning in 1966 profits declined. By 1968, when prices of fertilizers were very low, profits vanished for three years and remained below the average for all manufacturing corporations until 1974. During the 1968–1970 period of extensive oversupply, the average fertilizer firm showed net losses as high as 8.5% in 1969. Lack of profits discouraged additional investment; in fact, several production units were closed. This curtailment of new investment and plant closures eventually resulted in the United States having an inadequate capability to supply farmers' needs for fertilizers in 1973 and 1974. The price of fertilizer is compared with prices for other farm items in Figure 1.

Nomenclature. Discussing the fertilizer industry requires a number of special terms peculiar to fertilizers, some of which are traditional.

Analysis denotes the concentration of plant nutrients. The term high analysis is used to denote fertilizers of relatively high nutrient concentration.

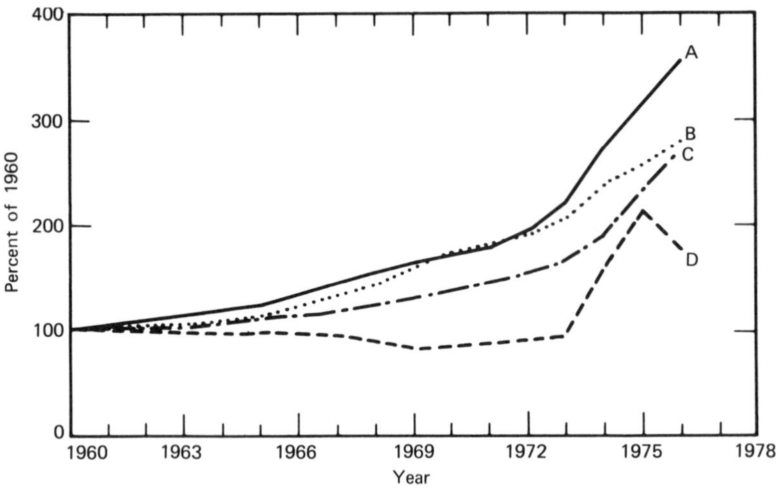

Figure 1. Prices of selected farm inputs: A, real estate; B, wage rates; C, machinery; D, fertilizer. Courtesy of USDA.

Grade defines the nutrient content expressed in percentage. For example, 10-10-10 usually denotes a fertilizer containing 10% N, 10% P_2O_5, and 10% K_2O. Three values are generally given, even if one or more nutrients are missing; eg, 46-0-0 for urea and 18-46-0 for diammonium phosphate. Other nutrients are rarely indicated but occasionally a grade such as 4-10-10-4-2-1 is given, the last three numbers indicating secondary or micronutrients—in this case, Mg, Mn, and Cu, respectively, expressed as the oxides. In contrast to the above, some countries express grade on an elemental basis and some use both oxide and elemental bases.

Unit refers to 1% of a ton of nutrient. In the metric system, 1 unit is 10 kg, and a 10-10-10 fertilizer contains 10 units or 100 kg N, 10 units or 100 kg P_2O_5, and 10 units or 100 kg K_2O. When the units are in short tons, as in the United States, a 10-10-10 fertilizer contains 91 kg each of N, P_2O_5, and K_2O.

BPL (*bone phosphate of lime*) is the phosphate content of phosphate ore expressed as tricalcium phosphate, $Ca_3(PO_4)_2$.

APA (*available phosphoric acid*) is the phosphate content (as P_2O_5) available to plants as determined by an empirical solubility test. Several empirical methods are used.

Muriate denotes potassium chloride.

Straight fertilizer is usually a fertilizer containing only one nutrient that is applied to the soil uncombined with other materials. An exception is ammonium phosphate, which contains both nitrogen and phosphorus, but which is considered a straight fertilizer when applied directly.

Mixed fertilizer is a fertilizer containing two or more nutrients. The term has traditionally been applied to a fertilizer containing two or more nutrients that has been made either by simple mixing of single-nutrient materials or by treating superphosphate with ammonia and adding supplementary materials. The term is not adequate for modern fertilizer practice and its increased complexity.

Complex fertilizer is a multinutrient fertilizer such as ammonium phosphate or nitric phosphate, made by processes based mainly on neutralization of an acid or other chemical interaction of ingredients. *Compound fertilizers* tend to be physical mixtures of materials. These terms are often used interchangeably so they are vague and often confusing.

Materials denotes nutrient carriers used in making mixed fertilizers.

Formulation refers to the amounts of materials used for a given weight of finished multinutrient fertilizer of specific grade.

Direct application is the application to the soil of a major primary fertilizer, such as diammonium phosphate, superphosphate, or ammonia, without first combining it with other fertilizer materials.

Conditioning is the treatment of fertilizer to reduce hygroscopicity and caking in storage. This usually involves coating the fertilizer particles with a finely divided material such as clay, oil, or wax, or with mixtures of these or other parting agents. Sometimes conditioner is incorporated into the product.

Fertilizer Evaluation and Control. Like other producers of consumer goods, the fertilizer producer is required by law to supply the quantity of fertilizer the bag or invoice indicates. In addition the producer must indicate and guarantee the nutrient content of the fertilizer and the quality of the nutrient. The fertilizer producer is checked continually by state authorities and is fined if the check analysis does not meet the guarantee.

Each state has a control group that samples fertilizer shipments and analyzes them for nutrient content. In the case of phosphate, the guarantee must be based on solubility (determined by a specified procedure) in a neutral ammonium citrate solution of specified composition. If not all the nitrogen is readily water soluble, the quality of the insoluble portion is also determined by a specified procedure.

Such legal requirements are also in effect in most other major fertilizer-consuming countries but the methods of analysis vary. In England, phosphate is tested by solubility in water rather than in an arbitrary solution. In some parts of continental Europe, phosphate analysis requires an alkaline ammonium citrate solution rather than the neutral citrate used in the United States. Alkaline citrate solubility is a more rigorous criterion because basic calcium phosphates generally are less soluble in the alkaline solution.

Raw Material Resources

The major raw materials required for manufacturing fertilizers are hydrogen and nitrogen for ammonia synthesis, phosphate rock, and potassium minerals for phosphorus and potassium, and sulfur. Sulfur, in addition to being a secondary macronutrient (it often is required in as large amount as phosphorus) is a key raw material for the fertilizer industry because of its use in production of sulfuric acid, the material most often used to solubilize phosphate rock. Although nitric acid can be used to solubilize phosphate rock, sulfuric acid is highly preferred, some 85% of fertilizer phosphate being derived from phosphoric acid made by use of sulfuric acid. However, if sulfur were to become unavailable or uneconomical as it has in the past, nitric acid made from ammonia could be used instead of sulfuric acid. Nitrophosphate, about 11% of fertilizer phosphate, is made by use of nitric acid. At present, production of nitrophosphate fertilizer is of more importance in Europe and Latin America than in the United States.

The other secondary macronutrients, calcium and magnesium, are widespread

phosphorus to be potential sources for industrial use are called phosphate rock and, sometimes, phosphate ore (see Phosphoric acid and phosphates).

Phosphorus makes up about 0.12% of the earth's crust. About 150 known minerals contain at least 1% P_2O_5. Nearly all mineable deposits of phosphate are of the apatite group represented by the general formula $Ca_5(F,Cl,OH,\frac{1}{2}CO_3)(PO_4)_3$. A small percentage comes from secondary aluminum deposits derived from apatite by weathering (see Calcium compounds).

Definitions. The definitions of terms are those published by Cathcart (17) and widely used by the phosphate rock industry.

Phosphate rock designates either (*1*) an apatite-bearing rock containing enough P_2O_5 to be processed into fertilizer or phosphorus, or (*2*) a beneficiated apatite concentrate.

Phosphorite is a sedimentary rock in which a phosphate mineral is a major constituent. This category includes phosphatic sandstones and limestones.

Ore, or *matrix*, is material that can be mined at a profit.

Grade of ore is expressed in terms of phosphate content, either as P_2O_5, BPL (bone phosphate of lime), TPL (triphosphate of lime), or P. BPL and TPL are equivalent to $Ca_3(PO_4)_2$.

World Reserves and Resources. Many variables affect reserve and resource tonnages. Supply–demand and pricing are very important factors. With expected increasing production rates, reserves (profitable, mineable material) will decline, causing prices to rise and making recovery of minerals from previously uneconomic sources profitable.

Inaccessibility of geographic location, as in the Sechura desert of Peru and also in northern Saudi Arabia, can cause otherwise high-grade deposits to be considered as resource rather than reserve. In the case of the Savannah River deposit in Georgia, environmental considerations cause down-grading of the deposit. This deposit contains at least 1.8×10^9 t of recoverable phosphate rock containing about 30% P_2O_5 (18), yet it is classified as reserve because its mining is not permitted.

Taking the foregoing and many other factors into account in assessing phosphate deposits, Cathcart (17) assembled the most recent, well-defined estimate of world reserves and resources. An adaptation of Cathcart's data is presented in Table 1. The amounts shown are about 21×10^9 t reserves containing at least 30% P_2O_5 and 9.1×10^{10} t resources of lower grade or for other reasons not mineable at a profit today.

In addition, there are a few hundred thousand metric tons of reserves and resources as guano, and about 250×10^6 t of reserves derived from guano.

At the 1974 rate of phosphate rock production (110×10^6 t), the reserve as defined above is sufficient for nearly 200 years, and the resource for nearly 900 years. At expected increased rates of production, the reserves and resources are adequate for at least 150 years and 700 years, respectively.

At projected rates of consumption, the high-grade reserves in Florida will be exhausted by the end of this century (18). In 1975 production from the Polk County Florida reserves constituted about 80% of all United States production and about one third of world production (19). This rate of depletion is causing increased interest in western United States reserves which represent nearly 80% of present United States total reserves (Table 1).

United States phosphate deposits with potential for future use are of special interest. For example, Cathcart (18) estimated the phosphorite formation in Idaho,

Table 1. World Reserves and Resources of Phosphate Rock [a]

Location	Reserves[b], 10^6 t		Resources[c], 10^6 t	
	Igneous apatite	Marine phosphorite	Igneous apatite	Marine phosphorite
North Africa				
Algeria		500		600
Morocco		5,000		35,000
Tunisia		500		800
West Africa				
Angola		20		100
Senegal		190		3,000
Spanish Sahara		1,600		15,000
Togo		100		200
Middle East				
ARE (Egypt)		800		2,000
Iran		30		100
Iraq		60		600
Israel		100		200
Jordan		100		200
Saudi Arabia				1,000
Syria		400		400
Turkey				300
Europe				
U.S.S.R.	400	1,450	400	2,950
Finland	50		100	
others		15		30
Asia				
People's Republic of China		100		1,000
India		70		200
Mongolia		250		700
Socialist Republic of Viet-Nam		100		400
Pakistan				150
Democratic People's Republic of Korea	5		30	
North America				
Mexico		1,140		0.07
United States				
(eastern)		1,600		6,000
(western)		6,000		7,000
South America				
Brazil	237	200	775	520
Columbia				600
Peru (Sechura)				6,100
Venezuela		20		0.1
South Africa	100		1,300	
Uganda	40		160	
Rhodesia	10		10	
Australian		500		1,500
others	10		30	
Total	*852*	*19,705*	*2,805*	*87,810*

[a] Ref. 17.
[b] Rock containing at least 30% P_2O_5, known mineable at a profit; variable with economic factors.
[c] Some solid information. Cannot be mined and processed at a profit.

Wyoming, Montana, and Utah to contain 14.5×10^9 t of phosphorus, many billions of metric tons of phosphate more than is shown in Table 1 for the United States western reserves and resources. Most of this material is deeply buried, is not weathered, and will require new mining and processing techniques. Its production is likely to be very costly. Cathcart also estimated that the Hawthorn formation in Florida would contain hundreds of billions of metric tons of phosphate pellets containing 10×10^9 t of phosphorus. Use of this material will have to await new mining and processing methods.

Billions of metric tons of phosphate rock are present offshore in the modern oceans. Best estimates are that a billion tons of pellets that may contain about 30% P_2O_5 are present in the Baja California–Mexico deposit alone.

Other areas in the world that contain large but unevaluated amounts of phosphate include Australia, Alaska, Africa, Tennessee, the Near East, Peru, Colombia, Brazil, the People's Republic of China, Mongolia, and the U.S.S.R.

Through exploration and discovery, billions of metric tons of phosphate rock have been added to the world reserves and resources in the last 20 years, and the search by geologists goes on throughout the world. There is reason to believe, therefore, that additional billions of tons will be found and that future economics and technology ultimately will make much of it valuable reserves.

Potash. Potassium is the seventh most abundant element in the earth's crust. The raw materials from which potash fertilizer is derived are principally bedded marine evaporite deposits but other sources include surface and subsurface brines and, to a limited extent, by-products of the sugar and cement industries. Both underground and solution mining are used to recover evaporite deposits and fractional crystallization is used for the brines. The potassium salts of marine evaporite deposits occur in beds in intervals of halite (NaCl) which also contains bedded anhydrite ($CaSO_4$) and clay or shale. The K_2O content of such deposits varies widely (see Potassium compounds).

The common ores of potassium include (*1*) sylvinite, a mixture of sylvite (KCl) and halite; (*2*) hartsalz, composed of sylvite, halite, and kieserite ($MgSO_4.H_2O$) or anhydrite; (*3*) carnalitite, carnalite ($KCl.MgCl_2.6H_2O$), and halite; (*4*) langbeinite ore, langbeinite ($K_2SO_4.2MgSO_4$) and halite; and (*5*) kainite ore, kainite [$4(KCl.MgSO_4).11H_2O$], and halite. More than 90% of the estimated potassium reserves occur principally as sylvinite and carnallitite (20). Sylvite, the richest of the minerals (63% K_2O), is and will remain for many years the principal economically exploitable reserve.

In addition, there are four major insoluble potassium silicate minerals, glauconite, leucite, nepheline, and orthoclase–sanidine which range in K_2O content from 7 to 22% (20). These minerals are plentiful but for economic reasons may never be exploited for potassium.

An assessment of world potash resources adapted from recent data of the U.S. Bureau of Mines (21) is shown in Table 2. Note that the approximately 10×10^9 t of reserves represent material economically mineable at average 1974 domestic mine prices. It is significant that of the 67×10^9 t of total resources in Canada, nearly 5×10^9 t is recoverable by conventional mining methods and the remainder by solution mining. Solution mining is now practiced at one or more mines, suggesting that this large resource may be nearing the status of an economic reserve. Canada has about half of present world reserves and more than half of known world resources of potassium.

Table 2. Assessment of World Potash Reserves[a], 10^6 t of K_2O

Location	Reserves[b]	Other resources	Total resources
North America			
Canada	4,537	62,613	67,150
United States	182	182	364
South America			
Chile	9	9	18
others	0	15	15
Europe			
France	91	91	182
GDR	2,450	2,450	4,900
FRG	1,633	1,633	3,266
Italy	9	14	23
Spain	73	73	146
U.S.S.R.	726	907	1,633
United Kingdom	23	23	46
others	0	64	64
Asia			
Israel and Jordan	218	907	1,125
People's Republic of China	5	5	10
Africa			
People's Republic of the Congo (Brazzaville)	18	18	36
others	0	68	68
Oceania and Australia	9	9	18
World total	*9,983*	*69,089*	*79,064*

[a] Ref. 21.
[b] At average 1974 domestic (U.S.) mine price.

The British Sulphur Corporation recently reported (20) the world recoverable K_2O at approximately 90 × 10^9 t. In addition, there are the extensive insoluble minerals and ores, by-product K_2O, and the oceans which contain 3.9 × 10^5 t K_2O km^3 of sea water (20–21). Although these latter sources may never become economic, they could be used if required for survival, an eventuality not anticipated. The 10 × 10^9 t of currently economic reserves of K_2O alone are adequate for 455 years at the 1975 rate of world consumption. The known recoverable potash is sufficient for more than 1000 years at any foreseeable rate of consumption.

Sulfur. In 1975 40% of all sulfur consumed, or about 21 × 10^6 t, was used in the manufacture of fertilizers and an increase to 44% by 1980 is projected (22) (see Sulfur compounds). This consumption is in the form of sulfuric acid, used mostly to manufacture wet-process phosphoric acid and ordinary superphosphate. The phosphoric acid is then used to produce the major fertilizers, concentrated superphosphate, ammonium phosphates, nitrophosphate, and fluid fertilizers. About 83% of manufactured fertilizer is based on use of sulfuric acid. The weight ratio of sulfur to phosphorus in fertilizer manufacture has increased steadily from 1.32 in 1950 to 1.8 in 1977 (22). In 1974 world sulfur production was 56.3 × 10^6 t and consumption was 53.9 × 10^6 t.

Sulfur is the thirteenth most abundant element in the earth and one of the few that occurs naturally in the free state. Most sulfur, however, is combined with other elements. The identified world sulfur resources, adapted from 1975 data of the U.S.

Bureau of Mines (21), are shown in Table 3. Note that sulfur is widespread throughout the world. Reserve definition is complicated by the variety and nature of the sources. For elemental, pyrite and sulfate types, the reserve is sulfur that is recoverable at 1975 prices using then current technology. The reserve assessment of coproduct sulfur is more complex because of the need to remove the sulfur from the prime product prior to marketing or because of emission-related restrictions that enforce production regardless of cost. A detailed analysis has been published by Bixby (22).

World reserves are estimated as 2.2×10^9 t, excluding coal and oil shales. United States reserves are 254×10^6 t. A major portion of the world reserves is in the sour natural gas deposits of the Near East, Canada, and Western Europe. Total sulfur resources in metric tons, as adapted from the U.S. Bureau of Mines, are: world, about 6×10^9 t; United States nearly 700×10^6 t. Estimates by the Sulfur Institute are slightly lower. According to Bixby (22), about 91% of known sulfur in the United States (not presently regarded as a resource) is distributed among coal (30.7%) and oil shale (60.7%). Technologies are currently under development for recovering sulfur from these sources but they are not expected to have significant commercial impact until after 1985. These sources may supply as much as 30% of the sulfur consumed by the year 2000, and they will ultimately supply much more (22).

A factor that could hasten the shift to new sources of sulfur is the high energy requirement of Frasch sulfur production, an operation that uses 6.33–8.44 GJ/t (6.01–8.01 10^6 Btu/t) of sulfur mined.

Table 3. Identified World Sulfur Resources[a], 10^6 t

Location	Reserves	Other	Total
North America			
United States	254	442	696
Canada	453	1,217	1,670
Mexico	100	66	166
others	5.5		5.5
Total	*812.5*	*1,725*	*2,537.5*
South America	33	33	66
Europe			
U.S.S.R.	166	498	664
Poland	77	11	88
France	39	11	50
FRG	22	5.5	27.5
Spain	22	498	520
Italy	17	28	45
others	205	315	520
Total	*548*	*1,366.5*	*1,914.5*
Africa	22		22
Asia			
Japan	22	44	66
Near East	664	332	996
People's Republic of China	28	28	56
others	55	221	276
Total	*769*	*625*	*1,394*
Oceania	28	11	39
World total	*2,213*	*3,761*	*5,974*

[a] Ref. 21.

In summary, the world supply of sulfur resources is adequate, with identified reserves, excluding coal and gypsum, totaling about 2×10^9 t, an amount sufficient to sustain rates of production higher than present ones until by-product sulfur is abundantly available.

SECONDARY MACRONUTRIENTS

Calcium. The calcium content of soils ranges from <0.1 to ca 25%. Although calcium nutrient deficiency is relatively rare in plants, probably no more than one third of the calcium needed for pH adjustment and other soil conditioning is used. This deficiency of calcium use is not owing to scarcity of raw material (see Calcium compounds).

Primary sources of calcium are lime materials and gypsum, generally classified as soil amendments. Soil minerals are major sources of calcium and magnesium. Among the more important calcium amendments are: blast furnace slag, calcitic limestone , gypsum, hydrated lime, and precipitated lime. Fertilizers that carry calcium are: calcium cyanamide , calcium nitrate , phosphate rock, and superphosphates. In addition, there are several organic carriers of calcium.

Calcium is widely distributed in nature as calcium carbonate, chalk, marble, gypsum, fluorspar, phosphate rock, and also other rocks and minerals. Calcium is the fifth most abundant element in the earth's crust. Thus calcium is plentiful all over the world in vast deposits. There is no foreseeable lack of this resource as it is virtually unlimited.

Magnesium. The magnesium content of soils usually is less than calcium because magnesium is more soluble. Important sources of magnesium are: dolomitic limestone, epsom salt, calcined kieserite, magnesia, potassium–magnesium sulfate, and a few organic sources.

Magnesium is the eighth most plentiful element in the earth and, in its many forms, makes up about 2.06% of the earth's crust (21). Dolomite, seawater, and well and lake brines, the sources of most of the world's magnesium and magnesium compounds, are available in unlimited quantities in nearly all countries. Seawater, with a magnesium content of 0.13 wt %, is an inexhaustible resource. Total world magnesium reserves and resources (magnesite) are placed at 2.4 and 3.2×10^9 t (magnesium), respectively (21) (see Magnesium compounds).

MICRONUTRIENTS

Boron. Virtually all United States boron production and about three fifths of the world production comes from bedded deposits and lake brines in California. Although marketed borates became scarce in 1973–1974 because of the high demand for borate-containing glass wool for insulation, United States reserves are adequate to support very high production levels. Turkey is the only other boron-producing country of significance. About 5% of boron used in 1975 went into agriculture. World boron reserves in 1975 were placed at about 73×10^6 t (21), 87% being accounted for in the United States, U.S.S.R., Turkey, and China (see Boron compounds).

Copper. World copper (qv) reserves are estimated at 408×10^6 t copper; one fifth being in the United States. The world resources are placed at about 1.5×10^9 t (21). These quantities are exclusive of deep sea resources known to be extensive but not well quantified.

Iron. World production of iron ore in 1974 was more than 800×10^6 t. World reserves are placed at 236×10^9 t of ore containing 90×10^9 t of iron (qv); world resources amount to 180×10^9 t of iron (21).

Manganese. At current prices there are no United States reserves of manganese ore containing 35% or more of manganese (qv). United States resources of significance are estimated to be on the order of 67×10^6 t containing Mn. World reserves are about 1.8×10^9 t Mn, and resources 1.45×10^9 t Mn (21).

Molybdenum. United States reserves and resources are about 3 and 13×10^6 t, respectively. World reserves and resources are about 6 and 23×10^6 t, respectively (21) (see Molybdenum).

Zinc. Zinc (qv) deposits in the United States extend from Maine through the Appalachian Mountains, and west through the Mississippi Valley in the Rocky Mountain states. United States reserves are estimated to be 27×10^6 t Zn (21). World reserves and resources are 135 and 110×10^6 t, respectively.

Nitrogen

Nitrogen is the nutrient used in fertilizer in the largest amount, being present in amounts almost double that of each of the other primary nutrients. Since 1962 its tonnage has tripled and its share of the fertilizer market has increased from 40 to 50%. Nitrogen's compound annual growth rate in the United States has averaged 7.6% over the past 15 years. World annual gain for the same period was 9.8%. World consumption of fertilizer nitrogen reached 43.3×10^6 t in 1975–1976 (23). Projected world annual consumption by 1980 is $57 \pm 4 \times 10^6$ t (24). World annual ammonia production capacity of ca 91×10^6 t has been predicted for 1980 (see Ammonia; Ammonium salts; Nitrogen).

World trends in production of the major nitrogen fertilizers for 1955–1976 are shown in Figure 2, and trends in the United States in Figure 3. Probably the most significant are the contrasting trends for ammonium nitrate and urea. From 1955–1969, ammonium nitrate production ranged from 25 to 30% of total commercial nitrogen fertilizer. Since 1969, with rapidly growing production of urea, the percentage of ammonium nitrate has declined despite continued growth in its tonnage. Urea has increased rapidly from about 6% of nitrogen fertilizer in 1959 to about 25% in 1975. The percentage for ammonium sulfate has also declined since 1956.

One of the striking features of nitrogen fertilizer usage in the United States is the large direct application of anhydrous ammonia, some 36% of nitrogen fertilizers used. This, together with compound fertilizers, ammonium phosphates, and nitrogen solutions, represents 76% of the nitrogen fertilizer used in the United States during the mid 1970s (25). The same group of materials represented only about 40% of world use—solid urea and ammonium nitrate both being stronger on a percentage basis than in the United States.

Ammonia. Nearly all commercial nitrogen fertilizer is based on ammonia, making it the most crucial basic material for the fertilizer industry. On a world basis, about 85% of all industrial ammonia is used in fertilizer. In 1980 more than 40% of the expected world ammonia capacity (ca 90×10^6 t/yr) resides in the United States and the U.S.S.R. In the United States, fertilizer consumption is 75% of total ammonia production. For the properties of ammonia, its production, storage, and economics, see Ammonia.

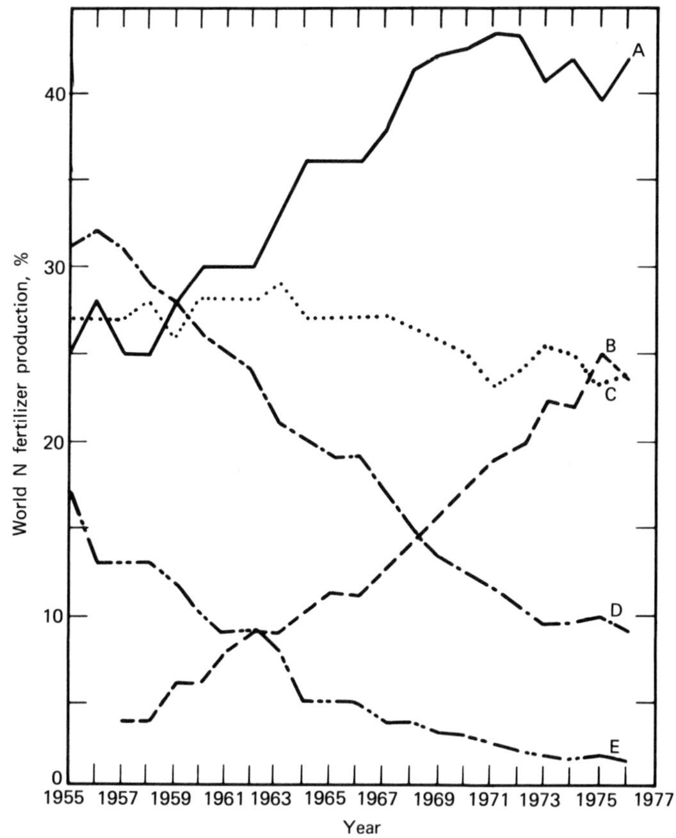

Figure 2. World trends in types of nitrogenous fertilizer materials: A, anhydrous ammonia, nitrogen solutions, ammonium phosphates, complex fertilizers, and others; B, urea; C, ammonium nitrate; D, ammonium sulfate; E, other low-analysis solid materials. Courtesy of TVA.

Ammonium Nitrate. Broadly defined, fertilizer ammonium nitrates include high-grade or straight ammonium nitrate, AN (33–34% N), ammonium sulfate–nitrate, ASN (26% N), and calcium–ammonium nitrate, CAN (20.5–28% N). The estimated total of world production of these materials was approximately 12×10^6 t N in 1976 (26), about one half high-grade AN (27), and most of the balance CAN. The two lower grades are used primarily because of their greater safety and better storage and handling qualities. Additional ammonium nitrate is a component part of nitrophosphate but this complex fertilizer is treated as a separate material in a later section of this article and the ammonium nitrate of such complex fertilizers is not included in the above statistics.

Ammonium nitrate has been the leading nitrogen fertilizer worldwide for most of the last three decades, it remained the world's leading N fertilizer through 1975, and its tonnage use in all fertilizer grades was still increasing, though at a slower rate, in 1976. As a percentage of total nitrogen fertilizer production, however, ammonium nitrate has been declining since 1967 (see Fig. 2), owing primarily to the rapid growth in the use of urea which now accounts for most of the nitrogen fertilizer used in the Asian and other developing countries. The overall importance of ammonium nitrate

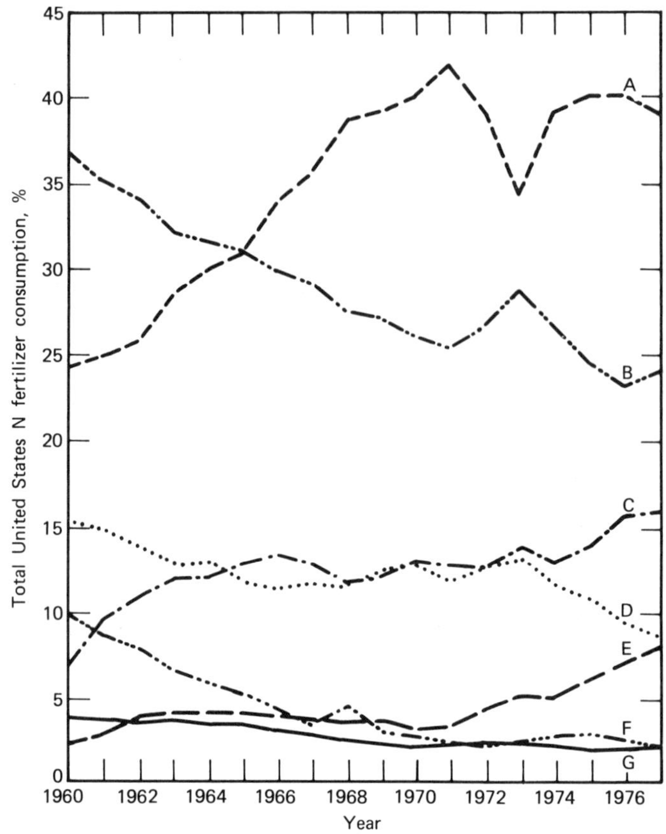

Figure 3. United States trends in types of nitrogenous fertilizer materials: A, ammonia; B, mixture; C, solutions; D, solid ammonium nitrate; E, solid urea; F, other; G, ammonium sulfate.

fertilizer is expected to continue to decline, with urea accounting for the bulk of increasing agricultural nitrogen usage.

About 70% of world ammonium nitrate production is located in Europe, being about equally divided between Eastern and Western European countries, and some 16% is produced in North America (26). The U.S.S.R. and Poland are the leading producers of 33–34% N grades in Eastern Europe, with France and UK the leaders in Western Europe.

Total United States production of fertilizer AN in 1975 was 5.34×10^6 t containing 1.83×10^6 t N (28). Of this total, some 56% was solid and 44% was liquid. About 80% of the materials were used in direct application (DA), continuing the strong trend to DA materials in the United States (28).

On January 16, 1978, the United States Gulf Coast fob price of ammonium nitrate was $121/t, and the delivered price across the nation ranged $120–138/t.

Fertilizer ammonium nitrates have a wide range of application in agriculture. Only for irrigated rice and where sulfur is a limiting factor are other nitrogen sources more useful.

Properties and manufacture of ammonium nitrate are discussed in the article Ammonium compounds.

Calcium–ammonium nitrate (CAN) is produced ranging in grade from 20 to 26% N. Addition of finely ground limestone to a concentrated solution of ammonium nitrate produces a material that can be prilled or granulated to yield a safe product with good storage and handling properties. When ammonium nitrate is already available, the only extra equipment needed to produce CAN is a limestone grinder and a mixing vessel where the two materials are combined.

Ammonium sulfate–nitrate (ASN) has a much lower explosion risk than ammonium nitrate, and its nitrogen content, 26–28% N, is substantially higher than that of ammonium sulfate. It is less hygroscopic than ammonium nitrate, but is subject to caking. It is produced by several methods (29).

In the Victor process ammonium sulfate is mixed with nitric acid and the mixture is ammoniated in an evaporator. Hardness promoters are added. The product contains 26% N.

In the Uhde-Hibernia process sulfuric acid is added to 80% ammonium nitrate solution, and the mixture is ammoniated at 160°C to form the double salt ammonium sulfate nitrate. Traces of ferrous sulfate and phonolite are added to improve the quality of the finished product which requires no coating.

Fisons produces ASN in an ammoniation granulation process wherein heat of ammoniation dries the product. Producing good quality prills of ASN is more difficult than with AN but prilling has been developed by both BASF and by Haltenback in Germany (26).

Most ammonium nitrate fertilizer is produced in solid form, usually prills or granules. Practice in the United States is somewhat different in that about one half is marketed as liquid such as urea–ammonium nitrate (UAN 28–32% N) containing 44–45% AN and 35–36% urea. The outlook is for continuation of this trend in the United States.

Prilling. Ammonium nitrate prilling is done in a relatively short tower, 21 m fed with 99.5–99.8% solution. The usual prilling operation and the design of prill towers have been described (30). Prills made from 99.5% or higher concentration generally contain less than 0.5% moisture, require no further drying, and little or no conditioning for good storage quality under usual storage practices. Under adverse storage conditions, as in parts of the United States, about 2% of kaolin clay and 1% of Petro Ag (sodium methylnaphthalene sulfonate) are added as conditioners.

A potentially important advance in simultaneously prilling and conditioning ammonium nitrate is represented by the Nitro-Top process developed by Fisons (31). The method involves prilling the ammonium nitrate droplets (0.3% moisture) in a fluidized cloud of a mixture of montmorillonite and talc dusts. A commercial tower is only about 7 m tall, with cross-section area much less than that of a standard tower with similar capacity.

Standard size prills segregate badly when blended with the larger granules of other fertilizer materials. For this reason, and because granulation is more susceptible to pollution control, granulation processes have been developed, and their expanded use is projected despite a marginally higher production cost for granules.

The Spherodizer granulation process developed by Cominco Ltd. (Canada) and C&I/Girdler, Inc. is one of the more advanced granulation processes. Very concentrated ammonium nitrate solution is sprayed onto a rolling bed of solid particles in a rotating drum, forming granules by a layer-solidifying process. The granules contain 0.5–0.1%

moisture and are said to require no conditioning (32). Oversize granules are crushed and recycled to the granulator along with fines to serve as nuclei for forming more product-size granules. The product size is controlled by choice of screen size. Air pollution is virtually eliminated by wet scrubbing in a manner that is simple and economical compared to that required for a standard prilling operation. The granules are much harder than prills.

Pan granulation of ammonium nitrate has been developed commercially (33–35). Essentially water-free melt (≤99.5% moisture), pH about 5.5, is sprayed onto a cascading bed of fines (microprills, crushed recycle, or other fine solid) in a tilted, rotating pan. Production of 18–20 t/h is common. The Norsk Hydro product is said to require no conditioning (35). The pan is an excellent size classifier, making it easy to produce granules of any desired size from 1 to 11 mm diameter (35). Major advantages of pan granulation are ease of granule size control, hardness of granule, and simplicity of pollution control.

Air and water pollution control in ammonium nitrate production are discussed in the section on Environmental Aspects of Fertilizer Production and Use, and elsewhere (36).

Fire and Explosion Hazard. Ammonium nitrate (AN) is a strong oxidizing agent. Moreover, certain materials, including fertilizer potassium chloride, induce self-heating and combustion, and others such as organic materials (1% or less) can sensitize the material to detonation. Serious fires and explosions have occurred. The worst of the explosions was that at Texas City, Texas, in 1947 where two shiploads of ammonium nitrate conditioned with petrolatum–paraffin wax exploded, killing 500 people and injuring 3000. Extensive studies of ammonium nitrate self-heating and detonation have since been carried out (37–39). Some of the more important findings are: (*1*) chemically pure ammonium nitrate does not decompose spontaneously at ordinary temperatures; (*2*) reducing agents such as carbonaceous matter, sulfur, certain metals when mixed with AN can lead to spontaneous heating; (*3*) the quantity of AN that can be detonated by impurities depends upon the geometry, confinement, temperature, packing density, and other factors; (*4*) AN is very difficult to detonate by ordinary impact or friction; (*5*) when heated unconfined AN does not detonate; and (*6*) heated under confinement AN may explode, the temperature and pressure required depending upon the type and amount of sensitizers present (see also Explosives).

The U.S. Bureau of Mines concluded that ammonium nitrate can be safely produced, stored, and handled if simple and reasonable precautions based on the findings listed above are taken. The thermal stability of fertilizers containing ammonium nitrate has been evaluated and discussed at length (31) and means for avoiding spontaneous burning are well developed.

Conditioning. The production of low moisture (0–0.3%) granules or prills greatly alleviates moisture absorption and caking problems. In the absence of preventive measures, the volume change accompanying the crystal transition at 32.3°C causes degradation of crystals, prills, and granules which increases moisture absorption and caking. Additives used to shift the transition temperature away from the usual storage temperature include potassium nitrate, magnesium nitrate, or nucleating agents like clay, talc, and other silicates. Numerous combinations of additives have been used to eliminate the phase transition as a major problem in storage of ammonium nitrate.

Coating dusts such as kieselguhr and clays are used as parting agents, sometimes also with small amounts of sulfonated hydrocarbon to reduce caking. These may also slightly retard moisture absorption, thus also reducing caking.

Materials such as petrolatum, paraffin, waxes, and oils are not used as conditioners as these mixed with ammonium nitrate can cause explosion.

Packaging and Storage. Multiwall paper or polyethylene bags, or combinations of paper and polyethylene, are used for packaging ammonium nitrate. In the United States these carry precautionary labels. Any spilled material may become contaminated and should not be rebagged. Bagged materials should not be stored within 76 cm of building walls or partitions. Because pressure enhances caking, bagged material should not be stacked too high. Ammonium nitrate should not be placed in storage while its temperature is above 54°C.

Ammonium nitrate is extensively stored in bulk. It should be placed on a clean floor and the building should be well ventilated. The height of piles should not exceed 6 m and in no case should the material be closer than 0.9 m to the roof or rafters. It must be kept separate from other materials and not close to sources of heat. It is preferable to store the bulk material in dehumidified areas but good granules and prills are often stored without dehumidification. Covering the floor and pile with plastic sheet reduces moisture absorption.

Material caked in storage should never be dynamited or blasted. No smoking signs are to be posted and observed, and all open flames kept away from the material.

Transportation. Ammonium nitrate is safely transported by rail, road, and water. Despite this, its transportation on United States navigable waterways is not permitted. Good ventilation must be provided and precautions taken against leakage and contamination. The material must be completely isolated from other cargo and must be kept free of extraneous combustible materials.

Urea. About 75% of all urea is used as fertilizer. It recently became the largest volume nitrogen fertilizer. Other uses for urea are as animal feed and industrial chemicals, representing ca 10 and 15%, respectively, of total urea production (see Urea).

The rapid growth in fertilizer urea production is shown in Figures 2 and 3. World production in 1973–1974 was about 9.1×10^6 t N (40); world consumption that year was about 8.4×10^6 t N. An indication of the magnitude of anticipated growth in urea demand for all uses is reflected by the following world capacity estimates of N: 1973–1974, 16.6×10^6 t; 1975–1976, 20.8×10^6 t; and 1980–1981, $30–40 \times 10^6$ t (40–41). Fertilizer usage of urea is expected to constitute about 31% of total N usage by 1980 (41). Some 47% of the projected 1980–1981 capacity is indicated for Asia–Oceania (40), the major rice-producing area of the world. World fertilizer urea production and consumption were about in balance at 13.7×10^6 t N in 1975–1976.

Consumption of fertilizer urea in the United States more than doubled during 1973–1977. Of this urea, 56% was solids and 44% was liquid. Combined United States and Canadian urea capacity is expected to increase from 5.5 to 8×10^6 t in 1975–1981. This area has 58 plants in production or under construction. Nearly 2×10^6 t (78%) of this projected growth is expected to be for granular urea, a grade largely unavailable prior to 1974. Granular urea annual capacity of the United States and Canada is expected to be 3×10^6 t by 1981. Use of liquid grades of urea is also increasing but at a slower rate than that for granular. The reasons for the spectacular growth in urea granulation are its use in bulk blends, as forest and rice fertilizer, and for coated fertilizer (lower surface per unit weight), and its generally simpler storage and handling.

The main reasons for the strong growth in urea fertilizer usage are: (*1*) high analysis (46% N), a strong economic factor in its storage, handling, and shipping; (*2*) ease and safety of applications, especially in comparison to anhydrous ammonia; (*3*) safety in production and handling (nonexplosive); (*4*) ease of pollution control in its manufacture; (*5*) its use in compound fertilizers to increase grade; (*6*) efficient production technology; and (*7*) its inherent suitability as a rice fertilizer (in contrast to the unsuitability of nitrates).

Disadvantages of urea are its high potential for loss of N by volatilization, especially when surface-applied; and the fact that urea, like other materials that release ammonia, can damage germination and seedlings when applied too near the seed. The latter problem is easily avoided by proper placement, and the former can be minimized by suitable timing of the applications and by soil incorporation. The complex granule-site chemistry of urea in soil is under study (42).

Properties. Urea properties of particular importance to its use as fertilizer include: N content (pure, 46.6%), 45–46%; solubility at 25°C, 119 g/100 g H_2O; melting point, 132.7°C; hygroscopic point (critical relative humidity at 30°C), 75.2%; specific gravity d_4^{20}, 1.335; and heat of solution, −60 J/kg (−14.34 cal/kg).

Commercial urea generally contains a small amount of biuret ($NH_2CONHCONH_2$). For fertilizer use other than foliar application, biuret content of 1% or slightly more, a level easily achieved in commercial production, is satisfactory. For most foliar applications the level should not exceed 0.1%.

Production. The technology of urea production is highly advanced. The raw materials for urea synthesis are ammonia and the by-product of ammonia production, carbon dioxide. For this reason urea plants are located at the site of ammonia production.

The ammonia and carbon dioxide are fed to a high pressure reactor (up to 30 MPa or 300 atm) and at temperatures of about 200°C where they react to form ammonium carbamate, urea, and water according to the following reactions:

$$2\,NH_3 + CO_2 \rightleftharpoons NH_4COONH_2 + \text{heat} \qquad (5)$$

$$NH_4COONH_2 \rightleftharpoons NH_2CONH_2 + H_2O - \text{heat} \qquad (6)$$

The fundamental chemistry and engineering of the process have been studied and described in great detail (43).

Because an excess of ammonia is fed to the reactor and the reactions are reversible, ammonia and carbon dioxide exit the reactor. Several major process variations have been developed to deal with the efficiency of the conversion and with serious corrosion problems (44). These variations deal primarily with ammonia utilization and recovery and with control of corrosion. The three main differences in ammonia handling are once through, partial recycle, and total recycle.

Urea plants with capacity of 275,000 t/yr are commonplace, and single-train units of almost 550,000 t/yr will be on-stream by 1980. Most recent advances have dealt with reduction of energy requirements in the total recycle process.

The economics of urea production is most strongly influenced by cost of the raw material, ammonia. When the ammonia cost is representative of production cost in a new plant (1977), it can amount to more than 50% of urea cost. The estimated price of prilled urea produced in a new (1977) plant, according to TVA, is $181/t.The estimate is summarized in Table 4. Note the estimated capital cost of $40,000,000 for a plant with capacity of 1000 t/d. The cost for producing urea in such a plant is sub-

Table 4. Estimated Price of Prilled Urea Produced in a New (1977[a]) Plant

	Units/t urea	Cost, \$/unit	Cost, \$/t urea
raw materials			
anhydrous ammonia, t	0.54	167[b]	90
carbon dioxide, t[c]	0.75	0	0
operating costs			31
capital charges, 20% of capital investment[d]			40
sales and administrative costs, 10% of sales			20
Estimated price			*181*

[a] Construction commenced in 1973, with escalation clause in contract. Capacity: 1,000 t/d; 330,000 t/yr. Capital investment: \$40,000,000 grassroots (include battery limits plant and 90-d storage, plus auxiliary units, site purchase and preparation, and support facilities). Working capital: \$15,000. Courtesy of TVA.

[b] Based on estimated price anhydrous ammonia in a new 1977 plant, deduct sales and administrative cost.

[c] By-product from NH_3 plant.

[d] Includes insurance and taxes, depreciation, interest, and pretax return on investment; higher percentage than for anhydrous ammonia because of higher ratio of working capital to plant investment capital.

stantially higher than recent fob list prices for urea. Those prices ranged from a low of about \$45/t in 1970 to a high of \$370/t in 1974. The very high prices of 1974 persisted to about mid-1975 when they decreased to \$115–170/t. Prices were in this range in 1976 and 1977. Much urea is still produced in plants with comparatively low capital investment and using ammonia produced at costs well below the cost in new plants. The higher costs for new plants reflects the expected trend for the future.

Prilling and Granulation. Prilling is the most widely used method of producing solid urea, some 70% of solid urea was produced in this form in 1976, but the use of granulation is increasing rapidly. In prilling, anhydrous, molten urea is forced through spray heads at the top of a tower to produce droplets which fall through a countercurrent stream of air where they solidify to form prills. The height of towers ranges 21–52 m. The temperature and rate of flow of air also affect the size and quality of prills.

Urea prilling without pollution control is an economical method for finishing but the prills have low strength and are generally too small for use in blending with other granular materials (DAP and TSP). The prilling operation is a serious polluter (see Environmental Aspects of Fertilizer Production and Use), the abatement of which is costly because of the very large volume of dust-laden air that must be treated. For these reasons there is a strong trend toward granulation.

Urea granules of preferred size are produced in a drum granulator (spherodizer) process developed by Cominco Inc. in conjunction with C&I Girdler (32), and by pan granulation methods developed by TVA (34,45–46) and by Norsk Hydro (35). The equipment and general operation are similar to that described for granulating ammonium nitrate. Ease of control and variation of granule size and ruggedness of granules with attendant lower dust formation in storage and handling are hallmarks of these methods.

Several estimates comparing the costs for granulation and prilling have been published in recent years (46–48). Indications are that drum granulation may be most expensive, and that when pollution control is included for both methods, pan granulation costs slightly more than prilling.

Ruskin's (48) estimates of the costs for producing granules in four types of plants with production capacity of 590 t/d is shown below:

Type of plant	*Estimated cost, $/t*
drum with scrubbing	13.86
pan with scrubbing	9.56
prilling with scrubbing	8.70
prilling without scrubbing	6.88

The drum granulation plant is a two-train plant because of limitations on the size of drum (4.3 m dia × 1.5 m long) that can be fabricated and transported. The other plants are single-train. Under these conditions the estimates indicate that drum granulation would cost nearly 50% more than pan granulation. Estimates for 1000 t/d plants would favor prilling even more, as single-train 1000 t/d prilling plants are feasible. A pan-granulation plant of that capacity would require two trains, and the drum-granulation plant would require three trains.

Conditioning. Notwithstanding the more favorable critical relative humidity of urea compared to ammonium nitrate (72.5 vs 59.4% at 30°C), some form of conditioner is often used on urea. Caking occurs as a result of pressure in storage, changes in temperature, and moisture absorption. Coating with a parting agent such as clay reduces caking tendency and coating with oil–wax reduces moisture absorption. Another very effective treatment is the incorporation of a few tenths of one percent of conditioner such as formaldehyde, or surface-active agents, into the urea melt prior to prilling or granulation. In addition to stabilizing the urea against caking and moisture absorption, granule and prill hardness are markedly increased. When 0.2–0.4% of formaldehyde is incorporated, no other conditioner is needed and 46% N urea is easily produced.

Handling and Shipping. Controlled humidity is often beneficial in bulk storage of urea. Use of controlled humidity, however, is not widely practiced, as bulk-stored urea covered with plastic sheet usually can be managed. Generally, bagging and bulk storage practices are similar to those for ammonium nitrate but handling problems are much less severe with urea because it is less likely to absorb moisture and to cake. It undergoes no crystal changes and it presents no fire and explosion hazard.

Urea must not be mixed with ammonium nitrate, either in storage or in compound fertilizers, as the mixture is extremely hygroscopic and liquefies.

Nitrogen Solutions. Nitrogen solutions are an important class of fertilizer in the United States and they are increasing in popularity abroad. The high-analysis solutions are of two forms, nonpressure and low pressure, 170–212 kPa (10–16 psig) at 40°C; the pressure forms contain free ammonia. The most popular is the nonpressure material made from urea and ammonium nitrate, UAN.

The largest use for nitrogen solutions is in direct application, but those solutions containing free ammonia are also very convenient sources of supplemental nitrogen in formulating complete liquid fertilizers and in granulation processes.

The basis for nitrogen solutions is the far greater solubility of mixtures of ammonium nitrate and urea as compared to that of either material alone. Solutions with nitrogen content almost as high as that of solid ammonium nitrate can be made. With the optimum ratio of ammonium nitrate and urea, 9 kg of water will dissolve 36 kg of the mixture.

The number of possible combinations is very large, requiring the adoption of a system of nomenclature making simple identification possible. The three digit number preceding parenthesis shows the nitrogen content to one decimal place but omitting the decimal, eg, 320 signifying 32% N. Inside the parenthesis the percentages of ammonia, ammonium nitrate, and urea are given in that order; 320 (0-45-35) signifying a 32% N solution containing no ammonia, 45% ammonium nitrate, and 35% urea; the balance is water. The compositions available are too numerous to list. Those listed are representative of the types available: 320 (0-45-35); 210 (0-60-0); 230 (0-0-50); 410 (22-66-0), vapor pressure at 40°C, 170 kPa (10 psig); and 454 (37-0-33), vapor pressure at 40°C, 494 kPa (57 psig).

Attempts to make 36–37% N fluids by suspending urea or ammonium nitrate in UAN solutions have been encouraging but the problem of excessive crystal growth during winter storage has not been overcome and use of this suspension has not yet been adopted. As problems of handling suspensions are overcome, this high-analysis material is likely to find use.

Nitrogen solutions can cause serious corrosion. Ammonium thiocyanate, sodium arsenite, Sulfonate 0A5, and a trace of free ammonia are the usual corrosion inhibitors. Aluminum is one of the more resistant equipment materials.

The consumption of nitrogen solutions in the United States reached about 2.0 $\times 10^6$ t N in 1975. Two thirds was nonpressure solutions and one third contained free ammonia. This is 35% of the N in the principal materials—nitrogen solutions, ammonium sulfate, solid ammonium nitrate, solid urea, and ammonium phosphates—used domestically in 1975 (28). Direct application accounted for 59% of this nitrogen solution usage and 41% went into liquid and solid mixed fertilizers. Future growth in nitrogen solution use will be influenced by competition from solid urea and potential supply limitations on ammonium nitrate.

Production. The usual manufacturing process is simple. The ingredient solutions are blended in a mixing tank in either a batch or continuous process. The usual feed solutions used contain about 83% ammonium nitrate and 72–85% urea. The plants usually are part of a nitrogen fertilizer complex that makes one or more of the required ingredients. The ammonium nitrate solution is withdrawn from the production line and combined in the desired ratio with urea solution taken directly from the concentrator. Mixing, transfer, and storage constitute the process steps. The process can be integrated with ammonia, urea, and ammonium nitrate processes.

Economics. Process economics is attractive in comparison to production of solids because evaporation is decreased and granulation, drying, and conditioning are avoided, greatly lowering investment cost.

Average prices paid by U.S. farmers for nitrogen solutions in 1976 were: 28% N, \$109/t; 30% N, \$113/t; and 32% N, \$121/t.

Ammonium Sulfate. Ammonium sulfate, $(NH_4)_2SO_4$ (21.2% N, 24.3% S), is used almost exclusively as fertilizer. It represented about 10% of total world fertilizer nitrogen consumption in 1977, this fraction having decreased steadily from 26% in 1960 (see Fig. 2). Estimated world production of ammonium sulfate in 1975 was 3.7–4.0 $\times$ 10^6 t N, about 75% of world capacity (49). Its low nitrogen content places ammonium sulfate at an economic disadvantage which is largely responsible for its low growth trend line (1.5%/yr) in the noncommunist world. It remains a useful plant food, however, especially where both nitrogen and sulfur are needed.

United States production of ammonium sulfate, as millions of metric tons of nitrogen, grew from 0.281 in 1960 to 0.539 in 1975. As a percentage, it decreased in the same period from 15.9 to 10.5% of the United States nitrogen fertilizer materials, excluding direct-applied ammonia. It represents 6% of total U.S. nitrogen fertilizer (28). Tonnage production during the decade 1965–1975 varied only slightly (see Ammonium compounds; Sulfur recovery).

During the period 1970–1974 list prices for fertilizer ammonium sulfate ranged from $16.50–51.80/t, and the price reached $104 during the nitrogen shortage in 1975. Delivered prices for granular ammonium sulfate, across the United States, on January 16, 1978, ranged $73–101/t. U.S. Gulf Coast fob prices were a few dollars lower.

Other Nitrogen Fertilizers. Several other nitrogen materials are used to a small extent. Individually, the usage of the several water-soluble forms (calcium nitrate, sodium nitrate, calcium cyanamide, and ammonium chloride) ranges from 0.4% to 0.6% of total world N use; collectively, they amount to about 2% of world N. Tonnage production for the individual materials ranges from 100,000 to 200,000 t N/yr. Their low analysis, often poor physical quality and poor compatibility with other materials in mixtures are leading reasons for their status.

Another form of material in limited but slowly increasing use is controlled-release fertilizers. Other chemicals are used to affect the rates of nitrification, denitrification, and urea hydrolysis as means for improving the efficiency of nitrogen utilization.

Calcium Nitrate. This fertilizer (15.5% N) is seen in two forms, $Ca(NO_3)_2.4H_2O$, and more frequently, the double salt $[Ca(NO_3)_2]_5.NH_4NO_3.10H_2O$. Originally much of this material was produced by treating limestone with nitric acid and neutralizing the resulting solution with ammonia. Much of the material produced today is byproduct of nitrophosphate processes.

Calcium nitrate is the most hygroscopic of fertilizer salts; its critical relative humidity at 30°C is 46.7%. The material is generally stored and shipped in moisture-proof bags. Before bagging it is partially dehydrated and either prilled or flaked. It then does not become sticky until after it rehydrates. The material must be used quickly after its container is opened.

Because of the large proportion of lime it contains and its quick-acting nitrate nitrogen, calcium nitrate is agronomically an excellent fertilizer.

Sodium Nitrate. Sodium nitrate (16% N) is mostly recovered from natural deposits in Chile but it can be produced by the reaction of nitric acid with sodium carbonate. Like calcium nitrate, sodium nitrate is alkaline, a desirable fertilizer property. Its physical properties are far superior to those of calcium nitrate; it can be used in mixtures but is used mostly in direct application. Its critical relative humidity at 30°C is 72.4%, a value between that of ammonium nitrate and urea. There appears to be no reason to expect its resurgence as a major fertilizer.

Calcium Cyanamide. Fertilizer calcium cyanamide, $CaCN_2$, in granular form is guaranteed to contain 21% N and in powdered form, 22% N. The fertilizer grade is about 60% $CaCN_2$ and contains about 5% oil, 12% graphite, 10% calcium hydroxide, and 6% calcium carbonate. The small amount of calcium cyanamide used is largely for direct application but some is used as plant defoliant and as herbicide. The material is made commercially by heating lime and coke in an atmosphere of nitrogen at 2200°C. Most of the cyanamide used in the United States is imported from Niagara Falls, Canada (see Cyanamides).

Ammonium Chloride. NH_4Cl, produced by neutralizing hydrochloric acid with ammonia, contains 26% N. Generally, however, it is a coproduct of soda ash made by the Solvay process. Little or none of the material is used as fertilizer in the United States. It is used largely in Japan and India, primarily as a rice fertilizer (see Ammonium compounds).

Controlled-Release Materials. Most nitrogen fertilizers are highly water soluble and tend to be leached, producing cycles of luxury and deficient feeding. Altogether they are inefficient: on average only about 50% of applied N is recovered by crops. To use multiple applications to feed crops more efficiently is labor intensive, thus costly. Another means for improving the efficiency of N use is through use of fertilizer whose release of N is in some way regulated or slowed to lengthen the period of feeding and decrease leaching and other losses. Such materials are known as controlled-release fertilizers.

Some properties and mechanisms that provide a measure of controlled release of nutrient are (*1*) low solubility, (*2*) resistance to attack by soil bacteria (urea–aldehyde reaction products), (*3*) coating soluble materials to provide a barrier to rapid dissolution, and (*4*) application with the fertilizer of other materials that inhibit nitrification, denitrification, or urease activity.

Controlled-release fertilizers are available but they have achieved little use on major farm crops, primarily because of insufficient benefit–cost ratio. One material still under intensive development and testing, sulfur-coated urea (SCU), shows promise for some long-feeding farm crops but not for such major nitrogen consumers as corn, small grain, and cotton. Both rising cost of fertilizer and further research on economic benefits may yet lead to more use of controlled-release fertilizers on farm crops.

Use of controlled-release fertilizers in the United States in 1975 was 180,000–220,000 t, mostly on nonfarm crops. This use represents only about 0.5% of United States fertilizer consumption in all markets and no more than 10% of nonfarm fertilizers. The market growth rate since 1960 has been some 8–10%/yr.

Urea Formaldehydes (Ureaforms). Urea formaldehydes containing 35–38% N are products of condensation of urea and formaldehyde. These materials have dominated the controlled-release fertilizer market in the United States through 1977, representing about 60% of all consumption. Despite their low efficiency—only about 50% of their nitrogen is available for crops—they do provide slow-release, nonburning nitrogen and have other advantages for specialty crops.

Isobutylidene Diurea. Isobutylidene diurea (IBDU), produced by condensation of urea with isobutyraldehyde, contains 31% N (fertilizer grade). Its United States use is only about 2.5–3% of controlled-release fertilizers—some 5000–5500 t/yr. Japanese output was 35,000 t in 1975.

The rate of mineralization of IBDU in soils is a function of moisture, pH, temperature, and particle size. Under average conditions about one half the N is released in 1.5 mo, and 80–90% in 3–6 mo. The material is also used in a variety of specialty mixed fertilizers.

Crotonylidene Diurea. Crotonylidene diurea (CDU), derived from reaction between crotonaldehyde and urea (theoretical N, 32.5%), is used straight and in mixtures, primarily for horticultural and other specialty applications. CDU is, however, a component of Triabon and horticultural substrates for which total sales amounted to about 100 t in 1974. About 3600 t of CDU was produced in Japan in 1973.

Overall nitrogen release by CDU is controlled by bacterial action, temperature, moisture and pH. Up to 10% of its N is released in 7–8 wk.

Sulfur-Coated Urea. Sulfur-coated urea (SCU) is a developmental material that shows promise of acceptance for certain farm crops as well as for the usual nonfarm uses of specialty materials. TVA has been developing a process for manufacturing SCU as a controlled-release fertilizer (50). Sulfur was selected as the least expensive coating agent with potential for controlling release of N from urea. (Because of explosion hazard, sulfur cannot be used on ammonium nitrate.) A demonstration-scale plant with capacity of 200 t/d began operation in 1978. In evaluations of the developmental program so far, the farm crops that show the most potential for favorable benefit–cost ratio are: rice under delayed and intermittent flooding, pineapple, sugarcane, fruits, and vegetables on sandy soils, turf, and forage crops (50).

Other developers of modifications of the TVA SCU process are Imperial Chemical Industries Limited (ICI), UK; and Canadian Industries Limited (CIL), Ontario, Canada. The former has produced and marketed Gold-N brand of SCU since 1972, produced in a small plant at Teeside. This product contains 68% urea, 32% N, 30% S, and 2% wax. The material has been used mostly in greenhouse applications and land reclamation projects. Sale price in England and in the United States has been about $550/t delivered to users.

CIL built a small plant (about 2.5 t/h of SCU) in Ontario, Canada, based on the TVA development. The product is being marketed in mixed lawn and garden fertilizer blends. Cost for the blends is thought to be no more than $300/t of blend to the distributor.

Estimated production cost in a 200 t/d (66,000 t/yr) plant, under three sets of conditions, escalated to 1978 costs, have been published (50). The estimated costs range from $156/t produced from prilled urea from an existing prilling unit, to $168/t produced from granular urea made in an existing pan-granulation system, and $177/t produced with granular urea made from an add-on pan-granulation system. These data indicate that SCU would cost from 31 to 41% more per unit of N than uncoated urea valued at $170/t. Important variables are the cost for the urea used and the size of granules, as the latter affects the amount of sulfur required.

Among the many other materials tested for controlled release, a few that show promise but that have not been produced commercially because of their high cost are: oxamide, 31.8% N; glycoluril, 39.4% N; guanylurea nitrate, 37.1%; and a wide variety of urea aldehydes.

Several controlled-release materials containing N, P_2O_5, and K_2O are used commercially in small amounts. Among the more important of these are: metal, ammonium and potassium phosphates, Osmacote, Preuse, Sunfresh, and tree spikes. These materials are not designed for use on general farm crops and their prices are much too high for such use.

Nitrification inhibitors are chemicals toxic to the soil bacteria that oxidize ammonia. These are used to delay nitrification and thus to slow the movement of nitrogen from the plant root zone as highly mobile nitrate. Urease inhibitors can be used to reduce the rate of urea hydrolysis. Denitrification inhibitors are used to slow volatilization of nitrogen from nitrates.

Phosphate

World phosphate fertilizer production grew at the rate of 6.5%/yr over the period 1963–1974, a rate slightly lower than that for nitrogen. World production reached 24.87 $\times 10^6$ t in 1976 (51), with production of 35.52 $\times 10^6$ t predicted for 1981–1982 (52). Most of the anticipated growth in consumption will occur in Eastern Europe and developing countries.

North American supply capability at 8.39 $\times 10^6$ t P_2O_5 in 1976–1977, already in some excess, is expected to grow only very slowly over the next 6 years, reaching 8.8 $\times 10^6$ t P_2O_5 in 1981–1982, some 26% of expected world production at that time. The United States will remain a major exporter of both phosphate rock and finished fertilizer for some years. A substantial amount of the United States export could go to the U.S.S.R. through the 20 billion dollar Occidental–U.S.S.R. transaction which calls for 0.7 $\times 10^6$ t P_2O_5/yr as superphosphoric acid over the next 20 years (53).

Trends in production of major phosphate fertilizer materials are shown in Figures 4 and 5 for the world and the United States, respectively. The most significant of the world trends is the strong growth in combined production of phosphates from 10% of world production in 1955 to almost 60% in 1976, and the simultaneous decrease for normal superphosphate from 63 to 21%. Ammonium phosphate is the strongest single growth material, surpassing normal superphosphate in 1976 when production of the two was approximately 6.4 and 5.3 $\times 10^6$ t P_2O_5, respectively, as estimated from FAO data, surpassing normal superphosphate in P_2O_5 production for the first time. The trends for these two leading materials in the United States are similar to, but even more extreme than, their world trends.

A noteworthy dissimilarity between nitrogen and phosphate fertilizer materials, especially in the United States, concerns their use as straight (direct application) and mixed goods. About 78% of nitrogen materials is used straight, whereas ca 82% of phosphates go into mixtures.

Use of electric furnace acid as fertilizer has virtually ceased because of its high cost and energy requirement compared to materials based on wet methods. For this reason electric furnace methods are not discussed in this article.

The principal routes for the production of phosphate fertilizers are sketched in Figure 6. Discussion of the individual phosphate fertilizers and of the processes for their production are presented in subsequent sections.

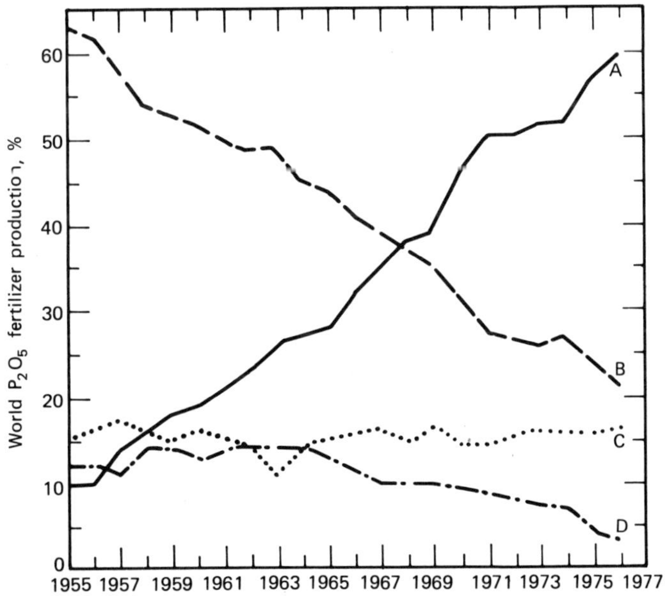

Figure 4. World trends in types of phosphatic fertilizer materials: A, ammonium phosphates complex fertilizers, other phosphates; B, normal superphosphate; C, concentrated superphosphate; D, basic slag.

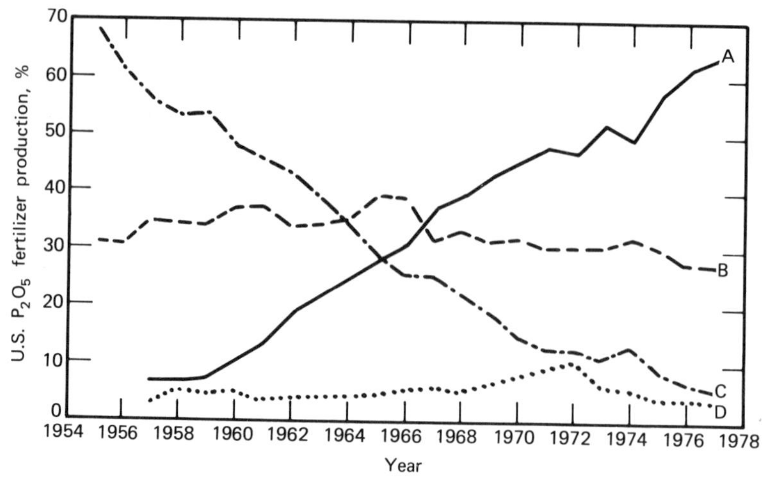

Figure 5. Trends in types of phosphatic fertilizer materials produced in the United States: A, ammonium phosphate; B, concentrated superphosphate; C, normal and enriched superphosphate; D, all other.

Phosphate Rock. The extent and production of phosphate rock as a fertilizer raw material has already been discussed (see also Phosphoric acid and phosphates).

Mining and Beneficiation. The phosphorite deposits of Florida and North Carolina (and those in northern Africa) are ideal for mining by use of large electric draglines with buckets ranging in capacities of 15–55 m^3. After mining, the ore is broken up hydraulically and the matrix is washed, and 840-μm (+20-mesh) pebble, 840–297-μm (−20 to +50-mesh) flotation feed, and <97 μm (−150-mesh) slime are produced by screening, scrubbing, grinding, and classifying. The >840 μm (+20-mesh) Florida pebble is a salable product, but because this fraction of the North Carolina deposit is quartz and carbonate grains, it is discarded. The <97-μm (−150-mesh) slime containing 10–12% P_2O_5 (up to one third of the P mined) is stored in earth-filled dams at high cost because it cannot be discharged into rivers. The 840–97-μm (−20 to +150-mesh) fraction is treated in flotation cells to separate the quartz and a high-grade phosphate product. The quartz tailings are used in land reclamation.

Smaller draglines are used in Tennessee, and the ore is hauled by truck or rail to the treatment plant. The <74-μm (−200-mesh) silt is removed by washing and is ponded, as in Florida. The >74-μm (+200-mesh) fraction is sintered or otherwise agglomerated and used as furnace feed for the production of phosphorus.

Western phosphate is mined both by open pit and underground methods, the latter being much more expensive. Much of the ore currently mined is high grade (31% P_2O_5) and requires only grinding and calcining to remove organic matter for use in production of wet-process acid. The mined shales (18–25% P_2O_5) are concentrated by froth flotation (qv) to separate the phosphate from calcite and quartz.

Energy consumed in phosphate (32% P_2O_5) production in the United States is 0.425 GJ/t (0.403×10^6 Btu/t) for western rock, 0.725 GJ/t (0.689×10^6 Btu/t) for Florida and North Carolina rock, and 1.839 GJ/t (1.745×10^6 Btu/t) for Tennessee rock (54).

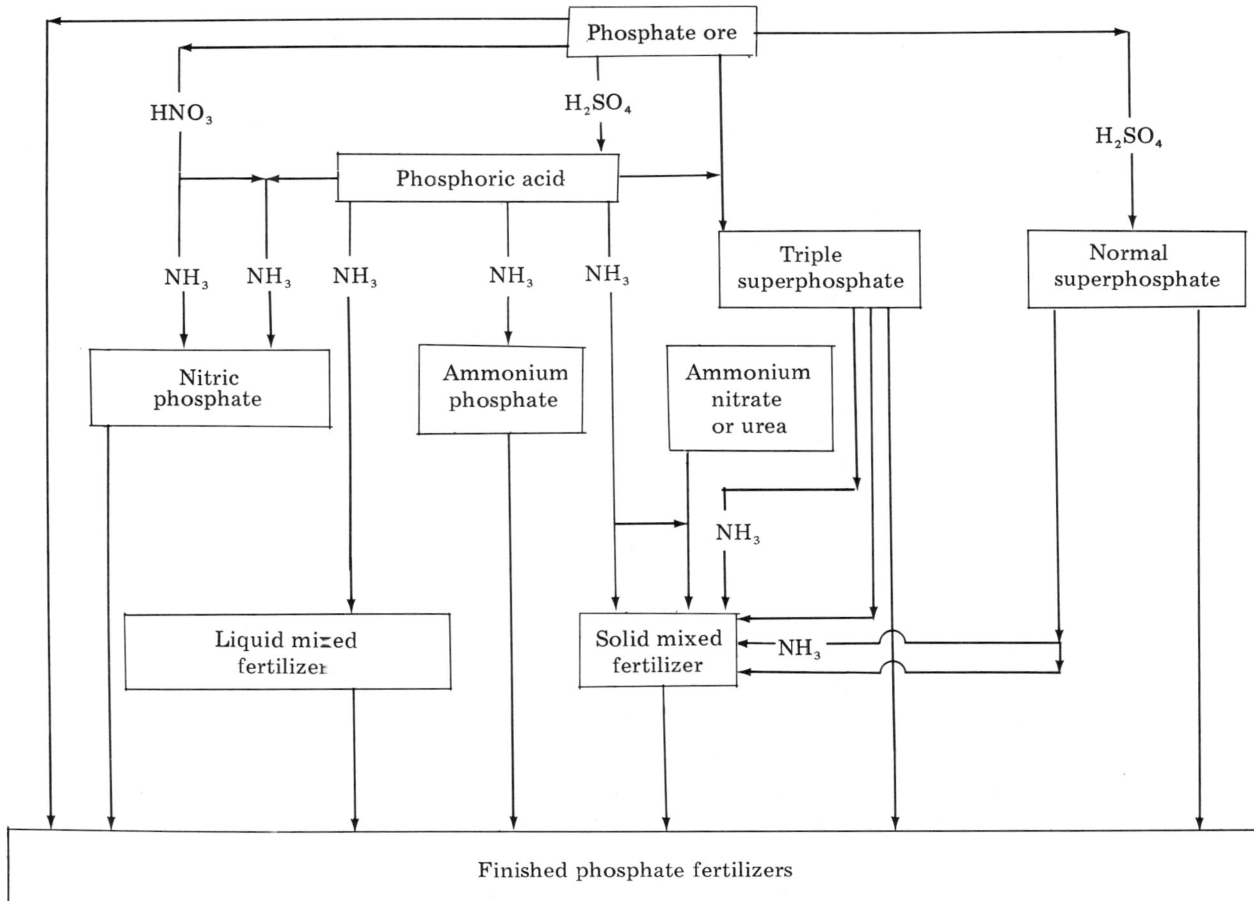

Figure 6. Main routes for producing phosphate fertilizers.

Quality Factors. Through selective mining of the higher grade and quality of rock, remaining reserves are becoming of progressively lower quality. This is especially true of the Florida reserves. For some fertilizer processes high P_2O_5 content of rock is of less importance than other quality factors. Important chemical factors are the presence of iron, aluminum, magnesium, heavy metals (Pb, Zn, Hg, Cd), toxic elements (Se, As, Cr, V), and radionuclides (U, Th, Ra, Pu) (55).

Means of using phosphate rock of marginal quality are beginning to emerge, as in the case of Christmas Island phosphate (55). It is quite possible, however, that the grade of fertilizers made from low quality rock will decrease because of increased amounts of impurities, many of which affect the solubility of the fertilizer phosphate (55). These problem areas do not appear to be insurmountable but they will lead to increased cost of fertilizer phosphate.

Structure of the World Phosphate Mining Industry. Of the total production of phosphate rock in 1974 (110×10^6 t), 75% was from three countries—United States, 38%; U.S.S.R., 20%; and Morocco, 17%. Other north African sources (Algeria, Tunisia, Syria, Jordan, Egypt, and Spanish Sahara) produced about 9% of the total. The remaining 16% was spread among 30 countries throughout the world.

Morocco, the United States, and the U.S.S.R. account for about 75% of international export trade. In the decade from 1964 to 1973, a total of about 345×10^6 t was exported—112×10^6 t (32%) by Morocco, 99×10^6 t (29%) by the United States, and 51×10^6 t (15%) by the U.S.S.R. Patterns of world trade in phosphate rock are unlikely to change drastically over the next decade but by the end of this century Australia may become a major producer and exporter.

Use as Fertilizer. Phosphate rock has had limited use as fertilizer for many years, with widely varying results. In recent years direct application of phosphate rock amounted to about 8% of total fertilizer phosphate used, primarily in the U.S.S.R., France, Brazil, Sri Lanka, Malaysia, and Indonesia.

The agronomic effectiveness of an apatitic rock depends strongly on its reactivity and the acidity of the soil; performance being better on more acid soils. Probably more than half of the potentially productive tropical soils are acidic, some with pH as low as 3.5–4.5. Certain phosphate rocks, thus, may become increasingly important as fertilizer. The International Fertilizer Development Center, with assistance from TVA, is active in research in this area (56).

Recent research (57) has elucidated the principles underlying the variable reactivity of phosphate rocks and has developed reliable laboratory methods for evaluating the reactivity of rocks. Broadly, the reactivity of apatitic rocks increases with increase in lattice substitution of carbonate and fluoride for phosphate and of sodium, magnesium, rare earths and other metals for calcium (57–58). These substitutions correlate well with changes in the unit cell a-axis dimension. Using this guideline, together with chemical methods, an absolute citrate solubility (ACS) value was defined, and a method for its determination was developed (59). An excellent correlation between ACS and plant response of several crops, including rice in greenhouse and field trials, was demonstrated. Thus, a method for selecting reactive rocks is now available. Among the more reactive are certain rocks from Algeria, Israel, U.S.S.R. (Akenbin, Egorier, Sarator, Tronbnia, and Viatka-Kama, but not Kohla), and North and South Carolina.

In addition to agronomic effectiveness, the relative economic effectiveness (REE) must be considered.

Wet-Process Phosphoric Acid. Wet-process phosphoric acid is the key intermediate for the phosphate fertilizer industry; over 90% of all the acid produced is used in fertilizer. In the United States, some 85% of all fertilizer phosphate is based on phosphoric acid. On a world basis 60% of phosphate fertilizer is based on phosphoric acid, but projections indicate that this proportion will reach 66% by 1981–1982 (52). Indications are that the trend of increasing phosphoric acid production has not yet peaked, in view of plans in North Africa, Mexico, and the United States to produce phosphoric acid and other high-analysis materials at the rock site as a means of reducing freight costs.

World production of wet-process phosphoric acid was 13.27×10^6 t of P_2O_5 in 1975–1976, 18.42×10^6 t in 1976–1977, with projected production of 24.35×10^6 t in 1980–1981 (52). About 36% of world capacity in 1977–1978 was in North America, mainly the United States, 20% in Western Europe, and 18% in Eastern Europe. World capacity is substantially in excess of production, 26.33×10^6 t P_2O_5 in 1976–1977, with 33.6×10^6 t projected for 1981–1982. With new units in operation in 1975, United States wet-process acid production reached 6.25×10^6 t P_2O_5, nearly half of world production, with increases continuing into 1976 (28).

Chemistry and Properties. The main chemical reaction in the acidulation of phosphate rock with sulfuric acid is as follows:

$$Ca_{10}F_2(PO_4)_6 + 10\ H_2SO_4 + (10\ x)\ H_2O \rightarrow 6\ H_3PO_4 + 10\ CaSO_4.xH_2O + 2\ HF \qquad (7)$$

where $x = 0$, 0.5–0.7, or 2.0. This acidulation process is strongly exothermic but with temperature control it can be operated to produce either the calcium sulfate dihydrate (gypsum), the hemihydrate ($CaSO_4.{}^1\!/\!_2H_2O$), or anhydrite ($CaSO_4$). These variations are discussed under Process Technology.

Part of the fluorine is volatized, the proportion depending upon temperature, and must be scrubbed to abate air pollution. Some of the fluorine reacts with silica and other impurities in the rock and contaminates the phosphoric acid with fluosilicic acid. In a recent fundamental study (60), 12 different fluorine compounds were identified as possible precipitates in the production of wet-process acid. Up to 77% of the fluorine can be immobilized as insoluble compounds, with possible improvement in acid quality and reduction in fluorine emissions from gypsum ponds.

In addition to the main acidulation reaction, other reactions occur also. Free calcium carbonate in the rock reacts with the acid used to produce additional by-product calcium compounds and CO_2 which causes foaming. Other mineral impurities (containing Fe, Al, Mg, U, and others) and organic matter dissolve, so that the wet-process acid is highly impure. Although the black acid is suitable for most fertilizer production, its impurities are the source of problems and added cost for some processes.

When the acidulent is hydrochloric acid, the product contains dissolved calcium chloride. Israel Mining Industries developed a process based on hydrochloric acid, including a solvent extraction method for separating the phosphoric acid from the calcium chloride.

Nitric acid acidulation produces phosphoric acid together with dissolved calcium nitrate. Separation of the phosphoric acid for use as an intermediate in other fertilizer processes has not been developed commercially. Solvent extraction is less effective in the phosphoric–nitric system than in the phosphoric–hydrochloric system. Instead, the nitric acidulate is processed to produce nitrophosphate fertilizer, see Nitrophosphates.

Filter acid from a dihydrate process contains 28–32% P_2O_5. It is usually concentrated to 40–42% P_2O_5 when used at the site of fertilizer production. For shipping, the acid is concentrated to 52–54% P_2O_5. At all of these concentrations the product is orthophosphoric acid, H_3PO_4.

On aging, many complex compounds (called sludge) precipitate from the acid. Examples of the major compounds are (61): $(Fe,Al)_3KH_{14}(PO_4)_8.4H_2O$; $FeNaH_5(PO_4)_3.H_2O$; $(Fe,Al)_3(K,H_3O)H_8(PO_4)_6.6H_2O$; $(Na,K)_2SiF_6$; $Na_xMg_xAl_{2-4}(F,OH)_6.H_2O$; $Fe(H_2PO_4)_2.2H_2O$; $CaSO_4.2H_2O$; $CaSO_4.5H_2O$; and $MgSiF_6.6H_2O$.

The partial compositions of several desludged shipping-grade phosphoric acids produced in the United States in 1976 and 1977 are shown in Table 5. As a result of lowering rock quality, the impurity content of the acid has increased by about one third over the last decade.

The superacid of fertilizer technology is a mixture of ortho- and polyphosphoric acids. The latter are based on open-chain arrangements of PO_4 tetrahedra (62). They are formed as water of constitution is removed from the ortho- or other member species of the series, usually by heating. The chemistry is illustrated by equations 8 and 9.

$$\underset{\text{orthophosphoric acid}}{2\,H_3PO_4} \overset{\Delta}{\rightleftharpoons} \underset{\text{pyrophosphoric acid}}{H_4P_2O_7} + H_2O \quad (8)$$

$$H_4P_2O_7 + H_3PO_4 \overset{\Delta}{\rightleftharpoons} \underset{\text{triphosphoric acid}}{H_5P_3O_{10}} + H_2O \quad (9)$$

The reactions proceed similarly up to linkage of some 10 to 14 phosphorus atoms. The general formula for the linear polyphosphoric acids is $H_{n+2}(P_nO_{3n+1})$. In the pure system, pyrophosphoric acid first appears at about 68% P_2O_5 and reaches its maximum at 77.5% P_2O_5. Triphosphoric appears at 71.7% P_2O_5, and tetraphosphoric at about 75.5% P_2O_5. The superacid of fertilizer contains little material polymerized beyond the tetramer, although additional species usually can be identified. High poly- and metaphosphoric acids, not described above, are present only at acid concentrations above 82% P_2O_5, concentrations not encountered in wet-process superacid.

Note that polymerization is reversible. This is very important for the storage of all polyphosphate fertilizers and especially for liquids. Also, hydrolysis of polyphosphate fertilizer in soil is necessary for its use by plants.

The presence of impurities in wet-process superacid is important in several respects. The polyvalent metals, particularly magnesium, cause the viscosity to increase sharply, virtually limiting the production and handling of superacid to concentrations

Table 5. Partial Composition of United States Commercial Wet-Process Phosphoric Acid

Constituent	Acid A, wt %	Acid B, wt %	Acid C, wt %
P_2O_5, total	53.2	51.8	54.4
Fe_2O_3	1.3	1.3	1.4
Al_2O_3	1.2	1.8	1.8
MgO	0.58	0.4	0.77
SO_4	3.1	4.8	3.1
F	0.7	1.0	1.0
organic	0.5	0.9	0.6
water insoluble solids	1.5	2.8	1.7

below about 73% P_2O_5. Compositions of some wet-process super acids produced in 1975–1977 are shown in Table 6. Moreover, at a given P_2O_5 concentration the ratio P_2O_5:H_2O in the impure acid is greater than that for the same P_2O_5 concentration in pure acid, so the polyphosphate content of the impure acid is higher. The polyphosphate content of a 70–72% P_2O_5 wet-process acid corresponds roughly to that of pure acid containing 76% P_2O_5.

An adaptation of an ascending paper chromatographic method provides a means for determining the phosphate species distribution in impure phosphoric acid and fertilizers (63–64). For most fertilizer analytical needs, however, a simpler determination of total P_2O_5 and nonorthophosphate P_2O_5 is sufficient.

Production Technology. The raw materials for production of wet-process acid are phosphate rock and, most often, sulfur (for sulfuric acid). Processes for extraction of P_2O_5 from phosphate rock by sulfuric acid vary widely but all produce a phosphoric acid–calcium sulfate slurry requiring solids–liquid separation (usually by filtration), countercurrent washing of the solids to improve P_2O_5 recovery, and concentration of the acid. Volatilized fluorine compounds are scrubbed, and calcium sulfate is removed in a variety of ways.

Phosphoric acid processes are dominated by the chemistry and crystallography of calcium sulfate because of its large amount and because its crystal form controls filtrability and often plant capacity, as well as the P_2O_5 recovery efficiency of the process. P_2O_5 is always incorporated in the crystal lattice of the calcium sulfate as it forms in the phosphoric acid solution, regardless of the crystal form precipitated. The most important process variations are those that affect crystal growth and crystal form. The different types of processes are named for the crystal form that predominates, as follows: dihydrate process, $CaSO_4.2H_2O$; hemihydrate process, $CaSO_4.0.5H_2O$; anhydrite process, $CaSO_4$; dihydrate–hemihydrate process: the dihydrate is formed first, then dissolved, and the calcium sulfate is recrystallized as hemihydrate; and hemihydrate–dihydrate process: the hemihydrate is formed first then dissolved, and the calcium sulfate is recrystallized as dihydrate. In the latter two processes, the purpose of the crystal transformation processes is to free the lattice P_2O_5 and recrystallize the calcium sulfate under conditions—low P_2O_5 and relatively high sulfate concentrations—where P_2O_5 will not be a serious contaminant in the new crystals. The main process variables are temperature of the extraction and recrystallization steps.

A broad comparison of the main types of processes, the strength and quality of phosphoric acid, and the form and quality of by-product calcium sulfate is summarized in Table 7. As the dihydrate process is most widely used, the quality of its acid and

Table 6. Composition of Some Wet-Process Superacids

Constituent	Acid E, wt %	Acid F, wt %	Acid G, wt %
P_2O_5, total	69.8	66.5	66.5
Fe_2O_3	1.3	1.6	2.6
Al_2O_3	1.0	0.25	2.2
MgO	0.3	0.87	1.1
F	0.3	0.6	0.3
SO_4	3.2	4.7	5.0
water insoluble solids	<0.1		

Table 7. Comparison of Wet-Acid Processes and Products

Type of process	Principal developers	Operating temperature, °C		Acid concentration, % P_2O_5	Acid impurity level vs dihydrate acid	Suitability of calcium sulfate for wallboard or cement	P_2O_5 recovery, %
		Extraction	Crystal conversion				
dihydrate	Prayon, Dorr and others[a]	71–85	none	28–32		no	95
hemihydrate	Fisons	91–99	none	45–50	ca same	no	91–94
anhydrite	Nordengren	102–238	none	40–50	lower	no	91
	Central Glass of Japan						
dihydrate–hemihydrate	Société de Prayon	62–68	93–99	33–38	higher	yes	97
	Nissan, NKK Mitsubishi, Fisons, Singmaster			30–35[b]	depends on process variation		
hemihydrate–dihydrate	Breyer/Heurty, Dorr HYS[c]	91–99	60–19	40–50[d]	can be lower	yes	96–98

[a] St. Gobain/Rhone Progil, Fisons, Swenson Isothermal, Kellog-Lopker.
[b] Without filtration in crystal transition (Nissan, NKK, Mitsubishi).
[c] The Dorr HYS process works poorly on Khola rock.
[d] With filtration in crystal transition (Fisons).

calcium sulfate, and P_2O_5 recovery are taken as reference for performance comparisons in the table. Illustrative flow diagrams of the principal variations in process types were published recently by Robinson (65). Numerous other variations in process details are also used (66–68). The majority of plants use a dihydrate process and some of these have production capacity up to 1150 t of P_2O_5/d.

The single-stage hemihydrate process is energy-efficient, producing 50% P_2O_5 acid directly, and therefore, is attractive for some situations. P_2O_5 recovery is low, however, leaving the serious disposal problem of impure calcium sulfate.

The only processes that yield high quality calcium sulfate and give high P_2O_5 recovery are those that carry out a crystal-phase transition to release lattice phosphate. Because of local demand for domestic sources of high-quality gypsum, certain Japanese developers offer three variations of hemihydrate–dihydrate processes, Table 7. All of these give high P_2O_5 recovery and high-quality calcium sulfate for commercial use but they produce only dilute acid (30% P_2O_5) because there is no filtration step in the crystal transition operation.

The Fisons hemihydrate–dihydrate process, Table 7, produces high concentration acid and high P_2O_5 efficiency (98%) as well as high-quality, market-grade gypsum. This combination of results gives the process attractive potential if commercial results, only now beginning to develop, prove successful.

The selection of a process can be complex, requiring careful evaluation of the many variables for each application. The hemihydrate process is energy efficient but this may not be an overriding consideration when energy is readily available from an on-site sulfuric acid plant. The energy balance in the total on-site complex may be the determining factor.

Dry grinding, handling, and feeding of phosphate rock has formerly been the most common practice but the trend in new plants is to wet grinding and feeding, with elimination of dust. The H_2SO_4 concentration fed to dihydrate reactors is increased when wet grinding is used.

The reactor of a phosphoric acid plant normally consists of a series of vessels connected by underflow or overflow parts. A typical Prayon dihydrate process reactor is a lined concrete shell that contains up to 9 compartments.

Many different filter types are used, including horizontal rotary table, belt, in-line pan, and horizontal rotary tilting pan.

The dihydrate acid containing 28–32% P_2O_5 is concentrated further for most uses, usually in single-effect vacuum evaporators. A variety of types are used, including forced circulation, natural circulation, and falling film. Direct-heated spray towers and submerged combustion have also been used but these are no longer preferred.

Because corrosion is a serious problem in wet-process acid plants, carbon brick linings and rubber linings are used extensively. Filters, pumps, and agitators use stainless steel, and piping is made of rubber-lined steel and a variety of plastics.

The abatement of fluorine emissions and disposal of by-product calcium sulfate from phosphoric acid plants is discussed below.

An estimate of the price (production selling costs, and return on investment) for 54% P_2O_5 wet-process acid produced in a new (1978) plant at a Florida–Gulf Coast location is summarized in Table 8.

The fob Tampa, Florida, sale price for 52% P_2O_5 acid in January 1978 was near \$330/t P_2O_5. The higher cost for acid produced in a new plant is owing primarily to increased capital costs.

Table 8. Estimated Price of 54% Wet-Process Phosphoric Acid Produced in a New (1978[a]) Florida–Gulf Coast Plant

	Units/t P_2O_5 in product	Cost, \$/unit	Cost, \$/t P_2O_5
raw materials			
phosphate rock (68 BPL), t	3.193	24.30	77.60
sulfuric acid (100%), t	2.613	53.03	138.56
antifoam agent, kg	2.45	0.86	2.11
operating costs			37.85
capital changes, 34% of capital investment[b]			43.28
sales and administrative costs, 10% of sales			33.27
Estimated price			*332.67*

[a] Plant completed Feb. 1978. Capacity: 1,000 t/d P_2O_5; 330,000 t/yr. Capital investment: \$42,005,000 grass roots (includes battery limits plant, 1.5-d rock storage, 8-d phosphoric acid storage, site purchase and preparation, and support facilities but not including sulfuric acid plant). Working capital (30-d value of raw material plus 8-d value of raw materials and operating costs): \$8,597,000.

[b] Includes insurance and taxes—3% depreciation, 6.67% interest—8% on one half of capital investment, and pretax return on capital investment—20% of capital investment.

Wet-Process Superphosphoric Acid. Production of wet-process superphosphoric acid (SPA) was about 540,000 t P_2O_5 in 1973. Growth in the production of SPA has been in the range of 7–9% over 1973–1980. About 95% of the acid is used in liquid fertilizers. A small additional amount of thermal superacid, ca 20,000–30,000 t P_2O_5/yr, is also used in fertilizer production. SPA fertilizer phosphate represents about 10% of the United States fertilizer phosphate consumption. Data on world production are not available but by far the bulk of this material is produced and used in North America, mostly in the United States.

SPA is produced by two methods, vacuum evaporation and submerged-combustion direct heating. About 70% of United States production is by vacuum evaporation. Both types of evaporators are fed with 50–54% P_2O_5 acid, which is usually desludged. When the feed acid is made from calcined rock, the product acid is green, but otherwise it is black.

Purification. The impure state of WPA (wet-process acid)—sludge, dissolved metals and F, and dissolved and colloidal organic compounds—is well known. The sludge is troublesome in storage, shipping, and handling of the acid; these aspects of the technology are considered below. Most shipping-grade acid has been at least partially clarified by settling or other means for removal of most of the sludge. If made from uncalcined rock, this acid retains the black organic matter.

Purification of WPA for fertilizer use is needed only to produce clear solution fertilizers that can be stored for extended periods without forming precipitates. Liquid fertilizers are made from SPA because of its capacity for sequestering metallic impurities but some of these metals accelerate hydrolysis of the polyphosphate. As this advances, precipitates form and shorten storage life and impair the liquid quality.

Until recently, purification of WPA for liquid fertilizer production has been largely limited to partial removal of magnesium, a particularly troublesome impurity. Allied Chemical Company forms an insoluble magnesium aluminum fluorophosphate and separates the solid before concentrating the acid to SPA (69). One manufacturer crystallizes an insoluble magnesium pyrophosphate compound in the course of manufacturing SPA. Precipitation of magnesium fluorosilicate has been reported also (70).

Many solvent extraction schemes for purifying WPA have been studied. Some are in commercial use for producing fertilizer- and chemical-grade acids, and others are under further development.

Israel Mining Industries developed a process for producing WPA by acidulating phosphate rock with an excess of hydrochloric acid and purifying the acid by solvent extraction and distillation (71). The acidulate contains phosphoric and hydrochloric acids, dissolved calcium chloride and other impurities. The acids are extracted from the bulk of the impurities by use of *n*-butanol or isoamyl alcohol. A water wash separates the mixed acids from most of the organic solvent. After further treatment for complete elimination of calcium, the hydrochloric acid is separated from the phosphoric acid by multiple-effect evaporation which yields pure 68% P_2O_5 phosphoric acid and recovered HCl for recycle. The phosphoric acid is much more pure than is required for fertilizer production. This mixed acid system is highly corrosive, requiring expensive materials of construction. Making the process economical and practical for fertilizer production requires a source of cheap hydrochloric acid and an environmentally acceptable disposal of calcium chloride—two conditions met in Israel but seldom met elsewhere.

Similar solvent extraction methods have been tested for nitric acid acidulates. Because separation coefficients generally are much less favorable in the nitrate system, phosphoric acid intermediate has not been produced commercially by this method.

An effective solvent extraction method in industrial use is the Israel Mining Industries acid cleaning process (72–74). The method is based on the large temperature-dependence of the solubility of phosphoric acid in some organic solvents: ethers, ketones, esters, and glycol ethers. The IMI process uses isopropyl ether. The method can reduce the cationic impurities in the acid by two orders of magnitude and the impurity anions by one order of magnitude (74). In the variation most relevant for fertilizer use, the acid is extracted into the solvent at 5–10°C and is released in purified form by adding water at about 30°C. From 60–70% of the feed P_2O_5 appears in the cleaned acid containing 50% P_2O_5 acid of a quality suitable for liquid fertilizer. The balance of the P_2O_5 and most of the impurities remain in the residual 30–40% of the acid, which is suitable for use in manufacturing solid fertilizers.

A method developed by USS Agri-Chemicals and tested commercially uses *n*-heptanol as solvent. The process is fed with 54% P_2O_5 impure acid, and the product, 80% of the input acid, contains 44% P_2O_5, 10–20% of the input Mg, 20–30% of the Al, 40–50% of the Fe, and 15–25% of the F (75).

The Occidental S-X process has also been operated on a large scale. The process uses sulfoxylic acid dissolved in kerosene to selectively extract magnesium and calcium by liquid ion exchange (75). This process largely avoids production of more impure acid as a by-product. The loaded extractant is regenerated by treatment with sulfuric acid. The spent sulfuric acid containing the impurities is limed and discarded—a potentially troublesome disposal problem.

In a different kind of process, urea is added to impure WPA to yield purified crystals of urea phosphate. Through ammoniation–pyrolysis of the urea phosphate, a high-quality ammonium polyphosphate base, liquid fertilizer (15-28-0) is produced. The method has been studied by several investigators, and TVA is developing the process in a pilot plant.

In a modification of the urea phosphate method, the purified crystals are treated with concentrated nitric acid to form purified phosphoric acid and urea nitrate. Ammoniation of the latter crystals yields urea–ammonium nitrate fertilizer solution. This method has been developed only on a laboratory scale.

Removal of organic matter from WPA by treatment with a variety of oxidizing agents, including nitric acid and ammonium nitrate, has been shown to be technically feasible but is not practiced on a plant scale, largely for air pollution reasons.

IMI have demonstrated on a commercial scale a method for precipitating the color-forming organic matter in the phosphate rock acidulation step (73). An acid-catalyzed phenol–formaldehyde condensation product is formed *in situ* and the organic matter is coprecipitated with the polymer. Only 20–50% of the organic matter is precipitated but the action is said to be selective for the species that causes color formation, increased viscosity, and interference with solvent extraction purification processes. The method is not yet widely used.

The most effective method currently in commercial use for producing clear acid is calcination of the rock before acidulation. For energy and cost reasons, calcination is practiced only with rocks very high in organic matter, such as those from North Carolina and the western United States.

Clarification of liquid fertilizers made from black acid is discussed below (Liquid Mixed Fertilizers).

TVA estimated the cost for purified acid produced by the methanol, acetone, and isopropyl ether solvent extraction processes, and for high-grade fertilizer base solution made by the urea phosphate method and by the pipe reactor-liquid fertilizer clarification method. Purification costs \$15–20/t P_2O_5. Analysis of the detailed results (76) indicates that within the probable accuracy of the estimates the in-plant unit nutrient prices of products from all processes are economically competitive, and other considerations would dictate choice of process. Such other considerations could be local conditions, energy costs, distribution costs, or a variety of intangible items for specific situations. The urea phosphate process, which suits the needs for N–P base solutions in the United States, is the simplest and least energy intensive of the five processes and has been chosen by TVA for study in greater detail.

Extraction of uranium from wet-process acid is of increasing interest. The acid made from Florida and North Carolina rocks contains 150–220 ppm of uranium, and its recovery could boost United States production of yellow cake (U_3O_8) by 25–30%. Estimates of potential recovery of U_3O_8 from the six major Gulf Coast–Florida acid producers run as high as $8\text{–}9 \times 10^3$ t.

The Oak Ridge National Laboratory has developed a process for extracting uranium from phosphoric acid based on a series of extractions with several complex solvents (77). Extractants used are mono- and dioctylphenyl phosphoric acids (10 *M* H_3PO_4), and a combination of di(2-ethylhexyl) phosphoric acid (DEPA) plus trioctylphosphine oxide (TOPO). High-grade U_3O_8 (>98%) is recovered (see Extraction, liquid–liquid; Uranium).

Freeport Minerals Company, Wyoming Minerals Corporation, and International Minerals & Chemical Corporation are using such methods to recover uranium from phosphoric acid in Louisiana and Florida. Earth Science Resources announced plans for an 8 million dollar solvent extraction plant for recovery of 45.4 t/yr of uranium oxide from phosphoric acid in Calgary, Alberta.

According to TVA the cost for recovering uranium by the ORNL process was

$33–44/kg U_3O_8 in mid 1977. Yellow cake was selling for $88/kg. In collaboration with the Oak Ridge group, TVA is attempting to develop the DEPA-TOPO technology for ores containing high amounts of impurities. The Power Reactor and Nuclear Fuel Corporation of Japan is reported to have developed a simpler process for recovering uranium from phosphoric acid at a cost of less than $22/kg U_3O_8 (79).

Storage, Handling, and Shipping. Properties that most affect the storage and handling of wet-process acid are sludge formation, corrosiveness, and viscosity (especially with SPA). These properties vary widely between the two usual shipping grades of acid, 50–54% P_2O_5 and 70–74% P_2O_5.

Often 54% P_2O_5 acid is aged for about a month before shipping to permit formation and settling of sludge. One month unassisted settling of 100 t P_2O_5/d as 54% P_2O_5 acid requires a minimum of 2830 m^3 (7.5×10^5 gal) of storage capacity. Acid made from calcined rock requires only about a fourth as long for postprecipitation and settling. Settling can be accelerated by mild movement, by flocculents, and by addition of a few percent of H_2SO_4. Sludge densities up to 40% solids can be produced in cyclones, and up to 70% in a conveyor-type centrifuge.

The viscosity of 50–54% P_2O_5 acid ranges from 30 to 300 mPa·s (= cP) over 4.4–38°C. The viscosity of SPA as a function of rock source, polyphosphate content, and temperature is shown in Figure 7. Satisfactory handling of SPA requires that it be warm (at least 30–50°C) so that its viscosity is not substantially above 1 Pa·s (10 P).

The container material most frequently used for storing and shipping phosphoric acid is rubber-lined mild steel (78). Both butyl and natural rubber are used. Type 316 stainless steel is also satisfactory for storage tanks and transfer lines. PVC plastics are sometimes used as tank liners but temperature-strength limitations must be considered. Below ground-level pits lined with plastic sheet are also used (79). Fiberglass tanks have not proven fully satisfactory.

Heating and insulation are not required for orthophosphoric acid tanks but they are generally used for SPA.

Wet-process acid is shipped in railroad tank cars, trucks, river barges, and ocean-going tankers.

Phosphoric acid is not classified by the ICC as a hazardous chemical but in contact with any part of the body it may cause burns. In appropriate situations the usual chemical plant safety equipment should be used—safety goggles, safety spectacles, plastic face shields, breathing apparatus and masks, gas masks, respirators, and rubber gloves.

All aspects of environmental controls in the production of wet-process phosphoric acid are discussed elsewhere in this paper.

Normal Superphosphate. From its beginning as the first commercial phosphate fertilizer, normal superphosphate (NSP), also called ordinary and single superphosphate, has continued among the top fertilizers of the world, Figure 4. While decreasing steadily on a percentage basis because of growing production of more concentrated materials, it grew on a P_2O_5 tonnage basis to a maximum of 6.7×10^6 t in 1967. Its production was 6.1 and 5.3×10^6 t P_2O_5 in 1975 and 1976, respectively.

In the United States, the impact of more concentrated triple phosphates was greater; normal superphosphate production decreased steadily over the entire period of 1955–1977 both in tonnage and as a percentage, amounting to 350×10^3 t P_2O_5 and 5.2% of total P_2O_5 production in 1977, see Figure 5.

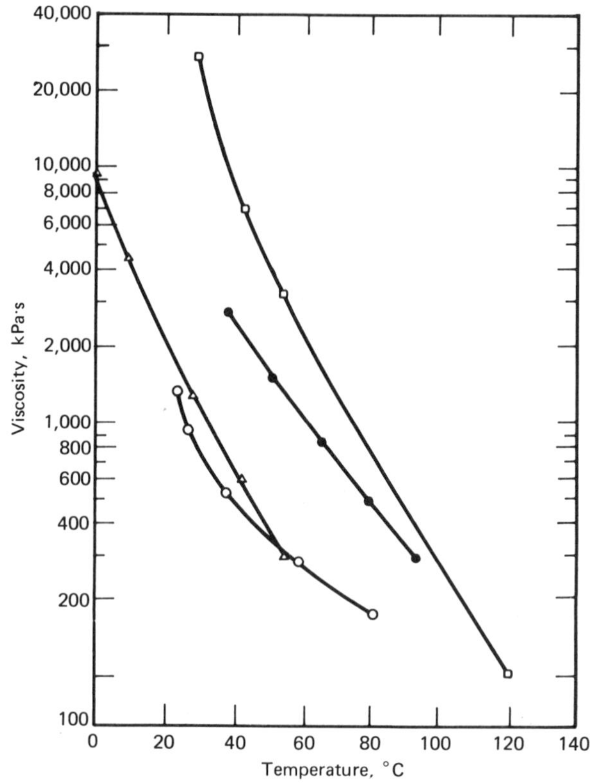

Figure 7. Viscosities of superphosphoric acid as related to polyphosphate content and phosphate rock source: □, western rock, calcined, 50% polyphosphate, 70.6% P_2O_5; ●, North Carolina rock, calcined, 50% polyphosphate, 71.3% P_2O_5; △, Florida rock, uncalcined, 45% polyphosphate, 70.5% P_2O_5; ○, North Carolina rock, calcined, 26% polyphosphate, 70.2% P_2O_5. Courtesy of TVA. To convert kPa·s to P, multiply by 10^4.

The sustained world popularity of NSP is owing to the simplicity of its production and its high agronomic quality as a good carrier of available P_2O_5, calcium, and sulfur. In terms of agronomic value for large numbers of crops, no phosphate fertilizer has been shown to be superior. It is likely to remain in strong demand in parts of the world where simplicity of production and sulfur needs have high priority and where transportation costs are not prohibitive.

Chemistry and Properties. NSP is produced by the reaction of phosphate rock with sulfuric acid. The method quickly yields a solid mass containing monocalcium phosphate monohydrate and gypsum, according to the following simplified equation:

$$Ca_{10}(PO_4)_6F_2 + 7\ H_2SO_4 + 3\ H_2O \rightarrow 3\ CaH_4(PO_4)_2.H_2O + 7\ CaSO_4 + 2\ HF \qquad (10)$$

Normally, a slight excess of sulfuric acid is used to bring the reaction to completion. There are, of course, many side reactions involving silica and other impurity minerals in the rock. Fluorine–silica reactions are especially important as they affect the nature of the by-product and of recovery methods. Thermodynamic and kinetic details of the chemistry have been described (80). Important properties of NSP are: free acid content, as H_2SO_4, 1–2%; moisture content, 5–8%; citrate soluble P_2O_5, in neutral citrate solution, 20–21%; hygroscopicity at 30°C, 94% relative humidity; and bulk density, nongranular, 800 kg/m^3, and granular, 970 kg/m^3. Porosity of the material is important for its absorption of ammonia, a requirement for much of its use in mixtures. Its ammoniation is considered below with multinutrient fertilizers. Analytical methods used in the United States are those of the Association of Official Analytical Chemists (A.O.A.C.).

Production Technology. A moderately high-grade of phosphate rock (33.5% P_2O_5, 73 BPL) is required for the production of a product containing 20% available P_2O_5. Major process variables in the manufacture of NSP are listed in Table 9.

A wide variety of mixers and curing dens have been used over the long history of NSP production (80) but the more advanced technology is illustrated in Figure 8. The concentrated acid and ground rock are combined by swirling in a cone mixer. The mixer discharges into a continuous den where a slat conveyor moves the solid to a disintegrator and thence by a conveyor to the storage pile. The product is known as run-of-pile normal superphosphate.

Table 9. Process Variables in the Manufacture of Normal Superphosphate

Variable	Range	Average
H_2SO_4 concentration, %	68–75	71
acid temperature, °C	28–95	54
particle size of rock, %		
through 149 μm (100 mesh)	80–95	90
through 74 μm (200 mesh)	50–95	70
acid:rock ratio, kg 100% H_2SO_4/100 kg rock	55–65	60
curing time	several weeks	

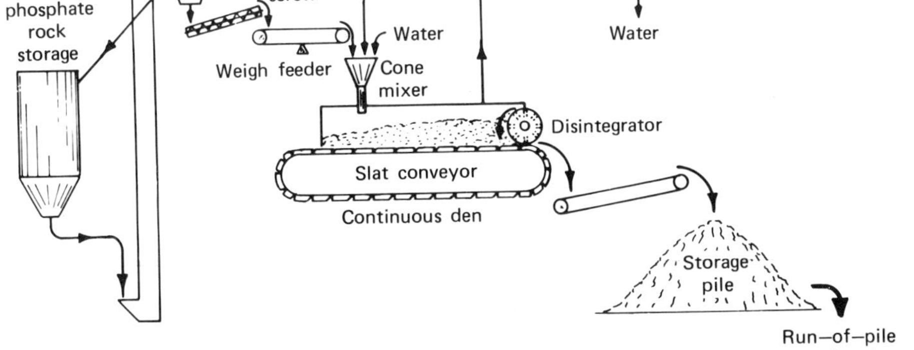

Figure 8. Continuous process for manufacture of normal superphosphate. Courtesy of TVA.

Sometimes a mixture of sulfuric and phosphoric acids is used to yield a product containing 22–40% but usually about 27% P_2O_5, called enriched superphosphate.

Production Economics. As the cost for delivered raw materials, phosphate rock, and sulfuric acid often accounts for more than 90% of the cost of producing NSP, the actual cost varies considerably with plant location. Because the rock is richer in P_2O_5 than the low analysis product, NSP need not be produced near the mine, but delivery of sulfuric acid and shipment to market are important factors. Most United States plants are located east of the Mississippi River, with concentration in the southeastern and extreme southern parts of the country where the major use of the product occurs. Use of the product is also high in California.

Raw materials requirements are 0.610 t 75% BPL rock and 0.460 t (sp gr 1.70 or 60° Bé) H_2SO_4/t of NSP.

A potentially advantageous variation in NSP production for wet, tropical climates is the material, Biosuper, a mixture of phosphate rock, sulfur, and bacteria. After application, the bacteria convert sulfur to sulfuric acid which solubilizes the phosphate into a form usable by plants. The cheap, simply made phosphate fertilizer has performed well in Australian field trials. The Commonwealth Scientific and Industrial Research Organization (CSIRO) estimates production cost at half to two thirds that of NSP. Such a process may prove of much interest and value to developing countries with suitable climate and acidic soils.

Triple Superphosphate. Although triple superphosphate (TSP), also known as concentrated superphosphate, was first made in Europe almost as early as NSP, it did not receive widespread use until after World War II.

The recent importance of TSP is shown in Figures 4 and 5. TSP was the first high-analysis phosphate fertilizer, containing 45–47% P_2O_5 compared to 20% P_2O_5 in NSP. World production of the material has remained at about 15–18% of total phosphate fertilizer production over 1955–1977, and world annual tonnage values have increased from 2.8 to 4.1×10^6 t P_2O_5 over 1970–1976. United States production of TSP ranged from 1.27 to 1.62×10^6 t P_2O_5/yr throughout 1965–1977.

There are 16 plants in the United States with annual production capacities of 30,000–340,000 t P_2O_5. Their combined capacity is 2.2×10^6 t/yr. Eleven of these are located in Florida, and one each in Missouri, Mississippi, North Carolina, Idaho, and Utah.

Simplicity of production, high analysis, and excellent agronomic quality are reasons for the sustained high production and consumption of TSP.

Chemistry and Properties. TSP is essentially impure monocalcium phosphate monohydrate ($CaH_4P_2O_8.H_2O$) made by acidulating phosphate rock with phosphoric acid according to the following reaction:

$$Ca_{10}(PO_4)_6.F_2 + 14\ H_3PO_4 + 10\ H_2O \rightarrow 10\ CaH_4P_2O_8.H_2O + 2\ HF \quad (11)$$

The complete chemistry of TSP production has been studied and reported in great detail (80). As in the production of NSP there are also reactions with impurity minerals. In fact, the increasing amounts of such impurities in United States commercial phosphate rocks, especially those from Florida, is beginning to be reflected in somewhat lowered amounts of citrate soluble P_2O_5 in the product (81). The range of constituents in commercial TSP made in 1965 from wet-process acid and phosphate rock are as follows (80): $CaH_4P_2O_8.H_2O$, 63–73%; $CaSO_4$, 3–6%; $CaHPO_4$, Fe and Al phosphates, 13–18%; silica, fluorosilicates, unreacted rock, organic matter, 5–10%; and free moisture, 3–6%.

The average citrate solubility of the P_2O_5 in best quality TSP is 98–99%, but products with citrate solubility values a few percentage points lower are not uncommon. The P_2O_5 citrate solubility of TSP made by a quick-cured process is ca 96%.

Other common properties of CSP are: bulk density, nongranular, 879 kg/m^3, and granular, 1116–2000 kg/m^3; and hygroscopic point at 30°C, 94% relative humidity.

Production Technology. Phosphate rock and wet-process phosphoric acid are the raw materials for manufacturing TSP. The grade of rock can be a little lower than that needed for NSP production. Over the years a large number of process modifications, both batch and continuous, have been used. Leading continuous processes are Broadfield, TVA cone mixing, Kuhlmann, Dorr-Oliver granular, Chemieban S.I.A.P.E., TVA rotary drum, mono-dicalcium phosphate, and high analysis.

For the production of nongranular TSP, 52–54% P_2O_5 acid is used without dilution or heating. The P_2O_5:CaO mole ratio, including the P_2O_5 in the rock, is 0.92–0.95. The rock is ground to 70% < 74 μm (−200 mesh). Pile curing for a few weeks is typical, as for NSP.

Granular TSP is preferred for direct application and is used in bulk blend fertilizers. A widely used slurry granulation process, such as the Dorr-Oliver process, is illustrated in Figure 9. The ground rock is mixed with 38–40% P_2O_5 acid in a series of tanks. The slurry from the tank is then mixed with a large proportion of recycled granules in either a pugmill or drum mixer. The granules are dried and screened, the product-size material being about 1 part in 13, making the recycle ratio 12:1.

TVA developed a one-step process for acidulation and granulation but the process has not been widely used in industry (82).

Economics. In contrast to NSP, the high nutrient content of TSP makes its shipment preferred over shipping the raw materials. Plants, thus, are located at or near the rock source. The phosphoric acid used, and the sulfuric acid required for its manufacture are produced at the site of the TSP plant.

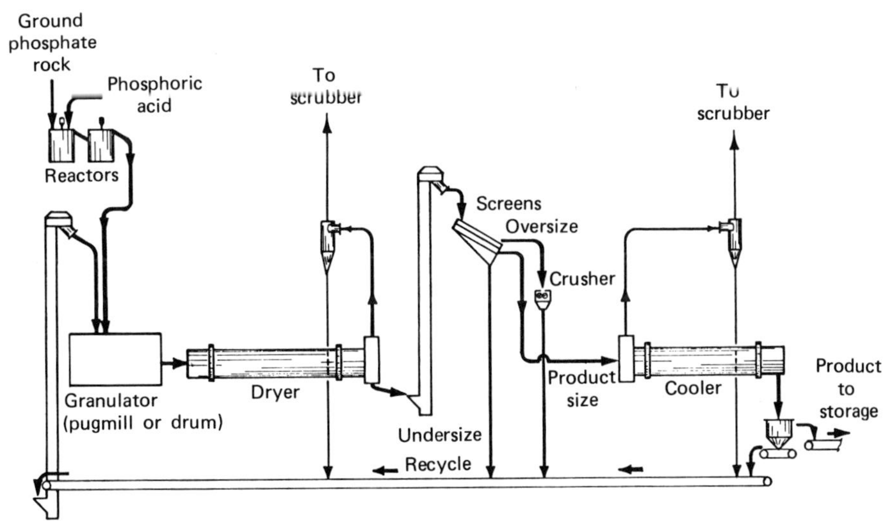

Figure 9. Continuous process for manufacture of granular triple superphosphate. Courtesy of TVA.

As for NSP, the cost for raw materials accounts for more than 90% of the total cost, most of this being the cost for acid. In January, 1978, the fob Tampa and North Carolina price for TSP was \$91–108/t, and delivered prices in Central California were \$130–154/t.

Fluorine emissions and recovery in the several TSP manufacturing processes are discussed under Environmental Aspects.

Basic Slag. Basic slag is produced only as a by-product of the steel industry, its yield and composition being determined by the phosphate and other impurity content of the iron ore used. It is formed by dephosphorization through aeration with air or oxygen and absorption of the phosphoric oxide in the basic siliceous slag produced in the furnace.

The principal producing countries are Belgium, France, the FRG, Luxembourg, and the UK. The material produced in the FRG and UK, basic Thomas or Bessemer converter slag, contains 15–20% P_2O_5, about 40–50% CaO and 0–13% Fe. Open hearth basic slag, containing only 8–10% P_2O_5, and blast-furnace slag, usually produced from low-phosphorus iron ores, are considered only liming agents and are used near the production site.

Basic slag is not a major fertilizer on a worldwide basis but it is important in the UK and Western Europe where its use ranges from 10–25% of the total fertilizer phosphate but is decreasing. World production of fertilizer basic slag was only 8×10^5 t P_2O_5 (51). Its use as percentage of total fertilizer phosphate is shown in Figure 4.

The phosphate in basic slag is water-insoluble; it is composed mainly of silicocarnotite, $5CaO.P_2O_5.SiO_2$; nagelschmidtite, $7CaO.P_2O_5.2SiO_2$; and sometimes tetracalcium phosphate, $4CaO.P_2O_5$ (83). The solubility of the phosphate in 2% citric acid is widely accepted as a measure of its availability for plants, although the correlation is only general. The slags usually contain small amounts of micronutrients, B, Co, Cu, Zn, and Mo.

Basic slag usually is ground to a finess of at least 80% $< 140\ \mu m$ (-100 mesh); most of the particles are 10–100 μm. Granulation has been attempted but has not been very successful. Mixtures of basic slag with potassium chloride are increasingly used, such salts sometimes being used as agglomeration aids.

The near-term outlook is for little or no growth in basic slag production, but if iron ores of sufficient phosphorus content continue to be processed, basic slag will remain a significant phosphate fertilizer.

Other Phosphates. Other phosphate materials that are used as fertilizer to a limited extent include bone meal, guano, and the thermal phosphates, Rhenania phosphate made in Germany and magnesium silicophosphate made in Japan. These materials are produced in decreasing relative amounts; together they represent some 3% of total phosphate fertilizer produced in 1976.

Recently, new tests of the effectiveness of thermal phosphates in tropical soils were initiated by the International Fertilizer Development Center. Production and use of such materials might become attractive in certain of the developing countries because of the simple technology of their production and of their suitability for use in acidic soils. The outcome of these tests is not yet available.

Potash

World fertilizer potash production, about 95% of all commercial potash, grew from 20.2×10^6 t K_2O in 1972–1973 to 23.5×10^6 t in 1975–1976 (51). Annual world consumption over the same period averaged about 2×10^6 t below production, with an average annual growth of 5.8% in consumption over the period 1972–1977. The Potash and Phosphate Institute estimates that world annual consumption is $28–29 \times 10^6$ t K_2O in 1980.

The potash industry is essentially a mining and beneficiation industry; the two main fertilizer materials, KCl and K_2SO_4, are produced by beneficiating ores at the mine site. The upgraded salts are then shipped to distributors and manufacturers of mixed goods.

Some 90% of all potash fertilizer is KCl; the proportion is slightly less in the United States. Most of the remainder of fertilizer potash is sulfate, either as K_2SO_4 or the double salt, langbeinite, $K_2SO_4.2MgSO_4$. A small but increasing amount of KNO_3 is used, mostly as a specialty fertilizer. This salt does not occur in significant quantities in nature; it is generally produced by the reaction of KCl and HNO_3. Test marketing of potassium phosphate fertilizer in the United States was begun by Pennzoil in 1976. Properties of the three principal fertilizer salts are shown in Table 10.

For most of the decade prior to 1965, the United States was the world's largest single producer of potash, more than 90% of this production coming from New Mexico. As the grades of these deposits have lowered and production costs risen, expanding production in Saskatchewan, Canada, has increasingly met the growing United States market. Today the United States is by far the largest single importer of potash.

The outlook for production at the 10 potash mines in Canada is in a confused state. Despite political disturbances, and takeover of some of the mines by the government of Saskatchewan, Canadian production is expected to continue at a very high level.

The Canadian problems have led to further explorations and rationalization of production in the United States. Drilling in Montana and North Dakota, extensions of the Saskatchewan deposits, have been intensified but the deposit in these regions is nearly 3000 m deep and would require solution mining which would be very costly at that depth (84) but less so than shaft mining. Renovations in progress in New Mexico should improve economy of production there, and reserves in the Carlsbad area are being reevaluated. Mining may continue there for several decades.

Table 10. Properties of the Principal Potassium Fertilizer Salts

Property	KCl	K_2SO_4	KNO_3
crystal form	cubic	rhombic	rhombic
K content, %	52.44	44.89	38.7
K_2O content, %	63.17	54.08	46.6
specific gravity (20°C)	1.988	2.662	2.11
refractive index	1.4904	1.494	1.5056
melting point, °C	790.0	588.0	334.0
hygroscopic point, critical rh at 30°C, %	84.0	96.3	90.5
solubility in water parts/100 parts water,			
at 0°C	27.6	7.35	13.3
at 100°C	56.7	24.1	246.0

Potassium Chloride. The three beneficiation methods for producing fertilizer grades of KCl are: thermal dissolution, flotation, and electrostatic refining. The choice between methods depends on factors such as grade and type of ore, local energy sources, amount of clay present, and local economic factors of primarily fuel and water costs.

The thermal method is based on the much higher solubility in water (100–110°C) of KCl compared to NaCl. The KCl is recovered in vacuum crystallizers, filtered or centrifuged, dried, and sometimes granulated by compaction, crushing, and screening.

The flotation process depends upon the fact that most potash deposits are mechanical mixtures of salts that can be separated after crushing. The process uses collectors such as long-chain fatty acid amines or acetates and frothing agents such as pine oil in a cell containing saturated liquor through which fine air bubbles are forced. KCl floats to the top of the bath where it is recovered from the foam (see Flotation).

In electrostatic beneficiation, the sylvinite ore is heated at 300–700°C to achieve greater retention of electrostatic charge. The ground ore is then electrostatically charged in a vibrating trough where the KCl becomes negatively charged and the NaCl becomes positively charged. The two are separated between plates with a potential difference of several thousand volts (see Extractive metallurgy).

The economics of new production facilities is of some concern to the industry. A facility of 9.07×10^4 t KCl/yr for which construction began in 1977, required an investment of 200 million dollars (84). Depreciation (20-year straight line) and interest (10% on one half of capital investment) charges alone would amount to about $20 per annual metric ton. To this must be added the operating, selling, and administrative costs and return on equity. The 1977 fob selling price of a ton of KCl was about $40.

Potassium Sulfate. Potassium sulfate is a preferred form for crops that have a low tolerance for chloride. It is produced from langbeinite by the metathetical reaction in aqueous solution:

$$K_2SO_4.2MgSO_4 + 4\ KCl \rightarrow 3\ K_2SO_4 + 2\ MgCl_2 \qquad (12)$$

The langbeinite, also called Sul–Po–Mag and K–Mag, is also used directly as a fertilizer. The beneficiated material contains 22% K_2O and 18% MgO. Production capacity for potassium sulfate and potassium-magnesium sulfate in the United States is ca 1.3×10^6 t/yr. Production of the double salt in 1976 was about 5.4×10^5 t, and United States demand for potassium sulfate in 1977 was about 8.5×10^5 t (some 4.25×10^5 t K_2O). Demand for these products is still increasing.

The price for potassium sulfate in September, 1977, was $193/t, carload lots, standard 50% K_2O, and $220/t for granular bagged. Price for the potassium-magnesium, standard, was $36/t and $47/t for granular.

Potassium Nitrate. Potassium nitrate has been known, but little used, as a fertilizer for many years. It may be reclaimed as a by-product of the production of sodium nitrate from natural deposits of caliche in Chile. It also has been produced by the double decomposition reaction between sodium nitrate and potassium chloride:

$$NaNO_3 + KCl \rightleftharpoons KNO_3 + NaCl \qquad (13)$$

The product salts were separated by fractional crystallization. The decline of natural saltpeter mining has virtually eliminated these processes as sources of KNO_3.

The current United States domestic commercial potassium nitrate contains 13.9% N, 44.1% K_2O, 0–1.8% Cl, 0.1% acid insoluble, and 0.08% moisture. The material is manufactured by Vicksburg Chemical Co. using a process developed by Southwest Potash Division of AMAX. The process uses highly concentrated nitric acid to catalyze the oxidation of by-product nitrosyl chloride and hydrogen chloride to more valuable chlorine (85). The much simplified overall reaction is:

$$2\,KCl + 4\,HNO_3 \rightleftharpoons 2\,KNO_3 + Cl_2 + 2\,NO_2 + 2\,H_2O \qquad (14)$$

The NO_2 is absorbed in nitric acid to produce the highly concentrated acid required for the overall process. Serious corrosion and other engineering problems appear to have been controlled, and production is increasing.

Israel Mining Industries developed a process in which KCl is converted to KNO_3 in one step at an ambient temperature:

$$KCl + HNO_3 \rightleftharpoons KNO_3 + HCl \qquad (15)$$

An organic liquid extractant is used to remove the HCl and shift the reaction in the desired direction. After concentration the by-product hydrochloric acid is suitable for production of phosphoric acid by the IMI process above.

Potassium nitrate is being used increasingly on intensive crops such as tomatoes, potatoes, tobacco, leafy vegetables, citrus, and peaches. The properties that make it particularly desirable for these crops are low salt index, nitrate nitrogen, favorable $N:K_2O$ ratio, negligible Cl content, and alkaline residual reaction in soil. Its low hygroscopicity (Table 10) favors its use in direct application and in mixtures. It is an excellent fertilizer but the high cost of production still limits its use to specialty fertilizers.

Potassium Phosphates. Because of their very high analysis—KH_2PO_4, 0-50-33; $(KPO_3)_x$, 0–60.13–39.87—their freedom from chloride, and their high solubility in liquid fertilizers, methods for economical production of potassium phosphates have been sought for many years. These methods have generally been based on the reaction of phosphoric acid or P_2O_5 with KCl, with formation of a variety of potassium phosphates and hydrochloric acid often contaminated with fluorine and other impurities. For products made from fertilizer KCl to be competitive, the added production cost must be offset by demonstrably superior value. The value and the means for disposal of the by-product HCl are critical components in the overall economic evaluation. The savings in shipment costs of high-analysis materials, the stability of high-potash liquid fertilizers, and the possibly improved fertilizing quality (Cl-free, low salt index) do, of course, justify some margin for increased cost. None of the earlier developments, however, has attained sustained large-scale production.

The most recent development to reach at least semicommercial scale is a process developed jointly by Goulding Fertilizer, Limited, of Ireland, and Pennzoil Company of Texas (86). The initial product is a mixture of H_3PO_4 and KH_2PO_4, which can be precipitated and heated to produce polymers with a range of grades and solubilities. Alternatively the H_3PO_4–KH_2PO_4 mixture may be ammoniated to produce an 8-48-12 fertilizer. A small plant for manufacture of the monomer material, sometimes ammoniated, as above, has been operated by Pennzoil at Hanford, California. An 8-48-12 fertilizer has been test-marketed on the United States west coast for crops benefitting from Cl-free fertilizer.

Other Potassium Fertilizers. A large proportion of fertilizer potash is used in mixtures—bulk blends, liquids, and granular N–P_2O_5–K_2O grades. These combination grades are discussed under Mixed Fertilizers (see Potassium compounds).

Secondary Nutrients

Calcium, magnesium, and sulfur have long been called secondary nutrients, as they are generally required at lower levels than the primary, macronutrients but above the trace levels of micronutrients. Because they are often found in plants at levels similar to that of the primary nutrients, they are often classified as macronutrients. The classification, secondary or secondary macronutrients is used here.

Calcium. Soil minerals are a main source of calcium for plants, making nutrient deficiency of this element in plants rare. As a soil amendment for pH adjustment and other soil conditioning, calcium is frequently under used in many acidic soils. Although correction of these conditions is important for agriculture, and liming agents often are sold by fertilizer distributors, this function is not one of fertilizer manufacture.

One major source of calcium fertilizer is phosphate rock. Acidulation of the rock to produce phosphate fertilizer provides large amounts of calcium in the fertilizer as sulfate, nitrate, and sometimes indirectly as carbonate. As calcium-containing superphosphates have given way to ammonium phosphates, the overall level of calcium in fertilizers has decreased substantially. Calcium cyanamide is another little-used calcium-containing fertilizer. Although the calcium content of fertilizers may be decreasing, it is not yet seen as a serious plant nutrient problem, and should such a problem develop it can readily be corrected by applying to the soil lime or limestone instead of adding such diluent materials to high analysis fertilizers.

Magnesium. Like calcium, magnesium is present in soil minerals, as well as being widespread in massive mineral forms. Magnesium minerals, however, are more soluble than those of calcium and its level in soils is lower than that of calcium. Dolomite, calcium–magnesium carbonate, is often used as a liming agent and to provide nutrient magnesium. Soils are occasionally magnesium deficient. An example of magnesium deficiency in a crop is often seen in the lush, spring growth of grass which produces the sickness grass tetany in cattle. Magnesium deficiency is also sometimes seen in citrus crops grown on calcareous and high-sodium soils. It, therefore, is not unusual to add magnesium compounds to fertilizers.

Most supplemental magnesium is supplied as dolomitic limestone but other important sources are magnesium sulfate, potassium–magnesium sulfate (langbeinite), as well as a few organic compounds.

Magnesium can undergo reactions with phosphorus compounds, sometimes lowering phosphate solubility. But the addition of magnesium to fertilizers, and the correction of magnesium deficiency generally presents no insurmountable problems. Sometimes, however, it is necessary to feed a magnesium supplement in the diet for animals.

Sulfur. Sulfur occurs in plants at levels as high as that of phosphorus. It is absorbed by plants as the sulfate anion (SO_4^{2-}), an important constituent of many fertilizers. Normal superphosphate contains much calcium sulfate. Wet-process phosphoric acid often contains a few percent of sulfur as sulfuric acid and calcium sulfate, and this sulfur appears in fertilizers made from the acid.

Other sources of sulfur in soils are decreasing. In industrialized, humid regions sulfur deposits in rain range from 28 to 168 kg/hm^2 yearly. Soils in such regions have not been deficient in sulfur, but as environmental goals of eliminating sulfur oxide emissions from industrial plants are met, sulfur deficiency in soils can be expected to increase. All indications are that fertilizers will need to contain increasing amounts of sulfur. In fact, symptoms of sulfur deficiency in soils of the United States have been increasing for some years. Sulfur deficiency had been identified in 13 states in 1960, and the number of such states grew to 31 by 1973.

The average sulfur concentration of fertilizers in the United States has been decreasing since the 1950s as the NPK grade of fertilizers has increased. At the same time, sulfur consumption in fertilizer production has increased from 9.6×10^6 t in 1969–1970 to 11×10^6 t in 1975–1976, but the increase has been largely for the production of wet-process acid, with disposal of the sulfur as by-product gypsum.

Some of the principal forms in which sulfur is incorporated in fertilizers are as sulfates of calcium, ammonium, potassium, magnesium, and as the element. Ammonium sulfate, normal superphosphate, and sulfuric acid frequently are incorporated in ammoniation granulation processes. Ammonium phosphate–sulfate is an excellent sulfur-containing fertilizer, and its production seems likely to grow. Some common grades made by this process are: 12-48-0-5S, 12-12-12-12S, and 8-32-8-6.5S.

Granular sulfur-containing fertilizers used in bulk blends include ammonium sulfate, urea–ammonium sulfate, ammonium phosphate–sulfate, elemental sulfur, and pelletized lignosulfonate. Sulfur-coated granular fertilizer can be used but this expensive form of sulfur is generally used to achieve other advantages such as controlled release of the coated nutrient.

Elemental sulfur and sulfur compounds can be added to suspension fertilizer, and ammonium thiosulfate is a common addition to clear liquid fertilizer.

Micronutrients

The generally recognized micronutrient elements are: boron, copper, iron, manganese, molybdenum, and zinc. Sometimes cobalt is considered a micronutrient, as it plays a role in the leguminous fixation of nitrogen.

Attention to meeting the micronutrient needs of crops has greatly increased during the last decade. Evidence of progress and need in the field was analyzed in a joint TVA-Soil Science Society of America symposium in 1972 (1).

These essential nutrients for plants are required only in micro amounts and some can be harmful to plants when used in excess. Herein lies the problem in micronutrient fertilization: adequately identifying the need, reducing this to a prescription, compounding the fertilizer, and distributing it evenly in the soil. Prescription compounding is being used increasingly as deficiencies are better identified but the shotgun approach is still used also. In this method a range of micronutrients is added to fertilizers at levels believed low enough to avoid harmful effects but adequate to prevent deficiencies. Fertilizers of this kind usually are produced in ammoniation-granulation plants. When these are sold as premium fertilizers because of the added micronutrients, the minimum amounts of guaranteed micronutrients allowed by law are: 0.02 wt % B, 0.05 wt % Cu, 0.10 wt % Fe, 0.05 wt % Mn, 0.0005 wt % Mo, and 0.05 wt % Zn.

Generally, soluble materials are more effective than insoluble ones, especially insoluble oxides. For this reason many soil minerals containing micronutrient compounds are ineffective for plants. Main micronutrient sources and uses are summarized below (see Mineral Nutrients).

Boron. The major materials used are borax, sodium pentaborate, sodium tetraborate, partially dehydrated borates, boric acid, and boron frits. Soil application rates for vegetable crops and alfalfa are usually in the range of 0.5–3 kg B/hm^2. Lower rates are used for more sensitive crops. Both soil application and foliar application are practiced but soil applications remain effective longer. Boron toxicity is not often observed in field applications.

Copper. Some 15 copper compounds have been used as micronutrient fertilizer. These include copper sulfates, oxides, chlorides, and cupric ammonium phosphate, and several copper complexes and chelates. Recommended rates of application range from a low of 0.2 to as much as 14 kg Cu/hm^2. Both soil and foliar applications are used.

Iron. As with copper, some dozen or more materials are used as fertilizer iron sources. These include ferrous and ferric oxides and sulfides and ferrous ammonium phosphate, ferrous ammonium sulfate , frits, and chelates. In many instances organic chelates are more effective than inorganic materials. Recommended application rates range widely, both with type of micronutrient used and with crop. Quantities range from as low as 0.5 kg Fe/hm^2 as chelates for vegetables to as much as a few hundred kg Fe/hm^2 as ferrous sulfate for some grains.

Manganese. Commonly used manganese fertilizer materials are manganous and manganic sulfates, chlorides, carbonates, and oxides and frits, and chelates. Soil application rates range from about 2 to 150 kg/hm^2 of Mn.

Molybdenum. The commonly used molybdenum materials are sodium molybdate, ammonium molybdate, molybdenum trioxide, molybdenum sulfate, and frits. Molybdenum is used in smaller amounts than any of the other micronutrients, ranging from a few grams to 3 kg/hm^2 of Mo.

Zinc. Zinc, one of the most widely used micronutrients, is applied as sulfates (both basic and normal hydrates), carbonate, sulfide, phosphate, oxide, chelates, and other organic materials. Rates of application range from 0.2 to 22 kg Zn/hm^2.

Incorporation in Fertilizer Granules. The surest way to achieve even distribution of micronutrients in fertilizer granules, and to achieve even field distribution with solid fertilizers, is to incorporate the micronutrient material as one of the feeds in the granulation process. A problem with this method, however, is that granulation processes are most efficient and economical when operated continuously to produce large tonnages of the same or similar composition. The granulation process is not adaptable to quick changes in product composition and to storage of a wide range of grades. These factors seriously limit the practice of prescription micronutrient formulation in ammoniation-granulation processes. These granulation processes more often use the shotgun approach. Granular grades that have been produced in this way without causing problems are: 6-12-12-10 S–2 Mg–0.10 B–0.10 Fe–0.45 Mn–0.35 Zn and 4-12-24-6 S–2 Mg–0.10 B–0.10 Fe–1.0 Zn.

Incorporation in Bulk Blends. Prescription formulation of micronutrient fertilizers is more practical in the production of bulk blends and fluid fertilizers, as these processes are more adaptable to rapid changes in grades, and thus, to custom formulation. The problems with formulation and distribution in bulk blends are in achieving even distribution and avoiding segregation of the nutrients in storage, transportation, and application. The fundamental concepts are the same as those already described in avoiding segregation of primary nutrients in bulk blends.

One way to achieve this is to blend properly-sized granular premium fertilizer with the other granular components of the blend. Granular micronutrient materials can also be blended. A potential problem with this method is that the micronutrient-containing granules may be too few to yield acceptable distribution of micronutrient in the field. Spaces between micronutrient-affected volumes of soil may be too large.

Coating granular fertilizers with powdered micronutrients overcomes the disadvantages of using granular micronutrient additives. When the coating is done well, segregation is not a problem because the micronutrient becomes essentially an integral part of each granule. The process generally consists of (*1*) dry mixing the granular blend with powdered micronutrient materials, (*2*) adding a liquid binder—a fertilizer solution or an oil—during mixing, and (*3*) continuing the mixing for a short period. The total cycle, requiring about 3 min, has been described in detail (87).

Incorporation in Fluid Fertilizers. In terms of homogeneity and even distribution, fluid fertilizers are probably the best micronutrient carriers. Fluid carriers of micronutrients usually are nitrogen solutions, clear liquid mixtures, or suspensions. Some micronutrients, however, are applied as simple water solutions or suspensions. Foliar micronutrient sprays often contain only the micronutrient material.

Several zinc and copper micronutrient compounds are soluble in a variety of nitrogen solutions. Ammonia–ammonium nitrate solutions containing 2.5% Zn and 1% Cu can be prepared (87). Moderate amounts of micronutrients are soluble in ammonium nitrate-water solutions. Micronutrients are not very soluble in urea–ammonium nitrate solution unless the pH is raised to 7 or 8 by adding ammonia, whereupon zinc and copper become much more soluble.

Orthophosphate liquid mixtures are ineffective as micronutrient carriers because of the formation of metal ammonium phosphates such as $ZnNH_4PO_4$. However, micronutrients are much more soluble in ammonium phosphate solutions in which a substantial proportion of the phosphorus is polyphosphate. The greater solubility is owing to the sequestering action of the polyphosphate. The amounts of Zn, Mn, Cu, and Fe soluble in 11-16-0 base solution with 70% of its P as polyphosphate are ten to sixty times their solubilities in ammonium orthophosphate solution. When a mixture of several micronutrients is added to the same solution, the solubility of the individual metals is lowered significantly. In such mixtures the total micronutrient content should not exceed 3% and the storage time before precipitates appear may be much shorter than when only one micronutrient is present.

Suspension fertilizers offer a simple and effective way of preparing homogeneous products containing high concentrations of macro- and micronutrients because the solubility of the micronutrient is immaterial, except as it affects agronomic response. Incorporation of the micronutrient may be carried out by adding the desired amount to the suspension either while it is mixed or after. As with clear liquids, it is important to provide good agitation to disperse the powder.

Synthetic chelates and natural organic complexes are sometimes more effective agronomically per unit of micronutrient than most inorganic forms. The organic materials are more expensive. The chelates can be used with both orthophosphate and polyphosphate liquids and suspensions.

Chemical Reactions in Mixtures. The chemical systems that can be created in mixtures of six micronutrients in NPK mixed fertilizers are many and highly complex (88). Many chemical reactions occur, and the solubility of the reaction products determines the micronutrient solubility.

The six micronutrients fall into two groups on the basis of their tendencies to undergo adverse reactions in carrier fertilizers. Boron and molybdenum undergo few reactions that lower their solubility. Copper, iron, manganese, and zinc react to form more insoluble compounds. Of these adverse reactions those of copper are least detrimental.

Use of micronutrient fertilizers is expected to become increasingly important. To achieve their effective use will require cooperative work by soil chemists and agronomists to identify and quantify needs, and of fertilizer production technologists for effective production.

Multinutrient Fertilizers

These materials, generally called compound fertilizers in Europe and other places, represent a very important class of fertilizers, both in terms of amount and in the large and varied technologies for their production. They contain more than one primary nutrient, eg, N–P, N–K, P–K, or N–P–K, in whatever form—liquid, solid, granular, or nongranular—combined in a single chemical compound or as several compounds.

The production and use of mixed fertilizers has grown steadily over many years. On a world basis, mixtures accounted for about 46% of all phosphate fertilizer produced in 1970, the proportion increasing to 59% in 1976, Figure 4. The trend in the United States is similar, as is indicated by the line for ammonium phosphates in Figure 5. More significantly, the contribution by class of fertilizers in the United States in 1975 indicates that 43% of all fertilizer was dry mixtures (bulk and bagged), and about 8% was liquid mixtures, for a total of 51% of all fertilizers used (28). Of the mixtures, 51% was dry bulk, 34% was dry bagged, and 15% was liquid. A trend of major economic importance is the increasing nutrient content of mixtures, from about 29% N–P_2O_5–K_2O in 1957 to 43% in 1978.

Four major types of mixtures are discussed: (*1*) those made generally by ammoniation–granulation (includes ammonium phosphates, ammonium phosphate-based materials, and superphosphate-based materials and some granular materials made without ammoniation), (*2*) nitrophosphates, (*3*) blends of granular materials, and (*4*) fluid mixed fertilizers (clear liquids and suspensions).

Granulation. The granulation of fertilizers represents one of the most significant advances of fertilizer technology. The proportion of mixed fertilizers produced by ammoniation–granulation in the United States passed through a maximum of about 80% during the early 1960s and has followed a decreasing trend line through 1975 as the production of bulk blends and liquid mixtures increased, Figure 10 (89). Recent data indicate a probable stabilizing of the market distribution at a little over 40% each for granular compounds and bulk blends and a little under 20% for liquid mixtures. The production of bulk blends and liquid mixtures are not yet major industry segments in most other parts of the world. In the United States, the number of granulation plants decreased from about 300 to 150–160, the tonnage of granular compounds decreased by 25.6%, and the average annual tonnage per plant increased from about 40,000–60,000 t (89).

There are many variations of granulation processes, the more significant differ-

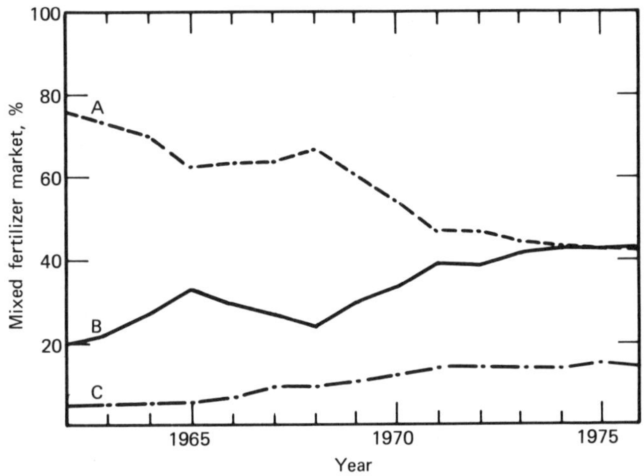

Figure 10. Share of United States mixed fertilizer market by type of product (89): A, granulated compounds; B, bulk blends; C, liquid mix. Courtesy of the British Sulphur Corporation Limited.

ences are based on the nature of the feed materials, how water is removed, and the granulation device used. Attention here is focused on granulation of three types of feed materials: primarily solids, slurries, and melts. Although the distinction between solid feed- and slurry-granulation is not sharp—liquids and solids may be feeds in both—the predominance of solids or of slurry in the feed affects the mechanism and control of granulation. Process design and economics can be greatly influenced by these variables.

Granulation of Solid Feeds. Granulation requires the presence of liquid and solid phases and the ratio of these determines whether granulation is effected by agglomeration or layering. Granulation of solid feeds usually occurs by agglomeration; layering occurs when larger amounts of liquid phase are present and much solid phase is recycled as in slurry granulation. Control of the process depends on the granulation efficiency of the formulation. When a formulation has a high granulation efficiency, the proportion of material recycled through the granulator is small, usually no more than one part per part of product. Thus the production capacity of a plant is high compared to systems requiring high recycle ratio.

Superphosphates have long been important feeds for these processes. When water solubility of the phosphate is a primary consideration, ammoniation of superphosphate is limited to neutralization of free acid. Further ammoniation forms water-insoluble dicalcium phosphate and can form even less soluble calcium phosphates. With NPK mixtures containing ammonium nitrate, precautions must be taken to avoid formation of calcium nitrate, a very hygroscopic material.

The push toward higher analysis granular products makes the use of urea in granulation highly desirable. Precautions must be taken when urea is granulated with superphosphate, as urea and monocalcium phosphate form an adduct [$CaH_4P_2O_8$-$.4CO(NH_2)_2$]. The water of crystallization that is freed from the superphosphate ($CaH_4P_2O_8.2H_2O$) can result in very wet products. Studies of this problem and the diverse practices around the world in dealing with it have been described (90–91). The reaction can be avoided or controlled by raising the pH of the feed materials to at least

4.5 by use of ammonia or chalk. Where products are not sold on the basis of water-soluble P_2O_5, light ammoniation of the superphosphate is beneficial. Over ammoniation interferes with granulation. The presence of ammonium sulfate or potassium sulfate in the formulation decreases the liquid phase through the formation of double salts such as $(NH_4)_2SO_4.CaSO_4.H_2O$ and $K_2SO_4.CaSO_4.H_2O$, which tie up water.

Because it is higher in P_2O_5 than triple superphosphate and also carries nitrogen, monoammonium phosphate (MAP) is a desirable solid feed for granulation processes. It can be stored, shipped and handled more conveniently and cheaply than phosphoric acid. Moreover, MAP does not react with urea as does TSP, making this combination of materials popular and useful in the production of high analysis granular products.

Economics. The flexibility and economy of solid granulation processes make them attractive. According to a recent cost comparison of granulation of solids and slurries, the plant investment cost per annual ton of product is one-third higher for slurry granulation, and operating and utility costs are said to be substantially higher also. Disadvantages of granulation of solid feeds are poorer granule appearance and quality (not of agronomic significance), and the requirement of a separate plant for producing the phosphate intermediate (MAP or superphosphate). The cost advantages mentioned above can, in some instances, outweigh the cost of producing the phosphate intermediate.

Cost comparisons for specific plants doubtless vary widely. Nevertheless, over the past 10 years with the upsurge of use of MAP as a granulation intermediate, solids-granulation processes have regained a substantial part of the market that they formerly lost to slurry granulation.

Slurry Granulation. In slurry granulation, a thin film of slurry is spread over a large surface area of recycled particles, and the particles are dried to form solid deposit. This cycle is repeated many times to produce granules of the desired size. This layering process is controlled by adjusting the slurry concentration (usually 8–12 wt % water) and by the amount of solids recycled. The recycle ratio may range up to 5 or more for some formulations but for others it can be as low as 2. Advances in means of removing water in the granulation-drying operation and of reducing the amount of water in the slurry have lowered the recycle ratio and improved the economics of slurry granulation.

For reasons of economy and flexibility it became common practice to ammoniate increasing amounts of acids (H_3PO_4 and H_2SO_4) in the granulator. The amounts of liquid phase and of recycle are adjusted appropriately. Frequently, economics favors use of increasing amounts of acids up to the point that the amount of solids that must be recycled becomes prohibitive—a recycle ratio of 6 or more. Another limitation on acid feed to the granulator is a combination of liquid phase and temperature in the granulator (92). Preneutralization makes it possible to use up to about 550 kg of combined 54% P_2O_5 phosphoric acid and concentrated sulfuric acid per ton of granular ammonium phosphate–sulfate NPK product (92).

A modern drum ammoniator-granulator showing the arrangement of spargers in the granulator for addition of ammonia, sulfuric acid, phosphoric acid, and preneutralized slurry, and the means for solids addition, is shown in Figure 11. These usually range in size from 5 to 6 m in length, the diameter being about half the length. The granulator is rotated at speeds of 40–50% of critical speed—that speed at which the contents of the drum are held against the shell by centrifugal force.

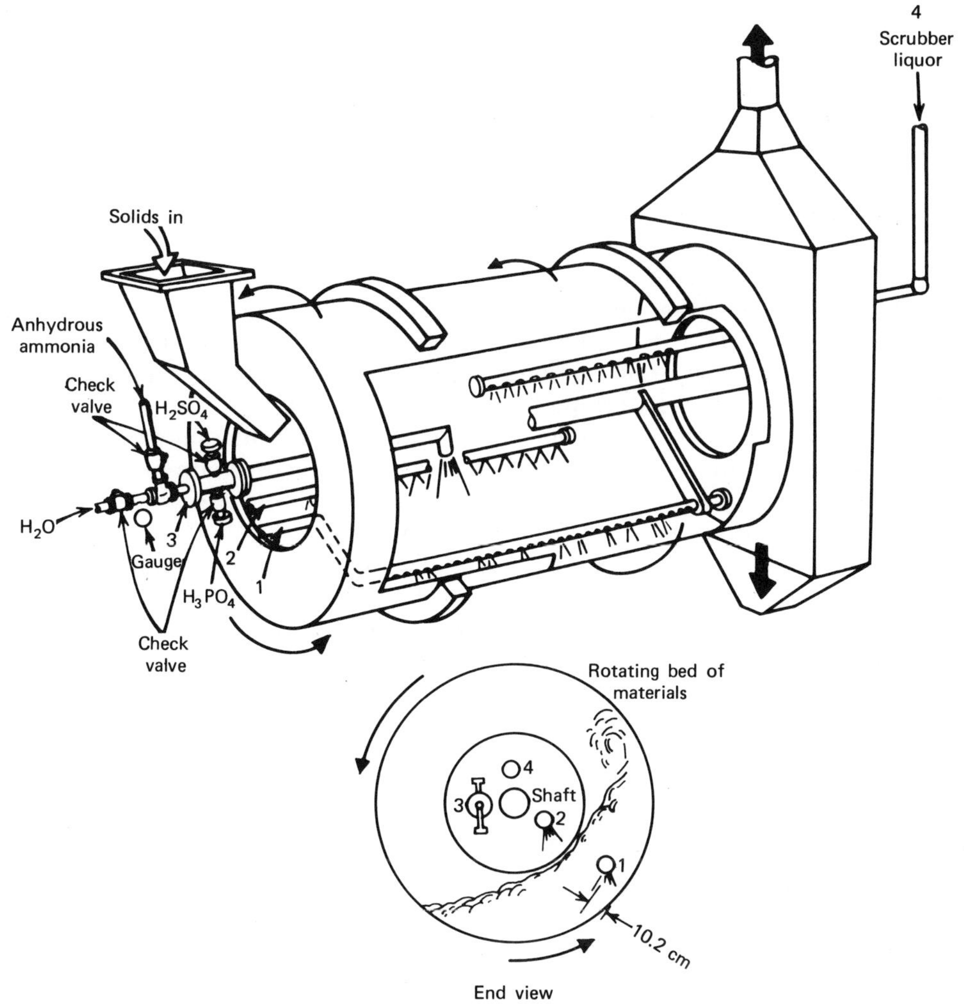

Figure 11. Location of pipe-cross reactor and spargers in ammoniator–granulator. 1, ammonia sparger, located at the 4 o'clock position 11.4 cm from granulation shell with holes facing the rotating stream; 2, phosphoric acid sparger, located to discharge phosphoric acid onto the top and near the center of the rotating bed of materials; 3, pipe-cross reactor; 4, scrubber liquor, located above the bed in granulator to dribble scrubber liquor onto bed.

The equipment required for drying wet granules and the cost of the energy for the dryer provide incentive for finding a more efficient means of water removal in slurry granulation processes. The most recent advance in ammoniation-granulation technology uses a pipe-cross reactor to replace the tank-type preneutralizer and to eliminate the dryer. This system is described under Melt Granulation.

Water Removal During Granulation. By spraying concentrated solution or slurry onto small, hot, aerated solid particles in rapid motion, granulation and drying can be carried out simultaneously in the same vessel. The Scottish Agricultural Industries (SAI) developed an internal recycle process for simultaneous granulation and drying, but the method was not widely adopted.

Chemical and Industrial Corporation (now C&I Girdler) developed a granulator dryer called the Spherodizer. Use of the method to granulate melts of ammonium nitrate and urea has already been described. The method, operated as a hot spherodizer, has found wide acceptance for granulating nitric phosphate and other NP and NPK slurries. Preneutralized slurry is sprayed under pressure onto a dense curtain of recycled material cascading from lifters in a rotating drum (93). A stream of heated air flows through the drum cocurrently with the solids and the sprayed slurry causing flash evaporation of water and deposition of dense concentric layers of solids around each of the particles (see also Drying).

Many plants using this method are now operating throughout the world. Typical drum sizes are 3–4.5 m in diameter, with lengths up to 9.2 m. Typical capacities are 250–500 t/d. Even larger drums are now planned, with single-train capacities up to 1,000 t/d, making them more competitive with prill towers.

The third and newest development in combination granulation-drying units is the spouting-bed granulator (94). A stream of hot gas, together with preneutralized slurry, is caused to flow upward through a mass of granular material in an inverted cone. No bed support is required for the rapidly moving fluidized solids. Water is rapidly evaporated and solids deposit, onion-skin fashion, on the granules. The exit gas flows through a cyclone dust collector, and the dust is recycled to the granulator. Recycle ratio is about 2:1. Several such plants have been built in Europe and a few small ones granulate ammonium sulfate in the United States. Although attractive advantages are claimed for the method—simple construction, absence of rotating parts, easy collection of effluents, exceptionally strong and regular granules—it has seen very slow adoption by the industry.

Melt Granulation. In melt granulation essentially all the water is removed before the granulation. The granules require no drying and usually the recycle ratio is small. Multicomponent fertilizers usually have higher melting points and high viscosities, making their granulation difficult. An exception is a mixture of monoammonium phosphate (MAP) and ammonium nitrate with N:P_2O_5 weight ratio of 2:1. This mixture melts at about 140°C and has a low viscosity. The melt is granulated in a drum granulator, using a recycle ratio of about 1:1. To form NPK products, fine crystalline potassium chloride is added to the granulator. Water content of finished products is 0.3–0.4%. Usually a coating agent is used to ensure good storage properties. Some typical grades at 24-24-0, 30-10-0, 17-17-17, and 23-11-11.

Although certain compositions containing ammonium nitrate and potassium chloride can cause burning, mixtures with component ratios 1:2:2, 1:1:1, and 1:2:1 can be granulated and handled safely (93).

The newest technology for melt granulation of multicomponent fertilizers was developed and demonstrated by TVA (95) and by Norsk Hydro (96–97). In the TVA method ammonium phosphate melt is produced by the reaction of preheated (66°C) 54% P_2O_5 wet-process acid with anhydrous ammonia in a pipe T reactor at 218°C (Fig. 12). After disengaging steam and ammonia vapor from the frothy melt in a rotating blade mechanism, the melt can be granulated in a pug mill to produce ammonium phosphate granules (11-55-0 to 12-57-0 grade) in which up to 25% of the P_2O_5 is polyphosphate. Recycle ratio is about 3.5:1. The granules require no drying or conditioning.

Alternatively, highly concentrated urea solution (97–99% urea) or prills can be cogranulated with the ammonium phosphate melt to produce urea–ammonium

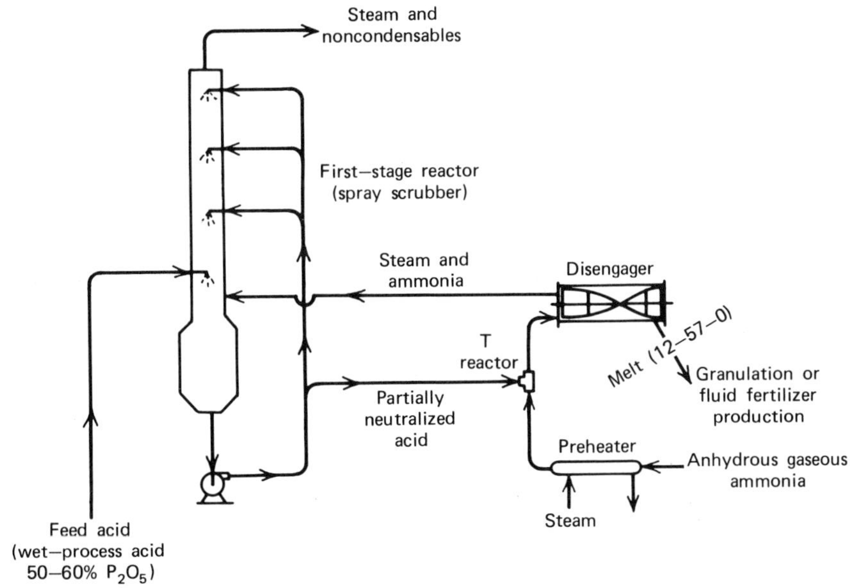

Figure 12. TVA ammonium polyphosphate process. Courtesy of TVA.

phosphate grades 28-28-0 and 35-17-0. Potassium chloride can also be added to the granulator to produce 19-19-19. In a pilot plant, the 19-19-19 grade has been granulated by spraying the phosphate melt onto the rolling bed of solids in a rotary granulator similar to that shown in Figure 11.

TVA estimated the energy savings for a new 50 t/h 11-55-0 melt-granulation plant to be about 1.055 GJ/t P_2O_5 (10^6 Btu/t) (98). This lowered energy requirement together with investment and operating savings gives an estimated savings of \$3.65/P_2O_5 as of September 1975. In the case of a new 30 t/h 19-19-19 melt-granulation plant, the estimated saving was about \$5.10/t of nutrient.

In a pilot plant at Porsgrunn, Norway, Norsk Hydro has developed the prilling of ammonium phosphate melt produced in a TVA pipe reactor, as well as urea-ammonium phosphate (UAP) and UAPK mixtures (96). The scale of the operation was 100 kg/h MAP and 300–900 kg/h of UAP and UAPK formulation. The prilling tower, equipped with a centrifugal distributor, was 28 m tall.

Norsk Hydro estimates the capital cost for a prilling plant to be 30–50% lower than that for a comparable granulation plant. Nutrient losses are near zero, and effluent problems are said to be less than those of granulation plants.

Another adaptation of a pipe-reactor melt-production system is the pipe-cross reactor (92,99). This reactor represents a significant advance in the production of ammonium phosphate–sulfate granular products in that it eliminates both the usual tank-type preneutralizer and the dryer (92,99). The use of this reactor also reduces the production of fumes and dust and facilitates their abatement. Several commercial plants have adapted the method by discharging the reactor melt into the granulator as shown in Figure 11. Widespread adoption of this method can be expected.

High-polyphosphate (50% of the P_2O_5) solid ammonium phosphate, such as that formerly produced from electric furnace acid (15-62-0), is a highly desirable shipping intermediate for use in the production of liquid and slurry fertilizers. Melts of such materials made from wet-process acid, however, are highly viscous and no way of granulating these has yet been found. Use of such a material in fluid fertilizer does not require that it be in the form of spherical granules. TVA is currently studying a flaking proces for such high polyphosphate melts.

Compacted Fertilizers. Compaction is used on a large scale for granulating the single-nutrient fertilizer, potassium chloride, but its use with multinutrient materials is limited mainly to producing specialty materials for lawns, shrubs, ornamental plants, fruits, etc. The method is well suited for producing the great variety of specialty materials, as the shift from one formulation to another can be made quickly without loss of material or efficiency and without time consuming modifications.

A compactor consists of two rolls rotating at the same speed but in opposite directions. The nip between the rolls is variable, as is the pressure (3.6–6.3 t/cm width of roll). The powder or grains of material to be compacted is fed into the nip and exits as a sheet or flakes. The flakes are crushed and sized.

Not all solids can be granulated by compaction. Potassium chloride, diammonium phosphate and concentrated superphosphate compact readily but compaction of potassium sulfate is less effective (100). Particle size of the material to be compacted is an important variable, and optimum moisture content is 0.5–1.5%, depending on the composition of the mix.

Economics. The number of variables in granulation methods and frequently overriding site-specific considerations preclude any meaningful overall economic comparison of methods. There are strong indications that frequently the formulation and grade have more effect on investment and operating costs than does the granulation method. On the other hand, the outlook for availability and cost for energy for the foreseeable future, together with investment savings, give strong incentive to the consideration of melt-granulation processes for new installations.

Production capacity is an important economic factor, and in this respect prilling has some economic advantages. Single-tower prilling units with capacity up to 3,500 t/d are now possible (97), giving this granulation method substantial advantage for some type of products. A single-drum slurry-granulator (hot Spherodizer) has not yet exceeded a daily capacity of 500 t, and drums of this size frequently must be fabricated at the installation site.

Because of their low recycle requirement (frequently <1:1) and process simplicity, solids granulation processes usually have unit production capacities higher than those of slurry granulation in units of similar size, sometimes by as much as a factor of 2 or 3. Cost comparisons (mid 1977) for slurry and solid granulation of 17-17-17 product are reflected in the following ratios: investment, 0.75; operation, 0.69; and utilities, 0.55. In some cases this difference may more than compensate for the added cost for the solid intermediates (usually TSP or MAP).

The foregoing are among the more important economic factors that require careful evaluation for specific situations. Further cost comparisons for ammonium phosphate processes, and for nitric phosphate processes vs sulfur-based NPK processes are provided in sections discussing these particular materials.

Ammonium Phosphates. The production of ammonium phosphate in the United States, mostly DAP, increased tenfold during 1955–1966 and tripled again in the next decade, reaching a total of 3.97 million t P_2O_5 in 1977—63.9% of all United States phosphate fertilizer (see Fig. 5). As of 1977 there were 47 ammonium phosphate plants in the United States and 10 in Canada.

On a world basis, ammonium phosphate production grew from 15% of all fertilizer P_2O_5 in 1970 to 26%—some 6.3 million t P_2O_5—in 1976 (see Fig. 4). In contrast to the heavy emphasis on DAP in the United States, MAP was the predominant form produced in Canada and Europe. In addition to DAP and MAP, ammonium polyphosphate (APP) is used extensively in liquid fertilizers, and recent development of economical methods for producing it in granular form opens the way for its increasing use in solids applications.

Because of the economy of their production and distribution, excellent physical properties, and excellent agronomic performance, these materials are applied in all phases of fertilizer production and use.

The properties of monoammonium phosphate (MAP), $NH_4H_2PO_4$, and diammonium phosphate (DAP), $(NH_4)_2HPO_4$, are shown in Table 11.

Because of impurities in wet-process acid, industrial grades of MAP range from 10-50-0 to 11-56-0. No set grade has been established for MAP. The established grade for industrial DAP is 28-46-0, and some difficulty is being experienced in making this grade because of the lowering quality of phosphate rock.

Diammonium Phosphate. The most widely used process for making DAP is illustrated by the flow diagram in Figure 13. The process takes advantage of the change in solubility of ammonium phosphates with change in mole ratio NH_4:H_3PO_4 to assist in control of granulation (101–102). This is done by operating the preneutralizer near the mole ratio 1:4.5 which gives maximum solubility, and the ammoniator–granulator is operated at mole ratio 2, a point of low solubility. This method of operation lowers the amount of recycle material required. The recycle ratio required in the production of 18-46-0 is about 5 kg/kg of product. The temperature in the preneutralizer is about 118°C, evaporating much water without excessive loss of ammonia. Use of the phosphoric acid feed to recover ammonia volatilized from the preneutralizer and granulator is common practice. The cost of producing DAP is compared with that for APP in a later section.

Monoammonium Phosphate. Within the last 10 years demand for MAP has grown steadily (103–104), and of the new ammonium phosphate plants announced in the last 5 years MAP plants have outnumbered DAP plants by about 2:1. This growing demand for MAP is attributed to the simplicity and economy of its production, storage, and transportation, its high analysis, its great versatility in use (granulation aid, bulk blends, suspensions), and its excellent physical properties. Since it is higher in grade

Table 11. Properties of Pure MAP and DAP

Property	MAP	DAP
N, %	12.2	21.2
P_2O_5, %	61.7	53.8
specific gravity	1.803	1.619
ammonia vapor pressure, kPa[a]		
73°C	insignificant	107.6
100°C	insignificant	140.0
125°C	0.004	300.0
solubility at 20°C, g/L	299.0	569.0
critical humidity at 30°C, %	92.0	83.0

[a] To convert kPa to psi, multiply by 0.145.

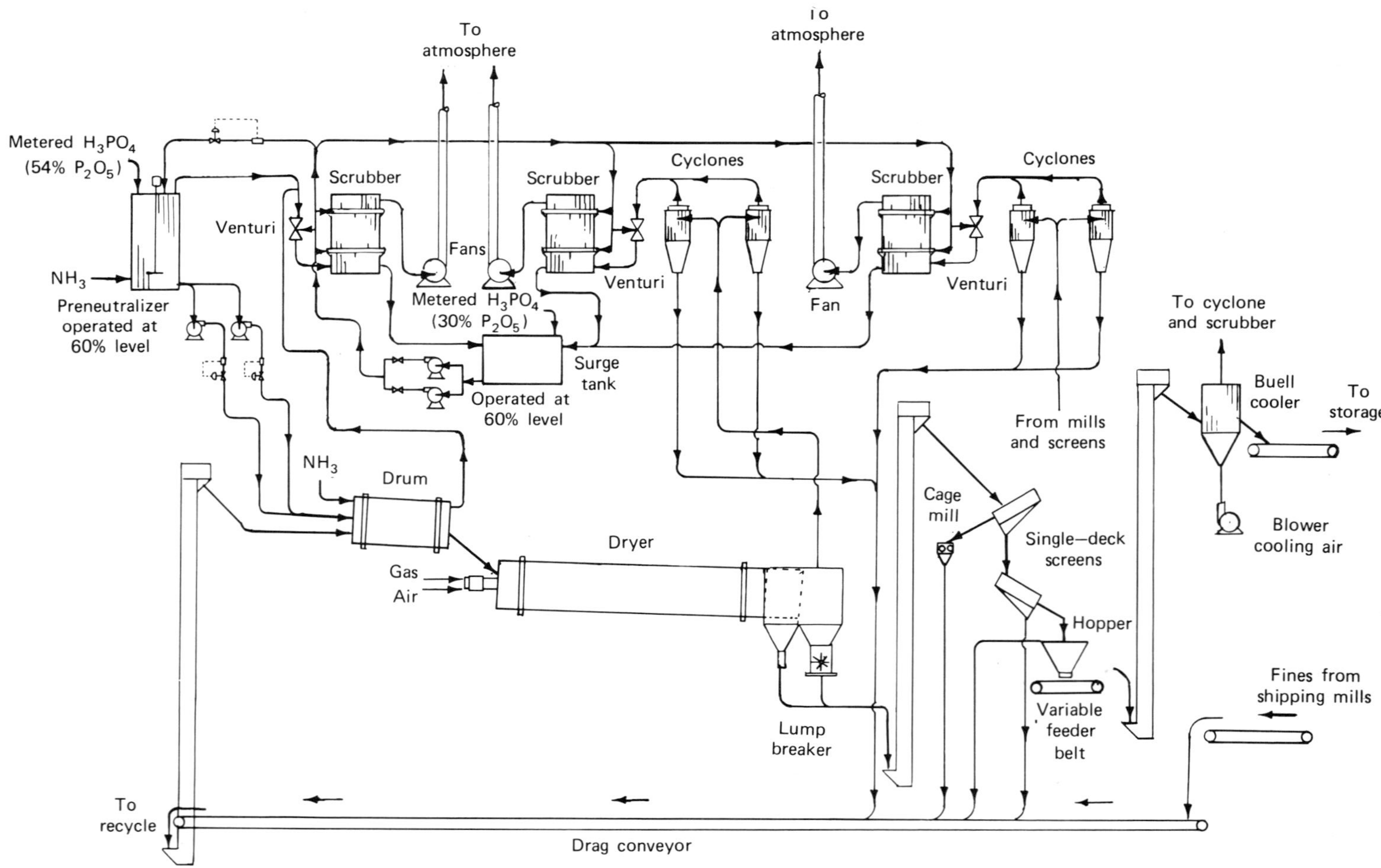

Figure 13. Flow diagram of granular diammonium phosphate plant.

than concentrated superphosphate and yet capable of ammoniation, it is widely used as a replacement for superphosphate in the production of high analysis granular materials. As a source of low polyphosphate APP (11-55-0), MAP's usefulness is expanded still further.

MAP is produced in granular, minigranular, and nongranular (powder) forms, and by several methods. An excellent review of all methods for its production was published recently (105).

Granular MAP (all orthophosphate) can be produced in the DAP system illustrated in Figure 13 or in a variety of other ammoniation–granulation systems. Acid concentrations of 36–54% P_2O_5 can be used and the ammoniation can be done completely in the preneutralizer or divided between the preneutralizer and the granulator. To obtain a product of higher nitrogen, sulfuric acid is added along with the phosphoric acid. Ammonia absorption is more complete than is achieved in producing DAP, reducing the magnitude of the scrubbing operation for ammonia recovery.

Pipe reactors are also used to produce MAP (all orthophosphate) by adjusting the acid concentration and temperature in the reactor to minimize or prevent polyphosphate formation.

Melt granulation of low-polyphosphate MAP, also called APP, is a recent and very promising development. Usually 52–54% P_2O_5 acid is ammoniated in a pipe reactor at about 218°C to produce a melt in which 15–25% of the P_2O_5 is polyphosphate. The melt can be granulated in a pug mill, rotary granulator, or by prilling. The usual product grade is 11-55-0. The properties of this material as produced by TVA are as follows: total N, 11–21%; ammonium N, 100% of total; total phosphorus, 54–58% P_2O_5; available phosphorus, 54–58% P_2O_5; water-soluble phosphorus, 99–100% of total P_2O_5; polyphosphate, 15–25% of total P_2O_5; moisture content, 2%; critical relative humidity at 30°C, 70–75%; bulk density (loose pour), 0.898 g/cm^3; and bulk density (tapped), 0.962 g/cm^3. This material is a mixture of MAP and diammonium dihydrogen pyrophosphate, $(NH_4)_2H_2P_2O_7$. When 20% of the phosphorus is polyphosphate, the compounds are in the ratio 3.5 mol MAP/mol pyrophosphate. The principal use of the material has been in the production of suspension fertilizers, but it is an excellent material for use in bulk blends and for direct application.

The first to produce nongranular MAP commercially as Phos SAI was Scottish Agricultural Industries, Ltd., their product being named Phos SAI. Gaseous ammonia and wet-process acid are allowed to react in a tank to produce a slurry with N:P mole ratio 1.3:1. Acid is added to the slurry in a pin mixer to adjust the mole ratio to 1.0 and to disengage water vapor (103). The powder product contains 6% moisture (a claimed advantage over smaller amounts) and a gel impurity phase covering the surface of the crystals which is claimed to have beneficial effect on the granulation quality of the material.

In the Fisons process, ammonia vapor and 48–52% P_2O_5 wet-process acid are mixed in a reactor at 308 kPa (30 psig) in an N:P mole ratio of 1. The reaction product containing 9–10% water is sprayed through a nozzle into a tower where water flashes off and powder MAP collects. The product contains about 6% moisture. At least 15 such plants have been built.

The characteristic feature of the Swift process is a 2-fluid nozzle which gives instantaneous mixing of liquid ammonia and 49–51% P_2O_5 wet-process acid. A reactor pipe discharges the slurry at about 127°C into the top of a tower where water is flashed and powder MAP collects. A typical product grade is 10.5-53-0 and moisture content is 4%.

The heart of the Gardinier process is a novel pipe reactor. Wet-process acid and ammonia are introduced tangentially into the reactor through diametrically opposed inlet lines. The reactants flow in a spiral pattern to the nozzle end of the pipe where they are ejected into a tower. The product, 11-53-0 grade, contains 2–4% moisture. Phosphoric acid containing up to 30% solids can be used in this process. A single reactor can produce up to 550 t/d of product.

Formation of Citrate-Insoluble P_2O_5. When wet process acid is ammoniated some of the dissolved inpurities react to form water-insoluble phosphates. The water insoluble P_2O_5 in a Florida wet-process acid during its ammoniation at 99°C over the pH range from 2.5–7.5, ranged 2–7%, the maximum being in the pH range 5–6.5. Much of this material is citrate soluble. The citrate-insoluble residues contain 35–41% P_2O_5, 12–15% Al_2O_3, up to 10% Fe_2O_3, and sometimes more than 10% MgO. Fluorine is also a major constituent (105).

Economics. The costs for producing DAP and APP (low polyphosphate MAP produced in a pipe reactor and pug-mill granulator) are illustrated by the TVA estimates in Table 12.

The estimated plant investment for the APP process is 1.9 million dollars less than that for the DAP process. The estimated total production cost for the APP process is \$1.94/t less than that for the DAP process. The main differences in the production costs of the two processes are the cost of fuel for drying the product in the DAP process, the higher investment cost for the DAP process, and the cost of the greater amounts of ammonia used in the DAP process. If the cost of ammonia were excluded to adjust for the difference in nitrogen cost, the APP would cost 5.2¢/kg P_2O_5 less than the DAP.

The market price for DAP fob Florida for April, 1977, was from \$128–135/t as compared with \$124.46 obtained in this estimate.

Ammonium Phosphate-Based Materials. ***Ammonium Phosphate–Nitrate.*** A detailed review of ammonium phosphate–nitrate NPK granulation fundamentals and technology was published recently (106). An attractive and increasingly used procedure

Table 12. Summary of Investment, Product Cost, and Inplant Price for APP and DAP

	DAP (18-46-0)[a]	APP (11-55-0)
nominal capacity		
t/d	982	982
t/yr	324,000	324,000
plant investment, \$	13,120,000	11,211,000
working capital, \$	12,912,000	12,893,000
Total capital investment, \$	*26,032,000*	*24,104,000*
cost of raw materials, \$/t		
ammonia	27.88 (0.23 t)	16.47 (0.13 t)
wet-process acid	66.09 (1.15 t 40% P_2O_5)	79.34 (1.02 t 54% P_2O_5)
cost, \$/kg P_2O_5	0.298	0.246
production cost (excluding ammonia cost), \$/t	95.58	106.05
cost per unit P_2O_5 (excluding ammonia cost), \$/kg P_2O_5	0.229	0.212

[a] TVA estimate, Sept., 1976.

is to feed solid nongranular MAP and concentrated ammonium nitrate liquor to the granulator. When the ammonium phosphate has good granulation properties, the recycle ratio can be as low as 1.2:1. Dutch State Mines, Albright & Wilson, and Norsk Hydro are the main producers of ammonium phosphate-nitrate by use of prilling.

Ammonium Phosphate–Sulfate. See Slurry Granulation and Melt Granulation.

Urea–Ammonium Phosphate. Following TVA pilot-plant studies of cogranulation of urea and DAP in a standard slurry-granulation process, a commercial-scale plant was built at Coromandel, India, to produce 28-28-0 grade UAP (107). Drying is a critical step for stability of the urea. The dryer product temperature must be around 71–77°C, with dryer inlet air temperature limited to 200°C. The plant operated well in excess of design capacity. Several other commercial plants of this type have been built, two in India and one or more in the United States. Both drum granulators and pug mills are used in these plants. A variety of grades such as 17-17-17, 14-28-14, 28-28-0, 24-12-12, and 10-26-26 are produced at rates ranging 35–50 t/h.

TVA operates a demonstration-scale plant for producing UAP by cogranulating urea with ammonium phosphate melt (11-55-0) in a pug mill to produce 28-28-0 and 35-17-0 grades. This melt-type process offers significant reductions in energy consumption, in investment and operating costs, and in dust and fume abatement problems (107).

Nitrophosphates. There are some 80 nitrophosphate plants in the world, two thirds of which are in Europe. Their total annual capacity is about 3.18×10^6 t P_2O_5, about 11% of the estimated 1975 world capacity for chemically produced phosphate fertilizer. Of this world total, 1.8×10^6 t is located in Western Europe and 1×10^6 t in Eastern Europe (substantial additional capacity is being installed there).

Chemistry and Technology. Nitrophosphates are made by acidulating phosphate rock with nitric acid followed by ammoniation, addition of potash as desired, and granulation or prilling of the slurry. The acidulate contains calcium nitrate and phosphoric acid or monocalcium phosphate according to equations 16–17:

$$Ca_{10}F_2(PO_4)_6 + 20\ HNO_3 \rightleftharpoons 10\ Ca(NO_3)_2 + 6\ H_3PO_4 + 2\ HF \qquad (16)$$

$$Ca_{10}F_2(PO_4)_6 + 14\ HNO_3 \rightleftharpoons 3\ CaH_4(PO_4)_2 + 7\ Ca(NO_3)_2 + 2\ HF \qquad (17)$$

Process variations include the use of phosphoric or sulfuric acid in addition to nitric acid and removal of part of the calcium nitrate by cooling, crystallization, and centrifuging. The purpose of those variations is to balance the amount of calcium available for combining with phosphate, thus reducing or preventing the formation of calcium phosphate that is less soluble than dicalcium phosphate and eliminating hygroscopicity caused by calcium nitrate. To achieve this, at least 40% of the calcium must be removed or combined with anions such as phosphate, sulfate, or carbonate. To the extent that water-soluble phosphate is required (dicalcium phosphate is citrate soluble but water insoluble), the ratio of available CaO to P_2O_5 is adjusted to less than 2. When this adjustment is made by use of phosphoric or sulfuric acid, the process loses its independence of sulfur. Process variations thus provide a wide range of possibilities which affect the phosphate solubility and, to some extent, uses that can be made of the product.

There is an extensive literature on nitrophosphates, so these materials will not be discussed at length here. Ewell (108) and Slack (109) have described the principal processes, and the newest developments have been described in a recent publication (110). Bibliographies on Nitric Phosphate Manufacture and Use, and Acidulation of

Phosphate Rock With Nitric Acid are available on request from the TVA Technical Library, Muscle Shoals, Alabama.

Agronomically, nitrophosphates are excellent fertilizers for many uses. Those products with phosphate water solubility below about 20% can be used effectively only on acidic soils or on those that are already highly phosphated. Like other nitrates, nitrophosphates are not used in submerged rice culture. One reason that they are little used in much of the United States and other countries where direct application of anhydrous ammonia supplies most of the nitrogen required is that nitrophosphate nitrogen cannot compete with this practice. Another is the extensive use of bulk blends, for which ammonium phosphates but not nitrophosphates are well suited.

Economics. Hignett (111) and Grundt and co-workers (112) recently published extensive comparisons of the estimated cost of producing nitrophosphates and some major sulfur-based fertilizers. Hignett estimated the 1975 United States gate sale price for nitrophosphates to be \$365 and \$360/t N + P_2O_5 for products containing 80% and 30% water-soluble P_2O_5, respectively. His estimated cost for equivalent amounts of DAP + urea was \$393/t N + P_2O_5. The differences between the sulfur-based and nitrophosphate materials were \$33 and \$28/N + P_2O_5. Grundt found similar differences for high water-soluble nitrophosphate and the Norsk Hydro granular ammonium phosphate-nitrate (17-17-17), and prilled UAPK (19-19-19). The nitrophosphates clearly have significantly lower production costs than DAP plus urea.

Outlook. The two major reasons why production of the more economical nitrophosphates is unlikely to expand as rapidly as their low cost would suggest are:

(*1*) The major phosphate exporters, especially in North Africa and the United States, have moved to ship upgraded materials, wet-process acid and ammonium phosphates, in preference to phosphate rock. The abundant, even excess, supply of these materials will keep them in a strong competitive position for at least the short-range future, eg, for five years.

(*2*) The developing countries, where nitrophosphates would seem to be appealing for most crops except rice, have already strongly committed to urea production, a material that combines readily with sulfur-based phosphates but not with nitrates.

Despite this general outlook, production and use of nitrophosphates in Europe are continuing to grow. The U.S.S.R. recently announced plans to build five new Norsk Hydro nitrophosphate plants with combined production capacity of 5×10^5 t P_2O_5/yr.

Chloride-Free Fertilizer. Some crops (eg, tobacco and potatoes) and a few soils are adversely affected by chloride ions. For these situations the potash is often supplied as the more costly K_2SO_4. Superfos A.G. (Denmark) has developed a modified nitrophosphate process that produces a chloride-free NPK material, starting with the conventional raw materials, ammonia, nitric acid, phosphate rock, and potassium chloride (113). The unwanted calcium and chlorine are removed as $CaCl_2$ by use of ion exchange.

The usual calcium nitrate–phosphoric acid solution is produced by acidulating phosphate rock with nitric acid. By the reaction of this ionized solution with a K-loaded cation-exchange resin, and by regenerating the cation exchanger with KCl solution, the desired result is obtained. The phosphate solution is then ammoniated to produce a Cl-free NPK solution which is subsequently concentrated and granulated. Normally about 2% of the nitrate and phosphate and 7% of the potash fed to the exchanger are lost in the regeneration effluent. The practice has been to dispose of the $CaCl_2$ by pumping it out to sea; a practice that could not be used at some locations.

Superfos has operated a plant with design capacity of about 400 t/d of 21-9-12 grade since December 1971 and the plant was enlarged in 1976. Other NPK grades, 17-17-17 and 18-11-14, have been produced also.

According to Knudsen (113) the chloride-free fertilizer can be produced at costs approximately equal to those of conventional chloride-containing NPK grades.

BULK BLENDS

A modern bulk-blend fertilizer is a mechanical mixture of granular fertilizer materials. The bulk blender often mixes grades to the farmer's order or to need recommendations based upon soil analysis. The mixtures, which may contain micronutrients and pesticides, is commonly marketed through bulk spreading services, usually within a radius of about 40 km. Prescription formulation and custom spreading service are the main reasons for the rapid growth of this segment of the fertilizer industry.

Bulk blending grew at a rapid pace in the United States, becoming the largest single system of marketing mixed fertilizer in the United States. Bulk blending is practiced much more widely in the United States than elsewhere, for a unique combination of reasons detailed by Hignett (114). The industry grew from some 440 plants in 1960 to about 5450 plants in 1976. Over the period 1967–1976 production of bulk blends increased from 4.0×10^6 to 6.7×10^6 t (89). The average annual capacity of United States bulk blending plants (all fertilizer materials) is about 5000 t/yr but the greatest number of plants range in capacity from 1000–3000 t/yr. A few plants have capacities up to 10,000 t/yr.

Raw Materials. Bulk blending works best with well-granulated, closely-sized, dry materials that do not deteriorate in storage. Materials commonly used are ammonium nitrate, granular (in preference to prilled) urea, ammonium sulfate, diammonium phosphate, concentrated superphosphate, and potassium chloride. Urea and ammonium nitrate cannot be used in the same mixture, as this combination is extremely hygroscopic. Diammonium phosphate has long been the leading phosphate source, with triple superphosphate also of considerable importance. However, monoammonium phosphate (including low polyphosphate APP, 11-55-0) is used increasingly because of its high analysis and stability in mixtures, and its high ratio of P_2O_5:N.

Urea has not been much used in blends with superphosphates because of the wetting reaction discussed under granulation. This combination may be used if certain precautions are taken. The superphosphate must be low in hydrate water, preferably less than 4%, and it must be protected from excessive atmospheric exposure. Drying the TSP at 121°C is an effective way of improving the compatibility of TSP with urea but the dried TSP must be protected from atmospheric exposure to prevent reabsorption of moisture. Urea blends well with ammonium phosphates but ammoniation of TSP does not fully prevent the urea–TSP reaction.

New materials for use in bulk blends, most of which are now available, are: granular urea, urea ammonium sulfate, urea ammonium phosphate, sulfur-coated urea, ammonium polyphosphate, and possibly in the foreseeable future, potassium phosphates.

Plant Design. Bulk-blend plants consist of bulk storage area for raw materials, facilities for weighing and transferring materials, a simple mixer (usually rotary drum) and a discharge and loading system. The plant usually operates spreader trucks for delivery and distributing the fertilizer on the farm.

To avoid serious segregation of the blended materials during storage and handling, the materials must be closely matched in granule size, or the plant must be designed to handle granules of considerable degree of size mismatch. These design features require use of a horizontal rotary drum mixer, use of egg-crate baffles in the holding bin, and either use of similar baffles or of a flexible fill pipe that is moved in a circular path while the spreader trucks are filled.

Adding Pesticides and Micronutrients. Because of the flexibility and simplicity of the blending process, micronutrients and pesticides can be added readily but steps must be taken to prevent their segregation from the mixtures. Recommended field application rates for micronutrients are usually less than 10 kg/hm^2, and it is important that the additive be well distributed on or in the fertilizer. In some instances the additives are granular and their incorporation is a simple dry blending along with the major fertilizer ingredients. Often, the materials are fine powders that must adhere to the surface of the granular fertilizers. The powder is first blended with the fertilizer in the mixer and a suitable sticking agent is then sprayed on. The blending and coating operation has been described (115).

Economics. Bulk blending plants vary so much in size that average costs have little meaning. Walkup made an in-depth analysis of costs for 18 plants whose annual volume of sales ranged from 391–7333 t (116). On average for these plants, the cost of buildings for storage and other purposes and the blending and associated equipment each required about 25% of the total investment. Spreading equipment represented the other large cost, amounting to about one third of investment cost. The remaining 17% consisted of the site, office equipment, scales, automotive vehicles, shop, and miscellaneous equipment. The estimated capital investment for a plant of average capacity for the group studied, 3,100 t product/yr escalated to December 1977 costs is about $150,000. The total cost per metric ton of blend of grade 8.5-24.6-22.1 (average for the plants analyzed) was distributed as follows: raw materials fob supplier, 52%; administrative and blending, 17%; freight on raw materials, 18%; handling to farm, 3%; and spreading, 10%.

The gross margin for bulk blends over 1972–1975 (116–117) was from 18–35% of sale price. Generally, the cost to the farmer of nutrient in bulk blends is a few dollars per metric ton below the cost for the same amount of nutrient in granular compounds.

LIQUID MIXED FERTILIZERS

The liquid mixed fertilizers discussed here are multinutrient fluids, both clear solutions and solutions carrying suspended solids (suspensions).

The liquid mixed fertilizer industry is most widespread in the United States, where there are more than 3000 plants. Data from a recent review (118) show rapid growth until 1970, and a lower growth rate since then. The production of liquid mixed fertilizers in the United States is compared with production of other liquid fertilizers, ammonia and other nitrogen solutions, in Figure 14. Of the 3.3×10^6 t of liquid mixed fertilizers produced in the United States in 1976, about 40% is estimated to be suspensions.

In recent years this segment of the fertilizer industry has shown significant growth in other countries, especially France, the U.K., and Canada, and more limited growth in Spain, Japan, New Zealand, and Yugoslavia. Recently, interest in liquid fertilizers has been shown in the FRG and the U.S.S.R.

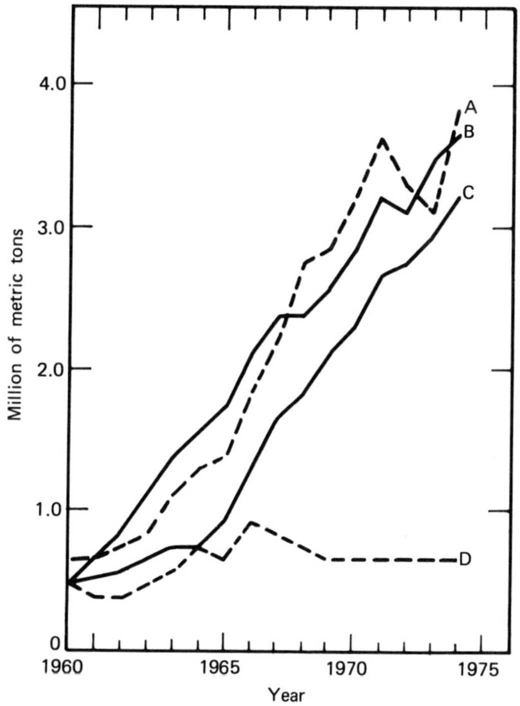

Figure 14. Consumption of fluid fertilizers in the United States. A, Anhydrous ammonia; B, nitrogen solutions; C, fluid mixed fertilizer; D, aqua ammonia. Courtesy of TVA.

Like bulk-blend plants, liquid-mix plants purchase the basic ingredients used, and many of them market other materials as well as liquid mixes. Most plants are small, directly serving farmers in limited areas with prescription-blended liquids, spreading services, soil-needs advice, and other services. A recent survey of the United States industry showed an average annual materials throughput per plant to be 2400 t represented by 1050 t of liquid mixtures, 650 t of anhydrous ammonia, 630 t of nitrogen solutions (UAN), and 570 t of liquid direct application materials such as 10-34-0 and 8-20-0. Sixty-five percent of the fluid mixers produced suspension fertilizers.

Major advantages of clear liquid fertilizers are ease of handling, homogeneity, accuracy of application, water solubility, freedom from dust, and ability of many blends to carry micronutrients. Their chief disadvantages are comparatively low analysis and generally higher cost of their raw materials.

Suspensions have most of the advantages of clear liquids and, in addition, can compete in nutrient content and cost with high-analysis granular fertilizers and bulk blends. They are compatible with most pesticides, and usually micronutrients can be added (dissolved and suspended) in the amounts desired. High purity, high cost raw materials are not required for their successful production.

Chemistry. Nutrient solubility is the most important single criterion for clear liquid fertilizers. Initially these liquids were made from ammonia, ammonium nitrate, urea, orthophosphoric acid, and potassium chloride. The total nutrient concentrations for such solutions was limited to 20–35% when the crystallization temperature was held to 0°C or lower. Ammonium polyphosphates are more soluble than the orthophosphates, and polyphosphates hold impurity metal ions in solution by sequestration.

Solubility curves for ammoniated polyphosphoric acids at 0°C, and for ammoniated mixtures of the acids, are shown in Figure 15 (119–120). Note that for the single acids the solubility increases with increase in chain length, but the highest solubilities occur at low degree of ammoniation (0.22 weight ratio $N:P_2O_5$). These are regions of high hydrolysis rates. As the superacid of the fertilizer industry is a mixture of poly acids, the solubilities for the ammoniated mixtures in the figure are more significant. The upper limit in grade for a mixture of ortho- and pyrophosphoric acids is 11-37-0. The highest solubility is obtained when all four acids are present but the practical ammoniation range is narrow (weight ratio $N:P_2O_5$ 0.27–0.28), where a nutrient solubility of 56–57% can be obtained (12-44-0 grade). However, at this low degree of ammoniation the product must be kept cool to minimize hydrolysis.

Hydrolysis of polyphosphates must be considered when these materials are used in liquid fertilizers. The various species in a mixture hydrolyze simultaneously but the longer chains hydrolyze more rapidly. A computer program is available to determine the constants for equations that describe the hydrolysis rates for each species (121). Low pH and increase in temperature promote hydrolysis. Thus the acid should be ammoniated as much as is feasible without reducing its solubility too much, and ammoniation should be done at as low a temperature as is feasible. Aluminum and iron accelerate the hydrolysis of all species above pyrophosphate. Magnesium suppresses the hydrolysis of tripolyphosphate but not the higher species.

Sequestration is a process by which polyphosphates form soluble complexes with metal ions and hold them in solution (see Chelating agents). Pyrophosphates are more effective than the higher species in sequestering iron and aluminum, and the longer chains are better sequestrants for calcium and magnesium (120). These mechanisms make polyphosphate-based fertilizers good carriers for the micronutrients Cu, Fe, Mn, and Zn.

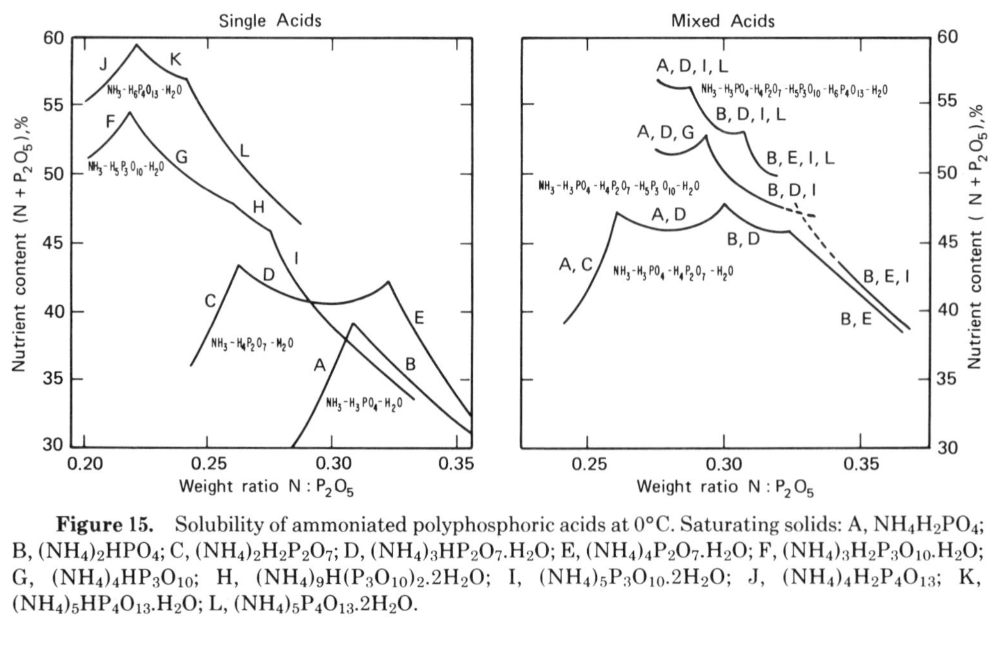

Figure 15. Solubility of ammoniated polyphosphoric acids at 0°C. Saturating solids: A, $NH_4H_2PO_4$; B, $(NH_4)_2HPO_4$; C, $(NH_4)_2H_2P_2O_7$; D, $(NH_4)_3HP_2O_7.H_2O$; E, $(NH_4)_4P_2O_7.H_2O$; F, $(NH_4)_3H_2P_3O_{10}.H_2O$; G, $(NH_4)_4HP_3O_{10}$; H, $(NH_4)_9H(P_3O_{10})_2.2H_2O$; I, $(NH_4)_5P_3O_{10}.2H_2O$; J, $(NH_4)_4H_2P_4O_{13}$; K, $(NH_4)_5HP_4O_{13}.H_2O$; L, $(NH_4)_5P_4O_{13}.2H_2O$.

Superphosphoric acids often contain more than enough polyphosphate to sequester their own impurities, therefore, fertilizer producers often use a combination of orthophosphoric acid (eg, 54% P_2O_5 acid) with superacid or ammonium polyphosphate solutions.

When fertilizer solution compositions are not such that impurities are fully sequestered, sludge-forming solids can precipitate (122) such as: $Al(NH_4)_2P_2O_7.3H_2O$, $Mg(NH_4)_2P_2O_7.4H_2)$, $Mg(NH_4)_6(P_2O_7)_2.6H_2O$, $Fe(NH_4)_2P_2O_7.2H_2O$, $MgAl(NH_4)_5(P_2O_7)_2F_2.6H_2O$. There is good laboratory evidence that F can be used to sequester Mg, Al, and Fe and avoid formation of some of these sludge-forming compounds but commercial application of the method has not yet been very successful.

Raw Materials. Electric furnace superphosphoric acid (white acid) was once a crucial material for use in producing clear liquid fertilizers because of its purity and high polyphosphate content. Because of its short supply and high cost, and because of advances in converting cheaper wet-process acid into polyphosphate-containing liquids, electric furnace acid is today used little or not at all.

Wet-process acids of three grades are used: 52–54% P_2O_5 (orthophosphoric acid); 64–66% P_2O_5 (low-conversion superacid with 15–25% of its phosphate as polyphosphate); and highly concentrated superphosphoric acid (up to 50% of the P_2O_5 as polyphosphate). Limited amounts of metal pickling phosphoric acid are used also, usually in production of suspensions.

Because of increases in freight rates and other high shipping costs for liquid-base materials normally used in fluid fertilizers, solids are being used increasingly in producing suspensions. The principal phosphate solid raw materials are: monoammonium phosphate, ammonium polyphosphate, diammonium phosphate, and urea ammonium phosphate. Dicalcium phosphate (DCP) is also used to a limited extent. Urea, UAP, and ammonium sulfate are the solid nitrogen raw materials that seem most likely to increase in use.

Cheaper (red) potash containing insoluble materials can be used in suspensions; however, white, fully-soluble potash is most often used for suspensions and it is always used in clear starter solutions.

The handling of dry clay in suspension plants causes dust problems and increases batch cycle time. Pregelled clay has been used to a limited extent. Two new clay suspensions that have recently become available may offer advantages to the suspension producer. One of these is a dispersed clay slurry containing 9% N and 25% attapulgite clay, made by mixing 75% urea solution with tetrapyrophosphate solution, clay, and water. The other, a UAN fluid containing 31% N and 2% attapulgite clay, is made by blending dispersed clay slurry with UAN (32% N) solution.

Production Methods. Many small liquid fertilizer plants in the United States use cold-mix batch processes. A base liquid such as 8-24-0, 10-34-0, or 11-37-0 is blended with other raw materials such as UAN solution, potash, and water to make the desired product such as 7-21-7. Clear liquids of this kind are used mostly as starter fertilizers.

A hot-mix plant provides for the reaction of phosphoric acid and ammonia to furnish the ammonium phosphate of the NPK fertilizer. A typical mixer–reactor system, equipped for cooling in the reactor and for cooling the product before storing, is shown in Figure 16. Continuous hot-mix plants are among the larger plants of the industry, and their principal products usually are NP-base liquids such as 8-24-0, 10-34-0, or 11-37-0 but they can produce NPK mixtures as indicated in the figure.

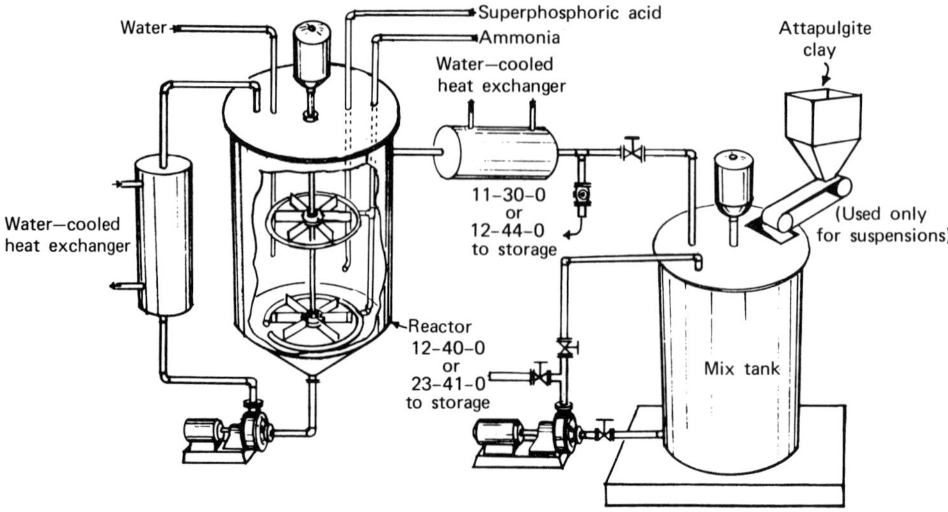

Figure 16. Reactor–mixer for fluid fertilizer production.

Because of the value of polyphosphate in liquid fertilizers and the wish to avoid the intermediate production of superacid, much research and development has been done on the production of polyphosphate-containing liquids by ammoniation of 54% P_2O_5 wet-process acid in a T reactor under conditions that convert 25–50% of the phosphate to polyphosphate. To produce fluid fertilizer directly, the melt from the disengager is dissolved in solution circulating from a fluid-fertilizer reactor. Supplemental ammonia and water are added to maintain the desired grade. Indirect cooling holds the temperature at about 82°C in the reactor, and the product is cooled to about 38°C before storing.

Because of its low polyphosphate content, 50% or less of the total phosphate, the 10-34-0 liquid made by this method frequently forms sludgy precipitates within a few days. One commercial plant has been built (123) but its product is either applied without storage for more than a few days, or is distributed as a suspension fertilizer.

The most advanced and now widely used NP base solution (11-37-0) process is based on a pipe-reactor method (124). By feeding low conversion wet-process superacid (20–30% of the P_2O_5 as polyphosphate) and ammonia vapor through the T ports of a pipe reactor the heat generated in the reactor (315–400°C) converts some 90% or more of the phosphate to polyphosphate (Fig. 17). The melt (10-65-0 grade) is discharged from the pipe reactor either below or above the surface of liquid in a tank from which a liquid of grade 11-37-0 (sometimes 10-34-0) is discharged. Usually 70–80% of the phosphate in the liquid fertilizer is polyphosphate. Such a material has excellent storage and use qualities. More than 100 of these pipe-reactor plants have been built in the United States in the last decade.

The pipe reactors in most commercial plants are made of 10.16-cm schedule 40 type 316 stainless steel. A pipe of this size can produce 23 t of liquid fertilizer per hour. A few plants use 15.24-cm diameter pipes and these have production capacity up to 36 t/h.

Liquid fertilizers containing acid produced from uncalcined rocks are generally

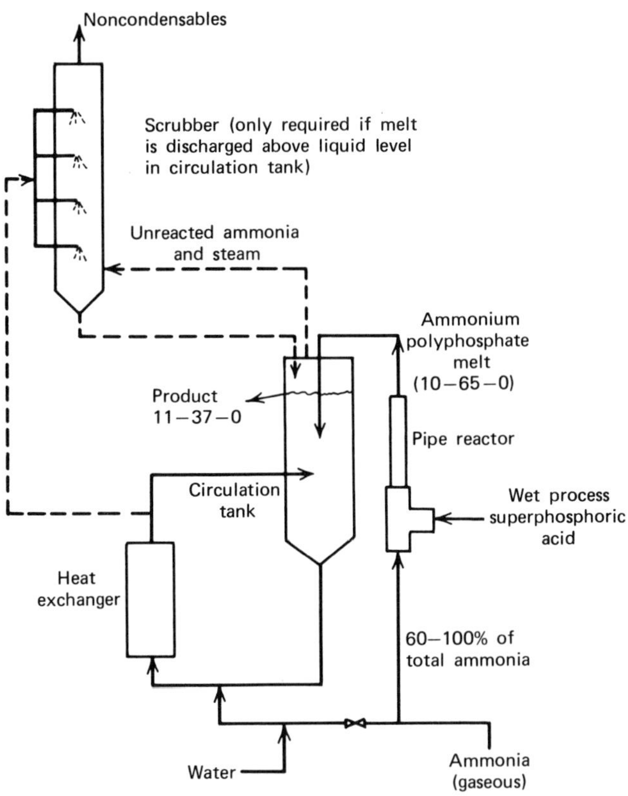

Figure 17. Production of 11-37-0 liquid fertilizer of high polyphosphate content from low-conversion, wet-process superphosphoric acid.

black owing to finely divided carbonaceous matter. The black matter does not harm the fertilizer but it can mask other solids that may clog application nozzles. The black matter can be removed by oxidation and by filtration but these methods have not proven very practical. A method for flocculating the black matter has been developed. In the TVA process a combination of two aliphatic compounds is used—a tallow amine and a quaternary ammonium chloride. When mixed properly with the black 11-37-0, these cause the black matter to rise rapidly to the surface where it can be removed. The flocculents are mixed with the fertilizer under high shear at 52°C. The mixture is then held quiescent for the flocculated material to rise to the top of the liquid. About 80–90% of the liquid fertilizer is recovered as clarified product. The top, containing the black matter and 10–20% of the fertilizer is recovered for use in suspension or solid fertilizer (see Flocculating agents).

The formation of high-quality liquid fertilizer (15-28-0) by pyrolysis of purified urea phosphate is described under purification of wet-process acid.

Based on mid-1975 costs, TVA estimated the costs for producing 450 t/d of P_2O_5 as 15-28-0 liquid fertilizer by the urea phosphate process, and the same amount of clarified 11-37-0 liquid by the pipe-reactor–clarification process. The estimated in-plant prices per metric ton of nutrients ($N + P_2O_5$) were $347 and $351, respectively. This difference is not considered significant within the precision of the estimates.

Suspensions. Production of suspension fertilizer by the conventional cold-mix method is widely practiced. Because of their favorable implications for the future, however, emphasis here is placed on more recent process developments.

The use of solids in the production of suspensions was described recently by Agrico (125) for producing high-analysis NPK suspension. MAP, the sole phosphate raw material, is ammoniated to an $N:P_2O_5$ mole ratio of 1.3:1.7, accelerating disintegration and dissolution of the MAP. Attapulgite clay is then added at the level of 1–2 wt %. After gel formation is complete, red potash and micronutrients are added. Costs of suspension grades 20-10-10, 14-14-14, and 5-15-15 are said to be slightly lower than the nearest comparable grades of dry bulk blends and were from \$0.18–0.25/kg of plant food, mid-1977.

DPA can be used in the same way except that wet-process acid and ammonia are added to produce heat and the liquid phase needed to hasten disintegration of the granular DAP. An 11-33-0 base grade is produced from which grades 16-8-8, 12-12-12, and 8-16-16 can be made.

Although suspensions can be made without use of polyphosphate, granular ATP (12-54-0, 15–25% of P as polyphosphate produced by TVA) is preferred as the phosphate source by many producers. A base suspension of grade 11-33-0 which stores well up to 3 months has been produced. This storage life exceeds that of most orthophosphate-based materials and eases the manufacturer's production scheduling. Less vigorous agitation is required for producing the APP-based material because the polyphosphate dissolves quickly, hastening disintegration of the granules.

TVA recently developed a process for production of orthophosphate suspension. It is made from wet-process orthophosphoric acid (52–54% P_2O_5) and anhydrous ammonia with an attapulgite gelling clay-suspending agent. Acid is ammoniated in three stages. Small crystals of diammonium phosphate are formed and suspended in a saturated orthophosphate solution. This is accomplished by ammoniation of the phosphoric acid to a $N:P_2O_5$ weight ratio of 0.350:0.360 and rapid cooling of the solution in an evaporative-type cooler, followed by a slight ammonia adjustment in the third-stage ammoniation tank containing cooling pipes. Sedimentation of crystals in the suspension during shipment is about 2%. The suspension is cold-mixed with urea–ammonium nitrate solution (28–32% nitrogen), potash, and water to produce various NPK suspension mixtures.

A recent TVA estimate of the cost of producing this all-orthophosphate 13-38-0 base suspension is summarized in Table 13.

Suspension Application Equipment. Other major advances, without which suspension fertilizer use would not have grown so rapidly, are in equipment for storing, transporting, and applying suspensions.

Cone-bottom tanks are preferred for storage. All storage tanks should be equipped with an air-sparging system for frequent mixing of the product, at intervals of one day to one week.

Modern transport tanks (truck-drawn) are equipped with a perforated pipe sparger and large capacity hydraulically driven centrifugal pump is used to circulate and transfer material. Some tankers are equipped for air-pressure unloading. Other transport tanks dump directly into application equipment.

Latest application equipment is of the high-flotation type. Broadcast applicators have 18-m booms, travel at speeds up to 24 km/h, and can cover up to 0.7 hm^2/min. As much as 648 kg/hm^2 of suspension is applied in some cases. The largest applicators have tank capacities up to 15 t.

Table 13. Estimated Price of 13-38-0 Ammonium Phosphate Suspension Produced in a New (1978[a]) Florida–Gulf Coast Plant

	Units/t product	Cost, \$/unit	Cost, \$/t 13-38-0
raw materials[b]			
anhydrous ammonia, t	0.145	109.45	15.87
wet-process phosphoric acid (54%), t	0.635	185.15	117.57
water, process, L	257,800	38,800	0.01
clay, t	0.014	60.00	0.84
operating costs			6.49
capital charges, 34%[c] of capital investment			6.68
sales and administrative costs, 10% of sales			16.38
Estimated price			*163.84*

[a] Plant completed February 1978 Florida–Gulf Coast area. Capacity: 720 t/d; 237,600 t/yr. Capital investment: \$4,718,000 grass roots (includes battery limits plant, 90-d product storage, site purchase and preparation, and support facilities. Working capital (30-d value of raw materials plus 90-d value of raw materials and operating costs): \$12,023,000. Courtesy of TVA.

[b] Raw materials purchased (fob Tampa).

[c] Includes insurance and taxes—3%, depreciation, 6.67%, interest—8% on one half of capital investment, and pretax return on investment—20% of capital investment.

The speed and reliability of application of suspensions by such modern equipment are unmatched by any other fertilizer application system.

Environmental Aspects of Fertilizer Production and Use

The potential for pollution of air and water is inherent in the production and use of fertilizers. Legislation to protect and preserve the environment and to protect workers in industrial plants is implemented in the United States by the EPA and the Department of Labor, respectively. There is a worldwide trend toward increased regulation of pollution (126). United States technology and regulations are taken as the principal basis for this discussion, with occasional reference to differing practices in other countries (see Air pollution; Air pollution control methods; Water pollution).

Of the six major industrial air pollutants deemed injurious to human health, three are directly involved in fertilizer manufacture: sulfur oxides (from sulfuric acid production), nitrogen oxides (from nitric acid and ammonium nitrate production), and particulates. National primary and secondary ambient air quality standards for these are shown in Table 14.

There are in addition designated or noncriteria pollutants which, although not shown to impair human health, may damage the environment. Controls for these are at various stages of development and implementation. Fluorides emitted in phosphate fertilizer manufacture have been designated and standards of performance issued in 1975 (127).

The goal of aqueous effluent guidelines is zero pollutant discharge by 1985; this was accelerated to 1977 for mixed and blend fertilizer processes and for ammonium sulfate production.

In addition to ambient air quality standards and emission standards, fertilizer manufacturers are required to control the environment in the working area in accor-

Table 14. National Primary and Secondary Ambient Air Quality Standards Affecting Fertilizer Industry

	Primary standard[a]		Secondary standard[b]	
Pollutant	Time period	Concentration, g/m^3	Time period	Concentration, g/m^3
sulfur oxides	annual average	80		
	24-h average	365		
			3-h average	
nitrogen oxides	annual average	100	same as primary	
particulate matter	annual average	75	annual average	60[c]
	24-h average	260[d]	24-h average	150[d]

[a] To protect human health.
[b] To protect human beings and the environment.
[c] Nonenforceable guideline.
[d] Not to be exceeded more than once per year.

dance with OSHA regulations. Table 15 gives OSHA standards for air contaminants associated with fertilizer manufacture.

The fertilizer industry is moving ahead on pollution control. Capital investment per year for control equipment in the United States fertilizer industry alone increased from 10.3 million dollars in 1972 to 99.7 million dollars in 1976 (126) while operating costs for pollution control for basic fertilizer producers and integrated companies more than quadrupled in that period, rising from 9.7 million dollars to an estimated 44 million dollars. Capital expenditures are expected to decline as much of the equipment to meet current air and water deadlines is now installed. The capital cost for environmental controls at fertilizer plants is 5–6% of total investment. The Fertilizer Institute estimated the operating costs of environmental control in United States fertilizer plants in 1976 to be about \$3.30/t N and P_2O_5.

Table 15. Industrial Hygiene Threshold Limit Values for Fertilizer Manufacture

Fertilizer material	Threshold limit values[a], mg/m^3 (ppm)
ammonia	35[b] (50[b])
fluoride (as F)	2.5
HF	2 (3)
H_3PO_4	1
CO	55 (50)
H_2SO_4	1
SO_2	13 (5)
HNO_3	5 (2)
NO_2	9 (5)
dusts	
silica	10
nuisance or inert, total	15
respirable	0.5
noise (8-h)	90 dB[c]

[a] 8-h time-weighted average value.
[b] Values probably will be lowered to 18 mg/m^3 and 25 ppm, respectively.
[c] Higher noise levels are permitted when exposure times are less than 8 h daily.

The technology for control of pollution at fertilizer plants is generally available, although further improvements may be expected in some areas. Under present conditions in the United States, large amounts of solid wastes—gypsum from wet-process phosphoric acid plants, sediments produced in the treatment of gypsum pond waters for discharge, and low-grade phosphate rock refuse (slimes from beneficiation plants)—must be stored. This storage of solid wastes, and the air emissions from gypsum ponds, represent the major remaining unsolved environmental problems for the industry. Under suitable economic conditions, not now existent in the United States, gypsum waste could be converted to useful products.

Pollution of groundwater and streams by fertilizer plant nutrients has caused some concern but there is strong evidence that, with caution in its use, fertilizer supplying the needed nutrients can be applied without risk of serious pollution in nearly all instances. Precautions, of course, are in order.

Air Pollution Control in Fertilizer Production. The EPA has identified fifteen major fertilizer processes as potential polluters. Standards for five categories of fluoride-emitting processes have been promulgated (127). Standards for particulate emission from phosphate rock, urea, and ammonium nitrate operations are under review and currently are under state regulation, as are those for other fertilizer process operations. Regulations as of 1977, and methods for controlling, are summarized briefly below.

Nitric Acid Plants. The allowable limit for NO_x (as NO_2) is 1.5 kg/t of HNO_3 produced (max 2-h av); opacity 10%. Methods developed for control of NO_x emissions include thermal and catalytic reduction with hydrocarbons or ammonia (see Exhaust control, industrial), reaction with urea, extended absorption, and adsorption on molecular sieves (qv). The most suitable method depends on type and size of plant and the uses to be made of the acid. All of these methods can meet or exceed EPA standards for new plants. Costs range from about 1% of production cost to net savings of $0.15/t of acid (128).

Sulfuric Acid Plants. The allowable limit for SO_2 is 2 kg/t of H_2SO_4 produced (max 2-h av); acid mist, 0.075 kg H_2SO_4/t H_2SO_4 produced (max 2-h av); and opacity, 10%.

The most advanced control technology for sulfuric acid (qv) plants involves double contact or double absorption (DL/DA). This extended catalytic control and absorption achieves 99.7% recovery of sulfur oxides, compared to 97–98% with single absorption. Extended absorption is now standard for most new plants. Stack gas from such plants contains 200–300 ppm of SO_2. In January, 1976, a new 1814 t/d sulfur-burning DL/DA acid plant cost approximately $16,000,000, some 10% more than a single-absorption plant.

Ammonia scrubbing is still used in some acid plants for SO_2 control. Overall economics depend on recoveries and ammonia costs. Old sulfuric and nitric acid plants are under state regulation and may be allowed somewhat higher emissions when it can be demonstrated that the concentrations of NO_2 or SO_2 in the ambient air resulting from the emission are well below the ambient standard (see Sulfur recovery).

Fluorides. Fluoride, a pollutant from several fertilizer operations, can be recovered in useful forms where the market and economic demands can be met. Whether the material is to be recovered or discharged determines the scrubbing process used for its removal. Fluorosilicic acid can be used in water fluoridation and to manufacture aluminum fluoride, synthetic cryolite, and hydrogen fluoride. The markets for these materials, however, are so uncertain that increased use of this by-product is not expected in the near-term future.

Scale formation in scrubbers for fluoride control can be a problem. For this reason spray-chamber scrubbers with little or no packing are often used. These may be either vertical or horizontal. Scrubbers tend to be more efficient than can be projected from the apparent effluent composition, the fluoride vapor pressure data available for scrubber design is apparently inapplicable.

EPA (129) estimated that costs for control of fluoride emissions under the proposed standards would be less than 1% of the 1975 wholesale price for wet-process acid, superphosphoric acid, and diammonium phosphate. Estimated costs for control of fluoride for run-of-pile triple superphosphate was about 2% of 1975 sale price, and for granular triple superphosphate, 4%.

The standards and the best methods for control of the six fluoride emitters are summarized below.

Wet-Process Phosphoric Acid Plants. No more than 10 g total fluoride per metric ton of equivalent P_2O_5 input to the process is allowed. The best demonstrated emission control consists of scrubbing offgas with gypsum pond water for removal of 98–99% of fluorides.

Gypsum ponds, devices used more extensively in the United States than elsewhere, are possibly the industry's greatest emitter of fluorides. Estimates of fluoride emitted are 0.03–0.92 kg/(hm^2·d). Because there is no accepted technology for measuring or controlling fluoride emissions from the ponds, standards for these are not expected in the near future. The only way known to prevent fluoride emission from gypsum ponds is by precipitating the fluoride with lime. Liming of entire gypsum ponds is impractical.

Superphosphoric Acid Plants. No more than 5.0 g total fluoride per metric ton of equivalent P_2O_5 input to the process is allowed. In the vacuum-evaporator process the fluorides are absorbed by water in the barometric condenser and no further control device is required. For the submerged-combustion process, packed scrubbers are considered the best available system for emission reduction.

Diammonium Phosphate Plants. No more than 30 g total fluorides per metric ton of equivalent P_2O_5 input to the process for new plants is allowed. Fluorides are removed by secondary scrubbing. Best results are obtained with a spray crossflow packed-bed scrubber. The best demonstrated control for ammonia consists of scrubbing emissions with phosphoric acid solution.

Run-of-Pile Triple Superphosphate Plants. No more than 100 g of total fluoride per metric ton of equivalent P_2O_5 input is allowed. The best demonstrated control of fluoride consists of scrubbing emissions with water. No visible emission is observed from storage facilities with this type of control.

Granular Triple Superphosphate Plants. No more than 100 g of total fluoride per metric ton of P_2O_5 input to the process is allowed. The best demonstrated control consists of scrubbing offgas with water.

Granular Triple Superphosphate Storage. No more than 0.25 g/h of total fluoride per ton of equivalent P_2O_5 in storage is allowed. The best demonstrated control of fluorides consists of scrubbing emissions with water.

Phosphate Rock Processing. Although federal standards for the various operations in phosphate rock processing were still under development in 1977, states have imposed regulations (130) relating allowable emissions to the hourly tonnage of material processed.

Nearly all the phosphate rock mined contains enough moisture so that dust emission in the mining operation and in the usual wet conveyance of the rocks is not a problem. The beneficiation also is a wet operation and does not emit dust.

Cyclone separators, wet scrubbers, electrostatic precipitators, and fabric filters are the devices usually used for particulate control.

Drying. The rock is dried either in fluidized-bed or rotary dryers. Dryer emissions from United States plants range from a low of 0.125 kg/t of rock, where a combination of impingement scrubbers and electrostatic precipitators is used, to a high of 2.45 kg/t where cyclone separators are the only dust collecting device. Wet scrubbing is most commonly used. High-pressure-drop scrubbers can have removal efficiencies of 96–99.9% for particulates of 1–10 μm in diameter and 80–96% for particles less than 1 μm in diameter. Wet electrostatic precipitators give similar performance, and a suitable bag filter is somewhat more efficient with particles in the lower size range (130).

Calcining. Fluid-bed calciners and a variety of control devices are used—cyclone, cyclone scrubber, electrostatic precipitator, impingement scrubber, and venturi scrubber. Particulate emission rates range 0.065–0.24 kg/t for venturi scrubbers, 0.25–0.90 kg/t for impingement scrubbers, 0.27–0.29 kg/t for cyclone scrubbers and electrostatic precipitators, and 2.5–5.5 kg/t for cyclones alone (130).

Grinding. Grinding is done with either roller or ball mills. The most common control device is the baghouse, although a few wet scrubbers, impingement scrubbers, and venturi scrubbers are used. Particulate emissions range from a low of about 0.002–0.1 kg/t for baghouse, and 0.05–0.15 kg/t for other devices (130). These variations are functions of exhaust air volume.

Handling and Storage. Material handling systems range from front-end loaders and other manual means to automated pneumatic equipment. Much of the emission potential for these operations is fugitive dust which is difficult to quantify. Bucket elevators and conveyor belts, covered or enclosed, are frequently used. Transfer points usually are hooded. Emissions from vents to storage silos frequently are controlled by fabric filters. The emission potential for a typical materials handling and storage system was estimated (130) to be about 1 kg/t of rock handled.

Wet grinding and feeding the wet rocks to acid reactors would avoid much of the particulate emission in phosphate rock operations. A few companies have overcome difficulties inherent in use of this system, and this practice appears to be the trend in the industry. One of the most serious problems to be overcome is the change in water balance in the plant. However, the potential savings through elimination of dryers and grinders is a substantial incentive.

The EPA (130) has estimated costs for particulate control for the several phosphate rock operations. These range widely with specified conditions, and careful analysis of each situation is required for dependable guidance.

Mixed Fertilizers. An in-depth study of emission control from these processes (excluding diammonium phosphate) was made for EPA (131), but because of a number of complexities in this segment of the industry, control standards have not yet been developed. Some states, however, have imposed regulations. A brief look at the emissions and technologies for their control follows.

Ammoniation–Granulation Plants. These plants are potential emitters of ammonia and particulates. Particulate emissions may be a problem, especially ammonium chloride fumes. Substitution of superphosphoric acid for sulfuric acid in ammoniation–granulation plants is partially effective in reducing emissions but the practice has not been adopted widely.

Among the control devices most frequently used are single-stage and two-stage cyclone scrubbers, venturi cyclones, packed-bed scrubbers, and baghouses. Using proper gas velocities through the cleaning equipment and providing seals on cyclones and conveyors effectively reduces dust emissions. Wet scrubbers have efficiencies of 85–99% for packed beds and venturis. Recovered material is recycled in the process.

Installation costs for baghouses are high and the houses must be operated at temperatures above 100°C because of the high moisture content (30–50%) of exhaust gases from dryers. They easily provide efficiencies exceeding control regulations. Their 1975 costs are reported (132) to be as follows:

	Cost, \$/annual m^3 of air capacity
capital	141–281
maintenance	21
operating	6.71

Other granulation processes also require control. The extensive experience of the TVA National Fertilizer Development Center in pollution control for a wide variety of granulation processes (drum, pug mill, melt, and pan granulation) and for reuse of the recovered materials has been described (133–134). Production of firm, closely-sized granules reduces particulate emission. With production of such granules the quantity of wastewater is more easily controlled so that it can be reused in the process. Where a conditioner is required, using oil–wax mixtures instead of clay simplifies the recycling of recovered material. These methods are generally applicable to a wide variety of granulation processes.

Although nitric phosphate production in the United States is small, these plants are widespread in Europe. Nitrogen oxides and fluoride-containing gases are their principal gaseous effluents. Although operating the reactors at low temperature reduces emissions, emission treatment is still necessary. The usual practice is to return scrubber liquor to the reactors but scrubbing with a medium to decompose the NO_x (probably a medium containing urea) has been practiced in Rumania.

Bulk-Blend Plants. These often use no special control equipment because their emissions can be kept below local state emission standards. Most of the particulate emission is fugitive dust that must be controlled at transfer points. Liquid fertilizer (10-34-0) and lightweight oil added in the mixer have been used effectively as dust suppressants. Dust discharges from some mixers have been reduced from 25 to 12 mg/m^3 and emissions from bagging machines from 282 to 11 mg/m^3. Occasionally, when the problem is severe, baghouses are used.

Screen conveyors and telescopic chutes improve control in transfer and distribution of materials. Product discharge elevators with sealed boots prove more effective than centrifugal discharge.

Liquid-Mix Plants. Plants of modern design usually produce no serious air pollution problems. The pipe reactor eliminates hot-mix reactor tanks, a major source of pollution in earlier plants. These modern plants are equipped with wet scrubbers which use water or partially cooled product, or both, to remove about 95% of the ammonia, fluoride, ammonium chloride, and particulate emissions. Usually a demister is used above the scrubber.

Prilling Towers. Particulate emission from prilling towers for ammonium nitrate and urea is a serious problem. The tower emissions are especially difficult to control because of their small size and the large amount of air in which they are entrained. Several methods for control of emissions from ammonium nitrate prill towers have been developed (135–137) and may serve as models for new source standards being considered by EPA.

In one method the volume of air to be treated in ammonium nitrate prilling towers is reduced and is limited by channeling air to initially cool the prills to a temperature at which vapor pressure of the ammonium nitrate is low and fume formation greatly reduced. Only the smaller air stream need be treated for particulate removal. This air is passed through an impingement scrubber and then through a Brink filter. The system meets both mass and opacity regulations.

Capital costs range from $50,000 for 227 t/d plant to $1,500,000 for a 907 t/d plant (1974 cost). The principal operating cost is for power, and as of 1975, net savings of $0.17–0.66/t of product has been claimed, owing largely to the value of recovered ammonium nitrate.

Another method uses Beco Engineering Company scrubbers to treat prill tower emissions. These scrubbers are designed to agglomerate and separate particles with low energy loss. This system reduced emissions to 3.3–6.6 t from a 454 t/d tower, an amount well below the maximum permissible level. Installed cost for this particular plant is reported as $725,000 (1974).

In a third system, the melt-spray head, was modified and surrounded by a metal shroud to provide a quiescent zone. The tower modification cost about $25,000 (1975). This system decreased both fume and microprill emissions. Ammonium nitrate emission was reduced by a factor of 5.5, and ammonia emission by a factor of 2.2. Plume visibility, however, was not affected.

The dust-prilling process for ammonium nitrate, developed in the UK, gives no fume in the effluent gas. Fume is trapped by the dust cloud and emitted dust particles are then removed in cyclones and bag filters.

A method of controlling fume emission from ammonium nitrate neutralizers has been developed. Control is said to be achieved through the method of operation of the neutralizer and scrubbers, and mist eliminators are not required. The critical feature of operation is pH control (138–139).

Defluorination of Phosphate. Defluorinated phosphate rock is used as animal feed, but the process usually is operated by fertilizer producers. Defluorination is one of the smaller operations of the fertilizer industry. About 95% of the fluorine is volatilized from the rock. The evolved fluoride is scrubbed with water in multiposition spray chambers and spray towers in series with high energy wet scrubbers which can yield removal efficiencies of 99.5% (140).

Water Pollution Control in Fertilizer Production. Polluted water from fertilizer plants originates in cooling water, process effluents, rainwater runoff, domestic effluent, and boiler blowdown. United States federal regulations prohibit discharge of any process wastewater pollutants from phosphate plants except under unusual and well defined weather conditions such as chronic or catastrophic rainfall events. When discharge is permitted, the effluent must comply with the guidelines given in Table 16. The amounts of nutrients that can be discharged from nitrogen fertilizer plants are shown in Table 17 (141–143).

Table 16. Effluent Guidelines for Phosphate Plants

	Concentration of pollutant, mg/L	
	Max for 1 d	Average of daily values for 30 consecutive days shall not exceed
total phosphorus as P	105	35
fluoride	75	25
total suspended solids	150[a]	50[a]
pH	6.0–9.5	

[a] Waived if wastewater is treated for P and F removal. Not applicable to nonprocess wastewater.

Phosphate and NPK Plants Control Technology. Aside from phosphate mining, production of wet-process phosphoric acid is the largest single phosphate process activity. Frequently acid concentrators and phosphate fertilizer processes—ordinary and triple superphosphate, diammonium phosphate, granulation—are operated in the same area as the acid plant. Where this is the case, water from gypsum ponds is used for scrubbing pollutants in all of the plants, and these scrubber effluents circulate back through the gypsum pond. Control of gypsum pond water, therefore, is of prime importance. Release of pond water is allowed only after it has been treated to lower phosphate and fluoride to the levels shown in Table 15. Slaked lime prepared from burned lime and fresh water is used for this purpose.

Gypsum pond waters vary widely in composition as shown by the following composite data from EPA tests on nine ponds (144): F/P, 0.2–2.5; SO_4/P, 0.6–2.0; and Ca/P, 0.1–0.6. The amounts of lime required for treatment vary widely. Liming is done in two stages using 10% calcium hydroxide slurry and allowing settling after each stage. In the first stage the lime is added continuously to maintain pH 3.5–4.0. Clear overflow from the settled first stage is treated with additional slaked lime to pH 8.0–9.5 and allowed to settle. After settling for 24 h, the sludge from the second stage, about one half the volume of pond water treated, is pumped to the center of the gypsum stack for disposal.

The cost of neutralization varies from \$0.53–1.59/m^3 (\$2–6/1000 gal) of water treated (144). Treatment costs for a 1000 t/d acid (P_2O_5) plant in central Florida often exceed \$550,000/yr. Water balance in the pond–plant system is extremely important and fresh water is used as sparingly as possible. The effluent quality of double-limed pond water often fails by a wide margin to meet EPA guidelines. It appears that in some instances water may have to be limed to pH >9.5 to reduce fluorides to the allowable 25 mg/d. Recent advances in the fluorine chemistry of gypsum ponds may point the direction for advances in this troublesome area (145).

In the UK where such plants are located in coastal areas, gypsum and its associated aqueous medium are discharged into tidal waters or directly into the sea. Seawater is buffered at about pH 8 which neutralizes the acidic effluent. Rapid dilution of the effluent quickly raises its pH above 4 and the gypsum dissolves. Each situation is judged on its merits and these are reviewed periodically by authorities.

Where granulation and NPK plants are not operated in conjunction with phosphoric acid plants, emphasis must be on recovery of nutrients and other pollutants and reuse of the water. The scrubber water is either recycled in the process or disposed of as low-analysis liquid fertilizer.

Table 17. Effluent Guidelines for Nitrogen Fertilizer Plants[a]

Type of plant	Parameter	BPCTA[b] limitation, kg/t product		BATEA[c] limitation, kg/t product		New source standard, kg/t product	
		Max 1 d	30 d av	Max 1 d	30-d av	Max 1 d	30-d av
ammonia	ammonia (as N)	0.1875	0.0625	0.05	0.025	0.11	0.055
urea solution	ammonia (as N)	0.95	0.48	0.53	0.27	same as BATEA	
	organic N (as N)	0.61	0.33	0.45	0.24	same as BATEA	
	pH	6–9		6–9		6–9	
prilled urea	ammonia (as N)	1.18	0.59	0.53	0.27	same as BATEA	
	organic N (as N)	1.48	0.80	0.86	0.46	same as BATEA	
	pH	6–9		6–9		6–9	
ammonium nitrate all forms	ammonia (as N)	0.73	0.39	0.08	0.04	same as BATEA	
	nitrate (as N)	0.67	0.37	0.12	0.07	same as BATEA	
nitric acid							
using gaseous ammonia	ammonia (as N)	0.007	0.0007	0.0045	0.00045	same as BATEA	
	nitrate (as N)	0.33	0.044	0.17	0.023	same as BATEA	
using liquid ammonia	ammonia (as N)	0.08	0.008	same as BPCTA		same as BPCTA	
	nitrate (as N)	0.33	0.044	0.17	0.023	same as BATEA	
	pH	6–9		6–9		6–9	
noncontact cooling water	pH	6–9		6–9		6–9	
ammonium sulfate		no discharge					
mixed and blend fertilizers		no discharge					

[a] Refs. 141–143.
[b] BPCTA = best practicable control technology available as of July 1, 1977.
[c] BATEA = best available technology economically available. To be obtained by July 1, 1984.

Nitrogen Plant Control Technology. *Ammonia Plants.* Effluent guidelines for ammonia plants are shown in Table 16. Because condensate from ammonia plants contains 800–1100 ppm of ammonia and substantial amounts of methanol and carbon dioxide, it cannot be released to streams without treatment. Condensate is steam stripped to remove 90–95% of the ammonia to meet effluent guidelines, with the volatile contaminants discharged to the atmosphere. The trend in modern plants is toward complete abatement of air and stream pollution and reuse of condensate. Raw water treating cost would thus be reduced. Current research is directed toward this end.

Urea Plants. A 1000 t/d urea plant produces 290–375 L/min of condensate, the amount depending on type of plant. The condensate contains from 2–6% NH_3 and 0.3–0.5% urea. These contaminants can be largely removed by a combination of urea hydrolysis and ammonia stripping.

The cooling water from urea plants is too warm (43.3°C) for release into streams. Cooling towers can alleviate this problem but these are expensive. Such warm water can often be used to advantage in evaporative scrubbing operations where the heat is dissipated in evaporating water. This is done in some urea and ammonium nitrate plants.

Ammonium Nitrate Plants. Evaporative scrubbing and water handling methods in ammonium nitrate plants are similar to those used in urea plants. The possibilities for this are expanded when the ammonium nitrate or urea plants are part of a complex involving nitric acid production and neutralization and the production of both explosive and fertilizer grades of ammonium nitrate.

An ion-exchange process has been used successfully to treat waste streams ranging in solute concentration from 1,700 to 11,000 ppm (146). The process uses a classical demineralization system.

After credit is taken for recovered ammonium nitrate and demineralized water, the operating cost is $0.55–0.77/t of ammonium nitrate, ca 0.5% of the value of a metric ton of ammonium nitrate.

Potash Production. The high water solubility of most minerals encountered in potash mining sometimes leads to environmental problems in disposing of tailings and waste solutions. Mines located in arid regions, as in New Mexico, have few problems because there is little rainfall and evaporation rates are high. In areas of high rainfall, seepage from tailing piles is collected and returned to the process. Solutions are disposed of either by evaporating to dryness in the plant or in ponds, or by injection into porous underground strata (21). In the UK, tailings often are discharged into the sea.

Solid Wastes. ***Gypsum.*** In the United States and in some other countries, most of this by-product gypsum is disposed of as solid waste. A limited amount is used to treat saline soils in the western United States, mainly in California. Some is disposed of in rivers and in the ocean but this practice in the U.S. and in some other countries is being phased out for environmental reasons. Sea disposal is still permitted in Europe provided accumulation in estuaries is avoided. The most widespread practice is to stack the material near the production plant. For each daily metric ton of P_2O_5 produced, the land required for gypsum stacks is 1357 m^3 (1.1 acre·ft)/yr—27.14×10^6 m^3 (22,000 acre·ft) for a 20-yr life of a 1000 t/d (P_2O_5) plant. The urgent need to find other means for disposal or use for by-product gypsum is obvious.

A variety of processes have been developed for recovering sulfuric acid from gypsum and for using gypsum in cement production but neither practice is widespread for economic reasons.

The production of sludge by liming of gypsum pond water has already been discussed; the sludge produced is another solid waste. The sludge volume is 50–60% of the volume of pond water treated. This sludge, a mixture of complex compounds of phosphates and fluorides, is disposed of on the gypsum stacks, further compounding the problem of gypsum disposal.

Phosphate Mining and Beneficiation. Increasing water demand in Florida, falling groundwater levels, mounting water quality problems, development of shorelands, decline of esthetic quality, and disorderly developments have combined to bring about more extensive planning and reclamation legislation in Florida than in any other state. This was essential in Polk County where nearly one third of the world supply of phosphate rock was produced in 1975. In Florida, mining plans and reclamation standards must be submitted for approval by a board of commissioners before new mining activities can begin. The mining company's environmental impact statement must provide means for and guarantee of reclamation of the mineral land and must describe how the reclamation will be accomplished. Reclamation and water protection regulations are also strong in North Carolina.

There is great concern among phosphate producers in the western United States that legislation will be passed requiring all mined areas to be completely restored to their original condition. Because of the nature of the terrain in that region, such restoration would be impossible, although useful restoration is possible. The passage of such legislation could preclude use of the largest phosphate resource in the United States.

Slime ponds. The water pollution aspect of phosphate slimes has already been mentioned. The slimes, containing 30–35% solids, are put into 160–240 hm^2 ponds where the solids settle over decades. There is 20,240 hm^2 of slime ponds in Florida alone. Occasional breaking of pond dikes has caused serious pollution of streams, with extensive fish kills. Research is in progress to find ways to improve phosphate recovery and lower the impact of slime deposits on the environment (147).

Radioactivity. Solid wastes, slimes, and tailings from phosphate rock beneficiation contain small amounts of radioactivity; some slimes contain fifty times the amount of radioactivity of typical soils of the United States (148).

The radioactivity of reclaimed phosphate lands sometimes is enough to cause exposures slightly greater than the present federal guidelines for maximum exposure of uranium miners. Guidelines for disposal of such radioactive wastes have not yet been developed.

Other Solid Wastes. There are several other minor solid wastes in fertilizer production. In ammonia plants using potassium carbonate for carbon dioxide removal, a small amount of arsenic oxide (As_2O_3) sludge is produced. A 1000 t/d plant produces about 20 t/yr of sludge containing 20% As_2O_3. The sludge is either stored for special treatment in the plant or is shipped to an arsenic producer for recovery.

Sulfuric acid plants that burn sulfur produce about 0.3 kg of a sludge from filtering molten sulfur per t of H_2SO_4 produced. This sludge is usually dumped on slag piles where it sometimes ignites and results in atmospheric pollution.

Some European liquid-fertilizer producers accumulate a sludge (4–20 kg/t of liquid fertilizer) containing phosphates and fluorides together with the sawdust filter aid. Usually this is recovered and added to solid fertilizers.

Fertilizer bags containing small amounts of residual fertilizer constitute a significant waste disposal problem. These can be burned, provided they are not excessively contaminated, or they can be disposed of in land fill.

Environmental Implications of Inorganic and Organic Fertilizer Use. Surface and groundwaters receive pollutants from a wide variety of sources, including fertilizers applied to soils. With the increasing use of fertilizer, there is a tendency to regard them as a major source of pollution. The situation is complex, however, and when used properly commercial fertilizers may be only minor contributors to pollution. The subject is under intensive study in search of the facts, and means for control where needed (149–150).

According to the EPA, United States waters are improving with respect to coliform bacteria and oxygen demand, but the concentrations of nitrogen and phosphorus are increasing, largely from agricultural runoff and sewage. Most United States surface waters contain at least 0.3 ppm of N and 10 ppb of P. In terms of the level of nutrients required for control of algae growth (<0.3 ppm of nitrogen and <10 ppb of phosphorus), most United States surface waters are polluted, the concentration of nitrogen and phosphorus nearly always exceeds these amounts. Rainwater contains on the average 0.7–1 ppm of nitrogen.

Fertilizer nitrogen is removed from soil by plants, surface runoff, leaching and erosion. The main factors involved in phosphorus removal are crop uptake and erosion; leaching is significant only in highly siliceous sands and organic soils.

High values of NO_3–N in drinking water may cause methemoglobinemia in very young infants but the critical level is not well defined. The United States federal standard is 10 ppm but the California standard is 20 ppm. Animals too may be adversely affected by excessive concentrations of NO_3–N in drinking water.

It should be noted that surface and groundwater receive plant nutrients from many sources other than the fertilizer used in agriculture. Other major sources from agriculture include decaying plant matter, soil erosion (ca 3×10^9 t of sediment/yr in the United States carrying ca 1.8×10^6 of total phosphorus), animal and other livestock wastes. Manufactured fertilizer usually is not the major source of nutrients. For example, some 90×10^6 t/yr of nitrogen worldwide is fixed by biological and other natural means (151), compared to about 42×10^6 t of N in manufactured fertilizer used in 1975.

Storage and soil disposal of animal and poultry wastes and disposal of municipal sewage are very serious sources of pollution and are receiving major efforts toward control.

Nitrogen. Ammonium nitrogen from whatever source (anhydrous ammonia, inorganic salts, urea, controlled-release fertilizers, animal waste) can be volatilized when applied to the surface of soils. When incorporated in the soil, it is not readily leachable, since it is held by ion exchange as are other cations. In most agricultural soils, however, ammonium–N is usually converted to nitrate–N in a few weeks when soil is suitable for crop growth. Although the rate of this conversion is variable with source, level of aerobic activity, temperature, and other soil conditions, most ammonium fertilizer is converted to nitrate in the soil. For flooded rice culture, conditions are different, and ammonium–N persists.

Nitrates are water soluble, leachable, penetrate soil to considerable depths and can result in high accumulations, or loss to groundwater in uncropped soil. Generally, however, nitrate does not accumulate under native grasslands because during much of the crop year it is utilized by the crop (150). Nitrate added as fertilizer or produced in the soil while plants are not growing may be leached below the root zone where it will either accumulate or move into groundwater. In high rainfall areas, the latter is very likely.

Water moves downward in soils when precipitation or irrigation exceeds loss of water by evapotranspiration. Nitrate may also move downward in the absence of a growing crop, and hence leaching is most likely to occur in late fall, winter, and early spring. Situations most conducive to loss of nitrogen from runoff and erosion occur when high rates of nitrogen are applied on sloping lands (an unlikely occurrence as such lands are usually in pasture) subject to intensive rainfall, when heavy rainfall occurs immediately following surface application of fertilizer, when fields are inadequately protected by plant cover, or when soils have low rates of infiltration.

Despite their limitation as analytical tools, runoff plots can yield data to illustrate the importance of crop cover, and also to compare treatments (152). Comparison of nutrient losses from fertilized and unfertilized plots in a small watershed shows no appreciable differences in nutrient content of runoff (153). Additions of N, P, K, and S in precipitation exceeded the losses in surface runoff and groundwater flow. Additions of these nutrients to surface waters by precipitation is often sufficient to promote algal growth.

Animal wastes and sewage sludges applied to the surface of the land as fertilizers, or stored on the surface, can yield runoff hazardous to surface waters. When placed on soil, manures and sludges should be incorporated as soon as possible to prevent groundwater pollution through runoff. Leaching of N from these organic materials can be a serious problem.

Degradation of ground water with soluble salts and nitrate from animal waste and sewage sludges is one of the factors limiting their use on the land. The pollution of groundwater by the nitrate and soluble salts will be most serious where the amounts of nitrate greatly exceed the needs of the crop, an increasingly serious problem in the area of large feedlots. Cattle in a 60,000-animal capacity feedlot excrete approximately 10^6 t of manure containing about 4000 t N annually within an area of 150 hm^2. Accumulation and storage of the waste can present serious problems of air and water pollution. To spread this amount of manure on the land at the rate of 23 t/hm^2, not an uncommon rate, requires 4000 hm^2, an area not usually accessible to most feedlots. Such large amounts of waste applied to the soil contain N far in excess of crop needs. Development of safe handling and use of this waste is receiving extensive study.

One main factor limiting the long-range use of sewage sludge on agricultural land is the potential hazard associated with the presence of certain heavy metals sometimes introduced by industrial effluents. The elements of greatest concern are zinc, copper, nickel, and cadmium. These elements, which are toxic to animals and man at high levels, are taken up by plants. Since they accumulate in the soil, the fear is that long-term heavy use of sewage sludge on soil will lead to hazardous amounts. Wide dispersion of sewage sludge, as by combining it at appropriate levels with commercial fertilizers, may be useful in this regard. Long-term experiments needed to evaluate the delayed effects are in progress. Economic factors also have limited the large-scale use of sewage sludge as fertilizer.

The loss of nitrogen and nitrogen oxides (mostly N_2O) from soil was mentioned earlier. Concern has arisen about the role of N_2O in depleting the ozone layer in the stratosphere. Important aspects of this phenomenon are yet to be clarified—the extent to which N_2O actually destroys the ozone, what fraction of fertilizer nitrogen undergoes denitrification to produce N_2O, how long N_2O remains in the atmosphere, and whether the oceans contribute most of the N_2O found in the atmosphere. Estimates of the extent of destruction of the ozone layer by N_2O have varied widely (154) (see Air pollution; Ozone).

The best hope of unraveling the complexities of the fate of fertilizer nitrogen in soil lies with research in progress using fertilizers labeled with ^{15}N-depleted nitrogen. Using this tool, scientists at TVA's National Fertilizer Development Center in cooperation with Los Alamos Scientific Laboratory, USDA, and university scientists across the United States are conducting field studies of the fate of fertilizer nitrogen in a variety of soils, climates, and types of agriculture (155). The method offers promise of leading to improved efficiency in use of fertilizer nitrogen and protecting the environment.

Phosphate. Phosphorus behaves entirely differently than nitrogen in soils, since it is strongly retained by most soils except siliceous sands and organic soils. The quantities of fertilizer P transported to surface and groundwaters is very difficult to determine under field conditions. However, many research reports show losses of P in solution to be about 1 kg(hm^2·yr). The situation is confused by the fact that the concentration of phosphorus in drainage waters is directly related to the phosphorus content of the rocks from which a soil is formed; the phosphorus content of rivers in high phosphatic limestone areas may range up to 0.22 ppm whereas those in sandstone and shale areas may contain only a trace. Moreover, whether soil is unfertilized or heavily fertilized, the concentration of phosphorus in the soil solution is usually within one order of magnitude of 0.1 ppm. Streams draining the most intensively cultivated watersheds carried concentrations of phosphorus below 0.03 ppm and the general levels were unchanged between 1921 and 1972 despite an eight- to tenfold increase in phosphorus fertilization (156).

Water-soluble phosphorus applied to fine-textured mineral soils is rapidly converted to solids only sparingly soluble in water. Generally, therefore, phosphorus is not leached from these soils. There is little question, however, that downward movement of P in highly siliceous sands may constitute a pollution hazard in localized areas. Similarly, organic colloids have little capacity to absorb P because of their low concentrations of Fe, Al, and Ca. This is important in choosing soluble or water-insoluble phosphate sources for organic forest soils.

The chemistry of soil phosphorus is different in flooded soils. Ferrous phosphate, a compound more soluble than the ferric phosphates of nonflooded soils, is formed. Moreover, the pH is increased and free iron oxide coatings are reduced, resulting in the release of occluded phosphate. These reactions occur and are important in paddy rice culture.

Sediment from soil erosion and other sources carries about 1.8×10^6 t of total phosphorus into United States surface waters annually (157). This sediment acts as a scavenger for dissolved phosphorus in streams, absorbing phosphorus from solutions containing more than 0.01 ppm. A practical consequence of this is that concentrations of 0.10–0.13 ppm of soluble phosphorus in runoff water from fertilized fields may be reduced to 0.009 ppm by adsorption on suspended sediment, a mechanism with large capacity for keeping phosphorus in natural waters near or below environmentally acceptable levels (156).

Reduction of runoff phosphorus in problem areas such as soils of high phosphate status, animal feedlots, and in urban runoff can be more readily achieved by water management to reduce runoff and erosion rather than by chemical treatment to remove phosphorus from the water. Farm management practices, thus, are of major importance for control of runoff and erosion. Schuman and co-workers (158) reported that the combined dissolved and sediment phosphorus lost from contour-plowed corn was nearly ten times higher than those losses from level-terraced corn when the two were fertilized at the same rate. Phosphorus losses from brome-grass pasture were substantially below those from contour-plowed corn fertilized at the same level.

In general, the trends for phosphorus pollution from animal wastes parallel those for nitrogen. Storage and application practices are crucial. Fall application of manure on alfalfa gave the highest losses, whereas fall application to fall-plowed cornfields gave much smaller losses (159).

Perspective and Corrective Measures. The production and use of fertilizers are vital for adequate production of food and fiber for human needs. To continue to meet these needs will require further increase in use of fertilizer. Although the impact on the environment of fertilizer nutrient losses from agricultural soils is difficult to quantify, losses to surface and groundwater do occur, erosion losses probably exceeding those of solution. It is not unusual to find that the combined pollutants from nonhuman sources and domestic sewage far exceed those from farming. Moreover, the farming inputs are only partially owing to use of commercial fertilizer. As important as control of soil erosion is for reducing losses of biologically active phosphorus to streams, on a national (U.S.) basis the effect of reducing soil erosion is small compared to the reduction that would result from tertiary treatment of municipal sewage.

USDA and EPA evaluated control of pollution from cropland (160) and listed the following practices for control of nutrient loss from agricultural applications: elimination of excessive fertilization; proper timing of nitrogen application; using crop rotations; using animal wastes for fertilizer; plowing under green manure crops; using winter cover crops; controlling fertilizer release; incorporating surface applications; controlling surface applications; using legumes in haylands and pasture; and timing fertilizer plow-down.

Energy Requirements for the Production and Distribution of Chemical Fertilizers in the United States

Production and distribution of fertilizers in the United States consumed only about 0.7% of the total national energy consumption at the 1976 level of production. The total national energy consumption of fertilizers in the United States agricultural system from extraction of raw materials to the application of fertilizer to the soil is about 5.67×10^8 GJ (538×10^{12} Btu), according to Davis and Blouin (161). About 88% of the total is for production, 1% is for transportation of raw materials, and 11% is for transportation, storage, handling, and application of the fertilizer. Energy requirements for production of the fertilizers are summarized in Table 18.

The estimated energy for raw material and product transportation is 38.5×10^6 GJ (36.5×10^{12} Btu), that for field application is 9.92×10^6 GJ (9.41×10^{12} Btu), and that for storage and transfer is 22.79×10^6 GJ (21.62×10^{12} Btu).

Although the energy consumed in fertilizer production and use is a very small percentage of total national energy consumption, the shortage of energy and its high cost make it a matter of concern, and serious efforts by the industry toward lowering energy consumption are underway. According to a recent report of The Fertilizer Institute (162), the average energy consumption per unit of fertilizer produced was reduced by 12% over the period 1972–1977.

Table 18. Total Energy Requirements for Production of Fertilizers in the United States[a]

Materials and intermediates	Quantity, 10^6 t	Energy requirements, 10^6 GJ[b]	
		Per t	Total
nitrogen[c]			
ammonia	8.26	48.84	403.42
solid urea[d]	0.64	8.14	5.21
urea solution[d]	0.64	3.49	2.23
solid ammonium nitrate[d]	1.18	9.30	10.97
ammonium nitrate solution[d]	0.64	4.65	2.98
Total nitrogen			*424.81*
phosphate[c]			
wet-process acid	3.99	11.63	46.40
triple superphosphate[d]	1.54	1.16	1.79
granular ammonoum phosphate[d]	2.45	1.16	1.79
ammonium polyphosphate solution[d]	0.45	5.81	2.61
normal superphosphate	0.27	2.32	0.63
Total phosphate			*53.22*
potassium[c]			
potassium chloride	2.27	4.65	10.56
Total potassium			*10.56*
mixtures[d,e]			
granular homogenous	8.62	0.70	6.01
fluid	3.18	0.12	0.37
bulk blend	8.62	0.12	1.03
Total mixtures			*7.41*
Total			*496*

[a] Ref. 161.
[b] To convert GJ to 10^6 Btu, divide by 1.054.
[c] Per t N, P_2O_5, K_2O.
[d] Does not include the energy required to produce intermediates.
[e] Per metric ton material.

The Future. Indications (discussed in more detail under ammonia production) are that new coal gasification technology now under development could decrease energy consumption for ammonia production from coal by about 20% compared with the old coal-based technology. Production of ammonia from coal could relieve the pressure on the United States natural gas shortage but it would not decrease total energy consumption; it would substitute an abundant feedstock for a scarce one. Studies of other routes for producing hydrogen for ammonia synthesis are still needed. Indeed, studies of nitrogen fixation by means other than reaction with hydrogen should also be pursued (see Nitrogen fixation).

Opportunities for producing wet-process acid with greater energy efficiency have been discussed. Trends toward greater utilization of chemical heat in fertilizer processes have already started (eg, melt granulation). The long-standing trend toward higher fertilizer analysis can be extended still further, with reduction in energy required for storage, transportation, and application of fertilizer. Technology in advanced stages of development in TVA (161) could result in an increase of 25–30% in product nutrient concentration.

BIBLIOGRAPHY

"Fertilizers" in *ECT* 1st ed., Vol. 6, pp. 376–452, by H. B. Siems, Swift & Company; "Fertilizers" in *ECT* 2nd ed., Vol. 9, pp. 25–150, by A. V. Slack, Tennessee Valley Authority.

1. J. J. Mortvedt and co-eds., *Micronutrients in Agriculture,* Soil Science Society of America, Madison, Wis., 1972.
2. F. E. Khasawneh, E. C. Sample, and E. J. Kamprath, eds., *The Role of Phosphorus in Agriculture,* American Society of Agronomy, Madison, Wis., 1979.
3. *Fertilizer Manual, United Nations Publication 67, II. B.1, ST/CID/15,* United Nations Industrial Development Organization, New York, 1967, p. 6.
4. *Worldwide Study of the Fertilizer Industry: 1975–2000,* International Centre for Industrial Studies, UNIDO/ICIS, United Nations Industrial Development Organization, Vienna, Austria, 1976.
5. *International Fertilizer Development Center, C-1,* TVA, Muscle Shoals, Alabama, May 1978.
6. *Chem. Week,* (Sept. 5, 1979).
7. L. B. Nelson, *National Association of Colleges and Teachers of Agriculture* 27 (June 1975).
8. *FAO Soil Bulletin No. 27, MI/F. 6588/E/275/1/1500,* FAO, New York, 1975, p. 356.
9. P. W. Arnold, *Proc. No. 115, The Fertilizer Society,* (Mar. 12, 1970).
10. Evaluation of TVA's National Fertilizer Introduction Program, *TVA Circular Z-79,* Tennessee Valley Authority, Muscle Shoals, Alabama, Apr. 1977.
11. N. L. Hargett and L. G. Sills, *Proc. 27th Fert. Ind. Round Table,* 148 (1977).
12. *Farm Chemicals Handbook,* Ware Brothers Co., Philadelphia, Pa., 1977, p. 1360.
13. D. E. Nichols, P. C. Williamson, and D. R. Waggoner, *paper presented at The Steenbock-Kettering International Symposium on Nitrogen Fixation, Madison, Wisc., June 12–16, 1978.*
14. *Paper presented to Commission on Natural Resources, National Academy of Sciences, Washington, D.C., 1975.*
15. H. Brown in J. M. Hollander, ed., *Annual Review of Energy,* Vol. 1, Annual Reviews, Inc., Palo Alto, Calif., 1976, pp. 1–36.
16. *Eur. Chem. News* **31**(802), 31 (Sept. 9, 1977).
17. J. B. Cathcart in ref. 2, Chapt I.
18. J. B. Cathcart in ref. 2, Chapt. II.
19. E. A. Imhoff, *U.S. Geological Survey Open File Report No. 76-648,* 1976.
20. *World Survey of Potash Resources,* 2nd ed., The British Sulphur Corporation Limited, London, Eng., 1966.
21. "Mineral Facts and Problems," *U.S. Bur. Mines Bull.* **667,** (1975).
22. D. W. Bixby in ref. 2, Chapt. V.
23. J. T. Shields and E. A. Harre, *paper presented at Florida Fertilizer and Lime Conference, May 1977.*
24. E. A. Harre, O. W. Livingston, and J. T. Shields, *World Fertilizer Market Review and Outlook, Bulletin Y-70,* National Fertilizer Development Center, TVA, Muscle Shoals, Alabama, 1974.
25. Derived from U.S. Department of Agriculture Annual Statistical Reports.
26. *Ammonium Nitrate—Supply/Demand 1957–1977,* The British Sulphur Corporation Limited, London, Eng., 1970.
27. *Annual Fertilizer Review, 1976,* Food and Agriculture Organization of the United Nations, Rome, Italy, 1977.
28. E. A. Harre, M. N. Goodson, and J. D. Bridges, *Fertilizer Trends 1976, Bulletin Y-111,* TVA, Muscle Shoals, Alabama, Mar. 1977.
29. *Nitrogen* (53), 26 (May–June 1968).
30. R. W. R. Carter and A. G. Roberts, *Proc. No. 110, The Fertilizer Society (London),* (Oct. 1969).
31. G. Perbal, *Proc. No. 124, The Fertilizer Society (London),* (1971).
32. *Hydrocarbon Process.* **56**(11), 130 (Nov. 1977).
33. R. D. Young and I. W. McCamy, *Can. J. Chem. Eng.* **45,** 50 (Feb. 1967).
34. R. S. Meline and co-workers, *J. Agric. Food Chem.* **16,** 235 (Mar.–Apr. 1968).
35. O. Skauli, *paper presented at the 1974 ACS Meeting, Atlantic City N.J., Sept. 8–13, 1974.*
36. *Chemical Economics Handbook,* Stanford Research Institute, Menlo Park, Calif., 1976, pp. 756, 4002A-D.
37. G. S. Scott and R. L. Grant, *U.S. Bur. Mines Inf. Circ.* **7463,** (June 1948).
38. J. J. Burns and co-workers, *U.S. Bur. Mines Rep. Invest.* **4994,** (Aug. 1953).

39. *Report of the Ammonium Nitrate Working Party,* His Majesty's Stationery Office, London, Eng., 1951.
40. *Urea—World Analyses and Prospects, 1970–1980,* The British Sulphur Corporation Limited, London, Eng., 1975.
41. *Urea—Chemical Economics Handbook,* Stanford Research Institute, Menlo Park, California, Mar. 1976, pp. 756–2007D.
42. R. D. Hauck and H. F. Stephenson, *J. Agric. Food Chem.* **13,** 486 (Nov.–Dec. 1965).
43. S. M. Lemkowitz, M. G. R. T. deCooker, and P. J. van den Berg, *Proc. No. 131, The Fertilizer Society (London),* (Dec. 14, 1972).
44. A. V. Slack and G. M. Blouin, *Urea Technology: A Critical Review, Tennessee Valley Authority Circular Z-4,* TVA, Muscle Shoals, Alabama, Dec. 14–16, 1969.
45. D. R. Waggoner and co-workers, *Proc. 24th Fert. Ind. Round Table 1974,* 70 (Dec. 3–5, 1974).
46. I. W. McCamy and M. M. Norton, *Farm Chemicals* **140**(2), 61 (Jan. 1977).
47. G. M. Blouin, *Urea Costs, Granulation vs Prilling, Tennessee Valley Authority Bulletin Y-92,* TVA, Muscle Shoals, Alabama, Apr. 1975.
48. R. P. Ruskin, *Chem. Eng.* **83,** 1114 (June 7, 1976).
49. E. M. Klapproth, *Chemical Economics Handbook,* Stanford Research Institute, Menlo Park, Calif., 1976, pp. 756, 6005A.
50. E. O. Huffman and co-workers, *Economic Commission for Europe, Seminar on Technical and Economic Aspects of the Fertilizer Industry, Helsinki, Finland, Aug. 11–15, 1975, TVA Circular Z-59,* TVA, Muscle Shoals, Alabama, 1975.
51. *Annual Fertilizer Review,* Food and Agriculture Organization of the United Nations, Rome, Italy, 1977.
52. W. F. Sheldrick, *Proc. 27th. Fert. Ind. Round Table 1977,* 13 (Oct. 1977).
53. *Green Markets,* 8 (Jan. 9, 1978).
54. W. C. White and K. T. Johnson in ref. 2, Chapt. IX.
55. J. R. Lehr and J. Hoar in ref. 2, Chapt. IV.
56. T. P. Hignett and co-workers in L. J. Carpentier, ed., *Proc. Tech. Conf. ISMA, Ltd.,* The Hague, The Netherlands, Sept. 13–16, 1976, pp. 273–288.
57. J. R. Lehr and co-workers, *Colloque International Sur Les Phosphates Mineraux Solides,* Toulouse, Fr., Mar. 16–20, 1967.
58. G. H. McClellan and J. R. Lehr, *Am. Mineral.* **54,** 1374 (1969).
59. J. R. Lehr and G. H. McClellan, *A Revised Laboratory Reactivity Scale for Evaluating Phosphate Rock for Direct Application, Tennessee Valley Authority Bulletin Y-43,* TVA, Muscle Shoals, Alabama, Apr. 1972.
60. A. W. Frazier, J. R. Lehr, and E. F. Dillard, *Chemical Behavior of Fluorine in the Production of Wet-Process Phosphoric Acid. Bulletin Y-113,* TVA, Muscle Shoals, Alabama, May 1977.
61. J. R. Lehr, A. W. Frazier, and J. P. Smith. *J. Agric. Food Chem.* **14,** 27 (1966).
62. R. F. Jameson in A. V. Slack, ed., *Phosphoric Acid,* Vol. I. Part 1, Marcel Dekker, Inc. New York, 1968, pp. 985–1006.
63. T. C. Woodis, *Anal. Chem.* **36,** 1682 (1964).
64. T. C. Woodis, J. R. Trim, and R. D. Duncan, *Anal. Chim. Acta* **65,** 469 (1973).
65. N. Robinson in ref. 2, Chapt. VI.
66. A. V. Slack in ref. 62, pp. 157–404.
67. *Eur. Chem. News* **21,** 526 (Mar. 1972).
68. R. L. Somerville, *paper presented to Div. Fert. Soil Chem., ACS, Chicago, Ill., Aug. 27, 1973.*
69. U.S. Pat. 3,642,439 (Oct. 15, 1969), W. P. Moore and co-workers (to Allied Chemical Corporation).
70. *Phosphorus Potassium* **79,** 25 (Sept.–Oct. 1975).
71. A. Baniel and R. Blumberg in A. V. Slack, ed., *Phosphoric Acid,* Vol. I, Part II, Marcel Dekker, Inc., New York, 1968, pp. 887–913.
72. *Israel Mining Industries Staff Report,* International Solvent Extraction Conference, The Hague, The Netherlands, 1971.
73. R. Blumberg, *Proc. No. 151, The Fertilizer Society Proc. (London),* (Nov. 13, 1975).
74. Institute for Research and Development, Israel, *ISMA Technical Conference (Seville),* 20 (Nov. 1972).
75. J. F. McCullough, *Proc. 26th Fert. Ind. Round Table,* 49 (Oct. 26–28, 1976).
76. Private communication, TVA, 1978.
77. F. J. Hurst and D. J. Krouse, *paper presented at 166th ACS National Meeting, Chicago, Ill., Aug. 27–29, 1973.*

78. *Eng. Mining J.* **178**(5), 9 (May 1977).
79. F. P. Achorn, B. Wright, Jr., and H. L. Balay, *Fertilizer Solutions Mag.* **20**(3), (June 1976).
80. *Superphosphate: Its History, Chemistry, and Manufacture.* U.S. Department of Agriculture and Tennessee Valley Authority, Washington, D.C., Dec. 1964.
81. J. R. Lehr in ref. 73, pp. 637–686.
82. A. B. Phillips, R. D. Young, J. S. Lewis, Jr., and E. G. Heil, *J. Agr. and Food Chem.* **6,** No. 7, 585 (July 1958).
83. G. G. Brown and K. F. J. Thatcher, *Proc. No. 96, The Fertiliser Society (London),* (Feb. 23, 1967).
84. R. C. Smith, Amax Chemical Corporation, "The Potash Industry," *paper presented to the American Association of Plant Food Control Officials, El Paso, Tex., Aug. 5, 1977.*
85. *Phosphorus Potassium,* (2), 32 (1962).
86. G. R. Hagstrom, *Fertilizer Solutions* **17**(2), 76, 78, 80, 86, 88 (1973).
87. J. Silverberg, R. D. Young, and G. Hoffmesiter, Jr., in ref. 1, pp. 431–458.
88. J. R. Lehr in ref. 1, pp. 459–503.
89. J. E. Reynolds, Jr., in A. I. More, ed., *Granular Fertilizers and Their Production,* Nov. 1977, British Sulphur Corporation, Limited, London, Eng., 1978, pp. 181–197.
90. R. D. Young and C. H. Davis, *Proc. 23rd Annual Meeting of the Fert. Ind. Round Table, Washington, D.C.,* (Nov. 6–8, 1973).
91. G. C. Hicks, *Tennessee Valley Authority Bulletin Y-108,* TVA, Muscle Shoals, Alabama, Oct. 1976.
92. F. P. Achorn and H. L. Balay, *Proc. 24th Fert. Ind. Round Table, 1974,* 53 (Dec. 3–5, 1974).
93. F. E. Steepvoeinkle and J. M. Hoogendonk, *Proc. No. 109, The Fertilizer Society (London),* (Apr. 24, 1969).
94. Y. F. Berguin in ref. 89, pp. 296–303.
95. R. G. Lee, M. M. Norton, and H. G. Graham, *Proc. 24th, Fert. Ind. Round Table, 1974,* 79 (Dec. 3–5, 1976).
96. T. Freng and C. H. Lee, *paper presented at Annual Meeting ACS, Atlantic City, N.J., Sept. 8–13, 1974.*
97. T. Grundt, T. Heggebo, and J. Ryan in ref. 89, pp. 44–58.
98. R. D. Young and R. G. Lee, *paper presented at Technical Conference of International Superphosphate and Compound Manufacturers Association, Ltd., and The Fertiliser Association of India, New Delhi, India, December 8–10, 1975, TVA Circular Z-70,* TVA, Muscle Shoals, Alabama, 1975.
99. F. P. Achorn and D. G. Salladay, *Proc. 28th, Fert. Ind. Round Table, 1975,* 196 (Nov. 4–6, 1975).
100. S. Maier in ref. 89, pp. 283–295.
101. R. D. Young, G. C. Hicks, and C. H. Davis, *J. Agric. Food Chem.* **10,** 442 (Nov.–Dec. 1962).
102. U.S. Pat. 3,153,574 (Oct. 20, 1964), F. P. Achorn, R. D. Young, and G. C. Hicks (to TVA).
103. I. S. E. Martin and F. J. Harris in ref. 89, pp. 253–270.
104. H. L. Balay and D. G. Salladay, *Proc. 26th, Fert. Ind. Round Table,* 93 (1976).
105. G. C. Hicks, *Tennessee Valley Authority Bulletin Y-119,* TVA, Muscle Shoals, Alabama, Nov. 1977.
106. J. L. Hawksley in ref. 89, pp. 110–124.
107. I. W. McCamy and G. C. Hicks, *Proc. 26th, Fert. Ind. Round Table,* 67 (1976).
108. R. Ewell, *paper presented at UNIDO, 2nd Interregional Fertilizer Symposium, Kiev, U.S.S.R., Sept. 1971.*
109. A. V. Slack and co-workers, *Farm. Chem.* **130,** (Apr.–July 1967).
110. *Phosphorus Potassium,* (39), Part I, 18 (Jan.–Feb. 1969); Part II, 29 (Mar.–Apr. 1969).
111. T. P. Hignett, *paper No. 8 presented to ANDA/ISMA Technical Seminar, Sao Paulo, Brazil, Apr. 22–24, 1975.*
112. T. Grundt, T. Hegglbo, and J. Ryan in ref. 89, pp. 44–58.
113. K. C. Kundsen in ref. 89, pp. 59–67.
114. T. P. Hignett, *Proc. No. 87, The Fertilizer Society (London),* (Mar. 25, 1965).
115. *TVA Circular Z-83,* TVA, Muscle Shoals, Alabama, Sept. 1977.
116. H. G. Walkup, *Fert. Prog.* **3,** 9–12, 34 (July–Aug.); 12–13 (Sept.–Oct.) (1972).
117. *Fertilizer Progress,* The Fertilizer Institute, Washington, D.C., 1975, pp. 14–16.
118. W. C. Scott, J. A. Wilbanks, and L. C. Faulkner, *Fertilizer Solutions* **20,** 54 (July–Aug. 1976).
119. T. D. Farr, J. W. Williard, and J. D. Fleming, *paper presented at 164th National Meeting of the ACS, New York, Sept. 6, 1972.*
120. J. W. Williard and E. F. Dillard, *J. Chem. and Eng. Data* **20,** 436 (1975).
121. A. V. Slack and F. P. Achorn, *Proc. No. 133, The Fert. Soc. (London),* (Feb. 15, 1973).

122. A. W. Frazier, R. M. Scheib, and R. D. Thrasher, *J. Agric. Food Chem.* **20,** 138 (1972).
123. National Fertilizer Solutions Association Staff Feature, *Fert. Solutions* **21**(6), 137 (Nov.–Dec. 1977).
124. R. S. Meline, *TVA Circular Z-60,* TVA, Muscle Shoals, Alabama, Sept. 1974.
125. *Phosphorus Potassium,* (89), 31 (May–June 1977).
126. K. T. Johnson, *Proc. FAI-IFDC Fertilizer Seminar 1977, Trends in Consumption and Production, Dec. 1–3, 1977, Tech-III/2-1-2-28,* The Fertilizer Association of India, New Delhi, India, 1977.
127. *Fed. Reg.* **40,** 33152 (Aug. 6, 1975).
128. M. L. Brown, *Environmental Symposium, New Orleans, La., Jan. 1976,* The Fertilizer Institute, Washington, D.C., pp. 137–177.
129. *Background Information of Performance: Phosphate Fertilizer Industry, EPA-450/2-74-019A,* Vol. 1, EPA, Washington, D.C., Oct. 1974.
130. L. L. Beck and W. Vataruk, *An Investigation of the Best Systems of Emission Reduction for the Phosphate Rock Processing-Industry,* EPA, Washington, D.C., Feb. 1976.
131. G. D. Rawlings and R. B. Reznik, *Source Assessment: Fertilizer Mixing Plants, EPA-600/2-76-032C,* EPA, Washington, D.C., Mar. 1976.
132. G. H. Powers in ref. 128, pp. 383–391.
133. J. C. Barber in ref. 128, 351–382.
134. R. G. Lee and D. R. Waggoner, *paper presented at ACS meeting, Chicago, Ill., Aug. 24–29, 1975.*
135. J. C. Stoner in ref. 128, pp. 251–286.
136. W. Unruk in ref. 128, pp. 287–292.
137. M. L. Brown in ref. 128, pp. 293–300.
138. U.S. Pat. 3,758,277 (Sept. 11, 1973), T. M. Cook, G. L. Tucker, and M. L. Brown (to Mississippi Chemical Corporation).
139. U.S. Pat. 3,870,782 (Mar. 11, 1975), T. M. Cook, G. L. Tucker, and M. L. Brown (to Mississippi Chemical Corporation).
140. T. M. Robinson and co-workers, *Engineering and Cost Effectiveness Study of Fluoride Emissions Control for Environmental Protection Agency, PB207 506,* EPA, Washington, D.C., Jan. 1972.
141. *Fed. Reg.* **39,** 2832 (1974).
142. *Fed. Reg.* **43,** sections 418.42, 418.43, 418.45 (Apr. 26, 1978).
143. *Fed. Reg.* **42,** sections 418.51, 418.52, 418.55, 418.56 (Mar. 25, 1977).
144. G. F. Palm in ref. 128, pp. 309–332.
145. A. W. Frazier, J. R. Lehr, and E. F. Dillard, *Bulletin Y-113,* TVA, Muscle Shoals, Alabama, Mar. 1977.
146. J. R. Brennen in ref. 128, pp. 217–250.
147. *Chem. Week* **120**(11), 44 (Mar. 1977).
148. *Preliminary Findings Radon Daughter Levels in Structures Constructed on Reclaimed Florida Phosphate Land, EPA technical note No. ORP/CSD-75-4,* EPA, Washington, D.C., Sept. 1975, p. 14.
149. K. S. Porter, *Nitrogen and Phosphorus, Food Production, Waste and the Environment,* Ann Arbor Science Publishers, Inc., Ann Arbor, Mich., 1975.
150. R. W. Simonson in ref. 149, pp. 21–45.
151. W. D. P. Stewart, *Science* **158,** 1426 (1967).
152. A. W. White and co-workers, *Crops Soils* **19**(4), 28 (Jan. 1967).
153. V. J. Kilmer and R. T. Joyce, *Proc. Natl. Fert. Solutions Assoc. Roundup,* 25 (July 25, 1972).
154. H. S. Johnson, *Science* **195,** 658 (Feb. 18, 1977).
155. J. Worth Wilkerson, ed., *Nitrogen Fertilizer—A Potential Polluter, Tennessee Valley Perspective,* TVA, Muscle Shoals, Alabama, 1976, pp. 15–19.
156. V. J. Kilmer and A. W. Taylor in ref. 2, Chapt. XIX.
157. C. N. Wadleigh, *Wastes in Relation to Agriculture and Forestry, Miscellaneous Publication No. 1065,* USDA, Washington, D.C., 1968.
158. G. E. Schuman, R. G. Spome, and R. F. Piest, *Soil Sci. Soc. Am. Proc.* **37,** 424 (1973).
159. R. A. Young and C. K. Mutchler, *J. Environ. Qual.* **5,** 174 (1976).
160. *Control of Water Pollution from Cropland,* Vol. I, U.S. Department of Agriculture and EPA, Washington, D.C., Nov. 1975.
161. C. H. Davis and G. M. Blouin, *paper presented at Conference on Energy and Agriculture, Center*

for the Biology of Natural Systems, Washington University, St. Louis, Missouri, June 16–19, 1976.
162. *TFI Fertilizer News Release*, The Fertilizer Institute, Washington, D.C., May 5, 1978.

E. O. HUFFMAN
Consultant

FIBERBOARD. See Laminated and reinforced wood.

FIBER OPTICS

Fiber optics have been in existence for a number of years. Previous uses have included medical endoscopes, cathode-ray-tube faceplates, and flexible light-distribution bundles. In these applications, transmission of light over a few meters is all that is needed. Telecommunication, in contrast, requires transmission over distances that are orders of magnitude longer. This article is principally a description of the development of fiber optic lightguides which has made communication by optical means practical.

Light has been used as a medium for transmission of messages for centuries. Exchange of messages in morse code by light flashes has been a common practice for the last century. An optical telegraph system operated in France from 1790 until replaced by the electric telegraph. The system reputedly could transmit an entire message a distance of 200 km in 15 min.

In 1880 Alexander Graham Bell invented the photophone by which speech could be transmitted on a light beam. Sunlight was bounced from a reflector through a lens to a mechanism that vibrated in response to speech. This caused the light beam to vary in intensity. At the receiving end, a selenium detector transformed these variations into electrical current to recreate speech through a telephone receiver.

The emerging technology of lightwave communications makes a different and more sophisticated use of light. Electromagnetic radiation in the visible or infrared region has frequencies ten thousand times the frequencies used in radio communications. Since information-carrying capacity increases with frequency, the potential for lightwave communications is enormous. For instance, it can be shown that by using only a small fraction of the bandwidth available from visible or infrared radiation, a signal could simultaneously carry the telephone conversations of every inhabitant of North America.

However, before the promise of this communication medium could be realized, several technical obstacles had to be overcome. Principal among them was the development of a suitable light source, modulator, and transmission medium. The laser, first demonstrated in 1960, was found to be capable of producing a light beam of appropriate intensity that could be modulated at sufficiently high rates. Although early lasers were cumbersome and technically undesirable, the later double-heterostructure gallium arsenide laser showed adequate potential and had a projected lifetime of decades. This device, which is only a few tenths of a millimeter on each side, can produce a beam of several milliwatts of optical power.

However, it was recognized from the beginning that transmission through the atmosphere would be unsatisfactory: fog, smog, rain, and snow attenuate the signal. Therefore, various kinds of light pipes were studied. These consisted of tubes with optical means of redirecting and refocusing the beam of light. These devices, which included both glass lenses and gas lenses, were expensive compared to optical fibers fabricated from glass or plastic. Such fibers had already been used for some years to carry light short distances, eg, for illumination of automobile instrument panels, and for medical applications as endoscopes. However, for long-distance communication, a much-improved transmission capability was required. For example, light intensity is reduced by half during passage through 3 cm of ordinary window glass. For lightwave communications to be competitive with existing systems, such a reduction in intensity can be tolerated only after the light has traversed 1 km of glass fiber.

Lightwave Communications System

A lightwave communications system consists of three major components: light source, detector, and transmission medium.

Light Sources. Light sources are of two types, light-emitting diodes (LEDs) and injection lasers. For an LED to be useful for lightwave communications, it must have a sufficiently high brightness (radiance) and the ability to be modulated directly at sufficiently high rates, ie, it must have a fast response time (see Light-emitting diodes and semiconductor lasers).

LEDs specifically for use in lightwave communications systems combine small area and high current density. Early GaAs diodes (1970) had attained radiance values of 20–35 W/(sr·cm^2) (see Color) with current flows of ca 300 mA (1). Improved design can yield a radiance of 100 W/(sr·cm^2) (2). On the other hand, LEDs of comparatively low brightness can be used with specially designed fibers. The advantage of these LEDs is that reliability improves as brightness is reduced (see also Digital displays).

Injection lasers (qv) are attractive as optical sources because they provide adequate launched power for either single-mode or multimode fibers. They are small, rugged, efficient, and can be modulated directly by means of the injected current. The most promising design at present is the double-heterostructure laser (3–4) which consists of $Ga_xAl_{1-x}As$ layers. The characteristics of these lasers are reviewed in ref. 5. Injection lasers usually had a short lifetime. However, extensive work to identify failure mechanisms has resulted in devices that now can operate continuously >12 yr (6).

Neodymium-doped yttrium–aluminum–garnet, a possible candidate for optical sources, emits a very narrow laser line at ca 1060 nm. In addition to its narrow line width, its output wavelength coincides with a low loss region of optical fibers. Single-frequency, single-mode output can be achieved easily, and the laser rods can be pumped with LEDs (7). However, use of the garnet requires a separate external modulator, which adds to both the complexity and the cost of the system. More attractive LED candidates are a series of indium–gallium–arsenide phosphides (8). These operate at wavelengths between 1200 and 1500 nm, where high silica fibers exhibit even lower losses than at 1060 nm. Also of importance to LED use, at these wavelengths the material dispersion of the glass becomes zero.

Optical Detectors. Optical detectors must have high sensitivity at the wavelength of the carrier and a fast enough response time to accommodate the data rate being used. For these and other reasons only solid state PIN (p-type silicon-intrinsic layer-n-type silicon sandwich) photodiodes and silicon avalanche photodiodes are adequate for optical communications use. The PIN structures have no avalanche gain, and the sensitivity of a receiver that employs such a device is usually determined by the thermal noise in the front-end circuitry of the receiver.

In an avalanche photodiode, the photocurrent generated by the incoming photons is internally amplified. The photocarriers are accelerated by a high field region at the junction to energies high enough to create additional electron-hole pairs through impact ionization. The avalanche multiplication process results in internal amplification or gain. The advantage of an avalanche device with gain, compared to a no-gain PIN structure, is 10–20 dB (9) when operating at a few hundred megahertz (see also Semiconductors).

Transmission Medium. A number of transmission media have been considered: the atmosphere, the lens waveguide, and the gas waveguide. These all have proven unsatisfactory or too expensive compared with the glass-fiber lightguide (see also Glass). This lightguide consists of a core of very high transparency, which transmits the light. The core is surrounded by a transparent cladding layer of lower refractive index.

When light is launched into the core region of clad-fiber structure, the portion falling within a certain solid angle is guided by total internal reflection. Snell's law defines the critical angle for total internal reflection based on the refractive indexes of the core and cladding. This process of light collection and guidance is illustrated in Figure 1.

There are two types of clad fibers, single-mode and multimode. Whether a fiber is single- or multimode is determined by the diameter of the core, the difference in refractive index between core and cladding, and the wavelength of the light being used. If the refractive index difference is reasonably large, eg, 1%, then for single-mode operation the core diameter must be small, on the order of the wavelength of the propagating light.

Step-index fibers are operating in single-mode when (10):

$$\frac{2\,\pi a}{\lambda}\cdot\sqrt{2\,n\Delta n} \leqslant 2.41$$

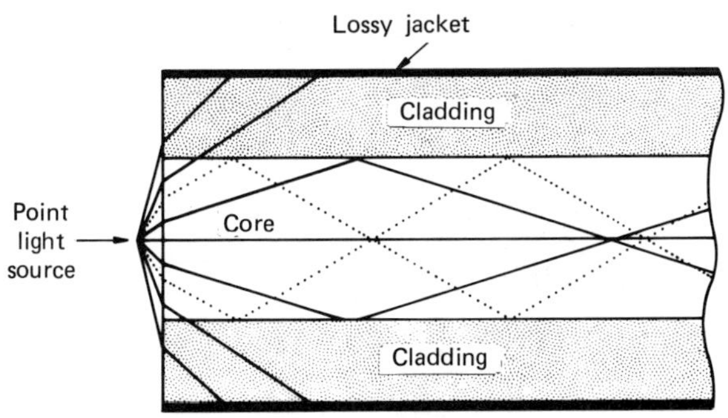

Figure 1. Light from a point source collected and guided by a glass-fiber lightguide. Courtesy of Academic Press, Inc.

where a = core radius, λ = wavelength of the transmitted light, n = core refractive index, and Δn = difference in refractive index between core and cladding. Thus for single-mode propagation of 800 nm wavelength light in a fused-silica-core lightguide, if the refractive index difference between core and cladding is 0.01, the maximum permissable core diameter is ca 3.5 μm. The cross section of a single-mode fiber is illustrated in Figure 2**a**.

A fiber is multimode if either the diameter of the fiber core or the core-cladding refractive index difference is larger than the limits for single-mode propagation. Light rays travelling at various angles within the core will be guided. An illustration of a multimode fiber cross section is shown in Figure 2**b**.

It is assumed in the preceding discussion that the core of the fiber has a uniform refractive index, and that the refractive index of the cladding, although lower than that of the core, is also uniform. The change in refractive index at the core-cladding interface is assumed to be an abrupt one. To decrease mode dispersion of multimode lightguides, the so-called graded-index fiber configuration is important. In such a structure, the refractive index of the core increases gradually from its periphery to the center as shown in Figure 2**c**.

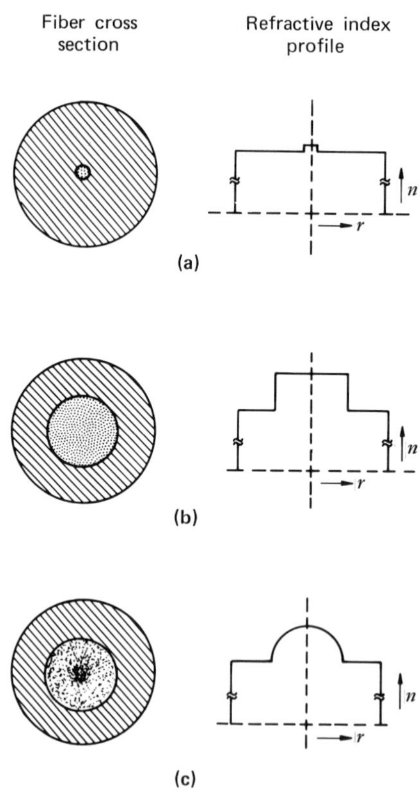

Figure 2. Three types of fiber lightguide: (**a**) single mode; (**b**) multimode step index; (**c**) multimode graded index. Courtesy of Academic Press, Inc.

Properties of Clad-Glass Fibers

Optical Characteristics. The most important characteristic of optical fibers is the attenuation or loss that the structure manifests. The amount of loss that can be tolerated depends on the particular use to which the fibers are put. In short-range systems such as interconnections within buildings, airplanes, or ships, lengths up to a few hundred meters are required, and losses of 50–100 dB/km are permissible. Systems of medium range, such as between the central switching offices of a telephone network, require spans of up to ca 12 km. Long-distance, city-to-city links would run for many tens or hundreds of kilometers. In medium- and long-distance systems, very low losses (<5 dB/km) would be required to eliminate or minimize the use of repeaters.

The second most important characteristic of optical fibers is the impulse response of the fiber. Mode dispersion and material dispersion broaden the light pulse, which causes pulses in consecutive time slots to overlap after transmission over a certain distance. This overlap limits the bandwidth of the transmission line. Use of single-mode fibers and laser light sources gives the lowest pulse dispersion and allows the highest data rate. However, because of the small core size and small refractive index difference, difficulties are encountered in splicing and connecting, and with bending losses and input coupling. Therefore, it is likely that early systems will operate at intermediate data rates and will use multimode low-loss fibers.

There are other important properties of optical fibers. The difference in the core-cladding refractive index of the fiber should be as large as reasonably possible in order to have a large light-acceptance angle for efficient coupling of the light source and to minimize microbending loss. The diameter of the core should be large enough to facilitate splicing and coupling to the source. Tensile strength and chemical durability should be maximized.

Particular specifications for an optical fiber lightguide depend on its use. For instance, a system operating at 50 megabits per second, with repeaters spaced 10 km apart, might use fibers with: loss, <4 dB/km; dispersion, <2 ns/km; core-cladding refractive index difference, 0.015; core diameter, 50 μm; cladding thickness, 25 μm; and tensile strength, ca 700 MPa (ca 10^5 psi).

Optical Attenuation. Optical attenuation, or loss, in a clad-glass fiber is caused by three distinct mechanisms: material absorption, material scattering, and lightguide scattering.

Material absorption is primarily owing to the presence in the glass of transition metal ions: iron, cobalt, nickel, copper, chromium, manganese, and vanadium. The valence state of these ions is important. For instance, ferrous iron absorbs in the near infrared region, with deleterious effects on lightguide transparency, whereas ferric iron absorbs in the ultraviolet and is, therefore, not important to transmission in the infrared. It has been estimated (11) that a few tens of parts per billion of these metals dissolved in a soda–lime silicate glass would cause sufficient absorption to render the glass unsuitable for lightguide use. Water can also be a serious contaminant. The fundamental absorption is ca 2800 nm in most silicate glasses with first and second overtones occurring at ca 1400 and 930 nm, respectively. The second overtone gives rise to an absorption of ca 1.2 dB/km for each 1 ppm of water present in fused-silica glass. Since this absorption occurs in the wavelength region of interest, it is important that water content be minimized.

Material scattering occurs because inorganic oxide glasses are not homogeneous structures. Compositional fluctuations and density fluctuations are inherent on the atomic scale. Macroscopic inhomogeneities such as bubbles and crystallites often occur. Those fluctuations that occur on a scale smaller than the wavelength of the light being used give rise to Rayleigh scattering, which varies in intensity with wavelength as $1/\lambda^4$ (12). In fused silica at 1000 nm, Rayleigh scattering amounts to about 0.7 dB/km. Those fluctuations that occur over distances that are long compared to the wavelength give rise to Mie scattering (12). Examples of such irregularities would be bubbles, crystallites, and phase separation. Mie scattering can be much larger than Rayleigh scattering and, whereas the latter is radially isotropic, Mie scattering can be strongly anisotropic.

Lightwave transmission is affected by the structure of the lightguide. For example, in a step-index lightguide, guidance takes place by total internal reflection at the core-cladding interface. The angle at which a particular ray of light strikes this interface is all important. If the angle is greater than the critical angle the ray is guided; if it is less, the light passes through the interface and is lost in the cladding. Irregularities at this interface must therefore be avoided or they might result in some of the light striking the interface at an angle such that it is lost into the cladding.

As mentioned above, mode dispersion and material dispersion cause pulse broadening in any multimode step-index fiber. In Figure 3**a** three rays of light are shown travelling down a step-index multimode fiber. These represent the fundamental mode, a medium-order mode, and a high-order mode. It is easy to see that the fundamental mode travels the shortest distance, and higher-order modes must travel a longer path through the core to reach the other end. If the refractive index of the core is uniform, then all three rays travel at the same speed. If all of the rays are injected into

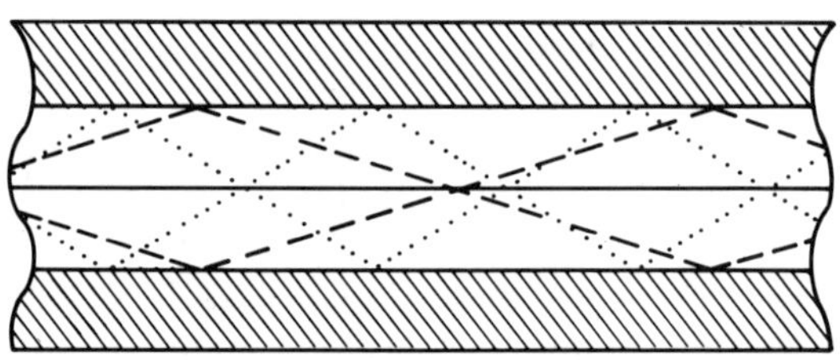

(a)

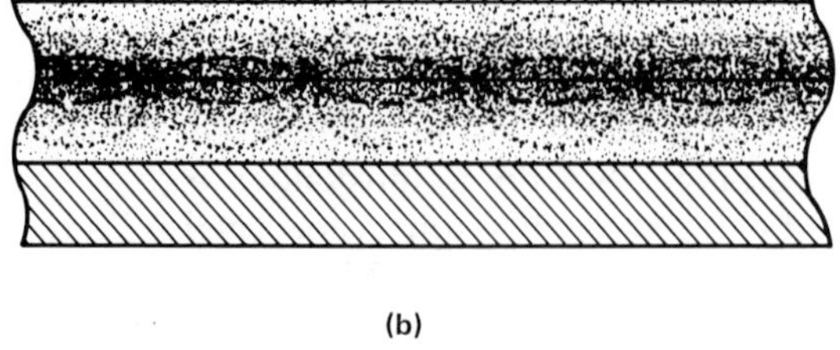

(b)

Figure 3. Paths of light rays in fiber lightguides: (**a**) multimode step index; (**b**) multimode graded index, the darkest shading represents the highest refractive index. Courtesy of Academic Press, Inc.

the fiber core as part of a short pulse of laser light, as in pulse-code modulation schemes, then the time–width of the pulse increases as the light travels down the fiber core. Eventually this broadening effect would cause the trailing edge of the pulse to overlap the leading edge of the next pulse, and the information being transmitted would be lost.

Such mode-dispersion problems can be minimized by a lightguide in which the refractive index is highest at the center of the core and gradually decreases radially until the cladding is reached. This is illustrated in Figure 3**b** where three rays, representing the fundamental mode and two higher-order modes are shown. The fundamental mode travels directly down the core axis, but the higher-order modes follow the quasi-helical path shown. The fundamental ray always travels along the axis, where the refractive index is highest and its speed is slowest. In the higher-order cases, the light travels mostly in the lower index regions of the core where its speed is higher. Thus under proper conditions, the modes can arrive at the end of the fiber almost simultaneously. Reductions in pulse width of more than one hundredfold have been observed, and reductions up to four hundredfold are possible.

Fabrication of Clad-Glass Fibers

The various types of fibers are all drawn on machines that are basically similar. The very refractory compositions, such as the high silica glasses, are normally drawn from solid rods or preforms, whereas the softer glasses, which are fluid at temperatures around 1200°C, can conveniently be drawn into fiber from special crucibles. Both types of glasses are drawn by the basic fiber-drawing machine illustrated in Figure 4. In the illustration, the machine is set up to draw fiber from a preform. A feed mechanism, A, at the top of the machine lowers the preform through the hot zone of the furnace or heat source, B, where the glass softens. The glass is pulled down into a thin fiber which is attached to the take-up drum, C. This drum provides the drawing force for the fiber-drawing operation, and stores the fiber neatly by moving horizontally as it revolves.

Elaborate fiber-drawing machines can include diameter monitors whose output can be coupled with the take-up drum drive to regulate the diameter of the fiber to a constant value. A capstan drive mechanism is frequently used. Here, the pulling force for the fiber-drawing operation is provided by two electrically driven rollers between which the fiber passes, not the drum on which the fiber is wound. Frequently, coating devices are incorporated into the fiber-drawing machine between the furnace and the take-up drum or capstan. These apply a protective coating to the pristine glass surface of the fiber before it is wound on the take-up drum to preserve its tensile strength. Present-day optical fiber-drawing machines produce 20–100 m of fiber/min, and scale-up to higher rates is envisaged for commercial production.

Crucible-Drawn Fibers. The heat source in the fiber-drawing machine is provided by a conventional wire-wound furnace, since the multicomponent glasses for optical lightguide use usually soften at 800–1200°C.

The double-crucible or double-bushing method (11,13) employs two crucibles that have concentric orifices in their bottoms as shown in Figure 5. The core glass is placed in the central crucible and the cladding glass is put into the outer crucible. The crucibles and their contents are heated to a sufficiently high temperature so that the glasses ooze from the concentric orifices at the bottom of the assembly. The glass is

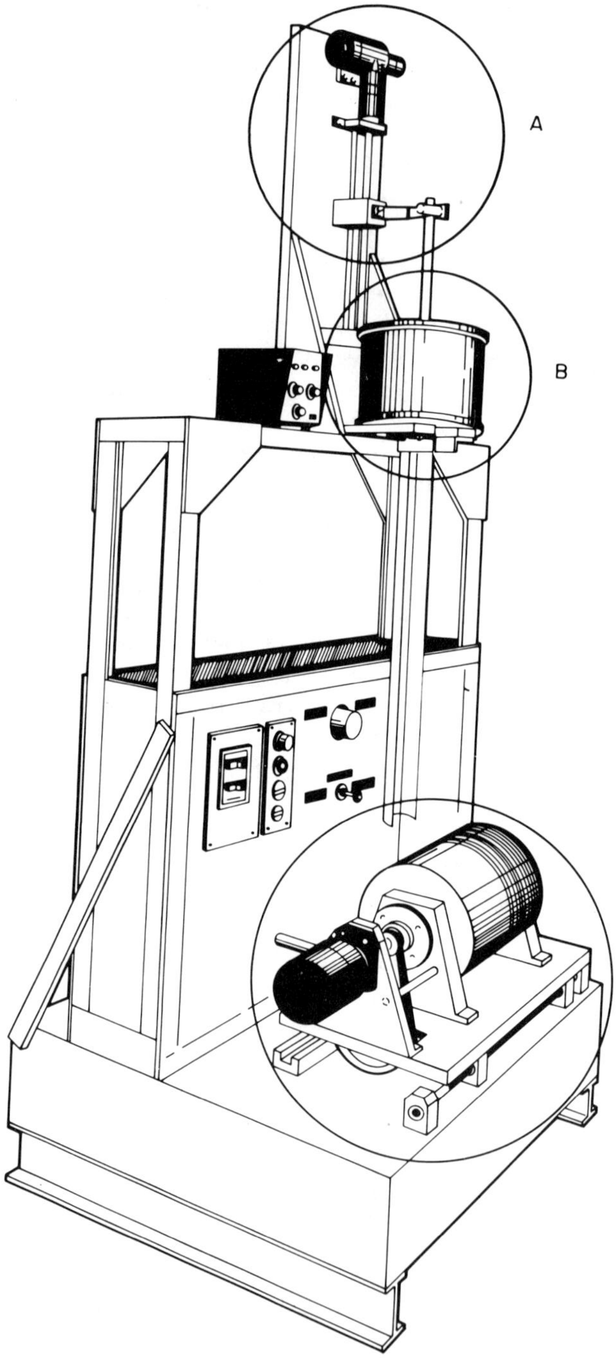

Figure 4. Laboratory fiber-drawing machine: A, feed mechanism; B, heat source; C, winding drum. Courtesy of *Bell Laboratories Record* (11).

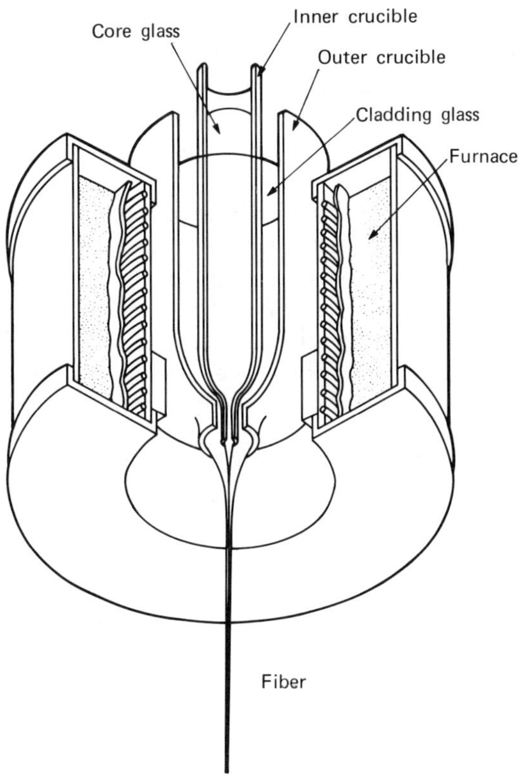

Figure 5. Double crucible for drawing clad-glass lightguides. Courtesy of *Bell Laboratories Record* (11).

pulled into a thin filament, which is wound on the take-up drum. By proper choice of geometry, glass composition, and temperature, diffusion of ionic species can occur across the core-cladding interface during the drawing operation to produce graded refractive indexes and low mode dispersion (14). The method is also attractive because it is relatively simple to feed the crucible with rods of glass from the top in a continuous fashion as fiber is being drawn from the bottom. Thus the fiber-drawing operation can be a continuous process.

Preform Drawn Fibers. Refractory glasses are usually drawn into fiber from specially prepared solid-rod preforms (see below). Since temperatures in excess of 1900°C are required to soften the high silica compositions sufficiently for fiber drawing, special heat sources are required. Special oxy-hydrogen torches can be used (15) but a more stable heat source is provided by high temperature furnaces with electrically heated carbon or tungsten elements. More recently, induction-heated furnaces have become popular (see Furnaces, electric).

For direct heating of high-silica preforms, carbon dioxide lasers have been used (16). The method has two main advantages: temperature response is rapid and the technique allows the drawing of fiber in any atmosphere.

Preparation of Bulk Multicomponent Glasses and Fibers. Multicomponent glasses are composed of one or more glass-formers, such as silica or boron oxide, modified by the addition of other metal oxides which themselves are not glass-formers (see Glass;

Boron compounds; Silica). The metal oxides modify the network of the glass-formers and reduce the softening temperatures. In contrast, the high silica glass compositions are modified with small quantities of other glass-formers such as boron oxide, germanium dioxide, or phosphorus pentoxide. The multicomponent glasses most studied for potential lightguide applications are the soda–lime silicate- and the sodium borosilicate-glasses. These glasses are usually prepared by melting together powdered silica and the modifying materials. The modifier oxides are usually introduced in the form of compounds such as carbonates or nitrates which decompose to form the oxides during the glass-melting process. In some cases the oxides themselves are used.

For a multicomponent glass to exhibit low light loss, it is essential that it contain only very small concentrations of those impurities that absorb in the wavelength region of interest for lightwave communications. There are no methods available at present for purifying a multicomponent glass. The only way to obtain such a material is to prepare it from raw materials that are themselves ultrapure, and to do so by a process that introduces either minimal or no contamination during the melting procedure. Preparation of the ultrapure starting materials required for such glass-melting processes is in itself a very difficult task, since the allowable concentrations of transition metal ion impurities lie in the range of a few tens of ppb (eg, ca 20 ng transition metals/g fiber).

The techniques used to prepare the starting materials vary depending on the particular material to be purified. They include ion exchange, solvent extraction, electrolysis, and distillation. Careful chemical analysis of trace transition metal impurities in both starting materials and glasses is also very important, and in most cases very difficult at the low concentration levels encountered. To control purification, reliable information on impurity content is essential. Techniques used to obtain impurity data include neutron-activation analysis, chemical coprecipitation followed by x-ray fluorescence analysis, atomic absorption spectrometry, and spark-source mass spectrometry (see Analytical methods).

In addition to preserving the purity of the material, the methods used to melt the glass must also ensure that the final product is of good optical quality. In other words, it must be chemically and optically homogeneous and free from bubbles and inclusions. These methods are relatively straightforward but must be carried out with great care. Homogeneity is obtained by stirring the melt, or by bubbling a stream of gas through it during the melting process. The main sources of contamination are the crucibles in which the glass is melted, the furnace refractories, and airborne dust particles that might be incorporated into the glass.

Several methods for melting soda–lime silicate glasses for optical waveguide applications have been described. In one example, premixed silica, sodium carbonate, and calcium carbonate of high purity was melted in platinum crucibles using r-f power to heat the crucible. This operation was carried out in a class-100 clean hood (11). When the batch was melted, the crucible and its contents were placed in a bottom-loading fining furnace. The glass was first homogenized by stirring it with a platinum paddle at 1500°C. It was then allowed to soak at the same temperature to allow large bubbles to float to the surface and burst. Slow stirring followed to ensure chemical and optical homogeneity. The temperature was reduced to 1300°C, and a second soaking was done during which microscopic gas bubbles were reabsorbed into the melt. The crucible and its contents were then allowed to cool to room temperature (a period of a few hours) in an insulated enclosure. When cool, the glass was removed from the crucible and annealed on a 24-h cycle. The lowest bulk glass loss reported from this technique was 20–30 dB/km over the spectral range 700–800 nm (17).

In most cases, the relatively high loss values obtained in the finished glasses are the result of contamination during the melting procedure. The most likely source of this contamination is the platinum crucibles in which the glasses were fused. For this reason many workers have begun to use high purity silica crucibles.

In some cases, r-f power has been used to fuse the glasses (18). The electrical conductivity of many multicomponent glasses in the molten state is high enough for r-f power in the megacycle frequency range to couple directly into the melt. If the crucible is cooled either by a stream of gas or cold water during the melting operation, attack by the molten glass on the silica crucible walls can be minimized. Thus it is possible to prepare glasses by this technique that have essentially the same purity as the starting materials from which they were melted. This technique has been used to prepare soda-lime silicate glasses with losses of 27–50 dB/km at a wavelength of 1060 nm (18–19). Although extremely high purity glasses should be preparable by this kind of technique, the attainment of good optical homogenity is not easy. The hottest parts of the molten batch are in the neighborhood of 1500°C, and the portions of the batch near the crucible walls are very much lower in temperature. Strong convection currents make it difficult to remove bubbles from the melt and to attain chemical homogeneity.

Various workers have drawn clad-glass fibers from soda–lime silicate compositions, and the best optical losses reported in the early work are 40–60 dB/km (17–18,20). In one case, total light losses of both the clad-glass fibers and the core material from which they were drawn are almost identical (17). These fibers, prepared by the double-crucible method, evidently had a near-perfect, core-clad interface, and negligible contamination had been added during drawing.

In recent years more attention has been paid to borosilicate systems than the soda–lime glasses. The rate of contamination from the crucible is reduced because the sodium borosilicate compositions can be melted and drawn into fiber at considerably lower temperatures than the soda–lime glasses. The second factor is that the optical loss coefficients (21) of the transition metal ions are generally lower in the sodium borosilicate glasses than those of the same ions in the soda–lime systems. The most striking example is the case of iron where the coefficient in borosilicate glasses is only about one tenth of its value in soda–lime silicate systems.

Of the losses reported for borosilicate glasses, among the lowest are those obtained on Selfoc fibers (21). The loss of the bulk glass has not been given but only the loss of the fibers that are drawn from it. The bulk glass from which the core of Selfoc fibers is drawn is a sodium borosilicate glass doped with thallium oxide. It has been successfully prepared in both silica and platinum crucibles. The cladding of the Selfoc fibers is formed from an undoped sodium borosilicate. The fibers are drawn by a double-crucible technique, and the geometrical arrangement of the crucible nozzles is such that the molten core and cladding glasses are in contact with each other for some time. Ion exchange of the sodium and thallium takes place across the interface between the two glass compositions, producing a graded refractive index.

A minimum loss of 20 dB/km in the wavelength range of 800–850 nm was reported for the Selfoc fibers (14). However, losses of 14 dB/km and 17 dB/km over this same wavelength range were reported (21–22). Sodium borosilicate glasses have been prepared with bulk absorption losses of <10 dB/km at 800 nm (23). This was done by carefully controlling the oxygen content of the atmosphere in which the glasses were melted. Under strongly reducing or strongly oxidizing conditions, the glasses showed relatively high losses. Between these two extremes, conditions were found which gave the minimum loss. Clad fibers drawn from these glasses by the double-crucible method showed total losses of ca 12 dB/km at 800 nm (23).

The early work described above strongly suggested that even lower loss multicomponent glass fibers could be prepared from ultra high-purity starting materials, provided that great care was taken to avoid contamination during processing. In addition, the oxidation states of residual impurities would have to be controlled. Such low-loss glasses and fibers have been produced by using these general principles (24–28). The fibers show minimum losses of 4–7 dB/km in the spectral range 800–850 nm. Their loss curves are shown in Figure 6.

Preparation of High Silica Fibers. Early in the development of optical fiber lightguides it became clear that some commercially available synthetic fused silicas showed low optical attenuations in the red and near-infrared regions of the spectrum (29). These synthetic silicas were prepared by vaporization of silicon tetrachloride and subsequent hydrolysis or oxidation. Synthetic silica produced in this manner is of very high purity, particularly with regard to the transition metal ions, because the impurities

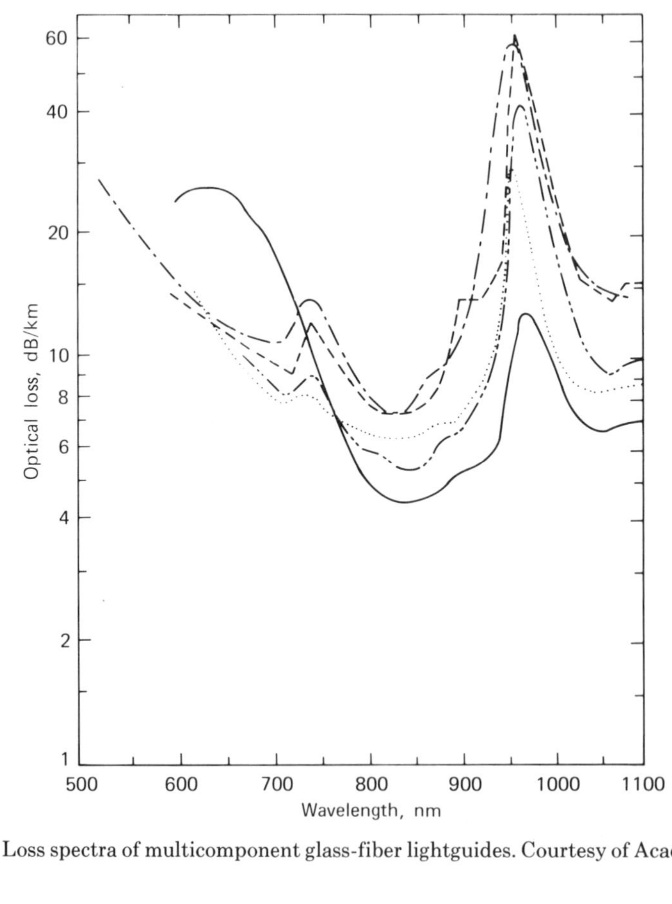

Figure 6. Loss spectra of multicomponent glass-fiber lightguides. Courtesy of Academic Press, Inc. (24–28).

are very much less volatile than the silicon tetrachloride starting material from which the silica is made. Purification occurs during evaporation of the silicon tetrachloride vapor from the liquid. Further, since the formation of such fused silica does not require a crucible or furnace for heating, contamination from such sources is avoided.

Soot Deposition. The very high working temperature of fused-silica glasses necessitates novel methods of fabrication. One technique was based on a discovery made by Hyde in 1942 (30). He found that submicrometer size silica particles (sometimes known as white soot), which are formed when gaseous silicon compounds are decomposed in the presence of water at high temperatures, would vitrify or sinter to bubble-free glasses at comparatively low temperatures. Application of these ideas gave rise to the soot process by which low-loss glass-fiber lightguides were made at Corning Glass Works.

Hyde's discovery was used in two ways to produce low-loss glass fibers (31–33). In the early work, a mixture of silicon tetrachloride and titanium tetrachloride was hydrolyzed in a gas-oxygen burner to produce a stream of fine particles (31). This stream of particles was passed through a silica tube where some of them adhered to the inside walls as illustrated in Figure 7. Subsequently, the tube was strongly heated to fuse the particles into the tube wall. The resultant film on the inside of the tube had a higher refractive index than the original silica, because it was doped with titanium oxide. The tube, with its inner coating, was drawn into fiber in such a way that the central hole closed and a solid fiber was obtained.

During the fiber-drawing process, a small amount of the titanium dopant was reduced from Ti^{4+} to Ti^{3+}. The fibers were lossy in that Ti^{3+} absorbs in the red and

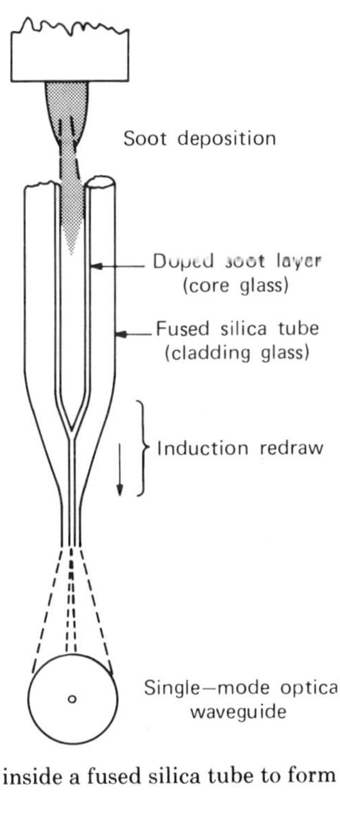

Figure 7. Deposition of soot inside a fused silica tube to form a single-mode lightguide. Courtesy of Corning Glass Works.

ir spectral regions. A subsequent oxidation step was required to reoxidize the Ti^{3+}. After this step, losses of ca 20 dB/km were obtained (31).

The second technique is known as the outside soot process (32–33). In this method, a stream of silicon tetrachloride mixed with either titanium tetrachloride or preferably germanium tetrachloride is oxidized in a gas-oxygen burner as previously described. The doped-silica particles so produced are directed at a mandrel which is rotated by a lathe. The particles adhere to the mandrel surface and can be built up to a thick layer by passing the burner repeatedly back and forth. After a sufficiently thick deposit is obtained, a second layer is deposited. The composition of this layer is chosen to give a lower refractive index and thus serve as cladding for the eventual fiber. The mandrel is removed and the "soot-form" sintered in a special furnace. The resulting product is a two-layer cylinder of bubble-free glass which is drawn into fiber. Using this technique, fibers have been obtained with losses of 4 dB/km at 820 nm and 2 dB/km at 1060 nm, and numerical apertures as high as 0.23 (32). Since many burner passes are used for the deposition step, the method affords the possibility of finely grading the refractive index of the structure.

In a technique that involves the direct deposition of the cladding layer, a pure fused silica rod is used as a mandrel for the deposition of borosilicate powder by a flame process similar to the Corning soot process (34). After sufficient thickness has been deposited, the whole preform is sintered under a restricted range of temperatures and atmospheres to produce a bubble-free glass. The preform can then be drawn into fiber. The original silica rod forms the core, and the deposited layer forms the cladding. Fibers of this type have shown losses as low as 4.5 dB/km at 820 nm and 3 dB/km at 1060 nm. Numerical apertures of about 0.17 have been obtained.

Chemical Vapor Deposition (CVD). Another technique that has been employed to produce low-loss optical fibers is chemical vapor deposition (CVD). A glassy layer or layers are deposited on the inside wall of a glass tube. The tube is heated strongly until it collapses uniformly to form a solid rod. The solid rod produced by the collapsing process is the preform, and from it the final fibers are drawn directly.

Two distinct variations of the CVD technique have been employed. Both use fused-quartz tubes for the deposition process. In the first method, the material of the tube itself can finally become the cladding of the fiber, so it is only necessary to deposit the core composition. The deposited core glass must of course have a higher refractive index than that of the fused-quartz tube. When fused silica is doped with other oxides, the refractive index almost always increases, thus there are a variety of chemical systems available as potential cores.

In the early stages of this work, the core material was composed of germanium oxide-doped silica (35–36). Germanium oxide was chosen as the dopant because germane (GeH_4) was available as a gas in high purity form and was easily mixed with silane (SiH_4). The two were allowed to react simultaneously with oxygen inside a fused silica tube by repeatedly passing the tube in and out of an electric furnace at ca 1100°C. The repeated passes were necessary in order to deposit a uniform film of SiO_2–GeO_2 glass on the inside wall of the tube. When the film deposit was sufficiently thick, the tube was placed in a glass-working lathe and collapsed to form a solid-rod preform, which was drawn into fiber with reasonably low losses.

Pure fused silica has excellent transparency and is an attractive material for fiber cores. However, because of its low refractive index, the choice of materials with which to clad it is very limited. Glasses of SiO_2–B_2O_3 show lower refractive indexes than fused silica when they are quenched from high temperatures (37).

The second adaptation of the CVD technique used cladding derived from layers of material that were laid down on the inside surface of the fused-quartz tube (38). These were followed by pure silica-core layers, and the tube was then collapsed to a solid rod. Reactants to produce the cladding layers were silane and diborane (B_2H_6) together with oxygen, and the core was formed by the reaction of silane and oxygen alone. These fibers also showed low losses.

The basic CVD method is attractive for three reasons: the raw materials used are available in high-purity form, they are transported as vapors, and they are allowed to react in a closed system, which eliminates external contaminants. However, for commercial production of optical waveguides, the method suffers from having a low deposition rate. This arises because the glassy layers are produced by a heterogeneous reaction at the surface of the tube. This heterogeneous reaction occurs only if dilute gas streams are used. The use of more concentrated reactants results in the formation of powdery deposits that will not yield clear glassy layers. Fabrication of a preform that can be drawn into a multimode fiber 1-km long typically requires 24–36 h of deposition time. An additional complication is that the hydride reactants produce water as a by-product, some of which becomes incorporated into films as hydroxyl ions and results in absorption loss in the fiber.

Modified Chemical Vapor Deposition (MCVD). To overcome the disadvantages attendant on low deposition rates and incorporation of water, the basic CVD technique was modified by use of more concentrated reactants and higher temperatures (1300–1600°C) (39).

At these temperatures chlorides, such as silicon tetrachloride ($SiCl_4$) and germanium tetrachloride ($GeCl_4$), react directly with oxygen, thus eliminating the need to use hydrides. However, the concentrated chloride–oxygen reactant streams produce not only a direct (heterogeneous) reaction at the wall but also a gas-phase (homogeneous) reaction that produces particles. Some of these particles collect on the tube walls and can be vitrified at the higher temperatures employed.

Because temperatures of 1300–1600°C are above the softening temperature of the substrate tube, the tube is rotated between the synchronous chucks of a glass-working lathe, as illustrated in Figure 8. The rotating tube is heated by a burner which traverses its length. Reaction takes place in the hot zone of the burner, and a glassy film, from heterogeneous reaction, is deposited in this region. In addition, particulate material obtained from homogeneous reaction is deposited downstream from the hot zone. The particulate layer is sintered to a clear vitreous deposit on a layer-by-layer basis as the torch moves along the tube. The torch is repeatedly traversed until sufficient deposit thickness is obtained. This procedure differs from the flame hydrolysis process where vitrification occurs by a separate sintering step after deposition is completed.

This deposition technique has several advantages. The increased reactant concentrations permit increased deposition rates over that of CVD by more than a hundredfold. Preforms that can be drawn into multikilometer lengths of multimode fiber can be fabricated in a few hours. These fibers show losses of ca 4 dB/km and 2 dB/km at 850 and 1060 nm, respectively (39), and numerical apertures up to 0.38 have been achieved (40), although the losses increase about 1 dB/km at the high doping levels required for such high numerical apertures.

Another use of the modified chemical vapor deposition (MCVD) technique has been to prepare fiber preforms with silica cores and borosilicate claddings (41). The

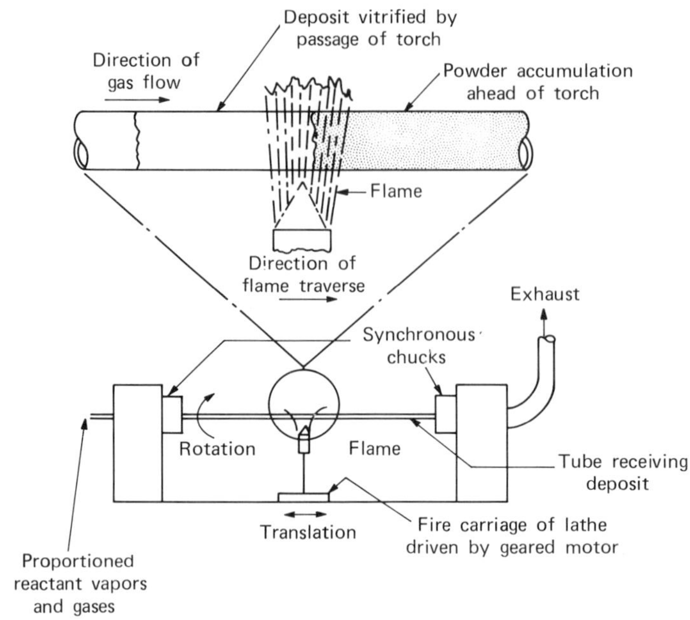

Figure 8. Modified chemical vapor deposition method for preform fabrication on a glass-working lathe. Courtesy of *Annual Review of Materials Science.*

cladding layer is formed first by the reaction of silicon tetrachloride vapor, boron trichloride, and oxygen. This is followed by the core layer produced by the reaction of silicon tetrachloride and oxygen. Fibers drawn from these preforms have shown very low losses: 1.9 dB/km and 1.1 dB/km at 860 and 1060 nm, respectively (41). Numerical apertures are not as high as those obtained in the germania-doped system. The maximum value achieved was 0.17 (41).

Another chemical system used to produce low-loss fibers gives cores that are composed of a phosphosilicate glass produced by a CVD process similar to those already described (42). The starting materials are phosphorus oxychloride, silicon tetrachloride, and oxygen. Cladding is provided by the original silica tube. Losses of 3 dB/km at 830 nm and numerical apertures of 0.18 have been achieved.

The lowest-loss fibers reported to date were made using MCVD with phosphosilicate cores and borosilicate cladding. The losses were reported to be as low as 0.5 dB/km at 1200 nm (43). Fibers with germania-doped silica cores and borosilicate cladding had a loss of 0.5 dB/km at 1500 nm (43). These low losses were owing in part to exceptionally low OH content.

The spectral attenuation of the lowest-loss fibers described are shown together for comparative purposes in Figure 9.

Fibers in which the refractive index changes gradually from the cladding towards the center of the core in a precise predetermined way would be expected to show much lower pulse dispersion than step-index fibers. All of the fibers produced by chemical-vapor deposition or soot-process techniques can be prepared with graded refractive indexes. This is accomplished by varying the concentration of reactants as the layers of material are deposited in the preform preparation process (41,44–49). When the resulting preform is drawn into fiber, the core has essentially the same graded refractive

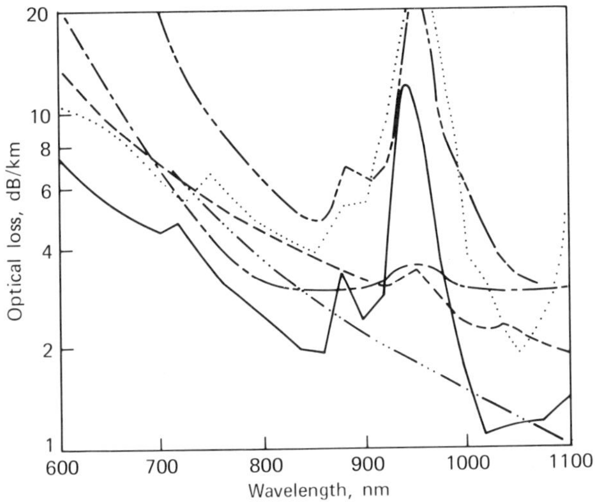

Figure 9. Loss spectra of high-silica glass fiber lightguides (32,34,40,43–45).

index profile as the preform. Fibers with this graded refractive index profile can be made with bandwidths of a few gigahertz.

Several fluorinated polymers have lower refractive indexes than fused silica and thus can be used as cladding for fibers with pure silica cores. Unfortunately, these polymers are lossy, and this loss affects that of the fibers produced with them (50). However, silica-core fibers clad with silicone resin can be made low-loss (51). The minimum loss is 3.2 dB/km at 840 nm, with a pulse dispersion of 19 ns/km. This is much less than the theoretical value (200 ns/km) calculated from the core-cladding index difference. Alternatively, plastic can be used as a loosely fitting tube enclosing the fiber core. By this means, the fiber is essentially clad with air. The points of contact between the core and the plastic tube are infrequent enough that losses are low.

In the use of perfluorinated ethylene–propylene as a cladding, the higher-order modes show higher losses than the low-order modes, so that the overall loss of the fiber depends upon the numerical aperture of the light-launching system (52). The loss of the fiber can be almost as low as the loss of the basic core material itself, such as 2.5 dB/km at 1060 nm if Spectrosil WF is used as the core (53). The theoretical maximum numerical aperture of such fibers is 0.58, but for most applications the effective numerical aperture is limited to values below this number because of increased loss in the high-order modes (see Fluorine compounds, organic).

Alkali-doped fibers can be prepared from fused silica tubes and metals that have high vapor pressures compared to such deleterious contaminants as iron and copper. In this technique the dopant metal is evaporated into a fused-silica tube under vacuum (54). After sealing, subsequent heating causes the metal (eg, potassium) to vaporize and react with the silica walls to form a silicide layer. Further heating in oxygen then forms potassium oxide which reacts with the silica tube to produce a potassium oxide–silica glass layer. The tube is collapsed and drawn as in the MCVD process.

Fibers made by this method have shown losses of 10 and 11 dB/km at 650 and 800 nm, respectively. Fibers of this type automatically have graded refractive indexes because of the diffusion of the dopant metal ions during the oxidation and collapsing step, but pulse dispersions have not been reported.

The earliest single-mode lightguides having reasonably low losses were composed of titanium dioxide-doped silica cores with silica cladding (31). The diameters of the cores were 2 or 3 μm. A theoretical formulation has been developed for the propagation of a pulse-modulated carrier wave through a single-mode lightguide (55). More recently, single-mode lightguides with large cores (up to 22 μm) of pure silica surrounded by borosilicate cladding have been fabricated (56). The fibers were made by the MCVD process already described. Large core diameters are desirable in order to facilitate splicing. The lowest loss obtained was 2.1 dB/km at 900 nm (54) and the pulse dispersion was 0.014 nm/km with a GaAs laser source (57). This value is equal to the material dispersion of fused silica for the 5–7.5 nm spectral width of the laser.

Finally, single-material, single-mode, fused-silica fibers have minimum losses of ca 50 dB/km at 1060 nm (58).

Strength of Clad-Glass Fibers

Glass fiber lightguides must be strong enough to undergo fabrication into cables. It is necessary also that the overall tensile strength of the cable be high enough to allow it to be subjected to handling and installation without fracturing its light-conducting fibers.

Fracture of fibers under tensile stress is usually the result of microscopic flaws in the glass surface. A stress–corrosion mechanism (chemical attack by atmospheric water vapor at the head of minute surface cracks where the tensile stress is concentrated) results in crack propagation and ultimate failure of the fiber (59).

The literature on the nature of glass failure and the strength of glass shows considerable variation in test results, presumably because the strength of a particular fiber being measured depends on the number and severity of the flaws in its surface (60–63).

When measured under ambient conditions with a loading time of ca 10 s, very carefully drawn glass fibers have strengths up to almost 7 GPa (10^6 psi) in the case of fused silica. However, even if a fiber does not break immediately on loading, it may do so at a later time. Surface cracks of a size smaller than the critical size for brittle fracture gradually grow in the presence of atmospheric water vapor until the critical size is reached, then a fracture occurs. The process is known as fatigue. Although the rate of fatigue varies with many factors, a strength reduction of ca 80% will occur in 20 years.

In addition to this fatigue effect, the strength of a glass fiber is also dependant on its length. This is because the probability of finding a flaw or weak link in the fiber surface increases as its length increases. Therefore, as the fiber length increases, the measured strength decreases. With imperfect production conditions, this effect is emphasized.

In order to obtain high fiber strength over kilometer lengths, the critical drawing conditions must be controlled. These include preform surface and bulk quality, drawing temperature, and drawing environment. If the high fiber strength is to be maintained, the newly formed fiber surface must be protected from abrasion. This can be accomplished by in-line coating with one of several organic polymers. However, none of the polymers so far used affords protection from fatigue-producing water vapor.

Each fiber produced is generally screened by proof testing. Some multiple of the use stress is applied to the fiber before incorporation into a cable, thus ensuring a safe minimum strength. An alternative technique is that of tempering. In this method an axial compressive stress is produced in the surface of the fiber. This stress is then the desired fatigue limit and proof stress level.

By employing careful cleaning and handling procedures for the preform both before and during fiber drawing, it has been possible to make fibers in kilometer lengths with strengths in excess of 4.1 GPa (600,000 psi) (64).

Cabling, Splicing, and Handling

Fibers to be used in telecommunications systems will seldom be used singly but will usually be formed into cables containing many fibers. In addition, techniques had to be developed for splicing the individual fibers within a cable to light sources, detectors, processing circuitry, or to other fibers. The fibers also had to be protected so that the cables could be handled by installation personnel without damage to the fibers.

Clad-glass fibers show their lowest loss at a given wavelength when they are microscopically straight. A fiber wound tightly on a rigid drum is forced to conform to imperfections on the drum surface. Such imperfections may include dust particles, metallic burrs, and surface scratches. Bending the fiber over these bumps forces short sections into small radius-of-curvature bends. Radiation losses occur at these bends and result in increased overall attenuation (46). For laboratory measurements one solution to the problem is to cool the drum, which has a higher coefficient of thermal expansion than the fiber. This allows the fiber to hang loosely and unstrained by the drum irregularities. An alternative solution is to wind the fiber on a soft drum.

The kind of induced loss just described is usually called microbending loss. The assembly of fibers into a cable inevitably causes such loss unless precautions are taken. A model of six fibers, stranded around a central member with bumps on its surface, and encapsulated in a thick layer of elastic material has a probable induced loss of 0.15 dB/bump for an rms (root mean square) bump height of 12 μm for a fiber with a 76-μm core, overall cladding diameter of 126 μm, and relative core-cladding index difference of 0.01 (65). Thus only 20 bumps over a 1-km length would result in a loss of 3 dB, which is ca equal to the total loss per kilometer of the best low-loss fibers.

A fiber 455 m long was wound on a cast acrylic drum, half at 7 MPa (1000 psi) tensile stress and the remaining half at 70 MPa (10^4 psi) (66). The microbending loss was 15 dB/km in the low-stress part and 145 dB/km in the high-stress portion. The measurements were made with an He–Ne laser, although the microbending loss was almost independent of wavelength. The loss decreased with increasing NA (numerical aperture) as predicted in ref. 67. The microbending loss decreased when a protective coating was applied to the fiber, because the coating cushioned the fiber against the effects of the drum irregularities described above (66). Such a jacket should have high longitudinal flexural rigidity in order to resist microbending (68). At the same time, it should have good lateral compressibility so that external forces are not transmitted to the fiber.

Rather than combining individual fibers into a cable, it was suggested that small groups of fibers would first be formed into units which would then be cabled. Two types of units were suggested, one circular and the other flat. In the unit with circular cross section, six coated-fibers were stranded around either a central coated steel wire or around a seventh similar fiber (69). Another unit consisted of a ring of coated fibers helically stranded around a central core rod and jacketed with an outer protective polymer layer (70). In either case, the assembly was encapsulated in urethane and jacketed with poly(vinyl chloride).

In the flat array, or fiber ribbon, parallel arrays of fibers are embedded in some other material to form a ribbon or tape (70–72). Close control of the geometry and fiber-to-fiber spacing in such ribbons would promote ease of splicing one length to the next in finished cables. At present there are many cable designs produced by various manufacturers. Some of these use the unit structure, and others strand individual fibers into the cable.

For two optical fibers to be joined without incurring excessive loss at the joint, it is desirable that their ends be flat and perpendicular to the fiber axes. A fiber or thin rod could be broken by scoring and stressing it. Satisfactory fiber ends were produced in this way, but the results were somewhat random. A tool was developed that could bend the fiber and place it under tension at the same time (73). Scoring was done with either a diamond or carbide scorer.

Preparation of multiple fiber ends for splicing proved to be more difficult than for single fibers since all the ends must be in one cross-sectional plane. Grinding and polishing was successful but the process was time-consuming, tedious, and not easily adaptable to field work. A method for fracturing the ends of fibers contained in flat arrays or tapes was developed that is a modification of the technique just described for single fibers (74). It utilizes a small simple hand tool, which can be operated in confined spaces.

Low-loss splices can be made between fibers with properly prepared ends (73). With proper alignment and use of index-matching liquid, splicing losses were always less than 0.04 dB for high silica multimode low-loss fibers. Two optical fiber connectors have been described, one that holds the fibers in metal sleeves and another that employs V-guides (75). The average connecting loss of 20 of these devices was 0.9 dB, without the use of index-matching liquid. The connectors could be disconnected and reconnected at will. A splicing technique used a high-precision glass sleeve wherein a transparent adhesive served to secure the fibers in place and also acted as an index-matching liquid (76). The average coupling loss observed was 0.29 dB.

Another method of splicing is to fuse the ends of the fibers together. Two techniques that use an electric arc to effect the fusion of high silica fibers showed average splice losses of 0.26 dB (77) and 0.14 dB (78).

Splicing tapes (ribbons) or other structures containing a number of fibers involves three operations: end preparation, alignment, and connection. A method that combines the last two steps makes use of a tool that first embosses grooves and prealignment slots into a plastic substrate (79). Fiber ribbons with prepared ends (74) could, therefore, be automatically aligned without the need for a microscope. The tool then applies a coverplate which is chemically bonded to the substrate. Out of 120 splice joints, the lowest splice loss was 0.0 dB; and the highest 0.53 dB. The average loss was found to be 0.11 dB. A technique was developed that combined all three operations to make splices (80). The two tapes to be spliced were placed in a specially grooved hinged holder. They were then scored simultaneously and broken simultaneously to prepare the fiber ends. The holder was then folded over, which automatically aligned all the fibers. Index-matching epoxy resin was used to make the splice permanent.

A method that does not employ V-grooves is described in ref. 81. The fiber tape is prepared for splicing by dissolving the plastic over a 1-cm distance. The tape is placed in a mold that holds the fibers in position and plastic is molded around the fibers. After curing, the assembly is removed from the mold, inverted, and scored at the position previously occupied by a spacer plate. The entire piece is then broken by bending and tension, and since the plastic breaks in the same place as the fibers, the end is now ready to be spliced. Two tapes, prepared as described, are placed in an alignment channel with an index-matching fluid to complete the splice. The average splice loss is <0.1 dB.

Uses

The research and development effort on the use of optical fibers for communications purposes began in the late 1960s. Since that time, when available fibers were very lossy, extraordinary progress has been made. Fibers with adequate bandwidths can now be produced in a routine fashion, with losses of about 3 dB/km in the wavelength range of 800–900 nm. In addition, fiber drawing and coating techniques have been improved to the extent that kilometer lengths of fiber can now be made with tensile strength well in excess of that required for practical use. Techniques have also been developed to form the fibers into cables, and to join the fibers and cables to each other and to transmitter and detector devices.

The work has resulted in a number of trials in the field. An example is the 1977 installation in Chicago by the Bell System. Fiber lightguide cables connect two central offices and a customer location, the Brunswick Building, for a total distance of 2.4 km. The laser and LED transmitters and avalanche photodiode detector receivers are contained in the three buildings. The system transmits voice, video, and data signals. The fiber lightguide cables were installed and used with zero breakage of fibers, and the system, carrying real traffic, exceeded all performance objectives.

Future Prospects

The potential for lightwave communications lies in providing almost limitless communication capacity. At present the economic position of optical fibers can be likened to that of the transistor in 1954. Then vacuum tubes could handle all the demands for hi-fi, radio, and TV. However, the potential of the transistor was realized about a decade later when its existence made possible the computers (qv) and microprocessors that have since revolutionized electronics. Thus, too, the almost limitless capacity that fiber optics can give telecommunications will open new services, such as electronic mail and subscriber access to computers and picture telephones.

At present the market for lightwave communication is rather limited. It is economically competitive at intermediate frequencies. First applications are planned at the DS3 rate in the North American digital hierarchy which operates at 44.7 megabits and is most used for trunk cable connections. At lower frequencies twisted copper pairs provide the least expensive transmission medium. The current cost of fibers will have to drop appreciably to compete. Extension of fiber optics to higher frequencies requires either single-mode fibers or improved multimode fibers and development of sources and detectors to allow operation at wavelengths in the 1.3–1.5 μm region. Another area where fibers can compete now is for data bussing. Here data transmitted at tens of megabits for distances under a kilometer require coaxial cable whose cost and weight make it disadvantageous compared to fiber. Exploratory engineering is currently being pursued toward an undersea lightwave system. Here low loss and high bandwidth can be expected to reduce repeater spacings and increase capacities over those of current technology.

BIBLIOGRAPHY

1. C. A. Burrus and R. W. Dawson, *Appl. Phys. Lett.* **17,** 97 (1970).
2. C. A. Burrus and B. I. Miller, *Opt. Commun.* **4,** 307 (1971).
3. M. Panish, I. Hayashi, and S. Sumski, *Appl. Phys. Lett.* **16,** 326 (1970).
4. Zh. I. Alferov and co-workers, *Sov. Phys. Semicond.* **3,** 1107 (1970).
5. H. C. Casey, Jr., and M. B. Panish, *Heterostructure Lasers,* Parts A and B, Academic Press, Inc., New York, 1978.
6. R. L. Hartman and R. W. Dixon, *Appl. Phys. Lett.* **39,** 239 (1975).
7. R. B. Chesler and D. A. Draegert, *Appl. Phys. Lett.* **23,** 235 (1973).
8. A. Dentai and co-workers, *Electron. Lett.* **13,** 484 (1977).
9. T. Li, *IEEE and Electronic Industries Assoc., Electronic Components Conference Record* (May 1975).
10. R. D. Maurer, *Proc. IEEE* **61,** 452 (1973).
11. A. D. Pearson and W. G. French, *Bell Lab. Rec.* **50,** 103 (1972).
12. S. E. Miller, E. A. J. Marcetili, and T. Li, *Proc. IEEE* **61,** 1703 (1973).
13. N. S. Kapany, *Fiber Optics,* Academic Press, Inc., New York, 1967, chapt. 5.
14. K. Koizumi and co-workers, *Appl. Opt.* **13,** 225 (1974).
15. A. R. Tynes and A. D. Pearson, unpublished work, 1971.
16. R. E. Jaeger, *Am. Ceram. Soc. Bull.* **55,** 270 (1976).
17. A. D. Pearson, *Proc. Xth Int. Congr. on Glass* **6,** 31 (1974).
18. B. Scott and H. Rawson, *Opto-electronics* **5,** 285 (1973).
19. C. E. E. Stewart and co-workers, *Electron. Lett.* **9,** 482 (1973).
20. C. E. E. Stewart and P. W. Black, *Electron. Lett.* **10,** 53 (1974).
21. Y. Ikeda, M. Yoshiyagawa, and Y. Furuse, *Proc. Xth Int. Congr. on Glass* **6,** 82 (1974).
22. Y. Ikeda, M. Yoshiyagawa, and Y. Furuse, "Low-Loss Glasses for Light-Focusing Fiber Waveguides," *Xth Int. Congr. on Glass* **6,** 82 (1974).
23. K. J. Beales, W. J. Duncan, and G. R. Newns, *Proc. 1st European Conf. on Opt. Fib. Communication, London, U.K., 1975,* p. 27.
24. Y. Ikeda and M. Yoshiyagawa, *Proc. 2nd European Conf. on Opt. Fiber Communication, Paris, Fr., Sept. 1976,* p. 27.
25. H. Imagawa and N. Ogino, *Tech. Digest, Int. Conf. on Integr. Opt. and Opt. Fiber Communication, Osaka, Japan, July 1977,* p. 613.
26. G. R. Newns, K. J. Beales, and C. R. Day, *Tech. Digest, Int. Conf. on Integr. Opt. and Opt. Fiber Communication, Osaka, Japan, July 1977,* p. 609.
27. A. D. Pearson and W. R. Northover, *Am. Ceram. Soc., Bull.* **57,** 1032 (1978).
28. S. Takahashi and T. Kawashima, *Tech. Digest, Int. Conf. on Integr. Opt. and Opt. Fiber Communication, Osaka, Japan, July 1977,* p. 621.
29. K. C. Kao and G. A. Hockham, *Proc. IEEE* **113,** 1151 (1966).
30. U.S. Pat. 2,272,342 (Feb. 10, 1942), J. F. Hyde (to Corning Glass Works).
31. U.S. Pat. 3,711,262 (Jan. 16, 1973), D. B. Keck and P. C. Schultz (to Corning Glass Works).
32. P. C. Schultz, *Am. Ceram. Soc. Bull.* **52,** 383 (1973).
33. U.S. Pat. 3,737,292 (June 5, 1973), D. B. Keck, P. C. Schultz, and F. Zimar (to Corning Glass Works).
34. F. W. Dabby and co-workers, *Appl. Phys. Lett.* **25,** 714 (1974).
35. J. B. MacChesney and co-workers, *Am. Ceram. Soc. Bull.* **52,** 704 (1973).
36. P. W. Black and co-workers, *Electron. Lett.* **10,** 239 (1974).
37. L. G. Van Uitert and co-workers, *Mat. Res. Bull.* **8,** 469 (1973).
38. W. G. French and co-workers, *Appl. Phys. Lett.* **23,** 338 (1973).
39. J. B. MacChesney and co-workers, *Proc. Xth Int. Congr. on Glass* **6,** 40 (1974).
40. P. B. O'Connor, J. B. MacChesney, and C. M. Melliar-Smith, *Electron. Lett.* **13,** 170 (1977).
41. W. G. French, *Proc. Xth Int. Congr. on Glass* **6,** 46 (1974).

42. D. N. Payne and W. A. Gambling, *Electron. Lett.* **10,** 289 (1974).
43. H. Osanai and co-workers, *Electron. Lett.* **12,** 549 (1976).
44. D. N. Payne and W. A. Gambling, *Opt. Commun.* **13,** 422 (1975).
45. G. W. Tasker and W. G. French, *Proc. IEEE* **62,** 1282 (1974).
46. J. B. MacChesney, P. B. O'Connor, and H. M. Presby, *Proc. IEEE* **62,** 1280 (1974).
47. L. G. Cohen and co-workers, *Appl. Phys. Lett.* **26,** 472 (1975).
48. R. D. Maurer, *Proc. Xth Int. Congr. on Glass* **6,** 59 (1974).
49. L. G. Cohen and co-workers, *Appl. Phys. Lett.* **28,** 391 (1976).
50. Y. Suzuki and H. Kashiwagi, *Appl. Opt.* **13,** 1 (1974).
51. S. Tanaka and co-workers; *Electron. Lett.* **11,** 153 (1975).
52. P. Kaiser, A. C. Hart, Jr., and L. L. Blyler, Jr., *Appl. Opt.* **14,** 156 (1975).
53. P. Kaiser, *Appl. Phys. Lett.* **23,** 45 (1973).
54. S. R. Nagel, A. D. Pearson, and A. R. Tynes, *J. Am. Ceram. Soc.* **59,** 47 (1976).
55. F. P. Kapron and D. B. Keck, *Appl. Opt.* **10,** 1519 (1971).
56. W. G. French and G. W. Tasker, *Tech. Digest. Topical Meeting on Optical Fiber Transmission, Williamsburg, Va., Jan. 1975,* TuA2-1.
57. L. G. Cohen, P. Kaiser, and H. M. Presby, in ref. 56, TuD2-1.
58. P. Kaiser and H. W. Astle, *Bell Syst. Tech. J.* **53,** 1021 (1974).
59. A. A. Griffith, *Phil. Trans. Roy. Soc.* **221A,** 163 (1921).
60. E. U. Condon, *Am. J. Phys.* **22,** 224 (1954).
61. E. B. Shand, *J. Am. Ceram. Soc.* **37,** 52 (1954).
62. E. B. Shand, *Am. Ceram. Soc. Bull.* **38,** 653 (1959).
63. C. J. Phillips, *Am. Scientist* **53,** 20 (1965).
64. H. Schonhorn and co-workers, *Appl. Phys. Lett.* **29,** 712 (1976).
65. R. Olshansky, *Appl. Opt.* **14,** 20 (1975).
66. W. B. Gardner, *Bell Syst. Tech. J.* **54,** 457 (1975).
67. D. Marcuse, *Theory of Dielectric Optical Waveguides,* Academic Press, Inc., New York, 1974, p. 235.
68. D. Gloge, *Bell Syst. Tech. J.* **54,** 245 (1975).
69. R. A. Miller in ref. 56, WA1-1.
70. M. I. Schwartz in ref. 56, WA2-1.
71. R. D. Standley, *Bell Syst. Tech. J.* **53,** 1183 (1974).
72. D. L. Bisbee and P. W. Smith, *Bell Syst. Tech. J.* **54,** 479 (1975).
73. D. Gloge and co-workers, *Bell Syst. Tech. J.* **52,** 1579 (1973).
74. E. L. Chinnock and co-workers, *Bell Syst. Tech. J.* **54,** 471 (1975).
75. K. Miyazaki and co-workers in ref. 56, WA4-1.
76. H. Murata, S. Inao, and Y. Matsuda in ref. 56, WA5-1.
77. Y. Kohanzadeh, *Appl. Opt.* **15,** 793 (1976).
78. D. L. Bisbee, *Appl. Opt.* **15,** 796 (1976).
79. A. H. Cherin and P. J. Rich in ref. 56, WB3-1.
80. E. L. Chinnock and co-workers in ref. 56, WA6-1.
81. P. W. Smith and co-workers, *Bell Syst. Tech. J.* **54,** 971 (1975).

General References

H. C. Casey, Jr., and M. B. Panish, *Heterostructure Lasers,* Academic Press, Inc., New York, 1978.
H. Krassell and J. K. Butler, *Semiconductor Lasers and Heterostructure LED's,* Academic Press, Inc., New York, 1977.
D. Marcuse, *Theory of Dielectric Optical Waveguides,* Academic Press, Inc., New York, 1974.
H. Melchior, "Detectors for Light-Wave Communications," *Phys. Today,* 32 (Nov. 1977).
W. S. Boyl, "Light-Wave Communications," *Sci. Am.* **237**(2), 40 (1977).
M. Zief, "Talking Through Glass Fibers—the Chemist's Role," *Chemtech* **8,** 610 (1978).
J. E. Midwinter, *Optical Fiber for Transmission,* John Wiley & Sons, Inc., New York, 1979.

A. DAVID PEARSON
JOHN B. MACCHESNEY
Bell Laboratories, Inc.

WILLIAM G. FRENCH
3M Company

FIBERS, CARBON. See Carbon—Carbon and artificial graphite.

FIBERS, CHEMICAL

The chemical fiber industry, a lineal descendant of prehistoric use of natural fibers in textiles, is now an important industry in its own right, a large segment of the chemical process industry, and heavily dependent on petrochemicals. Its history begins with the patent of an artificial silk by Count Hilaire de Chardonnet in France in 1885. Chardonnet silk was made by two of the three major spinning processes in use today. It was a regenerated cellulose fiber (now called rayon) made by converting a nitrated cellulose into fiber form and then chemically regenerating the cellulose to avoid the unacceptable flammability of cellulose nitrate. Initially, it was produced by extruding an alcohol solution of cellulose nitrate through small holes into water to effect fiber coagulation. In a later process it was generated by similarly extruding an alcohol—ether polymer solution into air, forming filaments that hardened as a result of solvent evaporation.

It is appropriate to introduce some definitions here. The word spinning as applied to chemical fibers refers to the process by which they are formed from polymeric substances. Chardonnet's first process is representative of wet spinning, in which a solution of a polymer is forced through one or more holes in a device called a jet or spinneret into a suitable coagulating bath. The individual thread-like strands are called filaments. Chardonnet's second process is called dry spinning, wherein a polymer solution is forced through the spinneret holes into a chamber where the solvent is evaporated from the filaments.

Chardonnet's fiber was derived from a naturally occurring long-chain polymeric material, cellulose. As the chemical fiber industry grew steadily from the 1880s to the 1930s, it continued to be based on use of naturally occurring polymers, primarily cellulose. Research in the 1920s and 1930s, particularly that of Staudinger and Carothers, laid the foundation for further advances, especially the development of the synthetic fibers. Staudinger, a 1953 Nobel prize winner, proved polymers were molecules of very high molecular weight. Carothers clearly defined the two major types of polymerization, condensation and addition; showed how condensation polymers of high molecular weight could be synthesized; and demonstrated that certain of these could be formed into filaments that exhibited the property of cold drawing and crystallization, resulting in oriented, strong fibers (1). Carothers' elegant research in DuPont's laboratories led in 1939 to two major innovations in the chemical fiber field. One was the commercialization of nylon fiber, the first truly synthetic fiber based on a polymer itself. The second was the technique of melt spinning. Instead of being dissolved as for wet or dry spinning, the nylon polymer was melted by heating, and the viscous liquid was forced through the spinneret holes to form filaments that hardened as they cooled.

The generally used term man-made, referred to here as chemical fibers, as applied to fibers differentiates between the natural fibers and those produced by chemical and physical means in manufacturing processes. The advent of nylon, and the subsequent proliferation of fibers from synthetic polymers, has brought about the subdivision of chemical fibers into two broad classes: those made from natural polymers such as the cellulosics (see Cellulose acetate and triacetate fibers; Rayon; Fibers, vegetable), and those derived from synthesized polymers. The latter, which are called synthetics (or sometimes noncellulosics), now represent the majority of chemical fibers. In this category fall the nylons, the polyesters, the acrylics, the polyolefins, and many others (see Acrylic and modacrylic fibers; Olefin fibers: Polyamides; Polyester fibers).

By 1977 world production of chemical fibers closely approached world production of natural fibers, and the production of synthetic fibers had grown to well over twice that of the cellulosics, with polyester taking the lead. Table 1 shows the changes in volume of various fibers between 1967 and 1977.

For polymers, whether naturally occurring or synthesized, to be suitable for fiber production they must meet several conditions. They must be capable of conversion into fibrous form and must be of sufficiently high molecular weight to permit development of adequate fiber properties. In order to develop such adequate fiber properties they must be capable of being oriented so that the long polymer molecules tend to lie parallel to the fiber axis.

The fiber is normally drawn (stretched) either during or subsequent to spinning to effect orientation. Drawing frequently imparts or enhances crystallinity resulting in enhanced fiber tensile strength. The choice of spinning process is primarily dictated by the nature of the polymer used and secondarily, in some cases, by the nature of the products desired. If the spinneret has a single hole, a monofilament is produced. Such monofilaments are desirable for certain uses, eg, fishing lines, sheer hosiery. Usually, however, the spinneret has many holes. For textile use multifilament yarns are produced, each thread line or "end" containing from fewer than 10 to 100 or more individual filaments depending on end-use, eg, apparel, upholstery, carpeting. Yarns of this type are usually referred to simply as filament yarns. Except for silk, however, natural fibers occur in relatively short lengths, usually measured in centimeters. These are called staple fibers, and yarns spun from them tend to have different aesthetic

Table 1. World Textile Fiber Production, 1967 and 1977[a], Thousands of Metric Tons

	1967	1977
Natural fibers		
raw cotton	10,180	14,138
raw wool	1,562	1,390
raw silk	32	49
Total	*11,774*	*15,577*
Chemical fibers		
cellulosics[b]	3,428	3,535
synthetics		
polyester	754	4,233
nylon	1,317	2,939
acrylic	542	1,765
olefin	150	949
other[c]	110	110
Sub total	*2,873*	*9,996*
Total	*6,301*	*13,531*

[a] Adapted from Textile Economics Bureau Data (2–3).
[b] Rayon and acetate, including acetate cigarette filter tow.
[c] Predominantly synthetic.

qualities from continuous-filament yarns. Chemical fibers are, in many cases, desired in staple form, either for use by themselves or for blending with other fibers. In the production of staple chemical fibers, polyester being preeminent, the spinneret contains hundreds or thousands of holes, and a very large bundle of tens of thousands of filaments, called a tow, is produced by combining output from several spinnerets. Such tows, after stretching, are crimped, ie, mechanically given a wavy form, and then cut into staple fiber of any length desired.

The physical, mechanical, and chemical properties vary widely among the various chemical fibers. They can also be varied to a degree in a given type of fiber by the method of spinning used and by the post-treatment, particularly extent of drawing. Key physical properties include the strength of the fiber, the degree of elongation or extensibility at the breaking point, and the modulus, or degree of stiffness. These are generally expressed relative to the fiber tex (denier times 0.1111), a measure of size defined as the weight in grams of 1000 (9000 for denier) meters of yarn or filament. The tex and the decitex, a unit ca 11% greater than denier, have come into use in Europe. Fibers of different densities thus have different cross-sectional areas for a given tex. Most of the chemical fibers in use today, except for glass and certain specialty products, have densities varying from slightly below 1.0 to ca 1.5 g/mL. The tenacity of a fiber or yarn is its breaking strength in grams divided by the dtex (or denier). Elongation at break is measured as a percentage of original length extended, and the modulus (or initial modulus), expressed in grams per dtex (or denier), is the force required to extend the fiber initially, divided by the degree of extension. As the properties of most fibers are affected by absorbed moisture and also by temperature, physical testing is done under controlled and defined conditions; ie, 21°C and 65% relative humidity (rh). Chemical fibers range widely in moisture regain from the dry state, from near zero for polypropylene and below 1% for polyester to ca 13% for rayon. Other properties significant to various uses include abrasion resistance, elasticity, resistance to drycleaning solvents and other chemicals, ease of dyeing and types of dyestuffs accepted, and dimensional stability, or resistance to shrinkage and wrinkling. Strength is important in industrial applications such as tire cord (qv). Abrasion resistance is significant in carpet-face yarns, but not particularly so in women's apparel. Though some of these properties influence what is called hand, the latter is a somewhat elusive, yet very important, property of a fabric. It denotes the tactile and aesthetic sensations experienced in feeling a fabric.

In addition to the several broad-based synthetic fibers introduced in the 1940s and the 1950s, there are now many high-performance specialty fibers introduced as a result of continuing research and development. For reference, Table 2 summarizes generic names and definitions established by the U.S. Federal Trade Commission for a number of chemical fibers. Several of them are discussed in more detail below.

Chemical Fibers Based on Natural Polymers

Rayon. Two other processes have long since superseded Chardonnet's method of regenerating cellulose, the cuprammonium process and the viscose process. They are based on two different methods of solubilizing the relatively intractable cellulose molecule. The cuprammonium, or Bemberg process, is currently of minor present significance. It is based on dissolving relatively purified cellulose, such as that from cotton linters or woodpulp, in an aqueous solution of ammonia, caustic soda, and basic

Table 2. Relevant F.T.C. Generic Names for Manufactured Textile Fibers

Generic name	Definition of fiber-forming substance[a,b]
acetate	cellulose acetate; triacetate where not less than 92% of the cellulose is acetylated
acrylic	at least 85% acrylonitrile units
aramid	polyamide in which at least 85% of the amide linkages are directly attached to two aromatic rings
azlon	regenerated naturally occurring proteins
glass	glass
modacrylic	less than 85% but at least 35% acrylonitrile units
novoloid	at least 85% cross-linked novolac
nylon	polyamide in which less than 85% of the amide linkages are directly attached to two aromatic rings
nytril	at least 85% long chain polymer of vinylidene dinitrile where the latter represents not less than every other unit in the chain
olefin	at least 85% ethylene, propylene, or other olefin units
polyester	at least 85% ester of a substituted aromatic carboxylic acid, including but not restricted to substituted terephthalate units and para-substituted hydroxybenzoate units
rayon	regenerated cellulose with less than 15% chemically combined substituents
saran	at least 80% vinylidene chloride
spandex	elastomer of at least 85% of a segmented polyurethane
vinal	at least 50% vinyl alcohol units and at least 85% total vinyl alcohol and acetal units
vinyon	at least 85% vinyl chloride units

[a] All percentages are by weight.
[b] Except for acetate, azlon, glass, novoloid, and rayon, the fiber-forming substance is described as a long chain synthetic polymer of the composition noted.

copper sulfate, followed by wet spinning into water. The regenerated cellulose fibers are stretched, extracted with sulfuric acid solution, washed, and dried. The process, first commercialized in Germany in 1899, was based on work by a Swiss chemist, Schweizer, in 1857 (see Cellulose; Rayon).

In 1891 the British workers Cross, Bevan, and Beadle found that cellulose is soluble in a mixture of caustic soda and carbon disulfide. The viscose process based on this finding was commercialized in 1903 in the United Kingdom (1910 in the United States) and is today predominant in rayon manufacture worldwide. It is a very complex process and its commercial utilization represented a triumph of ingenuity and technology. Basically, relatively purified wood-pulp cellulose, called dissolving pulp or chemical cellulose, is treated first with caustic soda, then with carbon disulfide, to form a xanthate which is, in turn, soluble in aqueous alkali solution. Control of conditions in the spinning solution, called viscose, and of the bath into which the viscose is extruded, importantly affect final fiber properties. The concentrations of sulfuric acid, sodium sulfate, zinc sulfate and other components of the spinning bath are especially important. The coagulated fiber formed in the spin bath is stretched, washed, and dried. For staple fiber production, the process sequence includes crimping and cutting. Hydraulic and coagulation circumstances limit the linear speed at which such wet spinning can be effected to not much over 100 m/min. Because of this, the process is best suited to the production of large total tex (denier) continuous filament yarns or of very large tows for production of staple fibers.

Rayon fibers, particularly staple fibers, serve in a wide variety of uses, alone and in combination with other fibers. High-tenacity, high total tex (denier) filament is used in tire cord and other industrial uses, although rayon's once-dominant tire cord position has long since yielded to nylon and polyester fibers.

Rayon staple is produced in a variety of types, including polynosic and high wet-modulus types, these retaining greater strength and dimensional stability when wet. Rayon staple is used in apparel, household goods, and various nonwoven fabrics (see Nonwoven textiles).

Despite the maturity of rayon, new product variants continue to be developed. Among the most recent developments are a highly crimped high wet-modulus item (4) closer to cotton in fabric hand and cover (opacity) either by itself or in blends with polyester staple, a higher absorptive capacity variant (5) for use in certain nonwoven products, and a hollow viscose rayon variant (6) that also provides more cotton-like hand and cover.

Cellulose Acetate. Unlike rayon, cellulose acetate [*9004-35-7*] is dry-spun, predominantly in continuous filament form. The fiber-forming base polymer is made by acetylating dissolving pulp with acetic anhydride usually in acetic acid solvent with sulfuric acid as catalyst. The fully acetylated cellulose initially produced is partially hydrolyzed to produce secondary acetate. Following catalyst removal, precipitation of the acetate flake, washing, and drying, the flake is dissolved in acetone containing a small amount of water. This spinning "dope" is subjected to extensive filtration and then dry-spun to evaporate the acetone solvent. The filament yarn so produced is wound at the bottom of the chamber on a bobbin or other suitable package for such further handling as end-uses dictate. Linear spinning speeds up to or exceeding 1000 m/min can be achieved.

Acetate fabrics have attractive appearance, pleasant hand, and excellent whiteness before dyeing. A wide variety of textile fabrics, both knit and woven, are produced. Women's apparel in which the relatively poor abrasion resistance of acetate fibers is not a particular problem, draperies, and upholstery are major use areas. Although acetate textile filament yarn is still an important chemical fiber, its production appears to have peaked ca 1970–1971. Since then acetate production has declined accompanied by closing of some plants and abandonment of acetate production by some manufacturers (7). A relatively small amount of acetate staple is also produced.

Cigarette filters represent a major and still growing use for acetate tow. The tow is a continuous-filament product containing several thousand filaments and made by combining the requisite number of multifilament yarns from several spinning chambers into a tow that is then crimped and baled for use in cigarette-making. Acetate filters predominate worldwide because of their combination of appearance, taste, and economics (see Cellulose acetate and triacetate fibers).

Cellulose Triacetate. Cellulose triacetate [*9012-09-3*] filament became a useful specialty fiber in the 1950s, initially in the United States and the United Kingdom. Its production is similar to that of acetate, differing in that the initial reaction product is not hydrolyzed to a product with less than 92% of the hydroxyl groups esterified, and the spinning solution for the triacetate flake is made with a mixture of methylene chloride and a small amount of methanol or ethanol. Its importance is that, unlike regular cellulose acetate, it can be heat-treated to induce a degree of crystallization, which produces dimensional stability and related "ease-of-care" characteristics permitting, for example, water washing. As a result heat-treated triacetate resembles such fully synthetic fibers as nylon or polyester.

Other Fibers Based on Natural Polymers. Natural polymers other than cellulose that are capable of producing fibrous structures in nature suggested themselves as bases for chemical fibers. Two types of such noncellulosic chemical fibers reached commercialization in the 1930s and 1940s. Although neither is any longer produced in the United States or of much commercial significance elsewhere, these fibers deserve mention.

Alginate Fibers. Both monofilament and multifilament fibers can be produced by wet spinning an aqueous sodium alginate solution based on alginic acid from seaweed (see Gums). The spin bath is acid to produce alginic acid fibers or a bath containing calcium ion is used to produce calcium alginate [*9005-35-0*] fibers. The latter was commercialized in the United Kingdom during World War II. These fibers are unsuitable for ordinary textile use, as they are very sensitive to weak soap or alkali solutions. Quite small amounts have been made for special uses in which the fiber is dissolved out of a textile structure. Production is understood never to have exceeded a hundred metric tons annually.

Protein Fibers. Just as cotton (qv) and flax (see Fibers, vegetable) are natural cellulose fibers so wool (qv) and silk (qv) are protein fibers in which polymeric amino acids provide the fibrous structure. These polypeptides are natural polyamides, analogous to the synthetic nylons, particularly nylon-6, though differing in that they are polymers of alpha amino acids (see Biopolymers).

Regenerated protein fibers have been made commercially since the 1930s. Several have been based on casein from milk, the fibers being produced by wet spinning (see Milk products). Most of these fibers are treated with formaldehyde to improve wet strength and reduce alkali sensitivity. The casein fibers have relatively high moisture absorption, similar to that of wool, and have been primarily used in staple form for blending with wool. A small amount is still made outside the United States.

The other regenerated protein fibers, based on wet-spinning proteins from peanuts, corn meal, and soybeans are chiefly of historical interest.

Synthetic Fibers

Polyamide Fibers. The importance of the development of nylon can hardly be overstated. It opened the way to a great number of synthetic fibers having a wide variety of chemical and physical properties. It freed the chemical fiber industry from the restrictions inherent in the use of natural polymers, permitting controlled tailoring of polymers.

Nylon-6,6. Nylon-6,6 [*9011-55-6*] is so designated because both the adipic acid and the diamine from which the polymer is synthesized contain 6 carbon atoms. Although coal tar intermediates were used, these starting materials are now derived from other intermediates based ultimately on petroleum hydrocarbons. The fiber-forming polymer is formed by heating nylon-6,6 salt under pressure to remove water as the adipic acid and diamine condense to form a long-chain, linear polymer in what is called a melt polymerization. When the desired average molecular weight is reached, pressure is reduced, and the polymer is usually extruded in spaghetti-like strands, cooled, and cut into chips. In some cases, the molten polymer is conveyed directly to melt-spinning equipment.

In spinning, the polymer chips are re-melted, either on a grid or in the melting zone of a screw extruder, metered via a gear pump, passed through a filter (usually sand) to remove solid impurities, and extruded vertically downwards through spinnerettes. The filaments are cooled and solidified, lubricated with a material designed to eliminate static and provide desired frictional characteristics, and wound on a rotating tube to form a "package." At this stage, the fiber is essentially unoriented, weak, and unsuitable for textile use.

Fiber strength is developed by subjecting the as-spun yarn to a drawing step, in which orientation and crystallization of the polymer molecules occur. This stretching is achieved by passing the yarn between two sets of rolls, the second of which operates at a surface speed of up to ca 5 times that of the feed roll, so that the yarn is drawn to several times the original length. In the draw zone, the yarn is usually passed over a pin to localize the draw point and enhance uniformity. Drawing is usually effected at ambient temperature, heat actually being generated in the individual filaments by the stretching process. Textile nylon is usually stretched and twisted simultaneously, the multifilament yarn being given a degree of twist to aid in subsequent processing. For certain uses, hot drawing (at temperatures, however, well below the polymer's crystalline melting point of about 263°C) or multiple drawing, or both, may be used to produce high tenacity with reduced extensibility.

Nylon-6,6 is a strong, tough, abrasion-resistant fiber that is relatively stable and relatively readily dyeable. It has good knot strength (ie, high percentage retention of tenacity when knotted) and good elasticity. By heat setting after being drawn, nylon fiber can be dimensionally stabilized at temperatures up to that of the annealing process, permitted the "boarding" of hosiery to preset shapes, pleat setting in garments, and importantly, providing ease-of-care characteristics.

Because of its broad range of characteristics, nylon-6,6 has a wide spectrum of uses. Textile nylon is used in hosiery, apparel, and home furnishings. Most of this fiber is multifilament, but monofilament nylon is used in sheer women's stockings. The high strength of nylon permits manufacture of lightweight and very sheer fabrics. Because of good abrasion resistance and resilience, nylon dominates the U.S. carpet market (8). A particular product for this area is BCF, or bulked continuous filament nylon. The bulking is accomplished by mechanically crimping multifilament yarn while it is heated, then cooling it in the bulky configuration. Nylon staple is also important in carpets.

Strong, durable nylon monofilaments are used in fishing lines and fishnets; high-strength nylon multifilament has a variety of automotive and industrial uses ranging from belting to filter fabrics. Tire cord (qv) is a major use, nylon having displaced high strength viscose rayon in this application.

Nylon-6. Although Carother's research in the late 1920s and 1930s touched on fiber-forming polymers from aminocaproic acid, commercialization of nylon-6 [*25038-54-4*] fibers, so named because of the six carbons in the caprolactam monomer, first occurred in Europe shortly after commercialization in the United States of nylon-6,6. Manufacture is based on the use of caprolactam, the internal cyclic amide of aminocaproic acid, derived petrochemically (see Polyamides).

Unlike nylon-6,6, caprolactam is polymerized by ring opening, followed by repetitive addition until the desired molecular weight is achieved. This reaction is also a melt polymerization in which an equilibrium content of monomer exists, the latter being removed after polymerization, usually by water washing. Nylon-6 polymer is also melt-spun, and the manufacture, fiber properties, and uses are very similar to those of nylon-6,6. Schlack at I.G. Farbenindustrie is credited with establishing appropriate conditions of converting caprolactam to a useful fiber-forming polymer.

Nylon-6 production has grown steadily, and has long since spread from Europe to the United States and elsewhere.

Qiana. In 1968 DuPont introduced a polyamide textile fiber different from either nylon-6 or 6,6. This fiber, named Qiana, is a condensation polymer containing alicyclic rings (9). As with other nylons, melt polymerization and melt spinning are used. The fiber is very silk-like in appearance and behavior, particularly with respect to fabric resilience after crushing. Qiana fiber is used mainly in apparel, particularly dresses, blouses, and shirts.

Aromatic Polyamides. Also introduced in the 1960s was a series of polyamides (10) in which the intermediates are wholly aromatic (see Aramid fibers). By virtue of the aromatic ring structures coupled with the relative regularity of the polymolecular chain, they are very high melting, very high in thermal stability, and possessed of generally high performance properties.

These polymers do not melt, for all practical purposes, other than at temperatures involving decomposition, and they are nearly insoluble. Consequently, conventional melt polymerization and melt spinning are not feasible. Polymers can be prepared from the dihalides of the dibasic acid by interfacial condensation with diamines at relatively low (<100°C) temperature or by reaction in suitable solvents (see Polymerization). In some cases sulfuric acid may be used as the solvent for these rather intractable polymers.

The intermediates, the polymerization, the fiber spinning and aftertreatment (drawing to orient, crystallizing and heat treatment) are all more costly than for the aliphatic nylons such as 6,6 or 6. The aramid fibers thus are higher in price, and find use in applications where exceptionally high strength, very high modulus (stiffness) or high resistance to heat, or both, are required.

Commercial aramid fibers (qv) are trademarked Nomex and Kevlar by DuPont. Nomex, based on meta-linked isophthalic acid and *m*-phenylenediamine, is used in specialty papers (11) for electrical insulation and aircraft structures, in protective garments, and in other applications requiring high thermal stability. Kevlar is an aromatic polyamide containing para-oriented linkages. It melts at over 500°C, is exceptionally high in strength with a tenacity more than twice that of high strength nylon or polyester, and a very high modulus. It finds use in heavy duty conveyor belts, and in composite structures with casting resins such as epoxies (see Embedding). It is contending against steel in radial tire reinforcement (12).

Polyester Fibers. Although Carothers had demonstrated that fiber-forming polyesters could be synthesized from glycols and dibasic acids, or from hydroxy acids, the laboratory work focused on polyamides for development. The original, and still growing, commercial success of polyester fibers is attributable to the work of Whinfield and Dickson in the United Kingdom (13). They demonstrated that the presence of aromatic rings in a linear polyester of sufficient molecular weight led to high melting points, good stability, and strong, tough fibers. Commercial production was initiated in the late 1940s and early 1950s (see Polyester fibers).

Polyester, now the largest-volume synthetic fiber in the world, surpassed nylon worldwide in 1972. The polymer, like nylon, is made by melt polymerization. Ethylene glycol reacts with either dimethyl terephthalate (DMT) or terephthalic acid (TPA) under heat and high vacuum. In the former case, methanol is eliminated in the condensation/transesterification reaction; in the latter, water is evolved. The intermediates are again petrochemicals-based. The polymer has a crystalline melting point of ca 250°C, similar to that of nylon-6,6.

Melt-spinning of polyester fibers is also similar to that of nylon. The as-spun fiber is amorphous and comparatively weak. Drawing, carried out hot because the glass transition temperature is higher than that of nylon, causes orientation of the polymer chains along with crystallization, and fiber strength is developed. The fiber may be heat-set (annealed) under tension to relax strains in the molecule and induce some additional crystallization. Like nylon, the extent and nature of the drawing can be used to control tenacity and elongation, these essentially being interchangeable. In more recent years, fiber production processes involving a combined spin-draw, or a combined drawing and texturing step, have proliferated.

Eastman Chemical Products, Inc. produces not only poly(ethylene terephthalate) [*25038-59-9*] under the trade name Kodel, but also a polyester based on condensing terephthalic acid with 1,4-dimethylolcyclohexane (14). This fiber is in many respects similar to the ethylene glycol-based polyester but melts higher and has a lower specific gravity and excellent recovery from stretch. It is designated Kodel II.

Polyester fibers are produced in both staple and continuous filament form. The former predominated in earlier years, but by 1977 polyester filament volume was 80% that of staple. The fiber is remarkably versatile, probably more so than any other chemical fiber. It is strong, abrasion resistant, relatively stable, higher in modulus than nylon, and of lower moisture regain. Its structure is such that polyester is more difficult to dye than most textile fibers, requiring somewhat specialized techniques.

Uses range widely over the apparel, home furnishings, automotive, and industrial fields. Enormous quantities of polyester staple are blended with cotton, rayon, or wool in spun yarns used for apparel. Good abrasion resistance also suits polyester to use in carpeting. Continuous filament yarn finds wide use in apparel; much of it is textured to produce bulk and opacity for this use.

In the automotive field, just as nylon tire cord (qv) to a great extent displaced rayon, high-tenacity polyester tire cord has made inroads against nylon, particularly in passenger car tires (15). Other industrial uses include belts, ropes, and filter fabrics.

Acrylic Fibers. Synthetic fibers based on acrylonitrile (qv) appeared in the United States in the 1940s (16). Subsequently, fibers of lower acrylonitrile unit content were introduced. These modacrylics present some use problems because they are lower in melting point and less resistant to the effects of heat than are acrylics, but they are relatively flame-resistant and useful in applications where this property is desirable (17) (see Flame retardant textiles; Acrylic and modacrylic fibers).

Acrylic and modacrylic fibers combined are now produced worldwide and represent the third largest in volume of the synthetic fibers produced. They are derived from petrochemical-based intermediates. Because acrylonitrile polymers cannot be melted without decomposition, melt spinning is not applicable. Highly polar solvents, such as dimethylacetamide, have been found to dissolve the acrylic polymers for either dry or wet spinning, both of which are used commercially (see Acetic acid). Virtually all acrylic fiber is produced in staple form. Earlier, filament acrylics were produced, but they proved very difficult to dye uniformly. A small amount of textile filament yarn is still produced in Japan, and acrylonitrile homopolymer filament is produced for carbonization in the manufacture of carbon fibers (see Carbon).

The fiber-forming acrylic polymers are high in molecular weight and are produced primarily in aqueous medium by free radical-initiated addition polymerization. The regular acrylics are usually copolymers with minor amounts of one or more comonomers, such as methyl acrylate, used to provide accessibility or sites for dyestuffs. Like other synthetics, acrylic fibers are stretched to develop orientation and fiber strength.

Acrylic fibers have good tenacity, although less than polyester or polyamides, excellent stability to sunlight, good dye acceptance as a result of the copolymer system used for this purpose, and a soft, pleasing hand of wool-like characteristics. Abrasion resistance, although well below that of nylon or polyester, is nevertheless good, and superior to that of wool. For these reasons, the acrylics find widespread use in both indoor and outdoor furnishings, including awnings and draperies, and in blankets, sweaters, and carpets (see Textiles). Blending low- and high-shrinkage yarns and shrinking the blend produces a bulky yarn. The levels of shrinkage are produced by different degrees of stretching in fiber production.

A variety of modacrylic fibers are produced in the United States and abroad. These are acrylonitrile copolymers or terpolymers with significant comonomer contents. Among the comonomers used are vinylidene chloride, vinyl chloride, and apparently acrylamides (qv). The fiber-forming modacrylic polymers are more soluble than the acrylics, some of them in acetone, for example, and are made commercially both by dry and wet spinning. The modacrylic fibers have various textile uses, including fake fur pile fabrics (see Furs, synthetic) and protective garments, among others. Modacrylics, because of their self-extinguishing or flame-resistant characteristics are also used in wigs (qv) and doll's hair.

Polypropylene. The discovery that catalyst systems could convert certain olefins, notably ethylene and propylene, to crystalline, ordered, or stereoregular polymers provided a basis for ordered polyolefin fibers of defined melting points (see Olefin fibers). Large-scale availability of low cost monomers from petrochemical operations, polymer specific gravities less than one, and melt-spinnability were key factors fostering the major development and commercialization of polyolefin fibers.

World polyolefin fiber production, most of it in the United States and Western Europe, is fourth in volume among the synthetics. Polypropylene [*9003-07-0*] is thought to constitute the majority of commercial polyolefin fiber. In stereoregular form, polypropylene has a crystalline melting point of about 165°C, compared to ca 125°C for low-pressure (regular) polyethylene [*9002-88-4*] and ca 110–120°C for high-pressure polyethylene (see Olefin polymers).

These fibers are melt-spun and produced in monofilament, multifilament, and staple forms. Both conventional melt spinning and a type of film extrusion are used. In the latter the film is mechanically slit, split, or fibrillated to produce coarser yarns where tex (denier) uniformity is not critical (18). In all cases the as-spun or extruded material is stretched to develop orientation and strength.

Polypropylene fiber is relatively low melting for a chemical fiber, limiting its uses. Offsetting this and other relative deficiencies is low cost, high strength, very great chemical inertness, and, because of the low density, high yardage of fiber of a given tex per kilogram (den/lb).

Because of the chemical inertness, dye acceptance is inadequate. For uses where color is requisite, either the base polymer is modified to provide dyeability or the fiber is spun-colored; that is, produced with pigments incorporated in the polymer and fiber during melting. Because of oxygen and light sensitivity, suitable stabilizers are also incorporated (19).

Textile uses include upholstery and carpeting, particularly indoor-outdoor carpets. In addition to face yarns for carpeting, polypropylene fiber, particularly that produced from film, is used for woven carpet backings. The low melting point, which prevents easy ironing, and the fabric hand have largely precluded polyolefins from application in apparel fabrics. Other important uses include rope and cordage, fishnets, and filter media. Negligible moisture absorption, resistance to decay by organisms, and low density, which causes the fiber to float make polyolefins particularly suited to such uses.

Vinyon and Other Vinyl Fibers. In addition to vinyl addition polymers based on acrylonitrile (ie, vinyl cyanide) others have been commercialized. Indeed, fibers known as Vinyon were commercialized in 1939 closely on the heels of the introduction of nylon-6,6, as a result of the joint efforts of the Union Carbide and American Viscose Corporations (now Avtex Fibers, Inc.) (see Vinyl polymers).

Poly(vinyl chloride) [*9002-86-2*] by itself tends to decompose on heating and is difficultly soluble in common solvents. However, a small amount of vinyl acetate copolymerized with it acts as an internal plasticizer, promotes solubility in solvents such as acetone, and yields a copolymer capable of producing useful fibers. Commercially the copolymers are about 85–90% vinyl chloride and 10–15% vinyl acetate units. Although textile filament was produced earlier, U.S. Vinyon production as of 1978 is entirely in staple form by Avtex. It is made by a dry-spinning process like that for cellulose acetate fiber. The fiber is temperature sensitive, starting to soften, shrink, and become tacky below the boiling point of water. It finds specialty use in applications requiring bonding and heat sealing, in conjunction with other fibers.

Other Vinyon fibers from vinyl chloride homopolymers, rather than copolymers, are produced abroad, notably in France, Italy, and Japan (20). They are produced primarily by spinning from solution, carbon disulfide being one of the solvents used. A variety of products are made but all share relatively limited thermal stability, tending to shrink at temperatures in the neighborhood of boiling water. They are, however, resistant to moisture and rotting, and inherently nonflammable, suiting them to specialty uses such as filter cloths, nonflammable garments, fishnetting, and felts (qv) for insulating purposes. Strength and residual shrinkage can be varied to fit the application.

Saran. Saran fibers based on vinylidene chloride polymers (21) are produced in several countries under various trade names. They are based on vinylidene chloride copolymerized with small amounts of vinyl chloride, and still smaller amounts of acrylonitrile.

Saran fibers are melt-spun and then stretched. They are characterized by a pale straw color and by resistance to water, fire, and light, and bacterial and insect attack. Their relatively low melting points require excessively low ironing temperature. They are difficult to dye, requiring pigmentation in the melt. The fibers find specialty use in certain types of upholstery, filter cloths, and fishnets. Both staple fibers and filament yarns are produced, the latter in both heavy tex (denier) monofilaments and in multifilament form. The relatively low cost of these fibers is one of their attractions (see Vinylidene polymers).

PeCe. PeCe fiber, a post-chlorinated vinyl chloride polymer, was developed in Germany and was of some significance there during World War II. The post-chlorination raises the chlorine content of the polymer from 57 to 64% and confers acetone solubility, permitting ready wet spinning. Like many other vinyl fibers, it is low melting and restricted to special applications.

Vinal. Vinal is the U.S. term applied to poly(vinyl alcohol) [*9002-89-5*] fibers. These fibers originated in Japan and considerable production continues there. Although vinyl alcohol does not exist as a monomer, its polymer can be obtained by the hydrolysis of the acetyl groups from poly(vinyl acetate). Vinyl acetate, a relatively abundant and modest-cost monomer, is polymerized and then saponified to poly(vinyl alcohol), which is wet-spun from hot water. In this form, the fiber is water soluble. To make it commercially useful, the fiber is treated with formalin and heat, rendering it water insoluble. As with other fibers, the relative tensile strength and extensibility can be varied by varying the degree of stretch used. The fiber has reasonable strength, a moderately low melting point (222°C), limited elastic recovery, good chemical resistance, and resistance to degradation by organisms. Both staple and filament forms, including monofilament, are produced. Vinal is used in bristles, filter cloths, sewing thread, fishnets, and apparel. Although other producers have closely studied this fiber, production has remained largely confined to Japan. Elsewhere, it would appear, the overall cost-performance ratio has not justified commercialization.

Polychlal. Polychlal fibers (22), unique to Japan, are closely related both to the Vinyon and Vinal types. They are produced from an emulsion of poly(vinyl chloride) and a matrix of poly(vinyl alcohol). This is essentially wet-spun into aqueous sodium sulfate and formaldehyde-treated. Production on a modest scale continues in Japan, and some of this fiber has been exported to the United States for use in flame-retardant apparel.

Other Synthetic Fibers. Synthetic fibers other than those already described have been commercialized. Although production is, in most cases, relatively small, these products are frequently important in specific applications.

Benzoate Fibers. A melt-spun polyester fiber is produced in Japan from the self-condensation polymer derived from *p*-(β-hydroxyethoxy)benzoic acid, the aim being to provide a silk-like fiber (22). Although limited production of this fiber continues, it does not appear likely to grow significantly.

Poly(hydroxyacetic Acid Ester) Fibers. Although most of the aliphatic polyesters, as distinct from the aromatic acid-based type, are low melting, poly(glycolic acid) [*26124-68-5, 26009-03-0*], also known as poly(hydroxyacetic acid) or polyglycolide, is an exception, melting at over 200°C (23). Fibers can be melt spun from homopolymers derived from glycolide or from copolymers derived from glycolide and a minor amount of lactide. The fibers are crystalline, and can be oriented by stretching. Because they are absorbable in the body, they have become important in surgical suture applications, where they replace catgut sutures in many such applications (see Sutures). Compared to the textile fibers, these are relatively high-priced, low-volume specialty items.

Spandex Fibers. Synthetic elastomeric fibers, developed to compete with or improve on natural rubber fiber, differ from ordinary textile fibers in having very high extensibility to break (500–600%) and high recovery from such stretching. The segments of the polyurethane are based on low molecular weight polyethers, such as poly(ethylene glycol), or polyesters, which then react with an aromatic diisocyanate to give a three-dimensional, elastomeric product. Though the chemistry is complex, basically the elastomeric products depend on the presence of hard and soft segments and cross-linking. The fibers produced are white and dyeable and are stronger and lighter than rubber, making them particularly suited for use in foundation garments, bathing suits, support hose, and other elastic garments (see Fibers, elastomeric; Urethane polymers).

Fluorine-Containing Fibers. Polymers based on polytetrafluorothylene [*9002-84-0*], DuPont's Teflon in the United States, have uniquely high chemical stability and inertness, no water absorption, low frictional characteristics, and high melting points with decomposition preceding and accompanying melting (see Fluorine compounds, organic). These properties clearly make conversion to fiber form difficult. Nevertheless, the problem was solved in the 1950s by emulsion spinning. This process appears basically to comprise sintering or fusing extremely small polymer particles that are themselves fibrous and have been aligned by passing a colloidal dispersion through a capillary. The fused fiber is subsequently drawn. Teflon fibers in both filament and staple forms are high priced and find, as would be expected, highly specialized applications, such as packings, special filtration fabrics, and other uses where corrosion resistance, lubricity, and temperature resistance are requisite.

Carbon Fibers. Carbon fibers (24), the subject of a great deal of research and development in the past decade, are characterized by extremely high strength and modulus along with high temperature resistance. These characteristics suggest themselves for particular applicability to high performance, reinforced composite structures where strength, stiffness, and lightness of weight are at a premium, eg, in special aerospace, industrial, and recreational applications. The composite structure may include other fibers, such as glass, along with a matrix resin, such as polyesters, epoxies, or polyimides.

Carbon fibers are made by controlled carbonization of an already formed fibrous structure based on an appropriate organic polymer. In particular carbonization has been applied to filament yarns of homopolymers of acrylonitrile, to high tenacity viscose rayon multifilament yarns, and to pitch for lower strength carbon fibers (see Ablative materials; Carbon; Composite materials).

Phenolic-Type Fibers. The Carborundum Company developed a fire-resistant fiber based on the technology of the well-known phenol–formaldehyde resin condensates. The latter, long used for moldings, laminates, and similar thermoset resin applications, are cross-linked, three-dimensional polymers made by curing a non-cross-linked precursor condensate of formaldehyde with phenolic compounds. Special processes were developed to produce fibers. The result, trade-named Kynol, is presently produced in Japan (Nippon Kynol Co., U.S. distributor, American Kynol Co.) and used for protective garments (see Novoloid fibers; Phenolic resins).

Polybenzimidazole Fibers. Poly(benzimidazole) [*26986-65-9*] fibers (25) have been produced semicommercially from condensation polymers based on the reaction of 3,3′-diaminobenzidine and diphenyl isophthalate. A process in which the fibers are dry-spun and then drawn was developed by the Celanese Corporation, partially supported by the government. The fibers are of the high performance type, of very high melting point, good strength and extensibility, nonflammable, and surprisingly for a synthetic fiber, of high moisture regain: 13%. These fibers are of interest for aerospace applications and potentially for industrial applications as well subject to incompletely defined economics (see Heat resistant polymers; Polyimides).

Glass. The world production of textile glass fiber listed at ca 800,000 metric tons for 1977 does not include substantial production of glass fiber batting materials. In many ways glass fiber is unique among the chemical fibers. It is an inorganic fiber, in contrast to the numerous organic, natural, and synthetic fibers, and the fibers are neither oriented nor crystalline. The fibers are strong, nonflammable, and rather heat-resistant, as well as highly resistant to chemicals, moisture, and attack by organisms. Not surprisingly, glass fibers are quite low in extensibility and higher in density than the organic fibers.

The fibers are produced by melt spinning. To offset their brittleness, they are spun into fine filaments. After lubrication, these are wound as a multifilament strand unless insulating batting is being produced; in the latter case the filaments are attenuated while still molten and are deposited in the form of a thick bat (see Insulation, thermal).

Glass fibers, in forms ranging from filament yarn to mats, woven rovings, and staple, find a variety of important uses. In addition to use as insulating bats, they are used in fabric form for draperies where the nonflammability, inertness, and resistance to the effects of sunlight are assets. Glass fiber cloth, as well as batting, is used in a variety of insulating applications and for filtration.

One of the most important uses for glass fiber is in reinforced plastics, particularly reinforced thermosetting polyester resins (26). Since the 1950s the construction of pleasure boats, both sail and power, has been shifted from hulls of wood to glass-reinforced polyester. Such composites are widely used in industrial and automotive applications. Glass-filled thermoplastic resins have also grown in volume. The glass fibers are used in nylon, polyacetal, and poly(butylene terephthalate) molding resins, for example (see Laminated and reinforced plastics). Glass fibers are also used in both radial and bias-ply automotive tire reinforcement (see Fiber optics; Glass).

Metal and Other Inorganic Fibers. Fine steel wire is, of course, well-known for radial tire reinforcement. However, a number of metallic and other inorganic fibers have been investigated for special applications. Metallic fibers have been used in small quantities with organic fiber carpet yarns to minimize static problems (see Antistatic agents). Fibers of boron, boron–tungsten, steel, beryllium, boron and silicon carbides, boron and silicon nitrides, alumina, zirconia, and other inorganics have received attention (27), particularly for high performance specialty uses. Most are quite costly, and the quantities employed quite small.

Fiber Modifications

It is axiomatic that the applicability of a given fiber is a function of the overall price-performance considerations for that fiber. Although certain fibers are barred from a particular use by some inherent limitation, there is much overlapping and inter-fiber competition. Moreover, as the chemical fiber industry has grown, modifications of individual fiber types have been developed to broaden uses.

These modifications may be chemical, physical, or a combination of both, to improve performance in a given use or to adapt it to uses for which it was not previously satisfactory. Since the preceding sections have only alluded briefly to such modifications, it is in order to review them in somewhat greater depth.

Physical Modifications. As noted, when chemical fibers are stretched, their tensile strength generally increases and their extensibility decreases. Use is made of this in producing highly drawn, high tenacity nylon and polyester for tire cord and industrial uses. By adjusting the draw ratio the fiber modulus can be altered, and polyester staple of a modulus suitable for blending with cotton is made this way.

By decreasing the filament tex, and therefore increasing the number of filaments in a yarn of given total tex, the suppleness of the yarn and the softness of hand in fabrics made from it are increased. This is the trend in polyester filament yarns for apparel (28).

Texturing or bulking of filament yarns is used to produce yarn with more of the characteristics of yarn spun from staple fibers while eliminating several steps involved in producing the latter. Texturing can also be used to produce stretch yarns, as from nylon. Most texturing of chemical fibers is based on the fact that if the yarn is distorted while heated to render it more plastic, then cooled, the yarn "remembers" its distorted structure and seeks to restore it. Several methods are briefly summarized in Table 3.

False-twist texturing has been used extensively particularly with polyester filament yarns. Basically, a running thread line is temporarily highly twisted between two points and heat-set. It is then untwisted, whereupon it kinks or bulks. Nylon of a high degree of stretch is thereby produced. Nonstretch bulked polyester yarn is also made this way; a second heater is used to set the bulked configuration. Initially, a spindle rotating at very high speeds and through which the yarn passes was used to insert the false twist. In recent years, the use of friction false-twist texturing has been growing. In this method, the running thread line proceeds over the edge of several disks in series, the twist being imparted by friction.

False-twist texturized polyester filament yarn has become a major item of commerce, with huge quantities being used in apparel. Originally, fully drawn filament yarn was textured. In recent years, it has been found advantageous in many cases for the fiber producer to make a partially drawn yarn with only some degree of orientation. This is known as POY (partially oriented or preoriented yarn). The texturer, who may also be the fiber producer, then simultaneously draws and textures this yarn.

Staple fiber bulk is enhanced by crimping before cutting into staple lengths and by using differential shrinkage fibers as described for acrylics. Rather elegant use of this principle is made in bicomponent acrylic staple fibers in which each fiber consists of two components fused together. These components are formulated to have different shrinkage potential. When subjected to hot water, a helical crimp develops.

Changing the cross section of the individual filaments in a yarn may affect fabric appearance and hand. In melt spinning the use of spinneret holes of special shapes

Table 3. Filament Yarn Texturing/Bulking Methods for Fibers

Means of yarn distortion	Comments
knitting, fabric heat-setting, de-knitting	called knit-de-knit; cumbersome, currently used little or not at all
passage over hot knife edge	Agilon process; spiral crimp; nylon stretch yarn an example
passage between gear teeth	gear-crimping process; zig-zag configuration; used for BCF carpet yarn
passage into tightly enclosed space	stuffer-box crimping; saw-toothed configuration; Ban-lon yarns; tows for crimped staple fiber
twisting, particularly false-twisting	stretch and bulky nonstretch yarns; textured polyester apparel yarn so produced
air turbulence	Taslan yarns; filaments looped randomly by air jet while over-feeding; not dependent on heat-setting; used for blends of nylon and polyester with acetates

can produce multilobal filament cross sections. Polyester yarn, for example, that is so spun produces a fabric without the sometimes objectionable shiny or plastic appearance. Alternatively, special cross sections can be used to provide "sparkle" yarns.

Chemical Modifications. Modified fibers can be produced by chemical alteration of the polymer, fiber, or fabric, or by the use of nonreacting additives. Improved dyeability is usually effected by use of a suitable comonomer. Polyester fibers ordinarily dyed with disperse dyestuffs are made cationically dyeable by use of a comonomer providing acidic sites in the base polymer.

Spun-colored fibers can be produced by incorporating a pigment before spinning. Acetate, triacetate, polyester, polyolefin, and acrylic fibers have used this approach for certain applications. In related processes, white pigments, particularly titanium dioxide, are used to dull the natural luster of chemical fibers, or optical brighteners may be added, such as to polyester, to increase the perceived whiteness of the fiber (see Brighteners, fluorescent).

Chemical approaches were of particular importance in seeking to meet U.S. Government regulations regarding flame-retardant garments for such uses as children's sleepwear. A halogenated phosphate compound known as Tris, tris(2,3-dibromopropyl) phosphate, used as an additive in acetate and triacetate spinning, and as a fabric additive with polyester, enabled fabrics from these fibers to meet federal requirements, but its use was dropped in light of work suggesting possible carcinogenicity. More recently, Hoechst Fiber Industries has introduced in the United States a flame-resistant polyester variant trademarked Trevira 271 based on copolymerization involving a phosphinic acid derivative (29) (see Flame retardants, phosphorus compounds).

The introduction of permanent-press treatment of polyester cotton staple fiber blends in fabric form was a major factor in the growth of polyester usage. Wash-and-wear characteristics can be imparted to fabrics by treatment with suitable thermosetting (cross-linking) resin systems, but an accompanying strength loss made this approach unsuitable for all-cellulosic fibers such as cotton or rayon. Blending the strong polyester fiber with the cellulosic overcomes this problem and yields a blend widely used in shirting and slacks fabrics.

Economic Aspects

Chemical fiber economics are complex, varying not only with the specific fiber and process, but also with the item (ie, staple or filament, tex, etc) of a given fiber, and with geography, scale, capacity utilization, and other factors. Certain generalities, however, provide guideposts. Of the noncapital-related costs, raw materials are the single largest item, followed by labor and energy. Both capital and production costs tend to be higher for filament yarn than for staple-fiber plants, and higher for low tex (denier) filament than for higher tex (denier) material. Because solvent and solution handling tend to increase energy consumption, labor, and environmental problems, both capital and operating costs for wet- and dry-spun fiber tend to be higher than those for melt-spun fibers.

High capital costs have for some years prevented construction of new rayon plants, except in special circumstances in certain Communist and developing countries. Rayon textile filament yarn has virtually disappeared because of high costs of wet-spinning fine tex items, despite the fact that cellulose is a renewable resource. Among the major synthetics, those with most favorable raw materials costs are polyester, polyolefins, and acrylics, with nylon-6 and 6,6 being significantly higher. Capital and energy costs, however, are less favorable for the acrylics. All these factors, plus versatility in use, have conspired to favor polyester in both staple and filament form.

Though highly profitable in much of their history, chemical fibers have been severely buffeted in recent years. In addition to the traditional cyclicality of the textile industry, during the 1970s overcapacity in both the fiber and petrochemicals industries and increases in energy and raw materials costs have significantly affected profitability for both U.S. and European fiber producers.

Table 4 lists typical market prices for several important fibers for 1970 and 1978, with 1973 and 1974 also included to reflect sharp increases in petroleum prices in that period. It must be emphasized that these prices reflect external factors as well as actual underlying production costs, that there are scores of other items at different prices, and that certain markets do not command these prices. All these economic pressures combine to place emphasis on the research and development of process improvements and cost reduction.

Present Status and Future Possibilities

Today there is one or more fibers to fit virtually any use, and there is an enormous range of properties in the broad spectrum of commercially available fibers. In addition, the direct conversion of synthetic polymers to nonwoven fabric form (30) has grown over the last few years. Polyethylene, polypropylene, and polyester resins, for example, can be converted to what are called spun-bonded fabrics in processes that combine melt-extrusion with a means of laying the filaments in a web and bonding the thermoplastic fibers at cross-over points to form fabrics useful in a variety of applications. Polypropylene carpet backing is just one example (see Nonwoven textiles).

Concurrent with these product developments has been a steady, continuing increase in uniformity of fiber products regarding properties, dyeability, and the like.

Scientific and technological possibilities are still plentiful, but in the future their exploitation is likely to be overshadowed by economic and political developments,

Table 4. Typical U.S. Market Prices Selected Chemical Fibers, 1970–1978[a], $/kg

Fiber and type	1970	1973	1974	1978
rayon staple, regular[b]	0.59	0.72	1.06	1.25
acetate filament, 8.3 tex (75 den)	1.61	1.47	2.02	2.42
16.7 tex (150 den)	1.23	1.10	1.63	2.11
polyester staple	0.90	0.84	0.97	1.19
polyester filament, 16.7 tex (150 den), drawn	2.99	1.94	1.87	1.43
nylon filament, 4.4 tex (40 den)	2.31	2.77	3.12	3.52
95.5 tex (860 den)	1.76	1.71	1.98	2.55
acrylic textile staple[c], low tex (denier)	1.96	1.71	1.76	1.63
polypropylene textile staple, natural[d]	1.01–1.08	0.79–0.86	1.03–1.10	1.14–1.21
177.8 tex (1600 den) BCF, colored filament	2.09	2.20	2.20	2.20

[a] Data from various industry sources, 1978.
[b] High wet modulus rayon staple ranges from $0.05–0.09 higher.
[c] Modacrylics tend to be higher, as do specialty items such as bicomponent acrylics.
[d] Natural refers to nonpigmented fiber.

particularly the increasing shortage of oil, increased cost of energy, environmental regulations, and other regulatory restrictions.

Overall, the chemical fiber industry can be expected to continue to grow, but at a rate lower than that of the 1950s and 1960s. Growth in the cellulosic sector is unlikely except for acetate cigarette filter tow. Growth will be in the synthetic fibers sector, and further inroads against the natural fibers are likely. Many chemical possibilities are in principle capable of exploitation. In the condensation polymers, eg, nylon or polyesters, the dibasic acid, the diamine, or the glycol can be varied widely to affect the properties of the fiber. For example, use of longer-chain glycols than ethylene glycol can impart greater resilience to polyester fibers based on terephthalic acid. Copolymerization in general, including block and graft copolymerization, offers seemingly endless possibilities for both addition and condensation polymers. Certain monomers not presently used commercially are known to produce interesting fibers; a notable case is pyrrolidone from which, by ring-opening polymerization, is obtainable nylon 4, a fiber of higher moisture regain than other synthetic fibers. Apparently, technical problems associated with polymer stability in melt-spinning and/or economic factors have prevented commercialization of nylon 4 fibers despite much work over several decades (see Polyamides).

Nevertheless, for the remainder of this century and perhaps indefinitely, it seems unlikely there will be any new, broad-based chemical fibers or that those relatively small-volume textile fibers already in existence will grow rapidly or become more widespread than at present. Thus, among the synthetics, the predominance of polyester, nylon, acrylic, and olefin fibers is likely to continue with continuing development of very specific modifications for specific uses.

BIBLIOGRAPHY

"Fibers" in *ECT* 1st ed., Vol. 6, pp. 453–467, by H. F. Mark, Polytechnic Institute of Brooklyn; "Fibers, Man-Made" in *ECT* 2nd ed., Vol. 9, pp. 151–170, by H. F. Mark, Polytechnic Institute of Brooklyn and S. M. Atlas, Bronx Community College.

1. H. Mark and G. Whitby, *Collected Papers of W. H. Carothers,* Interscience Publishers, Inc., New York, 1940.
2. *Textile Organon* **39**(6), 89 (June 1968).
3. *Ibid.,* **49**(6), 73 (June 1978).
4. *Mod. Text.* **52**(4), 19 (1976).
5. J. W. Schappel, F. R. Smith, and E. A. Zawistowski, *Textile Research Institute 48th Annual Research and Technology Conference, Atlanta Georgia, April 5–6, 1978.*
6. *Mod. Text.* **58**(6), 18 (1977).
7. *Ibid.,* **58**(3), 6 (1977).
8. *Chem. Eng. News.* **56,** 11 (Dec. 4, 1978).
9. G. Clayton and co-workers, *Text. Prog.* **8,** 1, 27 (1976).
10. *Ibid.,* pp. 98–115.
11. *Properties of Nomex Types E-54 and E-55 Aramid Paper,* DuPont Preliminary Information Memo 401, E. I. du Pont de Nemours & Co., Inc., April 1978.
12. R. E. Wilfong and J. Zimmerman, *J. Appl. Polym. Sci.* **31,** 1 (1977).
13. J. R. Whinfield, *Text. Res. J.* **23,** 289 (1953).
14. E. V. Martin, *Text. Res. J.* **32,** 619 (1962).
15. M. E. Denahm, *Mod. Text.* **59**(11), 13 (1978).
16. R. Meredith, *Text. Prog.* **7,** 21 (1975).
17. Ref. 9, pp. 61–69.
18. Ref. 9, p. 81.
19. Ref. 9, pp. 78–79.

20. R. W. Moncrieff, *Man-Made Fibers,* 6th ed., John Wiley & Sons, Inc., New York, 1975, pp. 522–532.
21. H. Mark, S. M. Atlas, and E. Cernia, *Man-Made Fibers Science and Technology,* Vol. 3, Wiley-Interscience, New York, 1968, pp. 303–326.
22. *Chemical Fibers of Japan 1970/1971,* Japan Chemical Fibers Association, Tokyo, 1970, p. 50.
23. Ref. 9, p. 50.
24. E. Fitzer and M. Heym, *Chem. Ind.* (16), 663 (1976).
25. R. H. Jackson, *Text. Res. J.* **48,** 314 (1978).
26. M. E. Dunham, *Mod. Text.* **59**(11), 16 (1978).
27. R. S. Goy and J. A. Jenkins, *Text. Prog.* **2,** 1, 31 (1970).
28. Ref. 8, p. 10.
29. W. S. Wagner in ref. 5.
30. R. A. A. Hentschel, *Chem. Technol.* **4,** 32 (1974).

General References

Refs. 9, 20, 21, and 27.
R. Hill, *Fibers from Synthetic Polymers,* Elsevier Scientific Publishing Company, Amsterdam, 1953.
J. J. Press, ed., *Man-Made Textile Encyclopedia,* Textile Book Publishers, New York, 1959.
E. M. Hicks and co-workers, "The Production of Synthetic Fiber," *Text. Prog.* **8,** 1 (1976).

WILLIAM J. ROBERTS
Consultant

FIBERS, ELASTOMERIC

An elastomeric fiber can be made from any natural or synthetic polymeric material that has high elongation and good recovery properties; however, at present only natural rubber (qv) and urethane polymers (qv) are raw materials for commercially successful elastomeric fibers. Previously, attempts have been made to introduce commercially fibers based on elastomeric polyamides (qv), polyesters (qv), and polyacrylonitriles (see Acrylonitrile polymers) (1–3). These efforts did not meet with success, owing primarily to economic problems and mechanical deficiencies in the experimental fibers. During times of natural rubber shortage, eg, World War II, neoprene (polychloroprene) has been used as a raw material for elastomeric fibers but, because of higher cost and less desirable performance characteristics, it is not used under normal supply conditions (see Elastomers, synthetic).

Elastomeric fibers based on urethane polymers and techniques for their experimental production were first reported in the early 1950s by Farbenfabriken Bayer, a pioneer in urethane and diisocyanate chemistry (4–5). It was not until the late 1950s, however, that elastomeric polyurethane fibers from pilot plant production facilities of the U.S. Rubber Co. (now Uniroyal, Inc.) and E. I. du Pont de Nemours & Co., Inc. were first introduced to the trade on a semicommercial scale. The fibers were given the generic designation spandex (6). The term spandex has come into common trade usage throughout the world in referring to fibers based on elastomeric urethane polymers.

Natural rubber thread has been in wide use for many years. If made by slitting rubber sheets, it is referred to as cut rubber thread. If made by extruding a natural rubber latex into an acid coagulation bath followed by washing, drying, and curing, it is referred to as extruded latex thread or fiber. Both production methods result in a thread or fiber that is a continuous monofilament. Cut rubber thread has a square or rectangular cross section, and extruded latex thread is substantially round.

Thermoplastic, hard (ie, inelastic) fibers, such as nylon and polyester, may be provided with a springlike, helical or zigzag structure by special processing procedures. Such fibers exhibit high elongations as the helical or zigzag structure is stretched but have very low recovery power. This apparent elasticity results from the geometric form of the filaments and not from polymer structure. Only fibers with elastic properties resulting primarily from entropy forces inherent within the basic polymer structure are described below.

Hard fibers, such as polyolefins, polyamides, and polyesters, can be converted into a state exhibiting relatively high degrees of elasticity by the use of critically controlled drawing and annealing techniques (7). Such fibers have been prepared with elongations in excess of 100% and good recoveries. The elastic recovery forces, however, result primarily from energetic effects (ie, ΔE) and not from entropy changes (ie, ΔS) as is the case in the rubberlike elasticity exhibited by natural rubber fibers and spandex. Hard, elastic fibers have not been produced commercially.

In the mid 1970s Monsanto Co. attempted to introduce commercially a nylon–spandex bicomponent fiber trade named Monvelle (8–10). Work on the development of nylon–spandex bicomponent fibers has also been reported by Toray Industries (11–14). These side-by-side (Monsanto) and sheath-core (Toray) bicomponent fibers were produced by extruding both nylon-6 and a polyurethane simultaneously through the same spinnerette hole. The resulting bicomponent fiber was then drawn under very carefully controlled conditions so that the nylon component was permanently deformed into a new, more highly oriented structure whereas the spandex was only elongated. On relaxation, recovery forces in the spandex component pulled the bicomponent monofilament back into a very tightly wound helical structure. An applied tensile force would tend to pull the helix and be resisted by the spandex component. Monvelle met with some customer acceptance as a replacement for nylon-covered spandex in stretch hosiery for women but was withdrawn from the market in 1976. Toray Industries has apparently made no attempts to commercialize its product.

Properties

In both rubber thread and spandex, mechanical properties may be varied over a relatively broad range. In rubber, variations are made by changes in the degree of cross-linking or vulcanization which is accomplished by changing the amount of vulcanizing agent, generally sulfur, and accelerants used. In spandex, many more possibilities for variation are available. By definition spandex is made from a urethane polymer (6). Again by definition, however, any polymer containing urethane linkages in the repeat structure:

$$\text{-}\!\!\left(\text{R}\text{—O}\overset{\text{O}}{\overset{\|}{\text{C}}}\text{NH—R}'\text{—NH}\overset{\text{O}}{\overset{\|}{\text{C}}}\text{O}\right)\!\!_n\text{-}$$

may be classified as a urethane polymer. The number of polymers in this classification is obviously very large. Most urethane polymers in current use for the manufacture of spandex are made by the reaction of a ca 2000 molecular weight polyester or polyether glycol with a diisocyanate at a molar ratio of ca 1:2, followed by reaction of the resulting isocyanate-terminated prepolymer with a diamine to produce a high molecular weight urethane polymer. The average repeat unit of the polymer molecule contains a polyester or a polyether chain of ca 2000 molecular weight, two urethane linkages, two residues from the diisocyanate, and a urea linkage. Mechanical properties may be affected by changing the particular polyester or polyether glycol, diisocyanate, and diamine used, and they also may be markedly modified by changing the molecular weight of the glycol and by changing the glycol–diisocyanate molar ratio used (15–18).

The long-chain urethane polymer molecules in spandex fibers are substantially linear block copolymers comprising relatively long blocks in which molecular interactions are weak, interconnected by shorter blocks in which interactions are strong (see Copolymers). The weakly interacting blocks, commonly referred to as soft segments, are from the polyester or polyether glycol component whereas the blocks having strong interactions result from the diisocyanate and chain extender and are referred to as hard segments. Hard segments from neighboring chains interact strongly, principally through hydrogen bonding, to form tie points between chains. As there are many hard segments in each chain, a network structure results. Any deformation of the network, either extension (stretching) or compression, results in an increase in the degree of order (ie, more uniform molecular alignment) and is opposed by entropy forces within the soft segments (19). Although in rubber the effect is similar, the network structure is formed not by strongly interacting hard segments but by covalent bonds between chain molecules that result from vulcanization with sulfur. For both spandex and rubber, modulus is directly, but not linearly, related to tie-point density, eg, in the case of spandex, hard-segment density. Similarly, the relationship for maximum elongation is an inverse function of tie-point density. In rubber threads, tie-point density is controlled by the amount of vulcanizing agent (sulfur), accelerants, and the reaction conditions used. In spandex, tie-point (hard segment) density is controlled by soft segment (glycol) molecular weight and by glycol–diisocyanate molar ratio.

Although rubber and spandex fibers can be produced with widely varying sets of mechanical properties, there is a relatively narrow range of properties for all commercial elastomeric fibers owing to common end-use requirements, practical limitations in raw materials availability and equipment limitations. Typical properties for spandex and rubber fiber are shown in Table 1. In Figure 1, stress–strain curves are shown for spandex and rubber fiber in comparison to hard fibers and hard fibers with mechanical stretch properties.

Manufacture

Cut Rubber. In producing cut rubber thread, smoked rubber sheet or crepe rubber is milled with vulcanizing agents along with stabilizers and pigments. The milled stock is calendered into sheets of thicknesses of 0.3–1.3 mm, depending on the final size of rubber thread desired. Multiple sheets (usually 20 or 40) are layered, heat-treated to vulcanize, then slit to form ribbons containing the same number of threads as layered sheets (see Fig. 2). Individual filaments for textile uses are square in cross section and may be readily separated from the ribbon.

Table 1. Physical Properties of Elastomeric Fibers

	Spandex	Extruded latex	Cut rubber
size range available[a]	2.2–360 tex (20–3,200 den)	16–610 tex (140–5,500 den)	2.5–21 μm dia (1,200–10,000 gauge)
tenacity, N/tex[b]	0.05–0.09	0.02–0.03	0.01–0.02
elongation, %	500–600	600–700	600–700
modulus (first cycle stress at 300% elongation), N/tex[b]	0.013–0.022	0.004–0.005	0.002–0.004
uv, ozone, and oxides of nitrogen resistance[c]	good (slowly yellows on uv exposure)	degrades rapidly	degrades rapidly
active chlorine resistance[c]	good (yellows on continual exposure)	poor	poor
oil (body oils, cosmetics, etc) resistance	good	poor	poor
dyeability	readily dyeable to broad range of colors	not dyeable	not dyeable
abrasion resistance (eg, nylon on elastomer)	very good	poor (fine sizes must be protected by covering yarns)	poor

[a] Spandex size is usually expressed in denier which is the weight in grams of a 9000 m length. However, the SI units is tex, the weight in g/1000 m. Rubber size is expressed as gauge which is the reciprocal of the diameter or side in inches (eg, extruded latex thread with a diameter of 0.01 in. is 100 gauge).

[b] To convert N/tex to g/den, multiply by 11.33.

[c] Both spandex and rubber threads are normally compounded with antioxidants and other stabilizers.

Extruded Latex Thread. In the manufacture of extruded latex thread, a concentrated (ie, containing up to ca 50% solids) natural rubber latex from the *Hevea Brasiliensis* tree is blended with aqueous dispersions of vulcanizing agents, pigments, and stabilizing agents. The compounded latex is held under carefully controlled temperature conditions for 8–36 h in order to allow some initial vulcanization to occur. This has the effect of increasing wet strength and thus processability of the extruded threads. The matured latex is extruded at constant pressure through precision-bore glass capillaries into an acetic acid bath where coagulation into thread form occurs. Threads are removed from the coagulation bath by transfer rollers, washed free of excess acid, and conducted through a dryer, after which a finish is applied and the threads are wound on individual bobbins or formed into multi-end ribbons and wound on ribbon spools. The thread is finally vulcanized by heating the ribbon spools or bobbins in vulcanizing ovens for up to 18 h. Recently, many producers have been changing from batch to continuous, in-line vulcanization. After vulcanizing, the multi-end ribbons are packed without support in boxes for shipment to the customer. Individual ends are shipped on bobbins. A typical extruded latex thread production line is illustrated in Figure 3.

The basic method for producing extruded latex thread has remained virtually unchanged since the process was first developed in the 1930s. In the past twenty-five years, however, there have been three modifications to the process that have significantly improved economics and convenience of use:

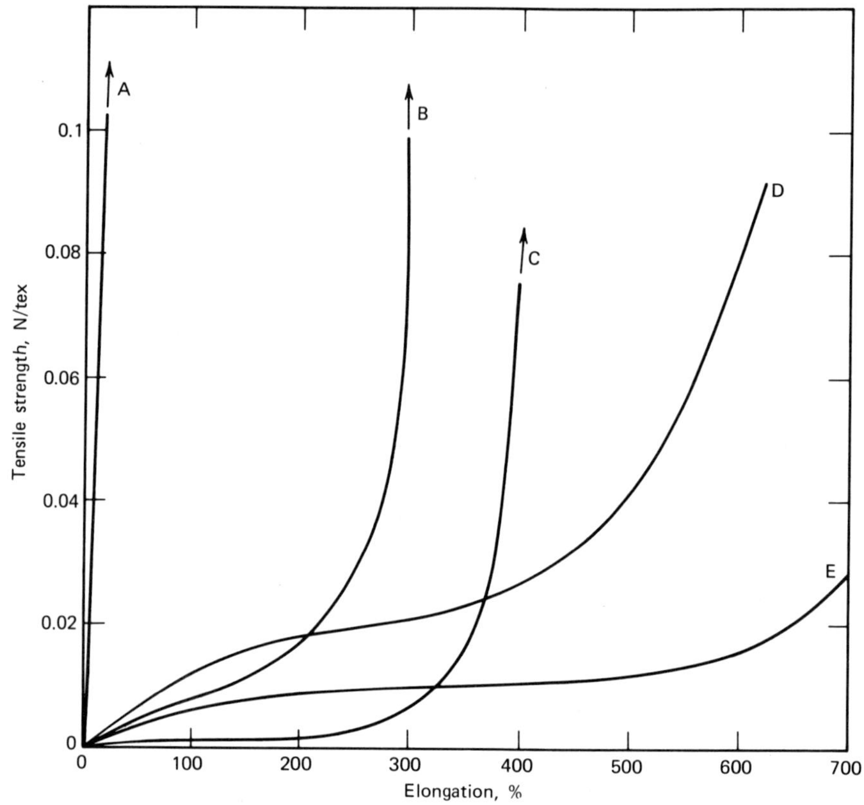

Figure 1. Stress–strain curves: A, hard fiber, eg, nylon, 0.35–0.79 N/tex; B, bicomponent nylon–spandex, 0.26–0.35 N/tex; C, mechanical stretch nylon, 0.26–0.44 N/tex; D, spandex; E, extruded latex. To convert N/tex to g/den, multiply by 11.33.

Ribbon-Forming Techniques for Latex Thread in the 1950s. Previously all latex thread was wound and sold on single-end bobbins. Most producers now form the threads into multi-end ribbons continuously on the production line by bringing groups of the dried and finished but unvulcanized ends into contiguous side-by-side configuration and applying carefully controlled pressure. The use of ribbons greatly reduces packaging expense and eliminates the need for returnable bobbins. It also simplifies operations in the customer's plant.

Talc-Free Rubber Thread. Until recent years the finish used for all extruded latex thread was talc powder (a problem in the pollution of the atmosphere of the working area and, because of its abrasive nature, acceleration of wear in production and processing machinery). Most producers have shifted from talc to newly developed silicone oil-based finishes.

Continuous Cure (Vulcanization). Until the early 1970s, the vulcanization process was conducted as a separate batch operation. Improvements in compounding and in oven design permit major producers to change to continuous in-line curing with savings in manpower and space as well as improved uniformities.

Extruded latex thread production lines vary widely. A typical line extrudes 100–200 ends (threads) into the coagulation bath in closely spaced side-by-side rela-

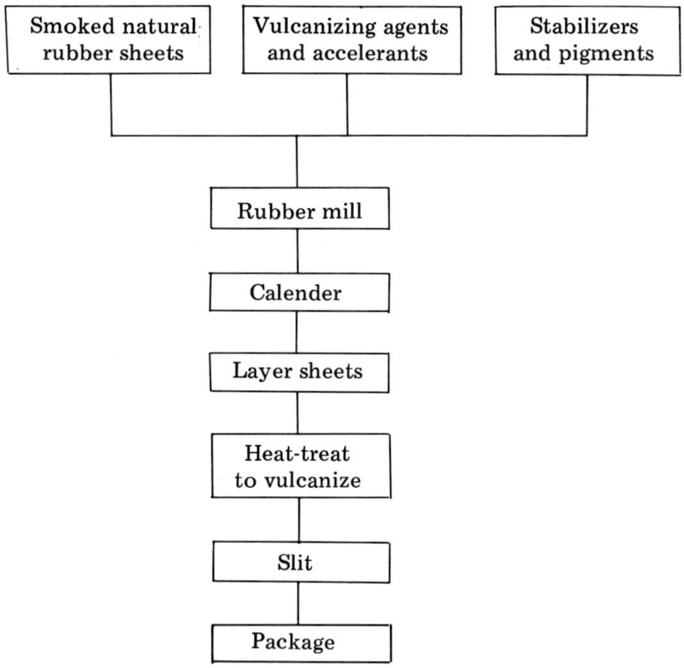

Figure 2. Cut-rubber thread manufacture.

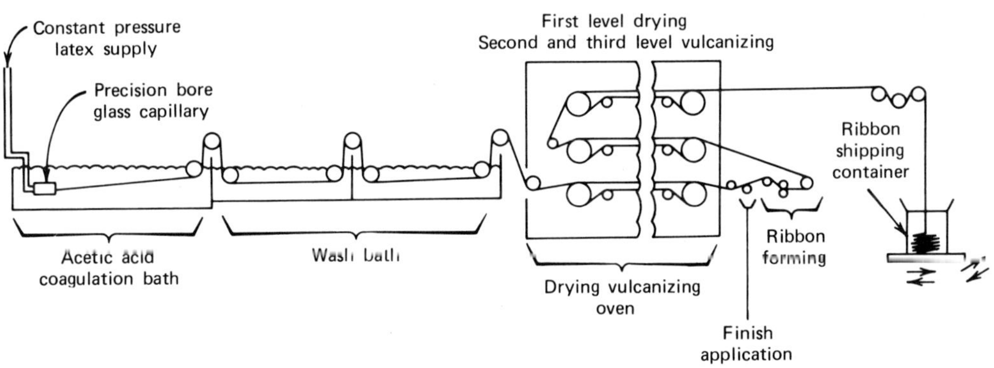

Figure 3. Extruded-latex thread production.

tionship. One or more wash baths follow the coagulation bath. In modern lines the oven is multipass with drying in the first pass followed by finish application and ribbon formation. The ribbons are continuously returned to other dryer levels for curing. Production speeds vary with thread size and equipment but, owing to hydrodynamic drag and the weak nature of the coagulating thread, maximum line take-up speeds are < 30 m/min.

Spandex. Spandex is produced commercially by four very different processes—melt extrusion, reaction spinning, solution dry spinning, and solution wet spinning. These are illustrated in Figure 4. Other approaches, such as emulsion spinning and sheet slitting, have been proposed but are not now in any commercial use (20). As is obvious from Figure 4, all processes involve different practical applications of basically similar chemistry. A block copolymer is formed by the reaction of a 1,000–3,000 molecular weight polyester glycol, polyether glycol, mixed polyester–polyether glycol, hydroxyl-terminated polycaprolactone, or any mixture of the preceding with a diisocyanate at a glycol–diisocyanate molar ratio of ca 1:2 to form a prepolymer which subsequently reacts with a low molecular weight diol or diamine at near stoichiometry. If the diol or diamine reaction with the prepolymer is carried out in a solvent, the resulting urethane block copolymer solution may be wet- or dry-spun into fiber. Alternatively, the prepolymer may be reaction-spun by extrusion into an aqueous or nonaqueous diamine bath to form a fiber and begin polymerization simultaneously, or the prepolymer may be permitted to react in bulk with a diol and the resulting block copolymer melt-extruded in fiber form.

Melt Spinning. Although considerable development work has been done in melt-extrudable urethane polymer formulation, particularly by Monsanto Co. and Toray Industries, there is currently only one commercial spandex (Mobilon) made by this process. Mobilon is produced in Japan by the Nisshin Spinning Co., and used primarily in waist bands and hosiery. Although many special techniques are required

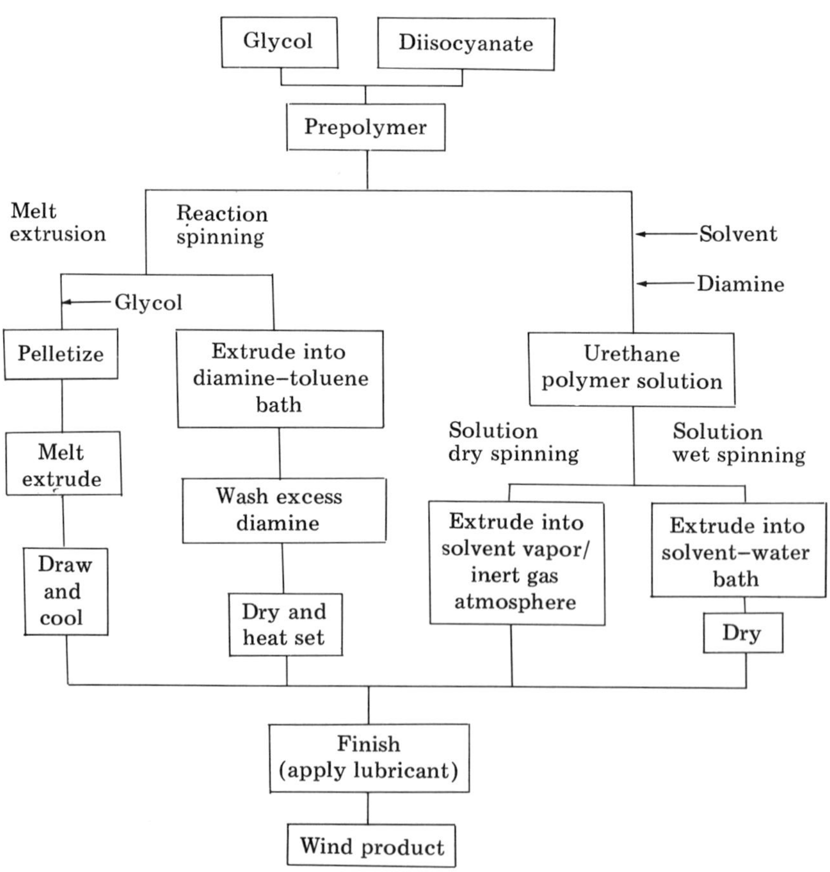

Figure 4. Spandex fiber production methods.

to accommodate heat sensitivity of the polyurethane, low modulus of the filaments, and inherent elastomeric tackiness, the process is similar to the melt extrusion of polyolefins.

Reaction Spinning. One of the earliest commercial spandex fibers, Vyrene, was produced by the U.S. Rubber Co. (now Uniroyal, Inc.) from a reaction-spinning process (21). Subsequently, both Globe Manufacturing Co. and Firestone Rubber, Inc. introduced spandex fibers, tradenamed Glospan and Spandelle, respectively, that were produced by reaction-spinning techniques. U.S. Rubber Co. and Firestone discontinued spandex production in the mid 1960s. The Firestone process and product was licensed to Courtaulds Ltd. which has developed the process further and produces a spandex trade named Spanzelle. Globe Manufacturing continues to produce Glospan and an unpigmented spandex trade-named Clearspan.

In the manufacture of spandex by reaction-spinning, a 1000–3000 molecular weight polyester or polyether glycol reacts with a diisocyanate at a molar ratio of ca 1:2 to produce an isocyanate-terminated prepolymer. After adjusting viscosity by the addition of minor amounts ($<10\%$) of inert solvent, the prepolymer is extruded into a diamine bath where filament and polymer formation occur simultaneously. The filaments are washed free of excess diamine, dried, cured, finished, and wound on tubes or bobbins. To minimize sticking filaments, the take-up tube, or bobbins are generally rewound before shipment to customer mills.

In general, some trifunctional hydroxy compound (eg, glycerol or trimethylolpropane) is added with the polyester or polyether glycol to produce covalent cross-links in the reaction-spun spandex. Additional covalent cross-links may result from allophanate and biuret formation during curing by the reaction of free isocyanate end groups with urethane or urea hydrogen along polymer chains. The early reaction-spun spandex fibers were extruded as monofilaments. Currently, a multiplicity of filaments are extruded from each spinnerette at individual filament tex values of 1.1–3.3 (10–30 den), then collected in bundles of the desired tex (den) at the exit of the reaction bath. This approach makes the surface area to mass ratio and diamine diffusion into the prepolymer cross section substantially constant in the diamine bath irrespective of final tex produced, thus minimizing condition changes required in changing tex. Since the individual filaments have reacted incompletely and are in a semiplastic state at the exit of the diamine bath, they interbond quite tightly into coherent bundles. The spandex fiber resulting from such production techniques is generally referred to in the trade as a fused multifilament.

As is the case with extruded latex thread, in reaction spinning production speeds are limited by filament weakness in the diamine bath and hydrodynamic drag. By combining multiples of relatively fine tex monofilaments to form one filament bundle, however, thread line strength develops much faster than with latex thread so that take-up speeds of as much as 100 m/min are possible, especially for the fine tex.

A master batch of delusterants (eg, TiO_2) and stabilizers usually is prepared by grinding the additives with the same polymeric glycol to be used in the prepolymer reaction. After grinding, the master batch is vacuum-dried and added to the polymeric glycol charge, prior to diisocyanate addition and the prepolymer reaction, in sufficient amount to provide the desired level in the finished fiber. Additional diisocyanate is added to the charge to account for the polymeric glycol introduced with the master batch. In selecting additives, care must be taken to avoid compounds that react significantly with diisocyanates or diamines under the conditions used in the process.

Further, additives chosen should have no adverse catalytic effect on the prepolymer or chain-extension reaction.

Reaction-spinning equipment is typically quite similar to that of solution wet spinning (see Fig. 6, p. 177). It differs principally in the use of fewer wash baths (only one or two as extruded latex) and in the use of a belt-type dryer rather than heated cans.

Solution Spinning. As in the case for both melt and reaction spinning, the first step in the preparation of a urethane polymer for wet or dry solution spinning involves reaction of a 1000–3000 molecular weight polyester or polyether glycol and a diisocyanate at a molar ratio of ca 1:2. In order to avoid solubility problems, great care must be taken to avoid any trifunctionality (or higher) in any of the reactants. Reaction conditions must be carefully selected and controlled so as to minimize side reactions, eg, allophanate and biuret formation, which can result in covalent cross-linking on chain extension (ie, reaction of the isocyanate prepolymer with diamine or diol to produce a high molecular weight urethane polymer). For the prepolymer reaction, most current commercial producers of solution-spun spandex fiber choose poly(tetramethylene ether) glycol (PTMEG) and methylenebis(4-phenyl isocyanate) (MDI) as the polymeric glycol and diisocyanate, respectively. Polyester glycols are lower in cost and, if properly selected, can yield substantially equivalent properties in the finished fibers. Solutions based on polyester glycols, however, are generally more difficult to handle in dry-spinning equipment, primarily owing to lower solvent removal rates.

In the polymerization reaction, commonly referred to as chain extension, the prepolymer, dissolved in a solvent, reacts with diamine or diol to form a urethane polymer in solution. In all commercial processes, the solvent of choice is dimethylformamide (DMF) or dimethylacetamide (DMAc). The chain extender is generally hydrazine or a diamine containing two primary amine groups. The diamine may be added to the prepolymer in solution or a solution of prepolymer may be added to a solution of the diamine. Stoichiometry is normally adjusted to provide a urethane polymer solution at 18–30 wt % solids and viscosity of 10–80 Pa.s (100–800 P) at 30°C. For dry spinning, viscosities and the percentages of solid are generally higher.

Additives, including pigments and stabilizers, are milled in the solvent of choice, generally with small amounts of the urethane polymer to improve dispersion stability, in a ball or sand mill, then blended to the desired concentration with the polymer solution after chain extension. Polymerization as well as additive preparation and addition of additives may be in either batch or continuous equipment. Some producers combine prepolymerization, chain extension, and additive addition and blending into one, integrated, continuous production line.

Dry Spinning. A majority of spandex manufacturers have chosen solution dry spinning as their production method. On a worldwide basis, ca 80% of all spandex produced is by various adaptations of this method. Although special mechanical techniques are required to handle dynamically the very low modulus, tacky filaments, dry spinning of spandex is very similar to the production of acrylic fiber by dry spinning. In both cases, DMF and DMAc are the solvents of choice (see Formic acid derivatives; Acetic acid derivatives).

The solution dry spinning process is illustrated by Figure 5. The spinning solution of urethane polymer is metered at a constant temperature by a precision gear pump through a spinnerette into a spinning cell of 30–50 cm dia and length of 3–6 m. Heated

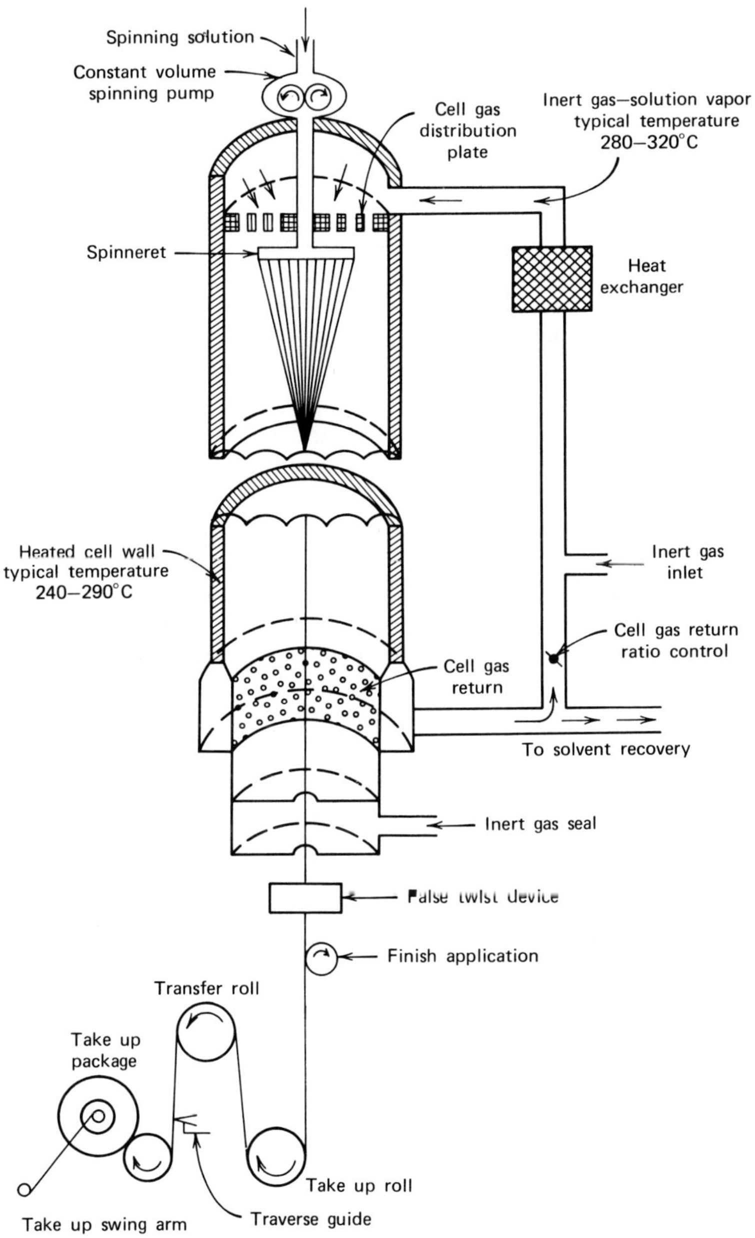

Figure 5. Spandex production, solution dry spinning process.

cell gas, made up of solvent vapor and an inert gas (eg, N_2) is introduced at the top of the cell and passes through a distribution plate behind the spinnerette pack. Because both cell gas and cell walls are maintained at high temperatures, solvent evaporates rapidly from the filaments as they travel down the spinning cell. Individual filament tex is normally maintained in the range of 0.6–1.7 (5–15 den) to maximize, within operable limits, surface to mass ratio and solvent removal rate. Individual filaments are grouped into bundles of the desired final tex at the exit of the spinning cell by a coalescence guide. A commonly used guide employs compressed air to create a mini-vortex which imparts a false twist and rounded cross section to the filament bundle. Depending on the tex produced and the manufacturer, each spinning cell may produce 1–8 filament bundles. Solution dry-spun spandex is normally referred to as continuous multifilament or coalesced multifilament. The terms coalescence and coalesced are misused, as the individual monofilaments remain discrete, ie, they adhere to one another only by a natural elastomeric tack and do not coalesce into larger structures.

After coalescence, a finish is applied to the multifilament bundle before it is taken up onto a tube. In addition to serving as a lubricant in subsequent textile processing, the finish must prevent bundle-to-bundle adhesion on the package. The white mineral oil normally used is modified with silicone oils and antistatic agents. Take-up speeds are ca 300–1000 m/min, depending on tex and producer.

Solvent is condensed from the cell gases, purified by distillation, and returned to polymer making for reuse. In the spinning cell and on distillation at atmospheric pressures, DMF hydrolyzes to a small extent to dimethylamine and formic acid. These compounds must be substantially removed (eg, by an ion exchanger) before the recovered DMF may be used in polymer making, because both dimethylamine and formic acid act as chain terminators. Some producers use vacuum distillation in order to minimize such hydrolysis and avoid the need for ion exchange. DMAc is more stable at atmospheric pressure distillation conditions and requires no further purification to remove hydrolysis products.

Wet Spinning. In general, any urethane polymer that may be solution dry-spun may also be wet-spun. In contrast, however, many wet-spinnable formulations may not be successfully spun on high-speed commercial dry-spinning equipment. Urethane polymer formulations utilizing polyester glycols are often used in wet spinning but only rarely in dry spinning.

A typical wet-spinning line is illustrated by Figure 6. Spinning solution is pumped by precision gear pumps through spinnerettes into a solvent–water coagulation bath. As in dry spinning, individual filament tex is maintained at ca 0.6–1.7 (5–15 den) in order to optimize solvent removal rates. At the exit of the coagulation bath, filaments are collected in bundles of the desired tex. A false twist may be imposed at the bath exit to give the multifilament bundles a more rounded cross section. After the coagulation bath, the multifilament bundles are countercurrently washed in successive extraction baths to remove residual solvent, then dried and heat-relaxed, generally on heated cans. Finally, as in dry spinning, a finish is applied and the multifilaments wound on individual tubes. A typical spinning line may produce 100–300 multifilaments at side-by-side filament spacings of less than 5 mm.

Water is continuously added to the last extraction bath and flows countercurrently to filament travel from bath to bath. Maximum solvent concentration of 15–30% is reached in the coagulation bath and maintained constant by continuously removing the solvent–water mixture for solvent recovery by distillation. Solvent recovery from

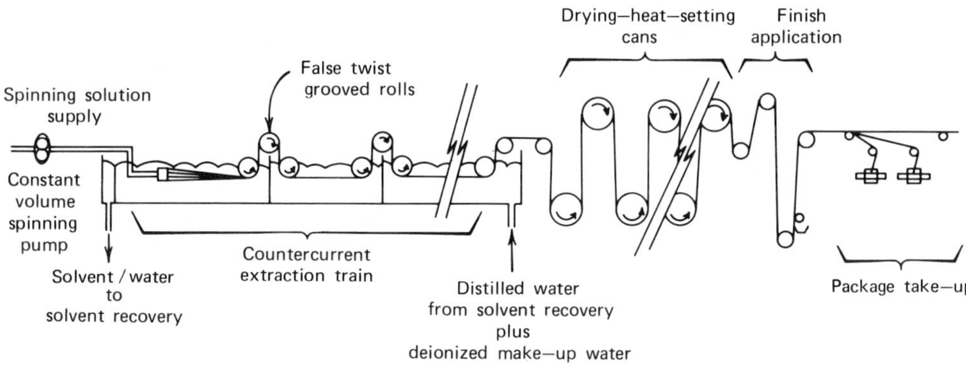

Figure 6. Wet-process spandex production.

wet spinning is generally a two-stage process, since three to five parts of water must be removed for every part of DMF or DMAc recovered. In the first stage, water vapor from the top of the column is condensed and either returned to the extraction train or sent to waste. DMF or DMAc is removed from the column reboiler as a liquid. The crude liquid DMF or DMAc, containing minor amounts of low molecular weight urethane polymer, extracted stabilizers, and other high or nonboiling contaminants, is transferred to a second column, vaporized, condensed, and stored for reuse in polymer making. In order to avoid accumulation of contaminants, the DMF or DMAc column reboiler must be purged periodically or continuously.

In wet-process synthetic fiber production, speeds are generally limited by hydrodynamic drag of the bath medium. For wet-spun spandex, such a consideration limits bath exit speeds to ca 100–150 m/min. It is, perhaps, because of this apparent limitation of production rate that most spandex producers have chosen dry-spinning techniques. More recent work, however, has substantially minimized this limitation by subjecting the spandex filament to drawing as much as three to four times after the spinning bath (22–23). Temperatures and residence times are selected so that the filaments are brought to temperatures above their second-order transition points (ie, the hard-segment melting points). This allows the molecular chains to move freely to relieve stresses and results in filaments of fine tex but with the same mechanical properties as the heavier tex feed. Thus it is possible to take up from a wet spinning process at speeds in excess of 300 m/min by continuously drawing and heat-relaxing the filaments after drying.

Producers

Since spandex was introduced commercially in the late 1950s, many companies have attempted production. Most were unsuccessful owing to a variety of technological and economic problems, and the list of producers on a worldwide basis, shown in Table 2, is now relatively small. With the exception of Bayer's Dorlastan, original work on all spandex processes in current use was done in the United States or Japan. A large percentage of worldwide capacity is controlled by DuPont, either directly or through subsidiaries and joint ventures. These include two plants in North America; one in South America, two in western Europe; and one in Japan.

Table 2. Spandex Fiber Producers

Company	Process	Trade name
North America		
Canada		
DuPont of Canada, Ltd.	solution dry spinning	Lycra
United States		
E. I. du Pont de Nemours & Co., Inc.	solution dry spinning	Lycra
Globe Manufacturing Co.	reaction spinning	Glospan and Clearspan
Western Europe		
Germany, Federal Republic of		
Farbenfabriken Bayer AG	solution dry spinning	Dorlastan
Italy		
Fillattice, S.p.A. (process licensed from Union Carbide Corp.)	solution wet spinning	Lynel
The Netherlands		
DuPont de Nemours, N.V.	solution dry spinning	Lycra
Spain		
Le Seda de Barcelona, S.A. (process originally licensed from Globe Manufacturing Co.)	reaction spinning	Enkaswing
United Kingdom		
Courtaulds, Ltd.	reaction spinning	Spanzelle
DuPont Company, Ltd.	solution dry spinning	Lycra
Eastern Europe		
USSR		
one plant licensed from and constructed by Toyobo Co., Ltd. of Japan	solution dry spinning	
Asia		
Japan		
Asahi Chemical Industry Co., Ltd.	solution dry spinning	Asahi Kasei Spandex
Fuji Spinning Co., Ltd. (process originally licensed from Ameliotex, Inc.)	solution wet spinning	Fujibo Spandex
Nisshin Spinning Co., Ltd	melt extrusion	Mobilon
Toyo Products Co., Ltd. (joint venture between E. I. du Pont de Nemours & Co., Inc. and Toray Industries, Inc.)	solution dry spinning	Opelon
Toyobo Co., Ltd.	solution dry spinning	Espa
South America		
Brazil		
DuPont Brazil, S.A.	solution dry spinning	Lycra

Economic Aspects

Except for some special purpose braids, no textile constructions consist entirely of elastomeric fibers. Use levels vary from a low of ca 5% in filling stretch cotton or cotton–polyester shirting fabrics (spandex core spun yarn in the fill) to a high of ca 40% for some warp-knitted tricot fabrics. The most commonly used raschel fabrics in the principal market area, foundation garments, contain ca 20–30% spandex.

Table 3 provides a summary of elastomeric fiber pricing in 1978 for the popular

Table 3. Elastomeric Fiber Pricing in the United States, 1978

Spandex		Extruded latex thread		
Tex[a]	Price[b], $/kg	Diameter, μm (gauge)	Approximate tex[a] equiv.	Price[c], $/kg
2.2	40.00	750 (34)	480	3.02
4.4	28.30	690 (37)	400	3.15
7.8	22.45	580 (44)	290	3.31
15.6	18.65	510 (50)	220	3.53
23.3	16.85	420 (60)	160	3.68
31.1	15.50	340 (75)	100	4.34
46.7	14.60			
62.2	12.35			
93.3	7.75			
124	5.95			
187	5.62			
249	5.40			

[a] To convert tex to den, divide by 0.111.
[b] On nonreturnable tubes.
[c] Unsupported multiend ribbons.

sizes. Cut rubber is not shown but prices, if adjusted for weight, are comparable to extruded latex thread. Raw material costs for spandex are on the order of $2.20–2.90/kg of finished product depending on the particular raw materials (eg, polyester or polyether glycol) used and conversion efficiencies. Similarly, rubber thread raw material costs are about $1.40–1.70/kg for both the cut and extruded processes. In comparing raw material costs to selling prices summarized in Table 3, it is obvious that the differences are much higher, especially for spandex, than for other chemical fibers such as nylon, polyester, acrylic, and acetate (see Fibers, chemical). Principal reasons for the higher differences are: the relatively small size of the total spandex market, which now totals worldwide ca 14,000 metric tons as compared to the multimillion ton levels enjoyed by the other chemical fibers, limits scale and thus efficiency of production units; and remaining technological problems limit maximum take-up speeds to much less than 1000 m/min and conversion efficiencies to under about 90%; and the relatively small number of producers continues to limit competition and allows higher gross profit margins.

In the early years of spandex commercial usage, many well-established, experienced synthetic fiber manufacturers projected near term markets on the order of 45,000 t. Such predictions were based on the following assumptions: (*1*) spandex would completely replace natural rubber; (*2*) major market areas for elastomeric fibers would continue to grow with the economy and population; (*3*) the superior performance characteristics of spandex would lead to many large volume end-use areas, eg, in outerwear, and shirtings, and sportswear; and (*4*) end-use areas for spandex staple fibers would develop rapidly.

Experience has shown that the first three assumptions were only partially correct and the fourth completely erroneous. Cost factors, improvements in rubber thread (eg, ribbon packaging and talc-free finishes) and adequate performance for specific areas have left considerable business to natural rubber thread. Dress styling and a freer life style have had a great negative effect on the use of foundation garments to the detriment of spandex growth. The rapid growth of lower cost and performance-adequate mechanical stretch yarns (eg, texturized nylons) have, instead of spandex, assumed a large volume potential in outerwear and sportswear areas, and the shift of consumer interest to wash-and-wear shirts retarded entry into shirting areas. After encountering many practical processing and construction problems in spandex staple research and development, all manufacturers have delayed introduction of a spandex staple fiber.

For the future, the original assumptions shown above may still be considered. Spandex raw material costs are not significantly higher than natural rubber fiber raw material cost. Cost differences between rubber and spandex can be expected to narrow, therefore, to the point where superior spandex properties will eventually mandate use of spandex over rubber in substantially all end-use areas. Despite the slow start, spandex is now beginning to penetrate sports and general outerwear areas and, as performance characteristics become more generally known, can be expected to grow at an increasing rate. Spandex staple is still commercially unknown but research and development work continues. Potentially, this area shows promise of large volume, as a staple spandex offers the possibility of making any staple yarn and staple yarn-based fabric elastomeric without change in the basic aesthetic qualities. In addition to the above, the Inmont Corporation has had a spandex spun-bonded fabric in semiworks-scale production for about three years that shows promise of broadening spandex use areas.

Uses

Cut Rubber and Extruded Latex. The manufacturing technology for cut and extruded rubber thread is much older and more widely known than that for spandex. Further production lines may be installed with a relatively modest capital investment. As a result, the manufacture of rubber thread is fragmented and more widely distributed with relatively few major and many minor producers. On a worldwide basis, Fillattice, S.p.A., of Italy with modern extruded latex plants in Italy, Spain, Malaysia, and the United States, is the largest producer of rubber thread. Second in production capacity is the Globe Manufacturing Co., Fall River, Mass., with production operations in the United States and England. Both of these firms also produce spandex (see Table 2).

Finer size extruded latex thread is generally wrapped with two covering yarns: one in a clockwise (S) direction and one in a counterclockwise (Z) direction. This covering yarn serves to protect the easily abraided rubber thread and to limit its performance to a selected area of the stress–strain curve.

Spandex. Both spandex and natural rubber fibers may be used in any area where extendability with good recovery is required. Initially, spandex was introduced as a replacement for natural cut rubber and extruded latex thread. Although spandex can indeed replace rubber with at least equivalent performance at lighter weight in almost all applications (excepting, primarily, golf balls where spandex is at a disadvantage owing to a slower recovery rate), such has not occurred to the degree originally expected mainly because of rubber's lower cost. Natural rubber fibers still find extensive use in narrow fabrics, braids, surgical hosiery, strip lace, and some circular knits. Spandex, however, has replaced almost all rubber in foundation garments and swimwear.

Spandex is processed into fabrics in three basic forms. These are: (*1*) bare (ie, as

Table 4. Spandex Uses

Spandex form	Fabric types	Uses
bare	warp knits, circular knits, narrow fabrics (wovens, knits, and braids), and hosiery (knit)	foundation garments, swimwear, control tops of panty hose, elastic gloves, waist and leg bands, and upholstery
covered	warp knits, circular knits, and hosiery (knits)	support hosiery, foundation garments, elastic bandages, sportswear, and upholstery
core-spun	wovens, circular knits, and men's hosiery (knits)	shirting, brassieres, and sportswear

supplied by the maker on tubes or beams without further processing), (*2*) covered (ie, with one or, generally, two covering yarns, and (*3*) core-spun yarns in which a staple fiber such as cotton or cotton-polyester is spun around a fine tex spandex core. Use areas for the various forms are not clearly defined and overlap in many cases. A general summary is provided in Table 4 (see Textiles).

Spandex is sold in tex as low as 2.2 (20 den), whereas the finest diameter (gauge) at which extruded latex thread has been supplied is 140 μm (180) equivalent to ca 16 tex (140 den). The availability of spandex in such fine size and its unique properties compared to rubber (eg, dyeability, high modulus, abrasion resistance, and resistance to degradation) have led to new use areas (eg, support hosiery, lightweight circular and warp-knitted fabrics, core-spun yarns for woven and knitted fabrics, and elastic surgical dressings).

For textiles, cut rubber thread is used primarily in braids and narrow fabrics and as cut ribbons for waist, leg, and arm bands. Cut rubber is also used in all wound-core the same areas as cut rubber.

BIBLIOGRAPHY

"Spandex" in *ECT* 2nd ed., Vol. 18, pp. 614–633, by Robert A. Gregg, Uniroyal, Inc.

1. E. W. H. Becker, R. Hontz, and W. Watkins, *Ind. Eng. Chem.* **40,** 875 (1948).
2. A. Nishimura and H. Komagata, *J. Macromol. Sci. A.* **1,** 617 (1967).
3. S. Melamed and R. Minton, "Anidex Fibers" in N. M. Bikales, ed., *Encyclopedia of Polymer Science and Technology,* Vol. 15, Wiley-Interscience, New York, 1971, pp. 161–168.
4. Ger. Pat. 826,641 (Jan. 3, 1952), E. Windemuth (to Farbenfabriken Bayer).
5. Ger. Pat. 886,766 (1951), W. Brenschede (to Farbenfabriken Bayer).
6. *Textile Fibers Products Identification Act U.S. Public Law 85-897,* U.S. Federal Trade Commission, Washington, D.C., effective Mar. 3, 1960.
7. S. Cannon, G. McKenna, and W. Statton *J. Polym. Sci. Macromol. Rev.* **II,** 209 (1976).
8. R. Dunbar, W. Nunning, and D. Martin *Lenzinger Ber.* **43,** 96 (1977).
9. A. Bruner, N. Boe, and P. Byrne, *J. Elastoplast.* **5,** 201 (Oct. 1973).
10. J. Saunders and co-workers, *J. Appl. Polym. Sci.* **19,** 1387 (1975).
11. Jpn. Pat. 75 95,516 (July 30, 1975), T. Hidaka, S. Mizutani, and A. Tsuchimoto (to Toray Industries, Inc.).

12. Jpn. Pat. 75 138,124 (Nov. 4, 1975), T. Hidaka, K. Ikawa, and S. Mizutani (to Toray Industries, Inc.).
13. Jpn. Pat. 78 49,121 (May 4, 1978), T. Takeka, S. Mizutani, and M. Shirido (to Toray Industries, Inc.).
14. Jpn. Pat. 78 52,718 (May 13, 1978), T. Hidaka and co-workers (to Toray Industries, Inc.).
15. D. Allport and A. Mohajer in D. Allport and W. H. Janes, eds., *Block Copolymers,* John Wiley & Sons, Inc., New York, 1973, pp. 443–492.
16. R. Bonart, *Angew. Makromol. Chem.* **58–59**(1), 259 (1977).
17. H. Oertel, *Chem. Ztg.* **98,** 344 (1974); *Chemiefasern Text. Anwendungstech. Text. Ind.* **27,** 1090, 1095 (1977); *ibid.,* **28,** 44 (1978).
18. A. Bleijenberg and co-workers, *Br. Polym. J.* **4**(2), 125 (1972).
19. E. Hicks, Jr., A. Ultee, and J. Drougas, *Science* **147,** 373 (1965).
20. Jpn. Pat. 72 39,717 (Dec. 8, 1972), K. Kosonoi, K. Ichikawa, and Y. Nakahara (to Asahi Chemical Industry Co., Ltd.).
21. U.S. Pat. 2,953,839 (Sept. 27, 1960), R. Kohan, D. Slovin, and F. Bliven (to U.S. Rubber Co.).
22. U.S. Pat. 4,002,711 (Jan 11, 1977), T. V. Peters.
23. Jpn. Pat. 76 04,313 (Jan. 14, 1976), Y. Ikeda, T. Hirukawa, and Y. Ishiki (to Fuji Spinning Co., Ltd.).

TIMOTHY V. PETERS
Consultant

FIBERS, MAN-MADE. See Fibers, chemical.

FIBERS, VEGETABLE

Natural fibers of vegetable origin are constituted of cellulose (qv), a polymeric substance made from glucose molecules, bound to lignin (qv) and associated with varying amounts of other natural materials. A small number of a vast array of these vegetable fibers have industrial importance for use in textiles (qv), cordage, brushes and mats, and paper (qv) products. Cotton fiber makes up about three-quarters of the world vegetable-fiber tonnage (see Cotton).

Vegetable fibers are conveniently classified according to the part of the plant where they occur and from which they are extracted as shown in Table 1 (1–2): (*1*) leaf fibers are obtained from the leaves of monocotyledonous plants, mostly tropical, which are part of their fibro-vascular systems. Commercially important examples are abaca (Manila hemp), sisal, and henequen. These are the hard fibers used for cordage. The long, multicelled fibers do not readily split apart and are wire-like in texture. (*2*) Bast fibers are obtained from the bast tissue or bark of the plant stem. The group includes flax, hemp, jute, and ramie which are the so-called soft fibers that are converted into textiles, thread, yarn, and twine. The long, multicelled fibers in this case can be readily split into finer cells which are manufactured into textile and coarse yarns. (*3*) Seed-hair fibers, eg, cotton, kapok, and the flosses, are obtained from seeds, seedpods, and the inner walls of the fruit. These fibers are short and single-celled. (*4*) Miscellaneous

Table 1. Selected Vegetable Fibers of Commercial Interest[a]

Commercial name	Botanical name	Geographical source	Use
Leaf (hard) fibers			
abaca	*Musa textilis*	Borneo, Philippines, Sumatra	cordage
cantala	*Agave cantala*	Philippines, Indonesia	cordage
caroa	*Neoglaziovia variegata*	Brazil	cordage, coarse textiles
henequen	*Agave fourcroydes*	Australia, Cuba, Mexico	cordage, coarse textiles
istle (generic)	*Agave* (various species)	Mexico	cordage, coarse textiles
Mauritius	*Furcraea gigantea*	Brazil, Mauritius, Venezuela, tropics	cordage, coarse textiles
phormium	*Phormium tenax*	Argentina, Chile, New Zealand	cordage
bowstring hemp	*Sansevieria* (entire genus)	Africa, Asia, South America	cordage
sisal	*Agave sisalana*	Haiti, Java, Mexico, South Africa	cordage
Bast (soft) fibers			
China jute	*Abutilon theophrasti*	China	cordage, coarse textiles
flax	*Linum usitatissimum*	north and south temperate zones	textiles, threads
hemp	*Cannabis sativa*	*all temperate zones*	cordage, oakum
jute	*Corchorus capsularis; C. olitorius*	India	cordage, coarse textiles
kenaf	*Hibiscus cannabinus*	India, Iran, USSR, South America	coarse textiles
ramie	*Boehmeira nivea*	China, Japan, United States	textiles
roselle	*Hibiscus sabdarifa*	Brazil, Indonesia (Java)	cordage, coarse textiles
sunn	*Crotalaria juncea*	India	cordage, coarse textiles
cadillo	*Urena lobata*	Zaire, Brazil	cordage, coarse textiles
Seed-hair fibers			
cotton	*Gossypium* sp.	United States, Asia, Africa	all grades of textiles, cordage
kapok	*Ceiba pentranda*	tropics	stuffing
Miscellaneous fibers			
broom root (roots)	*Muhlenbergia macroura*	Mexico	brooms, brushes
coir (coconut husk fiber)	*Cocos nucifera*	tropics	cordage, brushes
crin vegetal (palm leaf segments)	*Chamaerops humilis*	North Africa	stuffing
piassava (palm leaf base fiber)	*Attalea funifera*	Brazil	cordage, brushes

[a] Ref. 2.

fibers are obtained from the sheathing leaf-stalks of palms, leaves, stem segments, stems, and fibrous husks, eg, piassava and coir. These strawlike, woody, and coarse fibers are used for brush and broom bristles, matting, and stuffing.

Other classifications are applied according to different points of view such as commercial (hard and soft fibers), practical (textile, cordage, brushes and mats, stuffing and upholstery, papermaking and baskets, etc) (3), bast fiber crops for fine textiles, packaging, and soft cordage (4), morphological (hairs, bast, leaves, woody, and others) (3), and crop cultivation (cotton, jute, sisal, flax and hemp; and other fiber crops) (5) (see also Pulp; Bagasse; Sugar; Wood). Leaf fibers are also classed as structural fibers (6).

Cordage and fabrics were important to ancient man for fishing and trapping and applying motive power (7–8). The making of ropes and cords started in the paleolithic age (ca 20,000 BC); a mesolithic cave drawing (Spain) shows ropes partway down a cliff to recover honey (9). Evidence of predynastic Egyptian (ca 4000 BC) use of ropes and cords from reeds, grasses, and flax has been found (10). The vegetable fibers mostly used for rope were date palm fibers and papyrus. Matting was another important use of vegetable fibers in ancient Egypt, rushes and reeds, as well as papyrus grasses bound with flax string were used. Brushes from vegetable fibers were also used in ancient Egypt.

Flax, jute, ramie and sedges, rushes, and grasses have a long history of use in textiles and basketing.

Properties

Chemical Composition. Chemical analyses or compositions of various vegetable fibers are given in Tables 2 and 3. Cotton is at the high end of the range in cellulose content with >90%; jute and kenaf (bast fibers) and abaca (leaf fiber) have a relatively low cellulose content of 63–65%. The chemical compositions vary greatly between plants and within specific fibers depending on genetic characteristics, the part of the plant used and growth, harvesting, and preparation conditions.

Fiber Dimensions. Leaf fibers are multicelled, and not readily split into the component cells, whereas the bast fibers are readily broken down, permitting spinning. Strand and cell dimensions are given in Table 4. The cells themselves are composed of microfibrils which are comprised of groups of parallel cellulose chains.

Physical Properties. Mechanical properties of certain vegetable fibers are shown in Tables 5 and 6. The vegetable fibers are stronger but less extensible than cotton, ie, they have a higher breaking length and elasticity modulus with a lower strain (extensibility) and work modulus. The vegetable fibers approach glass in stiffness (resistance to deformation) and are considerably stiffer than man-made fibers, but have lower toughness (ability to absorb work) (see Fibers, chemical) (2). Kapok and other seed-hair fibers are relatively low in strength but have great buoyancy.

Vegetable fibers have spiral molecules that are highly parallel to one another. The spiral angles are low in flax, ramie, hemp, and other bast fibers, considerably less than cotton, accounting for the former fibers' low extensibility (12).

Table 2. Chemical Analysis of Vegetable Fibers[a], wt %

Fiber	Cellulose[b]	Moisture	Ash	Lignin[c] and pectins	Extractives
Leaf fibers					
abaca	63.72	11.83	1.02	21.83	1.6
bowstring hemp	69.7	9.7	0.7	13.7	6.2
caroa	60	10		12	18
cebu maguey (cantala)	75.8	5.5	1.4	14.1	3.2
henequen	77.6	4.6	1.1	13.1	3.6
phormium	63	11.61	0.63	23.07	1.69
piteira (Mauritius)	75.6	5.2	1.6	17.2	0.4
sisal	77.2	6.2	1.0	14.5	1.1
tula istle	73.48	5.6	1.65	17.37	1.9
Bast fibers					
Congo jute (cadillo)	75.3	7.7	1.8	13.5	1.4
hemp	77.07	8.76	0.82	9.31	4.04
jute	63.24	9.93	0.68	24.41	1.42
kenaf	65.7	9.8	1.0	21.6	1.9
ramie	91			0.65	
sunn	80.4	9.6	0.6	6.4	3.0
Seed and hair fibers					
cotton	90	8.0	1.0	0.5	0.5

[a] Ref. 2.
[b] See Cellulose.
[c] See Lignin.

Table 3. Chemical Composition of Various Vegetable Fibers[a], wt %

Type of fiber	Cellulose	Hemi-cellulose	Pectins	Lignin	Water-soluble compounds	Fats[b] and waxes[c]
cotton	91.8	6.4			1.1	0.7
flax (retted)	71.2	18.6	2.0	2.2	4.3	1.7
flax (nonretted)	62.8	17.1	4.2	2.8	11.6	1.5
Italian hemp	74.4	17.9	0.9	3.7	2.3	0.8
jute	71.5	13.4	0.2	13.1	1.2	0.6
Manila	70.2	21.8	0.6	5.7	1.5	0.2
New Zealand cotton	50.1	33.4	0.8	12.4	2.4	0.1
ramie	76.2	14.6	2.1	0.7	6.1	0.3
sisal	73.1	13.3	0.9	11.0	1.4	0.3

[a] Ref. 11.
[b] See Fats and fatty oils.
[c] See Waxes.

Plant Descriptions, Preparations, and Grades of Fibers

The vegetable fibers are grouped below partly according to their plant source and partly based on the importance of use within source groups (3–5,13).

Table 4. Dimensions of Fibers and Cells of Vegetable Fibers[a]

	Cells or ultimate fibers						Fibers or strands	
	Length, mm			Diameter, mm			Length, cm	Width, mm
Fiber	Min	Max	Av	Min	Max	Av	Range	Range
cotton	10	50	25	0.014	0.021	0.019	1.5–5.6	0.012–0.025
flax	8	69	32	0.008	0.031	0.019	20–140	0.04–0.62
hemp	5	55	25	0.013	0.041	0.025	100–300	
ramie	60	250	120	0.017	0.064	0.040	10–180	0.06–9.04
sunn	2	11	7	0.013	0.061	0.031		
jute	0.75	6	2.5	0.005	0.025	0.018	150–360	0.03–0.14
sisal	0.8	7.5	3	0.007	0.047	0.021	75–120	0.01–0.28
Manila (abaca)	2	12	6	0.010	0.032	0.024	180–340	0.01–0.28
phormium	2	11	6	0.005	0.025	0.013		
Mauritius	2	6		0.015	0.024			
kapok	15	30	19	0.010	0.030	0.018		
coir	0.3	1	0.7	0.010	0.024	0.020		
kenaf	2	11	3.3	0.013	0.034	0.023		
pineapple fiber	2	10	5.5	0.003	0.013	0.006		
nettle	4	70	38	0.020	0.070	0.042		
sansevieria	1	7	4	0.013	0.040	0.022		

[a] Ref 2.

Table 5. Mechanical Properties of Vegetable Fibers[a]

Fiber	Breaking length, km[b]	Ultimate strain at break, %[c]	Work of rupture modulus, kN·m/N	Initial linear elasticity modulus, N/tex[d]
kapok	16–30	1.2	0.1	1300
jute	27–53	1.5	0.3	1700–1800
hemp	38–62	2–4	0.6–0.9	180
flax	24–70	2–3	0.9	1800–2000
ramie	32–67	2–7	1.1	1400–1600
henequen	27–34	3.5–5		
manila hemp	32–69	2–4.5	0.6	
sisal	30–45	2–3	0.7–0.8	2500–2600
coconut	18	16	1.6	430

[a] Ref. 11.
[b] Tensile strength is reported on the basis of breaking length which measures strength per unit area.
[c] Elongation (stretch) at rupture.
[d] To convert N/tex to g/den, multiply by 11.33.

LEAF (HARD) FIBERS

Abaca. This fiber is extracted from the plant *Musa textilis*, native to the Republic of the Philippines. The fiber is also known as Manila fiber or Manila hemp. When mature, the individual abaca plant consists of 12 to 30 stalks radiating from a central root system. In appearance it is very similar to the banana plant (*Musa sapientum*). Each of the stalks is 2.7–6.7-m tall with a trunk 10–20-cm wide at the base. The sheaths forming the stalk expand into the overhanging leaf structure. The sheaths

Table 6. Coarseness and Breaking Length of Leaf Fibers[a]

Fiber	Coarseness, km/kg	Breaking length, km
Manila hemp		
mean	32	41.5
sisal	40	35.0
cantala	58	20.5
henequen	32	20.0
New Zeland flax	38	26.0
istle	34	27.0

[a] Ref 11.

before expanding are 2–4-m long, 13–20-wide, and ca 10-mm thick at the center. The fibers run lengthwise in the sheaths. The sheaths vary in length and width, and also in color. The mature stalks are cut off at the roots and at a point just below where they begin to expand. In the Philippines, the stripping method is generally used, ie, the fiber layer from the cut sheath is separated by inserting a knife just under the layer and a fiber strip or tuxy, 5–7-cm wide, is pulled off the sheath. Modern plantations have replaced this hand-pulling operation with power-driven pulling machines to accelerate production. After stripping, the fiber is air-dried; rain and sunlight need to be avoided to preserve color and luster.

A small proportion of Philippine abaca fiber is produced by mechanical decortication. The sheaths are fed into a decorticating machine in which the pulpy material is scraped from the fiber, then the fiber is washed and dried. The abaca fiber so produced is lower in quality than hand-cleaned fiber, ie, it has less sheen and is harsher in texture. Abaca fiber grown in Sumatra (Indonesia) and Central America is machine-decorticated exclusively.

The Philippines is the main source of abaca, contributing ca 90% of the total supply. The remainder is produced in Saba (North Borneo), Indonesia, and Central America. Philippine abaca is carefully graded by the government of the Philippines. The long cordage fiber is classified into four groups, determined by the fineness of the fiber and degree of separation of the individual fibers. Each group is further subdivided according to the range of color and associated fiber length. There are 18 grades, designated AB, CD, E, F, S2, S3, I, J1, G, H, H2, K, L1, L2, M1, M2, DL, and DM. These grades range in color from almost white to dark brown. Philippine abaca is longer and stronger than the corresponding grades produced elsewhere, and is generally lighter in color but bolder in texture.

Abaca is produced in two grades in Central America (introduced during World War II): clear and streaky. In Indonesia it is supplied in superior, good, fair, fair X, and B grades.

Sisal. This fiber is extracted from the leaves of the plant *Agave sisalana* which is widely cultivated in the Western Hemisphere, Africa, Asia, and Oceania. Other *Agave* sp. that furnish commercial vegetable fibers are henequen, cantala and istle (described below). However, only *A. sisalana* provides true sisal, which is further identified as African, Indonesian, Brazilian, and Bahama sisal.

The agaves have rosettes of fleshy leaves, usually long and narrow, which grow out from a central bud. As the leaves mature they gradually spread out horizontally and are 1–2-m long, 10–15-cm wide, and ca 6 mm thick at the center. The fibers are embedded longitudinally in the leaves, and are most abundant near the leaf surfaces. The leaves contain ca 90% moisture, but the fleshy pulp is very firm and the leaves are rigid. The fiber is removed when the leaves are cut because dry fibers adhere to the pulp. The fiber is removed by scraping away the pulpy material, generally by a mechanical decortication process.

In decortication, the leaves are fed through sets of fluted crushing rollers. The crushed leaves are held firmly at their centers and both ends are passed between pairs of metal drums on which blades are mounted to scrape away the pulp, and the centers are scraped in the same way. The fiber strands are washed and either air- or artificially dried. In Indonesia and Africa, for choice grades, the dried fiber is held against a revolving metal drum to remove remnants of dry adhering pulp.

Sisal fiber is graded according to the country and district of growth and further subgraded according to color, cleanliness, and length. There are eight Indonesian grades, designated, A, B, C, X, Y, X, D, and L; seven East African grades designated 1, 2, 3 long, 3 short, A, UG, and SCWF; five Haitian grades, designated A, B, X, Y, and S; eight West African grades, designated Extra, 1, 2, 2SL, 3 3L, A, and R; three Philippine grades, designated SR-1, SR-2, SR-3; and five grades of Brazilian sisal, designated Tipo 1, Tipo 3, Tipo 5, Tipo 7, and Tipo 9.

Henequen. This fiber is extracted from *Agave fourcryodes* and is also called Mexican henequen from its source, the state of Yucatan, Mexico. There are also El Salvador and Cuban henequens from *A. letonae*. These plants are harvested, decorticated, cleaned, and prepared for marketing by procedures similar to those used for sisal.

Mexican henequen is classified into seven grades according to color, cleanliness, and length. The grades are AA, A, B, B1, M, C, and M1.

Cantala. This variety of agave fiber is derived from *Agave cantala* and is also called Manila maguey from its Philippines source (also Indonesia). The fiber is prepared in the Philippines by retting the leaves in sea water followed by cleaning by hand or by decorticating as described for sisal. The Indonesian industry employs mechanical decortication without the retting step.

Catalpa is graded according to the method of decortication, retting, or machine decortication, into grades 1, 2, and 3.

Istle. This is the term for short, coarse fibers from the leaves of agaves and related plants growing wild in central and northern Mexico. There are three commercial varieties: tula istle from *A. lophanfu*, jaumave istle from *A. funhana*, and palma istle from *Samuela carnerosana*. The agave istle resembles small plants, whereas the palma istle looks like a small palm tree with leaves radiating from the top of the plant.

The istle fibers are extracted from the leaves by hand-scraping with subsequent sun-drying. The palma leaves are very gummy and are steamed before scraping.

Mauritius. This fiber is obtained from the giant cabuya plant, *Furcraea gigantea*, native to Brazil where it is called piteira or pita. It is cultivated commercially on the island of Mauritius and is known as Mauritius hemp or aloe. The cabuya plant resembles the agaves except that the leaves are heavier and larger.

The Mauritius fiber is extracted from the leaves by simple mechanical decortication by which the crushed leaves are scraped free from pulpy material, washed, steeped in a soapy solution and sun-dried.

There are five standard grades of Mauritius fiber: extra prime, prime, very good, good, and fair. Brazilian pita is graded according to standards for Brazilian sisal.

Phormium. This fiber is extracted from the plant *Phormium tenax*, native to New Zealand. The fiber is commonly called New Zealand flax or hemp, though it has no bast fiber characteristics. The plant is a perennial with a fan-shaped cluster of leaves 1.6–4.3-m long, and 6–10-cm wide. The leaves are green with a red midrib and red margins. The fibers are obtained by mechanical decortication of the cut leaves in a similar manner to sisal processing and then washed and sun-dried.

The phormium fibers are graded by New Zealand hard fiber standards according to color, strength and cleanliness and then subgraded into six classes: A, superior or superfine; B, fine; C, good; D, fair high-point, fair low-point; E, common; and F, rejections.

Sansevieria. This fiber is extracted from the perennial plant *Sansevieria* native to Africa, Arabia, India, and Ceylon. The fiber is also called bowstring hemp because of its primitive bowstring application. The plant grows wild or is cultivated in many countries. It is propagated with cultivated leaf cuttings. Extraction by mechanical decortication is similar to sisal processing.

Caroa. This leaf fiber is obtained from the plant *Neoglaziovia variegata*, which belongs to the pineapple family and is native to eastern and northern Brazil. The leaves for fiber production are collected from plants growing wild and are sword-shaped, 1–1.3-m long, and 2.5–5-cm wide. The fiber is extracted by hand-scraping after beating the leaves to break up the pulpy tissue, or after a retting process that partially ferments and softens the leaves. Caroa is graded by standards similar to those specified for Brazilian sisal. Its color and texture resemble sisal although it is apt to be incompletely cleaned.

Piassava. This fiber is extracted from *Attalea funifera*, a palm indigenous to Bahia, Brazil, where it grows wild and under cultivation. Bahia piassava and bass fiber are other terms. The plant is a tall feather palm with erect leaves 12 m in length. The fiber is taken from the sheathing leaf vases. The leaf sheaths generally need to be treated in water to free the fibers from the pulp by scraping. The mature lower leaves also may be cut at the base, the leaves crushed, and the fibers combed out with a knife. The fibers are sometimes hand-pulled from the leaf stem. The piassava fiber is classified into premium and second grades.

Broomroot. This fiber is extracted from the roots of the bunchgrass, *Muhlenbergia macroura*, a plant growing wild from Texas to Central America but produced commercially in Mexico. The plant grows 1–2-m tall, although the fiber is obtained from the roots by beating or rubbing the chopped roots to remove the bark-like covering. The fibers are bleached with sulfur fumes at the processing factory and graded into four qualities.

BAST (SOFT) FIBERS

Fine Textile Fibers

Flax. This fiber is extracted from the plant *Linum usitatissimum* L. which is grown chiefly in the USSR, Poland, France, Rumania, Czechoslovakia, Belgium, and Ireland. The plant is cultivated mostly for its oil-bearing seed (linseed), although it is also an important source of a vegetable fiber (see Vegetable oils). Flax is an annual plant with a slender, greyish-green stem growing to a height of 90–120 cm.

Several methods of harvesting are used and the correct time for harvesting for highest fiber quality and seed production is important. The plants are pulled by hand or machine for highest yield and quality, although mowing is practiced for some grades. After deseeding, the straw is retted, ie, the fibers are liberated through enzymatic action on pectinous binding material in the stem; dew- or water-retting or variations are employed. In the commonly used dew-retting, the straw is spread thinly on the ground and subjected to atmospheric precipitation with turning at intervals. Fungi enter the stem to cause the retting action. Water-retting is performed on bundles of dried straw which are immersed in rivers, pools, or ditches. A modern variation is a controlled warm-water procedure and, more recently, aerated retting is being used for reducing pollution of the effluents, water usage, and odors. The retted straw is dried before the scutching or fiber separation and cleaning step. This breaking of the straw was first done by hand-beating, and eventually in turbine-type machines consisting of multiple pairs of corrugated roll breakers and scutching wheels for cleansing and polishing. The fiber is then hackled for alignment and final cleaning. The short fibers from these operations (tow) and the longer fibers (line) are used for linen.

The qualities considered in grading flax are fineness, softness, strength, density, color, uniformity, luster, length, handle, and cleanliness. The line fiber from sources in different countries varies in grades and buyers have their own grading systems.

Hemp. This fiber is extracted from the annual plant *Cannabis sativa* which originated in Central Asia. The plant grows readily in temperate and tropical climates, but its commercial production for fiber is chiefly in the People's Republic of China and eastern Europe.

The hemp plant is grown for fiber from the stem, for oil from the seeds or for drugs from the flowers or leaves. Marijuana (see Hypnotics, sedatives, and anticonvulsants) is the narcotic alkaloid (cannabin) derived from a related hemp plant, which is not a commercial fiber-producing plant. The Marijuana Tax Act of 1937, nevertheless, includes *C. sativa.*

When mature, the stalks are 5–7-m tall and 6–16-mm thick. They are generally smooth and without branches or foliage except at the top. The stem structure is hollow. Thin-walled tissue adjoins the hollow center and outside of this is a layer of woody substance. The next layer consists of gummy tissue that cements the fiber layer to the woody layer. The fiber or bast layer is enveloped by a thin bark that constitutes the outside of the stalk.

The mature hemp stalk is harvested at the proper time to ensure highest quality and yield by hand-cutting and spreading or by a harvester-spreader for dew-retting. When water-retting is used, bundles of straw are dried in the field. Subsequently the leaves and seeds are removed by beating the tops on the ground, and they are then stacked before the water-retting. Water-retting is done in Italy, Spain, Hungary, Poland, and to a small extent in the USSR. Dew-retting is used mostly in the USSR. The retted and dried straw is further treated by either hand-breaking and hackling to remove the woody stem portion (hurd) or by mechanical breaking followed by turbine scutching. A wooden breaking device augments the hand-breaking.

The grading is by color, luster, density, spinning quality, cleanliness, and strength. Quality varies with country and regions within countries. The national classifications may include a number of grades and subgrades.

Ramie. This fiber is produced from the stems of *Boehmeria nivea*, a nettle native to central and western China but growing in regions varying from temperate to tropical, including the People's Republic of China (largest grower), the Philippines, Brazil,

Taiwan, Japan, and others. Another name for the raw fiber is China grass. The plant grows 1–2-m high or higher with stems 8–16 mm in diameter. It has perennial roots and yields three crops annually.

The ramie fiber is contained in the bark and is usually extracted by hand stripping and scraping. The plant is cut green and defoliated manually and bast ribbons are stripped from the woody stem. These are scraped in the field or stored in water for scraping in a central location. In Brazil, the Republic of China (Taiwan), and other countries the ramie ribbons are decorticated mechanically in small operations. Sometimes the ribbons are bleached by sulfur fumes.

Ramie fibers are graded according to their length, color, and cleanliness. Each country has its own grading system.

The raw ramie fibers contain 25–35% plant gums (xylan and L-arabinan) and small quantities of parenchyma cells which must be removed before the fibers can be spun. Degumming by boiling in aqueous alkaline solutions is sometimes done under pressure and at times with additives, followed by washing, bleaching with an oxidizing agent (eg, a hypochlorite), washing, neutralizing, and oiling to facilitate combing.

Packaging Fibers

Jute. This fiber is obtained from two species of the annual, herbaceous plant *Corchorus, C. capsularis* of Indo-Burma origin and *C. olitorius* from Africa. The major jute production area is India and Bangladesh and secondarily in other Asian countries and Brazil. The two sources of jute are differentiated by their seeds and seed pods. The *C. capsularis* is round-podded and is called white jute; the *C. olitorius* is long-podded and is known as tossa or daisee. The jute plants of both species have cylindrical stalks 2–4-m tall and 10–20 mm in diameter.

The plants are harvested by hand at an early seed stage using knives and the stems are left on the ground several days to promote defoliation. The defoliated stems are bundled and taken for wet-retting in canals, ditches, or ponds (slowly running water to carry off colored and acidic products) for periods of 10–20 d. The retting time for *C. insularis* is shorter than for *C. olitorius*. The retting can also be conducted on ribbons in concrete tanks. Stripping of the fibers follows immediately after stem-retting with sun-drying; the ribbon-retted material also is sun-dried.

The jute is made into crude bales for transport to a baling and grading center. Here, the fiber is designated as to species (white, tossa, or daisee), and source and grade according to the grades being used. The jute is then baled for shipment to domestic users or for export.

Kenaf and Roselle. These fibers are extracted from the stems of two closely related plants, *Hibiscus cannabinus* L. and *H. sabdariffa* L. var. *altissima*, known respectively as kenaf, Deccan hemp or mesta, and roselle. The two plants are native to Africa but are grown commercially in greatest quantity in the People's Republic of China, the USSR and Egypt for the kenaf, and India and Thailand for the roselle. The plants are herbaceous annuals growing in single stems to heights of 1–4 m for kenaf and 1–5 m for roselle with stem diameters of 10 to 20 mm. Kenaf is harvested as flowering begins whereas with roselle this takes place somewhat earlier. The kenaf can be hand-cut or mowed or pulled; the roselle may be hand-pulled. Both species may be field-dried for defoliation. Hand or mechanical ribboning followed by tank-retting is used on occasion, but the kenaf is often stem-retted followed by hand or mechanical stripping and washing. The washed fiber is finally dried on racks.

Kenaf is graded into three grades: A, good; B, medium; C, poor. The grading is based on color, uniformity, strength, and cleanliness. Roselle may be hackled to improve quality by removing dirt and increasing softness and luster. In Thailand roselle is classified into seven grades as to softness, color, and impurities.

Urena. This jutelike fiber is obtained from *Urena lobata* L., commonly called urena with many local names because of its widespread growth, although it is grown commercially mostlv in Brazil and Zaire. This perennial shrub grows under cultivation as an annual with few branches and it grows to a height of 4–5 m with stems 12–18 mm in diameter. The stems are cut 20 cm above ground level because of the highly lignified base. The plants are defoliated green or after piling in the field; retting is similar to that for jute or kenaf and requires 8–10 d. The retted material is then stripped, washed, and sometimes hand-rubbed to remove impurities.

Urena fiber is officially graded in Brazil and Zaire into four or five grades according to color, uniformity of color, luster, strength, and cleanliness.

China Jute. This jute-type fiber is extracted from the stem of a mallow, *Abutilon theophrasti* Medic or *A. avecennae Gaetn.*, growing chiefly commercially in the People's Republic of China and the USSR, although also in Japan, Korea, and Argentina, and known as China jute or Tientsin jute, and by many other local names. The plant is a herbaceous annual growing to a height of 3–6 m with stems 10–18 mm in diameter. China jute is harvested by hand in the early flowering stage, the leaves are removed, and bundles are water-retted similar to jute operations. The harvesting, retting and fiber extraction may be mechanical in certain regions. In the hand operation the fiber is hackled, baled, and graded into two classes.

Soft Cordage Fiber

Sunn Hemp. This fiber is extracted from the stem of the legume, *Crotolaria juncea* L., indigenous to India and known as sunn hemp, brown hemp, Indian hemp, Benares hemp, and other national references. India is the largest producer followed by Bangladesh and Brazil. Sunn hemp is an erect herbaceous shrub growing to a height of 1–5 m with a thin, cylindrical, branch-free stem. The plant is harvested in the seed-pod stage by hand-cutting or pulling and left in the field for defoliation. The stems are water-retted, washed, stripped, and dried by methods similar to those for stem fibers described previously. For export, sunn-hemp fibers are hackled or dressed and graded by a system with many classifications according to length, strength, firmness, color, uniformity, and percentage of extraneous matter.

Seed and Fruit-Hair Fibers

Coir. This fiber is contained in the fruit or husk of the coconut palm *Cocos nucifera* L., which is positioned beneath the outer covering of the fruit and envelops the kernel or coconut. Sri Lanka(Ceylon) and India are important producers. The fruits are gathered just short of ripeness and the husks are broken by hand or by use of a bursting machine. The main supply of fiber is obtained as a residue for copra production. The extraction of the fiber involves retting at the edges of rivers and also in pits or, in modern operations, in concrete tanks. The retted husks are beaten with sticks to remove extraneous matter and the dried fiber is suitable for spinning. Rougher fibers require less retting and the fibers are extracted from the husks mechanically. The fibers may be washed before drying and may be hackled before direct use or baling for shipment. The fiber can also be removed from the husk by a decorticating machine in connection with copra production.

Kapok. This fiber is obtained from the seed pod of a tree, *Ceiba pentranda*. The tree grows to a height of ca 35 m, and is indigenous to Africa and Southeast Asia; it is also known as the silk-cotton tree. The fiber is exported from Thailand, India, and Indonesia. The pods are picked short of opening, hulled or broken open, and the floss is dried in an area enclosed with cloth or wire netting to contain the floss. The seeds are removed by hand or separated mechanically and further cleaned by an air separation method. The fiber is finally baled for shipment. The seeds contain up to 25% of a nondrying oil resembling cottonseed oil in properties and uses.

Miscellaneous. A number of seed-hair fibers have some minor uses, including East Indian balsa fiber (*Ochroma pyramdale*), Indian kumbi (*Cochlospermum gossypium*), American milkweed floss (*Asclepias* sp.) and cattail fiber (*Typha* sp.).

Production

The world production of jute and hard fibers is given in Table 7 for the period 1963 to 1976 (14).

The growth since 1970 has been at a zero or negative rate.

The distribution for the production of the major hard cordage fibers in 1976 is shown in Table 8.

U.S. imports of unmanufactured vegetable fibers have shown serious decreases in the period of 1963 to 1976 (Table 9).

The consumption of paper pulps made from vegetable fibers in the United States in 1972 is summarized in Table 10 with sources of pulp and types of paper products.

In addition, 216,000 metric tons of U.S. sugar cane bagasse pulp were consumed in making wall board and fine papers in 1972.

Vegetable Fiber Yields. The yields of prepared vegetable fibers are generally low, 1–6%, as shown in Table 11. Linen flax fiber and kapok are exceptions.

Table 7. World Production of Jute and Hard Fibers in Thousands of Metric Tons

	1963	1970	1976
jute	2421	2051	2034
abaca	119	93	67
sisal-henequen	826	763	628

Table 8. Percentage Distribution of Production of Major Hard Cordage, 1976[a]

	Thousands of metric tons	%
abaca	84.0	13
sisal	421.3	66
henequen	130.0	21
Total	*635.3*	*100*

[a] Ref. 15.

Table 9. U.S. Imports of Unmanufactured Vegetable Fibers[a]

	1963	1970	1975
	Metric tons		
flax (Belgium, Luxembourg)	2,246	1,581	411
hemp (Canada)	35	0	0
jute and jute butts (Thailand, Bangladesh)	78,811	20,625	20,953
kapok (Thailand)	12,157	11,703	7,499
abaca (Philippines, Ecuador)	28,174	21,492	22,239
sisal and henequen (Brazil, Tanzania, Kenya, Mozambique, Mexico)	92,448	59,677	14,465
istle or tampico (Mexico)	7,074	29	0
others[b]	4,753	7,551	12,655
Total	*225,698*	*122,658*	*78,222*

[a] Ref. 14.
[b] Crin vegetal, phormium, broom root, rice straw, coir, sunn.

Table 10. U.S. Consumption of Nonwood Paper Pulps in 1978[a]

Vegetable fiber	Source	Pulp consumption, t	Paper products
cotton linters	United States	100,000	rayon and fine papers
rags, textile residues	United States	70,000	absorbent felt and fine papers
seed flax straw	United States, Europe	60,000	cigarette and lightweight papers
abaca	United States (old ropes), Philippines	25,000	specialty papers, including tea bags
miscellaneous		30,000	
Total		*285,000*	

[a] Ref. 16.

Table 11. Yields of Prepared Vegetable Fibers[a]

Plant	Fiber form	Yield[b], %	Plant	Fiber form	Yield[b], %
Leaf fibers					
abaca	decorticated	2–3	hemp	dry-line fiber	3.5
sisal	decorticated	3[c]		dry tow	1.0
henequen	decorticated	2–3	ramie	decorticated	3.5
Seed-hair fiber			jute	dry-retted fiber	6.0
kapok		17[d]	kenaf	dry-retted fiber	4.8
Bast fibers			roselle	dry-retted fiber	4.4
flax	scutched line fiber	12–16	sunn hemp	dry-retted fiber	3.4
	scutched clean tow	4–8			

[a] Ref. 3–5.
[b] Green plant basis.
[c] Leaf basis.
[d] Pod basis (600 pods per tree).

Economic Aspects

Prices. The prices of vegetable fibers for cordage and noncordage uses have varied widely over the years depending on the supply and demand, world economic conditions, weather, and starting in the 1960s, competition from man-made fibers particularly polypropylene fibers (see Olefin fibers). In 1970, polypropylene had a price of ca $400/t on the U.S. market and was projected to fall by 1980 to $275/t, making vegetable fibers uncompetitive. In 1978, however, polypropylene had risen to $1200/t because of the oil situation (17–18) (see also Olefin polymers).

The 1961–62 prices for unprocessed fibers imported into the United States for the hard fibers, abaca and sisal, were $33.00 and $17.82/100 kg, respectively. Recent prices for these fibers and a synthetic fiber are shown in Table 12.

Projections. Vegetable fibers, excluding cotton and flax, have been relatively unimportant in the total fiber supply and will become less important in relation to converted cellulosic and noncellulosic fibers as projected in Table 13.

Estimated world demands for hard fibers, as shown in Table 14, is expected to remain static (zero growth) for ropes, cables, nets and other cordage, paper and padding, and other uses.

Cordage goods from vegetable fibers accounted for 87% and synthetics for 13% of the total of 21,000 t estimated for the U.S. market in 1964–1966. The vegetable fiber

Table 12. Recent Prices for Various Fibers[a]

Fiber	Grade	Date	Price, U.S. $/100 kg	Market
abaca	Davao I	av 1977	44.80	ex (export) ship, New York
		Dec. 1978	37.90	
sisal	British East Africa No. 1	av 1977	102.30	landed, New York
		Oct. 1978	102.80	
jute	raw Bangladesh White C	av 1977	46.90	CIF (cost, insurance, and freight),
		Oct. 1978	48.90	United Kingdom
flax	water-retted	av 1977	185.30	United Kingdom
		Sept. 1978	215.80	
polyester		av 1977	123.50	
fiber[b]		July 1978	110.90	wholesale, United States

[a] Ref. 18.
[b] See Polyester fibers.

Table 13. Percentage of Total Fiber Supply in the United States[a]

	Cotton	Wool	Rayon and acetate	Non-cellulosic	Other[b]
1950	64	6.7	25.3	3.5	0.2
1960	55.9	4.4	19.5	19.5	0.3
1980 (estd)	46	2.5	10.8	40.5	0.2
2000 (estd)	44	1.9	5.4	48.6	0.1

[a] Ref. 19.
[b] Including sisal, jute, kapok, abaca.

Table 14. Estimated World Demand for Hard Fibers, 1000 Metric Tons[a]

	1964–66	1968–70	1980
sisal-henequen	772	765	695
abaca	120	95	95
Total	*892*	*860*	*790*

[a] Ref. 20.

share slipped to 71% of a total of 927,000 t in 1968–1970 and is projected to drop to 60% of a total of 965,000 t in 1980. The influence of higher oil prices on the cost of the polypropylene resins used in the synthetic cordages may well change the percentages and trend shown above.

Although there was a zero or negative annual growth rate for nonwood plant fibers in U.S. papermaking in the 1970s, it is expected that these fibers will hold their small percentage of under 1% of the total U.S. pulp consumption through the year 2000 (21).

Uses

Information on the broad end uses of vegetable fibers is given in Table 1 and consumptions by general products are discussed subsequently. Uses by certain categories are described as follows (1–3,5,22).

Textiles and Woven Goods. Clothing, sacks and bags, canvas and sailcloth, and fabrics are made from the bast fibers flax, hemp, ramie, jute, kenaf, roselle, and nettle. Coarse sacks, coffee and sugar bags, floor coverings, and webbing are made from the leaf fibers sisal, henequen, abaca, and Mauritius hemp. Phormium, nettle, and cantala also are used in coarse bagging.

Cordage and Twines. Industrial ropes, hoisting and drilling cables, nets, and agricultural twines are made from abaca, sisal, and henequen. Abaca serves particularly for hawsers and ship cables. Cantala, phormium, and Mauritius hemp are also used for cordage. Hemp, flax, and jute are used for string and yarns, and ramie is used for thread. Hemp and kenaf are used for nets as well.

Brushes. Fibers suitable for bristles in scrubbing and scraping brushes and brooms include coir, piassava, istle and broomwort.

Stuffing and Upholstery Materials. Stuffing for mattresses and pillows and furniture is made from kapok, crin vegetal, sisal tow, and flosses. Kapok is also used in life preservers and coir for door mats.

Paper. In addition to the production of bond and fine writing papers from cotton and linen textile residues and roofing felt from low-grade rags, heavy-duty, multiwalled bags for industrial packaging of flour, cement, chemicals, fertilizers, and hardware are made containing salvaged vegetable fibers or residues from primary product manufacturing of sisal, jute, abaca hemp, henequen, phormium, caroa, and Mauritius hemp. Other specialty papers using these fibers are abrasive papers, and gaskets. Abaca fibers are used specifically in tea bag paper and flax tow and sunn in cigarette paper. Esparto (*Stipa tenacissima*) fibers are made into highest-grade printing papers. Kenaf is proposed for newsprint (see Paper).

Miscellaneous. Tie material, basketry and furniture are made from raffia (*Raphia raffia*), a palm leaf segment, and rattan (*Calamus* sp.), a stem fiber.

BIBLIOGRAPHY

"Fibers, Vegetable" in *ECT* 1st ed., Vol. 6, pp. 467–476, by David Himmelfarb, Boston Naval Shipyard; "Fibers, Vegetable" in *ECT* 2nd ed., Vol. 9, pp. 171–185 by David Himmelfarb, Boston Naval Shipyard.

1. J. Cook, *Handbook of Textile Fibers,* 4th ed., Morrow Publishing Co., Ltd. Watford, Hertz, England, 1968.
2. M. Harris, *Handbook of Textile Fibers,* Harris Research Laboratories, Inc., Washington, D.C., 1954.
3. R. H. Kirby, *Vegetable Fibers,* Interscience Publishers, Inc., New York, 1943.
4. J. M. Dempsey, *Fiber Crops,* The University Presses of Florida, Gainesville, 1975.
5. J. Berger, *The World's Major Fibre Crops: Their Cultivation and Manuring,* Centre d'Etude de l'Azote, Conzett and Huber, Zurich, 1969.
6. W. Von Bergen and W. Krauss, *Textile Fiber Atlas,* American Wool Handbook Company, New York, 1942.
7. K. R. Gilbert, "Rope Making" in C. Singer, ed., *History of Technology,* Vol. I, Oxford University Press at Clarendon, 1954.
8. J. Grant, "A Note on the Materials of Ancient Textiles and Baskets," in ref. 7.
9. G. E. Linton, *Natural and Man-made Textile Fibers,* Duel, Sloan and Pearce, New York, 1966.
10. J. H. Harris, *Ancient Egyptian Materials and Industry,* Edward Arnold Publishers Ltd., London, 1962.
11. T. Zylinski, *Fiber Science,* Office of Technical Services, U.S. Dept. of Commerce, Washington, D.C., 1964.
12. W. E. Morton and J. W. S. Hearle, *Physical Properties of Textile Fibers,* 2nd ed., John Wiley & Sons, Inc., New York, 1975.
13. "Wood, Paper, Textiles, Plastics and Photographic Materials" in *Chemical Technology: An Encyclopedic Treatment,* Vol. VI, Barnes and Noble Books, 1973.
14. *U.S.D.A. Agricultural Statistics,* U.S. Government Printing Office, Washington, D.C., 1977.
15. H. Jiler, ed., *1977 Commodity Yearbook, Hard Fibers,* Commodity Research Bureaus, Inc., New York, 1977.
16. J. E. Atchison, *An Update on Utilization of Nonwood Plant Fibers, Recent Developments and Some Observations on a Visit to China,* paper presented at the 1979 Pulp and Fiber Fall Seminar, American Paper Institute, Charleston, S.C., Nov. 4, 1979.
17. "General Outlook" in *FAO Agricultural Commodity Projections 1970–1980,* Vol. I, Rome, 1971, pp. 264–2822.
18. *FAO Monthly Bulletin of Agricultural Economics and Statistics* **1**(12), 45, 47, 49 (Dec. 1978).
19. G. W. Thomas, S. E. Curl, and W. F. Bennet, *Food and Fiber for a Changing World,* The Interstate Printers and Publishers, Inc., Danville, Ill. 1976.
20. R. Grilli, *The Future of Hard Fibers and Competition from Synthetics,* World Bank Staff, The John Hopkins University Press, Baltimore, 1975.
21. J. N. McGovern, *Non-Wood Plant Fiber Pulping,* TAPPI, CAR Report No. 67, 1976.
22. T. F. Clark in R. G. Macdonald, ed., *Pulp and Paper Manufacture,* Vol. II, McGraw-Hill Book Company, New York, 1969.

General References

J. E. Atchison, "Agricultural Residues and Other Nonwood Plant Fibers," *Science* **191,** 768 (1976).

D. Lapedes, ed., *Encyclopedia of Science and Technology,* Vol. 5, McGraw-Hill Book Company, New York, 1977.

G. E. Linton, *THE Modern Textile and Apparel Dictionary,* 4th ed., Textile Book Service, Plainfield, N.J., 1973.

JOHN N. MCGOVERN
The University of Wisconsin

FIBRINOGEN. See Blood fractionation.

FIELD-ASSISTED SEALING. See Electrostatic sealing.

FILLERS

A filler is a finely divided solid that is added to a liquid, semisolid, or solid composition to modify the composition's properties and reduce its cost. Fillers can constitute either a major or a minor part of a composition. The structure of filler particles can range from irregular masses to precise geometrical forms such as spheres, polyhedrons, or short fibers. Fillers are used for nondecorative purposes, although they may incidentally impart color or opacity to a composition. Although additives that supply bulk to drugs, cosmetics, and detergents are often referred to as fillers, they are more properly termed diluents, since their purpose is to adjust the dose or concentration of a composition, rather than to modify its properties or reduce its cost. Fibers and whiskers are not discussed here because they are generally regarded as reinforcements, not fillers (see Carbon; Composite materials; Fibers).

Although the use of simple diluents and adulterants almost certainly predates recorded history, the use of fillers to modify the properties of compositions can be traced at least as far back as early Roman times when artisans used ground marble in lime plaster, frescoes, and pozzolanic mortar. The use of fillers in paper and paper coatings made its appearance in the mid 19th century. Functional fillers (introducing new properties into a composition, rather than modifying preexisting properties) were commercially developed early in the 20th century when Goodrich added carbon black to rubber, and Baekeland formulated phenol–formaldehyde plastics with wood flour.

Fillers can be classified according to their source, function, composition, or morphology (see Table 1). None of these classification schemes are entirely adequate owing to overlap and ambiguity of their categories. Yet if fillers are to be systematically

Table 1. Classification of Fillers

By particle morphology	By composition	
Crystalline	*Inorganic*	*Organic*
fibers	carbonates	celluloses
platelets	fluorides	fatty acids
polyhedrons	hydroxides	lignins
irregular masses	metals	polyalkenes
Amorphous	oxides	polyamides
fibers	silicates	polyamines
flakes	sulfates	polyaromatics
solid spheres	sulfides	polyesters
hollow spheres		proteins
irregular masses		

studied, some method of classification is necessary. Here the morphological scheme is used for the discussion of general filler properties, and the compositional scheme for the compilation of data on specific fillers (see Table 8, p. 211).

Properties

General Properties. ***Particle Morphology, Size, and Distribution.*** Many fillers, such as diatomaceous earth, pumice, and calcium carbonate, have morphological characteristics that allow these materials to be identified microscopically with great accuracy, even in a single particle. Photomicrographs, descriptions, and other aids to particle identification can be found in ref. 1.

Filler particle morphology affects rheology and loading of filled compositions. When filled compositions flow, it becomes necessary for some adjacent particles to be lifted over, pushed under, or detoured around other particles. Thus, particles with fibrous, acicular (needlelike), or irregular morphology yield compositions more resistant to flow than compositions filled with spheres, polygons, or other fillers with regular morphologies. Similarly, fillers with regular shapes can be used in compositions at higher loadings than fillers with irregular shapes, all other properties being equal.

Because of the diversity of filler particle shapes, it is difficult to clearly express particle size values in terms of a particle dimension such as length or diameter. Therefore, the particle size of fillers is usually expressed as a theoretical dimension: the *equivalent spherical diameter,* the diameter of a sphere having the same volume as the particle. The degree to which a particle diverges from sphericity can be expressed as its *sphericity coefficient* ψ.

$$\psi = \frac{\text{surface area of sphere with same volume as particle}}{\text{surface area of particle}}$$

Devices for measuring the particle size of fillers usually assign particles to one of several size ranges. The values of these ranges are then collected and ordered to yield size distributions which are frequently plotted as cumulative weight percent vs diameter. However, collected as the number of particles in size ranges, the data can be converted to a weight distribution by the following equation:

$$W_i = \frac{N_i \bar{d}_i^3}{\sum_i N_i \bar{d}_i^3}$$

where W_i = weight fraction of particles in a size range; N_i = number of particles in this size range; and $\bar{d}_i$ = mean diameter for the size range.

Particle size distributions can be described in a concise manner by values that indicate properties or tendencies of the distribution, such as the mean, mode, or median diameter, and the spread, or standard deviation:

$$\sigma w = \left[\frac{\sum_i W_i d_i^2}{\sum_i W_i} - \bar{d}w^2 \right]^{1/2}$$

However, standard deviation only applies to normal distributions that seem to occur only for narrow size ranges of classified particles. Most particle size distributions are skewed toward their larger particle size classifications (2).

The size distribution of fillers whose particles are larger than ca 40 μm and which have at least moderate sphericity can be conveniently measured by sieving. For fillers whose particles range from 4 to 40 μm, the Coulter technique is preferred. Filler particle size distributions smaller than 4 μm are measured by microscopy, sedimentation, permeametry, or radiation-scattering methods (2–3) (see Size measurement of particles).

Surface Area and Energy. Surface area is the available area of fillers, be it on the surface or in cracks, crevices, and pores. The values obtained from different methods for measuring the surface area of a given filler may vary significantly. Despite these variations, surface area is a very important property. It helps determine the ease of dispersion, the rheology, and the optimum loading of filled compositions.

The external surface of the filler can be calculated from its equivalent spherical diameter and its sphericity coefficient, but this method provides little information on the actual area of the filler that influences physical and chemical processes. In practice, surface area is usually determined (4) from the quantity of a gas that adsorbs in a monolayer on the surface of the filler, ie, the monolayer capacity D_m. From this value the specific surface area can be determined (5), which is an area per unit weight and is usually expressed in m^2/g.

At the phase boundary between a filler particle and a liquid composition, there can be an energy barrier to the liquid's wetting the filler, or the filler's dispersing in liquid. This energy barrier, often termed the free surface energy of the system, results from each phase's greater affinity for molecules of its own substance rather than those of the other phase.

The free surface energy of fillers is a function of surface area and composition. It can be derived from the contact angle of liquids on filler surfaces, the heat of immersion of fillers and liquids, or by adsorption methods (6). Many commercial fillers are surface-coated or treated to modify their free surface energy.

Loading and Packing. The relative amounts of filler and matrix (binder) in a filled composition are usually termed the loading and are always expressed quantitatively; however, the quantitative indexes vary from industry to industry. Formulators in the plastics industry use phr = parts filler per 100 parts resin, % wt = weight percent filler, and % vol = volume percent filler. In the paint industry, pvc = volume percent pigment (filler) in the dry paint film, or pv = the volume ratio of pigment to binder are commonly used. In the paper industry, filler loading is usually expressed as ar = the weight of adhesives per 100 parts by weight of filler.

The optimum degree of filler loading in a composition is usually determined by balancing the functional physical properties of the overall composition against its cost over a range of loadings. This determination can be greatly aided by the use of predictive models based on the packing geometry of filler particles. Some experimental determinations are always necessary because of error introduced by simplifying assumptions in the models but, at the very least, the models can be used to establish minimum and maximum values for filler loading.

For large amounts of fillers, the maximum possible loading with known particle size distributions is estimated. This method (7) assumes that the interstices (voids) between particles are occupied by smaller particles, and that the voids between these particles are occupied by still smaller particles. Thus, a wide filler particle size distribution results in minimum void volume or maximum packing. To get from maximum packing to maximum loading, it is only necessary to express maximum loading in terms of the minimum amount of binder that fills the interstitial voids and becomes adsorbed on the surface of the filler. This adsorption value can be estimated from the surface area of the filler or determined empirically by an oil adsorption test (ASTM D 281 or equivalent).

For small amounts of filler, eg, in order to alter the rheology of the composition, the same methods used to estimate maximum packing and loading can be used to estimate minimum packing and loading. For instance, if maximum loading is obtained with a filler with a wide particle size distribution and a low surface area, then a narrow particle size distribution and a high surface area yields a filled system at minimum loading.

In modern industrial practice, compositions frequently contain pigments, reinforcements, and other materials in addition to fillers. Hence, more complex models are needed to predict the effect of filler loading. Optimum filler loading for paints and filled paper can be calculated (8). An excellent discussion of filler loading in reinforced plastics is given in ref. 9.

Functional Properties. The functional properties of fillers are those that fillers impart to compositions to enhance their performance or economic utility. Although the properties that are required by compositions vary from one application to another, a given physical or chemical property of the filler may or may not be functional and depends on the requirements of the application in which it is used. A quantification of functional properties per unit cost, therefore, provides a valid criterion for filler comparison and selection.

Specific Gravity. In applications where fillers are used in a liquid system and it is necessary to control the volume and/or the weight of the filled composition, the specific gravity (true density) of the filler is one of its functional properties. The specific gravity of fillers composed of relatively large, nonporous, spherical particles is usually determined with an air comparison pycnometer (ASTM D 2840). The specific gravity of finely divided, porous, or irregular fillers is determined with a liquid pycnometer (ASTM D 153).

Bulk Density. Bulk density (apparent density) refers to the total amount of space occupied by a given weight of a dry powder. It includes the volume of the particles plus the volume of the interparticle voids of the powder. Bulk density is a functional property of fillers in powdered compositions, and it also has to be considered in the storage and shipping of fillers.

A reliable method for determining bulk density, such as ASTM D 1895, distinguishes the loose bulk density from the tamped bulk density so that allowances can be made for the compaction of the powder during shipping or processing.

Optical Reflectance. In applications where esthetic appeal is important, fillers with controlled optical reflectance (color and brightness) preserve the color uniformity of the finished composition. Reflectance is usually determined by instrumental methods such as tristimulus green filter reflectivity (ASTM D 97) (see Color).

Refractive Index. The refractive index of a filler must be controlled in order to maximize or minimize its opacifying effect on the composition (10). Microscopic methods of determination are available (11).

Free Moisture. The free moisture of fillers is water present on the surface of the particles where it can contribute to interparticle bonding (reinforcing) or filler–matrix interaction (binder adsorption or catalysis). A standard method for the determination is given in ASTM D 280.

Thermal Stability. Fillers for high temperature applications must not decompose under conditions of use. Even in low temperature applications, thermal stability of fillers can be an important functional property if the composition is molded or otherwise subjected to heat during its processing. Determination of thermal stability varies from industry to industry depending on conditions. The paper industry uses ASTM D 776, the paint industry ASTM D 2485, and the adhesives industry ASTM C 792.

Thermal Expansion. In order to preserve the structural strength of filled compositions over a wide temperature range, all the components of the composition should have similar coefficients of thermal expansion. This prevents the stress-induced damage that occurs as components expand at different rates. A typical method for quantifying the expansion of solid fillers is ASTM D 176.

Applications

Elastomers. In the ANSI/ASTM standard D 1566, a filler is defined as a solid compounding material, usually in finely divided form, which may be added in relatively large proportions to a polymer for technical or economic reasons. However, at present in the rubber industry these materials are invariably termed pigments, even though they are almost never used solely as colorants. The use of the term is probably owing to the long record of the industry of compounding with pigment-grade carbon black (see Elastomers).

Most rubber technologists distinguish reinforcing pigments from nonreinforcing or inert pigments but would agree that these terms are loosely defined and often used only in a comparative context. The standard definition (ASTM D 1566) for reinforcing agents is a material, not basically involved in the vulcanization process, used in rubber to increase the resistance of the vulcanizate to mechanical forces. In practice, these materials impart abrasion resistance, tear resistance, tensile strength, and stiffness to an elastomer after vulcanization. Whether or not a particular pigment is reinforcing or nonreinforcing often depends on its composition, its physical form, and the type of elastomer in which it is compounded. For instance, carbon black in isoprene rubbers and pyrogenic silica in silicone rubbers are more active reinforcing agents in these respective elastomers than other pigments with equivalent particle size distributions. In general, however, most pigments become increasingly reinforcing as their particle size diminishes.

The nomenclature used here to identify grades of carbon black is in conformance with ASTM D 2516 (see Carbon, carbon black). In this system, carbon black grades are identified by four letters: the first one classifies the cure rate as slow (S) or normal (N), the second one is a digit classifying the typical particle size, and the last two characters are arbitrarily assigned digits. The particle size range indicated by the second digit of the identifying codes is given in Table 2. The ASTM abbreviations used for elastomers are given in Table 3, and Table 4 lists typical elastomer fillers and their uses.

Since most fabricated elastomer products contain 10–50 vol % of filler, their physical properties and processing characteristics depend to a great extent upon the

Table 2. Carbon Black Particle-Size Identification Numbers

Size number	Average particle size, nm
0	1–10
1	11–19
2	20–25
3	26–30
4	31–39
5	40–48
6	49–60
7	61–100
8	101–200
9	201–500

Table 3. Elastomer Abbreviations

Elastomer	ASTM abbreviation
acrylate–butadiene	ABR
butadiene	BR
chlorinated polyethylene	CM
chloroprene	CR
ethylene–propylene–diene	EPDM
ethylene–propylene	EPM
isobutylene–isoprene	IIR
synthetic isoprene	IR
nitrile–butadiene	NBR
nitrile–chloroprene	NCR
nitrile–isobutylene–isoprene	NIR
natural isoprene	NR
pyridine–butadiene	PBR
styrene–butadiene	SBR
styrene–chloroprene	SCR
styrene–isoprene	SIR

nature and quantity of the fillers. The rubber technologist has to manipulate the formula so as to optimize a large number of properties and keep costs down.

Latex Stock. *Activation.* In rubber technology, an activator increases the effectiveness of an accelerator, ie, a material that accelerates vulcanization of an elastomer. Zinc oxide is an effective activator in NR, IR, and other isoprene rubbers. It apparently reacts with a fatty acid in the rubber or with an added fatty acid to form a soluble zinc salt which in turn reacts with the accelerator to form a zinc–accelerator complex. This complex reacts during vulcanization to release nascent sulfur which rapidly vulcanizes the elastomer (12). Magnesium oxide is an effective activator in CM, CR, NCR, and other chlorinated elastomers. Magnesium oxide seems to control the pH during vulcanization and act as an acceptor for the chlorine which is liberated during vulcanization (13).

Recovery (Nerve). Uncured latex stock tends to recover its previous shape after being rolled or extruded during processing. This tendency can result in processing difficulties and reduce dimensional accuracy of finished products. The recovery of

Table 4. Typical Elastomer Fillers and Their Uses

Filler	Specific gravity	Typical compatible elastomers	Uses
alumina	2.7	NR, CR, SRs	hose, mats
asbestos	2.4	NR, SRs	mats, tile
barium sulfate	4.0–4.5	NR, CR, SRs	O-rings, belts
carbon blacks	1–2.3		
S212		IR, NR	tires
S315		IR, NR	cable jackets
N110		IR, NR	tires, pads
N219		IR, NR	tire carcasses
N220		NR, SBR	tire treads
N231		NR, IR, SBR	tire treads
N234		NR, SBR	tire treads
N242		NR, BR	tire treads
N293		NR, SRs	belts
N326		NR, IR	tire carcasses
N356		NR, SRs	tire retreads
N375		NR, SRs	tire treads
N472		NR, SRs	electrical goods
N539		NR, SRs	tire sidewalls
N550		NR, SRs	extruded goods
N660		BR, NR	tubes
N762		NR, SRs	foot wear
N774		NR, SRs	tire carcasses
N990		CR, EPDM	extruded goods
calcium carbonate	2.7–2.9	NR, SRs	footwear, mats
clay, hard	2.6	NR, SBR, EPM, EPDM	flooring, footwear
soft	2.6		molded goods
mica	2.82	NR, SR	molded goods
resins	1.1–1.3	NBR, CR, NR, SBR	footwear, coatings
silicas			
colloidal sol	1.3	NR, SBR	sponge
diatomaceous	2.0–2.35	NR, IRR	carpet backing
novaculite	2.65	NR, SRs	molded goods
wet process	1.93–2.2	IIR, CR, NBR, NR	hygienic goods
pyrogenic	2.2	silicone rubber	molds, electrical goods
surface treated		NR, SRs	specialty goods
talc	2.7–2.9	NR, CR, IIR, EPM	molded goods
natural materials[a]	1.0–1.1	NR, SBR, CR	tape, extruded goods
wood and shell flour	0.9–1.6	CR, NR, SRs	footwear

[a] Such as chlorinated or vulcanized vegetable oils, or rice hulls.

latex stock can be reduced by adding large-particle-size fillers, or fillers with a high degree of particle aggregation (agglomerated or structured fillers), such as carbon black or pyrogenic silica, or by increasing the loading of other fillers.

Tack causes two layers of material to adhere when they are pressed together. This property can be reduced by employing fillers with finer particle size distributions, or by dusting the stock with a laminar filler such as mica.

Retardation (*Scorch Resistance*). Scorch is the premature vulcanization of an elastomer during processing. The tendency of an elastomer to scorch can be reduced by compounding it with a filler with a lower pH or a larger particle size. The type of filler also has an effect on the tendency of a latex stock to scorch: highly reinforcing fillers, such as carbon black, silica, and zinc oxide, can promote scorching, whereas most clay fillers tend to retard scorching.

Vulcanized Elastomers. *Abrasion resistance* is the reduced tendency of a rubber particle to undergo surface attrition when subjected to a frictional force. The abrasion resistance of rubbers is a function of both filler type and form. The following fillers are listed in the order of their decreasing effectiveness in preventing abrasion: carbon black $\gg$ silica $\gg$ calcium silicate > zinc oxide > clay > calcium carbonate. Although, in general, abrasion resistance of fillers increases with increasing sphericity and decreasing particle size, fillers should be compared by evaluating their abrasion resistance using samples prepared at equal volume loadings and by a procedure such as ASTM methods D 1630 or D 2228.

Elongation is the extension produced by a tensile stress applied to an elastomer. It is reduced by reinforcing fillers and fibrous fillers such as wood flour, cotton flock, asbestos, or wollastonite. Regardless of what type filler is used, elongation decreases with increasing loading.

Hardness. The resistance of a fabricated rubber article to indentation, its hardness (qv), is influenced by the amount and shapes of its fillers. High loadings increase hardness. Fillers in the form of platelets or flakes, such as clays, impart greater hardness to elastomers than other particle shapes at equivalent loadings.

Modulus is the tensile force required to stretch a uniform cross section of an elastomer to a given elongation. Reinforcing fillers result in higher-modulus composites than nonreinforcing fillers at equivalent loadings. In general, modulus increases with decreasing particle size of fillers.

Permanent Set. When an elastomer is stretched and then allowed to relax, it will not completely recover its original dimensions. This divergence from its original form is called its permanent set. It is principally affected by the affinity of the elastomer for the filler surface and is, therefore, primarily a function of the surface energy of the filler.

Resilience is the ratio of energy output to energy input in a rapid full recovery of a deformed elastomer specimen. It is, therefore, a reliable index to the energy that is lost through internal friction of viscous flow. In general, fillers with the least effect on resilience are those that are least reinforcing. Zinc oxide is an exception because it has good resilience and also gives good reinforcement.

Tear Resistance. The resistance of an elastomer to tearing is affected by the particle size and the particle shape of the fillers it contains. It generally increases with decreasing particle size and increasing sphericity of fillers.

Tensile strength of a rubber compound is increased by fillers of small particle size and large surface area. For most fillers, tensile strength increases with loading to an optimum value after which it decreases with increased loading.

Paper. Paper is typically prepared by depositing cellulose pulp fibers on a continuous wire screen from a dilute water suspension (see Paper). Fillers or loading materials are finely divided solids that are incorporated into a paper sheet by adding them to the pulp slurry and then depositing them on the wire along with the pulp. Finely divided solids dispersed in water containing an adhesive and then coated on the paper after it is formed are usually termed pigments. Here the term filler is used for both loading materials and pigments. Table 5 lists typical paper fillers and their uses.

Table 5. Typical Paper Fillers and Their Uses

Filler	Specific gravity	Refractive index	Uses
alumina trihydrate	2.4	1.57	printing paper coating
barium sulfate	4.4	1.64	photographic paper
calcium carbonate	2.6	1.66	printing paper
calcium silicate	2.1	1.50	printing paper
clay	2.7	1.57	printing paper
polystyrene	1.0	1.59	printing paper coating
satin white[a]	1.5–1.8	1.4	art paper
silica, diatomite	2.0–2.4	1.4	paper board
precipitated	2.1	1.4	printing paper
sodium aluminosilicate	2.1	1.55	printing paper
talc	2.75	1.57	rotogravure paper
titanium dioxide[b]	3.9	2.55	bond, printing paper
zinc oxide	5.6	2.01	electrophotographic paper

[a] Calcium sulfoaluminates.
[b] Anatase.

Most paper used today in the graphic arts industry contains 1–40 wt % of filler. The optical and mechanical properties of filled paper are superior to those of unfilled paper.

Optical Properties. *Brightness* of paper can be defined as the reflectance of a layered stack of paper such that no change in reflectance occurs if the backing material is changed from black to white. The brightness of fillers is usually defined as the diffuse blue reflectance factor obtained by testing a pressed plaque of filler in accordance with TAPPI standard T 534 dm-76.

The ability of fillers to improve the paper brightness increases with their surface area and refractive indexes. It is maximum at an optimum filler particle size, about 0.25 μm in most cases, where the filler particle size is approximately one-half the wavelength of the light used for observation.

Gloss is the property of a surface to reflect light specularly. It is typically associated with such phenomena as shininess, highlights, and reflected images. The gloss of paper is usually quantified with a spectrophotometer and a standard procedure such as TAPPI T 442 su-72.

Paper gloss is influenced by the size and shape of filler particles at the surface. A roughness of the order of one fourth the wavelength of light can produce a perceptible reduction of specularly reflected light (14). Since each surface particle protrudes only partially above the surface, the average height of surface irregularities does not exceed one half of the particle diameter. Thus gloss reduction does not take place until filler particles exceed roughly 3 μm in diameter (15). Filler particles with a laminar, platelike shape increase gloss more effectively than spherical or irregular particles. This is because platelike particles tend to orient parallel to the surface of the sheet, thus reducing its roughness more than would be predicted for their particle size distribution.

Opacity of paper is a function of its light absorption and the amount of light scattering that occurs at the pigment–air, pigment–fiber, and fiber–air interfaces. Light scattering is the principal cause of opacity in paper containing white fillers, whereas in colored paper opacity depends on light absorption.

The filler properties that influence opacity are color, particle size, surface area, and refractive index. Dark pigments generally provide better opacity than light pigments because they absorb more light. Light scattering, like brightness, is a reflective phenomenon; the optimum particle size is again approximately one-half the wavelength of the light that is used for observation. The opacity of fillers increases with both their surface area and their refractive index.

Mechanical Properties. *Ink Absorbency.* Ink penetrates into paper during the printing process. The more the ink penetrates, the less glossy the print. The degree of ink absorption in paper can be controlled by the particle size and shape of the fillers. It is monitored by TAPPI method RC19.

Since the void volume of fillers decreases with decreasing particle size, ink absorbency can be reduced by using small particle fillers. Loading or coating paper with platelike fillers, such as clays, also reduces ink absorbency because these materials tend to overlap and thus reduce porosity (see Ink).

Retention. After filler is added to the cellulose pulp slurry in the course of paper making, it has to remain in the sheet when the pulp is deposited on the wire. If the filler is not well retained, the sheet has a two-sided character, ie, the wire side of the paper contains less filler than the top side. Poor filler retention increases costs, since the filtrate containing the lost filler (white water) may have to be processed to recycle the filler or upgrade the water quality for environmental reasons.

Retention is reduced by increasing filler solubility, specific gravity, and sphericity. It is increased by increasing the filler's particle size or by flocculating smaller particle size fillers.

Sheet strength of paper is measured as bursting strength (TAPPI T 403 os-76), internal tear resistance (TAPPI T 414 ts-65), tensile breaking properties (TAPPI T 494 os-70), internal bond strength (TAPPI T 506 su-68), and folding endurance (TAPPI T 511 su-69). Fillers generally reduce the paper sheet strength since they dilute the pulp and reduce the number of fiber-to-fiber interactions. However, this is not necessarily detrimental. It often significantly increases flexibility and stretchability and thus reduces the number of tears that occur when the paper is run on high-speed equipment (16).

Plastics. In the plastics industry, the term filler refers to particulate materials that are added to plastic resins in relatively large (over 5%) volume loadings (see Laminated and reinforced plastics).

At present, the performance specifications for most compounded plastic products are less strict than for other filled products, such as elastomers and paints, and cost reduction is the primary objective. Hence, plastic compounders tend to compound with the objective of optimizing properties at minimum cost rather than maximizing properties at optimum cost. Table 6 lists typical plastic fillers and their uses.

Resins. *Curing.* Fillers with high surface areas can retard the curing of plastics by adsorbing catalysts and/or promoters. Other fillers can accelerate curing if they contain active sites or trace quantities of catalytic materials. Low density fillers cause high temperatures during curing and subsequent long cool-down periods because of their insulating effect. High loadings of dense solid fillers can appreciably increase the curing time needed to fully cure thermosetting compounds.

Viscosity of resin–filler mixtures is affected by particle shape, loading, and degree of dispersion. It decreases with increasing sphericity and degree of dispersion but increases with increasing loading.

Table 6. Typical Plastic Fillers and Their Uses

Filler	Specific gravity	Typical compatible resins	Uses
alumina trihydrate	2.4	polyesters	fire resistant filler
carbon blacks	0.2–2.3	epoxies	electrical goods
calcium carbonate			
mineral	2.7	most resins	tile, molded goods
precipitated	2.6	most resins	pipe, putty
clays	2.6	most resins	flooring tile
feldspar	2.6	thermoplastics	plastisols
metals[a]	2.5–11.5	epoxies	radiation shields, solders
mica	2.82	most resins	sheet molded goods
polymers[a]			
solid spheres	1.1–1.3	most resins	molded goods
hollow spheres	0.2–0.5	thermosets	molded goods
silica			
diatomite	2.0–2.4	polyethylene	films
novaculite	2.65	thermosets	electrical goods
quartz flour	2.65	thermosets	molded goods
tripolite	2.65	thermosets	molded goods
wet process	1.9–2.2	thermoplastics	sheets, films
vitreous	2.18	epoxies	electrical goods
silicate glass			
solid spheres	2.5	most resins	molded goods
hollow spheres	0.22	thermosets	molded goods
flakes	2.01	thermosets	electrical goods
talc	2.7–2.9	PVC, polyalkenes	extruded and molded goods
wood and shell flour	0.19–1.6	most resins	molded goods

[a] See Table 8.

Cured Plastics. *Tensile strength* of filled plastic compositions is affected by filler particle shape and size, size distribution, surface area, and interfacial bonding. In general, it increases with the decreasing sphericity of its fillers. At equivalent volume loadings, small filler particle sizes and narrow size distributions give better tensile strengths than larger particle sizes and broad size distributions. Higher filler surface areas and, in general, stronger filler-to-matrix bonding also result in higher tensile strength compositions. Tensile strength of plastics is measured by ASTM D 638-76.

Compressive strength of filled composition is normally governed by the strength of the weakest component of the system, be it filler, matrix, or the bond between filler and matrix. Thus weak compressive fillers, such as celluloses, reduce the compressive strength of composites, whereas the reverse is true for strong, rigid fillers such as mineral oxides. Compressive strength of plastic composites is measured by ASTM D 621-64.

Fire Resistance. Many fillers, particularly inorganic oxides, are noncombustible and provide a measure of passive fire resistance to filled plastics since they reduce the volume of combustible matter in the filled composition and, depending on their density, may also serve as insulation (see Flame retardants). Fillers that contain combined water and carbon dioxide, such as alumina trihydrate or dawsonite

[*12011-76-6*], increase fire resistance by liberating noncombustible gases when they are heated. These gases withdraw heat from the plastic and can also reduce the oxygen concentration of the air surrounding the composition.

Electrical Resistance. Most fillers are mainly composed of nonconducting substances that should, therefore, provide electrical resistance properties comparable to the plastics in which they are used. However, some fillers contain adsorbed water or other conductive impurities that can greatly reduce electrical resistance. Some of the standard tests for the electrical resistance of filled plastics are dielectric strength (ASTM D 149), dielectric constant (ASTM D 150), arc resistance (ASTM D 495), and d-c resistance (ASTM D 257) (see Insulation, electric).

Paint. The liquid phase of paint formulations, usually termed the *vehicle,* typically contains volatile and nonvolatile fractions. When paint is applied, the volatile fraction of the vehicle evaporates and the nonvolatile fraction polymerizes to become a film matrix in which prime pigments and fillers are embedded (see Paint). Prime pigments are coloring and opacifying pigments. Fillers in paints are variously referred to as inerts, extender pigments, and supplemental pigments. Table 7 lists typical paint fillers and their uses. Compared to prime pigments, fillers have lower refractive indexes and tinting strength and cost less (see Pigments).

Fillers contribute to application, durability, protection, and decoration. The properties that they impart to paints are similar to the optical and mechanical properties they impart to paper.

Optical Properties. *Gloss.* As in paper technology, gloss is the spectrally reflected light typically associated with such phenomena as shininess, highlights, and reflected images. It is usually evaluated by comparing a paint sample with a highly polished, black glass surface (ASTM D 523-67).

The degree of gloss of the painted surface depends mainly on the smoothness of the reflecting surface. For this reason, filler size, shape, and surface energy influence the glossiness of filled paints. High gloss paints are formulated with fillers and pigments that have small particle dimensions. An average particle diameter of less than 0.3 μm is necessary to provide high gloss surfaces (17). Low-gloss paint films are obtained by using fillers with a large average particle size. As was the case in paper, laminar, platelike fillers increase the gloss of paints more effectively than spherical or irregularly shaped fillers of equivalent particle size. Pigments that are completely wetted by the vehicle permit higher gloss than pigments that are incompletely wetted (18).

Table 7. Typical Paint Fillers and Their Uses

Filler	Refractive index	Use
barium sulfate	1.6	sanding primers
calcium carbonate	1.6	most paints
clay	1.6	exterior paints
mica	1.6	exterior paints
silica, diatomite	1.5	traffic paints
precipitated	1.5	flatted paints
talc	1.6	exterior paints
zinc oxide	2.0	exterior paints

Hiding Power. Like opacity in paper, the hiding power of fillers in paint is determined by the difference between their index of refraction and the index of refraction of their surrounding medium. The indexes of refraction of common paint vehicles are ca 1.5. Air has a refractive index of 1.0. Since most fillers that are used in paint have refractive indexes of from 1.4–1.7, it is readily apparent that they contribute little to the hiding power of paint when they are completely embedded in the vehicle. However, if filler particles extend above the vehicle film into the air, they contribute hiding power. Hence, hiding power can be expressed by wet hiding power (ASTM D 2805-70) and dry hiding power (ASTM D 344). Dry hiding and high gloss are mutually exclusive.

Mechanical Properties. *Stain resistance* of paints is directly related to their porosity. Therefore, fillers that reduce porosity, ie, with low surface area, wide particle size distribution, and laminar particle shape, contribute to stain resistance. Stain resistance of paints is evaluated by ASTM D 3023.

Viscosity. The ease with which paint is applied and the uniformity of the paint film depends largely on the paint viscosity. If the viscosity is too high, paint does not level and, therefore, brush, roller, or spray marks result. On the other hand, if a paint has too little viscosity, it sags or runs. High surface area (high vehicle demand), a high degree of filler agglomeration, and low sphericity contribute to paint viscosity. The viscosity of filled paints is evaluated by ASTM D 1200 and ASTM D 562.

Weathering Resistance. The principal causes of paint weathering are moisture, abrasion, uv radiation, and microorganisms. Fillers that tend to yield porous coatings increase water resistance since water damage occurs most often when water becomes trapped between the substrate and the paint film and blistering results. Hard pigments such as silica and clay contribute abrasion resistance to paints. Since uv radiation oxidizes prime pigments and paint vehicles in the presence of oxygen and water vapor, fillers that absorb uv radiation and limit the paint porosity reduce this type of degradation. Alkaline fillers, such as zinc oxide and calcium carbonate, inhibit the growth of mildew and other microorganisms.

Economic Aspects

Sales of fillers in the United States was estimated at 380×10^6 dollars for 1977, an increase of 48% over 1972 (19). During 1978 the industry experienced a growth in excess of inflation of 3–4%. Mineral fillers represent over 80% of the filler market, primarily because of their lower cost. They have been upgraded by the introduction of new grinding, beneficiation, and surface-treatment technologies. Thus mineral fillers can meet the competition of the highly engineered synthetic fillers in many markets. On the other hand, synthetic fillers have grown during the 1970s at a rate far above the 7%/yr average growth rate for fillers in general (20).

Table 8 gives 1979 price information on specific fillers, including some physical properties and manufacturing processes.

Projections. The market for fillers has been projected to reach 3,000,000 metric tons at a value of 901×10^6 dollars by 1986 (20). This projection is based on the increased cost of petroleum-derived matrix materials which they will replace.

Elastomers. World markets for tires and nontire products are predicted to grow over 50% by 1987 (21). However, carbon black is forecast to grow only 2–3% annually during the same period. This lag is based on the fact that the feedstock for carbon black is decant oil, which is also the source of fuel oil.

Table 8. Specific Fillers

Fillers	CAS Registry Number	Manufacturing process	Specific gravity	Refractive index	Oil adsorption, wt %	Approximate price, 1979, $/kg
Inorganics						
carbonates						
barium carbonate	*[513-77-9]*	precipitated	4.43	1.6	14	0.30
calcium carbonate						
precipitated	*[471-34-1]*	precipitated	2.65–2.68	1.63	15–65	0.11
limestone	*[1317-65-3]*	mined	2.71	1.60	10.5–15	0.02–0.04
whiting	*[13397-26-7]*	wet ground	2.71	1.60	6.5–15.0	0.04–0.06
magnesium carbonate	*[546-93-0]*	precipitated	2.20	1.5–1.71	80	0.48
fluorides						
calcium fluoride	*[7789-75-5]*	mined	3.18	1.43		0.24
sodium aluminum fluoride	*[15096-52-3]*	precipitated	2.95	1.3		0.55
hydroxides						
aluminum hydroxide	*[21645-51-2]*	precipitated	2.4	1.58		0.15
metals						
aluminum	*[7429-90-5]*	atomized	2.55			3.08
bronze	*[12597-70-5]*	reduced	8.0			3.30
lead	*[7439-92-1]*	atomized	11.4			1.32
zinc	*[7440-66-6]*	atomized	7.0			0.99
oxides						
aluminum oxide	*[1344-28-1]*	furnace	3.8	1.6	13	0.29
magnesium oxide	*[1309-48-4]*	calcined	3.5		70	0.33
silicon dioxide	*[7631-86-9]*					
colloidal sol		ion exchange	1.3	1.4		
diatomaceous		mined	2.65	1.4	120	0.09
novaculite		mined	2.65	1.55	20	0.11
pyrogenic		oxidation	2.2	1.4	150	3.41
quartz flour		ground	2.65	1.55	32	0.04
tripolite		mined	2.65	1.55	31	0.06
vitreous		furnace	2.18	1.4	15	0.55
wet process		precipitated	1.9–2.2	1.4	160	0.35

Table 8 (*continued*)

Fillers	CAS Registry Number	Manufacturing process	Specific gravity	Refractive index	Oil adsorption, wt %	Approximate price, 1979, $/kg
titanium dioxide	[*1317-70-0*]	oxidation	3.9	2.55	24	0.99
zinc oxide	[*1314-13-2*]	oxidation	5.6	2.0	13	0.88
silicates						
asbestos	[*1343-90-4*]	air flotation	2.56	1.57	34	13
clay						
kaolin	[*1327-36-2*]	mined	2.60	1.56	36	3
calcined kaolin	[*1327-36-2*]	calcined	2.63	1.62	25	17
calcium silicate	[*1344-96-3*]	precipitated	2.33	1.4	375	15
feldspar		mined	2.6	1.53	19	6
glass						
ground glass		milled	2.5	1.5	15	11
flakes		milled film	2.5	1.5		176
hollow spheres		heat expanded	0.2	1.5	35	143
solid spheres		furnace	2.5	1.5	30	44
mica	[*1327-44-2*]					
muscovite		mined	2.75	1.59	47	26
phlogopite		mined	2.75	1.60	24	26
vermiculite		heat expanded	2.25	1.60		33
nepheline	[*14797-52-5*]	mined	2.6	1.6	20	6
perlite		heat expanded	0.17			33
pyrophyllite	[*12269-78-2*]	mined	2.85	1.59	36	6
talc	[*14807-96-6*]	mined	2.85	1.59	27	4
wollastonite	[*13983-17-0*]	mined	2.9	1.59	26	6

sulfates						
barium sulfate						
barytes	*[13462-86-7]*	mined	4.47	1.64	6	15
blanc fixe	*[7727-43-7]*	precipitated	4.35	1.64	14	75
calcium sulfate						
gypsum	*[7778-18-9]*	mined	2.35	1.53	21	4
anhydrite	*[14798-04-0]*	mined	2.95	1.59	25	6
precipitated	*[23296-15-3]*	precipitated	2.95	1.59	50	15
sulfides						
lithopone	*[8006-32-4]*	precipitated	4.2	1.8	14	68
zinc sulfide	*[12402-34-5]*	precipitated	4.0	2.37	13	66
Organics						
cellulose						
cork	*[61789-98-8]*	ground	0.25			1.21
corn cob		cut	1.2			0.22
flock		ground				0.55
shell flour		ground	1.4		15	0.13
wood flour		ground	0.65		25	0.11
fatty acids and esters						
sulfur-chlorinated vegetable oil		sulfur chlorination	1.04			1.54
vulcanized vegetable oil		vulcanized	1.0			1.54
polymers						
polystyrene	*[9003-53-6]*	FR catalysis	1.05	1.59		1.21
phenol–formaldehyde	*[9003-35-4]*	condensation	1.1			1.21
mineral rubber		petroleum residue	1.0			1.21

Paper. Throughout the 1970s writing, printing, and coated papers enjoyed an average annual growth rate of 4–5.2%. This growth rate is projected to decline through the 1980s to 2.5–3.0% (22). Fillers marketed to the paper industry are expected to grow faster than the industry itself because fillers enjoy an ever-increasing cost advantage over TiO_2, and at the same time improve brightness.

Plastics. Historically, sales of plastics have grown at about twice the growth rate of sales of fillers (20). The future growth of fillers for plastics, however, is expected to outstrip the plastics industry, at least on a volume basis because increased filler-to-resin ratio accompanies resin price increases owing to the increased cost of petrochemical feedstocks.

Paint. The sales of fillers to the paint market parallels industry trends. From the late 1960s to the late 1970s the industry grew at about 2%/yr, and it is forecast to grow at the rate of 2.8%/yr throughout the 1980s (23). Although the cost of paint vehicles will undoubtedly increase in the future, paint manufacturers will not respond by increasing their ratio of filler to vehicle because vehicle loading in paints is determined primarily by performance rather than economic considerations.

Health and Safety Factors

The major hazard involved in the handling and use of fillers is the inhalation of airborne particles in the respirable size range (less than 10 μm aerodynamic diameter). At the present time, filler dusts are classified into carcinogens, fibrogens, and nuisance particulates. Nuisance particulates are dusts that have a long history of little adverse affect on lungs and do not produce significant organic disease or toxic effect when exposures are kept under reasonable control.

The American Conference of Governmental Industrial Hygienists established TLVs for the airborne concentration of many fillers in workroom air (24). These range from a proposed 0.5 fiber/cm^3 for amosite asbestos (a carcinogen) through 10 mg/m^3 divided by the percent respirable quartz + 2 for dust containing quartz (a fibrogen) up to ca 1.1×10^9 particles/m^3 of air (3×10^7 particles/ft^3 air) for calcium carbonate (a nuisance dust). Suppliers usually have information on the safe handling of these products.

BIBLIOGRAPHY

1. W. C. McCrone, R. G. Draftz, and J. G. Delly, *Particle Atlas Two,* Ann Arbor Science Publishers, Ann Arbor, Mich., 1973.
2. T. Allen, *Particle Size Measurement,* Chapman and Hall, London, Eng., 1974, p. 93.
3. R. R. Irani and C. F. Callis, *Particle Size: Measurement Interpretation and Application,* John Wiley & Sons, Inc., New York, 1963.
4. G. J. Gregg and K. S. W. Sing, *Adsorption, Surface Area, and Porosity,* Academic Press, Inc., New York, 1967.
5. S. Brunauer, P. H. Emmett, and E. Teller, *J. Am. Chem. Soc.* **60,** 309 (1938).
6. A. W. Adamson, *Physical Chemistry of Surfaces,* 2nd ed., John Wiley & Sons, Inc., New York, 1960.
7. C. C. Furnas, *Ind. Eng. Chem.* **23,** 1052 (1931).
8. P. B. Mitton, "Opacity Hiding Power and Tinting Strength" in T. C. Patton, ed., *Pigment Handbook,* Wiley-Interscience, New York, 1973, p. 289.
9. J. V. Milewski, "Packing Concepts on the Utilization of Filler and Reinforcement Combinations" in H. S. Katz and J. V. Milewski, eds., *Handbook of Fillers and Reinforcements for Plastics,* Van Nostrand Reinhold, New York, 1978, p. 66.

10. A. N. Winchell and H. Winchell, *The Microscopical Characters of Artificial Inorganic Solid Substances,* Academic Press, Inc., New York, 1964.
11. N. H. Hartshorne and A. Stuart, *Crystals and the Polarizing Microscope,* 4th ed., Arnold, London, Eng., 1970.
12. R. B. Sucher, "Reinforcement" in *Basic Rubber Technology,* ACS Rubber Group Lecture, Philadelphia, Pa., 1955, p. 13.
13. *Ibid.,* p. 15.
14. R. S. Hunter, *ASTM Bull.* **186,** 48 (1952).
15. M. P. Morse, "Surface Appearance" in ref. 8, p. 289.
16. H. C. Schwalbe, *Paper Web Transactions of the Cambridge Symposium,* Vol. 2, British Paper and Board Makers Association, London, Eng., 1966, p. 692.
17. N. F. Miller, *Off. Dig. Fed. Soc. Paint Technol.* **34,** 465 (1962).
18. E. Singer, "Additives" in R. Myers and J. Long, eds., *Treatise on Coatings,* Vol. 3, Part I, Marcel Dekker, Inc., New York, 1974, p. 32.
19. C. H. Kline Co., *Am. Paint Coat. J.,* 12 (Aug. 29, 1977).
20. Business Communications Co., *Chem. Mark. Rep.,* 5 (June 20, 1977).
21. International Institute of Synthetic Rubber Producers, *Rubber World,* 34 (Jan. 1978).
22. K. E. Lowe, *Pulp Paper,* 60 (May 1976).
23. Predicasts Inc., *Chem. Week,* 53 (Nov. 15, 1978).
24. American Conference of Governmental Industrial Hygienists, *TLVS® Threshold Limit Values for Chemical Substances in the Workroom Environment with Intended Changes for 1978,* ACGIH, Cincinnati, Ohio, 1978.

JOHN G. BLUMBERG
JAMES S. FALCONE, JR.
LEONARD H. SMILEY
PQ Corporation

DAVID I. NETTING
ARCO Chemical Co.

FILM AND SHEETING MATERIALS

A film, as defined by the *Modern Plastics Encyclopedia* and as dealt with for the purposes of this article, is a flat section of a thermoplastic resin or a regenerated cellulosic material which is very thin in relation to its length and breadth and has a nominal thickness not greater than 0.25 mm (see also Film deposition techniques). The same materials in similar configuration but greater thicknesses are classified as sheets. Film and sheet may be used alone, in combination with another thermoplastic through coating or coextrusion, or as a laminate in combination with paper, aluminum foil, or other film. Packaging in containers is discussed in the articles on Packaging materials.

To choose film and sheet materials to perform specific functions requires that many factors be considered. The diversity of physical, chemical, mechanical, and electrical properties which, theoretically, can be built into film and sheet is almost endless. Numerous film- and sheet-forming materials are currently available in various densities, melt indexes, copolymers, and blends, with additives for plasticizing, coloring, impact modification, slip, antifogging, ultraviolet stabilization, fire retardance, and biodegradability. The principal basic materials for film and sheet manufacture consist of about 49 thermoplastic resins and regenerated cellulose.

Whereas paper, which has been replaced for many uses by film materials, has its origin in antiquity, the history of commercial films in the United States dates only from 1924 when the first cellophane was produced (see Cellulose; Cellulose derivatives). Cellophane production today is far exceeded by polyethylene production which, at 1,703 million metric tons, accounts for 80% of total yearly film consumption (see Olefin polymers). The most important material for sheet is polystyrene because of its ease of thermoforming and low cost (see Styrene plastics).

Properties and Test Methods

The properties of film and sheeting materials cover a vast range. Thicknesses vary from 2.54×10^{-5} mm for films to 1.27 cm for sheets; widths from 3.2 mm for tear tapes to 12.2 m for tarpaulins; and tensile strengths from a film that is not even intended to have any appreciable tensile strength (and which completely dissolves in water in a few seconds) to one that is half as strong as steel; from materials that are virtually no barrier to moisture vapor or gases to some that are nearly as impermeable as metal or glass; from highly flammable to fire-resistant products; and from materials having nearly 100% light transmission to complete opacity. Film materials offer various degrees of weatherability, chemical resistance, flexibility, thermal and electrical response, bondability, printability, and thermoforming properties.

Tables 1–5 show a number of properties reported for the most important film and sheeting materials. Test methods have been standardized to a great extent by ASTM. Proper preparation of the test specimen is essential because temperature and moisture content of plastics affects properties. The most common test methods for film and sheeting are listed in Table 6.

Tear strength is the force required to propagate an initiated tear in a film under controlled conditions. The Elmendorf tear tester is used in the same manner as for paper, except for modifications to allow for softer film. Another tear test, the PPT-tear, measures the dynamic tear resistance of a film to snagging.

Impact strength measures the force required to rupture a film or sheet and can be measured in various ways. *Tensile impact* measures the energy extracted from a pendulum-type hammer rupturing the film. In the free-falling dart method, widely used for polyolefin films, a specially constructed dart of variable weight is dropped a measured distance onto a film or sheet sample. The *Izod impact test* determines the strength of materials that, because of thickness or toughness, fail to break or rupture in the dart or tensile impact test. The Izod test measures the energy extracted from a pendulum-type hammer in breaking standard compression-molded specimens with one pendulum swing. Results from one impact test cannot be correlated with those from any other test, nor can results from film or sheet of one gauge be applied to other gauges of the same material. Temperature has a great influence on impact strength.

Stiffness of film and sheeting can be reported as the *tensile modulus of elasticity* and measured (as are the tensile strength and elongation) on the Instron tensile tester. Typical values range from 69 MPa (10,000 psi) for limp polyethylene copolymer to 3,450 MPa (500,000 psi) for stiff cellophane.

The *moisture–vapor transmission rate* (MVTR) and the *gas permeability* are important properties in packaging applications, since moisture exclusion or retention is necessary for many packaged products (see Barrier polymers). Some products, paticularly foods, require low permeability to oxygen and carbon dioxide. Products such as coffee, cheese, meats, and potato chips need a good gas barrier to prevent oxidation of fats or oils. Other products such as lettuce and tomatoes require a highly permeable material that permits the escape of carbon dioxide as respiration takes place. The permeability of a film can be reduced by increasing its thickness, by coating or coextruding it with another material, or by laminating it to another film, paper, or foil.

Rockwell hardness is a measure of resistance to indentation from a round steel ball, and is reported as a whole number with a scale symbol. The latter represents the indenter size, load, and dial scale used. The higher the number on a given scale, the harder the material (see Hardness).

Among the optical properties, *transparency* refers to the percentage of light transmitted by the sample, whereas *haze* is the percentage of transmitted light scattered. *Gloss* (ASTM D 523) is the amount of light reflected from the film surface toward the viewer.

Characteristics important in packaging applications are referred to as service properties, eg, printability of films. Some materials can be printed without special preparation, provided the proper ink is selected, but others, such as polyolefins, must be treated by flame, chemical, or electrical corona discharge (the most common) before printing inks adhere to the surface. The method of sealing, whether by heat, adhesive, or solvent, has to be carefully considered. A film's machinability, which is a function of the stiffness, must be adequate to ensure proper feeding and folding on automatic packaging equipment. In sheeting, the ease of thermoforming is an important service property. Ease of thermoforming is a function of several properties including the temperature range that will allow drawing (forming) melt strength, ie, the material's stiffness at temperatures near its melt point, and the tackiness of the sheet near its melt point. Furthermore, the sheet has to be consistent with good gauge uniformity.

Heat shrinkability arises from an elastic memory imparted to some thermoplastic

Table 1. General and Thermal Properties[a]

Material	CAS Registry No.	Abbreviation	Forms available[b]	Thickness range, mm	Max width, cm	Film heat sealing-range, °C	Continuous service temperature, °C Max	Min	Flammability, cm/s	Method of manufacture[c]	Thermal conductivity, kW/(m·K)
acrylonitrile–butadiene–styrene	[*9003-56-9*]	ABS	S	0.25–12.7	267		97		2.0	EX, Cl	4.1–6.6
acrylonitrile–methylacrylate copolymer, rubber modified	[*31668-18-5*]		F, S	0.03–1.27	152	121–177	79	−29		EX	
cellophane (regenerated cellulose)	[*9005-81-6*]		F	0.02–0.04	119–132	82–177	177	−17	15		
cellulosic, ethyl cellulose	[*9004-57-3*]		F, S	0.05–0.38	107		121	−57		Ca	
cellulose acetate	[*9004-35-7*]		F, S	0.02–6.35	152	177–232	79	−26		Ca, EX	
cellulose acetate butyrate	[*9004-36-8*]		F, S	0.03–8.89	122		60			Ca, Ex	
cellulose propionate	[*9004-48-2*]		F, S	0.08–0.76	122	191–218	79	−34		EX	2.5
cellulose triacetate	[*9012-09-3*]		F, S	0.05–0.51	119		175			Ca	
fluoroplastics, ethylene–tetrafluoroethylene copolymer	[*26770-96-4*]	ETFE	F, S	0.01–2.1	122	274			nil	EX	
fluorinated ethylene–propylene copolymer	[*25067-11-2*]	FEP	F, S	0.01–2.41	122	282–371	249	−254	nil	EX	2.5
polytrifluorochloroethylene copolymer	[*9002-83-9*]	PCTFE	F, S	0.01–0.76	137	232–260	135	−196	nil	Ca, EX	2.5
polytetrafluoroethylene	[*9002-84-0*]	PTFE	F, S	0.01–3.18	122		260	−253	nil	Ca, EX	2.5
poly(vinyl fluoride)	[*24981-14-4*]	PVF	F, S	0.01–0.10	351	204–218	113	−73	nil	EX	2.5
ionomer	[*25608-26-8*]		F, S	0.03–0.25	152	93–260	66	−73	2.5	EX	2.5
polyamides, nylon-6	[*25038-54-4*]		F, S	0.01–0.76	213	193–232	150	−73	nil	EX	0.7–1.3
nylon-6,6	[*32131-17-2*]		F	0.01–0.03	76	238–249			nil	EX	2.3–2.9
nylon-11	[*25035-04-5*]		F, S	0.025–1.27	71	166–199	120		nil	EX	2.9–3.3
nylon-12	[*24937-16-4*]		F, S	0.01–1.52		160–193	110	−79	nil	EX	2.2
polybutylene	[*9003-29-6*]		F, S	0.001 and up	610	135–204	135	−46	2.5	EX	
polycarbonate	[*24963-68-3*]		F, S	0.006–12.7	137	204–221	132	−101	3.0	Ca, EX	1.9
polyester (poly(ethylene terephthalate))	[*25038-59-9*]	PET	F, S	0.003–0.36	305	218–232	149	−79	2.5	EX, O	1.5

polyethylene, low density	[9002-88-4]	LDPE	F, S	0.0008 and up	1219	121–204	88	−57	2.5	EX	8.0
medium density		MDPE	F, S	0.008 and up	610	121–204	104	−57	2.5	EX	3.4–4.2
high density		HDPE	F, S	0.01 and up	152	135–204	121	−46	2.5	EX	4.6–5.2
ultra-high molecular weight		UHMWPE	F, S	0.05 and up	152	135–204	121	−184 to −46	2.5	EX	
ethylene copolymers, vinyl acetate	[108-05-4]		F, S	0.019 and up	1219	66–149	71	G	2.5	EX	
methyl acrylate	[96-33-3]		F, S	0.019–0.25	1219	60–140	70	−75		EX	
polyimide	[25036-53-7]		F	0.008–0.13	91		399	−268			1.0
poly(methyl methacrylate), standard	[9011-14-7]		F	0.13–0.25	109		77	G	3.0	EX	1.0–2.2
type A (Korad acrylic)			F, S,	0.05–0.15	274	121–177			3.0	EX	
polypropylene, extrusion cast	[25085-53-4]	PP	F, S	0.02–0.25	152	141–204	140	−18	2.0	EX	0.9–2.5
biaxially oriented			F, S	0.01–0.03	203	uncoated 149–160, coated 88–154	143	−51	2.0	O	0.9–2.5
polystyrene, oriented	[9003-53-6]	PS	F, S	0.006–0.51	193	121–177	88	−63	3.8	O	0.9–1.5
foam			F, S	0.18–3.18	122		F	G		EX	
sulfone polymers, polysulfone	[25135-51-7]		F, S	0.003–6.35	132		177	−73		EX	0.7–2.7
polyethersulfone	[25667-42-9]		F, S	0.003–6.35	132	260–288	218	−73		EX	1.9
polyurethane elastomer	[27475-68-9]		F, S	0.013 and up	173	135–191	88	−73		Ca, EX	1.3–4.4
poly(vinyl alcohol)	[9002-89-5]	PVA	F	0.013–0.3	137		G		2.5	Ca	
poly(vinylidene chloride)	[9002-85-1]	PVDC	F	0.01–0.15	173	121–149	143		nil	EX	
poly(vinyl chloride), nonplasticized	[9002-86-2]	PVC	F, S	0.015–1.91	213	177–216	79		nil	Ca, EX, Cl	1.5–2.1
plasticized			F, S	0.013–2.54	203	157–182	79	−46	nil	Ca, Cl, EX	1.3–1.7
rubber hydrochloride	[25987-37-5]		F	0.01–0.064	152		89	−29	nil	Ca	
vinyl chloride–acetate copolymer,	[9003-22-9]										
nonplasticized			F, S	0.02–1.02	183	135–182	F	G	nil	Cl, Ca, EX	1.5–2.1
plasticized			F, S	0.02–0.25	213	149–157	79	G	nil	Cl, Ca, EX	1.3–1.7
acrylonitrile–rubber			F	0.025–0.08	137	107–135	G	G		Ca, EX	

[a] Much of this data was obtained from ref. 1.
[b] F = film, S = sheeting.
[c] EX = extrusion, Cl = calendering, Ca = casting, O = oriented.

Table 2. Physical and Mechanical Properties[a]

Material	Yield, $m^2/(kg{\cdot}mm)$	Specific gravity	Tensile strength, MPa[b]	Elongation, %	Tear strength, g/mm	Bursting strength, 2.54×10^{-2} mm thickness, Mullen points	Impact strength	Tensile modulus, MPa[b]	Rockwell hardness
acrylonitrile–butadiene–styrene	32.16	1.04	35.2–70.3	10–50			134–641[c]	897–2600	R50–120
acrylonitrile–methylacrylate copolymer, rubber modified	28.88	1.15	63.3–77.3	50–200		27			
cellophane (regenerated cellulose)	31.2	1.40–1.50	49.2–126.5	10–50	0.8–7.9	30–50	427–801[d]	3450	R35–100
cellulosics, ethyl cellulose	28.8	1.15	56.2–70.3	20–30	2.8–14.2	50–85	107–454[d]	1380	R50–115
cellulose acetate	25.2–26.9	1.28–1.31	49.2–115.3	15–55	1.6–3.9	30–60	106–454[d]	3450	R32–118
cellulose acetate butyrate	22.6–28.0	1.19–1.20	35.2–63.3	50–100	2.0–3.9	40–70	106–454[d]	1380	R35–113
cellulose propionate	28.0	1.20	28.1–35.2	60–80	39.4	50–70	106–454[d]	1380	R41–110
cellulose triacetate	25.2–26.9	1.28–1.31	63.3–112.5	10–40	1.6–3.9	50–70	106–454[d]		
Fluoroplastics, ETFE	19.6	1.7	49.2–56.2	300	600–900		2–15[e]		R50–70
FEP fluoroplastic	15.5	2.15	17.6–21.1	300	125	11	2–15[e]	345	D60[f]
polytrifluorochloroethylene copolymers	15.6	2.08–2.15	35.2–70.3	50–150	2.5–40	23–31	2–15[e]	1380	
polytetrafluoroethylene	15.4	2.1–2.2	10.6–31.6	100–350	10–100		2–15[e]	414	D50[f]
poly(vinyl fluoride)	20.6–24.0	1.38–1.57	49.2–126.5	115–250	12–100	19–70	2–15[e]	828	D80[f]
ionomer	35.3	0.94	35.2	250–450	30		30–90[e]	276	D60[f]
polyamides, nylon-6	29.4	1.13	63.3–126.5	250–550	50–90	elongates	4–6[e]	2070	R60–120
nylon-6,6	29.0	1.14	63.3–84.4	200	35–40	no break	4–6[e]	2070	R60–120
nylon-11	32.0	1.03	63.3–77.3	250–400	400–500	60	4–6[e]	1242	R80–115
nylon-12	32.6	1.01	49.2–84.4	290–330			4–6[e]		
polybutylene	36.0	0.908–0.917	26.7–30.9	300–380			20–30[e]	276	D55[f]
polycarbonate	27.7	1.20	59.1–61.9	85–105	7.9–9.8	25–35	641–961[d]	2208	M70–100
polyester (PE terephthalate)	23.8–27.1	1.38–1.41	281[g]	50[g]	4.7–10.6	55–80	25–30[e]	3450	M94–101
polyethylene, LDPE	36.0	0.91–0.925	10.6–21.1	100–700	19.7–118	10–12	7–11[e]	97–262	R10
MDPE	35.4	0.926–0.940	14.1–24.6	50–650	19.7–118		4–6[e]	172–380	R15
HDPE	34.8	0.941–0.965	16.9–42.9	10–650	5.9–118		1–3[e]	414–1242	R20
UHMWPE	35.5	0.940	21.1–38.7	300			no break[d]	600	

ethylene copolymers, vinyl acetate	34.8	0.924–0.940	7.0–21.1	400–800	19.7–118	10–12	100–200[c]	14–69	D17–45[f]
methyl acrylate	34.8	0.93	7.0–21	400–700	120	20	250[c]	14–50	
polyimide	31.2	1.42	17.6	70	2.5	75	37[d]	1297	E99
poly(methyl methacrylate), standard	28.1	1.18–1.19	57.7–61.9	4–12			37[d]	2760	M35–95
type A (Korad acrylic)	28.8	1.14	35.9	75		35–40			
polypropylene, extrusion cast	37.1–37.6	0.885–0.905	31.6–49.2	550–1000	236TD–10MD		1–3[e]	1550	R80–110
biaxially oriented	36.7	0.902–0.907	52.7–281	35–475	1.2–3.9		10–25[e]	3174	
polystyrene, oriented	31.6	1.05–1.06	56.2–84.4	3–40	2	16–35	1–5[e]	2760	M85
foam	variable		2.1–9.1	2–8					
sulfone polymers, polysulfone	26.8	1.24–1.25	59.1–74.5	64–110	2.5–4.7	0.41 MPa[b] at 7×10^{-3} Pa[h]	64[d]	2480	M70–116
polyethersulfone	24.0	1.37	70.3–84.4	20–150	2.8–6.3		85[d]	3.5	M80
polyurethane elastomer	27.6	1.11–1.24	35.2–84.4	200–700	87–279		no break[d]		D50–90[f]
poly(vinyl alcohol)	26	1.26	27.6–69	180–600	250–800		300–500[e]		
poly(vinylidene chloride)	20.1	1.65	256–113	30–80	10–90	25–39	40–20[e]	600	
poly(vinyl chloride), nonplasticized	24.0–27.6	1.2–1.5	49.2–70.3	25–50	10–700	30–40	12–20[e]	850–4140	D65–80[f]
plasticized	24.0–27.6	1.2–1.8	98.4–112.5	100–500	60–1400	20	12–20[e]	variable	A50–100[f]
rubber hydrochloride	30.0	1.11	24.6–35.2	200–800	60–1600	stretches	6–15[e]		
vinyl chloride–acetate copolymer,									
nonplasticized	24.0–27.6	1.3–1.4	38.7–56.2	3–100	10–30		12–20[e]	850–4000	D65–80[f]
plasticized	24.0–27.6	1.2–1.35	17.6–35.2	100–500	30–1400		12–20[e]	variable	A50–100[f]
acrylonitrile–rubber alloy	27.6	1.1–1.3	17.6–28.1	250–500	70	20			

[a] Much of this data was obtained from ref. 1.
[b] To convert MPa to psi, multiply by 145.
[c] 66 cm dart drop, g/0.025 mm.
[d] Izod test, 2.5 × 2.5 cm notched bar, (N·m)/m (to convert to ft·lbf/in., divide by 53.38).
[e] Pendulum impact on 3.2 mm specimen, kg/cm.
[f] Shore hardness.
[g] Tensilized.
[h] To convert Pa to μm Hg, multiply by 7.5.

Table 3. Chemical Properties[a]

Material	Water absorption, %	MVTR[b] at 37.8°C for 24 h, (g·mm)/m^2	Gas permeability cc/(mm^2·mm thickness), 24 h			Resistance to[c]			
			Oxygen	Carbon dioxide	Nitrogen	Strong acids	Strong alkalies	Greases	Organic solvents
acrylonitrile–butadiene–styrene	0.6–1.0		20–28	59–79	2–4	G to F	G	G	F to P
acrylonitrile–methylacrylate copolymer, rubber modified	0.28		0.3	0.6	0.1	P	F	G	G
cellophane (regenerated cellulose)	45–115	very high	0.25	0.2–2.4	0.4	P	P	G	coating attacked
cellulosics, ethyl cellulose	2.5–7.5		788	1970	236	F to P	G	G	P
cellulose acetate	3.0–8.5		46–59	339–394	12–16	P	P	G	P
cellulose acetate butyrate	1–2		374	2364	99	P	P	G	P
cellulose propionate	1.5–2.5					P	P	G	P
cellulose triacetate	2.0–4.5		59	347	12	F	P	G	F to P
fluoroplastics, ETFE	<0.02	0.02	39	99	12	G	G	G	G
FEP fluoroplastic	<0.01	0.02	296	658	126	G	G	G	G
polytrifluorochloroethylene copolymers	nil	0.01–0.02	3–6	6–16	1	G	G	G	G
polytetrafluoroethylene	nil	0.02				G	G	G	G
poly(vinyl fluoride)	<0.5	0.02	1.2	4	0.1	G	G	G	G
ionomer	0.4	1.6	600			G	G	G	G
polyamides, nylon-6	9.5		1	4–5	0.4	P	G	G	G
nylon-6,6	8.5		2	3.6	0.3	P	G	G	G
nylon-11	0.27		13	60	1.3	P	G	G	G
nylon-12	0.25	1.6	21–36	62–132	5–7	P	G	G	G
polybutylene	<0.01					G to P	G	G	
polycarbonate	0.35	6.7	118	424	20	G	P	G	G to P
polyester (PE terephthalate)	<0.8		1.2–1.6	6–10	0.3–0.4	G	P	G	G

polyethylene, LDPE	<0.01	0.51	197	1064	71	G	G	P	G
MDPE	<0.01		99–211	394–985	34–124	G	G	G	G
HDPE	nil		73	228	17	G	G	G	G
UHMWPE	nil					G	G	G	G
ethylene copolymers, vinyl acetate	<0.01		331	2364	158	F	F	P	F
methyl acrylate	nil								
polyimide	2.9		10	18	2	P	P	G	G
poly(methyl methacrylate), standard	0.35	0.5				G	G	P	
type A (Korad acrylic)	1.45					G	G	G	P to G
polypropylene, extrusion cast	<0.005	0.27	59–95	197–315	16–19	G	G	G	G
biaxially oriented	<0.005	0.14	63	213	8	G	G	G	G
polystyrene, oriented	0.04–0.10	2.76	99–138	355		G	G	G to P	G to P
foam						G	G	F	P
sulfone polymers, polysulfone	0.3		91	374	16	G	G	G	P
polyethersulfone	2.1		36	160	16	G	G	G	G to P
polyurethane elastomer	0.55–0.77		30–129	177–650	16–47	P to G	P to G	G	G
poly(vinyl alcohol)	soluble	0.1	59	98		P	P	G	G
poly(vinylidene chloride)	nil	0.4	0.3–2.7	1.5–17	0.05–0.6	G	G	G	G
poly(vinyl chloride), nonplasticized	nil	0.35–2.0	2–8	8–20		G	G	G	G to P
plasticized	nil	2.0–30	12–788	39–1182		G	G	G	G
rubber hydrochloride	5	0.0007–0.23	15–886	114–5319		G	G	G	G
vinyl chloride–acetate copolymer,									
nonplasticized	nil	1.46	6–8	16–28	3–4	G	G	G	P to G
plasticized	nil	0.23	8–59	28–315	3–24	G	G	F to G	P to G
acrylonitrile–rubber alloy	nil		20	95	2	G	G	G	G

[a] Much of this data was obtained from ref. 1.

[b] Moisture vapor transmission rate.

[c] P = poor, F = fair, G = good.

Table 4. Optical and Electrical Properties[a]

	Refractive index	Transparency, %	Haze, %	Dielectric constant, kHz	Dissipation (power) factor, kHz	Dielectric strength, kV/mm	Volume resistivity, Ω·cm
acrylonitrile–butadiene–styrene	1.53	33	100	2.75		14.3	10^{16}
acrylonitrile–methylacrylate copolymer, rubber modified				4.6	0.03	9.1	3×10^{15}
cellophane (regenerated cellulose)				3.2	0.015	79–99	10^{11}
cellulosic, ethyl cellulose				3.1	0.002–0.02	59	10^{15}
cellulose acetate	1.5	88	<1	3.6	0.013	126–197	10^{10}–10^{15}
cellulose acetate butyrate	1.47	88	<1	2.9	0.013	122	10^{11}–10^{15}
cellulose propionate	1.47	88	<1			76	3×10^{15}
cellulose triacetate				4.0	0.016	146	10^{13}
fluoroplastics, ETFE				2.6	0.0008	138	10^{16}
FEP fluoroplastic	1.34	<90	4	2.25	<0.002	276	10^{19}
polytrifluorochloroethylene	1.43			2.6	0.023	39–146	10^{18}
copolymers							
polytetrafluoroethylene	1.35			2.1	0.0002	17	10^{18}
poly(vinyl fluoride)				8.5	1.6	138	3×10^{13}
ionomer				2.4	0.002	39	10^{16}
polyamides, nylon-6				3.7	0.016	50	
nylon-6,6				4.0		34	4.5×10^{13}
nylon-11				3.7	0.03	34	7.8×10^{13}
nylon-12				3.8	0.05	39	2.5×10^{13}
polybutylene	1.5			2.3	0.0005		

polycarbonate	1.59	85–91	0.5–2	2.9	0.0015	59	10^{16}
polyester (PE terephthalate)				2.8	0.005	296	10^{18}
polyethylene, LDPE	1.51	0–75	4–50	2.2	0.0003	19	10^{16}
MDPE	1.52	10–80	4–50	2.2	0.0003	20	10^{16}
HDPE	1.54	0–40	10–50	2.3	0.0005	20	10^{18}
UHMWPE	1.54				2.3×10^{-4}	51	10^{18}
ethylene copolymers, vinyl acetate	1.54	0–80	2–40		0.015	18–22	2.4×10^{9}
methyl acrylate	1.54	0–40	10–50		0.015	18–22	2.4×10^{9}
polyimide				3.5	0.003	276	10^{18}
poly(methyl methacrylate), standard	1.5	92	1	3.75	0.04	16	10^{15}
type A (Korad acrylic)	1.5	92	1	4.4	0.046	95	10^{16}
polypropylene, extrusion cast	1.49	55–90	1.0–3.5	2.2	0.0003	118–177	3×10^{15}
biaxially oriented				2.2	0.0002	276–394	3×10^{16}
polystyrene, oriented	1.6	87–92	0.1–3.0	2.5	0.0005	197	10^{7}–10^{16}
foam							
sulfone polymers, polysulfone	1.63		5.0	3.1	0.0008	296	5×10^{16}
polyethersulfone				3.5	0.0035	95	10^{17}–10^{18}
polyurethane elastomer	1.55			6.3	0.045	24–51	2×10^{11}
poly(vinyl alcohol)							
poly(vinylidene chloride)	1.61	60–80	2–5	3.5	0.07	118–197	10^{12}–10^{15}
poly(vinyl chloride), nonplasticized	1.53	76–82	8–18	3.0–3.3	0.013	17–51	10^{16}
plasticized				4.0–8.0	0.11	10–39	10^{11}–10^{14}
rubber hydrochloride				3.0			10^{13}
vinyl chloride–acetate copolymers							
nonplasticized				4.0–8.0	0.0009–0.017	17–51	10^{16}
plasticized					0.07–0.16	10–39	10^{11}–10^{14}
acrylonitrile–rubber alloy							

[a] Much as of this data was obtained from ref. 1. Optical properties measured using white light. Electrical properties measured at 22°C.

Table 5. Permanence and Service Properties[a]

	Dimensional stability at 100°C, % change	Resistance to uv radiation[b]	Sealing method[c]	Heat-shrinkable	FDA approved grades
acrylonitrile–butadiene–styrene		G to F	H, A	no	yes
acrylonitrile–methylacrylate copolymer, rubber modified		P	A		yes
cellophane (regenerated cellulose)	−0.7 to −3.0	G		no	yes
cellulosics, ethyl cellulose	0 to −0.7	G to F	A, S	no	
cellulose acetate	+0.2 to −3.0	G	H, S	no	yes
cellulose acetate butyrate	+0.1 to −3.0	G	A, H	no	
cellulose propionate	−2	G	H, S	no	no
cellulose triacetate	0 to −0.7	G	A, S	no	no
fluoroplastics, ETFE			H, A, S	no	
FEP fluoroplastic	<1	G	H, A, S	no	yes
polytrifluorochloroethylene copolymers	+2 to −2	G	H, A, S	no	yes
polytetrafluoroethylene		G	H, A, S	no	yes
poly(vinyl fluoride)	1	G	H, A, S	no	no
ionomer		F to G	H	no	yes
polyamides, nylon-6		F to G	H, A	no	yes
nylon-6,6	+0.2 to −3.0	P	H, A	no	yes
nylon-11	0 to −0.8	G	H, A	no	yes
nylon-12	+0.1 to −3.0	F to G	H, A	no	yes
polybutylene		P	H, A	yes	yes
polycarbonate	nil	F	H	no	yes
polyester (PE terephthalate)	<0.5	F	H, A	yes	yes
polyethylene, LDPE	−2	F to G	H	yes	yes
MDPE	0 to −0.7	F to G	H	yes	yes
HDPE	−0.7 to −3.0	F to G	H	yes	yes
UHMWPE		F	H	yes	yes
ethylene copolymers, vinyl acetate	<1	G	H	yes	yes
methyl acrylate	<1	G	H	yes	yes
polyimide	0.3	G	A	no	no
poly(methyl methacrylate), standard		G	A	no	
type A (korad acrylic)		G	H, A	no	no
polypropylene, extrusion cast		F	H	no	yes
biaxially oriented	<2 (heat set)	F	A	yes	yes
polystyrene, oriented		F	H, A	yes	yes
foam		P	H, A	no	yes
sulfone polymers, polysulfone		P	A	no	yes
polyethersulfone	0.06	P	H, A	no	pending
polyurethane elastomer		P to G	H	no	
poly(vinyl alcohol)	−3.2	G	H, A	no	no
poly(vinylidene chloride) (PVDC)	10–35	F	H, A	yes	yes
poly(vinyl chloride), nonplasticized	−7 to 4	G	H, A	no	yes
plasticized	−7 to 15	G	H, A	yes	yes
rubber hydrochloride		F	H, A	yes	
vinyl chloride–acetate copolymer					
nonplasticized	<5	G	H, A	no	yes
plasticized	<5	G	H, A	yes	yes
acrylonitrile–rubber alloy		P	H, A		yes

[a] Much of this data was obtained from ref. 1.
[b] G = good, F = fair, P = poor.
[c] H = heat, A = adhesive, S = solvent.

Table 6. Test Methods

Property	SI units reported	Test method
general		
thickness	mm	ASTM D2103
thermal		
heat sealing range	°C	
continuous service temperature	°C	
thermal conductivity	W/(m²·K)	ASTM C177
flammability, burning rate	cm/min	ASTM D635
physical and mechanical		
yield	m²/kg	ASTM D2103
tensile strength	MPa	ASTM D882
elongation	%	ASTM D882
tear strength, Elmendorf	g/mm	ASTM D1922
burst strength	Mullen points	ASTM D774
impact strength, dart drop	g/0.025 mm	ASTM D1709
tensile impact	(N·m)/m	ASTM D1822
Izod impact	(N·m)/m	ASTM D256
stiffness, tensile modulus	MPa	ASTM D638
Rockwell hardness	scale reading	ASTM D785
chemical		
water absorption	%	ASTM D570
water vapor transmission rate for 24 h	g/m² [a]	ASTM E96
gas permeability for 24 h	m¹/m² [a]	ASTM 1434
optical		
refractive index, n_D		ASTM D542
transparency	%	ASTM D1746
haze	%	ASTM D1003
gloss	unitless	ASTM D523
electrical		
dielectric constant		ASTM D150
dissipation (power) factor		ASTM D150
dielectric strength	kV/mm	ASTM D149
volume resistivity	Ω·cm	ASTM D257
permanence		
dimensional stability at 100°C for 30 min	%	ASTM D1204
outdoor weatherability [b]		ASTM D1435

[a] For 0.025 mm film thickness.
[b] Relative resistance.

films during their manufacture by either stretch orientation or by cross-linking induced through irradiation. Shrinkage takes place when heat is applied to the film, and it tends to revert to its original unoriented state. It is important for such applications as taut, contoured-fit overwraps or sleeves.

Materials

Acrylonitrile–Butadiene–Styrene (ABS). ABS is only available in sheet form and has balanced engineering properties with high impact resistance, toughness, and rigidity. Its good hot-tensile strength allows large, deep-draw thermoforming over a wide melt temperature range. Limitations are poor solvent resistance and low continuous service temperature. Because ABS is mildly hygroscopic, it must be dried before sheet forming (see Acrylonitrile polymers; Engineering plastics).

Acrylic Polymers. Most acrylic polymer film and sheeting is produced from poly(methyl methacrylate). It has the optical clarity of glass and the best weatherability of any thermoplastic. Owing to these properties, in addition to good impact strength, rigidity, and scratch and stain resistance, it is widely used in the sign industry and in safety glazing for windows and large area enclosures. Abrasion and solvent resistance may be improved by various chemical coatings. Stretched sheet, with its increased toughness, is coated and used in aircraft canopies and windshields. Disadvantages of acrylic polymers are poor solvent resistance, susceptibility to stress cracking, combustibility, and a fairly low continuous service temperature (93°C). Flexible grades are not available (see Acrylic ester polymers; Methacrylic polymers).

Cellophane. Cellophane, once the most important packaging film material, is rapidly being replaced by the cheaper biaxially oriented polypropylene film. Cellophane is regenerated cellulose and, when uncoated, is extensively sensitive to humidity conditions and cannot be heat sealed. For these reasons, nearly all cellophane sold today is coated with cellulose nitrate (referred to as lacquer), poly(vinylidene chloride), or polyethylene. Cellophane has excellent transparency, gloss, stiffness, and tensile strength, and transmits a high percentage of uv radiation. It is used primarily in high performance packaging where high stiffness and abrasion resistance are of primary concern. It is still the most easily processed film in large-scale printing, converting, and packaging in-line operations.

Cellulosics. Film and sheet are available in four cellulose ester plastics: cellulose acetate, cellulose acetate butyrate, cellulose acetate propionate [*9004-39-1*], and cellulose triacetate. In general, properties of cellulosics include crystal clarity, good scuff and grease resistance, weatherability, high gas permeability, intermediate MVTR, and good electrical insulation characteristics. Cellulose sheet, except for cellulose triacetate, can be formed by all conventional thermoforming methods. Cellulose acetate is easily shaped for blister and skin packaging and transparent rigid containers. It does not become brittle with age, a notable property (see Cellulose derivatives; Cellulose acetate).

The acetate–butyrate and acetate–propionate grades are tougher than cellulose acetate and lower in water absorption. This toughness allows them to be drawn tightly over sharp points in skin packaging, protecting the products during handling and shipment. The toughness and weatherability of butyrate and propionate sheeting are important properties for outdoor signs.

Fluoroplastics. Fluoroplastic film and sheet possess outstanding chemical, physical, thermal, and electrical properties. The family of fluoroplastic materials includes, in addition to those listed in Tables 1–5, poly(vinylidene fluoride) [*24937-79-9*] (PVDF). Fluoroplastics are inert to most chemicals and solvents, nonflammable, have a continuous service temperature range of −254 to 250°C, and are heat sealable and thermoformable. They have outstanding electrical insulation properties and unique nonadhesive properties with low coefficients of friction. The films transmit more light and uv and ir radiation than window glass and, therefore, have very good weatherability. Applications include lamination to building materials for outdoor weatherability and indoor stain resistance. The fluorohalocarbons have the best MVTR barrier of all materials. Many FDA approved grades are available (see Fluorine compounds, organic).

Ionomer Resins. Ionomer refers to a class of polymers in which ionized carboxyl groups create ionic cross-links in the intermolecular structure. Ionomer resins exhibit excellent optical properties and abrasion resistance. They have good flexibility at low temperatures, high elongation, and excellent impact strength. Electrical properties as well as chemical resistance are good. The low upper-use temperature of 71°C gives ionomer film excellent heat sealability. Ionomers are used as a heat-seal layer in coextrusions. They also provide good bonding to untreated aluminum foil in extrusion coating. Film and sheet are used in overwrap, skin, and thermoformed blister packaging because of their durability and puncture resistance (see Acrylic acid; Acrylamide polymers).

Polyamide. In the United States, polyamide (nylon) film production consists primarily of nylon-6 and nylon-6,6. The outstanding properties of nylon are high tensile strength and elongation, abrasion resistance, good grease and solvent resistance, a high melting point combined with good, low temperature flexibility and strength, good clarity, and impermeability to gas and odors. Limitations are a high moisture absorption with related dimensional instability and deterioration of electrical and mechanical properties. Nylon-6 and nylon-6,6 also have poor resistance to sunlight. In combination with a heat sealable polymer, such as ionomer, LDPE, or PVDC, polyamide film is used for the vacuum packaging of fresh and processed meats and cheeses (see Polyamides; Engineering plastics).

Polybutylene. Polybutylene (butene homopolymer), is one of the newest available polyolefins. Polybutylene film and sheeting exhibit high tear, impact, and puncture resistance. The film has a high yield, is heat sealable, and can be hot-filled. Polybutylene has good chemical and abrasion resistance, good barrier properties, electrical insulating characteristics, and resistance to elevated temperatures (up to 104°C). Major film applications are heavy-duty trash liners and shipping containers, and industrial sheeting (see Olefin polymers).

Polycarbonate. Polycarbonate film and sheet have outstanding impact strength, dimensional stability, transparency, heat resistance, and self-extinguishing properties. Polycarbonate can be processed with conventional fabricating equipment and can be cut, routed, sheared, punched, drilled, and thermoformed. Because of its optical clarity and toughness, polycarbonate replaces acrylic polymers in high performance aircraft windshields, window glazing, and outdoor signs. A polycarbonate laminate is used for bullet-proofing (see Polycarbonates).

Polyester. The only polyester film produced in quantity is biaxially oriented poly(ethylene terephthalate). It offers very high tensile strength, wide service temperature, good dimensional stability, clarity, and good barrier and electrical insulating properties. Unfortunately, PET is attacked by bases and is not heat sealable. It is extensively used in photographic film base, reprographic products, magnetic tape, electrical insulation, and as metallized film. Both uncoated and PVDC-coated PET are used extensively in heavy-duty packaging applications (see Polyesters).

Polyethylene. Polyethylene, the largest volume film material, is available in a multitude of homopolymer and copolymer grades modified by additives. There are four general classifications: low density (LDPE), medium density (MDPE), high density (HDPE), and ultra-high molecular weight (UHMWPE). Mechanical, chemical, and barrier properties of film and sheet generally improve with higher density, the exception being tear strength and impact strength. All of these properties improve with increase in molecular weight (decrease in melt index). The advantages of LDPE over other thermoplastics include cost, transparency, tear resistance, impact strength, and moisture and chemical resistance. It can readily be processed by all thermoplastic methods, including orientation for shrink-film applications. Among its disadvantages are poor grease resistance, high permeability to odors and gases, poor weatherability, flammability, and poor resistance to high temperature.

Low density polyethylene film applications range from trash liners and shrink wrap to high impact produce and garment bags. Air cap sheet (bubbles of air laminated between two sheets of polyethylene) is used for insulating swimming pool blankets (2). Medium density polyethylene is stiffer and used mostly as an overwrap. High density polyethylene film is used as paper replacement, merchandise bag, multiwall bag and, because of its excellent barrier properties, as a barrier film in glassine replacement. Small amounts of LDPE are extruded into sheet for construction and agriculture uses. Much larger quantities of HDPE is extruded into sheet for thermoforming institutional disposable articles, shipping containers, protective athletic gear, toys, and children's swimming pools.

High density polyethylene with an average molecular weight from 2 to 6 million is known as ultra-high molecular weight polyethylene. These polymers, extruded into thin gauge film, are tougher and have higher chemical and abrasion resistance than HDPE (see Olefin polymers).

Polyimide. Polyimide, available as film, is tough, flexible, and highly resistant to combustion (see Polyimides). It provides a very wide continuous use temperature (−269 to 593°C). Room temperature properties compare to those of poly(ethylene terephthalate) but high temperature properties are much better. At 400°C polyimide has a tensile strength of 31.6 MPa (4600 psi). It has good dimensional stability, high dielectric strength, and resistance to ionizing radiation. Primary applications are high performance electrical insulation, printed circuits, and backing for magnetic and pressure-sensitive electrical tapes (see also Aramid fibers).

Polypropylene. Polypropylene (PP) is currently experiencing one of the highest growth rates of any film and sheeting thermoplastic material. The film can be formed by extrusion casting or by a blown-water quench process. The sheet is usually extrusion cast. Biaxial orientation of PP film (BOPP) significantly improves its properties and BOPP is replacing cellophane in most packaging applications because it provides comparable properties at lower cost. Its high melt point sets it apart from other polyolefins. It demonstrates good chemical and grease resistance, relatively low gas permeability and MVTR, stiffness, and excellent electrical properties. The major disadvantage of cast polypropylene is its poor impact strength at moderately low temperature (0°C). Low temperature impact is improved in ethylene copolymers of polypropylene, available in random or block form. Random copolymers, used primarily in film grades, retain good optical properties but have poorer impact strength than the block copolymers (see Olefin polymers).

Extrusion cast PP film has sparkling clarity, high stiffness and heat sealability. Primary applications for extruded PP film are textile bags, bread bags, and snack food overwrap. Biaxially oriented PP film has better clarity, higher stiffness, and very good cold-temperature impact strength, but it is difficult to heat-seal unless coated. It is usually coextruded, coated, or modified with additives to achieve heat sealability and improve barrier properties. It is used extensively in capacitors, in snack-food packaging laminations, as a cellophane replacement in cigarette overwrap and for gas-flush packaging of coffee, cheese, and meats. Nonheat-set oriented polypropylene–ethylene copolymer film provides high-clarity, high-strength shrink film.

Polypropylene sheet, with a little difficulty, can be thermoformed. It can be hot stamped and die cut and is readily foamed yielding an extremely low density foam. Hot-stamped PP sheet is extensively used in automotive interiors and under-the-hood applications. New applications include corrugated PP board, produced by new extrusion techniques developed in Japan and Europe, and polypropylene solar collectors for swimming pool heating (see Solar energy).

Polystyrene. Polystyrene (PS) film and sheet ranks only behind LDPE and PVC in total production. It is one of the most widely used thermoforming film and sheeting materials because of its low cost and high melt strength over a broad temperature range. Polystyrene can be vacuum formed, pressure formed, hot stamped, and die cut. Its properties range from clear but brittle to translucent with high impact (achieved by blending butadiene in polymerization or by adding impact modifiers during extrusion). It has poor weatherability and is soluble in aromatic and chlorinated hydrocarbons. High volume applications of PS film and sheet are thermoformed cups, trays, food containers, furniture components, luggage, and entire bathroom assemblies.

Biaxially orienting PS film and sheet overcomes brittleness, imparts strength, and improves production yield. It has high gas permeability and intermediate MVTR which make it suitable for packing fresh produce.

Foamed polystyrene sheet is widely used in food packaging, food service, and product packaging applications. Among the more important characteristics of PS foamed sheet are rigidity, thermal insulation, low density, shock absorbance, and clean appearance and feel. The sheet is readily thermoformed and colored. The foamed sheet is also widely coextruded with a thin film of crystalline PS on each side which improves the appearance and extends the utility of the product, especially in single-service food containers (see Styrene plastics).

Sulfone Polymers. Sulfone polymers, consisting of polysulfone and polyethersulfone are strong, rigid, and tough, high temperature engineering thermoplastics. They are transparent, chemically resistant and have low flammability with a low smoke emission. Sulfone polymers are used in high temperature electrical applications such as capacitor film and film for transparencies (see Polymers containing sulfur).

Vinyls. The term vinyls covers a large family of homopolymers and copolymers having a broad range of properties. Principal members are poly(vinyl chloride), vinyl chloride–vinyl acetate copolymers, and poly(vinylidene chloride), in addition to poly(vinyl alcohol) and poly(vinyl fluoride). The physical and chemical properties of the vinyl polymers can be varied extensively by compounding with plasticizers, modifiers, and other additives (see Vinyl polymers; Fluorine compounds, organic). Physical and mechanical properties are also highly dependent upon the method of manufacture, ie, extrusion, casting, or calendering. Vinyl polymers may range from hard brilliant film and sheet approaching cellulose acetate in properties, to soft, flexible materials similar to polyethylene. The usefulness of vinyl polymers such as PVC is limited by thermal decomposition, which can evolve HCl, attack by several solvent types, poor thermal stability, and staining by sulfur compounds, and a comparatively lower yield of film per pound of resin.

Solution-cast vinyl films have very good dimensional stability, clarity, and high surface gloss, but poor gas permeability and MVTR. Outdoor weatherability, toughness, and puncture resistance can be built into the film making it useful for decals. Cast vinyl film is also biaxially oriented for high-shrink film in packaging of meat, poultry, and produce. Blown extruded vinyl film has many of the same properties as cast film but with slightly less dimensional stability and lower surface gloss. Extruded rigid vinyl films are self-extinguishing and have outstanding weather and corrosion resistance.

Most film and sheeting above 0.076 mm is produced by calendering. Calendered flexible (plasticized) vinyl film and sheeting have moderate tensile strength and chemical resistance, high elongation, low water absorption, and good film tear strength. They can be easily sealed and thermoformed.

Rigid (unplasticized) calendered vinyls have excellent tensile strength, chemical and flame resistance, and good dimensional stability. They can be cut, machined, punched, thermoformed, and sealed by means of heat, solvents, or cements.

Film and sheeting made from copolymers of vinylidene chloride, vinyl chloride, alkyl acrylates, and acrylonitrile have flexibility, low gas permeability and MVTR, sparkling clarity, and resistance to chemicals, oils, and grease. Materials containing larger amounts of vinylidene chloride have greater chemical resistance and are among the best in barrier properties.

Applications for vinyl film and sheet are in upholstery materials, consumer packaging, medical and hospital packaging, and industrial liners for tanks and ponds.

Polyurethane. Extreme abrasion resistance is the most outstanding characteristic of polyurethane, a tough, flexible elastomeric material. Polyurethane is easily foamed and has good, low temperature capability and grease resistance. Unfortunately, it has poor weatherability, is attacked by most solvents, and is made from toxic isocyanates. Flexible polyurethane foam is widely used in furniture and bedding. Rigid polyurethane foam is one of the most efficient thermal insulations commercially available (see Urethane polymers).

Water-Soluble Films. Films made from poly(vinyl alcohol) (PVA), methyl cellulose [*9004-67-5*], starch [*9049-76-7*], and poly(ethylene oxide) [*25322-68-3*] (PEO) are soluble in water; PVA and PEO films are soluble between 0.6 and 100°C; for methyl cellulose, between 0.6 and 52°C, and starch between 52 and 100°C. Methyl cellulose is the only material approved for food contact by the FDA. Water-soluble films are used for disposables such as detergent packages, dyes, and hospital laundry bags (see also Resins, water soluble).

Plastic Paper. Plastic paper consists of spunbonds, film papers, and synthetic pulps (see Synthetic paper). Spunbonds are produced by melt spinning long continuous fibers, randomly dispersing the fibers into a bulky mat, and compacting the fibers into a paperlike product. Spunbonds are very strong and are water and chemical resistant. Film papers are made from polypropylene, HDPE, and polystyrene film treated to obtain opacity and ink receptivity. The Synpulp process makes fibrous pulps from polyolefins. Paper is prepared from the fibrous pulp by a commercial wood pulp process (see also Nonwoven textiles, spunbonded).

Manufacture

Properties of film and sheeting materials vary with the chemical composition, density and molecular weight distribution, additives employed, and method of manufacture. Of the six fundamental processes for forming a basic polymeric composition into a film, four are commercially significant. Chemical regeneration from a solution and melt forming are by far the most important. The four largest-volume film materials currently in use are made by chemical regeneration (cellophane) and melt forming (polyethylene, vinyls, and polypropylene). Acrylic, cellulosic, and vinyl films are made by solution forming.

Emulsion forming (together with plastisol and organosol forming) is used mainly for coating substrates. A modest volume is made by fiber forming. Powder forming is at present commercially not important.

Sheeting is manufactured by the melt-forming process of extrusion and calendering and to a small degree by solution forming. The latter is important primarily in making sheet from cellulosics and acrylic polymers. Uniaxial and biaxial orientation, metallizing, and radiation cross-linking, although part of the manufacturing method for specialized film and sheeting, are generally considered postfinishing operations after film and sheet forming.

Chemical Regeneration. Chemical regeneration from a solution is the oldest commercially significant film-forming process. Its primary application is in the manufacture of cellulose film by regeneration of cellulose from viscose as illustrated in Figure 1.

In this process, wood cellulose is first steeped in caustic soda solution to convert it to alkali cellulose. Then it is pressed to remove excess caustic, and shredded into light, fluffy crumb form. After aging (from a few hours to several days, depending on the viscose viscosity desired), the alkali cellulose is converted by reaction with carbon disulfide to sodium cellulose xanthate, which is dissolved in dilute aqueous caustic solution forming viscose. Cellophane is made by extruding viscose through a long narrow slot into a coagulating regenerating bath consisting of sulfuric acid and sodium sulfate dissolved in water. The film is washed to remove acids, salts, and other products, desulfurized, washed again, and bleached if desired. After washing in cold water, the

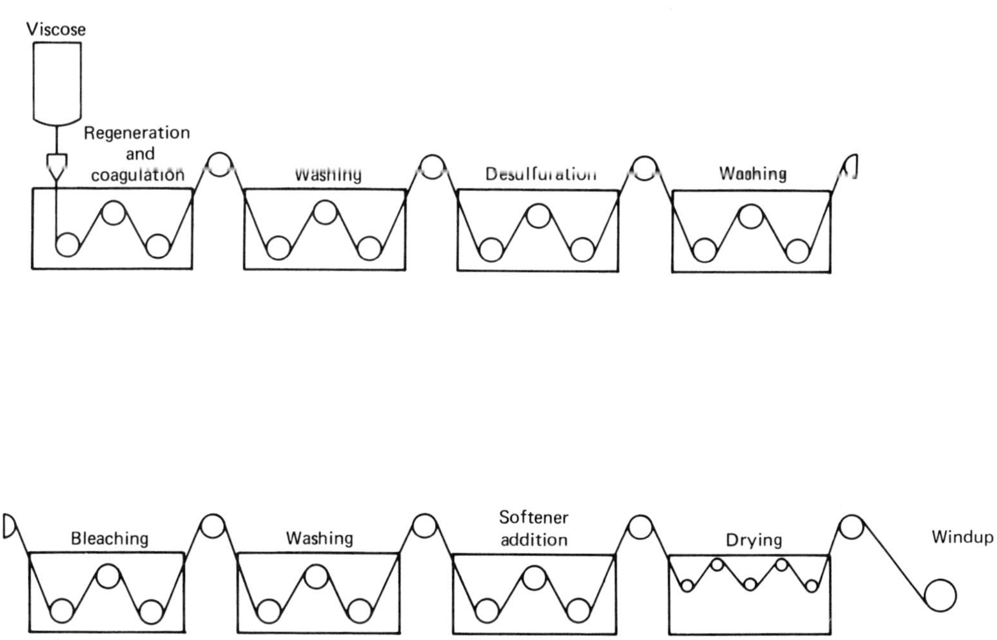

Figure 1. Manufacture of cellophane.

film goes through a softening bath, where plasticizer is added, and is then dried to the desired moisture level. Dried film is wound onto rolls for commercial applications.

Most varieties of cellophane go through an additional step during which a moisture-proof coating is applied to provide heat sealability and improved barrier properties. However, this complex process introduces many variables affecting the properties of the product. The major variables are molecular weight of the cellulose; rate of coagulation and regeneration; degree of stretch applied during the regeneration steps; amount of washing to remove impurities; amount of plasticizer added; rate and degree of drying; orientation developed; and type of colorant added.

Melt Forming. Melt-forming processes encompass calendering and three methods of extrusion: blown bubble, slot-die casting, and coating on a substrate.

Calendering. In calendering, a continuous film is formed by squeezing a thermoplastic material between two or more horizontal metal rolls, as shown in Figure 2. The polymers and other ingredients are compounded into a plastic mass and fed to the top rolls of the calender. The mass passes through successive nip rolls which mix it and reduce it in gauge (thickness). Further reduction in gauge may be effected by overdriving the film web take-off rolls (running those rolls faster than the calender rolls) to stretch the web before it cools. Surface finish of the film or sheet is controlled by the finish on the calender rolls or by the use of embossing rolls following the calender rolls. The film is then cooled, slit to the desired width, and wound on cores (see Coating processes).

Calendered film and sheet properties are influenced by (*1*) composition of the compound, ie, resin type and molecular weight, and type and amount of plasticizer and stabilizer; (*2*) degree of squeeze in each nip; (*3*) previous heat treatment of the plastic mass; (*4*) amount of stretching in the machine direction; and (*5*) type of finish applied to the surface.

The calendering process is employed chiefly to manufacture poly(vinyl chloride), vinyl chloride–acetate copolymer, and polyurethane film and sheeting. It is most economical for making relatively thick films. Vinyl chloride copolymer accounts for most calendering production today.

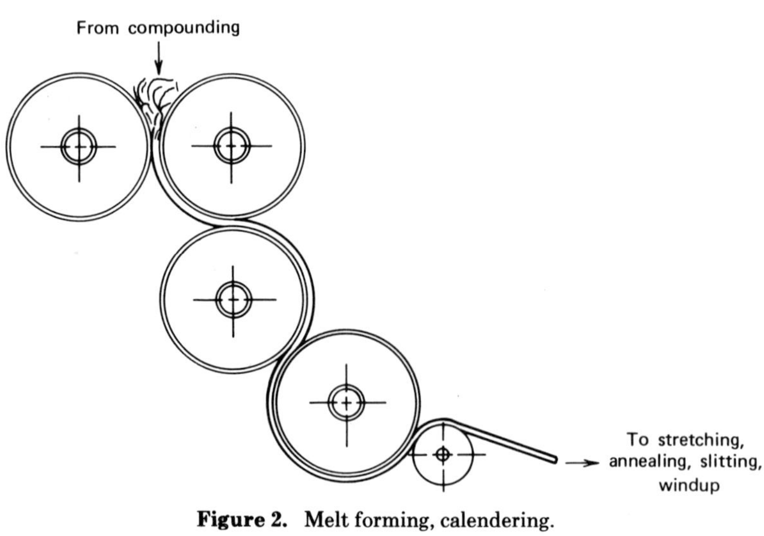

Figure 2. Melt forming, calendering.

Extrusion. In the blown-bubble extrusion process, shown in Figure 3, a tube is extruded from an upright annular die, inflated by air to a size dictated by the film properties desired and limited by feasibility, cooled with refrigerated blown air, collapsed to a flat tube, and wound into rolls of either slit or tubular film.

Common annular die openings range from 0.38 to 1.52 mm. Diameters of dies are on the order of three to six times the diameter of the extruder screw. Dies can be fed from either the bottom or the side. Tubing is generally extruded upward but it may be extruded horizontally or downward. Air is trapped in the bubble by the die at one end and by pinch rolls at the other.

As the melt emerges from the die and starts to cool, the tube is expanded to two or three times the die diameter. Air velocity, air temperature, and point of impingement of air from the cooling ring onto the outer surface of the bubble may be varied to give the properties desired in the film. The bubble may also be cooled by an internal air-cooling ring. The rate of cooling of the film determines the production rate, and is very critical in terms of optical properties obtained. Rapid cooling produces clear glossy

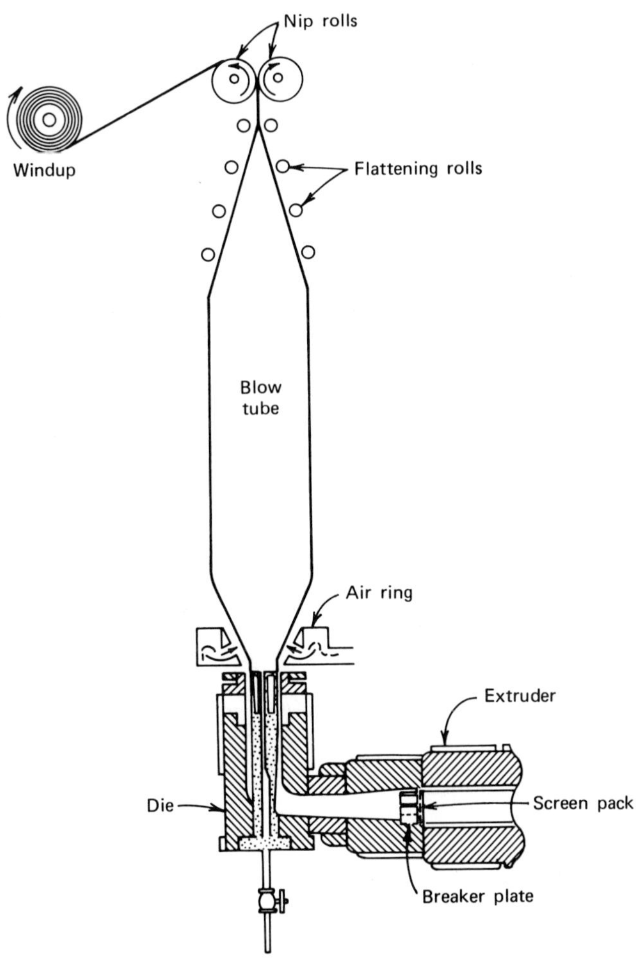

Figure 3. Melt forming, blown film extrusion.

films with fine crystalline structures. Impact and tensile strengths obtained in the film are directly related to the amount of stretch applied to the bubble.

The cooled bubble goes into a V-shaped frame which collapses the cylindrical tube into a flat tube. A set of nip rolls at the apex of the collapsing frame trap air in the bubble and draw the film away from the die. Surfaces of polyolefin films that are to be printed are treated by corona discharge before slitting and wind-up.

The blown-bubble process is simple, requires relatively low investment per kilogram of product, and can produce a film with balanced tensile and impact properties in machine and transverse directions. This process is used primarily to produce polyethylene films and to a lesser extent for vinyls, nylons, poly(vinylidene chloride), and other polymers.

The tubular water bath process is similar to the blown-bubble process except that the thermoplastic is quenched by a water-cooled mandrel inside and a water bath outside the tube of film. Subsequent operations remove the water and slit the film before winding. Polypropylene is the material most frequently processed by this method because of the crystal clarity achieved by the rapid quench.

In slot-die extrusion (see Fig. 4), a thin section of polymer melt is extruded from a slot die onto a chill or quenching roll that rapidly cools the melt. The quenching roll is overdriven relative to the linear rate of melt flow from the die to effect a drawdown, ie, a reduction in thickness from ca 0.51 mm to the range of 0.013–0.038 mm required for most applications. Alternatively, the melt is extruded directly into a liquid cooling bath. In either case, the basic objective is rapid quenching of the film, which generally results in excellent optical properties.

The quench roll is constructed internally to circulate chilled water so as to provide a uniform temperature. An air knife may be employed to lay the film onto the chill roll at a precise contact point. Edges of the film, which are usually thicker than the rest of the web because of polymer neck-in, are trimmed off prior to winding film onto rolls.

Critical factors in slot extrusion are die design for uniformity of gauge, rate of

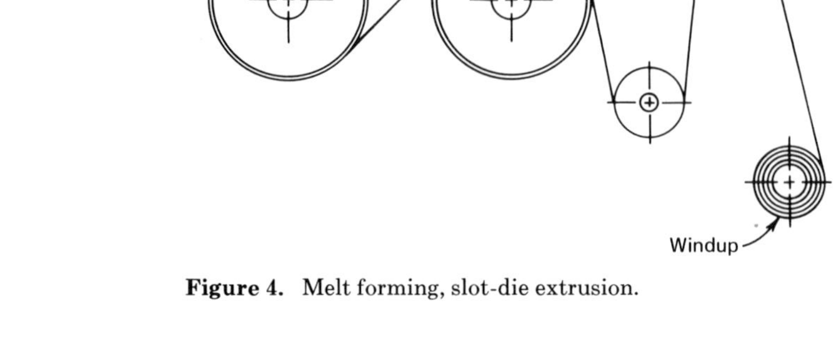

Figure 4. Melt forming, slot-die extrusion.

cooling, and degree of drawdown of melt. Since film made by this process is drawn in only one direction, strength properties are not balanced in machine and transverse direction. The film produced is used in applications in which balanced strength properties are not critical. Slot-die extrusion or casting is most commonly employed in making polyethylene and polypropylene films.

Sheet is made by an extrusion casting process similar to film extrusion casting except that three or more polished rolls are used in a vertical stack as shown in Figure 5. The polymer is either extruded into the nip formed by the top two rolls or cast down across the first roll, using an air knife to ensure roll contact. After the sheet is cast onto the first polished roll, subsequent rolls are used for heating or cooling, to relax or orient the film. Sheet extrusion into the nip of the roll stack is advantageous for 0.76 mm or thicker sheet. Sheet is extruded as close as possible to the desired product gauge directly from the die into the nip of the chill rolls. The chill rolls then give the final surface characteristics and cool the polymer. The extrudate is pulled either to a shear knife or to a turret winder where the sheet is wound or stacked.

The third extrusion method, coating on a substrate, is illustrated in Figure 6. In this process a high temperature melt is extruded from the lips of a flat die into the nip of two rolls, where it is bonded to the substrate under pressure. The roll that is in contact with the melt is chilled to solidify the melt; the roll that is in contact with the substrate is a rubber-surface pressure roll.

Extrusion coating requires a machine of high heat capacity to maintain the high melt temperature necessary for a good bond between melt and substrate. The melt should meet the substrate at the nip of the rolls or slightly over the rubber roll. If the extruded melt touches the chill roll first, a poor bond to the substrate is obtained.

The speed of the chill roll is adjusted to draw the extruded coating to the desired thickness. The pressure of the rubber roll is not critical. The bead that is formed along

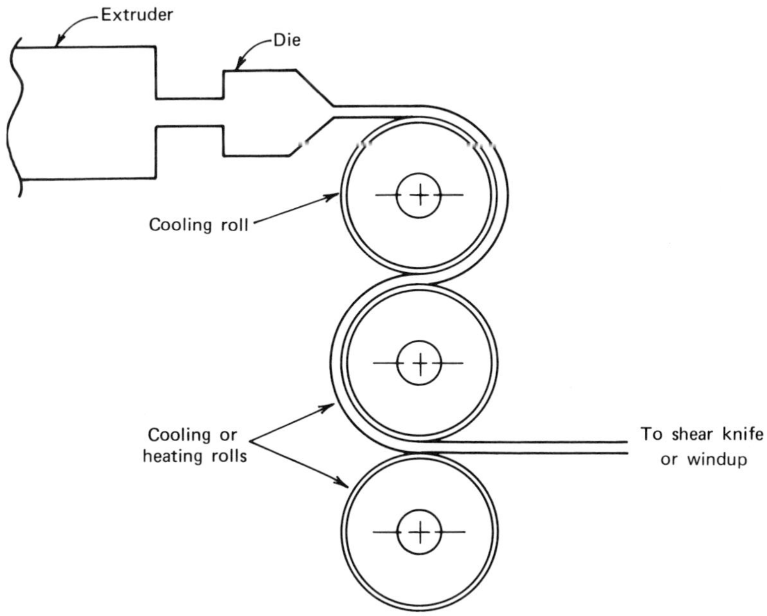

Figure 5. Melt forming, sheet extrusion casting.

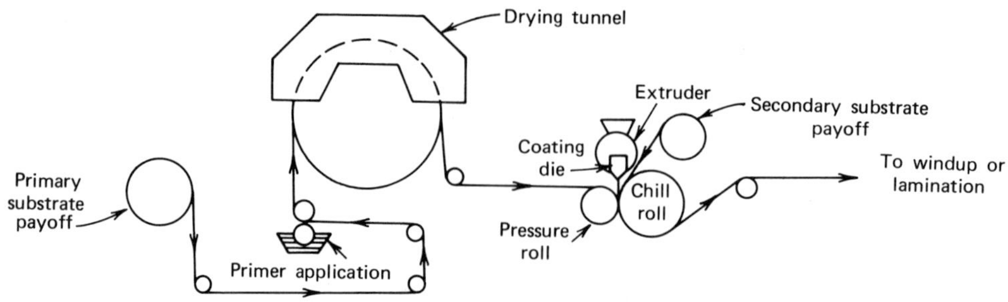

Figure 6. Melt forming, extrusion coating on a substrate.

each edge of the coated material must be trimmed before the film is wound on rolls.

Properties of materials produced by extrusion coating are generally governed more by the substrate than by the extruded layer. Coatings are generally employed to effect or improve heat sealability and barrier properties. This process is used most extensively for coating polyethylene onto cellophane, paper, polypropylene, and polyester substrates.

Coextrusion is widely used for producing multilayered film and sheet. In this process, two or more polymers are combined in the melt state either before, in, or immediately after exit from a die. It can be used in all major film and sheet extrusion processes. Only materials that have similar viscosity and thermal stability at the extrusion temperature can be combined via coextrusion. Poor interlayer bonding can be alleviated by use of an adhesive layer between the two dissimilar materials.

Coextrusion allows the use of thin layers of more costly polymers to reduce total gauge of a product while improving properties such as MVTR or gas barrier. It may add heat sealability to films that would not normally heat seal, as well as eliminate the cost of lamination. In some coextruded constructions cost is reduced by using a cheap material in the inner layer of a three-layer construction. The two outside layers provide surface properties and appearance. The inner layer may also be foamed.

Solution Forming (Solvent Casting). This method of film and sheet formation is not used extensively today for commercial production, since it presents serious problems of solvent recovery and safe handling of flammable solvents. In addition, the speed of the operation is limited by the migration rate of the solvent from the web. It is widely used in polymer research, however, because enough film for testing can be made from a very small sample of polymer. One of the most commonly employed solvents is xylene.

Unsupported Film and Sheet. The polymer plasticizer and other ingredients are dissolved and pumped through a filter and heat exchanger to a die. From the die the web is laid onto a belt or drum (Fig. 7), or poured in a suitable mold (or cell). Casting onto an endless belt is the more common commercial process. The problem associated with this method is the removal of solvent from the web. Solvent removal is not required in the case of acrylic materials where polymerization is completed by curing with heat.

The rotating-drum system is generally used in pilot plants or semiworks operations.

Film and sheeting made by solvent casting have excellent optical properties and

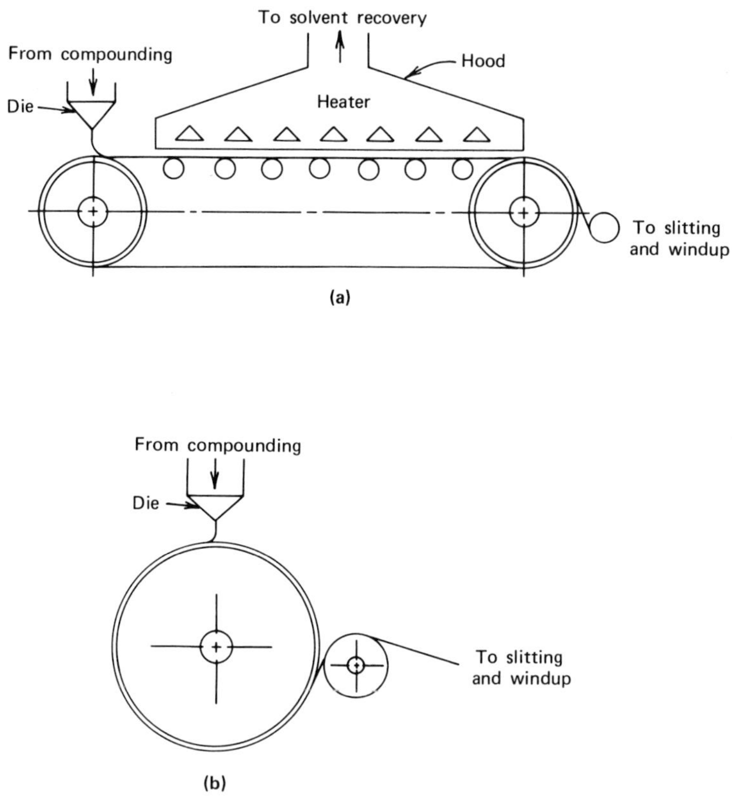

Figure 7. Solution forming: (**a**) on a belt; (**b**) on a drum.

very little orientation. This process is used commercially for making vinyls, acrylic polymers, cellulosics, polycarbonate, polyethersulfone, and some fluoroplastics.

Solution Casting on a Substrate. Solution casting techniques are employed extensively for applying water, MVTR, and gas barrier coatings on cellophane, biaxially oriented polypropylene, paper, cloth, and other substrates. Solutions are applied by flexographic roll, running the web through a bath of solution, and laying solution on the web and removing the excess with a doctor blade. Webs coated on only one side are dried on support rolls or on an endless belt in an oven. Webs coated on two sides are dried by passing them upward through a tower until the coating is tack-free. The coated films are then slit and wound onto rolls.

Transfer Coating. This process is a combination of casting and coating. Highly napped fabrics are transfer coated with urethanes (eg, for simulated suede leather) by casting polyurethane film on a release sheet, casting a second solution on the dried film, and then laminating the fabric to the wet film. The composite structure is dried and the release paper stripped away (see also Leather-like materials).

Emulsion, Plastisol, and Organosol Forming. This process is employed commercially for coating substrates but has little, if any, commercial significance in the production of unsupported or free films. The techniques shown in Figure 7 are used.

Emulsion coating is used in coating cellophanes and oriented polypropylene with vinylidene chloride polymers. Coating of textile substrates with plastisols or organosols

Table 7. Production of Major Film and Sheeting Materials, 1000 Metric Tons

	1967	1968	1969	1970	1971	1972	1973	1974	1975	1976	1977
LDPE, film	476	567	726	930	1057	1293	1455	1570	1261	1449	1657
polystyrene, film and sheet	168	204	238	40.4	434	442	556	614	500	415	455
foamed sheet	87.5	97.5	113	110	120	153	166	189	170	172	195
poly(vinyl chloride) and copolymers, film and sheet	410	410	465	454	451	542	482	453	370	383	393
cellophane film	180	176	163	153	147	144	147	151	142	135	99
polypropylene, film	34[a]	25[a]	25[a]	35[a]	45[a]	52[a]	65	63	64	79	91
sheet							17	26	34	36	42
HDPE, film	22.6[a]	36.3[a]	40.8[a]	34[a]	40.8[a]	21	33	42	40	43	59
sheet[b]						32	33	33	18	25	28
acrylic, sheet	32	35	38	43	45	50	55	60	59	61	65
polycarbonate, sheet	10.9	12.2	14	18	20.4	25	31	51	39	51	57
cellulosic, ≤0.76 mm	8.2	8.2	7.3	7.2	6.9	6.4	6.2	4.2	3.3		
>0.76 mm	21	24	27	25.3	23.2	21	22	39	40	48.5[a]	47[a]

[a] Film and sheet.

[b] Sheet over 0.3 mm.

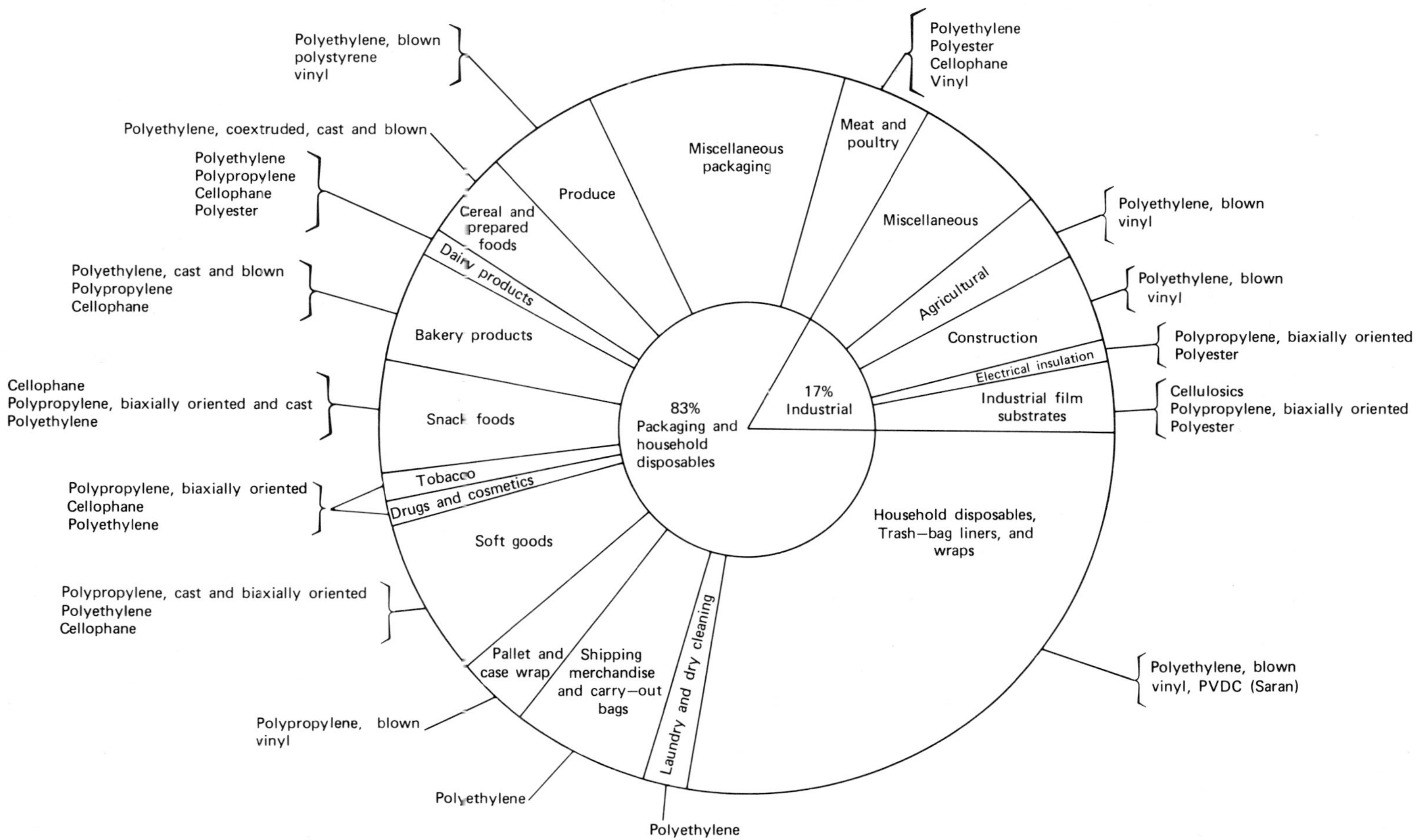

Figure 8. Major markets for film materials.

is employed in making fabrics for tents, tarpaulins, carpet backing, synthetic turf, and similar products. These coatings are applied by extrusion through a die, reverse-roll coating, cast coating, curtain coating, and spray coating.

Biaxial Orientation of Film and Sheet. Biaxially oriented film and sheeting is produced by tentering or bubble blowing. Tentering, which has been used extensively in the textile industry, consists of stretching a formed flat sheet first in the machine direction over heated rolls, and then in the transverse direction with a tenter-frame within an oven. In bubble-blowing orientation, the film can be stretched simultaneously in machine and transverse directions. In both processes, the film is subjected to a postorienting heating cycle for annealing (heat set) purposes. Shrink film is not annealed. Materials available in oriented film and sheet are polyester, polyethylene, polypropylene, poly(vinyl chloride), and polystyrene.

Postfinishing Operations. Film and sheet properties can be altered by treating for printability, cross-linking the polymer by ionizing radiation, and depositing a thin metal coating using vacuum metallizing (see Film deposition techniques). Many materials, especially polyolefins, need to be treated for ink adhesion. The most common method of treatment is corona discharge but flame and chemicals can also be used. In cross-linking by ionizing radiation, the radiation breaks chemical bonds creating highly reactive free radicals that form bonds with adjacent molecules. Cross-linking can impart shrink memory in heat-shrinkable thermoplastic products, improve crack resistance, abrasion resistance, and service temperature, and control cell structure and improve toughness in polyethylene foam.

In vacuum metallizing, thin films of metals or nonmetals are deposited under high vacuum on the surface of plastics. Aluminum is usually the deposited metal. It has been used in the automotive and appliance industries for many years, and has recently experienced a rapid growth in packaging films because of the decorative appeal. It substantially improves film barrier properties and is lower in cost than laminating with aluminum foil.

Forming a film by spraying a powder on the heated surface of a drum is not important commercially.

Economic Aspects

Table 7 gives the production of major film and sheet materials over the past decade. Cellophane production has decreased 41% since 1974 as it is replaced with biaxially oriented polypropylene. The markets for film materials depicted in Figure 8 show that the largest outlet is the packaging industry with more than 1.7×10^9 metric tons consumed in 1977. Polyethylene is the largest volume film material, and more than 56% of total LDPE production is consumed as film. Packaging uses for the major film materials are listed in Table 8, typical prices in Table 9, and trade names in Table 10.

Nonpackaging applications account for ca 45% of the usage of all film materials and cover a wide range, the largest segment being household trash liners, bags and wraps. Among industrial applications, agriculture and construction consume about half the volume of film materials. The relatively small-volume industrial application of film substrates is a high-value, high-performance market consisting of magnetic recording tape, electrical insulation, photographic film, microfilm, and pressure sensitive tape. Among these, the highest-volume materials are made of biaxially oriented polyester and polypropylene (see Magnetic tape; Insulation, electric).

Table 8. Packaging Uses (1979) for Cellophane, Polypropylene, Polyethylene and Poly(vinyl Chloride), Percentage of Total Material Consumed

		Polypropylene			
Packaging use	Cellophane	Cast	Biaxially oriented	Polyethylene	Poly(vinyl chloride)
meat and poultry	3			2	47
snack foods	47	12	34	3	
bakery products	12		10	6	
dairy products	3		6	1	
prepared foods	6			3	
produce	1			7	
cereal and mixes	5			1	
tobacco	11		17	<1	
drugs and pharmaceuticals	4		2	<1	
soft goods	2	88	1	8	5
pallet and case wrap				5	4
shipping, merchandise and carry-out bags				9	
laundry and dry cleaning				4	
household disposables, trash bags and wraps				39	
miscellaneous packaging	6		30	11	44

Table 9. Typical Average Prices of Film Materials, 1978

Material	Price, ¢/kg	0.025 mm thickness, ¢/m^2
low density polyethylene	25–29	2.9–3.3
high density polyethylene	28–34	3.4–3.9
ethylene-vinyl acetate copolymer	27–35	3.2–4.0
polypropylene, extrusion cast	32–34	6.5–7.4
biaxially oriented	59–78	7.0–12.9
polyester	79–95	13.1–19.8
nylon-6	76–84	10.7–12.3
poly(vinyl chloride)	38–67	5.5–10.6
cellophane	71–86	12.5–15
cellulose acetate	84–118	13–19
ionomer	73	8.5
polystyrene, oriented	26–36	3.5–4.8
polybutylene	39–46	4.4–5.2
poly(vinylidene chloride)	72–120	13.9–25
poly(vinyl alcohol)	104	16.3

Foamed sheet accounts for 28% of the sheet market. It is thermoformed into meat, egg, and poultry trays, and food and drink containers, polystyrene being by far the predominant material. Nonfoamed thermoformed applications, accounting for 40% of the total market, consist of blister and skin packaging, single-service food and drink containers, toys, appliance housings, and stamped automotive interior panels. Polystyrene is again predominant, with lesser amounts of polypropylene, vinyls, and HDPE being used. Lighting, outdoor signs, and window glazing account for 12% of sheet volume. Mostly used in this category are acrylic polymers, polycarbonate, and cellulosics. The balance of the sheet applications consist of vinyl apparel, wall coverings, upholstery, and rainwear.

Table 10. Trade Names of Film and Sheeting Material[a]

Material and trade name	Manufacturers
Acrylic polymers	
Acrylite	American Cyanamid Co.
Electroglas	Electro-Seal Glasflex Corp.
Gravoply	Hermes Plastics, Inc.
Koracryl H	Koro Corp.
Korad	Rohm & Haas Company
Lumasite	American Acrylic Corp.
Metallex	Hermes Plastics, Inc.
Plexiglass	Rohm & Haas Company
Swedcast	Swedlow, Inc.
Ultra-Glas	Fortin Plastics Co.
ABS	
Abselex	British Celanese, Ltd.
Boltaron	General Tire & Rubber Company
Gravo Flex	Hermes Plastics, Inc.
Nabutene	Convert, Ets. G.
Cellophane	
	various numerical and letter designations with manufacturer's name; E. I. du Pont de Nemours & Co., Inc.
	Olin
Cellulosics	
Cellonex	Dynamit Nobel, A.G.
Forticel	Celanese Plastics Company
Kodacel	Eastman Kodak Company
Rhodophane	M & B Plastics Company
Vuepak	Monsanto Company
Fluoroplastics	
Aclar	Allied Chemical Corp., Plastics Division
Coboflow	Cobon Plastics Corp.
Fluorofilm	Dilectrix Corp.
Gore-Tex	Gore, W. L. Associates, Inc.
Memoryline	Shamban, W. S., & Company
Tedlar	E. I. du Pont de Nemours & Co., Inc.
Teflon	E. I. du Pont de Nemours & Co., Inc.
Ionomer	
Sur-Flex	Flex-O-Glass, Inc.
Rubber hydrochloride	
Pliofilm	Goodyear Tire & Rubber Company
Nylon-6	
Capran	Allied Chemical Corp., Plastics Film Dept.
Trogamid T	Dynamit Nobel of America, Inc.
Polycarbonate	
Lexan	General Electric Company
Makrolon	M & B Plastics, Ltd.
Rowlex	Rowland Products, Inc.
Celanex	Celanese Plastics Company
Merlon	Mobay Chemical Company
Polyester	
Celanar	Celanese Plastics Company
Melinex	Imperial Chemical Industries, Inc.
Scotchpak	3M Company
Terphane	Rhodia, Inc.
Kodar	Eastman Chemical Products, Inc.
Mylar	E. I. du Pont de Nemours & Co., Inc.
Polyethylene	
Alathon	E. I. du Pont de Nemours & Co., Inc.
Ctyovac L	W. R. Grace & Company
Dynafilm	U. S. Industrial Chemicals Co.
Durethene	Sinclair Koppers Company
Fortiflex	Celanese Plastics Company
Polyethylene	
Clysar EH	E. I. du Pont de Nemours & Co., Inc.
Kordite	Mobil Oil Company
Nor Film	Northern Petrochemical Company
Poly-Star	St. Regis Paper Company
Polyfilm	Dow Chemical Company
Valeron	Van Leer Plastics, Inc.
VisQueen	Ethyl Corp.
Zeelon	Crown Zellerbach Corporation
Zendel	Union Carbide Corp.
Polypropylene	
Olephane	Crown Zellerbach Corporation
Propafilm	ICI America, Inc.
Bicor	Mobil Oil Corp.
Clysar EH C	E. I. du Pont de Nemours & Co., Inc.
Cryovac Y	W. R. Grace & Company
Dynafilm	U.S. Industrial Chemicals Company
Proprafilm	Imperial Chemical Industries
Udel	Union Carbide Corp.
Polystyrene	
Celatron	British Celanese, Ltd.
Cellofoam	United States Mineral Products Co.
Dri-Lite	Poly Foam, Inc.
Dylene	Sinclair Koppers Company
Santoclear	Mitsubishi Monsanto Chemical Co.
Kardel	Union Carbide Corp.
Polyflex	Monsanto Company
Trycite	The Dow Chemical Company
Polyurethane	
Artfoam	Strax Corp.

Table 10 (*continued*)

Material and trade name	Manufacturers	Material and trade name	Manufacturers
Nimbus	General Tire & Rubber Company	Flovic	Imperial Chemical Industries, Ltd.
Tufcote	Speciality Converters, Inc.	Gold Kast	Continental Plastics Company, Division CPI, Inc.
Tuftane	B. F. Goodrich Company		
U-Thane	Upjohn Company	Krystal	Allied Chemical Corp.
Vinyls		Plastifilm	Gross Plastic Film Corp.
Amer-Plate	Ameron Corrosion Control Div.	Profilm	Protective Lining Corp.
Astradur	Dynamit Nobel of America, Inc.	Rucoam	Hooker Chemical Corp.
		Saran	Dow Chemical Company
Bolta Flex	General Tire & Rubber Company	Vista Vinyl	Diamond Shamrock Corp.
		Reynolon	Reynolds Metals Company
Clopane	Clopay Corp., Plastics Film Div.	Cryovac	W. R. Grace & Company
		Vitafilm	Goodyear Tire & Rubber Company
Darvic	Imperial Chemical Industries, Ltd.	Visten	Union Carbide Corporation

[a] This is a representative list of major U.S. manufacturers only. Polyethylene film, for example, is manufactured by over 125 firms.

BIBLIOGRAPHY

"Film Materials" in *ECT* 2nd ed., Vol. 9, pp. 220–244, by Donald T. Surprenant, Union Carbide Corporation.

1. J. Agranoff, ed., *Modern Plastics Encyclopedia,* McGraw-Hill, Inc., New York, 1977, pp. 492–496.
2. *Chem. Week,* 36 (June 20, 1979).

General References

J. Agranoff, ed., *Modern Plastics Encyclopedia,* McGraw-Hill, Inc., New York, 1977; a very good reference for properties and forming methods of film and sheeting materials.

N. M. Bikales, ed., *Encyclopedia of Polymer Science and Technology,* John Wiley & Sons, Inc., New York, 15 volumes, 1964–1972.

J. H. Briston and L. L. Katan, *Plastic Films,* John Wiley & Sons, Inc., New York, 1974; discusses the technology, conversion, and application of plastic films.

J. H. Briston and L. L. Katan, *Plastics in Contact with Food,* Technomic Publishing Company, Westport, Conn., 1974.

J. Brandup and E. H. Immergut, ed., *Polymer Handbook,* 2nd ed., John Wiley & Sons, Inc., New York, 1975.

F. Bruins, ed., *Packaging with Plastics,* Gordon & Breach Science Publishers, New York, 1974.

E. G. Crosby and S. N. Kochis, *Practical Guide to Plastics Applications,* Cahners Publishing Company, Boston, Mass., 1972; discusses all phases of polymer processing.

J. H. DuBois, *Plastics,* 5th ed., Van Nostrand Reinhold Company, 1974; a handbook of plastics.

C. A. Harper, ed., *Handbook of Plastics and Elastomers,* McGraw-Hill Book Company, New York, 1975; an excellent textbook on the properties of plastics and elastomers.

C. J. Hilado, *Flammability Handbook for Plastics,* 2nd ed., Technomic Publishing Company, Westport, Conn., 1974; an introduction to the fire technology of plastics containing data on this subject.

M. I. Kohan, *Nylon Plastics,* John Wiley & Sons, Inc., New York, 1973.
R. P. Lukens and co-eds., *ASTM Standards, Part 35: Plastics—General Test Methods: Nomenclature,* American Society for Testing and Materials, Philadelphia, Pa., 1978.
R. P. Lukens and co-eds., *ASTM Standards, Part 36: Plastics—Materials Film, Reinforced and Cellular Plastics; High Modulus Fibers and Their Composites,* American Society for Testing and Materials, Philadelphia, Pa., 1978.
Y. L. Meltzer, *Foamed Plastics 1976: Recent Developments,* Noyes Data Corporation, Park Ridge, N.J., 1977.
J. G. Mohr and co-eds., *SPI Handbook of Technology and Engineering of Reinforced Plastics/Composites,* 2nd ed., Van Nostrand Reinhold, New York, 1973.
C. R. Oswin, *Plastic Films and Packaging,* John Wiley & Sons, Inc., New York, 1975; includes costs and benefits of various materials.
S. Sacharow, *Handbook of Packaging Materials,* Avi Publishing Company, Westport, Conn., 1976; a comprehensive discussion on all types of packaging materials.
S. Sacharow and R. C. Griffin, Jr., *Basic Guide to Plastics in Packaging,* Cahners Publishing Company, Boston, Mass., 1973.
B. C. Wendle, ed., *Engineering Guide to Structural Foams,* Technomic Publishing Company, Westport, Conn., 1977.
L. R. Whittington, *Whittington's Dictionary of Plastics,* Technomic Publishing Company, Westport, Conn., 1968; a reference containing more than 3110 basic terms with their scientific, commercial, and legal definitions.

The following periodicals are a very good source of information on recent developments in film and sheeting materials, forming processes, and end use markets:

Modern Packaging, McGraw-Hill, Inc., New York.
Modern Plastics, McGraw-Hill, Inc., New York.
Package Engineering, Cahners Publishing Company, Chicago, Ill.
Paper, Film, and Foil Converter, Peacock Business Press, Inc., Park Ridge, Ill.
Plastics Design Forum, Industry Media, Inc., Denver, Col.
Plastics Design and Processing, Lake Publishing Corporation, Libertyville, Ill.
Plastics Engineering, Society of Plastics Engineers, Greenwich, Conn.
Plastics World, Cahners Publishing Company, New York.

Trade associations that frequently publish bulletins concerning use, economics, manufacturing, and testing of film and sheeting materials:

American Society of Testing and Materials, Philadelphia, Pa.
National Flexible Packaging Association, Cleveland, Ohio.
The Packaging Institute, U.S.A., New York.
The Society of Plastics Engineers, Greenwich, Conn.
The Society of the Plastics Industry, New York; SPI has a number of publications, which are listed in *SPI Printed Matter;* statistical data is available in monthly *Statistical Report on Thermoplastic and Thermosetting Resins* and in the annual *Facts and Figures of the Plastics Industry.*
The Technical Association of the Pulp and Paper Industry, TAPPI, Atlanta, Georgia; TAPPI sponsors annual seminars in various phases of plastic processing.

E. LEA CRUMP
Gulf Oil Chemicals Company

FILM DEPOSITION TECHNIQUES

Thin films have been used for decorative purposes for well over a thousand years. One of the earliest thin films was "gold leaf" formed by controlled beating of gold to a thickness of 0.1 μm. As early as the seventh century, glass and ceramics were decorated by painting a pattern with a silver salt, then heating the painted object to cause reduction of the salt to a thin film of metallic silver.

From these early examples have evolved the unusually diverse types of thin films currently used. Included are the common electroplated films used for decorative and protective purposes as well as the conductive, resistive, magnetic, insulating, semiconducting, and even superconducting films used in microelectronics (see Electroplating; Magnetic materials; Semiconductors).

Thin films can have different properties than the bulk material. They generally exhibit higher resistivity than bulk material of similar chemical composition probably because of the higher concentration of structural defects generated by the deposition process. The most common thin film depositional anomalies are dislocations formed by stresses arising in film buildup, and from strains due to mismatch of thermal coefficients of expansion between substrate and deposited film.

Plasma and flame spraying methods for depositing coatings are not included in this article (see Plasma technology); the coatings produced by these techniques are normally in excess of the usual thin film thickness of ca 3 μm.

In addition to metallic and inorganic films, organic polymeric materials can also be applied as thin films. In order for a polymeric thin film to be considered for a thick film application, it should be able to form a homogeneous uniform layer less than 3 μm thick. This usually excludes the conventional thick polymeric coatings applied by spray, brush, or roller coat that require heat or catalytic curing, oxidation, or fusion (see Coatings; Paint).

Desirable properties of thin films include adherence, uniform thickness, impermeability, and absence of pinholes. Films of dielectric materials may also require inertness and thermal stability.

Deposition of Films From Solution

Electrolytic Deposition. ***Cathodic Films.*** Preparation of metal coatings by electrolytic deposition on the cathode of a cell, one of the oldest known methods for making films, was first employed by Jacobi in 1838 (1). The method involves the movement of metallic ions in solution to the electrodes under the influence of the applied electric field. At the cathode, the ions accept electrons, are neutralized, and incorporated into the metal lattice. The process, which is actually quite complex, is described in ref. 2 and Electroplating.

The boundary between cathode and electrolyte displays a very complicated structure during electrolysis and includes the adjacent region of the electrolytic bath, the cathode diffusion layer. Electrodeposition begins at cathode surface imperfections. During electrolysis, the cations migrate toward the cathode where they are deposited (with the exception of nonreducible cations, such as those of the alkali metals, magnesium and aluminum). The concentration of cations remains essentially unaltered within the electrolyte, and approaches zero immediately at the cathode. Thus, a concentration gradient is established with diffusion tending to restore the original, uniform concentration. The action of diffusion is superimposed on that of the electric field.

The supply of cations to the cathode is affected by the following parameters: (*a*) current density; (*b*) temperature and agitation of the electrolyte; (*c*) convection in the immediate vicinity of the cathode; (*d*) migration and diffusion velocity of the ions; (*e*) specific weight of the ions; and (*f*) geometric form of the cathode (distribution of current lines). All these factors influence the conditions of film formation (thickness of the diffusion layer) and the final film structure and properties.

The double layer at the surface of the cathode appreciably hinders penetration of metal ions, but contamination of the substrate (cathode) constitutes an even more severe impediment. Addition agents will reach the cathode and be adsorbed on it, frequently competing with cations for favorable lattice positions. These agents are organic or inorganic compounds added to the bath for buffering, as oxidizing agents, for brightening, complexing, bubble inhibition, reduction, etc. Thus their effect is of electrochemical as well as crystallographic nature (see Electroplating for a detailed discussion of the technology, and ref. 3 for nucleation and epitaxial growth of electrodeposited metal films).

Anodic Films. Anodized aluminum coatings have been known for ca 150 years. The first practical coating, that of a "duraluminum" seaplane, employed chromic acid which was subsequently replaced with sulfuric acid as the anodizing electrolyte. On becoming the anodes of electrolytic cells, many metals form a protective oxide film, frequently an extremely hard, compact well-adhering coating.

Metals on which anodic films can be formed are known as valve metals because of the rectifying characteristic of their anodic oxides. Among them are: aluminum, antimony, beryllium, bismuth, germanium, hafnium, magnesium, niobium, silicon, tantalum, tin, titanium, tungsten, uranium, and zirconium.

In addition to these metals, alloys of molybdenum and vanadium with low proportions of valve metals have also been anodized (4). The electrolytic cell used for the production of anodic oxide films contains an electrolyte in which a valve metal anode and an inert cathode are immersed while constant current and voltage are maintained. When the positive terminal is connected to the anode and the negative to the cathode, electrolysis and simultaneous growth of the oxide film on the anode result. Increased thickness of oxide is accompanied by an increase in resistance, and the rate of growth drops to a negligible value. The maximum thickness of anodic films is limited by the physical makeup of the film surface characteristics of substrate and the electrolyte composition. Film breakdown is characterized by oxygen evolution and sparking. The anodization mode involves the migration of oxygen, metal, or both, through the existing oxide film, the detailed mechanism of film growth is of some complexity (5). The basic requirement is a diffusion mechanism for ions through the film. It is generally assumed that it is the metal ions that travel through the oxide and that the metal oxide interface during growth is in dynamic equilibrium.

The rate of increase of film mass is proportional to the current under conditions where the electronic leakage current is negligible, so that dl/dt is constant where l = film thickness. The voltage/time relation is linear if the structure and composition of the film are assumed to be constant during formation and a constant differential field strength dV/dl is required (V = applied voltage) to maintain a constant ionic current.

$$\frac{dV}{dt} = \frac{dV}{dl} \cdot \frac{dl}{dt}$$

In practice, a slight dependence of dV/dl on current is observed. For tantalum at 18°C, the value of dV/dt for a current density of 10 mA/cm^2 is ca 3.6 V/s (6).

An electrolytically polished parent metal surface is required to produce well-adhering oxide films without pinholes because anodized surfaces can incorporate water and OH ions into the films. These can affect dielectric behavior and be a latent cause of corrosion.

The use of the anodic oxidized products includes aluminum protective coatings, aluminum and tantalum electrolytic capacitors and electronic tunnel devices. An unusual use of anodized aluminum film is as a substrate for transmission electron microscopy. After removal from the aluminum metal, the oxide is self-supporting in thickness down to less than 10 nm and is amorphous, so that no appreciable detail is seen in the micrograph or diffraction pattern. Heat treatment of amorphous anodic films can yield single-crystal films (7–8).

Tantalum oxide films exhibit unusual, intense interference colors and have been used to estimate thickness of dielectric films. However, aluminum oxide film interference colors are rather flat because of high reflectivity of underlying aluminum metal.

Chromate Conversion Coatings. Chromate conversion coatings, which are applied in thickness of 2.5 to 5 μm, are used for protection of aluminum and its alloys. The coating is applied by immersion or is rolled on rather than being electrodeposited. Although these coatings offer good chemical resistance to weather, they are not wear-resistant. They may be colorless or dyed. Some proprietary chromate coatings such as Alodine and Iridite employ a special coating and rinsing procedure.

Electroless Plating. Electroless plating, or chemical reduction plating, is a process that requires neither electrodes nor any external source of electricity (see Electroless plating). Also known as autocatalytic plating and occasionally referred to as chemical plating, it was invented in 1946 in an attempt to electrodeposit a nickel–tungsten alloy on the inside of a tube by adding a hypophosphite salt to decrease anodic oxidation (9). In the absence of electric current, hypophosphite ion reduced the nickel salt and plated the tube both inside and outside.

The electroless plating process is based on a chemical process involving the catalytic reduction of metal salts. The metal salt and reducer react in presence of a catalyst. The chemical reducers most often employed are sodium hypophosphite, formaldehyde, sodium borohydride, and amine boranes. Initially, the base material can be itself catalytic or be activated by use of a gold, platinum, or palladium compound. After the reaction is initiated, the deposited metal surface serves as the catalyst, thus ensuring continued buildup of metals such as nickel, palladium, and gold.

The most common electroless coating is electroless nickel plating which can be applied virtually to any substrate material including metals, plastics, ceramics, and other insulators (10–12). Electroless nickel plating is autocatalyzed, requiring no external electric current, and is accomplished by the chemical reduction of nickel ions in solutions containing either phosphite, boron, or hydrazine compounds. The nickel deposits obtained from hypophosphite solutions consist of nickel alloys containing 3 to 17% of phosphorus, depending on operating conditions of the plating bath. The wear resistance of electroless nickel produced with the phosphorus system is improved by heat treatment at 400°C and best properties are obtained when the phosphorus content is increased to 6.5–7.5%. The porosity of electroless nickel is very low compared to that of electrode-deposited nickel.

Nickel–boron plated alloys containing 0.02 to 12% boron can be produced using boron-containing chemicals as reducing agents. An interesting observation of this latter system is that the percentage of boron of the plated alloy is not a function of the bath life. In the nickel–boron alloy, boron is embedded in a nickel matrix with ca 5 to 10% of it as crystalline nickel, the remainder amorphous. Heating this nickel–boron alloy to 600°C causes a chemical reaction and changes the nickel state to a crystalline Ni_3B containing additional crystalline nickel (13). Maximum strength and ductility of nickel–boron alloys are obtained by decreasing the boron content. It has been demonstrated that a 0.0025 mm nickel–boron coating provides better corrosion protection than a coating of electrodeposited nickel ten times as thick.

Electroless plating permits plating in select areas and inside holes, making it very popular in the electronics industry for printed circuit board patterns, connector contacts, semiconductor devices, ceramic-metal integrated circuit packages, aluminum connector shells, and lead frames (14).

Polymeric Coatings. The simplest method of forming a polymer film is to dissolve the polymeric material in a solvent, paint or spread the solution over the surface to be coated, and allow the solvent to evaporate. Films produced by this technique often contain pinholes, are often porous and are not used as protective coatings. This approach suffers from another disadvantage, ie, the limited solubility of many polymers in volatile solvents. "Lacquer film" capacitors have, however, been successfully manufactured by this technique.

Vacuum Deposition of Films

The five major vacuum deposition techniques of sputtering, glow discharge, evaporation vapor plating, and ion plating are summarized diagrammatically in Figure 1. The starting materials (in square boxes) may be either solid or gas, and the type of energy applied either plasma or thermal (in the diamonds). Sputtering, for example, employs a solid target and a plasma energy source; glow discharge, a gas feedstock, and a plasma energy source.

Evaporation of Inorganic Materials. In the last century it was observed that "tarnished" lenses had higher transmittance than the clean lenses. Practical use of this phenomenon was demonstrated during World War II in the removal of unwanted sky reflections from aircraft instrumentation by application of a thin coating such as magnesium fluoride, aluminum oxide, or silicon oxide. Today, reflecting films prepared by vacuum evaporation are used to regulate the temperature of satellites by controlling solar absorbancy and thermal emittance. The vacuum evaporative process is a direct thin film deposition technique in which material is deposited by vaporization in a vacuum. Advantages of vacuum deposition are its simplicity, economy, and efficiency.

The heat required to evaporate metals and metal oxides can be introduced by resistance heating, electron bombardment, or audio-frequency induction. The rate of deposition by vacuum evaporation is generally higher than by sputtering. This may lead to structural disorder in the deposited thin film. To obtain evaporative films having well-defined chemical and physical properties, certain guidelines must be observed: (*1*) The greater the vacuum, the better chance that the film will resemble the starting material. Vacuums of the order of 1 mPa (ca 10^{-5} mm Hg) are often used. (*2*) Substrates should be clean, free of any grease or dust, and preferably subjected

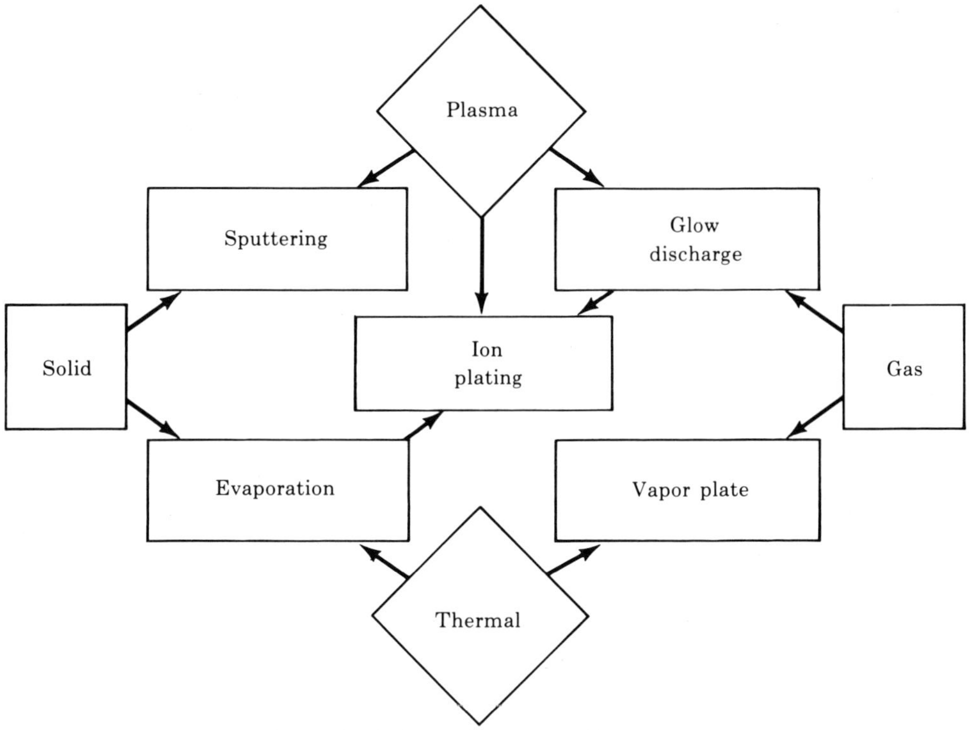

Figure 1. Relationship of deposition processes.

to a final glow-discharge cleaning. (*3*) The entire vacuum chamber and substrate should be baked at ca 400°C to remove adsorbed material such as water that would prevent good adhesion of the film. This last process eliminates the well-known water cycle in which tungsten filament reacts with water vapor to produce tungsten oxide and atomic hydrogen. The tungsten oxide then evaporates and is subsequently reduced to tungsten by the hydrogen, thus releasing water for process continuation.

The first criterion in the selection of a material for evaporation is its compatibility with processing in a vacuum system. For example, five forms of silicon oxide are described in literature (15) all containing silicon and oxygen in a ratio of 1.1. However, the most suitable for vacuum deposition is the black amorphous coke-like SiO which is normally very stable in air and easily handled. Next, the physical form or particle size must be suited to use with the heating system. Too fine a particle leaves the boat or crucible when energized by an electron beam; too large a particle results in poor heat transfer between particles and uneven melting. With resistive heating, particles of less than 1 mm are preferred; for electron beam heating, 2- to 4-mm particles are usual. Larger particulates in general have less surface area and minimize water adsorption and formation of oxides and carbonates at the surface.

In the selection of particle size, another consideration is the metal of which the boat or crucible is fabricated. Common refractory metals for this purpose are tungsten, tantalum, and molybdenum. With chamber pressures of 1 Pa (7.5×10^{-3} mm Hg), contamination by the boat material in the ppm range is evident, even though the solubility of the boat material in the added charge is practically nil. In addition, compatibility of the charge and crucible may be a problem. Silicon, for example, attacks many crucible materials, but this can be overcome by utilizing an electrostatically focused electron beam on the water-cooled silicon charge. Other heating techniques such as a high energy laser can be also applied.

The recommended evaporation conditions for commonly used materials are given in ref. 16. Most evaporation materials are metal alloys or simple inorganic compounds such as oxides, halides, sulfides, selenides, and tellurides. Because complex inorganic compounds as well as sulfates, nitrates, and phosphates decompose, they are not suitable for evaporation. The general rule is that inorganic compounds having radicals ending in -ite or -ate will decompose (16).

Vacuum deposition techniques are used extensively for the thin-film metallization of glass and ceramic substrates. In the fabrication of microelectronic circuitry the process is conducted in a vacuum system that may be as simple as a bell-jar system or an advanced in-line system of multiple vacuum chambers that allows interchamber substrate transfer without breaking vacuum and that contains mask changers, evaporation sources and substrate heater. In a typical deposition, the procedure is conducted without breaking vacuum to avoid film defects that arise from dust contamination. The substrates are placed in holders and cleaned by heating. Materials are evaporated sequentially through their respective masks. Nickel–chrome alloys, rhenium, nickel, titanium, and other resistive materials have been vacuum deposited for thin-film resistors. For dielectric coatings silicon monoxide is very common. Magnesium fluoride has been used. Thin-film conductor coatings employ aluminum, silver, copper and other good metallic conductors.

The process begins with the circuit design from which suitable masks are prepared defining the areas of deposition for the various materials. A typical circuit may require four masks. These are normally fabricated by photoetching openings in thin copper–nickel alloy sheets. These masks may include a resistor pattern, bottom capacitor plate, dielectric coating, top capacitor plate, and interconnections. Masks can be mounted in a changer which moves them sequentially in front of the substrate and positions them, aligning holes in the mask with location pins on the substrate. Alternatively, the masks can be positioned and the substrate moved.

Source heaters such as a stranded tungsten coil filament are used for materials such as aluminum. Silicon monoxide, the most common evaporated dielectric utilizes an electrically powered evaporation source fabricated from refractory metal. The design prevents line-of-sight evaporation from solid silicon monoxide to the substrate and thus avoids the ejection of particles that would cause defects in the film. The resistors fabricated from nickel–chrome alloy are evaporated from coiled tungsten wire or by direct current passage through the alloy wire. Electron-beam bombardment in a 0.01 Pa (ca 10^{-4} mm Hg) vacuum system has also been utilized.

Typical substrates include glass, alumina, and beryllia.

Evaporative Coating With Polymers. The wide variety of plastics available has lead to considerable interest in the possibility of coating materials by simple evaporation. Vacuum is the key element, for in the absence of vacuum most polymers heated in air decompose, char, or break into fragments incapable of recombination as films of the original polymer. In theory, it should be possible simply to evaporate a polymer and to condense it as a coating in a vacuum, as shown in Figure 2. Materials that do not decompose on heating should be capable of transportation through a vapor phase by vacuum evaporation. The evaporation of a polymer appears at first to be a process of depolymerization on heating, followed by repolymerization of the fragmented polymers and oligomers on the substrate surface. However, it should be noted that

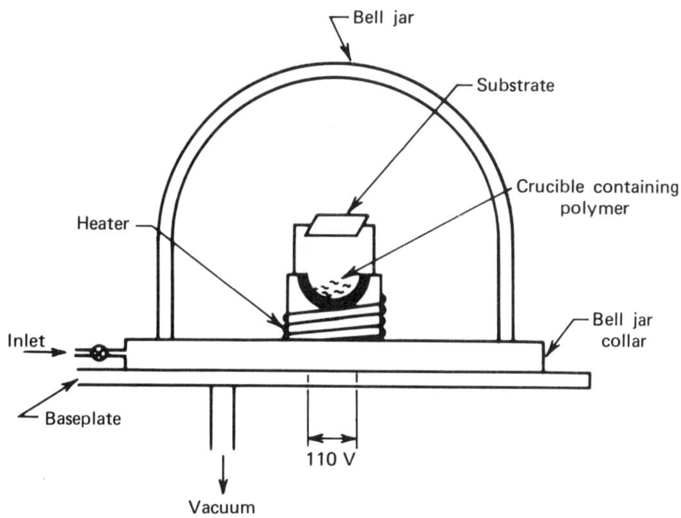

Figure 2. Vacuum deposition of polymers.

vacuum deposition of organics is not entirely analogous to metals in that vaporized polymer molecules are ca 25 times larger than vaporized metal atoms and have major differences in bond energies, reactivities, and crystalline structures.

Certain polymers have been shown capable of being evaporated in a vacuum and condensed directly to yield a polymeric coating. Success has been reported using Teflon (PTFE) (17), polyethylene (18), and nylon (19). Coatings resulting from these polymers have been deposited by electrically heating the powders to a full red heat. In a typical example (20), Teflon heated for only a few minutes in a crucible or tungsten boat deposited films of the order of 1 μm. A problem in using simple evaporation-condensation techniques is the quality of the adhesive bond between the thin film and substrate and the cohesive makeup of the film. Good quality Teflon film can be produced by evaporation in a hydrogen atmosphere (21). Other disadvantages of this method include the inability of many polymers to deposit thin films and the reported high dielectric loss of these coatings (22).

Vapor-Phase Polymerization. Vacuum-deposited polymers have certain capabilities that suit them uniquely for thin polymeric coatings. One such polymer is Parylene, which is produced by vapor-phase polymerization and deposition of *p*-xylylene (or its substituted derivatives). The polymers are highly crystalline, straight-chain compounds with excellent dielectric characteristics and molecular weight of ca 500,000. Parylene has been successfully applied in the vapor phase via a thermal decomposition free radical mechanism in the absence of either high temperature or high vacuum (23). The starting material can be a purified dimer called di-*p*-xylylene, a white stable crystalline solid vaporized at ca 133 Pa (1 mm Hg) and 200°C and then conducted through a pyrolysis or cleavage zone at 600°C to quantitatively yield monomeric *p*-xylylene, a diradical as shown in Figure 3. At room temperature this stable diradical, when allowed to contact a solid surface, instantaneously polymerizes quantitatively without the formation of any intermediate phase, yielding high purity, ultrathin, pinhole-free coatings. Parts to be coated are rotated by a controlled variable speed motor to allow even distribution of the polymeric coating. The

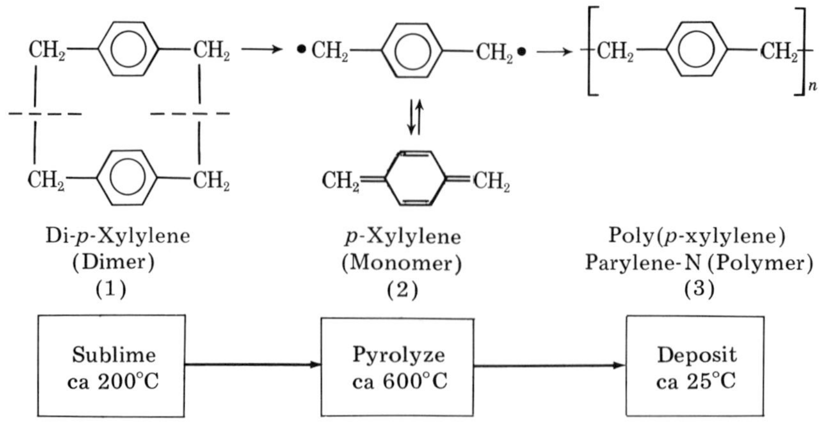

Figure 3. Vacuum deposition of Parylene.

process is not "line-of-sight" as in vacuum metallizing; the vapor coats evenly over edges, points, and internal areas. The vapor is pervasive, but it coats without bridging so that holes can be jacketed evenly. Moreover, with Parylene, the object to be coated remains at or near room temperature eliminating the risk of thermal damage. Coating thickness is controlled easily and very accurately simply by regulating the amount of dimer vaporized. Parylene coatings are usable at temperatures as high as 120°C for extended periods in air, and at even higher temperatures (greater than 220°C) in the absence of oxygen.

Deposition chambers of virtually any size can be constructed. Large parts (currently up to 152 cm long and 46 cm high) can be processed. In current commercial applications, Parylene is deposited in thicknesses ranging from 6 to 38 μm in a single tumble-coating operation for use as a capacitor dielectric material in extended foil Flat Kap capacitors (Union Carbide) (24), as a circuit-board conformal coating for high reliability applications (25), as a method for immobilizing particulate matter in hybrid microcircuits (26–28), as a semiconductor device discrete-component protective coating (29), in thermistors, thermocouples, sensing probes, and photocells, and for ferrite toroids. Ultrathin unsupported membranes (pellicles) are used in radiation, optical applications, and as media supports of extremely low mass.

Ion Plating. The technique used for the evaporation of a coating material in the positive glow region of a gas discharge is known as ion plating. The ionized evaporant is accelerated to the negative electrode surface under influence of electrical fields in low pressure d-c glow discharge (usually of argon). In the process the deposition of coating on a substrate is maintained at a faster rate than the sputtering removal of material from the substrate surface. As a result, the film grows in thickness with time (see also Ion implantation).

The technique involves placing the part in the vacuum chamber as shown in Figure 4. The pressure is initially reduced to ca 0.7 mPa (5×10^{-6} mm Hg) and then an inert gas or gas mixture is introduced while the pressure is held at 0.6–53 mPa (5 to 40×10^{-6} mm Hg) during the coating cycle. A negative bias of several thousand volts with respect to the plasma is applied to the part being coated (cathode). The surface to be coated is first cleaned for ca 10–30 min by bombardment with the ionized gas which removes the oxide film and forms nucleating sites for the coating. Then evap-

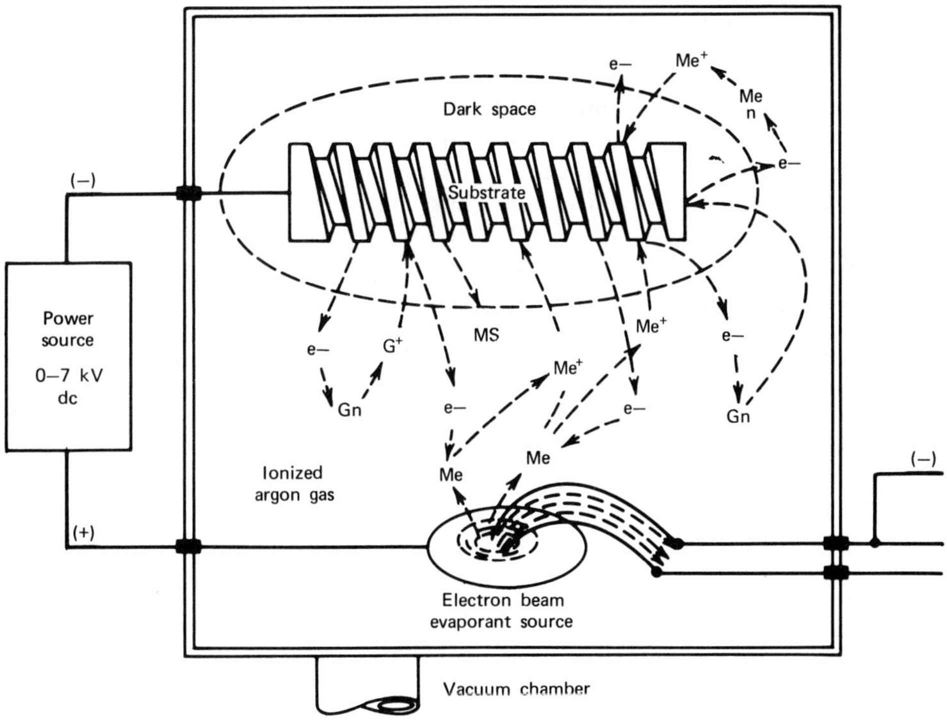

Figure 4. Ion-plating process. e– = electrons; Gn = neutral gas atoms; G^+ = positively charged gas ions; MS = substrate atoms; Me = evaporant atoms; Me^+ = positively charged evaporant ions. Courtesy of Hohman Plating and Manufacturing, Inc.

oration of the coating material is initiated and continued until the desired thickness is reached.

Some of the advantages of this technique include:

(*1*) Excellent throwing power. The technique is not limited to a line-of-sight deposition path because of the effect of gas and the electrical lines of force on the ions. Holes as small as 0.05 mm diameter and 2 mm depth can be ion-plated.

(*2*) The coatings obtained demonstrate superior adhesion because of the high energy imparted to the material, which results in some diffusion at the interface, and even, under certain conditions in ion implantation.

(*3*) The parts need not be rotated and uniform coatings are obtained without build-up on the corners.

(*4*) Unusual substrate materials such as titanium, which is hard to coat, have been ion-plated with copper.

Disadvantages include the difficulty of masking parts owing to the high throwing power of the plasma process and, compared to electroplating, the expense of the process.

Sputtering. Sputtering is a process in which material is removed from a source or target (the cathode), carried by a plasma and deposited on the substrate (the anode). It was first described more than 100 years ago as cathode disintegration, and in the late 1800s sputtering was used for metallizing mirrors commercially. One common application is in the manufacture of records; the wax master disk is made conductive by sputtering gold, allowing replication of the record surface by copper electroforming.

Sputtering is the main competition to the vacuum evaporative process and yields more consistent sheet resistivity values and an improved temperature coefficient of resistance. For the deposition of complex metal oxides and alloys, sputtering is generally preferred to vacuum evaporation. The disadvantages of sputtering as compared to vacuum evaporation are the higher cost of equipment and operation, more complex equipment, and slower deposition rates. In addition, when bias sputtering is not employed or when gas pressure is excessive, gas entrapment becomes a factor.

Sputtering is considered inefficient because as much as 90% of the supplied energy is being transformed into heat. Poor energy transfer between substrate, substrate holder, and anode table raise the substrate temperature as high as 250–300°C. This may be an advantage because of annealing of structure defects in the film.

Diode sputtering employs a glow discharge with an inert gas such as argon as shown in Figure 5. Material is removed from a water-cooled target by momentum transfer owing to the attraction of positively charged ions in the inert gas in an electrically energized plasma. The impact of the positively charged ions dislodges atoms of the cathode material which are then accelerated by an applied electric field toward the substrate held on a water-cooled anode table.

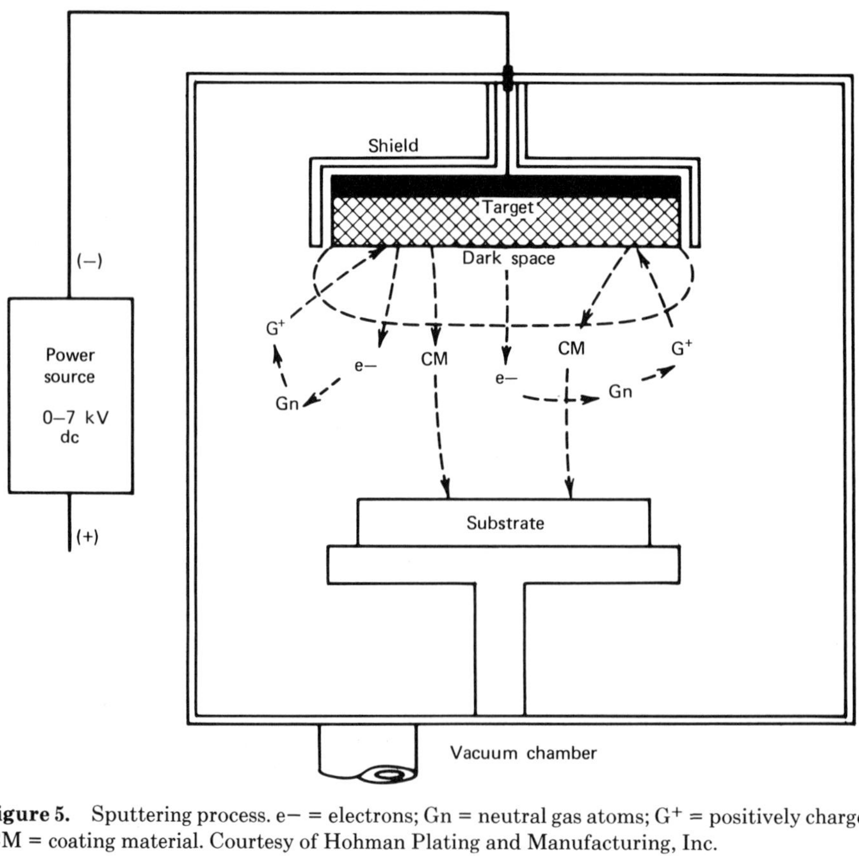

Figure 5. Sputtering process. e− = electrons; Gn = neutral gas atoms; G+ = positively charged gas ions; CM = coating material. Courtesy of Hohman Plating and Manufacturing, Inc.

Ions formed by sputtering strike the substrate surface with energies between ten and several hundred electron volts; electroplating, by contrast, usually develops only a few tenths of a volt. Some of the sputtered atoms develop sufficient energy to penetrate several atomic layers of substrate surface, resulting in adhesion anchoring and films of excellent adhesion. Sputtering is suitable for depositing multilayer films and films of complex alloys and dielectrics.

Surface cleanliness is most important in adhesion. Normal cleaning techniques are incapable of removing monolayers of surface contaminants. However, the substrate surface can be cleaned by reversing the sputtering process and bombarding the substrate with inert gas ions. This technique, called sputter etching, also creates nucleating sites that promote adhesion. Adhesion can also be increased by putting a negative electrical bias on the substrate; this in effect causes a partial back-sputtering of weakly bonded atoms. The technique also minimizes inert gas entrapment and removes oxygen, resulting in a purer and denser coating and an increase in substrate temperature.

The character of the bond between film and substrate depends upon the film and substrate materials. It may be a chemical or metallic bond or a van der Waals interaction.

The sputtering rates are dependent upon the material and the heat of sublimation. The following materials are ranked in order of increasing sputtering rate (30): (*1*) silicon; (*2*) titanium; (*3*) niobium; (*4*) tantalum; (*5*) tungsten; (*6*) aluminum; (*7*) chromium; (*8*) nickel; (*9*) gold; (*10*) silver.

Diode or conventional deposition employing radio-frequency (r-f) equipment is the sputtering technique used for most applications involving conductive and insulating materials. To ionize the gas at lower pressures than are used in the diode system, a modified technique called triode sputtering can be employed. This method enables the accelerating voltage to be varied over a wider range than is possible with the diode system. As a result, better control of physical properties of the coating can be achieved, sputtering efficiency is increased, and diffusion of sputtered atoms is decreased.

Several individual target materials inside the chamber can provide a variety of combinations or utilize different power levels for sputtering at different rates. The sequential use of different target materials can produce a graded coating or overcome coating incompatibility by providing a buffer coat. Targets having several material combinations have also been used to provide heterogeneous coatings.

One improved sputtering technique is planar magnetron sputtering, a process first patented in 1939 (31–33). Basically, the planar magnetron uses a closed magnetic field loop (Fig. 6) to confine and compress the plasma causing the ionized gas to sputter more efficiently. Under the influence of the two fields, the electrons travel in a cycloidal path rather than in straight lines. This extended path, and its formation into an endless loop, greatly increases the probability of ionizing collisions and produces a highly dense plasma near the target surface. The net result is to create a higher ion flux with deposition rates of 5 to 20 times those of conventional d-c or r-f diode methods. Several other process improvements have resulted including the ability to operate at a lower pressure and temperature, and modification of the cathode to a long rectangular shape which is less affected by plasma and gives excellent overall evenness of coating.

Another modification of r-f sputtering is reactive sputtering which has provided an extended means of compound preparation. If small amounts of oxygen or nitrogen are introduced into the chamber with the argon, the gas mixture reacts with the target

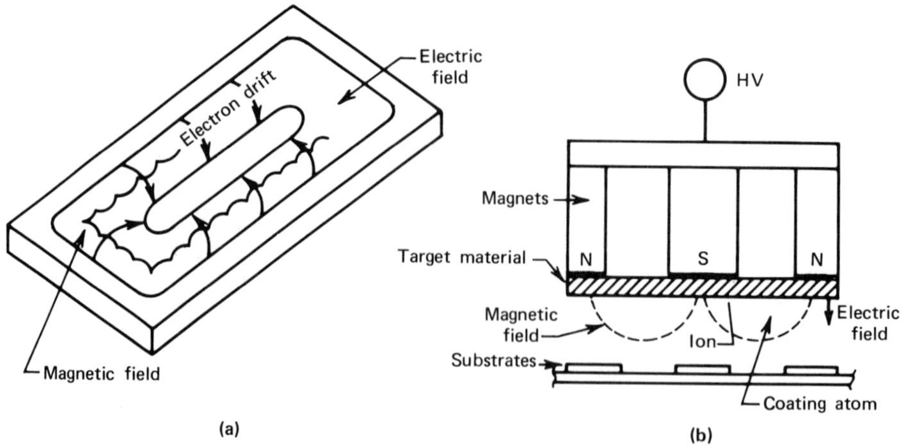

Figure 6. Planar magnetron cathode courtesy of Materials Research Corporation. HV = high voltage.

atoms and converts the normal coating to an oxide or nitride of the original target material. For example, aluminum may be sputtered in an argon–oxygen gas mixture to form an aluminum oxide film.

The commercial uses of sputtered inorganic materials is extensive and includes:

(*1*) Electronic applications such as formation of thin film conductors, resistors and capacitors, passivation, insulative and magnetic films, and solderable coatings on insulators.

(*2*) An automatic technique for coating plastic automotive grills for decorative purposes. This process, developed by Varian, employs multiple S-Gun Sources (SGS), and is widely used in the semiconductor industry to cover the large surfaces. Each SGS contains its own solid metal target which is bombarded with argon ions causing sputtering of metal atoms and coating the parts. The resultant metal coating is uniform with a typical thickness of 0.03 μm.

(*3*) Application of carbides, nitrides, borides, and disilicides to metal tool surfaces.

(*4*) Formation of soft lubricating films such as molybdenum disulfide.

(*5*) Application of antiglare coatings on optics and filters.

(*6*) Coating of molds to increase mold release characteristics.

(*7*) Coatings of highly stressed aircraft and machine parts to prevent corrosion and hydrogen effects.

R-f Sputtering of Polymers. For sputtering conductive materials, the power supply may be d-c or r-f as discussed. However, a polymeric insulating material used as a cathode takes on a positive charge during d-c glow discharge and does not attract ions. As a result, it resists the starting or maintenance of a d-c glow discharge. R-f energy is therefore used instead. The luminance of the glow discharge is generated by an r-f field between the cathode located at the top and the substrate situated on the anode base. Subjecting the target to r-f power operating at 13.56 MHz gives it a negative bias, causing it to become a cold cathodic electron emitter. Free electrons leaving the target ionize part of the gas, creating positive ions that are accelerated toward the polymeric target, strike it, and dislodge atoms. The dislodged matter impacts the substrate at high energy creating atomic bonding between the deposited coating and the substrate. During the first half of the discharge cycle, the target is bombarded with ions; neutralization of the surface charges by the electrons follows. Sputtering occurs because the insulator is subjected alternately to ion and electron bombardment. The positive cycle portion and the target material is always ready for bombardment.

Sputtering polymers requires basically simple equipment, a cathode and anode for establishing a gas discharge similar to that in a fluorescent tube, a source of solid sputtering material, the target, a deposition site, and a substrate, all of which are housed in an evacuated chamber which is backfilled with an inert gas (Fig. 7). Targets and electrical feed-throughs must be shielded to prevent electromagnetic interference (EMI) and power losses, and relatively high self-bias applied to the metal cathode behind the target must be maintained (34). A photograph of a simple laboratory setup for sputtering organic materials by the diode technique is shown in Figure 8. This is contrasted to the three-electrode, or triode, system (sometimes called hot cathode) which employs a thermionic cathode, an anode, and a target (34–35). In the triode process the ionization of gas and sputtering of material are separated. The energized filament emits electrons and ionizes the sputtering gas, and the negatively biased target material attracts the positive ions. The advantage of the triode system is that the glow discharge can be maintained at 0.01 Pa (ca 10^{-4} mm Hg) as compared to 1 Pa (10^{-2} mm Hg) required for the diode method. The triode method also produces higher deposition rates of materials, owing possibly to larger mean-free paths at lower pressures. These result in less backscattering of sputtered atoms with more atoms contacting the substrate. Deposition rates are low because in ionic bombardment of a surface most energy is manifested as heat, only a small portion is available for kinetic energy of the ejected atoms. Typical growth rates are of the order of 1 μm/h. Increasing the sputter-coating rate becomes a problem of decreasing the target temperature. Increasing the voltage increases the sputtering rate but only to a point; too high a voltage results in gas ions penetrating the target surface, without dislodging atoms. Before deposition of the film begins, a potential is applied to the substrate to sputter-etch the target (reverse sputtering) and remove surface oxides and contaminants and produce good adherence of the sputter coat.

Ideally and under optimum conditions, sputtering of materials can result in a

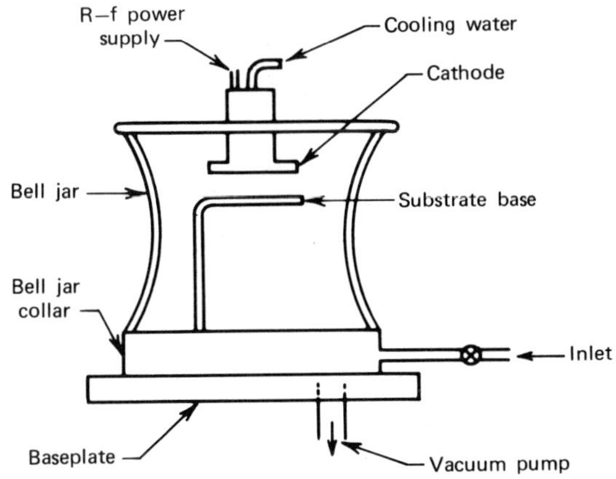

Figure 7. R-f polymer sputtering.

Figure 8. Polymer sputtering.

coating having the same composition as the original target. However, with organic polymeric targets marked differences are encountered. For example, DuPont commercial Teflon (PTFE) and sputtered Teflon (35) have different colors. This is usually observed in sputtered polymers and has been ascribed to disorder (36). Evidence of amorphous structure and increased hardness appears to indicate that sputter coatings contain a greater amount of cross-linking and molecular disorientation. The infrared spectra made of three Teflon coatings (Fig. 9), the commercial material, a coating obtained by a simple evaporation condensation, and a sputtered coating, confirm the structural changes. A comparable situation is apparent in the sputtering of Parylene dimer and polymer (37–38), with only a rudimentary similarity between the infrared spectrum of Parylene produced by the standard vapor-phase synthetic technique and that of sputtered Parylene from targets fabricated from dimer and the polymeric material. Even though there appears to be differences in electrical and/or physical properties between the original and sputtered polymeric coatings, these variations have not precluded use of these coating materials as unique substitutes, eg, the use of sputtered Teflon as a lubricant (35) and of sputtered Parylene as a coating (37–39).

A simple technique for evaluating the electrical qualities of a thin polymeric film is by fabrication of a thin-film capacitor. This can be accomplished by sputtering a thin film of nickel–chromium onto both sides of the thin polymeric coating producing a sandwiched metal film that can be subjected to capacitance, dissipation factor, and dielectric constant measurements for the determination of its film condition and electrical properties (40).

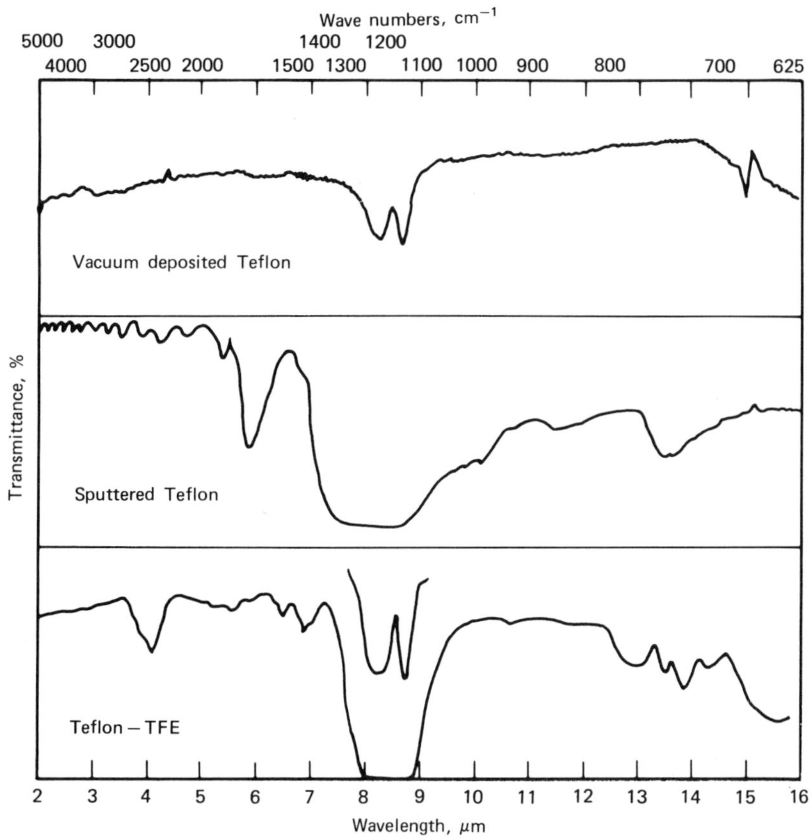

Figure 9. Comparative infrared spectra of Teflon.

Ultraviolet Irradiation, Photopolymerization. Over 40 years ago it was observed that a polymer film formed on walls of a flask containing methyl methacrylate that was irradiated with ultraviolet light (41–42). The mechanism postulated excitation of the monomer in the gas phase resulting in vinyl addition polymerization.

The equipment required for gas-phase photopolymerization includes a vacuum system capable of evacuation to ca 100 μPa (ca 10^{-6} mm Hg), provisions for introducing the gaseous monomer or vaporizing the liquid monomer and an ultraviolet source between 0.2 and 0.3 μm (usually a mercury arc lamp with filter). Gaseous monomer is introduced and maintained at ca 400 Pa (3 mm Hg) during radiation.

Contact with ultraviolet light initiates film formation at rates up to several hundred micrometers per minute. With the substrate maintained at a low temperature, the process has yielded ultrathin pinhole-free adherent films of only 0.05 μm thickness (43–44). The deposited polymers are generally flexible and contain little cross-linking. However, extensive side reactions have been observed in some instances to yield low molecular weight materials and other products (43) (see Photochemical technology; Radiation curing).

The polymerization mechanism involves a two step process: the gaseous monomers are first attached to the vessel surface by chemisorption or physical adsorption, and then activated with ultraviolet. The ultraviolet polymerization process utilizes low energies of about 8.0×10^{-19} J (5 eV) causing free radical formation. The resultant free radicals couple to form polymers at a rate that generally increases with increasing light intensity. A variety of monomers, including tetrafluoroethylene and hexachlorobutadiene, have been polymerized to yield pinhole-free films. If a polyfunctional monomer such as divinylbenzene is activated, cross-linked polymers can result.

Substrates can be coated by ultraviolet polymerization without the necessity of a mask by using quartz optics to direct the radiation to provide a suitable pattern of ultraviolet light at the surface.

Deposition of Films In Gaseous Discharge

R-f Glow-Discharge of Inorganic Materials. At ambient pressures and temperatures, gases are poor conductors of pressure and temperature. However, at high temperatures or under reduced pressures, gases will conduct electricity when subjected to a strong electrical field. At this time, the gas becomes partially ionized and contains free electrons and ions that carry current through the gas. In a glow or gaseous discharge, the free electrons in the gas acquired from the electrical field collide with gas molecules, and transfer energy to them. Electron-molecule collisions result in molecular excitation and in dissociation and ionization of the gas molecules. Free radicals, positive and negative ions are formed, and ultraviolet, visible and infrared radiation emitted.

Deposition of thin inorganic films by the low temperature plasma process is becoming important to the semiconductor and microelectronics industry (45). Thin inorganic films of silicon and of silicon nitride have been prepared in an r-f plasma by allowing silane to react with water or ammonia, respectively (46). Boron nitride has been prepared from B_2H_6 and NH_3 (47–48), and the glow discharge of tetraorthosilicate has yielded silicon oxide films (49).

Gaseous Anodization. An extension of the conventional wet anodizing technique uses an oxygen plasma electrolyte maintained by a glow discharge at 1 Pa (7.5×10^{-3} mm Hg). The mechanism of film growth is essentially similar to electrochemical anodization where the rate of oxide growth is a function of anodizing voltage and oxygen pressure. The metal surface to be anodized is subjected to an oxygen plasma which results in the formation of pinhole-free oxide films of about 0.0025 μm (50), with a glow rate of 0.0022 μm/V for aluminum oxide (51). Films have also been prepared of tantalum pentoxide, hafnium oxide, and lanthanum titanate. The disadvantages of this method of anodization lie in its low current efficiency and low throwing power; the advantages are the avoidance of entrapping water or hydroxyl ions in the oxide film.

Glow-Discharge Polymerization. Glow-discharge polymerization has been shown to be useful for preparation of thin organic films (52). When an organic substance in the vapor phase is excited to luminescence by an electric discharge, a solid film tends to deposit on all surfaces exposed to the luminous plasma. In the last thirty years over one hundred patents and publications have appeared on glow-discharge studies in organic chemistry; polymerization of as many as forty different organic monomers is reported in a single reference (53).

The use of the glow-discharge reactions to initiate polymerization is a rather violent method leading to fragmentation of the original monomer and incorporation of these fragments in the polymer. Low temperature plasmas can be generated using an electrodeless technique at the radio-frequency level. This process utilizes excitation throughout the reactor walls at reduced pressure for internal electrodes.

Glow-discharge polymerization can be initiated with internal electrodes or by electrodeless equipment (r-f). The advantages of the electrodeless method are that contamination by consumable or erodable electrodes is avoided and that the large volume of the reactor can be filled with a uniform plasma, thus allowing for engineering flexibility in experimental design. In addition, the substrate to be coated may be either a conductor or a nonconductor.

The equipment for preparing glow-discharge films with electrodes consists of an evacuated chamber with a source of organic monomer vapor and electrodes wired to an external power supply. The electrodes are pairs of flat plates of any conductive material, placed ca 1 cm apart. With a vapor pressure of ca 133 Pa (1 mm Hg), the glow discharge is usually initiated at 300–400 V and sustained with a current density of the order of 1–3 mA/cm^2.

The basic equipment for r-f discharge polymerization consists of an evacuated chamber, a source of organic monomeric vapor, and a power supply generating 13.56 MHz with a 300-W maximum output to transfer electrical energy in the microwave frequency range through the coil into the bell jar (Fig. 10). In a typical glow-discharge polymerization procedure, the system is first evacuated to vacuum of ca 1 Pa (7.5×10^{-3} mm Hg). Argon is then allowed to enter at about a flow rate of 0.028 m^3/h (1 ft^3/h) while the pressure increases to ca 13 Pa (0.1 mm Hg). The r-f generator is turned on and the power adjusted to 250 W to give a brilliant violet glow discharge. The monomer gas flow is started and adjusted until the glow is quenched just at the surface of the substrates to be coated. At this point the glow turns blue, and the system pressure is ca 50 Pa (0.4 mm Hg) as shown in Figure 11.

A dynamic glow-discharge polymerization system is usually employed in the electrodeless system because of the advantages of operating with a monomer flowing through the discharge zone rather than under static conditions. The use of this approach ensures that the concentration of unwanted reactive species will not be allowed to build up. In addition, the use of a flowing gas has been shown to exert a cooling effect on the substrate (54).

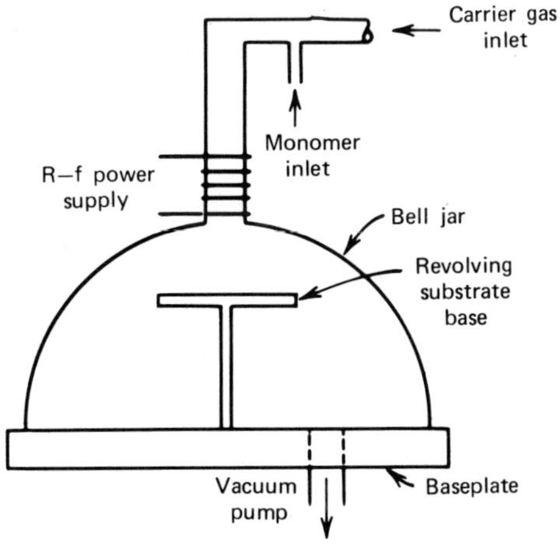

Figure 10. Glow-discharge polymerization.

Figure 11. Glow-discharge polymerization equipment.

When discharge conditions are optimized, most organic vapors will deposit films (55–56). Unsaturated and aromatic compounds polymerize at the fastest rate, ie, with the most efficient conversion of the energy dissipated in the discharge into new chemical bonds; the slowest, or most inefficient, materials are the saturated aliphatics, particularly perhalogenated hydrocarbons (57–61).

The mechanism of formation of solid films does not require the uniting of functional groups in a repetitive chain. For example, methane and benzene will yield solid films consisting of a complex macromolecule having a fairly well-defined structure. Although the physical appearance and properties of the coating in general depend on the conditions employed, the product is highly branched, cross-linked and unsaturated regardless of the starting vapor (62). It has been reported that under conditions of relatively high monomer vapor pressure and low current density, films are usually clear and transparent but soft and tacky, suggesting incomplete polymerization (63); under conditions of low vapor pressure and high current density, the films produced are mechanically hard and often discolored (yellow or brown) because of cross-linking. In addition, when the current density is relatively high and the deposition time long, films may become degraded and appear blistered or loosely adherent.

Other disadvantages of glow-discharge polymerization reactions are the difficulty of obtaining high resolution patterns and precise control of film composition owing to the variety of recombining species and the possibility of contamination or sputtering. In addition, relatively high gas pressure is required and considerable substrate heating results.

Plasma polymerization has been used to overlay an organic protective layer on vacuum-deposited aluminum to achieve high reflectivity using the same equipment (64). Other commercial uses include the production of a film with composition and properties similar to Teflon (trade name Glofilm) which is reported in use as a capacitor dielectric.

Deposition of Films At Atmospheric Pressure

Metallo-Organic Deposition. The thermal decomposition of some organometallic materials (65–68) can produce films of thickness 0.1 to 0.2 μm. The chemical structures of these organometallic starting materials are dissimilar to those used in vapor plating as described in the next section. In metallo-organic deposition, the metal atom is attached to either an oxygen, sulfur, nitrogen, or phosphorus atom; chemical vapor deposition utilizes an organometallic in which the metal atom is directly bonded to a carbon as in nickel carbonyl.

Metallo-organic deposition is a simple process compared to sputtering or vacuum evaporation and requires simple equipment. The chemical compounds used are organo-metallic halides, hydrides, or oxides that are readily dissociated or reduced below the temperature of the substrate material. They are mixed in liquid or paste form in a homogeneous single-phase solution called a formulated resinate. In the actual process the formulated metal resinate of an organometallic is applied by brushing, spinning, dipping, spraying or screen printing, and drying and firing in a furnace. This treatment decomposes or oxidizes the organic portion, depositing the adherent thin film of metal oxide on the substrate. The process is limited to the deposition of thin films on inorganic substrates of glass, ceramics, sapphire, ferrites, quartz, and salt-enameled metal plates as well as metals (Al, Mg, Ti, Mo, stainless steel and alloys).

The resultant film thicknesses range between 0.1–0.2 μm, similar to the thickness obtained by vacuum techniques. When fired, the thin films obtained by this technique exhibit superior adherence to those obtained by vacuum evaporation techniques. The fired thin films produced by metallo-organic deposition can be used for conductors, resistors, dielectrics, magnetics, and photomasks. Firing of metallo-organic resinates in air produces thin films containing negligible carbon. However, if firing is performed under a reducing or inert atmosphere, a carbon residue (0.1% or higher, depending upon firing conditions) is produced in the film.

Chemical Vapor Deposition (Cvd). Chemical vapor deposition (cvd), also called vapor plating or vapor forming, is a process in which a surface is coated with vapors of volatile stable chemical compounds at a temperature below the melting point of the surface. The compound is reduced or dissociated (69–72), resulting in an adherent coating of material. An example of this process is titanium vapor plating by the reduction of titanium tetrabromide with hydrogen at an elevated temperature. Solid $TiBr_4$ is vaporized and the vapor mixed with hydrogen. The mixture is conducted over the surface which is heated to about 1300°C. There it decomposes, depositing a coating of titanium metal.

In another variation of this technique, nickel carbonyl vapor is passed over a mandrel in a closed chamber at ca 175°C. Upon contacting the mandrel surface, the nickel carbonyl decomposes into nickel and carbon monoxide. A minute amount of nickel is deposited, and the carbon monoxide vapors are recirculated, yielding a controlled growth of nickel at a rate of 0.025 mm/h. This produces a uniform precise nickel coating on the mandrel without additional buildup on edges or projections and has proven commercially useful for preparation of nickel shells and molds.

The cvd technique is also applicable to the decomposition of surface coating of refractory metals and compounds such as the carbides of tantalum and silicon, tin oxide, and carbon. The advantages of the process include deposition of coatings varying from the ultrathin up to a thickness of many millimeters; the main disadvantage is the necessity of heating the surface to be coated.

Commercially, cvd is used to produce epitaxially grown single crystal silicon by the reduction of silicon tetrachloride with hydrogen and to make epitaxial compounds, polysilicon, silicon nitride, silicon dioxide, and both doped polysilicon and silicon dioxide. Each coating can be deposited ideally in a selected temperature range (73) as shown in Table 1.

Several different types of equipment have evolved to meet select requirements of chemical vapor deposition. Among them are horizontal, vertical, cylindrical, tubular, cold-wall, and hot-wall systems. In the horizontal system, the stationary silicon wafers are processed lying flat or on a moving belt in a continuous process furnace. The vertical reactor employs a bell jar in which wafers are placed on a susceptor disk and rotate continually for uniform coverage. In the cylindrical reactor the wafers are peripherally mounted. The tubular system configuration is similar to the horizontal type with multitiered carriers being used. A typical low pressure cvd configuration is shown in Figure 12.

Table 1. Typical Deposition Reactions—Atmospheric Pressure Cvd Reactors

Film	Carrier gas	Reaction	Deposition temperature range (°C)
Si_3 epitaxy	H_2	$SiCl_4 + H_2 \rightarrow Si + HCl$	1125–1200
Si_3 epitaxy	H_2	SiH_4 + heat $\rightarrow Si + H_2$	1000–1075
Si_3N_4	H_2	$SiH_4 + NH_3 \rightarrow Si_3N_4 + H_2$	900–1100
poly Si	H_2	$SiH_4 + NH_3 \rightarrow Si_3N_4 + H_2$	600–700
poly Si	H_2	SiH_4 + heat $\rightarrow Si + H_2$	850–1100
SiO_2	H_2	$SiH_4 + CO_2 \rightarrow SiO_2 + H_2 + CO$	600–900
SiO_2	H_2	$SiH_4 + CO_2 \rightarrow SiO_2 + CO$	500–900
SiO_2	H_2	$SiCl_4 + CO_2 \rightarrow SiO_2$	800–1000
SiO_2	H_2	$SiH_4 + O_2 \rightarrow SiO_2 + 2H_2$	200–500

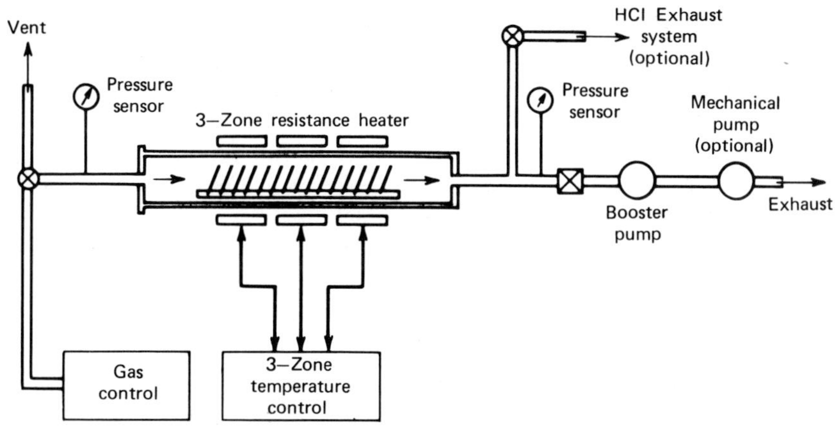

Figure 12. Typical low-pressure cvd configuration.

Two types of systems are used for heating the equipment. The cold-wall system contains either an external r-f induction coil that couples energy through the quartz chamber wall to the susceptor which in turn heats the wafers, or an external radiant heater in contact with the clear quartz wall of the chamber. In the hot-wall system the outside of the chamber wall is wrapped with resistance heaters which heat the wafers through internal-directed radiation.

Other commercial uses of chemical vapor deposition are deposition of carbon thin films by pyrolytic decomposition of organic materials and manufacture of capacitors of titanium dioxide with dielectric constants as high as 82. These capacitors are produced by the hydrolysis of titanium tetrachloride.

Electron-Beam Polymerization. Electron beams can provide the necessary energy to cause polymerization of organic monomers. Monomeric materials that have been irradiated to produce films include silicones (74), butadiene (75), styrene (76), methyl methacrylate (77), divinylbenzene (78), and epoxy resins (79–80). However, the lack of selectivity of this energy source results in formation of unwanted contaminant materials (81–82) (see Radiation curing).

Film properties can be tailored somewhat by varying the voltage used to accelerate the electrons. The lower the voltage, the less cross-linked and softer the polymer film. Use of voltages greater than 1000 V cause crazing of films owing to excessive cross-linking. Typical growth rates are ca 0.5 μm/h.

Electron-beam polymerization, like ultraviolet polymerization, is useful for forming films without a mask (80). It is capable of yielding film pattern resolution as low as 1 μm. Film patterns can be obtained by irradiating a large surface area with a defocused electron beam or by scanning with a small electron beam.

Gamma Radiation. Gamma radiation can be used for formation of thin polymeric films via a mechanism of free-radical polymerization similar to that occurring in ultraviolet and electron-beam polymerization. The condensation polymerization reactions effected by gamma irradiation can be as specific as most chemical reactions because the initial energy is rapidly degraded in matter and much of the chemical change produced by radiation is a consequence of the action of electrons of less than 100 eV. The transient species produced by such electrons do not give final products immediately, but take part in various transfer processes in such a way that the molecules finally transformed to products are not necessarily the ones that were initially affected.

Equipment for effecting gamma polymerization consists of a source of gamma radiation, such as, cobalt-60, cesium-127, or iridium-192. Although gamma radiation seems to have great potential as a technique, its commercial applications have been limited.

Uv Solid Polymerization. Solid materials, such as the photosolubilizable lacquers which first appeared in the early 1950s, also undergo a useful conversion on exposure to light. These lacquers were used for photomechanical purposes by first coating a substrate, then exposing the coating to light through a mask having the desired pattern. Exposure to light changes the solubility of the coating, and washing away the soluble portion of the coating leaves a relief image of the remaining pattern on the substrate. The pattern formed is usable as a printing surface or stencil to resist chemical etchants (see Photoreactive polymers; Radiation curing).

Photopolymerizable coatings in use today include, photoresists, photocurable coatings, inks, and photopolymer printing plates. The most important use of liquid photoresists is in the manufacture of semiconductor devices. The properties of these materials that suit them to this application are their high resolving power, chemical resistance and formation of adherent coatings on silicon wafers.

Photoresists are categorized by their response to actinic radiation, negative, or positive acting. Negative-acting resists are those in which the portions of film exposed to ultraviolet light are rendered less soluble than the unexposed portions; the converse is true for the positive-acting resists, ie, the areas exposed to radiation react, becoming soluble in the developing solution. The five basic methods for applying photoresists include spraying, roller coating, dipping, whirling, and flow coating.

The negative-acting resists function most satisfactorily with a thickness of 2.75 to 3.75 μm; positive-acting resists are coated slightly thicker, to ca 5 μm. After drying, it is recommended that the coated substrates be processed through the image-forming procedures as soon as practical because of the possibility of unwanted polymerization owing to metal surfaces, air oxidation, smog, moisture, and visible light.

Properties

The ideal thin-film coating adheres well and is chemically inert to the substrate and atmosphere. However, the film is not a separate entity and must also be considered together with the attached substrate. Among the properties of interest are its crystallographic orientation, the density of dislocations or imperfections, and the distribution of adsorbed gases.

Other properties of interest are related to the intended use of a film. For example, for optical films the main physical properties of interest are refractive index and the transmission value at a particular wavelength. Laser components require a high transmittance, necessitating low impurity concentrations of metals (83) that absorb in the desired wavelength. Thin polymeric coatings considered for modern microelectronic application must be pure and homogeneous solid layers of uniform thickness. When in intimate contact with semiconductors, the films may influence device performance through a number of mechanisms that may be discernible only under conditions of combined electrical and environmental stresses. Good electrical properties, including high intrinsic breakdown voltage, low dielectric constant, low electrical conductivity, and small dissipation factor, can be enhanced or otherwise affected in use. For example, the electric field in a polymeric film may be large because of the relatively small thickness of insulation (84), and the measured and predicted geometric capacitances may be different (85). Thin-film insulators can also function as conductors, a phenomenon that has been ascribed to field emission, electron tunneling, impurity, and space-charge limited conductions.

Adhesion. The adherence of thin films is governed by the substrate surface, crystallographic orientation, if any, density of dislocations or imperfections, gases adsorbed, and by the bonding type and spatial distribution of surface atoms. Thin films by their nature are fragile and rely upon the substrate for structural support. This skeletal support in turn is normally most effective when good adhesion exists between the thin film and substrate. Metallic thin films depend almost entirely on an adhesion-enhancing layer of a chemically active metal deposited directly on the substrate. If foreign materials cover the surface, bonding is prevented.

The bonding forces are of two types: chemical and physical. Each is further subdividable. Chemisorption processes usually lead to maximum bonding. If there is a reaction between the arriving film material and the substrate, a chemical bond is formed. Some reactions form undesirable intermetallics that are quite brittle and do not promote bonding. The amount of chemical reduction that occurs in reactive bonding depends upon the free energy of formation of the compound formed by the arriving metal and the substrate material and upon the kinetics of the reaction. Complete reduction rarely occurs, but adequate bonding can result from partial reduction of the substrate. For example, tantalum cannot reduce Al_2O_3 to aluminum metal under conditions of thermodynamic equilibrium, but Ta does adhere very well to Al_2O_3 substrates because partial reduction of the substrate occurs.

Physisorption alone occurs in some systems. The bonding forces involved (van der Waal's and electrostatic) are weak compared to those described above. Film-substrate systems that involve only physical bonding (eg, gold on glass) have few practical uses, though they are of some laboratory significance. In Figure 13, the ad-

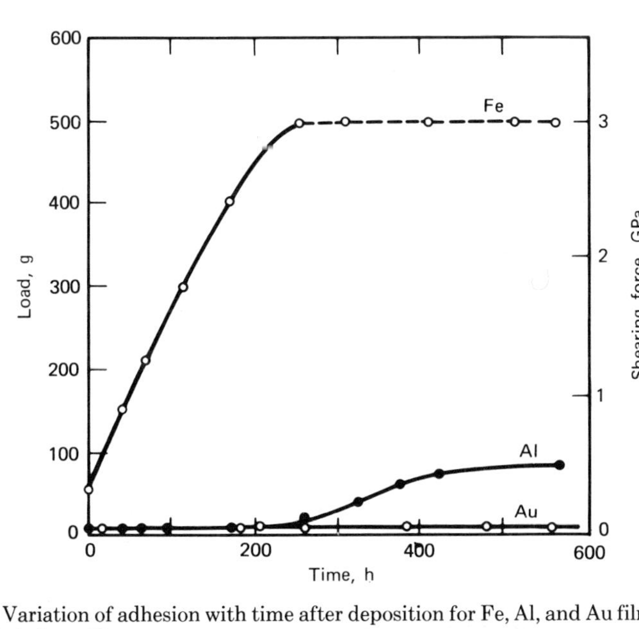

Figure 13. Variation of adhesion with time after deposition for Fe, Al, and Au films. To convert GPa to dyn/cm^2, multiply by 10^{10}.

hesion of iron, aluminum, and gold films on glass is plotted as a function of time after deposition (86). In the case of gold, the adhesion remains weak and is probably caused by van der Waals forces. For aluminum, the adhesion increases with time, and iron, which has stronger adhesion from the beginning, attains very high values that remain constant after a relatively short period of time. Benjamin and Weaver established (87) that the increase of adhesion with time is due to the formation of an oxide layer between film and substrate, resulting in a very strong chemical bond.

The major internal forces acting to detach films from substrates are stresses: internal and interfacial. Interfacial stress is largely due to thermal expansion mismatches between film and substrate. In the case of ceramic substrates, a large thermal expansion mismatch exists between the ceramics and the metals commonly deposited on them as conductive films. This lateral expansion mismatch is compensated elastically by changing film thickness. Substantial temperature changes in a system result either in the strain being borne elastically by the entire system, in cracking of the metal layer or in plastic deformation of metal. If the metal is brittle and its elastic limit in tension is exceeded, the metal cracks (crazes) or delaminates from the substrate. Pure metal systems, especially face-centered-cubic conductors (eg, gold, silver, copper, and aluminum), are more likely to accommodate the thermal expansion mismatch by deforming plastically. Although internal stress has been the subject of numerous experimental investigations, its origins are still not well understood. It is known that the magnitude of the detachment force resulting from internal stress can be extremely large. For example, molybdenum films do not adhere to clean molybdenum substrates if they are evaporated by electron-beam techniques or sputtered at low bias voltages. Such films exhibit very large tensile stresses. However, films sputtered at relatively high bias voltages (−125 V) exhibit compressive stress. Generally, films deposited at zero stress or in slight compression by the use of intermediate bias voltage adhere, and those that are in substantial tensile stress do not.

Thickness. The single most significant film parameter is probably film thickness. Thickness can be continually monitored during the deposition process or after deposition is complete. The choice of technique is dictated by the film thickness and accuracy required, and by the film use. The latter is related to the uniformity of film thickness, hardness, transparency, substrate size, and physical properties. Film thickness measurement techniques utilize light wavelength, mechanical, electrical, and magnetic approaches. These techniques have been explored in several texts (88–98). In the following discussion, techniques for film thickness measurement will be grouped according to the general method.

Film Thickness Measurement

METHODS

Weight Measurement. Probably the oldest and most direct method is the weighing of a measured film section on a microbalance having a sensitivity of 1 to 2 micrograms. The accuracy of the gravimetric method is of the order of ±1%. The limits of film measurement are ca 0.0009 μm of silicon dioxide, 0.0008 μm of aluminum, and 0.0001 μm of gold and platinum. The disadvantage is that the same substrate must be weighed before and after deposit. The film density used for calculating the weight is reaction-dependent and is not always known accurately. Actually, this technique is best utilized for measuring the density rather than thickness of a film.

The gravimetric method has been extended to provide a continuous measure of the vapor deposition rate and total weight by utilizing the microbalance monitor (99–102), which contains moving coil meters that operate in the 0.0001 to 0.002 μm range. To increase the sensitivity, a lightweight, large-area vane with counterweight on the other side of the meter has been employed. Another modification measures the vapor momentum transfer rate using a simple torsional pendulum (103) with a sensitivity of 0.00016 μm/min per degree rotation for tin vapor.

Oscillating Quartz Crystal. Using the thickness shear mode of a piezoelectric quartz crystal, an oscillating quartz crystal placed near a coated substrate can be used to monitor film thickness during deposition. The crystal is exposed to the vapor stream, and the deposited film causes a frequency shift which is measured (104–113). Since the shift is proportional to the number of atoms deposited, the device indicates the film thickness, and the rate can be derived by differentiation. This technique is probably the single most important method of monitoring thin-film deposition rates of metal, nonmetal, and multicomponent films.

One of the main sources of error in the crystal oscillator is the frequency shift that results from changes of the crystal temperature during deposition. Such changes can be reduced by enclosing the crystal in a water-cooled holder (105). If, in addition, the crystal temperature is stabilized prior to exposing the substrate to the vapor stream, then this source of error is virtually eliminated (106). The device has also been used for automatic control of the deposition rate. For example, in an evaporator in which an alloy film was produced by simultaneous evaporation of the two components, rates of ca 0.0001 μm/s could be controlled to ca $\pm 1\%$ (108). This corresponds to a detection of mass changes of ca 10^{-9} g/cm^2 which is about the limit of useful sensitivity under conventional conditions. Mass changes in the 10^{-12} g/cm^2 region have been detected with special precautions (110). This corresponds to a coverage of less than 10^{-4} of a monoatomic layer of a film material like iron.

Stylus. A direct read-out technique for film thickness measurement employs a stylus connected to a transducer that is moved over the step of the film edge and then across a film surface (114–115). Accuracies of ca 0.002 μm can be attained. This is a rapid method requiring a step and a film sufficiently hard to resist damage by the diamond stylus. Vertical movement of the delicately balanced stylus is converted to an electrical signal, amplified ca 10^6 times, and then fed into a recorder.

Stylus instrumentation is one of the most popular techniques for the quantification of surface geometry and can be used for planar as well as radial surfaces. A recently introduced system permits three-dimensional surface analysis instead of only the two-axis (x and z) measurement previously available. Besides providing surface-related information such as waviness, average roughness, and profile, the newer instrumentation yields data on peak count, average roughness height, and bearing factors (macro and micro) (116).

Sectioning. Sectioning techniques are a simple destructive means of determining thickness of films (117). With films of 1 μm thick, microscopic measurement can be performed on a cross section; with thinner ones, the film is beveled at a small angle using a technique called angle lap stain. Thickness measurement of epitaxial or diffused semiconductor films utilize this approach. The lapping angle is critical and must be known accurately. The method is applicable to transparent tapered films on reflective substrates as well as to opaque films.

OPTICAL METHODS

Photometric. A variety of optical techniques can be adapted to measuring film thickness. By determining the optical constants for a film, the thickness can be calculated. Reflectance can be reduced or enhanced depending on the relative values of the indexes of the film and of the substrate material. The relatively sharp reflectance changes that occur form the basis for monitoring and controlling deposition of multilayer dielectric films by the use of beam photometers (118–125).

Wedge Technique. Film thickness can be simply measured by depositing a thin film on a glass slide using a wedge arrangement of a metal shim in contact with the edge of the slide (29,49,123). The deposited film surface and the glass make an appreciable angle when the shim is removed. The interfering light rays do not enter the eye parallel to each other but appear to diverge from a point near the film. The resulting fringes resemble the localized fringes in a Michelson interferometer. They appear to be formed in the film itself. The film is wedge shaped with the fringes practically straight following the lines of equal thickness.

In estimating the thickness of films less than 1 μm thick, the 0.5461 μm green-line wavelength, at a quarter wavelength, and the index of refraction are employed. Each green line appears at N (number of fringes) times 0.084 μm. Thus, to estimate the film thickness, the number of green lines up to the wedge of film are counted (under a fluorescent light), one is subtracted, and the result is multiplied by 0.084.

Interferometry. Interference fringes are produced when two reflecting surfaces are close together. A direct determination of film thickness by measurement of interference fringes was developed by Tolansky (95,97). Two types of fringe measurement can be utilized, Fizeau fringes and fringes of equal chromatic order (feco). In the Fizeau technique the interferometer consists of two inclined optical flats, one of which supports a film. Bringing the second optical flat in contact with the film surface while illuminating with a monochromatic beam causes the appearance of dark fringes separated by $\lambda/2$. The displacement is expressed as a fraction of $\lambda/2$ fringe spacing which gives the film thickness. In the feco technique (95,97,125), an interference-objective with beam splitter is used to produce two-beam Fizeau interference fringes which are more difficult to obtain, but more accurate.

Light that is split into two beams which traverse different optical paths can be recombined with a resulting interference pattern. The Michelson interferometer method separates the light into two beams by amplitude splitting; the Nomarski polarization interferometer splits a linearly polarized beam reflected from the sample into two of equal intensity with a Wollaston prism. Various modifications of the Michelson interferometer have been used for the measurement of film thickness (90,92,97). Microscope manufacturers have marketed instruments utilizing the Michelson interferometer principle for film thickness measurement; Nomarski polarization interference attachments are commercially available for polarized metallurgical microscopes.

Variable-Angle Monochromatic Fringe Observation (Vamfo). By rotating a stage on an axis normal to the optical axis of the microscope, reflected light can be observed at various angles. The film thickness can be obtained by rotating the stage and sample and observing the bright and dark fringes on the film surface (126–128).

Constant-Angle Reflection Interference Spectroscopy (Caris). This nondestructive technique is useful for the measurement of thickness of transparent films on a reflective substrate. The method is based on the interference of radiation reflected from the dielectric-air interface with that from the dielectric-substrate interface (129). A spectrophotometer is used to observe fringes formed by a sample film as a function of wavelength.

Electron Impact Emission Spectroscopy (Eies). Vacuum deposition of alloy films often requires monitoring and control of deposition rate of each material in order to exactly maintain the film composition. The monitoring devices normally used, such as the quartz crystal microbalance and ionization gauge, cannot distinguish between the different components of a film. Eies allows the multicomponent evaporant to enter a miniature sensor and then pass through a region of excitation (130–131). There it is cross-bombarded by an electron beam. The collision process excites some fraction of the atoms or molecules causing optical emission. The wavelengths of the emitted radiation generally fall in the ultraviolet light region and are characteristic of each atomic or molecular species. The intensities of the emission lines are linearly proportional to the vapor densities of corresponding species, and thus also to the evaporation rate. In Figure 14, a scanning monochromator was used to determine sensor output signal during evaporation of an Al/Cu binary alloy. Strong copper and aluminum emission lines are evident; the line amplitude represents light intensity or evaporation rate.

Spectrophotometric Method. The double-beam recording spectrophotometer is widely used for measuring thickness of epitaxial semiconductor films with a precision of better than 1% for SiO_2 films as thin as 0.04 μm (132) (see Analytical methods).

Ellipsometry. A nondestructive method sometimes referred to as polarimetry and polarization spectroscopy is used to measure both thickness and refractive index of a transparent film. This more complicated technique involves the phase shift and intensity of two components of reflected polarized light and measures the change in the state of light polarization reflected from a substrate. The subject is fully covered in standard optics texts under the heating of elliptically polarized light (93,98,133).

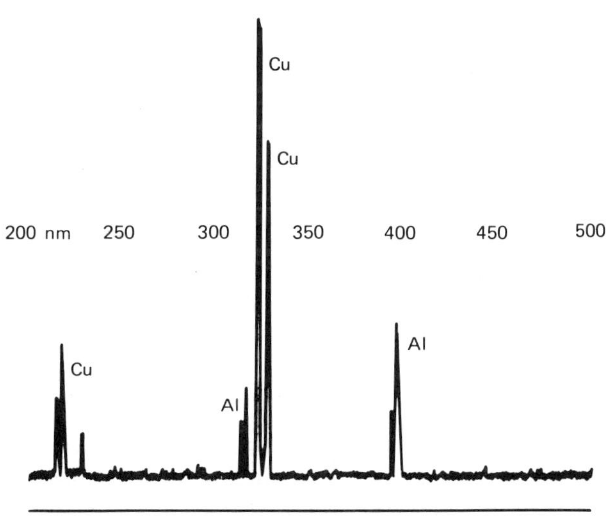

Figure 14. Typical binary alloy scan. Courtesy of Inficon.

ELECTRICAL METHODS

Capacitance and resistance as well as conductivity and isothermal Hall coefficient have been used for measurement of film thickness. Capacitance is inversely proportional to the film thickness and directly proportional to the area of the electrode deposited on the dielectric film (edge effects minimized) and to the dielectric constant of the film. Resistivity measurements, which are applicable to metallics, have also been employed to measure thickness of epitaxial semiconductor films (134). The four-point probe technique is a simple nondestructive method for measuring the thickness of conducting films and nonconducting substrates (135). Film thicknesses of metallics have been determined with accuracy to 0.02 μm by combining conductivity and isothermal Hall coefficient measurements (136–138) (see Nondestructive testing).

Capacitance Monitoring. Dielectric film thickness can be obtained by measuring the electrical capacitance of the deposited film. This method has been used to measure evaporation rates by monitoring changes in capacitance of a parallel plate condensor as a result of dielectric constant changes (139). In another procedure a deposited film is photoetched to obtain a line pattern from which capacitance changes can be measured and related to changes in thickness (140).

Ionization-Gauge Monitor. The ionization-gauge monitor is a modification of the ion gauge commonly used for pressure measurements (105,141–146). Atoms of the evaporant are ionized as they traverse the gauge and the resulting ion current is measured. Since this current is proportional to the number of atoms of the element being deposited, the device indicates the incidence rate, and film thickness is derived by integration over the deposition time. The thickness obtained will be correct only if the sticking coefficient on the substrate is unity, as it is for most metals. If it is necessary to interrupt deposition upon attaining a certain film thickness, either an integrating network is necessary or the rate must be kept constant during the period of deposition. Uniform rates are desirable in any case, and it is advantageous if the measuring device can simultaneously act as an automatic rate control. The ionization-gauge monitor has been used for this purpose, and rates of less than 0.001 μm/s have been controlled to $\pm 5\%$, with special care to $\pm 2\%$ (147). At higher rates control to $\pm 1\%$ has been obtained (148).

CHEMICAL METHODS

A fairly accurate measure of film thickness can be made by the reaction of a thin film with an excess of chemical reagent and back-titrating the unreacted reagent. In a simple instance, a known section of aluminum thin film can be completely dissolved with a known amount of standardized hydrochloric acid. The excess acid can then be determined by titration, and the amount of hydrochloric acid consumed found by difference. This in turn can yield the weight of aluminum and the film thickness.

RADIATION METHODS

X-Ray Emission. By measuring the attenuation of the characteristic radiation of the substrate material by the film, the film thickness can be ascertained. For a particular wavelength, attenuation is an exponential function of the film thickness and is dependent on the mass-absorption coefficient of the film material. If the film material is excited with x-rays and the intensity of a selected wavelength of the characteristic radiation emitted by the film material is measured, the thickness of the film is linearly proportional to the intensity of emitted radiation (see X-ray technology). This approach is excellent for measuring metallic thin films of 0.002 to 1 μm thickness. It is nondestructive but requires rather expensive equipment and calibrated thickness standards. A lower-cost, compact and stable radioisotope x-ray source has also been used (149).

X-Ray Interference. The use of x-ray interference fringes for calculating film thickness has been reported (150–152). X-ray interference can be observed by allowing an incident beam of monochromatic x-rays to strike a thin film in the region of the critical angle for total reflectance. Film thicknesses as small as 0.03 μm have been determined by separation of interference fringes. The use of a secondary interference pattern has been reported to give a more accurate determination of film thickness (153).

Beta Backscatter. If the metallic film and the substrate have large differences in atomic number, the film thickness can be measured from the amount of backscatter from the film of beta particles emitted from a radioactive source. Intensity of the backscattered particles is a function of film thickness and increases with atomic number of the film material. The technique requires thickness standards, but is inexpensive, simple, quick, nondestructive, and applicable to opaque and transparent films. Films of 0.1–50 μm can be measured on ca a square centimeter with an accuracy of ±5%.

Pinholes

One principal concern in the production or the deposition of uniform polymeric thin films is that they be continuous and pinhole-free. Pinholes, which usually originate from the presence of dust particles on the surface of the substrate, cannot be tolerated because they allow moisture and ambient contaminants to penetrate to the substrate, resulting in corrosion, electrical and physical degradation, or other harm. The inherent physical dimensional makeup of thin films dictates that the thickness is very small compared to the other dimensions. This can result in ease of pinhole formation.

Thin polymeric films prepared by the glow-discharge method on metallic substrates do not generally contain pinholes. A gaseous molecule at interaction temperature striking a metallic surface has a sticking coefficient greater than that for impingement upon the polymer surface (44). Therefore, any holes are quickly filled owing to the greater chance of a molecule sticking to the metallic substrate than to a polymer layer. In sputtering of organic materials, pinhole-free coatings are also readily attainable because of the relatively high operational gas pressure; the atoms arrive from many different angles and can penetrate under any dust particle present. This is in contrast to the formation of thin films by evaporation techniques where film formation begins at nuclei and continues with growth about these nuclei. Here complete surface coverage will not take place until these nuclear islands have coalesced. This results in vertical as well as horizontal growth leading to films with variable thicknesses. In addition, particles of dust on the surface of the growing film can cast shadows and give rise to tiny areas where deposition is restricted or absent. Sensitive detection methods for the location of pinholes as small as 0.1 μm in organic polymeric films include the electrographic method (154) which uses benzidine-impregnated filter paper which is anodically oxidized to a colored product only at conductive pinhole sites. Hydrogen gas evolution (152) can also provide a rapid and quantitative method of developing decorative effects in a thin film.

Film Composition Analysis

Instrumental surface analysis is the ultimate technique for the examination of thin films. Photons, electrons, atoms, and molecules or ions can be used as probes in a variety of combinations with an imposing array of instrumental methods. Each instrumental technique for surface analysis contains its own complex of capabilities and corresponding specificity. Surface analysis identifies the chemical elements present and shows their concentration, distribution, surface depth, and the type of bonding involved. All of the following techniques are discussed in more detail in the article on Analytical methods.

Transmission Electron Microscope (TEM). The electron has a wavelength several hundred times shorter than light and the use of electrically charged particles permits focusing as with a light beam. This is the principle of the transmission electron microscope (TEM) which has achieved magnification up to one million times with a 0.2 nm resolution. The TEM is one of the best systems for molecular and crystallographic studies and is most useful for forming images of lattice defects such as stacking faults, microtwins, dislocations, vacancy clusters, and gain domain boundaries. However, it suffers from the following limitations: (*1*) Image intensifers cannot allow use of thick specimens because the electrons penetrating a thick specimen lose varying amounts of energy owing to scattering and are no longer monochromatic. (*2*) Sample preparation is difficult and requires ultrathin slicing or embedment.

Scanning Electron Microscope (SEM). Unsurpassed three-dimensional high magnification photographs can be obtained showing microtopography and indicating possible surface anomalies (Fig. 15). The SEM does not contain any image forming lenses and is not truly a microscope.

Used for surface analysis the SEM represents only a first step for elemental

Figure 15. SEM photograph of Al, 1.3 μm on SiO_2 and Si. Courtesy of International Plasma Corporation.

compositional analysis. As the electron beam scans the surface, the emerging x-rays are fed into the x-ray analyzer which has been set for the element of interest. Whenever an x-ray of that energy is detected, a dot is brightened at the corresponding point in the display tube of the SEM. By positioning the SEM electron beam on a spot of interest, a simultaneous semiquantitative analysis of all elements can be performed in one minute by plotting counts vs energy (keV). The heights of the peaks are not an exact measure of concentration because all elements are not excited with the same efficiency. For a more quantitative measurement, known elemental standards must be employed.

Scanning Auger Microprobe (SAM). A nondestructive analytical technique for surface analysis based on the ejection of secondary, or Auger, electrons is a phenomenon discovered by Pierre Auger in the 1920s (155). The SAM is a highly complex instrument ideal for the determination of the composition of very thin surface films down to 0.1% of a monolayer and films as thin as 0.001 μm.

A technique that has added dramatically to SAM allows for sputter etching a specimen surface and simultaneous analysis giving the profile thickness. Data on specimen composition as a function of depth into the specimen are obtained. This technique involves bombardment of the specimen surface with a beam of argon or xenon ions which sputter-etch the surface at a rate of ca 1 μm/h. Simultaneously, bombardment of the specimen with an electron beam yields Auger analytical data. It is important that the sputter-etched area be much larger than the area of the electron beam so that the region of analysis is in the center of a relatively flat-bottomed crater. This precludes analysis of the sides of the crater where the depth resolution would be seriously impaired. In typical present-day equipment, the ion beam diameter is on the order of a few millimeters, and the electron beam can be focused to as little as 25 μm (Fig. 15).

Secondary Ion Mass Spectrometer (SIMS). One of the newest surface analytical instruments is the secondary ion mass spectrometer (SIMS) which directly analyzes the mass of the positive and negative ions that are sputtered from the surface as a result of ion bombardment (133,156–157). Measuring both the positive and negative ions gives two different types of mass spectra. A typical negative spectrum of stainless steel is shown in Figure 16. Positive SIMS are especially sensitive to low-atomic-number elements that have a low electronegativity and ionization potentials; the negative SIMS is most sensitive to the low-atomic-number elements with high electronegativity, hetero atoms and halides.

Ion Scattering Spectrometry (Iss). Ion scattering spectrometry (iss) can yield information about the atoms present in the first monolayer of a surface (158–160). Noble gas ions of 200–5000 eV suffer an energy loss upon striking the surface atoms. The iss spectrum shows a single peak of scattered ion intensity at an energy loss that is specific to the mass of surface atom involved (161). Chemical bonding information can be obtained by observing the characteristic pattern of scattered ion yields as a function of the primary ion energy. The high sensitivity of iss permits detection of elements at ppm levels in the top monolayer.

Electron Spectroscopy for Chemical Analysis (Esca). A unique technique for surface analysis that employs x-rays for bombarding the surface is known as electron spectroscopy for chemical analysis (162–163). In esca a specimen under vacuum is irradiated with monochromatic x-rays and emits electrons from inner orbitals with kinetic energies that are characteristic of the parent atoms and the incident x-ray energy. The

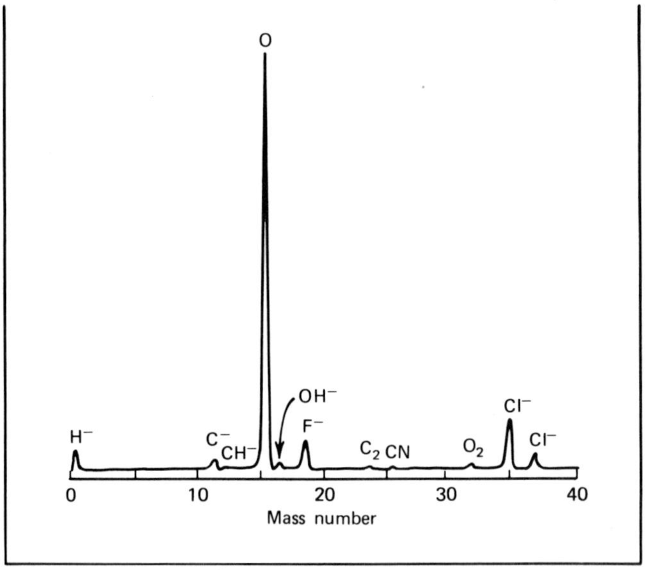

Figure 16. Negative sims spectra of stainless steel. Courtesy of Physical Electronics Industries, Inc.

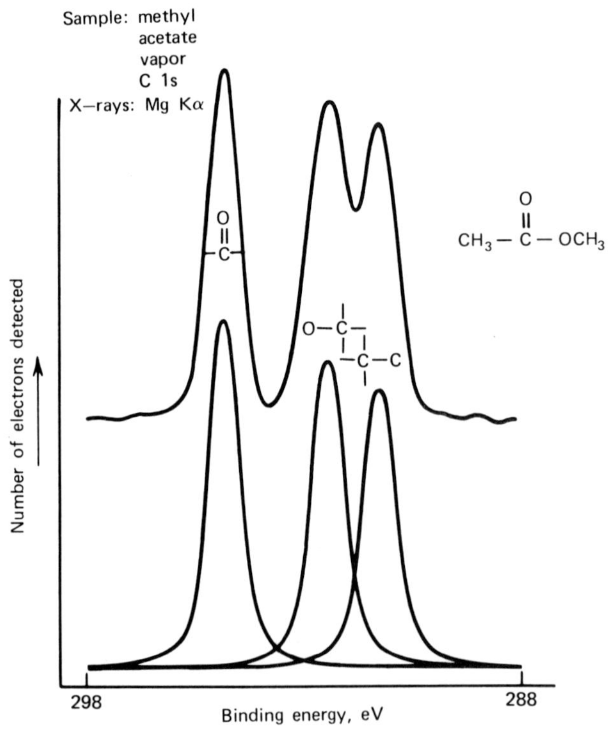

Figure 17. Esca spectra of methyl acetate. Courtesy of Physical Electronics Industries, Inc.

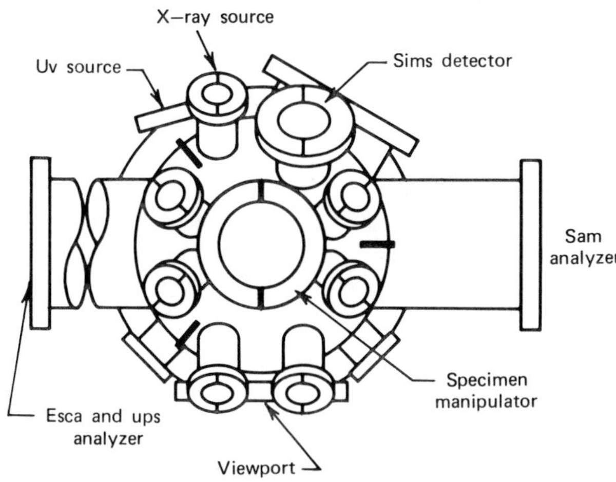

Figure 18. Multiple instrumentation design. Courtesy of Physical Electronics Industries, Inc.

spectrum thus obtained contains distinguishing peaks for each constituent element. Esca can detect elements above helium and analyze films down to a surface depth of 2 nm.

Esca is the only spectroscopic technique that is useful for nondestructive analysis of organic materials and that can also yield molecular structural data. For example, a film of methyl acetate shows three characteristic carbon peaks owing to the different bonds present (Fig. 17). Oxidation of most metals shifts peaks to a higher binding energy, and such inorganic shifts are also observed with esca.

Ultraviolet Photoelectron Spectroscopy (Ups). One of the newest of surface analytical techniques, ultraviolet photoelectron spectroscopy, typically uses helium in the uv source. The radiation excites photoelectron emission from electronic orbitals with low electron binding energies. Ups is extremely sensitive to surface bonding interactions which redistribute electron density.

Multi-Instrumentation. Each surface analysis technique has its own capabilities and definitive application areas, and it may also be complementary to other techniques. Frequently, the information required cannot be provided by a single technique, therefore, and two or more techniques are combined in a single instrument. This multi-instrumentated surface analysis achieves a synergistic combination of capabilities in one instrument and provides information not readily attainable by other methods. Equipment for these multiple instrumentation techniques allows the investigator to perform several types of analysis in a short period on a specimen without removing the specimen from the ultra-high vacuum environment. Geometrical configuration is a design factor that determines adequate specimen access to the various stations as in the multiple-instrumental system in Figure 18 which contains sam, sims, esca, and ups. Simultaneous Auger and sims analyses can be performed, providing two complementary surface composition measurements during ion bombardment, and the iss technique produces sputtered ions from the surface which can be used as a source of secondary ion mass spectrometry (sims).

BIBLIOGRAPHY

"Film Deposition Techniques" in *ECT* 2nd ed., Vol. 9, pp. 186–220, by Klaus H. Behrndt, Bell Telephone Laboratories.

1. *Athenaeum,* (April 5, 1839).
2. A. Brenner, *Electrodeposition of Alloys—Principles and Practice,* Vols. I and II, Academic Press Inc., New York, 1963 and 1964.
3. K. R. Lawless, *J. Vac. Sci. Technol.* **2,** 24 (1963).
4. G. C. Wood and co-workers, *J. Electrochem. Soc.* **110,** 1167 (1963).
5. L. Young, *Anodic Oxide Films,* Academic Press Inc., New York, 1961.
6. O. S. Heavens, *Thin Film Physics,* Chapman and Hall, London, 1970, p. 7.
7. R. L. Schwoebel, *J. Appl. Phys.* **34,** 2784 (1963).
8. P. E. Doherty and R. S. Davis, *J. Appl. Phys.* **34,** 619 (1963).
9. A. Brenner and G. Riddel, *J. Res. Nat. Bur. Stand.* **39,** 385 (1947); *Proceedings Am. Electroplaters Soc.* **34,** 156 (1947).
10. N. Feldstein, *Solid State Technol.,* **16,** 87 (1973).
11. *The Engineering Properties of Electroless Nickel Deposits,* International Nickel Co., New York, 1971.
12. G. Mallory, "Recent Developments in Applications of Electroless Nickel Plating," Paper 690646, *SAE National Aeronautic and Space Engineering and Manufacturing Meeting, Los Angeles, 1969.*
13. *Mater. Eng.* **81,** 50 (1975).
14. N. Feldstein, *RCA Engineer* **19**(1), (1973).
15. V. F. Stetter and M. Friz, *Chem. Ztg.* **97,** 138 (1973).
16. J. Feldman, M. Friz, and F. Stetter, *Res. Dev.* **27,** 49 (1976).
17. J. H. Gorrell, *Plast. Technol.* **10,** 45 (1964).
18. M. White, *Vacuum* **15,** 449 (1965).
19. U.S. Pat. 3,121,177 (Feb. 11, 1964), R. H. Davis.
20. J. H. Gorrell, *Plast. Technol.* **10,** 45 (1964).
21. L. V. Gregor, "New Materials In Electronics—Polymer Dielectric Films," *Proceedings Congress of Canadian Engineers,* IBM Report MP 22.0116, Montreal, June 2, 1967.
22. L. V. Gregor, *IBM J.,* 140 (March 1968).
23. W. F. Gorham, *J. Polym. Sci.* **4,** 3027 (1966).
24. "Development of Thin Organic Rolled Film Capacitor," *Bureau of Ships, Code 606/681A2, Contract NOGSR-89519,* Final Report, Union Carbide, 1968.
25. R. Holbrook, "Evaluation of Parylene C as a Conformal Coating For Electronic Devices," *Hughes Aircraft Company Internal Reports Nos. 2748.40/2714 Oct. 1969 and 2748.40/2580, Feb. 1969.*
26. V. S. Kale and T. J. Riley, *IEEE Trans. Parts Hybrids Packag.* **13,** 273 (1977).
27. D. M. Wylie, *Development of Parylene Passivation For Centaur Hybrid Microelectronics,* NASA CR135071, NASA, Washington, D.C., Aug. 1976.
28. F. W. Oberin, *Development For Application of Parylene Coating,* NASB29940, NASA, Huntsville, Alabama, June 1974.
29. S. M. Lee, J. J. Licari, and I. Litant, *Trans. Metall. Soc.* **1,** 701 (1970).
30. *Bulletin SL-1002,* Hohman Plating & Manufacturing Inc., pp. 2–3.
31. U.S. Pat. 2,146,024 (Feb. 7, 1939), F. M. Penning (to Darco Corp.).
32. E. Kay, *J. Appl. Phys.* **34,** 760 (1963).
33. U.S. Pat. 3,325,394 (June 13, 1967), E. Kay and A. P. Pensich (to IBM).
34. G. S. Anderson and co-workers, *J. Appl. Phys.* **33,** 2991 (1962).
35. T. Spalvins, "NASA TMX-52819," Presented at *Conference and School on Industrial Applications of Sputtering,* Materials Research Corporation, Tarrytown, N.Y., June 1970.
36. S. M. Lee, "RF Coating with Polymers—What Does It Offer," Presented at *Electronic Insulation Conf., Electronic Materials and Processing Symposium,* IEEE, Chicago, Oct. 3, 1973.
37. U.S. Pat. 3,663,265 (May 16, 1972), S. M. Lee and W. A. Bailey (to North American Rockwell Corporation).
38. U.S. Pat. 3,666,533 (May 30, 1972), S. M. Lee (to North American Rockwell Corporation).
39. D. T. Morrison and A. Taylor, "RF Sputtering of Plastics" in *Proceedings of The Second German Sputtering School and Congress,* Tegernsee, Germany, May 3–5, 1971, p. 134.
40. D. S. Alam and C. T. H. Stoddert, *Chem. In Britain* **1,** 410 (1965).
41. H. W. Melville, *Proc. Roy. Soc. London* **A163,** 511 (1937).

42. *Ibid.*, **A167,** 99 (1938).
43. *General Electric Research and Development Center Thin Film Polymers,* Schenectady, N.Y., July 1967.
44. R. White, *Electrochem. Technol.* **4,** 468 (1966).
45. J. R. Hollahan and R. S. Rosler, "Plasma Formation of Inorganic Thin Films" in J. L. Vossen and W. Kern, eds., *Thin Film Processes,* Academic Press, New York, 1978.
46. H. F. Stierling and R. C. G. Swan, *Solid State Electron.* **8,** 653 (1965).
47. S. B. Hyder and T. O. Yep, *J. Electrochem. Soc.: Solid State Sci. and Technol.* **123,** 1721 (1976).
48. S. Veprek, *J. Phys. Chem. Solids* **37,** 544 (1976).
49. L. L. Alt, S. W. Ing, and K. W. Laendle, *J. Electrochem. Soc.* **110,** 465 (1963).
50. L. S. Palatnik and Yu. F. Komnik, *Dokl. Akad. Nauk SSSR* **124,** 308 (1959); *Sov. Phys. Dokl.* **4,** 196 (1959).
51. G. J. Tibol and R. W. Hull, *J. Electrochem. Soc.* **111,** 1368 (1964).
52. F. K. McTaggert, *Plasma Chemistry in Electrical Discharges,* Elsevier Scientific Publishing Co., New York, 1967.
53. A. Bradley and J. P. Hammes, *J. Electrochem. Soc.* **110,** 15 (1963).
54. A. M. Mearns and C. Catsibiris, *Thin Solid Films* **3,** 204 (1969).
55. G. Smolinsky and J. H. Heiss, "Thin Organic Films Found by Glow Discharge Technique," *Proceedings ACS 155th Meeting,* San Francisco, Vol. 28, No. 1, April 1968.
56. J. P. Redmond and A. F. Pitas, "Investigation of Polymeric Coatings by Electrodless Discharge," *NASA Final Report W-1625,* May 1968.
57. H. Pagnia, *Phys. Status Solidi* **1,** 90 (1960).
58. R. A. Connell and L. V. Gregor, *J. Electrochem. Soc.* **112,** 1198 (1965).
59. J. Goodman, *J. Poly. Sci.* **44,** 551 (1960).
60. Private Communication from F. A. Sliemers, Batelle Columbus Laboratories, June 19, 1978.
61. S. M. Lee, *Insul. Circuits* **17,** 33 (1971).
62. K. Jesch, J. E. Bloor, and P. L. Kronick, *J. Polym. Sci.* **4,** 1487 (1966).
63. A. Denero and co-workers, *Eur. Polym. J.* **4,** 93 (1968).
64. I. Gonde Dittman, *Ind. Res.* **20,** 169 (1978).
65. H. C. Kaufman, *Handbook of Organo-Metallic Compounds,* D. Van Nostrand, New York, 1961.
66. H. Zeiss, *Organo-Metallic Chemistry,* Reinhold Publishing Corp., New York, 1960.
67. J. H. Harwood, *Industrial Applications of Organo-Metallic Compounds,* Reinhold Publishing Corp., New York, 1963.
68. R. T. Hopper, *Ceram. Ind.* **80,** 65 (1963).
69. C. F. Powell, J. L. Oxley, and J. N. Blocher, *Vapour Deposition,* The Electrochemical Society, John Wiley & Sons, Inc., New York, 1966.
70. C. F. Powell, I. E. Campbell, and B. W. Gonser, *Vapor-Plating,* John Wiley & Sons, Inc., New York, 1955.
71. *International Conferences on Chemical Vapor Deposition,* 1968–1972 (1st, 2nd, and 3rd), The Electrochemical Society, New York.
72. H. E. Podall and N. M. Mitchell Jr., *Ann. N.Y. Acad. Sci.* **125,** 218 (1965).
73. H. W. Markstein, *Electron. Packag. Prod.* **17,** 31 (1977).
74. R. W. Christy, *J. Appl. Phys.* **31,** 1680 (1960); U.S. Pat. 3,119,707 (Jan. 21, 1964), R. W. Christy (to Space Technology Laboratories).
75. I. Haller and P. White, *J. Phys. Chem.* **67,** 1784 (1963).
76. R. A. Fotland and W. J. Burkhard, *Proceedings 121st Electrochem. Soc. Mtg., Los Angeles, May 7, 1962.*
77. E. M. Da Silva and E. Klokholm, *Proceedings Electrochem., Soc. Mtg. Electric Insulation Division,* Abstract No. 14, San Francisco, May 9, 1965.
78. A. E. Brennemann and L. V. Gregor, *J. Electrochem. Soc.* **112,** 1194 (1965).
79. L. V. Gregor in G. Hass and R. Thun, eds., *Physics of Thin Films,* Vol. 3, Academic Press, New York, 1966, p. 131.
80. R. F. Thornley and T. San, *J. Electrochem. Soc.* **112,** 1151 (1965).
81. A. E. Ennos, *Brit. J. Appl. Phys.* **5,** 27 (1954).
82. K. M. Poole, *Proc. Roy. Soc.* **167A,** 99 (1938).
83. M. Sparks, "Theoretical Studies of Materials For High Power Infrared Coatings," *Contract DAHC 15-73-C-0127,* Defense Supply Service, Washington, D.C., 1973.
84. J. J. O'Dwyer, *Theory of Dielectric Breakdown of Solids,* Clarendon Press, Oxford, 1964.
85. H. Y. Ku and F. G. Ullman, *J. Appl. Physics* **35,** 265 (1964).

86. P. Benjamin and C. Weaver, *Proc. Roy. Soc. London* **A254,** 177 (1960).
87. *Ibid.,* **261,** 516 (1961).
88. H. E. Bennett and J. M. Bennett in G. Hass and R. E. Thun, eds., *Physics of Thin Films,* Vol. 4, Academic Press Inc., New York, 1967.
89. P. S. Flint in B. Schwartz and N. Schwartz, eds., *Measurement Techniques for Thin Films,* The Electrochemical Society, New York, 1967, p. 141.
90. M. Francon, *Optical Interferometry,* Academic Press Inc., New York, 1966.
91. D. J. Gillespie in B. Schwartz and N. Schwartz, eds., *Measurement Techniques for Thin Films,* The Electrochemical Society, New York, 1967, p. 102.
92. D. S. Heavens in G. Hass and R. E. Thun, eds., *Physics of Thin Films,* Vol. 2, Academic Press, Inc., New York, 1964, p. 193.
93. O. S. Heavens, *Optical Properties of Thin Solid Films,* Butterworth Scientific Publications, London, 1955.
94. H. Mayer, *Physik Dunner Schichten,* Wisenschaftliche Verlagsgesellschaft m.b.H., Stuttgart, 1950.
95. S. Tolansky, *Multiple-beam Interferometry of Surfaces and Films,* Oxford University Press, London, 1948.
96. S. Tolansky, *An Introduction to Interferometry,* Longmans, Green & Co., Ltd., London, 1955.
97. S. Tolansky, *Surface Microtopography,* Interscience Publishers, Inc., New York, 1960.
98. A. Vasicek, *Optics of Thin Films,* North Holland Publishing Company, Amsterdam, 1960.
99. D. S. Campbell and H. Blackburn, *Trans. 7th Natl. Vacuum Symposium,* Pergamon Press, New York, 1960, p. 313.
100. A. R. Beavett, *J. Sci. Insti.* **43,** 182 (1965).
101. R. E. Hayes and A. R. V. Roberts, *J. Sci. Insti.* **39,** 428 (1962).
102. A. L. Houde in *Vacuum Microbalance Techniques,* Vol. 3, Plenum Press, 1963, p. 109.
103. C. A. Neugebauer, *J. Appl. Phys.* **35,** 3599 (1964).
104. P. Lostis, *Rev. Opt.* **38,** 1 (1959).
105. K. H. Behrndt, "Thickness and Rate Measuring Devices and Methods for Obtaining Films of Uniform Thickness" in G. Hass and R. F. Thun, eds., *Physics of Thin Films,* Vol. 3, Academic Press Inc., New York, 1966.
106. K. H. Behrndt and R. W. Love, *Vacuum,* **12,** 1 (1962).
107. S. J. Lins and H. S. Kukuk, *Trans. 7th Natl. Vacuum Symp.,* p. 333, Pergamon Press, New York, 1960.
108. K. H. Behrndt, *Trans. 7th Natl. Vac. Symp., 1960,* Pergamon Press, London, 1961, p. 137.
109. D. McKeown, *Rev. Sci. Instrum.* **32,** 133 (1961).
110. A. W. Warner and C. D. Stockbridge, *Vacuum Microbalance Techniques,* Vol. 3, Plenum Press, New York, 1963, p. 55.
111. A. W. Warner and C. D. Stockbridge, *J. Appl. Phys.* **34,** 437 (1963).
112. W. Steckelmacher and co-workers, *Trans. 10th Natl. Vacuum Symp.,* Macmillan Company, New York, 1963, p. 415.
113. W. P. Mason, *Piezoelectric Crystals and Their Application to Ultrasonics,* D. Van Nostrand Inc., Princeton, N.J. 1956.
114. R. E. Reason in H. W. Baker, ed., *Modern Workshop Technology,* Part 2, Cleaver-Hume Press, Ltd., Bristol, England, 1960, Chapt. XVII.
115. N. Schwartz, and R. Brown, *Trans. 8th AVS Symp. and 2nd Intern. Congr. Vacuum Sci. Technol.,* 1961, Pergamon Press, New York, p. 836.
116. D. W. Freyberg, *Res. Dev.* **25,** 5, 25, 41 (1974).
117. A. Kutzelnigg, *Testing Metallic Coatings,* Robert Draper Ltd., Teddington, England, 1963.
118. K. Chopra, *Thin Film Phenomena,* McGraw-Hill Book Company, New York, p. 99.
119. T. Sawaki and co-workers, *J. Physique* **25,** 2158 (1964).
120. A. C. Traub, *J. Opt. Soc. Am.* **45,** 999 (1956).
121. P. Giacomo, *Rev. Opt.* **35,** 317, 422 (1956).
122. A. Andant, *J. Phys. Radium* **11,** 351 (1958).
123. D. L. Perry, *Appl. Opt.* **4,** 987 (1965).
124. W. Stecklelmacher and co-workers, *Vacuum* **9,** 171 (1959).
125. A. F. Bogenschutz and co-workers, *Z. Angew. Phys.* **14,** 469 (1962).
126. W. A. Pliskin and E. E. Conrad, *IBM J. Res. Dev.* **8,** 43 (1964).
127. W. A. Pliskin and R. P. Esch, *J. Appl. Phys.* **36,** 2011 (1965).

128. W. A. Pliskin and H. S. Lehman, *IBM Tech. Report TR 22.279,* Phoenix, 1966.
129. W. A. Pliskin in E. M. Must and W. G. Guldner, eds., *Physical Measurement and Analysis of Thin Film,* Plenum Publishing Corp., New York, 1969.
130. C. Lu and co-workers, *J. Vac. Sci. Technol.* **14,** 103 (1977).
131. C. Lu, *Thin Solid Films* **45,** 481 (1977).
132. N. Goldsmith and L. A. Murray, *Solid State Electron.* **9,** 331 (1966).
133. A. W. Czandera, ed., *Methods of Surface Analyses,* Elsevier Scientific Publishing Co., New York, 1975.
134. E. E. Gardner and co-workers, *Measurement Techniques For Thin Films,* The Electrochemical Society, New York, 1967, pp. 240, 258.
135. L. P. Valdes, *Proc. IRE* **42,** 420 (1954).
136. W. F. Leonard and R. L. Ramey, *J. Appl. Phys.* **35,** 2963 (1964).
137. *Ibid.* **37,** 2190 (1966).
138. V. A. Marsoci and P. Siegal, *J. Appl. Phys.* **39,** 29 (1968).
139. G. C. Riddle, *Proc. 4th Symp. Electron Beam Technol.,* Alloyd Electronics Corp., Cambridge, Mass., 1962.
140. F. Z. Keister and R. Y. Scapple, *Trans. 9th Natl. Vacuum Symp.,* The Macmillan Company, New York, 1962, p. 116.
141. M. H. Perkins, *Trans. 8th Natl. Vac. Symp., 1961,* Pergamon Press, London, 1962, p. 1025.
142. O. Haase, *Z. Naturforsch.* **12a,** 941 (1957).
143. F. Metzger, *Helv. Phys. Acta* **16,** 323 (1943).
144. G. R. Giedd and M. H. Perkins, *Rev. Sci. Instrum.* **31,** 773 (1960).
145. H. Schwartz, *Trans. 7th Natl. Vacuum Symp.,* Pergamon Press, New York, 1960, p. 326.
146. H. Schwartz, *Rev. Sci. Instrum.* **32,** 194 (1961).
147. M. H. Perkins, *Trans. 8th Natl. Vac. Symp., 1961,* Pergamon Press, London, 1962, p. 1025.
148. M. Beckerman and R. E. Thun, *Trans. 8th Natl. Vac. Symp., 1961,* Pergamon Press, London, 1962, p. 905.
149. J. F. Cameron and J. R. Rhodes in G. L. Clark, ed., *Encyclopedia of X-Rays and Gamma Rays,* Reinhold Publishing Co., New York, 1960, p. 150.
150. H. Kiessig, *Naturwissenschaften* **18,** 847 (1930).
151. W. Hink and W. Petzold, *Z. Angew. Phys.* **10,** 135 (1958).
152. Y. Fujike, *J. Phys. Soc. Jpn.* **14,** 1308 (1959).
153. J. Sauro, I. Fankuchen, and N. Wainfan, *Phys. Rev.* **132,** 1544 (1963).
154. S. M. Lee, J. J. Licari, and J. P. McCloskey, *Insulation* **15,** 40 (1969).
155. P. Auger, *J. Phys. Radium,* **6,** 205 (1925).
156. C. W. White and N. H. Tolk, *Phys. Rev. Lett.* **26,** 486 (1971).
157. R. L. Park and co-workers, *Rev. Sci. Instrum.* **41,** 1810 (1970).
158. D. D. Smith, *Surf. Sci.* **25,** 171 (1971).
159. D. J. Bell and co-workers, *Surf. Sci.* **30,** 69 (1972).
160. D. G. Armour and G. Carter, *J. Phys. E.* **5,** 2 (1972).
161. F. W. Karasek, *Res. Dev.* **24,** 24 (1973).
162. K. Siegbahn and co-workers, *Nov. Acta Regiae Soc. Sci. Ups.* **20,** 282 (1967).
163. T. A. Carlson, *Phys. Today* **25,** 30 (1972).

STUART M. LEE
Ford Aerospace & Communications Corp.

FILM THEORY OF FLUIDS. See Mass transfer.

FILTERS, OPTICAL. See Optical filters.

FILTRATION

Filtration is the separation of particles of solids from fluids (liquid or gas) by use of a porous medium (see also Air pollution control methods; Gas cleaning; Ultrafiltration). This article discusses only separations of solids from liquids.

The two main mechanisms of filtration are cake filtration and depth filtration. In cake filtration, solids form a filter cake on the surface of the filter medium. In depth filtration, solids are trapped within the medium using either cartridges or granular media such as sand or anthracite coal. Cartridges are usually disposable, with media of various kinds of fibers or porous structures, and generally mounted in pressure enclosures. In depth filtration, also called bed filtration, gravity flow as well as pressure operation are used (see Water, industrial water treatment, municipal water treatment, water pollution, water reuse).

Filtration-thickening can be surface filtration or a modification of cake filtration. Slurry cross-flow, mechanical action, or electrical energy in the form of electrophoresis prevent or limit cake buildup on the medium surface. An alternative operation is cyclic with automatic intermittent cake formation, discharge, and reslurry.

Filtration can also be accomplished by centrifugal force (see Centrifugal separation), or by ultrafiltration, which involves a cross-flow filtration that is actually a classification based on molecular size. In straining and screening, closely related to filtration, the objective is to separate only a proportion of the solids (see Size classification).

An operation where any successive layers of solids are accumulated, even though cake thickness is negligible, is properly called a cake filtration. If no successive layers are formed, then it is either a straining or a filter-thickening operation. Both might be termed surface filtration. In the filtration of extreme fines, such as bacteria, with absolute-rated cartridges, both depth and surface filtration are involved (see under Cartridge Filtration).

There is no sharp dividing line between the need for depth and cake, or surface, filtration. However, cartridge depth filtration is seldom used when solids recovery is required. The finer the solids and the more dilute the suspension (conc. <0.01%) the more likely that depth filtration will be economically preferable.

Cake Filtration

In the last 25 years, several developments have shaped the course of evolution of filtration. (*1*) Polyelectrolyte flocculants permit the nature of feeds to be modified, increasing equipment possibilities. (*2*) Synthetic fibers and membranes vastly enlarged the choices of filter media and consequently a suitable filter material is available for most applications. (*3*) Endless-belt filters with a variety of configurations are of increasing importance. Their key feature is the inclusion of continuous medium washing. (*4*) The filter press, after steadily losing to continuous filters since their introduction, is receiving renewed interest because of automated discharge, grooved rim recessed plates, and its inherent capability for dewatering to the lower moisture content necessary for landfill or conserving fuel in cake drying and incineration. Membrane development further extends its performance capability.

Theoretical Aspects. Cake filtration is unique in that filter testing provides directly the rates, cake moistures, washing effectiveness, etc, without application of theory. Theoretical considerations, however, contribute to the understanding of the filtration mechanism, estimating effects of variables, and analyzing data.

In filtration theory, there are two ideal types of materials: incompressible and compressible. In a filter cake, a mechanical force is transmitted between the particles by interparticle contact. This mechanical force is called the effective pressure, a term originating in the field of soil mechanics. Incompressible cake has a rigid structure that does not deform under varying levels of the effective pressure. For practical purposes, filter cakes of nonflocculent (particulate) solids, such as sand or salt crystals, are usually incompressible. On the other hand, filter cakes of flocculent solids such as waste sludge, metal hydroxide, or polyelectrolyte floc, have a deformable structure and are called compressible cakes. The concentration of the solids in a compressible cake varies through the cake, responding to the local levels of the effective pressure. The four filtration operations shown in Figure 1 are all used with compressible materials and all but expression with incompressible material.

Cake formation (Fig. 1**a**) is the process of removing liquid from a suspension by maintaining a hydraulic pressure difference between the feed slurry supplied from one chamber and the filtrate received by another chamber, the two chambers being separated by a filter medium. During the process, feed slurry flows towards the medium, and solids are deposited on the medium. Under the applied hydraulic pressure, filtrate is forced to flow through the growing cake and the filter medium.

Cake washing (Fig. 1**b**) is the process of removing a dissolved solute from the pores of a filter cake by forcing a wash liquid to flow through the cake. The objective may be either to increase the purity of a solid product or to improve the recovery of the liquid product, or even to reduce the liquid viscosity to aid in dewatering. Dewatering before or after washing always interacts with the washing.

In gas displacement dewatering (Fig. 1**c**), a gas (usually air or steam) is forced through the cake and the liquid phase is substantially displaced by the gas. However, the flow channels in the cake may be of such a small diameter that capillary forces prevent the gas at the applied pressure differential from displacing the liquid significantly.

Dewatering by expression (Fig. 1**d**) is the process of applying mechanical pressure to compress a cake and squeeze out the liquid. The basic mechanism is operable during dewatering in equipment such as the filter belt press and filter presses equipped with movable diaphragms. When the original liquid–solid mixture in the cylinder is a suspension and not a cake, expression would then involve two processes, ie, cake formation followed by dewatering.

Formation. The theory of filtration is based on the concept of the specific resistance (1–4), with the following assumptions: (*1*) flow is one dimensional; (*2*) growth of cake is unrestricted; (*3*) only solid and liquid phases are present; (*4*) the feed is sufficiently dilute such that the solids are freely suspended; (*5*) the concentration of the solid phase in the filtrate is zero; (*6*) pressure losses in feed and filtrate piping are not considered; and (*7*) flow is laminar, a valid assumption in most cake formation operations of practical interest. These assumptions must always be considered carefully. For instance, in filter-press operations (*1*) and (*2*) would not hold once cakes from the two sides of a chamber meet.

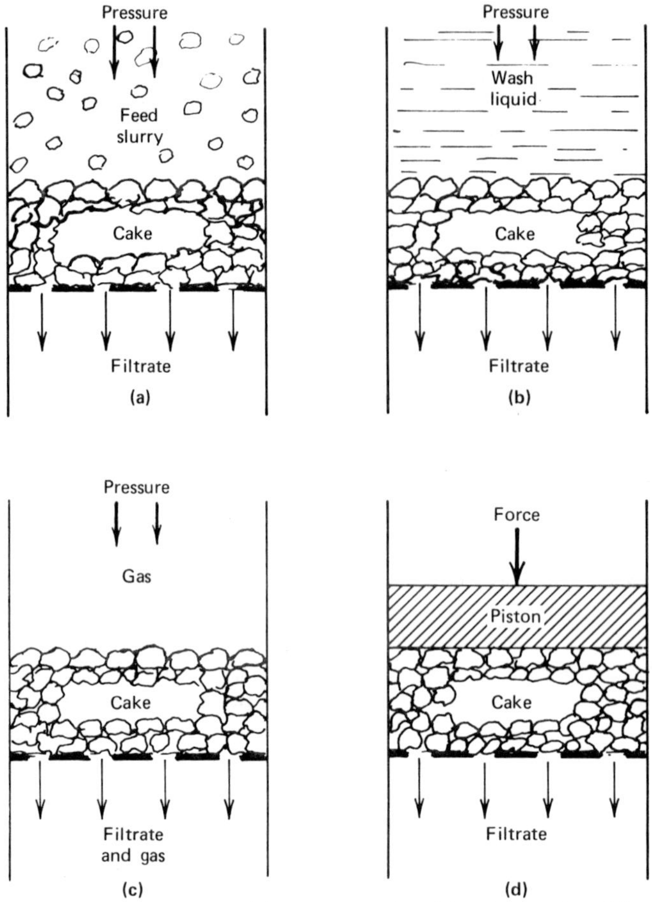

Figure 1. Basic filtration processes: (**a**) cake formation; (**b**) cake washing; (**c**) cake dewatering by gas displacement; (**d**) cake dewatering by expression.

The differential equation of cake formation is:

$$\frac{dv}{dt} = \frac{\Delta p}{\mu(\overline{\alpha}C^*v + R_m)} \tag{1}$$

where v is the volume of filtrate per unit area of filtration surface, t the filtration time, Δp the filtration pressure differential across the cake–medium system, μ the viscosity of the liquid phase, C^* the mass of dry solids in the cake per unit volume of filtrate, $\overline{\alpha}$ the average specific resistance of the filter cake, and R_m the hydraulic resistance of the medium. The variables $\overline{\alpha}$ and C^* are introduced in the derivation of an expression for the hydraulic resistance of the cake R_c, which is represented by the term $\overline{\alpha}C^*v$ in equation 1.

Specific resistance, an index of the filterability of a slurry, is the hydraulic resistance of the cake per unit mass of dry solids in the cake per unit area of filtration surface. If the cake is incompressible, the concentration of the solids and, therefore, the specific resistance is independent of position within the cake, the time, and the applied pressure differential. For a compressible material, on the other hand, the concentration of the solid phase and the specific resistance vary through the depth of the cake. Furthermore, they vary at each local position with the filtration time and applied pressure differential. Then, an average specific resistance is defined (4) and used to determine R_c, as follows:

$$R_c = \overline{\alpha} w_0 \tag{2}$$

where w_0 is the total mass of dry solids in the cake per unit area. For practical purposes, the average solids concentration and average specific resistance of a compressible cake depend primarily on the filtration pressure differential independent of the overall thickness or mass of the cake. They are increasing functions of the pressure differential; ie, when Δp is increased across a compressible cake, a higher average solids concentration and thus higher average specific resistance result.

The mass of dry cake solids w_0 can be expressed in terms of the filtrate volume v:

$$w_0 = \frac{C_s \rho_l v}{1 - C_s/\overline{C}_c} = C^* v \tag{3}$$

where C_s is the mass fraction of solids in the feed slurry; $\overline{C}_c$ is the average mass fraction of solids in the cake; and ρ_l is the density of the liquid phase (filtrate). For compressible material $\overline{C}_c$ and, therefore, C^* depends on Δp. Combining equations 2 and 3 yields the desired expression for R_c.

A characterization of the cake formation process for two important classes of filtration operations, constant pressure filtration and constant rate filtration, can be obtained readily through application of equation 1.

In constant pressure filtration, equation 1 is integrated and rearranged to yield:

$$v^2 \left(\frac{\mu \overline{\alpha} C^*}{2} \right) + v(\mu R_m) = \Delta p t \tag{4}$$

In constant rate filtration:

$$q = \frac{dv}{dt} = \frac{v}{t} = \text{constant} \tag{5}$$

Equation 1 can be solved to relate the filtration pressure differential to the time, eg:

$$\Delta p = q\mu(\overline{\alpha} C^* q t + R_m) \tag{6}$$

If the cake is compressible, $\overline{\alpha}$ and C^* in equation 6 depend on the instantaneous value of Δp, which increases during the filtration. For the general case of variable pressure and variable rate filtration, the equations become more complicated (4).

The specific resistance is usually determined in a filtration test conducted at a constant pressure differential. The volume of filtrate is recorded as a function of time. The data are plotted as t/v vs v. According to the following rearranged form of equation 4 this plot should be a straight line:

$$\frac{t}{v} = \left(\frac{\mu \overline{\alpha} C^*}{2 \Delta p} \right) v + \frac{\mu R_m}{\Delta p} \tag{7}$$

Let a equal the slope of this line and b the intercept on the t/v axis. Then the specific resistance and the resistance of the medium can be calculated from:

$$\bar{\alpha} = \frac{2a\Delta p}{\mu C^*} \tag{8}$$

and

$$R_m = \frac{b\Delta p}{\mu} \tag{9}$$

The average specific resistance is correlated with filtration pressure differential by the following power function:

$$\bar{\alpha} = \alpha_0(\Delta p)^s \tag{10}$$

where α_0 is an empirical constant and s is known as the coefficient of compressibility. Coefficient s is zero for incompressible materials and positive for compressible materials.

Regarding application of the above ideas for compressible materials, two possible pitfalls (2) are noted: $\bar{\alpha}$, $\bar{C}_c$, and C^* depend on the fraction of the total pressure drop taken across the cake Δp_c, not on the total Δp. Hence, an implicit assumption is that $R_c \gg R_m$, in order to simplify the theoretical analysis. This assumption is violated in the early stages of a filtration but is considered valid when the cake reaches an appreciable thickness. Equations 4–6 applied to compressible materials are correct strictly only when R_m is negligible. Another qualification is that for concentrated slurries of compressible materials the specific resistance depends on the slurry concentration as well as on the pressure differential; $\bar{\alpha}$ decreases as slurry concentration increases (5–7). The explanation of this phenomenon is based on the existence of solids fluxes as well as liquid fluxes in a compressing cake.

Washing. Cake washing (Fig. 1**b**) is the process of removing one liquid from the pores of a filter cake by directing a desired substitute liquid to flow into and through the cake. If cake washing is preceded by a stage of gas displacement dewatering, considerable pore liquid may be removed before the washing. In the descriptions following, the cake is assumed completely saturated with feed liquid at the start. If the wash liquid were to displace this original liquid by pure plug flow, 100% removal would be achieved using wash liquid in an amount equal to one pore volume. However, in practice this ideal situation never occurs. Typically, two pore volumes of wash liquid may be needed to remove, eg, 85–95% of the original solute.

The washing process takes place in three stages (8):

Hydraulic Displacement. As in pure plug flow, optimum washing efficiency is obtained. However, owing to the distribution of sizes of flow channels within a cake, wash liquid flows more rapidly in the larger channels and breakthrough occurs before one pore volume of wash liquid is used.

Intermediate Stage. Hydraulic displacement continues from the smaller flow channels. The breakthrough from the larger channels contains feed liquid that entered the wash stream by diffusional mechanisms. Hence, both diffusion and hydraulic displacement are operative in this stage.

Mass Transfer Stage. It is not possible to remove all the feed liquid by hydraulic displacement. Some remains either as a film on particles or trapped in pores or stagnant pockets, such as the ones situated inside of floccules or at the point of contact of particles. This liquid is transferred to the wash streams by diffusional mechanisms.

Dewatering. *By Gas Displacement.* In gas displacement (Fig. 1**c**), both liquid and gas phases flow simultaneously in the filter cake. There are distinct interfaces between the two fluid phases within the pores of the cake. Flows of this general type (9) are also encountered in other fields, such as oil reservoir engineering, hydrology, and soil science.

At the start of the process the pores of the cake are filled completely with liquid. A certain pressure difference must be reached at the gas–cake interface to initiate displacement of the liquid. This pressure is the capillary pressure. For a capillary tube of radius r_c the capillary pressure p_c is defined by:

$$p_c = \frac{2\sigma \cos \phi}{r_c} \qquad (11)$$

where σ is the surface tension between the gas and liquid phases and ϕ is the contact angle (assumed to be less than 90°). If it is assumed for illustrative purposes that tube radius r_c represents typically the radius of the pores of the cake, equation 11 demonstrates the strong effect of pore size on the pressure requirement. The analysis suggests that in dewatering cakes of very fine particles, relatively high pressures may be required to initiate gas displacement (8).

After displacement of the liquid by the gas has started, liquid continues to flow from the cake until the residual saturation level is reached. The residual saturation S_r is the fraction of the pore volume containing liquid at equilibrium. This liquid is considered trapped in the smaller interstices of the cake by capillary forces.

Theoretical models to describe the time-wise reduction of liquid saturation (fraction of pores containing liquid) from an initial value of one to the final value of S_r have been developed (10–11) with one model (10) relying on empirical correlations or direct experimentation to derive necessary coefficients. Another model (11) requires experimental determination of a capillary pressure curve. It has been proposed (12–14) that cake moisture is a function of an empirical correlating factor:

$$F_c = \frac{u_{\text{air}} \Delta p_d \theta_d}{\mu w_0} \qquad (12)$$

where u_{air} is the air velocity measured at downstream pressure, Δp_d the pressure drop, θ_d the dewatering time, μ the viscosity of the liquid, and w_0 the cake weight of dry solids per unit area.

By Expression. The theory of dewatering by expression is based on the configuration depicted in Figure 1**d** in which a piston is used to compress a filter cake held in a cylinder having a permeable bottom. During expression the effective pressure (the pressure transmitted by interparticle contact) and solids concentration increase and the liquid pressure decays; the sum of these two pressures at any position in the cake at any time equals the applied pressure. At the end of the process, flow ceases and the effective pressure becomes equal to the applied pressure for all points in the cake.

A theoretical description of dewatering by expression (15–19) has been derived by solving a differential equation:

$$\frac{1}{C_e}\frac{\delta p_s}{\delta t} = \frac{\delta^2 p_s}{\delta w^2} - \frac{1}{\alpha}\frac{d\alpha}{dp_s}\left(\frac{\delta p_s}{\delta w}\right)^2 \qquad (13)$$

which considers the dewatering processes occurring locally in the cake as a function of time. Equation 13 is based on a continuity-of-mass equation and a force-balance equation. In addition, it is assumed that the local values of solids concentration and specific resistance are functions of the local effective pressure. In equation 13 the dependent variable is the effective pressure p_s and the independent variables are the time t and the mass w of solids per unit of area lying above the medium. In deriving equation 13, it is more convenient to use material coordinate w instead of the actual position in the cake. The consolidation coefficient C_e is defined by:

$$C_e(p_s) = \frac{\rho_s}{\mu \alpha m_v (1 + e)} \tag{14}$$

In equations 13 and 14, two fundamental parameters, the local specific resistance α and the coefficient of volume change m_v, characterize, respectively, the fluid flow and solids deformation properties of the cake. The specific resistance (see eq. 2) is now considered to vary locally. The new parameter, the coefficient of volume change, is defined by:

$$m_v(p_s) = -\frac{1}{(1 + e)} \frac{de}{dp_s} \tag{15}$$

Here and in equation 14 the term $e = \epsilon/(1 - \epsilon)$ is the local void ratio and is used instead of porosity, solids concentration, etc, to represent the relative proportions of the solid and liquid phases. The functional dependency of the specific resistance and the coefficient of volume change on the effective pressure can be determined experimentally using compression-permeability cell (16) data.

Using numerical procedures (15) equation 13 can be solved for a specific material for various specified boundary and initial conditions. By making the approximation that the consolidation coefficient is constant during mechanical expression, explicit solutions are obtained (16).

Practical Aspects. ***Cake Formation.*** *Pressure Differential.* Increased pressure differential on incompressible cakes is productive, its benefit on rate calculable. However, on compressible cakes it increases cake resistance to an extent determinable only empirically. (Eq. 10 modifies the Δp effect by the coefficient of compressibility.)

High initial Δp tends to drive fine particles into and through the filter medium, increasing the medium resistance and the solids content of the filtrate. The same force can add to the resistance of initial cake layers and certainly can contribute to an extension of solids bleed to the filtrate. These considerations are important reasons for starting cake formation with reduced pressure differentials, and are justifications for variable rate, variable pressure filtration, which is the normal practice in batch pressure filtration. The main reason for variable rate and pressure filtration is the practicality and economics of centrifugal pump selection, dictated by their performance curves. Sizing the pump for a final rate of 0.7–0.8 of the average rate usually gives good overall results (20). This contrasts with a final constant pressure rate approximately one half the average. Some filter cakes formed from flocculated solids have critical pressures at which the cakes densify significantly, decreasing rates substantially. It is then desirable to operate below such pressures or modify the feed with filter aids to maintain porosity.

Formation Time. Increased length of cake formation time at constant pressure usually affects the rate by the square root function. Thus any doubling of cake thickness or liquor throughput per cycle takes about four times as long. A logarithmic plot of form time versus cumulative filtrate or cake solids normally gives close to the theoretical straight line with 0.5 slope. A plot of cumulative solids or filtrate vs the reciprocal rate (eq. 7) also gives a useful straight line function.

Liquor viscosity is changed by temperature and its effect on rate is often unexpectedly large. Doubling viscosity doubles the time to form the same cake thickness, thus cutting the rate in half.

Cake specific resistance is mostly the result of the basic feed nature. It can be improved, however, in at least four ways (in addition to changes in Δp and C^*): (*1*) constant rate rather than constant pressure filtration initially gives lower resistance nearest the filter medium and minimizes any increase in medium resistance caused by solids penetration; (*2*) flocculation of feed solids directly affects the specific resistance (measurement by the capillary suction test (CST) (21) simplifies evaluation of feed treatment, especially on municipal waste sludges); (*3*) admixture of filter aid materials also decreases the specific resistance and reduces cake compressibility; (*4*) slight modification of feed can improve cake resistance, such as desliming an essentially coarse feed.

Filter medium resistance can be maintained at a minimum by: (*1*) low initial pressure differential; (*2*) precoating; (*3*) cloth cleaning by reverse flow of liquor or air, especially air blow under submergence on continuous filters; (*4*) regular cloth washing; and prevention of (*5*) excessive dosage of polyelectrolyte flocculants and (*6*) very fine solids reaching the cloth first. The medium selected should always be as coarse as the operation permits. It is generally best not to have one piece of equipment attempting both optimum cake processing and liquor clarification.

Solids Concentration. The rate of filtration depends on the difference of solids concentration between feed and filter cake (eq. 3). The C^* factor is the mass of dry cake solids per unit volume of filtrate; hence, as feed concentration approaches cake concentration, the filtrate rate approaches zero and cake formation rate approaches infinity. This emphasizes the benefits of thickening before a filtration where cake rate is the first consideration. Where existing feed concentration might be increased by thickening or process changes or where dilution might enhance cake washing or cake moisture, the approximate rate changes that might be expected are worth calculating.

With cake thickness held constant to assure comparable dischargeability, washing, or dewatering, cake formation time will be directly proportional to the change in filtrate volume. For constant formation time, the only remaining variable in equation 4 other than C^* is filtrate volume. With R_m considered negligible:

$$v^2 \propto \frac{1}{C^*} \text{ and } w_0^2 \propto C^* \tag{16}$$

This calculation does not allow for unpredictably large decreases in specific cake resistance with increasing solids concentration (22).

Feed Modification. Filtration is affected by particle size distribution and is extremely sensitive to very small changes in extreme fines or slimes. Less than 1% shift in such fines content can make a critical change in filtration performance. Such extreme fines often tend to form a restrictive layer near the cloth, on the cake surface, or even within the cake. Similarly, any shift in natural flocculating tendencies or in conditions of flocculant addition may create a large shift in effective particle size. Natural flocculation may be very sensitive to pH changes. In addition, aging of slurries affects particle or floc size or nature.

Agitation. Any shear action on flocculated solids or soft, friable solids can break down flocs or particles and totally alter the filterability. This harmful agitation can originate from a pump, the feed storage tank, or in the filter itself.

Within a filter vessel or tank, agitation can inhibit cake formation. The action is greater on larger particles. Thus the solids concentration in the filter tank increases and its particle size distribution shifts to a coarser one. This dynamic phenomenon is called filter thickening.

Without adequate agitation or with a wide range of particle size, segregation of solids can occur, giving nonuniform cakes that cannot be washed or dewatered effectively on all vertical filter surfaces. Drum filters can counteract adverse effects by starting cake formation at bottom center. Top-fed filters may need a surface raking action to counteract fines sealing the surface.

Cake Washing. The nature of the cake determines whether the high efficiency hydraulic displacement or low efficiency diffusion mechanism predominates in a given situation. Displacement action is at a maximum when cake pores are roughly the same size and the cake is uniform. A compressible cake may shrink when a less dense or less viscous fluid phase moves in—whether air or liquid. This can create nonuniformities, often appearing as cracks. Spray washing, vs flood washing, can help counteract the effect of nonuniformities.

Wash efficiency is best interpreted graphically by relating wash displacements and the soluble solids content of a washed cake, expressed as a percentage of that in an unwashed cake (see Fig. 2). Pure diffusion washing is often uneconomical, especially on continuous filters, so washing is often terminated while the efficiency is in the transitional stage, after only two to four displacements. Washing efficiency should include that which results from dewatering after washing.

Countercurrent displacement washing (excluding reslurrying) on continuous horizontal filters, partly acts as though each stage starts a new displacement and partly reflects the loss of efficiency owing to mother liquor remaining in the smallest pores and the channeling developed in the prior stages. Countercurrent washing requires at least 1.10 to 1.25 displacement volumes. More than three stages are seldom effective.

The rate of wash passage and the compatibility of wash and original liquor significantly influence wash efficiency. The higher the velocity of passage and the more divergent the wash properties, the lower the efficiency. Thus a hot wash does a faster but less efficient job. Thicker cakes reduce the wash flow rate (in inverse proportion), improving efficiency unless the thicker cake is subject to channeling. The cake nearest the filter medium is the least washed and least dewatered; so a remaining cake heel is often beneficial, as is the case on some horizontal and drum filters.

Cake Dewatering. Removal of liquor from filter cake (dewatering) operates independently of cake formation but the result can be significantly affected by the cake structure, which, in turn, depends on slurry characteristics, feed preparation, and cake formation variables.

Whether dewatering can be benefited most by exclusion of liquor, expression, or gas displacement also depends on cake structure and characteristics. Forming a more dense cake minimizes void volume, thus excluding liquor. Forming a more porous cake structure, thus reducing capillary pressure, may permit better drainage by gas displacement. However, most flocculants which improve filtration rates by forming a more porous but compressible cake contribute to higher cake moistures.

Dewatering by expression involves pressure differential as an independent variable, with its effectiveness being related to the cake thickness, specific resistance, time, consolidation properties, and shear forces. In filter belt presses cake shearing contributes to effective dewatering by changing cake structure, modifying specific resistance and consolidation properties.

The empirical correlating approach factor (eq. 12) relates the rate of dewatering by gas displacement to the pertinent variables. Plotting the approach factor vs percent moisture would show the expected relationship of rapid liquor drainage tapering off quickly to slow flow, with limited benefit of significant extension. Individual variables are best correlated separately to avoid obscuring significant variables by a dominating one.

Pressure differential, the most important variable in dewatering, is only constrained by equipment limitations, uneconomic gas usage, and cake collapse.

Air flow rates of 700 $m^3/(h{\cdot}m^2)$ (2300 $ft^3/(h{\cdot}ft^2)$) are often justifiable in dewatering coarse solids on continuous horizontal and top feed filters. The energy usage, though quite significant, is less than thermal dewatering.

Increased *cake thickness* (or cake weight per unit area) reduces the rate of dewatering approximately in an inverse proportion, ie, dewatering time and air flow rates are most meaningful with regard to unit mass of cake solids.

The *viscosity* of the liquor in the filter cake has a significant effect on the drainage rate and may affect the residual saturation level (10,14). A hot water wash or steam are most effective in reducing.

Reduction of *surface tension* of the liquor by adding surfactant to the feed or cake wash may reduce the cake moisture of some materials (23).

With compressible cakes *cake shrinkage* occurs during cake dewatering. As liquor is removed from pores, cake structures may tend to collapse, especially those with flocculated solids. Cakes with a high content of fiber generally shrink only toward the filter medium, giving a thinner cake. Nonfibrous cakes also shrink in other dimensions, forming cracks. After a short period of crack formation, further cake dewatering is not significant. Cake compression devices may reduce moisture significantly.

Filter Media. Selection of a filter medium has to be based on many factors, with overall economics being the first criterion. The most permeable medium consistent with process requrements permits the highest output. The filter medium should retain the first layer, with the cake then acting as its own medium. Obtaining ideal particle retention often conflicts with optimizing cake filtration; hence, priorities have to be set including consideration of separate steps for cake filtration and clarification (24–25).

The important properties of media (excluding precoats) are: particle retention on surface vs depth; minimum flow resistance; good cake release; resistance to blinding and ease of cleaning vs disposable media; flexibility for the intended use, eg, caulking or sewing bags; strength and durability for the particular equipment design and process conditions; dimensional stability under tension or pressure; and suitability as a gasket between filter press plates.

Permeability and porosity are prime qualities to be considered in cloth selection. The tightness of a cloth is identified by its Frazier permeability rating through a dry cloth, which correlates with percent open area. At present, this rating is expressed as $cm^3/(cm^2{\cdot}s)$ at an assumed pressure of 125 Pa (0.5 in. H_2O) (ASTM, tentative). Particle retention ratings would not be meaningful for cake filtration and have not been developed. Retention can be estimated knowing the yarn, weave, and permeability. Calendering imparts a shiny finish giving improved cake release, lower permeability, and better particle retention.

Woven Fabrics. Wool has been all but eliminated from the filter media market and cotton is limited to only a small share, being replaced by synthetics with better cake release, nonblinding characteristics, long life, and resistance to attack from chemicals, bacteria, or fungi. Polypropylene is moderately priced and the most versatile and popular woven synthetic fabric. Nylon is a close second. Other synthetics such as polyethylene, polyesters, Dynel, Kynar, Orlon, Saran, and Teflon are also available.

The filtering characteristics of woven fabrics depend mostly on the type of yarn and weave (26). Yarns can be monofilament, multifilament, spun from staple fiber, or a combination of the latter two. A high twist can make a multifilament behave more like a monofilament. The effects of weave and yarn on filtration are compared in Table 1.

Undercovers are desirable for lightweight cloths, and where drainage support may be rough or have reduced open area, where a coarse drainage screen allows significant cloth sag and stretching, and where discharge action applies a pressure that is relieved by an undercushion.

Wire Cloths and Miscellaneous Porous Structures. Metal wire cloths perform similarly to monofilament synthetics. They are used in pressure leaf and continuous filters where solvent or temperature resistance are important (eg, for sterilization in food processing), or where their resistance to wear gives enough longer life to warrant the higher cost. On the other hand, flexing may shorten their life compared to that of synthetics.

Metal coils are used on the Komline-Sanderson coil filter (continuous vacuum drum) as the filter medium and conveyer to the discharge point, as in the belt discharge. Long life and ease of cleaning have made coils a popular medium for waste sludge filtration.

Perforated plates with small holes serve as direct filter media for the filtration of some coarse solids and particularly for the cachaza mud filtration in cane sugar mills.

Wedge-wire screens are used on some tubular filters as precoat support. A wide variety of porous metals including diffusion-bonded laminates, composites, and air-laid nonwovens, are available for specialty processing.

Table 1. Relative Effects of Weave and Yarn on Filtration[a]

Fabric	Flow rate	Retention	Cake discharge	Moisture residue in cake	Resistance to blinding
Weave					
plain	3	1	3	3	3
twill or chain	2	2	2	2	2
satin	1	3	1	1	1
Yarn					
monofilament	1	3	1	1	1
multifilament	2	2	2	2	2
spun staple	3	1	3	3	3

[a] Rating: 1 = best, 3 = worst.

Porous ceramics have traditionally been used in the nutsche as well as in some cartridge filters, where their resistance to strong acids and alkalis is especially valuable. Because of their low cost compared to metal filters, ceramics are widely used in tubular-pressure precoat filter applications (27).

Nonwoven Media. The term nonwoven is used in a dual context to describe a wide variety of fabrics (see Nonwoven textiles). It includes all dry-laid materials such as felts (qv) and bonded sheets as well as wet-laid materials such as paper. It also refers to fabrics constructed from loose assemblages of fibers arranged in sheet form and physically bonded into stable structures by bonding systems such as resins, latexes, or regenerated cellulose (24).

Felts. Today the term felt applies exclusively to needle felts, which comprise about 60% of the total nonwoven filter market and 80–90% of the dry-laid nonwoven filters (including cartridge filtration applications) (28). Felts are being used more and more in all wet-filtration areas, including continuous filters. They are available in most of the same fibers as woven cloths (excluding cotton and glass only in limited applications) with polypropylene predominating. By using a very open-mesh monofilament woven cloth called scrim or base media near the middle of the felt, high tensile strength and stability are obtained, extending the range of applications. Since production of these cloths involves several layers of fiber webs, they can be varied by changing fiber characteristics, dimensions, and orientation in the different layers.

Needle felts combine high permeability with high retention of solids. Selecting the proper top web layer and calendering or singeing can provide a particle-retaining surface with good cake release properties.

By varying fiber denier and diameter and felt density, the felt can be tailored to a particle retention range suitable for straining applications in the range of 2–200 μm particle diameter (29). This property is utilized in bag straining of slurries containing a few oversized particles such as paints, and viscous polymers for film and fiber production.

Because costs are comparable, the choice between needle felts and woven cloths is one of performance in a given application. However, wear qualities of felts are generally inferior.

Papers and Other Nonwovens. Paper (qv) is used frequently in horizontal leaf filters, filter presses, and plate filters, usually as a disposable item with a suitable cloth as a back-up (27). Thus it substitutes for a precoat. It may preferably be impregnated with neoprene or melamine–formaldehyde resin to add wet strength (see Amino resins).

The bonded fiber of nonwovens, which are varied in type (28), are lighter and thinner than felts and stronger and more permeable than papers (24). They are the primary fabrics for milk filtration, both in the barn and the processing plant.

Spunbonded nonwovens (24) are made from continuous filaments of synthetic or cellulosic fibers, such as polyester, nylon, polyolefin, and rayon, which are self-bonded following melt extrusion.

Rolls of paper, bonded nonwovens, or spunbonded webs are fed onto a variety of continuous or semicontinuous types of filters, serving as a consumable medium. In the handling of so-called coolants, which comprise a wide variety of aqueous solutions, oils and emulsions from machining, grinding, and general metal-working and rolling mills, paper, bonded nonwovens and spunbonded webs, especially the dry laid, are used as horizontal belts, their movement being indexed.

Precoats. In conjunction with filter cloths, precoats provide a renewable filter medium system that can accomplish optimum clarification with negligible risk of blinding. Batch operations use precoats 1.5–2 mm thick, consisting of diatomaceous earth, perlite, cellulose fibers, or sometimes asbestos, carbon, fly ash, calcium carbonate, gypsum, etc. Continuous filters use layers, 5–15 cm thick, mostly of diatomaceous earth or (see Diatomite) perlite (30–31).

Only diatomaceous earth, perlite, and cellulose fibers are marketed extensively. They are available in a wide range of particle sizes with varied flow rates and clarification capabilities. Perlite is less dense and less compressible than diatomaceous earth, and has less particle retention capability, so is not generally useful for thin batch precoats. Cellulose is most frequently used with diatomaceous earth or perlite to gain precoat stability and improve clarification or permit a more open medium. It is more resistant to caustic and avoids silica bleed-through.

A precoat material for batch operations is selected by Buchner funnel screening tests, looking for the highest flow rate that gives the desired clarity. The precoat should be more retentive than the supporting medium (30).

For continuous-rotary-drum precoat filters, medium size ranges of diatomaceous earth or perlite are mostly used, in order to form a 5–15 cm precoat in 30–60 min and retain feed particles with little penetration.

Feed Pretreatment. Filterability may be affected by changes in the following feed characteristics or feed treatments: average particle size of solids; ratio of slimes to coarser particles; agitation and pumping, solids degradation; age, pH, and temperature of slurry; flocculation or dispersion; heat treating or freeze-thawing; viscosity and vapor pressure of liquid; dissolved materials; aeration; surface tension of liquid; filterable solids concentration; filter thickening; and admix of filter aid.

Particle Modification. Many processes offer the possibility of improving particle size, or reducing the proportion of extreme fines or slimes, of which even 2–4% can completely dominate filterability. Crystallization or precipitation conditions can be controlled for particle growth and minimum nucleation. Grinding circuits can be closed and controlled to minimize overgrinding. When fines are unavoidable, they can be removed by classification in, eg, hydrocyclones.

Many materials form dispersed colloids in water because of the predominance of the repulsive force of the electric double layer (resulting in a zeta potential) over the attracting London-van der Waal's forces (32). Change of pH or addition of electrolytes, such as salts of aluminum, iron, and calcium, can reduce or cancel the normally negative zeta potential. The higher the cation valence, the greater the effect (33). Activated silica sols act in conjunction with these cations as coagulation aids providing negatively charged sites to grow agglomerates by a bridging action. Aluminum and ferric iron can also function by a totally separate mechanism, ie, the formation of voluminous hydroxide flocs that physically entrap other suspended solids.

Particles can also be flocculated by a bridging action using natural or synthetic polymers. The natural polymers, such as treated starch, glues, and gelatins, mostly act through hydrogen bonding, but their effectiveness depends on the pH and ionic strength of the solution (32). The synthetic polyelectrolyes function through a bridging action from active sites, whose nature varies with anionic, cationic, and nonionic varieties, and with pH, and specific molecular configurations. A combination of treatments is frequently most effective (see also Flocculating agents).

The efficient use of polyelectrolytes requires (32): (*1*) a dilute solution of the flocculant (ca 0.01%) for good dispersion; (*2*) quick and uniform dispersion of the flocculant through the slurry; (*3*) after dispersion, enough agitation to promote only particle collision without break-up; and (*4*) once flocculated, avoidance of turbulence. A highly concentrated feed tends to hinder good dispersion. Low efficiency of usage or excessive dosage can restabilize a colloidal suspension or add a gelatinous component to the feed, inhibiting cake formation and dewatering, and blinding the filter medium.

The objective of flocculation for gravity settling is the formation of large flocs; for centrifugal settling, shear-resistant flocs; for filtration, small, dense flocs with moderate strength and shear resistence. The polyelectrolytes are tailored to such objectives.

Flocculated solids formed into a filter cake usually encompass increased moisture; often they collapse above a certain critical pressure.

Liquid Modification. Viscosity affects cake formation rate and the rate of approach to final moisture as an inverse function. It is most significantly affected by temperature, and may be decreased by dilution.

High vapor pressure of the liquid at operating temperature, especially with a saturated solution, may necessitate the use of pressure filters. The surface tension of the liquid has little effect on cake formation rate but it can significantly affect the residual level of cake moisture, which can be modified by the addition of surfactants. However, some cause foam in a vacuum system.

Solids Concentration Modification. Increasing filter feed concentration by thickening or process change is not always beneficial. Some filter cakes may become too thick to dewater or wash well. Process objectives must be the guideline.

Admix. Another alternative for improving low filtration rates is filter aid admix, also called body feed, if contamination of the solids is acceptable. Diatomaceous earth or perlite are used primarily (31). The admix is used to maintain adequate porosity in the continually forming cake to provide an acceptable filtration rate. It extends a batch filter cycle by preventing a rapid pressure increase and an uneconomically low rate. Since it also increases the volume of cake solids, it shortens filtration cycles when determined by cake capacity.

The interaction of admix with flocculants and coagulants varies, depending on the nature of the flocs. Dense flocs of discrete particles may need less and coarser admix, if any, whereas flocs of colloids or gelatinous material need more admix especially if filtering pressures are high enough to collapse the flocs. For solids having discrete particle size, admix should be one to two times the solids weight. For amorphous and gelatinous materials, admix may range from four to ten times the solids weight. The admix particle size should be about the same as that of the solids being filtered.

Testing and Scale-up. The aim of filtration testing is to learn the filtration characteristics of a particular slurry, its expected variations, any desirable modifications by pretreatment, and the response to meaningful variables (34). Matching this behavior to the process objectives and equipment capabilities permits sizing commercial equipment and predicting its performance.

In some applications special test results can be correlated with a body of experience to predict size and performance, such as the CST (capillary suction test) data on sewage sludge and fiber freeness (usually CSF, Canadian standard freeness test). The former have been correlated with specific resistance, and the latter, in combination with other factors, with filtration rates.

In sizing any commercial filter from test data, consideration must be given to a scale-up factor which includes safety considerations of: normal and exceptional plant feed variability; how representative the tested samples may be including age and handling; the extent of testing; whether design flow includes catch-up capability for downtime of the filter or other equipment; and the risks of possible undersizing. True technical scale-up considerations include: feed segregation in larger units, relative completeness of cake discharge, the degree of cloth conditioning, potentially poorer precoat coverage, feed inlet velocity effects, agitation effects.

Batch Pressure Filters. The three types of batch pressure filters are suited to handling different types of cake (34). Pressure vessel filters (leaf-type) handle incompressible or slightly compressible cakes. Filter presses handle both compressible and incompressible cakes, especially with the flexibility potential of membranes. Variable volume presses cannot be used on incompressible materials. Tubular, or candle type of pressure vessel filters are used for clarification applications, using precoat, and often admix, resulting in cakes that are often slightly compressible.

Pressure Vessel Filters. Preliminary testing and sometimes sizing of pressure leaf filters is generally done on pressure filter cells (bomb filter tests). Tests of 20–30 min duration are made at constant pressure and results are extrapolated to a commercial cycle, using either a logarithmic plot of time vs accumulated filtrate or a linear coordinate plot of reciprocal rate (t/v) vs accumulated filtrate (v), both of which usually give straight lines or with a slight downward curve. (See eqs. 4 and 7.) Tests at three different pressures are suggested. If the slopes of the plotted lines are different, compressible cakes are indicated. If the plots are significantly curved, the feed slurry may have changed because of segregation or cooling, or more filter aid may be desirable (35).

A major use of these cell tests is to check filter aid admix type and dosage. Test techniques with precoat are not practical. Such short tests can give only preliminary information on cake washing or dewatering. Cycle length may be dictated by cake thickness or low rate at the maximum pressure. The commercial operation is usually at a variable rate and pressure, which requires interpretation from the constant pressure extrapolation. Since a variable rate and pressure operation involves specific filter and pump sizes and interactions, a constant rate basis gives the most conservative overall rate using the following relationship:

$$\frac{t_{cp}}{t_{cr}} \times \frac{P_{cp}}{P_{cr}} = \frac{1}{2}\left(\frac{V_{cp}}{V_{cr}}\right)^2 \tag{17}$$

where t is the filtering time, V the cumulative filtrate volume, P the maximum pressure, cp the constant pressure, and cr the constant rate (34–35).

Constant rate and constant pressure should be compared at the same cake thickness and the same cumulative filtrate volume. If the maximum pressures are the same, the time required at constant rate is double that at constant pressure; therefore, the filtering area is doubled. Since the true objective is a correction to variable rate and pressure, an estimate of 50% more area is a good compromise.

Pilot tests at full cycle for leaf and tubular types are necessary for critical washing or drying applications, and desirable to verify cake discharge, preliminary selection of the filter medium, any precoat, and admix. Constant rate tests are most conservative. Increasing pressure in increments, simulating commercial operation, is often preferable. Either basis is better than constant pressure because it reveals whether there is a critical pressure when the cake collapses, sharply increasing the pressure or decreasing the rate. The feed slurry is drained at reduced vessel pressure, before introducing wash liquor. Washing is continued until the filtrate shows constant or tolerable levels of feed liquor. Removal efficiency is determined by analyzing the cake for residual feed liquor.

Cake dewatering data consist primarily of the relation of time vs resultant cake moisture at various pressures of displacement gas or of membranes. These data can be roughly correlated with the diminishing rate of filtrate flow to guide cycle termination. Sizing based on any constant rate tests needs no scale-up factor for variable rate and pressure operation. If an incremental pressure run closely approximates a commercial run, then a 0.8 factor on the rate is desirable for safety.

Filter Belt Presses. The difference between pressure leaf filters and filter presses is that filter press cakes are confined and compressed as part of the cake-forming cycle, or alternatively by a membrane. Pressure cell tests are thus not representative of filter press action.

Filter press pilot tests should follow a sequence such as outlined below (34) for a nonwashing application: optimize (*1*) the feed rate by several tests run at constant rate or incremental pressure (or both) up to the tentatively desired terminal pressure, then at constant pressure until the chamber is adequately packed or the rate is uneconomical; (*2*) the cake thickness to give the desired solids content and maximum output; and (*3*) maximum pressure. Then verify the filter medium selection.

Variable-Volume Filters. To simulate the elastomeric membrane function, tests are performed on a compression cell, also called a compression–permeability cell or piston press. A free piston transmits hydraulic pressures in the range of 0.7–35 MPa (100–5000 psi). Tests are performed at constant pressure for the short cycle used on commercial units. First the desired pressure differential is determined, then the rate, cake thickness, and cycle.

Capabilities of the various tube or diaphragm types of presses must be known. In order to interpret the data (34) the cake permeability is calculated using a plot of t/v vs v and the Darcy equations (effectively the same as eq. 7, substituting cake permeability for its reciprocal specific resistance), assuming medium resistance is negligible compared to the moderately thick cake. The optimum output rate, cake thickness, and cycle are calculated from the permeability and process criteria knowing the relative downtime to on stream time (assumed equal for the tube press).

Continuous Pressure Filters. *Filter Belt Presses.* The two distinct stages in filter-belt press operation (36) and testing are (*1*) gravity drainage of a superflocculated sludge and (*2*) progressive compression dewatering.

In superflocculation, the liquor not contained within the flocs is almost unrestricted in its flow away from the flocs. In the CST test recommended for evaluating flocculation, the value is close to that for pure water. If superflocculation is not obtained on the commercial unit, the sludge cannot dewater to the extent necessary for further compression. Sludge flows away from the compression rolls and impregnates the filter medium. On the other hand, if excess polyelectrolyte flocculant is used, the filter medium may become blinded.

Polyelectrolyte flocculants and their dosage are evaluated by performing the CST test with timed agitation of a flocculated sample with a standard stirrer.

Superflocculated sludge is tested for gravity drainage rate using a funnel or test leaf which can use the filter-belt medium and an unrestricted drainage and where the dewatering can be initiated at a precise time without introducing any significant kinetic energy to the charge of slurry sample. The rate and extent of drainage is tested with a range of sludge volume samples, allowing drainage rate to drop almost to zero, because the commercial machine gravity section must obtain that extent of drainage before the first compression roll.

The effect of compression can be reasonably approximated using the compression test cell as used for variable-volume filters. With gravity-dewatered sludge feed the cell pressure is gradually increased at the same rate and to the same extent as in the commercial unit. This does not take the cake shearing action into account which is believed important for best filter-belt press dewatering. Tests results do, however, give a good indication of final solids concentration.

Rotary units in pressure vessels are the same as the continuous vacuum filters, and are used because vacuum is not compatible with the process. Generally they are tested by substituting another liquid with comparable filtration properties (especially viscosity) at conditions that can be vacuum-leaf tested, in order to get preliminary data to guide pilot operations.

Batch Vacuum Filters. Batch vacuum filters are either nutsche or vertical leaf or tube. They are tested with bench-scale-sized units of comparable configuration or of a standard vacuum test leaf. The nutsche is tested using typical top-feed techniques, forming cakes of whatever thickness desired and proceeding through cycles of washing and dewatering as needed, with complete flexibility for cycle. The leaf, such as the Moore, or tubular types of filters, is tested by putting the small leaf through the same sequence of immersions and air exposure as planned for the larger unit, being careful to use a vertical leaf surface and observe solids settling.

Continuous Vacuum Filters. Most rotary continuous vacuum filters are constructed with the filtering surface divided into sections with each having its own medium support, drainage, and piping connected to a filter valve, where section functions are controlled.

These filters are tested with 93-cm^2 circular test leaf, which is successively moved through the same sequence of operations as a section of a filter. A Buchner funnel is not an adequate substitute, because it lacks sufficient drainage. In the top-feed test the slurry is fed onto a horizontal (or nearly so) filtering surface as in the commercial filter. In the dip or bottom feed test the leaf is submerged in the slurry to form a cake, just as the drum or disc filter is partly submerged in slurry to form a cake. These tests have a high validity and can reliably be used either to size filters or predict performance or to indicate the need for piloting.

If the cake is not washed, the operations on most filters are cake formation, dewatering (also called drying), discharge, possible cloth cleaning, and dead time. If the cake is washed, it is preceded by a short period of air drying and followed by dewatering. Washing, if applied countercurrently is interrupted by short air drying times.

For proper test planning and execution, the proportion of time available for each of the successive filter functions should be known. Table 2 gives these data for a variety of filters.

Adequate cake thickness for discharge and the normal maximum is important for testing and equipment selection. Table 3 gives the important range for most filters.

One approach to test planning that minimizes testing is a progressive determi-

Table 2. Time Available for Each Function as Percent of Total Cycle Time

Filter	Cake formation	Air dry	Cake washing	Final drying	Minimum discharge, cloth cleaning, nonfunctional, etc
drum[a]	10–30[b]			50	20
	10–30	10	25–5	15–35	20
belt or string discharge	10–30			40	35
	10–30	10	15–5	15–25	35
paper pulp	55–70			5	40–25
	35	10	20	2–3	33–32
disc					
metallurgical	12–33			40	27
paper pulp	40–50			5–10	55–40
internal drum	10–20			40–35	50–45
	10–20	5	15–10	20	50–45
horizontal pan or table	5–50			70–25	25
	5–40	5	50–5	15–25	25
horizontal belts	5–70			95–30	not included
	5–50	5	65–10	25–35	not included

[a] With scraper, wire, blow, roll, precoat discharge, and also single-compartment drum.
[b] Larger diameter filters can use 33%, smaller ones and all belt discharge 25% max.

Table 3. Cake Thickness Range for Effective Operation

Filter type	Min, mm	Max, mm
drum		
scraper discharge	3–5	25–75
wire discharge	2–3	6
string discharge	5–6	12
belt discharge	3–5	25
roll discharge	skin–1	3
precoat discharge	skin	3
roll or air-doctor discharge (paper-pulp washer)	3–5	25
blow discharge (single compartment)	2–3	12
internal	9–12	25–50
disc		
scraper or snap blow (metallurgical)	6–9	25–75
sluice discharge (paper pulp)	3–5	9
horizontal		
table (scroll discharge)	20–25	100
belt	4–6	100
tilting pan	20–25	100

nation of the desired levels of variables, promptly setting those that need no further change. Thus most testing of key variables is done at conditions close to the final operating basis, reinforcing the validity of conclusions.

The order of establishing the desired levels of variables depends on the application and necessarily overlaps. Each test adds information on the effect of several variables.

A typical sequence of variable evaluation on top-feed tests for a simple dewatering on a horizontal filter would be: (*1*) any feed variables, especially solids concentration; (*2*) filter media; (*3*) cake thickness (note formation time is not an independent variable on top-feed tests but depends on the slurry volume, feed concentration, vacuum, etc); (*4*) cake heel thickness; (*5*) vacuum level and flow; and (*6*) dewatering time.

If a washing operation is included, preliminary evaluation is made while exploring the cake thickness and continued through evaluation of vacuum level and dewatering time. Wash efficiency is tested by three significant volumes at two cake thicknesses and analyzed using the semilog plot of percent mother liquor remaining vs wash displacements shown in Figure 2.

The dip test program for a drum or disc filter follows a similar approach but with the cake formation time as an independent variable and the vacuum level is more significant than flow. Both are determined early in the program, to get the desired cake thickness for discharge. When moisture is important and affected by several variables, a graph of moisture vs cake capacity can be used (see Fig. 3) for visualizing performance.

Alternatively, the test procedure might develop successively sufficient data on each of the three main operations: cake formation, washing, and dewatering. Graphical interpretations can then provide the basis for sizing calculations (34,37–38). Sizing would then be done by sequential analysis of the pertinent graphs with appropriate consideration of the interdependence based on the filter design. When sizing in this manner, verify that an actual test was made at the conditions selected.

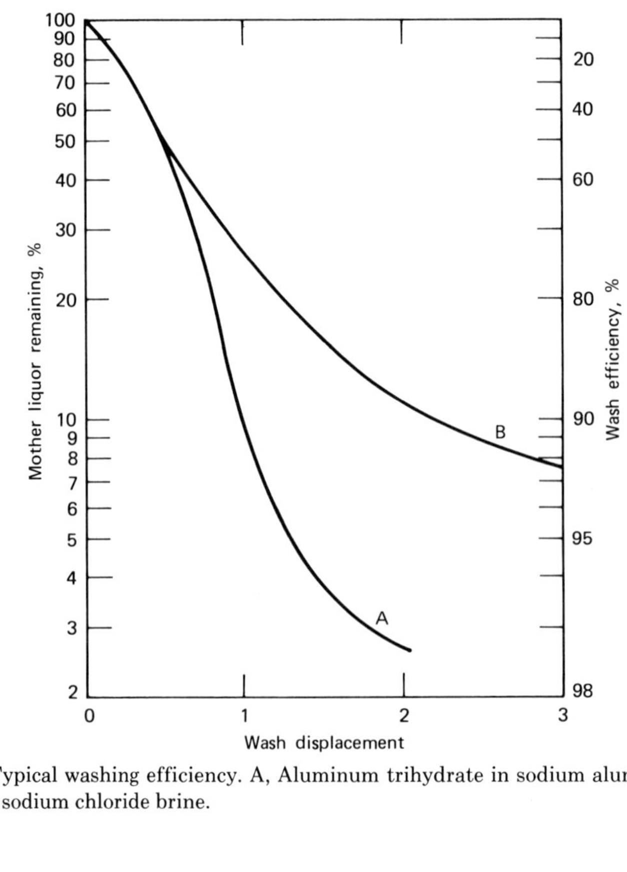

Figure 2. Typical washing efficiency. A, Aluminum trihydrate in sodium aluminate solution; B, phosphate rock in sodium chloride brine.

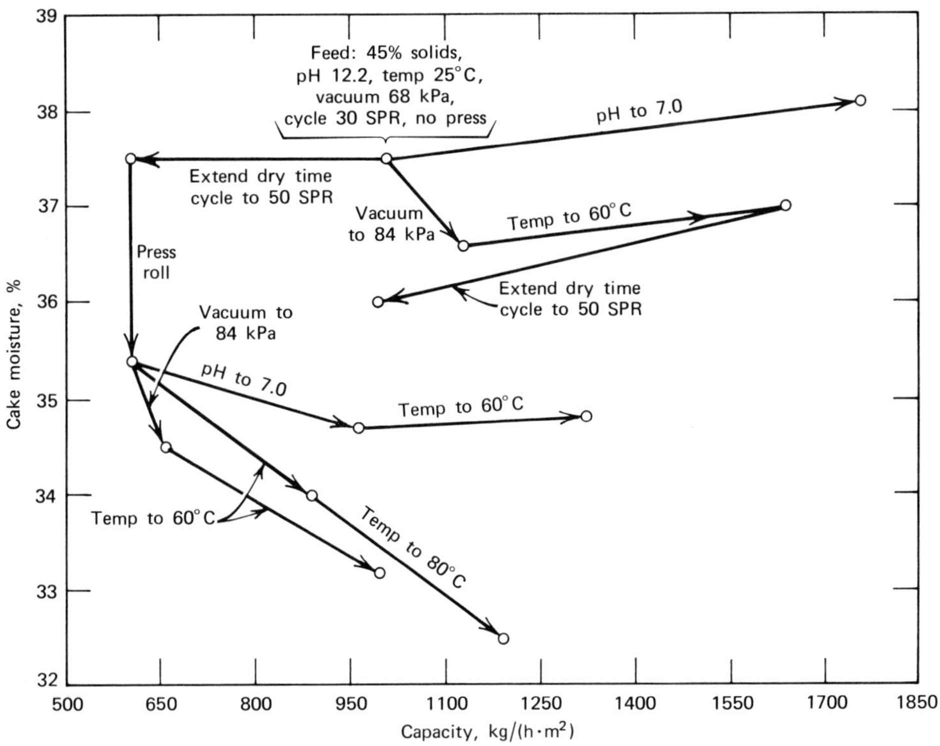

Figure 3. Flue dust cake moisture (SPR = seconds per rotation). To convert kPa to psi, divide by 6.9.

On continuous filters, before the advent of synthetic cloths, a scale-up factor of 0.65 times the rate was normal, reflecting cloth conditioning probability. Now a factor lower than 0.75 is unusual. Filters with effective cloth cleaning systems or using open monofilament cloths have the highest factors of 0.8 to possibly 0.9. An effort has been made to make the scale-up more precise by identifying a factor for each individual function or aspect (34,38).

Cake Filtration Equipment

A preliminary guide to matching filterability with suitable types of equipment is presented in Table 4. Table 5 illustrates the capabilities of many filter types. Approximate costs of many filters are indicated in Figure 4.

The distinction between batch and continuous filters is not clear-cut because many automated-cyclic batch operations become almost as continuous as sectionalized rotary drum filters. Units with a continuous output of both filtrate and cake are treated here as continuous.

Batch Pressure Filters. ***Pressure Vessel Filters.*** The filtering elements are vertical or horizontal leaves, stationary or rotating; vertical cylindrical tubes; or even cartridges. Pressure vessels are usually cylindrical in design either for horizontal or vertical operation and for operating pressures to about 500 kPa (75 psi). They can be jacketed or insulated.

Pressure vessel filters are employed in clarification operations, generally using

Table 4. Slurry Filterability vs Filter Capability[a,b]

Slurry characteristics	Fast filtering	Medium filtering	Slow filtering	Dilute	Very dilute
Filterability aspects					
cake formation	cm/s	cm/min	1–5 mm/min	<1 mm/min	no cake
normal solids concentration	>20%	10–20%	1–10%	<5%	<0.1%
dry solids rate, kg/(h·m²)[c]	>2500	250–2500	25–250	<25	
filtrate rate, m³/(h·m²)[d]	>10	1–10	0.1–1	0.02–5	0.02–5
solids settling rate	difficult to suspend	fast	slow	slow	
Filter application					
continuous vacuum					
horizontal table, scroll discharge	++	++			
tilting pan	++	++			
horizontal belt	++	++	+		
top feed drum	++				
internal drum	++	++	+		
single-compartment drum	++				
multicompartment drums					
most dischargers	+	++	++		
roll discharge		+	++		
rotary precoat				++	++
disc		++	++		
filter belt press			++		
batch nutsche	++	++	++	++	++
batch vacuum leaf		+	++	++	
batch pressure					
filter press		++	++	++	++
horizontal leaf	++	++	++	++	++
tubular				++	++
vertical leaf		++	++	++	++
tube press		++	++		
automated diaphragm press		++	++		
cartridge, disc, edge					++
filter thickeners					
high shear or cross flow		++	++		
cycling backflush				++	++

[a] Excluding pulp and paper which is fast filtering but dilute giving high filtrate loadings.
[b] Crosses indicate degree of efficiency.
[c] To convert kg/(h·m²) to lb/(h·ft²), multiply by 0.205.
[d] To convert m³/(h·m²) to gal/(h·ft²), multiply by 24.5.

filter aid, or for cake filtration where solids concentrations are too low or particle size too fine for continuous filters.

A cake thickness of 10–15 mm is desirable in a 4–24 h cycle. Pressure vessel filters compete primarily with filter presses. The vertical leaf and tubular types have higher capacities because of faster cycles, especially when a concentrated slurry can be discharged from a closed vessel. The closed vessel discharge is also important when the cake should not be exposed to air.

Vertical leaf filters with uniform cakes (no slurry segregation, cracking, or sloughing) and horizontal leaf filters have a washing advantage over nonmembrane

Table 5. Special Filter Capabilities—Main Criteria for Selection

← Coarsest, most concentrated slurries			Finest, most dilute slurries →	
Continuous filters		Batch filters		Miscellaneous
Horizontal filters (high filtering rates; effective countercurrent washing; high hydraulic and pneumatic capacity) table-scroll discharge (cake heel usual advantage in lower moisture and solubles discharged; air blow under feed cleans cloth, reslurries heel, distributes feed) tilting pan (large filter areas; wash containment; cloth cleaning one side, adds wastewater) horizontal belt (also good for washing finer feeds, competes two stage drums; cloth washing both sides, adds wastewater) *Top feed* (dewaters and dries coarse crystals at high capacities) *Internal drum* (for fast settling materials not uniformly coarse; feed pool helps handle fines) *Single-compartment drum* (very high capacity on crystals, good washing)	*Multicompartment drum* (flexible, moderate wash and capacity, interchangeable discharge devices) scraper and wire discharge with blow (simplest, cheapest drum; effective workhorse) string (discharge ideal for feed to apron dryers; if cake too soft or thin discharge fails) belt discharge (cloth washing gives lowest risk of cloth conditioning; loses drying and/or washing area; poorer filtrate solids; adds operator attention and cloth rinse waste stream) coil discharge (nonblinding, long-life media on waste sludges; poor filtrate clarity) roll discharge requires consistently sticky cake, then gives high capacity with thin cakes; lacks flexibility) *Disc* (low floor space; cheapest straight dewatering filter; cloth maintenance moderate)	*Batch vacuum leaf* (for high efficiency washing on slow filtering slurries) *Batch vacuum nutsche* (flexible for most feeds where small batches or special wash preclude continuous filters) *Rotary vacuum precoat* (used for polishing or slow filtering thin cakes where liquid is valuable; usually saves filter aid vs batch pressure admix) *Filter belt press* (gentle dewatering of superflocculated waste sludges containing fibers; low power; incremental Δp)	*Pressure vessel filters* vertical leaf (good polishing or cake filtering; sluice or dry discharge; often good washing; less expensive than horizontal leaf; slurry heel disadvantage; rotating discharge best for thin, sticky cakes) horizontal leaf (good back-up polishing; good with intermittent feed, varying pressure; efficient washing; high labor unless discharge by rotation) tubular (polishing normal purpose using precoat and admix; high filter area per volume; high hydraulics) *Filter presses* (flexible-widely varied applications; gives low moisture on compressible cakes, especially with membranes; washing slow, inefficient except with membranes) *Variable volume* (for compressible cakes only; low moistures. Short cycles give moderate capacity; use on fine products of high value; tube press uses high Δp; automated diaphragm press has high filter area per floor area)	*Granular media* (for polishing very dilute streams where area is available) *Cartridge filters* (for clarifying a wide variety of very dilute slurries) absolute rated (mostly membrane types, can sterilize filtrates) *Bag filters* (train fine oversize particles) *Edge filters* (strain coarser oversize) *Filter thickeners* cyclic backflush (of tubular and specialty elements clarifies quite dilute feeds giving low concentration slurries (1–10%)) cross-flow shear (concentrates more dilute slurries, as the cyclic backflush) continuous disc filters (submerged discharge builds solids concentration from low (1–10%) to medium (30–50%)) high mechanical shear (with cake heel gives very high conc paste (60–80% solids))

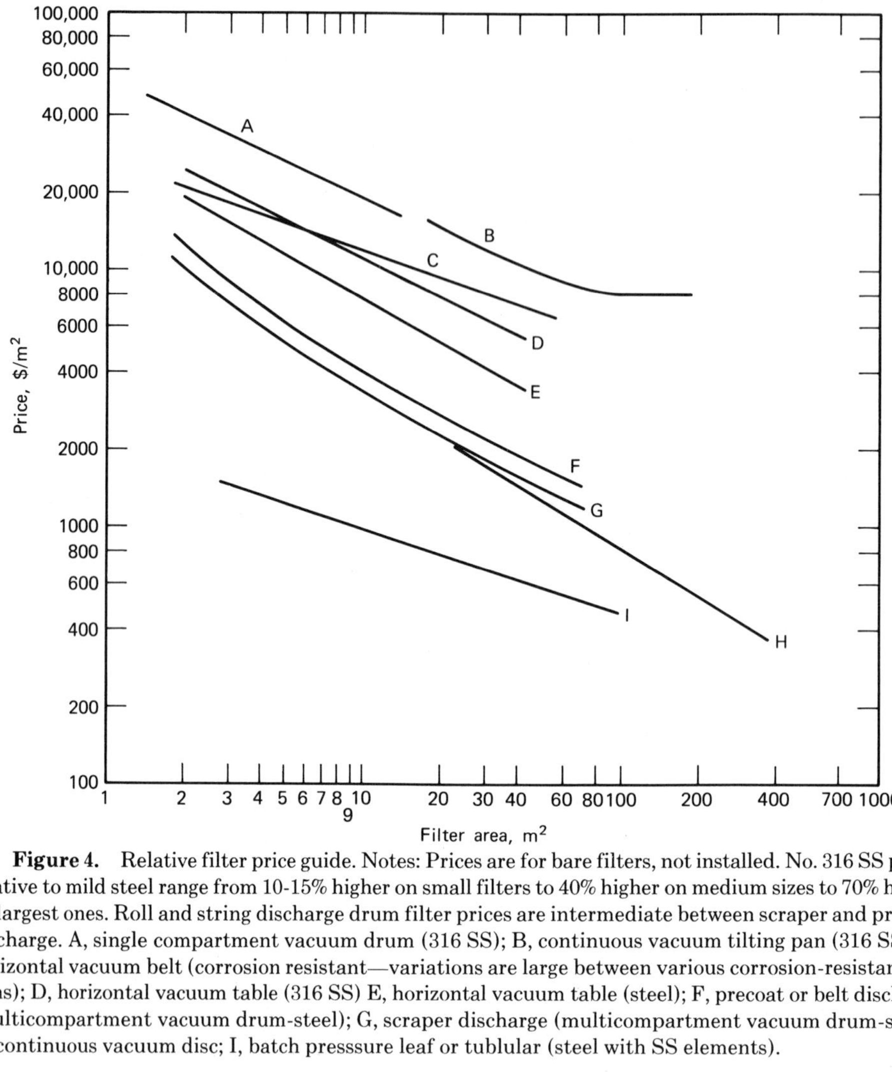

Figure 4. Relative filter price guide. Notes: Prices are for bare filters, not installed. No. 316 SS prices relative to mild steel range from 10-15% higher on small filters to 40% higher on medium sizes to 70% higher on largest ones. Roll and string discharge drum filter prices are intermediate between scraper and precoat discharge. A, single compartment vacuum drum (316 SS); B, continuous vacuum tilting pan (316 SS); C, horizontal vacuum belt (corrosion resistant—variations are large between various corrosion-resistant options); D, horizontal vacuum table (316 SS) E, horizontal vacuum table (steel); F, precoat or belt discharge (multicompartment vacuum drum-steel); G, scraper discharge (multicompartment vacuum drum-steel); H, continuous vacuum disc; I, batch presssure leaf or tublular (steel with SS elements).

presses. Labor economics and the cleanliness of the filter station depend on the operation, a sluice discharge being best. However, leaf filters have a slurry heel, and a further potential disadvantage vs the press on moistures of compressible cakes.

Stationary vertical leaf designs concentrate more area into less volume and are less expensive than horizontal leaf designs. The type of discharge desired and the particular cake characteristics generally dictate the choice between the horizontal or vertical tank design. With both designs, discharge could be assisted by sluicing, leaf shock or vibration, and reverse air blow, and media could be cleaned by sluicing showers.

The horizontal tank (see Fig. 5) is favored for dry cake discharge because the shell or leaf assembly is usually retracted exposing the leaves for inspection and manual completion of discharge. In addition, the horizontal tank has smaller leaves of uniform size and hence uniform media size, less cake weight per leaf, shorter sluicing distance, and better cloth cleaning.

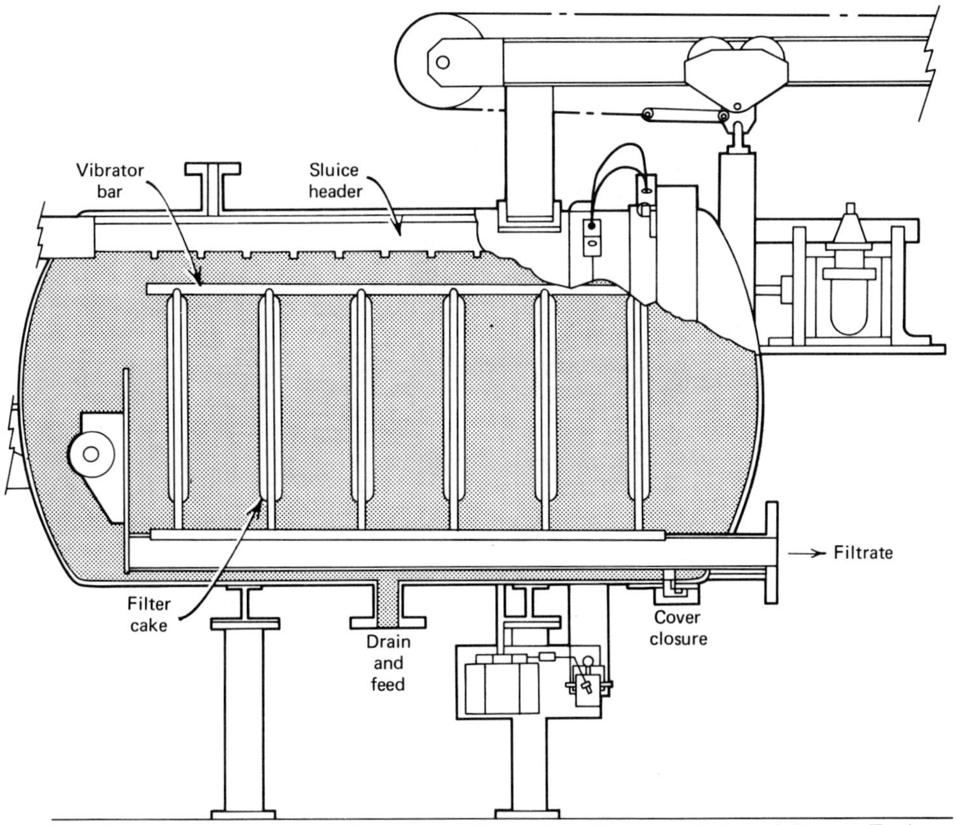

Figure 5. Vertical pressure leaf filter and horizontal tank. Courtesy of Ametek Process Equipment Co.

For sluice discharge with fixed or oscillating showers, the vertical tank is preferred, because it discharges into a compact area where reslurry and removal is assured. Since leaves are taller, however, sluicing showers have to reach farther, making it difficult to dislodge some cakes completely. Other factors favoring the vertical tank are less floor space needed, fewer leaves, less media waste on rectangular leaves, and higher upflow rate counteracting segregation when fed from bottom center.

Horizontal stationary leaf designs have the primary benefit of cake stability even with intermittent feed and variable pressure. Excellent cake washing is another advantage because feed can be drained and wash introduced with negligible pressure; therefore, no cake cracking occurs. Since only the top half of the leaf is used, these filters are more expensive. Manual discharge is high in labor costs. They are thus used for long cycles and are good polishing or guard filters. In some designs leaves tilt for discharge.

The pressure nutsche is actually a single-leaf design, finding use in small batch operations to dewater the slurry heel from a bigger filter.

Rotating leaf filters may also be called pressure disc filters. The vertical leaf filter may rotate throughout the cycle to improve cake uniformity, or just at discharge, moving the leaf against a fixed scraper or to close proximity of a sluice nozzle. The latter provides the shortest possible sluicing distance, hence the most effective of any leaf filter for removing thin, sticky cakes.

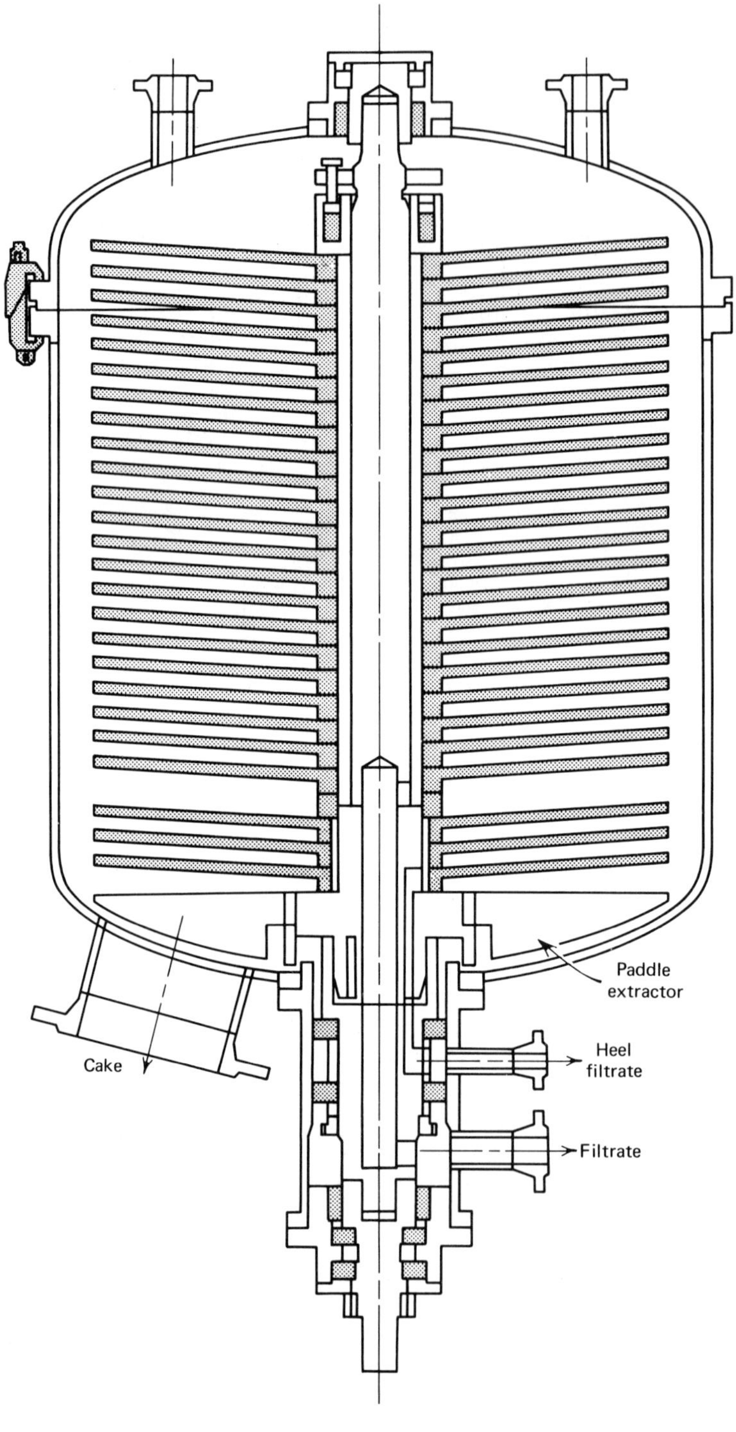

Figure 6. UHDE horizontal pressure leaf filter. Courtesy of Dorr-Oliver Inc.

The horizontal leaf filter (see Fig. 6) rotates for discharge purposes only, using a scraper or centrifugal force, often in conjunction with a sluice, filtrate back flow, or with slurry in the vessel.

Tubular pressure filters (*candle filters*) can concentrate more filtering area into a fixed volume and have higher hydraulic capacity than leaf filters, and thus are effective for high-rate clarification especially with very low solids. Most designs anticipate use of a filter aid both as precoat and admix. Some designs aim specifically at applications not requiring filter aids and, hence, minimize tube clearances. The flexing of cloth media at discharge sometimes gives them an advantage over wire media in dislodging fines.

Tube and media designs are available in a wide variety of forms; some are rigid and some flex on discharge. They are made of metal, ceramic, plastic, wire cloth, or fabricated socks of woven or felt media. They are used mainly in water clarification and for water recirculation in cooling systems. Miscellaneous applications are in the production of beer, fruit juices, sugar syrups, pharmaceutical products, oils, and waxes.

Specialized Designs. Several leaf filter designs may incorporate scavenger plates to filter the slurry heel, when it is important to maintain batch identity. The scavenger plates must have sufficient area in order to keep the heel filtration time to a minimum. Some of the horizontal leaf units recycle the heel to all the leaves.

Leaf or tubular filters may also function only as filter thickeners by sequential backflush mechanisms which may be by timer-controlled cycling, or by use of a filter valve, rotating either the valve or the filter elements.

Filter Presses. The basic filter press consists of either a series of recessed plates (see Fig. 7) or an alternating combination of flush plates and cake-holding frames, a framework to support these elements, and a system for closing the elements together with enough force on the sealing faces to counterbalance the forces during filtration. Both plate designs are available in automated versions and with diaphragms, or membranes, but recessed plates give better discharge. A rectangular plate shape is most common, making fullest use of filter media, but circular shapes are also used.

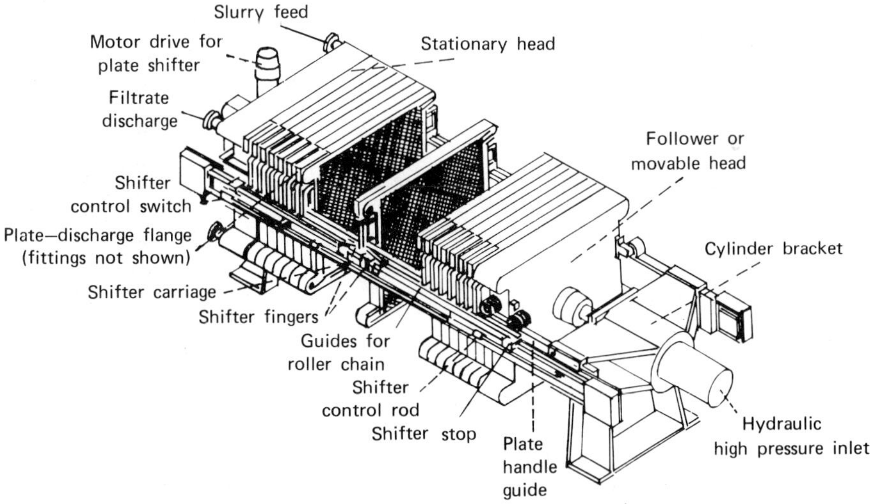

Figure 7. Filter Press. Courtesy of Shriver Operations, Eimco PMD Division, Envirotech Corp.

Materials of construction include metals, with or without rubber or plastic coatings, wood, polypropylene, and glass-filled polyester (39). Polypropylene has performed outstandingly as both plate and frame material and has largely displaced wood. Approval by the FDA made it popular in the food and pharmaceutical industries (40). Operating pressures depend on the materials of construction but often go up to 1.5–2 MPa (220–290 psi).

Automation includes opening and closing the press, shifting plates (and frames) in a manner to dump cakes, and a photoelectrical curtain to stop the mechanism when operators move in to complete a faulty discharge.

Membranes, or diaphragms, molded of thick elastomeric material can be installed on each plate or every other plate (see Fig. 8). A drainage pattern is molded into the membrane surface, and the membrane is securely fixed to the outer periphery of the

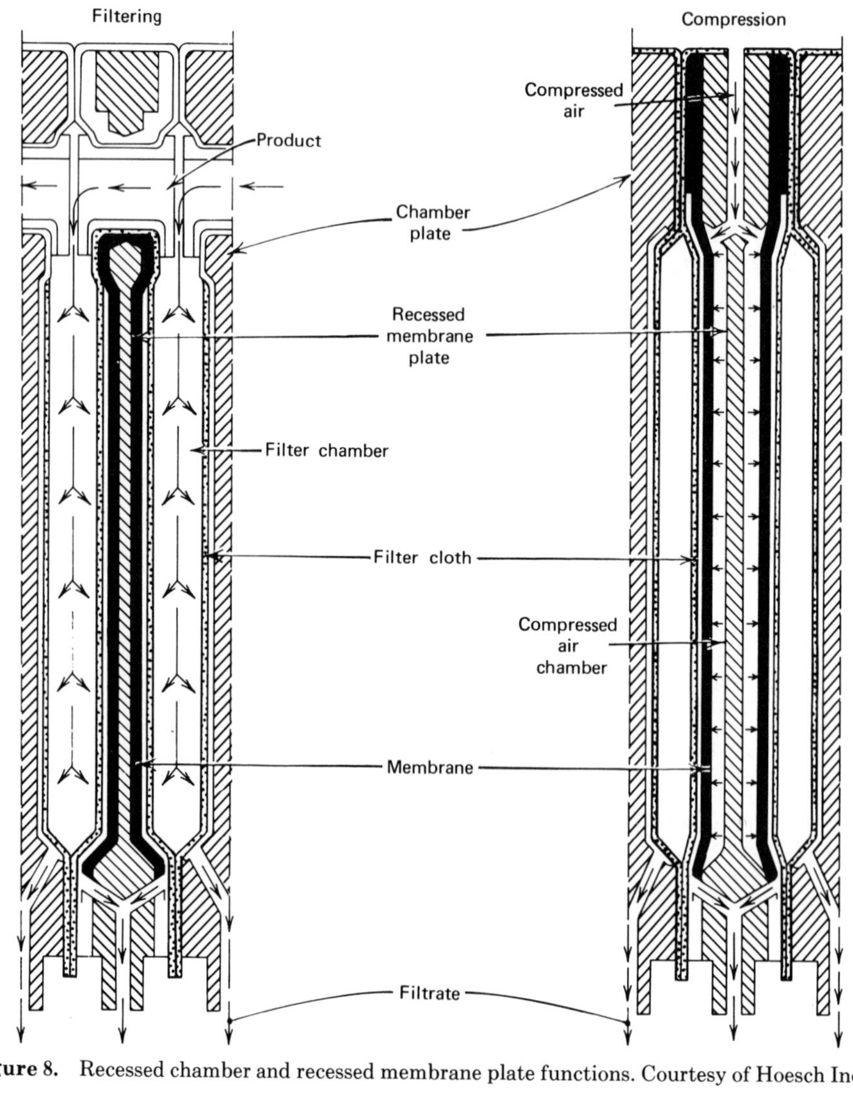

Figure 8. Recessed chamber and recessed membrane plate functions. Courtesy of Hoesch Industries, Inc.

support plate. Pneumatic or hydraulic pressure introduced behind the membrane then creates a variable-volume chamber permitting expression of additional filtrate by compaction of compressible cakes, often including completion of cake formation. On compressible cakes the membrane usually gives lower moistures than air blowing. On incompressible cakes the membrane is of no use for dewatering. Gas blowing provides the lowest moistures on incompressible cakes (especially if heated).

Applications involve a wide range of cake filtration and clarification operations but are best on compressible cakes (eg, sewage sludge, some antibiotic mycellia, fullers earth in oil). Filter presses are exceptionally flexible in being able to filter almost any material. Filtering area is easy to add or subtract by adding or removing filter elements, or lengthening element support rails, if need be. Washing efficiency is low on filled presses, because cakes are not uniform, but can be good using membranes to complete cake formation. For good efficiency, wash must flow across the cake and not follow the feed. Washing can often be satisfactory when plenty of time and wash liquor are practical.

The plate-and-frame type generally has frames for 25–50 mm thick cakes but can use much thicker cakes. With thinner frames and sheet filter media, it may be called a sheet filter. Feed and wash entry and filtrate removal is via corner holes.

The recessed plate type may have center or corner feed. Media installation varies widely: some cloths are hung over the plates and serve as gaskets; some are caulked to avoid press leakage and media damage; some are mounted externally to enhance discharge or incorporate cloth washing; paper media cannot be used. Cloth life is best when caulked, next best on the plate-and-frame, and worst on the standard recessed plate.

Variable-Volume Filters. In variable-volume filters the pressure differential for most of the cake formation, as well as dewatering, is transmitted to the slurry and cake by a diaphragm or membrane. Dewatering is by expression as one continuous action with cake formation. They differ from filter presses in the very short cycles and relatively thin cakes, and hence relatively high capacity. As with diaphragms on filter presses, the benefits of lower cake moistures are obtainable only with compressible cakes. Economics generally limit application to fine size feeds of relatively high value where the reward for lower moistures is sizable.

In addition to plate press filters, two other types of variable volume filters are currently marketed: the tube press and the automated-diaphragm pressure filter.

The tube press consists of two concentric hollow tubes, the inner one being perforated and covered with the filter cloth. Attached to the inside of the outer tube is a cylindrical elastomeric bladder providing the compression by hydraulic pressure between the bladder and outer tube.

The automated diaphragm pressure filter consists of rectangular modules mounted in a vertical arrangement providing multiple horizontal filter faces. The continuous filter cloth is threaded in a zig-zag fashion between the horizontal filter faces and finally through the wash chamber for continuous, effective cleaning. A programmed sequence forms cake, washes, dewaters, opens the unit, and discharges cake by indexing the movement of the medium.

Continuous Pressure Filters. Most continuous rotary vacuum filter designs have been converted to pressure designs to handle materials not suited to vacuum operations or where preceding and following operations are under pressure. Designs are generally limited to about 200–400 kPa (30–60 psi) for sizable units for cost reasons. However, at least one filter, the Bird-Young, operates up to 1000 kPa (145 psi). These units cannot usually be justified only for the benefit of increased pressure differential. Since operations are similar to their vacuum counterparts, except for the problem of dry cake removal, they are discussed under vacuum filters. Hot oil and solvent clarifications on continuous pressure precoat filters and oil dewaxing with propane on drum filters are typical applications.

Filter Belt Presses. The principle of the filter belt press (or belt filter press) is a gradual increase of pressure differential across a sludge starting with gravity or low vacuum and shifting to compression between belts and rolls (41).

The sludge must be superflocculated either naturally or with flocculants (usually polyelectrolytes) to make it free draining by gravity or light vacuum. With significant sludge variability a set dosage or even a single flocculant may not provide the superflocculation needed to drain the sludge adequately for effective compression (42). Initial compression is usually provided by a second belt. Further compression is provided by two belts with the sludge between passing through a series of rolls in a zig-zag manner such as to give combined compression and shear to the cakes. The number of belt reversals and the maximum belt tension are of great importance in producing the driest cakes at the highest throughput. Some units impart shear by a slight differential speed between belts. All units incorporate belt-cleaning sections to counteract the natural blinding tendencies of sludge and most flocculants. Recycled filtrate is generally used.

Applications are almost exclusively on waste sludges, especially sewage and paper mill wastes that contain a significant proportion of fiber contributing cake cohesiveness under compression (see Wastes).

The variety of about 40 filter-belt press configurations is only barely suggested by the two units outlined in Figure 9. Tough service requirements include exposure of roll bearings to abrasive slurries, belt-tracking systems with appreciable belt tension, and belt puncturing by sharp objects.

Batch Vacuum Filters. The three main types of batch vacuum filters are the horizontal single-compartment nutsche, the Moore-type vertical leaf assemblies, and tubular units.

The nutsche may be a relatively small porous ceramic unit, a sizable homemade flat drainage surface with a vacuum box underneath, or a fabricated metal design capable of tilting for solids discharge. They are finding increased usage because of the development of automatic discharge mechanisms (27). They are used for moderately fast-draining solids where batch processing, low capacity, or special wash requirements preclude continuous filters. Labor costs for discharge may be significant.

In Moore filters vertical leaves, generally of wood or possibly plastic to minimize weight, are assembled in units of, eg, 50 leaves with the filtrate manifold and structural support at the top. These assemblies are moved by crane to a slurry vat, a wash solution vat, a dewatering area, and a discharge area. Overall cycles might be 2–3 h.

Continuous Vacuum Filters. Continuous vacuum filters are grouped into three categories, based on the orientation of the filtering surface: disc, drum, and horizontal. Most filters are made up of sections or sectors each of which go through a sequence of functions for which the filter valve programs selection of vacuum, venting, airblow, or no function, and diverts filtrate to the desired receiver.

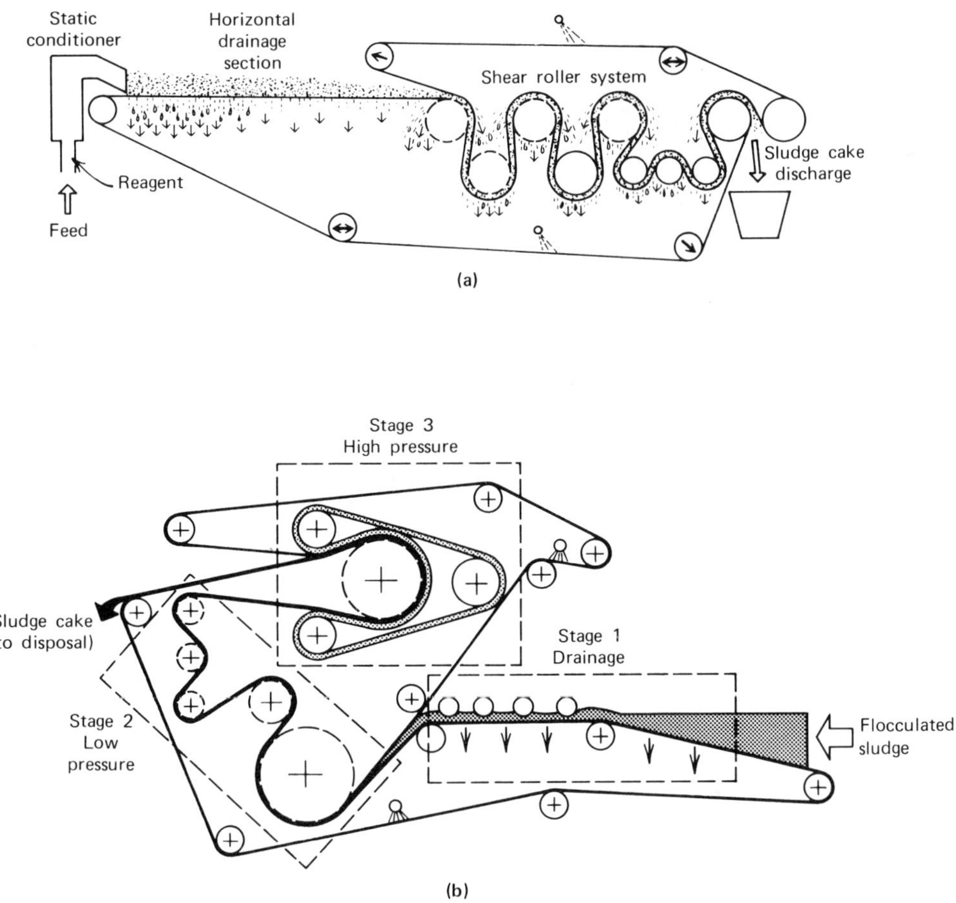

Figure 9. Filter belt presses. (**a**) Courtesy of Simon-Hartley. (**b**) Courtesy of Parkson Corp.

Disc Filters. The disc filter, as applied in the metallurgical field, is employed for straight dewatering applications (see Fig. 10). It concentrates more filtering area, all of it vertical, into less space, and is the least expensive of the continuous filters. Each disc is made up of 8 to 16 sectors fabricated of metal, wood, or plastic, each with the filter medium installed as a bag or caulked. In most designs, each row of sectors discharge filtrate into a manifold pipe which is either contained within or nested around the outside of the central shaft. Easily replaceable external pipes compete with thick-walled cast-iron center shafts in an effort to gain better performance and minimize overall cost in highly abrasive applications such as taconite. High vacuum capacities, ca 180 $m^3/(h{\cdot}m^2)$ (10 gfm/ft^2) on taconite filtration, favor large external pipes to reduce pressure drop, enhancing performance and reducing abrasive wear.

Most filter tank designs are open on the rising side with indentations on the descending side to pass discharged cakes. These tanks usually provide 40% diametric submergence and include a paddle agitator to counteract solids settling. Alternatively, each disc can be housed in individual tanks, eliminating the agitator and its submerged stuffing box. Here limited slurry volume results in rapid turnover supplementing the

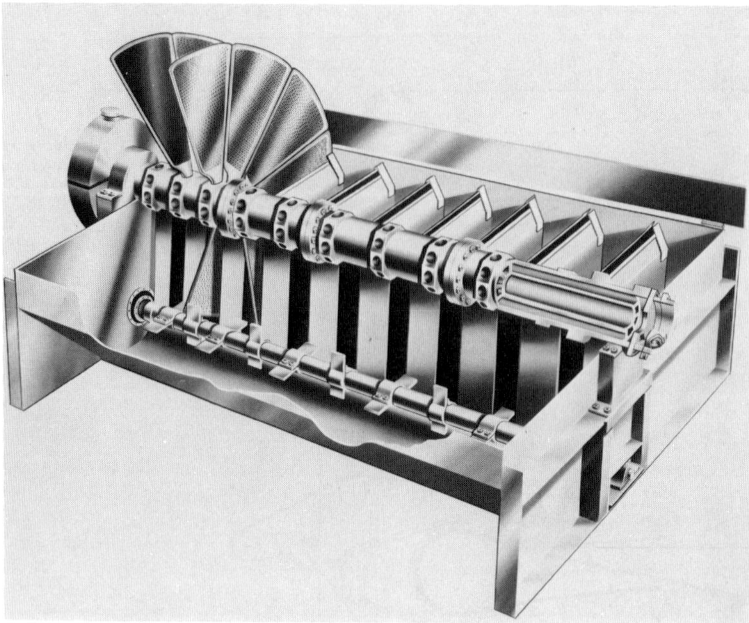

Figure 10. Disc filter. Courtesy of Dorr-Oliver Inc.

agitation by the disc itself. Dry cake discharge is generally by scraper assisted by a soft (low pressure) air blow, or, for some firm cohesive cakes, by a snap blow and deflector blades (no scrapers), using a timed quick influx of high pressure air to inflate bags with a snap.

Applications of the disc filter in metallurgy include the removal of fine iron ore, coal fines and refuse, flue dust, magnesia, many metallurgical concentrates, and alumina trihydrate seed.

In the pulp and paper industry, a modification of the disc filter is used as a save-all and for pulp deckering. Discharge is by jet sluice on the rising side. The applications are essentially filter thickening (see Paper; Pulp).

A completely submerged disc filter is used for filter thickening of miscellaneous slurries and a pressurized disc unit for sugar refinery applications.

A relatively new design has several more sectors per disc, permitting larger diameter discs with somewhat more effective submergence; each sector of each disc has a separate filtrate outlet providing more hydraulic and pneumatic capacity per sector at a sacrifice in the weight of the center shaft and limiting a filter to three discs per valve; cake washing is included.

Drum Filters. The drum filter has proliferated into many variations since the original Oliver continuous vacuum filter. The basic piped drum, submerged in a filter tank, has many discharge devices, each extending its range of application. Some paper pulp filters with ring or cone valves use part of the drum as a receiver. The single compartment filter (Bird-Young) uses the entire drum as a receiver. Top feed versions are also available, including filter dryers and internal drum designs (Dorrco).

In the standard drum filter (multicompartment drum) the drum surface is divided into sections separated by division strips (see Fig. 11). Each section has one or more

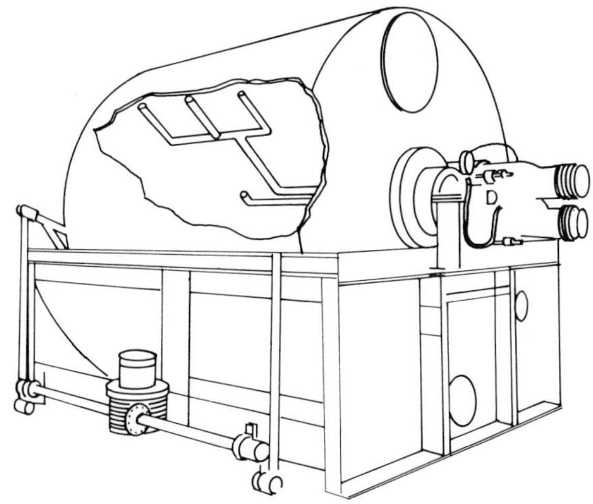

Figure 11. Drum filter. Courtesy of Dorr-Oliver Inc.

radial pipes draining the section to a trunnion pipe and the filter valve. Piping may be inside or outside the drum and may be located on the leading or trailing edge, or both, as dictated by the application. Drum filters are mostly fabricated from a wide variety of metals, including titanium, but small and medium sizes are available in various plastics.

The various discharge devices available for the standard drum filters are scraper, wire, strings, coils, belt, roll, and precoat (see Fig. 12). A reverse air blow is always used in conjunction with scraper or wire discharge and sometimes with the roll discharge.

The normal floating scraper and wire discharge are used with cloths caulked in panels. A soft air blow at less than 5 kPa (37.5 mm Hg) billows the cloth, which deflects as it passes the scraper or wire, giving a cover-cleaning action and effective discharge down to 2–3 mm thick cakes. The wire can better handle the thinnest cakes, because sticky cakes do not cling. A fixed scraper discharge can be used with cloths (usually metal or heavy abrasion resistant) installed by wire winding, to handle 12 mm or thicker cakes. The scraper and wire discharges are the least expensive, easiest to operate, and can usually give good performance with one of the myraid of available filter media.

With the string discharge, strings on ca 1.3 cm centers lift the cake off the medium, support it to the discharge roll, then pass through a comb to maintain alignment and remove any clinging cake before returning to the drum. It requires a cake that is thick enough (5–7 mm, minimum) and cohesive enough to hold together until final discharge. The filter medium must be very smooth to release the cake well.

The coil filter uses about 10 mm diameter helical coil springs both as the filter medium and for conveying the cake to the discharge roll. Before the advent of filter belt presses, the coil filter was one of the major continuous filters handling sewage and other fibrous waste sludges.

The belt discharge for the drum has the filter medium as a continuous belt conveying the cake to the discharge roll and then being thoroughly washed with water sprays before returning to the drum. Its advantage over other discharge devices is in

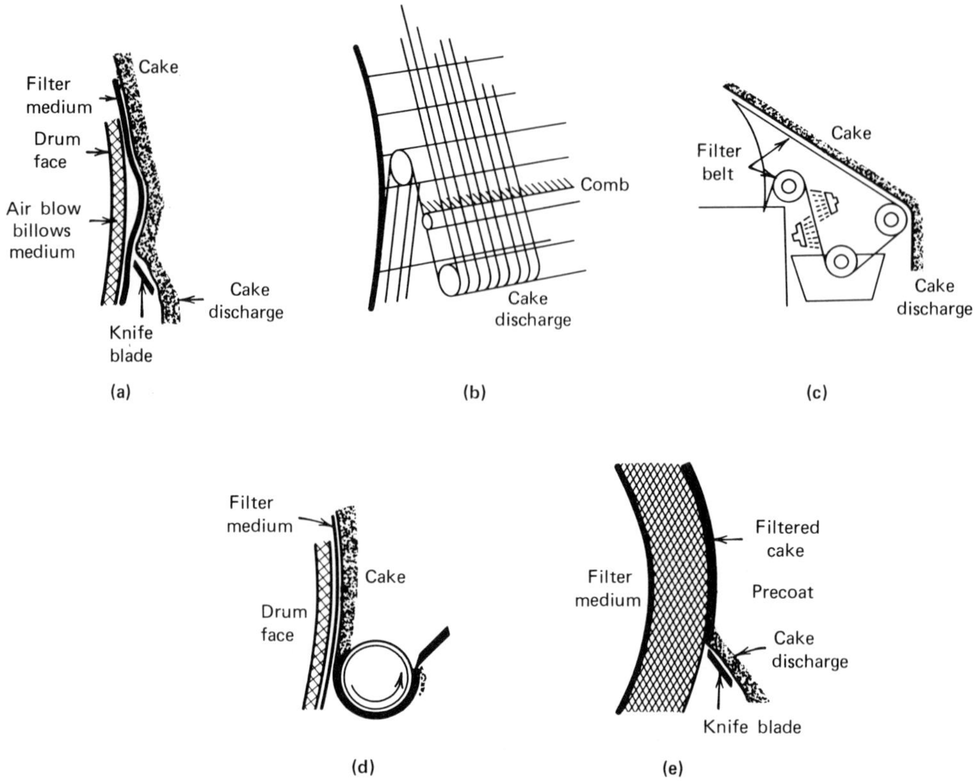

Figure 12. Drum filter discharge devices: (**a**) scraper; (**b**) strings; (**c**) belt; (**d**) roll; (**e**) precoat.

reducing the risk of cloth conditioning, assuring more consistent performance. Furthermore, the belt is faster to replace than media caulked at each section. However, there is about 10% loss in functional area affecting both dry time and cake washing capability; poor filtrate clarity vs media anchored in place; the need to reprocess cover cleaning waste amounting to 3–6 $m^3/(h{\cdot}m^2)$ of face (4–8 gpm/ft^2); and belt maintenance and tracking.

The roll discharge is specific for those materials where the cake is thin and tends to stick to itself or a bare roll in preference to a medium. The cake transfers either to a bare roll that is then scraped clean, or to a roll carrying about 10–30 mm of a cake heel left by the scraper. Since no minimum cake thickness is needed, rotational speed and, hence, capacity can be high. It is a low cost method to dewater clay and the red mud produced in bauxite processing.

Precoat discharge makes the drum filter a totally different operation with different objectives, generally liquor clarification with feed concentration in the range of 0.1–1%. A 50–150 mm layer of precoat, usually consisting of diatomaceous earth or perlite, is first formed on the drum using a filter medium suited only to precoat retention. Process slurry filtration then takes place on top of the precoat with the steadily advancing precoat scraper shaving off a few hundredths of a millimeter of precoat per revolution along with process cake. When the precoat layer is used up, reprecoating is necessary. The entire drum is usually kept under vacuum and the filter valve is not needed.

If the precoat is protected by the cake against erosion, the cake can be washed. The wash filtrate usually cannot be separated from mother liquor because large amounts of filtrate are retained in the precoat. Typical applications are in processing antibiotics, wine, slop oil, and sugar syrups in corn processing. It usually competes with batch pressure filters using appreciable filter aid admix and is justified by reduced usage of filter aid.

In the recausticizing system in the pulp and paper industry, lime mud (calcium carbonate) has proven to be an excellent precoat material both for its own and dregs dewatering and washing. Used as its own precoat heel, it has contributed to lower cake moistures and smooth operation (see Paper; Pulp).

The pulp and paper industry uses quite specialized drum filters (see Fig. 13) for brownstock and bleach washing, for thickening ("deckering") dilute pulp slurries, and for fiber recovery ("savealls"). The design specialization stems from the vast water flows, the free filtering nature of the pulps, and the cohesive sheet nature of most cakes. Discharge devices include air and steam "doctors" and couch rolls.

The single-compartment drum filter is a special tank type unit where the inside of the drum is machined so that stationary dividers riding against the inside of the drum replace the filter valve functions. The drum then serves as a filtrate receiver. Speed of rotation is high, up to 26 rpm, and coarse crystalline or granular materials filter at exceptionally high rates, forming thin cakes suited to fast washing and discharge by jet pulses of air from the very narrow (50–60 mm) sections. The unit competes with top-feed and horizontal types of filters.

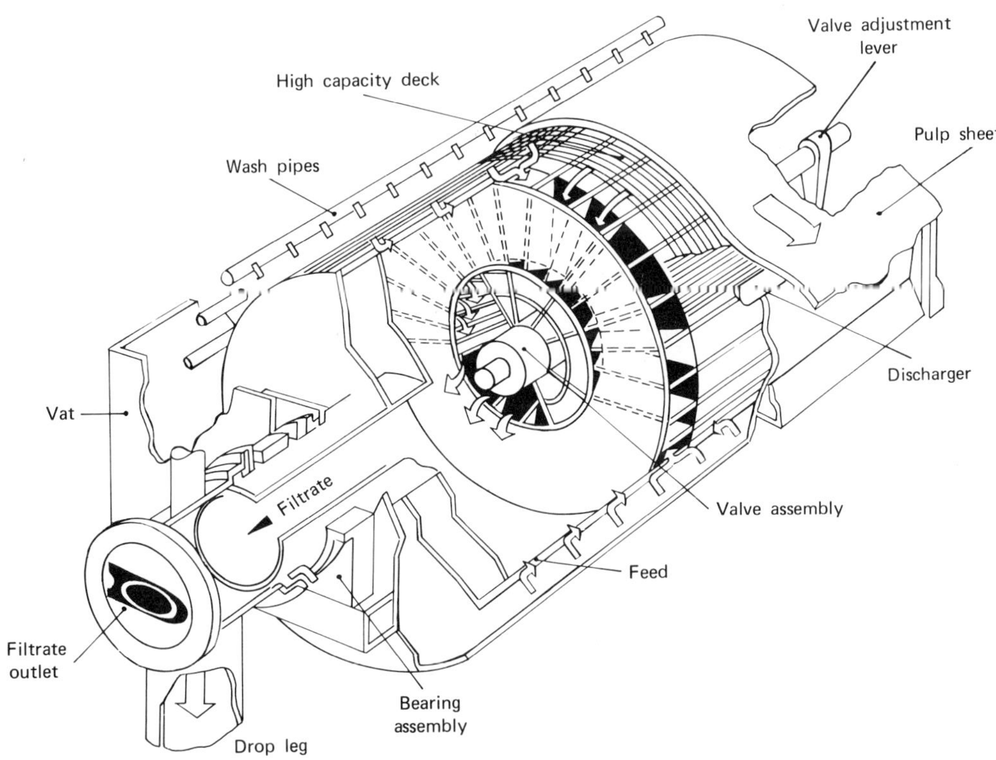

Figure 13. Center valve pulp filter. Courtesy of Dorr-Oliver Inc.

Top feed drum filters are designed with hydraulic and pneumatic capability to handle crystalline and granular material too coarse and free filtering to handle in a tank type unit. As filter dryers (see Fig. 14) they use hot air to dewater and dry coarse crystalline material such as sodium chloride and ammonium sulfate. With multiple compression rolls they are used to dewater synthetic rubber crumbs.

The internal drum filter (see Fig. 15) forms the cake from a slurry pool contained within the drum up to the rim of the partial front drum head, providing an effective 20% maximum submergence. Discharge is by reverse blow, often pulsating, at the top. Cake discharges down a chute through the open front head or can be conveyed out. It is applied to fast-settling, crystalline or granular metallurgical slurries unsuited to any tank type units and not coarse enough for most horizontal filters. The cake washing and cloth cleaning capabilities are moderate. Nickel ore processing, iron oxide concentrates and tailings, and alumina trihydrate are typical applications.

Horizontal Filters. The horizontal table with scroll discharge, the tilting pan, and several horizontal belts (not including the filter-belt presses covered under pressure filters) are the common horizontal vacuum filter designs.

The horizontal table filter (see Fig. 16) is a circular unit with a sectionalized horizontal filter deck, an outer rim, and an inner island. The unique use of a reverse air blow under the feed section is effective in cleaning the filter medium and distributing the feed while reslurrying the heel. The very high capacity sloped drainage of each section and the filter valve provide quick liquor drainage, hence sharp separation of wash liquors, which makes it an excellent countercurrent washing filter. Owing to the ability to handle high air flows at any vacuum needed, it efficiently dewaters coarse

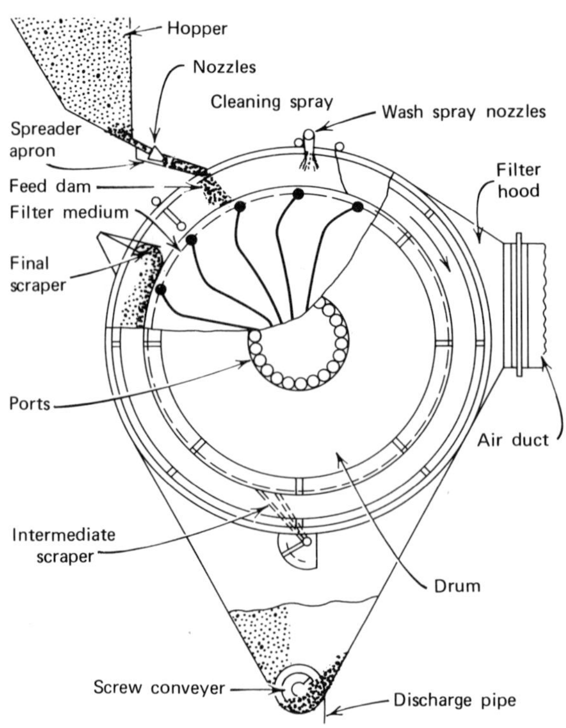

Figure 14. Top-feed filter dryer. Courtesy of Swenson Div., Whiting Corp.

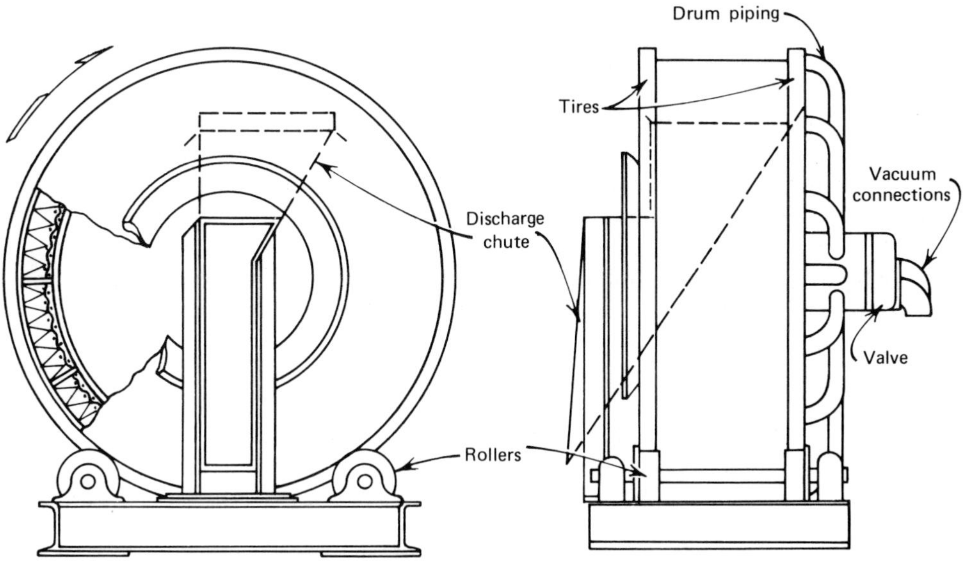

Figure 15. Internal drum filter. Courtesy of Dorr-Oliver Inc.

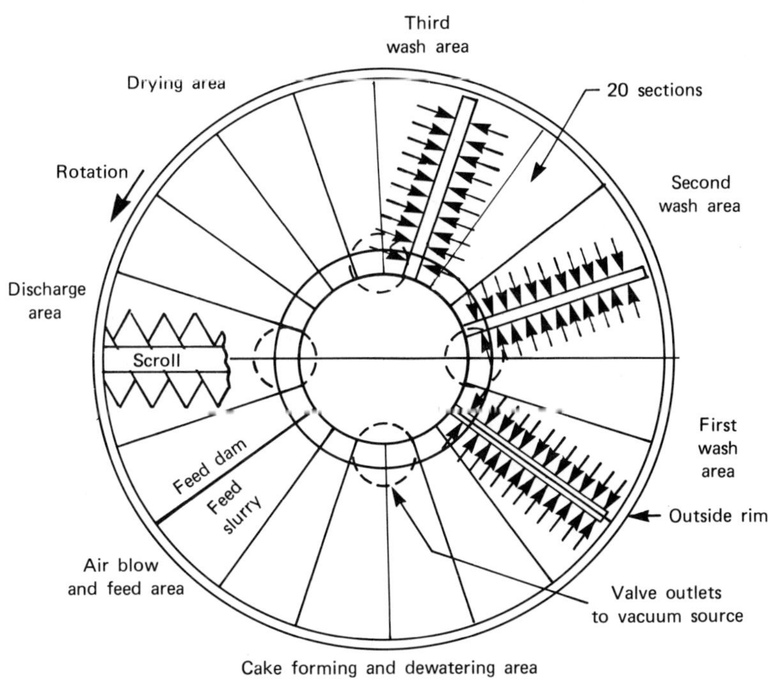

Figure 16. Horizontal filter. Courtesy of Dorr-Oliver Inc.

solids to minimum moisture. Its effectiveness is enhanced by leaving a significant cake heel which contains a disproportionally high moisture and mother liquor content. Major dewatering applications include glass and foundry sands, phosphate rock, silica gel, and potash. Washing applications include alumina trihydrate, gypsum from phosphoric acid, phosphate rock in brine, caustic salt, cotton linters, solvent extraction of cottonseed, alfalfa, etc.

The tilting pan filter (see Fig. 17) has individual pans rotating around a vertical axis, with large pipe connections to the filter valve. The pans invert to discharge cake and the medium is then washed by jet sprays. The pan rims automatically confine wash liquors permitting flood washing, if desired. It is made in large sizes and is used for the countercurrent washing recovery of phosphoric acid from gypsum.

Horizontal belt machines (see Fig. 18) may or may not have a rubber support belt. All provide for medium belt washing on both sides. The design with rubber belt

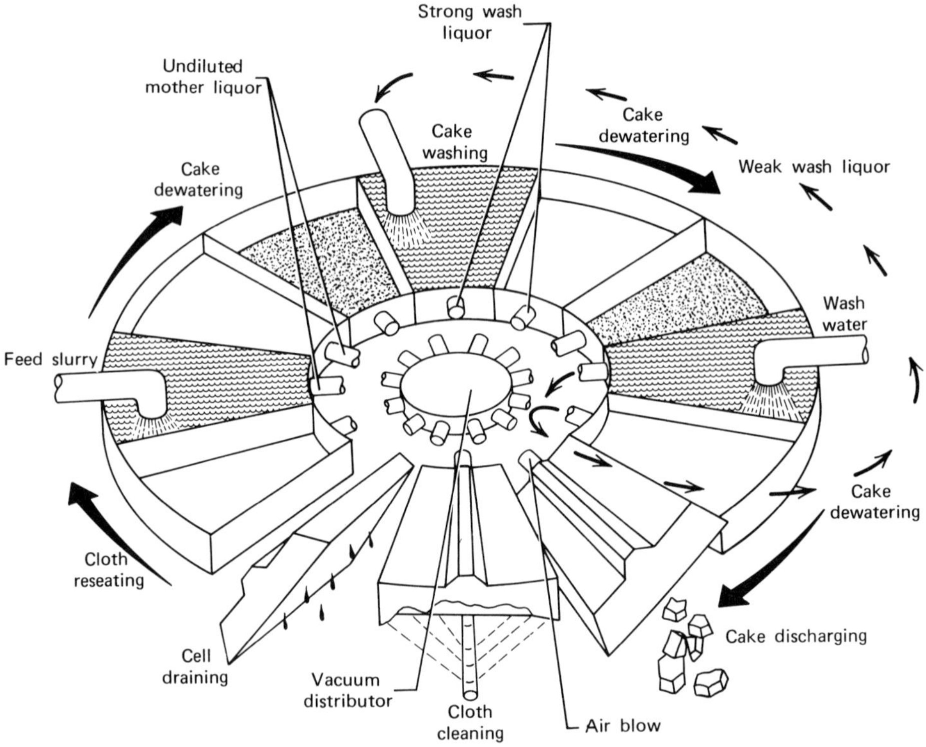

Figure 17. Tilting pan filter. Courtesy of Bird Machine Co.

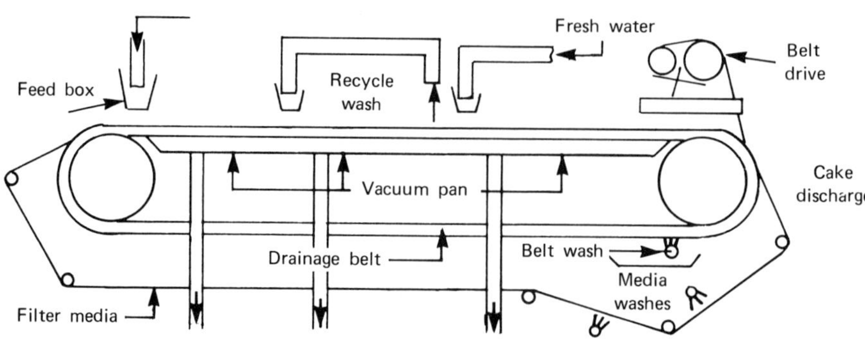

Figure 18. Horizontal-belt filter. Courtesy of Eimco Process Machinery Div., Envirotech Corp.

support incorporates the drainage in the belt, which has a low friction wearing surface with tracking guides that rides along the vacuum trough. Such a unit can operate effectively, though possibly not economically, with cakes down to 3 mm, so it competes with washing drum filters as well as other horizontals.

In the designs without support belts, vacuum is intermittent. The medium belt may be advanced without vacuum or moveable drainage trays may retract without vacuum. Either way, with heavy cakes, the medium is subjected to moderate wear. A major advantage over the belt-supported unit is in handling chemicals such as solvents where elastomeric belt materials are not suitable.

Machine tool coolants and cutting oils are clarified by a variety of vacuum or pressure filters, such as an indexing disposable horizontal belt with drag conveyer.

Specialized Functions. ***Compression.*** Physical compaction of compressible cakes has long been practiced to reduce cake moistures. It was always an integral benefit in filter press operation, and now, with membranes, it is even more so. It is, of course, the essential driving force in operation of the filter-belt press, and vital to performance of the tube press and automated diaphragm press.

Various devices have been used on continuous filters, especially drum types. Press rolls, often in conjunction with press belts to prevent cake couching or sticking to rolls, are most common. The top feed synrub filters use several rolls to squeeze liquor from synthetic rubber crumb. Rotating flappers, vibrating devices, and leno drag nets are also employed.

Steam Drying. Lowering filter cake moisture economically with heat requires a high transfer rate of heat input. Slightly superheated steam is ideal, functioning synergistically by transferring the high heat of vaporization to the cake while condensate is removed by the pressure differential. In cakes of moderately low porosity the condensate line moves through the cake as a front, giving high efficiency. In batch filters it can substitute effectively for air displacement. In dewatering some fine iron ores, on disc filters to meet critical moisture levels for pelletization; it is a vital part of the operation, though not as efficient as expected.

Vapor-Tight Hooding. Most continuous vacuum filters can be placed in vapor-tight hooded enclosures for operation with volatile liquids. Vacuum systems must then recycle vapors to the hoods, excluding air. Solvent extraction is practiced on drum and horizontal table vacuum filters.

Cartridge Filtration

Cartridges are removable elements of cylindrical shape incorporating a filter medium surrounding a drainage core with top and bottom seals. They may be disposable or cleanable. They may filter through the depth, on the surface, or at the edge of the medium. Cleanable cartridges function like tubular filters, except that they are removed from their enclosure for cleaning.

Cartridges are generally used for clarification of liquors rather than for solids recovery (see Water; Waste). Solids concentrations below 0.01% are normal. Particle size removal is usually within the range of 0.02–100 μm, thus covering the 0.02–10 μm range, often referred to as microfiltration. For sterile filtration, a 0.2 μm absolute removal is required.

Cartridge filters compete with tubular and leaf-type pressure filters, which usually need filter aid both as precoat and admix. They are also used as prefilters before ultrafiltration or other fine cartridge filters, or as polishing filters backing up various types of macrofilters.

Theory of Depth Filtration. Depth filtration theory has been developed for granular-media beds but most aspects apply as well to cartridges (43–48). The three mechanisms involved in the removal of particles from a slurry by depth filtration are particle transport, attachment, and simultaneous detachment.

The transport step, relatively insignificant, is a physical-hydraulic process that provides forces to cause particles to leave the flow stream lines and approach media surfaces where attachment forces can be effective. At work within the media pores are gravity, direct interception, diffusion by Brownian motion, and particle rotation caused by hydrodynamic drag forces interacting with angular particle shape. Gravity is not as significant in cartridges as in granular media. Direct interception does not involve much inertial force in liquids, in contrast to air filtration. Brownian motion is the major transport factor for particles less than 1 μm. Transport mechanisms function efficiently even though they are insignificant in filter design.

The mechanism of attachment is comparable to that of flocculation. The double-layer repulsive forces and London-van der Waal's attractive forces interact, in addition to bridging and chemical bonding. Some particles are also removed by interstitial straining. Attachment forces are often vastly increased by small amounts of previously deposited particles (49).

Hydrodynamic drag forces contribute to particle detachment. As pores gradually fill with solids, flow velocity increases as flow rate is maintained, introducing an increasing detachment force. The coincidental increase in pressure differential and compression of the medium compounds the process.

Flocculants can contribute in a major way to filtration as well as to sedimentation. They can, however, also clog filter the media. Cationic flocculants behave the same in both, and overdoses are detrimental to both. Flocculant selection is very important, as are pH and zeta potential. Alum is an effective and economical agent for color removal from water but is ineffective in resisting shear forces. Some polyelectrolytes are essential for efficient removal of colloidal matter. Flow velocity is crucial since a high velocity can prevent deposition, whereas low velocity may cause solids to deposit on the surface (see Flocculating agents).

Theoretical models attempting to predict performance of clean depth filters only are based on particle trajectory calculations and assume some porous media model such as cylinders, spheres, or constricted tubes (43). However, an effective simulation procedure has been developed (50) for predicting dynamic behavior over the entire practical range of a granular-bed filter operation by viewing deposition as a two-stage process, ie, a smooth coating mode and a constriction clogging mode. Current models do not apply when feed contains flocculants or when medium porosity decreases with depth. Mathematical expressions for rate of deposition and pressure are available, but are based on at least four adjustable parameters that lack proper physical meaning and are obtained by curve-fitting experimental data.

Practical Aspects. Cartridge performance is judged by particle retention, flow rate maintenance, and solids holding capacity. Particle retention is normally reported in terms of absolute or nominal ratings. Absolute refers to 100% removal of the specified size particle but a downstream concentration of 10^{-6} of the challenge level may also be considered absolute (51). This is of special concern in the pharmaceutical field where cartridges with absolute 0.2 μm ratings are used for positive sterilization removing all bacteria. In the cartridge manufacturing industry, absolute ratings are found to be reliable under uniform test conditions. Nominal ratings refer to a retention point of 98, 95, and occasionally 90%, and are an approximate guide to a manufacturer's range of retention. The conditions at which the nominal rating is to be determined are not yet standardized within the cartridge industry. Thus cartridges by two manufacturers with identical ratings may well show different retentions.

Mean pore size, efficiency, and filtration ratio are alternative retention rating methods (52): mean pore size for quick screening of media for an application or for quality control; efficiency as a weight percent removal of a given contaminant on less critical applications; and filtration ratio in the industrial hydraulic area.

Retention capability is tested by the bubble-point or forward-flow test, both nondestructive, whereas testing with known sizes of particles or bacteria is destructive. The bubble-point test (51) measures the pressure at which the first bubble shows through a wetted medium and correlates with pore size by the following relationship:

$$P = \frac{K^4 \sigma \cos \phi}{d} \tag{18}$$

where P = bubble point pressure; σ = surface tension of liquid; ϕ = liquid to membrane contact angle (0–40° hydrophilic, 90°+ hydrophobic); d = pore diameter; and K = shape correction factor.

The forward-flow test measures air flow at fixed pressure drops on a wetted medium, indicating total porosity.

In testing a medium with specific sizes of particles, the removal efficiency may be determined gravimetrically or by analyzing the filtrate by turbidimeter, particle-size counter, or microscope. The ISO multipass filter test method uses an AC (Arizona Coarse) dust to determine the β ratio of the number of particles greater than a specified size in the influent to the number in the effluent.

Flow rate depends directly on the cartridge permeability and the pressure differential (51):

$$Q = A\Delta p / R\mu \tag{19}$$

where Q is flow rate, R is resistance to flow, A is filter area, Δp is differential pressure, and μ is liquid viscosity.

Cartridges are usually rated for this performance aspect. If a cartridge is properly matched to the application, flow rate is maintained with only a gradual pressure increase until the cartridge is almost fully loaded, and then pressure rises rapidly.

Solids capacity directly affects the run length or cartridge life and is a key measure of how well the cartridge is matched to the solids being removed. With a depth medium, the solids must build up throughout the depth at a rate that is about equal in percentage of full load at all depths, or some depth will prove the limiting one and start a cake filtration. The inefficient use of depth filters is most commonly caused by surface filtration blocking. On the other hand, surface filters are used inefficiently with feed of high solids concentration; in this case, a prefilter should be employed. Prefilters, usually depth types, may be separate elements for maximum flexibility, or incorporated within a single cartridge.

In the food and pharmaceutical industries and in certain applications, such as metal-finishing baths, cartridge make-up must be carefully evaluated for extractable content. Some resins or polymer media contain extractables that can poison the bath or contribute taste, odor, or color (53–54).

Bacteria growth within the cartridge may foul the medium and decrease its capacity. Bacterial growth progressing to the downstream side is called grow-through. Filled and depth cartridges are most vulnerable. In sterile filtration with absolute-rated cartridges, grow-through cannot be tolerated. In applications where it is a potential problem, direct steam, uv irradiation, chemical sterilization, or cartridge replacement every eight hours is necessary.

Equipment and Media. Cartridges are almost exclusively enclosed by pressure vessels capable of holding one or more elements assembled into tubes (52). Some of the smaller housings are made of plastic, but most are fabricated of suitable metals. It is necessary that the cartridges can be easily and safely replaced. In one unit, multiple tubes are mounted in a removable, disposable cage for radioactive waste filtration.

The cartridges are generally made of metal or plastic hardware. The filter media are formed (discs are stacked) around the drainage core to pack a maximum of filtering area or capacity. Some media are strong enough that a separate core is not necessary. Media are often categorized as depth or surface but many types involve both features, depending, of course, on the particle size being removed. Cartridges depending primarily on depth filtration are wound types, felts, some porous ceramics, sintered metal or metal fibers, and bonded fibers. Surface filtration cartridges are made of fiber sheets in the form of stacked discs or pleated cylinders, generally resin impregnated, some membranes, some porous metal or plastic, and some edge filters using discs or wire winding. Some membranes function in depth as well as on the surface. However, they do not present the problems of compressible depth filters in migration of media and sloughing of particulates with increasing pressure differential (51).

Disposable Cartridges. Disposable cartridges include those with the following media: absolute-rated membranes (plastic, inorganic microfilament, and nuclear); thin-walled ceramics; wound types; fiber sheets, usually resin impregnated as pleated cylinders or as flat discs; hollow discs of porous metal, plastic, or impregnated paper; bonded fibers, powders. The absolute-rated membranes are used for sterile filtration with absolute ratings of 0.2 μm. Essential requirements in addition to absolute ratings and fluid compatability, are strength and rigidity to prevent the unloading of entrained particles of microorganisms and the migration of the filter medium itself.

The plastic membrane is a microporous structure (about 150 μm thick) composed of a wide selection of polymeric materials but mostly cellulose esters. It retains in depth as well as on the surface but relies on direct interception for particle removal with excellent integrity (52). It is used both as flat sheet discs often stacked with alternating drainage plates, and pleated cylindrical elements. When pleated (Fig. 19), it is generally combined into 2–4 layers for support and prefiltration. The inorganic microfilament membrane is structured by controlled dispersion and synthetic resin bonding of a synthetic inorganic crystalline microfilament. The result is functionally comparable to the plastic membrane in providing a uniform pore structure with 80–85% void volume, and with the highly important resistance to bacteria grow-through. It is available in pleated cylindrical cartridge form. The Nuclepore (track etch) membrane has nearly straight-through pores created by nuclear bombardment followed by the etching of holes selectively through a polycarbonate or other suitable polymer film. The number of holes is controlled by the residence time in the nuclear reactor and the pore diameter by the etching time. The membrane has a lower void volume but is thinner, resulting

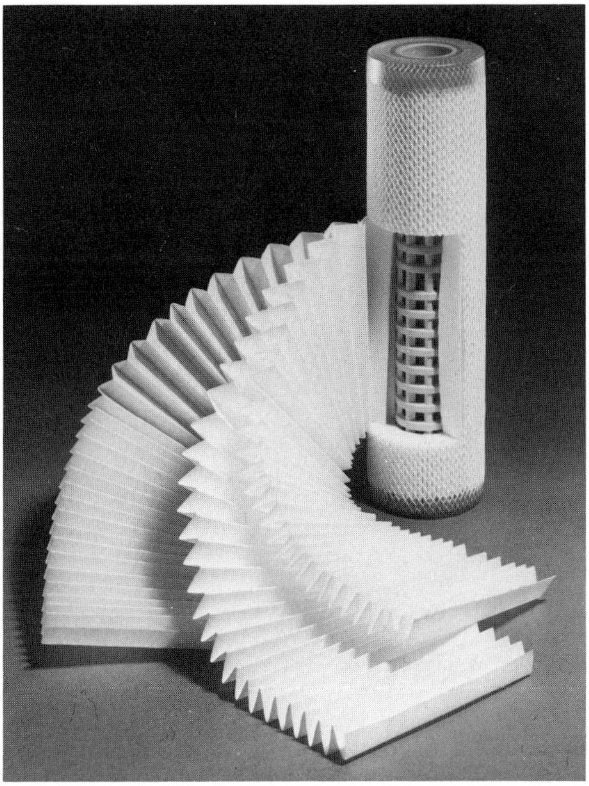

Figure 19. Pleated cylindrical cartridge. Courtesy of Gelman Sciences, Inc.

in equivalent flow rates per unit area. The durable film (ca 10 μm thick) has an extremely smooth surface especially suited to microscopic analytical work, in contrast to the coarse surface of the microfilament type. It functions purely as a surface filter or screen with 0.03–10 μm standard pore sizes and is available in flat sheets and pleated cartridges (see Particle-track etching). Another relatively new disposable medium is the thin-walled, small-diameter microporous ceramic cylinder having absolute ratings from 0.8 to 0.12 μm.

All absolute-rated cartridges lack significant dirt-holding capacity, making prefiltering frequently desirable. Some absolute cartridges include a built-in prefilter.

Wound cartridges (Fig. 20) are by far the most common with nominal ratings down to 0.5 μm. The entire medium may consist of brushed yarn wrapped spirally around the core, or a sheet medium may first be wrapped around the core. Winding pitch, tension, fiber length, fiber characteristics, and quantity affect removal ratings (52).

Porous fibrous sheets, usually resin impregnated, are made from cellulose, glass, plastic, and many other fibers. Multiple layers are common, adding depth. Pleating is normal for cylindrical elements to maximize surface area, solids capacity, and element rigidity. Nominal ratings range from 1–100 μm; they are generally used in coarse filtration applications such as oil filters.

A unique disc stack arrangement designed for oil filtration alternates filter sheets

Figure 20. Wound cartridge. Courtesy of Filterite, a Brunswick Co.

with plates which provide alternate feed entry plus storage capacity for solids removed and passage for filtrate removal (see Fig. 21).

Another design compacting large surface area into small volumes is a hollow disc stack using media such as resin impregnated paper, porous metal or plastic, or a combination of cellulose fibers and inorganic filter aids designed to give a positive zeta potential (Fig. 22). Porous plastics such as polyethylene and polytetrafluoroethylene (PTFE) are formed by varying techniques and are also available in cylindrical cartridges.

Many types of loose fibers bonded with a wide variety of resins have been formed into light but strong porous filter media, often with graded density to enhance dirt-holding capacity. Long polypropylene fibers without resins have been formed into a comparable structure, as have layers of needle felts, without resin bonding.

For low-flow applications, specialty designs are available filled with dry powders to combine filtration with adsorption by carbon, fuller's earth, or ion-exchange resins.

Cleanable Cartridges. Some cartridges are designed to be cleaned either in place or after removal. Edge filters made from discs or wire wound are, in effect, strainers and are easily cleaned by backflushing or by rotating the slots past a cleaning blade. Coarse, plain weave woven wire functions comparably. Finer, more complicated woven wire weaves, porous ceramics, and porous sintered metal powders cannot generally be cleaned as effectively by backflushing, and at intervals may require special cleaning treatments with acids, solvents, detergents, or high heat.

Sintered metal fiber or felt structures have unique properties. Advantages over other porous metal structures are flexibility for pleating, elasticity to absorb pulsating

Figure 21. Paper disc stack. Courtesy of Hilliard Corp.

pressure differentials, weldability, and good retention and dirt-holding capacity. These structures can be graded in depth by superimposing different porosity webs. Absolute ratings of 3–40 μm are available. They are highly effective in critical hydraulic systems (55) and in the filtration of gels from polymer melts at 300°C (56).

Bag Filters. Bag filters are open-top bags or socks made from varied needle felt, multifilament, or monofilament media (see Fig. 23). Most frequently they are used with a restrainer basket within a pressure housing, the basket providing necessary support. These bags have diameters to 200 mm and are 450 or 800 mm long. Larger bags are also available for tieing on to a pipe outlet. Particle retention capabilities cover a wide range, as low as 5 μm with several media and even lower.

Bags function primarily as strainers, even with felt where depth is limited. They are used when oversized particles must be removed but smaller particles must be passed reliably. Typical applications include adhesives, cutting oils, paints, pigments, phosphate coatings (57), printing inks, polymers, pharmaceuticals, and vegetable oils.

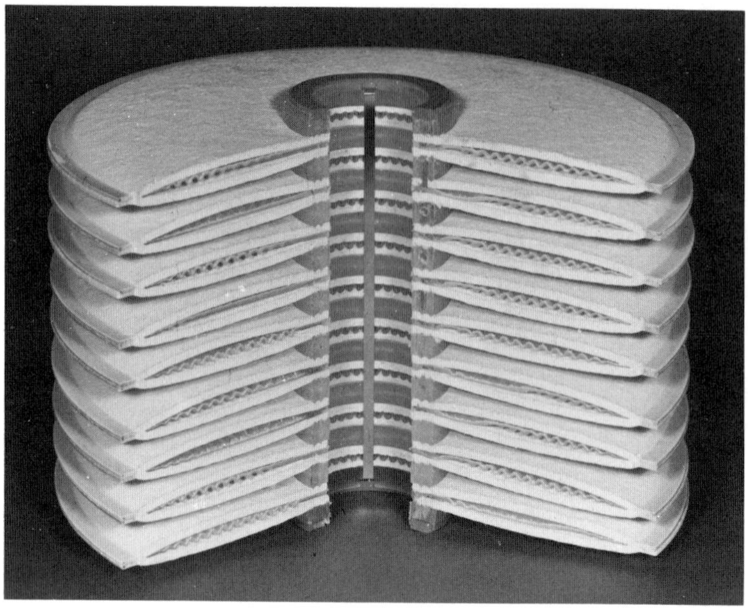

Figure 22. Hollow disc filter assembly. Courtesy of AMF Cuno Div., AMF Corp.

Filtration Thickening

Any filtration can be considered as a thickening operation, ie, solids concentration is increased. Filter cakes are often discharged by sluicing or backwashing from filters such as pressure leaf and tubular, or even continuous drum or disc filters, the net effect being an increase in slurry concentration (and possibly a change in liquid).

Filtration thickening may be an alternative or a supplement to other solids concentration. For example, for the clarification of a low-solids concentration slurry where a dewatered cake is not needed or is best obtained from a much higher slurry concentration.

Filtration thickening may provide a more efficient overall dewatering operation; eg, where space for gravity sedimentation is unavailable (see Gravity concentration), where specific gravity is inadequate for sedimentation, or where enclosed or pressurized operation is desirable. It may even permit dewatering to a higher solids concentration than cake filtration. In most types of equipment cake build-up is limited, in order to obtain the high flow rates associated with negligible or thin cakes.

Backwashing batch-type designs operate on dilute slurries giving high liquor clarification rates from several elements operating in parallel with intermittent or cycling backwash. Continuous drum and disc types of filters with high or complete submergence operate with backwashed cake being reslurried. Alternatively, the initial submergence of a drum or disc filter might involve intermittent backwash in conjunction with cake shearing from scrapers or agitation to give slurry prethickening before final cake formation, termed delayed cake formation (58). Filtration thickening can also take place by limiting or almost eliminating cake formation by high shear generated either by high velocity slurry cross-flow or mechanical shear. A totally different mechanism is used in electrically augmented vacuum filtration where electrophoretic action limits cake formation while vacuum removes the liquid.

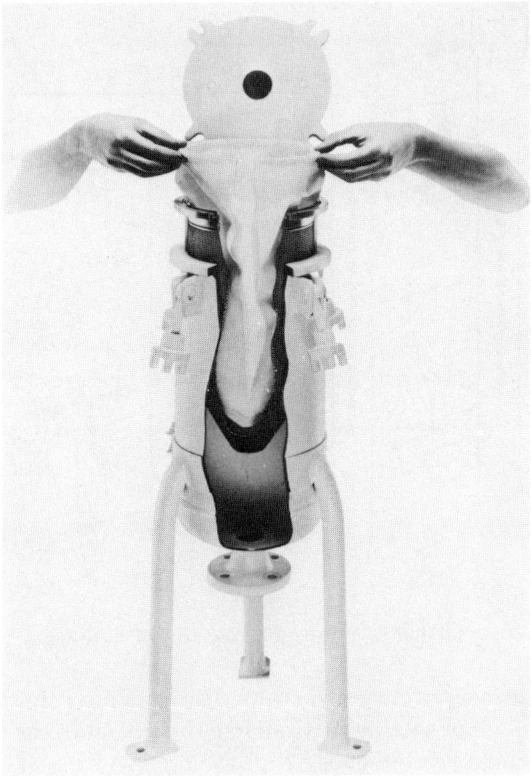

Figure 23. Bag filter. Courtesy of American Felt and Filter Co.

Backwashing Designs. The cost effectiveness of backwashing units involves: the required frequency of backwashing (thus the dirt-holding capacity in relation to feed solids and flow rate); the amount of backwash liquor in proportion to filtrate; the percentage of cake removal by backwashing (minimum channelling); the value of liquor lost in backwashing or the expense of recovering it by another solid–liquid separation step; and the frequency of replacement of the filter media. Intermittent backwashing is practiced on many tubular and special leaf filters.

The most common cycling backwash design uses several tubular filters, each in a separate housing manifolded together for parallel operation and sequential backwashing (see Fig. 24). For internal (filtrate) backwash, three manifolds are used; for external, four. Continuous liquid output is common. Features include vibrators or air shocks assisting filtration or backwashing. Typical filter media are wire screens, perforated or slotted metal screens, and synthetic fabrics.

Another device uses pleated metallic media formed around the periphery of thick disc elements, stacked around a perforated center core. Dual pistons moving within the core control backflushing of one element at a time. Filtration is from inside out.

In yet another device, filtering takes place alternately on both sides of the same medium, usually a filter cloth between elements of a stack of frames similar to a plate-and-frame press. Feed enters alternate frames, with those in between carrying backflushed solids initially and then filtrate.

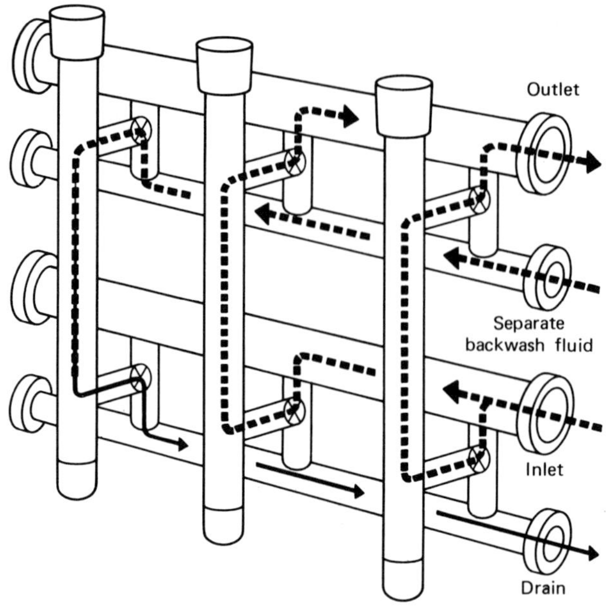

Figure 24. Multiple tubular filter function. Courtesy of Albany Engineered Systems.

A completely submerged rotary vacuum disc or drum filter uses filtrate reverse flow on the bottom sector or section to discharge the cake into the conical tank bottom from where a screw conveyor removes it.

Steady-State High-Shear Designs. In several different equipment designs mechanical or cross-flow-velocity shearing forces limit cake formation and continuously produce thickened slurries and filtrates.

The plate-and-frame thickener uses flow velocity through channels to limit cake thickness to an equilibrium amount. The slurry flow steadily thickens as it progresses through the channels formed into the frames. Filtrate is steadily withdrawn from the plates. Flow control into and out of the channels controls velocity and pressure, and hence concentration increase. The primary application areas have been pigments and catalyst recovery.

The Mott Inertial Filter uses porous metal pipe filter elements with pressurized slurry flowing at approximately 6 m/s inside the pipes, minimizing cake buildup, comparable to ultrafiltration. Tubes rated at 0.5 μm retention provide best overall performance. Periodic backflushing contributes to maintaining higher average rates.

Delayed cake formation describes the filtration process in two different types of equipment. A mathematical analysis (58) of benefits obtainable from an operation where the initial submergence of a continuous filter is devoted to filter thickening by preventing cake formation demonstrates that the overall filtration rate is higher than would be for straight filtration. The reaping of such potential benefits depends on effectively delaying cake formation without adversely affecting particle size or flocculation characteristics.

Figure 25 shows the operational features of staged, high pressure, thin-cake filters

(59–60). Filter cake on the recessed plates is limited to a thin heel primarily by the mechanical action of the rotating turbines and partially by the hydraulic forces from the slurry passage. The filtering elements are all in series, thus steadily thickening the slurry as it passes from stage to stage until it exits as a very thick slurry, termed extrudate. Depending on the rheological properties of the material, this extrudate may be more concentrated than a filter cake from conventional units.

Electrofiltration. ***Electrokinetic Separations.*** An electric field maybe imposed on a suspension in compression by passing a current through the suspension. A colloidal particle with its gegenion (counter ion) cloud experiences a force couple (61–65); for example, a negatively charged particle is driven toward the anode, while the positively charged medium holding the gegenion cloud experiences an equal and opposite force toward the cathode.

The migration, under the influence of such an electrical field, of a solid suspended in a liquid in a closed container is called electrophoresis (66). If a barrier causes accumulation of solid, the continuing force couple expels fluid from between the particles, causing a densification. For a negatively charged colloid (as in the above example) this process is called the electrophoretic deposition of an anode cake (67). The expulsion of the fluid is called electroosmosis. As a phenomenon electroosmosis has been proposed and has long been used on an ad hoc basis for, eg, dewatering moist soils (68–69). It has also been proposed and indeed was originally developed for dewatering peat moss (70).

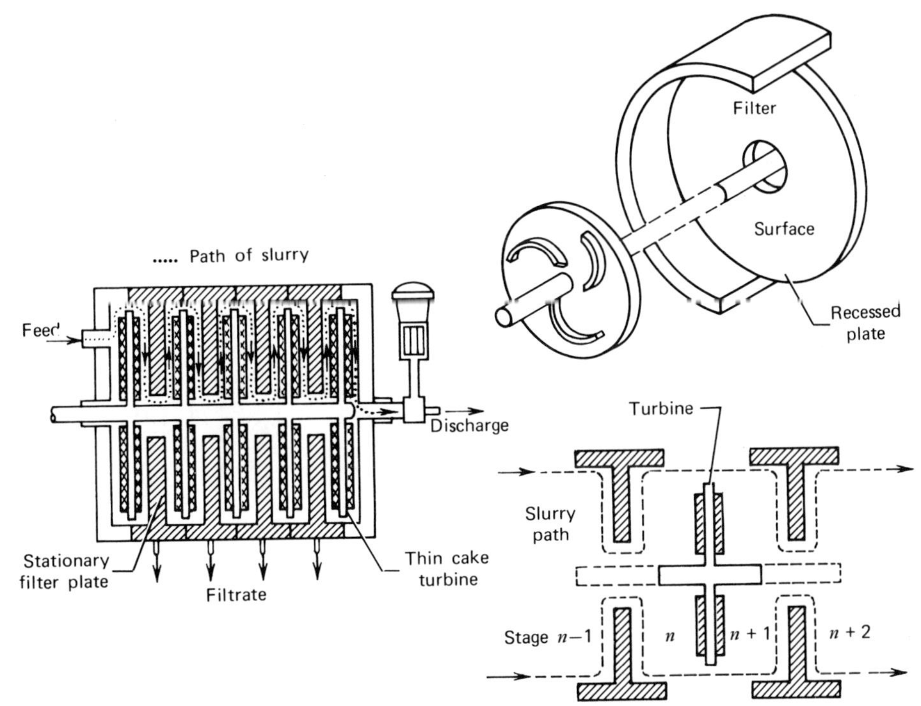

Figure 25. Staged, thin-cake, filter thickener. Courtesy of Artisan Industries, Inc.

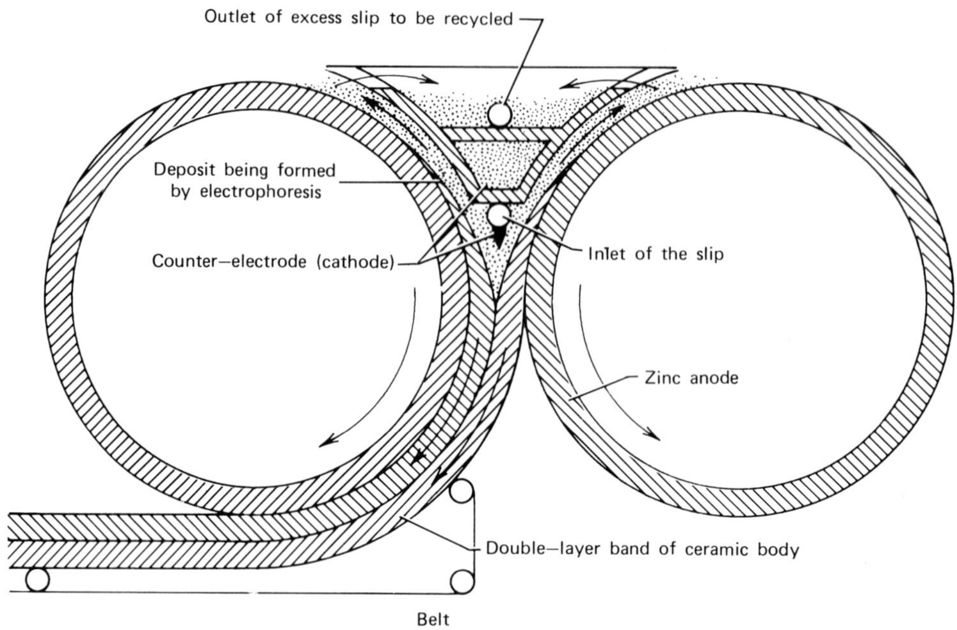

Figure 26. The Aliphat, a device used for continuously casting a uniform thickness of ceramic sheet by electrodeposition.

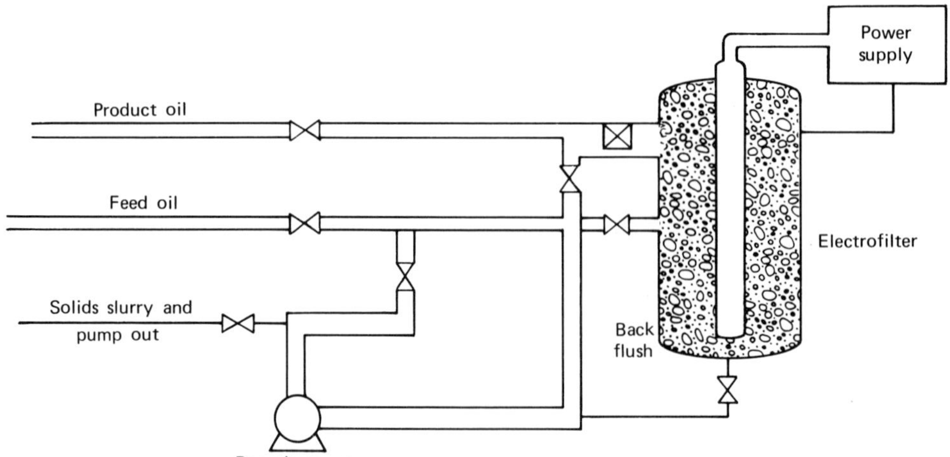

Figure 27. Flow chart used in the electrofiltration of catalyst particles from vegetable oils by dielectrophoresis. Courtesy of Petreco, Div. of Petrolite Co.

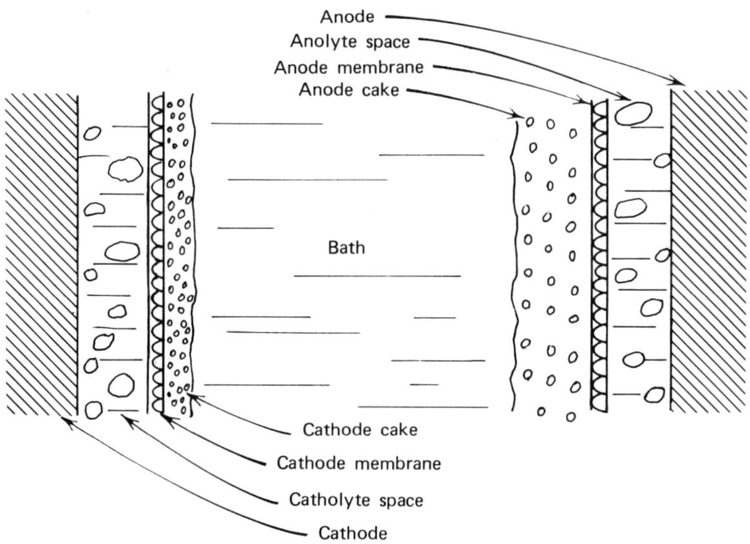

Figure 28. The nine important zones in electrically augmented vacuum filtration.

In recent years, a method has been developed for dewatering clay slurries by electroosmosis through a membrane. Reagents have to be added to keep the fluid removed acid to neutral to preclude hydroxyl ions otherwise released in cathodic chemistry from migrating into the membrane and reversing the body force on the fluid. The electrophoretic deposition of a cake from a suspension is currently being used for making uniform clay sheets from which ceramic objects can be formed by die cutting (64), see Figure 26. Electrophoresis has also been used since 1914 to refine kaolin clay at the Chodau plant of the Carlsbad Electro-Osmose Company, Czechoslovakia (65).

Dielectrophoresis. Electroosmosis and electrophoretic deposition both take place from suspensions in compression, or nearly so, where the force couple on the colloidal particle and the suspending medium exert a direct effect. However, in a dilute suspension of a solid in a gas or low-dielectric-strength liquid an electric field causes the same force couple as before but without net movement. The particle shifts one way, the gegenion cloud the other, and a contrary field results exactly canceling out the applied field. However, in a field gradient it may be shown that net movement will occur. The migration of colloidal particles in a field gradient is called dielectrophoresis. In the case of, eg, removal of Raney nickel catalysts from hydrogenated vegetable oil, a filter is used that has a packed bed of high dielectric strength pebbles. A high voltage may be directly applied across the bed. Localized regions of high field gradient are produced; these cause the (positively charged) catalyst particles to migrate to and adhere to the matrix of pebbles. Figure 27 shows electrofiltration of catalyst particles from vegetable oil.

A number of schemes for the electrical flocculation of suspensions have been proposed by means of pulsed or alternating uniform fields (66–68). It seems likely that this technique would be limited to cases of marginal stability.

Electrically Augmented Filtration. Electrically augmented (or aided) conventional filters combine the best features of both conventional filtration and electrokinetic separation to give unique separation capability. They may be categorized according to the driving force for the conventional filtration step, namely, pressure (electrically augmented pressure filter, EAPF), vacuum (EAVF), and gravity (EAGF). The unifying principle is that the filtration step occurs at the cathode, which is the filtrate receiver, just as in conventional filtration. As might be expected, a cake forms as filtration proceeds with attendant decrease of flux. However, if a current is being passed through the cake, cake buildup stops when the flux has been slowed to a value somewhat slower than the electrophoretic speed of migration of colloidal particles relative to the slurry liquid. The cake thickness and porosity are then constant as long as the pressure differential and current remain constant. Water flows out of the system at a rate depending in a complicated way on the volume fraction of water in the slurry but linearly on the current density. Thus the EAPF (69) and EAGF (70), because they are batch devices, show gradual decreases in flux as the slurries thicken. However, the EAVF, a continuous device, has a constant slurry composition depending only slightly on the rate of throughput and hence, for a given feed composition, has a constant flux approximately proportional to current density. The nine physically identifiable zones of the EAVF are shown in Figure 28.

Nomenclature

a = slope of experimental plot of t/v vs v, s/m^2
A = filtration area
b = intercept on t/v axis of experimental plot of t/v vs v, s/m
cp = constant pressure
cr = constant rate
C^* = mass of dry solids in filter cake per unit volume of filtrate obtained, kg/m^3
$\overline{C}_c$ = average concentration of dry solids in the filter cake expressed as a mass fraction, kg/kg
C_e = consolidation coefficient, kg^2/(m^4·s)
C_s = concentration of dry solids in feed slurry expressed as a mass fraction, kg/kg
d = pore diameter
e = local value of void ratio
F_c = empirical factor used in correlating gas displacement dewatering data, m^3/(kg·s)
K = shape correction factor for pore size
m_v = coefficient of volume change, Pa^{-1}
p_c = capillary pressure, Pa
p_s = effective (interparticle) pressure, Pa
Δp = pressure differential across cake–medium system applied during cake formation, or depth filtration, Pa
Δp_d = pressure differential applied during cake dewatering by gas displacement, Pa
P = bubble point pressure; maximum pressure
q = superficial filtrate velocity, m^3/(m^2·s)
Q = flow rate per unit area
r_c = radius of a capillary tube, m
R = total resistance to flow through a medium
R_c = resistance of the filter cake, m^{-1}
R_m = resistance of the filter medium, m^{-1}
s = coefficient of compressibility
t = time, s
u_{air} = superficial gas velocity measured at downstream pressure, m/s
V = cumulative filtrate volume
v = volume of filtrate obtained during cake formation per unit area, m

w = material coordinate, ie, mass of solids per unit area lying above the medium at various positions, kg/m^2
w_0 = total mass of solids contained in a filter cake per unit area, kg/m^2

Greek symbols

α = local specific resistance of filter cake, m/kg
α_0 = empirical factor
$\bar{\alpha}$ = average value of the specific resistance for a filter cake, m/kg
ϵ = porosity of a filter cake, ie, volume fraction of the pores, m^3/m^3
μ = dynamic viscosity, Pa·s
ϕ = contact angle, degrees
ρ_l = density of liquid phase, kg/m^3
ρ_s = density of solids, kg/m^3
σ = surface tension, N/m
θ_d = dewatering time, s
ϕ = liquid-to-membrane contact angle

BIBLIOGRAPHY

"Filtration" in *ECT* 1st ed., Vol. 6, pp. 506–530, by L. E. Brownell, The University of Michigan; "Filtration" in *ECT* 2nd ed., Vol. 9, pp. 264–286, by L. E. Brownell, The University of Michigan.

1. B. F. Ruth, *Ind. Eng. Chem.* **27,** 708 (1935).
2. R. S. Gale, *Water Pollut. Control* **66,** 622 (1967).
3. P. Kos and D. D. Adrian, *J. Sanit. Eng. Div. Am. Soc. Civ. Eng.* **101,** 947 (1975).
4. F. M. Tiller and J. R. Crump, *Chem. Eng. Prog.* **73** (11), 65 (1977).
5. F. M. Tiller and H. R. Cooper, *AIChE J.* **6,** 595 (1960).
6. F. M. Tiller and H. R. Cooper, *AIChE J.* **8,** 445 (1962).
7. M. Shirato and co-workers, *AIChE J.* **15,** 405 (1969).
8. R. J. Wakeman, *Filtration Post Treatment Processes,* Elsevier Scientific Publishing Co., Amsterdam, The Netherlands, 1975.
9. J. Bear, *Dynamics of Fluids in Porous Media,* American Elsevier Publishing Co., Inc., New York, 1972, pp. 439–519.
10. L. E. Brownell and D. L. Katz, *Chem. Eng. Prog.* **43,** 537, 601, 703 (1947).
11. P. J. Lloyd and J. A. Dodds, *Filtr. Sep.* **9,** 91 (1972).
12. D. A. Dahlstrom, *Chem. Eng. Prog.* **74**(3), 69 (1978).
13. R. J. Piros, R. A. Brusenback, and D. A. Dahlstrom, *Min. Eng.* **4,** 1200 (1952).
14. C. E. Silverblatt and D. A. Dahlstrom, *Ind. Eng. Chem.* **46,** 1201 (1954).
15. M. Shirato and co-workers, *J. Chem. Eng. Jpn.* **4,** 263 (1971).
16. M. Shirato and co-workers, *Filtr. Sep.* **7,** 277 (1970).
17. M. Shirato, T. Murase, and T. Aragaki, *DECHEMA Monogr.* **74,** 9 (1974).
18. M. Shirato, T. Murase, and N. Hayashi, *Proceedings, The Second Pacific Chemical Engineering Congress (PACHEC '77),* AIChE, New York, 1977, pp. 667–674.
19. M. Shirato and co-workers, *J. Chem. Eng. Jpn.* **10,** 154 (1977).
20. C. A. Jahreis, *Chem. Eng.* **70,** 237 (1963).
21. R. C. Baskerville and R. S. Gale, *Water Pollut. Control* **67,** 233 (1968).
22. A. Rushton and M. Hosseini, "Concentration and Velocity Effects in Filtration" in *Proceedings, International Symposium, 198th Event of the European Federation of Chemical Engineering, June 6–7, 1978, Antwerp, Belgium,* pp. 49–60.
23. C. E. Silverblatt and D. A. Dahlstrom, *Ind. Eng. Chem.* **46,** 1201 (1954).
24. W. Shoemaker, *AIChE Symp. Ser. 171,* **73,** 26 (1977).
25. D. B. Purchas, *Filtr. Sep.* **2,** 465 (1965).
26. R. C. French, *Chem. Eng.* **70** (21), 177 (1963).
27. A. S. Ward, "Filter Media Development and Innovation," *paper presented at Filtration Society Symposium, Oct. 3, 1972, Manchester, UK.*

28. R. L. Westerfield, "Food Filtration Applications Utilizing Dry Lay Nonwovens," *paper presented at Idea 78 Conference Workshop on Nonwovens in Filtration Sponsored by INDA, Rosemont, Ill., Oct. 23–26, 1975.*
29. A. C. Wrotnowski, "Final Filtration with Felt Bag Strainers," *paper presented at AIChE 84th National Meeting, Atlanta, Ga., Feb. 26–Mar. 1, 1978.*
30. F. B. Hutto, Jr., *AIChE Symp. Ser. 171* **73,** 50 (1977).
31. A. F. Alciatore and E. L. Neu, "Comparison of Diatomite and Perlite Filter Aids," *paper presented at 69th National Meeting AIChE, Cincinnati, Ohio, May 16–19, 1971.*
32. C. M. Thomas, *AIChE Symp. Ser. 171* **73,** 18 (1977).
33. *Dorr-Oliver Standard Test Procedure I-D-2,* Dorr-Oliver Inc., Stamford, Conn..
34. D. B. Purchas, ed., *Solid–Liquid Separation Equipment Scale-Up,* Uplands Press Ltd., Croydon, Eng., 1977.
35. *Instruction Manual Bomb Filter Filtration Tests,* Minerals Div., Grefco, Inc., Los Angeles, Calif.
36. R. C. Baskerville, A. M. Bruce, and M. C. Day, *Filtr. Sep.* **15,** 445 (1978).
37. C. E. Silverblatt, H. Risbud, and F. M. Tiller, *Chem. Eng.* **81,** 127 (1974).
38. D. A. Dahlstrom, *Chem. Eng. Prog.* **74**(3), 69 (1978).
39. C. Orr, *Filtration Principles and Practices,* Part I, Marcel Dekker, Inc., New York, 1977.
40. C. A. Jahreis, *Chem. Eng.* **83**(4), 80 (1976).
41. E. P. Austin, *Filtr. Sep.* **15,** 320 (1978).
42. A. F. Cassel and B. P. Johnson, "Evaluation of Dewatering Units to Produce High Sludge Solids Cake," *paper presented at 51st Annual Conference of WPCF, Oct. 2, 1978.*
43. C. Tien and A. C. Payatakes, *AIChE J.* **25,** (1979).
44. L. A. Spielman, "Particle Capture Mechanisms," *Symposium on Deposition and Filtration of Particles from Gases and Liquids at Loughborough University, Eng., Sept. 6–8, 1978,* Luton, Ltd.
45. C. S. Oulman, A. M. Saatci, and E. R. Baumann, *AIChE Symp. Ser. Water 1978* **75,** (1979).
46. A. Adin, E. R. Baumann, and J. L. Cleasby, *JAWWA* **71**(1), 17 (1979).
47. C. R. O'Melia and W. Stumm, *J. Am. Water Works Assoc.* **59,** 1393 (1967).
48. K. J. Ives, *Capture Mechanisms in Filtration: The Scientific Basis of Filtration,* Noordhoff-Leyden, The Netherlands, 1975.
49. C. Vreeken, P. M. Heertjes, and J. A. Wesselingh, "The Effect of Particle Coagulation on the Performance of Filter Beds," in ref. 44.
50. C. Tien, R. M. Turian, and H. Pendse, "Simulation of the Dynamic Behavior of Deep Bed Filters," *paper presented at 84th National Meeting AIChE, Atlanta, Ga., Feb. 1978.*
51. H. W. Ballew and co-workers, *Basics of Filtration and Separation,* Nuclepore Corp., 1978.
52. N. Nickolaus in ref. 24, pp. 38–49.
53. R. R. Read, *Filtr. Sep.* **16,** 79 (1979).
54. P. G. T. Evans, *Prod. Finish.* **43**(11), 20 (1972).
55. R. J. Akers and A. S. Ward in ref. 39, p. 191.
56. W. Kastle, "Economical Differences in Polymer Filtration with Different Filter Media," in ref. 22, pp. 203–210.
57. W. Meier, *Filtr. Sep.* **15,** 142 (1978).
58. F. M. Tiller and J. R. Crump, *Chem. Eng.* **84**(12), 183 (1977).
59. F. M. Tiller, *Filtr. Sep.* **15,** 204 (1978).
60. A. Bagdasarian and F. M. Tiller, *Filtr. Sep.* **15,** 594 (1978).
61. L. Casagrande, *Review of Past and Current Work on Electro-Osmosis Stabilization of Soils,* Harvard Soil Mech. Series, No. 45, Harvard University Press, Cambridge, Mass., Dec. 1953.
62. U.S. Pat. 894,070 (July 12, 1908), B. Schwerin.
63. U.S. Pat. 4,110,189 (Aug. 29, 1978), A. C. Kunkle, W. F. Abercrombie, Jr., and C. J. Akins, Jr.
64. M. S. Chronberg, *Ind. Ceram.* (706), 318, (May, 1977).
65. C. E. Curtis, *J. Am. Ceram. Soc.* **14,** 219 (1931).
66. F. E. Stuart, *Public Works N.Y.* **78**(4), 27, 36 (1947).
67. D. O. Herer and F. E. Woodard, *Tappi* **59**(1), 134 (Jan. 1976).
68. N. C. Lockhart, *Filtr. Sep.* **15,** 211 (May–June 1978).
69. U.S. Pat. 3,773,640 (Nov. 20, 1973), L. M. Burrage.
70. R. H. Sprute and D. J. Kelsh, *Electrokinetic Consolidation of Slimes in an Underground Mine, RI8190* U.S. Bureau of Mines, Washington, D.C., 1976.

General References

R. J. Wakeman, *Filtration Post Treatment Processes,* Elsevier Scientific Publishing Co., Amsterdam, The Netherlands, 1975.
D. B. Purchas, ed., *Solid–Liquid Separation Equipment Scale-up,* Uplands Press Ltd., Croyden, Eng., 1977.
A. Tomiak and M. A. Lanzon, *Tappi* **60**(9), (1977) (has an extensive bibliography).
E. B. Fitch, "Countercurrent Filtration Washing," *Chem. Eng.* **69,** 119 (1962).
J. Hermia and D. Taeymans, "Considerations on the Choice of a Washing Process," *Proc. Intl. Symp. 198th Event of the European Federation of Chemical Engineering, June 6–7, 1978, Antwerp, Belgium.*
G. R. S. Smith, "How to Use Rotary Vacuum Precoat Filters," *Chem. Eng.* **83**(4), 84 (1976).
Filtration Leaf Test Procedures, Dorr-Oliver Inc., Stamford, Conn., 1979.
Filtration Manual No. 95, Dorr-Oliver Inc., Stamford, Conn., 1979.
Celatom Diatomite Filter Aids, Eagle-Picher Industries, Inc., Cincinnati, Ohio.
Kenite Diatomite, Kenik Corp. Div., Witco Chemical, New York.
Celite Filter Aids, Johns-Manville Corp., Denver, Col.
Diatomite, Minerals Div., Grefco, Inc., Los Angeles, Calif.
Perlite, Minerals Div., Grefco, Inc., Los Angeles, Calif.
Solka-Floc Filter Aids, Minerals Div., Grefco, Inc., Los Angeles, Calif.
AIChE Equipment Testing Procedures on Batch Pressure Filters, AIChE, N.Y., 1967.
C. Orr, *Filtration Principles and Practices,* Part I, Marcel Dekker, Inc., New York, 1977.
R. H. Perry and C. H. Chilton, *Chemical Engineers' Handbook,* 5th ed., McGraw Hill-Book Co., New York 1973.
D. B. Purchas, *Industrial Filtration of Liquids,* Leonard Hill, London, Eng., 1967; (extensive equipment descriptions).
L. Svarovsky, *Solid–Liquid Separation,* Butterworths, London, Eng., 1977.
J. M. Chalmers, L. R. Elledge, and H. F. Porter, *Chem. Eng.* **62**(6), 191 (1955).
J. E. Flood, H. F. Porter, and F. W. Rennie, *Chem. Eng.* **73**(13), 163 (1966).
R. C. Emmett and C. E. Silverblatt, "When to Use Continuous Filtration Hardware," *Filtr. Sep.* **12,** 577 (1975).
J. M. Regan, "Filter Presses—Still Alive and Kicking," *Filtr. Sep.* **14,** 485 (1977).
J. Bouvier, "Selection of SLS Methods as a Function of Liquid and Process Characteristics: the Fermentation Broth in Antibiotics Industry" in ref. 22, pp. 171–189 (good bibliography).
S. M. Moos, and R. E. Dugger, "Vacuum Filtration: Available Equipment and Recent Innovations," *paper presented at SME-AIME meeting at Lake Buena Vista, Florida, Sept. 11–13, 1978.*
M. G. Howard, "The Economics of Horizontal Leaf Pressure Filter Design and Operation," *Filtr. Sep.* **16,** 150 (1979).
R. F. Madsen, "Vacuum Pressing, a New Unit Operation for the Production of Low-Moisture-Content Filter Cake," *paper presented at A.S.S.B.T. Meeting,* San Diego, Calif., Feb., 1978.
N. W. Dyson "Cartridge Filtration for Processing Edible Oils," *Filtr. Sep.* **16,** 166 (1979).
Bag Filters, bulletins from American Felt & Filter Co., Carborundum FSI.
G. Burton, "A New Process for Reducing the Moisture Content of Filter Cake," *paper presented at the 4th International Coal Preparation Congress, Eng., 1962.*
C. S. Simons and D. A. Dahlstrom, "Steam Dewatering of Filter Cakes," *Chem. Eng. Prog.* **62**(1), 75 (1966).
R. C. Emmett, Jr. and D. A. Dahlstrom, "Steam Drying of Filter Cake," *Chem. Eng. Prog.* **68**(1), 51 (1972).
E. R. Cunningham, "Strainers, Screens, Filters, and Separators—Mechanical Methods for Removing Solids from Process Liquids," *Plant Eng.* **31**(24), 80 (1977).
B. Schiele and C. Alt, "Dynamic Pressure Filtration of a Highly Dispersed Suspension by Means of Microporous Membranes," in ref. 22, pp. 75–88; (excellent bibliography on filter-thickening).

ROBERT M. TALCOTT
CHARLES WILLUS
M. P. FREEMAN
Dorr-Oliver Incorporated

FINE ART AUTHENTICATION AND PRESERVATION. See Supplement Volume.

FINE CHEMICALS

Fine chemicals generally are considered to be chemicals that are manufactured to high and well-defined standards of purity, as opposed to *heavy chemicals* which are considered as being made in large amounts to technical levels of purity. Although fine chemicals usually are thought of as being produced on a small scale, this is by no means always the case. The production of some fine chemicals is in tens or hundreds of kilograms per year, others are in thousands of metric tons (especially fine chemicals used as drugs or food additives). For example, annual production in the United States of aspirin is of the order of 14,000 t.

This article discusses standard-setting for drug, food additive, laboratory reagent, and electronic chemicals, and chemical standards used in analytical procedures. In Table 1 are standards for sodium hydroxide, ranging from the technical grade through the drug, food, and reagent grade, to the electronic grade. The advances in purity represented by these various grades of chemicals are based upon the special uses of the chemicals.

Standards for drugs are established by the United States Pharmacopeial Convention, Inc. (USPC), and have been published in 20 revisions of the *United States Pharmacopeia* (USP). In the past, standards for many drugs that were not in the USP were established by the American Pharmaceutical Association, and were published in the *National Formulary* (NF). The last edition of the NF, *NF XIV* (1), was published in 1975. In 1974 the USPC acquired the NF and the continuing revisions of *USP XIX* and *NF XIV* were published in consolidated annual supplements. *USP XX* and *NF XV* now have been published in a combined volume (hereafter referred to as USP)

Table 1. Typical Specifications for Various Grades of Sodium Hydroxide, Wt %

	Technical	NF	FCC	Reagent	Electronic
assay (NaOH)	94.0 min	95.0 min	95.0 min	97.0 min	98.0 min
Na_2CO_3		3.0 max	3.0 max	1.0 max	0.4 max
heavy metals (as Pb)		0.003 max	0.003 max		
heavy metals (as Ag)				0.002 max	0.001 max
lead (Pb)			0.001 max		
mercury (Hg)			0.00001 max	0.00001 max	0.00001 max
arsenic (As)			0.0003 max		
potassium (K)		to pass test		0.02 max	0.01 max
iron (Fe)				0.001 max	0.0003 max
nickel (Ni)				0.001 max	0.0005 max
ammonium hydroxide ppt				0.02 max	0.02 max
chloride (Cl)				0.005 max	0.001 max
sulfate (SO_4)				0.003 max	0.0005 max
phosphate (PO_4)				0.001 max	0.0002 max
nitrogen compounds (N)				0.001 max	0.0003 max

(2). In this compendium, drug substances, and dosage forms of drug substances, are designated USP, and pharmaceutic ingredients are designated NF. The latter substances are used to make the active ingredient(s) into a suitable dosage form for use by the patient.

Standards for food-grade chemicals in the United States are set by the Committee on Food Chemicals Codex of the NAS which publishes them in the *Food Chemicals Codex* (FCC) (3) (see Pharmaceuticals).

Standards for laboratory reagents are set by the ACS Committee on Analytical Reagents, and are published in *Reagent Chemicals—ACS Specifications* (4).

Standards for electronic-grade chemicals are quite new, and a number have recently been published in *The 1979 Book of SEMI Standards* (5).

The publications mentioned above include both specifications and methods of analysis for the substances for which they set standards. The establishment of standards of quality for chemicals of any kind presupposes the ability to set numerical limits on physical properties, allowable impurities, and strength, and to provide the test methods by which conformity to the requirements may be demonstrated. Published tests are considered applicable only to the specific requirements for which they were written. Modification of a requirement, especially if the change is toward a higher level of purity, often necessitates revision of the test to ensure its validity.

One of the greatest tasks of those who provide compendial standards is to obtain, adapt, or develop test methods for determining compliance with the standards. Such methods must be capable of routine use in many laboratories with different personnel and equipment. There is a vast difference between a method that can be used in one laboratory by specialists, and one that can be used in many laboratories to determine whether chemicals pass or fail the established specifications. Such determination must be reliable, because the results obtained may determine whether a product is safe, or legal.

There are no established specifications for the standard reference samples used in general chemical analysis. Many such substances, however, are analyzed and certified by the NBS. Specific reference standards are required for many of the analyses included in the USP and NF specifications for drugs. The USP and the NF are recognized in the *Food, Drug, and Cosmetic Act* as establishing legal drug standards that the FDA is responsible for enforcing. The FCC is recognized by FDA regulations as defining food grade. The ACS reagent specifications have no special legal status per se, but they are used by the USP, the FCC, and in government procurement, and are referenced in FDA regulations, and in ASTM methods.

In recent years federal regulation of drugs, food additives, and reagents used in *in vitro* diagnostic tests has increased markedly. With increasing consumer concern about the safety of fine chemicals, especially those used in foods and drugs, the United States government has become increasingly sensitive to the manner in which standards are set. Freedom of information legislation has confirmed the public's right to know, and this has introduced the objective of due process into the development of standards. The USP and the FCC have mechanisms that make possible public participation in setting the standards with which they are involved. The ACS does not have such a mechanism, but industry, government, and academia are represented on the reagent committee.

Fine chemicals are produced by a wide spectrum of manufacturers, largely because the distinction between different kinds of chemicals is not sharp. There are specialty producers of fine chemicals. Many companies that manufacture drugs in dosage form also manufacture the chemical drug substances that are used in the dosage forms. A number of companies manufacture drug chemicals and food chemicals. Some fine chemicals are made by manufacturers of heavy chemicals, and either may be simply a segment of their regular production, or some of that production which has been subjected to additional purification steps. Many fine chemicals are imported into the United States from countries such as Japan, the Federal Republic of Germany, and the Netherlands.

Drug Chemicals

Standards for drug chemicals are published, as has already been mentioned, in USP–NF. Examples of drug substances are: acetaminophen [*103-90-2*], ampicillin [*69-53-4*], aspirin [*50-78-2*], powdered ipecac, riboflavin [*83-88-5*], stannous fluoride [*7783-47-3*], and thyroid. Examples of pharmaceutical ingredients are: ethylparaben [*120-47-8*], lactose [*63-42-3*], lanolin [*8006-54-0*], magnesium stearate [*557-04-0*], sodium hydroxide [*1310-73-2*], starch [*9005-25-8*], and vanillin [*121-33-5*].

Drug substances are chemicals that have therapeutic or diagnostic uses, whereas pharmaceutical ingredients provide preservative action, flavoring, or fulfillment of a function in the formulation of dosage-form drugs. Dosage forms include acetaminophen tablets (see Analgesics), ampicillin capsules (see Antibiotics), aspirin tablets (see Salicylic acid), ipecac syrup (see Expectorants), riboflavin injection (see Vitamins), and thyroid tablets (see Hormones). Stannous fluoride (see Tin compounds) is used in fluoride-containing toothpastes (see Dentifrices). Ethylparaben (see Salicylic acid) is a preservative, lactose (see Sugars) is a tablet filler, lanolin (see Wool) is an ointment base, magnesium stearate (see Carboxylic acids) is a tablet lubricant, sodium hydroxide (see Alkali and chlorine products) is a neutralizing agent, starch (qv) is a tablet binder, and vanillin (qv) is a flavor (see also Flavors and spices).

The United States Pharmacopeial Convention (USPC). The USP was established in 1820 by the assembly of the first United States Pharmacopeial Convention (USPC) in Washington, D.C. Among the objectives of the meeting was the inclusion in a pharmacopeia of those drugs most fully established and best understood and standards of pharmaceutical quality for them.

A brief history of the USPC is included in the front of *USP XX–NF XV* (12). An excellent discussion of the USPC is given in ref. 6, and recent descriptions are available in refs. 7 and 8.

The early USPC was dominated by physicians, and they were the people who selected the best drugs. This prevented inclusion in the USP of a large number of drugs that were widely used, particularly elixirs, a popular dosage form in the late 19th century. To fill this gap, the American Pharmaceutical Association, in 1888, published the first NF, which provided standards for drugs in wide use but not included in the USP.

When the *Federal Food and Drug Act* was passed in 1906, both the USP and the NF were recognized in the act as setting the legal standards of strength, quality, and purity for the drugs that were described in them. The publications of USP and the NF between them never included standards for all drugs manufactured in the United States. Since the acquisition of the NF, the USP has changed its policy so that its objective now is to provide names and standards of pharmaceutical quality for all

United States drugs. The inclusion of such standards for all drugs in the USP will be done as rapidly as possible.

From the perspective of modern science, the tests described a hundred years ago in the compendia were simple analyses. They could be performed by a pharmacist with a minimum of laboratory equipment. Present-day tests employ complex instrumentation such as uv and ir spectrophotometers, various kinds of chromatography, x-ray, nmr, and atomic absorption (see Analytical methods). Thus they are no longer performed by pharmacists but are largely used by manufacturers and the regulatory agencies (eg, the FDA and comparable state agencies).

The USP Committee of Revision. The work of wholly revising the USP every five years and making revisions by supplement when needed is performed by the 70-person Committee of Revision. The committee members represent the fields of medicine, pharmacy, chemistry, microbiology, etc.

The committee members, supported by the Drug Research and Testing Laboratory, write the general chapters and product monographs. The present *USP XX–NF XV* contains monographs for 2510 substances.

The USP Book. As has been noted earlier, in addition to defining the quality of the reagents used in testing, the USP sets the legal standards of strength, quality, purity, and requirements for packaging and labeling for the articles they include. Thus much of the text in the USP represents enforceable legal requirements.

The main sections of the book are as follows:

The General Notices. These are the basic requirements for the application and interpretation of the tests and specifications that follow in the USP. They define many of the terms used in the text, and answer the majority of procedural questions that may arise with the monograph of each substance.

The Monographs. A separate monograph is provided for each substance included in the USP. Each monograph describes a substance and includes general physical properties.

In the past the monograph has included nonenforceable information such as the category of use, the usual dose, and sizes available. However, starting with *USP XX*, this dispensing information has been greatly expanded, and has been published in a book separate from the USP quality standards. This dispensing information is designed to be of assistance to the physician, the pharmacist, and the patient. It is particularly designed to guide the pharmacist in advising those who take the dispensed drug of precautions that may be necessary, possible adverse effects, and other information the patient should report to the prescriber.

The body of the monograph includes a statement of assay, followed by a packaging and storage statement, a labeling statement where applicable, identification tests, and then a variety of tests to establish the strength, quality, and purity of the drug. The tests either provide the methods to be followed, or refer to other tests which are in the General Tests section of the book.

The General Tests. This section of the USP gives methods for tests that are general in nature, and which apply to a number of substances; thus repetition is avoided in the monographs. Tests are included for such characteristics as heavy metals, melting point, chloride, sulfate, sterility, pyrogens, etc. Also included are descriptions of various analytical techniques, such as spectrophotometry and chromatography, and descriptions of tests to be used on glass or plastic containers, rubber closures, etc.

Reagents, Indicators, and Solutions. This section includes the specifications and testing methods for reagents to be used in the tests specified in the USP, and directions for making the various indicator, buffer, colorimetric, test, and volumetric solutions used in the testing. Reagents for which ACS specifications exist are referenced to the ACS book.

Reference Standards. Many of the identification tests and assays, in particular, described in USP (and NF) monographs require the use of reference standards. The USP provides reference standard material for all of the USP (and NF) tests that call for their use.

The Standard-Setting Process. Setting USP standards is a continuing, and by no means unilateral, process. The Committee of Revision not only develops monographs for new substances but also is continually reviewing the monographs, specifications, and testing methods for existing substances. The results of the continuous review are published in annual supplements. A complete review of the entire book is done every five years and a revision is published.

When a new substance is admitted to the USP, the first step is to locate producers and to obtain their specifications and testing methods. The person to whom the monograph is assigned examines these, decides on the appropriate specifications, writes the monograph in USP language and style, and submits it to USP headquarters. If there are no objections, the new monograph is published by the USP for comment in *Pharmacopeial Forum* which is published every other month, and includes proposed changes in existing monographs, the general notices, the general tests, or other sections of the USP, proposed new monographs, and proposals relative to policy or philosophy. The public may comment, protest, or make suggestions, and their views receive due consideration. If significant revisions are necessary after this process, a revised proposal is published. Through *Pharmacopeial Forum,* USP standards are openly arrived at.

Food-Additive Chemicals

The FCC is to food-additive (qv) chemicals what the USP is to drugs. In fact, many chemicals that are used in drugs also are food additives, and thus may have a drug monograph in the USP, and a food-additive monograph in the FCC. Examples of food-additive chemicals are: ascorbic acid [*50-81-7*] (see Vitamins), butylated hydroxytoluene [*128-37-0*] (see Antioxidants), calcium chloride [*10043-52-4*] (see Calcium compounds), ethyl vanillin [*121-32-4*] (see Vanillin), ferrous fumarate [*7705-12-6*] (see Iron compounds), ferrous sulfate [*7720-78-7*] (see Iron compounds), niacin [*59-67-6*] (see Vitamins), sodium chloride [*7647-14-5*] (see Alkali and chlorine products), sodium hydroxide [*1310-73-2*] (see Alkali and chlorine products), sodium phosphate, dibasic [*7558-79-4*] (see Phosphoric acids and phosphates), spearmint oil [*8008-79-5**] (see Oils, essential), tartaric acid [*133-37-9*] (see Hydroxy dicarboxylic acids), tragacanth [*9000-65-1*] (see Gums), and vitamin A [*11103-57-4*] (see Vitamins).

Although the USP predates by many decades the *Food, Drug, and Cosmetic Act,* positive action to compile standards for food-grade chemicals did not take place until the years following enactment of the *Food Additive Amendment* to the Act in 1958 (9). This amendment stated that substances added to foods should be of food-grade quality, but it contained no criteria by which such quality could be determined (see also Colorants for food, drugs, and cosmetics). The Food Protection Committee of the NAS-NRC undertook the project of producing a Food Chemicals Codex.

The objective of the FCC is to define food-grade chemicals in terms of the characteristics that establish their identity, strength, and quality. It provides specifications in monograph form for some 650 food additives, together with analytical test procedures by which compliance with the specifications can be determined. The third edition is being published in 1980.

Food Chemicals Codex specifications were recognized in 1971 by the FDA as establishing food grade under the terms of *Title 21, Code of Federal Regulations.* FCC specifications have also been adopted by the governments of Australia, Canada, the UK, and New Zealand. There is extensive international activity in the field of food additives. The FCC is represented at meetings of the Joint FAO/WHO Expert Committee on Food Additives and the Food Additive Commission of the IUPAC.

The Food Chemicals Codex. The committee on codex specifications has the responsibility for the development and revision of the FCC. It relies upon the expertise and laboratory work of its members in setting methods and specifications. It also contacts manufactures, trade associations, etc, to obtain their comments and criticisms of monographs proposed by the committee.

Reagent Chemicals

Reagent Chemicals—ACS Specifications, now in its 6th edition (4), is to reagent chemicals what the USP and the FCC are to drug and food-additive chemicals. The ACS Committee on Analytical Reagents, and its activity relative to specifications for reagents, has a history dating back to 1917, and more tenuously, to 1903 (10). Examples of reagent chemicals are acetone [*67-64-1*] (qv), arsenic trioxide [*1327-53-3*] (see Arsenic compounds), barium chloride [*10361-37-2*] (see Barium compounds), bromine [*7726-95-6*] (qv), bromothymol blue [*76-59-5*] (see Hydrogen-ion concentration), cupferron [*135-20-6*] (see Copper; Iron), ethyl ether, anhydrous [*60-29-7*] (see Ethers), hexanes (see Hydrocarbons), hydrochloric acid [*7647-01-0*] (see Hydrogen chloride), perchloric acid, 70% [*7601-90-3*] (see Chlorine oxygen acids and salts), silver diethyl dithiocarbamate [*38351-46-1*] (see Silver compounds), silver nitrate [*7783-99-5*] (see Silver compounds), sodium hydroxide [*1310-73-2*] (see Alkali and chlorine products), and uranyl acetate [*541-09-3*] (see Uranium).

The ACS Committee on Analytical Reagents, made up of 20 members representing industry, government, and academia (both manufacturers and users of reagents), meets twice a year, and carries on much of its work by correspondence.

Requirements and details of tests are based on published work, on the experience of committee members in the examination of reagent chemicals, and on studies of test procedures made by committee members.

When a specification for a reagent is first prepared, it generally is based on the highest purity level that is competitively available in the United States. If a higher level of purity subsequently becomes available on a competitive basis, the specification is revised accordingly. An exception to this approach is a need for higher purity in a reagent than is competitively available. In this case the committee sets the standard based on need, presumably stimulating manufacturers to meet that need.

Although the book on reagent chemicals contains many tests for the determination of trace impurities in reagents, it is not intended to be a text on the techniques of trace analysis but rather to provide tests that are reproducible in various laboratories, and which are accurate, economic, and feasible (see Trace and residue analysis).

The usual reagents suffice for many purposes but the committee recognizes uses for reagents that require a higher purity than that defined by existing ACS specifications. Sometimes reagents must be further purified with respect to one or more impurities. However, when there is a reasonably widespread use, the committee may define a special grade, such as "Suitable for Use in Spectrophotometry."

The committee takes into account the practical matters involved in the manufacture of reagents. In general, the higher the purity requirement, the greater the cost. In the combination of high-purity requirements with low usage, economics may become an important factor.

The ACS Book. The ACS book, *Reagent Chemicals—ACS Specifications,* establishes a standard of quality for reagents to be used in precise analytical work, for which purpose it contains both specifications and testing methods for over 250 reagent chemicals.

The book is similar to the two compendia already described. It is published approximately every 5 years, and supplements are published between editions of the book. In addition, notices of changes that need to be publicized after their adoption at meetings of the committee are published promptly in the ACS journal, *Analytical Chemistry*.

The Standard-Setting Process. The committee has three main lines of endeavor—the improvement of existing limits for impurities, the improvement of present test methods, and the development of specifications and testing methods for additional compounds.

The ACS committee has a general rule: if two manufacturers meet a given specification, the specification is so defined. Usually the methods are checked by several laboratories with samples from various sources. Because of the many different matrices in which impurities are determined, even the simplest impurity tests need to be checked for accuracy in the laboratory.

ACS reagents are specified for tests in the USP and the FCC, and in many ASTM methods of analysis. ACS specifications also are used by government agencies for procurement purposes, and are mentioned in FDA regulations pertaining to the labeling of *in-vitro* diagnostic reagents. Although they do not have legal status, as is the case with USP and FCC specifications, they are generally recognized, in the United States and elsewhere, as *the* standards for reagents.

Most of the common reagent chemicals already are covered by ACS specifications. However, as tests become more highly instrumented and complex, it becomes difficult to write instructions that cover them, and to obtain agreement on tests and results among laboratories. It also may be necessary to develop tests for specialized uses for existing reagents, eg, tests on selected solvents to assure that they are satisfactory for spectrophotometric use or for use in the determination of pesticide residues by glc. Large quantities of such solvents are used to extract pesticides from natural products. This means that the solvents themselves must show little or no response to the test, if the reagents are not to overwhelm the substance sought. Testing of such solvents involves a many-fold concentration of possible impurities, followed by a glc examination for peaks that would interfere with the analysis for pesticide residues (see Analytical methods).

Electronic Chemicals

Although chemicals used in the electronics industry frequently are required to be of especially high purities or are made to exacting physical specifications, there is no complete compendium of standards. The requirements have been set by agreement between the user and the supplier on an item by item basis. However, in the case of certain solvents and acids and bases used in processing semiconductors this was done by specifying ACS reagent grade plus additional requirements for heavy metals and/or other trace elements. Examples of electronic chemicals are: acetic acid, glacial [*64-19-7*] (qv), acetone [*67-64-1*] (qv), ammonium fluoride [*12125-01-8*] (see Ammonium compounds), ammonium hydroxide [*1336-21-6*] (see Ammonium compounds), dichloromethane [*75-09-2*] (see Chlorocarbons), hydrofluoric acid [*7664-39-3*] (see Fluorine compounds, inorganic), hydrogen peroxide, 30% [*7722-84-1*] (qv), methanol [*67-56-1*] (qv), nitric acid [*7697-37-2*] (qv), 2-propanol [*67-63-0*] (see Propyl alcohols), sodium hydroxide [*1310-73-2*] (see Alkali and chlorine products), tetrachloroethane [*127-18-4*] (see Chlorocarbons), toluene [*108-88-3*] (qv), and xylenes (qv).

The ASTM has published a few selected specifications for materials used in the electronics industry, such as gold wire for semiconductor lead bonding, but it does not provide a comprehensive set of standards (see Electrical connectors).

The SEMI Book. In 1970 the Semiconductor Equipment and Materials Institute (SEMI) was founded as a nonprofit organization of firms, corporations, and individuals engaged in supplying fabrication equipment, materials, or services to the semiconductor and microelectronics industries (see Integrated circuits; Liquid crystals; Semiconductors). This organization recognized the need for minimum standards of quality for key materials and advanced electronic devices. The Subcommittee on Standards for Chemicals began its work in 1975, and now has published specifications and analytical procedures for a number of acids and organic solvents, as well for ammonium hydroxide and hydrogen peroxide in *The 1979 Book of SEMI Standards* (5). Products meeting these requirements can be described as "meeting SEMI specifications."

Chemical and Other Standards Used in Analysis

National Bureau of Standards (NBS). The NBS is the source of many of the standards used in chemical and physical analysis in the United States, and throughout the world. The standards prepared and distributed by the NBS are used to calibrate measurement systems and to provide a central basis for uniformity and accuracy of measurement. At present, nearly 900 Standard Reference Materials are available and are described in *NBS Special Publication 260,* which can be obtained from the Superintendent of Documents (11). Included are many steels, nonferrous alloys, high-purity metals, primary standards for use in volumetric analysis, microchemical standards, clinical laboratory standards, biological material certified for trace elements, environmental standards, trace element standards, ion-activity standards (for pH and ion-selective electrodes), freezing and melting point standards, colorimetry standards, optical standards, radioactivity standards; particle-size standards, and density standards. Certificates are issued with the standard reference materials showing values for the parameters that have been determined.

Some of the standards are fine chemicals in themselves, and others, such as filters for checking spectrophotometers, are of interest in the testing and control of fine chemicals (see Optical filters).

United States Pharmacopeia (USP). Reference standards are required in many USP and NF tests, and in a few FCC tests. The USPC distributes about 730 such standards. The USP reference standards include several for melting points, and also specimens of narcotics and other controlled substances.

Food and Drug Administration (FDA). Reference standards for biologicals are prepared and distributed by the Bureau of Biologics of the Food and Drug Administration.

Impact of the *Food, Drug, and Cosmetic Act* on Fine Chemicals

FDA Quality Standards. Although standards for many drugs are included in the USP, and for many food additives in the FCC, the FDA also establishes some specifications of its own.

In the drug field, specifications and testing methods for biologicals and antibiotics currently are set by the FDA. Also specifications and testing methods are prescribed for colorants. Many food additive (qv) petitions are granted with the requirement that certain specifications are met.

Device Legislation. Regulations covering medical devices define reagents used in *in vitro* diagnostic tests as devices (see Medical diagnostic reagents; Prosthetic and biomedical devices).

Regulations Concerning Good Manufacturing Practice. Chemicals that are drugs, as defined in the *Food, Drug, and Cosmetic Act,* are subject to the requirement of the act that they be made under conditions of Current Good Manufacturing Practice (CGMP). Specific GMP regulations for such chemicals have not been published but the regulations which have been published for dosage form drugs include many points which should be considered (12).

In general, GMP requirements introduce more record-keeping and paperwork into the manufacturing process than has been typical in most chemical plants. The primary thrust of GMP is that it is not enough merely to make chemicals to meet USP or other applicable specifications. The chemicals must be made under clean and sanitary conditions, procedures and processes must be documented, and processing and packaging must be carried out under conditions that preclude mixup and mislabeling. Records must be kept of complaints, and the manufacturer must know enough about the storage properties of the products to specify storage conditions, and if necessary, expiration dates on the label.

The effects of these regulations on drug chemical producers should not be large. However, a manufacturer of drug chemicals is required to register with the FDA, and is subject to FDA inspection at least once every two years. Some manufacturers who make chemicals that incidentally are drugs are impelled to drop the drug designation from their labeling in order to avoid the exposure to inspection that registration entails.

Food-additive chemicals are subject to food GMPs which mainly deal with sanitation (13). They are not subject to GMPs similar to those of drugs. However, guidelines similar to those mentioned above for drug chemicals can be applied with little modification to food-additive chemicals.

Reagent chemicals per se are not subject to any GMP regulations because normally they are not considered to be drugs or devices. However, *in vitro* diagnostic reagents are considered to be devices, and many reagent chemicals can be considered to be *in vitro* diagnostic reagents. Such reagents, to which the *in vitro* diagnostic regulations apply, are subject to the GMPs for *in vitro* diagnostic products (14). Thus GMP may impinge upon certain reagents, unless the manufacturer specifically declares that they are not *in vitro* diagnostic reagents, and does not sell them for this use (15).

Since electronic chemicals are not drugs, food additives, or devices, they are not subject to any GMP regulations.

BIBLIOGRAPHY

1. *The National Formulary XIV,* The American Pharmaceutical Association, Washington, D.C., 1975.
2. *The United States Pharmacopeia XX, (USP XX–NF XV),* The United States Pharmacopeial Convention, Inc., Rockville, Md., 1980.
3. *Food Chemicals Codex,* 3rd ed., National Academy of Sciences–National Research Council, Washington, D.C., 1980.
4. *Reagent Chemicals—American Chemical Society Specifications,* 6th ed., American Chemical Society, Washington, D.C., 1980.
5. *The 1979 Book of SEMI Standards,* Semiconductor Equipment and Materials Institute, Mountain View, Calif., 1979.
6. L. C. Miller and others, *Anal. Chem.* **32,** 19A (1960).
7. A. Hecht, *FDA Consumer,* 24 (Oct. 1977).
8. W. M. Heller, *Hosp. Formul.* **12,** 883 (1977).
9. *Public Law No. 618,* 86th U.S. Congress, Washington, D.C., 1960.
10. S. M. Tuthill, *Anal. Chem.* **42,** 30A (1970).
11. *NBS Special Publication 260,* 1979 ed., National Bureau of Standards, Washington, D.C., 1975.
12. *Title 21, Code of Federal Regulations,* Parts 210 and 211, Government Printing Office, Washington, D.C.
13. *Ibid.,* Part 110 (Sections 110.1–110.99).
14. *Ibid.,* Part 820 (Sections 820.1–820.198).
15. M. C. Bracken and E. J. Thompson, "Chemical Substances Legislation" in N. I. Sax, ed., *Dangerous Properties of Industrial Materials,* 5th ed., Van Nostrand, New York, 1979.

SAMUEL M. TUTHILL
JOHN A. CAUGHLAN
Mallinckrodt, Inc.

FIREBRICK. See Refractories.

FIRE CLAY. See Refractories.

FIRE EXTINGUISHING AGENTS. See Plant safety.

FIRE PREVENTION AND EXTINCTION. See Plant safety.

FIRE-RESISTANT TEXTILES. See Flame-retardants for textiles.

FIREWORKS. See Pyrotechnics.

FISCHER-TROPSCH PROCESS. See Fuels, synthetic.

FISH LIVER OILS. See Aquaculture.

FISH OILS. See Fats and fatty oils.

FLAME PHOTOMETRY. See Analytical methods.

FLAMEPROOFING. See Flame retardants.

FLAME RETARDANTS

AN OVERVIEW

Most of the immediate surroundings are combustible—clothes, furnishings, and much of the construction material in dwellings and work places, the interiors of cars, buses, airplanes, etc. Wood, paper, textiles, and synthetic polymers all burn under the right conditions; many burn vigorously and ignite readily. Humans have always been plagued by unwanted fire. So it is not surprising that they have sought ways to reduce the combustibility of their surroundings. Early chemists tried a variety of concoctions to reduce the flammability of textiles (1). By the 19th century enough was understood to enable Gay-Lussac to make a systematic study of available chemicals. He found that ammonium salts of sulfuric, phosphoric, and hydrochloric acids were effective as well as certain mixtures of these with borax. Later the dye chemist Perkin discovered that certain complex heavy metal ions (stannates and tungstates) helped make the treatment with ammonium salts water-resistant. In the 1930s the effect of mixing antimony oxide with organic halogen compounds was discovered. These three efforts represent the major discoveries on which modern flame-retardant chemicals are based. The technology has become far more sophisticated in recent years but for the most part it represents variations on these earlier themes.

Terminology

The application of such terms as fireproof, flameproof, self-extinguishing, nonburning, noncombustible, etc, has often led to ambiguity about the relative flammabilities of different materials. These ambiguities eventually led the Federal Trade Commission to take action in the case of cellular plastics and to restrict the use of such terminology (2). This action, in addition to the prohibition placed on the use of certain terminology, requires the use of a caveat whenever the results of burning tests are cited. Much of the older literature, as well as some of the recent publications, uses this restricted terminology.

Fire retardant (*flame retardant*) is used to describe polymers whose basic flammability has been reduced by some modification as measured by one of the accepted test methods.

Fire-retardant chemical is used to denote a compound or mixture of compounds that when added to, or incorporated chemically into a polymer serves to slow up or hinder the ignition or growth of fire. A flame retardant chemical technically is one that has the foregoing effect primarily in the vapor phase.

Terms such as self-extinguishing, nonburning, and noncombustible are no longer used by the fire research community because they have been found to be misleading.

Means of Achieving Fire Resistance

The materials to be made resistant to fire are invariably polymers—natural and synthetic. Some are inherently harder to ignite and burn than others. Graphite is very difficult to burn, whereas nitrocellulose burns explosively. One way to reduce fire hazard is to use polymers especially tailored to resist burning. By making a polymer rich in ring structures and low in readily oxidizable side groups, low flammability may be achieved (3). An example is an aromatic polyimide that contains no hydrogen atoms:

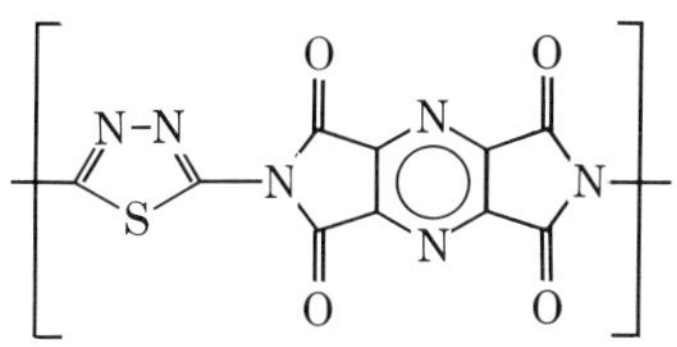

The aromatic polyimides have excellent fire resistance, but they are expensive and therefore are limited to specialty uses such as aerospace applications. Some of the many candidate aromatic polymers will gradually enter markets now held by more flammable polymers treated with flame-retardant chemicals (see also Aramid fibers; Novoloid fibers).

The more common approach to obtain fire resistance is to add one or more of the commonly used fire-retardant elements to a flammable polymer system. This may be done either with an inert additive (eg, surface treatments of fibers and films or unreactive admixtures to molding or extrusion mixes) or by copolymerizing a moderate amount of a monomer containing the requisite element. Occasionally, a polymer can be treated reactively after polymerization as in chlorinating polyethylene (see Olefin polymers).

As shown in the following articles, there are many ways to incorporate the elements antimony, boron, chlorine, bromine, phosphorus, and mixtures of these into polymer systems, sometimes with other metals or with certain nitrogen compounds. The concerns of the chemist are many as he or she formulates the system (eg, performance in the flammability test, cost, and ease of processing). Often the most difficult task is to achieve the desired safety level and to retain the functional properties of the parent polymer, properties for which the system is sold. Achieving good fire resistance often requires 10–20 wt % of the fire-retardant element in the compound containing that element. At such a high level the effect on polymer properties is substantial and considerable ingenuity is required to maintain functional performance. Thus, for most systems the fire retardant must be carefully tailored; for many the result is a compromise between safety and functional performance. Sometimes the margins are small, eg, the fire retardants for ABS resins (4). ABS is processed at ca 250°C, a temperature at which many fire-control chemicals are unstable (see Acrylonitrile polymers). Bromine compounds are reasonably effective in ABS but stability at 250°C limits the choice to aromatic bromine compounds. These are generally less effective fire retardants than aliphatic bromine compounds but the aliphatics do not survive the processing. A compromise must be struck, ie, a higher level of the more stable aromatic bromine compound must be used thereby increasing the cost and the effects on polymer properties.

Requirements For Fire Resistance

Nearly all markets for fire-retardant chemicals are created by fire regulations. The markets are as varied as the regulations. Textile products are covered by federal standards for apparel, carpets, and mattresses, as well as by local building codes for textile coverings used in places of public assembly (see Textiles). The building codes control the flammability of wall and ceiling coverings for the public parts of most buildings; in some areas of the United States the codes are extended to furnishings, eg, the seat coverings in airport waiting rooms.

Some of the regulations are voluntary or at least nongovernmental, eg, the ease of ignition of appliance housings is specified by UL. To obtain UL certification the plastic used in the housing must pass a prescribed fire test.

Traditionally in the United States fire test methods have been written by consensus standard groups in the private sector. After a particular hazard has been identified, a committee, usually in either the ASTM or the National Fire Protection Association (NFPA), prepares and publishes a method of testing. For building products the next step is incorporation of the method in a model building code. The method becomes law when the model code is adopted by municipalities. In recent years the federal government has played an increasing role in establishing standards. The Department of Housing and Urban Development (HUD) has authority to set minimum performance standards for its insured housing. In 1974, it was given authority over all mobile homes manufactured or sold in the United States. Although HUD takes suggestions from every quarter, the process of setting a federal standard is not the same as that in a voluntary consensus standards group. It is unnecessary for HUD to reach full consensus, merely to provide an opportunity for all to be heard. The U.S. now has fire standards on materials and products for housing, automobile interiors, buses, trains, aircraft interiors, mines, etc. The impact of these regulations has been felt most by the plastics industry although a few federal standards have affected natural products, eg, the children's sleepwear standard issued in the early 1970s by the Department of Commerce evidently caused a considerable shift away from cotton (qv) to polyester fibers (qv).

The rate at which new standards are introduced reached a peak in the early 1970s and has declined somewhat since. Whereas at least one new mandatory federal fire standard was published each year in the early 1970s, the rate is roughly half that now. Most recently the standard on mobile homes has taken effect and a federal standard on cellulosic home insulation has been mandated by the Congress.

A recent survey provides estimates of the market for chemicals used as fire retardants in plastics and related products (see Table 1). About two thirds of these chemicals are used in plastics (see the succeeding Flame retardants articles).

Table 1. Use by Product Type[a]

Chemical types	Estimated volume, thousands of metric tons 1976	1986
phosphate esters	[23][b]	[32][b]
halogenated phosphate esters	10	20
chlorinated hydrocarbons	20	32
brominated hydrocarbons	13	20
antimony trioxide (Sb_4O_6)	14	22
borates	4	7
polyols containing phosphorus	5	12
brominated bisphenol A	5	7
others	5	7

[a] Ref. 5. Does not include $Al_2O_3.3H_2O$ which is extensively used as a filler and heat sink, particularly in carpet backing (see Antimony and other organic compounds).
[b] Reduced to eliminate the functional fluids believed to have been included.

Fire Testing And Fire-Retardant Effects

Some of the reasons for the slowdown in standards-setting have to do with changing attitudes towards federal regulations. In recent years research has shown that the basis for many of the older fire tests is doubtful and that test methods for fire performance can produce misleading results if the tests are applied too broadly. Results of the old tests must be used with great care; new tests must be far more sophisticated and complex. Two needs must be addressed: (*1*) tests on isolated materials must cover a range of exposure conditions and, (*2*) interactions of materials must be taken into account. The first problem is not new to science and engineering. The pitfalls of the *one-point* measurement are well known, eg, measurement of the viscosity coefficient of a non-Newtonian liquid. The common mistake is to make a viscosity measurement

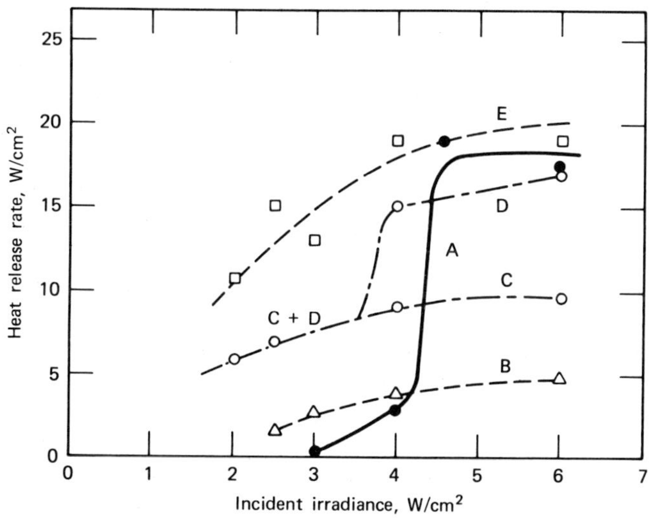

Figure 1. Fire tests of five materials, heat release rate vs incident irradiance. Courtesy of W. Parker, National Bureau of Standards.

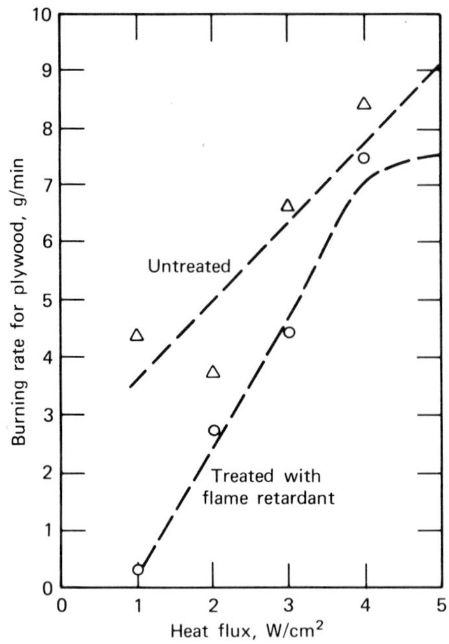

Figure 2. Burning rate for plywood as a function of external heat flux (6) (see also Laminated and reinforced wood).

at one shear stress, derive the coefficient of viscosity and then use it to predict behavior at other shear stresses. Because the liquid behaves in a nonlinear manner, the prediction will be wrong.

The situation in a fire test is more complex. The variables are burning rate, or some function of it, plotted against heat flux at the sample surface. Such plots are not linear in any case. Some results are shown in Figure 1.

The graph illustrates both the nonlinear behavior of materials in fire tests and also that rank ordering of materials changes, sometimes dramatically, when the exposure is changed. At 2 W/cm^2 material A has the lowest rate of heat release; at 5 W/cm^2 this material has the next-to-highest rate. The important lesson is that it is far better to define experimentally the fire response curve than to try to estimate it from one or two points. Much of the difficulty with unexpected flammability of synthetic polymers has arisen from failure to perform testing over a wide range of exposure conditions.

Many fire tests begin with an exposure to a fire source until ignition is obtained, then the source is withdrawn and observations of continued burning or extinguishment are made. This has been the most common *single-point* test. The exposure is the energy feedback from the flame of the sample. (Another such test is the Oxygen Index (OI) in which the O_2 content of the gas stream is varied until the sample just fails to support flame and combustion. The result is reported as LOI, Limiting Oxygen Index). Addition of a fire retardant is usually sufficient to produce self-extinguishing behavior. The material would be expected to resist burning rather well in isolation. But in most fires there are many materials involved, and some, eg, newspapers, are very flammable. The effect of the other materials is to raise the heat flux seen by the particular material.

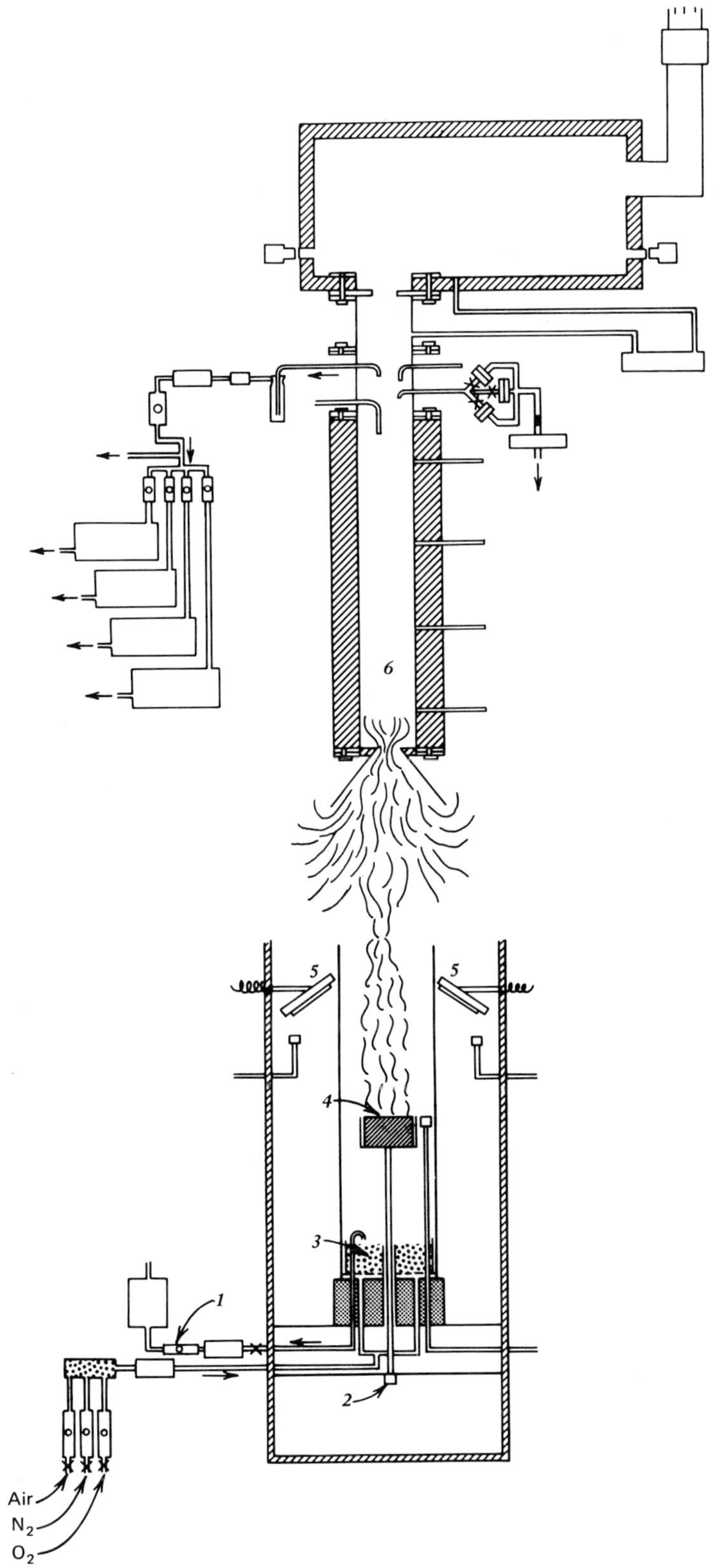

Figure 3. Recent experimental apparatus. For description of numbers, see text.

One needs to know whether addition of the retardant produces fire-resistant behavior at heat fluxes characteristic of large fires, ie, whether the presence of a particular material will increase substantially (ie, $\geqslant$100%) the rate of fire growth in a room or decrease substantially the escape time as measured from the time a sustained ignition has occurred (see Materials reliability). The effect of raising the heat flux is that most fire-retarded polymers will burn if the external flux is high enough. A typical result is shown for plywood in Figure 2. In Figure 2 the sharp lowering of the burning rate by the chemical treatment at 1 W/cm^2, as well as the return to rapid burning at an increase in exposure to ca 3 W/cm^2, are shown (the 3 W/cm^2 heat flux is not uncommon in a room fire). There is a small reduction in burning rate of the treated material at the highest fluxes measured. This offset occurs with other cellulosics but is not always seen with synthetic polymers, ie, for many polymers the retardant effect is entirely removed at high heat fluxes.

There is no doubt that flame-retardant additives affect the response of many materials to fire. The following articles provide abundant testimony to the improved, often dramatic, resistance to ignition of treated materials. However, materials rich in carbon will burn, despite the presence of additives, if the exposures to heat and oxidizer are severe enough. The flame retardants increase the threshold of ignition of the treated materials and thereby render the materials safer in many uses. The product developer and users should realize that the flammability of the material is decreased but not eliminated. The description of this flammability behavior in terms of response to increasing heat flux portrays the limitations of fire-retardant materials, and permits them to continue to play an important role in fire-safety engineering (see Plant safety).

The newest test equipment proposed for flammability tests includes provision for varying the exposure. The heat-release-rate calorimeters are typical of this trend. Figure 3 is a schematic of one of the newest devices. In this apparatus the sample (*4*) is exposed to a gas stream of O_2 content regulated at (*1*), and passed through a bed of glass beads (*3*) to smooth the flow. External heat is supplied through radiant heaters (*5*). Weight loss is recorded by a suitable transducer (*2*); gas temperature, particulates, and chemical composition are measured in the overhead channel (*6*). The device in Figure 3 is a research instrument which probably would be simplified for purposes of routine testing.

BIBLIOGRAPHY

1. J. W. Lyons, *The Chemistry and Uses of Fire Retardants,* Wiley-Interscience, New York, 1970, Chapt. 5.
2. *Fed. Regist.* **40**(12), 255 (1975).
3. W. DeWinter, *J. Appl. Polym. Sci., Appl. Polym. Symp.* **9**(195), (1969).
4. Ref. 1, p. 328.
5. *Am. Dyest. Rep.* **67,** 14 (June 1978); *Chem. Eng. News* **56,** 22 (April 24, 1978).
6. J. W. Rowen and J. W. Lyons, *J. Cell. Plast.* **14**(1), 25 (1978).

JOHN W. LYONS
National Bureau of Standards

ANTIMONY AND OTHER INORGANIC COMPOUNDS

In many polymers, the high concentration of halogenated organic compounds, needed to impart flame retardancy, adversely affects their physical properties. In practice, halogen-containing flame retardants are formulated with inorganic compounds that behave synergistically with the halogen. This enables formulators to use less additives without diminishing flame retardance. Indeed, in many instances, flame retardancy is improved when inorganic halogen synergists are used.

Antimony Compounds

Antimony Trioxide. In 1979 approximately 15,900 metric tons of antimony trioxide [*1309-64-4*] (commonly referred to as antimony oxide) was used to impart flame retardance to a variety of plastics. Antimony trioxide is manufactured by oxidizing molten antimony sulfide ore and/or antimony metal in air at 600–800°C (1). Typical properties for antimony trioxide are listed in Table 1 (see Antimony compounds).

Antimony trioxide is a white pigment (qv). Its pigment strength is a function of the average particle size and the particle size distribution. Particle size can be controlled during its manufacture to produce either a high tint or a low tint product. The difference in the particle size and particle size distribution between high tint and low tint antimony trioxide is illustrated in Figure 1. Both grades have the same flame inhibiting efficacy but have different effects on pigmentation and physical properties.

Domestic products and trade names of antimony trioxide are summarized in Table 2.

Special grades are available at higher costs. They include White Star S15 from the Harshaw Chemical Company (2) and ultra-fine antimony oxide from PPG Industries (3).

Antimony Trioxide in Cellulosics. Antimony trioxide can be used as a condensed-phase flame retardant in cellulosic materials (4). In these substrates, it reacts endothermically with the hydroxyl groups and forms a variety of products. The endothermic reaction absorbs heat needed to propagate the flame. The products formed are difficult to ignite and shield the underlying cellulose from the flame, minimizing pyrolytic and oxidative degradation.

Antimony Pentoxide. Antimony pentoxide [*1314-60-9*] is manufactured by the oxidation of antimony trioxide with nitrates or peroxides (5–9). For Sb_2O_5 the wt % of antimony is 72.8 and the specific gravity is 3.8.

When the pentoxide is heated above 380°C, it disproportionates into antimony tetroxide [*1332-81-6*] with the evolution of oxygen.

Table 1. Properties of Antimony Trioxide

Physical properties	High pigmenting	Low pigmenting
specific gravity	5.3–5.8	5.3–5.8
average particle size (Fisher subsieve method), μm	0.8–1.8	0.9–2.5
oil absorption, g/100 g Sb_2O_3 (ASTM 332-64)	9–12	9–12
refractive index, n_D^{20}	2.087	

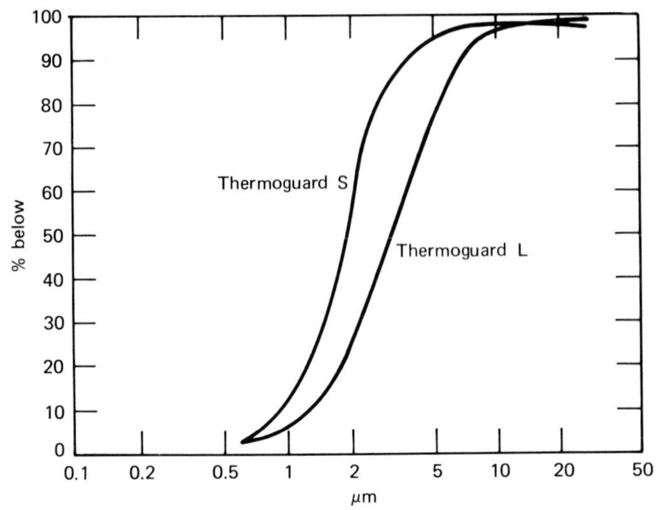

Figure 1. Particle size distribution of antimony oxides.

Table 2. Trade Names of Antimony Trioxide (Sb_2O_3)

High tint	Low tint	Company
HP	LP	ASARCO, Inc.
KR		Harshaw Chemical Co.
White Star	KR-LTS	
White Star M		
Thermoguard S	Thermoguard L	M & T Chemicals, Inc.
H Grade	L Grade	McGean Chemical Co.
Fire Shield H	Fire Shield L	PPG Industries, Inc.
Montana Brand		U.S. Antimony Corp.

Commercially, antimony pentoxide is primarily available as a stable colloid (Nyacol, Inc.) (10) or as a redispersible powder (Nyacol, Inc.; PPG Industries, Inc.) (11). It is significantly more expensive than antimony trioxide and is designed primarily for highly specialized applications. Antimony pentoxide manufacturers suggest fiber and fabric treatment applications as a potential area for its use. The redispersible powder form of antimony pentoxide, which is also recommended for plastics, contains 88% antimony pentoxide, and 12% dispersing agents. Care must be exercised when this product is incorporated into plastic since the dispersing agents can adversely affect the thermal stability and physical properties.

Sodium Antimonate. Sodium antimonate [*15593-75-6*], $Na_2OSb_2O_5 \cdot \frac{1}{2}H_2O$, is a free-flowing white powder made by the oxidation of antimony trioxide in a basic medium (12). A few of its properties are shown in Table 3.

The pigmenting strength of sodium antimonate is less than antimony trioxide. It is recommended for formulations in which deep tone colors are required. Because it contains 62 wt % antimony, somewhat higher concentrations are needed to make it as effective as antimony trioxide which has 83 wt % antimony.

Table 3. Some Properties of Sodium Antimonate

Property	Value
physical form	white powder
antimony content	61.7 wt %
bulk density, kg/m^3	39.4–46.4
specific gravity	4.8
fineness	99.75% through 74 μm (200 mesh) sieve

Mixed Metal Antimony Compounds. Recent developments in inorganic flame-retardant synergists have centered on mixed products that contain antimony and other metals which reportedly give excellent performance at reduced cost.

Thermoguard CPA (12) (M & T Chemicals, Inc.) appears to be as effective as antimony trioxide in most flame retardant applications, and has a significantly lower price. Although it contains a lower level of antimony compared to antimony trioxide, other metals contained in the product significantly boost its flame retarding properties.

NL Industries has developed a series of antimony–silico complexes under the trade name Oncor (13). These products contain up to 50% antimony trioxide. They are less opacifying than either high- or low-tint antimony oxide. Generally, antimony–silico complexes are less effective flame retardants than antimony trioxide. Therefore, although the cost per kilogram is less than antimony trioxide, the cost-effectiveness of the antimony–silico complexes can be higher.

Antimony–Halogen Mechanisms. Antimony trioxide is used almost exclusively with heat-labile halogen compounds. Most of the mechanisms proposed indicate that antimony trioxide is activated by reaction with halogens, forming antimony trihalides or antimony oxyhalides (14–15). This can be shown simply by the following reactions:

$$Sb_2O_3 + 6\,HCl \xrightarrow{\Delta} 2\,SbCl_3 + 3\,H_2O$$

$$Sb_2O_3 + 2\,HCl \xrightarrow{\Delta} 2\,SbOCl + H_2O$$

Antimony trichloride and antimony oxychloride work primarily as flame-phase flame retarders. The type of antimony halide formed depends on the concentration of the hydrogen halide and the temperature of the reaction. The interaction of two halogenated organic compounds with antimony trioxide is reported in ref. 16.

In this study, a typical aliphatic chlorinated paraffin containing 70 wt % chlorine (Chlorowax, Diamond Shamrock) was heated alone at a rate of 20°C/min. A 67% weight loss was noted at 250–360°C (see Fig. 2). The loss is equivalent to 93 wt % of the theoretical stoichiometric quantity of hydrogen chloride.

When an equal weight of antimony trioxide was added to the chlorinated paraffin and the mixture was heated at the same rate, a 76% weight loss at 310–400°C was noted (see Fig. 3). If there were no reaction, the loss would have been only 37.5% since only half of the mixture was the chlorinated paraffin, and antimony trioxide does not volatilize below 656°C. The higher weight loss indicates that some reaction between either the decomposition products of the chlorinated paraffin or the chlorinated paraffin itself and antimony trioxide have taken place. The gas generated by the re-

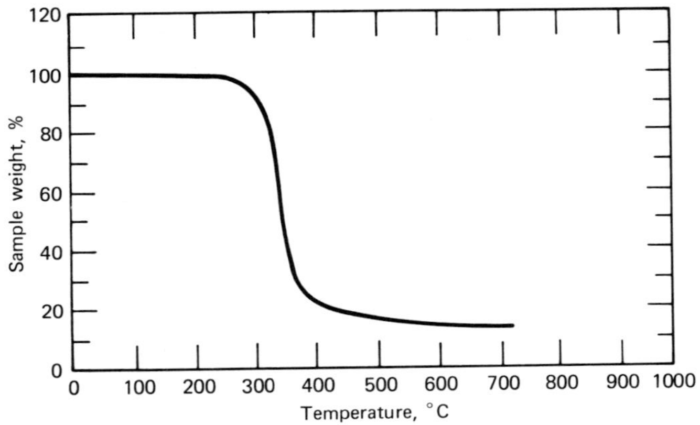

Figure 2. Thermogravimetric analysis (tga) of 70 wt % chlorinated paraffin.

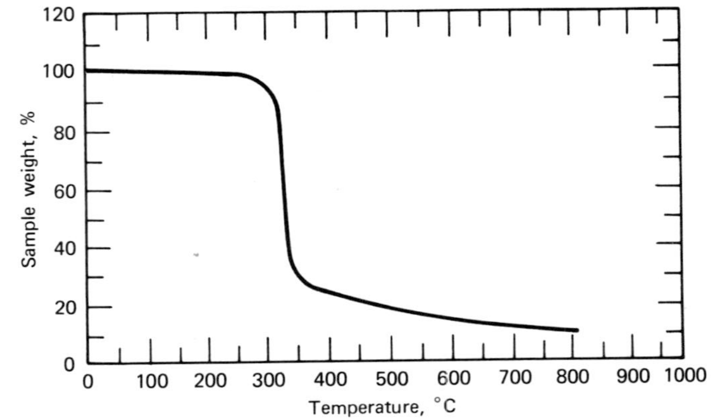

Figure 3. Tga of a 50:50 mixture of a 70 wt % chlorinated paraffin and antimony oxide.

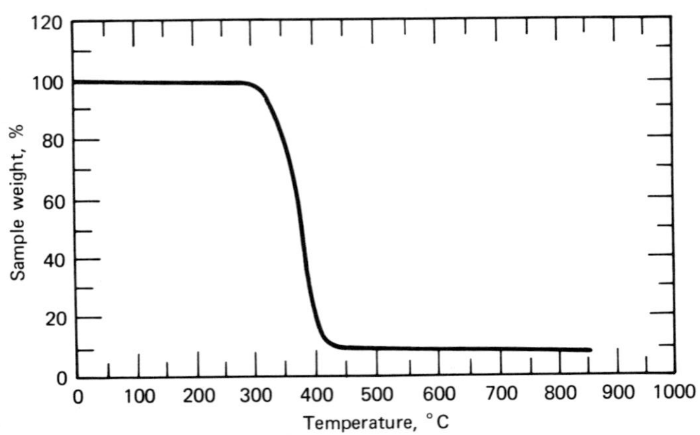

Figure 4. Tga of Dechlorane, a Diels-Alder adduct of hexachlorocylopentadiene.

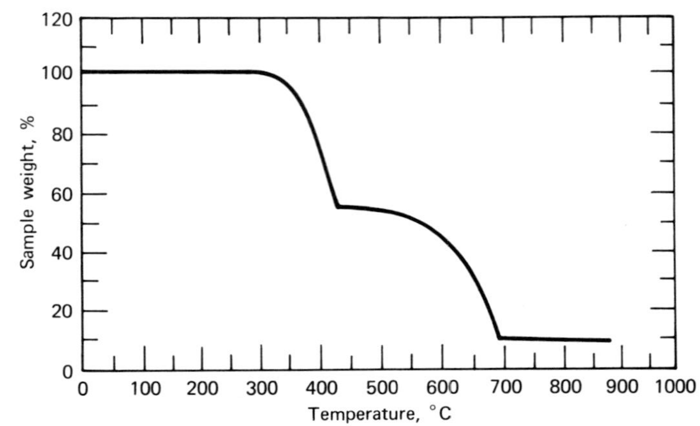

Figure 5. Tga of a 50:50 mixture of Dechlorane and antimony oxide.

action has been analyzed and identified as antimony trichloride. The weight loss is equivalent to 90% of the theoretical quantity of antimony trichloride that can be formed from the mixture. From this thermal analysis, it is apparent that antimony trichloride is the predominate antimony species formed from combinations of antimony trioxide and aliphatic chlorine compounds that generate high concentrations of hydrogen chloride upon thermal degradation.

When a cyclic halogenated organic compound, Dechlorane 5–10 (Hooker), that contains 77% chlorine was heated at a rate of 20°C/min, 90% of its weight was lost between 280 and 400°C. It does not generate hydrogen chloride directly upon decomposition (see Fig. 4). When equal weights of Dechlorane 5–10 and antimony trioxide were heated at the same rate, a different weight loss pattern was noted (see Fig. 5).

Instead of the smooth continuous decomposition pattern observed for either the chlorinated paraffin-antimony trioxide mixture, or Dechlorane 5–10 itself, a two-stage decomposition pattern was observed. There was a 45% weight loss between 305 and 410°C and another between 490 and 680°C. The decomposition pattern was remarkably similar to the decomposition for antimony oxychloride illustrated in Figure 6 (17).

$$5\ SbOCl(s) \xrightarrow{245\text{–}280°C} Sb_4O_5Cl2(s) + SbCl_3\uparrow$$

$$4\ SbOCl_2(s) \xrightarrow{410\text{–}475°C} Sb_3O_4Cl(s) + SbCl_3\uparrow$$

$$3\ Sb_3O_4Cl(s) \xrightarrow{475\text{–}565°C} 4\ Sb_2O_3(s) + SbCl_3\uparrow$$

$$Sb_2O_3(s) \xrightarrow{658°C} Sb_2O_3\uparrow$$

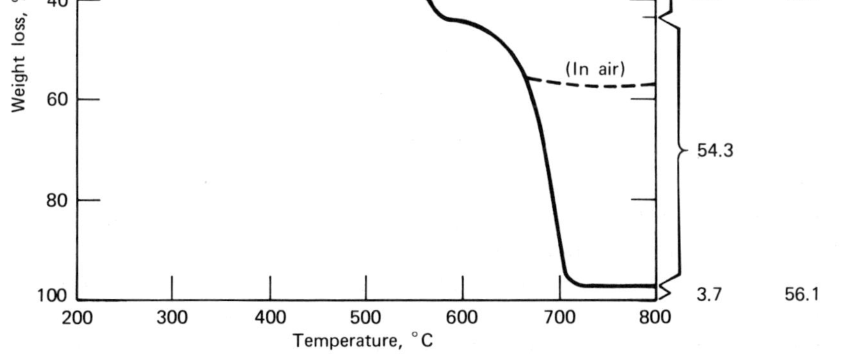

Figure 6. Tga of antimony oxychloride (SbOCl). Scan rate = 10°C/min; flow rate = 80 mL argon/min or 80 mL air/min.

It appears that antimony oxyhalides are the primary antimony compounds formed when organic halogen compounds, which do not generate hydrogen chloride directly upon thermal exposure, and antimony trioxide are heated together.

Antimony trihalides are the flame-retarding species whether they are generated directly from the starting antimony–halogen mixture or from antimony oxyhalide (18). They inhibit combustion by altering the manner and type of decomposition products formed by the plastic and by modifying the reactions in the flame to make them less exothermic. In the condensed phase or molten polymer just beneath the flame, antimony trihalide promotes reactions that form carbonaceous chars instead of highly volatile reactive gases (19–20). The chars act as heat shields, which deflect the heat of the flame, and slow down the thermal and oxidative decomposition of the polymer. The chars also form a seal around the polymer preventing potentially flammable gas from escaping and entering the flame.

Once in the flame, the antimony trihalides decompose into various antimony oxides and halogen compounds. The decomposition mechanism has not been completely determined.

A decomposition mechanism has been proposed in which antimony trihalides such as antimony tribomide decompose in a stepwise manner and participate directly in flame-quenching reactions (21):

$$SbX_3 + H\cdot \rightarrow HX + \cdot SbX_2$$

$$SbX_3 \rightarrow X\cdot + \cdot SbX_2$$

$$SbX_3 + \cdot CH_3 \rightarrow CH_3X + \cdot SbX_2$$

$$\cdot SbX_2 + H\cdot \rightarrow \cdot SbX + HX$$

$$\cdot SbX_2 + \cdot CH_3 \rightarrow CH_3X + \cdot SbX$$

$$\cdot SbX + H\cdot \rightarrow Sb + HX$$

$$\cdot SbX + \cdot CH_3 \rightarrow Sb + CH_3X$$

The stepwise decomposition also releases the halogen slowly, enabling it to participate in flame quenching reactions for longer periods:

$$X\cdot + \cdot CH_3 \rightarrow CH_3X$$

$$X\cdot + H\cdot \rightarrow HX$$

$$X\cdot + HO_2\cdot \rightarrow HX + O_2$$

$$HX + H\cdot \rightarrow H_2 + X\cdot$$

$$X\cdot + X\cdot + M \rightarrow X_2 + M$$

$$X_2 + \cdot CH_3 \rightarrow CH_3X + X\cdot$$

M is a third body that absorbs energy.

The antimony oxides formed also participate directly in reactions with the hydrocarbons to give water and molecular hydrogen instead of flame propagating radicals. Since the formation of the nonpropagating molecules is less exothermic than the formation of flame propagating radicals, less heat is generated.

When the heat output is diminished sufficiently, the combustion reaction stops and the flame is extinguished:

$$Sb + O\cdot + M \rightarrow SbO\cdot + M$$

$$\cdot SbO + 2\,H\cdot \rightarrow \cdot SbO + H_2$$

$$\cdot SbO + H\cdot \rightarrow SbOH$$

$$SbOH + \cdot OH \rightarrow \cdot SbO + H_2O$$

Boron Compounds

Approximately 2700 metric tons of borates was used as flame retarders for poly(vinyl chloride), cellulosics and unsaturated halogenated polyesters in 1979. Zinc borate is, by far, the most widely used of this class of compounds (see Boron compounds). There is a variety of zinc borates available that vary in zinc, boron, and water content (22).

Manufacturers and trade names of commercially available borate flame retardants are shown in Table 4.

Zinc borate is rarely used alone (23). It acts synergistically with antimony oxide, enabling compounders to extend antimony trioxide in some formulations.

Zinc borate is also used with high levels of alumina trihydrate in some halogenated unsaturated polyester resins (24).

Boric Acid–Sodium Borate. Boric acid [*10043-53-3*] and sodium borate [*1303-94-4*] (borax) are two of the oldest known flame retardants (25). They are used primarily to flame-retard cellulosics such as cotton (qv) and paper (qv). Both products are inexpensive and fairly effective in these applications. Their use is limited to products for which nondurable flame retardancy is acceptable since both are very water soluble.

Boron Mechanism. Boron compounds function as flame retardants in both the flame (26) and condensed (27–28) phases. Flame-phase-active boron compounds are generated from combination of borates and halogenated organic compounds. These compounds usually generate boron trihalides, which have been used to reduce the flame volatility of air–hexane mixtures (29).

Boric acid and borax are effective condensed-phase flame retardants in polyhydroxyl compounds, especially in cellulosic fibers. When these compounds are exposed to a flame, they melt and form a glasslike coating around the fibers. Prolonged exposure causes the coating to dehydrate, generating water which cools the flame and cause it to extinguish. The boron residue also reacts with the hydroxyl groups of the cellulose to generate additional quantities of water and form an inorganic char that is difficult to ignite and burn. The char is an insulator that slows down the rate of polymer degradation and fuel formation.

Boron compounds that also contain other metals are active in both phases. Although zinc borate is not used alone to flame-retard PVC, it does inhibit flammability

Table 4. Manufacturers and Trade Names of Borate Flame Retardants

Trade name	Chemical formula	CAS Registry No.	Manufacturer
Firebrake ZB	$2ZnO.3B_2O_3.3.5H_2O$	[*12513-27-8*]	U.S. Borax
X Fire			Pearsall Chemical Co.
Zinc Borate	$Zn(BO_2)_2$	[*13701-59-2, 14720-55-9*]	Atomergic
ZB	$ZnO.B_2O_3.2H_2O$		Humphrey Chemical Co.
Busan M-11	$Ba(BO_2)_2$	[*27043-84-1*]	Buckman Labs.

in the condensed and flame phases. Upon exposure to the flame, the PVC generates hydrogen chloride which can react with the zinc borate to form nonvolatile zinc compounds as well as volatile and nonvolatile boron compounds.

$$2\ ZnO.3B_2O_3 + 12\ HCl \xrightarrow{\Delta}$$

$$Zn(OH)Cl + ZnCl_2 + 3\ BCl_3 + 3\ HBO_2 + 4\ H_2O$$

The nonvolatile zinc compounds and boric acid promote char, reducing fuel formation, and the boron trichloride and water cool and extinguish the flame.

Ammonium Fluoroborate. Ammonium fluoroborate [*13826-83-0*], NH_4BF_4, is another boron-containing compound that has some utility as a flame retardant. It can decompose to yield both halogen and boron functionalities to the flame-retarding process. Flame-retardant plastic formulations recently published (25) suggest that ammonium fluoroborate should be used primarily in combination with antimony trioxide. Manufacturers propose that the following reaction describes functionally what takes place when the two products are exposed to flaming conditions.

$$6\ NH_4BF_4 + Sb_2O_3 \xrightarrow{\Delta} 6\ NH_3 + 6\ BF_3 + 2\ SbF_3 + 3\ H_2O$$

The products formed contribute to extinguishing the flame by the mechanisms proposed in preceding paragraphs (16–17,21,25–27).

Principal producers of ammonium fluoroborate include Allied Chemical Corporation, Marcus Hook, Pa.; Cabot Corporation, Boyertown, Pa.; and Harshaw Chemical Company, Cleveland, Ohio.

Alumina Hydrates

Approximately 159,000 metric tons of alumina trihydrate [*21645-51-2*] (ALTH) was used to flame-retard unsaturated polyesters and foam carpet backing in 1979. ALTH is made either from bauxite by the Bayer process (see Aluminum) or from recovered aluminum by the sinter process (30) (see Aluminum compounds). Physical properties listed in Table 5 and principal suppliers in Table 6.

Alumina trihydrate is the only aluminum compound of commercial significance as a flame retardant. It functions as a flame retardant in both the condensed and flame phases.

When alumina trihydrate is exposed to temperatures above 250°C, it forms water and alumina.

$$Al_2(OH)_6 \xrightarrow{\Delta} Al_2O_3 + 3\ H_2O$$

The evolution of water absorbs heat. The water cools the flame and dilutes the flam-

Table 5. Physical Properties of Commercial Alumina Trihydrate

density, g/cm^3	2.42
refractive index, n_D^{20}	1.579
Mohs hardness	2.5–3.5
color	white to off-white

Table 6. Principal Suppliers of Alumina Trihydrate

Company	Hydrate type	Particle size
Olean Aluminum Corp.	Bayer	15 μm (97% through 44 μm(325 mesh) screen)
Aluminum Company of America	Bayer	6.5 μm (98% through 44 μm (325 mesh) screen)
	Sinter	6.5 μm (65% through 44 μm (325 mesh) screen)
	Submicron	0.4–0.8 μm
	Bayer	6 μm (97% through 44 μm (325 mesh) screen)
Reynolds Chemicals	Bayer	98% through 44 μm (325 mesh) screen
	Sinter	60% through 44 μm (325 mesh) screen
Great Lake Foundry Sand Co.	Bayer	1.3–15.5 μm
	Sinter	1.3–15.5 μm
Industrial Mineral Chemicals	Bayer	4–20 μm
Solem Industries, Inc.	Bayer	1.5–25 μm
	Sinter	1.5–25 μm

mable gases and oxidant in the flame. The alumina residue, an excellent heat conductor, increases removal of heat from the flame zone.

Although ALTH is an inexpensive compound, it is a comparatively inefficient flame retardant. High add-on levels, up to four times as much as the plastic (26,30) itself, are needed to impart acceptable flame retardance. It is used alone only in polymers in which large amounts of filler can be tolerated and increased weight (or density) is desired. The major application areas for ALTH are filled thermoset polyesters and styrene–butadiene rubber latex rug backing.

Alumina trihydrate is also used as a secondary synergist to improve the flame retardance of polymer systems that already contain antimony trioxide, zinc borate or some phosphorus flame retardants (31).

Molybdenum Oxides

Molybdenum compounds have been used as flame retardants of cellulosics for many years (22) (see Molybdenum compounds). Recently, they have found some use in other polymers. Molybdenum compounds appear to function as condensed-phase flame retarders (32). After ignition of PVC formulations containing molybdenum oxide [*1313-29-7*] (MoO_3) and antimony oxide, 90% of the molybdenum remained in the ash and only 10% of the antimony was found.

Since most of the molybdenum remained in the ash and the formulation did have flame-retardant properties, molybdenum is probably a condensed-phase flame retardant that promotes char (32). The precise mechanism of action has not been sufficiently defined to warrant further speculations.

Applications

Poly(vinyl Chloride). Unplasticized PVC is considered a self-extinguishing plastic because of the high levels of chlorine contained in the polymer. Flexible PVC is not self-extinguishing. Generally, plasticizers commonly used to flexibilize PVC significantly increase its flammability (see Vinyl polymers).

Bis(2-ethylhexyl) phthalate (DOP) is the most commonly used plasticizer in the vinyl industry. Table 7 demonstrates the effect of DOP concentration on the oxygen index of vinyl formulations. The higher the DOP concentration, the lower the oxygen index.

Antimony Trioxide. In many applications, a low oxygen index is undesirable. The addition of antimony trioxide to vinyl formulations significantly increases the flame resistance. Figure 7 shows how antimony trioxide increases the oxygen index of PVC systems at various plasticizer levels.

Mixed-metal flame retardants have shown significant utility in PVC applications as replacements for antimony trioxide. Table 8 shows the flame retardant performance of Thermoguard CPA (Sb_2O_3) vs that of antimony trioxide in a typical vinyl wire-coating formulation.

In this system, Thermoguard CPA offers performance equivalent to antimony trioxide at substantially reduced cost. Physical properties and processing characteristics of the two formulations shown are also equivalent.

Table 7. Oxygen Index of Flexible PVC

Dioctyl phthalate (DOP), wt %	Cl, wt %	Oxygen index
0	57.0	41–49
16.6	48.5	31.5
28.5	41.0	36.4
37.5	36.4	22.2
44.4	32.0	21.2
47.3	30.0	19.2

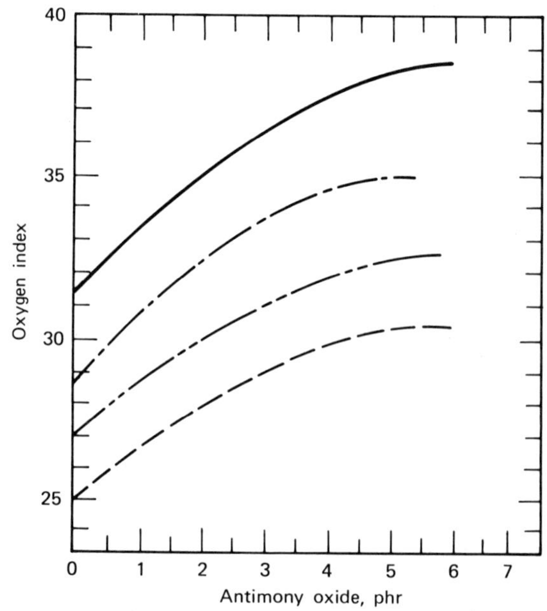

Figure 7. The effect of antimony oxide on the oxygen index of poly(vinyl chloride) plasticized with: —, 20 phr dioctyl phthalate; — - —, 30 phr dioctyl phthalate; – - - –, 40 phr dioctyl phthalate; and - - - -, 50 phr dioctyl phthalate. phr = parts per hundred resin.

Table 8. Performance of Thermoguard[a] CPA vs Antimony Trioxide[b] Synergists in PVC Wire Coating

Component	Parts by wt	
Formulations		
PVC	100.0	100.0
trimellitate ester	54.0	54.0
lead stabilizer	7.0	7.0
filler	10.0	10.0
lubricant	0.5	0.5
antimony oxide	3.0	
Thermoguard CPA		3.0
Test results		
UL-94, classification	V-0	V-0
flame-out time, s	2.3	2.4
oxygen index	28.9	28.6

[a] M & T Chemicals, Inc.
[b] Ref. 13.

Zinc Borate. Zinc borate has also been used to increase the flame resistance of flexible vinyl. It is not as effective as antimony oxide. The performance of zinc borate vs that of antimony trioxide in a 50 phr, DOP-plasticized vinyl formulation is shown in Figure 8 (24).

Zinc borate can be used as an extender for antimony trioxide in some systems. The effect of antimony trioxide and zinc borate combinations on the oxygen index of flexible PVC is shown in Table 9. The base formulation (50 phr DOP) has an oxygen index of 22.7%. The addition of 1.5 phr of antimony trioxide raises the oxygen index to 27.5%, and 1.5 phr of zinc borate raises the oxygen index to 23.1%. Addition of 1.5 phr of each of these materials raises the oxygen index to 28.2%, showing the additive effect of this combination.

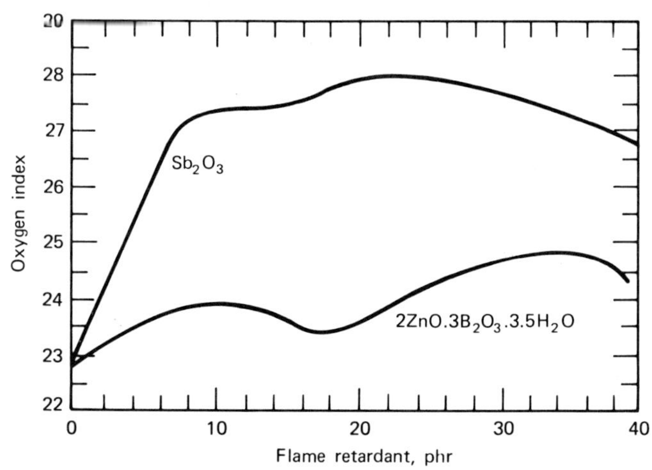

Figure 8. A comparison of the effect of antimony oxide and Firebrake ZB on the oxygen index of PVC with 50 phr DOP.

Table 9. Effect of Zinc Borate and Antimony Trioxide on the Oxygen Index of Flexible PVC (50 phr[a] DOP[b])[c]

Zinc borate, phr	Antimony trioxide, phr	Oxygen index
0	3	28.7
0	1.5	27.5
0.5	2.5	28.6
1.0	2.0	28.4
1.5	1.5	28.2
3.0	0	24.1
1.5	0	23.1
0	0	22.7

[a] phr = parts per hundred resin.
[b] DOP = dioctyl phthalate.
[c] Ref. 24.

Molybdenum Oxide. Molybdenum oxide is also being used as a flame retardant in flexible PVC. Its flame-retarding efficiency vs antimony trioxide is shown in Figure 9 (32).

Although molybdenum oxide is not as effective a flame retardant as antimony trioxide, it can be used to replace some antimony trioxide in vinyl formulations that yield lower smoke generation. The effect of such combinations is equal to the additive effect of the individual components (see Table 10).

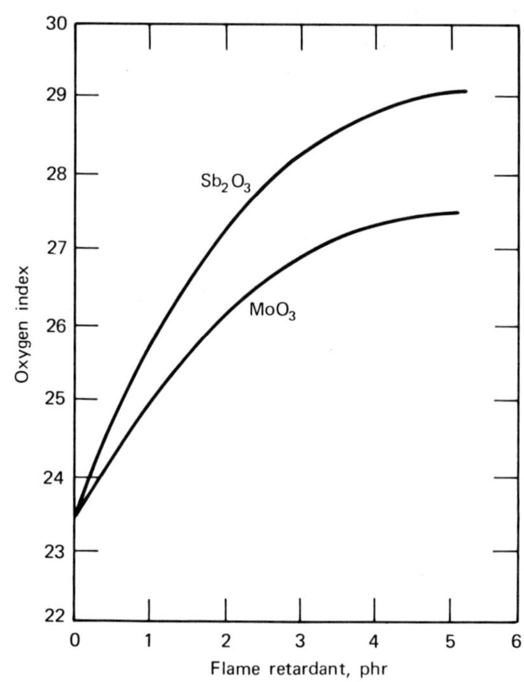

Figure 9. A comparison of the effect of antimony oxide and molybdenum oxide on the oxygen index of PVC that contains 45 phr DOP.

Table 10. Effect of Molybdic Oxide and Antimony Trioxide on the Oxygen Index of Flexible PVC[a]

Additive	Concentration, phr	Oxygen index
none	0	24.5
antimony trioxide	2	27.5
	3	28.3
	4	29.0
molybdic oxide	2	26.0
	3	27.0
	5	27.0
antimony trioxide	2	28.0
molybdic oxide	1	
antimony trioxide	1	27.0
molybdic oxide	2	
antimony trioxide	2	29.5
molybdic oxide	2	

[a] Ref. 32.

Unsaturated Polyesters. The flame retardancy of unsaturated polyesters (qv) can be increased by the addition of antimony trioxide and a halogen. The halogen may be incorporated directly into the polyester during manufacture or final formulation. The effect of antimony trioxide and chlorine on the oxygen index of a cured polyester is shown in Table 11.

Oxygen indexes may be increased by the addition of antimony trioxide or maintained at a desired level by varying the chlorine–antimony ratio. In this manner, a formulator can balance the physical properties of the polyester by carefully controlling the amount of flame retarder in a formulation.

Alumina Hydrate. Combinations of zinc borate and either antimony oxide or alumina hydrate also may be used to increase the flame retardancy of halogenated unsaturated polyesters. Table 12 shows how zinc borate and antimony trioxide affect the flame-spread index (ASTM E 162) and oxygen index of a halogenated polyester.

In Table 13, the flame-spread index of various halogenated polyester resins containing zinc borate and alumina hydrate are shown.

Both general purpose and halogenated unsaturated polyesters may be made flame resistant by incorporation of alumina hydrate. Table 14 shows that polyester formu-

Table 11. Oxygen Index of Polyesters[a]

Cl, wt %	Antimony trioxide, wt %	Oxygen index
25	0	30.2
25	1	33.4
25	3	38.7
20	1	30.2
12.5	2.5	30.0
0	0	18.2

[a] Ref. 33.

Table 12. Comparative Fire Tests for Laminates (26% Fiberglass) Containing Firebrake ZB–Antimony Oxide Mixtures[a]

Resin	Antimony trioxide, phr	Firebrake ZB[b], phr	Flame spread index	Oxygen index
Hetron resin[c]	0	0	15.0	26.0
Hetron resin	2	0	9.1	37.3
Hetron resin	2	6	3.7	40.2
Hetron resin	1	6	12.5	36.8
Hetron resin	0.5	6.5	5.4	34.1

[a] Ref. 34.
[b] Zinc borate from U.S. Borax Corporation.
[c] Halogenated polyester resin from Ashland Chemical Company.

Table 13. Comparative FireTests for Laminates (26% Fiberglass) Prepared from Various Halogenated Polyesters Containing Zinc Borate and Alumina Hydrate Mixtures[a]

Resin	Firebrake ZB[b], phr[c]	Alumina hydrate, phr[c]	Flame spread index	Oxygen index
Hetron 92[d]			23.0	30.4
Hetron 92	5	15	10.0	38.5
Dion 6692[e]			28.0	32.0
Dion 6692	4	15	8.6	36.3
Dion 6692	10	15	14.0	39.1
Dion 6692	10	10	11.0	38.0
Atlac A711-050[f]			29.0	28.1
Atlac A711-050	5	5	24.0	34.5
Atlac A711-050	5	10	10.0	38.3
Stepan C4717[g]			12.5	34.1
Stepan C4717	5	15	2.5	37.0
Stepan C4717	10	15	3.5	39.0
Roblen resin[h]			28.0	31.1
Roblen resin	5	15	13.0	49.6

[a] Ref. 35.
[b] $2\,ZnO.3\,B_2O_3.3.5\,H_2O$, U.S. Borax Corp.
[c] Parts per hundred of resin.
[d] ICI Americas Inc.
[e] Koppers Chemical Co.
[f] Atlas Chemical Co.
[g] Stepan Chemical Co.
[h] Robuco Chemical Co.

lations must contain ca 40–60% alumina hydrate to achieve a significant improvement in the oxygen index. Halogenated systems require significantly lower levels of ALTH to provide equivalent oxygen indexes.

The effect of molybdenum oxide and antimony oxide on the oxygen index of polyesters is shown in Table 15 (18). Molybdenum oxide is shown to work effectively with antimony trioxide to reduce flammability of polyesters and in many instances reduce smoke generation.

Table 14. Oxygen Index of Polyester Resin Castings[a]

	Filler		
Polyester resin	Sb_2O_3[b], wt %	$Al(OH)_3$[c], wt %	Oxygen index
general-purpose			18.5
general-purpose		20	21.5
general-purpose		40	26
general-purpose		60	37
general-purpose		65	53
chlorinated			27
chlorinated	5		38
chlorinated		20	29
chlorinated		40	36
chlorinated		60	55
brominated			25
brominated	5		35
brominated		20	28
brominated		40	33
brominated	3	40	44
brominated		60	62

[a] Polyester resin, alumina trihydrate and antimony trioxide equal 100%. No glass-fiber reinforcement (24).
[b] Sb_2O_3, antimony trioxide, used at 5 phr.
[c] $Al(OH)_3$, alumina trihydrate, C-331 grade, 7–9 μm median particle size.

Table 15. Effect of Molybdenum Oxide and Antimony Trioxide on Oxygen Index of Polyester[a,b]

Additive	Concentration, wt %	Oxygen index	Smoke reduction, %
none	0	33.0	
antimony trioxide	1	34.5	+67
	3	42.0	+96
	5	43.0	−26
molybdic oxide	5	36.0	−36
	7	39.0	
antimony trioxide	1	39.5	−54
molybdic oxide	1		
antimony trioxide	2	41.0	−20
molybdic oxide	1		
antimony trioxide	1	40.0	−15
molybdic oxide	2		

[a] Ref 32.
[b] Atlac 711-05A polyester (ICI Americas Inc.) with 25% glass-fiber catalyzed with 1% Lupersol (Pennwalt).

Polyurethanes. Flame retardancy is usually imparted to polyurethane foams (see Urethane polymers) by incorporating phosphorus into the formulation, either as part of the polyol or an additive. Polyurethane formulations have also been flame-retarded with antimony trioxide and a halogen. The effect of combinations of antimony trioxide and chlorine and antimony trioxides, chlorine, and zinc oxide on the burn rate of a typical flexible foam is illustrated in Table 16.

Table 16. Flexible Urethane Foam Formulations [a]

Component	Parts by wt A	B	C
3000 mol wt poly(oxypropylene) glycerol	100.0	100.0	100.0
surfactant	1.0	1.0	1.0
amine blowing agent	0.15	0.15	0.15
M & T catalyst T-9[b]	0.40	0.40	0.40
deionized water	4.0	4.0	4.0
toluene diisocyanate (TDI)	52.0	52.0	52.0
PVC resin	0	22.0	22.0
antimony trioxide	0	7.0	14.0
zinc oxide (French process)	0	3.0	0
ASTM D-1392 flame test results			
self-extinguishing time, s		25.0	70.0
distance burned, cm	15.2	4.1	6.1

[a] Ref. 36.
[b] M & T Chemicals, Inc.

Antimony trioxide is superior to either antimony tetroxide or antimony pentoxide as a halogen synergist in flexible foams. A comparison of all three oxides in a formulation that contains 6.3 wt % chlorine is shown in Table 17.

Polyolefins. The most expeditious flame retardant for polyolefins is a combination of antimony trioxide and a halogen-containing additive. The amount of antimony oxide necessary depends upon the polymer (polypropylene or polyethylene, etc) and the type of halogen (chlorine vs bromine, aromatic vs aliphatic) (see Olefin polymers). Typical formulations are shown in Table 18.

Polystyrene–ABS. Flame retardance can be imparted to styrenics (see Acrylonitrile polymers; Styrene plastics) in much the same manner as to polyolefins. Typical formulations containing a halogen source and antimony trioxide are shown in Tables 19 and 20.

Table 17. Flammability of Flexible Urethane Foams Containing 6.3 wt % Cl [a,b]

Oxide	Parts by wt added	Rating[c]	Extent of burn, cm	Burning rate, cm/min
O	0	B	12.7	13.5
Sb_2O_3	5	SE	8.3	7.4
Sb_2O_3	10	SE	5.3	7.6
Sb_2O_3	15	SE	4.8	7.6
Sb_2O_3	20	SE	4.6	7.6
Sb_2O_4	15	B	12.7	9.7
Sb_2O_4	20	B	12.7	10.4
Sb_2O_5	10	B	12.7	9.4
Sb_2O_5	15	B	12.7	10.7
Sb_2O_5	20	B	12.7	9.7
Sb_2O_5	40	B	12.7	8.1

[a] Ref. 23.
[b] Initial mixture parts by wt: oxypropylated glycerol, 75; diol containing 45.5% Cl, 25; TDI 80 (Olin Corp.), $C_9H_6N_2O_2$, 51; water, 4; DC 190 (Dow Corning), 1.5; DABCO 33 LV (Houdry), $C_6H_{12}N_2$, 0.6; stannous octanoate, 0.2.
[c] B = burning, SE = self-extinguishing.

Table 18. Polyethylene–Polypropylene

Component	Wt % A	Wt % B	Wt % C
Flame-retardant polyethylene			
polyethylene	85.0	60.0	82.0
antimony trioxide	7.5	13.0	6.0
chlorinated paraffin (70 wt % Cl)	7.5		
Dechlorane Plus[a], ($C_{18}H_{12}Cl_{12}$)		27.0	
decabromobiphenyl oxide			12.0
Flame test results			
oxygen index	24.0	24.0	27.0
UL-94 class	V-2	V-1	V-0
Flame-retardant polypropylene			
polypropylene	60.0	93.0	
antimony trioxide	10.0	2.5	
Dechlorane Plus[a], ($C_{18}H_{12}Cl_{12}$)	30.0		
Citex BN-451[b]		5.0	
Flame test results			
oxygen index	27.0	27.0	
UL-94 class	V-1	V-2	

[a] Hooker Chemical Company.
[b] Saytex Chemical Corporation.

Table 19. Flame-Retardant ABS

Component	Wt % A	Wt % B	Wt % C
ABS	85.0	72.0	75.0
antimony trioxide	5.0	7.2	7.5
PVC	10.0		
Firemaster 680[a], ($C_{14}H_8Br_6O_2$)		20.8	
Dechlorane Plus[b], ($C_{18}H_{12}Cl_{12}$)			17.5
Flame test results			
oxygen index	24.0	28.0	28.0
UL-94 class	V-2	V-0	V-1

[a] Velsicol Chemical Corp.
[b] Hooker Chemical Company.

Table 20. Flame-Retardant Polystyrene

Component	Wt % A	Wt % B	Wt % C
polystyrene	75.0	85.0	80.0
antimony trioxide	1	3.0	5.0
PVC	15.0		
Citex BC-93[a]		12.0	
Dechlorane Plus[b]			15.0
Flame test results			
oxygen index	24.0	25.0	24.5
UL-94 class	V-2	V-0	V-1

[a] Saytex Chemical Corporation.
[b] Hooker Chemical Company.

BIBLIOGRAPHY

1. I. Touval and H. H. Waddell, *Handbook of Fillers and Reinforcements for Plastics,* Van Nostrand, Reinhold, New York, 1978, 13, p. 219.
2. D. A. LeBlanc, *Proceedings of the 1978 Symposium on Textile Flammability,* LeBlanc Research Institute, 1978, p. 270.
3. *Fire Shield/Flame Retardant Bulletin, Ultra-Fine Antimony Oxide,* Chemetron Chemical Corporation, Chicago, Ill., 1976.
4. J. E. Ramsbottom, *Fire-proofing of Fabrics,* His Majesty's Stationery Office, London, Eng., 1947.
5. U.S. Pat. 3,873,451 (Mar. 25, 1975), C. C. Cumbo and P. C. Yates (to E. I. du Pont de Nemours & Co., Inc.).
6. U.S. Pat. 3,994,825 (Nov. 30, 1976), C. E. Crompton (to Chemetron Chemical Corporation).
7. U.S. Pat. 4,017,418 (Apr. 12, 1977), C. E. Crompton and A. M. Z. Kazi (to Chemetron Corporation).
8. U.S. Pat. 4,022,710 (May 10, 1977), T. Kobashi and H. Shiota (to Japan Exlan Corp.).
9. U.S. Pat. 4,051,064 (Aug. 10, 1977), C. E. Crompton and A. M. Z. Abdulla (to Chemetron Corporation).
10. Colloidal Antimony Oxide, Bulletin, Nyacol, Inc.
11. *Bulletin T-P2,* Chemetron Chemicals, Inc., 1976.
12. *Thermoguard Flame Retardant Bulletin,* M & T Chemicals, Inc., 1978.
13. *Bulletin F-2,* NL Industries, Inc.
14. *Little RW Flame-proofing Textile Fabrics, American Chemical Society Monograph Series 40,* Reinhold, New York, 1947.
15. J. A. Rhys, *Chem. Ind. London* **187,** (1969).
16. I. Touval, *Am. Chem. Soc. Prepr. Org. Coat. Div.,* (1971).
17. J. J. Pitts, P. H. Scott, and J. Powell, *Cell Plast.* **6,** 35 (1970).
18. C. P. Fenimore and F. J. Martain, *Combust. Flame* **10,** 135 (1966).
19. I. Einhorn, *The Flammability of Polymeric Materials,* lecture series, University of Utah, Salt Lake City, Utah, 1970.
20. G. S. Learmouth and D. G. Thwaite, *J. Br. Polym.* **1,** 54 (1969).
21. J. W. Hastie, *Combust. Flame* **21,** 49 (1973).
22. J. J. Pitts, *Flame Retardancy of Polymeric Materials,* Marcel Dekker, Inc., New York, 1973, p. 149.
23. R. W. Sprague, *Systematic Study of Firebrake 2B as a Flame Retardant for PVC,* Parts III, IV, VI, U.S. Borax Research Corp., Anaheim, Calif., 1976.
24. P. V. Bonsignore and H. P. Hsieh, "Alumina Trihydrate and Other Inorganic Additives for Polyester Resins," *Flame-Retardant Chemicals Association (FRCA) Meeting, Houston, Tex., March 1978.*
25. J. A. Finn, "Ammonium Fluoroborate in Flame Retardant Thermoplastics," *6th International Symposium on Flammability and Fire Retardants, Nashville, Tenn., May 1979.*
26. C. J. Hilado, *Flammability Handbook for Plastics,* 2nd ed., Technomic Publications, Westport, Ct., 1974.
27. G. A. Byrne, D. Gardiner, and F. H. Holmes, *J. Appl. Chem.* **16,** 81 (1966).
28. P. H. Kemp, *Chemistry of Borates,* Part 1, Borax Consolidated Limited, London, Eng., 1956.
29. M. E. Morrison and J. Scheller, *The Effect of Burning Velocity Inhibitors on the Ignition of Hydrocarbon, Oxygen, Nitrogen Mixtures,* AeroSpace Research Laboratory, Wright Patterson Air Force Base, Apr. 1969.
30. E. A. Woychesin and I. Soboleu, *Handbook of Fillers and Reinforcements for Plastics,* Van Nostrand-Reinhold, New York, 1976, p. 237.
31. S. Kaufman and M. M. Yocum, "Balancing Flame Retardancy and Low Temperature Properties in PVC," *FRCA Meeting, New York, Oct. 1975.*
32. F. W. Moore, *Molybdenum Compounds as Smoke Suppressant for Poly(vinyl chloride) Formulations,* Society of Plastics Engineers (SPE ANTEC), Montreal, Can., Apr. 1977.
33. I. Touval, "Antimony Compounds as Flame Retardants for Plastics," *NYU Flame Retardancy Conference, Miami, Fla., 1973.*
34. *Technical Data Sheet* IP-22, U.S. Borax Corp.
35. E. A. Woychesin and I. Soboleu, "Effect of Particle Size on Performance of Aluminum Trihydrate in Glass Reinforced Polyesters," *Society of Plastics Engineers Technology Conference, 1975.*
36. U.S. Pat. 3,574,149 (Apr. 6, 1971), J. T. Harrington (to General Tire & Rubber Company).

JOHN M. AVENTO
IRVING TOUVAL
M & T Chemicals, Inc.

HALOGENATED FLAME RETARDANTS

The development and extensive use of synthetic polymers in both old and new types of applications has intensified the concern for combustibility. Although these new polymers are not necessarily more flammable than natural polymers, they are more readily used in forms, eg, foams, electrical applications, etc, that can result in an increased fire control problem.

Along with the development of many synthetic polymer systems during the 1930s and 1940s, a significant advance in the science of imparting flame resistance occurred, ie, the use of halogenated organic materials to impart ignition resistance to these new polymer systems.

In early plastics applications, the small size of fabricated articles and the relative scarcity of these articles made fire retardancy a secondary consideration. Advances in plastics technology have led to increasingly large-scale applications, especially in the construction industry. Since many polymers have fuel values (heats of combustion) comparable to common fuels, eg, wood, oil, alcohol, etc, it is readily understandable that they contribute to the burning process in a typical fire.

Commercial halogenated products used as flame-retardants for plastics currently in use are mainly compounds containing high (50–85 wt %) levels of either chlorine or bromine, ie, decabromodiphenyl oxide [*1163-19-5*], chlorendic acid [*115-28-6*], tetrabromophthalic anhydride [*632-79-1*], etc. These materials fall into two distinct types: additives and reactives. The additives have the advantage of being readily added to a polymer by mechanical means with a minimum of reformulation being required. The reactives, on the other hand, require the development of essentially new polymer systems.

Only massive polymer forms are considered, though the materials and concepts discussed are almost similarly applicable to fibers, fabrics, coatings, and elastomers. Halogenated phosphorus compounds are included under Flame retardants—phosphorus compounds.

Principles of Developing Flame-Retardant Polymers

Any discussion of the principles of developing flame-retardant polymer systems must acknowledge the chaotic situation that exists at present. This situation has arisen for a variety of reasons: technical, economic, legal, and semantic.

The semantic problem is the worst in that it is at the root of most of the other problems and is caused by the fact that the term fire or flame retardant may be perceived in a variety of ways depending upon the user's viewpoint. The term, as defined above, means simply that some change has been made in a polymer system so that it will pass one or more of at least a hundred different flammability tests. These tests are normally designed to minimize, but not eliminate, the fire risk associated with the use of a polymer in some specific use or product. As a consequence, a modification of a polymer that makes it suitable for one use does not necessarily make it suitable for others. There is no single fire-retardant chemical or method that is applicable to all polymer systems or even to all uses of a single polymer.

It is, therefore, necessary that early in the development of a flame-retardant polymer system the question "Why?" is answered before much effort is put into answering the question "How?".

It is not unusual to see many compounds proposed as flame-retardant chemicals that are clearly unusable in any practical sense, but that allow a polymer system to pass a specific flammability test. A polymer system can be easily modified so that it can be called "flame-retardant" by some "test". It is difficult, however, to do so and keep a polymer system that is low-cost, environmentally and physiologically acceptable, and also mechanically and esthetically not too dissimilar from nonfire-retardant counterparts.

One of the most common approaches used to modify the burning properties of polymers at the present time is by incorporation of halogen into the polymer matrix, either directly or through the use of halogenated additives. The usual rationale for the use of the halogens as flame retardants is based on the theory that they function in the gas phase as radical traps (1–3). It is generally agreed that the combustion of gaseous fuels is a high temperature process which proceeds via a free radical mechanism. The overall reaction, ie,

$$\text{fuel} + O_2 \rightarrow CO + CO_2 + H_2O + \text{heat} + \text{light} \qquad (1)$$

involves a great variety of discrete reactions depending on the nature of the fuel. However, a relatively small number of propagation and chain-branching reactions are common to most fuels and are critical for maintaining the combustion process. These reactions are of the type:

$$\cdot OH + CO \rightarrow CO_2 + H\cdot \qquad (2)$$

$$\cdot OH + RCH_3 \rightarrow RCH_2\cdot + H_2O \qquad (3)$$

$$RCH_2\cdot + O_2 \rightarrow RCHO + \cdot OH \qquad (4)$$

$$H\cdot + O_2 \rightarrow \cdot OH + \cdot O\cdot \text{ (chain branching)} \qquad (5)$$

Halogenated compounds have been shown to enter into flame chemistry via a series of reactions (4):

$$HBr + \cdot OH \rightarrow H_2O + Br\cdot \qquad (6)$$

$$HBr + \cdot O\cdot \rightarrow \cdot OH + Br\cdot \qquad (7)$$

$$HBr + H\cdot \rightleftharpoons H_2 + Br\cdot \qquad (8)$$

$$HBr + RCH_2\cdot \rightleftharpoons RCH_3 + Br\cdot \qquad (9)$$

$$RBr \rightleftharpoons R\cdot + Br\cdot \qquad (10)$$

In the radical-trap theory of flame inhibition it is thought that equations 6–10 effectively compete with equations 2–5 for those radical species that are critical for flame propagation, ie, ·OH and ·O·, thereby slowing the rate of energy production and resulting in the extinction of the flame. Hydrogen fluoride does not significantly enter into the flame chemistry, thus fully fluorinated compounds are generally considered to be ineffective as flame-retardant agents (1). The radical-trap theory of flame inhibition, although attractive in that it can be adapted to any situation, tends to lead to the belief that the simple inclusion of small amounts of halogen into a polymer system will render the system flame retardant.

A recent physical theory of flame suppression by the halogens, although conceding that the halogens enter into flame chemistry, suggests that this participation *per se* cannot be the primary mechanism by which the halogens function (5). Rather, it is postulated that the halogens act by altering the physical properties, ie, the density and mass heat capacity of the gaseous fuel–oxidant mixture so that flame propagation is effectively prevented. The physical theory is primarily based on the observations that any gaseous mixture of fuel and halogenated agent generally propagates flame when mixed with air as long as the mass fraction of halogen in the mixture is less than ca 0.7, and the relative effectiveness of the halogens is directly proportional to their atomic weights, ie, F:Cl:Br:I = 1.0:1.9:4.2:6.7. The halogenated agents probably act by the same basic mechanisms as the inert gases, ie, CO_2, N_2, etc, and their suppressant effects are additive to those of the inert gases (6).

A functional relationship (7) has been shown to exist between a polymer burning rate and the Spalding (8) mass transfer number, B, where:

$$B = \frac{\frac{M_O}{r} H_c - C_p(T_s - T_a)}{H_G} \qquad (11)$$

M_O is the mass fraction of oxygen in the combustion zone; H_c is the net heat of combustion of the sample (J/g); r is the stoichiometric mass oxygen/fuel ratio; C_p is the specific heat of the gases in the combustion zone; T_s is the surface temperature of the sample (°C), T_a is the ambient temperature (°C); and H_G is the apparent heat of gasification (J/g). The B number contains the fundamental properties of the polymeric materials. Thus the mass burning rate or burning intensity can be related to the fundamental properties of the material.

Flammability peak determinations show that the minimum mass fraction of inert gas required to prevent flame propagation through all fuel–oxygen-inert mixtures in a premixed gas system is ca 0.8, ie,

$$M_i = \frac{m_a + m_N}{m_f + m_O + m_a + m_N} \qquad (12)$$

where M_i is the mass fraction of the inert components of the mixture and m_a, m_N, m_f and m_O are the weights of agent, nitrogen, fuel and oxygen, respectively (6). If all of the terms in the B number remain constant, an increase in the mass of inert gas in the combustion zone ($m_a \neq 0$) results in a lower oxygen mass fraction, m_O, a lower B number, and a corresponding reduction in the polymer burning rate.

When applied to liquid fuels, the Spalding B number in its simplest form can be visualized as the ratio of the heat of combustion and the heat of vaporization ($\Delta H_c/\Delta H_v$). Table 1 shows the significance of this ratio applied to several halogen-containing fuels (9). In Table 1 the flash and fire points are expressed both in °C, as normally reported, and as the weight of compound present in the gas phase over the surface of the liquid at this temperature (mg/L). The introduction of halogen has a lesser effect upon ΔH_v per milligram of compound evaporated. The ratio $\Delta H_c/\Delta H_v$ decreases with added halogen indicating that less energy is available from the flame for gasification, and, in order to keep the flame burning, additional heat from some outside source is required. ΔH_v is the amount of heat required to vaporize the weight of fuel (latent heat of vaporization) present in the gas phase at the appropriate flash and fire points after the fuels have been raised to these temperatures by the outside source. Note the large increase in mass that must be vaporized in order to obtain sustained burning in the case of bromobenzene, at least 100 times the mass that must be vaporized in the case of benzene itself.

The physical theory apparently accounts for the effects seen when halogenated

Table 1. Flash and Fire Points of Halobenzenes

Fuel	wt % X	Flash point, °C	Fire point, °C	$\Delta H_c{}^a$, J/mg	$\Delta H_v{}^a$, J/mg	$\Delta H_c/\Delta H_v$	Flash, mg/L	Fire, mg/L	Δ mg/L
C_6H_6	0	13	ca 14	42	0.39	106	47	ca 50	nil
C_6H_5Cl	31.5	40	51	27	0.33	83	63	250	187
o-$C_6H_4Cl_2$	48.2	68	103	19	0.27	70	134	600	466
C_6H_5Br	50.9	52	154	ca 19	0.24	ca 80	152	5900	5700
$C_6H_3Cl_3$	58.6	99		ca 17	0.22	ca 75	185		

[a] To convert J to cal, divide by 4.184.

agents are used as flame retardants. In view of the fact that the halogen content of a typical plastic is generally ca 1–30 wt %, it is obvious that if the typical polymer were totally vaporized the gases given off would be quite capable of flame propagation.

In order to visualize the role of halogen it is necessary to examine the heat balance that occurs at the surface of the polymer. Figure 1 shows a schematic of this balance (10). Heat received by the polymer surface may arise either as a heat flux from the flame ($\dot{Q}_T$) or as an externally applied heat flux ($\dot{Q}_E$) derived from another source. Heat is lost either as the heat required for gasification ($\dot{Q}_G$) of the polymer or as heat lost ($\dot{Q}_L$) through radiation, conduction, convection, dripping, etc. $\dot{Q}_T$ and $\dot{Q}_G$ are agent dependent, whereas $\dot{Q}_E$ is obviously agent independent, except in char-forming systems. $\dot{Q}_L$ may be agent dependent if the agent acts by increasing the drip rate of the burning polymer. Halogenated agents affect the heat balance through $\dot{Q}_T$, $\dot{Q}_G$, and $\dot{Q}_L$. Although phosphorus may act in the gas phase, it appears to be the most important element affecting $\dot{Q}_G$ and $\dot{Q}_E$ through char formation (11).

Qualitatively, the burning process involves heating of the substrate to a temperature high enough to drive off flammable vapors. When the rate of vapor evolution becomes high enough to generate a flammable mixture, the mixture ignites. If the rate of vapor or gas evolution becomes sufficiently high, the heat produced by the combustion process may return enough heat to the substrate so that the evolution of fuel becomes self-sustaining.

When a flame retardant that acts in the vapor phase is added to the system, part of the vapor that distills from the polymer does not contribute to the heat of combustion but results only in a reduction in the mass fractions of the oxygen and fuel in the combustion zone. Hence, there is an increase in the total mass of material that must be vaporized per unit time in order to keep the fire burning. A corresponding increase in the amount of energy must be added to the system from an external heat source ($\dot{Q}_E$, Figure 1) in order to vaporize the extra material.

Both dripping and char formation interfere with the energy feedback cycle ($\dot{Q}_T$

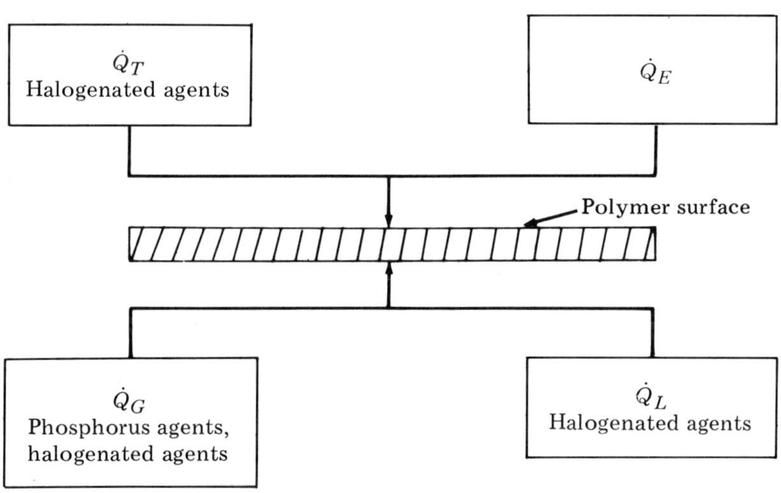

Figure 1. Heat balance at the surface of a burning polymer. $\dot{Q}_T$ = the total heat flux from the flame; $\dot{Q}_E$ = the heat flux from any external heat source; $\dot{Q}_G$ = the heat flux required for gasification of the polymer; $\dot{Q}_L$ = total heat flux lost owing to convection, radiation, dripping, etc. Active halogen participation occurs in $\dot{Q}_T$, $\dot{Q}_G$, and $\dot{Q}_L$.

and $\dot{Q}_E$) and consequently cause an increase in the intensity of the external heat flux required to balance the energy–fuel cycle.

Where the flame is actively spreading over the surface of a material, the elemental composition of the vapor being evolved ahead of the moving flame is not necessarily the same as the elemental composition of the polymer. The composition of the vapors may vary considerably between the temperature at which the material first begins to evolve vapors and the temperature at which the rate of evolution supports the flame. With this type of dynamic burning condition, changes in the substrate and the structure of the agent are more important than they are under steady-state conditions.

There are five fundamental methods used to fire-retard both natural and synthetic polymer systems. They are:

Raise the decomposition temperature of the polymer. This is generally accomplished by increasing the cross-linking density of the polymer, as with ladder polymers (increase $\dot{Q}_G$).

Reduce the fuel content of the system. This approach generally involves halogenating the polymer backbone, adding halogenated additives, adding inert fillers, or by resorting to inorganic systems (increase $\dot{Q}_G$, decrease $\dot{Q}_T$).

Induce polymer flow by selective chain scission. This approach is generally applicable to thermoplastic polymer systems where interrupting the polymer backbone results in reduction of the viscosity of the polymer and promotes dripping (increase $\dot{Q}_L$).

Induce selective decomposition pathways. This method is most applicable to cellulosics where the introduction of phosphorus compounds generates phosphorus acids which catalyze the loss of water and the retention of the carbon as char (increase $\dot{Q}_G$, decrease $\dot{Q}_T$).

Mechanical means include (*1*) bonding a nonflammable skin on the polymer, (*2*) covering the polymer with an intumescent coating, (*3*) design of the system, and (*4*) the use of sprinklers (decrease $\dot{Q}_E$).

Antimony–Halogen Synergism. Antimony oxide, a commonly employed fire retardant adjunct for halogen-containing polymer systems, is usually employed as a means of reducing the halogen levels required to obtain a given degree of flame retardancy with the polymer system. This reduction is often desirable since the required halogen content for the system may be so high that it affects the physical properties of the system. In other cases, the antimony oxide is used simply to give a more cost-effective system.

Antimony–halogen systems have been widely studied in attempts to explain the apparent synergistic effects obtained with this combination of elements. No completely satisfactory theory is available as yet, but it is generally agreed that the active agents, antimony trihalides or antimony oxyhalides, act principally in the gas phase (12–13). As with the halogens, it is generally postulated that the antimony halides act as radical traps.

Small scale tests show that the optimum halogen (Cl,Br)/antimony atom ratio in most systems is 3/1 (14), corresponding to the atom ratio found in the antimony trihalides, ie, $SbCl_3$, $SbBr_3$. On the usual weight basis, this corresponds to a ratio of ca 0.9/1 for the chlorine–antimony system and ca 2/1 for the bromine–antimony systems.

The formation of the antimony trihalides results from a complex series of reactions that are discussed in detail in refs. 13 and 15.

Although the antimony halides appear to act principally in the gas phase, some effect on the condensed-phase chemistry cannot be ruled out. Antimony–halogen flame-retardant compositions usually produce a carbonaceous residue, even in polymers such as polypropylene, which produces none in the absence of fire retardants. The production of the carbonaceous residue probably results from the antimony trihalides, strong Lewis acid catalysts, which are capable of promoting the dehydrohalogenation of organic halides, and coupling and rearrangement reactions in organic systems.

Phosphorus–Halogen Systems. A large number of phosphorus-containing compounds have been used in halogen-containing polymer systems as a means of improving their ignition resistance. In many of these cases both the phosphorus and the halogen reside in the same molecule, although there is little if any evidence to indicate that having both elements in the same molecule has any particular advantage. Because there is no fixed optimum phosphorus–halogen ratio, in contrast to the antimony–halogen system, it is frequently easier to optimize the ignition resistance when the phosphorus and halogen are adjusted separately (see Flame retardants, phosphorus compounds).

Generally, phosphorus appears to act as an acid precursor in the solid phase to induce selective decomposition pathways that result in a reduction in the rate of fuel formation and an increase in charring (16). This mode of action is most applicable to cellulosics, but may also be important in other oxygen- or nitrogen-containing polymers such as polyesters (qv), polyamides (qv), and polyethers (qv).

In polymers, such as polyolefins and polystyrene, the formation of acids has little affect on the mode of polymer decomposition and much of the phosphorus may be volatilized, in some cases as much as 50–99% (16). Even in these cases some of the phosphorus may end up as polyphosphoric acids which serve to protect the substrate from the heat produced by the burning gases. The phosphorus that volatilizes will show some beneficial flame-retarding effects in that it has gas-phase flame suppressing activity similar to the halogens (5).

Whether phosphorus can be considered to be synergistic or simply additive to the halogens depends on the model employed (17).

Testing

Flammability tests are employed to determine the response of a product or material when exposed to an ignition source. The responses usually measured are: ease of ignition, surface spread rate, fuel contribution, fire endurance resistance, and smoke evolution (18). Factors often affecting the apparent flammability of a material, in addition to the material itself, are geometry of the sample, test apparatus, type of ignition source, the size of ignition source, drafts, age and conditioning of sample, atmosphere, temperature, etc.

Common test methods used for plastics are listed below, with a brief description of several of the most commonly cited tests.

Screening Tests. ***Match Test.*** A trivial test giving much information about the burning characteristics of a plastic, the match test consists of simply igniting a test specimen with a common paper match. Some materials that appear to be difficult to ignite with a Bunsen burner flame can be ignited by a match.

ASTM D 2863-76, Oxygen Index Test. This test has become one of the most widely employed screening tests because it is one of the few available test methods that gives reproducible numerical values (19).

In the oxygen index test, a polymer sample (6 × 3 × 80 mm) is placed vertically in a chimney with a controlled atmosphere, and the upper end is ignited with a gas flame. By varying the oxygen:nitrogen ratio, the ratio that just supports a flame is determined. The oxygen index is defined as:

$$\mathrm{OI} = \text{oxygen index} = O_2/(O_2 + N_2) \tag{13}$$

A number of variants have been developed for use with films and fibers. For a given polymer substrate, the higher the index values by this test, the more flame retardant the polymer. The OI of a polymer is primarily sensitive to the elemental composition of the polymer, ie, agent loading. The polymer's flame spreading characteristics (ASTM E-84) are related not only to the elemental composition but also to the cross-link density, degree of cure, presence of volatile compounds, etc (20).

The tendency for some polymers to drip or flow under the conditions of the OI test can result in erroneously high OI values. This effect can be mitigated to a large degree by incorporating a suitable wick, eg, fiber glass, in the test specimen.

UL-94, Test for Flammability of Plastic Materials for Parts in Devices and Appliances. This is a group of simple test methods used with plastic materials to provide a preliminary indication of their suitability with respect to flammability for electrical appliance applications (21). UL-94 specifies material specimens, material conditioning, apparatus, test method, and a classification system for six tests: (*1*) horizontal burning test for classifying materials, 94 HB; this test is a modification of ASTM D 635-76 (22); (*2*) vertical burning test for classifying materials, 94 V-0, 94 V-1, or 94 V-2; (*3*) horizontal burning test for classifying foamed materials, 94 HBF, 94 HF-1, or 94 HF-2; (*4*) vertical burning test for classifying materials, 94-5V; (*5*) flame spread index test using radiant panel; this test cites use of ASTM E 162, test for surface flammability of materials using a radiant heat energy source (23); (*6*) ignition temperature test.

In much of the literature prior to 1972 the results of UL-Subject 94 test, Burning Test for Plastics Classed as Self-Extinguishing (24), refer to a variant of the present Vertical Burning Test for Classifying Materials, 94 V-0, 94 V-1, or 94 V-2. Those results that classified the material as self-extinguishing (Group I) correspond roughly to 94 V-0, and those that classified the material as self-extinguishing (Group II) correspond roughly to 94 V-2. (See Overview article for discussion of these terms.)

ASTM E 84-77, Test for Surface-Burning Characteristics of Building Materials (Tunnel Test). This test, commonly cited in building regulations, is designed to determine the surface flame-spreading, fuel contribution, and smoking characteristics of construction materials (25).

The test apparatus consists of a tunnel 7.62 m long by 44.5 × 30.5 cm in cross section, one end of which contains two gas burners. The test specimen (7.62 m × 50.8 cm) attached to the roof is exposed to a gas flame for 10 min while the maximum flame spread, temperature, and smoke development down the tunnel are determined.

Other Tests. A large number of other tests and standards are used to determine the flammability of materials (26). Many of these are applicable to specific uses of the materials and are seldom cited in the literature.

A few of the more common small- and medium-scale tests are given, along with their characteristics and the type of plastic with which they are most commonly used in Table 2.

Table 2. Medium and Small-Scale Tests for Flammability of Plastics

Designation	Description	Characteristic measured	Type of plastic	Reference
ASTM D 568-77	film and sheeting	flame spread	polyolefins	27
ASTM D 635-77	horizontal rigid bars	flame spread	PVC, unsaturated polyester	22
ASTM D 1692-76	cellular plastics, horizontal	flame spread	flexible polyurethane foams, rigid polyurethane foams, styrenics	28
ASTM D 3014-76	rigid cellular plastics	flame spread	rigid polyurethane foams, styrenics	29
ASTM E 162-76	radiant panel	flame spread, fuel contribution, smoke	engineering thermoplastics, polyolefins, PVC, styrenics, unsaturated polyester	23
ASTM E 119-76	building materials	fire endurance	engineering thermoplastics, PVC, rigid polyurethane foams, styrenics, unsaturated polyester	30
ASTM E 108-75	roof coverings	flame spread	engineering thermoplastics, PVC, rigid polyurethane foams, styrenics, unsaturated polyester	31
Factory Mutual calorimeter	calorimeter	fuel contribution	engineering thermoplastics, PVC, rigid polyurethane foams, styrenics, unsaturated polyester	32
corner-wall fire test		flame spread	engineering thermoplastics, PVC, rigid polyurethane foams, styrenics, unsaturated polyester	33
UL-44 (1975)		ease of ignition	engineering thermoplastics, polyolefins	34
UL-796 (1975)		flame spread	epoxy resins, polyolefins, unsaturated polyester	35
Hooker HLT-15	rigid reinforced plastics	flame spread	unsaturated polyester	36
MVSS 302		flame spread	engineering thermoplastics, flexible polyurethane foams, PVC, rigid polyurethane foams, styrenics, unsaturated polyester	37

Polymer Classes

Synthetic polymers (qv) can be grouped into two broad classes: thermoplastics and thermosets. Thermoplastics are generally linear polymers that melt readily and are molded with heat and/or pressure. Thermosets are generally handled as liquids, set by cross-linking the polymer chains, and do not melt after they are cured. Flexible polyurethane foams are an intermediate class of polymers that are lightly cross-linked and flow, or shrink, upon the application of sufficient heat.

Thermoplastics such as polystyrene are frequently flame-retarded by adding a small amount (1–4%) of an aliphatic bromine compound having a low (<250°C) decomposition temperature. Upon heating, the bromide decomposes, promotes chain scission, and allows the polymer to melt and drip away from the ignition source.

The addition of any material, eg, glass, rubber, fillers, etc, to a thermoplastic that relies on dripping for its flame retardancy retards the dripping and allow ignition to occur. In applications where dripping is prevented or retarded, a considerably increased loading of agent is generally required. For example, high-impact (rubber-filled) polystyrene requires the use of ca 20 wt % bromine (38–39) vs the 4–5 wt % required for neat polystyrene in order to obtain a UL-94 V-2 rating (40).

Thermoset polymers generally require ca 20–30 wt % chlorine or 12–15 wt % bromine in order to show a significant degree of flame retardancy in moderately severe tests, such as the HLT-15 test (41). To meet the most severe requirement (ASTM E 84, Flame Spread Classification (FSC) of <25) unsaturated polyesters may require >20 wt % bromine (42) or chlorine with an adjunct agent such as antimony oxide.

Chlorine and bromine compounds are the only halogens commercially significant as flame-retardant chemicals. Fluorine compounds are expensive and, except in special cases, ineffective. Iodine compounds, although effective, are expensive and too unstable to be used.

The number of chlorinated compounds used as fire retardants has decreased in recent years, and most of the newer agents are based on bromine. This shift has occurred at least partially as a result of the increasingly severe flammability standards imposed on plastic products. These standards normally require levels of chlorine that are difficult to obtain without destroying the desirable properties of the polymer. In addition, the decreased use of lead in gasoline has freed large quantities of bromine that were previously used in the production of ethylene dibromide. The anticipated large excess in bromine capacity prompted new uses for bromine, among them a number of flame-retardant chemicals has been developed.

Bromine compounds have a number of inherent properties that make them excellent flame retardants. For example, bromine compounds generally are ca twice as effective as chlorine compounds having the same halogen content on a weight basis and are ca 1.7 times as dense as chlorine compounds.

It is desirable for the agent to decompose with the liberation of halogen at a somewhat lower temperature, ca 60°C, than the decomposition temperature of the polymer. Bromine-containing compounds generally decompose in the range of 200–300°C because of the relatively low carbon–bromine bond energy and consequently have decomposition temperatures that mesh quite well with the decomposition temperatures of most common plastics.

Table 3 shows a comparison of two typical bromine and chlorine compounds in ABS formulated to give about the same degree of fire retardancy: UL-94 Group 1

Table 3. Comparison of Typical Bromine- and Chlorine-Containing Agents in ABS[a]

Component	Density (g/cm^3)	Parts			
ABS	1.05	100[b]	100[b]	100[c]	100[c]
COD-diHex[d]	1.80	0	25		
DBDPO[e]	3.04			0	10
Sb_2O_3	5.7	0	10	0	5
		Properties			
tensile strength, MPa[f]		39.77	33.82	37.37	40.65
UL-94		fail	group 1	fail	group 1 (VO)
agent in system (estd), vol %		0	15	0	4
density of plastic (estd), g/cm^3		1.05	1.27	1.05	1.21

[a] Refs. 38 and 42.
[b] Blendex-211, Marbon Chemical Company.
[c] Dow ABS 500, Dow Chemical Company.
[d] Dechlorane Plus, Hooker Chemical Company.
[e] FR-300-BA (decabromodiphenyl oxide), Dow Chemical Company.
[f] To convert MPa to psi, multiply by 145.

(39,43). The data, although not strictly comparable because of differences in the properties of the ABS, are nevertheless informative. The high loading required with 1,5-bis(chlorendo)cyclooctane [*13560-89-9*] (COD-diHex or Dechlorane Plus (**1**))-Sb_2O_3 results in a system that is only ca 85 vol % polymer as compared to 96 vol % polymer for the bromine-containing system and probably accounts for much of the loss in tensile strength with the COD-diHex-Sb_2O_3 system. The brominated system has lower density. Such lower densities often exist when bromine is used in place of an equivalent flame-retarding amount of chlorine. This decrease in density may result in 5–10% additional volume for the same weight of plastic. The lower density of the bromine-containing systems should not be overlooked since plastics are almost always sold on a volume basis, not a weight basis.

Table 4 lists the relative amounts of either chlorine or bromine required for equivalent degrees of flame retardancy for a number of plastics (40). The specific polymer and other additives may affect the test performance; therefore, the exact level should be established for the specific system.

Table 4. Relative Halogen Requirements for Equivalent Degree of Fire Retardancy of Various Polymers[a]

Polymer	wt % Cl	wt % Br
polyolefins	40	20
acrylates	20	16
polyacrylonitrile	10–15	10–12
styrene	10–15	4–5
ABS	23	3
polyurethane	18–20	12–14
unsaturated polyester	25	12–15
epoxies	26–30	13–15

[a] Ref. 40.

Additive Flame Retardants

Nonreactive additives have commonly been used for flame-retarding polymers. The ideal additive is inexpensive, colorless, easily blended, compatible, heat and light stable, efficient, permanent, and has no deleterious effects upon the properties of the base polymer. The most important limitations are incompatibilities that affect the physical properties of the polymers, and the tendency for additives to be fugitive.

Owing to environmental and toxicity problems as well as the tendency for additives to be fugitive, several excellent halogenated flame retardants have been abandoned during the development stage or have been removed from the market. Notable examples include polychlorobiphenyls, polybromobiphenyls, tris(2,3-dibromopropyl) phosphate and several derivatives of hexachlorocyclopentadiene.

The principal commercially available additive flame retardants are listed individually below. A number of other agents are listed by their manufacturers as available on a commercial basis but few specific details are available. Some of these agents are tetrabromoxylene [*23488-36-2*] (44), 1-(1,2-dibromoethyl)-3,4-dibromocyclohexane [*3322-93-8*] (44), and 1,2-bis(tetrabromophthalimido)ethane [*32588-76-4*] (45).

Newer agents that are listed by their manufacturers as available on a developmental basis are pentabromodiphenyl oxide [*32534-81-9*] (46), hexabromobenzene [*87-82-1*] (46), bis(2,4,6-tribromophenyl)carbonate [*67990-32-3*] (46), and pentabromotoluene [*87-83-2*] (46).

Polychlorinated Paraffins. Among the first of the halogenated organics to be developed as flame retardants were the resinous chlorinated paraffins which reached commercial importance during World War II as a treatment for cotton duck for the military. They are still by far the biggest sales item of the halogen compounds in the flame-retardant chemicals market.

The chlorinated paraffins as a family are produced by several manufacturers by the controlled chlorination of normal paraffinic hydrocarbons. The chlorination is controlled to yield mainly liquid products containing 20–70 wt % chlorine. Several solid resinous products containing 70 wt % chlorine are also available (see Chlorocarbons). The products are available in a variety of formulations, ranging from neat oils and solids to water-based emulsions (qv). A number of the products are marketed with heat stabilizers (qv) and/or antimony oxide present.

Chlorinated paraffins are generally used to replace conventional resins and plasticizers, and provide the desired degree of flame retardancy. The organic or polymeric portion of a typical formulation relying solely on chlorinated paraffins for flame retardancy contains 25–90 wt % of one or more chlorinated paraffins. These compounds tend to be plasticizers (qv) and the amount that may be used is generally limited because of their lack of compatibility at higher concentrations (47–48).

Relative to the aromatics, the chlorinated paraffins are thermally unstable and restricted to applications where the processing temperature does not exceed ca 175°C. In applications requiring higher processing temperatures, the addition of stabilizers of the types normally used for processing PVC may be used (see Vinyl polymers). Polychlorinated paraffins are principally employed as secondary plasticizers in PVC for wire and cable coatings, calendered products, and profile extrusions; in unsaturated polyester resins; in polyolefins for paints, canvas, and adhesives, and in intumescent coatings.

The polychlorinated paraffins are classified as nontoxic, both dermally and by ingestion, according to the Federal Hazardous Substances Labelling Act of 1961. They are all somewhat biodegradable and insoluble in water. New tests have been recommended by the ITC (Interagency Testing Committee), authorized under TSCA (49).

COD-diHEX (Dechlorane Plus). A variety of additive type flame-retardant chemicals based upon hexachlorocyclopentadiene have been offered previously, but most of these products have been discontinued during recent years, eg, Dechlorane [*2385-85-5*].

The principal remaining product from this family of compounds is (1) COD-diHEX or Dechlorane Plus. COD-diHEX is readily prepared by a Diels-Alder reaction of two equivalents of hexachlorocyclopentadiene (HEX) and 1,5-cyclooctadiene (COD), as indicated in the following reaction:

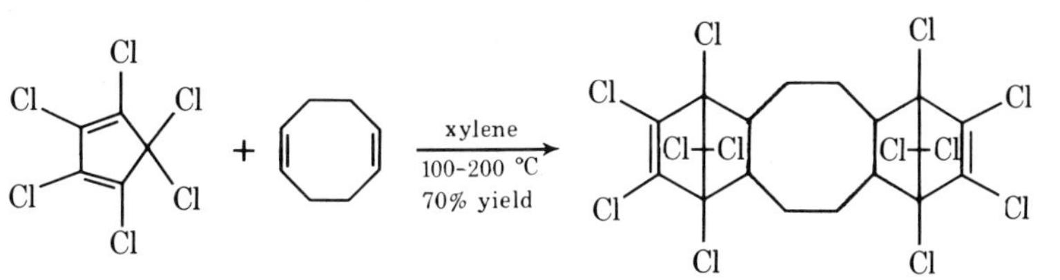

(1)

COD-diHEX is prepared by heating a 2:1 molar ratio of hexachlorocyclopentadiene (HEX) and 1,5-cyclooctadiene (COD) in xylene at 100–200°C, without catalyst, for ca 12 h (see Cyclopentadiene). The reactor slurry is cooled and the solid product recovered. Xylene and unreacted HEX and COD are recovered by vacuum distillation and recycled. The yield is ca 70% based on monomers charged to the reactor (50). Properties of COD-diHEX are shown below.

chlorine content, wt %	65
melting point, °C	350
specific gravity, 25°C	1.80

COD-diHEX, a very stable flame-retardant additive, is especially suited for applications that require high temperature processing characteristics, no weight loss, no plasticization, and good moisture resistance. Its primary markets are in polypropylene, polyethylene, nylon, and thermosetting resins, where it is often used with antimony oxide.

Decabromodiphenyl Oxide (DBDPO). Decabromodiphenyl oxide [*1163-19-5*] is produced and sold by several manufacturers (39,51). It is a free-flowing white powder that usually contains more than 85 wt % of decabromodiphenyl oxide, and less than 15 wt % of other brominated diphenyl oxides, principally nanobromodiphenyl oxide [*68928-79-0*]. Properties of DBDPO are shown below (40).

molecular weight	960
bromine content (theoretical), %	83
melting range, °C	295–305
decomposition point, °C (dta)	ca 425
specific gravity	3.04

DBDPO is prepared by the bromination of diphenyl oxide in the presence of a Friedel-Crafts catalyst (see Friedel-Crafts reactions). The exact details of manufacture, ie, raw materials, catalysts, solvents, temperatures, by-products, etc, have not been disclosed by any of the several manufacturers. Process differences are probably not significant with respect to the efficacy of the product as a flame retardant. However, differences in compatibility with the polymer during processing are observed in specific applications, eg, wire and cable coatings, fibers, etc.

Manufacturing differences may also affect the nature and amounts of impurities in the product, and thus the toxicological and environmental properties. The toxicological and environmental properties of DBDPO have been evaluated in great detail (52–54), and although the results of these studies may be superficially applied to any DBDPO, regardless of the manufacturing process, care should be exercised in extrapolating these data to materials prepared by processes other than that employed in obtaining the product used in the actual toxicological tests.

DBDPO is a very thermally stable bromine compound and can be employed with polymer systems that require high processing temperatures. DBDPO is widely employed in conjunction with antimony oxide in high-impact polystyrene polymers that are used in the manufacture of television cabinets.

DBDPO is also recommended for use in polycarbonates, polyester resins, polyolefins, ABS, polyamides, PVC, and rubber, although its acceptance in many of these systems is limited to special applications. DBDPO has limited utility in polymers that have relatively low decomposition temperatures since the high thermal stability of the aromatic bromine tends to make the agent inefficient.

DBDPO has relatively poor uv stability in styrenic systems and causes the formation of a red–brown coloration. The use of uv stabilizers can minimize, but not eliminate, this effect. This photochemical instability accounts to a large extent for the environmental acceptability of this material.

Pentabromochlorocyclohexane (PBCCH). Pentabromochlorocyclohexane [*25495-99-2*] is a dense white, high-melting solid (mp >190°C) containing 77.9 wt % bromine and 6.9 wt % chlorine (55). No manufacturing details have been published. The compound is employed extensively in foamed polystyrene to modify its flammability characteristics. Other potential uses have been suggested, eg, rubber, but the agent's relatively low decomposition temperature (235°C) tends to limit its use to those polymer systems that are not exposed to excessive processing temperatures.

Hexabromocyclododecane (HBCD). Hexabromocyclododecane [*25637-99-4*] was first proposed as a flame retardant for polystyrene in 1963. The compound, a white crystalline solid having a bromine content of 74.7 wt % and a melting point of 170–180°C, is prepared by the bromination of cyclododecatriene, a butadiene trimer (46). The major market for the compound is in foamed polystyrene, although it has been proposed as a flame retardant for polyolefins.

2,3,4,5,6-Pentabromoethylbenzene (PBEB). 2,3,4,5,6-Pentabromoethylbenzene [*85-22-3*] is a solid of high bromine content (80 wt %) with a melting point of 136–138°C and a decomposition temperature of ca 210°C. The compound is prepared by the Friedel-Crafts-catalyzed bromination of ethylbenzene in an inert solvent (51).

Few details are available with respect to its use, although one of its manufacturers recommends its use in epoxy resins, ABS, and urethane foams.

1,2-Bis(2,4,6-tribromophenoxy)ethane (BTBPE). This additive [*37853-59-1*] is a white crystalline powder having the high bromine content and thermal stability required to make it an excellent flame retardant for those polymer systems requiring high processing temperatures (56). The compound also has the light stability needed for many plastic applications.

BTBPE (**2**) can be prepared by the reaction of tribromophenol with ethylene dibromide in the presence of a base.

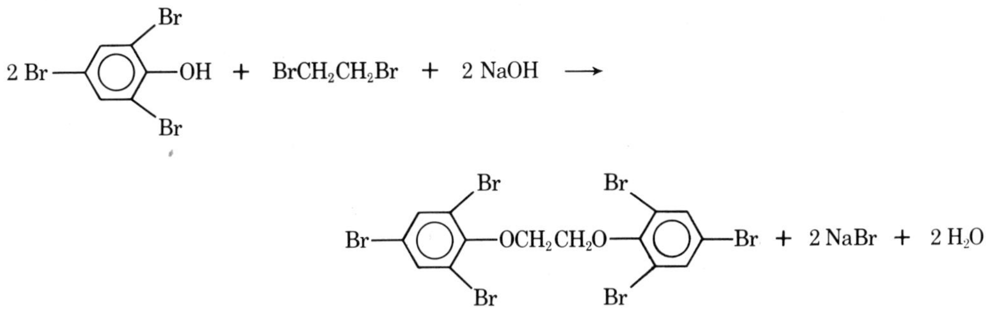

(**2**)

Properties of BTBPE are shown below.

bromine, wt %	70
melting range, °C	223–225
density, g/cm^3	2.58
TGA, 1 % wt loss, °C	244

BTBPE has an acute oral LD_{50} greater than 10 g/kg of body weight (male rats) and an acute dermal LD_{50} greater than 10 g/kg of body weight (rabbits). Within the meaning of the Federal Hazardous Substances Act as amended January, 1971, BTBPE is nontoxic orally or dermally and is not a skin or eye irritant. Subacute feeding studies conducted on white rats showed minimal effects and low accumulation in fat, liver, and muscle tissues. The small amounts that accumulated rapidly decreased on withdrawal of the test feed.

The principal market for BTBPE at present is in ABS polymer systems, although other polymer systems, such as high impact polystyrene, polyesters, etc, represent potential markets.

Reactive Flame Retardants

Reactive halogenated flame-retardant agents usually consist of compounds that contain, in addition to the halogen, a chemically reactive moiety that can enter into the polymerization reaction. The halogen-containing monomer can be: a vinyl group, for agents intended for use in vinyl polymers, eg, polystyrene, polyacrylates, etc; or an anhydride, carboxyl, or hydroxyl group, for agents intended for use in condensation polymers, eg, polyesters, polyurethanes, epoxies, etc.

Reactive agents generally lose their chemical identity and become a permanent part of the polymer backbone. As a consequence, many of the problems associated with additive agents do not apply to reactive agents. The development of polymer systems based upon reactive agents is more time-consuming and more expensive because in using these agents the manufacturer is developing new copolymer systems. It is no longer a question of avoiding the loss of some properties of an existing polymer but a question of developing an inherently flame-retardant polymer system having the desired fire retardancy as well as the physical properties required for the desired application.

Two of the oldest vinyl monomers, vinyl chloride and vinylidene chloride, are prime examples of the use of reactive flame-retardant monomers that are extensively employed to produce inherently flame-retardant polymer systems, eg, PVC and Saran-type resins (see Vinyl polymers). These polymer systems have become so numerous and economically important that they are not normally included in estimates of the amount of halogen consumed in the flame-retardant market.

The principal reactive flame retardants are listed individually below along with some of their properties and uses. In some cases, the agent listed forms the basis for other agents, eg, tetrabromobisphenol-A (TBBP-A) is employed to prepare a variety of other commercially important agents such as the diglycidyl ether of TBBP-A for use in epoxy resins and vinyl ester resins, and ethoxylated and propoxylated TBBP-A for use in polyester resins and polyurethanes.

A number of developmental reactive agents are presently available, including: epibromohydrin [*3132-64-7*] (51), tri- and pentabromophenol [*118-79-6*] and [*608-71-9*] (46), 3-bromo-2,2-bis(bromomethyl)propanol [*1522-92-5*] (57), 2,3-dibromopropyl methacrylate [*3066-70-4*] (46), and *trans*-2,3-dibromo-2-butene-1,4-diol [*3234-02-4*] (58). In certain cases, elemental bromine and hydrogen bromide have been employed as reactive agents in the production of inherently flame-retardant polyester resins.

A postbrominated, unsaturated polyester resin has achieved commercial importance. This resin is prepared by brominating a methylene chloride solution of a maleic anhydride–tetrahydrophthalic anhydride–glycol resin (59). The resin solids are dissolved in styrene after the methylene chloride is stripped. The bromine reacts selectively with the double bond in the tetrahydrophthalate moiety. Polyester resins containing up to 23 wt % bromine can be prepared in this manner.

A developmental, unsaturated polyester resin prepared by the simultaneous hydrobromination and polymerization of a mixture of pentaerythritol, maleic anhydride and phthalic anhydride has been recently introduced (60). This process allows the preparation of resin solids containing up to 52 wt % bromine, although the material as currently marketed contains ca 45 wt % bromine in the resin solids or 33 wt % bromine in the styrenated (25 wt % styrene) resin (61).

Tetrachlorophthalic Anhydride (TCPA). The reactive compound (**3**) [*117-08-8*], has long been available as a flame retardant. The compound is manufactured by the ferric chloride catalyzed chlorination of phthalic anhydride.

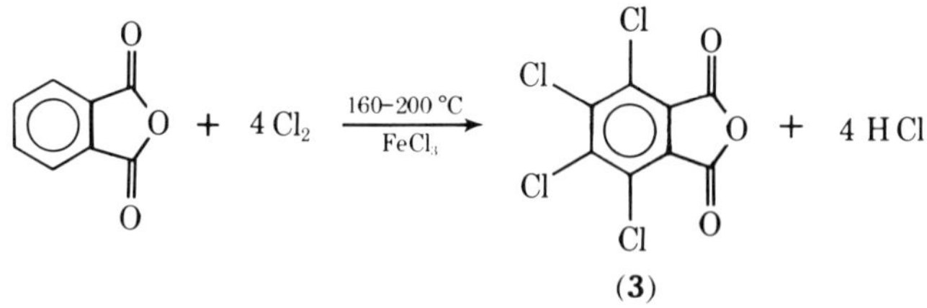

(**3**)

The relatively low chlorine content (49.6 wt % Cl) and the lower flame-retardant efficiency of the aromatic halogens limits their use to unsaturated polyester resin formulations that do not require a high degree of flame retardancy.

Chlorendic Anhydride and Chlorendic Acid. Chlorendic acid [*115-28-6*] (**4**) (hexachlorobicyclo[2,2,1]-hept-5-ene-2,3-dicarboxylic acid) and its corresponding anhydride [*115-27-5*] (**5**), are among the most important reactive flame retardants and flame-retardant precursors in use today (see Cyclopentadiene).

Chlorendic acid (**4**) is synthesized by a Diels-Alder reaction of maleic anhydride and hexachlorocyclopentadiene (HEX) in toluene followed by hydrolysis of the anhydride with aqueous base (62).

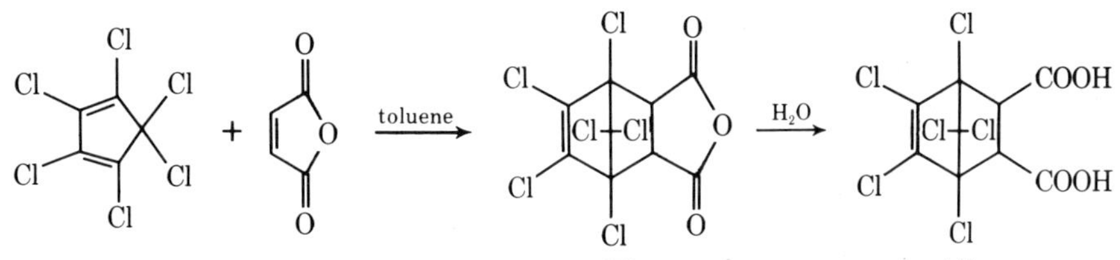

(5) (4)

As described in the patent literature, chlorendic acid can be prepared by the reaction of equimolar quantities of HEX and maleic anhydride in chlorobenzene at 140–150°C for 8–10 h. The hot anhydride solution is then added to water to effect the hydrolysis and the insoluble acid is isolated as the monohydrate in better than 90% yields. Drying at 100–105°C yields an essentially anhydrous product of 99.6% purity.

The anhydride can be isolated directly from the reaction mixture or can be prepared in a very pure form by dehydration of the acid. Properties of chlorendic anhydride and chlorendic acid are shown below (63).

Property	*Chlorendic anhydride*	*Chlorendic acid*
molecular weight	370.9	388.9
chlorine content, wt %	57.4	54.7
appearance	crystalline solid	crystalline solid
melting point, °C	240–241	decomposes to the anhydride

The principal use of chlorendic anhydride and chlorendic acid has been in the manufacture of unsaturated polyester resins. Because the esterification rate of chlorendic anhydride is similar to that of phthalic anhydride, it can be used in place of phthalic anhydride in commercial polyester formulations. The double bond in chlorendic acid is not reactive as a cross-linking site, hence reactive monomers, such as maleic anhydride, must be included in the polyester backbone to achieve cross-linking.

The relatively low halogen content of this monomer limits the amount of chlorine that can be introduced to ca 28 wt %. In order to formulate resins with ASTM E 84 flame-spread ratings in the under 25 class, antimony oxide is usually employed as a synergist.

Chlorendic anhydride also can be used in oil-modified alkyds (see Alkyd resins), and as flame-retardant curing agent for epoxy resins. The acid has found some use in polyurethane foams when allowed to react with nonhalogenated glycols to form halogenated polyols.

Tetrabromobisphenol-A (TBBP-A). This versatile brominated bisphenol [*79-94-7*] (**6**) is the primary raw material for several other widely used fire-retardant agents (57,64). TBBP-A is a high melting (mp 179–181°C), dense white solid containing 58.8 wt % of bromine. TBBP-A is manufactured by the bromination of bisphenol-A according to the equation:

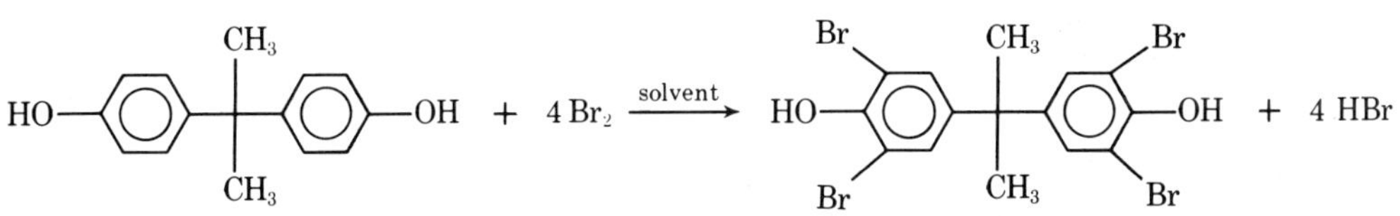

(**6**)

Details of the commercial process are not available; however, use of an alcohol as the solvent allows the preparation of a very high quality of TBBP-A, as well as the recovery of much of the by-product HBr as a marketable alkyl bromide (65).

TBBP-A is marketed in two grades, the specifications of which are given below. The higher quality polycarbonate grade is employed in polymer systems that require high thermal stability during processing, and the epoxy resin grade is usually suitable as a chemical intermediate or in polymer systems that are processed at lower temperatures. TBBP-A specifications (51) are listed in Table 5.

Although TBBP-A has achieved commercial importance as a flame retardant for polycarbonates, its primary use is as an intermediate in the manufacture of epoxy resins. For use in epoxy resins, TBBP-A reacts with epichlorohydrin and caustic to give the diglycidyl ether of TBBP-A (**7**) (see Epoxy resins).

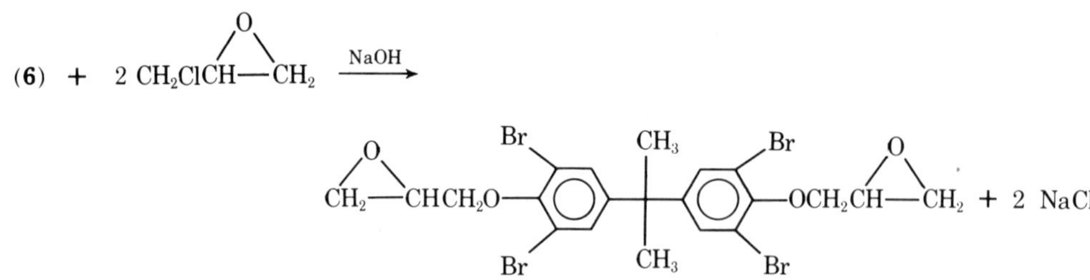

(**7**)

The diglycidyl ether of TBBP-A may react further with methacrylic acid to yield vinyl ester resins.

TBBP-A may react also with alkylene oxides to yield glycols which are employed in the manufacture of unsaturated polyester resins and thermoplastic polyesters.

Dibromoneopentyl Glycol (DBNPG). The commercial form of DBNPG is a mixture containing 80–82 wt % DBNPG, 5–7 wt % monobromoneopentyl triol [*19184-65-7*] and 13–15 wt % tribromoneopentyl alcohol [*1522-92-5*]. The principal compound in the mixture, 2,2-bis(bromomethyl)-1,3-propanediol [*3296-90-0*] $(BrCH_2)_2C(CH_2OH)_2$ is DBNPG.

Properties of the pure compound and the commercial product are shown in Table 6.

DBNPG is prepared by the stepwise hydrobromination of pentaerythritol to yield a product in which about half of the pentaerythritol hydroxyl functions have been replaced (see Alcohols, polyhydric). One process consists of the acetic acid-catalyzed reaction of pentaerythritol, slurried with a solvent, and anhydrous hydrogen bromide until the proper composition is obtained (67). Pure DBNPG can be readily obtained by crystallizing the crude product from an appropriate solvent, eg, water, methanol, toluene, etc.

Table 5. TBBP-A Specifications

Specification	Polycarbonate grade	Epoxy resin grade
APHA color[a], max	25	50
freezing pt, °C min	179.5	178
iron, ppm max	1.0	3.0

[a] As measured on 10 g in 50 mL reagent methanol.

Table 6. Properties of Dibromoneopentyl Glycol[a]

Property	DBNPG (pure) White crystalline solid	DPNPG (commercial) Small white flakes
melting point, °C	109–110	75–95
bromine content, wt %	61.01	59–61
hydroxyl content, wt %	13.0	11.5–12.5
color (20% in methanol) APHA	20	60
decomposition point, °C	235	

[a] Ref. 66.

The principal use of DBNPG is in unsaturated polyester resins, where it can be used to replace part of the regular glycol to yield a resin having the desired bromine content. A preferred method of using the compound is to prepare a resin using DBNPG as the sole glycol. This technique gives a resin solid containing ca 45 wt % bromine which may be blended with either a nonhalogenated or halogenated resin to give a product having the desired degree of flame-retardant properties (68).

DBNPG has been successfully used in both rigid and flexible urethane foam systems (see Foams). For rigid systems DBNPG can be used by itself to produce Class II foams or with phosphorus co-agents to produce Class I foams as evaluated by the ASTM E 84 test (69).

DBNPG has been employed to prepare a variety of derivatives that are useful in other polymer systems. Its diglycidyl ether is found to be an unusual epoxy resin. The di-DBNPG ester of adipic acid is employed in urethane foams (69). A number of reaction products of DBNPG and various phosphorus compounds have been shown to be of value in other polymer systems.

DBNPG can in certain systems be employed as an additive to increase the halogen content of polymer systems that do not require high processing temperatures.

Tetrabromophthalic Anhydride (TBPA). This compound [*632-79-1*] is a widely used reactive flame retardant in unsaturated polyesters as well as the precursor to a number of other fire retardants, eg, [*20566-35-2*] (**8**) which are designed for use in other polymer systems, such as polyurethane foams (70).

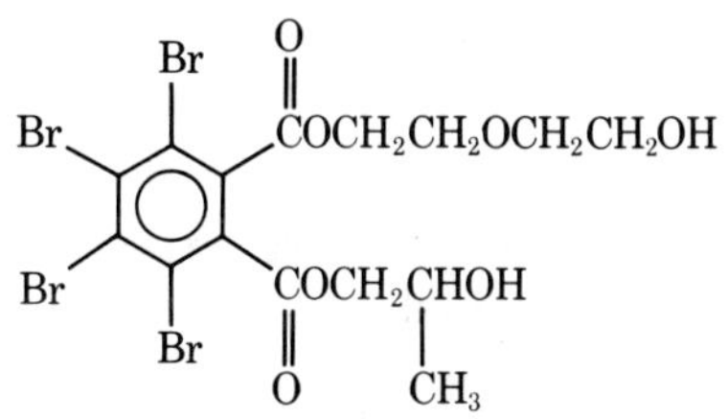

(**8**)

TBPA is prepared by the bromination of phthalic anhydride in 60% oleum in 95% yields (71). The use of oleum as the bromination solvent results in some sulfonation of the aromatic ring (36). Sulfonated material is removed as far as possible by hydrolyzing the anhydride with dilute NaOH, filtering, and acidifying with dilute HCl. The precipitated acid is washed several times with hot water and reconverted to the anhydride by heating at 150°C for several hours. Residual sulfonic acid groups are normally neutralized by adding a small amount of a basic substance, eg, sodium acetate or alkanolamine, to the reaction mixture when TBPA is employed in unsaturated polyester reactions.

TBPA is an off-white, crystalline, high-melting (mp 276–280°C) solid with a bromine content of 68.8 wt %. Polyesters prepared from this compound have relatively poor photochemical stability and tend to discolor upon exposure to light. This tendency to discolor can be reduced somewhat, though not eliminated, by the use of uv stabilizers (qv).

Vinyl Bromide (VBR). Vinyl bromide [*593-60-2*], a colorless gas at ambient temperature and pressure, is a reactive flame-retardant monomer (72). This monomer, through its vinyl functionality, is capable of homopolymerization, or copolymerization with other reactive monomers. The homopolymer, because of its thermal and photochemical instability, is of very little interest *per se;* however, various copolymers are suitable for the manufacture of fibers, films, and extruded or molded objects.

Vinyl bromide is manufactured by the reaction of 1,2-dibromoethane (ethylene dibromide, EDB) with hot, aqueous sodium hydroxide. The monomer, which has a bromine content of 74.7 wt %, boils at 15.85°C at atmospheric pressure. The monomer has flammability limits in air of 9–15 vol %, although it does not have a flash point by the Tag Closed Cup method.

The monomer is shipped and stored as a liquid under pressure, and is available in cylinders, tank trucks, and tank cars.

Vinyl bromide has a relatively low acute toxicity, but because of its chemical similarity to vinyl chloride it would be expected to show similar chronic effects at similar exposure levels. Potential users should check with the manufacturers for current recommended exposure levels.

Vinyl bromide is used primarily as a flame retardant in the manufacture of modacrylic fibers (see Acrylic and modacrylic fibers). It is copolymerized with vinyl acetate and maleic acid or anhydride to produce finely divided granular products, or with vinyl chloride to yield relatively low molecular weight polymers. The vinyl chloride–vinyl bromide copolymers, because of their solubility in organic solvents, are suitable for preparing films, for impregnating or laminating fibers, and for use as rubber substitutes.

Economic Aspects

The consumption of halogenated flame retardants in the United States in 1976 was ca 30.5 thousand metric tons valued at ca 60 million dollars, excluding the halogenated organophosphorus compounds (73). The totals include compounds ranging in price from ca 50¢/kg for some of the chlorinated paraffins to ca \$3/kg for some of the more expensive bromine-containing agents. The growth rate for the total flame-retardant chemical market, which includes the halogenated compounds, phosphorus compounds, alumina trihydrate, antimony oxide, etc, is estimated to be greater than 10%/yr over the next 5 yr (74). The future growth of flame retardants in general will be greatly affected by both the rate of introduction of new standards and the strictness of those standards.

The future adoption of low-smoke and noncorrosive requirements for plastic products could force the development of new flame-retardant technology completely independent of the halogen technique. This, however, will require a massive breakthrough in the development of low-cost thermally-stable polymers, and in sensitive areas, it may require the abandonment of organic polymers entirely.

The introduction of new halogenated flame retardants may be significantly slowed owing to the expense of toxicology testing now required. Most flame-retardant research appears to be directed toward developing methods of expanding the usefulness of the present commercial agents in the larger volume polymer systems. This could be highly advantageous in that the high price of many of the agents is due more to the low production volumes than to the price of the raw materials.

BIBLIOGRAPHY

"Halogenated Fire Retardants" in *ECT* 2nd ed., Supplement, pp. 467–488, by V. A. Pattison and R. R. Hindersinn, Hooker Chemical Corp.

1. J. W. Lyons, *The Chemistry and Uses of Fire Retardants,* Wiley-Interscience, New York, 1970, pp. 14–17.
2. E. C. Creitz, *J. Res. Natl. Bur. Stand.* **74A,** 521 (1970).
3. G. Avondo, C. Vorelle, and R. Delbourgo, *Combust. Flame* **31,** 7 (1978).
4. E. R. Larsen and R. B. Ludwig, *J. Fire Flammability* **10,** 69 (1979).
5. E. R. Larsen, *J. Fire Flammability/Fire Retardant Chemistry* **1,** 4 (1974).
6. E. R. Larsen, Paper INDE-054, *presented at 166th National Am. Chem. Soc. Meeting, Chicago, 1973.*
7. A. Tewarson and R. F. Pion, *Final Tech. Report, FMRC No. 22524,* Factory Mutual Research, Norwood, Mass., 1977, pp. 56–57.
8. D. B. Spalding, *Fourth Symposium (International) on Combustion,* The Combustion Institute, Pittsburgh, 1953, p. 847.
9. E. R. Larsen, *Proceedings Annual Conference Reinf. Plast./Compos. Inst.,* Soc. Plast. Ind., Vol. 32, Section 17-B, 1977.
10. A. Tewarson, *Technical Progress Report,* FMRC No. 22429, Factory Mutual Research, Norwood, Mass., 1975, p. 8.
11. S. K. Brauman and N. Fishman, *J. Fire Retardant Chem.* **4,** 93 (1977)
12. S. K. Brauman and A. S. Brolly, *J. Fire Retardant Chem.* **3,** 66 (1976); S. K. Brauman, *ibid.,* **3,** 117, 138 (1976).
13. J. W. Hastie and C. L. McBee in R. G. Gann, ed., *Halogenated Fire Suppressants,* ACS Symposium Series 16, American Chemical Society, Washington, D.C., 1975, p. 118.
14. J. J. Pitts in W. C. Kuryla and A. J. Papa, eds., *Flame Retardancy of Polymeric Materials,* Vol. 1, Marcel Dekker, Inc., New York, 1973, pp. 163–181.
15. J. J. Pitts, P. H. Scott, and D. G. Powell, *J. Cell. Plast.* **6,** 35 (1970).
16. S. K. Brauman, *J. Fire Retardant Chemistry* **4,** 18 (1977).
17. E. D. Weil in W. C. Kuryla and A. J. Papa, eds., *Flame Retardancy of Polymeric Materials,* Vol. 3, Marcel Dekker, Inc., New York, 1975, pp. 185–243.
18. C. J. Hilado, *Flammability Handbook for Plastics,* 2nd ed., Technomic, Westport, Conn., 1974, pp. 36–37.
19. *ASTM D 2863-76, ASTM Standards,* Vol. 35, 1977, p. 721.
20. E. R. Larsen, *J. Fire Flammability* **9,** 156 (1978).
21. *UL-94 Standard for Tests for Flammability of Plastic Materials for Parts in Devices and Appliances,* 2nd ed., Underwriters Laboratories, Skokie, Ill., 1976.
22. *ASTM D 635-76,* ref. 19, p. 209.
23. *ASTM E 162-76, ASTM Standards,* Vol. 18, 1977, p. 790.

24. *Underwriters Bulletin Subject 94 Burning Test for Plastics Classed as Self-Extinguishing,* Underwriters Laboratories, Skokie Ill., 1965.
25. *ASTM E 84-77,* ref. 23, p. 690.
26. Ref. 18, p. 91.
27. *ASTM D 568-76,* ref. 19, p. 187.
28. *ASTM D 1692-76,* ref. 19, p. 536.
29. *ASTM D 3014-76,* ref. 19, p. 743.
30. *ASTM E 119-76,* ref. 23, p. 739.
31. *ASTM E 108-75,* ref. 23, p. 727.
32. *Fire Hazard Standard, Class I Building Materials,* Factory Mutual Research, Norwood, Mass., 1970.
33. *Factory Mutual Building Corner Fire Test Procedure,* Factory Mutual Research, Norwood, Mass., 1972.
34. *Std. UL 44,* Underwriters Laboratories, Skokie, Ill., 1975.
35. *Std. UL 796,* Underwriters Laboratories, Skokie, Ill., 1975.
36. R. C. Nametz and R. S. Nulph, *Proc. 20th Anniv. Tech. Conf., Soc. Plast. Ind., Reinf. Plast. Div., Sect. 11-C,* Chicago, 1965.
37. *Fed. Regist.* **36**(5), 289 (1971).
38. R. V. Petrella in M. Lewin, S. M. Atlas, and E. I. Pearce, eds., *Flame Retardant Polymeric Materials,* Plenum Publishing Corp., New York, 1978, p 159.
39. *FR-300-BA (Decabromodiphenyl Oxide),* Form No. 101-92-75, Dow Chemical, Midland, Mich., 1975.
40. Ref. 1, p. 23.
41. E. R. Larsen, *J. Fire Flammability/Fire Retardant Chemistry* **2,** 209 (1975).
42. E. R. Larsen and W. C. Weaver, *Proc. Annu. Conf. Reinf. Plast. Compos. Inst.,* Soc. Plast. Ind., Vol. 28, Sect. 2A, 1973.
43. I. Touval and H. H. Waddell, *What You Should Know About Flame Retardants,* M & T Chemical, Rahway, N.J., 1971.
44. *Emery 9345 TBX Flame Retardant,* Data Sheet 812, Emery Industries, Cincinnati, Ohio, 1977.
45. *Plast. Technol.* **23**(8), 75 (1977).
46. *Development Products—Flame Retardants,* Great Lakes Chemical, West Lafayette, Ind., 1976.
47. *Built-In Flame Retardancy with Chlorowax Chlorinated Paraffins,* EC-CW-48B, Diamond Shamrock Chemical, Cleveland, Ohio, 1972; *Chlorowax and Delvet,* EC-CW-28E, Diamond Shamrock Chemical, Cleveland, Ohio, 1977.
48. *Chlorinated Paraffins Tech. Data Sheets,* Dover Chemical, Dover, Ohio 1977.
49. *Fed. Regist.* **43,** 50, 134 (Oct. 26, 1978).
50. U.S. Pat. 3,403,036 (Sept. 24, 1968), R. R. Hindersinn and J. F. Porter (to Hooker Chemical); U.S. Pat. 3,392,136 (July 9, 1968), R. R. Hindersinn and H. W. Marciniak (to Hooker Chemical).
51. *Flame Retardants,* Great Lakes Chemical, West Lafayette, Ind., 1976.
52. *Ecotox Studies on FR-300-BA,* Form No. 101-129-78, Dow Chemical, Midland, Mich., 1978.
53. J. M. Norris and co-workers, *J. Fire Flammability/Comb. Tox.* **1,** 52 (1974).
54. R. J. Kociba and co-workers, *J. Comb. Tox.* **2,** 267 (1975).
55. *Flame Retardant Chemicals—Development—FR-651-A,* Form No. 101-113-78, Dow Chemical, Midland, Mich., 1978.
56. *Firemaster 680,* Michigan Chemical (Velsicol), Chicago, Ill., 1974.
57. *Fire Retardant Chemicals,* Form No. 101-13-74, Dow Chemical, Midland, Mich., 1974.
58. *Dibromobutenediol,* Tech. Bull. 9640-101, GAF, New York, 1975.
59. U.S. Pat. 3,536,782 (Oct. 27, 1970), U. Toggweiler and F. F. Roselli (to Diamond Shamrock Chemical).
60. *Fire Retardant Chemicals—FR-1540,* Form No. 101-99-76, Dow Chemical, Midland, Mich., 1976.
61. U.S. Pat. 3,891,605 (June 24, 1975), E. R. Larsen and E. L. Ecker (to Dow Chemical).
62. E. Prill, *J. Am. Chem. Soc.* **69,** 62 (1947).
63. *Velsicol Chlorendic Anhydride,* Tech. Bull. 50200-1, Velsicol Chemical, Chicago, Ill., 1968.
64. *Emery 9350 TBBA Flame Retardant,* Data Sheet 811A, Emery Industries, Cincinnati, Ohio, 1978.
65. U.S. Pats. 3,182,088 (May 4, 1965), H. E. Hennis (to The Dow Chemical Co.); 3,234,289 (Feb. 8, 1966), H. E. Hennis (to The Dow Chemical Co.).

66. *FR-1138,* Form No. 101-21-74, Dow Chemical, Midland, Mich., 1974.
67. U.S. Pat. 3,932,541 (Jan. 13, 1976), R. A. Davis and co-workers (to The Dow Chemical Co.).
68. U.S. Pat. 3,507,933 (Apr. 21, 1970), E. R. Larsen, B. R. Andrejewski, and D. L. Nelson (to The Dow Chemical Co.).
69. D. P. Miller, *Reactive Additive Flame Retardants for Urethane Foams, Polymer Conf. Series,* Univ. of Detroit, Detroit, Mich., 1978.
70. *Firemaster PHT4-DIOL,* Michican Chemical (Velsicol), Chicago, Ill., 1977.
71. U.S. Pat. 3,382,254 (May 7, 1968), H. Jenkner, O. Rabe, and R. Strang (to Chemische Fabrik Kalk).
72. *Vinyl Bromide—Properties, Characteristics, Uses,* Form 101-10-69R, Dow Chemical, Midland, Mich., 1969.
73. R. C. Kidder, *Plast. Eng.* **38,** (Feb. 1977).
74. *Chem. Mark. Rep.,* 5 (Mar. 3, 1978).

General References

J. W. Lyons, ed., *The Chemistry and Uses of Fire Retardants,* Wiley-Interscience, New York, 1970.

W. C. Kuryla and A. J. Papa, eds., *Flame Retardancy of Polymeric Materials,* Vols. 1–3, Marcel Dekker, Inc., New York, 1973–1975.

M. Lewin, S. M. Atlas, and E. M. Pearce, eds., *Flame Retardant Polymeric Materials,* Plenum Press, New York, 1975.

C. J. Hilado, *Flammability Handbook for Plastics,* 2nd ed., Technomic, Westport, Conn., 1974.

R. G. Gann, ed., *Halogenated Fire Suppressants,* ACS Symposium Series 16, American Chemical Society, Washington, D.C., 1975.

E. R. Larsen
Dow Chemical U.S.A.

PHOSPHORUS COMPOUNDS

The main fire retardants currently used in plastics and textiles fall into several distinct classes: (*1*) alumina trihydrate; (*2*) halogenated compounds, usually used in combination with antimony oxide; (*3*) borax and boric acid; and (*4*) the phosphorus, phosphorus–nitrogen, and phosphorus–halogen compounds.

Mechanism of Action of Phosphorus Flame Retardants

The overview article presents a broad discussion of flame-retardant mechanisms. The following discussion deals specifically with phosphorus flame retardants.

Condensed-Phase Mechanisms. The mode of action of phosphorus-based flame retardants in cellulose has been more extensively studied and is better understood than in most other polymer systems. Two alternative routes of cellulose (qv) pyrolysis are known to occur: one route proceeds first to a tarry depolymerization product called levoglucosan (**1**) which decomposes to volatile combustible fragments; the other route (catalyzed by acids) leads primarily to water and difficultly combustible char (1–2).

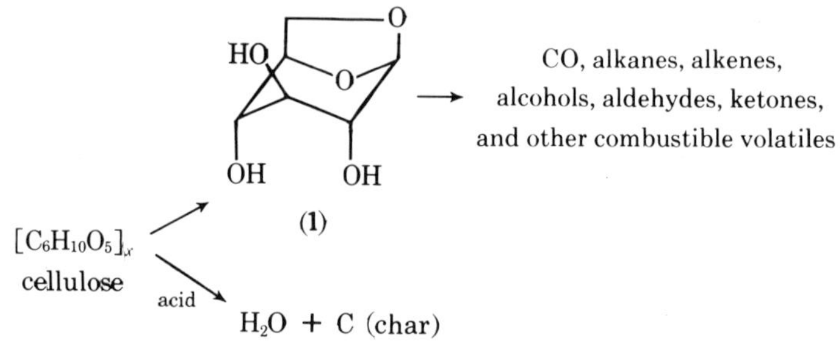

Although mineral acids in general catalyze the desired water- and char-forming pyrolysis route, phosphoric acid [*7664-38-2*] is particularly advantageous because of its low volatility. Also, when strongly heated, phosphoric acid yields polyphosphoric acid which is even more effective in catalyzing the desired dehydration reaction. The flame-retardant action of phosphorus compounds in cellulose is believed to proceed by way of initial phosphorylation of the cellulose. The phosphorylated cellulose then breaks down to water, phosphoric acid, and an unsaturated cellulose analogue, eventually char, by repetition of these steps. Certain nitrogenous compounds such as melamines, guanidines, ureas, and other amides appear to catalyze the cellulose phosphate-forming steps and are found to enhance or synergize the flame-retardant action of phosphorus on cellulose (3–4).

In poly(ethylene terephthalate) and poly(methyl methacrylate) the mechanism of action of phosphorus-based flame retardants has been shown to involve both a similar decrease in the amount of combustible volatiles and a similar increase in the amount of residue (aromatic residues and char) (5–7). The char thus formed also acts as a physical barrier to heat and gases (5).

In rigid polyurethane foams, the action of phosphorus flame retardants also appears to involve char enhancement (8–9).

The physical character of the char from rigid urethane foams was found to be affected by the retardant. The presence of a phosphorus-containing flame retardant caused a rigid urethane foam to produce a more coherent char, possibly serving as a physical barrier to the combustion process (10–13). There is evidence that a substantial fraction of the phosphorus may be retained in the char (14–15).

In polymers such as polystyrene that do not readily undergo charring, phosphorus-based flame retardants tend to be less effective, and such polymers are usually flame retarded by antimony–halogen combinations. However, even in noncharring polymers, phosphorus additives exhibit some activity that suggests at least one other mechanism of action. It has been proposed, and some evidence adduced, that phosphorus compounds may produce a barrier layer of polyphosphoric acid on the burning polymer (16–17).

There is evidence that phosphorus-containing additives can act in some cases by catalyzing thermal breakdown of the polymer melt, reducing its viscosity, and favoring the flow or drip of molten polymer from the combustion zone. In polystyrene, tris(2,3-dibromopropyl) phosphate [*126-72-7*] acts at least in part by this mechanism (18).

Several commercial polyester fabrics are flame-retarded with low levels of phosphorus additives or reactives which cause them to melt and drip more readily than fabrics without the flame retardant (19). This mechanism can be counteracted or completely defeated by the presence of nonthermoplastic fibers such as cotton, which can serve as wicks, or by silicone oils which can form pyrolysis products capable of impeding melt flow (20–21).

Vapor-Phase Mechanisms. In addition to the condensed-phase mechanisms discussed above, phosphorus flame retardants can exert vapor-phase flame-retardant action. It has been demonstrated that trimethyl phosphate [*512-56-1*] retards the velocity of a methane–oxygen flame with about the same molar efficiency as $SbCl_3$ (22–23). Both physical and chemical vapor-phase mechanisms have been proposed for the flame-retardant action of certain phosphorus compounds. Since tris(dibromopropyl) phosphate was found not to change the activation energy of thermo-oxidative degradation of polypropylene, although it raised the oxygen index, a vapor-phase physical-shielding action was postulated (24). Possibly this action may be produced by bromine-containing pyrolysis products rather than by the phosphate itself.

Triphenylphosphine oxide [*791-28-6*] and triphenyl phosphate [*115-86-6*] as model phosphorus flame retardants were shown by mass spectroscopy to break down in a flame to give small molecular species such as PO, HPO_2, PO_2, and P_2 (25). The rate-controlling hydrogen atom concentration in the flame was shown spectroscopically to be reduced when these phosphorus species were present. These data indicate the existence of a vapor-phase mechanism; however, the stable volatile compounds used in this study are not typical of many of the phosphorus-based flame retardants used commercially. Physical or chemical vapor-phase mechanisms may be reasonably hypothesized in cases where a phosphorus flame retardant is found to be effective in a noncharring polymer, and especially where the flame retardant or phosphorus-containing breakdown products are capable of being vaporized at the temperature of the pyrolyzing surface. In General Electric's engineering thermoplastic Noryl, which consists of a blend of a charrable poly(phenylene oxide) and a noncharrable polystyrene, experimental evidence indicates that effective flame retardants such as triphenyl phosphate act in the vapor phase to suppress the flammability of the polystyrene pyrolysis products (26).

A comparison of a variety of phosphorus additives at equivalent phosphorus loadings was made in poly(methyl methacrylate), which can be retarded by condensed phase action (7), but should also be subject to vapor-phase inhibition because it depolymerizes to monomer. Poor flame retardancy was found with trimethylphosphine oxide [*676-96-0*], a volatile stable species, whereas a much larger oxygen index elevation was observed with phosphoric acid; this result suggests that the condensed-phase mechanism is the more efficient one in poly(methyl methacrylate).

Poly(ethylene terephthalate) exhibits a higher oxygen index with 5 wt % phosphorus incorporated in the backbone of the polymer as phenylphosphinyl groups, as contrasted to 5 wt % phosphorus incorporated as a relatively volatile additive, triphenylphosphine oxide (27). This result suggests that a condensed-phase mechanism is more effective than a vapor-phase mechanism in this polymer.

The question as to whether a flame retardant operates mainly by a condensed-phase mechanism or mainly by a vapor-phase mechanism is especially complicated in the case of the haloalkyl phosphorus esters. A number of these compounds upon thermal degradation release volatile halogenated hydrocarbons which are plausible flame inhibitors. At the same time, their phosphorus content remains as relatively nonvolatile phosphorus acids which are plausible condensed-phase flame retardants (6,18,28). There is no evidence for the formation of phosphorus halides.

Interactions With Other Flame Retardants. Some claims have been made for a phosphorus–halogen synergism but, unlike the firmly established antimony–halogen synergism, phosphorus–halogen interactions are often merely additive and, in some instances, slightly less than additive (6,29–30). Cases of phosphorus–halogen synergism (ie, activity greater than that predicted by some additivity model) usually do not hold up to careful analysis and some supposed cases are artifacts of nonlinear response–concentration relationships (13,30). Nevertheless, combinations of phosphorus and halogen in separate compounds or in a single compound are often quite useful even if not truly synergistic.

Antagonism between antimony oxide and phosphorus flame retardants has been reported in several polymer systems, and has been explained on the basis of phosphorus interfering with the formation or vaporization of antimony halide. This phenomenon is also not universal and some useful commercial PVC formulations have been described for antimony oxide and triaryl phosphates (31).

An interesting case of synergism has been described involving a bisphosphine oxide, American Cyanamid's RF-699, and ammonium polyphosphate [*13566-20-6*] (5).

Phosphorus-Based Flame Retardants in Commercial Use

Since the original report of ammonium phosphate as a flame retardant by Gay-Lussac in 1821 (32) and the commercial introduction of tricresyl phosphate [*1330-78-5*] as a flame-retardant plasticizer for cellulosics early in the present century, many thousands of phosphorus compounds have been described as having flame-retardant utility. A broad sampling of these is covered in ref. 33. The more specialized topics of phosphorus monomers and polymers containing built-in phosphorus have been reviewed (34–36). This article is confined to the much more limited groups of compounds that have found commercial or semicommercial use.

Inorganic Phosphorus Compounds. ***Red Phosphorus*** **[*7723-14-0*].** This allotropic form of phosphorus is relatively nontoxic and, unlike white phosphorus, is not spontaneously flammable (although easily ignited). It is a polymeric form of phosphorus with thermal stability up to ca 450°C. In finely divided form, it has been found to be outstandingly effective as a flame-retardant additive (37). In Europe it has found commercial use in molded nylon electrical parts. Handling hazards, such as flammability, odor, partial reversion to toxic white phosphorus, and the imparting of color, have deterred broader usage. A product, Exolit 505, available from Hoechst (FRG) consists of red phosphorus treated with caprolactam, and is reported to be safer than the untreated material (38). Related products are marketed in Japan.

Ammonium Phosphates. These salts were recommended for treating theater curtains in 1821 (32). Their use in forest-fire control is well established. Monoammonium phosphate [*7722-76-1*] and diammonium phosphate [*7783-28-0*], or mixtures of the two, which are more water soluble and nearly neutral, are still used in large amounts for nondurable flame-retarding of paper, textiles, disposable nonwoven cellulosic fabrics, and wood products. Their advantage is high efficacy and low cost. Ammonium phosphate finishes are not resistant to laundering or even to leaching by water, but they are resistant to organic solvents such as dry-cleaning solvents. One important advantage of ammonium phosphates as flame retardants, and phosphorus flame retardants in general over borax (also used for nondurable cellulosic flame retardants), is their effectiveness in preventing afterglow (39).

The crystalline nature of ammonium phosphates may produce a gritty texture on the surface of some substrates. This characteristic is lessened by commercial ammonium phosphate formulations containing softening and penetrating agents.

Self-cross-linking acrylic latexes have been formulated with diammonium phosphate and organic phosphates to obtain flame-retardant textile backcoatings and nonwoven binders with a small but useful degree of durability to laundering and dry cleaning.

Insoluble ammonium polyphosphate. When ammonium phosphates are heated with urea, or by themselves under ammonia pressure, relatively water-insoluble ammonium polyphosphate (Phoschek P/30, Monsanto) is produced (40–41). These products are long chains having repeating units of the structure —$OP(O)(ONH_4)$—. This product, a finely divided solid, is a principal ingredient of intumescent paints and mastics (42). In such formulations, ammonium polyphosphate is considered to function as a "catalyst." Thus, when the intumescent coating is exposed to high temperature, the ammonium polyphosphate yields a phosphorus acid which then interacts with an organic component such as dipentaerythritol to form a carbonaceous char. A blowing (gas-generating) agent such as melamine or chlorowax is also present to impart a foamed characteristic to the char, thus forming a fire-resistant insulating barrier to protect the substrate. In addition, the intumescent formulations typically contain resinous binders, pigments, and other fillers. Mastics are related but generally more viscous formulations, intended to be applied in thick layers to girders, trusses, and decking; they generally contain mineral fibers to increase their coherence.

Ammonia/P_2O_5 products. The reaction of ammonia gas with phosphorus pentoxide at high temperature yields an amorphous colorless solid, slowly soluble in water to form a nearly neutral solution. The product consists of a mixture of ammonium salts of metaphosphorimidic acid [*12770-65-9*] (43). Analysis shows about two ammonium nitrogen atoms and one imide nitrogen atom for every two phosphorus atoms. Stauffer's Victamide is known to be a complex mixture but a typical component is believed to have structure (**2**) (44):

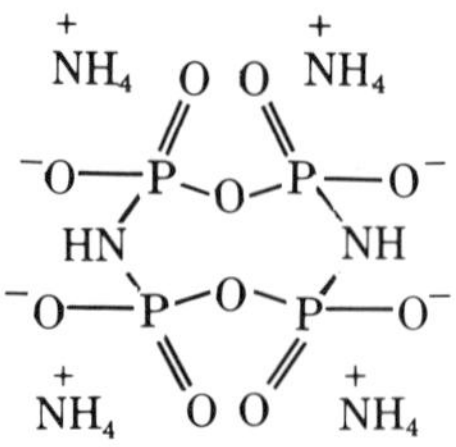

(**2**)

Victamide, as an aqueous solution, can be applied to paper, cotton cloth, cotton batting, and nonwovens. When dry, it produces a smoother surface texture than that produced by the crystalline ammonium phosphates. Proprietary formulations have been developed affording some degree of water resistance; presumably the Victamide acts therein as a phosphorylating agent. Ammoniation of Victamide by concentrated ammonia produces a product which, when applied to a cellulosic substrate and heated, yields a semidurable flame-retardant finish that withstands several aqueous washes (45).

Phosphoric Acid-Based Systems for Cellulosics. Semidurable flame-retardant treatments for cotton can be attained by phosphorylation of cellulose. This was originally accomplished by heating cotton or paper with phosphoric acid in the presence of basic compounds such as urea at ca 145–180°C. Commercial formulations have been developed utilizing as coreactants either phosphoric acid and cyanamide or phosphoric acid and a dicyandiamide–formaldehyde resin (46). The nitrogenous component catalyzes the phosphorylation of cellulose, retards the acid-degradation of the cellulose, and synergizes the flame-retardant action of the phosphorus. A fair degree of durability to laundering is achieved by such treatments. Typically, several launderings can be tolerated, and dry-cleaning resistance is good. A substantial part of the decline in flame retardancy during laundering is caused by ion exchange of the protons of the phosphoric acid groups by sodium, calcium, and magnesium cations that suppress the flame-retardant effectiveness of the phosphorus groups. Such finishes also have limited utility because of fabric damage during cure, although applications have been made on draperies, nonwoven fabrics, and paper products.

Commercial formulations have been developed based on phosphoric acid, urea–formaldehyde resins, and dicyandiamide as leach-resistant, clear, flame-retardant coatings for wood (47).

Organic Phosphorus Flame Retardants—Additive Types. ***Alkyl Acid Phosphates.*** The lower alkyl acid phosphates have found some limited use as additive flame retardants in cast thermoplastics and polyester resins. In cast poly(methyl methacrylate), methyl and haloalkyl acid phosphates are effective in combination with halogen-containing additives (48–51).

Trialkyl Phosphates. Triethyl phosphate [*78-40-0*] is a colorless liquid boiling at 209–218°C and containing 17 wt % phosphorus. It is manufactured from diethyl ether and phosphorus pentoxide via a metaphosphate intermediate (52). Triethyl phosphate has been used commercially as an additive for polyester laminates and in cellulosics. In polyester resins it functions as a viscosity depressant and as a flame retardant. The viscosity–depressant effect of triethyl phosphate in polyester resin permits high loadings of alumina trihydrate, a fire-retardant smoke-suppressant filler (53). Triethyl phosphate has also been employed as a flame-resistant plasticizer in cellulose acetate. Because of its water solubility, the use of triethyl phosphate is limited to situations where weathering resistance is unimportant. The halogenated alkyl phosphates are generally used for applications where lower volatility and greater resistance to leaching are required.

Trioctyl phosphate [*1806-54-8*] has been employed as a specialty flame-retardant plasticizer for vinyl compositions where low temperature flexibility is critical, eg, in military tarpaulins. It can be included in blends with general purpose plasticizers such as phthalate esters to improve low temperature flexibility.

Dimethyl Methylphosphonate. Dimethyl methylphosphonate [*756-79-6*] (DMMP) is made by molecular rearrangement of trimethyl phosphite. It contains 25 wt % phosphorus (near the maximum possible for a phosphorus ester) and it is therefore highly efficient on a weight basis as a flame retardant. DMMP is a low-viscosity, colorless liquid, bp 185°C. Because of its volatility, it has been useful mainly in thermoset systems. DMMP is an efficient viscosity depressant in polyester resins and epoxy resins. As a flame retardant it has somewhat greater efficiency than triethyl phosphate which is used in similar systems. DMMP is used commercially in mineral-filled and glass-reinforced polyester compounds where its viscosity–depressant effect permits use of higher filler loadings. The use of alumina trihydrate as filler and DMMP in the dual role of viscosity depressant and flame retardant affords reinforced polyester resin formulations with low flame spread suitable for bathtubs and shower stalls (54). DMMP has been used commercially for boosting the phosphorus content of flame retardants used in rigid foams. DMMP is also used as a chemical intermediate for the manufacture of several other flame retardants (Fyrol 76, Fyrol 51, Fyrol 58 and Antiblaze 19).

Halogenated Alkyl Phosphates and Phosphonates. This class encompasses a number of important and versatile flame-retardant additives. The presence of the halogen contributes to flame retardancy although this contribution is to some degree offset by the lowering of the percentage of phosphorus because of the heavy halogen atoms in the molecule. The halogens also depress vapor pressure and depress water solubility, thus aiding the permanency of these additives. Available are manufacturing processes that permit these compounds to be priced favorably. Individual compounds of commercial importance are discussed below.

CAS nomenclature is employed commercially in preference to the older tris(haloalkyl) phosphate names because of the unfortunately widespread use of the nickname "tris" for the compound tris(2,3-dibromo-1-propyl) phosphate.

2-Chloroethanol phosphate (3:1). Tris(2-chloroethyl)phosphate [*115-96-8*] (**3**) is a liquid containing 10.8 wt % P and 36.7 wt % Cl, made from ethylene oxide and phosphorus oxychloride (55).

This ester has found a wide variety of flame-retardant uses because of low color and low odor, favorable economics, high percent phosphorus, and compatibility with most polymers except polyolefins and polystyrene. Tris(2-chloroethyl) phosphate

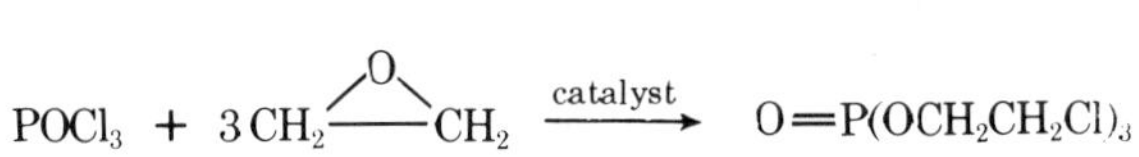

(**3**) (Fyrol CEF, Stauffer)

affords good permanence except under very hot and humid conditions. It is fairly stable to hydrolysis and its volatility lies between that of dibutyl phthalate and dioctyl phthalate. This phosphate additive is used commercially in polyurethane and polyisocyanurate foams, carpet backing, flame-laminated polyurethane foam, flame-retardant paints and lacquers, epoxy resins, phenolic resins, amino resins, poly(vinyl acetate) coatings and adhesives, urethane coatings, cast acrylic sheet, polyester resins, and wood-resin composites such as particle board.

Emulsions of tris(2-chloroethyl) phosphate, blended with a binder such as a vinyl or acrylic emulsion, can be used for applications such as the backcoating of upholstery.

Tris(2-chloroethyl) phosphate can be used as a secondary plasticizer in PVC to suppress the flammability resulting from plasticizers such as phthalates. Where a particularly high degree of flame retardancy is required, it can be used in combination with the aromatic phosphate plasticizers. Such formulations can serve as an alternative to the use of antimony oxide in plasticized vinyl polymers.

1,3-Dichloro-2-propanol phosphate (3:1). Tris(1,3-dichloro-2-propyl) phosphate [*13674-87-8*] (**4**) is a liquid made from epichlorohydrin and phosphorus oxychloride and contains 49 wt % chlorine and 7.2 wt % phosphorus. The resultant dichloropropyl groups are mainly 1,3-dichloro-2-propyl groups. The principal component is (**4**) although there are many literature references to this product under the erroneous name tris(2,3-dichloropropyl) phosphate. Compared to tris(2-chloroethyl) phosphate (3:1) (**3**), (**4**) has substantially reduced volatility, and much lower water solubility. It has an extremely low rate of hydrolysis, and it resists attack by bases. This stability probably results from steric hindrance which prevents reagents from attacking the phosphate's P—O—C linkages. Because of its good stability, this phosphate is a very useful additive flame retardant for flexible urethane foams (56). It does not interfere with the catalyst system, cause discoloration, nor adversely affect the tensile strength of flexible urethane foams. Foams containing it are able to meet the Motor Vehicle Safety Standard 302 requirements for automotive seating. It is also used in rigid urethane and isocyanurate foams. This phosphate additive is also useful as a flame retardant in styrene–butadiene rubber (SBR) and acrylic latexes for textile backcoating and binding of nonwoven fabrics.

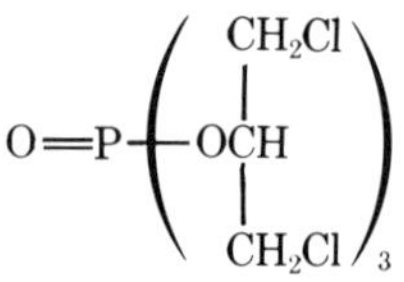

(**4**) (Fyrol FR-2, Stauffer)

1-Chloro-2-propanol phosphate (3:1). Tris(1-chloro-2-propyl) phosphate [*13674-84-5*] (Fyrol PCF, Stauffer) is made from propylene oxide and phosphorus oxychloride, and contains 33 wt % chlorine and 9.5 wt % phosphorus. The product has both 1-chloro-2-propyl and 2-chloro-1-propyl groups in the molecule, but the former group predominates. Its physical properties are rather close to those of the chloroethyl analog, but the chloropropyl compound has still lower reactivity to water and bases. Therefore, it has been found to function well in urethane foams, especially in rigid foam systems where good storage stability in the isocyanate or in the polyol/catalyst mixture is required (57). Formulations for low smoke evolution have been developed suitable for manufacture of sheathing foam. This compound is also effective in isocyanurate foams for reducing brittleness. Applications also have been found in flexible molded foam.

2,3-Dibromo-1-propanol phosphate (3:1). Tris(2,3-dibromo-1-propyl) phosphate [*126-72-7*] had been used widely as an additive in a variety of plastics and textiles. Its manufacture involved addition of bromine to allyl alcohol followed by reaction of the resultant 2,3-dibromo-1-propanol with phosphorus oxychloride (58). The product is a dense, nearly colorless, fairly viscous liquid, insoluble in water, very stable to hydrolysis, and stable to ca 200°C. This highly active flame retardant was used in clear cast acrylic sheet, cellulose acetate or acrylic fibers, polystyrene (especially, expanded foam), polyurethane foams, lacquers, SBR latexes, and cured unsaturated polyesters. Textile finishes having good durability to laundering were developed. For example, tris(dibromopropyl) phosphate was applied as an emulsion to polyester fabric and the fabric then heated to "thermofix" the flame retardant in the fibers (59). This flame retardant could also be exhausted onto the fabric in an operation similar to dyeing. These finishing techniques were used on 100% polyester to pass the federal requirements for children's sleepwear. This application was discontinued when mutagenic properties were demonstrated by a bacterial bioassay technique (60–61), and the compound is no longer manufactured.

Oligomeric chloroalkyl phosphonates. Oligomeric substances are those having several repeating units in their molecular structure. Where reduced vapor pressure and thus increased permanence of the flame-retardant additive is required, oligomeric phosphorus additives have been found useful. The first of this class to become commercial is Monsanto's Phosgard C-22-R (**5**), a chloroethyl phosphonate containing ca 27 wt % chlorine and ca 15 wt % phosphorus, which is made in several steps from PCl_3, ethylene oxide, and acetaldehyde (62).

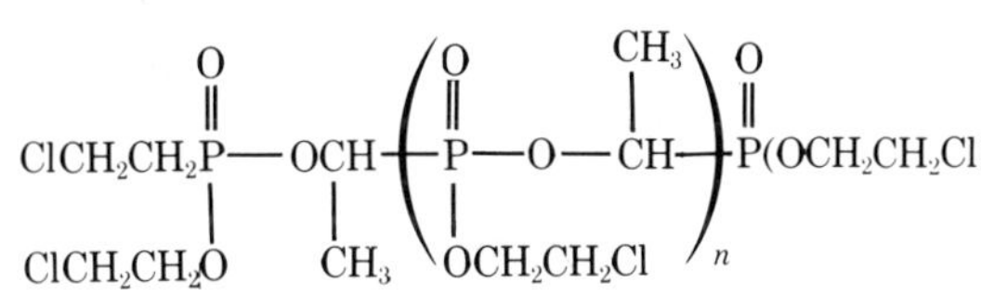

(**5**)

n = 1 [*4351-70-6*]

n = 2 [*71720-64-4*]

(Phosgard C-22-R)

Phosgard C-22-R has been used commercially in rigid polyurethane foam, rebonded foam, textile backcoatings, cast poly(methyl methacrylate), acrylic-impregnated wood flooring, resin-impregnated paper products, phenolics, and epoxy electrical parts. This additive has good permanency because of its low volatility and low water solubility. It has only limited utility in flexible urethane foams because of reactivity toward some of the catalysts used in these systems and a tendency to cause scorch.

Bis(2-chloroethyl) 2-chloroethylphosphonate and higher oligomers. The commercial product, Antiblaze 78 (Mobil), is a mixture of bis(2-chloroethyl) 2-chloroethylphosphonate [*6294-34-4*], and higher boiling oligomers, mainly bis(2-chloroethyl) [2-[[(2-chloroethoxy)(2-chloroethyl)phosphinyl]oxy]ethyl]phosphonate [*58823-09-9*] as made by the rearrangement of tris(2-chloroethyl) phosphite [*140-08-9*]

(63). Antiblaze 78 contains 12 wt % phosphorus and 32 wt % chlorine. It has been found useful as a flame-retardant additive for closed-cell rigid urethane foams. It has also been found useful in rebonded flexible foams, adhesives, and coatings. Bis(2-chloroethyl) 2-chloroethylphosphonate is also an intermediate to Stauffer's Fyrol Bis-Beta (bis(2-chloroethyl) vinylphosphonate [*115-98-0*]), and Fyrol 76. With bases, eg, amines, the 2-chloroethylphosphonates undergo ready dehydrochlorination, which imposes some limitations on the formulations.

Diphosphates. Three 2-chloroethyl diphosphates with a similar pattern of utility as low-volatility flame-retardant additives are sold commercially. As a class, these compounds have low volatility, low water solubility, and good-to-fair thermal stability; thus, in products such as open-cell foams they show resistance to dry and humid aging.

The first of these to be introduced was Monsanto's Phosgard 2XC-20 (**6**). It is manufactured in several steps from pentaerythritol, phosphorus trichloride, chlorine, and ethylene oxide (64). It contains 36 wt % chlorine and 10 wt % phosphorus.

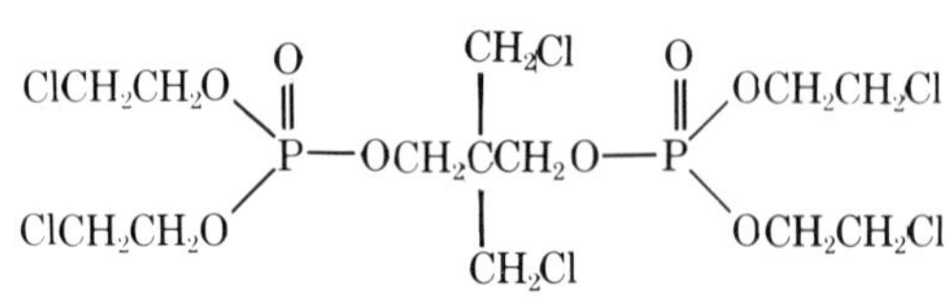

(**6**) [*38051-10-4*]

Phosgard 2XC-20 is recommended by the manufacturer as a nonvolatile flame-retardant additive for urethane foam, acrylics, and epoxies. The sterically bulky dichloroneopentyl structure is intended to enhance the chemical stability of the compound. The principal use for (**6**) is in flexible urethane foams, although it is useful also in phenolics, acrylics, and epoxy resins.

A more recently introduced diphosphate is Olin's Thermolin 101 (65) containing primarily compound (**7**). It is a liquid of rather low viscosity (242 mPa·s (= cP) at 25°C) containing 31 wt % chlorine and 13 wt % phosphorus. The synthesis described in the patent literature begins with the reaction of ethylene oxide and phosphorus trichloride to form tris(2-chloroethyl) phosphite, which is then chlorinated to bis(2-chloroethyl) phosphorochloridate. This acid chloride is treated with ethylene glycol in the presence of a base (66).

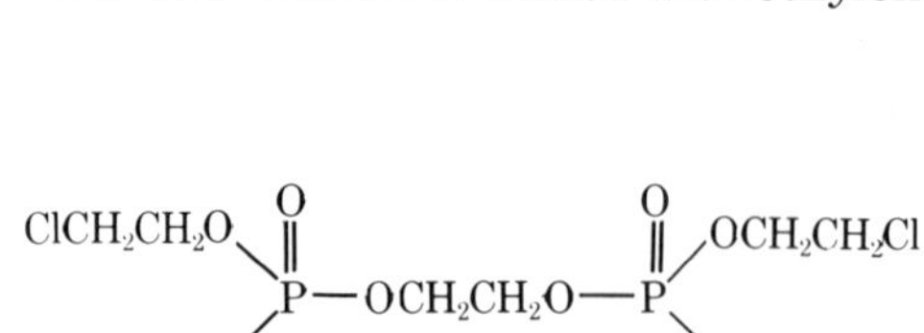

(**7**) [*33125-86-9*]

Another commercial diphosphate of related structure is Monsanto's Phosgard 1227 (**8**), containing 27 wt % chlorine and 12 wt % phosphorus. The compound was

(**8**) [*53461-82-8*]

described in an earlier USSR publication (67). The synthesis is believed to involve reaction of diethylene glycol with $POCl_3$, followed by reaction with ethylene oxide. A comparison of these diphosphates to related monophosphates has recently been published, showing the diphosphates to have better permanency in flexible foams subjected to accelerated aging (65).

Oligomeric phosphate. Fyrol 99 contains the ethylene glycol diphosphate described above together with a series of higher oligomers (**9**); it contains typically 14 wt % phosphorus and 26 wt % chlorine. A process of manufacture is shown in the following equation (68):

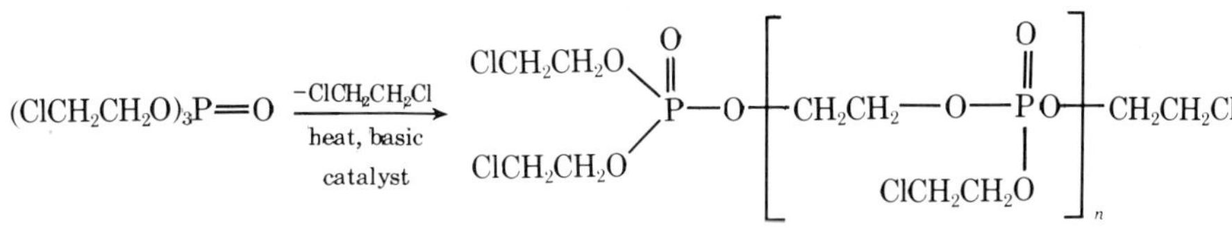

(**9**)

2-chloroethyl phosphate polymers [*28205-79-0*]

Fyrol 99 is low in volatility and has advantageous resistance to water and to solvents. It is useful in resin-impregnated paper automotive air filters, in flexible urethane foam, rebonded urethane foam, and in structural foam. It is also useful as an additive flame retardant in other thermoset polymer systems. Applications have been described for Fyrol 99 in textile backcoating compositions.

3,9-Tribromoneopentoxy-2,4,8,10-tetraoxa-3,9-diphosphaspiro(5.5)undecane-3,9-dioxide [*61090-89-9*]. Velsicol VCC48 (**10**), recently introduced, is a thermally stable, high-melting (218–225°C) solid, containing 53–54.5% bromine and 7.3% phosphorus. It is prepared from tribromoneopentyl alcohol (pentaerythritol tribromide), phosphorus oxychloride, and pentaerythritol by a two step reaction (69–70). This compound is intended primarily as a flame-retardant additive for polypropylene. It is melt-blendable under normal processing conditions and affords a UL-94 V-O rating at low loadings.

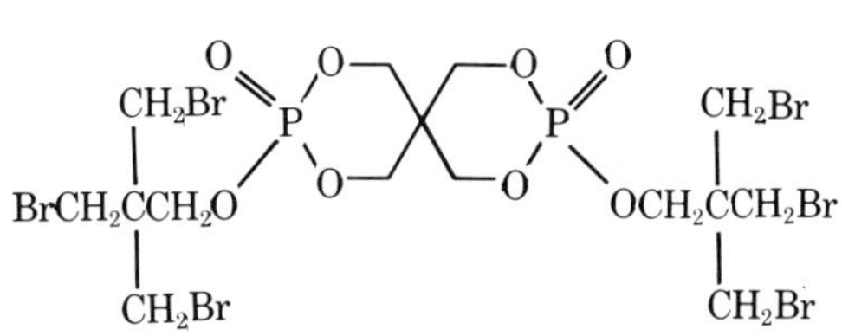

(**10**) [*61090-89-9*]

Oligomeric Halogen-Free Phosphorus Esters. Low-volatility oligomers of high phosphorus content (**11**) have been prepared by reactions such as the following (71):

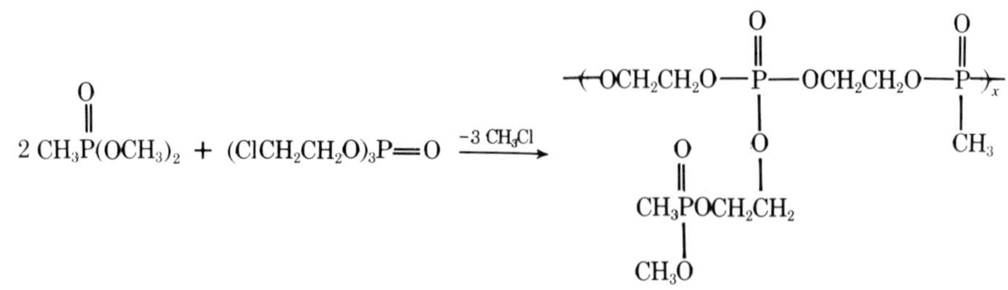

(**11**)

The first commercial product of this family, Fyrol 51 [*67326-33-4*] is a water soluble liquid, compatible with thermosetting resins. Diluted with a small percentage of alcohol to a conveniently low viscosity, it has been commercialized as Fyrol 58 for use in amino resin-impregnated automotive air filters. Reactive end groups, primarily alcohol groups, can be introduced to permit oligomers of this class to be cured with thermoset resin coreactants to produce water-resistant, resin-impregnated paper and textile backcoatings. These oligomers also show utility as reactive flame retardants in urethane foams.

Oligomeric Phosphite. A polymeric phosphite of 6–8 Pa·s (60–80 P) viscosity has been introduced as Stepan's Wiltrol HRF for use as an additive to impart relatively permanent flame retardancy to high-resiliency flexible urethane foams. The chemical structure has not been disclosed.

Oligomeric Cyclic Phosphonate. This flame-retardant additive is a very viscous, colorless, water-soluble syrup having a high phosphorus content (21 wt %). It is a mixture of cyclic phosphonate esters made by the following reaction sequence (72):

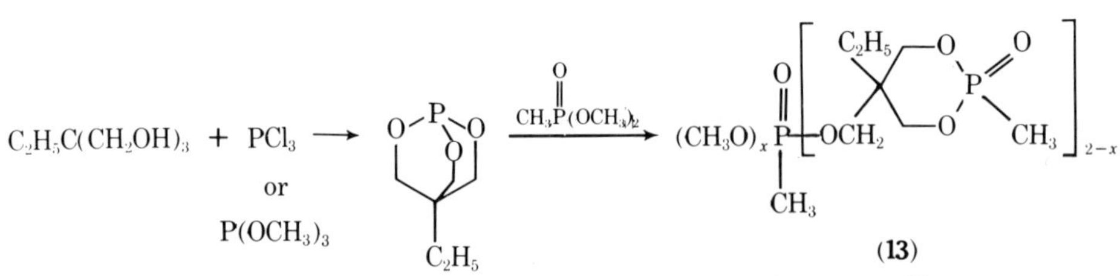

(**12**) (**13**)

$x = 0$ [*42595-45-9*], $x = 1$ [*41203-81-0*] (Antiblaze 19)

Although the intermediate phosphite (**12**) is very toxic (73), the final phosphonate product (**13**) has low mammalian toxicity, and it has excellent thermal and hydrolytic stability. Antiblaze 19 can be thermally fixed in polyester fabric (74), an application for which tris(2,3-dibromopropyl) phosphate was used until recently. Antiblaze 19 is also useful as an efficient flame-retardant additive in rigid urethane foams and thermoset resins.

Bis(5,5-dimethyl-2-thiono-1,3,2-dioxaphosphorinanyl) Oxide. An unusually effective flame retardant additive for rayon has been developed by Sandoz and introduced in Europe. This compound is believed to have structure (**14**) (75):

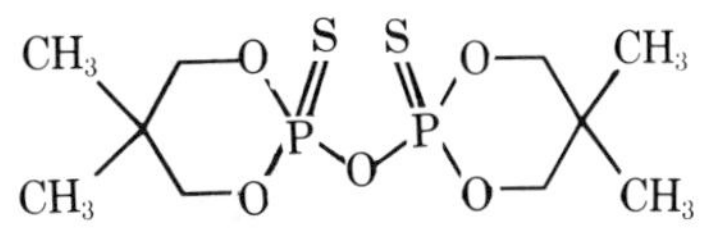

(**14**) [*4090-51-1*]

The compound is dispersed in the viscose before spinning and becomes entrapped in the spun fiber. Despite the anhydride linkage, the compound is surprisingly stable to hydrolysis, presumably because of steric hindrance resulting from the neopentyl structures together with the hydrophobic thiono groups. The flame-retardant action persists through numerous launderings.

Phenylphosphonates. Oligomeric phenylphosphonates of dihydric phenols have been found to have sufficient thermal stability and freedom from volatility to permit their use in thermoplastics processed above 250°C. In particular, the phenylphosphonate oligomer (**15**) from sulfonylbisphenol made by the following reaction:

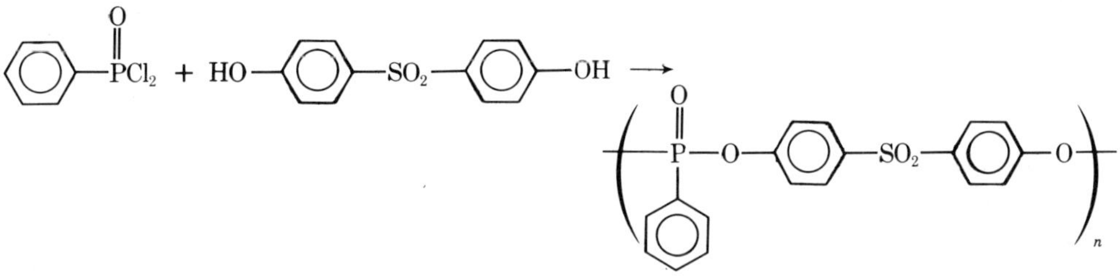

(**15**) [*35398-70-0*]

was used commercially in a polyester fiber HEIM developed by Toyobo in Japan (76) although it has now been superseded by Toyobo GH, a phosphorus-containing modified polyester fiber.

Triaryl Phosphates. The triaryl phosphates, as exemplified by tricresyl phosphate, have over a half-century of commercial history, having been originally introduced for cellulosic plastics. They are still used in cellulosics but their more important use is now in poly(vinyl chloride), commonly in admixture with nonphosphate plasticizers (77–79). The principal applications are in electrical insulation, automotive interiors, furniture upholstery, conveyor belts, and vinyl foams. The triaryl phosphates as a class have the following features: flame retardant properties, fast processing and gelation properties, good oil and gasoline extraction resistance, good microbial resistance, high extender tolerance, good high frequency heat sealing (welding) characteristics, high thermal stability, and relatively poor low temperature properties (79). The amount of aryl phosphate plasticizers employed in vinyls has been increased by the Motor Vehicle Safety Standard 302 flammability requirement for auto interiors. Stricter flame retardance requirement for electrical insulation and for vinyl film and sheet used in construction also have enhanced the growth of these flame-retardant plasticizers (qv). These compounds are also used in major quantities as hydraulic fluids and in lubricant compositions.

Triaryl phosphates have been traditionally produced by the reaction of mixtures of phenol, cresols, and xylenols (cresylic acids) in various proportions with phosphorus oxychloride $POCl_3$, generally in the presence of a catalyst. The esters are subsequently distilled and washed to a high degree of purity (78,80). Cresylic acids from petroleum and from coal tar have been employed as starting materials. To avoid the toxic *o*-cresyl phosphates, cresylic acids containing very little of the ortho isomer have been selected by the manufacturers.

The lowest-cost member of this group of products is cresyl diphenyl phosphate (CDP). CDP has advantages in regard to light stability, plasticizer efficiency, and low temperature flexural properties but has somewhat greater volatility than tricresyl phosphate (TCP).

A more recent development (81–82) is the use of C_3 and C_4 alkylphenols as synthetic cresylic acid substitutes, largely in response to the shortage and increased cost of the "natural" cresylic acids, particularly those cresylic acids low in cresol. The first synthetic substitutes to appear on the market were isopropylphenyl diphenyl phosphates (FMC's Kronitex 100, Stauffer's Phosflex 41P); the isopropylphenol is made by alkylation of phenol by propylene. Mixed tertiary-butylphenyl diphenyl phosphates have also reappeared in the marketplace, after earlier marketing for cellulosics by Dow; the current products are Monsanto's Santicizer 154 and Stauffer's Phosflex 51B. As was the case with TCP, these newer alkylphenyl phosphates are usually used in blends with conventional plasticizers.

Chlorine-containing blended aliphatic/aryl phosphates have been introduced by Stauffer principally for use in vinyls and rebonded urethane foam. These blends appear to have advantages over their chlorine-free components in regard to smoke and flame-retardant effectiveness.

Alkyl Diphenyl Phosphates. The use of triaryl phosphates in vinyls tends to lead to poor low temperature flexibility. Related phosphate plasticizers having two phenyl groups and one long chain alkyl group were designed to avoid this difficulty. The two principal commercial examples of this group are Monsanto's Santicizer 148 and 141, which are, respectively, isodecyl diphenyl phosphate [*29761-21-5*] and 2-ethylhexyl diphenyl phosphate [*1241-94-7*] (83). The latter has FDA approval for use in food packaging applications.

Triphenyl Phosphate. Triphenyl phosphate [*115-86-6*] is a colorless solid, melting at 48°C and boiling undecomposed at 370°C. It is made from phenol and phosphorus oxychloride and is generally purified by distillation. It has been used as a flame retardant in cellulosic polymers such as cellulose acetate.

More recently, triphenyl phosphate has been found effective as a flame retardant for engineering thermoplastics (qv), a use which imposes stringent requirements for thermal stability on additives. Because of its stability, triphenyl phosphate has also been found useful in rigid urethane foam systems where the flame retardant, catalyst, and other ingredients may be premixed and held for long periods before foaming.

Phosphonitrilics. A phosphonitrilic additive (**16**) for rayon represented the first commercial flame retardant application of the much investigated phosphonitrilic chlorides (which are also known as phosphazenes). The chemistry of the flame retardant additive, introduced by the former Avisco Division of FMC, is as follows (84):

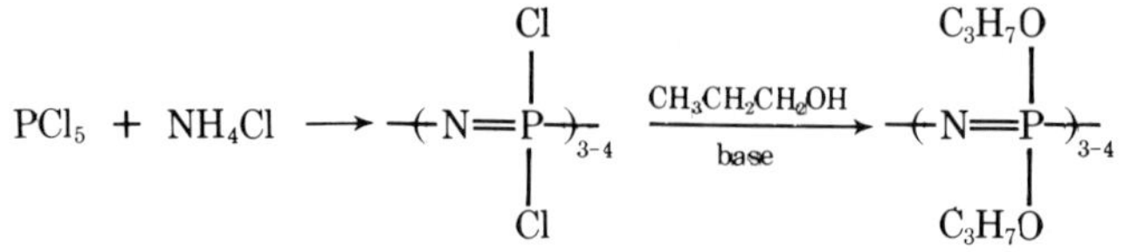

(**16**) [*28779-95-5*]

The additive is an odorless liquid of low toxicity, insoluble in water, and stable to acids and bases. It can thus be admixed into the highly alkaline viscose and spun into acid to obtain rayon (qv) fibers where the additive remains locked in as liquid droplets. The high cost has prevented significant commercial use of this product.

The phosphonitrilics have been further elaborated into a large and varied family of polymers (85–86) such as trifluoroethoxy-, and aryloxy-substituted phosphazene polymers, but to date none of these has been commercialized primarily as a flame retardant material (see Fluorine compounds, organic).

Phosphonium Bromides. These high-melting thermally stable solids, American Cyanamid's Cyagards, can be milled into thermoplastics such as polypropylene and are effective flame retardants at low concentrations (one-half or less of the concentration required with chlorine–antimony systems). Because they are insoluble solids, they do not plasticize the polymer. The synthesis of one member of this series is as follows (87):

$$PH_3 + 3\ CH_2{=}CHCN \longrightarrow P(CH_2CH_2CN)_3 \xrightarrow{BrCH_2CH_2Br} [(NCCH_2CH_2)_3PCH_2]_2^{2+}\ 2\ Br^-$$

(**17**) [*10310-38-0*]

Despite its ionic structure, this bis-phosphonium compound has low water solubility. These compounds are expensive, but can be used in combinations with cheaper halogenated hydrocarbons, or ammonium polyphosphate (88). This group of products appears to be in a semicommercial stage.

Phosphine Oxides. A novel phosphine oxide of the structure (18), American Cyanamid's RF-699, has been found highly effective in polypropylene, especially when synergized by ammonium polyphosphate (5,89).

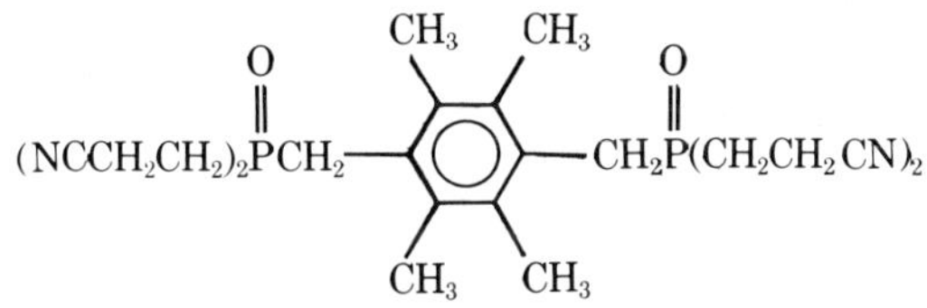

(**18**) [*53506-00-0*]

Organic Phosphorus Compounds—Reactive Types. ***Organophosphorus Monomers.*** Of the many vinyl monomers containing phosphorus which have been described in the literature (34–35), few have gone beyond the laboratory. Bis(2-chloroethyl) vinylphosphonate [*115-98-0*] (**19**) is a commercially available phosphorus-containing monomer made from ethylene oxide and phosphorus trichloride (90) by way of two other commercially available intermediates, tris(2-chloroethyl) phosphite [*140-08-9*] and bis(2-chloroethyl) 2-chloroethylphosphonate (discussed earlier as an additive).

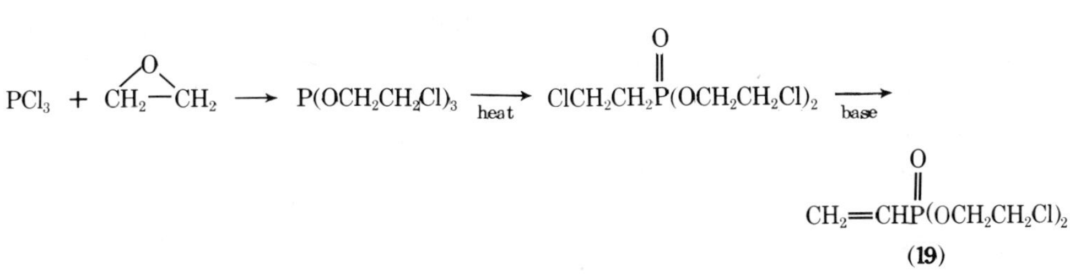

(**19**)

(Fyrol Bis-Beta, Stauffer)

Several applications have been found for bis(2-chloroethyl) vinylphosphonate. Emulsion copolymers with other vinyl monomers have been introduced as flame retardant latexes for textiles and paper applications (91–92). Fyrol Bis-Beta has also been found effective as a flame retardant cross-linking agent for cured unsaturated polyesters, especially those based on halogenated acids (90). The vinylphosphonates have been found to copolymerize particularly well with vinyl halides and vinylidene halides. Vinyl chloride–vinylphosphonate copolymers having inherent plasticity, low flammability, and low smoke generation properties have been described recently (93) (see Vinyl polymers).

Phosphorus-Containing Diols and Polyols. The commercial development of several phosphorus-containing diols occurred in response to the need to flame-retard rigid urethane foams used for insulation in the transportation and construction industry. A 1970 review (94) has 337 references mostly to patents on phosphorus polyols, but few have been used commercially. Perhaps the earliest product introduced (and still used) for this application was Mobil's Vircol 82, a diol obtained by the reaction of propylene oxide and dibutyl acid pyrophosphate. The neutral liquid product has an "OH number" of 205 mg KOH/g, and contains 11.3 wt % phosphorus (95–96). The phosphorus content and OH number correspond to the following structure (**20**):

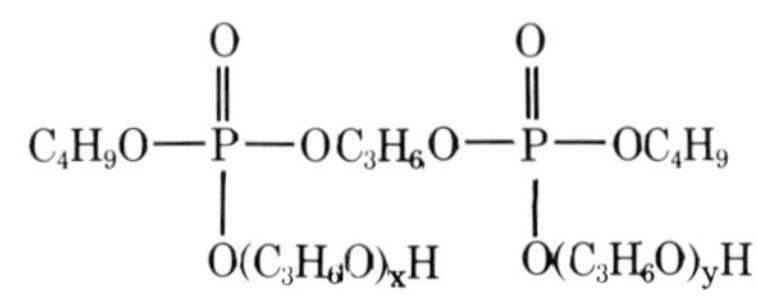

(**20**)

where x + y = 3.4 (avg). When a higher percentage of phosphorus is needed, Vircol 82 is used in admixture with dimethyl methylphosphonate [*756-79-6*].

Major commercial usage has been attained by a phosphonate diol containing a tertiary amine structure which may exert a synergistic effect on the flame-retardant action of the phosphorus. This diol, diethyl *N,N*-bis(2-hydroxyethyl)aminomethylphosphonate [*2781-11-5*] (**21**), is synthesized as shown (97):

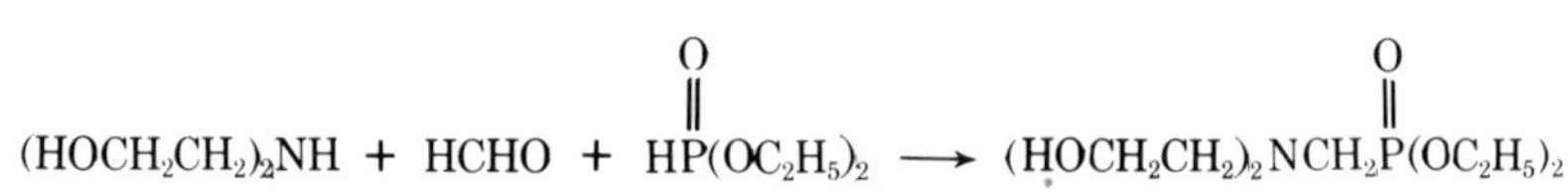

(**21**)

The product contains 12.6 wt % phosphorus, and has an OH number in the 450 mg KOH/g range.

Fyrol 6 is the leading flame retardant used for Class II (UL-94 flame-spread rating) rigid urethane foams for applications such as insulation for roofing and walls. Particular advantages of this diol phosphonate are its stability in polyol–catalyst mixtures and outstanding humid-aging resistance of foams in which it is used. Fyrol 6 is used in both spray foam and board stock.

A number of commercial phosphorus-containing polyols have been made by the reaction of propylene oxide with phosphoric or polyphosphoric acid (94). Because of the tendency of the propylene oxide to oligomerize, these products have relatively low phosphorus content, and some of them also have had stability problems.

Other phosphorus-containing polyols of undisclosed structure are available; an example is BASF Wyandotte's Pluracol 648, a high functionality polyol containing 4.5 wt % phosphorus for use in slab stock and E-84 class II rigid foams.

Tris(dipropylene glycol) phosphate [*51798-31-3*], Weston 430, has been used in rigid water-blown foams where low levels of phosphorus are achieved, sufficient to pass the small-scale ASTM D 1692 flammability test with retention of most of its physical properties.

Flame retardancy of flexible foams is still largely achieved by use of additives such as tris(1,3-dichloropropyl) phosphate [*13674-87-8*] or the diphosphates and oligomeric phosphates described above. The requirements for reactive flame retardants (polyols) for flexible foam have been generally more difficult to meet. Additives continue to be the principal means for flame-retarding flexible foams. In Europe, a polyol made from tris(2-chloroethyl) phosphate, polyphosphoric acid, and ethylene oxide or propylene oxide has been commercialized (98).

Three related polyols containing 24–36 wt % bromine and 2.3–2.7 wt % phosphorus were briefly on the market for use in semirigid foam; one of them, Brominex 160 P, had an OH number of 47 and was suggested for flexible foam. The probable chemistry of these products is indicated by a patent (99) that describes addition of bromine to castor oil and subsequent introduction of phosphonate groups by reaction with trialkyl phosphite.

Reactive Organophosphorus Compounds in Textile Finishing. Although synthetic fibers can be flame-retarded with additives or comonomers, the flame-retarding of cotton requires the application of a textile finish. Earlier markets for such finishes were in military goods, industrial protective clothing, and curtains. However, in recent years a large amount of cotton goods has been treated to pass federal requirements established for children's sleepwear under the Flammable Fabrics Act Amendment of 1967. Several durable flame retardant reactive cotton finishes have competed for this market which is now diminishing in favor of inherently flame-retardant synthetics (see Flame retardants in textiles). The industrial clothing market continues to employ these finishes.

Tetrakis(hydroxymethyl)phosphonium salts. The reaction of formaldehyde with phosphine in aqueous hydrochloric or sulfuric acid yields tetrakis(hydroxymethyl)phosphonium chloride [*124-64-1*] (**22**) or sulfate [*55566-30-8*], as shown in the following equation:

$$PH_3 + 4\ HCHO + HCl \rightarrow (HOCH_2)_4P^+Cl^-$$

(**22**)

These are water-soluble crystalline compounds sold as concentrated aqueous solutions. They are capable of being cured by reaction with amino compounds (ammonia, urea, melamine–formaldehyde resins, and the like) to form durable cross-linked resin finishes (100) (see Amino resins).

Recently, the chloride has been replaced in commercial usage by the corresponding sulfate, in order to avoid the possibility of formation of bis(chloromethyl) ether, a carcinogen, during manufacture or use.

Finishes for cotton based on tetrakis(hydroxymethyl) phosphonium salts plus melamines, ureas, or other coreactants were developed by Albright and Wilson in England, the USDA Southern Regional Laboratory, and Hooker Chemical Corp. Such finishes have been used extensively on industrial garments, hospital cubicle curtains, and military cotton goods.

Considerable effort has been expended in recent years on developing modifications of these finishes for lightweight cotton goods to try to overcome the excessive fabric stiffness caused by the earlier finishes. One improved finish developed by the USDA and Hooker Chemical Corp. is the so-called THPOH-ammonia finish, wherein the tetrakis(hydroxymethyl)phosphonium salt is first converted to a free organic base by adding a measured amount of an inorganic base, and then cured on the fabric by reaction with ammonia gas (100–101). The cloth is usually treated with hydrogen peroxide or perborate to convert all of the phosphine groups to more stable phosphine oxide groups. Durable flame retardancy with good tactile qualities (hand) and strength retention has been obtained. Because the ammonia gas necessitates special equipment, the commercialization of this process had to await the design of special ammoniation chambers, but the finish is now in commercial production.

Dimethyl 3-[(hydroxymethyl)amino]-3-oxopropylphosphonate [*20120-33-6*] (**23**) (Pyrovatex CP) developed by Ciba-Geigy, is based on the following chemistry (102–103):

$$(CH_3O)_2\overset{\overset{O}{\|}}{P}H + CH_2{=}CHCONH_2 \longrightarrow (CH_3O)_2\overset{\overset{O}{\|}}{P}CH_2CH_2CONH_2 \xrightarrow{HCHO} (CH_3O)_2P(O)CH_2CH_2CONHCH_2OH$$

(**23**) Pyrovatex CP (Ciba-Geigy)

Pyrovatex CP co-reacts on cellulose with an amino resin in the presence of an acid catalyst. Finishes based on Pyrovatex CP allow cotton to meet the federal standard for children's sleepwear and this finish was commercially available for several years for this use. It is also used for industrial garments and upholstery.

Oligomeric Vinylphosphonate. Bis(2-chloroethyl) vinyl phosphate [*7446-88-0*], polymer with dimethyl methylphosphonate (Stauffer's Fyrol 76) [*41222-33-7*], is a unique textile flame retardant in that curing is accomplished by a free-radical method.

Fyrol 76 contains a water-soluble vinylphosphonate oligomer of 22.5 wt % P. It has generally been applied to cotton in admixture with *N*-methylolacrylamide or an amine resin plus a free radical catalyst such as persulfate (104–105).

Cure temperatures of ca 145–180°C have been most commonly used. The finish has also been found to be curable by electron-beam radiation.

Cotton flannel treated with Fyrol 76 finish is able to pass the federal standard for children's sleepwear, and retains acceptable hand and strength. Perborate bleach can be used without adversely affecting flame-retardant properties after 50 wash-and-dry cycles.

Good results have been obtained with Fyrol 76 on cotton flannel knits, print cloth, sheeting, terry cloth, corduroy and heavy-weight fabrics, cotton–rayon blends, 100% rayon, and certain polyester–cotton blends and acetate–cotton blends.

Inherently Flame-Retardant Polyester Fibers Containing Phosphorus. Numerous patents describe poly(ethylene terephthalate) (PET) flame-retarded by reaction with phosphorus-containing dicarboxylic acids, esters, or diols. At least two such fibers have been commercially introduced.

Farbwerke Hoechst's Trevira 271 appears to be based on the following chemistry (106–107):

$$CH_4 + PCl_3 \longrightarrow CH_3PCl_2 + HCl$$

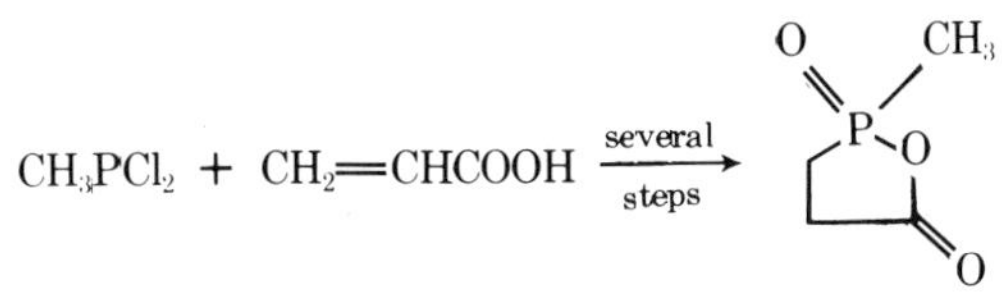

(**24**) [*15171-48-9*]

This phosphinic anhydride is then co-polycondensed with the usual precursors of PET thus introducing —$OP(O)(CH_3)CH_2CH_2COO$— units into the polymer chain. Trevira 271 is marketed for children's sleepwear and for work clothing.

Toyobo's GH is apparently based on the following reactions (108):

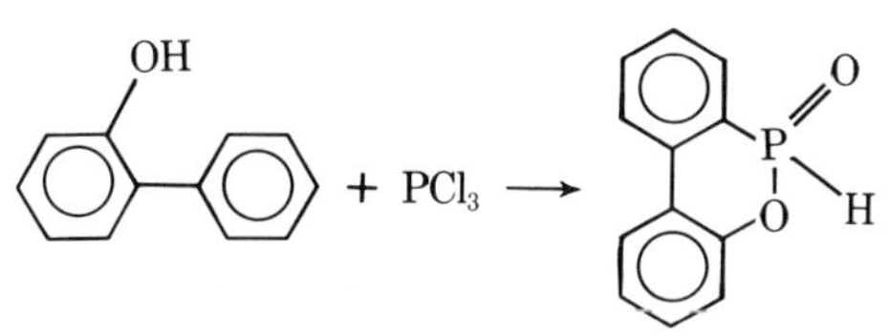

(**25**) [*35948-25-5*]

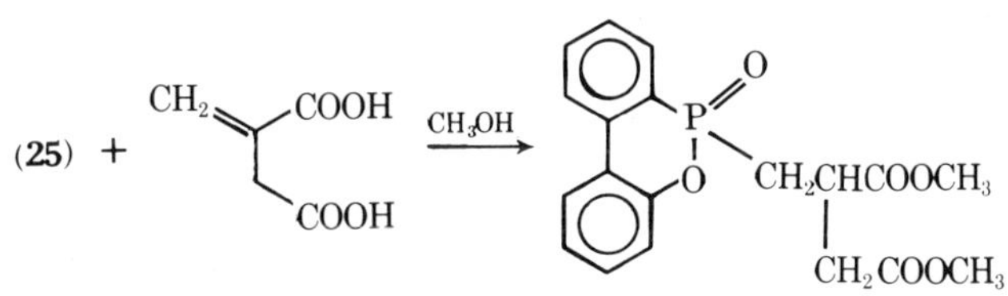

(**26**) [*63562-31-2*]

This dicarboxylic ester is then copolycondensed with the other reagents in PET manufacture to produce GH. The advantage of this rather unusual phosphinate structure is its high thermal and hydrolytic stability. The fibers have processing and dyeing characteristics similar to conventional PET. Toyobo's GH is offered for children's sleepwear in the United States and in curtains for public buildings in Japan.

Health and Safety Factors

Two factors must be considered when discussing the toxicology of flame-retarded materials; (*1*) the toxicology of the compounds themselves, and (*2*) the effect of the flame retardants on combustion-product toxicology.

Product Toxicology. No valid generalization can be made regarding the toxicity of phosphorus compounds as a broad class. Some phosphorus compounds are essential components of all living tissues, eg, inorganic phosphate, ATP, DNA, RNA, and phospholipids. Phosphoric acid and various salts of phosphoric acid and pyrophosphoric acid are on the GRAS (Generally Recognized as Safe) list of FDA-approved food additives. Unlike certain specific organophosphorus structures which have been deliberately designed (109) as biocides, the phosphorus flame retardants as a class have only moderate to low acute toxicity. Published data compiled by NIOSH (110) and EPA (111) show the following values (rat, oral) for representative commercial phosphorus flame retardants (Table 1).

Table 1. NIOSH and EPA Data for Representative Commercial Phosphorus Flame Retardants

	Dose, mg/kg	Lethality[a]	Reference
triethyl phosphate	1600	LD_{LO}	110
2-chloroethanol phosphate (3:1)	521, 1230, 1410 (3 reports)	LD_{50}	111
1,3-dichloro-2-propanol phosphate (3:1)	2830	LD_{50}	111
triphenyl phosphate	3000	LD_{LO}	110
tricresyl phosphate (commercial mixture)	4680	LD_{LO}	110

[a] LD_{50} = 50% lethal dose, LD_{LO} = lowest published lethal dose.

A particular mode of neurotoxicity was discovered during the Prohibition era (ca 1930) when tricresyl phosphate was used to adulterate bootleg liquor, causing a characteristic type of paralysis. This mode of toxicity was correlated with the presence of the *o*-cresyl isomer (or certain other specific alkylphenyl isomers) in the triaryl phosphate and many details of the chemistry and toxicology of the toxic process have since been elucidated (112). The use of low *o*-content cresols became the accepted practice in industrial production of tricresyl phosphate. Standard *in vivo* tests, usually conducted with chickens sensitive to this mode of toxicity, have been developed for premarket testing of new or modified aryl phosphate products (112–113). Commercial 2-ethylhexyl diphenyl phosphate has been cleared by FDA for food-wrapping uses.

Mutagenic properties were found with tris(2,3-dibromopropyl) phosphate by the recently developed *Salmonella* assay (58,59). This test assesses mutagenicity on special strains of *Salmonella* bacteria and is considered by some authorities as a plausible predictive test for carcinogenicity. As a result of these findings, confirmed by a number of *in vitro* tests, tris(2,3-dibromopropyl)phosphate, hitherto used to treat polyester fabric to meet the federal standard for children's sleepwear, was voluntarily removed from the market by its manufacturer. The same study showed mutagenic activity of certain related dibromopropyl compounds not containing phosphorus. Also, in the same study, 2-chloroethanol phosphate (3:1) and 1,3-dichloro-2-propanol phosphate (3:1) were found not to display mutagenic activity (60). The latter compound has been found inactive in several other mutagenicity tests (114).

Effect on Combustion Toxicology. At the present stage of knowledge, generalities are not possible on this topic; little or no empirical field data are available. There are no documented instances of any type of flame retardants contributing to human fire casualties. A survey (115) of data from small scale combustion or pyrolysis experiments revealed either decreases or increases in the yield of toxic gases (CO, HCN, and the like) with phosphorus flame retardants.

Laboratory experiments with rodents, or the use of gas analysis, tend to be confused by artifactual responses depending on burning configuration, equipment design, polymer substrate, temperature and toxicity criteria (ie, death vs incapacitation of various sorts) (116–117). Moreover, data from small-scale testing have doubtful relevance to real fire hazards.

One noteworthy neurotoxic response was demonstrated in laboratory pyrolysis studies with various types of phosphorus flame retardants in a rigid urethane foam, but the response was caused by a highly specific interaction with certain trimethylolpropane polyols, producing a toxic bicyclic trimethylolpropane phosphate [*1005-93-2*] (**27**) (118–120). Similar formulations based on other polyols avoided this neurotoxic product. Large-scale fire tests also failed to confirm (**27**) as a major contributor to hazard (115).

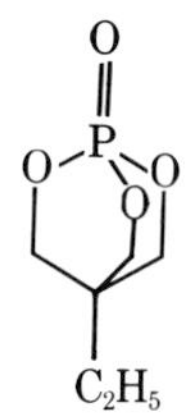

(**27**)

Several recent laboratory studies using rodents have demonstrated reduced toxicity of combustion products from urethane foams containing commercial chloroethyl phosphate flame retardants, compared to similar foams without the retardants (121). In a small-scale test with mice, forced burning of polyurethane foams containing diethyl *N,N*-bis(2-hydroxyethyl)aminomethylphosphonate (**20**) seemed to require significantly longer times to cause death than burning of polyurethane foams without flame-retardant additives (122).

Again, the relevance of such information to full-scale fires is questionable, but such laboratory tests do offer a logical basis for refuting unwarranted generalizations derived from other small-scale tests.

Effects on Visible Smoke. Although examples have been cited in which specific phosphorus flame retardants increased smoke relative to the nonflame-retarded polymer, this effect appears to be a complex function of burning conditions and the total formulation (115). Instances are reported in which the presence of the flame retardant lowered observed smoke, eg, in rigid urethane foams where the phosphorus additive increased char (123). In PVC, the addition of an alkyl diphenyl phosphate was observed to increase smoke when calcium carbonate was used as filler, and to decrease smoke when certain magnesium minerals were used as filler (124). In rigid urethane foams, the effect of a phosphorus flame retardant on smoke was found to depend on the polyol used, on the testing mode (ie, flaming, or smoldering) and on the criteria applied (ie, total smoke, maximum optical density, or rate of smoke evolution) (115).

Environmental Considerations. As with other large-volume industrial chemicals, the phosphate flame retardants and plasticizers are given close environmental scrutiny. Results published to date on acute toxicity to aquatic invertebrates and fish indicate, not surprisingly, large differences between one aryl phosphate mixture and another. However, preliminary results suggest a substantially lower degree of acute fish toxicity for even the more toxic members of this class than for the polychlorobiphenyls (PCBs) (125) (see Chlorocarbons).

Moreover, results of standardized biodegradation testing in river water and activated sludge showed that commercial isopropylphenyl diphenyl phosphate [*55864-04-5*], *tert*-butylphenyl diphenyl phosphate [*981-40-8*], cresyl diphenyl phosphate [*26444-49-5*], alkyl diphenyl phosphates, and trialkyl phosphates underwent primary degradation at moderate to rapid rates and were ultimately degraded to simple inorganic molecules (126).

Economic Aspects

Published production and sales figures are not generally available for individual flame retardants. One publication has estimated 1978 U.S. usage as 18,000 metric tons of nonhalogenated phosphate esters and 10,000 t of halogenated phosphate esters (127). Reactive polyurethane flame retardants used in rigid foams were estimated at 9000 t in 1978 (127), of which a major portion may be assumed to be comprised of phosphorus diols and polyols.

The growth rate of the aryl phosphate flame retardant plasticizers was reported in 1978 to have been 6% during the preceding two years, exceeding that of nonflame-retardant plasticizers (128). The principal polymer application for these materials is in vinyls where the aryl phosphates serve as both plasticizers and flame retardants.

The other major use areas for the phosphorus-based flame retardants are polyurethane foams (both rigid and flexible), cellulose and cellulosic polymers, certain engineering thermoplastics, polyester resins, other thermoset resins, and adhesives. Despite considerable research, phosphorus flame retardants have found only minor use in polyolefin and styrenic polymers, where flame retardancy continues to be dominated by halogenated additives and antimony oxide.

Trends in the commercial development of phosphorus flame retardants have been in the direction of less toxic, more stable or less volatile compounds, and built-in or oligomeric additive types. Increased attention is directed to the effect of polymer composition, including flame-retardant nature and quantity, on smoke and toxic fumes. At the same time, there is a growing realization that the results of arbitrary tests, especially small-scale tests, may correlate poorly with real fire hazards. Examples cited above illustrate the invalidity of broad generalizations concerning the effect of flame retardants on toxicity, smoke, and combustion products. The development of flame retardants in all likelihood will be conducted as part of a systems approach involving the total polymer composition and its physical form as well as its chemical structure. More attention will be given to fire barriers and coatings to protect flammable materials. Increased emphasis will also be given to questions of safety of flame retardants in handling, long-term exposure, disposal, and biodegradation.

BIBLIOGRAPHY

1. S. L. Madorsky, *Thermal Degradation of Organic Polymers,* Wiley-Interscience, New York, 1964, p. 238 ff.
2. M. Lewin and A. Basch in M. Lewin, S. M. Atlas, and E. M. Pearce, eds., *Flame-Retardant Polymeric Materials,* Vol. 2, Plenum Press, New York, 1978, pp. 1–3.
3. J. E. Hendrix, G. L. Drake, Jr., and R. H. Barker, *J. Appl. Polym. Sci.* **16,** 41, 257 (1972).
4. E. D. Weil in W. C. Kuryla and A. J. Papa, eds., *Flame Retardancy of Polymeric Materials,* Vol. 3, Marcel Dekker, Inc., New York, 1975, pp. 185–243.

5. A. Granzow, *Accounts Chem. Res.* **11**(5), 177 (1978).
6. G. Avondo, C. Vovelle, and R. Delbourgo, *Combust. Flame.* **31,** 7 (1978).
7. I. J. Gruntfest and E. M. Young, *ACS Div. Org. Coatings & Plastics Preprints,* Vol. 2, No. 2, 1962, pp. 113–124; C. Camino, N. Grassie, and I. C. McNeill, *J. Poly. Sci., Poly. Chem. Ed.* **16,** 95 (1978).
8. A. J. Papa in ref. 4, p. 31.
9. L. Zabski, W. Walczyk, and Z. Jedlinski, *Chem. Zvesti* (English translation) **30,** 311 (1976).
10. J. J. Anderson, *Ind. Eng. Chem. Prod. Res. Dev.* **2,** 260 (1963).
11. I. Einhorn, "Thermal Degradation and Flammability Characteristics of Polymeric Materials," *Polymer Conference Series, Univ. of Utah, Salt Lake City, June 15–20, 1970.*
12. M. A. Boult, R. K. Gamadia, and D. H. Napier, *Instn. Chem. Eng. Symposium Series,* (33), 56 (1972).
13. J. E. Kresta and K. C. Frisch, *J. Cell. Plast.* **11,** 68 (Mar./Apr. 1975).
14. H. Piechota, *J. Cell. Plast.* **1,** 186 (1965).
15. A. J. Papa and W. R. Proops, *J. Appl. Polym. Sci.* **16,** 2631 (1972).
16. A. E. Sherr, H. C. Gillham, and H. G. Klein, *Ad. Chem. Ser.* **85,** 307 (1968).
17. S. K. Brauman and co-workers, *J. Fire Flammability* **6,** 41 (1976); S. K. Brauman, *J. Fire Retardant Chem.* **4,** 18 (1977); S. K. Brauman and N. Fishman, *ibid.,* 93 (1977).
18. A. W. Benbow and C. F. Cullis, *Combust. Flame* **24,** 217 (1975).
19. Ref. 4, pp. 215–218; H. Mori, *Proceedings of the 1975 Symposium on Textile Flammability,* LeBlanc Research Corp., East Greenwich, R.I., 1975, pp. 124–154.
20. R. Sanders, *Text. Chem. Color.* **5**(3), 48 (1973).
21. T. J. Swihart and P. E. Campbell, *Text. Chem. Color.* **6**(5), 109 (1974).
22. W. A. Rosser, Jr., S. H. Inami, and H. Wise, *Combust. Flame* **10,** 287 (1966).
23. E. T. McHale, *Fire Research Abstracts and Reviews* **11**(2), 90 (1969).
24. D. Steutz and co-workers, paper presented at *the University of Utah Polymer Conference, June 8, 1970.*
25. W. Hastie and C. L. McBee, *National Bureau of Standards Final Report, NBSIR 75-741,* Aug. 1975.
26. J. Carnahan and co-workers, paper presented at *4th Intl. Conference on Flammability and Safety, University of San Francisco, Jan. 15–19, 1979.*
27. A. B. Deshpande and co-workers, *J. Appl. Poly. Sci. Appl. Polym. Symp.* **31,** 257 (1977).
28. Y. Okamato, N. Kimura, and H. Sakurai, *Bull. Chem. Soc. Jpn.* **47,** 1299 (1974).
29. J. W. Lyons, *The Chemistry and Uses of Flame Retardants,* Wiley-Interscience, New York, 1970, p. 23.
30. Ref. 4, pp. 207–214.
31. Ref. 4, pp. 203–207.
32. J. L. Gay-Lussac, *Ann. Chim.* **18,** 211 (1821).
33. Ref. 29, pp. 29–74.
34. E. L. Gefter, *Organophosphorus Monomers and Polymers,* Associated Technical Sources, Inc., Glen Ridge, N.J., 1962, English translation from Russian.
35. M. Sander and E. Steininger, *J. Macromol. Sci. Rev. Macromol. Chem.* **C1,** 1 (1967).
36. *Ibid.,* **C2,** 1 (1978).
37. E. N. Peters in W. C. Kuryla and A. J. Papa, eds., *Flame Retardancy of Polymeric Materials,* Marcel Dekker, Inc., Vol. 5, New York, 1979.
38. *Exolit 505 Product Information Bulletin,* Hoechst AG, Knapsack, Germany, July 1976.
39. Ref. 28, pp. 135 and 215.
40. C. Y. Shen, N. E. Stahlheber, and D. R. Dyroff, *J. Am. Chem. Soc.* **91,** 62 (1969).
41. U.S. Pat. 3,397,035 (Aug. 13, 1968), C. Y. Shen and N. E. Stahlheber (to Monsanto Co.).
42. H. L. Vandersall, *J. Fire Flammability* **2,** 97 (1971).
43. U.S. Pat. 2,122,122 (June 28, 1938), W. H. Woodstock (to Victor Chemical Works).
44. M. Becke-Goehring and J. Sambeth, *Z. Anorg. Allg. Chem.* **297,** 287 (1958).
45. U.S. Pat. 3,684,559 (Aug. 15, 1972), J. C. Chapin (to Ventron).
46. S. J. O'Brien, *Text. Res. J.* **38,** 256 (1968).
47. U.S. Pat. 3,887,511 (June 3, 1975), S. C. Juneja (to Canadian Patents & Development Ltd.); S. C. Juneja and L. R. Richardson, *For. Prod. J.* **24**(5), 19 (1974).
48. Brit. Pat. 996,914 (June 30, 1965), (Rohm and Haas G.m.b.H).
49. U.S. Pat. 3,347,818 (Oct. 17, 1967), K. L. Howe (to E. I. du Pont de Nemours & Co., Inc.).
50. Brit. Pat. 1,100,283 (Jan. 24, 1968), (to British Celanese).

51. U.S. Pat. 3,666,712 (May 30, 1972), E. D. Weil (to Stauffer Chemical Co.).
52. U.S. Pat. 2,430,569 (Nov. 11, 1947), D. C. Hull and A. H. Agett (to Eastman Kodak Co.); U.S. Pats. 2,407,279 (Sept. 10, 1946), and 2,508,389 (Sept. 30, 1975), D. C. Hull and J. R. Snodgrass (to Eastman Kodak Co.).
53. U.S. Pat. 3,909,484 (Sept. 30, 1975), A. N. Beavon (to Universal-Rundle Corp.).
54. P. V. Bonsignore and J. H. Manhart, *Proc. Ann. Conf. Reinf. Plast. Compos. Inst. Soc. Plast. Ind.* **23c,** 29 (1974).
55. U.S. Pat. 3,100,220 (Aug. 6, 1963), A. L. Smith (to Celanese Corp.).
56. U.S. Pat. 3,041,293 (June 26, 1962), R. Polacek (to Celanese Corp.).
57. J. W. Crook and G. A. Haggis, *J. Cell. Plast.* **5,** 119 (1969).
58. U.S. Pat. 3,046,297 (July 24, 1962), D. E. Overbrook and R. C. Nametz (to Michigan Chemical).
59. Brit. Pat. 1,292,878 (Oct. 18, 1972), (to Eastman Kodak).
60. M. J. Prival and co-workers, *Science,* **195,** 76 (1977).
61. A. Blum and B. N. Ames, *Science* **195,** 17 (1977).
62. U.S. Pats. 3,014,956 (Dec. 26, 1961), and 3,042,701 (July 3, 1962), G. H. Birum (to Monsanto).
63. M. I. Kabachnik and P. A. Rossiyskaya, *Izv. Akad. Nauk. SSR Ser. Khim,* 403 (1946); Brit. Pat. 1,503,537 (Mar. 3, 1978), R. E. Kinney and co-workers (to Mobil Oil Corp.).
64. U.S. Pat. 3,192,242 (June 29, 1965), G. H. Birum (to Monsanto).
65. J. S. Babiec and co-workers, *Plast. Technol.* **21,** 47 (June 1975).
66. U.S. Pat. 3,707,586 (Dec. 26, 1972), R. J. Turley (to Olin Corp.).
67. T. M. Moshkina and A. N. Pudovik, *Zh. Obschei Khim.* **32,** 1761 (1962).
68. U.S. Pats. 3,896, 187 (July 22, 1975), and 3,513,644 (May 26, 1970), E. D. Weil (to Stauffer Chemical Co.).
69. C. J. Kmiec and J. A. Albright, paper presented at *Fire Retardant Chemicals Assn. Meeting, Chicago, Ill., Oct. 30–Nov. 2, 1977.*
70. U.S. Pats. 3,978,167 (Aug. 31, 1976), and 3,997,505 (Dec. 14, 1976), J. A. Albright (to Velsicol Chemical Corp.).
71. U.S. Pat. 3,891,727 (June 24, 1975), E. D. Weil (to Stauffer Chemical Co.).
72. U.S. Pat. 3,789,091 (Jan. 29, 1974), J. J. Anderson, J. G. Camacho, and R. E. Kinney (to Mobil Oil Corp.).
73. E. M. Bellet and J. Casida, *Science* **182,** 1135 (1973).
74. U.S. Pat. 4,066,812 (Jan. 3, 1978), W. M. Kaupin (to Wm. Carter Co.).
75. Brit. Pats. 1,510,381 and 1,510,382 (May 10, 1978), (to Sandoz).
76. U.S. Pat. 3,719,727 (Mar. 6, 1973), Y. Masai, Y. Kato, and N. Fukui (to Toyo Spinning Co.).
77. L. G. Krauskopf in L. I. Nass, ed., *Encyclopedia of PVC,* Marcel Dekker Inc., New York, 1976, pp. 529–534.
78. T. W. Lapp, *The Manufacture and Use of Selected Aryl and Alkyl Aryl Phosphate Esters,* Final Report prepared for the EPA, Midwest Research Institute, distributed by N.T.I.S. as PB-251678, Feb. 1976.
79. D. L. Buszard, *Chem. Ind.,* 610 (Aug. 19, 1978).
80. F. A. Lowenheim and M. K. Moran, *Faith, Keyes, and Clark's Industrial Chemicals,* 4th ed., John Wiley & Sons, Inc., New York, 1975, pp. 849–853.
81. J. M. Heaps, *Plastics (London)* **33**(366), 410 (Apr. 1, 1969).
82. U.S. Pat. 3,919,158 (Nov. 11, 1975), D. R. Randell and W. Pickles (to Ciba-Geigy AG).
83. J. R. Darby and J. K. Sears in N. M. Bikales, ed., *Encyclopedia of Polymer Science and Technology,* Wiley-Interscience, New York, 1969, pp. 229–306.
84. L. E. A. Godfrey and J. W. Schappel, *Ind. Eng. Chem. Prod. Res. Dev.* **9,** 426 (1970).
85. H. R. Allcock, *Science* **193,** 1214 (1976).
86. *Chem. Week* **116,** 33 (Apr. 2, 1975).
87. L. Cipriani, *Tech. Papers Reg. Tech. Conf. SPE Palisades Section, Oct. 27–28, 1970,* pp. 37–47.
88. U.S. Pats. 3,422,048 (Jan. 14, 1969), and 3,513,119 (May 19, 1970), J. F. Cannelongo (to American Cyanamid Co.).
89. C. Savides, A. Granzow, and J. F. Cannelongo, *Tech. Pap. Reg. Tech. Conf., Soc. Plast. Eng., 1975,* pp. 18–28.
90. E. D. Weil, *J. Fire Flammability/Fire Retardant Chemistry* **1,** 125 (1974).
91. U.S. Pat. 3,489,706 (Jan. 13, 1978), B. Mikofalvy (to B. F. Goodrich).
92. U.S. Pat. 3,691,127 (Sept. 12, 1972), P. Kraft and R. Brunner (to Stauffer Chemical Co.).

93. Ger. Appl. 2,452,369 (May 5, 1975), P. Kraft and P. S. Yuen (to Stauffer Chemical Co.).
94. A. J. Papa, *Ind. Eng. Chem. Prod. Res. Dev.* 9, 478 (1970).
95. L. Naturman, *SPE J* **1,** 965 (1961).
96. Brit. Pat. 954,792 (Apr. 8, 1964), (to Virginia-Carolina Chem. Co.).
97. U.S. Pat. 3,235,517 (Feb. 15, 1966), T. M. Beck and E. N. Walsh (to Stauffer Chemical Co.).
98. U.S. Pat. 3,850,859 (Nov. 26, 1974), J. Wortmann, F. Dany, and J. Kandler (to Hoechst AG).
99. U.S. Pat. 3,534,073 (Oct. 13, 1970), M. Lewis (to Swift).
100. W. Reeves and co-workers, *Text. Chem. Color.* **1,** 365 (1969).
101. W. F. Herbes, K. H. Remley, and J. B. Trecek, *Am. Dyest. Rep.* **65** (9), 72 (1976).
102. R. Aenishanslin and co-workers, *Text. Res. J.* **39,** 375 (1969).
103. R. Aenishanslin and N. Bigler, *Textilveredlung* **3,** 467 (1968).
104. B. J. Eisenberg and E. D. Weil, *Text. Chem. Color.* **6**(8), 180 (1974).
105. J. P. Bruce, *Am. Dyest. Rep.* **62**(10), 68 (1973).
106. U.S. Pat. 4,033,936 (July 5, 1977), U. Bollert and co-workers (to Hoechst AG).
107. U.S. Pat. 3,941,752 (Mar. 2, 1976), H. J. Kleiner and co-workers (to Hoechst AG).
108. Ger. Appl. 2,646,218 (April 28, 1977), S. Endo and co-workers (to Toyobo Co. Ltd.).
109. R. D. O'Brien, *Toxic Phosphorus Esters,* Academic Press, New York, 1960; C. Fest and K. J. Schmidt, *The Chemistry of Organophosphorus Pesticides,* Springer-Verlag, New York, 1978.
110. *Registry of Toxic Effects of Chemical Substances,* DHEW (NIOSH) Publication No. 78-104-A, 1977.
111. S. S. Lande and co-workers, *Investigation of Selected Potential Environmental Contaminants: Haloalkyl Phosphates,* Report TR-513 prepared for EPA, distributed by N.T.I.S. as PB-257910, Syracuse Research Corp., Aug. 1976.
112. M. K. Johnson, *CRC Critical Rev. Toxicol.* **3,** 289 (1975); M. K. Johnson, *Arch. Toxicol.* **34,** 259 (1975).
113. F. R. Johannsen and co-workers, *Toxicology and Applied Pharmacology* **41,** 291 (1977).
114. D. Brusick and co-workers, *J. Environmental Pathology and Toxicology,* **3**(1), 207 (1980).
115. E. D. Weil and A. M. Aaronson, "Phosphorus Flame Retardants—Meeting New Requirements," in *Proceedings of 1st European Conference on Flammability and Fire Retardants, Brussels, July 14–15, 1977,* Technomic Publishing Co., Westport, Conn., 1978.
116. W. H. Lawrence and co-workers, *J. Combust. Tox.* **5,** 39 (1978).
117. C. Herpol, *Fire and Materials* **1**(1), 29 (1976).
118. J. H. Petajan and co-workers, *Science* **187,** 742 (1975).
119. K. J. Voorhees and co-workers, *J. Polym. Sci. Polym. Lett. Ed.* **13,** 293 (1975).
120. G. Kimmerle and co-workers, *Arch. Toxicol.* **35,** 149 (1976).
121. C. J. Hilado, *J. Combust. Tox.* **3,** 32 (1976); C. J. Hilado and A. Furst, *ibid.,* 425 (1976).
122. C. J. Hilado and A. M. Machado, *J. Combust. Tox.* **5,** 162 (1978).
123. I. N. Einhorn, "Effects of Fire Retardants on the Nature of Smoke," in *Polymer Conference Series, Flammability Characteristics of Polymeric Materials, Univ. of Utah, Salt Lake City, July 8–12, 1974.*
124. U.S. Pat. 3,869,420 (Mar. 4, 1975), T. C. Mathis and A. W. Morgan (to Monsanto Co.); T. C. Mathis and J. D. Hinchen, *31st Ann. Tech. Conf. SPE, Technical Papers,* Vol. 19, 1973, pp. 343–348.
125. M. J. Nevins and W. W. Johnson, *Bull. Environ. Contam. and Toxicol.* **7,** 250 (1978).
126. V. W. Saeger and co-workers, *Envir. Sci. Tech.* **13,** 840 (1979).
127. *Mod. Plast.* **55**(9), 64 (1978).
128. *Plast. Eng.* **34**(6), 35 (1978).

EDWARD D. WEIL
Stauffer Chemical Company

FLAME RETARDANTS FOR TEXTILES

Hazards associated with the ready combustibility of cellulosic materials were recognized as early as the 4th century BC, when Aeneas is said to have recommended treatment of wood with vinegar to impart fire resistance (1). The annals of Claudius record that wooden storming towers used in the siege of Piraeus in 83 BC were treated with a solution of alum to protect them against fire.

The technique of imparting flame resistance to textile fabrics is relatively new. Among the earliest references is an article by Sabattini published in 1638. Recognizing a need to prevent fire, he suggested that clay or gypsum pigments be added to the paint used for theater scenery to impart some flame resistance. Perhaps the first noteworthy recorded attempt to impart flame resistance to cellulose was made in England in 1735 (2) when Obadiah Wyld was granted a patent for a flame-retardant mixture containing alum, ferrous sulfate, and borax.

In France in 1821, Gay-Lussac (3) developed a flame-resistant finish by treating linen and jute fabrics with a mixture of ammonium phosphate, ammonium chloride, and borax.

The first successful, launder-resistant, flame-retardant finish for fabric was based on the work of Perkin (4) who precipitated stannic oxide within the fiber. This fabric was flame resistant but afterglow was severe and persistent enough to completely consume the fabric.

Flame retardants are mainly used on cottons and rayons. Fabrics made from wool (qv), silk (qv), and protein-like synthetic polymers are not considered sufficiently combustible, for the most part, to warrant the need for flame-retardant finishes (see Biopolymers; Textiles).

Since World War II the flammability of textiles of all types has received greatly increased attention, spurred by the Conference on Burns and Flame Retardant Fabric in 1966 (5), and by the 1967 amendment to the Flammable Fabrics Act of 1953 (6). Flammability standards were established by the Department of Commerce and enforced by the Federal Trade Commission. This responsibility was taken over by the Consumer Product Safety Commission when it was created in 1973.

The terms used in connection with flame-resistant fabrics are sometimes confusing. Fire resistance and flame resistance are often used in the same context as the terms fireproof or flameproof. A textile that is flame resistant or fire resistant does not continue to burn or glow once the source of ignition has been removed, although there is some change in the physical and chemical characteristics. Fireproof or flameproof, on the other hand, refer to material that is totally resistant to fire or flame. No appreciable change in the physical or chemical properties is noted. Asbestos is an example of a fireproof material.

Most organic fibers undergo a glowing action after the flame has been extinguished, and flame-resistant fabrics should also be glow resistant. Afterglow may cause as much damage as the flaming itself since it can completely consume the fabric. The burning (decomposition) temperature of cellulose is about 230°C, whereas afterglow temperature is approximately 345°C.

Chemical modification of cellulose with fire retardants gives products whose resistance to laundering and weathering is superior to that of finishes based on the physical deposition of the flame retardant within the fabric, yarn, or fiber. The reactions involved are either esterification or etherification. The latter is preferred because ether linkages are more stable to hydrolysis.

Flame Resistance

The flame resistance of a textile fiber is affected by the chemical nature of the fiber; ease of combustion; fabric weight and construction; efficiency of the flame retardant; environment; and laundering conditions.

Fire-resistant characteristics can change significantly when treated fabric is exposed to sunlight, followed by laundering, even though repeated washing and tumble drying of samples of the same specimen did not indicate any significant changes, especially in the durability of the finish (7). Dry heat alone, followed by laundering or autoclaving can also have a deleterious effect (8).

Fibers are classified into natural fibers, eg, cotton, flax, silk, or wool; regenerated fibers, eg, rayon; synthetic fibers, eg, nylons, vinyls, polyester, acrylics; and inorganic fibers, eg, glass or asbestos. Combustibility depends on chemical makeup and whether the fiber is inorganic, organic, or a mixture of both (see Fibers, chemical).

The weight and construction of the fabric affect its burning rate and ease of ignition. Lightweight, loose-weave fabrics usually burn much faster than heavier-weight fabrics; therefore, a higher weight add-on of fire retardant is needed to impart adequate flame resistance.

Phosphorus-containing materials are by far the most important class of compounds used to impart durable flame resistance to cellulose (see also Flame retardants, phosphorus compounds). They usually contain either nitrogen or bromine and sometimes both. A combination of urea and phosphoric acid imparts flame resistance to cotton fabrics at a lower add-on than when the acid or urea is used alone (9). Other nitrogenous compounds, such as guanidine, or guanylurea, could be used instead of urea. Amide and amine nitrogen generally increase flame resistance, whereas nitrile nitrogen can detract from the flame resistance contributed by phosphorus. The most efficient flame-retardant systems contain two retardants, one acting in the solid and the other in the vapor phase.

Bromine in flame-resistant fabric escapes from the tar to the vapor phase during pyrolysis in air. It appears to have little or no effect on the amount of phosphorus remaining in the char. Bromine contributes flame resistance almost completely in the vapor phase.

Nitrogen when used in conjunction with phosphorus compounds has synergistic effects (10–12). Phosphorus content can be reduced without changing the efficiency of the flame retardant (13).

The temperature of the environment influences the burning characteristics of fabric as measured by the oxygen index (OI). This is true for untreated as well as flame-retardant fabrics. For example, the OI value of untreated fabric is 0.18 when burned at 25°C and 0.14 when burned at 150°C. For flame-retardant fabric, an OI value of 0.35 at 25°C may be reduced to 0.27 when burned at 150°C. Sunlight and heat can also destroy some flame retardancy (13–15), especially when followed by laundering or autoclaving (7–8). The moisture content of fabric can also affect flame retardancy (16).

Fire retardancy of a treated cellulosic fabric is reduced when the retardant contains acid groups and the treated fabric is soaked or laundered in water containing calcium, magnesium, or alkali metal ions. Phosphate- and carbonate-based detergents affect durability of fire retardants (17). Soap-based detergents can result in a substantial loss of fire resistance because of deposit of fatty acid salts (18). Phosphorus-based flame retardants are adversely affected by water hardness and sodium hypochlorite (see Drycleaning and laundering).

Mechanisms. Cellulose, as such, has no appreciable vapor pressure and does not burn. However, on exposure to high temperatures it decomposes exothermically into flammable compounds causing further degradation and decomposition until complete disintegration has taken place. Numerous studies have been made on burning of untreated and flame-retardant-treated cellulose (19–24). Decomposition takes place in two stages. First, thermal decomposition causes cellulose to decompose heterogeneously into gaseous, liquid, tarry, and solid products (25,26). The flammable gases thus produced ignite, causing the liquids and tars to volatilize to some extent. This produces additional volatile fractions which ignite and produce a carbonized residue which does not burn readily. This process continues until only carbonaceous material remains. After the flame has subsided, the second stage begins: the residual carbonized residue slowly oxidizes and glowing continues until carbonaceous char is consumed.

Cotton (qv) treated with an effective flame retardant forms, in general, the same decomposition products upon burning as untreated cotton; however, the amount of tar is greatly reduced with a corresponding increase in the solid char. Consequently, as decomposition takes place, smaller amounts of the flammable gases are available from the tar, and greater amounts of nonflammable gases from the decomposition of the char fraction.

Char is essentially carbon and its oxidation causes afterglow. Phosphorus-containing compounds (in some cases, polymers) are particularly effective in inhibiting char oxidation. The oxidation of carbon takes place through either of the following reactions:

$$C + \tfrac{1}{2}\,O_2 \rightarrow CO \quad \Delta H = 110.45 \text{ kJ } (26.4 \text{ kcal})$$

$$C + O_2 \rightarrow CO_2 \quad \Delta H = 394.96 \text{ kJ } (94.4 \text{ kcal})$$

Effective glow-retardant chemicals, such as compounds containing phosphorus, cause the first reaction to be prevalent. Oxidation of carbon monoxide is not sufficiently exothermic to maintain afterglow of the char.

Imparting flame resistance to cellulose has been explained by the following theories:

Coating theory. As early as 1821, Gay-Lussac (3) suggested that fire resistance was due to formation of a layer of fusible material which melted and formed a coating thereby excluding the air necessary for the propagation of a flame. This was based on the efficiency of some easily fusible salts as flame retardants. Carbonates, borates, and ammonium salts are good examples of coating materials that produce a foam on the fiber by liberation of gases such as carbon dioxide, water vapor, ammonia, etc.

Gas theory. The flame retardant produces noncombustible gases at burning temperature which dilute the flammable gases produced by decomposition of the cellulose to a concentration below the flaming limit.

Thermal theory. Heat input from a source is dissipated by an endothermic change in the retardant and the heat supplied from the source is conducted away from the fibers so rapidly that the fabric never reaches temperature of combustion.

Chemical theory. Strong acids, bases, metal oxides, and oxidants that tend to degrade cellulose, especially under the influence of heat, usually impart some degree of flame resistance to cellulose (27). This is also true of the more efficient flame retardants, such as phosphoric and sulfuric acid, which are good dehydrating agents. When this happens, cellulose on combustion produces mainly carbon and water rather than carbon dioxide and water.

Flame retardants for cotton may possibly act through a dehydration process by Lewis acid or base formation through a carbonium ion or carbanion mechanism (21). This theory is being further investigated.

Earlier theories suggested that flame-retarded cellulose decomposed at high temperatures to *l*-glucosan which in turn broke down to form other volatile products which were highly flammable. However, if bases are present in the fabric during burning, dehydrocellulose is formed by a base-catalyzed dehydration followed by char formation (28). Base-catalyzed dehydration prevents formation of *l*-glucosan by propagating structural changes at an energy level below that required to convert the coformers of the glucopyranose ring (29).

Durability

Nondurable Finishes. Flame-retardant finishes that are not durable to laundering and leaching are, in general, relatively inexpensive and efficient (30). In some cases, mixtures of two or more salts are much more effective than any one of the components alone. For example, an add-on of 60% of borax $Na_2B_4O_7.10H_2O$ is required to prevent fabric from burning. Boric acid, H_3BO_3, by itself, is ineffective as a flame retardant even in amounts that equal the weight of the fabric. However, a mixture of seven parts borax and three parts boric acid imparts flame resistance to a fabric with as little as 6½% add-on.

The water-soluble flame retardants are most easily applied by impregnating the fabric with a water solution of the retardant, followed by drying. Adjustment of the concentration and regulation of the fabric wet pickup controls the amount of retardant deposited in the fabric. Fabric can be processed on a finishing range consisting of any convenient means of wetting the fabric with the solution, such as a padder or dip tank, followed by a drying on cans, in an oven, on a tenter frame, or merely by tumbling in a mechanical dryer. Water-soluble flame retardants also may be applied by spraying or brushing, or dipping fabrics, or as a final rinse in a commercial or home laundry (31). The water-soluble flame retardants most widely used for textiles are listed in Table 1. Less commonly used are sulfamates of urea or other amides and amines; aliphatic amine phosphates, such as triethanolamine phosphate [*10017-56-8*], phosphamic acid [*2817-45-0*] (monoimido phosphoric acid), $H_2PO_3NH_2$, and its salts; and alkylamine bromides, phosphates, and borates.

Semidurable Finishes. Semidurable fire retardants are those that resist removal for one to about 15 launderings. Such retardants are adequate for applications such as drapes, upholstery, and mattress ticking. If they are sufficiently resistant to sunlight or can be easily protected from actinic degradation, they can also be applied to outdoor textile products.

The principal disadvantage of the water-soluble flame retardants is their lack of durability. It can be overcome by precipitating inorganic oxides on the fabric; for example, $WO_3.xH_2O$ and $SnO_2.yH_2O$:

$$2\,Na_2WO_4 + SnCl_4 + (2x + y)H_2O \rightarrow 4\,NaCl + 2\,WO_3.xH_2O + SnO_2.yH_2O$$

Table 1. Water-Soluble Flame Retardants for Textiles

Component	CAS Registry Number	Formula	Composition
borax–	[1330-43-4]	$Na_2B_4O_7.10H_2O$	70
boric acid	[10043-35-3]	H_3BO_3	30
borax–			47
boric acid–			20
diammonium phosphate	[7783-28-0]	$(NH_4)_2HPO_4$	33
sodium phosphate dodecahydrate–	[10101-89-0]	$Na_3PO_4.12H_2O$	50
boric acid			50
boric acid–			50
diammonium phosphate			50
borax–			50
boric acid–			35
sodium phosphate dodecahydrate			15
ammonium sulfamate–	[7773-06-0]	$NH_4OSO_2NH_2$	75
diammonium phosphate			25
ammonium bromide	[12124-97-9]	NH_4Br	100
borax–			15
boric acid–			47
sodium phosphate–	[7601-54-9]	Na_3PO_4	18
sodium tungstate dihydrate	[10213-10-2]	$Na_2WO_4.2H_2O$	20

These codeposits add flame- and glow-resistance properties to textile fabrics. However, some insoluble deposits may also degrade the fabrics. Codeposits quite frequently improve glow resistance. They are usually more soluble than the deposit responsible for flame resistance and are more easily removed during the laundering process.

There are several methods for introducing the insoluble deposits into the fabric structure. Most generally used is the multiple-bath method, in which the fabric is first impregnated with a water-soluble salt or salts in one bath, and is then passed into a second bath which contains the precipitant.

Most semidurable retardants are used on cotton and are based on a combination of phosphorus and nitrogen compounds (32).

Durable Finishes. Earlier studies to produce durable flame retardants for cellulose were based on treatment with inorganic compounds containing antimony and titanium (33). Numerous patents were issued based on these types of treatments, eg, DuPont's Erifon process (34) and the Titanox FR process of the Titanium Pigment Corporation (35).

In the Erifon process titanium and antimony oxychlorides were applied from acid solution (pH 4) to fabric, which was then neutralized by passing through a solution of sodium carbonate, followed by rinsing and drying. Fabrics thus finished exhibited good flame resistance but also considerable afterglow. Some fabric characteristics were changed by this treatment. A large amount of tent fabric was treated by this type of process for the military service. However, it has now been replaced by flame retardants based on phosphorus.

The basic chemicals used in the Titanox FR process were titanium acetate chloride and antimony oxychloride. As with the Erifon process, it was difficult to process the fabric without dulling its appearance.

Excellent fire-resistant fabric was obtained by treating fabric with a suspension or emulsion of insoluble fire-retardant salts or oxides, eg, antimony oxide, with a chlorinated organic vehicle, such as chlorinated paraffin (36). Antimony oxide alone is only a poor flame retardant. When used, however, in conjunction with a chlorinated compound, which can form hydrochloric acid on heating, a very good flame retardant is produced.

The abbreviation, FWWMR, for fire, water, weather, and mildew resistance has frequently been used to describe treatment with a chlorinated organic metal oxide. A plasticizer, coloring pigments, fillers, stabilizers, or fungicides are usually added. However, hand, drape, flexibility, and color of the fabric are more affected by this type of finish than by other flame retardants. Add-ons of up to 60% are required in many cases to obtain adequate flame resistance. Durability of this finish is good and fabric processed properly retains its flame resistance after four to five years of outdoor exposure. This type finish is well suited for very heavy fabrics, eg, tents, tarpaulins, or awnings, but not for clothing or interior decorating fabrics. The metal oxides can be fixed to cotton by use of resins, eg, vinyl acetate–chloride copolymers (vinylite VYHH) or PVC (37).

A flame retardant has been developed based on an oil–water emulsion containing a plasticizer (PVC latex) and antimony oxide (38). High add-ons are necessary to impart adequate flame resistance but the strength of the fabric is little affected.

Test Methods

The important test methods are listed in Table 2.

Fire Resistance. The most widely used test is the standard vertical flame test, although it is only one of the many tests in which the fabric to be tested is suspended vertically. A strip of fabric (30 cm × 6.3 cm) is suspended vertically with its lower edge hanging just 1.9 cm above the top of a bunsen or tirrell burner. The burner is adjusted to give a 3.8-cm luminous flame, extending 1.9 cm over the fabric. After an ignition time of 12 s the flame is removed and the duration of the afterglow and the length of the charred area are measured. Most of the specifications allow an afterflaming of 2 s and an average char length (10 strips) of 8.9 cm, with a maximum of 11 cm for any one strip. These values are, of course, based on weights of the fabric. This method has been incorporated into a great many specifications, including CCCT 191a and CCCD 746 (Federal), 6-345 (Army), 24-C-20 (Navy), DOC FF 3-71, DOC FF 5-74, and others.

Refinements of this method include standard draft-free cabinets, conditioning the specimens, special fabric clamps and holders, manometers to control flow of gas to the burner, and techniques for measuring char length.

Flammability. The position of a fabric usually influences its burning rate. A fabric suspended vertically and ignited on the bottom burns considerably more rapidly than one held in a horizontal position. The speed with which vertically suspended fabrics burn makes distinctions among fabrics difficult and for that reason many flammability testers mount the fabric specimen at some angle from the vertical position.

The AATCC (American Association of Textile Chemists and Colorists) or inclined flammability tester was developed after World War II and has been more widely used than any other rate-of-burning device: A strip of fabric (15.2 cm × 5 cm) is clamped in a rack inclined at 45°, and the surface of the fabric near the bottom is exposed to the flame of a microjet burner for 1 s. If the sample is ignited by this flame, the time

Table 2. Summary of Test Methods For Textiles

Test standard	Instrument	Materials tested	Specimen size, cm	Angle of specimen	Ignition source	Properties measured
FT MS 191-5910	burn rate of	fabrics	2.5 × 15.2	30°	paper match	burn rate
CS 191-53 (Rev.), AATCC 33-1962, ASTM D1230-61, NFPA 702-1975, California Tech. Bulletin 117	45° flammability tester for wearing apparel	fabrics used for wearing apparel, bonded and nonbonded upholstery, filling materials and fabrics	varies with procedure	45°	microburner (butane gas)	ease of ignition, flame spread time, self-extinguishing time
FT MS 191-5908.1	burn rate tester (Method 5908)	fabrics	2.5 × 15.2	45°	microburner (butane gas)	ease of ignition, flame spread time
DOC FF-1-70 DOC FF-2-70, ASTM D2859 CPAI-84 DDD-C-95	carpet flammability tester	carpets, rugs, tent flooring	22.8 × 22.8	horizontal	burning methenamine tablet	extent of char area
DOT FMVS 302, California Recreational Vehicles	horizontal flammability tester	motor vehicle interior materials	10.1 × 35.5	horizontal	bunsen burner	flame spread time
CPAI-75	flammability tester	multicomponent sleeping bags	30.4 × 35.5	horizontal	bunsen burner (mfg gas type B)	burning rate over 25.4 cm
FTMS 191-5906, FAA Reg. Para. 25.853	burn rate tester (Method 5906)	fabrics-aircraft interiors	11.4 × 31.7	horizontal	bunsen burner	flame spread time
FT MS 372, NFPA 253	radiant panel tester for floor covering systems	floor covering systems	25.4 × 107	horizontal	radiant panel—gas/air mixture	critical radiant flux at flame out
NFPA 254-77	furnace tester for floor covering systems	floor covering systems	60.9 × 244	horizontal	gas flame	flame propagation index
NFPA-701-1976 (large scale), State of California Title 19-1237.3	vertical flammability tester (NFPA 701 large scale)	fabrics and films	in sheets—12.7 × 213 in folds—63.5 × 213	vertical	bunsen burner	afterflame, afterglow, char length
ASTM D568-74, NFPA 701-1976, (small scale), City of Boston Fire Code 1959, City of New York Fire Code 1940, State of California Title 19-1237.1	flammability tester for fabrics and plastics	fabrics, plastic sheet and film	varies with procedure	vertical	bunsen burner	burning rate, afterflame, afterglow, char length

FT MS 191-5903, FAA Reg. Para. 25.853 CPAI-84, California Tech. Bulletin 117, Sections A, B and F	vertical flammability tester	aircraft interior materials tenting-walls and top, resilient cellular materials and plastics	varies with procedure	vertical	bunsen burner	burn length, afterglow, char length
DOC FF3-71 DOC FF5-74 FAA proposed std flight attendants' uniforms	children's sleepwear flammability tester	fabrics and garments children's sleepwear, flight attendants' uniforms	8.9 × 25.4	vertical	special methane gas burner	residual flame time, char length
J. C. Penney modification of DOC FF3-71, FF3-71 proposed, ASTM method	vertical flammability tester (semirestraint method)	wearing apparel fabrics	15.2 × 38.1	vertical	special methane gas burner	burning time, weight loss, char length, char area
TAPPI-T461-5U-72	TAPPI flammability tester	treated paper and paper board	6.9 × 20.9	vertical	bunsen burner (natural gas)	flame time, afterglow, char length
CPSC proposed standard for the flammability wearing apparel	mushroom apparel flammability	clothing items, fabrics and related materials used in apparel	31.7 × 60.9	vertical-clamped to heat sensor assembly	microburner, methane gas	ignition time, maximum heat time
proposed FAA regulation	heat-flux resistance apparatus	fabrics for flight attendants' uniforms	15.2 × 15.2	vertical	elec. heated radiant panel	heat flux transmission
ASTM D2863-74	oxygen-index flammability tester	plastics, textiles, paper and other materials	0.6 × 0.3 × 12.7	vertical	gas flame, burning in atmosphere of variable oxygen concentration	oxygen index (oxygen content required to support combustion)
FTMS-191 Method 5920	Mackey apparatus	cloth or related materials	5 × 30.4	not applicable	not applicable	tendency of material to undergo self-heating at moderate temperatures

of flame travel over 12.7 cm of the specimen is used as an indication of relative flammability of the fabric (39). In July 1954 this method was adopted for the first federal legislation relating to combustible products.

With the passage of the amendment to the 1953 Flammable Fabrics Act in 1967, other federal flammability standards have been issued by the Department of Commerce (DOC). The test methods and textile products relating to these standards are as follows:

DOC FF 1-70 (40) and DOC FF 2-70 (41) Standards for the Surface Flammability of Carpets and Rugs (April 16 and December 29, 1971, respectively).

DOC FF 3-71 (42) and DOC FF 5-74 (43) Standards for the Flammability of Children's Sleepwear are a modification of the vertical flame test (July 29, 1973 and May 1, 1975, respectively). These standards have been amended by FF3-71 and FF5-74 (February 6, 1978). The former requires that no single sample could have flaming material on the bottom of the test cabinet 10 s after the ignition source was removed.

DOC FF 4-72 (as amended) Standard for the Flammability of Mattresses (44).

Motor Vehicle Safety Standard No. 302—Flammability of Interior Materials; (passenger cars, multipurpose passenger vehicles, trucks), tested in a horizontal position.

AATCC Test Method 34-1969—Fire Resistance of Textile Fabrics (45), based on a modification of the vertical flame test.

Oxygen Index Test. This test (OI) is based on the minimum oxygen concentration that supports combustion of the material (46). The OI values of various fibers are listed in Table 3.

Types of Retardants

Mesylated and Tosylated Celluloses. A study on imparting flame resistance to cellulose by chemical modification (47) established that the flame resistance of cellulose is improved by oxidation of —CH_2OH groups to —COOH. To correct some of the shortcomings of this treatment, mesyl or tosyl cellulose was prepared and the mesyl (CH_3SO_2) or tosyl ($CH_3C_6H_4SO_2$) group was replaced with bromine or iodine (47):

$$R_{cell}OSO_2CH_3 + NaBr \rightarrow R_{cell}Br + CH_3SO_2Na$$

This treatment produced a fabric that had durable flame resistance, good strength retention, but suffered from an undesirable afterglow which was eliminated by phosphorylation with diethyl chlorophosphate.

Table 3. Oxygen Indexes of Various Fabrics

Fabric	OI	Fabric	OI
acrylic	0.182	nylon	0.201
acetate	0.186	polyester	0.206
polypropylene	0.186	wool	0.252
rayon	0.197	flame-retardant cotton	0.270
cotton	0.201	aramid	0.282

Urea-Phosphate Type. Phosphoric acid [*7664-38-2*] alone imparts flame resistance to cellulose (25–26) resulting, however, in degradation. This can be prevented by incorporation of urea [*57-13-6*]. Phosphorylating agents for cellulose include ammonium phosphate [*7783-28-0*], urea–phosphoric acid, phosphorus tri- and oxychloride [*7719-12-2*] and [*10025-87-3*], respectively, monophenyl phosphate [*701-64-4*], phosphorus pentoxide [*1314-56-3*], and chlorides of partially esterified phosphoric acids.

Cellulose phosphate esters are also produced by treatment with sodium hexametaphosphate [*14550-21-1*] by the pad-dry-cure technique with a high retention of breaking and tearing strength (48). Phosphorus contents up to 3.5% have been reported. The reaction products contain more than 1.6% phosphorus and were insoluble in cupriethylenediamine, indicating that the cellulose was cross-linked:

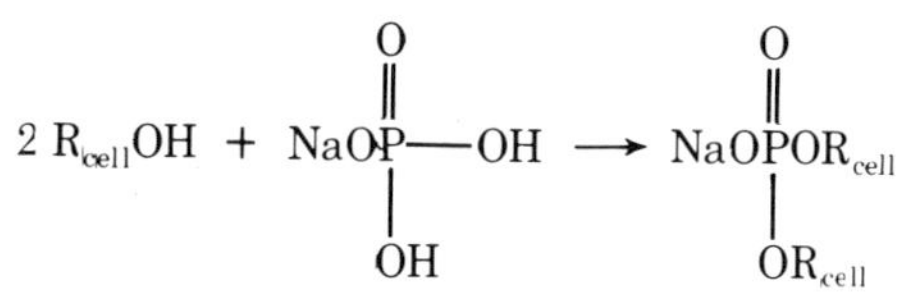

Phosphorylated cottons are flame resistant in the form of the free acid or the ammonium salt. Since these fabrics have ion-exchange properties, conversion to the sodium salt takes place easily in laundering if acid sours and basic tap water are used. However, flame retardation can be restored if the fabric is treated with a solution of ammonium hydroxide after the acid sour and dried.

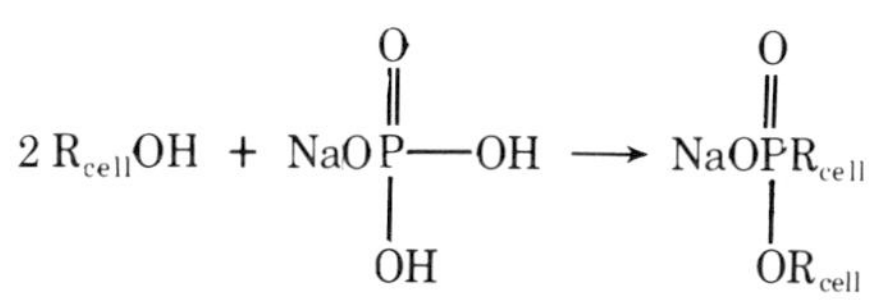

Phosphonomethylated Ethers. A phosphorus-containing ether of cellulose was prepared by the reaction of cotton cellulose with chloromethylphosphonic acid [*2565-58-4*] in the presence of sodium hydroxide by the pad-dry-cure technique (49). Phosphorus contents between 0.2 and 4% were obtained. This finish was durable but had high ion-exchange properties and was flame resistant only as the ammonium salt.

A durable flame-retardant cellulosic fabric with good hand is obtained by treating phosphorylated or phosphonomethylated cotton with titanium sulfate [*13825-74-6*] (50). The reactions are shown below.

Amide-Based Systems. ***Cyanamide.*** Pyroset CP (trade name for 50% cyanamide) solution produces a fire retardant that does not adversely affect the initial handle of treated fabric (51). Durability to laundering on industrial fabrics, blankets, draperies, and garments made from heavyweight fabrics is good; however, durability to laundering on lightweight fabrics remains limited (see Cyanamides). Treated fabrics also show an increase in wrinkle recovery. Drying and curing is accomplished in a single step, preferably at 149°C for 2–5 minutes. The finish is subject to ion exchange and, therefore, the hardness of the wash water has a decisive effect on durability of the flame retardancy. Fabric flammability fails after seven washings at a water hardness of 70

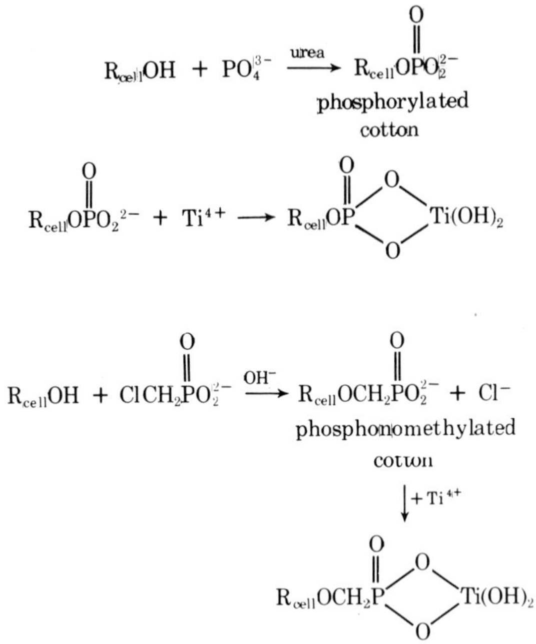

ppm whereas in soft water (0 ppm hardness) flame retardancy can withstand 50 wash cycles. Tensile strength losses of Pyroset CP treated fabric are 40–45%. The treated fabrics are dimensionally stable and have improved rot resistance.

Another system (52) is based on methylphosphoric acid [*993-13-5*] (MPA) and cyanamide [*420-04-2*]:

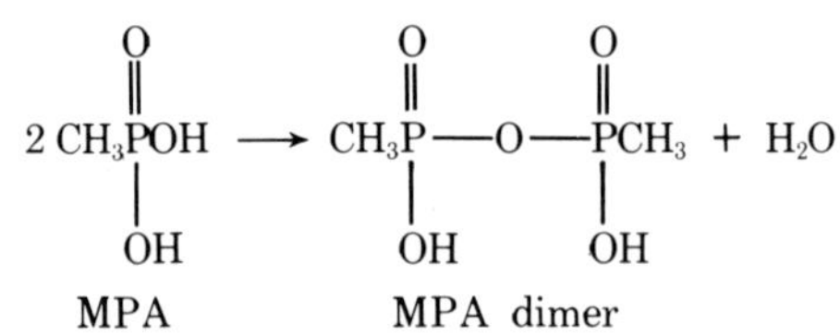

One or more of the hydroxyls in MPA or in its dimer react with cellulose:

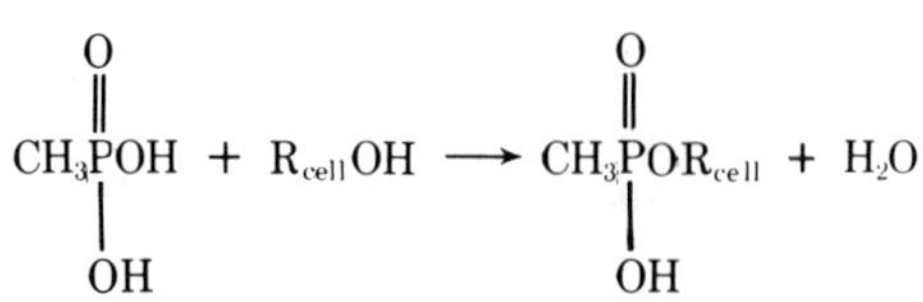

The water is taken up by the cyanamide, forming urea:

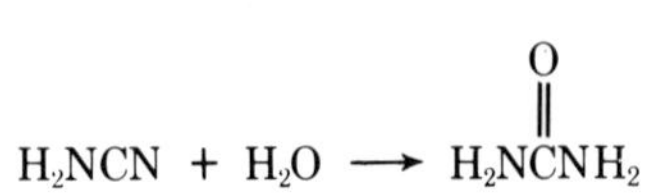

Fabrics are treated by a pad-dry-cure technique; however, smoke is evolved in the curing step. At an add-on of 10% this flame-retardant finish was durable to 40–50 laundry cycles, had dry wrinkle-recovery angles of 200–220°, and wet wrinkle-recovery angles of 220–270°. (The wrinkle-recovery angle reported is the sum of the recoveries in the warp and fill directions (W + F); thus, the maximum possible wrinkle recovery is 360°). The same finish gave tensile-strength retention of 60–80% and tear-strength retention of about 50%, and raised the moisture regain of the fabric about 3%. The system shows a high tolerance for calcium.

Acrylamide. Pyrovatex CP, is based on the reaction product of a dialkyl phosphite and acrylamide (qv) which is methylolated with formaldehyde (53).

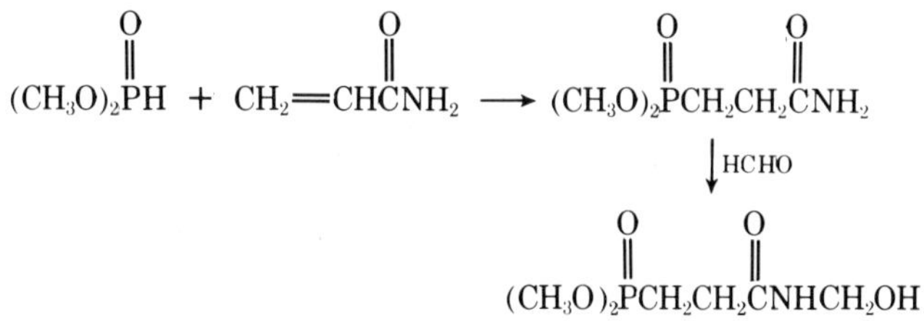

Excess formaldehyde reacts with cotton in the presence of an acid catalyst. A cross-linking agent, such as trimethylolmelamine [*1017-56-7*], TMM, improves the efficacy and durability of this finish. It is nonirritating to the skin, does not adversely affect the hand of the treated fabric, and is durable to 3–50 launderings and dry-cleanings. Tensile-strength losses are between 20 and 30% and tear-strength losses about 30%. Cotton textiles require add-ons of about 20–35%.

Dialkylphosphonopropionamides. Cellulosic derivatives very closely resembling those based on the dialkylphosphonopropionamides have also been prepared (54). The fabric was treated with *N*-hydroxymethylhaloacetamides (chloro, bromo, or iodo) in DMF solution by a pad-dry-cure technique with a zinc nitrate catalyst. It was then allowed to react in solution with trimethyl phosphite [*121-45-9*] at about 140–150°C; the reaction rates decreased in the order iodo > bromo > chloro.

With phosphorus contents above 1.5%, good flame resistance durable to laundering was obtained without noticeable loss in fabric-strength properties.

Triazines. When the dialkoxyphosphinyl group is attached to the triazine ring rather than to an alkyl group, 2,4-diamino-6-diethoxyphosphinyl-1,3,5-triazine [*4230-55-1*] (DAPT) is obtained (41,55,56):

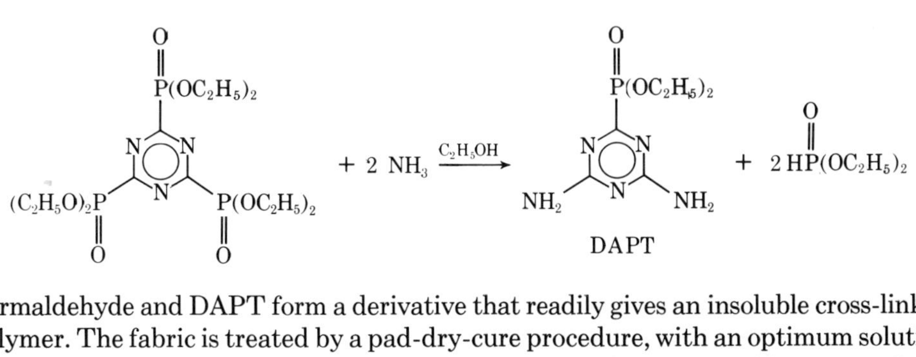

Formaldehyde and DAPT form a derivative that readily gives an insoluble cross-linked polymer. The fabric is treated by a pad-dry-cure procedure, with an optimum solution pH of 6.6 (no catalyst). Fabrics containing from 17.5–20% resin add-on passed the standard vertical flame tests even after 35 laundry cycles. Tear-strength losses were about 35% and tensile-strength losses about 18%. Although the treated fabrics did not yellow on chlorine bleaching and scorching, some strength loss resulted, indicating chlorine retention.

More recently a new effective flame retardant for cotton based on 2,4-diamino-6-(3,3,3-tribromopropyl)-1,3,5-triazine [*62160-38-7*] (DABT) was prepared from ethyl γ-tribromobutyrate and biguanide:

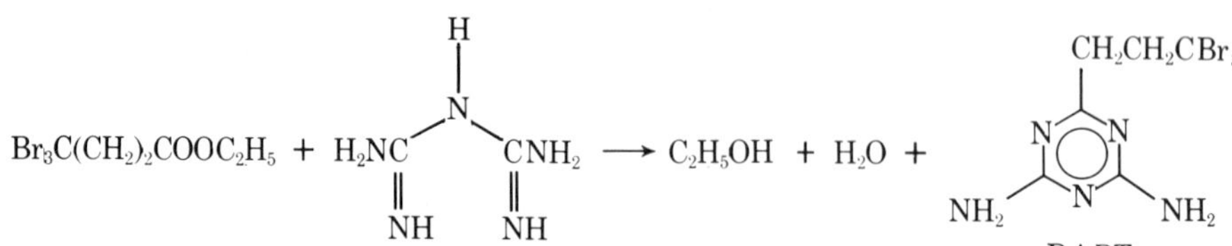

DABT

The tetramethylol derivative of DABT, prepared by reaction of DABT with alkaline aqueous formaldehyde, polymerized readily on cotton to impart excellent flame retardancy which was very durable to laundering with carbonate- or phosphate-based detergents and to hypochlorite bleach. This was accomplished at low add-on without use of phosphorus compounds or antimony oxide (57–59).

THPC Based. A flame-retardant finish for cotton (60) is based on tetrakis(hydroxymethyl)phosphonium chloride [*124-64-1*] (THPC), $[HOCH_2]_4P^+Cl^-$. It is water soluble and is prepared by the reaction of phosphine with formaldehyde and HCl. The fact that THPC reacts with aminized cotton to impart flame resistance led to further studies with materials such as melamine and urea, in an effort to eliminate the costly aminization step. THPC reacts and polymerizes with methylolmelamine [*937-35-9*], as well as with many other materials containing active hydrogens, eg, amines, phenols, and polybasic acids (61), to form insoluble polymers in the cotton fiber and produce good flame- and glow-resistant finishes (see also Amino resins).

The methylol groups in THPC react with amines (or amides) in the following manner:

$$(HOCH_2)_4PCl + RNH_2 \rightarrow RNHCH_2PCl(CH_2OH)_3 + H_2O$$

One or more methylol groups are replaced by the $RNHCH_2$ groups and the ionic chlorine is hydrolyzed:

$$(RNHCH_2)_3PCl(CH_2OH) + H_2O \rightarrow HCl + CH_3OH + (RNHCH_2)_3PO$$

Thus, the product contains no chlorine and the phosphorus is in the form of phosphine oxide, a structure that is very stable to hydrolysis. If the nitrogen compounds are di- or polyfunctional, the product is a highly insoluble polymer. THPC alone also reacts with cellulose, but only very slowly under processing conditions associated with the finishing of cotton textiles.

A typical application procedure is as follows: a 268 g/m^2 cotton twill is treated by padding it through a solution containing 15.8% THPC, 9.5% methylolmelamine, 3% triethanolamine, and 9.9% urea, then dry cured for 4–5 minutes at 140°C, and washed. A dry add-on of 18–20% on the dry fabric imparts flame and glow resistance and also some crease and rot resistance.

A durable flame-retardant finish is obtained from bromoform [*75-25-2*] and triallyl phosphate [*1623-19-4*] (62). Although this finish (known as BAP) is durable to laundering, it causes some stiffness and imparts color to fabric. The char lengths obtained with the standard vertical flame test were longer than those obtained with other durable flame retardants.

Another durable flame retardant (63) is based on the telomerization of poly(allyl phosphonitrilate) with bromoform and its application to cellulose by the pad-dry-cure technique. This finish, known as PNE-bromoform, is more efficient than BAP but has some of the same shortcomings (see also Allyl monomers and polymers).

The THPC-amide flame-retardant finish penetrates the fibers whereas BAP and PNE-bromoform have a coating effect. To achieve penetration and coating, the THPC-amide is combined with either PNE-bromoform (64) or BAP (65). These finishes produced good results on many fabrics; however, again the treated fabrics were somewhat stiff because of deposited telomer.

A highly effective and durable flame retardant for cotton cellulose was obtained by treating cotton with the reaction product of THPC with tris(1-aziridinyl) phosphine oxide [*545-55-1*] (APO) (66). A thermosetting resin forms inside the fiber with a pad-dry-cure treatment. However, owing to the toxicity of APO, this product is no longer made (see Chemotherapeutics, antimitotic).

Flame resistance can also be imparted to fabric by chemical fixation of a water-soluble precondensate of THPC and amides with ammonia and/or ammonium hydroxide (67–69). This technique (the ammonia cure method) improves the degree of flame resistance by adding additional nitrogen to the finish. Thus the treated fabric does not lose strength as is the case with the heat cure method. Even though very little cross-linking is obtained by this technique, it greatly improves durability to laundering. The English flame retardant, Proban, is based on this technique (68).

A new, highly durable flame-retardant finish for cotton is based upon THPC, cyanamide, and phosphoric acid (70). Fabric is padded through a water solution of the three components, then dried and cured at elevated temperature. This formulation imparts a high degree of flame resistance to fabric. Substituting Na_2HPO_4 [*13708-85-5*] for H_3PO_4 maintains fabric strength.

Modified THPC-amide formulations are available containing antimony oxide and organic chlorine compounds (71–72), generally PVC or a chlorinated paraffin. Although these compounds are cheaper than THPC, they are less effective on a weight basis (see Flame retardants, antimony and other inorganic compounds).

A formulation based on THPC, thiourea [*62-56-6*], and small amounts of sodium dihydrogen orthophosphate [*7558-80-7*] has been reported (73). A pad-dry cure procedure was used. The treated fabric retained its flame retardancy for 30 laundry cycles.

Before 1976, all flame retardants based on THP salts were made from THPC. After this date, however, both Hooker Chemical Co. and American Cyanamid Company replaced THPC with the corresponding sulfate, (THPS) because of the possible formation of the carcinogen bischloromethyl ether (BCME) from THPC. No evidence has been found, however, showing that such a compound is formed. Although neither THPC or THPS are currently manufactured in the United States, Albright and Wilson, Co. Ltd. in England produces both of these compounds, and markets them in many countries including the United States.

Flame retardants based upon THPC are predominantly used for apparel and household goods, and for certain military items, eg, tent-liner fabrics. Processes based upon THPC are employed in the United States, Europe, and other countries (74,75).

THPOH Based. Reactive methylol phosphorus compounds are made from 1 mole of THPC with 0.8–1 mole of sodium hydroxide. The aqueous solution of the product, referred to as THPOH (76), is not yet adequately characterized. Some researchers consider it to be mainly an equilibrium mixture of tetrakis (hydroxymethyl)phosphonium hydroxide [*512-82-3*], THPOH, and tris(hydroxymethyl)phosphine [*2767-80-8*], THP (78):

$$(HOCH_2)_4PCl + NaOH \rightarrow (HOCH_2)_4POH + NaCl$$

$$(HOCH_2)_4POH \rightleftharpoons (HOCH_2)_3P + HCHO + H_2O$$

Others have evidence that the solution consists of a mixture of THP and a hemiacetal of THP (79):

$$(HOCH_2)_4PCl + NaOH \rightarrow (HOCH_2)_3P + HCHO + NaCl + H_2O$$

$$(HOCH_2)_3P + HCHO \rightleftharpoons HOCH_2OCH_2P(CH_2OH)_2$$

In the reaction of THPC with sodium hydroxide (79) or other bases to produce THPOH, the pH of the solution must not exceed 7.1–7.5. In this range THPOH solution is stable. Like most other phosphines, THP can be oxidized at pH above 7.8 to tris(hydroxymethyl)phosphine oxide [*1067-12-5*] (THPO), which reacts only slowly with amides and amines.

Amides with two or more NH groups react with THPOH to produce insoluble polymers containing nitrogen and phosphorus. The phosphorus is in the form of a phosphine oxide; a methylol phosphine structure is probably an intermediate in these and similar reactions (see Phosphorus compounds).

THPOH–NH_3 Flame Retardants. This flame retardant is applied by padding fabric with a solution of THPOH and various auxiliaries, then partially drying the fabric and finally exposing the partially dried fabric to ammonia gas (80). A typical formulation for flameproofing a 268 g/m^2 cotton fabric contains 30% THPOH (5.4% P) and wetting and softening agents. The fabric is dried to about 10–20% moisture, exposed to ammonia, and passed through a bath of H_2O_2 solution, which extends the durability of the finish by oxidizing phosphorus to the most stable phosphorus oxide (P═O) form, and removes odors and discoloration. The process can be modified by the addition of copper salts (81) to produce a system stable in the presence of ammonium hydroxide.

Since THPOH–NH_3 flame retardant does not impart stiffness, the process is applicable to most cotton fabrics. Fabrics weighing as little as 67 g/m^2 (56 g/yd^2) have been satisfactorily flameproofed by this process. The treated fabric has high breaking strength, however tear strength is frequently reduced. With the proper amount of softening, tear strength can be restored. The phosphorus-to-nitrogen ratio in fabric treated by this process is about 3.5 to 1.

THPOH–TMM–NH_3 Flame Retardant. A third system contains typically 18.5% THPOH, 11.5% trimethylolmelamine (TMM), plus wetting and softening agents (82). Such a formulation is adequate for a 268 g/m^2 cotton fabric, and enables it to pass the standard vertical flame test. TMM can replace about one-third of the THPOH required in the THPOH–NH_3 process at a considerable savings (83).

THPOH-Amide Flame Retardant. This flame retardant is applicable to a wide variety of fabrics (84). The reagents are padded onto the fabric, dried, and then cured at about 150°C for 3 min. Fabrics retain 80–90% of their original breaking strength. Stiffness imparted is less than that imparted by the THPC-amide process, and like

THPC-amide, it gives a high degree of rot resistance. The treated fabrics have less tendency to yellow when exposed to sodium hypochlorite bleach.

Miscellaneous. Highly efficient flame retardants have been developed with reactive water-soluble compounds of low molecular weight and high phosphorus content. These compounds can be insolubilized by reaction with cellulose hydroxyls rather than *in situ* polymerization or copolymerizations (85). Included in this group are bis(aziridinyl)chloromethylphosphine oxide [*13846-34-9*] (86) and chloromethylphosphonic diamide [*6326-70-1*] (87–90). Cotton fabric treated with methylphosphonic diamide [*4759-30-2*] (MPDA) containing 12–14% add-on passes the children's sleepwear flammability standard FF3-71 (91).

Flame and crease resistance can be imparted to cotton textiles in one operation (52), using guanidine [*133-00-8*], dimethyl phosphite [*868-85-9*], and formaldehyde [*50-00-0*].

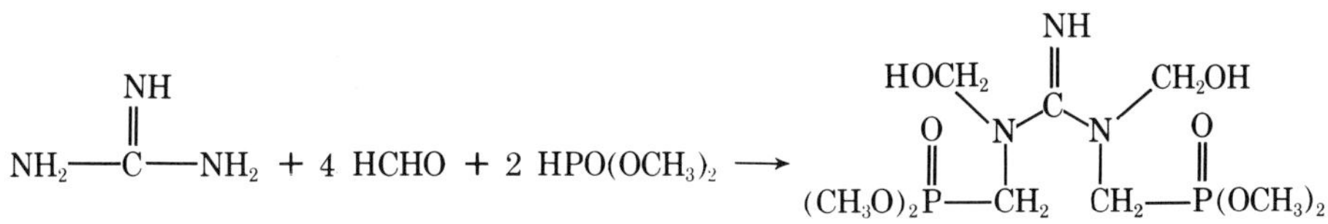

A durable finish is obtained by using methanol in place of water as the solvent; an application time of 45 min is required for cotton. Excessive strength losses and poor reproducibility diminish the usefulness of this finish.

Fyre-Fix. The Fyre-Fix finish is based on a fiber-reactive material (92), and is a stable water-soluble material which is easy to handle and compatible with most resins and softeners. It is applied by a pad-dry-cure technique. The treated fabrics have good flame resistance, a soft hand, and no ion-exchange properties.

Fyrol 76. This flame retardant is a reactive vinyl phosphorus ester containing 22.5% phosphorus. It can be used alone or with methylolacrylamide (NMA) and a free-radical catalyst, eg, potassium persulfate. Fabric is treated with the conventional pad-dry-cure technique; rapid steam curing or radiation curing also can be applied. Sleepwear fabric requires approximately 25–30% add-on to pass the DOC FF3-71 test. The fabric has an excellent hand and some permanent press properties. Approximately 80–100% tensile strength and about 65% tearing strength are retained. Chlorine bleach and some carbonate detergents reduce the durability of this finish, whereas perborate bleaches do not affect it. As with the THPC-type finishes, an oxidation step is generally used.

Pentamethylphosphorotriamide. Of the phosphoramide derivatives, pentamethylphosphorotriamide [*10159-46-3*] is the most effective finish when applied to fabric in conjunction with dimethylolmelamine and when using an amine hydrochloride as catalyst. The finished fabric passes the FF3-71 flammability test. Its main application is for use on heavyweight clothes since the finish imparts a harsh hand to lightweight fabrics (93).

Application Techniques

Use of radiation to effect fixation of some flame retardants is being investigated (94). Electron-beam fixation requires the selection of compounds that can be insolubilized inside or outside of the fiber with high yield in a short time. Polyunsaturated compounds, eg, Fyrol 76, are showing promise (see Radiation curing).

Flame retardants suitable for cotton are also suitable for rayon. A much better product is obtained by incorporating flame retardants in the viscose dope before fiber formation. The major classes of flame retardants used in the dope are listed in Table 4.

Thermoplastic Fibers. Thermoplastic fibers, eg, polyester and nylon, are considered less flammable than natural fibers because they possess a relatively low melting point and the melt drips rather than remaining to propagate the flame when the source of ignition is removed. Most common synthetic fibers have low melting points. Reported values for polyester and nylon are 255–290°C and 210–260°C, respectively.

Flame retardancy of the synthetic fibers is obtained by either mechanically building the retardant with the polymer before it is drawn into a fiber, or chemically modifying the polymer itself. Incorporation of chemicals in the dope before spinning the fiber has not been very successful. The most widely used technique for flame-retarding polyester acetate and triacetate is the application of Tris, tris(2,3-dibromopropyl) phosphate [*126-72-7*] (TDBP) [$(BrCH_2CHBrCH_2O)_3P{=}O$] by a Thermosol diffusion or an exhaustion technique (130–133). Polyester fiber can only retain about 4–5% TDBP calculated on the weight of the fiber. This was banned in 1977 by the Consumer Product Safety Commission (CPSC) as a potential carcinogen (see also Flame retardants, halogenated, and phosphorus compounds). Tris is not effective on cotton fabrics and was not used as a flame retardant for cotton textiles.

Table 4. Major Classes of Dope-Additive Flame Retardants for Viscose Rayon[a]

Active elements	Examples	Comments
phosphorus	alkyl and aryl phosphates, phosphonates and phosphites, polyphosphonates, Sandoflam 5060[b]	members of this class can be mixed with halogenated materials for synergistic effects
phosphorus–nitrogen	phosphazenes, phosphoryl or thionophosphoryl amides or ester amides, spirocyclotriphosphazenes, THPC-amines of condensates[c]	very efficient and effective; no unusual toxicological effects have been reported for a large number of these
phosphorus–halogen	halogenated alkyl or aryl phosphonates or polyphosphonates, halogenated alkyl or aryl phosphates, phosphites, or phosphazenes[d]	probably the largest class; severe toxicological problems for most useful members of this class
halogen	poly(vinyl or vinylidene halides) as latexes, poly(halogenated acrylate) latexes; emulsion or dispersions of alkyl or aryl halides, halogenated paraffins[e]	halogenated materials are not effective by themselves but are excellent synergists when combined with phosphorus compounds or colloidal antimony pentoxide; some latexes are not stable in viscose
antimony oxide	colloidal antimony pentoxide[f]	pigmentary antimony trioxide is not compatible with viscose; must be mixed with suitable halogenated synergist for necessary effectiveness

[a] Ref. 95.
[b] Ref. 96, 97.
[c] Ref. 98–109.
[d] Ref. 110–127.
[e] Ref. 127.
[f] Ref. 128, 129.

An alternative to the diffusion technique is the application of decabromodiphenyl oxide, on the surface of fabrics in conjunction with binders (134). Experimental finishes using graft polymerization, *in situ* polymerization of phosphorus-containing vinyl monomers or surface halogenation of the fibers also have been reported (132–133,135,136).

Mobil developed a flame retardant, Antiblaze 19, for polyester fibers (137). It is a nontoxic mixture of cyclic phosphonate esters. Antiblaze 19 is 100% active whereas Antiblaze 19T is a 93% active, low-viscosity formulation for textile use. Both are miscible with water and are compatible with wetting agents, thickeners, buffers, and most disperse dyes formulations.

Antiblaze 19/19T can be diffused into 100% polyester fabrics by the Thermosol process for disperse dyeing and printing. This requires heating at 170–220°C for 30–60 s.

Fire-Retardant Fiber Blends

The flammability behavior of thermoplastic and nonmelting polyester/cellulosic fiber blends cannot be predicted from the flammability of the single-fiber structures (138–141). The treatment of one of the fiber components with a flame-retarding agent specifically effective for that component does not necessarily render the two-component blend flame resistant (138,140) unless the treated component comprises at least 85% of the blend (74). An effective flame-retarding treatment must reduce the flammability of each component. It is sufficient to flameproof only one fiber component and the second one can be completely free of finish, if a flame-retarding agent is used that is effective on both types of fibers (142).

Considerable effort is being made to develop satisfactory flame retardants for blended fabrics. It is now feasible to produce flame-resistant blended fabrics that contain about 65% or more cellulosic fibers. Before additional significant advances can be made, it is necessary to develop flame retardants for thermoplastic fibers. Flame retardants are already available for the cellulosic fibers. The most practical approach may be to produce flame-resistant thermoplastic fibers by altering the chemical structure of the polymers and, after blending these fibers with cotton or rayon, treating the finished textile with an appropriate flame retardant for the cellulose.

In 1972, the White Chemical Company developed Caliban F/R P-44, a nontoxic flame retardant for textiles and textile blends (143). It consists of decabromodiphenyl oxide, an acrylic latex, and antimony oxide in emulsion form, which is applied by the pad-dry-cure technique.

A flame retardant based on THPC-amide plus poly(vinyl bromide) [*25951-54-6*] (144) is suitable for use on 50/50 polyester-cotton blends. It is applied by the pad-dry-cure process, with curing at 150°C for about 3 min. A typical formulation contains 20% THPC, 3% disodium acid phosphate, 6% urea, 3% trimethylolglycouril [*496-46-8*], and 12% poly(vinyl bromide) solids. Approximately 20% add-on is required to impart flame retardancy to a 168 g/m^2 50/50 polyester–cotton fabric. Treated fabrics pass the FF 3-71 test.

The THPOH-NH_3 finish, and the Fyrol 76 finish also impart flame retardancy to certain cotton–polyester blends if the blends contain at least 65% cotton.

The use of LRC-100 flame retardant for 50/50 polyester cotton blends has been reported recently (145). It is a condensation product of tetrakis(hydroxymethyl)-

Table 5. Flame-Resistant Synthetic Fibers [a]

Fiber	Company	Chemical nature	Composition of fabric	Applications
PFR Rayon	Avnet Fibers	rayon containing phosphazene ester	100% PFR, blends	safety clothing, aircraft upholstery and institutional drapery
Polyester Type G.H.	Toyobo Co.	polyester containing aromatic, sulfur-containing phosphonate	100% polyester, blends with certain fibers	apparel and interior furnishings
Extar FR [b]	Teijin	bromine-containing polyester	100% polyester, blends with certain fibers	developmental product; drapery
Trevira 271	Hoechst	polyester with comonomer		commercial product; apparel, interior furnishings
SEF [c]	Monsanto	modacrylic	100% SEF, blends with acrylic or polyester	apparel, drapery and industrial fabrics
Verel	Eastman	modacrylic	100% Verel, blends with rayon or polyester	interior furnishings
Kanekaron	Kanegafunci Chemical Co.	modacrylic		apparel, interior furnishings
Lufnen	Kanebo Acrylic Fibers Co.	modacrylic		apparel, interior furnishings
Treviron [d]	Teijin	vinyon	100% vinyon, blends	apparel, interior furnishing
Leavil [e]	Montedison	vinyon	100% vinyon, blends	apparel, interior furnishings
Clevyl T	Rhone-Poulenc	vinyon	100% vinyon, blends	apparel, interior furnishings
Cordelan I and II [f]	Kohjin Co.	polychlal	100% Cordelan, blends with up to 40% polyester	apparel, interior furnishings
Nomex 111	DuPont	aramid	100% Nomex, blends with Kynol, wool, etc	safety apparel, aircraft upholstery, specialty
Kevlar	DuPont	aramid	100% Kevlar, blends	safety apparel, industrial fabrics
Durette	Fire Safety Products, Inc.	modified aramid	100% Durette	space program, high oxygen areas, race drivers' gloves, specialty products
Fiberglass	Owens-Corning	glass	100% Fiberglass, blends	interior furnishing, safety apparel
Kynol	American Kynol	novoloid	100% Kynol, blends	safety apparel, specialty products

[a] From ref. 148, unless otherwise stated.
[b] Ref. 149.
[c] Ref. 150.
[d] Ref. 151.
[e] Ref. 152.
[f] Ref. 153.

phosphonium salt (THP salt) and *N,N′,N″*-trimethylphosphoramide [*6326-72-3*] (TMPA). The precondensate is prepared by heating the THP salt and TMPA in a 2.3 to 1 mole ratio for one h at 60–65°C and is applied in conjunction with urea and trimethylol melamine in a pad-dry-cure oxidation wash procedure. To pass the FF 3-71 test, 3.5–4.0% phosphorus is needed.

In February 1978 CPSC approved changes in the FF-3 and FF-5 standards for childrens' sleepwear, eliminating the melt-drip time limit and coverage for sizes below 1, and revised the method of testing the trim. This permits the use of untreated 100% nylon and 100% polyester for children's sleepwear.

Halogens are less effective flame retardants on nylon than on polyester, and most of the flame retardants effective on cellulosics or on polyester substrates are not effective on nylon. Thiourea and thiourea formaldehyde appear to be the most effective treatments for imparting flame resistance to nylon (146–147).

The burning characteristics of synthetic and natural fibers are very different. Synthetic fibers tend to melt, drip, and shrink away when in contact with a flame, whereas cotton does not distort when subjected to a heat source. When synthetic and natural fibers are combined, the cotton serves as a grid for the polyester, thus preventing it from dripping, and the polyester, which has a higher heat of combustion, accelerates the pyrolysis of cotton. Thus, when natural and synthetic fibers are combined, many of the resulting fabrics are more hazardous than fabrics made from the individual components. Table 5 lists the commercially available inherent flame-retardant fibers.

Mutagenicity

Because Tris—a polyester flame-retardant chemical—is a potential carcinogen (154–156), workers in this field have tested a number of commonly used chemicals for potential mutagenicity.

The THPOH–NH_3 finish was tested recently (157) by the Ames mutagenicity test (158–161) and neither the finish nor its extracts caused a significant systematic increase in mutations.

The Hooker Chemical Co. has just released the results of tests conducted by an independent laboratory which indicated no significant mutagenic potential from any of the company's proprietary textile flame retardants.

Fyrol 76 also is reported to be nontoxic and nonmutagenic.

Stauffer's substitute for Tris, Fyrol FR2, was accused of mutagenic activity by the Environmental Defense Fund, and was withdrawn from the market by the company.

Because of such uncertainties, several major suppliers of flame-retardant chemicals have reduced their operations in this area or have abandoned the market altogether. This has forced the apparel industry to use inherently flame-resistant fibers.

BIBLIOGRAPHY

"Fire-Resistant Textiles" in *ECT* 1st ed., Vol. 6, pp. 544–558, by G. S. Buck, Jr., National Cotton Council of America; "Fire-Resistant Textiles" in *ECT* 2nd ed., Vol. 9, pp. 300–315, by George L. Drake, Jr., U.S. Department of Agriculture.

1. M. M. Sandholtzer, *Natl. Bur. Std. (U.S.), Circ.* **C455** (1946).
2. Br. Pat. 551 (1735), Obadiah Wyld.

3. J. L. Gay-Lussac, *Ann. Chim.* **18**(2), 211 (1821).
4. W. H. Perkin, *J. Ind. Eng. Chem.* **5,** 57 (1913); *Met. Chem. Eng.* **10,** 636 (1912).
5. *Bull. N.Y. Acad. Med.* **43**(8), 615 (1967).
6. *Public Law 90-189, S 1003, 90th Congr., Sect. 4, Dec. 14, 1967,* U.S. Government Printing Office, Washington, D.C., 1967.
7. D. A. Yeadon and co-workers, in *Proc. 10th Ctn. Util. Res. Conf. USDA, ARS 72-83,* SRRC, New Orleans, La., p. 70.
8. L. W. Mazzeno, Jr., and co-workers, *Text. Chem. Color* **5**(3), 43 (1973).
9. R. W. Little, *Flameproofing of Textile Fabrics,* Reinhold Publishing Corp., New York, 1947.
10. G. C. Tesoro, *Textilveredlung* **2,** 435 (1967).
11. G. C. Tesoro, S. B. Sello, and J. Willard, Jr., *Text. Res. J.* **38,** 245 (1968).
12. *Ibid.,* **39,** 180 (1969).
13. J. E. Hendrix and co-workers, *J. Appl. Sci.* **14**(7), 1701 (1970).
14. J. E. Hendrix and co-workers, *J. Fire Flammability* **3,** 2 (1972).
15. D. A. Yeadon and co-workers, *Proc. 10th Ctn. Util. Conf. USDA, ARS 72-83,* SRRC, New Orleans, La., 1970, p. 70.
16. Gulf Coast Section, *Text. Chem. Color.* **4**(12), 287 (1972); J. E. Hendrix and co-workers, *Text. Res. J.* **41**(10), 854 (1971).
17. R. J. Brysson and co-workers, *Information Council on Fabric Flammability, Proceedings of the 5th Annual Meeting,* Dec. 9, 1971, New York, p. 138.
18. R. M. Perkins, G. L. Drake, and W. A. Reeves, *J. Am. Oil Chem. Soc.* **48**(7), 303 (1971).
19. J. M. Church, "Fundamental Studies of Chemical Reactants for Fire-Resistant Treatment of Textiles" in *U.S.Q.N.C. Textile Series, Report* **38,** U.S.Q.N.C., Natick, Mass., 1952.
20. R. W. Little, *Flameproofing Textile Fabrics, American Chemical Society Monograph No. 104,* Reinhold Publishing Corp., New York, 1947, p. 41.
21. H. A. Schuyten, J. W. Weaver, and J. D. Reid, *Adv. Chem. Ser.* **9,** 7 (June, 1954).
22. "A Fundamental Study of the Pyrolysis of Cotton Cellulose To Provide Information Needed for Improvement of Flame Resistant Treatments For Cotton," *Final Report 1959/1964 on PL 480 Project No URE-29-(20)-9; Grant No. FG-UK-108-59,* The Cotton, Silk and Man-Made Fibers Research Association, Didsbury, England, 1964.
23. J. B. Berkowitz-Mattuck and T. Naguchi, *J. Appl. Polym. Sci.* **7,** 709 (1963).
24. E. Heuser, *The Chemistry of Cellulose,* John Wiley & Sons, Inc., New York, 1944, p. 546.
25. R. W. Little, *Flameproofing Textile Fabrics, American Chemical Society Monograph No. 104,* Reinhold Publishing Corp., New York, 1947; U.S. Pat. 2,482,755 (Sept. 27, 1949), F. M. Ford and W. P. Hall (to Joseph Bancroft and Sons Co.).
26. J. D. Reid and L. W. Mazzeno, Jr., *Ind. Eng. Chem.* **41,** 2828 (1949).
27. M. Leatherman, *U.S. Dep. Agric. Circ.* **466** Washington, D.C., 1938.
28. F. J. Kilzer and A. Broido, *Pyrodynamics* **2,** 151 (1965).
29. C. H. Mack and D. J. Donaldson, *Text. Res. J.* **37,** 1063 (1967).
30. W. A. Reeves, G. L. Drake, Jr., and R. M. Perkins, *Fire Resistant Textiles Handbook,* Technomic Publishing Co., Inc., Westport, Conn., 1974, p. 45.
31. J. D. Reid, W. A. Reeves, and J. G. Frick, *U.S. Dep. Agric. Leaflet No. 454,* USDA, Washington, D.C., 1959.
32. J. R. W. Perfect, *J. Soc. Dyers and Colour.* **74,** 829 (1958).
33. U.S. Pat. 2,570,566 (Oct. 9, 1951), F. W. Lane and W. L. Dills; U.S. Pat. 2,607,729 (Aug. 19, 1952), W. L. Dills; U.S. Pat. 2,785,041 (March 12, 1957), W. W. Riches (to E. I. du Pont de Nemours & Co., Inc.); U.S. Pat. 2,658,000 (Nov. 3, 1953), W. F. Sullivan and I. M. Panik; U.S. Pat. 2,691,594 (Oct. 12, 1954), J. P. Wadington; U.S. Pat. 2,728,680 (Dec. 27, 1955), D. Duane (to National Lead Co.).
34. U.S. Pat. 2,570,566 (Oct. 9, 1951), F. W. Lane and W. L. Dills (to E. I. du Pont de Nemours & Co., Inc.); H. C. Gulledge and G. R. Seidel, *Ind. Eng. Chem.* **42,** 440 (1950); *J. Text. Inst.* **41**(7), 357 (July, 1950).
35. U.S. Pat. 2,658,000 (Nov. 3, 1953), W. F. Sullivan and I. M. Panik (to National Lead Corp.).
36. U.S. Pat. 2,229,612 (Oct. 20, 1942), E. C. Clayton and L. L. Heffner (to Wm. E. Hooper and Sons Co.).
37. R. Van Tuyle, *Am. Dyest. Rep.* **32,** 297 (1943); *J. Text. Inst.* **34**(10), 587 (Oct. 1943); N. J. Read and E. G. Heighway-bury, *J. Soc. Dyers Colour.* **74,** 823 (1958).
38. U.S. Pat. 2,518,241 (Aug. 8, 1950), J. F. McCarthy (to Treesdale Laboratories); U.S. Pat. 2,591,368 (April 1, 1952), S. H. McAllister (to Shell Development Co.).

39. U.S. Dept. of Commerce, *Commercial Standards 191-53* (revised), U.S. Government Printing Office, Washington, D.C., 1954.
40. *Fed. Regist.* **35**(74), 6211 (1970).
41. *Fed. Regist.* **35**(251), 19702 (1970).
42. *Fed. Regist.* **37**(141), 14624 (1972).
43. *Fed. Regist.* **39**(85), 15210 (1974).
44. *Fed. Regist.* **37**(110), 11362 (1972); *Fed. Regist.* **38**(78), 10111 (1973); *Fed. Regist.* **38**(81), 10482 (1973); *Fed. Regist.* **38**(110), 15095 (1973); *Fed. Regist.* **38**(117), 15990 (1973).
45. *Text. Chem. Color.* **2**(3), 49 (1970).
46. C. P. Fenimore and F. J. Martin, *Mod. Plast.* **43,** 141, 146, 148, 192 (1966); J. S. Isaacs, *J. Fire Flammability* **1,** 36 (1970).
47. E. Pacsu and R. F. Schwenker, *Text. Res. J.* **27,** 173 (1957); R. F. Schwenker and E. Pacsu, *Ind. Eng. Chem.* **50,** 91 (1958); U.S. Pat. 2,990,232–2,990,233 (June 27, 1961), E. Pacsu and R. F. Schwenker, Jr. (to Textile Research Institute).
48. D. M. Gallagher, *Am. Dyest. Rep.* **53,** 23 (1964).
49. G. L. Drake, Jr., W. A. Reeves, and J. D. Guthrie, *Text. Res. J.* **29,** 270 (1959).
50. R. B. LeBlanc and D. A. LeBlanc, *Proceedings of the Symposium on Textile Flammability,* LeBlanc Research Corp., East Greenwich, R.I., 1974, p. 1.
51. S. J. O'Brien, *Text. Res. J.* **38,** 256 (1968); *Pyroset DO Fire Retardant, Text. Fin. Bull. No. 130,* 2nd ed., American Cyanamid Co., Bound Brook, N.J., 1959.
52. W. A. Sanderson, W. A. Muller, and R. Swidler, *Text. Res. J.* **40**(3), 217 (1970).
53. Fr. Pats. 1,395,178 (April 9, 1965) and 1,466,744 (Jan. 20, 1967) (to Ciba Ltd.); U.S. Pat. 3,374,292 (March 19, 1968), A. C. Zahir (to Ciba Ltd.); R. Aenishanslin and co-workers, *Text. Res. J.* **39**(4), 375 (1969); *Technical Bulletin 081-4B (270M),* Ciba Chemical and Dye Co., Toms River, N.J.
54. G. C. Tesoro, S. B. Sello, and J. J. Willard, *Text. Res. J.* **38**(3), 245 (1968).
55. J. P. Moreau and L. H. Chance, *Am. Dyest. Rep.* **59**(5), 37, 64 (1970).
56. L. H. Chance and J. P. Moreau, Paper presented at the *American Chemical Society 159th Meeting, Division of Cellulose, Wood, and Fiber Chemistry, Feb. 22–27, 1970, ACS,* Houston, Texas, 1970.
57. L. H. Chance and J. D. Timpa, *J. Chem. Eng. Data* **22**(1), 116 (1977).
58. L. H. Chance and J. D. Timpa, *Text. Res. J.* **47**(6), 418 (1977).
59. J. D. Timpa and L. H. Chance, *J. Fire Retard. Chem.* **5,** 93 (1978).
60. W. A. Reeves and J. D. Guthrie, *Text. World* **104**(2), 176, 178, 180, 182 (Feb. 1954).
61. W. A. Reeves and J. D. Guthrie, *Ind. Eng. Chem.* **4,** 64 (1956).
62. J. G. Frick, Jr., J. W. Weaver, and J. D. Reid, *Text. Res. J.* **25,** 100 (1955).
63. C. Hamalainen and J. D. Guthrie, *Text. Res. J.* **26,** 141 (1956).
64. C. Hamalainen, W. A. Reeves, and J. D. Guthrie, *Text. Res. J.* **26,** 145 (1956).
65. J. D. Reid, J. G. Frick, Jr., and R. L. Arceneaux, *Text. Res. J.* **26,** 137 (1956).
66. W. A. Reeves and co-workers, *Text. Res. J.* **27,** 260 (1957); G. L. Drake, Jr., J. V. Beninate, and J. D. Guthrie, *Am. Dyest. Rep.* **50,** 129 (1961).
67. U.S. Pat. 2,722,188 (Nov. 27, 1956), W. A. Reeves and J. D. Guthrie (to U.S. Secy Agr.).
68. U.S. Pat. 2,983,623 (May 9, 1961), H. Coates (to Albright and Wilson Ltd.); Brit. Pat. 906,314 (Sept. 19, 1962), H. Coates (to Albright and Wilson Ltd.); Brit. Pat. 938,989–938,990 (Oct. 9, 1963), H. Coates and B. Chalkley (to Albright and Wilson Ltd.).
69. G. L. Drake, Jr., W. A. Reeves, and R. M. Perkins, *Am. Dyest. Rep.* **52,** 608 (1963).
70. F. L. Normand, D. J. Donaldson, and G. L. Drake, Jr., *Text. Ind.* **134**(6), 169, 176, 186, 188, Atlanta (1970).
71. R. B. LeBlanc, *Text. Ind.* **132**(10), 274, 294, 296, 298, Atlanta (1968).
72. U.S. Pat. 3,243,391 (March 29, 1966), G. M. Wagner (to Hooker Chemical Corp.).
73. J. K. Sharma, L. LaL, and H. I. Bhatnagar, *Indian J. Text. Res.* **2,** 116 (1977).
74. W. A. Reeves and J. D. Guthrie, *U.S. Dep. Agric. Ind. Chem. Bull.* (364), (1953).
75. J. D. Guthrie, G. L. Drake, Jr., and W. A. Reeves, *Am. Dyest. Rep.* **44**(10), 328 (1955).
76. J. V. Beninate and co-workers, *Text. Ind.* **131**(11), 110, Atlanta (1967).
77. N. Filipescu and co-workers, *Can. J. Chem.* **41,** 821 (1963).
78. W. J. Vullo, *J. Org. Chem.* **33**(9), 3665 (1968).
79. S. E. Ellzey, Jr., and co-workers, private communication, 1970.
80. J. V. Beninate and co-workers, *Am. Dyest. Rep.* **57**(25), 74 (1968).
81. D. J. Donaldson and D. J. Daigle, *Text. Res. J.* **39**(4), 363 (1969).
82. J. V. Beninate and co-workers, *Text. Res. J.* **39**(4), 368 (1969).

83. D. J. Daigle and D. J. Donaldson, *Text. Chem. Colour.* **1**(24), 34 (1969).
84. J. V. Beninate and co-workers, *Text. Res. J.* **38**(3), 267 (1968).
85. G. C. Tesoro, *Pure Appl. Chem.* **46,** 239 (1976).
86. G. C. Tesoro, W. F. Olds, and R. M. Babb, *Text. Chem. Color.* **6,** 148 (1974).
87. Dutch Pat. 661 5460 (May 3, 1967), (to Badische Anilin & Soda Fabrik Co.); Brit. Pat. 1,126,259 (Sept. 5, 1968).
88. C. E. Morris and L. H. Chance, *Text. Res. J.* **43,** 336 (1973).
89. G. Tesoro, E. I. Valko, and W. Olds in V. M. Bhatnagar, ed., *Proceedings of the Second International Conference on Flammability and Flame Retardants Montreal, Canada, May 22–23, 1975,* Technomic Publishing Co. Inc., Westport, Conn.
90. G. Tesoro, E. I. Valko, and W. Olds, *Text. Res. J.* **46,** 152 (1976).
91. G. Tesoro, *J. Appl. Polym. Sci.* **21,** 1073 (1977).
92. *Fyre-Fix,* Syntron, Inc., Pawtucket, R.I.
93. R. B. LeBlanc, J. H. Badger, and K. P. Clark, *AATCC 1974 National Technical Conference, Book of Papers,* AATCC, 1974, p. 462.
94. B. Court and co-workers, *Am. Dyest. Rep.* **67**(1), 32 (1978).
95. N. A. Portnoy and G. C. Daul, *AATCC Book of Papers, AATCC 1978 Nat. Tech. Conf.,* AATCC, Research Triangle Park, N.C., 1978, p. 269.
96. A. Granzow, *Acc. Chem. Res.* **11**(5), 177 (1978).
97. Brit. Pat. 1,156,588 (Oct. 24, 1967), H. Watanabe, K. Yagomi, and T. Tumori (to Kokoku Rayon & Pulp Co.).
98. Gen. Offen. 2,242,681 (March 15, 1973), H. Noehbur and A. Maeder (to Ciba Geigy A.-G.).
99. U.S. Pat. 3,947,276 (March 30, 1976), F. Sidan and P. Rossi (to Snia Viscosa Societa Nazionale Industria Applicapori Viscosa S.p.A.).
100. Fr. Pat. 2,310,376 (Dec. 3, 1976), (to Ethyl Corp.).
101. Jpn. Pat. 76,139,197 (Dec. 1, 1976), (to Ethyl Corp.).
102. U.S. Pat. 3,994,996 (Nov. 30, 1976), R. Borivo; and co-workers (to FMC).
103. U.S. Pat. 4,006,125 (Feb. 1, 1977), K. A. Regnard and A. H. Guber (to Horizons, Inc.).
104. U.S. Pat. 3,990,900 (Nov. 9, 1976), R. Borivoj, E. F. Orwoll, and J. F. Start, (to FMC).
105. U.S. Pat. 4,011,089 (March 8, 1977), J. T. Kao (to Ethyl Corp.).
106. Jpn. Pat. 74 124,323 (Nov. 28, 1974), I. Masuda and co-workers (to Omi Kenshi Spinning Co. Ltd.).
107. Jpn. Kokai 76 17,321 (Feb. 12, 1976), A. Kawai and co-workers (to Mitsubishi Rayon Co. Ltd.).
108. Jpn. Pat. 76 26,319 (March 4, 1976), H. Terasawa, T. Kondo, and I. Saito (to Dondo, Ltd.).
109. R. Wolf, "Flame Retardant Dope Additives for Regenerated Cellulose Fibers" in R. Bruce LeBlanc, ed., *Proceedings of the Symposium on Textile Flammability,* LeBlanc Research Corporation, East Greenwich, R.I., 1978, p. 248.
110. U.S. Pat. 3,266,918 (Aug. 16, 1966), J. W. Schappel and A. I. Bates (to FMC Corp.); U.S. Pat. 3,455,713 (July 15, 1969), J. W. Schappel and A. I. Bates (to FMC Corp.).
111. J. W. Shappel, *Mod. Text.* **49,** 54 (1968).
112. L. E. A. Godfrey, *Text. Res. J.* **40,** 116 (Feb. 1970).
113. Can. Pat. 769,630 (Oct. 17, 1967), M. P. Godsay (to Courtaulds, Canada, Ltd.).
114. Neth. Appl. 6,914,207 (Dec. 29, 1969), N. V. Aku.
115. Jpn. Kokai 76 17,325 (Feb. 12, 1976), K. Mimura, A. Kawai, and H. Ito (to Mitsubishi Rayon Co., Ltd.).
116. Jpn. Kokai 76 17,324 (Feb. 12, 1976), M. Inoue, K. Mimura, and K. Kagewa (to Mitsubishi Rayon Co., Ltd.).
117. Jpn. Kokai 76 17,322 (Feb. 12, 1976), M. Inoue, M. Mimura, and K. Kagewa (to Mitsubishi Rayon Co., Ltd.).
118. Jpn. Kokai 76 17,323 (Feb. 12, 1976), K. Mimura, A. Kawai, and T. Kafsuyama (to Mitsubishi Rayon Co., Ltd.).
119. Jpn. Pat. 72 23,612 (Oct. 13, 1962), Y. Oskubo, M. Ohki, and H. Terayawa (to Kaneyafuchi Co., Ltd.).
120. Jpn. Pat. 74 72,415 (July 12, 1974), Y. Kato and co-workers (to Toyobo Co. Ltd.).
121. Jpn. Pat. 74 125,625 (Dec. 2, 1974), K. Nagai and co-workers (to Toyo Rayon Co., Ltd.).
122. Jpn. Pat. 74 75,821 (July 22, 1974), Y. Koto and co-workers (to Toyobo Co., Ltd.).
123. Jpn. Pat. 75 59,512 (May 22, 1975), S. Nagai and co-workers (to Mitsubishi Rayon Co. Ltd.).
124. Jpn. Pat. 73 75,817 (Oct. 12, 1973), H. Kishida and co-workers (to Diawa Spinning Co., Ltd.).

125. Jpn. Pat. 73 91,312 (Nov. 28, 1973), K. Nagai, H. Okada, and I. Takequichi (to Toyo Rayon Co., Ltd.).
126. Jpn. Pat. 73 91,313 (Nov. 28, 1973), N. Yamomoto.
127. Jpn. Pat. 75 12,318 (Feb. 7, 1975), K. Nagai, H. Okoda, and I. Takeuchi (to Toyo Rayon Co. Ltd.).
128. U.S. Pat. 3,575,898 (April 20, 1971), R. L. McClure (to Beaunit Corp.).
129. U.S. Pat. allowed but unissued (Ser. No. 800,186), D. J. Unrau, N. A. Portnoy, and P. J. Hartmann.
130. E. Baer, in E. Greenwich, ed., *Proceedings of the Symposium on Textile Flammability,* LeBlanc Research Corp., East Greenwich, R.I., 1973, p. 117.
131. T. J. McGreehan and J. T. Maddock, *A Study of Flame Retardants for Textiles Report No. PB 251-441/AS,* National Information Service, Springfield, Va., 1976.
132. H. E. Stepniczka, *Textilveredlung* **10,** 188 (1975).
133. M. Lewin, S. M. Atlas, and E. M. Pearce, *Flame Retardant Polymeric Materials,* Plenum Publishing Corp., New York, 1975.
134. V. Mischutin, in R. B. Leblanc, ed., *Proceedings of the Symposium on Textile Flammability,* LeBlanc Research Corp., East Greenwich, R.I., 1975, p. 211.
135. R. Liepins and co-workers, *J. Appl. Polym. Sci.* **21,** 2529 (1977).
136. *Ibid.,* 2403 (1977).
137. B. E. Johnston, A. T. Jurewicz, and T. Ellison, *Proceedings of the Symposium on Textile Flammability,* LeBlanc Research Corp., East Greenwich, R.I., 1977, p. 209.
138. G. C. Tesoro and C. Meiser, Jr., *Text. Res. J.* **40,** 430 (1970).
139. W. Kruse and K. Filipp, *Mell.* **49,** 203 (1968).
140. W. Kruse, *Mell.* **50,** 460 (1969).
141. G. C. Tesoro and J. Rivlin, *Text. Chem. Col.* **3,** 156 (1971).
142. P. Linden and co-workers, *Textilveredlung* **6,** 651 (1971).
143. V. Mischutin, *J. Coated Fabr.* **7,** 308 (1978).
144. D. J. Donaldson and co-workers, *Am. Dyest. Rep.* **64**(9), 30 (1975).
145. R. B. LeBlanc, J. P. Dicarlo, and D. A. LeBlanc in ref. 138, p. 177.
146. D. Douglas, *J. Soc. Dyers Colour.* **73,** 258 (1957).
147. H. Stepniczka, *Ind. Eng. Chem. Prod. Res. Devel.* **12**(1), 29 (1973).
148. R. B. LeBlanc, *Text. Ind.* **142**(2), 37, 44, 50, 53, 55, 57, 59 (1978).
149. H. Mori in ref. 135, p. 124.
150. A. A. Dunham in ref. 50, p. 181.
151. R. Tsuzuki, A. Hashizume, and H. Mori in ref. 150, p. 283.
152. P. L. Susani in ref. 50, p. 268.
153. W. L. Morrison, *Proceedings of the Symposium on Textile Flammability,* R. I. LeBlanc Corp., 1976, p. 144.
154. R. W. Morrow and co-workers, *Am. Ind. Hyg. Assoc. J.* **37,** 192 (1976).
155. F. A. Danher, *Information Council on Fabric Flammability, Tenth Annual Meeting, New York, 10 December 1976.*
156. N. K. Hooper and B. N. Ames in *Regulation of Cancer-Causing Flame-Retardant Chemicals and Governmental Coordination of Testing of Toxic Chemicals, Serial No. 95-33,* Government Printing Office, Washington, D.C., 1977, p. 42–45.
157. L. W. Mazzeno, Jr. and N. Greuner, *Text. Chem. Color.* **9**(8), 38 (1977).
158. B. N. Ames, J. McCann, and E. Yamasaki, *Mutat. Res.* **31**(6), 347 (1975).
159. A. F. Kerst, *JFF/Fire Retard. Chem.* **1,** 205 (1974).
160. J. McCann and co-workers, *Proc. Natl. Acad. Sci. U.S.A.* **72,** 5135 (1975); J. McCann and B. N. Ames, *Proc. Natl. Acad. Sci.* **73,** 950 (1976).
161. B. N. Ames, J. McCann and E. Yamasaki, *Mutat. Res.* **31,** 347 (1975).

General References

W. C. Kuryla and A. J. Papa, eds. (1973–75), *Flame Retardancy of Polymeric Materials,* Vol. 5, Marcel Dekker, Inc., New York, 1979.

J. W. Lyons, *The Chemistry and Uses of Fire Retardants,* Wiley-Interscience, New York, 1970.

W. A. Reeves, G. L. Drake, Jr., and R. M. Perkins, *Fire Resistant Textiles Handbook,* Technomic Publishing Co., Inc., Westport, Conn. 1974.

Textile Flammability, A Handbook of Regulations, Standards and Test Methods, American Association of Textile Chemists and Colorists, Research Triangle Park, N.C., 1975.

GEORGE L. DRAKE, JR.
United States Department of Agriculture

FLAMETHROWERS. See Chemicals in war.

FLARES. See Pyrotechnics.

FLAVIANIC ACID. See Naphthalene derivatives.

FLAVOR CHARACTERIZATION

The aroma of fruit, the taste of candy, and the texture of bread are examples of flavor perception. In each case, physical and chemical structures in these foods stimulate receptors in the nose and mouth. Impulses from these receptors are then processed into perceptions of flavor by the brain. These are subjective psychological processes and as such have all the complexities of the human mind. The central questions posed in flavor characterization are: (*1*) what causes flavor? (*2*) how is it detected? and (*3*) how can human response to flavor be predicted? Their answers remain elusive and, at present, knowledge is extremely primitive.

The intricate nature of human perception itself creates the greatest difficulty in the study of flavor. Just as people perceive events differently, they often perceive flavors differently, and this is due to both psychological and physiological factors. For example, certain aryl thiocarbamates taste exceedingly bitter to some people and are almost tasteless to others (1). Not only is this difference genetically determined but the frequency of occurrence of the nontaster expression differs from one population to another. Among U.S. Caucasians there are 40% nontasters, whereas only 3% of the Korean population cannot perceive the strong bitter taste of the aryl thiocarbamates (2).

Flavor characterization is concerned with the similarities in human flavor perception using methods that are designed to average out the differences. A collection of people (a panel) tastes or smells the same material and reports the perceptions according to previously explained guidelines. Using appropriate statistical methods, the similarities, if any, in the panelists' perceptions can be isolated (3). Such a process, previously called organoleptic testing but now called sensory analysis, requires a large amount of time to design, execute, and analyze, and is therefore expensive. Consequently, manufacturers concerned with flavor are motivated to find less expensive instrumental procedures to predict the flavor perception. However, such methods are indirect and their accuracy can only be determined by direct sensory analysis.

Flavor is generally defined in terms of three components: odor, taste, and texture (4). Odor is a result of stimuli interacting with specialized receptors in the nose. Similarly, taste results from the interactions between stimuli and receptor organs on the tongue (5). There are, however, no single identifiable organs involved with the perception of texture. The presence of pain, the sense of touch, and the detection of sound all contribute to the sense of texture. Thus the texture of a food includes all perceptions of flavor detected in the mouth that are owing to neither odor nor taste, and flavor is the combined perception of odor, taste, and texture (see Flavors and spices).

The idea that a stimulus, applied to a sense organ, produces a response is easily understood since this phenomenon is observed as an external process. For example, the application of an electric current (a stimulus) to a frog leg will result in an observable muscle contraction (a response). However, the perception of a stimulus by an organism is not directly observable but is either inferred from behavior or reported with language. What then is perception and can it, in fact, be measured? It may be that the data collected in a sensory analysis are measures of responses to flavor and not quantifications of perception. Nevertheless, the analyst characterizing flavor operates under the belief that he or she is measuring perception, knowing that the resolution of this issue will not decrease the validity of the experiments but merely change the language used to describe them.

It is easy to understand that the taste of sucrose and the smell of geraniol are two different kinds of sensory perceptions and that each of these two kinds of sensations can have different intensities. This is because sensory perception is both qualitative and quantitative. Sweet, bitter, salty, fruity, floral, etc are different flavor qualities, each produced by different chemical compounds, and the intensity of a particular sensory quality is determined by the amount of the stimulus present. The saltiness of a sodium chloride solution becomes more intense if more of the salt is added, but its quality does not change. However, if hydrochloric acid is substituted for sodium chloride then the flavor quality changes from salty to sour. For this reason, quality is substitutive, and quantity, intensity, or magnitude is additive (6). The sensory properties of food are generally complicated, consisting of many different flavor qualities at different intensities. Thus the first task of sensory analysis is to identify the component qualities and then to determine their various intensities.

Psychophysics

The study of the relationship between sensory perceptions and the stimuli that produce them, called psychophysics, has produced some very useful concepts for flavor characterization. Over a century and a half ago E. H. Weber observed a mathematical relationship between the perceived intensity of a sensation and the physical intensity of the stimulus that produced it. He observed that the minimum detectable difference between the intensity of two stimuli (called the just noticeable difference, JND) is proportional to the total intensity of the stimulus. This relationship has been verified for a large number of sensory qualities, eg, the taste and odor of many different chemicals, the loudness of sounds, the brightness of lights, the pain produced by electric shock, etc (7). So general is this relationship between sensory intensity and stimulus intensity that it is called Weber's law.

For determination of JND, a discriminability test is used in which people are asked if they can discriminate between a stimulus at two different intensities. The JND is the value of this minimally detectable difference. Data from such experiments contain only physical variables, ie, stimulus intensities. The gradual acceptance of the existence of psychological variables corresponding to the perceived intensity of sensations led to a more refined form of Weber's law (6). This is generally referred to as Stevens' law:

$$\psi = k\phi^n$$

where ψ is a measure of the perceived intensity of the sensation and ϕ is a measure of the physical intensity of the stimulus which produced the response; k and n are constants. The constant k depends only upon the units chosen for the variables ψ and ϕ. However, the exponent n is characteristic of the sensory quality (sometimes called modality) and is independent of stimulus intensity. The relationship between quality and modality is sometimes confusing. For example, sweetness is considered a modality, and both sucrose and saccharin are considered sweet (8). However, the exponent for sucrose is ca 1.5 and the exponent for saccharin is 0.8. Furthermore, saccharin is partially bitter whereas sucrose is not. Since the taste of sucrose is qualitatively different from that of saccharin, they are different sensory qualities.

Table 1 shows a list of several different sensory qualities and their corresponding Stevens' law exponents. These values were determined by having people associate numbers with their perception of the intensity of a sensation, a process called magnitude estimation (6). A plot of the log of these numbers vs the log of the stimulus magnitude yields a straight line with a slope equal to the exponent n. Those sensory qualities having an exponent less than one, eg, the odor of n-hexanol, $n = 0.66$, are said to be compressive. A large range of stimulus magnitudes are compressed into a small sensory scale. This allows detection and quantification of a large range of odor concentrations but, at the same time, limits the ability to detect small changes or differences in odor intensities. In contrast, an expansive sensory quality (eg, the taste of sucrose, $n = 1.5$) has a small total range of detectable concentrations, but minute changes in concentration are easily detected (7).

The existence of many different sensory exponents may at first seem unnecessarily complicated. But there are some compelling arguments that may justify these complicated perceptual processes. Consider those female insects that use volatile chemicals (called pheromones) to attract males from great distances. During the male's flight

Table 1. The Stevens' Law Exponents For Different Sensory Stimuli

Stimulus	Exponent of power function ($\psi = k\phi^n$), n
electric shock	3.5
temperature	1.6
loudness of sound	0.6
brightness of light	0.3
sweetness of sucrose	1.5
bitterness of quinine	0.6
saltiness	1.0
sourness	1.0
odor of n-heptane	0.6

toward the female (see Insect control technology) the concentration of the pheromone that must be detected varies over many orders of magnitude. The insect can deal with this variation by compressing a large range of stimulus concentrations into a narrower range of neural responses (9). On the other hand, it is extremely useful to be able to detect small changes in skin temperature over the narrow range that most organisms can tolerate and, indeed, the exponent for the sensation of warmth is 1.5 (7).

Descriptive Sensory Analysis

The method used to determine the quality and quantity of a sensory perception is called descriptive sensory analysis, and a knowledge of psychophysics plays an essential role in its design and execution. In a descriptive sensory analysis, a person is asked to associate a name or a number with his sensory perceptions. The set from which these names or numbers are chosen is called a scale. There are four general types of scales (nominal, ordinal, interval, and ratio) each with different properties and allowable statistics (7).

A nominal scale is always used to determine the quality or qualities of a flavor. It consists of an unordered array of names that people are asked to associate with their perceptions of flavor quality, eg, sweet, floral, crispy, etc. Recently it has been demonstrated that the nominal scaling of odors requires only that the person (*1*) has frequently encountered the odor, (*2*) has a long-standing connection between the odor and the name, and (*3*) has aid in recalling the name (10–11). The first two requirements are satisfied by training and the third is satisfied by presenting the person with a familiar list of names from which to choose. Unfortunately, the nominal scaling of the flavor qualities of a real material is complicated by the fact that a number of different flavor qualities are perceived simultaneously.

The measurement of flavor intensity, unlike the evaluation of quality, requires an ordered scale, the simplest of which is an ordinal scale. A scale consisting of the words weak, moderate, strong, or the numbers 1, 2, 3, is an example of an ordinal scale with three points.

Although ordinal scales are used to evaluate flavor intensity, they are more often used to determine people's attitudes toward a sensation. Do they like or dislike a sensation; do they find it pleasant or unpleasant? This aspect of flavor, called hedonics, is an attempt to determine what is the psychological impact of flavor perceptions. It is distinct from flavor characterization which deals with the perceptual impact of a stimulus. In flavor characterization, ordinal scales are used to arrange a set of samples in their order of increasing intensity (ranking) but they are seldom used to estimate the magnitude of these intensities.

The mathematical form of Stevens' law implies that a ratio scale should be used in order to determine the intensity of a flavor (7). In order to explain the nature of a ratio scale, consider the evaluation of a pair of samples each with a different flavor intensity. A panelist is asked, "Assume that the first sample has a flavor intensity of 'X' what number would you associate with the flavor intensity of the second?" The answer to this question is Y. If the ratio of X to Y is always the same no matter what value of X is used in the original question, then the panelist is estimating the flavor intensity on a ratio scale. However, if the difference between X and Y is constant for different values of X the flavor is being estimated on an interval scale. The process called magnitude estimation mentioned previously is an example of ratio scaling.

There has been considerable discussion in the literature (12–14) of the many psychological scales used to quantitate sensory perceptions. This effort is motivated by the desire to devise analytical methods that are consistent with the demands of Weber's and Stevens' laws. At the moment, the method of magnitude estimation appears to be the most rigorous. However, the Arthur D. Little flavor profile (FP) and quantitative descriptive analysis (QDA) are two of the most popular methods used in the food industry (15–16). Both methods involve the training of people in the nominal scaling of the flavor qualities present in the food being studied, but they differ in their method for quantitation.

Table 2 shows examples of an FP and a QDA questionnaire. Clearly the FP procedure uses an ordinal scale to quantify each flavor quality. However, the type of scaling process employed in the QDA procedure is not obvious. The authors of the QDA system consider it an interval scale (17), but it has been described elsewhere as a ratio scale because there are no intervals marked off on the line (18). In addition, the three words, weak, moderate, and strong, form an ordinal scale above the line. This points out one of the most serious, practical problems in perceptual measurement: the responses that a panelist gives in a sensory analysis depend upon his or her perception of the questionnaire as well as perception of the stimulus. However, rigorous training and the analysis of stimulus–response data can minimize some of these difficulties. These comments are not meant to question the usefulness of the QDA procedure but to point out the conceptual problems present in all sensory methods.

The practical problems posed by psychological scaling have led to a novel approach called cross-modial matching or magnitude production (7,19). Rather than have panelists associate their sensory perceptions with numerical scales, instead they

Table 2. A Comparison of the Type of Scales Used in the Flavor Profile and the Quantitative Descriptive Analysis

Type 1	Flavor profile			
	Instruction: check the appropriate intensity			
Flavor quality	Threshold	Slight	Moderate	Strong
sweetness	( )	( )	( )	( )
fruitiness	( )	( )	( )	( )
mercaptan	( )	( )	( )	( )
crispness	( )	( )	( )	( )
dryness	( )	( )	( )	( )

Type 2	Quantitative Descriptive Analysis		
	Instruction: make a vertical line marking the intensity of each quality		
Flavor quality	Weak	Moderate	Strong
sweetness	|	|	|
fruitiness	|	|	|
mercaptan	|	|	|
crispness	|	|	|
dryness	|	|	|

are asked to associate their perception of flavor intensity with their perception of some other stimulus, eg, the loudness of a sound, which they produce by turning a dial. The association of the intensity of one kind of stimulus with that of another can be achieved with remarkable ease, probably because it requires no conception of numbers.

One such stimulus that could be used to quantify flavor perception by cross-modial matching is the length of a line. The magnitude estimation of the perceived length of a line has been shown to follow Stevens' law with an exponent of ca one (7). Thus, asking people to draw a line the length of which represents their perception of a flavor intensity is an example of cross-modial matching. The QDA procedure may, in fact, be this type of analysis and not involve any psychological scaling.

There are two aspects of flavor perception that are not accounted for in the Weber–Stevens' laws: threshold and saturation. For every sense there is a minimum detectable stimulus intensity called the threshold. When the concentration of the stimulus is below the threshold value there is no perception. Also, the threshold value is not absolute but is greatly affected by the presence of other stimuli. For example, the threshold for geosmin in both beet juice and fish flesh is ca 50 times higher than it is in water (20–21). Saturation is the concentration of a stimulus above which no increase in perception can be detected. It is true that the Weber–Stevens' law can predict the relationship between stimulus intensity and sensory response with some precision; however, they do not describe the very common situation of stimuli at or near the threshold or point of saturation.

Discriminant Sensory Analysis. Discriminant sensory analysis is used to determine whether or not a difference can be detected in the flavor of two or more samples. These differences can be qualitative, quantitative, or both. The nature and magnitude of the difference is not revealed; discriminant analyses merely do or do not detect differences. Although these procedures yield much less information about the flavor of a food than descriptive analyses, they are extremely useful; eg, a manufacturer might want to substitute one component of a food product with another safer or less expensive one without changing the flavor in any way. Several formulations can be attempted until one is found with flavor characteristics that cannot be discriminated from the original or standard sample.

In the most popular discriminant sensory analysis, called the triangle test, panelists are presented with three samples, two of which are the same. The panelists examine the three samples and try to identify the odd sample. Their probability of choosing the correct sample by chance is one in three, or 33%. Invariably, the samples analyzed in these tests are very similar and statistical procedures must be used to determine whether or not the panelists detected a difference, usually expressed as probability of randomness or level of significance.

The one-third chance of guessing correctly on a triangle test is a decided advantage over other methods such as the duo–trio and the paired comparison. In the duo–trio method, a standard sample is identified and the panelist must determine which of two other samples is the same as the standard; the probability of guessing correctly is one in two, or 50%. The paired comparison is usually a little more than a discriminability test since it contains an element of direction, ie, a person is asked to choose which of two samples is weaker or stronger in some particular flavor quality. Nevertheless, the probability of choosing any one of them is 50%. The higher probability of guessing correctly in the duo–trio and paired tests requires more trials in order to establish significance, thus increasing the cost of analysis. For both descriptive and discriminant sensory analyses, proper experimental design requires a clear idea of the question being asked and the factors that must be controlled. It requires appropriate panel selection, choice of test environment, method of sample preparation, type of statistical analysis, etc, and these are not simple tasks (22).

The development of precise and reproducible methods of sensory analysis is prerequisite to the determination of what causes flavor, or the study of flavor chemistry. Knowing what chemical compounds are responsible for flavor allows the development of analytical techniques using chemistry rather than human subjects to characterize flavor (23). Routine analysis, a practice that is now rare in most food production for the quality control of flavor, will be much more common in future. Once standards for each flavor quality have been synthesized or isolated then they can also be used to train people to do more rigorous descriptive analyses. Perhaps the development of a universal vocabulary or list of descriptors to characterize flavor perceptions is possible.

Flavor Chemistry. The chemical compounds having odor and taste number in the thousands. In 1969 a description of the odor characteristics of more than three thousand chemical compounds used in the flavor and perfume industry were described (24). By now the list of compounds known to have odor and taste is much larger, and the list of all possible flavor compounds seems limitless. Many different compounds have the same flavor character or quality, differing perhaps in their relative intensity but indistinguishable in the type of flavor they elicit. Exactly how many different flavor qualities exist is not known but it is certainly much less than the total number of compounds with flavor.

One of the more persistent ideas is that there is a very small number of flavor qualities or characteristics, called primaries, each detected by a different kind of receptor site in the sensory organ. It is thought that each of these primary sites can be excited independently but that some chemicals can react with more than one site producing the perception of several flavor qualities simultaneously (5). Sweet, sour, salty, and bitter qualities are generally accepted as the four primaries for taste; sucrose, hydrochloric acid, sodium chloride, and quinine, respectively, are compounds that have these primary tastes. Sucrose is only sweet, quinine is only bitter, etc; saccharin, however, is slightly bitter as well as sweet and its Stevens' law exponent is 0.8, between that for purely sweet (1.5) and purely bitter (0.6) compounds (18). There is evidence that all compounds with the same primary taste characteristic have the same psychophysical exponent even though they may have different threshold values (19). The flavor of a complex food can be described as a combination of a smaller number of flavor primaries, each with an associated intensity. A flavor could be described as a vector in which the primaries make up the coordinates of the "flavor space."

Table 3 shows examples of compounds with taste and their associated sensory qualities. Sour taste is primarily produced by the presence of hydrogen ion slightly modified by the types of anions present in the solution. For example, acetic acid is more sour than citric acid at the same pH or molar concentration (25). Saltiness is due to the salts of alkali metals, the most common of which is sodium chloride. However, such salts as cesium chloride and potassium iodide are bitter, and potassium bromide has a mixed taste (salty and bitter) (26). Thus saltiness, like sourness, is modified by the presence of different anions but is a direct result of a small number of cations.

Sweetness and bitterness are much more complicated because of the great variety of chemical compounds having these taste qualities. Furthermore, very different

Table 3. Compounds With Taste

Name	Structure	Taste quality
hydrogen ion	H^+	sour
sodium ion	Na^+	salty
potassium ion	K^+	salty
potassium iodide	KI	bitter
quinine	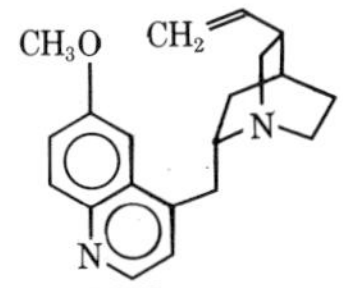	bitter
β-D-mannose	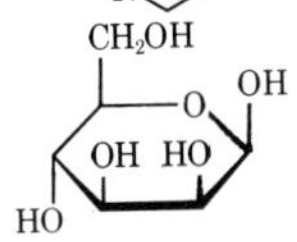	bitter
α-D-mannose	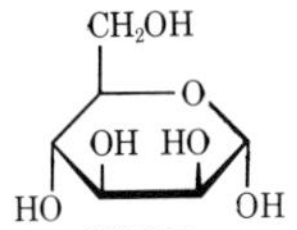	sweet
β-D-glucose	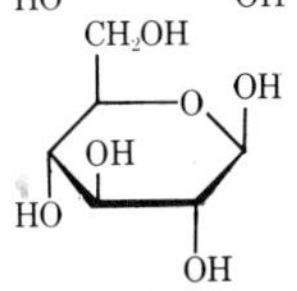	sweet

compounds can have the same flavor quality, such as quinine and β-D-mannose both of which are bitter. Conversely, two very similar compounds can have different flavor qualities, eg, α-D-mannose is sweet whereas β-D-mannose is bitter. Examination of the structures in Table 3 with sweet and bitter taste exemplifies the problem of identifying those molecular features that distinguish sweet- from bitter-tasting compounds. Many chemical and physical properties have been examined to determine the requirements for a chemical to elicit a particular sensory response (8). The AH-B (acid hydrogen-base) theory of sweetness postulates that a molecule must have a proton attached to an electronegative center located 0.3 nm away from another electronegative center in order to taste sweet (27). This theory has been extended to include a hydrophobic site near the AH-B system, but it is still unable to predict the taste of a particular structure (28). All sweet compounds have an AH-B system as demonstrated in Figure 1, but some bitter and some tasteless compounds also have such a system in their structures.

The relationship between molecular structure and sensory properties is even less clear for compounds with odor. It seems likely that there is a set of odors that could be called primaries, but a generally accepted list of such primary odor qualities has not been devised. Molecular size and shape have been used to describe the features that distinguish different odor qualities (29); eg, there are seven odor primaries including ethereal, camphoraceous, musty, floral, minty, pungent, and putrid. Like all theories attempting to relate chemical structure with sensory properties, these ideas have little predictive value. The complexity of this problem is exemplified by the compounds shown in Table 4 (see also Air pollution control methods). These compounds are not meant to represent odor primaries but are examples of naturally oc-

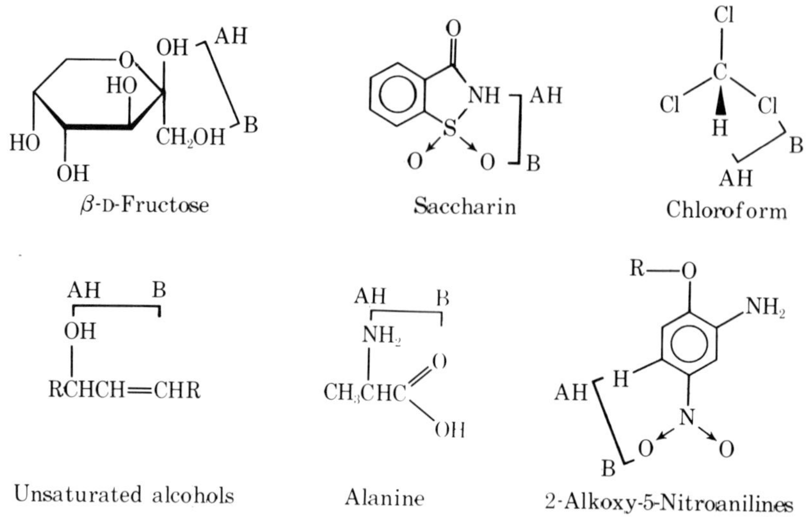

Figure 1. Compounds of various chemical classes and their common (*AH, B*) unit.

curring substances that have very low odor thresholds and diverse structures. For example, geosmin, the compound responsible for the characteristic odor of beets and of wet soil, can be detected in water at a concentration of 10^{-12} g/g (34–35,40–41). The isolation and identification of trace components such as geosmin that are responsible for the odor of natural products is one of the major challenges in flavor chemistry.

The methods used to isolate, concentrate, and identify odor components are not unique to flavor chemistry, but the use of sensory analyses to monitor their progress is unique. Once a particular compound has been identified and a standard unambiguously synthesized, its odor characteristic must be verified by sensory analysis. The complexity of natural products requires the use of separation techniques having the highest available resolution, such as gas and liquid chromatography. It is difficult to determine a compound occurring at a concentration of 10^{-9} to 10^{-12} g/g without concentrating it at least several thousandfold. Even after concentration there are often only a few nanograms available for structure analysis. The extraction of 100,000 kg of beets, the amount produced on 10 hm^2 (hectares) of land, would be required to obtain 20 mg of geosmin from beets for a standard pmr (proton magnetic resonance) spectrum (42). Therefore, in addition to high resolution, the most sensitive analytical techniques such as mass spectroscopy must be used.

Figure 2 shows a portion of a reconstructed chromatogram obtained from a system combining a capillary gas chromatograph with a mass spectrometer-data system (see Analytical methods). The sample was a fraction of odor components extracted from grape essence collected from a gel permeation column used for preliminary separation. Using gel permeation, 2-phenylethanol, a major constituent of all grapes, can be removed from the components of the grape essence responsible for its characteristic odor. Peak 3 in the figure is due to the trace amount of 2-phenylethanol remaining in this fraction and it is clearly separated from peak 2, a very important odor component of the grape. Without the use of gel permeation, this peak is completely obscured by the large 2-phenylethanol component. The structure of this component as well as that of peaks 1, 4, 5, and 6 has yet to be fully determined. Using sensory analysis of the gas

Table 4. Naturally Occurring Compounds With Odor

Compound	Structure	Odor	Source	Reference
isoamyl acetate	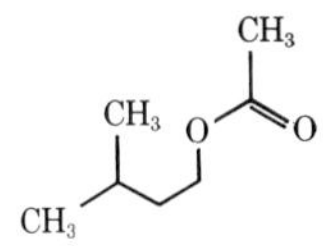	fruity	banana	24
geraniol	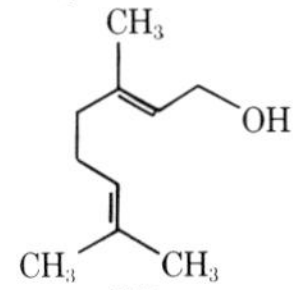	floral	rose	30
trans-β-sinensal	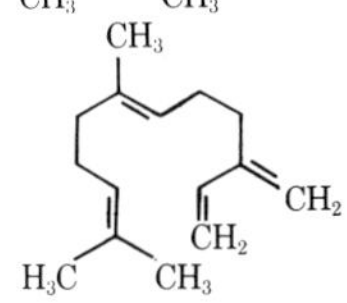	fruity	orange	31
nootkatone	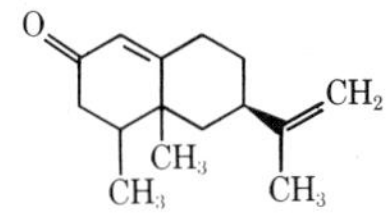	fruity	grapefruit	32
eremophilone	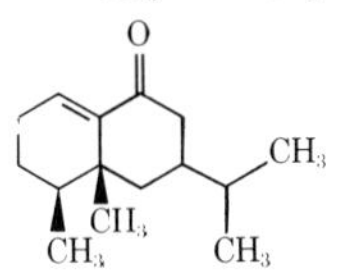	woody	sandalwood	33
geosmin	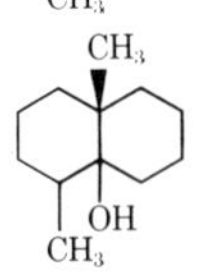	earthy	beets	34–35
muscone	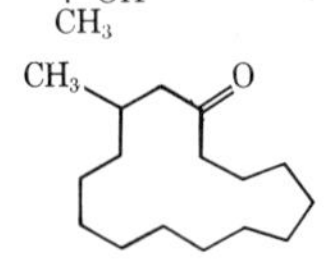	musk	musk deer	36
hydrogen sulfide	H_2S	putrid	rotten eggs	37
5α-androst-16-en-3-one	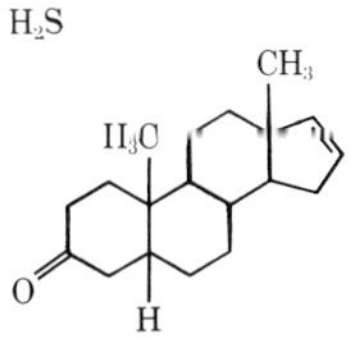	animal	boar urine	38
cadaverine	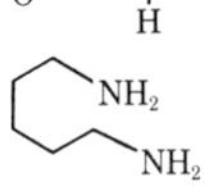	putrid	carrion	39

chromatographic effluent, they have been shown (43) to have strong characteristic odors. The amount of each of these compounds in the whole grape is ca 10^{-11} g/g. This fraction from the gel permeation column had the characteristic odor of the Catawba grape cultivar (*Vitus labruscana*, Bailey) from which it was taken, but it contained no methyl anthranilate, a compound often associated with the characteristic odor of this grape species (44).

The above example demonstrates only one of the challenging problems of flavor

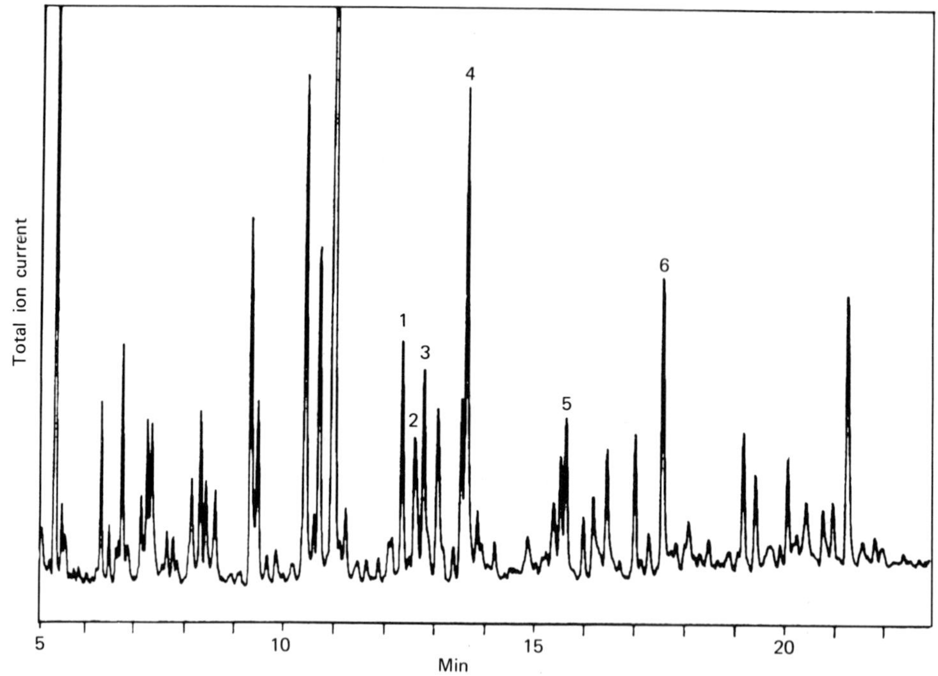

Figure 2. A reconstructed gas chromatogram from a gc-ms-data system. The sample is a fraction of grape essence with peaks 1–6 the important odor components.

chemistry, ie, each flavor requires its own analytical approach; however, a sensory analysis is always required. Most of the remaining problems, and they are many, demand the most sensitive and selective techniques, and the use of methods for concentration and isolation that preserve the sensory properties of these complex and often delicate flavors. The importance of chemistry to the study of flavor was recognized 160 years ago by a famous gourmet who wrote, "Men who will come after us will know much more than we of this subject; and it cannot be disputed that it is chemistry which will reveal the cause of the basic elements of taste" (45). Although we have learned much since Brillat-Savarin, his sentiments are still true.

BIBLIOGRAPHY

"Organoleptic Testing" in *ECT* 2nd ed., pp. 336–344, by Loren B. Sjöström, Arthur D. Little, Inc.

1. A. L. Fox, *Proc. Nat. Acad. Sci.* **18,** 115 (1932).
2. J. Cohen and D. P. Ogdon, *Psychol. Bull.* **46,** 490 (1949).
3. M. A. Amerine, R. M. Pangborn, and E. B. Roessler, *Principles of Sensory Evaluation of Food,* Academic Press, New York, 1965.
4. J. Brozek, *Am. J. Clin. Nutr.* **5,** 332 (1957).
5. R. W. Moncrieff, *The Chemical Senses,* 3rd ed., CRC Press, Cleveland, 1967.
6. S. S. Stevens, *Science* **127,** 383 (1958).
7. S. S. Stevens, *Am. Sci.* **48,** 226 (1960).
8. M. G. J. Beets, *Structure-Activity Relationships in Human Chemoreception,* Applied Science Publishers, London, 1978.
9. H. Arn, *Z. Naturforsch. Teil B.* **30,** 722 (1975).

10. E. Gibson, *Principles of Perceptual Learning and Development,* Appleton Century, New York, 1969.
11. W. S. Cain, *Science* **203,** 467 (1979).
12. S. S. Stevens, *Psychol. Rev.* **70,** 153 (1963).
13. H. Eisler, *Psychol. Rev.* **70,** 243 (1963).
14. H. R. Moskowitz and J. L. Sidel, *J. Food Sci.* **36,** 677 (1971).
15. S. E. Cairncross and L. B. Sjöström, *Food Technol.* **4,** 308 (1950).
16. H. Stone and co-workers, *Food Technol.* **29,** 24 (1974).
17. R. C. Singleton, J. Sidel, and H. Stone in ref. 16, in press.
18. J. J. Powers and H. R. Moskowitz, *Am. Soc. Test. Mater. Spec. Tech. Publ.* **594,** 35 (1974); R. A. Scanlan, ed., *Flavor Quality: Objective Measurement,* ACS Symposium Series No. 51, 1977.
19. H. R. Moskowitz, *Percept. Psychophys.* **9,** 51 (1971).
20. L. D. Tyler, T. E. Acree, and N. L. Smith, *J. Food Sci.* **44,** 79 (1979).
21. M. Yurkowski and J. L. Tabachek, *J. Fish Res. Board Can.* **31,** 1851 (1974).
22. J. L. Sidel and H. Stone, *Food Technol.* **30,** 32 (1976).
23. R. Teranishi and co-workers, *Flavor Research, Principles and Techniques,* Marcel Dekker, New York, 1971.
24. S. Arctander, *Perfume and Flavor Chemicals,* 2 Vols., Montclair, N.J., S. Arctander, 1969.
25. F. W. Fabian and H. B. Blum, *Food Res.* **8,** 179 (1943).
26. L. M. Beidler, *Handbook of Sensory Physiology,* Vol. IV, Springer-Verlag, Berlin, 1971, p. 256.
27. R. S. Shallenberger and T. E. Acree, *Nature* **216,** 480 (1967).
28. L. B. Kier, *J. Pharm. Sci.* **61,** 1394 (1972).
29. J. E. Amoore, *Proc. Sci. Sect. Toilet Goods Assoc. Special. Suppl.* **7,** 1 (1962).
30. T. E. Furia and N. Bellanca, *Fenaroli's Handbook of Flavor Ingredients,* Chemical Rubber Co., Cleveland, 1971.
31. M. G. Moshomas, *J. Agric. Food Chem.* **19,** 769 (1971).
32. W. D. MacLeod and N. M. Buigues, *J. Food Sci.* **29,** 565 (1964).
33. J. L. Willis in G. Ohloff and A. F. Thomas, eds., *Gustation and Olfaction,* Academic Press, New York, 1971, p. 169.
34. T. E. Acree and co-workers, *J. Agric. Food Chem.* **24,** 430 (1976).
35. K. E. Murray, P. A. Bannister, and R. G. Buttery, *Chem. Ind.* **26,** 973 (1975); G. Charalambous and I. Katz, eds., *Phenolics, Sulfur and Nitrogen Compounds in Food Flavors,* ACS Symposium Series No. 26, 1976.
36. L. Ruzicka, *Helv. Chim. Acta* **9,** 1008 (1926).
37. R. P. Pomeroy and H. Cruse, *J. Am. Water Works Assoc.* **61,** 21 (1969).
38. R. L. S. Patterson, *J. Agric. Food Chem.* **19,** 31 (1968).
39. D. Ackerman, *Z. Physiol. Chem.* **54,** 16 (1907).
40. R. G. Buttery and J. Garibaldi, *J. Agric. Food Chem.* **24,** 1247 (1976).
41. L. D. Tyler, T. E. Acree, and R. M. Butts, *J. Agric. Food Chem.* **26,** 1415 (1978).
42. L. D. Tyler and co-workers, *J. Agric. Food Chem.* **26,** 775 (1978).
43. T. E. Acree and co-workers, *Anal. Chem.* **48,** 1821 (1976).
44. R. R. Nelson and co-workers, *J. Food Sci.* **42,** 57 (1977).
45. J. A. Brillat-Savarin, *The Physiology of Taste,* 1825, translated, M. F. K. Fisher, Harcourt Brace Jovanovich, New York, 1978.

TERRY E. ACREE
Cornell University

FLAVORS AND SPICES

In this article, flavors and spices are presented separately, although they are employed for similar purposes in the food and flavor industries. Spices, of course, have many other applications. The emphasis here is on the use of spices in flavorings.

The quintessence of the spices, the essential oils, derived from spices by various methods, are described separately in Oils, essential. The application of essential oils to flavorings, however, is included in this section. Oleoresins, which are derived from spices by extraction, also have many applications; only their use in flavorings is considered here.

Flavoring materials are often classified as condiments, ie, materials used to give relish to food and gratify the taste (see also Carbonated beverages; Chocolate and cocoa; Food additives; Food processing).

FLAVORS

Flavor is that property of a substance (commonly a food or one used in food) that causes a simultaneous reaction or sensation of taste on the tongue and odor in the olfactory center in the nose. Thus defined, taste and odor are descriptive of a sensation, whereas taste in its colloquial meaning is practically synonymous with flavor.

Flavor has been defined as "a memory" and "an experience," and it is indeed both of these (1). These definitions have always included as part of the explanation at least two phenomena, ie, taste and smell (2). It is suggested that in defining flavor too much emphasis is put on the olfactory (smell) and gustatory (taste) aspects (3). It is further suggested that vision, hearing, and tactile senses also contribute to the total flavor impression. Flavor is viewed as a division between physical (appearance, texture, and consistency), and chemical senses (smell, taste, and feeling) (4). The preferred definition is proposed by the Society of Flavor Chemists: "Flavor is the sum total of those characteristics of any material taken in the mouth, perceived principally by the senses of taste and smell and also the general senses of pain and tactile receptors in the mouth, as perceived by the brain" (5).

Food Acceptance

Four quality factors of food are recognized as significant: (*1*) appearance, (*2*) mouthfeel, (*3*) flavor, and (*4*) nutritive value. Flavor is, therefore, an important factor in food acceptance, and research and development of flavors has expanded due to the increase in the use of prepared foods and the technical advances in instrumental analytical methods. When all four factors are in proper qualitative proportions, a food finds general acceptance. When all four are interdependent, the sense of sight (appearance) takes precedence over the others and nutritive value obviously is not an important consideration.

A food must have the expected or proper appearance and color before it will be readily consumed (6). There are a number of prepared foods in which artificial flavors and colors are used, whose flavor is sufficiently bland to make color essential for flavor identification, eg, margarine. The preservation of color in a natural food during processing or the development of color by processing are factors of primary importance in food acceptance.

Mouthfeel (7) is described as a texture or kinesthetic factor evaluated by the skin or muscles in the mouth. It includes smoothness, roughness, stickiness, slickness, and related surface characteristics, as well as brittleness and viscosity which are a food's physical characteristics. Texture is not only determined by mastication, but also by the ease of cutting or dividing before consumption. This property is closely allied to flavor since a particular flavor observed without the proper mouthfeel may not only be unappreciated, but go unrecognized. Thus a nutty flavor is associated with brittleness, a butter flavor with slickness or smoothness, etc.

The appreciation or enjoyment of certain foods varies with different races, nationalities, and social classes throughout the world, and can be ascribed to custom, training, and climate. The latter restricts the natural availability of foods. In regions where there is the monotony of a limited diet or of a bland diet such as rice, the use of flavorings and spices to impart a more pleasing variety to food is demanded. In an affluent society, since foods are available from all regions, a variety is provided that results in greater acceptability of foods, flavorings, and spices.

Highly seasoned foods appear to be more prevalent in hot climates. Whether this is due to the early use of spices as preservatives prior to refrigeration, or to a physiological reaction to temperature, is speculative. The acceptability of certain foods may also be due to availability and habit, which in turn affect certain customs. Thus the sugar maple, indigenous to the northern United States, provides a syrup generally enjoyed and familiar in this region; however, it is generally unacceptable in the southern states. Such differences in taste demand variety in flavorings. For this reason, although advances of foods and food processing are making increased use of science and technology (8), food acceptance depends significantly upon a proper blending of art and science, in order to satisfy individual flavor preferences and prejudices.

Taste

Certain basic principles are involved in the physiology of flavor perception. It is generally accepted that there are four tastes: salty (9), sour (10), bitter (11), and sweet, although at various times three more have been proposed, ie, alkaline, meaty, and metallic (12) (see Flavor characterization). It is well-established that the four basic tastes are perceived by certain sense organs (taste buds) located on the tongue. Theoretically, the taste buds are chemically stimulated and the reaction can be measured by electrical impulses of their nerve fibers to the brain. There is some evidence that the stimulation may be physical (specific receptor cells are shaped to fit the particular molecular structure of the substance, a lock-and-key theory). Taste in this instance would be a surface phenomenon (13). When several taste qualities are perceived practically simultaneously, the overall sensation depends upon the temporal and spatial patterns of the nerve activity.

There are several factors that affect the extent and character of the sensation. A hereditary or genetic factor may cause a variation between individual reactions. Thus phenylthiourea, which can cause a bitter taste sensation, may not be perceptible to certain individuals whose general ability to distinguish other tastes is not noticeably impaired (14). The variation of pH in saliva (which acts as a buffer) may influence the perception of acidity differently in individuals (15). Enzymes in saliva can cause rapid chemical changes in basic food ingredients, such as proteins and carbohydrates, with variable effects on the individual.

When a food contains both sweet and bitter substances, the temporal pattern of reception, ie, the order in which sweet and bitter tastes are perceived, affects the total qualitative evaluation. This temporal effect is caused by the physical location of taste buds. The buds responding to sweet are located on the surface and tip of the tongue, the bitter in grooves toward the rear. Therefore, the two types of taste buds can be activated progressively.

Texture also influences the evaluation of taste. Sweetness in a liquid is associated with body or viscosity. An artificially sweetened beverage that lacks body, therefore, may be rated qualitatively lower than one equally sweet but containing sucrose.

The metallic taste (12,16–17) is not ascribed to any special taste buds or mouth area. It can be described as a chemical sense, similar to the sense of pain, and may be generated by the salts of iron, copper, and tin, which are the salts most commonly found in foods. The concentrations of metal ions found to affect the taste of orange juice were 20 ppm for Cu^{2+}, 30–40 ppm for Fe^{2+}, and ca 100 ppm for Fe^{3+}, Al^{3+}, Cr^{3+}, Ni^{2+} and Sn^{2+} (18). Other similar reactions that are sometimes classed as taste sensations are pungency (eg, pepper) and cold (eg, menthol), both of which may be included among pain reactions.

Although the values cannot be considered absolute, approximate magnitudes or taste sensitivity have been measured (see Table 1).

Certain taste interrelationships should be considered in the evaluation of taste magnitude. The apparent sourness of citric acid is depressed by both sucrose and sodium chloride. Although the saltiness of sodium chloride is reduced by sucrose, it can be significantly enhanced by acid (20).

Stimulation of the tongue with the application of different taste stimuli almost simultaneously can produce an interaction, modification, or blending of the stimuli in some instances and not in others. Warm and cold sensations were reported to act similarly on the tongue in two groups: bitter, warm, and sweet; and sour, cold, and salty (21). Thus the physiology of taste is still unresolved. The theory of the specificity of the taste buds may be subject to modification (22).

There are several generalizations that can be made regarding taste (23):

(*1*) There are four, or perhaps five, taste qualities: salty, sour, bitter, and sweet, and metallic (12,16–17,24).

(*2*) The material must be in water solution (eg, saliva) to have taste. Water solubility is the first requirement of a taste stimulus (12).

(*3*) Only acids are sour. Sourness is not identical to chemical acidity, or pH, which

Table 1. Approximate Taste Thresholds[a]

Basic taste	wt%	Test standard
sour	0.007	hydrochloric acid
salty	0.25	sodium chloride
sweet	0.50	sucrose
bitter	0.00005	quinine

[a] Ref. 19.

is a function of the hydrogen ion concentration (qv), but appears to be a function of the entire acid molecule. A combination of pH and acid concentration determines the actual degree of sour taste. At the same pH, any of the organic acids (eg, citric acid) exhibits a far greater sourness than a mineral acid (eg, hydrochloric acid) (25–26).

(*4*) Only salts are salty; however, not all salts are salty, some are sweet, bitter, or tasteless. The salty taste is exhibited by ionized salts, and the greatest contribution to salty taste comes from the anions (24).

(*5*) Organic molecules are usually sweet or bitter, or a combination of these, or tasteless (probably owing to lack of water solubility).

Odor

The physiology of odor, which is the determining characteristic of flavor, is more complex and less understood than that of taste. A large number of odors are distinguishable, but it is not known how this is accomplished. Olfactory response is only observed when the substance contacts the olfactory membrane (27). This membrane is called the olfactory mucosa or olfactory epithelium and occupies an area of ca 2.5 cm^2 in each nostril. Above the nasal passages, the two olfactory clefts are separated by the nasal septum. For a substance to have an odor, it must be capable of reaching the olfactory epithelium high up in the nose, ie, it must possess sufficient volatility and low molecular weight so that, by inspiration, it may make contact in the nasal passages. However, a normally nonodorous, or nonvolatile substance placed in contact by means of a spray or mist may become odorous. Many materials now regarded as useless (27) may have practical odorant applications because of this phenomenon.

The odor of a substance is most logically attributed to its molecular structure. As in taste, its perception is theorized to be a chemical reaction with a receptor cell that excites a nerve center, giving a sensation. Other theories that have been ascribed to nonchemical reactions, such as molecular vibration or infrared absorption, seem to be without solid foundation (28–29). Although odor appears to be associated with structure, it has not been possible to predict odor types accurately on this basis (30–31). In general, compounds of carbon and hydrogen are odorless or of uncharacteristic odor until such atoms as oxygen, nitrogen, or sulfur are introduced. These atoms, combining to form certain functional groups such as alcohols, aldehydes, amines, esters, ketones, etc, make up what have been called osmophore, or odor-bearing groups. It is not only the atoms that determine the odor character, but the type of linkage also has a decided influence. Another factor that affects odor is the structural form of the molecule which depends on isomeric and spatial variations. Certain metals (especially arsenic and selenium) as their hydrogen compounds may exhibit intense odor.

The identification of odor (32) may not depend upon a single sensory reaction but upon several responses occurring almost simultaneously. If there are spatial patterns (on different areas inside the nose) and temporal patterns (depending on receptor sensitivity at a given time), and a number of specifically variable sensory cells, then these factors in combination with molecular structure would permit the distinction between a very large number of odors. The identification would result in a type of integration of the several sensory perceptions by the brain. A volatile substance disturbs the equilibrium in the odor receptor, thus starting or stopping a biological or metabolic process, possibly enzymatic in nature (33). The odor molecule may behave as a positive or negative catalyst causing a change in membrane potential. Whatever the physiology of odor perception may be, the sense of smell is keener than that of taste (19). If aromas are classed into odors and tastes as in common practice in science, it can be calculated that there are probably more than 10^6 possible sensations of odor and only a few, perhaps six, sensations of taste (17,34–35).

The detection-threshold values in Table 2 have been determined by different investigators and may vary by as much as 1000, depending on the test methods. The values give approximate magnitudes and are relatively significant when the same techniques for evaluation are used. As late as 1952 it was stated (38) that the vast field of the chemistry of foodstuffs remained unexplored. Since then, through new techniques of analysis, the chemistry of odorous materials has been the subject of intense research (see Table 3). Today many new compounds have been identified in natural products (36,37,39,40–47).

Table 2. Selected Odor Detection Thresholds of Some Organic Compounds [a]

Compound	Detection threshold[b]		
	ppm/air	mg/m^3 air	ppm/water
acetaldehyde	0.066		0.000688
acetic acid		0.025–76	24.3
acetophenone		1.5	65.0
amyl acetate		0.04–31.0	0.08
benzaldehyde	0.042		0.00425
butanol		0.158–1000	2.5
butyric acid		0.001	
1,8-cineol		0.055–0.19	0.012
ethanol	4400		100.0
ethyl acetate		3.6–1.120	5.0
ethyl acrylate		0.0010	0.067
ethyl benzoate	0.62		
ethyl lactate		8.0	1.4
ethyl 2-methylbutyrate		4.6	0.0001
ethyl vanillin		0.000007	0.1
hexanol		0.04–1.5	5.2
2-hexenal		0.034	0.017
isobutyl alcohol		1–500	7.0
isobutyric acid			8.1
isovaleric acid		0.005	0.7
α-ionone		0.0001–0.008	0.5
dl-menthone		42.9	0.17
methyl anthranilate	0.0094		
methyl salicylate		0.002–112	0.10
phenol	0.001		
pyridine	0.03		
skatole		0.0000004	
vanillin		0.000001–0.005	0.2

[a] Detection threshold is the minimum physical intensity detected by a subject (36) who is not required to identify but just detect the existence of the stimuli.

[b] Ref. 37.

Table 3. Odor Detection Threshold Levels in Water and Mineral Oil[a]

Compound	Concentration in water, ppm	Concentration in mineral oil, ppm
2-methylpyrazine	105	27
2,5-dimethylpyrazine	35	17
2,6-dimethylpyrazine	54	8
2,3,5-trimethylpyrazine	9	27
2,3,5,6-tetramethylpyrazine	10	38
2,5-dimethyl-3-ethylpyrazine	43	24
2,6-dimethyl-3-ethylpyrazine	15	24

[a] Ref. 39.

Flavor Materials

Materials for flavoring may be divided into several groups. The most common groupings are labeled natural and artificial. Under natural materials are found spices and herbs; essential oils and their extracts, concentrates, and isolates; fruit and fruit juices (qv); animal and vegetable materials and their extracts; and aromatic chemicals isolated by physical means from natural products, eg, citral from oil of lemongrass, and linalool from bois de rose. Under artificial materials are found aliphatic, aromatic, and terpene compounds that are made synthetically as opposed to those isolated from natural sources. Examples of each grouping are: benzaldehyde made synthetically (artificial) or isolated from oil of bitter almond (natural); and 1-menthol made synthetically (artificial) or isolated from oil of Mentha arvensis (natural).

Natural Flavorings. Natural flavorings include essential oils, fruit and fruit juices, and plant extracts. Many essential oils are used for flavoring and perfumery, eg, neroli (bitter orange blossoms), geranium, and ylang-ylang (see Oils, essential). The whole fruit, crushed fruit, and puree may be used directly in foods, ice cream, cakes, and confections. Fruit juices or their concentrates are more commonly employed (see Fruit juices).

Another group of natural flavorings comprises those obtained by extraction from certain plant products such as vanilla beans, licorice root, St. John's bread, orange and lemon peel, coffee, tea, kola nuts, catechu, cherry and elm bark, and gentian root. These products are used in the form of alcoholic infusions or tinctures, or of concentrations of alcohol or alcohol–water extractions termed fluid extracts or solid extracts. Official methods for their preparation and specifications for all products used in pharmaceuticals are described in the *U.S. Pharmacopeia* (48) and the *National Formulary* (49–50). There are many flavor extracts in use for which no official standards exist and the properties of which are based solely upon suitability for commercial applications (51).

Artificial Flavorings. The largest group of flavoring materials is composed of hundreds of isolates from natural materials and synthetic materials, eg, aldehydes, ketones, acids, esters, phenols, phenol ethers, lactones, organic derivatives of sulfur, and aliphatic, aromatic, and terpene alcohols. More than one of these chemically active or osmophoric groups may be contained in a single chemical compound, as in vanillin (qv) (4-hydroxy-3-methoxybenzaldehyde) with phenolic, aldehydic, and ether groups, and in methyl anthranilate (methyl 2-aminobenzoate) with ester and amine groups, and in ethylmaltol (2-ethyl-3-hydroxy-4*H*-pyran-4-one) with hydroxy and ketone groups, etc.

Progressive Use of Synthetics. Several manuals devoted, at least in part, to flavor formulation have been published (52–55). The increasing number of materials available today has resulted in the improvement of flavor characteristics and has permitted a closer approach to natural flavors. Often such materials in themselves bear a scant sensory relationship to the true natural flavor character, but when used as a component and judiciously applied may serve a useful purpose in a properly compounded flavor.

Typical Synthetics. Among the synthetics in common use for certain applications as indicated are: anethole (anise-licorice) for cough drops and chewing gum; benzyl acetate [*140-11-4*] (fruity raspberry and cherry) for candies and soft drinks; cinnamaldehyde [*104-55-2*] (cinnamon) for candies, chewing gum, and baked goods; methyl anthranilate (concord grape) for candies and soft drinks; methyl salicylate [*119-36-8*] (wintergreen) for chewing gum, candies, and oral preparations; citral [*5392-40-5*] (lemon) for soft drinks and candies; menthol [*89-78-1*] (mint) for candies, chewing gum, and oral preparations; allyl caproate (pineapple) for candies and soft drinks; γ-undecalactone (peach) for candies; ethyl methylphenylglycidate [*77-83-8*] (strawberry) for candies, soft drinks, and gelatin desserts; vanillin (vanilla) for ice cream, baked goods, and candies; ethyl butyrate (fruity) for soft drinks and candies; benzaldehyde (bitter almond, cherry) for candies, soft drinks, and baked goods; *trans*-α-ionone (raspberry) for candies; diacetyl (butter) for margarine and cheese; eugenol (clove) for candies and condiments; and isoamyl acetate (banana) for soft drinks, chewing gums, and candies). Many more synthetic flavors in current use are shown in Table 4. Although their principal character and purpose may be as indicated, they can be employed for other applications as well.

In commerce, several classifications of flavoring and compounded flavorings are listed according to composition in order that the user may conform to state and federal food regulations and labeling requirements, as well as to indicate their proper application. Both supplier and purchaser are subject to the control of the FDA and also of the Treasury Department. The latter regulates the alcoholic content of flavors and the tax drawbacks on alcohol (return of a portion of the tax paid on ethyl alcohol used in flavorings).

One class of flavorings known as True Fruit is composed of fruit juices and their concentrates.

A second group includes Fruit Flavors With Other Natural Flavors (WONF) in which fruit concentrates or extracts may be fortified with natural essential oils or extractives (isolates) of other naturally occurring plants (56). This class is employed when the manufacturer is compelled by regulation to use only natural products, as in the case of wines and cordials in the United States.

A third class includes Artificial Fruit Flavors, in which the fruit concentrates are fortified with synthetic materials. These may be subdivided into two or more groups according to price, use of the proportionate strengths of the natural fruit, and synthetic fortification.

Table 4. Selected Character Impact Items[a]

Flavor character	Flavor materials	CAS Registry Number
almond	5-methylthiophen-2-carboxaldehyde	[*13679-70-4*]
anise	anethole	[*104-46-1*]
	methyl chavicol (estragol)	[*140-67-0*]
apple	isoamyl acetate	[*123-92-2*]
	ethyl 2-methylbutyrate	[*7452-79-1*]
bergamot	linalyl acetate	[*115-95-7*]
blueberry	isobutyl-2-buteneoate	[*589-66-2*]
blue cheese	2-heptanone	[*110-43-0*]
butter	diacetyl	[*431-03-8*]
caramel	2,5-dimethyl-4-hydroxy-3(2*H*)-furanone	[*3658-77-3*]
caraway	*d*-carvone	[*2244-16-8*]
cassia	*p*-mentha-8-thiol-3-one	[*34352-05-1*]
celery	3-propylidene-1(3*H*)-isobenzofuranone	[*17369-59-4*]
cherry	benzaldehyde	[*100-52-7*]
	tolyl aldehyde	[*1334-78-7*]
chocolate	5-methyl-2-phenyl-2-hexenal	[*21834-92-4*]
	isoamyl butyrate	[*106-27-4*]
	vanillin	[*121-33-5*]
	ethyl vanillin	[*121-32-4*]
	2-methoxy-5-methylpyrazine	[*2882-22-6*]
coconut	γ-nonalactone	[*104-61-0*]
	α-octalactone	[*698-76-0*]
coffee	furfuryl mercaptan	[*98-02-2*]
	furfuryl thiopropionate	[*59020-85-8*]
cognac	ethyl oenanthate	[*106-30-9*]
clove	eugenol	[*97-53-0*]
coriander	linalool	[*78-70-6*]
cucumber	nona-*trans*-2-*cis*-6-dienal	[*557-48-2*]
garlic	diallyl sulfide	[*592-88-1*]
grape (concord)	methyl anthranilate	[*134-20-3*]
grapefruit	nootkatone	[*4674-50-4*]
green bell pepper	2-methoxy-3-isobutylpyrazine	[*24683-00-9*]
green leafy	*cis*-3-hexanol	[*928-96-1*]
hazelnut	(methylthio)pyrazine	[*21948-70-9*]
horseradish	1-penten-3-one	[*1629-58-9*]
jasmin	benzyl acetate	[*140-11-4*]
maple	2-hydroxy-3-methyl-2-cyclopenten-1-one	[*80-71-7*]
meat	methyl-5-(β-hydroxyethyl)thiazole	[*137-00-8*]
melon	2-methyl-3-*p*-tolypropionaldehyde	[*16251-78-8*]
	hydroxycitronellal dimethyl acetal	[*141-92-4*]
	2,6-dimethyl-5-heptenal	[*106-72-9*]
	2-phenylpropionaldehyde	[*93-53-8*]
	2-methyl-3-(4-isopropylphenyl)propionaldehyde	[*103-95-7*]
mushroom	1-octen-3-ol	[*3391-86-4*]
mustard	allyl isothiocyanate	[*57-06-7*]
peach	γ-undecalactone	[*104-67-6*]
	6-amyl-α-pyrone	[*27593-23-3*]
peanut	2,5-dimethylpyrazine	[*123-32-0*]
	2-methoxy-5-methylpyrazine	[*2882-22-6*]
pear	ethyl decane-*cis*-4-*trans*-2-dienoate	[*3025-30-7*]
peppermint	menthol	[*89-78-1*]
pineapple	allyl caproate	[*123-68-2*]
	methyl β-methylthiopropionate	[*13532-18-8*]
	ethyl butyrate	[*105-54-4*]
popcorn	methyl 2-pyridyl ketone	[*5910-89-4*]

Table 4 (*continued*)

Flavor character	Flavor materials	CAS Registry Number
potato	methional	[*3268-49-3*]
	2,3-dimethylpyrazine	[*5910-89-4*]
prune	benzyl-4-heptanone	[*7497-37-7*]
	dimethylbenzylcarbinyl isobutyrate	[*59354-71-1*]
raspberry	6-methyl-α-ionone	[*79-69-6*]
	trans-α-ionone	[*127-41-3*]
	p-hydroxyphenyl-2-butanone	[*5471-51-2*]
red currant	*trans*-2-hexenol	[*928-95-0*]
seafood	pyridine	[*110-86-1*]
	piperidine	[*110-89-4*]
	trimethylamine	[*75-50-3*]
smoke	guaicol	[*90-05-1*]
	2,6-dimethoxyphenol	[*91-10-1*]
spearmint	1-carvone	[*6485-40-1*]
tomato	isobutylthiazole	[*18640-74-9*]
	cis-4-heptenal	[*6728-31-0*]
vinegar	ethyl acetate	[*141-78-6*]
	acetic acid	[*64-19-7*]

[a] Character impact item is a material and/or combination of materials that, when tasted and/or smelled, is reminiscent of the named character, eg, vanillin is the flavor impact item for vanilla flavors.

Compounding

In the compounding technique constituents are selected or rejected because of (*1*) their odor, and (*2*) the results of flavor tests in water, syrup, milk, etc. After the compound is considered to be characteristic, it is tested as a flavor in the final product by an applications laboratory. Flavoring proportions used in high-fat foods must be greater than in low-fat foods because flavors are absorbed by fat thereby lowering flavor impact (57). A hard candy with a high glucose content relative to the cane sugar present requires either more flavoring or an increase in the more potent ingredients of the flavoring than one with a relatively low glucose content.

Specifications for many of the essential oils and artificial flavorings may be found in monographs compiled by the Essential Oil Association (EOA) of the U.S., and refs. 48, 52, and 58 (see also Fine chemicals). The EOA specifications include essential oils, natural flavor and fragrance materials, aromatic chemicals, isolates, general tests, spectra, suggested apparatus, and revisions adopted by the EOA. The purpose of these physical specifications is to encourage standardization and uniformity in basic flavor and perfume materials. Although compliance with specifications does not guarantee that the flavor quality standards will be acceptable, the specifications fill a need and provide a valuable reference for the flavor industry.

At present, the FDA lists 1161 flavorings and food adjuncts accepted, proposed, and under investigation (56) (for day-to-day status of these materials see the *Federal Register*). The Flavoring Extract Manufacturers' Association listed ca 1596 chemicals for food use which included nonflavoring agents as well; see also refs. 53 and 59. The Food Chemicals Codex lists ca 350 flavoring chemicals (58), and ca 300 additional compounds will be listed in its supplements.

The flavor chemist is responsible for the basic knowledge of sensory and application properties of each of this large number of raw materials. The tremendous number of possible combinations of these items to produce specifically flavored finished compounds is readily apparent. It is not uncommon to develop a flavor that combines essential oils, plant extractives, fruit juices, and synthetics. As mentioned previously, the choice of materials depends upon the type of product, conditions of manufacture, and intended use.

Although the terms synthetic, artificial, and chemical have aroused the doubts and suspicions of consumers in some instances (60), the fact that a flavoring has been synthesized should not be prejudicial to its proper use and ready application in foods since many such chemical components also occur naturally in foods (40,61) (see Food additives). To illustrate this point, a quote from a former Commissioner of the FDA, A. M. Schmidt, is offered: "Comments were submitted that the use of the term artificial misleads the public into believing that an artificial flavor is in some way inferior to a natural flavor. Comments pointed out that there is no available evidence to indicate any difference in safety or nutritional value between a naturally occurring flavor and its synthetic counterpart. The commissioner concurs with the comment that an artificial flavor is no less safe, no less nutritious, and not inherently less desirable than a natural flavor. The sole purpose for distinguishing between natural and artificial flavor is for economic reasons. In most instances, natural flavor is more expensive than the artificial flavor. Where a label creates an impression that a natural flavor is present, the consumer has the right to rely upon that implication or representation" (62). Since it would be economically impractical to isolate components from natural products in most instances, synthesis creates effective replacements.

The formulas in Table 5 demonstrate modern approaches to flavor compounds.

In Table 5(A), part (a) contributes basic sweet and lasting notes; (b) the spicy lift; (c) the citrus fruit note, which is commonly orange, but modified in this example by tangerine oil; (d) the fruity ester note, of which isoamyl acetate is the most familiar component. Ethyl acetate gives lift without affecting other flavor characteristics. It also serves as a plasticizer for the chewing-gum base. The proportions used depend upon the type of base used, whether chicle, latex, or a synthetic plastic.

Lemon with other natural flavors (Table 5(B)) has characterizing features that either simulate, resemble, or reinforce the lemon flavor (56). The title of Table 5(C) is similar to lemon, ie, apricot flavor with other natural flavors (a natural flavor with characterizing natural flavors added). In Table 5(C) the addition of the natural flavors are again either simulating, resembling, or reinforcing the apricot flavor (56,63–64).

Pineapple fortifier (Table 5(D)) is a mixture of several esters (eg, ethyl butyrate and ethyl acetate), carboxylic acids (eg, butyric, caproic, and acetic acids), and essential oils (eg, oil cognac and oil balsam Peru). Some of the esters do not occur in nature, ie, allyl caproate, although they produce pineapple aroma. The proportion of fortifier can be reduced in order to augment the natural fruit effects, but its reduction weakens the combined strength proportionally. Such a flavor as shown in Table 5(D) is suitable for use in cream centers, beverages, and frozen desserts.

Essential oil mixtures (see Table 6) are now commonly used to replace, in whole or in part, the dry spices from which they are derived. The latter now serve primarily to enhance the attractiveness or appearance of the products which are usually packed in glass containers. For example, a blend of 12.7 g of the essential oils of black pepper, ginger, allspice, cinnamon, nutmeg, and clove is equivalent to the flavoring strength of 1 kg of the dry seasoning blend.

Table 5. Modern Approaches to Flavor Compounds

	wt %
(A) Artificial Fruit Flavor for Chewing Gum	
(a) heliotropin	0.05
ethyl vanillin	1.00
vanillin	2.00
aldehyde C_{16}	0.55
(b) oil clove	1.00
oil peppermint	2.00
(c) oil orange	37.00
oil lemon	5.00
oil tangerine	5.00
(d) ethyl acetate	16.40
ethyl butyrate	15.00
isoamyl acetate	10.00
isoamyl butyrate	5.00
Total	*100.00*
(B) Lemon With Other Natural Flavors	
oil lemon	98.8
citral (isolated by distillation from oil of lemongrass)	1.0
orange aldehydes (isolated from orange oil by rectification)	0.2
Total	*100.0*
(C) Apricot With Other Natural Flavors	
apricot juice concentrate, 65 wt % sucrose (65 deg. Brix)	99.6
absolute carob	0.1
vanilla extract, tenfold	0.1
oil orange, tenfold	0.1
oil petitgrain	0.05
oil bitter almond	0.05
Total	*100.00*
(D) Fortified Artificial Pineapple Flavor	
pineapple juice concentrate	60.0
apple juice concentrate	23.0
tincture St. John's bread	0.5
malt syrup concentrate	0.2
vanilla extract, tenfold	0.1
citric acid	2.0
ethyl alcohol, 95%	14.0
pineapple fortifier	0.2
Total	*100.0*

Economic Aspects

Table 7 includes only those materials that are commonly used in the flavoring industry. Anethole, methyl salicylate, and ethyl butyrate are used mainly in flavorings, whereas ionones and heliotropin are used chiefly in perfumes. A recent report states that the flavor and flavor-enhancer market will almost double in volume by 1985, ie, from $498 million in 1976 to $869 million by 1985 (65).

A 1964 Census of the Food and Beverage Industry, issued by the publication *Food Industries*, has listed a total of 16,311 manufacturing plants in the United States, of

Table 6. Essential Oil, Herb and Spice Equivalents

Herb or spice	Essential oil equivalent to 100 kg of corresponding best quality, dry herb or spice, kg
allspice (pimental berries)	2.5
almond, bitter	0.5
angelica root	0.75
angelica seed	1
anise seed	2.5
basil, sweet reunion	0.15
calamus	2.5
caraway seed	2.5
cardamon seed	3
cassia cinnamon	1.5
celery seed	2
cinnamon Sri Lanka	0.5
clove	15
coriander seed	0.75
cumin seed	3
dill seed	4
dill weed	1
estragon	0.13
fennel seed	5
garlic	0.5
ginger	0.25
horseradish	1
laurel leaves (bay)	1
lovage root	0.5
mace	5
marjoram, sweet	0.5
mustard seed	0.25
nutmeg	5
onion	0.75
origanum	0.75
parsley seed	3
pepper, black	1
rosemary	0.5
sage	1.25
savory	0.25
thyme	2
valerian root	1

which 348 employ over 1000 workers. The plants engaged in the production of flavorings number 182. The number of employees probably engaged by the flavorings industry is ca 17,830. The production of foods and beverages has increased substantially to meet increased demands (see Food processing).

Flavor Precursors

The characteristic flavors of foods such as in fruits and vegetables are considered to result from enzymatic action upon certain more complex components during the normal developmental, or ripening, process (66). The formation of volatile flavor

Table 7. Flavor and Perfume Materials: U.S. Production and Sales, 1977

Flavor and perfume materials	CAS Registry No.	Production, metric tons	Sales: Quantity, metric tons	Sales: Value, thousands of dollars	Sales: Unit value[a], $/kg
Benzenoid and naphthalenoid					
4-allyl-2-methoxyphenol (eugenol)	*[97-53-0]*	135	126	1188	1.95
4-allyl-2-methoxyphenyl acetate	*[93-28-7]*	1.8	3.1	34	2.27
anisyl acetate	*[1331-83-5]*		1.8	27	2.86
benzophenone[b]	*[119-61-9]*	441	258	1112	0.88
benzyl acetate	*[140-11-4]*	699	633	1370	0.44
benzyl benzoate	*[120-51-4]*		329	643	0.40
benzyl propionate	*[122-63-4]*	12.7	11.7	45	0.77
cinnamaldehyde	*[104-55-2]*	911	553	1634	0.66
cinnamyl acetate	*[103-54-8]*	14.5	6.3	70	2.23
cinnamyl anthranilate	*[87-29-6]*		0.4	20	8.55
cinnamyl propionate	*[103-56-0]*		1.3	16	2.80
2-ethylhexyl salicylate	*[118-60-5]*		31	130	0.85
isobutyl phenylacetate	*[102-13-6]*		10.4	55	1.08
isobutyl salicylate	*[87-19-4]*		4.5	19	0.83
isopentyl salicylate	*[87-20-4]*	477	361	958	0.54
4′-methoxyacetophenone	*[100-06-1]*	9.9			
2-methoxy-4-propenyl-phenol (isoeugenol)	*[97-54-1]*	58	59	896	3.12
p-methylanisole	*[104-93-8]*	10.4	11.3	50	0.91
methyl anthranilate	*[134-20-3]*		101	401	0.81
α-methylcinnamaldehyde	*[101-39-3]*		3.1	16	1.06
methyl phenylacetate	*[101-41-7]*	12.7			
2-phenethyl phenylacetate	*[102-20-5]*	13.6	6.8	70	2.15
phenethyl propionate	*[122-70-3]*	9.9			
3-phenyl-1-propanol (hydrocinnamic alcohol)	*[122-97-4]*	24			
3-phenylpropyl acetate	*[122-72-5]*	3.1			
p-propenylanisole (anethole)	*[104-46-1]*	1099	1048	8603	1.69
p-toylaldehyde	*[104-87-0]*	12.7			
all other benzenoid and naphthalenoid materials			13,663	89,260	1.34
Total		*21,088*	*17,223*	*106,617*	*1.27*
Terpenoid, heterocyclic, and cyclic					
cedrol	*[77-53-2]*	15.8	13.6	196	2.93
cedryl acetate	*[77-54-3]*	145	78	728	1.92
dihydroterpinyl acetate	*[58985-18-5]*	62	35	144	0.83
guaiac wood acetate	*[134-28-1]*		20	199	2.03
α-ionone	*[127-41-3]*	33	26.8	494	3.81
β-ionone	*[14901-07-6]*	12.2	9.9	131	2.74
isobornyl propionate	*[2756-56-1]*		1.3	8	1.16
6-methyl-α-ionone	*[79-69-6]*	9.9			
6-methyl-β-ionone	*[79-70-9]*	310	201	2482	2.54
α-terpineol	*[98-55-5]*	1166	1085	1161	0.22
α-terpinyl acetate	*[80-26-2]*	1004	408	903	1.00
vetivenyl acetate	*[10486-25-6]*	16.7			
all other terpenoid, heterocyclic, and alicyclic materials		3186	2105	21,493	2.13
Total		*5425*	*4009*	*28,011*	*1.44*

Table 7 (*continued*)

Flavor and perfume materials	CAS Registry No.	Production, metric tons	Sales: Quantity, metric tons	Sales: Value, thousands of dollars	Sales: Unit value[a], $/kg
Acyclic					
allyl heptanoate	[*142-19-8*]	0.9	0.9	10	1.98
butyl butyryl lactate	[*7492-70-8*]	20	20	192	1.96
butyl 10-undecenoate	[*109-42-2*]	10.4			
citral dimethyl acetal	[*7549-37-3*]	17.6			
citronellyl acetate	[*150-84-5*]	24	12.7	103	1.65
citronellyl formate	[*105-85-1*]	9.5	8.1	89	2.29
citronellyl isobutyrate	[*97-89-2*]	3.1			
citronellyl propionate	[*141-14-0*]	2.2	0.9	12	2.22
3,7-dimethyl-*cis*-2,6-octadien-1-ol (nerol)	[*106-25-2*]		145	293	0.42
3,7-dimethyl-*trans*-2,6-octadien-1-ol (geraniol)	[*106-24-1*]		809	4553	1.16
3,7-dimethyl-*cis*-2,6-octadien-1-yl acetate (neryl acetate)	[*141-12-8*]	4.5			
3,7-dimethyl-1,6-octadien-3-ol (linalool, linalyl alcohol)	[*78-70-6*]	1485	1226	5139	0.86
3,7-dimethyl-1,6-octadien-3-yl acetate (linalyl acetate)	[*115-95-7*]	413	428	2546	1.22
3,7-dimethyl-6-octen-1-al (citronellal)	[*106-23-0*]	459			
3,7-dimethyl-6-octen-1-ol (citronellol)	[*106-22-9*]	1138	606	3690	1.25
ethyl heptanoate	[*106-30-9*]	2.2	2.7	17	1.32
ethyl hexanoate (ethyl caproate)	[*123-66-0*]	6.3	3.6	23	1.28
ethyl isovalerate	[*108-64-7*]	5.4			
ethyl myristate	[*124-06-1*]	9			
ethyl propionate	[*105-37-3*]		56	155	0.57
geranyl acetate	[*105-87-3*]		43	331	1.57
geranyl butyrate	[*106-29-6*]	2.2			
geranyl formate	[*105-86-2*]	8.1	53	88	2.24
geranyl propionate	[*105-90-2*]		0.4	7	3.33
2-hexanal	[*505-57-7*]	1.3	0.4	21	9.28
7-hydroxy-3,7-dimethyl-1-octanol (hydroxycitronellal)	[*107-75-7*]	767	283	3889	2.83
isopentyl butyrate	[*106-27-4*]	58	48	152	0.65
isopentyl formate	[*110-45-2*]	3.1	3.1	17	1.11
isopentyl isovalerate	[*659-70-1*]	11.3			
rhodinol	[*6812-78-8*]	5.9			
all other acyclic materials		37,247	23,854	51,146	0.44
Total		*41,715*	*27,559*	*72,473*	*0.54*
Cyclic		26,513	21,232	134,628	1.31
Grand total		*68,229*	*48,791*	*207,101*	*0.87*

[a] Calculated from the unrounded figures.
[b] Includes significant quantities having other uses.

components has been studied *in vitro* (67–69). Fruits and vegetables were deflavored by heat, usually by boiling, and their flavor was partially restored by the addition of an enzyme. Although many different enzymes develop from a precursor, only certain ones produce the characteristic flavor (67–68). Thus a true raspberry flavor is produced in a deflavored raspberry puree only by enzymes obtained from raspberry fruit (70). A vegetable product contains a number of different enzymes in varying proportions that cause the formation of volatile components in ratios different from the actual ratio of precursors present in that particular product.

The complexity of the total reactions in nature has hindered commercial exploitation of the type that would enable a normally prepared fruit preserve, for example, to regain the flavor lost in processing (71).

Enzymes not only produce characteristic and desirable flavor (72) but also cause flavor deterioration (73–74) (see Enzymes, industrial). The latter enzyme types must be inactivated in order to stabilize and preserve a food. Freezing depresses enzymatic action. A more complete elimination of enzymatic action is accomplished by pasteurization.

The creation of flavor by enzymes is utilized in the fermentation process which is used to prepare products such as alcoholic beverages, cheese, pickles, and sauerkraut (see Fermentation). A more complex flavor development occurs in the production of chocolate. The chocolate beans are first fermented to develop less complex flavor precursors that, upon roasting, give the chocolate aroma. The beans from unfermented cocoa do not develop the chocolate notes (75–79) (see Chocolate and cocoa).

The use of dry heat as in roasting (80–81), baking, and frying, develops flavor characteristics not found in the unheated product (82). The roasting of meats (44,83), coffee (84), peanuts (85), and sesame seeds (86), the baking of bread (87–88), the baking (89) or frying of potatoes (90), are methods by which desirable flavor is created. There is ample evidence to prove that the distinctive flavor development in breakfast foods, the crust of baked bread, the aroma of roasted coffee, etc, can be directly attributed to the chemical combinations brought about during the heat-treatment operation. The reactions of sugars and amino acids or proteins have been shown to be accomplished by a loss of free amino groups and increased acidity (91). Glucose combines with α-aminobutyric acid to give a product with a maple flavor, and glucose and/or fructose react with proline to produce a bread-like flavor (92). Many amino acids generate a typical aldehyde odor when heated with glucose, probably due to the Strecker degradation reaction resulting from the decarboxylation and deamination of the amino acid (93–94). Products such as dried skim milk, dried eggs, and dehydrated vegetables and fruits are particularly susceptible to deteriorative flavor changes ascribed to this reaction.

Flavor Regulations

The Pure Food and Drug Act of 1906 introduced federal regulations "to combat the atrocities of food and drug alteration and fraud" (95). The law was superseded by the Food, Drug, and Cosmetic Act of 1938 and the Amendments of 1954, 1958, and 1962 (96) (see Food additives).

Sensory Evaluation

For the evaluation of flavor materials it is evident, since the type of foods and its processing affect flavoring efficiency, that they must be taste-tested in the food itself. Because there has been a lack of standardization of testing techniques, a committee on sensory evaluation of the Institute of Food Technologists has offered a guide (97) which is designed to assist in developing standard procedures.

For each type of problem, appropriate taste tests are suggested, together with the type of panel, number of samples per test, and analysis of data. This useful summary of a field in which a great amount and variety of experimentation has occurred eliminates a large amount of data which is difficult to interpret. It contains a list of original references. Specific tests are outlined for new product development, product and process improvement, cost reduction, selection of a new source of supply, quality or rating selection of the best sample, market testing of a new or improved product, consumer preference tests, and a selection of trained panelists (98–101). Ten types of tests are explained (102–103), and certain ones are designed for trained, untrained, semitrained, and highly-trained panels. The methods of sample presentation include single-sample, paired-comparison, and triangle test (in which three samples are employed, two of them alike). Other methods are fitted to the attainment of specific objectives, and for these the type of panel and number of panelists are also suggested. Flavor evaluation is adequate only when the origin of flavor is known and differences can be measured by a scientific method based upon units or significant qualities of flavor.

The order of sample presentation has been shown to lead to errors in judgment (104). For example, the later samples in a series were rated lower; serving good samples first lowered the rating of poor samples and serving poor samples first lowered the rating of good samples. The significance of this finding is premised by the existence of an ability to determine what is a good flavor. One can determine what is an acceptable flavor; this decision rests with the consumer of the food product. The literature on sensory testing includes refs. 105–108 (see Flavor characterization).

SPICES

A spice may be any of various aromatic plant materials used to flavor foods, or a variety of dried plant products (sometimes used in the form of extracts) that exhibit a pronounced aroma and flavor, but these descriptions are incomplete. Spices are derived from vegetable substances that have flavor and aroma from which no volatile or flavoring principles have been removed.

Spices (109) can be divided into: (*1*) the so-called true or tropical spices, such as black and white pepper, cloves, cinnamon, ginger, nutmeg, and mace; these may be buds, fruit, bark, roots, or other parts of tropical plants; (*2*) herbs, such as sage, rosemary, marjoram, and oregano; these are usually the leafy parts of plants that grow in the temperate zones; (*3*) spice seeds, such as mustard, celery, caraway, dill, fennel, and anise; these grow in both tropical and temperate zones, and (*4*) dehydrated aromatic vegetables, such as onion, garlic, parsley, leek, and sweet pepper, many of which are now produced in the United States.

Condiments are sometimes classified as spices, although they are usually employed in a different manner. Spices are used during the cooking or processing of food, condiments are added to food after it has been served at table (chutney, piccalilli, catsup, and salad dressings might be mentioned in this category). Black pepper may be considered a spice or a condiment, depending on its use.

The essential oils and oleoresin derivatives of spices have already been mentioned. Other derivatives are used, but since extensive definitions would be required, the types are mentioned and references offered: absolutes, concretes, essences, extracts, tinctures, resinoids, and others (110).

History of Spices

The notion that foul odors are evil and pleasant odors are good has persisted since the beginning of recorded history. Chinese myths and legends report that Shen Nung "divine cultivator" founded Chinese medicine and discovered the curative powers of more than 100 plants having medicinal properties about 2700 BC. China was the earliest spice trader. Records of the use of cassia and ginger were written in the 5th century BC by Confucius.

References to spices in the Old Testament include a description of the land of Hevila, in the book of Genesis, through which flowed a branch of the river from the Garden of Eden and in which there was bdellium, an aromatic exudate from a shrub-like tree and common fragrant resins like myrrh and olibanum.

In Egypt a medical scroll of ca 1515 BC, described almost 800 drugs, herbs, and spices. Spices such as cinnamon and cassia were available only from China and other parts of Asia. Shortly after 1000 BC, the Queen of Sheba visited King Solomon with a gift of spices. Some of these spices were obtained on the "golden road to Samarkand," from the Orient to Babylon and Alexandria, through China and the Moluccas (the original Spice Islands, now part of Indonesia) to the Red Sea. China was the earliest spice trader but actual trading with the Middle East and Europe was done through the Arabs in the large cities along the route from about 900 BC until 1500 AD, when the Portuguese seized command.

In the 6th century BC Greek settlements and trading posts were established under Alexander the Great. The trading posts encompassed the spice trading routes to the East.

The Romans expanded the use of such items as pepper, clove, cinnamon, and nutmeg for the flavoring of food and wine.

At the end of the 13th century Marco Polo's expedition returned to Venice after traveling for 25 years in the Orient. Their stories and descriptions led to the opening of direct trade between Europe and the East.

In 1418 Prince Henry the Navigator, of Portugal, explored the west coast of Africa and discovered a pepper-like spice called grains of paradise, actually melaluca, related to ginger. In 1498 Vasco da Gama seized the Spice Islands and gained control of the eastern spice trade.

Christopher Columbus returned to Spain from the New World with red pepper and pimenta (allspice). In the 16th century the western spice centers were shifted from Italy and Alexandria to Portugal and Spain, largely because of trade by way of the Cape of Good Hope. In the 1620s the Dutch drove the Portuguese out of the Spice Islands. Spain was removed from competitive commercial shipping with the destruction of the Spanish Armada. The Dutch were left to regulate the spice trade.

In 1770 the French had plantations of nutmeg, clove, and cinnamon on many of the islands off the east coast of Africa.

By the 19th century the center of spice trade moved to London with British control of India, Ceylon, Singapore, and the Malay mainland.

By the 1820s New England privateers gave Salem, Massachusetts, at that time the sixth largest city in the United States, a virtual monopoly in the pepper trade with Sumatra.

A comprehensive history of spices is found in refs. 111–113.

Production and Economic Aspects

The quantities and monetary value of spices imported into the United States in 1964 and 1978 are shown in Table 8. The consumption of spices in the United States has continued to increase. The value of U.S. imports of specified condiments, seasonings, and flavoring materials in 1978 totaled a record $197.5 million, nearly 8% above 1977 imports of $183.5 million and 31% above the 1976 imports of $151.1 million. The

Table 8. U.S. Importation of Spices, 1964 and 1978

	1964		1978	
	Metric tons	10^3 $	Metric tons	10^3 $
allspice (pimenta)	514	780	830	1,301
anise seed	196	85	489	636
basil			479	447
capsicum (red pepper)	5,223	2,292	4,394	5,266
caraway seed	3,261	1,058	3,089	4,137
cardamom seed	83	272	169	1,531
cassia	6,161	2,861	7,711	5,538
celery seed	955	347	2,160	1,824
cinnamon	573	683	886	1,540
cloves	1,051	770	1,145	6,609
coriander seed	1,403	202	4,279	2,262
cumin seed	1,702	862	3,339	5,803
dill seed			557	462
fennel seed	270	85	906	703
ginger	1,795	1,083	3,556	4,703
laurel (bay) leaves	343	111	393	497
mace	295	534	253	681
marjoram			380	351
mustard seed	10,103	1,285	32,040	9,835
nutmegs	1,593	2,139	2,110	4,066
origanum leaves	773	363	2,385	3,741
paprika	4,576	3,435	5,005	7,569
parsley			132	181
pepper black	19,843	11,901	26,142	53,053
pepper white	1,735	1,397	2,382	6,522
rosemary			259	147
sage	650	319	1,309	2,168
savory			73	104
sesame seed	11,280	3,602	32,000	27,256
tarragon			22	127
thyme	404	152	603	785
turmeric	1,623	668	1,839	1,986
vanilla beans			1,185	25,281
Total			*142,501*	*187,112*

volume of imports was 149,510 t in 1978, 140,510 t in 1977, and 149,640 t in 1976. The increased import value of 1978 was attributed mostly to higher shipments of black pepper and sesame seed.

Vanilla bean prices soared during 1978 because of another poor crop in Madagascar. In New York, spot prices for bourbon vanilla were quoted at about \$62–66/kg, up from the earlier 1978 level of \$23/kg. Black pepper prices, however, were on the decline during 1978 for the first time in several years owing to an improved supply situation.

The United States continues as the world's largest importer of spices, accounting for approximately one fourth of the total produced. Imports supply ca 80% of U.S. consumption needs, the remainder coming from domestically produced spices such as paprika and other capsicum peppers, basil, mint, parsley, thyme, mustard seed, dill, fennel, sage, sesame, tarragon, marjoram, and fenugreek seed. U.S. spice sales at the retail level in 1975 were placed at about \$450 million. The port of New York received ca 75% of the spices, seeds, and herbs imported into the United States (114).

U.S. spice exports in 1978 totaled 11,360 t valued at \$20.3 million, compared with 1977 shipments of 5470 t valued at \$14.7 million. the U.S. export classification system was changed in 1978, thus a direct comparison for individual spices for the past two years is not possible. In addition, since January 1, 1979, import statistics for spice oleoresins have been available in a separate classification (see Oils, essential).

Recent Developments

The U.S. spice market can be divided into three sectors based on areas of applications, ie, industrial (food processing), retail and institutional (restaurants, hospitals, schools, military, etc).

Particularly in the industrial sector, interest continues in expanding the use of oleoresins and oils of the spices. The oleoresins and essential oils are made by solvent extraction or steam distillation of the spice. They have certain advantages over the ground spices for large volume use, such as consistency of quality, freedom from microorganisms, uniform dispersion throughout the product, easy handling and storage. The oleoresins consist of natural resins, any volatile or essential oil present, and other valuable extractives. Solvent selection can determine the natural components to be retained, the cellulosic material, tannins, minerals, starches, and sugars are usually discarded, and the flavor, odor, pungency, or color (if desired) of the spice is concentrated in the finished oleoresin.

The amount of oleoresin or essential oil varies according to the particular spice: from 0.1% in oil of garlic to 20% in oleoresin clove. These items can be used according to the yield as a replacement for the dried spice (115). In many cases, natural spices are preferred, eg, where the end product requires a natural filler, or the spice appeals to the eye, as nutmeg on an egg nog, cinnamon on a bun, or paprika on a salad.

In recent years, essential oils and oleoresins have been produced in the growing areas throughout the world. Fewer and fewer oils and oleoresins are being manufactured domestically from imported spices. Shipping costs are higher for bulk-dried spice than for the concentrated oil or oleoresin.

In the retail and institutional sectors, the consumption of highly spiced dishes has continued to increase over the last decade. The "nationality foods" are available more and more as convenience items. Dietary foods that are generally bland require the use of spice and flavors to make them more attractive.

Spices and certain oleoresins, eg, sage and rosemary, are antioxidants that prevent rancidity (oxidation) in fats and fatty foods. Analytical control has standardized spices and extracts for greater uniformity of the type and origin of the spice.

Some spice characters can be synthesized. Cinnamaldehyde, the main component of cinnamon (70–80%), can be manufactured synthetically for use in chewing gum, mouthwash, and toothpaste. Cinammon of Sri Lanka costs $2.31/kg (avg 1977 price) and contains 0.3% (max) oil. Therefore, the oil that contains 80% cinnamaldehyde costs in excess of $770/kg. Synthetic cinnamaldehyde 99.9% costs $2.75/kg.

Another prime example is *l*-carvone [*6485-40-1*] in spearmint oil at 80% concentration which can be manufactured from a by-product of the citrus industry, *d*-limonene. *l*-Carvone can be used in flavors where spearmint flavor is required. Vanillin (qv) is also a synthetic by-product of the wood-pulp industry. Lignin, the binder for wood fibers, is composed of vanillin precursors. Artificial vanilla extract using synthetic vanillin is a common household flavor.

In 1977, ca 2000 t of cinnamon spice of all types was produced, valued at $2,650,000. More than 450 t of cinnamaldehyde, the vital cinnamon flavoring principle, was manufactured in the United States. That much synthetic flavor would require an equivalent production of more than 22,725 t of cinnamon spice.

Specifications, Analysis and Quality

Since spices are natural products they exhibit a range of variations of their many characteristics. Climatic conditions can change flavor, quality, strength, and appearance. The qualities of the spice are influenced by soil conditions, rainfall, storms, blights, insects, growing and harvesting methods, and storage.

The USDA considers most spices GRAS (generally recognized as safe) and there are no standards of identity or legal definitions of spices (116). Spices used in drugs must meet the official standards of the *U.S. Pharmacopeia* (48). Advisory specifications may also be applied in commercial spice trading. The abbreviation SRA, occasionally used in conjunction with the spice, means Service and Regulatory Announcements promulgated under the old Food and Drug Act of 1906.

Another notation sometimes observed in descriptions of a particular spice is FAQ which means fair average quality. This term has meaning to experts who can evaluate spices by size, appearance and odor, or taste. Individual spices have specific grading descriptions; eg, cinnamon "quills" or "pipes," is graded 000, 0000, or 00000. The latter, five zeros, is prime grade of finest, smoothest, pale brown, and least broken quality.

In the second edition of the *Official Analytical Methods of the American Spice Trade Association* (117), twelve general methods for evaluating spices are described, eg, moisture, ash, and volatile oil. These are generally accepted methods but purchasers have established specifications for acceptance of particular lots, or even grades, by such determinations as amount of essential or volatile oil and a gas chromatographic evaluation of an oil experimentally produced.

The FDA has methods for the determination of residual solvents in spice extracts, such as oleoresins, and has limited the concentrations of specific solvents permitted (118). Solvents such as benzene, the chlorinated hydrocarbons, and other toxic solvents have been removed entirely from use.

The presence in spices of pesticides or herbicides and the like are rigidly controlled by the FDA (see Trace and residue analysis). The ASTA also has 22 procedures for the detection and enumeration of microorganisms in spices (118).

In order to answer questions raised about the nutritional value of spices, the Research Committee of ASTA has made available the nutritional values shown in Table 9 which indicates that spices *per se* are not high in nutrients.

Purchasing and Processing

Spices arrive in the United States usually in whole form (dried roots, pieces of roots, bark, buds, fruits, etc) cleaned, but not further processed. Control is strict, particularly with regard to adulteration and cleanliness. Lots containing impurities, dirt, insects, rodent droppings, etc, are rejected. If contamination is not too serious, the lot may have to be reworked and cleaned; for example, mold can be removed in some cases by passing the spice through brushing machines. If the lot on visual examination shows slight insect contamination, it can perhaps be cleaned by treatment for 24 h with methyl bromide or ethylene oxide (usually as a mixture with CO_2). This may also cause a reduction of the bacteria count, by as much as 90%; rejection depends upon the degree of contamination and the type of spice.

After release from the Customs House, the spice usually is sent to a spice-packaging house for grinding, or to an essential oil house for processing (by distillation or extraction). If the spice is to be stored for any length of time prior to usage, it should be kept in whole form.

Grinding or milling reduces the spice to a particle size desirable for a specific processing operation. It is accomplished in various types of comminuters, such as pulverizers, roller mills, choppers, and hammer mills, each type adapted to a special spice and a special purpose. The ground spice is finally sifted through silk or woven wire sieves, the degree of fineness expressed by a number indicating the number of holes per linear inch (2.54 × no. holes per centimeter), eg, Ginger 70, Black Pepper 40, and Allspice 50.

By grinding, the surface area of the spice is increased enormously and most of the oil glands or sacs are ruptured and exposed to the air. Obviously there is evaporation of the volatile essential oil, and in some cases loss of aroma by oxidation and resinification. Ground spices, therefore, do not retain their characteristic aroma. All spices, whether whole or ground, should be stored with certain precautions. The temperature should not exceed 17°C, and the humidity 65%. A dark, cool cellar is preferable, as direct sunlight may cause loss of color in the spice. Cleanliness is essential to prevent insect infestation.

Use of Spices

Only a few general indications on the use of spices in culinary preparations are given here. The essential oils carry the aroma of the spice (or aromatic plant) in concentrated form (see Oils, essential). The essential oils are volatile and do not contain the very high boiling or nonvolatile constituents that are responsible for the characteristic taste of certain spices. For example, piperine and zingerone, the biting principles of black pepper and ginger root, respectively, are completely lacking in the essential oils of these two spices. To isolate these compounds the spice has to be extracted

Table 9. Approximate Composition of Ground Spices, 100 g, Edible Portion[a]

Spice	Water, g	Food energy, kJ[b]	Protein, g	Fat, g	Total carbohydrate, g	Fiber carbohydrate, g	Ash, g	Calcium, g	Phosphorus, g	Sodium, g	Potassium, g	Iron, mg	Thiamine, mg	Riboflavin, mg	Niacin, mg	Ascorbic acid, mg	Vitamin A value, Int'l units
allspice	8.8	1589.92	6.0	6.6	74.4	21.6	4.2	0.8	0.11	0.08	1.1	7.5	0.10	0.06	2.9	39.2	1445
basil leaves	6.1	1359.80	11.9	3.6	61.7	20.5	16.7	2.1	0.47	0.04	3.7	42.8	0.15	0.32	6.9	61.3	290
bay leaves	4.5	1715.44	7.6	8.8	75.4	25.2	3.7	1.0	0.11	0.02	0.6	53.3	0.10	0.42	2.0	46.6	545
caraway seed	6.4	1945.56	21.4	23.1	43.5	10.4	5.6	0.7	0.50	0.02	1.9	8.5	0.38	0.38	8.1	12.0	580
cardamom seed	8.4	1506.24	9.8	2.9	74.2	9.9	4.7	0.3	0.21	0.01	1.2	11.6	0.18	0.23	2.3	12.0	175
celery seed	5.1	1882.80	18.1	22.8	43.8	12.9	10.2	1.8	0.55	0.17	1.4	44.9	0.41	0.49	4.4	17.2	650
cinnamon	9.9	1485.32	4.6	2.2	79.8	20.3	3.5	1.6	0.05	0.01	0.4	4.1	0.14	0.21	1.9	39.8	175
cloves	5.4	1799.12	6.3	14.5	68.8	11.1	5.0	0.7	0.11	0.25	1.2	9.5	0.11	0.04	1.5	80.9	175
coriander seed	6.3	1882.80	12.3	19.6	56.5	31.5	5.3	0.8	0.44	0.02	1.2	5.9	0.26	0.23	3.2	12.0	175
cumin seed	6.2	1924.64	17.7	23.8	44.6	9.1	7.7	0.9	0.45	0.16	2.1	47.8	0.73	0.38	2.5	17.2	175
dill seed	6.6	1820.04	13.1	17.9	56.4	20.7	6.0	1.6	0.21	0.01	1.1	11.8	0.42	0.28	2.8	12.0	175
fennel seed	6.3	1548.08	9.5	10.0	60.8	18.5	13.4	1.3	0.48	0.09	1.7	11.1	0.41	0.36	6.0	12.0	1040
garlic powder	5.4	1527.16	17.5	0.6	73.3	1.9	3.2	0.1	0.42	0.01	1.1	3.5	0.68	0.08	0.7	12.0	175
ginger	6.9	1589.92	8.6	6.4	72.4	5.9	5.7	0.1	0.15	0.03	1.4	11.3	0.05	0.13	1.9	12.0	175
mace	4.7	2363.95	8.1	38.8	46.1	4.8	2.3	0.2	0.11	0.07	0.5	11.3	0.37	0.56	1.2	12.0	175
marjoram	6.6	1527.16	12.5	6.8	64.4	16.7	9.7	2.5	0.23	0.11	1.4	72.7	0.29	0.32	4.1	51.5	550
mustard powder	3.2	2426.72	31.7	42.6	18.5	1.9	4.0	0.3	0.79	0.01	0.7	8.3	0.65	0.45	8.5	22.1	195
nutmeg	4.4	1363.95	7.4	38.9	47.3	3.1	2.0	0.2	0.20	0.01	0.4	2.2	0.36	0.25	9.4	12.0	175
onion powder	4.6	2548.08	10.6	0.8	80.5	6.4	3.5	0.3	0.29	0.04	1.0	2.2	0.42	0.06	0.6	14.7	175
oregano	8.0	1506.24	11.7	6.4	64.9	11.0	9.0	1.7	0.20	0.02	1.7	53.3	0.34	0.04	6.2	12.0	1010
paprika, imported	6.4	1610.84	15.8	10.1	58.0	19.4	9.7	0.2	0.27	0.02	2.4	23.1	0.40	1.17	15.4	48.6	3350
paprika, domestic	7.9	1631.76	13.8	10.4	60.3	19.2	7.6	0.2	0.30	0.02	2.4	23.1	0.60	1.36	15.3	58.8	4915
parsley flakes	4.1	1485.32	21.9	5.6	54.3	8.7	14.1	1.2	0.31	0.54	3.6	14.5	0.17	1.23	7.9	392.0	1090
pepper, black	8.3	1673.60	10.4	10.2	66.5	10.0	4.6	0.4	0.16	0.01	1.2	17.0	0.07	0.21	0.8	12.0	175
pepper, chili	6.5	1736.36	14.0	14.1	58.2	15.6	7.2	0.1	0.32	0.01	2.1	9.9	0.59	1.66	14.2	63.7	6165
pepper, red	6.2	1757.28	16.0	15.5	54.3	26.0	8.0	0.1	0.32	0.01	2.1	9.9	0.52	0.93	13.6	29.4	3530
pepper, white	9.6	1652.68	12.4	8.0	69.0	4.5	1.0	0.2	0.15	0.01	0.1	6.9	0.02	0.13	0.2	12.0	175
poppy seed	3.7	2217.52	23.1	35.5	30.2	8.7	7.5	1.6	0.88	0.01	0.8	9.9	0.69	0.18	0.9	12.0	175
rosemary leaves	5.7	1840.96	4.5	17.4	66.4	19.0	6.0	1.5	0.07	0.04	1.0	33.0	0.51	0.04	1.0	61.3	175
sage	5.7	1736.36	10.2	14.1	62.3	16.0	7.7	1.8	0.09	0.01	1.0	27.3	0.75	0.34	5.7	39.8	2395
savory	9.1	1485.32	7.1	5.2	69.9	15.3	8.7	2.2	0.14	0.02	1.1	37.8	0.37	0.04	4.1	12.0	865
sesame seed	3.2	2008.32	23.7	22.4	47.1	37.7	3.6	0.1	0.79	0.03	0.4	6.2	0.80	0.13	5.7	12.0	345
tarragon	4.7	1527.16	24.2	7.3	51.5	6.8	12.3	1.3	0.31	0.07	3.2	35.7	0.25	1.34	8.9	12.0	175
thyme	7.1	1422.56	6.8	4.6	68.3	24.3	13.2	2.1	0.20	0.08	0.9	135.0	0.51	0.40	4.9	12.0	175
turmeric	5.8	1631.76	8.6	8.9	69.9	6.9	6.8	0.2	0.26	0.01	2.5	47.5	0.09	0.19	4.8	49.8	175

[a] Analyses performed by Research 900 Laboratories, St. Louis, Mo.
[b] To convert kJ to kcal (cal for food energy, only), divide by 4.184.

with a volatile solvent (ethanol, acetone, hexane, etc); on removal of the solvent by careful evaporation of the solution *in vacuo,* a viscous, semisolid, or sometimes solid residue is obtained (oleoresin). These products have lately achieved great importance in the food industry; they offer many advantages over straight spices, eg, cleanliness and uniformity of flavor from lot to lot. This is especially important when foods are produced by an automated process; the use of oleoresins permits better distribution of flavor and a standardization of the flavor level that is difficult to obtain with straight spices. Moreover, oleoresins, when properly prepared, are almost free of bacteria, molds, spores, etc, which is a major advantage over crude spices and aromatic herbs, especially in prepared foods. Because of their high concentration, a few kilograms of oleoresins (or in many cases of essential oils, where the nonvolatile portion is of minimal significance to the total flavor) can replace many kilograms of natural spices. Oleoresins (see also Resins, natural) are often added to the food product in a diluted form; the most common method is distribution on a dry edible carrier, such as table salt or dextrose. These products, commonly known as dry solubles, are usually equivalent to the corresponding natural spice.

Individual Spices

Allspice (Pimenta). The dried, nearly ripe berry of the tree *Pimenta dioica* L. (formerly called *Pimenta officinalis* Lindley) (family *Myrtaceae*) is a tropical evergreen growing semiwild in Jamaica and other Latin American countries. It should not be confused with pimento or pimiento, a variety of red pepper. Pimenta is the only major commercial spice exclusive to the Western Hemisphere. Pimenta berries resemble black pepper corns (*Piper nigrum*); they are round, up to 0.65 cm dia. The berries are hand-picked while green, then slightly fermented and sun-dried, turning them reddish-brown. The aroma of pimenta berries resembles that of cloves, black pepper, cinnamon, and nutmeg, hence the term allspice or quatre épices. Pimenta is used to flavor vegetables, fruits, pickles, and spicy table sauces.

Anise Seed. The dried, ripe fruit of *Pimpinella anisum* L. (family *Umbelliferae*), is a member of the parsley family indigenous to Asia Minor, Greece, and Egypt. The bulk of supply comes from Spain, Turkey, and Egypt, but can be cultivated in most temperate climates. Anise seed resembles caraway seed, but has a sweet, licorice flavor very popular in Mediterranean and Latin countries, particularly in beverages and confectionary.

The star anise (*Illicium verum* Hook f.) is from a small evergreen tree native to Southwestern China. When ripe, the hard brown fruits of this tree open up into an eight-pointed star, hence the name. The flavor and aroma of this spice is very similar to that of *Pimpinella anisum*; the essential oils are of similar composition.

Basil (Sweet Basil). Basil, the brown dried leaves and tender stems of *Ocimum basilicum* L. (fam. *Labiatae*), is an annual, native to India, Africa, and Asia, and cultivated in southern France, the Mediterranean countries, and in the United States as well. Basil is one of the oldest herbs known. The flavor is warm, sweet, somewhat pungent, and peculiar (methyl chavicol and linalool). It is used with meats, fish, and certain cheeses, and particularly with tomato-based recipes. It is a main component of the liqueur Chartreuse.

Bay (Laurel) Leaves. Bay leaves, the dried leaves of *Laurus nobilis* L. (fam. *Lauraceae*), is also called sweet bay or laurel tree, an evergreen with shiny green leaves up to 7.6 cm length. It is not to be confused with the bay rum tree (*Pimenta racimosa* Mill (from *Myrtacea*)) from Puerto Rico and neighboring islands, or California bay laurel (*Umbellularia californica* Nutt. (fam. *Lauraceae*)). *Laurus nobilis* has been cultivated since antiquity in Mediterranean countries, and is now grown extensively in Turkey, Yugoslavia, France, and Central America. The odor is delicate, but distinctly aromatic, the flavor slightly bitter and burning. It is used extensively for meats, sauces, and fish dishes, for pickling spice, bouillabaisse, and stews.

Benne (Also Benni or Bene). See Sesame seed.

Capsicum. Several important condiments are derived from the dozens of spices and varieties of the genus *Capsicum* (fam. *Solanaceae*) (nightshade family), particularly from *C. annum* L., such as paprika, red pepper, and cayenne pepper. These plants are indigenous to Mexico, Central America, the West Indies and much of South America. Capsicums are very sensitive to climatic and soil conditions, and many horticultural varieties have been developed, differing in size, shape, taste, pungency, and color. In general, capsicum can be divided into two main groups, the sweet and the pungent.

(*1*) Paprika, also called sweet pepper or pimento (in Spanish, pimiento), is the ground dried ripe fruit (stemless pod) of chiefly *Capsicum annum* L. The varieties of this annual herb, grown mainly for color, not pungency, are mild and slightly sweet; the odor is pleasant and somewhat aromatic. This herb is grown mainly in the United States, Spain, Hungary, Bulgaria, and Morocco. It is a rich source of vitamin C and serves as a garnish and flavorant in many foods, such as meats, soups, poultry, salad dressing, etc, and is an important constituent of catsup and chili sauces.

(*2*) Red pepper, cayenne pepper, and chillies belong to the pungent group of capsicums derived from varieties of *C. annum* L. and *C. frutescens* L. They are usually smaller and exhibit a very hot, pungent flavor. They are cultivated mainly in the tropical and subtropical latitudes (southern United States, Mexico, China, Pakistan and India, Africa, and Japan). They occur in dozens of types (African chilies, Birdseye, tabasco, Jalapeño pepper, Louisiana Sports, Japanese Santaka, etc), that vary in degree of pungency and color. Often, ground peppers are blended to arrive at a required strength of flavor, pungency, and color. Red pepper serves in foods where pungency is essential, eg, in barbecue and spicy table sauces, chili powder, hot curries, and also in certain medicinal preparations.

Caraway Seed. The dried ripe fruit of *Carum carvi* L. (fam. *Umbelliferae*), is a biennial plant, cultivated extensively in the Netherlands and now also in Poland, Egypt, and North Africa. The seed is brown and hard, ca 0.48 cm long, curved, and tapered at the ends. It is perhaps the oldest condiment cultivated in Europe. The odor is pleasant and the flavor is aromatic, warm, and somewhat sharp (carvone). Caraway is used in dark bread, potatoes, sauerkraut, kuemmel liqueurs, cheese, applesauce, and cookies.

Cardamom Seed. The dried ripe seed of *Elettaria cardamomum* M. (fam. *Zingiberaceae*) is a large perennial herb, indigenous to and cultivated in southeastern India, Sri Lanka, and now cultivated mainly in Guatemala. The seeds are removed from the capsules (pods) and, when dried, are irregularly round and ca 0.23 cm long. The aroma and flavor are pleasant, aromatic and somewhat camphoraceous. It is a tropical spice, known since the 4th Century BC. Throughout Arab countries, cardamom is the most popular spice, where it is used in a coffee-like beverage, a symbol of hospitality. The spice is also used in baking pastries and pies in Nordic countries, as well as in ground meats. It is an ingredient of curry powder and pickle flavor.

Cayenne. See under Capsicum.

Celery Seed. The dried, ripe fruit of *Apium graveolens* L. (fam. *Umbelliferae*), is a biennial, sometimes an annual herb, native to southern Europe and grown extensively in France, India, China, and the United States. The seed is 0.42 cm long, and brown. The odor is characteristic and warm, the taste somewhat bitter. It is used in tomato catsup, sauces, soups, pickles, pastries, salads, and certain cheeses.

Chili Powder. Chili powder is a commercial blend of several spices, ie, chili peppers, oregano, cumin seed, onion and garlic powder, allspice, perhaps cloves, and others. Chili powder is the basic flavor for many highly spiced dishes, among them chili con carne, and also in cocktail sauces, ground meats, etc.

Chilies. See under Capsicum.

Cinnamon. There are two distinct types of cinnamon, each with its own varieties:

(*1*) *Cinnamomum cassia* Blume (fam. *Lauraceae*), the so-called cassia. This spice is native to southeastern China and is the dried bark of this evergreen tree, and, from this source, true cassia essential oil is produced. Another name for this variety is cassia lignea. Other varieties are called cassia and are identified as (*a*) Saigon cinnamon, C. *loureirii*, Nees; (*b*) from Sumatra in Indonesia (Korintje or Kerintji cinnamon, and Padang or Batavia cinnamon, C. *burmani* Blume); (*c*) C. *sintok* Blume of minor commercial importance and native to Malaysia. *The National Formulary XIV* (49) classified cinnamon as the dried bark of *Cinnamomum loureirii* Nees (fam. *Lauraceae*). The spice from Saigon and Indonesia arrives in the United States in the form of sticks or quills, which is the same form as true cinnamon from Sri Lanka (Ceylon). Saigon "thin quill" is the preferred quality of these spices since it has the highest essential-oil content, and next in preference is cassia lignea. Cassia oil is made from the leaves and twigs of the lignea type rather than the bark from which the ground spice is produced.

(*2*) The dried inner bark of shoots of *Cinnamonium zeylanicum* Nees, a moderately sized coppiced evergreen bush of the laurel family, cultivated in Sri Lanka. The bark is stripped and rolled into quills, and shipped in large bundles of quills. The aroma and flavor of zeylanicum is much milder than the cassia type.

Owing to the more pungent flavor, grinders in the United States generally prefer the product from Saigon to the Sri Lanka cinnamon. The spice varieties are used widely for the flavoring of baked goods and confectionary, in pickling, curry powders, and spicy table sauces.

Cloves. Cloves, the dried, unopened flower buds of an evergreen tree, *Eugenia caryophyllus* Thumb (fam. *Myrtaceae*), is also called *Syzyguim aromaticum* L. Other botanical names are used, but some discrepancies exist in the proper nomenclature. This tree is indigenous to the Molucca Islands, but in 1770 was smuggled out to the islands of Mauritius and Reunion, eventually being cultivated on extremely large plantings in Tanzania (Zanzibar and Pemba Islands) and Malagasy (Madagascar). The term clove is derived from the French clou (nail), the dried clove buds resembling round-headed nails about 1.3 to 1.9 cm long. The color is dark reddish-brown, the aroma is powerful, and the taste is hot, pungent, and spicy. Cloves, both whole and ground, are widely used in baked goods, confectioneries, puddings, desserts, sweet syrups, in pickling and pickled fruit, and spicy table sauces. About half of the world production of cloves is used, mixed with tobacco, to supply kretek cigarettes in Indonesia.

Coriander Seed. Coriander seed, the dried, ripe fruit of *Coriandrum sativun* L. (fam. *Umbelliferae*), is indigenous to southern Europe and the Mediterranean, now cultivated in Mexico, Romania, and Argentina as well. The seed is small (0.5 cm in diameter), yellowish to brown, and was one of the first herbs grown in the United States, in Massachusetts, in 1670. The foliage and green, unripe fruit have a fetid, "buggy" odor, described by the Greeks as koris or bedbuggy. As the fruit ripens, the disagreeable odor fades away and is replaced by a warm, fragrant, and spicy odor. It is used in ground meats (sausage and frankfurters), in cheeses, curry powders, and pickling spice. It has many medicinal uses as well.

Cumin Seed. Cumin seed is the dried, ripe fruit of *Cuminum cyminum* L. (fam. *Umbelliferae*). The seed-like fruit is elongated ca 0.31 to 0.63 cm and is yellowish-brown, native to upper Egypt and the eastern Mediterranean, now cultivated in Iran, Turkey, Lebanon, India, and Egypt; it possesses a strong, peculiar aroma and flavor, pleasing to some, offensive to others. Cumin seed is used in chutney and curry, chili powders, chili con carne, cheeses, and pickling of cabbage.

Curry Powder. Curry powder is a commercial blend of many spices. The composition of curries varies widely in the East according to region (Bombay curry, Madras curry, etc). It is employed in the flavoring of exotic dishes, particularly Indian, Indonesian, and Chinese specialties. Typical constituents of curry powder are coriander, ginger, nutmeg, clove, cinnamon, fenugreek, red pepper, onion salt, turmeric, etc.

Dill Seed. Dill seed, the dried, ripe fruit of *Anethum graveolens* L. (fam. *Umbelliferae*), is indigenous to the Mediterranean region and southern USSR, cultivated chiefly in India, Hungary, and Egypt. The seed is oval shaped, light brown, and up to 0.16 cm in length. The aroma resembles that of caraway seed, but is fainter. The herb is grown extensively in the United States, principally for the leaves, which must be harvested before the plant flowers. The leaves, known as dillweed, are used mainly for essential oil production.

Dillseed is used as a condiment and in soups, processed meats, sausages, spicy sauces, salads, sauerkraut, and particularly in pickling.

Fennel Seed. Fennel seed, the dry, ripe fruit of *Foeniculum vulgare* Mill. (fam. *Umbelliferae*), a biennial or perennial aromatic herb, is native to India, southern Europe, and the Mediterranean, particularly in the vicinity of the sea. The seed is oblong–oval, straight or slightly curved, up to 0.80 cm in length, and yellowish-brown. The flavor is aromatic and sweet, and resembles that of anise seed. It is very popular in Mediterranean countries for the flavoring of baked goods, soups, fish dishes, and sweet pickles. It is used in cordials and pharmaceutical preparations.

Fenugreek Seed (Foenugreek). Fenugreek seed, the dried, ripe fruit of *Trigonella foenum graecum* L. (fam. *Leguminosae*), is an annual plant of the pea family, native to western Asia and southern Europe and now cultivated in southern France, Morocco, Lebanon, India, and Argentina. The seed is irregularly oval, up to 0.48 cm in length and is a brownish-yellow. It has a slightly bitter, peculiar flavor and odor, somewhat pleasant, resembling maple or burnt sugar. It is used in curry powders, chutneys, and imitation maple flavors. Medicinal uses continue from ancient times.

Garlic Powder. Garlic powder, selected, dehydrated, and ground cloves of *Allium sativum* (fam. *Liliaceae*), is a widely cultivated, perennial herb, with an underground compound segmented bulb. The aroma is powerful and offensive, developing by enzymatic action and more pronounced by crushing and moisture. In diluted form it is very attractive and indispensable in certain Italian and French dishes, salad dressings, and sauces. It is widely used in the United States for meat packing, sausage, and frankfurters. One of the oldest of spices, it is indigenous to Egypt, China, and India, and is now grown throughout the world, extensively in the United States, People's Republic of China, Republic of China, and Japan.

Garlic Salt. Garlic salt is a mixture of garlic powder and table salt.

Ginger. Ginger is the washed or peeled sun-dried rhizome (underground stem) of *Zingiber officinale* Rose (fam. *Zingiberaceae*), a tuberous perennial. The color is an even, light buff, and the tubers are referred to as hands because of their size and shape. Cultivated by the ancient Chinese and Hindus, it is grown throughout the world, particularly in China, Fiji, India, Nigeria, Sierra Leone, Nicaragua, and Jamaica. The aroma and flavor is distinctively pleasant, aromatic, sweet, and pungent, with a slight bite. African ginger is camphoreaceous, Cochin ginger is lemon-like, and Jamaica ginger perhaps the finest combination. Ginger is used in soft drinks, baked goods (gingerbread), confectionery, curry powders, meats, pickling, and certain cordials.

Mace. Mace, the dried aril or skin surrounding the kernel of *Myristica fragrans* Houtt (fam. *Myristicaceae*), is extensively produced in Indonesia. The color is yellow-orange. The flavor is similar to nutmeg, but is softer and somewhat less pungent.

Marjoram (Sweet Marjoram). Marjoram, the dried leaves with or without small portions of the flowering tops of *Majorana hortensis* Moench (fam. *Labiatae*), is a perennial herb cultivated in Egypt, France, Turkey, Greece, Mexico, and the United States. It may be confused, sometimes purposely, with oregano. Marjoram is a grey–green color and the flavor is highly aromatic, warm, and somewhat bitter. It is used in salad dressings, vegetables, meat stews, sausages, soups, and poultry stuffings.

Mint. Mint, the dried leaves of *Mentha spicata* L. (spearmint) and *Mentha piperita* L. (peppermint), hardy perennial herbs, is indigenous to Europe and the Mediterranean, but now cultivated extensively in the United States Far West (Washington and Oregon) and Midwest (Indiana and Michigan regions). Although it has many medicinal uses, it is a desirable flavor in candies, sauces, desserts, salads, jellies, vinegars, teas, and particularly, in toothpaste, mouthwash, and chewing gum.

Mustard. Mustard is the dried, ripe seeds of several closely related genera, species, and varieties of herbaceous annuals of the fam. *Cruciferae*; *Sinapis alba* L. (syn. *Brassica hirta* Moench) known as white or yellow mustard. *Brassica nigra* (L.) Koch; black mustard and *Brassica juncea* Coss; and Czern, brown mustard. The first two are probably indigenous to southern Europe and the Mediterranean region; *B. juncea* to the northern Himalayan area. They are all of the mustard family and are cultivated widely in Canada, France, the United Kingdom, China, South America, and the United States. The seed is extremely small, globular, and hard; at one time considered the smallest seed, within one growing season the plant reaches a height of 3 m or more.

When ground and mixed with water, the seeds of black mustard ferment and yield a sharp and irritating odor and a pungent taste; white mustard does not react in this manner.

Mustard is marketed in three forms: (*1*) as seed; (*2*) as prepared mustard (a blend of ground mustard seed, vinegar, salt, sugar, and other spices); and (*3*) as powdered dry mustard, also known as ground mustard or mustard flour. The seed of the white or yellow mustard adds pungency to any preparation, and that of black or brown mustard is required for aroma. Mustard is by far the largest volume spice imported into the United States.

Nutmeg. Nutmeg is derived from an apricot-like fruit of *Myristica fragrans* Houtt (fam. *Myristicaceae*), an evergreen tree, indigenous to the Moluccas and other East Indian islands. It is cultivated as well in Indonesia and Grenada. The fleshy fruit is globose, lemon yellow to light brown which, when it ripens, splits in half exposing a beautiful scarlet reticulated arillole (mace) investing the thin, hard, brown, and shiny shell which contains the seed (nutmeg). The oval nutmegs range in size up to 4 cm in diameter.

The flavor of nutmeg is pleasant, warm, spicy, and aromatic. Nutmegs are used to flavor baked goods, processed meats, curries, and beverages.

Onion Powder. Onion powder consists of dehydrated and ground onions *Allium cepa* L. and other *Allium* species (fam. *Libiaceae*) that are cultivated extensively in many countries. The spice has a sharp and pungent odor and flavor, but is very desirable on dilution. It is widely used in meats, sausages, vegetables, salads, and many dishes.

Onion Salt. Onion salt is a mixture of onion powder and table salt.

Oregano. The dried leaves of *Lippia graveolens* (fam. *Verbenaceae*) and of *Origanum* mostly *vulgare* sp. (fam. *Labiatae*). The lippia species are grown in Mexico and the origanum in the Mediterranean countries, Greece and Turkey. The color of the leaf is light green, about 1.59 cm long. The odor is strong and aromatic, the flavor pungent, warm, and slightly bitter. The aroma resembles marjoram. It is used in chili powders, Italian specialties, gravies, salad dressings, vegetables, and various meats.

Paprika. See under Capsicum.

Parsley (Parsley Flakes). Parsley, the dried leaves of *Petroselinum hortense,* syn. *P. crispum* (Mill) (fam. *Umbelliferae*), is a hardy biennial native to the Mediterranean region and now cultivated in the United States and southern Europe. It is green and the flavor is pleasant, characteristic, and mild. It is used for the seasoning of fish, meats, soups, salads, sauces, and the like.

Parsley Seed. Parsley seed is the dried, ripe fruit of the same plant. The aroma and flavor are less pronounced than in the leaves.

Pepper, Black. Black pepper is the dried, immature berries (fruit) of *Piper nigrum* L. (fam. *Piperaceae*), a vine-like perennial, native to the Malabar coast of southern India, but now widely cultivated in the tropics of both hemispheres. There are many varieties of pepper, differing in appearance, size, and flavor (51). Dried black pepper consists of a light-colored kernel, surrounded by a dark-colored, shriveled, outer hull, with ca 3629 to the kg. It has a strong characteristic odor and a hot, pungent flavor, with a "bite." It is the most extensively used seasoning known.

Pepper, White. White pepper is derived from the same plant as black pepper, except that the berries are allowed to ripen fully. The dark hulls are removed by soaking and then fermentation, with removal of the outer hulls by rubbing and washing. It has generally the same flavor and aroma as black pepper; however, it is milder, with a slight fetid odor.

Poppy Seed. Poppy seed, the tiny dried seed of *Papaver somniferum* L. (fam. *Papaveraceae*), is an annual plant cultivated widely in Turkey, Iran, India, and China for the preparation of opium. It is also grown in Australia, the Netherlands, and Yugoslavia. The poppy seeds are not narcotic and are black, kidney-shaped seeds, about 0.12 cm long. Poppy seed is one of the most ancient of spices; it was found among the ruins of prehistoric lake dwellings in Switzerland. The seeds have an agreeable, nutty flavor. They are used in whole form as a topping for bread, rolls, and pastries, and also in pasta, salads, and some vegetables.

Red Pepper. See under Capsicum.

Rosemary. Rosemary, the dried leaves of *Rosmarinus officinalis* L. (fam. *Labiatae*), is an evergreen shrub growing wild in the Mediterranean countries. It is grown in Yugoslavia, Portugal, Spain, France, and the United States. The leaves resemble curved pine needles in shape and seldom exceed 2.5 cm in length. They are brownish–green, the odor is strong, fragrant, and aromatic, and the taste is somewhat bitter and camphoraceous. Rosemary is used for the flavoring of stuffings, poultry, beef, lamb, fish, stews, and soups.

Saffron. Saffron, the dried stigmas of *Crocus sativus* L. (fam. *Iridaceae*), is a bulbous perennial native to southern Europe and Asia Minor, and cultivated in the Mediterranean countries, particularly Spain. True saffron should not be confused with either meadow saffron (*Colchicum autumnale* L., fam. *Liliaceae*), also called safflower, or bastard saffron (*Carthamus tinctorius* L., fam. *Compositae*), which are occasionally used to adulterate true saffron.

Saffron is the world's most expensive spice. It is necessary to handpick the stigmas from 75,000 flowers to produce the 225,000 stigmas necessary to make one-half kilogram of spice. It is bright yellow–red, and the aroma is powerful, somewhat bitter, peculiar, and exotic. It is used in exotic dishes, particularly for the coloring and flavoring of rice in Spanish specialties: arroz con pollo (chicken and rice); paella (chicken, sausages, shellfish, rice, vegetables); and bacalao vizcaino (codfish Spanish style). It is important in bouillabaisse (a shellfish and fish mixture from the French Mediterranean coast) and is also used in cheese, pastry, and confectionery.

Sage. Sage, the dried leaves of *Salvia officinalis* L. (fam. *Labiatae*), is a perennial low shrub growing spontaneously in Mediterranean countries. Of the numerous sage species, the Dalmatian variety is considered the finest for flavoring. It is widely cultivated in Albania, Yugoslavia, Turkey, and Italy. The color is greyish–green or silver–gray, and the aroma is strong, fragrant, warm, slightly bitter, with no camphoraceous or piney off-odor. It is used for the flavoring of meats, poultry, fish, certain cheeses, salad dressings, and also sausages and poultry seasoning.

Savory (Summer Savory). Savory, the dried leaves and flavoring tops of *Satureia hortensis* L. (fam. *Labiatae*), is an annual plant cultivated in southern Europe, particularly Yugoslavia, France, and Spain. The dried leaves are brown–green and up to 0.95 cm in length. The odor is strong, warm, and highly aromatic, the flavor is sharp and somewhat resinous. It is used in poultry seasoning, meats, soups, eggs, salads, and sauces, and is sometimes substituted for parsley.

Sesame Seed (Benne, Benni, or Bene Seed). Sesame seed is the whole dried seed of *Sesamum indicum* L. (fam. *Pedaliaceae*), an annual plant now cultivated in Central America, although indigenous to Indonesia and tropical Africa. It may be the oldest condiment known. It is widely cultivated in China, India, Mexico, Guatemala, and Nicaragua. The seed is small, shiny, and oval-shaped, ca 0.32 cm long. The unhulled seeds are dark, and the hulled seeds are pearly white. Sesame seeds, when baked, have a pleasant roasted, nutty flavor. It is used on bakery goods and in confections, eg, halvah.

Tarragon (Estragon). Tarragon, the dried leaves and flowering tops of *Artemisia dracunculus* L. (fam. *Compositae*), is a perennial herb, indigenous to the southern USSR and western Asia, and cultivated throughout southern Europe, particularly France and Spain, and also in the United States. The aroma and flavor are warm, aromatic, and reminiscent of anise, but bittersweet. It is used in the flavoring of vinegar and in salads, soups, stews, and sauces such as bearnaise, tartare, etc.

Thyme. Thyme is the dried leaves and flowering tops of *Thymus vulgaris* L. and *T. zygis* L. (fam. *Labiatae*), perennial small shrubs native to the Mediterranean region and Asia Minor. It is grown extensively in France, Spain, Portugal, Greece, and the United States. The tiny, brownish-grayish-green, narrow, curled leaves rarely exceed 0.63 cm in length. The flavor is aromatic, warm, and pungent. It is used in flavoring clam and fish chowders, sausages, meats, poultry dressing, fish sauces, and in the liqueur, Benedictine.

Turmeric (Curcuma). Turmeric (curcuma) is the clean, sound, dried rhizome of *Curcuma longa* L. (fam. *Zingiberaceae*), a herbaceous perennial indigenous to southern Asia. It is cultivated in India, China, Kuwait, Indonesia, Sri Lanka, and the West Indies. Turmeric has a thick round rhizome or underground stem, with blunt "fingers." The flavor is distinctly aromatic, earthy, and somewhat bitter, but it is used as much for its orange–yellow color as for flavor. It is an important constituent of curry powder, and is used for flavoring poultry, seafood, and rice dishes, and as a coloring and flavoring agent in pickles.

Vanilla. Vanilla is the dried, cured, full-sized, but not fully ripe fruit pods (beans) of *Vanilla planifolia* And. (and of *V. tahitensis*, J. W. Moore) (fam. *Orchidaceae*). It is native to the tropical rain forests of southern Mexico, Central America, the West Indies, and northern South America. Plantings were started in Madagascar, Reunion, Java, Mauritius, and Zanzibar in 1840. Today, the Malagasy Republic (Madagascar), Comoros Islands, and Reunion grow about 90% of the world's supply of vanilla beans, Bourbon vanilla. The structure of the flower prevents self-pollination, and therefore, hand pollination has become necessary where insects are not prevalent.

A large portion of vanilla goes into the production of tinctures and extracts. These are used for ice creams, bakery goods, liqueurs, and many other items.

BIBLIOGRAPHY

"Flavors and Spices" in *ECT* 1st ed., Vol. 6, pp. 581–594, by Ernest C. Crocker, Arthur D. Little, Inc.; "Flavor and Spices" in *ECT* 2nd ed., Vol. 9, pp. 347–380, by Edmund H. Hamann and Ernest Guenther, Fritzsche Brothers Inc.

1. *Food Tech.* **23,** 11 (1969).
2. E. von Sydow, *Food Tech.* **25,** 40 (1971).
3. F. J. Hammerschmid, *Drogoco Report,* 244 (Oct. 1977).
4. J. F. Caul, *Cereal Sci. Today* **12**(1), 275 (1967).
5. *Soc. of Flavor Chemists Candy Ind.*, 5, 10 (Aug. 6, 1968).
6. G. Mackinney and A. C. Little, *Colors of Foods,* Avi Publications Co., Westport, Conn., 1962.
7. S. A. Matz, *Food Texture,* Avi Publications Co., Westport, Conn., 1962.
8. P. C. Paul and H. H. Palmer, *Food Theory and Applications,* John Wiley & Sons, Inc., New York, 1972.

9. R. W. Moncrieff, *Flavour Ind.*, 828, (Dec. 1971).
10. *Ibid.*, 84 (Feb. 1971).
11. *Ibid.*, 583 (Sept. 1970).
12. R. W. Moncrieff, "1967 Introduction to the Symposium" in H. W. Schultz, H. Day, and L. M. Libby, eds., *Symposium of Foods; The Chemistry and Physiology of Flavor,* Avi Publications Co., Westport, Conn.
13. D. R. Evans, *Proc. 1st Intern. Symp. Wenner Gren Center Stockholm, 1962,* New York, 1963.
14. *Flavor Research and Food Acceptance,* Reinhold Publishing Corp., New York, 1958.
15. J. J. Brennan and U. Leffler, *Proceedings Can. Inst. Food Technologists Conf., Montreal, May 1964.*
16. R. W. Moncrieff, *Perfumery Res. Oil Record* **55,** 205 (1964).
17. S. S. Schiffman and R. P. Erickson, *Physiology* **7,** 617 (1971).
18. D. K. Tessler and M. A. Joslyn, *Fruit and Vegetable Juice Processing Technology,* Avi Publications Co., Westport, Conn., 1961.
19. R. W. Moncrieff, *The Chemical Senses,* CRC Press, Cleveland, 1967.
20. R. M. Pangborn and R. B. Chrisp, *J. Food Sci.* **29**(4), 490 (1964).
21. G. vonBekesy, *Science* **145,** 834 (1964).
22. H. L. Meiselman, *Crit. Rev. Food Technol.,* 89 (Apr. 1972).
23. R. W. Bragg and co-workers, *J. Chem. Ed.* **55,** 281 (1978).
24. S. Price and J. A. Desiomone, *Chem. Senses and Flavor* **2,** 432 (1977).
25. K. Kulka, *Agric. Food Chem.* **15,** 48 (Jan./Feb. 1967).
26. S. Price and J. A. Desiomone, *Chem. Senses and Flavor* **2,** 448 (1977).
27. D. Ottoson, "Odor and Taste" in *Proc. 1st Intern. Symp. Wenner Gren Center, Stockholm, 1962,* New York, 1963.
28. F. Thomel, *Dragoco Rep.,* 191 (Sept. 1977).
29. *Chem. Ind.,* 8 (Mar. 25, 1967).
30. Y. R. Hayes, *Soc. Chem. Ind. (London),* 38 (1957).
31. M. Stoll, *Soc. Chem. Ind. (London),* 1 (1957).
32. R. H. Adrian "Odor and Taste" in *Proc. 1st Intern. Symp. Wenner Gren Center, Stockholm 1962,* New York, 1963.
33. L. Mullins, *Ann. N.Y. Acad. Sci.* **62,** 247 (1955).
34. H. R. Schiffman, *Sensation and Perception: An Integrated Approach,* John Wiley & Sons, Inc., New York, 1976, p. 123.
35. F. Drawert, *Proc. of the Intern. Symp. on Aroma Research, Zeist, The Netherlands, 1975,* p. 13.
36. W. H. Stahl, ed., *Compilation of Odor and Taste Threshold Values Data,* DS 48, American Society for Testing and Materials, Philadelphia 1973.
37. L. J. van Gemert and A. H. Nettenbreiger, *Compilation of Odor Threshold Values in Air and Water,* National Institute for Water Supply, Voolburg, the Netherlands, Central Institute for Nutrition and Food Science, TNO Zeist, the Netherlands.
38. E. C. Bate-Smith and T. N. Morris, *Food Science,* Cambridge University Press, England, 1952.
39. P. E. Koehler, M. E. Mason, and G. V. Odel, *J. Food Sci.* **36,** 817 (1971).
40. S. Van Straten, ed., *Volatile Compounds in Food,* 4th ed., Central Institute for Nutrition and Food Research, TNO, Zeist, The Netherlands.
41. J. A. Maga, *Critical Reviews in Food Science and Nutrition,* (July 1975).
42. *Ibid.,* (Sept. 1975).
43. *Ibid.,* (Jan. 1975).
44. B. M. Dwivedi, *Critical Reviews in Food Science and Nutrition,* (Apr. 1975).
45. I. Schutte, *Critical Reviews in Food Science and Nutrition,* (March 1974).
46. J. A. Maga and C. E. Sizer, "Pyrazines in Foods" in *Fenarolis Handbook of Flavor Ingredients* Vol. 1, 2nd ed., CRC Press, Cleveland, 1975, p. 47.
47. H. E. Nursten and A. A. Williams, *Chem. Ind.,* 487 (Mar. 25, 1967).
48. *U.S. Pharmacopeia XX, (USP XX-NF XV),* The United States Pharmacopeial Convention Inc., Rockville, Md., 1980.
49. *National Formulary XIV,* American Pharmaceutical Association, Washington, D.C., 1975.
50. "Food Flavorings" in *Code of Federal Register,* Part 22, Sections 22.1–22.7.
51. A. V. Saldarini, *Flavour Ind.,* 247 (Sept./Oct. 1974)
52. T. E. Furia and N. Bellanca, *Fenaroli's Handbook of Flavor Ingredients,* 2nd ed., CRC Press, Cleveland, 1975.

53. R. A. Ford and G. M. Cramer, *Perfumer and Flavorist* **2**(1), (1977).
54. M. B. Jacobs, *Synthetic Food Adjuncts,* D. Van Nostrand Co., Inc. New York, 1947.
55. J. Merory, *Food Flavoring (Composition, Manufacture and Use),* Avi Publications Co., Westport, Conn., 1968.
56. *Fed. Regist.* **42,** 14315 (Mar. 15, 1977).
57. W. Grab, E. Beck, and T. Bernegger, *Intern. Flav. Food. Add.,* 63 (Mar./Apr. 1977).
58. *Food Chemicals Codex,* 3rd ed., National Academy of Sciences—National Research Council, Washington, D.C., 1980.
59. F. Grundschober and co-workers, *Flavours,* (July/Aug. 1975).
60. N. J. Leinen, *Food Proc.,* 28 (March 1978).
61. B. M. Lawrence, *Perfumer and Flavorist,* 4 (Dec. 1976/Jan. 1977).
62. *Fed. Regist.* **38,** 33285 (Dec. 3, 1973).
63. P. J. Serritella, *Food Prod. Dev.,* 43 (Aug. 1978).
64. E. F. Oneil, *Proceedings Flavor Labeling Regulations Manf. Conf., Nov. 1974.*
65. *Food Additive Market Report No. 561,* Frost and Sullivan Inc., New York, 1978.
66. M. S. Konigsbacker and E. J. Hewitt, *Ann. N.Y. Acad. Sci.* **116,** 705 (1964).
67. E. J. Hewitt and co-workers, *Food Tech.* **10**(10), 487 (1956).
68. K. S. Konigsbacher, E. J. Hewitt, and R. L. Evans, *Food Tech.* **13**(2), 128 (1959).
69. D. A. M. Mackay and E. J. Hewitt, *Food Res.* **24,** 253 (1959).
70. C. Weurman, *Food Tech.* **15,** 531 (1961).
71. A. J. MacLeod, *Flavour Ind.,* 665 (1970).
72. B. C. deLumen and co-workers, *J. Food Sci.* **43,** 698 (1969).
73. G. Reed, ed., *Enzymes in Food Processing,* Academic Press, New York, 1975.
74. T. Godfrey, *International Flavours and Food Additives,* 163 (July/Aug. 1978).
75. T. A. Rohan, *J. Food Sci.* **29,** 457 (1964).
76. T. A. Rohan and M. Connell, *J. Food Sci.* **29,** 460 (1964).
77. T. A. Rohan, *Flavour Ind.,* 147 (1971).
78. G. A. Reineccius, P. G. Keeney, and W. Weisberger, *J. Agr. Food Chem.* **20,** 202 (1972).
79. G. C. Harlee and J. C. Leffingwey, *Tob. Int.* **46,** 407 (Mar. 9, 1979).
80. L. C. Maillard, *Compt. Rend.* **154,** 66 (1912).
81. J. Adrian, *Labo Pharma Prob. Tech.,* (244) (June 1975).
82. J. E. Hodge, *Chemistry and Physiology of Flavors,* Avi Publications Co., Westport, Conn., 1967.
83. R. A. Lawrew, *Flavour Ind.* 591 (1970).
84. C. G. Tassan and G. F. Russell, *J. Food. Sci.* **39,** 64 (1974).
85. J. A. Newell, M. E. Mason, and R. S. Matlo, *J. Agric. Food Chem.* **15,** 767 (1967).
86. C. H. Manley, P. P. Vallon, and R. E. Erickson, *J. Food Sci.* **39,** 73 (1974).
87. I. R. Hunter and co-workers, *Cereal Sci. Today* **11,** 493, 496 (1965).
88. C. E. Sizer, J. A. Maga, and K. Lorenz, *Lebensm Wiss. Technol.* **8,** 267 (1975).
89. S. R. Pareles and S. S. Chang, "Identification of Compounds Responsible for Baked Potato Flavor" in *Paper of the Journal Series,* New Jersey Agricultural Experimental Station, Rutgers University, N.J., Oct. 1, 1973.
90. S. S. Chang, R. J. Peterson, and C-T. Ho in M. K. Supran, ed., *Chemistry of Deep Fat Fried Flavor in Lipids as a Source of Flavor,* ACS Symposium Series 75.
91. T. F. Stewart, *A Survey of the Chemistry of Amino Acid-Reducing Sugar Reaction in Relation to Aroma Production,* Scientific and Technical Surveys No. 61, British Food Manufacturing Industries Research Association, Dec. 1969.
92. Brit. Pat. 1,223,796 (Mar. 3, 1971), E. Hitoshi and S. Okumura (to Ajinomoto Co. Inc.).
93. R. H. Walter and J. S. Fagerson, *J. Food Sci.* **33,** (1968).
94. W. J. Herz and R. G. Schallenberger, *Food Res.* **25,** 491 (1960).
95. W. W. Lawrence, *Of Acceptable Rise, Science and the Determination of Safety,* William Kaufman Inc., Los Altos, Calif., 1976.
96. R. L. Hall and B. L. Oser, *Food Tech.* **19**(2), (Feb. 1965).
97. *Institute of Food Technologists* **18**(8), 25 (1964).
98. E. Larmond, *Laboratory Methods for Sensory Evaluation of Food,* Pub. No. 1637, Canada Department of Agriculture, 1977.
99. W. H. Stahl and M. A. Einstein, "Sensory Testing Methods" in F. D. Snell and L. S. Ettre, eds., *Encyclopedia of Industrial Chemical Analysis,* Vol. 17, Wiley-Interscience, New York, 1973, pp. 608–644.

100. *A General Guide to Sensory Evaluation,* McCormick and Co., Hunt Valley, Md., 1978.
101. *Manual of Sensory Testing Methods,* ASTM Special Technical Publication 434, Philadelphia, 1968.
102. J. F. Caul, *Advan. Food Res.* **7,** 1 (1957).
103. S. E. Cairncross and L. B. Sjostron, *Food Tech.* **4,** 308 (1950).
104. J. Eindoven and co-workers, *J. Food Sci.* **29,** 520 (1964).
105. R. M. Pangborn, *Food Technol.* **9**(11), 63 (1964).
106. M. A. Amerine, R. M. Pangborn, and E. B. Roessler, *Principles of Sensory Evaluation of Food,* Academic Press, New York, 1965.
107. A. W. Williams, *Intern. Flav. and Food Add.* **80** (Mar./Apr. 1978); 131, 133, 135 (May/June 1978); (July/Aug. 1978).
108. H. W. Spencer, *Flavour Ind.,* 293 (May 1971).
109. M. W. Neale, *Canner/Packer* **32**(3), 41 (1963).
110. W. E. Dorland and J. A. Rogers, Jr., *The Fragrance and Flavor Industry,* Dorland Publishing Co., Mendham, N.J., 1977, p. 55.
111. F. Rosengarten, Jr., *The Book of Spices,* Pyramid Books, New York, 1973.
112. *A History of Spices,* American Spice Trade Association Inc., New York, 1960.
113. J. W. Parry, *The Story of Spices,* Chemical Publishing Co., New York, 1962.
114. *Cleanliness Specifications for Unprocessed Spices, Seeds and Herbs,* American Spice Trade Association, Englewood Cliffs, N.J., 1976, p.i (preface).
115. *Equivalents, Flavor Data Sheet—Essential Oils, Herbs and Spices,* Fritzsche Dodge Olcott, New York, 1968.
116. *Fed. Regist.,* (Jan. 5, 1960).
117. *Official Analytical Methods of the American Spice Trade Association,* 2nd ed., American Spice Trade Association, Inc., Englewood Cliffs, N.J., 1968.
118. *Microorganisms in Spices, Microbiological Methods,* American Spice Trade Association, Englewood Cliffs, N.J.

J. A. ROGERS, JR.
F. FISCHETTI, JR.
Fritzsche Dodge & Olcott, Inc.

FLAX. See Fibers, vegetable.

FLAXSEED OIL. See Drying oils (linseed oil); Fats and Fatty oils.

FLOCCULATING AGENTS

Utilization of natural substances as sedimentation aids for clarifying household water in various parts of the world apparently long predates written history. The first specific mention of lime and aluminous earth for this purpose appeared in the latter part of the first century AD (1). Use of such aids in large industrial plants appears to have started in the first part of the nineteenth century, and municipal use started during the later nineteenth and early twentieth centuries. A large variety of aqueous suspensions of solids were found to be susceptible to treatment. Lime and alum were the most generally useful settling aids but other products exhibited special utility in some applications. In particular, certain naturally occurring organic polymers, such as glue and starch (2), found wide use in the mineral processing area.

Introduction of synthetic, water-soluble polymers to the mining industry in 1951 ushered in a new era in solid–liquid separations by chemical reagents (3). They were the first of a wide range of flocculants tailored to meet many needs such as clarification of water (municipal and industrial), treatment of municipal sewage and industrial wastes (food processing, oil refining, metal finishing, pulp and paper mills, etc), mineral processing (beneficiation, recycle-water clarification, effluent treatment), and manufacturing processes (paper production, sugar refining, phosphoric acid production, etc). Although there are several hundred commercial synthetic flocculants currently available in the United States and Canada alone (4), the number of significantly different types of chemical structures represented is relatively limited. In the market place, selection of a flocculant depends on optimizing the cost-to-performance ratio, ie, achieving desired performance at minimum cost. Although a systems point of view predominates (including flocculant availability, reproducibility, handling, storage, tolerance to fluctuations in treatment-plant loading while meeting output specifications, equipment-in-place and necessary modification, etc), the delivered cost per unit weight of individual flocculants always enters as one factor. Consequently, a relatively few monomers, suitable for incorporation into water-soluble polymers and produced on a sufficiently large scale to have low cost, are the major building blocks for commercially important synthetic polymeric flocculants.

The flocculants currently in commercial use are conveniently classified as inorganic, synthetic organic, and those derived from natural products (naturally occurring organic polymers which may have been modified chemically). In 1978 the category with the largest sales value was synthetic organic, although inorganic flocculants had the largest sales volume on a weight basis (see Water; Sedimentation; Dispersants; Extractive metallurgy).

Flocculation

To provide perspective on the types of flocculating agents, and clarify conflicting terminology (5) regarding flocculation, a brief survey of general principles and practice in use of flocculating agents is presented. Several reviews (6–11) can be consulted for more detail.

Consider a suspension of solid particles in water, with a distribution in sizes of particles. This suspension is called the substrate. In the usual situation, if the suspension is allowed to stand quietly, the large particles settle out first. There is also a portion of very fine material that settles only very slowly or not at all. This resistance to settling by the very fine particles relates to the size and density of the particles, to their surface properties, and to the composition of the suspending medium. For a suspension of chemically homogeneous particles there is no fundamental distinction between the particles that remain in suspension and those that settle out. Any difference is in degree rather than in kind, and reflects the much larger surface-area-to-volume ratio of the smaller particles.

In water suspensions the solid surfaces tend to have a net electrical charge, balanced by ions of opposite charge in solution in the vicinity of the particle surface. The primary sources of this surface charge are (*1*) ionization of surface groups (such as OH groups for mineral oxides (12) or carboxyl groups for latex particles (13)), (*2*) isomorphic substitution in the solid lattice (14), and (*3*) preferential adsorption of ions (15) or ionizable species (16) from the suspending medium. This situation of a charge localized on the particle surface, with a diffuse distribution of counterions extending into the liquid medium, is referred to as an ionic double layer (17).

The surface charge generates a repulsive force between particles, tending to keep them apart. For the large particles, gravitational forces are dominant and make them unstable with respect to suspension, but for individual particles of colloidal dimensions the electrostatic and microhydrodynamic forces dominate and make their suspensions relatively stable. If these colloidal particles aggregate into larger entities for which gravitational forces dominate, suspension stability is lost. This formation of aggregates is termed flocculation; an aggregate thus formed is termed a floc or floccule, and any chemical agent that enhances the process is termed a flocculating agent or a flocculant.

The terms flocculation and coagulation are both used in reports in connection with formation of aggregates, frequently interchangeably (5) and sometimes with distinctions that vary among professional disciplines. In view of the more general (but not universal) acceptance of the equivalence of the words coagulation and flocculation, IUPAC recommends that any author wishing to make a distinction state the difference clearly in the publication (5). Although no distinction is made in this article, the more common types of distinction appearing in the literature are enumerated: (*1*) based on mechanisms for destabilization of a suspension and/or type of aggregate formed, coagulation implying formation of compact aggregates and flocculation implying formation of loose or open network aggregates (2,5): (*2*) based on chemical agents used, coagulation for inorganic materials and flocculation for organic polymers; (*3*) based on engineering process steps, coagulation representing conditioning the particles with the chemical agent, and flocculation representing the mechanical particle transport step (collisions between conditioned particles) leading to aggregates (7); and (*4*) based on another engineering usage, coagulation representing the overall aggregation process and flocculation again representing the particle transport step (8–9).

In the absence of a flocculating agent, the energy of interaction between a pair of particles (a factor in formation of aggregates) traditionally is formulated as arising from two components, a term arising from overlap of double layers leading to repulsion, and a term arising from London dispersion forces which lead to van der Waals attraction. Since dispersion forces are short range, the particles must be close together before a significant attractive force develops. The distance of closest approach for colloidal particles is strongly influenced by the magnitude of the double layer, which therefore, influences the stability of a suspension. Flocculating agents enhance aggregation either by modifying these energy terms or by introducing new interaction terms, thereby destabilizing the suspension and effecting solids–liquid separation.

Emphasis on electrostatic stabilization of suspended particles in an aqueous substrate does not imply the absence of other kinds of stabilization. The surface characteristics of particles range from hydrophobic to hydrophilic. Suspensions of hydrophobic particles, such as minerals, tend to be stabilized primarily by surface charge. Hydrophilic particles, such as sewage sludges, are stabilized by solvation as well as by charge. For example, the hydrophilic surface may result from a coating of firmly anchored polysaccharide or similar polymeric species with loops or tendrils of the polymer extending out from the particle surface into the aqueous phase. When two particles approach one another and the polymer segments interpenetrate, solvent molecules are forced out of the interpenetration volume. There is a free energy associated with the solvent exclusion called steric stabilization, which provides the thermodynamic barrier to close approach of particles (18).

Destabilization. There appear to be five general modes by which flocculating agents bring about destabilization of aqueous suspensions: double-layer compression, specific-ion adsorption, sweep flocculation–enmeshment, polymer charge patch, and polymer bridging. More than one mode of action may be involved at the same time in any real flocculating system.

Double-Layer Compression. Increasing ionic strength of the suspending medium reduces the thickness of the double layer (19), thereby reducing the range of interparticle repulsion. Ions of similar charge to the surface are repelled by the surface, and counterions are attracted to its vicinity. As the counterion concentration increases, the volume of the diffuse ionic layer necessary for electroneutrality decreases. Simple electrolytes whose effect is purely electrostatic in nature, those that have no specific chemical interaction with the surface, are termed indifferent electrolytes, and their effectiveness as flocculants increases with charge on the counterion (Schulze-Hardy rule (20)). Critical flocculation concentrations (cfc), the minimum electrolyte concentration necessary to develop measurable rate of flocculation, depend on the magnitude of the surface charge, but a common circumstance would involve ca 10^{-1}, 2×10^{-3}, and 2×10^{-4} mol/L of mono-, di-, and trivalent ions, respectively (19). Dosage is relatively independent of the level of suspended solids (ss) in the suspension.

Specific-Ion Adsorption. Some counterions can participate in a chemical reaction with the particle surface, losing at least a portion of their hydration sphere and complexing with other atoms in the surface structure (21). Counterion adsorption may be sufficient to reduce the net surface charge to zero, or even to reverse the sign of the net surface charge (charge reversal). When the net surface charge is reduced to zero by this mechanism, electrical repulsion between particles is minimized and the rate of flocculation should be high. Actually the surface charge need not be reduced to zero to effect flocculation at a measurable rate by this mechanism; it need only be lowered enough to allow close approach of particles (22). The cfcs for destabilization by specific-ion adsorption are considerably smaller than those for an indifferent electrolyte and are dependent on the available surface area of the substrate particles (23). Adsorption tends to be strong, and desorption relatively slow, if it is observed at all.

Hydrolyzable metal ions (such as aluminum and iron) frequently are considered to operate by this mechanism (22–23), although several mechanisms may be involved simultaneously, depending on experimental conditions. When partially hydrolyzed, these metal ions can form positively charged dimers, trimers, and multimers of metal ions bridged by hydroxyl or oxo groups, which exhibit specific adsorption onto many types of solid surfaces. When the surface charge of the suspended particles is negative, a typical situation in common substrates, these hydrolysis species can decrease the magnitude of the charge and even cause charge reversal (22,24).

Sweep Flocculation (Enmeshment). Hydrolyzing metal salts usually are applied as flocculants in a concentration/pH region that is oversaturated with respect to the neutral metal hydroxide. The rate of precipitation of the hydroxide, which is a function of temperature, other ions present, and nucleation sites, may be slow unless a certain degree of supersaturation is exceeded. Under appropriate conditions, however, a fluffy amorphous hydroxide forms rapidly, trapping and enmeshing colloidal particles as it settles, and giving rise to the term sweep flocculation (24–25). Dosage is relatively independent of concentration of substrate particles and does not involve double layer effects. Hence, sweep flocculation is applicable for all types of substrates in the pH range where it occurs. It is particularly appropriate for substrates with a very low level of suspended solids such as some water supplies.

Polymer Charge Patch (26–27). The particles in many common substrates have relatively fixed sites of negative charge on their surfaces. When organic polymer molecules containing positively charged sites are added to such a substrate, they are adsorbed onto the particle surface via electrostatic bonds. In most respects this is specific-ion adsorption, with the effects described previously, but for some situations there is a significant additional feature that justifies designation as a separate mode. One general type of cationic polymer has a high charge density along the chain (such as a cationic charge every few carbon atoms), providing a greater density of charge in the polymer molecule than is present on the surface of a particle. When this type of polymer molecule is adsorbed onto the surface, it not only neutralizes the negative charge within the geometric area where it is attached but it also provides excess cationic charge to compensate for other negative charge sites on the surface. The entire negative charge on a particle can be neutralized by a relatively small geometric surface coverage by polymer (28). Thus the polymer adsorbs onto a particle surface in patches of molecular size, forming positive patches surrounded by regions of negative surface sites. This mosaic of positive and negative surface areas then allows a direct electrostatic attraction between particles. That is, the flocculant introduces a positive element into the flocculation process instead of merely reducing the repulsive force between particles. Dosage for optimum flocculation, termed optimum polymer dosage (OPD), is essentially stoichiometric with respect to surface area of substrate particles (26,28). Dosages in excess of OPD can restabilize the suspension.

Polymer Bridging. Water soluble polymers can bond to a particle surface by a variety of mechanisms, of which the electrostatic bond is only one. Other types include the hydrophobic bond, van der Waals bond, and a group of rather specific physicochemical interactions ranging from hydrogen bonding to covalent bonding. Hydrogen bonding probably is the most common of this group but other chemical interactions can be crucial at times (29–31).

When very long polymer molecules (high molecular weight linear polymers) are adsorbed onto the surface of particles, they tend to form loops that extend some distance from the surface into the aqueous phase, and the ends may also dangle. These loops and ends may come into contact with, and attach to, another particle, forming a bridge between the two particles. This is the bridging mode of flocculation (6).

The particles must come close enough together for this bridging to occur but charge neutralization is not required. Nevertheless, the factors that reduce electrostatic repulsion described above are pertinent here also. The optimum polymer dosage can vary widely and need not increase stoichiometrically with suspended solids.

Chemical Precipitation. Flocculating agents participate in several commercially important precipitation reactions including not only those that remove a component from solution but those that flocculate the resulting suspended solids. These include precipitation of soluble phosphates (32–33) from waste streams with Al^{3+}, Fe^{3+}, or Ca^{2+} ions, removal of fulvic and humic acids from water supplies with hydrolyzing cations (Al^{3+}, Fe^{3+}) (34) or with high charge density cationic polymers (35), and removal of soluble anionic color components (36–37) generated in certain types of wood pulping operations (see Pulp). Frequently, the flocs formed under these conditions are small and settle slowly. If more rapid clarification is desired, a high molecular weight anionic polymer is added to form larger flocs for faster settling but this does not effect intrinsic removal of the soluble component (33).

Selective Flocculation. In most practical systems the substrate to be treated is very complex, heterogeneous in both size and composition of suspended particles. Although removal of all suspended solids is usually the goal, there are many situations, particularly in the mineral processing area, where selective removal according to composition is desired . For selective flocculation to be effective there must be some difference in surface characteristics between particles of different composition (38) and some manipulation of the suspending medium may be critical. Although a number of laboratory studies on synthetic mixtures of ground minerals have demonstrated significant enrichment in one component (39), the enrichment factor decreases when the suspension is prepared from components that have been ground as a mixture (39–40). During the grinding process, which is essential for liberating minerals from ores, surfaces of one component tend to be contaminated with portions of other components. As of 1978 there were two commercial applications of selective flocculation, one involving taconite ores (41) and the other involving sylvinite ores (42).

Flocculant Selection. Selecting the best flocculant for a specific job is a challenge, and only a few comments outlining complexities encountered and representative approaches used are given here. Although cost–performance in the treatment plant is the final criterion, this article emphasizes the performance aspect.

In connection with applications it should be mentioned that sometimes two flocculants are used sequentially. The first is termed the primary flocculant. A common term for the second is flocculant aid but it appears more appropriate to call it a secondary flocculant or coflocculant. The terms coagulant and coagulant aid are also used in the literature.

Although the discussion in previous sections has emphasized gravity settling of flocs, other techniques for separating flocs from the liquid phase are practiced. Common techniques are filtration and centrifugation but for some applications a technique termed microflotation (43) (also dissolved air and induced air flotation) is used. In this approach tiny air bubbles are generated in the substrate. Flocs become attached to them and are carried to the top of the liquid where they are skimmed off. The method is particularly adapted to low density flocs such as are encountered in paper wastes, sewage, and oily wastes.

Substrate Variations. The surfaces of mineral substrates are usually hydrophobic but they may have been involved in a processing step that increased hydrophilicity. Activated sewage sludge (44) involves polysaccharides and proteinaceous materials which contribute to steric stabilization. A raw water supply may have <100 mg of colloidal-size suspended solids per kilogram of substrate, whereas a mineral processing stream may have several hundred grams of suspended solids per kilogram of substrate.

Although emphasis in the summary of modes of action of flocculants above appeared to be on particles of colloidal size, a wide range of sizes are involved in practice. The principles apply equally to all sizes, including attachment of small particles to large ones. Since the general features stabilizing suspensions of small solid particles also apply to aqueous suspensions of small liquid droplets (emulsions), flocculating agents are effective in flocculation of droplets (45) and breaking emulsions (qv).

The pH of the suspending medium ranges from strongly acidic (as in uranium ore processing), through the medium pH range for many waste streams, to strongly basic (as in the Bayer process for bauxite ore (see Aluminum)). Ionic strength ranges from very high (as in potash processing) to relatively low (as in certain water supplies). Temperatures cover the range from near freezing (outdoor settling facilities in cold climates) to near boiling (Bayer process). And in many practical situations there are fluctuations in substrate parameters with time. This variety in types of substrates suggests the need for a variety of flocculants, thereby posing a problem in selecting the best one.

Performance Criteria. Although flocculation was defined in terms of formation of aggregates, it is the consequences of that aggregation that constitute performance of a flocculant in commercial applications. The many facets of performance are specified by descriptors that measure a given attribute of the treated substrate. Among these facets are the settling rate of the flocculated solids, clarity of the supernatant, volume of settled solids, fraction of color removed, filler retained on paper fiber, filterability of flocculated solids, residual moisture in filter cake, etc. The importance of individual descriptors varies with the application. Clarity may be of primary importance in a municipal water treating application and settling rate of greater importance in a mineral processing plant. Figure 1 displays a generalized plot of magnitude of descriptor versus dosage for a flocculant. OFD designates optimum dosage for this flocculant and this descriptor. (Not all descriptors exhibit the same functional dependence on dosage of the flocculant (46).) Dosages used commercially tend to be some fraction of OFD, sometimes only a few percent, for economic reasons.

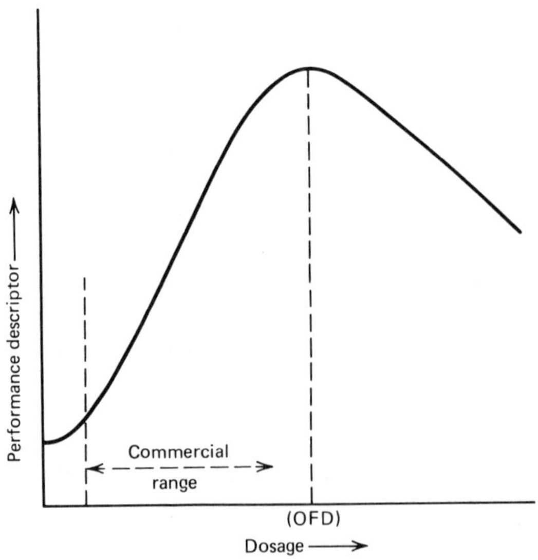

Figure 1. Schematic performance–dosage curve. OFD = optimum flocculant dosage.

Flocculant Addition. Flocculants are introduced into the substrate as aqueous solutions and are mixed with the suspended solids as rapidly and as uniformly as possible. Since correct preparation of flocculant solutions is important, for best results the manufacturer's recommendations should be followed. For example, there are special concentration–time factors for some inorganic flocculants. Usually organic flocculants should be introduced to the substrate in as high dilution as is practical, sometimes utilizing special techniques (8,47) for this purpose. Extended, intense agitation of the flocculated system should be avoided, since it breaks up flocs, but for some applications a moderate amount of agitation may be desirable (eg, sewage sludge dewatering) (48).

Laboratory Performance Testing. Selection of the flocculant for a given commercial application hinges on laboratory testing. Because a heterogeneous system is involved, rarely does a laboratory test accurately reflect the dynamics of a full-scale plant unit. Most laboratory testing is done on a batch basis, whereas most commercial units operate continuously. Nevertheless, the conventional laboratory tests developed over the years allow an experienced practitioner to select the better flocculants for a particular job. These laboratory tests should be made directly on the feed to the plant unit under consideration. Final determination of the operating dosage, however, must be made in the plant unit.

A wide variety in types of laboratory tests have been described in the literature (48–55); only a few representative ones are mentioned here. Information about the charge on suspended particles can be obtained from measurement of the zeta potential (49,56) which is useful in some situations (see Electrodialysis). In others, compositional heterogeneity of the substrate imposes difficulties in interpretation and utilization (57).

For suspensions having low solids concentration (a qualitative term, usually corresponding to less than ca 2 wt %), some form of the jar test (49–50,58) is commonly used. This involves an arrangement of 4–6 identical beakers (or jars, 500–1500 mL in size) of substrate to be treated simultaneously with identical paddle stirring in each beaker. Frequently, the stirring is rapid during flocculant addition (fast-mix), followed by a period of slow stirring for floccule growth, and finally a quiescent period to permit flocs to settle. Typical measurements include floc size and turbidity reduction (clarity) as a function of dosage for different flocculants. Qualitative observations of floc appearance and settling are also made. This method is used extensively for testing raw water supplies and low solids wastewaters, with attempts to simulate the mixing and flocculation sections of the treatment plant by varying the stirring regimes in the jar test.

For substrates that contain medium to high solids (up to ca 40 wt %) and with gravity settling of solids, a settling or thickening test (53) is used. The substrate, also referred to as pulp, is placed in a graduated cylinder (0.5–2.0 L) and agitated to obtain a uniform suspension. Then flocculant is added as a dilute solution with agitation to ensure uniform mixing with the substrate, agitation is stopped, and the time for the solids–liquid interface to settle to various levels is measured. Descriptors commonly checked are settling rate as a function of sediment height, sediment volume as a function of time, and supernatant clarity. Flocculant addition and agitation must be controlled very closely in order to obtain reproducibility.

For solids separation via vacuum filtration for substrates containing medium to high solids, a Buchner funnel test (51) is used. The funnel is provided with appropriate filter paper or filter cloth and is mounted in an arrangement allowing application of a prescribed vacuum while filtrate is collected in a graduated cylinder. Flocculant is mixed with substrate in a beaker, the mixture is poured into the Buchner funnel, and vacuum is applied for a prescribed time. The descriptors measured are rate of accumulation of filtrate, clarity of filtrate, and moisture in the filter cake. The test is used widely for waste streams, even when centrifugation is the mechanical separation technique.

Other common laboratory tests simulating some aspects of application areas for flocculants include drainage from paper pulp (59), retention of fines and fillers on paper fibers (Britt jar test) (55), solids concentration by microflotation (54), and as supplements to the Buchner funnel procedure, the filter-leaf test (51) and the capillary-suction time (CST) test (52).

Flocculating Agents

INORGANIC MATERIALS

Many soluble inorganic salts can function as indifferent electrolytes, following the Schulze-Hardy rule in general effectiveness, but few are significant commercially. More important are inorganic salts that have hydrolyzing ions and specific ion interactions with the substrate.

A convenient description (60) of hydrolysis reactions of cations involves denoting hydroxide ions as explicit ligands, other ligands being implicit, as in the generalized reaction:

$$x\ M^{3+} + y\ (OH^-) \rightleftharpoons M_x(OH)_y{}^{(3x-y)+}$$

where M is the cation (Al and Fe are of primary interest). Although a cation with 3 units of charge is indicated, by appropriate modification M could have been written with z units. A specific hydrolysis product of M is designated as the x,y species. The mole ratio of bound OH^- ions (bound to M) to total M(III) in solution, (OH/M), is termed the average ligand number $\bar{n}$, and is one of the important parameters of the solution. Hydrolysis products present in a solution are a function of M(III) concentration, pH, $\bar{n}$, age of solution, temperature, and to some extent, the anions present.

When the concentration of M(III) is very low ($\leqslant 10^{-5}\ M$), the stable species in solution tend to be monomeric, ie, $x = 1$, with the distribution between values of y changing with pH. Solubility of the 1,3 species is very small, although supersaturation is readily attained. When freshly precipitated at room temperature, the 1,3 species tends to be gelatinous and fluffy but becomes more crystalline with time. If the concentration of M(III) in solution is high ($\geq 10^{-2}\ M$), charged polynuclear hydroxides form. These may exist indefinitely in solution if $\bar{n} < 2.5$ but they are metastable with respect to ultimate precipitation of the neutral crystalline hydroxide. Thus in the use of this type of compound as flocculating agents, there frequently are specific instructions on solution preparation, dilution, and time factors (61).

Aluminum Derivatives. The hydrolysis behavior of the aluminum ion in aqueous solution has been the subject of many investigations. Although differences in detail are reported, there seems to be a consensus on general features (60,62–63). Kinetic data show that mononuclear products form rapidly and reversibly. Although formation of polynuclear species is relatively fast, breakdown is a slow process (62). For most effective use as a flocculant, aluminum solutions are prepared in concentrated form and then diluted at time of use (61). A schematic distribution among hydrolysis products is sketched in Figure 2 (60), where dashed curves represent supersaturation relative to $Al(OH)_3$, gibbsite [*14762-49-3*]. (The dashed curve labeled polymers represents a range of polynuclear species with $x = 6–20$ or more. These are nonlinear polymers, probably with hexameric units resembling those in crystalline $Al(OH)_3$ (62).) Minimum solubility of $Al(OH)_3$ occurs at pH = 6.2–6.8, and the concentration of soluble Al(III) species would be about $10^{-6.5}$ *M* but equilibration with polynuclear species in solution may be slow when $\bar{n} < 2.5$. Hydroxide precipitation occurs more rapidly at higher pH.

Alum. In the literature on flocculants the term alum [*7784-31-8*] refers to a commercial aluminum sulfate hydrate, $Al_2(SO_4)_3.XH_2O$ where X is about 14 (64). It also is called papermakers' alum or filter alum, and is available either in the dry form or in solution. Dry alum is available in several grades, with a minimum aluminum content of 17% expressed as Al_2O_3 (65). Liquid alum is about a 49 wt % solution of $Al_2(SO_4)_3.14H_2O$, or about 8.3 wt % aluminum as Al_2O_3. It can be stored indefinitely without deterioration (64). Alum is the most widely used inorganic flocculant. Its use in water treatment is estimated at ca 3.6×10^5 metric tons in the United States during 1978 (66). The growth rate for alum is slower than that exhibited by synthetic organic flocculants.

Sulfate ions appear to act as a catalyst in the formation of polynuclear complexes and their linkage to form a solid lattice (62). Thus precipitates form at lower values of $\bar{n}$ (and lower pH) in the presence of sulfate ions than for other anions, making alum effective in sweep flocculation under more acidic conditions than other aluminum salts (67).

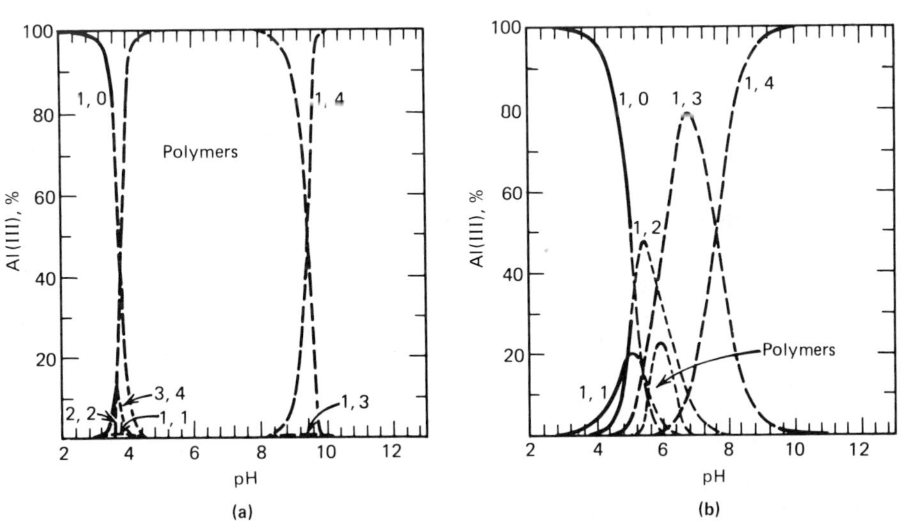

Figure 2. Approximate distribution of hydrolysis products (60). (**a**) 0.1 *m* Al(III); (**b**) 10^{-5} *m* Al(III). Dashed curves denote supersaturation relative to α-$Al(OH)_3$, and pairs of numbers designate the species $(Al^{3+})_x(OH^-)_y$.

Poly(aluminum chloride) and Related Materials. A partially hydrolyzed aluminum chloride [*7446-70-0*] solution ($\bar{n} \simeq 1.5$), which may incorporate a small amount of sulfate, has been introduced in Japan, England, and Australia as an alternative to alum (68). Known as poly(aluminum chloride), PAC, an approximate empirical formula for it is $Al(OH)_{1.5}(SO_4)_{0.125}Cl_{1.25}$, and it is available as a solution with an aluminum content of 10 wt % expressed as Al_2O_3. It is reputed to provide faster flocculation and stronger flocs than alum in some applications. Stabilizers and procedures for preparing related compositions are described in the patent literature (69).

Sodium Aluminate. Commercial sodium aluminate [*1302-42-7*] is available either in dry form or in solution, with an excess of base present. It provides a strongly alkaline source of water-soluble aluminum, particularly useful when addition of sulfate ions is undesirable. Sometimes it is used in conjunction with alum for pH control (see Aluminum compounds).

Iron Derivatives. Compared with aluminum, the hydrated ferric ion is more acidic (70), it forms stronger complexes with simple anions, and its amorphous hydroxide is less acidic but the two show a gross similarity in hydrolysis reactions (60,71). Aging characteristics of the polynuclear products of the ferric ion are more dependent on the anions (72). Minimum solubility of ferric hydroxide [*1309-33-7*] occurs in the pH range 6.8–8.4, where concentration of soluble Fe(III) species is about $10^{-11.5}$ *M*, but equilibration with polynuclear species in solution ($\bar{n} < 2.5$) may be slow. Ferrous ions form analogous mononuclear species but comparable data on tendency to form polymers are not available. Minimum solubility of ferrous hydroxide occurs near pH 10.7, where concentration of soluble Fe(II) species would be about $10^{-6.5}$ *M*, but a tendency for air oxidation to the ferric species complicates the system. Because of the color of iron compounds, they tend to be used in waste streams rather than in water supplies.

Liquid ferric chloride [*7705-08-0*] (73), a dark brown oily-appearing solution which is 35–45 wt% $FeCl_3$, is the customary form for flocculant use (73). Ferric chloride also is available in solid form.

Ferric sulfate [*10028-22-5*] (73) is marketed as dry granules, $Fe_2(SO_4)_3.XH_2O$, with $X \simeq 7$ [*35139-28-7*].

Ferrous sulfate [*7720-78-7*] (73) also known commercially as copperas, is generally available in dry form with the nominal composition $FeSO_4.7H_2O$ [*7782-63-0*]. It is a by-product of the production of titanium dioxide from ilmenite and of the pickling of steel (see Iron compounds).

Silicon Derivatives. Hydrolysis reactions of Si^{4+} usually are written in terms of the neutral species $Si(OH)_4$ and hydroxide ion ligands (60) (or of SiO_2 if the solid is involved). Solubility of SiO_2 in equilibrium with $Si(OH)_4$ is low (ca 10^{-3} *M*) below pH about 9, but in more alkaline solutions soluble anionic monomeric and multimeric silicate species form. When the pH of an alkaline sodium silicate solution is lowered into the insolubility range, polymerization occurs. With appropriate timing and subsequent dilution, this process gives rise to metastable polysilicates or colloidal aggregates that function as anionic flocculants (activated silica) (74–75) (see Silicon compounds; Silica).

Lime. The term lime refers to a number of chemicals that are principally calcium and oxygen, but which may contain a considerable amount of magnesium (dolomitic lime) (76). Hydrolysis reactions of calcium do not produce polymeric species, nor is there significant specific ion adsorption (by itself) under the usual conditions of commercial use. Lime is used primarily for pH control or chemical precipitation (including water softening) in water and wastewater treatment, or both. At the same time it assists flocculation, frequently functioning as a coflocculant. As the largest tonnage chemical involved (ca 2×10^6 t during 1978) it represents an important flocculating agent. The major types are quicklime (CaO [*1305-78-8*] or mixed CaO–MgO) and slaked or hydrated lime which may be $Ca(OH)_2$, $Ca(OH)_2/MgO$, or $Ca(OH)_2–Mg(OH)_2$. A dilute aqueous suspension of hydrated lime particles with the consistency of milk is termed milk-of-lime (see Lime).

Miscellaneous. Various by-product streams containing aluminum or iron salts (such as ferrous chloride pickle liquor) are used as flocculants on a local basis. Magnesium bicarbonate [*2090-64-4*] is used in a recycle system (77). Bentonite [*1302-78-9*] and other clays sometimes are added as aids in flocculation (see Clay). Main United States producers of inorganic flocculants, except lime, are listed in Table 1. Lime producers are omitted since they are so numerous (76).

SYNTHETIC ORGANIC MATERIALS

Practical synthetic organic flocculants are water-soluble polymeric substances with weight average molecular weights ranging from about 10^3 to greater than 5×10^6

Table 1. Main Producers of Inorganic Flocculants in the United States

Company	Products[a]					
Allied Chemical Corp.	a					f
American Cyanamid Company	a					
Associated Metals & Minerals Co.			c			
Burris Chemical, Inc.	a					
Catco, Inc.			c			
Cities Service Company, Inc.	a			d		
Conservation Chemical Co.			c			
The Cosmin Corp.					e	
Diamond Shamrock Corp.						f
The Dow Chemical Company			c			
E. I. du Pont de Nemours & Co., Inc.	a		c			
Essex Chemical Corp.	a					
Filtrol Corp.	a					
Philip A. Hunt Chemical Corp.			c			
Imperial West Chemical Co.	a		c			
Nalco Chemical Co.		b				
NL Industries, Inc.					e	
Olin Corp.	a					
Pennwalt Corp.			c			
Pfizer, Inc.					e	
Philadelphia Quartz Co.						f
Quality Chemicals, Ltd.					e	
Reynolds Metals Co.		b				
Southern California Chemical Co.			c			
Stauffer Chemical Co.	a					
K. A. Steel Chemicals, Inc.			c			
Vinnings Chemical Co.		b				

[a] a = alum; b = sodium aluminate; c = ferric chloride; d = ferric sulfate; e = ferrous sulfate; and f = sodium silicate.

(reported values as high as 20×10^6). If some subunits of the polymer molecule are charged, it is termed a polyelectrolyte (78). When the charge on the subunits is positive, as for polyamine salts, the polymer is termed cationic; when the charge is negative, as for poly(sodium acrylate) [*9003-04-7*], it is termed anionic. Polyelectrolytes (qv) containing both positive and negative charges in the same molecule are termed polyampholytes. Some water-soluble polymers do not contain any charged subunits (or at least no significant number). Two such polymers are pure polyacrylamide and poly(ethylene oxide). These are termed nonionic polymers. Although nonionic water-soluble polymers do not fall within the definition of a polyelectrolyte, they tend to be placed in the same category in the flocculant literature.

Synthetic organic flocculants are supplied to the user in a variety of physical forms—dry powder, granules, beads, aqueous solutions, aqueous gels, and water-in-oil emulsions (qv). Relatively concentrated aqueous solutions of low molecular weight polymers are convenient to handle, and many are marketed in this form. As the molecular weight increases, viscosity increases, and the polymer concentration must be reduced to retain ease of handling. Storage, shipping, and handling costs increase per unit of active flocculant as the concentration decreases. Thus, a combination of economic and technical factors dictates the form in which a polymer is manufactured and sold. High and very high molecular weight polymers tend to be sold as dry products or as oil-in-water emulsions. Each form has advantages. Requirements imposed at different treatment sites vary, and thus it is not unusual for essentially the same polymer to be sold in more than one physical form.

There is no comprehensive critical monograph on synthetic organic flocculating agents available. An understandable reticence on the part of manufacturers to make details of processes and products available to competitors precludes a thorough treatise, particularly on those flocculants for which a large commercial market exists. Various reviews that include selected flocculants have appeared in the literature (7,79) but the discussion has been relatively brief. For reports on patents see ref. 80 for recent developments and 81 for earlier ones.

In this article, synthetic organic flocculants are grouped according to chemical functionality, as cationic, anionic, ampholytic and nonionic compositions, and are dealt with in that order. Subdivisions of these broad groups are made somewhat arbitrarily to emphasize either synthesis similarities or chemical similarities. To assist in visualizing chemical composition, representative structural repeating units (SRUs) for the primary types of commercial flocculants are depicted schematically in Table 2. It is necessary to note the SRUs are an idealized depiction, believed to represent the major structural elements but omitting secondary features.

Cationic

Although a number of polysulfonium and polyphosphonium compounds have been reported (82), quadrivalent nitrogen is the charged site in all commercial cationic flocculants at the present time (see Ammonium compounds). Cationicity derives from the quadrivalent nitrogen, either via protonation of primary, secondary, or tertiary amines, or via generation of quaternary nitrogen groupings. Flocculants of this latter type sometimes are referred to as polyquats. Cationicity of the quaternary nitrogen is independent of pH, although other parts of the polymer molecule may exhibit sensitivity to pH, such as hydrolysis of ester linkages. Cationicity dependent on protonation, however, is a function of pH. Without getting enmeshed in details of polymer–particle surface interactions, the gross effects of pH on effective cationicity can be obtained from colloid titration behavior of the polymer at fixed levels of pH. Data on four representative polymer types, titrated with poly(potassium vinyl sulfate), are sketched in Figure 3. Results are given as fraction of maximum cationicity titrated for each flocculant, in buffered solutions. (The slight change in shape of curves with electrolyte content does not alter their general features (85).) It is clear why some function very well as cationic flocculants in acidic substrates, but very poorly in moderately basic substrates. A substrate example is municipal sewage sludge at pH ca 6.5, vs its lime-treated counterpart at pH ca 12.

Step-Growth Polymers. ***Poly(alkylene polyamines) and Poly(hydroxyalkylene polyamines).*** These are step-growth polymerization products of polyfunctional amines and alkyl dihalides or bifunctional alkyl epoxides and alkyl epoxide derivatives. As such they are of relatively low molecular weight ($\lesssim 10^5$) but have a high level of cationicity. In the flocculant literature they are referred to generally as polyamines. Typically, in the first case a poly(ethyleneamine) is produced by the reaction of ammonia, or low molecular weight alkylene polyamine, with ethylene dichloride in the presence of aqueous base (86). This is sometimes followed by reaction with 1-chloro-2,3-epoxypropane (epichlorohydrin) (87). Reaction variables have to be programmed carefully to meet molecular weight requirements (88). A relatively high content of piperazine rings in the chain of the product show in the infrared spectrum (89). Completely linear poly(ethyleneamine) is known (90) (Table 2) but is not available commercially.

The second family of polyamines is usually prepared from 1-chloro-2,3-epoxypropane and ammonia or monomethylamine (Table 2), or both, under basic conditions (86,91–92). The reaction involves a series of nucleophilic ring-opening additions and condensations. Depending on reaction conditions, these polyamines tend to be branched, particularly the higher molecular weight products. Completely linear model polymers, derived from 1-alkyl-2-hydroxyazetidines through ring-opening polymerization, have recently been described (93).

Poly(ethyleneimine) [*26913-06-4*] (PEI) is a special poly(ethyleneamine) produced commercially from ethyleneimine in an aqueous medium by ring-opening polymerization, using carbon dioxide, a mineral acid, or ethylene dichloride as an initiator (94) (see Imines, cyclic). This product (PEI or polyaziridine) is highly branched, with primary, secondary, and tertiary amine groupings. Its primary use is in the manufacture of paper (95) (qv). Since a series of products with relatively well-defined molecular weights (ca 10^3–10^5) have been available, PEI has been used in a number of model investigations of flocculation (96).

Polyamine Quaternaries. These cationic polymers are prepared either by quaternization of poly(alkylene polyamines) and poly(hydroxyalkylene polyamines) with alkyl halides or sulfates (86,97), or by step-growth polymerization from dialkylamines, tetraalkyl amines, or derivatives thereof (98), with suitable bifunctional alkylating agents, and with or without small amounts of polyfunctional primary amines (such as ammonia, ethylenediamines, and others) for molecular weight enhancement (98) (see Diamines). Polyamines produced from ammonia and ethylene dichloride, quaternized with methyl chloride, and polyquaternaries produced directly from dimethylamine and 1-chloro-2,3-epoxypropane (Table 2) are of primary commercial significance. The fact that polyquaternaries are relatively insensitive to pH variation broadens the spectrum of possible applications. Quaternaries are also much more

Table 2. Principal Commercial Synthetic Organic Flocculant Structures

Polymer	CAS Registry No.	Molecular formula	Charge	Comments
poly(ethyleneamine)	*[26913-06-4]*	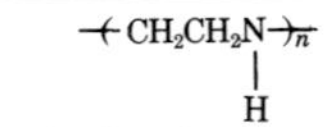 (idealized structure)	cationic	homopolymer; charge density function of pH
poly(2-hydroxypropyl-1-*N*-methylammonium chloride)	*[31568-35-1]*	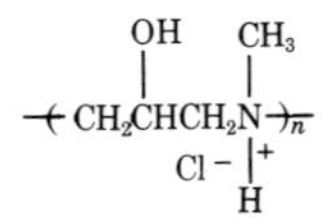 (idealized structure)	cationic	homopolymer; charge density function of pH
poly(2-hydroxypropyl-1,1-*N*-dimethylammonium chloride)	*[39660-17-8]*	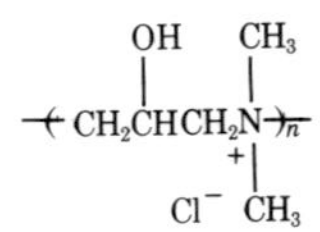	cationic	homopolymer; strongly cationic, pH-insensitive, chlorine resistant; major representative
poly[*N*-(dimethylaminomethyl)-acrylamide]	*[25765-48-4]*	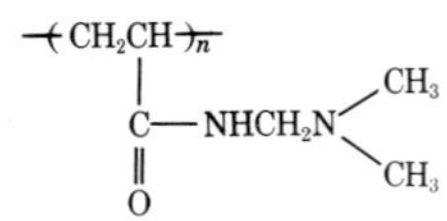	cationic	solution equilibria; charge density function of pH; major commercial importance
poly(2-vinylimidazolinum bisulfate)	*[0000-00-0]*	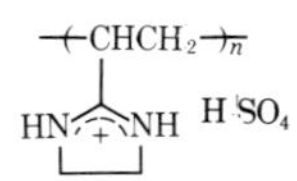	cationic	homopolymer; charge density function of pH (amidine structure)
poly(diallyldimethylammonium chloride)	*[26062-79-3]*	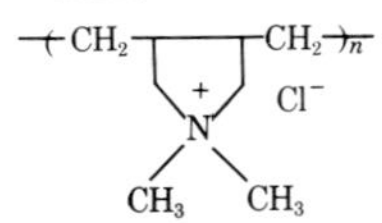	cationic	homopolymer; strongly cationic, pH-insensitive; chlorine resistant; major representative; copolymers with acrylamide also of significance

poly(*N*,*N*-dimethylaminoethyl methacrylate), neutralized or quaternized	[*25145-86-3*]	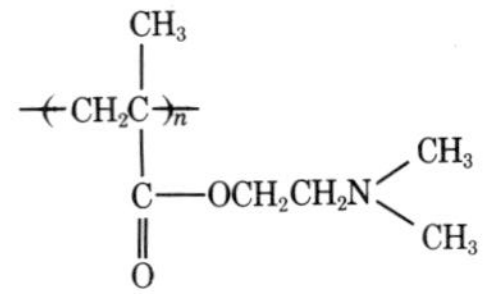	cationic	homopolymer; copolymer with acrylamide of more importance
poly[*N*-(dimethylaminopropyl)-methacrylamide]	[*64080-86-0*]	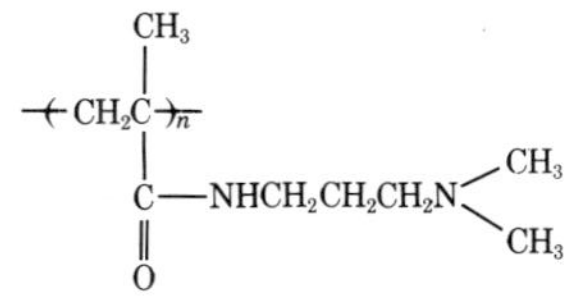	cationic	hydrolytically stable cationic acrylamide derivative
poly(sodium or ammonium acrylate)	[*7446-81-3*] [*10604-69-0*]	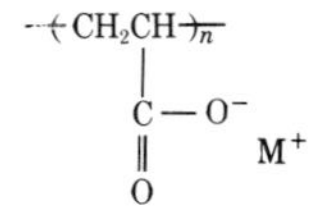	anionic	homopolymer and major anionic constituent with acrylamide
poly(sodium styrenesulfonate) (PSS)	[*9080-79-9*, *60495-06-9*]	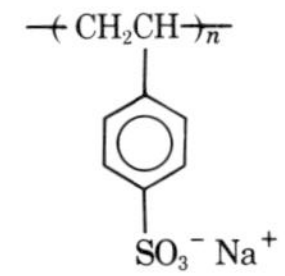	anionic	homopolymer of some significance as strongly anionic, pH-insensitive flocculant
polyacrylamide (PAM)	[*9003-05-8*]	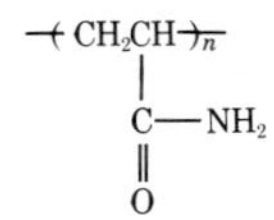	nonionic	principal nonionic flocculant and a major constituent in copolymers
poly(ethylene oxide) (PED)	[*25322-68-3*]		nonionic	effective nonionic flocculant for special applications
poly(vinylpyrrolidinone)	[*9003-39-8*]	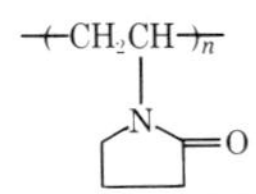	nonionic	flocculant for special applications

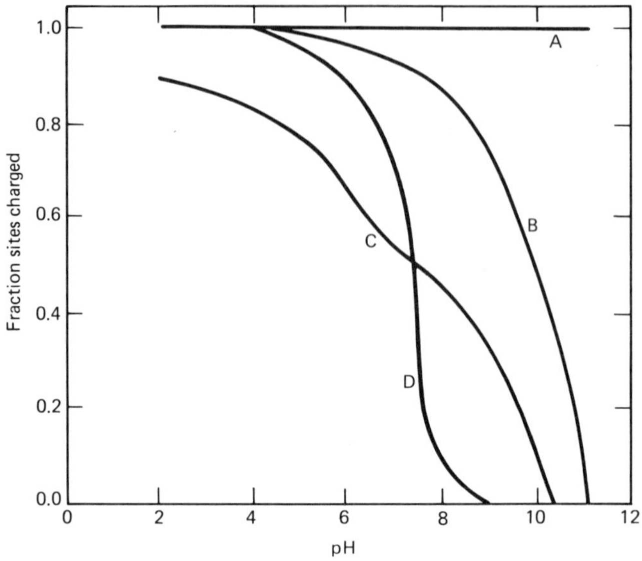

Figure 3. Fraction available sites with cationic charge. A, Quaternary nitrogen, poly(*N,N,N*-trimethylallylammonium methyl sulfate) (83); B, poly(vinylimidazoline); C, poly(ethyleneimine), Montrek 1000; D, Mannich reaction product of poly(acrylamide). Courtesy of American Cyanamid Company (84).

resistant to degradation by chlorine than are the simple polyamines, a fact that can be important in chlorinated raw water. The mechanism of degradation by chlorine is not established but probably is analogous to fragmentation of trimethylamine with hypochlorite (99).

Poly(epichlorohydrin) (100) reacts with lower trialkylamines (101) to generate a series of polymers with pendent quaternary nitrogen groups. Processing does not appear to be simple (see Chlorohydrins).

Cationic Cyanamide Derivatives. A diverse group of amino-aldehyde step-growth condensation products are included in this group. Amino building blocks such as dicyandiamide (102), melamine (103), urea (104), and guanidine (104) can be derived from cyanamide as a basic raw material. Formaldehyde is used for the aldehyde (105). The products obtained are water dispersible and may be modified with multifunctional amines (104) and amino alcohols (106). The molecular weight of these products is rather low (ca 1–5 $\times 10^3$) (107), and they have a tendency to set on heating (see Amino resins). Cationic cyanamide derivatives were initially developed for improving paper properties (108) but their action in destabilizing colloidal materials also was recognized (109) (see Cyanamides).

Vinyl Polymer Conversion Products. ***Cationic Carbamoyl Polymers.*** These cationic polymers are a versatile and commercially important group of synthetic organic flocculants. The homopolymer of acrylamide reacts with formaldehyde and a suitable amine (eg, dimethylamine) *via* the Mannich reaction to form a polytertiary amine (Table 2) (110). Starting with polyacrylamide of the appropriate molecular weight, highly charged flocculants with molecular weights much greater ($>10^6$) than those of step-growth cationic polymers can be produced. High viscosity mandates low active solids content ($<10\%$) for aqueous solutions, a disadvantage in shipping costs. New technology, however, involving high-solids hydrocarbon-based polymer dispersions (20–50%) is aimed at alleviating the problem (111). Recent patent literature deals with quaternized carbamoyl polyamines, which are remarkably stable. Both high (112) and lower (113) molecular weight products have been described.

Poly(2-vinylimidazoline). As a polycyclic amidine (Table 2), this polymer [*10577-96-5*] has a structure unique for a flocculating agent with very high cationicity and medium or high molecular weight. It is produced from polyacrylonitrile by a sulfur-catalyzed condensation with ethylenediamine. A mechanism for the reaction is discussed in the literature (114). A process in which polymer remains suspended throughout the conversion presumably is used commercially (115) (see Acrylonitrile polymers).

Polymers made by vinyl polymerization of 2-vinylimidazoline are known (116) but they have low molecular weight unless copolymerized with acrylamide. Step-growth polymerization products of acrylonitrile (qv) and an alkylene polyamine (117) contain cyclic amidine structures but they are also low in molecular weight because of the nature of the synthesis.

Vinyl Addition Polymers. ***Poly(diallyldimethylammonium chloride).*** Poly(diallyldimethylammonium chloride) is a linear polymer synthesized from a 50–70% aqueous solution of diallyldimethylammonium chloride by free-radical cyclopolymerization, initiated with peroxide catalysts (118–119). The polymers are in the low to medium molecular weight range ($<10^6$), chlorine resistant, and insensitive to pH variations (quaternary salt). Contrary to a long held belief, the predominant structural unit is the 5-membered pyrrolidine ring (120) not the 6-membered piperidine ring (Table 2).

Copolymers [*26590-05-6*] of diallyldimethylammonium chloride with acrylamide (qv) (119,121) also are of importance, and can be prepared with a wide range of molecular weights (119,122) (see Allyl monomers and polymers).

Dimethylaminoalkyl Acrylic and Methacrylic Ester Polymers. *N*-(Dialkylaminoalkyl)methacrylic ester polymers and *N*-(dialkylaminoalkyl)acrylic ester polymers, both as homopolymers and as copolymers with acrylamide, represent an important class of cationic flocculants. Although there are many amino ester monomers (123), polymers based on dimethylaminoethyl methacrylate (Table 2) have the largest commercial use, particularly as the methyl chloride or dimethyl sulfate quaternary salts. As derivatives of methacrylic acid and acrylic acid, the amino ester monomers copolymerize readily with acrylamide to yield high molecular weight polymers (123–124). The fact that the ester group is susceptible to hydrolysis, especially in neutral and basic media, influences treatment of monomer and polymer (125). Since the amino ester monomer is considerably more expensive than acrylamide, major flocculant usage involves copolymers containing about 1–40 mol % of the cationic monomer. Cost of the homopolymer limits its area of application. Flocculants are available in all the common physical forms. Most manufacturers of synthetic organic flocculants include a dimethylaminoethyl methacrylate product in their line. Technical advances in manufacture of the polymers are linked to those of polyacrylamide technology.

N-(Dialkylaminoalkyl)acrylamide Polymers. These polymers are of interest as flocculants since they are more stable toward hydrolysis than the corresponding methacrylate and acrylate structures. Many references to suitable cationic amide monomers exist in the literature (124,126). Recently *N*-(dimethylaminopropyl)-

methacrylamide (Table 2), which has been known for some time (127–128), has become available on a semicommercial basis, both as the tertiary amine and as a quaternary salt. As a derivative of acrylamide it copolymerizes readily with acrylamide to yield high molecular weight polymers.

Miscellaneous. Other flocculants have been synthesized from a number of commercial cationic monomers, such as 2-vinylpyridine, 5-vinyl-2-picoline, and *p*-vinylbenzyltrimethylammonium chloride, but their sales volumes are apparently small. References to them can be found in review literature (82,123). Many other cationic polyelectrolytes are described in the literature (82); selections in the previous paragraphs are based primarily on commercial availability.

Anionic

In anionic flocculants, 1–100% of the monomer units contribute to the charge. Molecular weights tend to be in the high or very high range. Low molecular weight, highly anionic polymers behave as dispersants (qv). Although several types of anionic sites in a polymer are possible, the major type in commercial flocculants is the carboxylate ion.

Polymers Containing Carboxyl Groups. Homopoly(acrylic acid) [*9003-01-4*], poly(acrylic acid-*co*-acrylamide) [*9003-06-9*], and hydrolyzed polyacrylamide (Table 2) dominate the anionic flocculant market. Hydrolyzed poly(acrylonitrile) plays a lesser role. Polyacrylamide can be hydrolyzed readily with alkali (usually sodium hydroxide or sodium carbonate) in an aqueous medium (129). The hydrolysis is relatively rapid up to about 50% conversion of amide groups and then slows down considerably, probably owing to neighboring group effects. High molecular weight anionic flocculants, with 2–50% of the amide groups hydrolyzed by stoichiometric amounts of alkali, are prepared conveniently in this way. Patents have been issued to several flocculant producers covering the use of a number of alkalies under various process conditions, including simultaneous polymerization and hydrolysis of acrylamide (130).

Addition copolymerization of acrylic acid with acrylamide covers the entire anionicity range, with various polymerization techniques described in the patent literature (131). The rate of copolymerization is pH dependent, and the reactivity ratio for acrylic acid copolymerizing with acrylamide decreases with increasing pH (132), important factors in the technology of production (see Acrylic acid and derivatives; Acrylamide polymers).

Although the carboxyl grouping is a relatively strong acid, the extent of ionization is a function of pH in acidic substrates. Both the inherent acidity of the carboxyl group and the effect of neighboring carboxyl groups in the same molecule (electrostatic and statistical) influence the ionization. Electrostatic effects are moderated by ionic strength of the medium, and all of these factors influence the conformation of the molecule in solution. A representative situation is plotted in Figure 4 (10) where the solid line is for an acrylic acid–acrylamide copolymer with 30 mol % acrylic acid, in a solution with moderate ionic strength. The dashed lines represent a very high density of ionizable sites (polyacrylic acid), and a very low density of sites. Plots of this type, based on data for a homogeneous medium (133), are useful as guides to the level of anionicity. For a heterogeneous system, however, conditions in the immediate vicinity of a particle surface may be somewhat perturbed relative to those in the bulk phase.

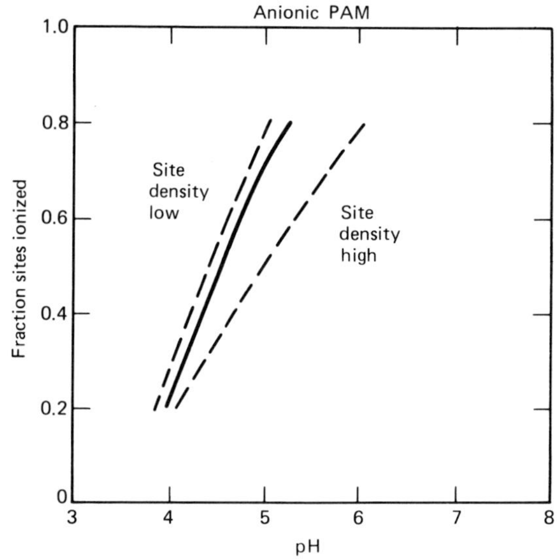

Figure 4. Carboxyl ionization in acrylic acid–acrylamide copolymers. Courtesy of Canadian Mineral Processors (10).

Polymers Containing Sulfonic Acid Groups. The sulfonic acid grouping is an inherently stronger acid than the carboxyl grouping. Polymers with sulfonic acid moieties, therefore, retain their anionic charge in media at low pH. The sodium salt of poly(styrenesulfonic acid) (Table 2) is the best known commerically available flocculant of this type. It can be derived from sodium styrene sulfonate by free radical polymerization (134) or from polystyrene by postsulfonation (135). Very high molecular weight products have been investigated as flocculants for so-called red muds in the Bayer process for bauxite, to demonstrate the effect of molecular weight on that substrate (136). Copolymers of sodium styrene sulfonate with acrylamide also have been patented as flocculants (137) (see Styrene).

Related poly(ethylenesulfonic acid) [*26101-52-0*] polymers, including copolymers with acrylamide, have been reported to be flocculants (138–139). A sulfonic acid monomer, 2-acrylamido-2-methylpropanesulfonic acid (140) has been recently test-marketed, and is projected to be available commercially by 1980. Utility as an anionic monomer for flocculants is cited in the patent literature (138,140).

Nonionic

For purposes of classifying flocculants, a polymer is considered nonionic if fewer than 1% of the monomer units are charged. In aqueous systems such polymers function as flocculants primarily by the bridging mechanism. Hence, they must be high or very high in molecular weight for practical applications. Polyacrylamide is the most prominent member of the class, followed by poly(ethylene oxide) and others are of little commercial significance as flocculants.

Polyacrylamide. Unique polymerization characteristics of acrylamide (129) allow production of very high molecular weight polymers (Table 2), (some having weights up to 20×10^6). During processing there may be some inadvertent hydrolysis, introducing small amounts of anionicity. Some manufacturers allow up to 4 mol % hydrolysis in the nonionic category, but since hydrolysis can be controlled to a few tenths of 1%, an upper limit of 1 mol % is reasonable. Various processes for producing acrylamide homopolymers by addition polymerization have been discussed extensively in the review literature (129–130,141) (see Acrylamide polymers). Chemical and physical properties of the polymers often are dependent on synthesis variables, an aspect dealt with in the reviews. One feature that has been the subject of many basic investigations is the stability of dilute aqueous solutions on aging; a variety of synthetic procedures have been tested (142).

Acrylamide Polymer Production. Since acrylamide-based polymers (PAMs) play such a dominant role in the flocculant market (major contributor to synthetic organic category in 1978), a few comments on recent developments in production methods are pertinent. Industrial processes include aqueous solution polymerization, mixed solvent polymerization (also called precipitation polymerization), and dispersed-phase polymerization (also referred to as suspension and inverse emulsion polymerization), using a variety of initiators. Low solids (8–10%) aqueous solution polymerization techniques have been practiced the longest, with newer technology aimed at improving the economics of drying (143). Continuous (belt) processes, at high solids levels, are aimed at improved overall process economics via high-capacity production and lowered water-removal costs (144). A process of this type, based on photopolymerization technology (actinic photoinitiation) (145) is operated on a pilot scale in France, and in a full-scale plant in Finland, for production of a range of cationic, nonionic, and anionic PAMs (146). In another approach, drying has been circumvented via introduction of a product in gel form with high solids, synthesized using cobalt-60 radiation (147).

Dispersed-phase polymerization represents a significantly different approach (148). Minute droplets of a concentrated aqueous monomer solution (up to 50% or more monomer) are dispersed in an organic medium, such as xylene or other hydrocarbon, providing a system with relatively low bulk viscosity, and still maintaining essentially aqueous polymerization conditions conducive to formation of high molecular weight polymers. The technology involved in producing flocculants this way is complex but it has resulted in recent introduction of several commercial products in emulsion form. Although some require addition of an emulsion breaker to develop dilute aqueous flocculant solutions, most of the recently introduced products are self-breaking on dilution with water in flocculant feed tanks. Since the emulsion must have good storage stability under extremes of ambient temperature conditions and still break quickly on dilution, this represents a significant feat (149) (see Emulsions; Latex technology).

Poly(ethylene oxide). Flocculant-grade ethylene oxide polymers (Table 2) have much higher molecular weights ($\geqslant 3 \times 10^6$) than that for other application areas (150). These polymers are made by a special catalytic process, probably using an alkaline earth catalyst (151) (see Polyethers).

Poly(vinylpyrrolidinone). Poly(vinylpyrrolidinone) (Table 2) is produced by addition polymerization of *N*-vinylpyrrolidinone for a variety of applications (152). Flocculant usage is relatively specialized, eg, for clarification of beverages, such as fruit juices, beer, wine, and other food products (152–153) (see Acetylene-derived chemicals; Vinyl polymers).

Poly(vinyl alcohol). Poly(vinyl alcohol) [*9002-89-5*] is produced by alcoholysis of poly(vinyl acetate) catalyzed either by alkali or acid. Molecular weight is rather low (ca 3×10^5 or less) for a nonionic flocculant, and it does not appear to have established a commercial position as a flocculant. On the other hand, it has been used in a number of fundamental studies on flocculation (154) since it is relatively easy to achieve a balance between hydrophobicity and hydrophilicity in the molecule by controlling the extent of alcoholysis of the parent poly(vinyl acetate) (see Vinyl polymers).

Polyampholytes

Zwitterion flocculants containing both cationic and anionic functional groups in the same polymer molecule are intriguing but little is known about them. Mannich reaction products of polyacrylamide, slightly hydrolyzed, have been claimed to have special properties for treatment of sewage sludge (155). Copolymers containing 2-aminoethyl methacrylate and methacrylic acid have been reported effective in treating anionic substrates at or below the isoelectric point of the polyampholyte (156) (see Methacrylic acid and derivatives).

Major Manufacturers and Brand Names

Most large worldwide producers of synthetic organic flocculants are listed in Table 3, along with their brand names and general classes of products. Principal manufacturers also sell flocculants to specialty firms who may resell under their own brand names. When combined with smaller manufacturers not listed in Table 3, the total number of flocculants offered for sale is in the hundreds.

FROM NATURAL PRODUCTS

The first flocculants known were derived from natural products, crushed almonds, beans, and other nuts, eg, were used to clarify drinking water. Current flocculants derived from natural products involve starch, starch derivatives, plant gums, seaweed extracts, cellulose derivatives, proteins, and tannins. Starch is the most important on the list by a considerable margin, followed by guar gum. An important point to note is that flocculant sales represent a very minor part of the market for natural products, and that other uses dominate the industry. For example, in some countries starch is reserved for food only.

Although the price per kilogram for natural products tends to be low relative to synthetic flocculants, dosage requirements tend to be high. Hence, on a cost–performance basis synthetic flocculants are replacing naturally-derived ones in a number of areas, such as PAMs replacing protein colloids in uranium processing. Furthermore, the composition of natural products tends to fluctuate, and they tend to be more susceptible to microbiological attack which creates storage problems. Nevertheless, they enjoy a considerable market.

Starch [*9005-25-8*]. Starches consist of two types of polysaccharides: linear chains of D-glucose units joined by α-D($1\rightarrow4$) linkages (amylose) [*9005-82-7*], and highly branched structures consisting of short chains of 20–25 units of the linear polysaccharide joined to each other by α-($1\rightarrow6$) glycosidic linkages (amylopectin) [*9037-22-3*]. Usually amylopectin is the major component and has a much higher molecular weight, ranging up to 10^6. The amylose fraction has been reported superior to amylopectin or native starch in red mud flocculation (157). Starches isolated from corn, wheat, sorghum, rice, potatoes, and tapioca are all used to some degree as flocculants but some work better with certain substrates than others (2).

Table 3. Worldwide Producers of Synthetic Flocculants[a]

Company	Brand names	Types[b]
Allied Colloids Ltd. (UK)	Magnafloc, Percol, Zetag	a
American Cyanamid Company (U.S.)	Magnifloc, Superfloc, Cyfloc	a, b, c
BASF A.G. (FRG)	Sedipur	a, b
Betz Laboratories, Inc. (U.S.)	Betz Polymer, Polyfloc	a, b
The Calgon Corp. (Merck & Co.) (U.S.)[c]	Calgon, Coagulant Aid, Catfloc	a, b, c
Chemische Fabrik Stockhausen & Cie (FRG)	Praestol	a, b, c
Dai-Ichi Kogyo Seiyaku Co., Ltd. (Japan)	Hi-Set	a
Dearborn Chemical (W. R. Grace) (U.S.)[c]	Aquafloc	a
Diamond Shamrock (U.S.)	Nopcofloc	a, b, c
The Dow Chemical Co. (U.S.)[c]	Separan, Purifloc, PEI, Montrek	a, b
Drew Chemical Corp. (United States Filter Corp.) (U.S.)	Amerfloc, Drewfloc	a, b, c
Floerger (Fr.)	Floerger	a, b, c
Hercules, Inc. (U.S.)	Hercofloc	a, b
KEMIRA Oy (Finland)	Fennofloc	a
Kyoritsu Yuki (Japan)	Himoloc	a
Mitsubishi Chemical Industries, Ltd. (Japan)	Diaclear	a
Mitsui-Cyanamid Ltd. (Japan)	Accofloc	a
Nalco Chemical Co. (U.S.)[c]	Nalco, Nalcolyte	a, b
National Starch Corp. (U.S.)	Versa-TL	c
Nobel-Hoechst (Fr.)[c]	Bozefloc	a
Rhone-Poulenc (Fr.)	Flocogil	a
Rohm G.m.b.H. (FRG)	Rohafloc	a, b
Rohm and Haas (U.S.)	Primafloc	b
Sandoz (U.S.)	Sandolec	b, c
Sankyo Chemical Industries, Ltd. (Japan)	Sanpoly	a, b
Sanyo Chemical Industries (Japan)	Sanfloc	a, b
Sumitomo Chemical Co., Ltd. (Japan)	Sumifloc	a, b
Toagosei Chemical Industry Co., Ltd. (Japan)	Aronfloc	a
Union Carbide Corp. (U.S.)	Polyox	c

[a] Data supplied by manufacturer unless otherwise indicated.
[b] a = nonionic, anionic, and/or cationic PAMs; b = polyamines, polyquaternaries; and c = others, PEO, PSS, cationic cyanamide derivatives.
[c] Data from Refs. 4, 81.

Dosage efficiency and dispersibility of starch flocculants are improved by introduction of cationic or anionic substituents. Cationic starch derivatives are produced by etherification with reagents such as *N,N*-diethylaminoethyl chloride (158) and 2,3-epoxypropyltrimethylammonium chloride (159). Degree of substitution (DS) is generally low (≤ 0.05). Another route to cationicity is through grafting of cationic monomers onto starch (160–161) (see Cyanoethylation).

Anionic starch usually is produced by causticization with alkali (2). Other techniques include treating mixtures of dry starch with mono- or disodium phosphate at a temperature in excess of 100°C for relatively short periods of time (162), and grafting (161). Large amounts of causticized starch currently are used in the mining industry but it is gradually being replaced by synthetic flocculants. One unique commercial application, however, is selective flocculation in a low-grade iron ore pulp (41) (see Starch).

Guar Gum [*9000-30-0*]. This water-soluble high molecular weight polysaccharide has been of interest since its commercialization in the early 1950s as a hydrocolloid (163). Although it is a legume seed extract, mechanical seeding and harvesting has made it available on a large scale. Chemically, guar gum is a galactomannan, more specifically it is a poly-(1→4)-β-D-mannose substituted on alternating mannose units with D-galactose through α-(1→6) links. It is nonionic and, hence, an effective flocculant over a wide range of pH and ionic strengths. A disadvantage of guar gum is relatively rapid biological decomposition in aqueous solutions. This can be alleviated by addition of chelating agents.

Cationic aminoethyl ether and quaternary ammonium derivatives are known. A commercial product of the latter type is available (DS ca 0.13), as is an anionic product, the carboxymethylated sodium salt (DS ca 0.06) of guar gum (164) (see Gums).

Protein Colloid. Animal glue or gelatin [*9000-70-8*] is derived from fibrous collagen, the major protein of skin, bones, and connective tissues (165). Collagen has a multistranded structure of helically coiled polypeptide chains, varying somewhat with origin of the collagen (see Biopolymers). The polypeptide chains contain a variety of amino acids, with a high content of glycine, proline, and hydroxyproline (see Amino acids). Gelatin is produced from collagen by partial hydrolysis, resulting in some dissociation of the polypeptide strands and some degradation of peptide bonds within the chains, and generating a water-soluble protein. The terms gelatin and glue represent essentially the same complicated mixtures of hydrolysis products, differing mainly in purity (edibility). A purified glue sometimes is sold as technical or inedible gelatin. The term protein colloid is used to emphasize the functional behavior of these products (see Glue; Gelatin).

The polymers are ampholytic owing to the amine and acid groups present. Acid hydrolysis yields products with an isoelectric point around pH = 9; alkaline hydrolysis yields products with an isoelectric point around pH = 5, primarily owing to formation of acid sites via deamidation of glutamine and asparagine moieties (166). Properties are a function of hydrolysis conditions. Animal glue is used as a flocculant in mineral processing (167) but is being supplanted by synthetic products. Edible gelatin is used for clarification of wine, beer, vinegar, and nonalcoholic beverages (168).

Miscellaneous. Other polymeric materials derived from natural products have been tested for use as flocculants, but do not appear to have developed a significant market. A few are listed below.

***Alginic Acid* [*9005-32-7*].** Sodium alginate [*9005-38-3*] (169), another polysaccharide-based flocculant, is produced from brown algae (seaweed) by alkaline extraction. The acid either is a mixture of linear homopolymers of (1→4)-β-D-mannuronic acid and of L-guluronic acid, or else is a copolymer of the two uronic acids. Although flocculation is a minor application, sodium alginate is offered commercially for this purpose. There are indications that microbial alginate produced by continuous culture processing (170) (with *Azobacter vinelandii*) could provide a source independent of the sea (see Gums).

***Chitosan* [*9012-76-4*].** This water-soluble cationic polysaccharide, is made from chitin, poly (1→4)-2-acetamido-2-deoxy-β-D-glucose, by partial deacetylation with alkali. Chitin is derived from crustacean shells (fishery waste), and, therefore, the raw materials supply is limited at present. Activity as a flocculant has been demonstrated (171).

***Dextran* [*9004-54-0*].** Dextran is a nonionic microbial extracellular polysaccharide made from sucrose, predominantly a (1→6) linked poly-α-D-glucose. It is reported to have a very high molecular weight (see Sugar).

***Sodium Carboxymethyl Cellulose* [*9004-32-4*].** This water-soluble gum is produced from alkali cellulose (alkali salt of poly-β-D-glucose) and monochloroacetic acid with DS in the range 0.4–1.5. Unlike many flocculants derived from natural products, it is relatively resistant both to biological and to hydrolytic degradation (see Cellulose derivatives).

***Tannins* [*1401-55-4*].** Tannins are a complex mixture of polyphenolics containing gallic acid, resorcinol, phloroglucinol, pyrogallol, and other phenol derivatives, often combined with sugars. Flocculant activity has been reported (172) (see Lignin).

***Xanthan* [*11138-66-2*].** This anionic polysaccharide, probably acetylated, is obtained through microbial action on D-glucose. It is a complex poly D-glucose–D-mannose carbohydrate that owes its anionicity to the presence of D-glucuronic acid and pyruvic acid (ketal) groups. Molecular weight is estimated to be in the $(2 - 10) \times 10^6$ range. Polycationic microbials may eventually become available when toxicity problems with cationic polysaccharide-producing organisms have been overcome (173) (see Gums; Microbial transformations).

Major Manufacturers. Most major United States manufacturers of flocculants derived from natural products are listed in Table 4 along with general classes of products.

Economic Aspects

The market for flocculating agents is large and growing. Surveys tend to separate markets into two areas: (*1*) water–wastewater treatment (174–175) (water supplies, and effluent streams requiring treatment before discharge for municipalities and industries other than mining) and (*2*) mining–mineral processing applications (176). Another area (*3*) is in manufacturing processes, such as paper, sugar, etc. Sources of uncertainty in survey data arise from reluctance of manufacturers, suppliers, and users to disclose information which may provide competitors with commercial intelligence since the variance in selling prices depends on quantity, location, handling conditions, etc, and the variance in fraction of active component in an as-sold flocculant (a few % to 100%).

Table 4. Key United States Producers of Flocculants Derived from Natural Products[a]

Company	Products[b]
CPC International, Inc.	a
Staley Manufacturing Corp.	a
Harvest Queen Mill and Elevator Co.	a
Celanese Corp.	b
General Mills Chemicals, Inc. (Henkel K.G.a.A.)	b
Hercules, Inc.	b
Swift & Co. (subs. Esmark, Inc.)	c

[a] Not a complete listing of all flocculant producers.
[b] a = starch (starch derivatives); b = guar gum (guar gum derivatives); and c = technical gelatin.

Production of aluminum sulfate (alum) and lime in the United States during the period 1968–1978 remained nearly constant (increase of ca 0.1% for the former and ca 1.2% for the latter) (177). About 35–40% of the aluminum sulfate is used for water and wastewater treatment, amounting to about 4×10^5 t (17% Al_2O_3) during 1978 (roughly 90% for water and 10% for wastewater) (66). At the quoted price, 16.06¢/kg on Dec. 24, 1979 (178), the value was about 64×10^6 dollars. Although production has remained constant, there has been a modest increase in the fraction consumed as a flocculant. Growth is retarded by steps to replace a portion or all of the alum in conventional treatment systems by synthetic flocculants (179).

It is impossible to attach a figure to the use of lime as a flocculant, since that function is usually incidental to its other uses in a treatment system. Approximately 10% of United States production is used in water and wastewater treatment (roughly twice as much in the former as in the latter) (180). Thus usage during 1978 is estimated to have been about 2×10^6 t, and at the quoted price, 3.4¢/kg on Dec. 24, 1979 (178), corresponds to 68×10^6 dollars. In addition, a rough estimate indicates sugar refining (carbonatation and phosphotation) (181) and mineral ore processing each consumed about (8–9) $\times 10^5$ t (see Sugar; Extractive metallurgy).

Production of ferric chloride in 1975 was about 75,000 t of which about 85% was used for sewage treatment (182). At 9.9¢/kg, this represents about 6.3×10^6 dollars for flocculant usage. Data on ferric sulfate are not readily available. It is doubtful that the total for all iron compounds as flocculants in 1975 exceeded 10×10^6 dollars, but a moderate growth rate in sewage treatment is expected.

Sales of synthetic organic flocculants for the water–wastewater area in the United States during 1977 were estimated at 100×10^6 dollars, with a projected annual growth rate of 8–10%/yr in constant dollars through 1985 (175). Specialty chemicals for the water–wastewater area, mainly synthetic organic flocculants are expected to grow at a comparable rate in Europe (183), Japan, and throughout most industrialized areas.

Reports on the mining industry frequently lump flocculants and other chemical reagents together. The 1975 *Minerals Yearbook* indicates that about 3000 t of synthetic organic flocculants was used in United States mineral processing areas surveyed in connection with froth flotation (qv) (184). Sales of United States produced PAMs during 1977 were estimated at (18–23) $\times 10^3$ t (175), a large portion going to the mineral processing area although not all in the United States. It is estimated that coal and mineral-related industries in Canada spent 20×10^6 dollars for synthetic flocculants in 1978 (185), mainly imported from the United States, Europe, and Japan. Use of synthetic flocculants in mineral processing is worldwide, with excellent growth prospects.

One major segment of the market for synthetic flocculants in manufacturing processes is the paper industry (drainage and retention aids (186)). A rough estimate of the United States market, made by combining production figures for grades of paper (187) in which retention aids might be used, with approximate levels of usage (188), suggests a current potential of $10–20 \times 10^6$ \$/yr (see Paper).

On a dry weight basis the price of synthetic flocculants was \$2.75–5.50/kg during 1977 (175) with a few in the \$6–9/kg range.

Since flocculant usage represents a very small part of the natural products market, similar data for naturally derived flocculants are not readily available (189). Although large amounts of starch are used in the mining industry, only a portion functions as a flocculant. Worldwide consumption for this purpose is estimated to be in the range 50–100 × 10^3 t/yr. Assuming a price of ca 18¢/kg, this corresponds to 9–18 × 10^6 dollars. World flocculant usage of guar gums is estimated to be 2–5 × 10^3 t during 1977, and at a price of ca $1.00/kg, this corresponds to 2–5 × 10^6 dollars. Flocculant usage of protein colloids has decreased rapidly to less than 10^3 t (estimated) during 1977. In general, flocculants from natural products are being displaced by synthetic products, except for specialty applications.

Analytical and Test Methods

Characterization techniques (190) pertinent to polymers generally (osmometry, viscometry, light scattering, gel permeation chromatography, ultracentrifugation, fractionation, spectroscopy, etc) apply to polyelectrolytes as well but with complications introduced by the presence of charged sites (191–192). For example, gel permeation chromatography of aqueous solutions of ionic polymers presents rather different problems than those encountered with organic solvents (193). Colloid titration (194) can be used to determine the nature and magnitude of the charge on many macromolecules, subject to appropriate calibration procedures and knowledge of the limitations. Conversely, if the charge on the polyelectrolyte is known, colloid titration can be used to determine the amount of polyelectrolyte in solution (195). In a variation of the procedure, zero electrophoretic mobility (19) can be used as the end point of the titration (196) in place of the dye color change. Relationships between zero electrophoretic mobility (ZEM), point of zero charge (PZC), dye end point, isoelectric point (IEP), and surface charge are submerged in the calibration procedure (see Electromigration). The carboxyl content of polymers can be determined by acid–base potentiometric titration (197), although endpoints are not sharp, and by conductiometric titration (198).

Quantitative determination of very small amounts of flocculant remaining in supernatant liquids after flocculation is difficult, and perhaps is accomplished most accurately by use of a flocculant containing a radioactive tag (26,199). However, other methods (200) have been used and appear satisfactory when conditions are favorable (see Analytical methods).

Health and Safety Factors

For the safety of personnel handling bulk flocculants, procedures recommended by the manufacturer should be followed. Since many flocculants can be considered to be practically nontoxic, most procedures merely represent good practice in industrial handling and use.

Although the polymer molecules themselves generally appear to exhibit low mammalian toxicity, this is not necessarily true for the corresponding monomeric species. An important example is polyacrylamide (201) compared with acrylamide monomer (202). Production procedures, however, can hold residual monomer in the polymeric product to a level considered a minimal hazard (see Acrylamide polymers).

The EPA has responsibility for establishing controls on water quality in a number of areas where flocculants are used. A detailed protocol exists (currently being revised in light of the *U.S. Safe Drinking Water Act*) for obtaining EPA approval for use of

Table 5. Representative Areas of Use for Common Types of Flocculants

Flocculant type		Representative applications
Inorganic		
aluminum compounds		principal flocculant for municipal water supply clarification, and for color precipitation–flocculation; industrial water supply clarification; settling industrial and municipal sewage sludge; phosphate precipitation–flocculation; general purpose flocculant
iron compounds		phosphate precipitation–flocculation; industrial and municipal sewage settling and dewatering
lime		pH adjustment in wide range of substrates (water, wastewater, ore pulps, etc); carbonate precipitation (softening) in water supplies; phosphate precipitation in sugar refining–coflocculant; phosphate precipitation in primary and secondary sewage treatment–coflocculant; coflocculant with ferric compounds in industrial and municipal sewage sludge settling and dewatering
Synthetic organic		
Ionicity[a]	Molecular weight[b]	
C-H, VH	h	industrial and municipal sewage sludge dewatering; microflotation of activated sewage sludge; clarification of black liquor in titanium dioxide processing
C-H, VH	m	thickening iron ore concentrates and tailings; thickening coal refuse (as coflocculant); industrial and municipal sewage sludge dewatering
C-H, VH	l	municipal and industrial water supply clarification; thickening iron ore concentrates and tailings
C-L, M	h	industrial and municipal sewage sludge dewatering; microflotation of activated sewage sludge; flocculation of fines and mineral fillers onto paper fibers (retention aid); microflotation of white water in paper mills
A-VH	vh	thickening red muds (iron oxides) in Bayer alumina process
A-M	h, vh	clarification of liquors in sugar processing; secondary flocculant after chemical precipitation; thickening and dewatering coal refuse; thickening clay slimes; dewatering alum sludge; clarification in phosphoric acid production; filtration aid for fine coal concentrates
A-L	h, vh	secondary flocculant after chemical precipitation; retention of fines and fillers on alum treated paper fibers; thickening potash tailings (clay slimes); general purpose flocculant for mineral processing
nonionic	h, vh	general purpose flocculant for mineral processing; thickening acidic ore pulps and concentrates; thickening pickle liquor sludge; settling acid or carbonate leach solutions (uranium processing)
Natural products		
starch		thickening red muds in Bayer alumina process; selective flocculation in taconite beneficiation; (use of starches as depressants in flotation of ores is not a flocculant application)
guar gums		settling carbonate leach solutions (uranium processing); filtration aid for gold and uranium ore leach residues; (use of gums as depressants in flotation of ores is not a flocculant application)
protein colloids		clarification of wines and beverages; settling acid leach solutions (uranium processing)

[a] A = anionicity; C = cationicity; L = low (1–10%); M = medium (10–40%); H = high (40–80%); and VH = very high (80–100%).

[b] l = low, 10^3–10^5; m = medium, 10^5–10^6; h = high, 1–5 × 10^6; and vh = very high, >5 × 10^6.

a flocculant in potable-water treatment (203). Without endorsing any products, EPA maintains a listing of products that can be used in water treatment without adverse physiological effects on those using the water, when applied in the concentration recommended by the manufacturers or distributors and provided the products continue to meet the quality specifications furnished (204). The list carries no implications regarding flocculants for other areas.

There are a number of discussions on health and safety aspects of flocculants in the literature (205). For specific commercial products the manufacturer's data sheets should be studied.

Uses

In the literature on uses, flocculants frequently are mentioned as aids for a specific performance descriptor, eg, filter aid, dewatering aid, drainage aid, retention aid, settling aid, etc, but the basic function is to promote flocculation. Sometimes several types of flocculating agents provide satisfactory technical performance, and the final selection is based on other factors. Overall cost-effectiveness, including effects on other aspects of the treatment or processing plant, is the dominant element. Because of the variety of flocculating agents and their wide spectrum of uses, only representative applications involving significant commercial quantities can be indicated, and then only for broad categories of flocculants. Information of this type can be obtained from flocculant suppliers, from publications (4,32–33,41–42,81,176,179,206) and from various economic analyses on chemical products (207). This information is summarized in Table 5.

For convenience in grouping synthetic organic flocculants for functional behavior, the following somewhat arbitrary designations will be used. Level of ionicity within a polymer molecule is categorized in terms of the monomer units that contribute charge—low, l (1–10%); medium, m (10–40%); high, h (40–80%); very high, vh (80–100%)—with A designating anionicity and C cationicity. Similarly molecular weights (weight average) are lumped into categories—low, l (10^3–10^5); medium, m (10^5–10^6); high, h ($1–5 \times 10^6$); very high, vh ($>5 \times 10^6$).

In connection with applications, mention of pellet flocculation should be made (208). Under certain conditions of substrate, flocculant, and physical treatment, large pelletlike aggregates can form when flocs are rolled over one another gently. This phenomenon has been exploited commercially in Japan (209) for dewatering various sludges using PAMs and sodium silicate (see Wastes).

BIBLIOGRAPHY

1. M. N. Baker, *The Quest for Pure Water,* American Water Works Association, New York, 1948, pp. 3–4.
2. V. K. LaMer in S. D. Faust and J. V. Hunter, eds., *Principles and Applications of Water Chemistry,* John Wiley & Sons, Inc., New York, 1967, pp. 254–256.
3. R. B. Booth, J. E. Carpenter, and H. Hartjens, *Ind Min. J.* (Special Issue), 335 (1957).
4. H. A. Hamza and J. L. Picard, *Index of Commercial Flocculants—1974, Canmet Report 77-8,* Canada Centre for Mineral and Energy Technology, Ottawa, Can., 1975.

5. D. H. Everett, *Pure Appl. Chem.* **31,** 579 (1972).
6. J. A. Kitchener, *Br. Polym. J.* **4,** 217 (1972).
7. R. Akers, *Flocculation,* Institute of Chemical Engineers, London, Eng., 1975; *Rep. Prog. Appl. Chem.* **60,** 605 (1976).
8. W. J. Weber, Jr., *Physiochemical Processes for Water Quality Control,* Wiley-Interscience, New York 1972.
9. R. F. Packham and J. Sheiham, *J. Inst. Water Eng. Sci.* **31**(2), 96 (1977).
10. F. Halverson, *Proc. Tenth Annual Meeting, Canadian Mineral Processors, Ottawa, Can., Jan. 1978,* pp. 404–450.
11. K. J. Ives, ed., *The Scientific Basis of Flocculation, NATO Advanced Study Institute Series,* Vol. E27, Sijthoff and Noordhoff, Alphenaan den Rijn, Amsterdam, The Netherlands, 1978.
12. T. W. Healy, *J. Macromol. Sci. Chem.* **A8,** 603 (1974).
13. J. Stone-Masui and A. Watillon, *J. Colloid Interface Sci.* **52,** 479 (1975).
14. H. vanOlphen, *An Introduction to Clay Colloid Chemistry,* 2nd ed., John Wiley & Sons, Inc., New York, 1977, pp. 18, 64, 92.
15. H. Hohl and W. Stumm, *J. Colloid Interface Sci.* **55,** 281 (1976); P. W. Schindler and H. Gamsjager, *Kolloid Z.* **250,** 759 (1972).
16. R. D. Vold and C. C. Konecny, *J. Phys. Colloid Chem.* **53,** 1262 (1949); T. M. Doscher, *J. Colloid Chem.* **5,** 100 (1950).
17. R. Aveyard and D. A. Haydon, *An Introduction to the Principles of Surface Chemistry,* Cambridge University Press, New York, 1973.
18. D. H. Napper in M. Keiker, A. C. Zettlemoyer, and R. L. Rowell, eds., *Colloid and Interface Science,* Vol. 1, Academic Press, Inc., New York, 1977, p. 413.
19. P. Sennett and J. P. Olivier, *Ind. Eng. Chem.* **57**(8), 32 (1965).
20. Ref. 8, pp. 68–69.
21. J. A. Davis, R. O. James, and J. O. Leckie, *J. Colloid Interface Sci.* **63,** 480 (1978); J. A. Davis and J. O. Leckie, *J. Colloid Interface Sci.* **67,** 90 (1978); W. Stumm, H. Hohl, and F. Dalang, *Croat. Chem. Acta* **48,** 491 (1976).
22. R. O. James, G. R. Wiese, and T. W. Healy, *J. Colloid Interface Sci.* **59,** 381 (1977).
23. E. Matijevic, *J. Colloid Interface Sci.* **43,** 217 (1973).
24. W. Stumm and C. R. O'Melia, *J. Am. Water Works Assoc.* **60,** 514 (1968).
25. R. F. Packham, *J. Colloid Sci.* **20,** 81 (1965).
26. D. R. Kasper, Ph.D. Thesis, California Institute of Technology, Pasadena, Calif., 1971.
27. J. Gregory, *J. Colloid Interface Sci.* **42,** 448 (1973).
28. A. Rembaum, "Flocculation of Model Waste Particles by Means of Ionene Polymers" in T. F. Yen, ed., *Recycling and Disposal of Solid Wastes,* Ann Arbor Science, Ann Arbor, Mich., 1974.
29. F. M. Fowkes and M. A. Mostafa, *Ind. Eng. Chem. Prod. Res. Dev.* **17**(1), 3 (1978).
30. Y. A. I. Attia and J. A. Kitchener, *Proc. Eleventh International Mineral Processing Congress, Cagliari, Italy, Apr. 1975,* pp. 1233–1248.
31. A. S. Teot and S. L. Daniels, *Environ. Sci. Technol.* **3,** 825 (1969).
32. *Process Design Manual for Phosphorus Removal, EPA 625/1-76-001a,* EPA, Washington, D.C., 1976.
33. A. Benedek and co-workers, *Assessment of Polyelectrolytes for Phosphorus Removal, Research Report No. 37,* Research Program for the Abatement of Municipal Pollution, Ontario Ministry of the Environment, Ontario, Can., 1976.
34. E. S. Hall and R. F. Packham, *J. Am. Water Works Assoc.* **57,** 1149 (1965); S. P. Sheton and W. A. Drewey, *J. Am. Water Works Assoc.* **65,** 627 (1973).
35. R. F. Packham, *TP 100,* Water Research Association, Buckinghamshire, U.K., 1973; N. Narkis and M. Rebhun, *J. Am. Water Works Assoc.* **69,** 325 (1977).
36. M. S. Nasr and D. G. MacDonald, *Can. J. Chem. Eng.* **56,** 87 (1978); J. Clarke and M. W. Davis, Jr., *Tappi* **52**(10), 1923 (1969).
37. T. C. Kisla and R. D. McKelvey, *Environ. Sci. Technol.* **12,** 207 (1978).
38. J. P. Friend and J. A. Kitchener, *Chem. Eng. Sci.* **28,** 1071 (1973).
39. A. D. Read and A. W. Whitehead, *Proc. 10th Int. Mineral Processing Cong., Institute of Mining and Metallurgy, London, Eng., 1974,* p. 949; A. D. Read and C. T. Hollick, *Miner. Sci. Eng.* **8,** 202 (1976).

40. J. A. Rubio and J. A. Kitchener, *Trans. Inst. Min. Metall.* **86,** C97 (1977).
41. U.S. Pat. 3,292,780 (Dec. 20, 1966), D. W. Frommer and A. F. Colombo (to U.S. Dept. of Interior); J. W. Villar and G. A. Dawe, *Min. Cong. J.* **61,** 40 (Oct. 1975).
42. U.S. Pat. 3,805,951 (Apr. 28, 1974), W. P. Brogiotti and F. P. Howald (to American Cyanamid Co.); A. F. Banks, *Proc. Eighth Annual Meeting, Canadian Mineral Processors, Ottawa, Can., Jan. 1976,* pp. 111–140.
43. E. A. Cassell and co-workers, *AIChE J.* **17,** 1486 (1971); K. Roberts and P. Barla, *J. Colloid Interface Sci.* **49,** 75 (1974); H. Lundgren, *Tappi* **53**(2), 287 (1970).
44. P. L. Busch and W. Stumm, *Environ. Sci. Technol.* **2,** 49 (1968).
45. R. G. Luthy, R. E. Selleck, and T. R. Galloway, *Environ. Sci. Technol.* **11,** 1211 (1977).
46. D. Dollimore and T. A. Horridge, *J. Colloid Interface Sci.* **42,** 581 (1973).
47. U.S. Pat. 3,528,218 (Oct. 6, 1970), L. J. Bloch von Blottnitz and B. S. Dykhouse (to American Cyanamid Co.).
48. R. C. Baskerville, A. M. Bruce, and M. C. Day, *Filtr. Sep.* **15,** 445 (Sept.–Oct. 1978).
49. R. J. TeKippe and R. K. Ham, *J. Am. Water Works Assoc.* **62,** 260, 594, 620 (1970).
50. R. F. Packham, *Br. Polym. J.* **4,** 305 (1972); A. Benedek and J. J. Bancsi, *J. Environ. Eng. Div. Proc. Am. Soc. Civil Eng.* **102**(EE1), 17 (Feb. 1976).
51. W. W. Eckenfelder, Jr., *Industrial Water Pollution Control,* McGraw-Hill Book Co., New York, 1966, pp. 240–243.
52. R. S. Gale and R. C. Baskerville, *Chem. Ind.* (9), 355 (Mar. 4, 1967); R. C. Baskerville and R. S. Gale, *J. Inst. Water Pollut. Control Fed.* **67,** 233 (1968).
53. J. R. Christian, D. Dollimore, and T. A. Horridge, *J. Phys. E. Sci. Inst.* **3,** 744 (1970); ref. 51, pp. 230–235.
54. Ref. 51, pp. 56–59.
55. K. W. Britt, Tappi **56**(3), 83 (Mar. 1973); J. E. Unbehend, *Tappi* **60**(7), 110 (July 1977).
56. *Process Design Manual for Suspended Solids Removal, EPA 625/1-75-003a,* EPA, Washington, D.C., 1975, pp. 4-11–4-15.
57. A. F. Colombo, *Eleventh International Mineral Processing Congress, Cagliari, Italy, April 1975,* (Special Volume), pp. 468–487.
58. Ref. 56, pp. 4-5–4-11.
59. *Tappi T227 m-58 Standard.*
60. C. F. Baes, Jr. and R. E. Mesmer, *The Hydrolysis of Cations,* Wiley-Interscience, New York, 1976, pp. 112–123.
61. J. D. Griffith and R. G. Williams, *J. Am. Water Works Assoc.* **64,** 825 (1972).
62. A. C. Vermeulen and co-workers, *J. Colloid Interface Sci.* **51,** 449 (1975); R. J. Stol, A. K. van Helden, and P. L. de Bruyn, *J. Colloid Interface Sci.* **57,** 115 (1976); H. de Hek, R. J. Stol, and P. L. de Bruyn, *J. Colloid Interface Sci.* **64,** 72 (1978); E. Strazdins, *Tappi* **46,** 432 (1963); *ibid.,* **48,** 157 (1965).
63. E. Matijevic and co-workers, *J. Phys. Chem.* **65,** 826 (1961); L. H. Allen and E. Matijevic, *J. Colloid Interface Sci.* **35,** 66 (1971); E. Matijevic, F. J. Mangravite, Jr., and E. A. Cassell, *J. Colloid Interface Sci.* **35,** 560 (1971).
64. *Alum, Technical Brochure ICP 400,* Industrial Chemicals and Plastics Division, American Cyanamid Co., Wayne, N.J., 1972.
65. Ref. 56, pp. 5-1–5-15.
66. *Chemical Economics Handbook,* Stanford Research Institute, Menlo Park, Calif., 1976, 702.5000D.
67. G. P. Hanna, Jr. and A. J. Rubin, *J. Am. Water Works Assoc.* **62,** 315 (1970).
68. S. Ban, *Nikkakyo Geppo* **26**(1), 27 (1973); *Jpn. Chem. Week,* (June 15, 1972); *LAPOFLOC PAC Technical Information Bulletin Ref. No. TI/P.27.1,* Laporte Industries Limited, Cheshire, Eng., 1977; *Chem. Eng. News* **57**(32), 19 (Aug. 6, 1979).
69. Jpn. Kokai 77 99994 (Aug. 22, 1977), M. Isao, H. Miyazaki, and H. Kawamura (to Toyo Soda Mfg. Co. Ltd.); Ger. Offen. 2,547,695 (Apr. 29, 1976), B. J. L. Kilby (to Laporte Industries Ltd.); Ger. Offen. 2,612,504 (Oct. 21, 1976), W. H. Redmayne and D. Berry (to Laporte Industries Ltd.); Ger. Offen. 2,634,126 (Feb. 10, 1977), T. Maryvonne and J. Grosbois (to Rhone-Poulenc S.A.).
70. R. N. Sylva, *Rev. Pure Appl. Chem.* **22,** 115 (1972).
71. J. Dousma and P. L. de Bruyn, *J. Colloid Interface Sci.* **56,** 527 (1976); **64,** 154 (1978).
72. P. J. Murphy, A. M. Posner, and J. P. Quirk, *J. Colloid Interface Sci.* **56,** 270, 284, 298, 312 (1976).
73. Ref. 56, pp. 5-16–5-24.

74. W. Stumm, H. Huper, and R. L. Champlin, *Environ. Sci. Technol.* **1,** 221 (Mar. 1967); J. G. Vail and J. H. Wills, *Soluble Silicates,* Vol. 2, Reinhold Publishing Corp., New York, 1952, Chapt. 9.
75. R. K. Iler, *Surface Colloid Sci.* **6,** 1 (1973); W. J. Weber, Jr., and L. H. Ketchum, Jr., *U.S.N.T.I.S., PB Rep. 1974, No. 232454/9GA.*
76. Ref. 56, pp. 5-24–5-34.
77. C. G. Thompson and A. P. Black, *AIChE Symp. Ser.* (*Water*) **71**(151), 170 (1975).
78. C. Tanford, *Physical Chemistry of Macromolecules,* John Wiley & Sons, Inc., New York, 1961, p. 5.
79. A. Audsley, *The Flocculation of Suspensions of Solids with Organic Polymers,* National Physical Laboratory, London, Eng., 1963; A. Audsley, *Mineral Processing Information Note No. 5.* Warren Spring Laboratory, 1965; K. A. Kini, *Note. No. 6,* 1970, obtainable from Ministry of Technology Reports Centre, Station Square House, St. Mary Cray, Kent, Eng.
80. S. Gutcho, *Waste Treatment with Polyelectrolytes and Other Flocculants,* Noyes Data Corporation, Park Ridge, N.J., 1977.
81. J. Vostrcil and F. Juracka, *Commercial Organic Flocculants,* Noyes Data Corporation, Park Ridge, N.J., 1976.
82. F. E. Bailey and E. M. LaCombe, *J. Macromol. Sci. Chem.* **A4,** 1293 (1970); M. F. Hoover, *J. Macromol. Sci. Chem.* **A4,** 1327 (1970).
83. U.S. Pat. 4,053,512 (Oct. 11, 1977), H. P. Panzer and D. F. Bardoliwalla (to American Cyanamid Co.).
84. Private communication, E. Strazdins and J. McClure, Stamford Research Center, American Cyanamid Company.
85. E. Kokufuta, *Macromolecules* **12**(2), 350 (1979).
86. Br. Pat. 454,723 (Oct. 6, 1936), (to I. G. Farbenindustrie, A.G.); U.S. Pat. 2,322,201 (June 15, 1944), D. W. Jayne, Jr., N. M. Day, and S. E. Erickson (to American Cyanamid Co.); U.S. Pat. 3,275,588 (Sept. 27, 1966), D. G. Garms (to The Dow Chemical Company); Fr. Pat. 1,449,204 (Aug. 12, 1966), K. G. Phillips (to Nalco Chemical Company); A. T. Coscia in N. M. Bikales, ed., *Encyclopedia of Polymer Science and Technology* Vol. 10, John Wiley & Sons, Inc., New York, 1969, p. 616.
87. U.S. Pat. 3,523,892 (Aug. 11, 1970), D. L. Schiegg (to Calgon Corp.).
88. T. J. Suen and co-workers, *J. Polym. Sci.* **45,** 289 (1960); E. Dyer and C. C. Anderson, *J. Polym. Sci. A-1* **5,** 1959 (1967); G. H. Olive and S. Olive, *J. Polym. Sci. A-1* **10,** 675 (1972).
89. H. L. Spell, *Anal. Chem.* **41,** 902 (1969).
90. T. Saegusa, H. Idoda, and H. Fujii, *Polym. J.* **3,** 35 (1972).
91. *Chem. Eng. News* **46**(3), 46 (1968).
92. U.S. Pats. 3,248,353 (Apr. 26, 1966), 3,493,502 (Feb. 3, 1970), A. T. Coscia (to American Cyanamid Co.); 3,567,659 (Mar. 2, 1971), D. E. Nagy (to American Cyanamid Co.).
93. A. K. Banthia, E. H. Schacht, and E. J. Goethals, *Makromol. Chem.* **179,** 841 (1978).
94. O. C. Dermer and G. E. Ham, *Ethyleneiminine and Other Aziridines,* Academic Press, Inc., New York, 1969.
95. H. Wilfinger, *Das Papier* **2,** 265 (1948); U.S. Pat. 2,943,013 (June 28, 1960), H. F. Arledter (to Hulbert Paper Co.); D. Horn, "Polyethyleneimine-Physicochemical Properties and Applications," *IUPAC International Symposium on Polymeric Amines and Ammonium Salts, Sept. 24–26, 1979, Ghent, Belg.*
96. M. Ku, *J. Colloid Interface Sci.* **56,** 596 (1976).
97. U.S. Pat. 3,725,312 (Apr. 3, 1973), H. P. Panzer and R. Rabinowitz (to American Cyanamid Co.).
98. U.S. Pat. 3,738,445 (June 12, 1973), H. P. Panzer and K. W. Dixon (to American Cyanamid Co.).
99. A. J. Ellis and F. G. Soper, *J. Chem. Soc.,* 1750 (1954).
100. L. C. Pizzini and J. T. Patton, Jr. in N. M. Bikales, ed., *Encyclopedia of Polymer Science and Technology,* Vol. 6, John Wiley & Sons, Inc., New York, 1967, p. 171; R. A. Briggs and E. E. Gruber, *ibid.,* p. 196.
101. U.S. Pat. 3,320,317 (May 16, 1967), W. A. Rogers (to The Dow Chemical Co.).
102. J. Hasek and co-workers, *Listy Cukrov.* **93,**(5), 104 (1977); Ger. Offen. 1,917,050 (Oct. 23, 1969), H. Stettler and K. Hofer (to Sandoz, Ltd.).
103. U.S. Pat. 2,345,543 (Mar. 28, 1944), H. P. Wohnsiedler and W. M. Thomas (to American Cyanamid Co.); U.S. Pat. 3,716,483 (Feb. 13, 1973), A. Renner (to Ciba-Geigy); Jpn. Pat. 5,0105,575 (June 14, 1978), (to Nippon Carbide Kogy KK).
104. Belg. Pat. 549,049 (1956); Ger. Offen. 1,243,646 (July 6, 1967), R. Schwarz and T. Hennig (to Zschimmer and Schwarz); U.S. Pat. 3,484,837 (Dec. 16, 1969), J. J. Odum, T. P. Shumaker, and P. R. Bloomquist (to Reichold Chemicals, Inc.).

105. R. W. Auten, *Tappi Spec. Tech. Assoc. Publ.* **31,** 468 (June 1948).
106. Jpn. Kokai 71 10,408 (March 16, 1971) (to Kuxita Ind., Co., Ltd.).
107. J. K. Dixon, G. L. M. Christopher, and D. J. Salley, *Tappi Spec. Tech. Assoc. Publ.* **31,** 412 (June 1948).
108. C. S. Maxwell, *Pap. Trade J.* **16**(19), 39 (1943).
109. U.S. Pat. 2,668,111 (Feb. 2, 1954), J. R. Lindquist (to Certain-Teed Prod. Corp.).
110. A. M. Schiller and T. J. Suen, *Ind. Eng. Chem.* **48,** 2132 (1956); U.S. Pat. 3,171,805 (Mar. 2, 1965), T. J. Suen and A. M. Schiller (to American Cyanamid Co.); U.S. Pat. 3,539,535 (Nov. 10, 1970), R. L. Wisner (to The Dow Chemical Co.); Jpn. Kokai 69 8,226 (Apr. 17, 1969), C. J. McDonald and R. H. Beaver, *Macromolecules* **12,** 203 (1979).
111. Ger. Offen. 2,333,927 (Jan. 23, 1975), K. Dahmen, W. H. Hubner, and E. Barthell (to Chem. Fabrik Stockhausen).
112. U.S. Pat. 3,897,333 (July 29, 1975), J. R. Field and G. Smalley (to Allied Colloids Mfg. Co.).
113. Ger. Offen. 2,546,102 (Apr. 27, 1978), A. T. Coscia and M. N. O'Connor (to American Cyanamid Co.).
114. N. Sawa, *Nippon Kagaku Zasshi* **89,** 780 (1968).
115. U.S. Pat. 3,406,139 (Oct. 15, 1968), M. J. Hurwitz and H. Aschkenasy (to Rohm and Haas Co.).
116. U.S. Pat. 4,006,247 (Feb. 1, 1977), H. P. Panzer, M. N. O'Connor, and L. J. Baccei (to American Cyanamid Co.).
117. U.S. Pat. 3,450,646 (June 17, 1969), R. R. Annand, D. Redmore, and B. M. Rushton (to Petrolite Corp.).
118. U.S. Pat. 3,288,770 (Nov. 29, 1966), G. B. Butler (to Peninsular Chem. Res., Inc.); C. E. Schildknecht, *Allyl Compounds and Their Polymers,* Wiley-Interscience, New York, 1973.
119. J. E. Boothe, H. G. Flock, and M. F. Hoover, *J. Macromol. Sci. Chem.* **A4,** 1419 (1970).
120. J. E. Lancaster, L. Baccei, and H. P. Panzer, *Polym. Lett.* **14,** 549 (1976).
121. W. H. Schuller and co-workers, *J. Chem. Eng. Data* **4**(3), 273 (1959); U.S. Pat. 4,077,930 (Mar. 7, 1978), S. K. Lim, A. E. Bloomquist, and R. J. Schaper (to Calgon Corp.).
122. U. S. Pats. 3,147,218 (Sept. 1, 1964), R. B. Booth and co-workers (to American Cyanamid Co.); 3,171,805 (Mar. 2, 1965), T. J. Suen and A. M. Schiller (to American Cyanamid Co.).
123. L. S. Luskin in R. H. Yocum and E. B. Nyquist, eds., *Functional Monomers,* Vol. 2, Marcel Dekker, Inc., New York, 1974, pp. 555–739.
124. H. Rauch-Puntigam and T. Volker, *Acryl- and Methacrylverbindungen,* Springer-Verlag, Berlin, Heidelberg, New York, 1967.
125. T. Matsumoto, M. Okubo, and S. Onoe, *Kobunshi Ronbunshu Eng. Ed.* **4**(3), 205 (1975); *Technical Brochure CM-42F/Cd,* Rohm and Haas Company, Philadelphia, Pa., Dec. 1970, p. 5.
126. U.S. Pat. 3,666,810 (May 30, 1972), D. I. Hoke (to The Lubrizol Corp.).
127. U.S. Pat. 2,595,907 (May 6, 1950), W. M. Thomas and J. A. Price (to American Cyanamid Co.); Ger. Offen. 1,095,467 (Dec. 22, 1960), A. Maeder (to Ciba Ltd.); Can. Pats. 583, 352 (Sept. 15, 1959), 595,642 (Apr. 5, 1960), G. T. Jones (to Imperial Chem. Ind. Ltd.).
128. U.S. Pat. 3,878,247 (Apr. 15, 1975), P. H. Moss and R. M. Gipson (to Jefferson Chemical Co.).
129. *Chemistry of Acrylamide,* American Cyanamid Company, Wayne, N.J., 1969.
130. Y.-C. Yen, *Acrylamide and Polyacrylamide, Stanford Research Institute Report No. 99, Patent Summary,* Stanford Research Institute, Menlo Park, Calif., 1976, pp. 203–204.
131. *Ibid.,* pp. 249–250.
132. J. Bourdais, *Bull. Soc. Chim. Fr.* **22,** 485 (1955); W. R. Cabanes, T. Y. C. Lin, and C. Parkanyi, *J. Polym. Sci. A-1* **9,** 2155 (1971).
133. M. L. Miller and C. E. Rauhut, *J. Colloid Sci.* **14,** 524 (1959).
134. Belg. Pat. 628,018 (Aug. 5, 1963), H. Volk and W. E. Smith (to The Dow Chemical Co.).
135. U.S. Pat. 3,336,271 (Aug. 15, 1967), T. E. Durocher (to The Dow Chemical Co.).
136. W. E. Walles, *J. Colloid Interface Sci.* **37,** 364 (1971).
137. U.S. Pat. 2,909,508 (Oct. 20, 1959), G. D. Jones (to The Dow Chemical Co.); Fr. Pat. 1,383,924 (Jan. 1, 1965), H. Pohlemann, H. Spoor, and E. Heil (to BASF).
138. K. Sakaguchi and K. Nagase, *Kogyo Kagaku Zasshi* **69,** 1335 (1966).
139. U.S. Pat. 3,617,572 (Nov. 2, 1971), D. J. Monagle (to Hercules, Inc.).
140. U.S. Pats. 3,478,091 (Nov. 11, 1969), D. L. Murfin and L. E. Miller; 3,506,707 (Apr. 14, 1970), L. E. Miller and D. L. Murfin; 3,692,673 (Sept. 19, 1972), D. I. Hoke (to The Lubrizol Corp.).
141. N. M. Bikales in N. M. Bikales, ed., *Water-Soluble Polymers,* Plenum Press, New York, 1973, pp. 213–225.

142. W. M. Kulicke and J. Klein, *Die Angew. Makromol. Chem.* **69,** 169, 189 (1977); K. L. Gardner, W. R. Murphy, and T. G. Geehan, *J. Appl. Polym. Sci.* **22,** 881 (1978); W. P. Shyluk and F. S. Stow, Jr., *J. Appl. Polym. Sci.* **13,** 1023 (1969).
143. U.S. Pat. 3,634,944 (Jan. 18, 1972), M. L. Zonis and co-workers (to American Cyanamid Co.).
144. Ger. Offen. 2,009,748 (Sept. 16, 1971), H. Willersinn and C. H. Krauch (to BASF); Ger. Offen. 2,136,887 (Feb. 8, 1973), D. Sander and H. Schenk (to BASF); U.S. Pat. 3,732,193 (May 8, 1973), J. Svarz (to Nalco Chemical Company); U.S. Pat. 4,138,539 (Feb. 6, 1979), P. H. Landolt and L. N. Allen (to American Cyanamid Co.).
145. Fr. Pat. 2,064,729 (Oct. 22, 1969), J. P. Communal, J. Fritz, and B. Papillon (to Rhone-Progil).
146. Private communication, J. P. Regeaud, Rhone-Poulenc Chimie Fine.
147. *Chem. Mark. Rep., 67* (Nov. 27, 1978).
148. U.S. Pats. 2,982,749 (May 2, 1961), R. E. Friederich, R. M. Wiley, and W. L. Carret (to The Dow Chemical Co.); 3,284,393 (Nov. 8, 1966), J. W. Vanderhoff and R. M. Wiley (to The Dow Chemical Co.); 3,624,019 (Nov. 30, 1971), D. R. Anderson (to Nalco Chemical Company); 3,734,873 (May 22, 1973), D. R. Anderson and A. J. Frisque (to Nalco Chemical Company).
149. U.S. Pats. 4,022,731, 4,022,736 (May 10, 1977), J. M. Schmitt (to American Cyanamid Company).
150. *Union Carbide Product Bulletin F-44029A,* Union Carbide Corp., New York, 1973.
151. F. N. Hill, F. E. Bailey, Jr., and J. T. Fitzpatrick, *Ind. Eng. Chem.* **50,** 5 (1958); U.S. Pats. 2,987,489 (June 6, 1961), F. E. Bailey and F. N. Hill (to Union Carbide Corp.); 3,271,328 (Sept. 6, 1966), F. N. Hill (to Union Carbide Corp.).
152. D. H. Lorenz in N. M. Bikales, ed., *Encyclopedia of Polymer Science and Technology,* Vol. 14, John Wiley & Sons, Inc., New York, 1974, pp. 239–251.
153. J. L. Azorlosa and A. J. Martinelli in R. L. Davidson and M. Sittig, eds., *Water-Soluble Resins,* Reinhold Book Corp., New York, 1968, p. 131.
154. G. J. Fleer, L. R. Koopal, and J. Lyklema, *Kolloid Z.* **250,** 689 (1972); G. J. Fleer and J. Lyklema, *J. Colloid Interface Sci.* **46,** 1 (1974); **55,** 228 (1976).
155. U.S. Pat. 3,323,979 (June 6, 1967), W. A. Foster and J. E. Stout (to The Dow Chemical Co.).
156. U.S. Pat. 3,947,396 (Mar. 30, 1976), D. A. Kangas and W. R. Neuendorf (to The Dow Chemical Company).
157. Belg. Pat. 626,712 (July 1, 1963), E. F. Paschall and D. D. Wolfe (to Corn Products Co.).
158. U.S. Pat. 2,813,093 (Nov. 12, 1957), C. C. Caldwell and O. B. Wurgburg (to National Starch Products, Inc.).
159. U.S. Pat. 2,970,140 (Jan 31, 1961), C. H. Hullinger and N. H. Yui (to American Maize-Products Company.
160. U.S. Pat. 3,669,915 (June 13, 1972), D. A. Jones, G. F. Fanta, and R. C. Burr (to U.S. Sec. of Agriculture).
161. R. C. Burr and co-workers, *Proc. Chem. Inst. Canada, Flocculation and Dispersion Symposium, Nov. 4–5, 1974, Toronto, Can.,* pp. 253–271.
162. J. Hollo and co-workers, *Die Starke* **16,** 118 (1964).
163. A. M. Goldstein and E. N. Alter, "Guar Gum" in R. L. Whistler and J. N. BeMiller, eds., *Industrial Gums,* Academic Press, Inc., New York, 1959.
164. *Jaguar® Products Bulletin,* Stein Hall and Company, New York.
165. H. H. Young in N. M. Bikales, ed., *Encyclopedia of Polymer Science and Technology*, Vol. 7, John Wiley & Sons, Inc., New York, 1967, pp. 446–460.
166. W. F. Harrington and P. H. von Hippel, *Adv. Protein Chem.* **16,** 1 (1961).
167. P. D. Wood and A. G. Ward and A. Courts, eds., *Science and Technology of Gelatin,* Academic Press, London, Eng., 1977.
168. *Gelatin,* Gelatin Manufacturers Institute of America Inc., New York, 1973.
169. K. B. Guiseley, "Seaweed Colloids" in A. Standen, ed., *Encyclopedia of Chemical Technology,* 2nd ed., Vol. 17, Wiley-Interscience, New York, 1968, p. 763.
170. L. Deavin and co-workers, *Symposium on Extracellular Microbial Polysaccharides of Practical Importance, 172nd ACS Meeting, San Francisco, Calif. Aug. 29–Sept. 3, 1976.*
171. R. A. A. Muzzarelli, *Chitin,* Pergamon Press, New York, 1977.
172. Br. Pat. 1,310,491 (Mar. 21, 1973), O. Berger (Davey, Taxman, and Co., Ltd.).
173. A. Jeanes and co-workers, *Nature (London) New Biol.* **233,** 259 (1971); private communication, Northern Regional Research Laboratory, U.S. Department of Agriculture, Peoria, Ill., 1975.
174. A. C. Gross, *Environ Sci. Technol.* **8,** 414 (1974).
175. W. J. Storck, *Chem. Eng. News* **56**(4), 9 (Jan. 23, 1978).
176. D. W. Roger and G. W. Poling, *CIM Bull.* **71**(793), 152 (May 1978).

177. *Chem. Eng. News* **57**(24), 35, 36, 40 (June 11, 1979).
178. *Chem. Mark. Rep.*, (Dec. 4, 1978).
179. D. L. Terrell, *J. Am. Water Works Assoc.* **69**(5), 263 (1977); *Inst. Water Eng. Sci. Mag. (U.K.)*, 138 (May 1976).
180. Ref. 66, *Limestone and Lime 746.1001 I-M,* Mar. 1978.
181. M. C. Bennett, *Chem. Ind.* **22,** 886 (Nov. 16, 1974).
182. Ref. 66, *Ferrous Products 738.5000 A-D,* Feb. 1977.
183. *Water Waste Eng.* **16**(7), 8 (July 1979).
184. *1975 Minerals Yearbook,* Vol. 1, U.S. Bureau of Mines U.S. Government Printing Office, Washington, D.C., 1977, pp. 107–124.
185. *Can. Bull. Res. Dev.* (75), 1 (June 1979).
186. *Retention of Fine Solids During Paper Manufacture, CA Report No. 57,* Technical Association of the Pulp and Paper Industry, Atlanta, Ga., 1975.
187. *Pulp Pap.* **53**(7), 25, 104, 108 (June 30, 1979).
188. W. E. Frankle and J. L. Sheridan, *Tappi* **59,** 84 (1976); J. A. Nelson, *Tappi* **61,** 108 (1978).
189. Private communication, A. R. Risdon, Celanese Corporation.
190. P. W. Allen, ed., *Techniques of Polymer Characterization,* Academic Press, Inc., New York, 1959; "Characterization of Macromolecular Structure," *Proc. Conference, Apr. 5–7, 1967, Warrenton, Va., National Acad. Sci., Washington, D.C., 1968;* M. J. R. Cantow, ed., *Polymer Fractionation,* Academic Press, Inc., New York, 1967.
191. N. M. Bikales, ed., *Water-Soluble Polymers,* Part III, Plenum Press, New York, 1973.
192. Ref. 78, Chapts. 4, 6–8.
193. F. A. Buytenhuys and F. P. B. Van der Maeden, *J. Chromatogr.* **149,** 489 (1978); U.S. Pat. 4,118,316 (Oct. 3, 1978), C. P. Talley, G. M. Bradley, and R. T. Guliana (to Calgon Corp.); A. R. Copper and D. P. Matzinger, *J. Appl. Polym. Sci.* **23,** 419 (1979); A. Domard, M. Rinaudo, and C. Rochas, *J. Polym. Sci. Phys. Ed.* **17,** 673 (1979).
194. S. Kawamura and Y. Tanaka, *Water Sewage Works* **9,** 348 (1966); H. J. Walther and D. Knaak, *Acta Hydrochim. Hydrobiol.* **2,** 471 (1974); I. Sato and T. Saito, *Ebara Infiruko Jiho* **70,** 26 (1977).
195. L. K. Wang and W. W. Shuster, *Ind. Eng. Chem. Prod. Res. Dev.* **14,** 312 (1975); D. Horn, *Prog. Colloid Polym. Sci.* **65,** 251 (1978).
196. Private communication, Mr. E. Strazdins, Stamford Research Center, American Cyanamid Co.
197. I. Tuktamysheva and A. Khaidas, *Metody Anal. Kontrolya Proizvod. Khim. Prom-Sti* **11,** 55 (1977).
198. L. YaSkvirskii, T. M. Gurkova, and I. M. Shelepugina, *Obogashch. Rud* **21**(1), 47 (1976); S. Crisp, B. G. Lewis, and A. D. Wilson, *J. Dent. Res.* **54,** 1238 (1975); E. Korosteleva and N. Dymarchuk, *Zh. Anal. Khim.* **30,** 11 (1975).
199. A. P. Black, F. B. Birkner, and J. J. Morgan, *J. Am. Water Works Assoc.* **57,** 1547 (Dec. 1965); N. Narkis and M. Rebhun, *J. Am. Water Works Assoc.* **67,** 101 (Feb. 1975).
200. W. B. Crummett and R. A. Hummel, *J. Am. Water Works Assoc.* **55,** 209 (1963); Y. A. Attia and J. Rubio, *Br. Polym. J.* **7,** 135 (1975); R. K. Iler, *J. Colloid Interface Sci.* **37,** 364 (1971); M. J. A. Vreugde and G. W. Poling, *CIM Bull.* **68**(764), 54 (Dec. 1975); L. Jankovics, *Polym. Lett.* **2,** 247 (1964).
201. D. D. McCollister and co-workers, *Toxicol. Appl. Pharmacal.* **7,** 639 (1965).
202. *Investigation of Selected Potential Environmental Contaminants: Acrylamides, EPA-560/2-76-008, TR 76-507 (PB 257704),* EPA, Washington, D.C., July 1976.
203. *Guidelines for the Information Needed for Evaluation of Acceptability of New Products and Formulations Intended for Use as Coagulant Aids for Water Treatment,* EPA, Washington, D.C., Jan. 1971.
204. *Report on Coagulant Aids for Water Treatment,* Environmental Protection Agency, Office of Water and Hazardous Materials, Cincinnati, Ohio, Feb. 1978.
205. S. L. Daniels, *Proc. 1974 National Conference on Control of Hazardous Material Spills,* AIChE, New York, 1974, pp. 31–37.
206. R. S. Gale and R. C. Baskerville, *Filtr. Sep.* **7,** 37, 47, 51 (1970); A. S. Teot, S. L. Daniels, and J. C. Hoffman, *Proc. Chem. Inst. Canada, Flocculation and Dispersion Symposium, Nov. 4–5, 1974, Toronto, Can.,* pp. 226–239; V. Stump, "Polyelectrolyte Selection for Direct Filtration," *Proc. 96th Ann. Conf. AWWA, June 20–25, 1976, New Orleans, Louis.;* E. Whayman and O. L. Crees, *Sugar J.* **38**(6), 20 (1974).
207. *Chemical Economics Handbook,* Stanford Research Institute, Menlo Park, Calif.
208. M. Yusa, *Int. J. Mineral Process.* **4,** 293 (1977).

209. M. Yusa, H. Suzuki, and S. Tanaka, *J. Am. Water Works Assoc.* **67,** 397 (1975).

General References

W. J. Weber, Jr. *Physicochemical Processes for Water Quality Control,* Wiley-Interscience, New York, 1972.

N. M. Bikales, ed., *Water-Soluble Polymers,* Plenum Press, New York, 1973.

J. Vostrcil and F. Juracka, *Commercial Organic Flocculants,* Noyes Data Corporation, Park Ridge, N.J. 1976.

C. F. Baes, Jr., and R. E. Mesmer, *The Hydrolysis of Cations,* Wiley-Interscience, New York, 1976.

J. K. Dixon, "Flocculation" in N. M. Bikales, ed., *Encyclopedia of Polymer Science and Technology,* Vol. 7, John Wiley & Sons, Inc., New York, 1967, pp. 64–78.

F. HALVERSON
H. P. PANZER
American Cyanamid Company

FLOTATION

Froth flotation is a process for separating finely ground valuable minerals from their associated gangue. The process is based on the affinity of properly prepared surfaces for air bubbles. A froth is formed by introducing air into a pulp of finely divided ore in water containing a frothing or foaming agent. Minerals with a specific affinity for air bubbles rise to the surface in the froth and are thus separated from those wetted by water. In preparation, the ore must first be ground to liberate the intergrown valuable mineral constituent from its worthless gangue matrix. The size reduction, usually to about 208 μm (65 mesh), reduces the minerals to such a particle size that they may be easily levitated by the bubbles (see Extractive metallurgy; Flocculating agents; Gravity concentration; Sedimentation; Size classification; Size reduction).

Froth flotation is usually used to separate one solid from another, for solid–liquid separations, as in dissolved air flotation, and for liquid–liquid separations, as in foam fractionation. The process also has the potential to make a particle size separation since fine particles are more readily flocculated and floated than are coarse ones.

Froth flotation is the principal means of concentrating copper, lead, molybdenum, zinc, phosphate, and potash ores, and a host of others. In the United States, nearly 400 million metric tons of ore are treated per year by this unit operation. Its chief advantage is that it is a relatively efficient operation at a substantially lower cost than many other separation processes. Separations by flotation also include widely divergent applications such as the separation of ink from repulped paper stock, peas from pea pods, oils from industrial wastes, and metal ions, bacteria, proteins, and colloidal particles from water. The first and most thoroughly studied and used application, however, is in the field of mineral engineering, and therefore, it is appropriate to consider the subject from this aspect.

It is essential to choose reagents to coat the desired minerals selectively in the presence of many other mineral species. This is a difficult undertaking since a pure mineral is rarely found in nature. Extensive atomic substitution in the crystal lattice, adsorbed and occluded ions, and intimate interlocking of various minerals usually occur. In addition, the same chemical compound may be found in several crystal forms often with varying surface properties. The same mineral may have a different chemical composition depending on its environment of formation. These complications require ingenuity of the mineral engineer and the applied surface chemist. Flotation was largely developed in the early part of the century as an empirical art, but in the past few decades great strides have been made toward understanding the basic physicochemical principles underlying the process.

Flotation, as applied to the concentration of minerals, is used in conjunction with other mineral engineering operations, such as comminution, classification, gravity concentration, thickening, and filtration (qv).

The recently published *Flotation—A. M. Gaudin Memorial Volume* (1) is an invaluable reference book as are several older but still useful books on the subject (2–7). Several other books and industrial bulletins given in the list of general references are also helpful. Professional journals, such as those published by the world's major mining and metallurgical societies, and governmental publications are useful sources of information.

Fundamentals

The flotation separation of one mineral species from another depends on the relative wettability of surfaces. Typically, the surface free energy is lowered by the adsorption of heteropolar surface-active agents. The hydrophobic coating then acts as a bridge so that the particle may be attached to an air bubble.

The basic principles of surface chemistry have contributed substantially to understanding flotation phenomena (8–11). Many of these physicochemical surface phenomena may be directly correlated with flotation. The contact angle is used as an investigative tool and the electrokinetic properties of surfaces and the adsorption density of certain ions are known to correlate well with flotation response (12).

Contact Angles. An air bubble when brought into contact with a clean mineral surface usually does not adhere to the surface. However, if a suitable reagent is added, the mineral acquires a hydrophobic coating and an air bubble may be attached quite readily. This is illustrated in Figure 1. A schematic diagram of this phenomenon is shown in Figure 2, which may be represented as the well-known Young equation,

$$\gamma_{SA} = \gamma_{SL} + \gamma_{LA} \cos \theta$$

where γ_{SA}, γ_{SL}, and γ_{LA}, are the interfacial tensions at the solid–air, solid–liquid, and liquid–air interfaces, respectively. This may be combined with the Dupré equation for the work of adhesion at the solid–air interface,

$$W_{SA} = \gamma_{SL} + \gamma_{LA} - \gamma_{SA} \text{ to give } W_{SA} = \gamma_{LA} (1 - \cos \theta)$$

Considering that γ_{LA} remains essentially constant for the concentrations of surfactant usually used in flotation, it is apparent that the contact angle is a good measure of the work to create the solid–air interface. For this reason the contact angle has proved to be a simple yet very important tool for flotation research.

Figure 1. Photomicrograph of bubble-particle attachment in flotation. Galena particles in ethyl xanthate solution clinging to air bubbles. Courtesy H. R. Spedden.

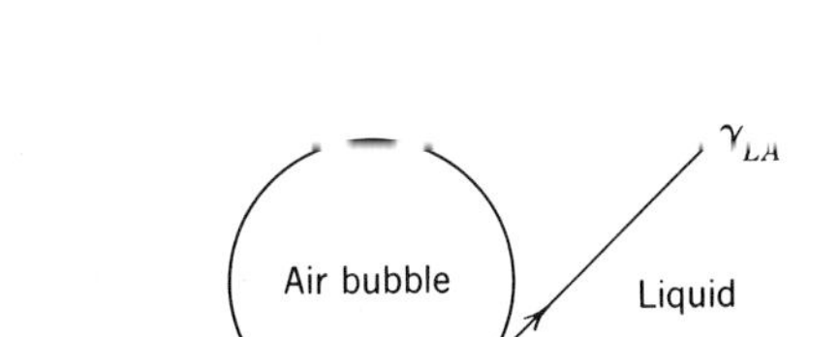

Figure 2. The contact angle formed by an air bubble on a solid submerged in water.

The Electrical Double Layer. Particulate matter in aqueous solution invariably has an electrical charge because an excess of cations or anions exists at the solid surface. This may be the result of, for example, the dissociation of surface groups, unequal dissolution of lattice ions, atomic defects occurring in the natural solid, or broken bonds between atoms resulting from comminution.

Figure 3 is a representation of the electrical double layer existing at a mineral surface showing the potential-determining ions. Attracted to this surface and extending

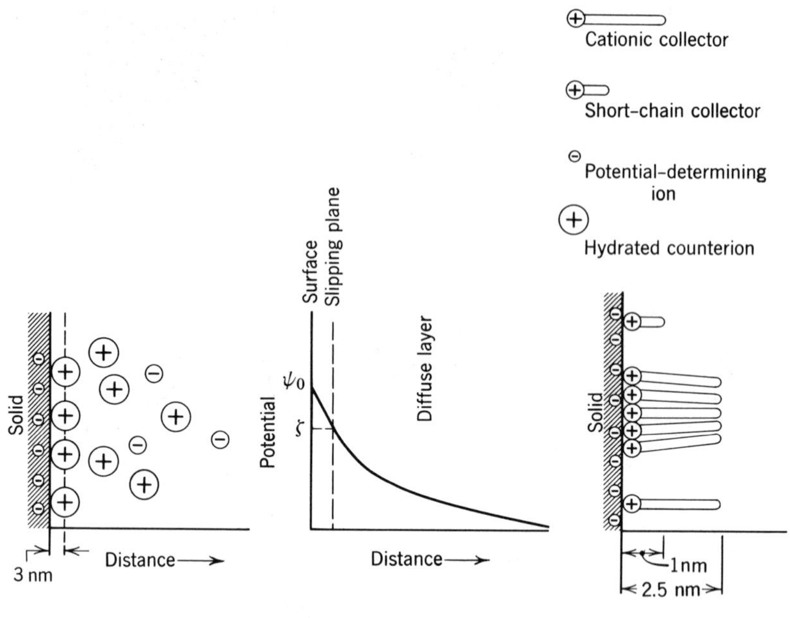

Figure 3. (**a**) Schematic representation of the double layer; (**b**) potential diagram; (**c**) adsorption of flotation collector (11). Courtesy American Institute of Mining, Metallurgical and Petroleum Engineers.

out into the solution for some distance are counterions, in this case hydrated cations which maintain electroneutrality. The center line of closest approach of the counterions is approximately 0.3 nm from the surface. The diffuse double layer extends into the bulk of the solution for varying distances, depending largely upon the concentration and valence of the counterions. In a 10^{-5} molar solution of a monovalent electrolyte, the median distance may be about 100 nm, but this distance is reduced sharply with increasing concentration. The potential frequently measured in moving a particle through a solution is the potential at the slipping plane, called the zeta potential, ζ. The zeta potential is not the surface potential, ψ, but is rather the potential at some distance from the surface. For this reason its qualitative significance in determining, for example, the sign of the charge on the surface is often of greater value than is its exact quantitative value.

The surface potential, ψ_0, and the concentration of potential-determining ions in solution C^+ or C^- are related as follows (8):

$$\psi_0 = kT/ve \ln (C^+/C_0^+) = -kT/ve \ln (C^-/C_0^-)$$

where k is the Boltzmann constant, T the absolute temperature, v the valence of the potential-determining ion, e the electronic charge, and C_0^+ and C_0^- the concentration of potential-determining ions at the point of zero charge. The concentration of potential-determining ions controls the sign and magnitude of the surface potential. The valence of the ion also makes a significant contribution. The ability of the potential-determining ions to change the surface charge is very important in flotation. An excellent, concise summary of double-layer computations is given in ref. 13.

The classic example of potential-determining ions is that of Ag^+ and Cl^- for a silver chloride sol (Table 1). With a point of zero charge, PZC (isoelectric point) at pAg = 4 and a solubility product constant (K_{sp}) of 2×10^{-10}, silver chloride is positively charged in solutions containing greater than 10^{-4} mol Ag^+/L and negatively charged in solutions containing Cl^- in excess of 2×10^{-6} mol/L. In like manner, for other slightly soluble minerals (K_{sp} ca 10^{-10}), the lattice ions are also potential determining. Barite, $BaSO_4$, is an excellent example of this class of minerals, and the concentration of Ba^{2+} and SO_4^{2-} determines its sign of charge and potential in solution. The PZC varies for each specific mineral.

For the insoluble oxide minerals, H^+ and OH^- are the potential-determining ions, and these minerals are positively charged at pH values below their PZC and negatively charged at higher pH values. As shown in Table 2, the PZC is highly variable for this class of minerals. The same mineral type (eg, chromite or asbestos) may show more than one PZC, depending on compositional or morphological differences, or a series of minerals with a common cation (eg, goethite, hematite, or magnetite) may have a common PZC. Ref. 16 presents a comprehensive review of the PZC values for heavy metal oxides and hydroxides and methods for the prediction of their PZC.

Other ions may also alter the surface charge of a mineral in solution. Polyvalent ions such as Al^{3+}, Fe^{3+}, and Co^{2+}, near their pH of precipitation in dilute solution, may become specifically adsorbed and decrease the potential or reverse the sign of the charge of a mineral in solution. This charge reversal process (17) has been applied to flocculation (18) and to activation phenomena (14) (see below).

Heteropolar surfactants of the colloidal electrolyte class are also known to adsorb onto oppositely charged minerals through electrostatic attraction and change the sign of the charge of a mineral in solution. These ions probably adsorb singly at dilute concentrations, but at greater concentrations (ca 10^{-4}–10^{-5} mol/L) (19) they may tend to form in patches called hemimicelles. At this concentration they may cause the charge on the mineral to be reversed.

Another class of ions, called indifferent ions, can influence the potential of a particle in solution. With increasing concentration and valence, these ions may adsorb onto the oppositely charged solid and reduce its zeta potential to a small value. They

Table 1. Points of Zero Charge of Slightly Soluble Minerals[a]

Mineral	Formula	Potential-determining ions	Sign of charge in saturated solution	K_{sp}[b]	PZC
ceragyrite	$AgCl$	Ag^+, Cl^-	−	2.0×10^{-10}	pAg 4.0
barite	$BaSO_4$	Ba^{2+}, SO_4^{2-}	−	1.5×10^{-9}	pBa 3.4
calcite	$CaCO_3$	Ca^{2+}, CO_3^{2-}	−	4.7×10^{-9}	
fluorite	CaF_2	Ca^{2+}, F^-	+	1.7×10^{-12}	
gypsum	$CaSO_4.2H_2O$	Ca^{2+}, SO_4^{2-}	−	2.4×10^{-5}	
scheelite	$CaWO_4$	Ca^{2+}, WO_4^{2-}	−	8×10^{-13}	pCa 4.8
anglesite	$PbSO_4$	Pb^{2+}, SO_4^{2-}	+	1.3×10^{-8} [c]	
celestite	$SrSO_4$	Sr^{2+}, SO_4^{2-}	+	7.6×10^{-7} [c]	

[a] Refs. 8 and 11.

[b] Solubility product constant.

[c] PZC near the stoichiometric ratio in saturated solutions.

Table 2. Points-of-Zero Charge of Insoluble Oxide Minerals[a]

Mineral	Formula	PZC, at pH
quartz	SiO_2	ca 2
feldspar, albite	$NaAlSi_3O_8$	2.0
clay,		
montmorillonite		<3.0
kaolinite		3.4
olivine, forsterite	Mg_2SiO_4	4.1
cassiterite	SnO_2	4.5
ilmenite	$FeTiO_3$	5.6
chromite	$MgO.Cr_2O_3$	5.6, 7.0, 7.2
rutile	TiO_2	6.0
magnetite	Fe_3O_4	6.5
goethite	$FeO.OH$	6.7
hematite	Fe_2O_3	6.7
iron hydroxide ppt	$Fe(OH)_3$	8.6
alumina	Al_2O_3	9.4
asbestos,		
lizardite		9.6
chrysotile		11.8

[a] Refs. 11, 14–16.

cannot, however, change the sign of the charge on the mineral as can the potential-determining ions. Their principal value is to cause spontaneously occurring flocculation when the zeta potential is reduced to some low value, eg, about 10 mV or below.

Phases. In flotation, gas, liquid, and solid phases have to be considered.

Gas Phase. The gas universally used is air, though often any gas will serve. Oxygen plays a special role in the flotation of the sulfide minerals (20–22). Theoretical analysis (23) indicates that the equilibrium adsorption density of the collector at the solid–gas interface exceeds its density at the solid–liquid interface. Experimental work with the adsorption of 1-hexanethiol by gold (24) and of sulfonate by mercury (25) appears to substantiate this thesis.

Liquid Phase. The universal liquid for froth flotation is water, though in a few instances saturated brine (3) or seawater (5) is used. To this phase various reagents are added for selective control of the wettability of various mineral surfaces and to achieve the desired frothing. There are three general classes of reagents: collectors (promoters), modifiers, and frothers.

Typical flotation collectors are shown in Table 3, though this is by no means an exhaustive list, and similar compounds often work equally well. For sulfide mineral flotation, both soluble and relatively insoluble sulfhydryl compounds are used. The hydrocarbon chain is relatively short (typically C_2–C_5) and only a small amount of collector is usually required (0.01 to 0.1 kg/t). For nonsulfide mineral flotation, the soluble colloidal electrolytes are the generally accepted collectors except in a few instances where the nonionized fatty acid is used. The hydrocarbon chain length ranges from C_{12} to C_{18} and a larger quantity of collector is required (ca 0.1 to 1 kg/t) than that used for sulfide mineral flotation. The amount of collector required varies considerably and depends not only on the nature of the collector but also on the nature of the mineral to be floated.

Chelation agents have also been given serious study as flotation reagents (11);

Table 3. Typical Flotation Collectors

Type	Formula[a]	Charge on the soluble ion
Sulfhydryl collectors[b]		
xanthate	ROCSSNa	anionic
dithiophosphate	$(RO)_2PSSNa$	anionic
monothiocarbamate	RHNCSOR	
thiol (mercaptan)	RSH	
dixanthogen	$(ROCSS)_2$	
thiocarbanilide	$(C_6H_5NH)_2CS$	
Colloidal electrolytes[c]		
fatty acids and their soaps	RCOOH, RCOONa	anionic
alkyl or aryl alkyl sulfonates	RSO_3Na	anionic
alkyl sulfate	$ROSO_3Na$	anionic
primary amine salt	RNH_3Cl	cationic
secondary amine salt	R_2NH_2Cl	cationic
quaternary ammonium salt	$RN(CH_3)_3Cl$	cationic

[a] $R = CH_3(CH_2)_n$.
[b] For sulfide minerals, $R = C_2–C_5$.
[c] For any mineral, the colloidal electrolyte collectors are generally straight chain C_{12} to C_{18}, or a benzene or naphthalene ring may also be incorporated into the R group.

however, their industrial use has been insignificant, largely because of their high cost.

Modifying agents are classed as pH-regulating agents, activators, depressants, dispersants, and flocculants. They function in a wide variety of ways depending on the system. Their specific purposes are explained below.

Frothing agents are used to provide a stable flotation froth, persistent enough to facilitate the mineral separation, but not so persistent that it cannot be broken to allow subsequent handling. The most commonly used frothing agents are short, branched-chain alcohols, such as MIBC (methyl isobutyl carbinol, 4-methyl-2-pentanol), poly(propylene glycol) methyl ethers, pine oil, and cresylic acid and creosote. The froth serves a secondary purpose in providing a zone in which the unfloated mineral particles that have been mechanically entrapped by the rising air bubbles may drain away from the flotation concentrate. When the collector is an amine, soap, or sulfonate, an additional frother is often not needed (see also Foams).

Solid Phase. Crystal chemistry (3,11) plays an extremely important role in flotation, and the nature and structure of the solid phase are of critical importance. In addition to the direct influence of crystal structure, minor elemental substitution or chance adsorption of ions is sometimes the dominant feature. The solid phase may be roughly categorized into the five groups given in Table 4: sulfides, slightly soluble, insoluble oxides, soluble salts, and naturally floatable. Since each class of mineral is floated in a different manner, this categorization is a convenient way to assess the potential floatability of any mineral. The ease of floatability roughly correlates with the surface energy. Naturally floatable minerals have low surface energies, whereas those more difficult to float have much higher surface energies. The sulfides occupy a somewhat anomalous position because they have relatively high energies, though in practice they are so readily oxidized in aqueous solution that they present a much lower energy surface.

In addition to the gross differences between minerals of these various classes, more

Table 4. Collector Systems Used to Float Various Types of Minerals

Mineral classification	Potential-determining ions	Typical mineral examples	Collector type	Typical collector	Mode of collector attachment	Typical collector concentration, kg/t	Remarks
sulfide	M^n, S^{2-}, HS^-, H^+, OH^-	chalcocite, Cu_2S chalcopyrite, $CuFeS_2$ galena, PbS	sulfhydryl	ethyl xanthate	chemisorption	0.02–0.05	hydrocarbon chain length C_2–C_5
slightly soluble	lattice ions	barite, $BaSO_4$ fluorite, CaF_2 hornsilver, AgCl	carboxylic acid	oleic acid/sodium oleate	chemisorption	0.25–0.5	hydrocarbon chain length C_{12}–C_{18}
insoluble oxide	H^+, OH^-	feldspar, (K, Na)$AlSi_3O_8$ hematite, Fe_2O_3 silica, SiO_2	colloidal electrolyte				
			anionic	aryl alkyl sulfonate	electro-static	0.5–1	direct adsorption of collector onto oppositely charged mineral surfaces
			cationic	tallow amine			
			colloidal electrolyte, anionic	aryl alkyl sulfonate	chemisorption, electrostatic	0.5–1	use of metal ion as activator
soluble salt		sylvite, KCl	amine	tallow amine		0.15–0.4	float KCl from NaCl in saturated brine
naturally floatable		coal molybdenite, MoS_2 sulfur, S_8	(oil)	(fuel oil)		0.1–1	often use frother only 0.05–0.25 kg/t, sometimes augmented with oil

subtle, but still important, changes in their floatability may come about because of naturally occurring or induced point defects. Such defect structures are typified by the semiconducting minerals. Point defects strongly influence flotation behavior (26–28) and recent experiments with a wide variety of minerals such as hornsilver, fluorite, nickel oxide, hematite, and galena have confirmed this (28).

Kinetics. Under a fixed set of conditions, different minerals have different flotation rates. This variation is a function not only of the type of mineral but also its size, the reagent system employed, and the hydrodynamic conditions within the flotation cell. A first-order rate equation generally expresses the data satisfactorily (29–31):

$$\frac{dc}{dt} = -kC^n$$

where C = concentration of floatable material, t = time, s, k = rate constant, and n = 1 (first-order).

The rate of flotation depends on the strength of collector adsorption. In general, a mineral with a strongly chemisorbed collector coating (eg, xanthate on galena) floats faster than a mineral that has a weakly adsorbed collector coating.

Particle size also influences the flotation rate of a mineral and the optimum rate for many minerals occurs for particles 10–200 μm in size (30). The rate decreases as the particle size is reduced below 10–20 μm and decreases precipitously for particles above 210 μm. The data obtained by a kinetic analysis (32) of an industrial copper flotation circuit are represented by:

$$W = W_o(1 - P)^n$$

where W = weight of a given mineral constituent remaining in the pulp after passing through n cells, W_o = weight of that constituent in the flotation feed, and P = probability of flotation of any mineral constituent in any one cell.

A graph of n vs log weight fraction remaining gives a straight line with slope of log $(1 - P)$. In fact, the analysis of a flotation circuit gives two such straight lines: one for the first few cells in line representing the fast floating constituents and one for the last cells representing the slower floating fraction.

Differences in the flotation rate of various minerals are not often intentionally used to achieve a mineral separation, though on a practical basis, the "cleaner" cells typically capture the fastest floating fraction, whereas the later cells in the "rougher" section recover the slower floating material. Recently, the flotation rate differential between coal and pyrite has been proposed as a means of rejecting pyrite from coal (33).

Flotation rate data are used to estimate the necessary retention time for a flotation circuit (34), to scale up flotation cells (29), and to model flotation circuits for process optimization (35–36).

Collectors

The collector may be adsorbed on a mineral by chemical or physical forces. Attachment by chemisorption to the surface is the preferred method in practice because it generally leads to greater selectivity in separation of the mineral species, and the flotation can be achieved at a lower reagent concentration and, hence, at a lower cost. Physical adsorption of the collector may involve van der Waals, hydrogen, hydrophobic bonding, etc, but the usual adsorption mechanism is electrostatic attraction between an ionic collector and a mineral of opposite charge (11,37). Selection of the collector depends on the mineral assemblage under consideration, and this is the basis of the methodology developed in Table 4.

Sulfide Minerals. The sulfides of copper, lead, nickel, and zinc are the largest class of minerals separated by flotation. These minerals and the native metals (eg, gold) are usually floated with a short-hydrocarbon-chain collector (C_2–C_5) of the sulfhydryl type. Only small amounts of reagent are required with typical levels of 0.01–0.05 kg of reagent per metric ton of ore. In simplified form, the driving force for the adsorption of these short-chain sulfhydryl collectors is related to the insoluble nature of many of the heavy metal xanthates (7). The K_{sp} of lead ethyl xanthate is about 10^{-17}, and lead sulfide (galena) floats readily with a short-chain xanthate (3). In contrast, the solubility product of zinc ethyl xanthate is 5.3×10^{-9}, and sphalerite, ZnS, does not float with ethyl xanthate unless the concentration of xanthate is raised to very high levels. The sulfide minerals appear to be very stable as indicated by the following equilibrium constant (38):

$$PbS \rightarrow Pb^{2+} + S^{2-} \qquad K_{sp} = 7 \times 10^{-29}$$

In practice, however, the sulfides are readily oxidized (20) in aqueous solution, and therefore the surface lead ions are not so tightly held to the solid and may thus react with the xanthate.

Recent studies have shown that metathetic reaction proposed for xanthate adsorption is much too simple (22,28,39–41). In many cases the adsorbed xanthate is converted to a dixanthogen on the surface through an electron transfer (22,40).

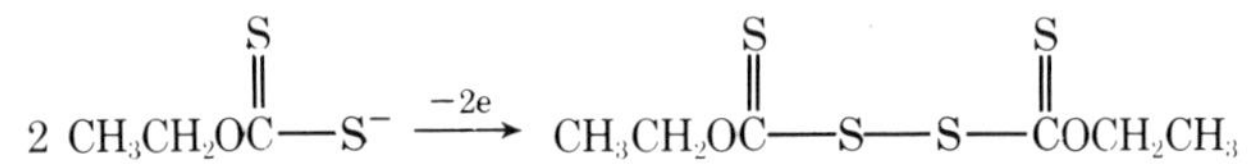

ethyl xanthate ion dixanthogen

Both xanthates and dixanthogen (or any similar thiolate or dithiolate) may be present on the surface of a sulfide material, and which species predominates depends on the rest potential of that mineral in solution (22). When dixanthogen is used as a collector for galena, the primary reaction product is lead ethyl xanthate produced by an electron transfer from the mineral surface to the dixanthogen (22,28).

Although the sulfide minerals are almost universally floated industrially with sulfhydryl collectors, they may also be floated by the appropriately charged colloidal electrolyte collector. For example, sphalerite that is negatively charged in solution may be readily floated with dodecylamine (42). By appropriate control of the various potential-determining ions listed in Table 4 for the sulfide minerals, either cationic or anionic collectors may be used. As sulfides are rendered strongly negative in solution by as little as 10^{-6} mol HS^-/L, they may thus be conveniently floated by cationic amines (43–44). There are no known commercial applications of this method, but the potential is considerable.

Sulfide minerals may also be floated without the use of any collector (45). Freshly produced sulfide mineral surfaces, before any substantial oxidation has taken place, are naturally hydrophobic and thus may be floated using only a frother. At present, this method is only of laboratory interest but it shows great promise.

Slightly Soluble Minerals. These minerals are defined here as those with solubility product constants of ca 10^{-10}. Because of their moderate solubility, their cations are readily available at the mineral surface for collector attachment, and accordingly, this class of minerals is generally floated industrially with fatty acid soaps. The driving force is the relatively low solubility of alkaline earth and other metal soaps in solution (K_{sp} ca 10^{-15}) (11,14,46). It was demonstrated (11,47) with infrared techniques that above the pK_a for fatty acids (ca 4.7) the fatty acid is chemisorbed onto minerals such as fluorite, whereas below their pK_a value, where the neutral molecule, RCOOH, predominates, the adsorption is principally physical.

The lattice ions of the slightly soluble minerals are potential determining, as shown in Table 4. Thus, in addition to their flotation by a chemisorbed fatty acid, they may be floated by a collector of the colloidal electrolyte class. A typical example would be that of barite, which is positively charged in strong Ba^{2+} solutions and may be floated with an anionic organic sulfonate or sulfate, whereas in strong SO_4^{2-} solutions it is negatively charged and is most readily floated with a cationic amine.

Other more insoluble minerals ($K_{sp} \ll 10^{-10}$) may sometimes also act as minerals of this class since oxidation and various dissolution reactions in solution may lead to a concentration of their lattice ions in solution approaching 10^{-5} mol/L. The mineral apatite, $Ca_5(PO_4)_3(OH,F)$ would be an example of such a mineral. Furthermore, CO_2 and OH^- may cause HCO_3^-, CO_3^{2-}, and OH^- to act as potential-determining ions for some minerals of the slightly soluble class. The carbonation reaction, under appropriate conditions, may give calcium minerals of this class a surface coating of carbonate so they would respond as would calcite (48). The present state of knowledge of this class of minerals has recently been summarized (49).

Insoluble Oxide Minerals. This class of minerals is normally floated with long-chain (C_{12}–C_{18}) fatty acids, sulfonates, and amines. In fact, fatty acids may be used to float nearly any mineral, given the appropriate conditions. The problem is then essentially one of how to achieve selectivity between the various minerals. Flotation of this class of minerals may be either by use of electrostatic adsorption of the oppositely charged collector or by use of an activator (Table 4). The latter procedure (see below) is more commonly used industrially.

When collector attachment is by electrostatic attraction, the data of Table 2 may be used to estimate the PZC, and thus the appropriate anionic or cationic collector for the pH in question may be chosen. The response of the system may be seen in Figure 4 (11,50). Below its PZC (pH 6.7), goethite (FeO·OH) is positively charged, and the anionic dodecyl sulfate or sulfonate is adsorbed and thus allows the mineral to be floated. Above the PZC, where the goethite is negatively charged, a cationic amine is used to achieve flotation.

The curves of Figure 4 were obtained with reagents having 12 carbon atoms in the chain. As the number of carbon atoms in the chain is increased to 18, the curves translate, so that with octadecyl sulfate goethite flotation is possible up to pH 10 (51). Selectivity between species is lost by use of higher homologues or of an excessive amount of collector. This is the result of an ever-increasing interaction energy between the hydrocarbon chains, masking many of the electrostatic effects. The interaction between hydrocarbon chains may cause the formation of patches or hemimicelles of collector (Fig. 3). This occurs at a concentration of about one-hundredth of that of the bulk critical-micelle concentration of the collector in question (52).

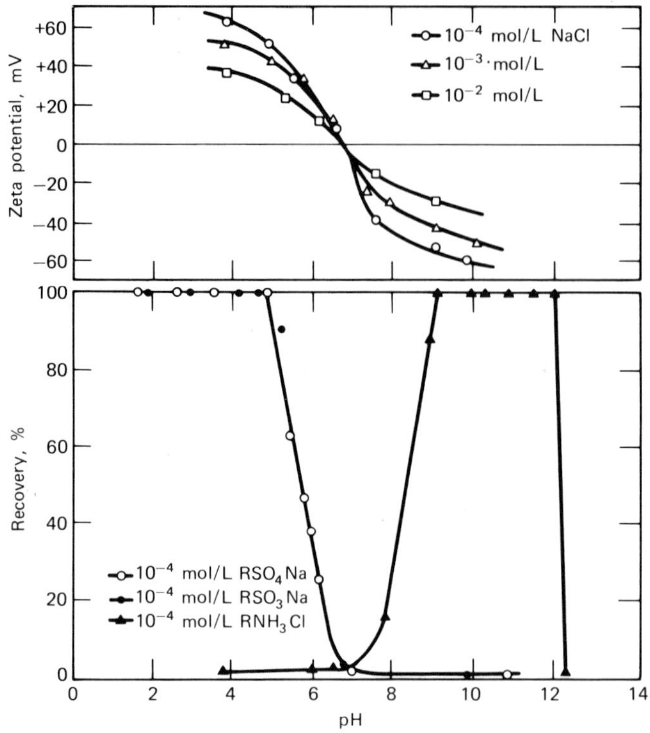

Figure 4. Effect of pH on the zeta potential and flotation recovery of goethite (50). Upper curves show the PZC at pH 6.7. Lower curves show flotation with the anionic collectors, dodecyl sulfate and sulfonate below the PZC, and with the cationic dodecylamine above the PZC.

Soluble Salt Minerals. Froth flotation of soluble salt minerals is used extensively in the concentration of potash ores. The flotation separation of KCl from NaCl particles is accomplished in a saturated brine with the use of an amine. A plausible explanation for the selectivity between these two alkali halides is one of size fit (11,53). The ionic radius of K^+ ion, 0.133 nm, is more nearly like that of the aminium ion than is the Na^+ ion at 0.095 nm. This theory has been extended with reasonable success to all the alkali halides. Others have postulated that if a soluble salt has a sufficiently negative heat of solution, flotation is possible. This would explain some of the exceptions to the size-fit hypothesis.

Naturally Flotable Minerals. These minerals do not constitute a class in the same sense as do the preceeding classes, since they have only their inherent floatability in common. Such natural floatability may arise from a variety of causes. The content of organic compounds of minerals, such as asphalt, coal and tar sands, provides a natural hydrophobicity and they may be readily attached to an air bubble. Minerals such as boric acid, graphite, molybdenite, and talc are characterized by a weak van der Waals bonding between layers. When these crystals are fractured, no strong chemical bonds are broken, hence the surface is essentially a nonpolar, nonwetted one, and the compound is naturally floatable (54).

Naturally floatable minerals are generally floated only with a frother or with a frother plus a small amount of fuel oil. An oily frother such as pine oil may also be conveniently used. For many of the naturally floatable minerals, the surface is not completely hydrophobic but contains a mosaic of hydrophobic and hydrophilic sites (28). Their natural hydrophobicity, and hence, their floatability is highly variable. If the use of an oil as collector proves inadequate, or if for reasons of selectivity it is undesirable, it is common to use an amine collector as these mineral surfaces usually have a negative charge. The minerals are so readily floated that either a short-chain amine (ca C_8) or a minute quantity (ca 0.1 kg/t) of a higher homologue (C_{12}–C_{18}) of the collector is adequate. At this reagent level, minerals of the other classes are generally not very floatable.

Attention has previously been called to the natural floatability of a freshly produced sulfide mineral (45), and in this case natural hydrophobicity is probably destroyed by subsequent oxidation of the surface.

Regulating Agents

pH Regulators. For those minerals whose potential-determining ions are H^+ and OH^-, the role of pH in altering the surface charge is quite obvious. In other systems, the relationship is not so simple. For instance, pH may indirectly affect potential-determining ions through chemical equilibria that alter the concentrations of the potential-determining ions. At lower pH values, CO_3^{2-} ion from calcite would react further with H^+, whereas at higher pH values Ca^{2+} ion would be precipitated as the hydroxide. In addition, pH controls the ionization of, especially, the fatty acid and amine collectors from molecular to ionic species, which in turn influences the collector adsorption and, hence, flotation. At high concentrations of H^+ or OH^-, these ions compete with the collector for the mineral surface.

Activators. Activators are added to a flotation system to permit better collector attachment to the mineral to be floated. The classic example is the addition of copper sulfate to a flotation pulp containing sphalerite, ZnS. The cupric ion attaches to the sphalerite, allowing the mineral to be floated with a xanthate. Such flotation is not possible with a short-chain xanthate in the absence of an activating agent. The activation of sphalerite with Cu^{2+}, Pb^{2+}, or Ag^+ is a chemical reaction in which these ions replace zinc in the sphalerite lattice through the formation of a more insoluble sulfide. In the case of copper activation, the exchange occurs rapidly until three layers of Zn^{2+} have been replaced and then continues more slowly (55).

The activation of most minerals occurs through a different mechanism. It has long been known that metal ions activate minerals for flotation near the pH of precipitation of that metal ion in dilute solution (56). More recent work (14,57–58) has demonstrated that this form of activation occurs by the adsorption of the first hydroxy complex (or in some cases, polynuclear complexes) of the metal ion onto the mineral surface. A sulfonate or fatty acid soap collector may then be attached to this activated surface. Delicate control of these activated systems is often required as the concentrations of collector, metal ion, and hydroxyl ion are all critical. Either an inadequate or an excessive quantity of any of the three can seriously reduce mineral floatability (58). Self-activation is often used. For example, manganese dioxide minerals are activated with SO_2 which dissolves some of the manganese and makes these ions available to serve as the activator (11).

For some systems the electrostatic model offers a means of explaining how activators function. With fused alumina (corundum), sulfate ion serves as an activator when using a cationic collector at pH values below the PZC (pH 9.4) where the alumina is positively charged (37).

Another method of activation is sulfidizing. In 1905, sodium sulfide as an aid in the flotation of oxidized copper and lead ores was patented. The reasons for its effectiveness are still not known, though it is probably connected with the formation of a new mineral surface simulating that of the metal sulfide. After sulfidizing, the oxidized mineral is floated with a xanthate or similar collector.

Depressants. The electrostatic model is of some help in understanding the function of depressants in achieving selectivity. Where H^+ and OH^- ions determine the potential, their use as a depressant assures that the mineral has the opposite charge to that of the collector. Sulfate ion depresses corundum when using an anionic collector (37). Barite depression by Ba^{2+} when using a cationic amine collector is another example. Chemical forces are undoubtedly also at work in depression phenomena, and mass action effects have been used to explain the use of zinc sulfate to prevent the activation of sphalerite by Pb^{2+} (59).

Actually, many depressant phenomena should more appropriately be called deactivation. The activating ion may be incorporated into the mineral lattice by minor elemental substitution, or may be on the mineral surface through chance adsorption. An elegant proof of this is offered in the case of the beryllium silicate (beryl), where a small amount of iron is responsible for the attachment of the sulfonate in the commercially used beryl flotation process (57). When the iron was leached from the mineral, flotation by the sulfonate was impossible. The deactivating type of depressant action is mostly accomplished by use of a complexing agent to sequester the offending ion. For sulfide mineral flotation, sodium cyanide is frequently used. Figure 5 (4), using the contact-angle method, shows that the flotation of copper and iron minerals is greatly influenced by the use of a small amount of cyanide. On the other hand galena, PbS, is unaffected since Pb^{2+} is not sequestered by cyanide. These curves are of great practical value in indicating selectivity between species. For example, chalcocite, Cu_2S, may be easily separated from pyrite, FeS_2. The importance of pH control is also clearly indicated by these curves.

For nonsulfide mineral flotation, the complexing agents are usually meta-, poly-, or pyrophosphate, quebracho, citric acid, carbonate, and fluoride ion. These are used because of their effectiveness and low cost. Numerous other reagents perform the same function (11).

In sulfide mineral flotation, a very powerful depressing action is achieved by the use of sodium sulfide which is even more effective than cyanide or hydroxyl ion (4). The exact reason for the depression is not known, but it has been demonstrated that the effective species is probably the HS^- ion.

Another important method of mineral depression is by a mechanism called surface blocking (60). In this method, sodium silicate or hydrocolloids, such as starch, guar, glue, proteins, and tannins, are used to prevent collector attachment probably by attachment of the hydrocolloid as a surface polymer. Sodium silicate probably also functions through other mechanisms as well, in view of its similarity to silicate minerals and its well-known properties as a dispersant. Surface blocking is typified by the use of the large polysaccharide, guar gum molecule in potash flotation. Here it blocks the surface of clay minerals, naturally present in the ore, to the adsorption of amine (3). In this way the relatively expensive amine collector can be used efficiently to float KCl.

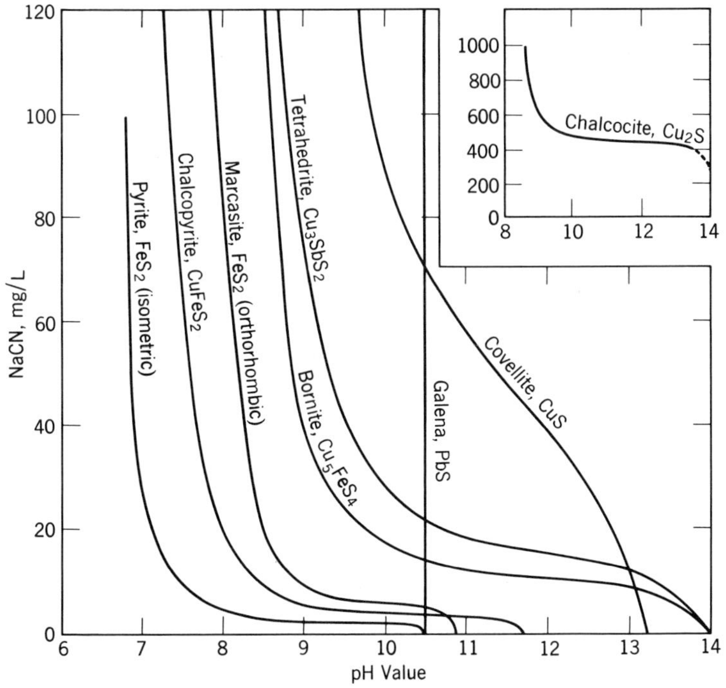

Figure 5. Contact angles for several sulfide minerals (4). Bubble contact and flotation possible to left of the curve (solution contains 25 mg/L potassium ethyl xanthate). Courtesy the Australian Institute of Mining and Metallurgy.

Depression is also sometimes achieved by the use of oxidizing agents to oxidize the mineral surface and prevent collector attachment. Coal may be depressed in this way (31). Acid and strong oxidizing agents such as sodium hypochlorite are also used to destroy collector coatings so that the floated mineral can be depressed in a subsequent flotation step. This procedure is used in copper–molybdenum separations and in phosphate flotation and is discussed below.

Dispersants. In sulfide mineral flotation, separation is ordinarily made in the presence of slimes, or very fine particles. However, for nonsulfide flotation where selectivity between species must be accomplished by subtle means, slime control is almost always necessary. In some cases, for example, the separation of clay from quartz, a dispersant can be added to the flotation system to ensure that coflocculation of the desired mineral and gangue does not occur (61). More often, however, the usual method is to disperse the pulp physically (eg, by the use of a hydrocyclone; see Size separation) and/or chemically to remove the slimes before flotation. The term slimes is ambiguous and usually refers to a particle size below which the particles cannot be removed commercially. This size was about 74 μm (200 mesh) 50 years ago, but the widespread use of hydrocyclones has moved the usual definition down to 10 μm and finer.

The electrostatic model can be successfully used to explain the deleterious effects of slimes (62–63). Consider, for example, the mineral flocculation to be expected in a pulp of iron oxide (goethite), kaolinite, and finely ground quartz at pH 5 (see Table 2). The goethite under these conditions is positively charged and would be expected to attract to its surface the very small particles of negatively charged kaolinite and quartz. These slime coatings seriously interfere with flotation. Even 50 mg bentonite clay slimes per liter can completely inhibit the flotation of goethite (63). In addition to preventing collector attachment, slimes may also adsorb the collector, decreasing its availability to the mineral species to be floated. Amines are particularly sensitive to clay slimes.

Dispersion is usually accomplished through pH regulation and by dispersants (qv), such as the pyro-, poly-, and hexametaphosphate, lignin sulfonates, sodium silicate, and the tannins.

Flocculants. Flocculating agents are normally added after flotation to aid in the filtration of the products or in thickening finely ground pulps. Lime and alum are commonly used as are natural and synthetic polymeric substances such as starches, gums, polyethylene oxides, polyacrylamides, and copolymers of polyacrylamide with acrylates or quaternaries. Anionic, nonionic, and cationic forms may be used (see Flocculating agents).

In some cases mineral particle flocculation is purposely induced to affect the flotation step. Examples of this are agglomerate flotation, flocculation flotation, and carrier flotation. In agglomerate or emulsion flotation, an oil in the amount of 0.5 to several kg/t is added, together with collecting and emulsifying agents, to cause agglomeration of only the desired particles. The floccules are then removed by froth flotation or other means. This process is especially effective for coarse particles (<14 mesh, ca 1.2 mm), naturally floatable particles, and particles of minerals that are difficult to float. A modification of this process is spherical agglomeration where 3–10% oil is added to the ground ore to produce compact spherical agglomerates that are then removed from the gangue by screening (64).

In a selective flocculation–flotation process (65), as applied commercially to hematitic iron ores, the ground ore is dispersed with sodium silicate and/or phosphate. The fine hematite particles are then flocculated with starch and allowed to settle in a thickener while other slime particles overflow the thickener to waste. This is followed by the removal of the coarser silica particles from the flocculated iron particles by flotation with a cationic collector in the presence of dextrine as an iron depressant. Such a process has considerable merit for the flotation of very fine (slime) particles that are otherwise difficult to float selectively.

In carrier or piggy-back flotation, an easily floatable inert mineral is added to the ground ore to become coated with slime particles of a mineral otherwise difficult to float. The inert mineral is then removed by conventional flotation, which carries with it the poorly floatable mineral. The process is used commercially to improve the brightness of coating-grade clays by removing color causing impurities (66).

Industrial Technology

Successful flotation requires a rather delicate balance of many operating variables, only some of which are under the direct control of the mineral engineer. The most important among these are ore, water, reagents, and machinery.

Most ore bodies consist of from one to five economically extractable minerals combined with a much larger amount of nonvaluable minerals referred to as gangue. The texture or physical association of the minerals is one of the most important considerations, for it determines whether or not the valuable constituents may be economically removed and the degree of treatment complexity required for the separation.

It also dictates the selection of auxiliary operations such as grinding and gravity separation. As a general rule, in the most economical treatment, separation takes place at as coarse a size as possible to secure liberation of the minerals. If this size is coarser than that at which flotation is practiced, other methods of concentration might be indicated. Likewise, if the size of liberation is too fine to come within the practical flotation range, chemical leaching might be the answer.

Perhaps no two ore bodies are identical and no single ore body is consistent in composition throughout. The extensive substitution of foreign elements into the mineral crystal lattice, the adsorption of interfering ions, the fine intergrowths of different mineral species, and the presence of mineral slimes that coat the surfaces of the minerals often make flotation separations in practice very difficult. The problems are not unlike those encountered by chemists dealing with natural products. One of the best methods of assessing the treatment possibilities of a new ore body is to compare it to the treatment schemes used for processing known ore bodies. Summaries of practice given in the general reference books are especially recommended. Once an ore body has been selected, the representative samples, collected by the geologist or mining engineer, are tested in the laboratory to determine processing feasibility (67–68). It is advisable for the test engineer to work closely with a mineralogist during this phase of the study so that the identification and association of the mineral species may be made.

Equipment. The most commonly used flotation machines are the Agitair (Galigher Co.), Denver D-R (Denver Equipment Co.), and Fagergren (Envirotech Corp.) machines. The basic elements of such machines are shown in Figure 6. The ore, usually ground to a nominal value below 210 μm (<65 mesh) is introduced into the cell in the form of a water slurry or pulp at about 25% solids by weight. The regulating, collecting, and frothing agents are added either before, or by stages to the bank of flotation cells, depending on the type of separation and the need for conditioning time of the various reagents. Air is introduced through the central shaft and is dispersed into the well-mixed pulp by the impeller. The mineralized bubbles rise to the surface of the cell to produce a froth which is then scraped over the lip to effect the separation.

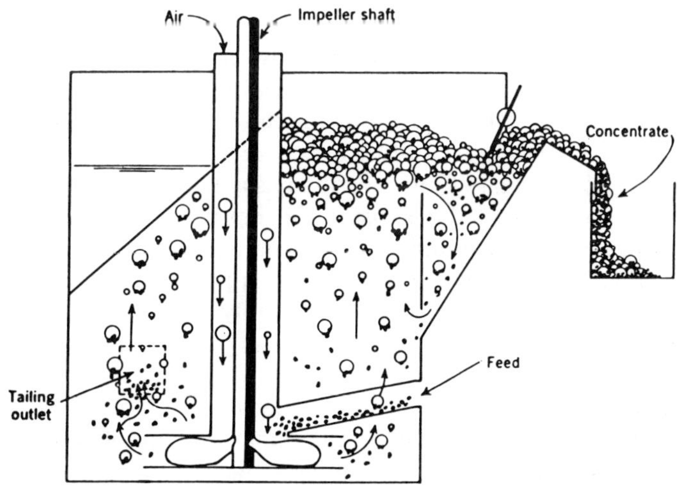

Figure 6. Froth flotation cell.

Industrial flotation machines usually range in size from 1.1 to 28 m^3 (40–1000 ft^3)/cell with cells having a volume of 5.7–14 m^3 (200–500 ft^3) each, commonly used in large concentrators. Since the flotation residence time may range from 3 to 40 min, it is necessary to use many cells in series to allow sufficient time for all the valuable mineral to float. Commonly, units of 12–18 individual cells are used in series for copper sulfide flotation. Design elements are given in refs. 29, 69–70.

Processes. It is impossible to state exactly the conditions for any one flotation separation because of the numerous variables involved and the constantly improving techniques. The following examples of typical operations illustrate conditions generally applicable.

Lead–Zinc Ores. These ores often consist of the minerals galena, PbS, sphalerite, ZnS, and a mixture of nonvaluable minerals, such as pyrite, FeS_2, quartz, SiO_2, and calcite, $CaCO_3$. The galena and sphalerite are usually separated by grinding in the presence of sufficient lime or sodium carbonate to maintain a pH of 8–9.5, with the addition of 0.05–0.25 kg sodium cyanide per metric ton of ore. About 0.2 kg zinc sulfate per metric ton of ore may also be added. The combination of pH, cyanide, and zinc sulfate is selected to prevent the flotation of the sphalerite, while the galena is being recovered by the use of a sulfhydryl collector such as ethyl xanthate. A small quantity of frother (0.01–0.05 kg/t) is used to provide only sufficient froth for the optimum recovery of a comparatively clean galena concentrate. The residue is further treated by the addition of sufficient copper sulfate to activate the sphalerite. A xanthate, frequently propyl or higher, is added as a collector, and the pulp is subjected to more vigorous frothing than before. Each of the concentrates from the first or rougher stage is usually retreated by further flotation steps, called cleaning, to aid in the elimination of entrapped minerals. The residues or cleaner tailings from these retreatment steps are recirculated to the head of their respective preceeding flotation stage. A flow sheet for flotation of a typical lead–zinc ore consisting of galena, sphalerite, and pyrite in a nonsulfide gangue is shown in Figure 7. Silver, if present, generally follows the galena during flotation.

Reagent usage and results obtained from a concentration process for lead–zinc–silver ores with a flow sheet roughly similar to that of Figure 7 are given for the Bunker Hill mine, Kellogg, Idaho, in Tables 5 and 6 (71).

Copper Sulfide Ores. Copper sulfide ores constitute the largest tonnage of ores treated by flotation. The minerals most commonly encountered include chalcopyrite, $CuFeS_2$; chalcocite, Cu_2S; bornite, Cu_5FeS_4; covellite, CuS; enargite, Cu_3AsS_4; and tetrahedrite, Cu_3SbS_2. The first two are by far the most significant. Sulfhydryl collectors are used, and, since copper sulfide ores nearly always contain an appreciable amount of pyrite, 0.15–0.3 kg/t of lime is added as a depressant. In some instances a small quantity of cyanide is added to assist in the depression of pyrite but its use must be carefully controlled since cyanide also has a depressing effect on copper-bearing sulfides as well (Fig. 5). Additions of collector of approximately 0.01–0.03 kg/t, and a comparable quantity of frother, are used successfully. If the pyrite contains sufficient gold or is to be used for acid manufacture or other purposes, it may be floated as a separate concentrate.

Many large porphyry copper deposits of the western United States contain, in addition to their copper content (ca 0.6% Cu), a small amount of molybdenite (ca 0.02 % MoS_2). This is an important by-product mineral that because of its ready floatability follows the copper minerals during the flotation process. It is separated from the copper

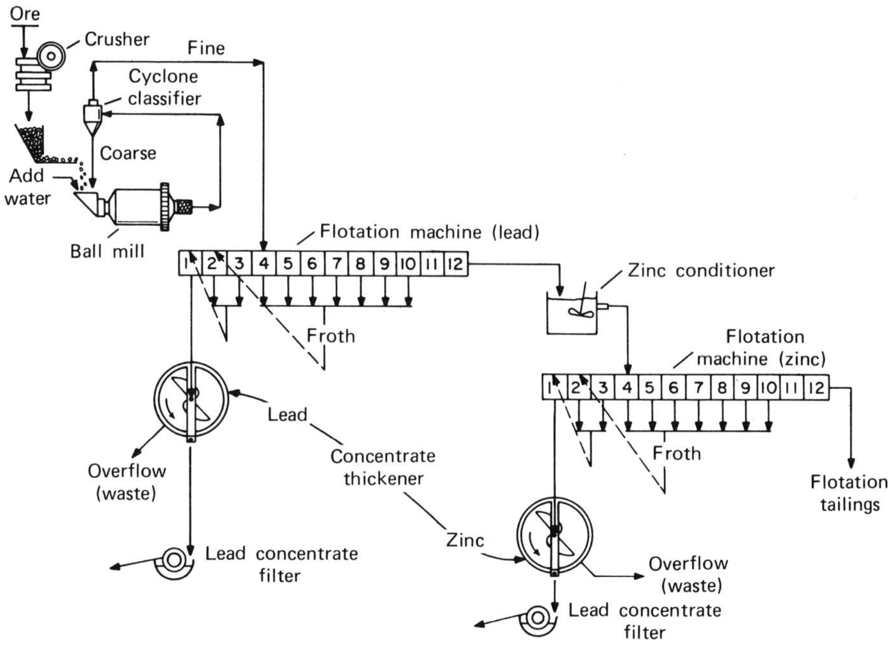

Figure 7. Simplified flow sheet for treating a lead–zinc ore.

Table 5. Reagent Practice for the Treatment of a Typical Lead-Zinc-Silver Ore[a]

Reagent	Point of addition	Quantity added, kg/t	Function of reagent
soda ash	ball mill and Pb flotation	0.156	depressant for pyrite and sphalerite
sodium cyanide	ball mill and Pb flotation	0.058	depressant for pyrite and sphalerite
zinc sulfate	ball mill and Pb flotation	0.122	depressant for sphalerite
sodium isopropyl xanthate	ball mill	0.020	collector for galena
	Zn conditioner	0.037	collector for sphalerite
MIBC[b]	Pb flotation	0.032	frother
	Zn conditioner	0.040	frother
lime	Zn conditioner and Zn flotation	0.708	depressant for pyrite
copper sulfate	Zn conditioner	0.172	activator for sphalerite
polyacrylamide	Zn concentrate thickener	0.0005	flocculant

[a] Based partly on data from ref. 71.
[b] MIBC = methyl isobutyl carbinol.

minerals in the final concentrate by selective flotation and is sold as a separate product. In a typical separation process, the bulk copper–molybdenite concentrate, containing ca 25% Cu and ca 0.5% MoS_2, is treated with steam or an oxidizing agent to destroy the collector coating on the minerals. A small amount of fuel oil and a frother are then used to float the molybdenite away from the copper minerals.

Table 6. Typical Metallurgical Results for Lead–Zinc–Silver Ore Using Reagent Practice Given in Table 5[a]

Product	Weight %	Assays Pb, %	Assays Zn, %	Assays Ag, g/t[b]	Distribution Pb, %	Distribution Zn, %	Distribution Ag, %
mill feed	100.00	6.19	4.90	136	100.00	100.00	100.00
Pb concentrate	8.83	66.59	4.98	1397	95.16	8.97	90.62
Zn concentrate	7.85	1.18	54.21	97.8	1.50	86.78	5.65
tailing	83.32	0.25	0.25	6.2	3.34	4.25	3.73
soluble Pb		0.08					

[a] Ref. 71.
[b] To convert g/t to troy oz/avoirdupois ton divide by 34.3.

Phosphate Ores. Phosphate ore containing about 25% BPL (bone phosphate of lime), $Ca_3(PO_4)_2$, is concentrated in very large tonnages from the Florida pebble phosphate deposits (72). The ore slurry from strip mining is sized at about 1.2 mm (14 mesh) and the coarser fraction, after scrubbing to break up mud balls, is a finished product. The 1.2 mm (<14 mesh) fraction is further sized at 0.42 mm (35 mesh) and 0.10 mm (150 mesh). The 0.10 mm material (<150 mesh) is discarded. From the sizing operation the 1.2 × 0.42 mm (14 × 35 mesh) material in thick slurry is conditioned with fatty acid (tall oil), fuel oil, and caustic, and the resulting agglomerates are separated by flotation or with spirals or spray belts. The 0.42 × 0.10 mm (35 × 150 mesh) fraction is conditioned with the same type of reagents and floated by the conventional froth flotation route. Not all the silica gangue is rejected by the fatty acid flotation, so the concentrate is blunged with acid to remove collector coatings, deslimed, washed of reagents, and subjected to an amine flotation with fuel oil at pH 7–8. This latter flotation removes additional silica and raises the final concentrate grade to about 75% BPL (see Phosphoric acid and phosphates; Fertilizers).

Coal. Coal is produced in increasing quantities by the flotation process (31) though the principal means of coal beneficiation remains gravity concentration (see Gravity concentration). Since coal is typically a naturally floatable mineral, often a frother only (ca 0.15 kg/t) or a frother with a small quantity of fuel oil (0.1–0.5 kg/t) suffices. The amount of fuel oil required is a function of the rank of the coal (see Coal). The higher rank bituminous coals require little or no oil as collector whereas the lower rank sub-bituminous and lignite coals will require rather large quantities that are often substantially in excess of 2 kg/t. Resistance to flotation because of surface oxidation can be sometimes overcome by the addition of small amounts of amine.

Ash removal has historically been the primary purpose of coal processing. More recently, a major objective has been the elimination of pyritic sulfur. Pyrite may be separated from coal by the use of differences in their respective flotation rates (31,33), variations in circuitry (31,73), pyrite depression (33), and coal depression with concurrent flotation of the pyrite (73).

Economic Aspects

Production. The U.S. Bureau of Mines periodically reports the statistical trends of froth flotation in the United States. Their most recent study indicated that 383 million metric tons of ore was concentrated by the flotation process in the United States

Table 7. U.S. Production Data Ores Treated by Flotation, 1960 and 1975[a]

	Ore treated, 10^6 t		Reagent cost, \$/t	
	1960	1975	1960	1975
Ore type				
copper, copper–molybdenum, and molybdenum	126.32	236.00[b]	0.068	0.142[b]
lead–zinc (including those containing some copper)	14.52	17.49	0.171	0.408
iron oxide	1.39	25.96	0.305	0.417
phosphate	19.03	68.11	0.225	0.364
potash	10.87	12.56	0.184	0.216
fluorspar	0.44	0.25	0.828	2.280
glass sand	0.73	6.67	0.227	0.371
feldspar, mica, quartz, spodumene	0.94	2.33	0.740	0.644
coal	3.73	11.87	0.098	0.201
other[c] (by difference)	1.94	2.06		
Total	*179.91*	*383.31*	*0.121*	*0.229*
number of flotation plants	194	252		
rated daily capacity, 10^6/d	0.662	1.462		
water used: m^3/t		5.30		
(gal/short ton)		(1270)		
power used[d], MJ/t[e]		62.7		
(kW·h/short ton)		(15.8)		

[a] Refs. 74–75.
[b] Estimated value. Individual data for molybdenum producers and those listed as "other" not reported separately.
[c] Includes garnet, manganese (1960 only), barite, bastnaesite, clay, gold, silver, ilmenite, kyanite, limestone–magnesite, mercury, talc, tungsten, and vermiculite.
[d] For total concentration process.
[e] To convert J to cal, divide by 4.184.

in 1975 (74). As shown in Table 7, this is more than twice the amount treated in 1960 (75); rated daily capacity had also more than doubled in this period. The largest category treated are copper and molybdenum ores, and sulfide ores of all types (Cu, Mo, Pb, Zn) accounted for 66%. Nonsulfide ores such as iron oxide, phosphate, glass, sand, and coal have shown a large increase. Table 7 also gives reagent cost in dollars per metric ton. On the average, the reagent cost nearly doubled in the 15-year period. In 1975 the average reagent cost per metric ton processed was \$0.23. Reagent costs vary widely, ranging from the treatment of simple sulfide ores of Cu and Mo (\$0.14/t) to fluorspar (\$2.28/t). The separation of complex sulfides and, especially, the nonsulfides generally requires large amounts of reagents with a corresponding increase in costs. The bulk of the energy used is for grinding the ore to flotation size. The average amount of water used to process a metric ton of ore (5.3 m^3 or 17 ft^3) is somewhat distorted by the large amount used to process a metric ton of raw phosphate rock (13.1 m^3 or 43 ft^3). Copper and copper–molybdenum ores, for example, require water in the amount of ca 3.3 m^3/t (10.8 ft^3). In a typical flotation plant, approximately 90% of the water requirement is supplied by recycling.

Table 8 gives the consumption and value of flotation reagents used in the United States. Whereas consumption doubled, the value of these reagents increased fourfold in the period 1960–1975. Collectors are, by far, the most expensive class of reagents used.

Table 8. Consumption and Value of Flotation Reagents Used in the United States in 1960 and in 1975[a]

Reagents	Consumption, 10^3 t		Value, 10^6 \$	
	1960	1975	1960	1975
collectors	99.85	181.77[b]	8.74	40.24
frothers	7.67	12.07	2.45	7.71
activators	5.65	5.32	1.47	2.55
depressants	7.19	26.70	1.64	7.27
modifiers	257.44	571.45[c]	5.98	25.32
flocculants	7.90	4.75	1.51	4.29
other		0.35		0.30
Total	*358.70*	*802.41*	*21.79*	*87.68*
kg/t ore treated	1.949	1.899		
$/t ore treated			0.121	0.229

[a] Refs. 74–75.
[b] Approximately 53% fuel oil and kerosene and 39% tall oil and other fatty acids.
[c] Approximately 82% lime and 10% sulfuric acid.

Costs. Estimating capital and operating costs is a difficult task, further complicated by recent problems of a serious inflation, substantial changes in the cost of borrowed money, and dramatic increases in energy and environmental costs. These complications have imposed a particularly heavy burden on the mining industry which must process large quantities of ore to recover a rather small amount of product and is a capital-intensive industry.

Capital construction costs for a flotation mill, including tailings pond, average $3600–4500/t/d for small mills (500–1000 t/d), and up to $1800–3600/t/d for large plants. The lowest costs per metric ton are for very large concentrators with capacities in excess of 50,000 t/d. Preliminary capital costs can be estimated (76–77). Operating costs for a very large copper sulfide treatment plant are currently ca $2.20/t. Note that the average reagent costs (Table 7) represent only about 10% of these direct costs. Typically, crushing and grinding accounts for more than 60% of the direct costs. As the plant size decreases or the processing step becomes more sophisticated, direct costs escalate substantially. Furthermore, as shown in Table 7, the reagent cost for the flotation of many commodities is ca $0.40/t which increases the operating costs accordingly. Of the direct costs approximately one-third is for labor, maintenance, and supplies, to which 30% or more must be added for indirect costs, including financing charges. Total operating costs for a large copper sulfide mill with all of its advantages of simplicity, ready floatability of the copper sulfide minerals, and the economies of scale thus approaches $3.00/t. A recent survey indicated that the cost of producing copper in a large plant increased 85% from 1972 to 1976 (78). The energy cost increase in this period was 183%.

Uses

Although the principal use of flotation has been in mineral processing, it has also considerable use elsewhere. For example: the removal of ink, pigments, and coatings from paper; the flotation of sewage and industrial waste; purification of water (qv); removal of oil from refining waste; concentration of naphthalene from coke-oven-gas cooling-tower water; flotation of ions and precipitates; separation of bitumens from

Table 9. Classification of Various Bubble Separation Techniques[a]

Size range	Natural surface activity	In association with surface-active species
molecular	foam fractionation: detergents from solution	ion flotation, molecular flotation: Sr^{2+}, Ag^{2+}, Pb^{2+}, Hg^{2+}, cyanides, phosphates
microscopic	foam flotation: microorganisms, proteins, dyes	microflotation: particulates in waste, microorganisms
macroscopic	froth flotation of naturally floatable minerals: sulfur	froth flotation: minerals, eg, silica precipitate flotation: ferric hydroxide

[a] Ref. 83.

tar sands (qv); flotation removal of seeds, pods, and other impurities from peas; the separation of the oil-rich fraction of seeds; removal of wax from wool (qv); flotation of dyes; flotation of microorganisms; and the separation of textile fibers from process streams (79–81). The separation of ion-exchange resins from unclarified hydrometallurgical liquors and pulps has been proposed (82).

Perhaps the best characterization of the wide variety of flotation methods is shown in Table 9 (83). For those substances that are naturally floatable, and can be concentrated or attached at a bubble wall, the terms foam fractionation, foam flotation, or froth flotation have been applied. For materials that are not naturally floatable, a surface-active agent is added as a collector and the terms ion or molecular flotation, microflotation, and froth flotation are used. An ion can be precipitated with a carboxylic acid that is removed by flotation. Sulfonates and other detergents have also been used for this purpose (80). More efficiently, the ion is precipitated first and subsequently the precipitate floated with a smaller quantity of collector. This is called precipitate flotation and can be carried in standard froth flotation equipment with slight modifications. Such separations are typified by the removal of copper precipitated as the hydroxide or sulfide from solution (43,84).

For pollution control and the removal of fine particulates, conventional froth flotation is not adequate. Instead, a dissolved-air flotation cell called a save-all is used (79). The liquid to be purified is pressurized with air, and upon pressure release, tiny bubbles nucleate and grow on the surface of the particles. The bubbles, with their attached particulates, then rise to the surface where they are removed as an overflow product. This has traditionally been used for clarifying water containing sewage sludge, paper mill white-water effluent, and packing-plant wastes where surfactants and oils naturally present may act as the collector. When absent, a small quantity of the appropriate collector should be added.

BIBLIOGRAPHY

"Flotation" in *ECT* 1st ed., Vol. 6, pp. 595–614, by H. R. Spedden, Massachusetts Institute of Technology; "Flotation" in *ECT* 2nd ed., Vol. 9, pp. 380–398, by Frank F. Aplan, Union Carbide Corporation, Mining and Metals Division.

1. M. C. Fuerstenau, ed., *Flotation, A. M. Gaudin Memorial Volume,* 2 vols., AIME, New York, 1976.
2. D. W. Fuerstenau, ed., *Flotation—50th Anniversary Volume,* AIME, New York, 1962.

3. A. M. Gaudin, *Flotation,* 2nd ed., McGraw-Hill, New York, 1957.
4. K. L. Sutherland and I. W. Wark, *Principles of Flotation,* Australian Institute of Mining and Metallurgy, Melbourne, 1955.
5. V. A. Glembotskii, V. I. Klassen, and I. N. Plaksin, *Flotation,* transl. by R. E. Hammond, Primary Sources Publishers, New York, 1963 (original title: *Flotatsiya,* Moscow, 1961).
6. V. I. Klassen and V. A. Mokrousov, *An Introduction to the Theory of Flotation,* translated by J. Leja and G. W. Poling, Butterworth & Co., London, 1963.
7. A. F. Taggart, *Handbook of Mineral Dressing,* John Wiley & Sons, Inc., New York, 1945.
8. H. R. Kruyt, *Colloid Science,* Vol. I, 1952; Vol. II, 1949, Elsevier Publishing Co., Amsterdam.
9. A. W. Adamson, *Physical Chemistry of Surfaces,* 3rd ed., John Wiley & Sons, Inc., New York, 1976.
10. G. E. Agar and P. L. deBruyn in ref. 2, Chapt. 5.
11. F. F. Aplan and D. W. Fuerstenau in ref. 2, Chapt. 7.
12. D. W. Fuerstenau, *Trans. AIME* **288,** 1365 (1957).
13. H. van Olphen, *An Introduction to Clay Colloid Chemistry,* Interscience Publishers, a division of John Wiley & Sons, Inc., New York, 1963.
14. M. C. Fuerstenau and B. R. Palmer in ref. 1, Chapt. 7.
15. F. F. Aplan, *Min. Eng.* **20,** 91 (Feb. 1968).
16. G. A. Parks, *Chem. Rev.* **65,** 177 (1965).
17. R. O. James and T. W. Healy, *J. Colloid Sci.* **40,** 53 (1972).
18. B. K. Parekh and F. F. Aplan, "Charge Reversal and Flocculation of Fine Particles with Co^{2+}," presented at *53rd Colloid and Surface Science Symposium, Rolla, Missouri, June 1979.*
19. A. M. Gaudin and D. W. Fuerstenau, *Trans. AIME* **202,** 958 (1955).
20. A. F. Taggart, T. C. Taylor, and A. F. Knoll, *Trans. AIME* **87,** 217 (1930).
21. I. N. Plaksin, *Trans. AIME* **214,** 319 (1959).
22. N. P. Finkelstein and G. W. Poling, *Miner. Sci. Eng.* **9,** 177 (1977).
23. P. L. deBruyn, J. Th. G. Overbeck, and R. Schumann, *Trans. AIME* **199,** 519 (1954).
24. F. F. Aplan and P. L. deBruyn, *Trans. AIME* **226,** 235 (1963).
25. C. A. Smolders, *Rec. Trav. Chim.* **80,** 699 (1961).
26. I. N. Plaksin and R. S. Shafeev, *Trans. IMM,* **72**(608), 715 (1962–63).
27. G. Simkovich, *Trans. AIME* **227,** 306 (1963).
28. E. Y. Spearin and co-workers, to be published by U.S. Bureau of Mines.
29. N. Arbiter and C. C. Harris in ref. 2, Chapts. 8 and 14.
30. W. J. Trahar and L. J. Warren, *Int. J. Min. Proc.* **3,** 103 (1976).
31. F. F. Aplan in ref. 1, Chapt. 45.
32. D. F. Kelsall, *Trans. Instit. Min. Metall.* **70,** 191 (1960–61).
33. F. F. Aplan in T. D. Wheelock, ed., *Coal Desulfurization,* ACS Symposium Series, No. 64, ACS, Washington, D.C., 1977, Chapt. 6.
34. A. C. Dorenfeld in ref. 2, Chapt. 15.
35. E. T. Woodburn and co-workers in ref. 1, Chapt. 24.
36. G. C. Thorne and co-workers in ref. 1, Chapt. 26.
37. H. J. Modi and D. W. Fuerstenau, *Trans. AIME* **217,** 381 (1960).
38. W. M. Latimer, *Oxidation Potentials,* 2nd ed., Prentice-Hall, New York, 1952.
39. S. A. Allison and co-workers, *Met. Trans.* **3,** 2613 (1972).
40. R. Woods in ref. 1, Chapt. 10.
41. P. E. Richardson and E. E. Maust in ref. 1, Chapt. 12.
42. H. H. Kellogg and H. Vasquez-Rosas, *Trans. AIME* **169,** 476 (1946).
43. J. W. Perez and F. F. Aplan in P. Somasundaran and R. B. Grieves, eds., *Advances in Interfacial Phenomena of Particulate/Solution/Gas Systems; Applications to Flotation Research,* AIChE Symposium Series 150, Vol. 71, AIChE, New York, 1975.
44. J. W. Perez, K. C. Thompson, and F. F. Aplan, unpublished data.
45. V. M. Lepetic, *Bul. Can. Inst. Min. Metall.,* **76**(746), 71 (1974).
46. C. Du Rietz in E. Ohman, ed., *Progress in Mineral Dressing, Trans. Int'l. Mineral Dressing Congress, 1957,* Almquist and Wiksell, Stockholm, 1958.
47. A. S. Peck, *U.S. Bur. Mines Rept. Invest. 6202,* 1963.
48. J. D. Miller and J. B. Hiskey, *J. Colloid Interface Sci.* **41,** 567 (1973).
49. H. S. Hanna and P. Somasundaran in ref. 1, Chapt. 8.
50. I. Iwasaki, S. R. B. Cooke, and A. F. Colombo, *U.S. Bur. Mines Rept. Invest. 5593,* 1960.
51. I. Iwasaki, S. R. B. Cooke, and H. S. Choi, *Trans. AIME* **217,** 237 (1960).
52. D. W. Fuerstenau, *J. Phys. Chem.* **60,** 981 (1956).

53. D. W. Fuerstenau and M. C. Fuerstenau, *Trans. AIME* **204,** 302 (1956).
54. A. M. Gaudin, H. Miaw, and H. R. Spedden, *Proc. Intern. Congr. Surface Activity, 2nd London,* Vol. III, Butterworth & Co., London, 1957.
55. A. M. Gaudin, D. W. Fuerstenau, and G. W. Mao, *Trans. AIME* **214,** 431 (1959).
56. G. Gutzeit, *Trans. AIME* **169,** 272 (1946).
57. M. C. Fuerstenau and R. B. Bhappu, *Trans. AIME* **229,** 164 (1963).
58. M. C. Fuerstenau and S. Atak, *Trans. AIME* **232,** 24 (1965).
59. D. W. Fuerstenau and P. H. Metzger, *Trans. AIME* **271,** 119 (1960).
60. A. F. Taggart, *Elements of Ore Dressing,* John Wiley & Sons, Inc., New York, 1951.
61. H. H. Kellogg, *Trans. AIME* **169,** 548 (1946).
62. D. W. Fuerstenau, A. M. Gaudin, and H. L. Miaw, *Trans. AIME* **211,** 792 (1958).
63. I. Iwasaki and co-workers, *Trans. AIME,* **223,** 97 (1962).
64. C. E. Capes, A. E. Smith, and I. E. Puddington, *Bul. Can. Inst. Min. Metall.* **67,** 115, 1974.
65. A. F. Columbo and D. W. Frommer in ref. 1, Chapt. 47.
66. F. A. Seeton, *Ultraflotation at Minerals and Chemicals Phillip Corp., DECO Trefoil,* Bull. M4-B117, Denver Equipment Co., Colorado Springs, Colo., Jan.–Feb. 1963.
67. R. D. MacDonald and R. J. Brison, in ref. 2, Chapt. 12.
68. F. F. Aplan in A. B. Cummins and I. A. Given, eds., *SME Mining Engineering Handbook,* Vol. II, Section 27.3, American Institute of Mining, Metallurgical and Petroleum Engineers, New York, 1973, pp. 15–27.
69. C. C. Harris in ref. 1, Chapt. 27.
70. N. Arbiter and N. Weiss, *Trans. AIME* **247,** 340 (1970).
71. N. J. Sather and F. L. Prindle in D. O. Rausch and B. C. Mariacher, eds., *AIME World Symposium on Mining and Metallurgy of Lead and Zinc,* Vol. 1, AIME, New York, 1970, Chapt. 20.
72. P. S. Smith, Jr. in ref. 2, Chapt. 46.
73. K. J. Miller, *Trans. AIME* **258,** 30 (1975).
74. *Minerals Yearbook,* 1975, Vol. 1, U.S. Bureau of Mines, U.S. Government Printing Office, Washington, D.C., 1977, pp. 96–124.
75. C. W. Merrill and J. W. Pennington in ref. 2, Chapt. 4.
76. A. Mular, *Mineral Processing Equipment Costs and Preliminary Capital Cost Estimations,* Special Vol. 18, Canadian Institute of Mining and Metallurgy, Montreal, 1978.
77. N. Weiss and S. D. Michaelson in ref. 2, Chapt. 22.
78. F. M. Lewis, C. K. Chase, and R. B. Bhappu, *Min. Eng.* **30,** 371 (April 1978).
79. W. W. Eckenfelder, Jr., *Industrial Water Pollution Control,* McGraw-Hill Book Company, New York, 1966.
80. R. Lemlich, ed., *Adsorptive Bubble Separation Techniques,* Academic Press, New York, 1972.
81. L. A. Roe, *Unusual Applications of the Flotation Process, DECO Trefoil,* Bulletin F10-B104, Denver Equipment Co., Colorado Springs, Colo., Sept.–Oct. 1961.
82. R. B. Bhappu, *Colo. School Mines Quart.* **56,** 13 (July 1961).
83. P. Somasundaran, *Sep. Purif. Methods* **1**(1), 117 (1972).
84. R. E. Baarson and C. L. Ray in M. E. Wadsworth and F. T. Davis, eds., *Unit Processes in Hydrometallurgy,* Gordon and Breach, New York, 1964, p. 656.

General References

Refs. 1–7, 9, and 60

A. M. Gaudin, *Principles of Mineral Dressing,* McGraw-Hill Book Co., Inc., New York, 1939.

F. B. Michell, *The Practice of Mineral Dressing,* Electrical Press, London, 1950.

Mineral Processing Flowsheets, 2nd ed., Denver Equipment Co., Denver, Colo., 1965.

Mining Chemicals Handbook, Mineral Dressing Notes No. 26, American Cyanamid Co., Wayne, N.J., 1976.

Flotation Fundamentals and Mining Chemicals, Dow Chemical Company, Midland, Mich., 1976.

Chemicals for Mining, Hercules, Inc., Wilmington, Del., 1977.

DECO Trefoil, Denver Equipment Div., Joy Mfg. Co., Colorado Springs, Colo.

FRANK F. APLAN
The Pennsylvania State University

FLOUR. See Bakery processes and leavening agents.

FLOWMETERS. See Fluid mechanics.

FLUIDIZATION

The term fluidization is used to designate the gas–solid contacting process in which a bed of finely divided solid particles is lifted and agitated by a rising stream of process gas. At low velocities, the amount of lifting is slight and the bed behaves as a boiling liquid, hence the term boiling bed. At high velocities, the particles are fully suspended in the gas stream and are carried along with it, and the terms suspension, suspensoid, and entrainment contact have been used to describe this effect.

The fluidization technique as it is known today began with the work of Standard Oil Development Co. (now Exxon Research & Engineering Co.), The M.W. Kellogg Co., and Standard Oil of Indiana, in their efforts to find a catalytic cracking process better than the fixed-bed process that was introduced commercially in 1937. The fixed-bed process was a major improvement over the earlier thermal cracking methods. It yielded more gasoline of higher octane rating and less low-value heavy fuel oil by-product.

Innovations to develop a superior process include the fixed-bed and alternating-bed methods. In the latter, the regeneration and reaction beds are in series and the catalyst is continuously moved from one to the other. In order to avoid severe erosion as well as attrition, lower gas velocities are required. This led to the observation that dense beds of powder-form catalysts could be maintained with relatively low carryover losses even at superficial gas velocities that were orders of magnitude greater than the calculated settling velocity of the individual particles making up the bed. At these gas velocities the particles were observed to be considerably agitated as gas bubbles passed upward through the bed in a manner analogous to the boiling of liquids. The pressure drop through such a boiling or fluidized bed was equal to the weight of the bed charge; the bed was in effect heterogeneously buoyed by the gas stream and thus took on effective flowing properties similar to liquids (1).

The first commercial fluid-bed catalytic cracking plant was put in operation in 1942 followed by 31 additional plants in the war years. During the past 38 years the application of the fluidization technique has spread rapidly to metallurgical ore roasting, limestone calcination, synthetic gasoline, petrochemicals, and even to the design of nuclear reactors (qv) (2). Table 1 gives a sampling of fluid-bed applications that have been investigated over the past 20 years (see Reactor technology). Each application in Table 1 may represent a large number of operating units. Table 2 indicates the extent of activity in fluidized-bed combustion (see Furnaces, fuel-fired). Equal if not greater activity exists in fluidized-bed gasification (see Fuels, synthetic) than in fluidized-bed combustion, and nearly 200 fluidized-bed catalytic cracking plants are in operation (see Petroleum). Fluidization as a unit operation has touched almost every process industry.

In most fluidized-solid processes, the solid material is handled in one or more steps and transferred from step to step through pipelines (109). To raise the material to a higher level, it is carried as a suspension in a gas stream; to take it to a lower level or to a region of higher pressure, the settled material is allowed to flow down a pipeline. This is illustrated schematically in Figure 1. Process gas enters a plenum below a bed-supporting grid and bubbles up through the bed of catalytic or reacting solids thus fluidizing the bed. The bubbles burst at the surface and entrain some particles into the disengaging height. The exciting gases pass through a cyclone which returns the entrained particles to the bed and exhausts clean product gases to downstream equipment. If bed particles require regeneration, they can be withdrawn through a standpipe and returned to the fluidized bed via a pneumatic lift line and separator. From the separator they again flow through a standpipe seal leg into the reactor. The host of variations in solids flow between processing vessels can be visualized best by reference to the so-called fluid-solids phase diagram (110–111).

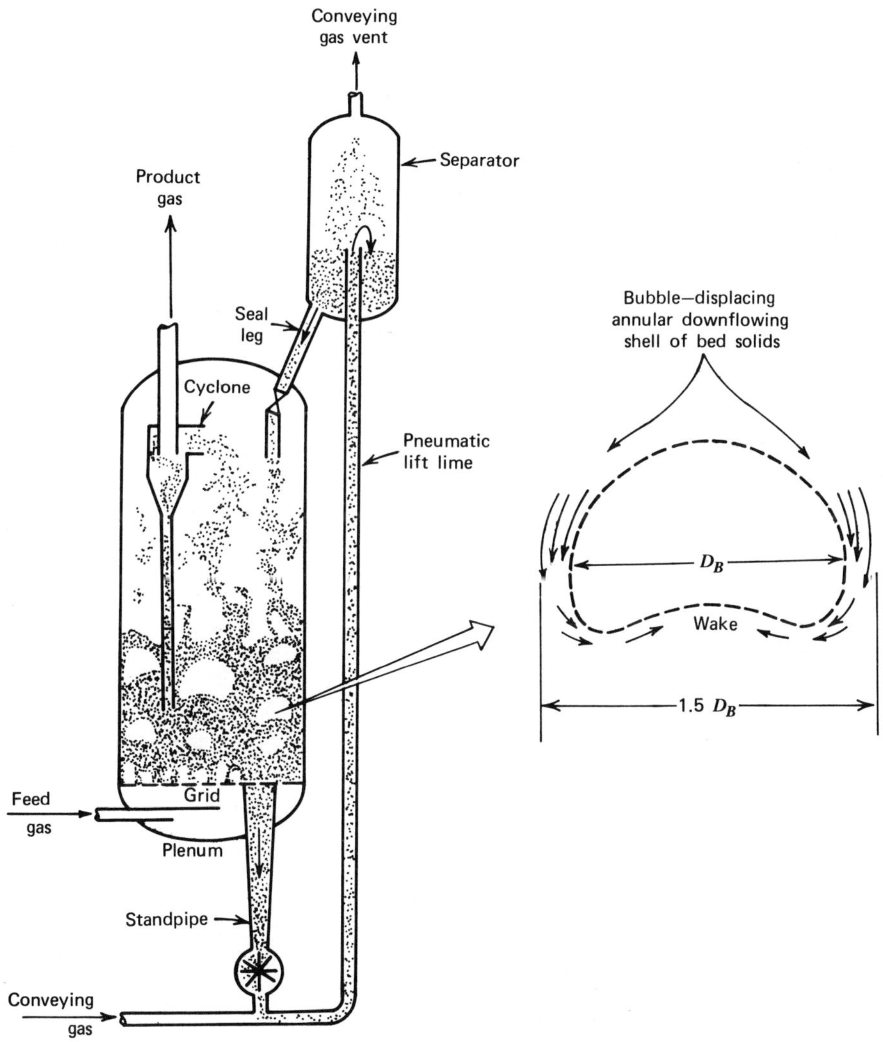

Figure 1. Bubble rise via displacement by inflow of a surrounding downflowing shell of bed solids.

Table 1. Selected Applications of Fluidization

Process	Investigator/user	Ref.	Process	Investigator/user	Ref.
acetone (qv) recovery	British Celanese Ltd., Wrexham, Wales	3	synthetic fiber heat stretching and heat setting	Dunlop Textiles; Courtaulds Ltd., Eng.	33
acetylation of polyformaldehyde	Sci. Res. Inst., Sofia, Bulgaria	4	electrolysis	Akzo Zout Chem., The Netherlands	34
activated carbon regeneration (see Carbon)	Battelle Columbus Labs., Ohio	5	plating and similar electrochemistry (see Electroplating)	Centre Nat. de La Rech. Sci., Nancy, France; Central Electrochem. Research Institute, India	35
agglomerates from fines	The Pillsbury Co., Krim-Ko Corp., Calif., and Hawaiian Sugar Refining Corp.	6	polyethylene, LD and HD (see Olefin polymers, polyethylene)	Union Carbide Corp.	36
alkylation and dehydrogenation of aromatic hydrocarbons	Mamedaliev Petrochemical Institute, USSR	7	fuel oil from polypropylene residue	Procedyne, New Brunswick, N.J.	36
production of AlF_3 (see Fluorine compounds, inorganic)	Montedison-Lurgi	8	fertilizers from oxidation and ammoniation of coal	Central Fuel Research Institute of India	37
calcining aluminum hydroxide (see Aluminum compounds)	Metals Research Institute, Budapest, Hungary	9	freezing fruits and vegetables; grilling meats and potatoes	Swedish Food Processor	38
			roasting chocolate beans (see Chocolate and cocoa)	General Goods Corp.	39
ammonium chloride and similar condensations of sublimable materials	Ivanov Chemical Engineering Institute, USSR	10	gasoline from methanol	Mobil Oil Corp.	40
			granulation and drying	Aeromatic, Inc.; Niro Atomizer	41
ammoniation of superphosphate	Inorg. Chem. Res. Inst., Czechoslovakia	11	hot air for peak shaving	Stal-Laval (UK) Ltd.	42
aniline via fluid-bed process	American Cyanamid Co.	12	hydrogen (qv) from steam	Institute of Gas Technology	43
aniline from nitrobenzene	M. S. Murthy and co-workers	13	HCl and Fe_2O_3 from waste pickling liquors	Nittetu Chemical Engineering Ltd., American Lurgi Corp.	44
roasting of arsenopyrites	Piritas Espanolas Auxini S.A., Madrid, Spain	14			
benzotrichloride from toluene	Western N.Y. Nuclear Research Center	15	regenerating HCl pickle liquor	Lurgi, G.m.b.H.; Hilgers A.G. at Rhein-Brohl, FRG	45
conversion of ethanol to butadiene (qv)	Indian Institute of Technology, Kharagpur	16	H-coal and H-oil (liquid-fluidized beds)	Hydrocarbon Research, Inc., Cities Service Oil, Co.	46
coke calcination (qv)	Aluminum Co. of Canada	17	CaO and phosphorus from tricalcium phosphate in a plasma bed at 1093°C	Battelle Memorial Institute	47
spent carbon recovery	Westvaco Corp.	18			

CS_2 adsorption on carbon in 11.6 m dia fluid bed (see Fuels, synthetic, gaseous)	Courtlauds, Ltd., Holywell, Wales	19
carbon tetrachloride from char	C.S.I.R.O., Clayton, Victoria, Australia	20
trichloroethane (methychloroform)	Ethyl Corp.	21
trichlor- and perchloroethylene (see Chlorocarbons)	PPG Industries	22
coal (qv) gasification	Philadelphia and Reading Corp., New York; Hydrocarbon Research, Inc., Inst. of Gas Tech., FMC Corp., M.W. Kellogg Co., Consolidation Coal Co., etc	23
fluid-bed boiler	CEGB, Eng.	24
coal combustion	Babcock and Wilcox	25
coating drug tablets	Abbott Laboratories	26
coating in fluid beds (see Coatings)	Knapsack-Griesheim A.G., Cologne; Whirlclad Div., The Polymer Corp.; General Electric Co.; Rockwell Manufacturing Co.	27
fluid coking of coal (see Coal conversion processes)	Atlantic Refining Co.	28
cristobalite from quartz in 1538°C fluid bed	A.D. Little, Inc.; Wedron Silica Co., Chicago, Ill.	29
oil decontamination of sand	University of California, Santa Barbara, Calif.	30
dimethyl terephthalate from toluene	Bergwerksverband A.G.; Toyokoatsu Ind , Japan	31
paper and textile drying (qv)	The Shirley Institute; Dunlop Textiles; Man-Made Fibers Research Association	32
phthalic anhydride (qv) from *o*-xylene	Badger, Cambridge, Mass.	48
prilling fertilizers	Fisons Fertilizers, Ltd.	49
fluidized solids propellants (see Explosives and propellants)	Bell Aerospace Div. of Textron Corp.	50
radioactive waste solidification	Newport News Ind. Corp.; Aerojet Energy Corp.; Atlantic-Richfield Hanford Co.	51
hydrocarbon reforming		52
scouring in sea water desalination	Brookhaven National Lab.	53
shale preheater	Oil Shale Corp. (TOSCO)	54
Si and Zr tetrachlorides	Stauffer Chemical Co.	55
$HSiCl_3$ from Si + HCl	K. A. Andrianov, USSR Academy of Sciences	56
styrenes from aromatic hydrocarbons	Mamedaliev Institute, USSR	57
combustion of S to SO_2	Celleco	58
SO_2 removal from gases	R. J. Best and J. G. Yates, Exxon-Esso, Abingdon, Eng.; Mitsubishi, Japan; Westvaco Corp., New York	59
sulfur and NH_3 recovery	Chemetics International, Canadian Industries, Ltd.	60
terephthalonitrile	Lummus, Bloomfield, N.J.	61
titanium dioxide production	LaPorte Titanium Ltd., London	62
reclamation of scrap metals	USI (Engineering) Ltd.	63
metal carbide and nitrides (see Carbides)	AERE (UK)	64
LPM process for fluidized bed methanation of H_2 + CO	Chem. Systems, Inc.	65
catalytic cracking of methylcyclohexane	S. Tone and co-workers	66

Table 1 (*continued*)

Process	Investigator/user	Ref.
recovery of metals from dilute solutions	Electronics Council Research Center, Capenhurst, UK; Rockwell International, Constructors John Brown	67
condensation of naphthalene	J. Ciborowski	68
neutralization of liquids	H. W. Gehm and L. T. Purcell	69
NiO and nickel chloride	Falconbridge Nickel Mines, Quebec, Can.	70
olefins from crude oil on fluidized coke	BASF, Ludwigshafen, FRG	71
oxidation of aromatic hydrocarbons in fluid beds	H. K. Pargal	72
generation of ozone (qv)	Iowa State University	73
paper mill black liquor recovery	Container Corp. of America; Copeland Process Corp.; Green Bay Packaging Inc.; Dorr-Oliver	74
continuous particle size separation	Baskakov and co-workers; SWECO, Inc.	75
perchlorethylene	Diamond Alkali; Columbia-Southern Chemical Corp. (Pittsburgh, Plate Glass Co.)	76
incineration of spent caustic and other refinery wastes	Amoco- Dorr Oliver; Nichols Engineering	77
Econ-Abator catalytic incineration of organic emissions	Air Resources, Inc., Harshaw Chem. Co., Nipro, Inc., Dutch State Mines	78
indole and benzofuran	SNAM Progetti, Italy	79
ion exchange (qv)	Himsley Engrs Ltd., Toronto, Can., Liquitech, Inc., Houston, Tex., Farbenfabri-ken Bayer (Mobay)	80
chlorination of pyrites cinder to FeO	Montedison S.p.A.–A.G. McKee	81
FeS to iron oxide in 4.3 dia bed at 1038°C	Lurgi, G.m.b.H.; Outokumpu Oy, Finland	82
iron ore reduction by the H-iron process	Hydrocarbon Research, Inc.; Bethlehem Steel Corp.	83
magnetic Fe roasting	French Iron and Steel Research, Institute	84
iron ore reduction (FIOR)	Exxon Research & Engineering Co.; A. G. McKee	85
isophthalonitrile	Badger, Boston, Mass.; Mitsubishi, Japan	86
reforming and desulfurization in a magnetically stabilized bed	Exxon Research & Engineering Co.	87
maleic anhydride	Shakhtakhtinskü and co-workers; Mitsubishi Chem. Industries	88
maleic anhydride from benzene and butylenes	Institute for General Chem., Warsaw, Poland, Petrochem. Processes Inst. USSR Academy of Science	89

melamine from urea	BASF, Ludwigshafen, FRG, U.S. Steel, Pitt. Coke Co., Chemico	90	uranium carbides and nitrides from UO_2	AERE, Harwell, Eng.	64
melamine from urea and ammonia	Osterreichische Stickstoffwerke AG; Power Gas Corp.;Uhde, G.m.b.H.	91	uranium carbides from uranium plus hydrocarbons	Argonne National Laboratories, Illinois Institute of Technology	100
titanium tetrachloride	Fabriques De Produits Chimiques, Belgium	92	vinyl chloride production (see Vinyl polymers)	M. G. Geiger, Jr., Pechiney-Saint-Gobain	101
chlorinating rutile to $TiCl_4$ at 980°C	Chemical and Metallurgical Research, Inc.	93	oxychlorination to vinyl chloride	B. F. Goodrich; Ethyl Corp.	102
tungsten from H_2 reduction of WF_6	Battelle Memorial Institute; Allied Chem. Co.	94	vibratory fluid-bed cooler	Rexnord, Inc., Alfa-Lavalthermal, Inc.	103
H_2 reduction of Nb, Ta and W halides	E. I. du Pont de Nemours & Co., Inc.	95	vulcanization of rubber (qv)	Rubber and Plastics Research Association of Great Britain	104
fluorination of U to UF_6 (see Fluorine compounds, inorganic)	General Electric Co., Union Carbide Corp.	96	hy-flo process for treating industrial and municipal wastewater	Ecolotrol-Dorr Oliver	105
direct conversion of UF_6 to UO_2 and UO_2 to UF_6	Argonne National Laboratories, E. I. du Pont de Nemours & Co., Inc.	97	fluid-bed distillation of wood (qv)	Georgia Institute of Technology	106
fluorination of UF_4 to UF_6 at 1090°C	Union Carbide Nuclear Co., Paducah, Ky.	98	regenerating spent NH_3-base pulping liquor	Copeland Systems, Inc.	107
UO_2 from UF_6	Royal Institute of Technology, Stockholm, Sweden	99	recovery of zinc chloride (see Zinc compounds)	Conoco Coal Development Co.	108

Table 2. Fluidized-Bed Coal Combustion Developments in the United States[a]

Company, location		Developments
Argonne National Laboratories, Argonne, Ill.	PFBC	designing a 11–43 t/h steam boiler as a CTIU to fire 5–6 t of coal/d, expected completion 1981; Stearns-Roger doing the actual design; bed velocity ca 2 m/s, pressure at 0.3–1.2 MPa (3–12 atm), size 0.9 m × 0.9 m to 1.2 m × 1.2 m; operated a 15 cm diam PFBC at up to 1 MPa (10 atm)
American Electric Power Co., New York	PFBC	sponsored feasibility study of combined cycle plant (105 MW steam turbine and 65–70 MW gas turbine) with Stal-Laval and Woodall-Duckham; sponsoring pilot-scale tests at CRE in Leatherhead
Babcock Contractors, subsidiary of Babcock & Wilcox Ltd.	AFBC	installing a 27 t/h boiler in an Ohio State mental hospital (at a cost of $4,300,000)
Babcock & Wilcox, Alliance, Ohio	AFBC	operating a 9 t/h steam 1.8 m × 1.8 m boiler under EPRI contract firing 20 t/d of coal; operates a 1 m × 1 m unit under EPRI contracts to study sorbent utilization firing 0.23 t/h of coal with 2.4 m/s fluidizing velocity; operates a 0.3 m × 0.3 m unit also studying sorbent and fuel characteristics
Babcock & Wilcox, North Canton, Ohio	AFBC	designed a product line of FBC boilers producing steam in 23–136 t/h
Battelle, Columbus, Ohio	AFBC	operated a 15 cm dia MS-FBC bench-scale unit with a combustor superficial velocity of 9.1 m/s; designing (with Foster-Wheeler A.G. and McKee) a 18 t/h steam boiler prototype to serve Battelle's facility wherein the combustor will be 1.83 m ID and the external heat exchange 1.5 m × 3.0 m
Burns & Roe Inc., Oradell, N.J.	AFBC	designing a 200–300 MW utility power plant under a DOE $1,300,000 study contract; designed a 570 MW power plant under a DOE contract
	PFBC	designed a 583 MW FBC combined cycle plant in association with Babcock & Wilcox and Pratt & Whiting div. of United Technologies, Inc.; unit will have 5 beds 22 m × 7 m
Cleaver Brooks Div. of Aqua-Chem. Inc., Milwaukee, Wis.	AFBC	a manufacturer of packaged boilers supplied the Alexandria, Va. test facility of Pope, Evans & Robbins
Combustion Engineering, Windsor Locks, Conn.	AFBC	designing a 23 t/h steam generator at the Great Lakes Naval Training Center scheduled to be in operation before the end of 1980; operating a 0.9 m × 0.9 m test unit delivering 1 t/h steam; also have a Plexiglas cold flow model of this unit; made a design study for a retrofit FBC unit for Con. Edison's 500 MW Arthur Kill #3 unit on Staten Island under EPRI and NYSERDA sponsorship
Combustion Power Corp., Menlo Park, Calif.	PFBC	operated a 3.7 m^2 bed (ca 2.1 m ID) at 0.4 MPa (4 atm), burning Illinois No. 6 coal (4 wt % sulfur)
	AFBC	operated a 0.7 m × 0.7 m combustor principally for corrosion testing
Curtiss-Wright Corp. Woodridge, N.J.; Stone & Webster, Boston, Mass.; Dorr-Oliver, Stamford, Conn.	PFBC	constructing a combustor capable of producing 59 t/h of steam from combustion of 99 t/d of coal; operating a bench-scale unit one fourth the diameter of the above 13 MW pilot plant to provide design data
The Ducon Co., Mineola, N.Y.	AFBC or PFBC	patented a modular combustor incorporating conceptual improvements in mechanical design, control, and operation

Table 2 (*continued*)

Company, location		Developments
Energy Products of Idaho, a subsidiary of Energy, Inc., Idaho Falls, Idaho	AFBC	sold a dozen FB incinerators (trade-named Fluid Flame) in sizes producing 4.5–13.6 t/h steam burning wood wastes, corn cobs, fruit pits, etc
Energy Resources Co., Cambridge, Mass.	AFBC	operated a 0.6 m × 0.6 m FBC unit at 1.2–4.9 m/s testing 16 different coals; is building a 1.8 m × 1.8 m unit for further test work and to establish credibility; offering to design and build industrial size AFBC boilers producing up to 227 t/h of steam
Exxon Research & Engineering Co., Florham Park, N.J.	AFBC	operates a Plexiglas cold flow model 30 cm × 2.3 m in cross section and 3.7 m high at superficial air velocities up to 4.6 m/s using horizontal simulated tubes 5.1 cm × 15.2 cm in dia at various C-C spacings
Exxon Research & Engineering Co., Linden, N.J.	PFBC	operates a 30.5 cm ID miniplant burning 172 kg of coal/h with a max operating pressure of 1 MPa (10 atm) at superficial velocities up to 3 m/s; operates a 11.4 cm ID bench-scale unit burning 13.6 kg coal/h
Fluidyne Engineering Corp., Minneapolis, Minn.	AFBC	constructing two 1.5 m × 3.4 m beds (capable of producing up to 29.5 GJ/h or 28 × 10^6 Btu/h) at the Owatonna Tool Co. in Minnesota to supply high temperature air for paint-curing ovens, plating lines, and space heat; are offering 15.8 GJ/h (15 × 10^6 Btu/h) modules on the open market; have 3 test units in operation for demonstration and customer's systems studies
Fluidyne Engineering Co., The City of Wilkes-Barre, and the Shamokin Area Industrial Corp.	AFBC	submitted proposals to DOE for construction of units to burn 832 mm of culm or anthracite wastes (equivalent to 2.3 × 10^8 t of coal) scattered in northeastern parts of Pennsylvania
Foster Wheeler Corp., Livingston, N.J.	AFBC	constructing a 45.4 t steam/h unit at Georgetown University; this is a 2 bed unit with over bed feed of coal and limestone to operate at 2.4 m/s fluidizing velocity burning 113 t/d of coal; operated a 1.8 m × 15.2 cm cold model to study scale-up correlations regarding tube spacing, solids distribution, etc; operates two test facilities in Livingston, one is 0.5 m × 0.5 m and the other a 10.9 m² bed; offering commercial warrantees on FBC boilers up to 272 t/h of steam; operated a 45.7 cm dia unit burning 45 kg/h of coal; this unit now at METC; designed a 32 t/h steam AFBC boiler for Ford Motor Co.
The General Electric Co., Schenectady, N.Y.	AFBC and PFBC	operates a 0.3 × 0.3 m and a 0.6 m × 0.6 m test bed to explore effects on in-bed exchanger tubes
General Electric and TVA	AFBC and PFBC	completed conceptual design and cost comparison study of a 750–925 MW FBC utility power plant
Grand Forks Energy Research Center, Grand Forks, North Dakota	AFBC	this DOE Laboratory has been testing various coals and lignites for bed sulfur retention in a 15 cm dia combustor
International Boiler Works, subsidiary of Comb. Equip. Assoc., Inc., East Stroudsburg, Pa.	AFBC	manufacturing an FBC from an Energy Resources Co. design

Table 2 (*continued*)

Company, location		Developments
Johnston Boiler Co., Ferrysburg, Mich.	AFBC	operates a 4.5 t/h FBC steam generator at its Michigan offices; offering 1.1–22.7 t steam/h units (under license from Britains Fluidfire Ltd. through CSL); constructing a 24.8 MJ/h (23,500 Btu/h) boiler for the Ohio Center Convention Complex in Columbus, based on a design developed by Britain's CSL
John Zink Co., Tulsa, Ohio	AFBC	is reported to be fabricating an in-house FBC
Morgantown Energy Technology Center, Morgantown, W. Va.	AFBC	designing a 27 t/h steam capacity unit to burn 9–82 t coal/d as a Component Test and Integration Unit with two beds each 1.8 m × 1.8 m stacked vertically surmounted by a 0.9 m × 1.8 m carbon burnup cell; the beds will operate at superficial velocities of 2.1–4.6 m/s through beds 0.6–2.4 m deep; this CTIU will be sited on the Medical Center Campus of West Virginia University and is scheduled for operation end of 1979; have an 45.7 cm dia unit for test purposes
WASA-Lewis Res. Center, Cleveland, Ohio	PFBC	operates a bench-scale PFBC for technical support studies; burns 4.5–36.2 kg/h of coal at bed pressure of 172–1379 kPa (25–200 psia) in a 22.6–50.8 cm conical bed
Oak Ridge National Laboratory, Oak Ridge, Tenn.	AFBC	carried out conceptual design of a unit geared to large apartment complexes; operated a 1.2 m × 1.2 m cold flow model to observe bubble flow past tube arrays; developed a concept for an atmospheric fluidized-bed coal combustor for cogeneration (ccc) to produce 300–500 kW of electricity plus some 2.6 GJ/h (2.47 × 10^6 Btu/h) of useful heat
Oregon State University, Corvalis, Oregon	AFBC	operates a 0.9 m × 0.9 m × 10.7 m high cold flow model to study tube spacing effects, mixing, etc
Pope, Evans and Robbins, New York	AFBC	operating a 0.5 MW (2.3 t/steam h) pilot plant at Alexandria, Va., 0.5 m × 1.8 m in cross section; operating a 30 MW plant at Rivesville comprising 3 modules each 3.1 m × 3.7 m rated at 30 MW plus a smaller carbon burnup cell; this unit produces 136 t steam/hr burning 299 t of coal/d; licensed Mitsubishi and IHI of Japan to build 136 t/h AFBC boiler to burn coke breeze waste
Process Equipment Modelling and Mfg. Co., Nelsonville, N.Y.	AFBC	designed and operated a Plexiglas cold model of a 15.2 cm combustor and 7.6 cm × 35.6 cm external heat exchanger for Battelle's MS-FBC development
Stone & Webster Engr'g. Co., Boston, Mass.	AFBC	designing a 500–600 MW utility power plant under a DOE study contract
The Trane Co., La Crosse, Wisc.	AFBC	has under development a proprietary design of FBC packaged boilers
University of West Virginia, Morgantown, W. Va.	AFBC	operates a 0.6 m × 0.6 m cold model for mixing studies and has under construction a 0.6 m × 0.6 m FBC boiler
Wormser Engr'g. Inc., Lynn, Mass.	AFBC	operating a demonstration unit in a Lynn, Mass. factory where it is used for space heating; expects to offer a line of FBC boilers in the 3–10 MW range

[a] PFBC = pressurized, fluidized-bed combustion; AFBC = atmospheric, fluidized-bed combustion; CTIU = Component Test and Integration Unit; CRE = Commonwealth Research Establishmen; EPRI = Electric Power Research Institute; MS = multisolid; NYSERDA = New York State Energy Research and Development Authority.

ADVANTAGES AND DISADVANTAGES OF THE FLUIDIZATION TECHNIQUE

Advantages

Temperature Control. The ability of the fluidized-solid bed to approach isothermal conditions is the outstanding advantage of this method over other methods of carrying out reactions. Of the several reaction variables, temperature is one of the most important, for reaction rates change exponentially with temperature (often doubling for a 10°C change). Commonly there are several competing reactions in which a temperature change of a few degrees may shift the balance of the several rates.

Three factors, in order of their importance, of the relatively close control of temperature in a fluidized-solid bed are:

(*1*) Turbulent agitation within the fluidized mass, which breaks and disperses any hot or cold spots throughout the bed before they grow to significant size. The catalytic activity differs somewhat for each particle, and those with greater activity accelerate the reaction in their neighborhood to a greater extent; consequently, their temperature is different than that of the surrounding particles of lower activity.

(*2*) High heat capacity of the bed is relative to the gas within it. This factor stabilizes the temperature of the bed, permitting it to absorb relatively large heat surges with only small temperature changes. For example, a bed of ordinary sand, fluidized with air at a solids concentration of ca 1120 kg/m^3 (70 lb/ft^3) would contain only about 0.8 kg air/m^3 (0.05 lb air/ft^3), corresponding to a mass ratio of ca 1400:1.

(*3*) A high-heat-transfer rate, which is possible because of the large amount of transfer surface per unit volume of the fluidized bed. This permits rapid leveling of any temperature surges either from the incoming gas or from reactions within the bed. Although the heat transfer coefficients are not unusually high, the amount of surface per unit volume is very large; eg, the surface area of a bed of ordinary sand would be in the range of 3280–16,400 m^2/m^3 (1000–5000 ft^2/ft^3) of bed.

The point-to-point variation of temperature in large fluidized-catalyst beds is less than 10°C when the feed–gas temperature is not greatly different from that of the bed, and particularly if the inlet gas is carefully distributed.

Continuity of Operation. The ability to handle the fluidized solid like a liquid permits the technique to be easily adapted to many continuous operations, thereby obtaining the advantages of lower labor requirements, precise and automatic control of process variables, and uniformity of product quality.

Heat Transfer. The fluidized-solid technique is a convenient method for transferring heat, either alone or in conjunction with other operations, such as catalysis, gas–solid reactions, and transport of solids and fluids incidental to these operations (see Heat exchange technology).

Catalysis. The fluidized-solid technique can be used for the contact of free flowing, nonsticky, granular solids with gases. It may be applied in catalytic gas reactions in which solid catalysts are used. The technique has been used most widely for the catalytic cracking of petroleum because of: (*1*) control of reaction temperature; (*2*) maintenance of uniform catalyst activity (as well as continuous catalyst regeneration); (*3*) continuous removal of solid by-products; (*4*) supply of heat to endothermic reactions; (*5*) simple equipment with few moving parts; and (*6*) continuous operation with automatic control.

Most catalysts gradually lose their activity during use because of poisoning or coating of the active surface with by-products. As the activity decreases, operating conditions in conventional catalyst contactors must be altered in order to maintain operating rate. Use of higher temperature is one expedient but it may increase the cost of the product. Lowering the operating rate is another expedient, but this also results in increased costs through higher investment because the plant must be large enough to manufacture at an average rate sufficient to compensate for the low-rate period. In contrast, the fluidized-solid technique makes possible the maintenance of a definite level of catalyst activity because partially spent catalyst can be continuously withdrawn and fresh catalyst added. A stable catalyst requires a lower catalyst withdrawal rate than one which loses activity rapidly (see Catalysis).

Gas–Solid Reactions. Advantages of the fluidized-solid technique for carrying out gas–solid reactions are the isothermal reaction bed, the easily varied time of contact, the effective contact (as compared with rotary kilns or tray-type reactors), the simple methods of handling solids (no moving parts) and transferring heat, and the ease of continuous, automatic operation.

Disadvantages

Solids that do not flow freely or tend to agglomerate cannot be processed in a fluidized-solid reactor; instead, rotary kilns and tray-type reactors are used. As a reaction proceeds, fine solid particles may be formed that become entrained in the gas leaving the fluidized bed, and recovery methods must usually be included in the design. The pressure drop in the gas system of a fluidized-solid boiling-bed reactor is larger than in kilns or tray reactors because the gas supports and fluidizes the solid. This pressure drop may sometimes be a serious objection to the fluidized-solid reactor because of the larger compressors required.

Lack of Temperature Gradient. Occasionally, it is desirable to obtain a temperature gradient in a catalytic process; for example, a higher temperature may be desired at the upper part of a reactor in order to effect a cleanup of residual reactant. In the simple boiling-bed fluidized-solid reactor this is not possible. However, temperature gradients may be secured with baffles or other zoning means plus internal heat exchangers, or by the use of series reactors.

Pressure Drop. The pressure drop through a fluidized bed may be larger than that through an ordinary heat exchanger. If the fluidized bed is used solely for heat transfer, the high pressure drop may be detrimental. The pressure drops must be balanced throughout the system, which may necessitate providing a gas purge at several points.

Gas–Solid Reactions. Gas bubbles rising through the fluidized solids contain little of the solids because the efficiency of contact in a boiling bed may be much less (in terms of the availability of the active surface of the solids) than in a fixed-bed reactor. This disadvantage is at least partially overcome by the fluidized catalyst of much smaller particle size (and, therefore, much greater surface area) than that used in fixed-bed reactors.

Without experimental reaction rates and conversions from an operating fluidized-bed pilot plant, predictions of what might occur in a commercial scale plant are doubtful; fixed-bed reaction kinetics are as yet not assuredly transferable to a bubbling bed.

OPERATING CHARACTERISTICS AND DESIGN PROCEDURES

Choice of Operating Gas Velocity

Fluidization occurs in a bed of particulates when an upward flow of fluid through the interstices of the bed attains a frictional resistance equal to the weight of the bed. At this point, an infinitesimal increase in the fluid rate lifts or supports the particles. Hence, the particles are envisioned as barely touching, or as floating on a film of fluid.

In order to avoid ambiguities, this condition can be considered as incipient buoyancy, since at such a fluid rate, the particles are still so close as to have essentially no mobility, whereas the usual desire in fluidization is to create bed homogeneity. Such homogeneity can only be achieved by violent mixing. This is brought about by increasing the fluid velocity to the point of blowing bubbles or voids into the bed which is mixed as they rise. The increased fluid velocity at which bubbles first form is called the incipient-bubbling velocity.

Fine powders exist over a wide range of bulk densities and therefore exhibit substantial differences between incipient buoyancy and incipient bubbling, as illustrated qualitatively in Figure 2.

There is no way to predict precisely a powder's range of bulk densities. However, this is a relatively simple physical measurement. For example, FCC (fluidized catalytic cracking) catalysts with particle-size distributions of ca 10–200 μm dia exhibit incipient-buoyancy velocities in ambient air on the order of 0.3–0.9 cm/s, and incipient-bubbling velocities of ca 3–9 cm/s. Many catalysts used in other processes consist of silica–alumina carriers (such as are used in FCC catalysts) impregnated with desired catalytic agents. Hence, these velocities are broadly representative. The incipient-bubbling velocity is more significant in reactor design.

The terms particulate and aggregative were coined (112) in the 1940s to differentiate bubbling beds (aggregative) and nonbubbling beds (particulate). In general, liquid-fluidized beds were nonbubbling, whereas gas-fluidized beds bubbled. Bubbling is related to fluid and particle properties in a manner permitting the prediction of a

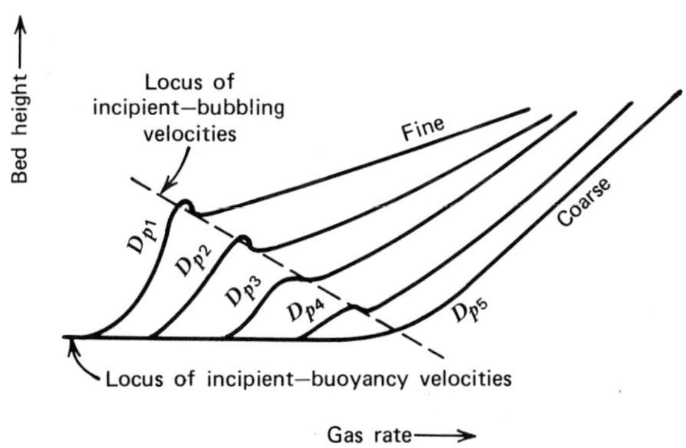

Figure 2. Particle size affects incipient-buoyancy and incipient-bubbling velocities. ρ_B = bed bulk density; $D_{P1} \ldots D_{P5}$ = increasing particle size.

system's maximum attainable (113) bubble size, which if negligible leads to the observation of so-called particulate fluidization. The maximum stable-bubble size for a particular system is described rather than the terms aggregative and particulate.

The optimum operating fluid velocity is not likely to have narrow limits. It generally represents compromises with entrainment, attrition, pressure drop, and costs. The lower limit is the bed's incipient bubbling velocity and the upper limit usually approaches the terminal or free fall velocity of the largest particle in the bed. A nearly identical analogy exists for the optimum physical dimensions of a distillation or absorption tower (see Distillation; Absorption).

The composition of the fluidizing gas are fixed by the process conditions. If the density and size distribution of the particles to be fluidized are not predetermined, they must be obtained from an incorporation of the cost of various degrees of grinding or, in the case of a high-conversion gas-catalytic reaction, be taken as the size which optimally leads to an incipient bubbling velocity of 3–9 cm/s under reaction conditions (111). Most frequently the optimum superficial fluidizing velocity is based on pilot-plant experience.

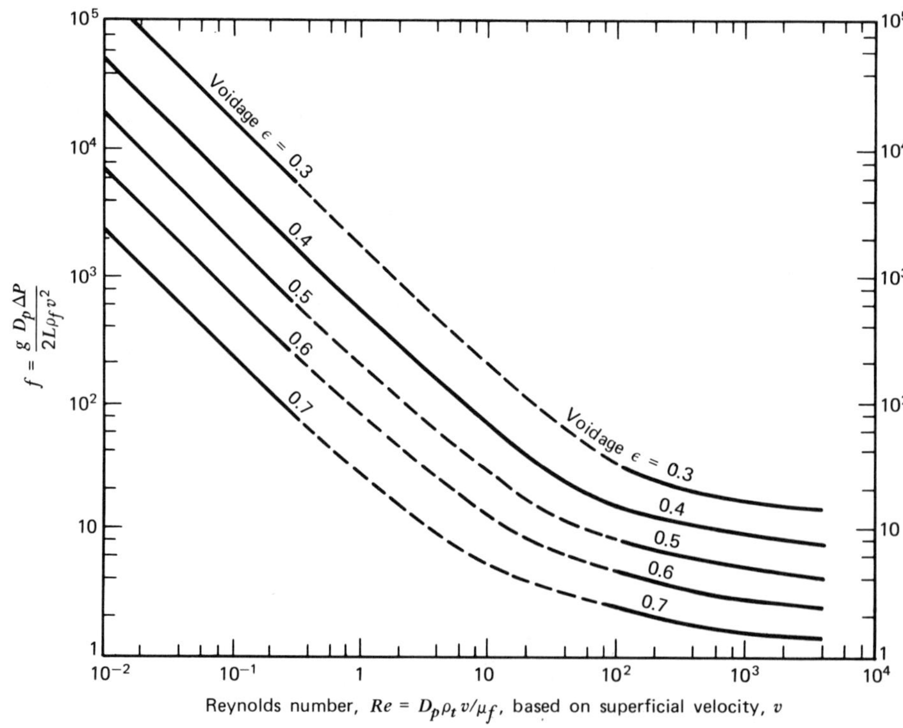

Figure 3. Friction factor for pressure drop through packed beds of uniform particles. Curves for $Re > 100$ based on ref. 115; curves for $Re < 0.1$ based on ref. 116 correlation; curve for $\epsilon - 0.4$ between $1 < Re < 100$ based on ref. 117. For spheres, D_p = diameter; for particles of other shape, $D_p = 6\ V_p/S_p = {}_jD_p$, where V_p = particle volume and S_p = particle surface; $j = V_p$/volume of a sphere having the same surface area as the actual particle, D_p is the diameter of a sphere having the same surface as the actual particle.

Calculation of Incipient Bubbling and Terminal Velocities

The superficial upward fluid velocity through a bed of particles that initiates incipient bubbling is definable as the minimum flow which in passing through the interstices of the bed at its loosest bulk density suffers a particle-to-fluid frictional resistance equal to the weight of the bed. Any additional fluid, therefore, lifts the bed (which must be unrestrained on its surface) and passes upwards in the form of a so-called bubble. Mechanistically, it pushes the bed upward at its point of entry through the supporting grid port and thereby creates a void into which the bed solids can slide and thereby displace the bubble-form void in the upward direction (114). The calculation of the incipient bubbling velocity makes use of a substitution of the bulk density for the pressure drop in the conventional correlations for flow through packed beds and the determination of the attendant velocity. Correlations suggested in Figure 3 (115–118) are far more reliable than many published equations (119–120) for direct calculation of incipient fluidization velocities since they empirically account for the voidage or loosest bulk density, which can have a very significant effect as shown in the curves.

The superficial velocity at which the largest particle in the bed is able to be blown out of the reactor is frequently the practical or near-optimum operating velocity. This is calculable from the conventional drag coefficient vs Reynold's number curves (121) for the free fall, or terminal, velocity of equivalent single spheres as represented in Figure 4.

It is obvious from Figures 3 and 4 that drag coefficient and friction factor differ solely in the constants 4/3 and 1/2 and are, therefore, interchangeable. Since both velocity and particle diameter appear in the abscissas as well as the ordinates, the prediction of incipient bubbling velocity or of terminal velocity involves a trial and

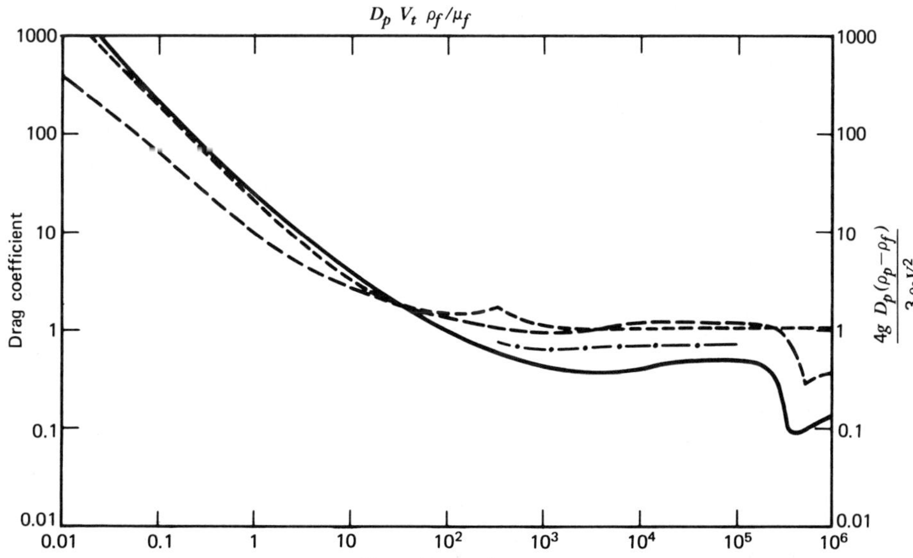

Figure 4. Drag coefficient for spheres, —; horizontal disks, - - -; infinite cylinders, – –; and cylinders of length = 5 × diam, -·-.

error procedure. This can be circumvented (122) by converting the friction factor to the drag coefficient, $C_D = (8/3)f$, and plotting $(C_D Re^2)^{1/3}$ vs $(Re/C_D)^{1/3}$, as shown in Figure 5. The denominators of the abscissa and the ordinate represent composites of the physicochemical properties of the fluidizing medium and the solids in which composites are constants for any given system. Figure 5 is a generalized correlation of velocity versus particle diameter at any known bed density or void fraction. The incipient bubbling velocity is found from the value of the ordinate corresponding to

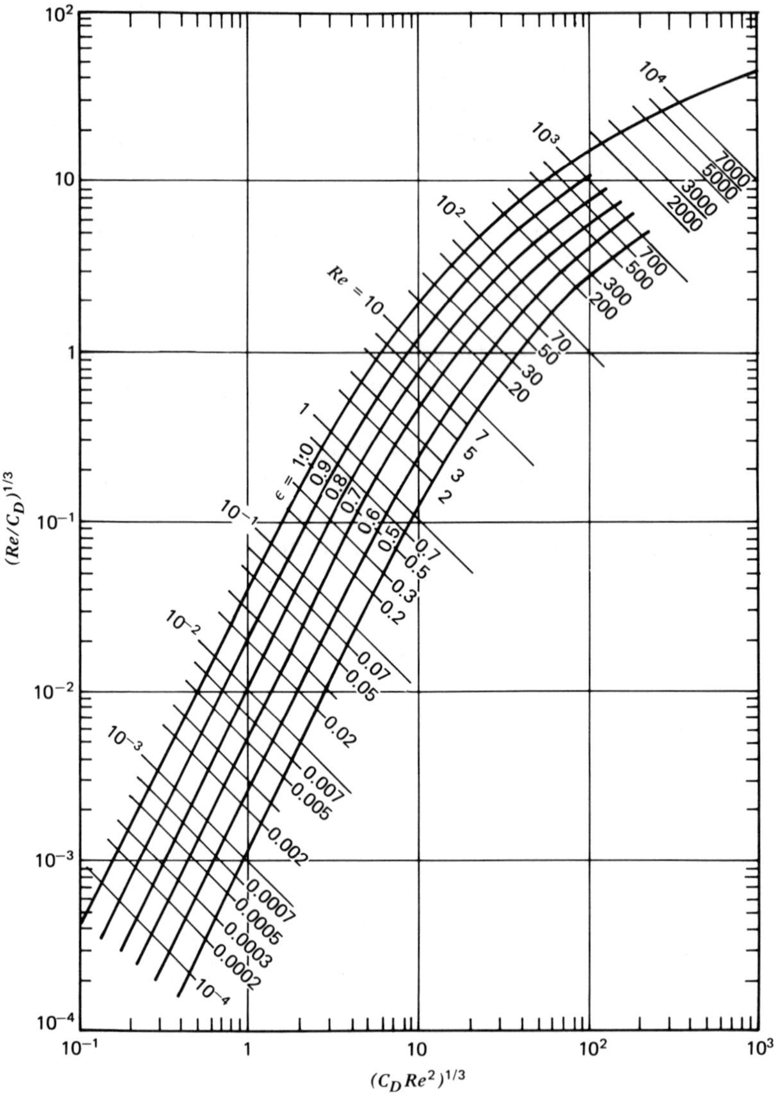

Figure 5. Smoothed correlation of particulate fluidization.

$$(Re/C_D)^{1/3} = \left(\frac{D_P v_\epsilon \rho_f}{\mu_f} \Big/ \frac{4gD_P(\rho_P - \rho_f)}{3\rho_f v_\epsilon^2}\right)^{1/3} = v_\epsilon \Big/ \left(\frac{4g\mu_f(\rho_P - \rho_f)}{3\rho_f^2}\right)^{1/3}$$

$$(C_D Re^2)^{1/3} = \left[\frac{4gD_P(\rho_P - \rho_f)}{3\rho_f v_\epsilon^2}\left(\frac{D_P v_\epsilon \rho_f}{\mu_f}\right)^2\right]^{1/3} = D_P \Big/ \left(\frac{3\mu_f^2}{4g\rho_f(\rho_P - \rho_f)}\right)^{1/3}$$

an abscissa determined by the geometric weight mean particle diameter of the bed and the bed's void fraction at its loosest bulk density. The value of the abscissa representing the largest particle in the bed, the ordinate determined from the $\epsilon = 1.0$ curve, represents the velocity which would blow the largest particle out of the bed. This is not an upper limit to bed operation because the particles entrained at any operating velocity must be separated from the exit gases and returned to the bed in order to maintain the bed. This particle recovery system usually consists of one or more cyclones in series whose dip legs are submerged in the bed to discharge as near the grid level as feasible (123). The greater the superficial operating velocity, the greater the entrainment rate and the greater the need to design efficient cyclone recovery systems (see Air pollution control methods).

Prediction of the Particle Entrainment Rate

The entrainment of particles from the surface of a bubbling bed is directly analogous to the entrainment of liquid droplets from a boiling or bubbling pool of liquid on a distillation tray. Since solid particles cannot coalesce in the manner of liquid droplets, the entrainment above a fluidized bed of solids, although declining with increasing height, inevitably reaches a constant rate representing all the bed particles with a free fall velocity less than the operating superficial velocity. The force of ejection is a function of the size and frequency of the bubbles erupting on the bed surface. The bursting bubbles represent intermittent higher-than-superficial velocity profiles, which dissipate with distance or so-called particle disengaging height above the bed. The so-called Transport Disengaging Height (TDH) is correlated empirically as shown in Figure 6 (124). A reactor height greater than TDH serves no purpose in decreasing the entrainment rate. The abscissa of Figure 5 is related to the frequency of surface bubble eruptions and the parameter to the physical size of the surface bubbles. The size of the surface bubbles is calculable from the grid characteristics, the bubble merger rate, and the maximum stable bubble size for the particle–fluid system under consideration.

The particle entrainment rate, at and above TDH, can be determined from the empirical correlation given in Figure 7 (124). The detailed design of the cyclone recovery system can best be found in ref. 123.

Grid Design and Initial Bubble Size

Bubbles form at the grid ports where fluidizing gas enters the bed because the velocity at the interface of the bed just above the hole represents a gas input rate in excess of what can pass through the interstices. The gas input has a frictional resistance less than the bed weight, hence the layers of solids above the holes are pushed aside until they represent a void through whose porous surface the gas can enter at the incipient fluidization velocity (125). As the void grows the interface velocity becomes insufficient to hold back the walls of the void and hence the bubbles cave in from the sides cutting off the void and presenting a new interface to the incoming gas (114). This sequence is illustrated in Figure 8. The depth of penetration of the grid gas jets is correlated empirically (126–128) in Figure 9. The diameter of the initial bubble re-

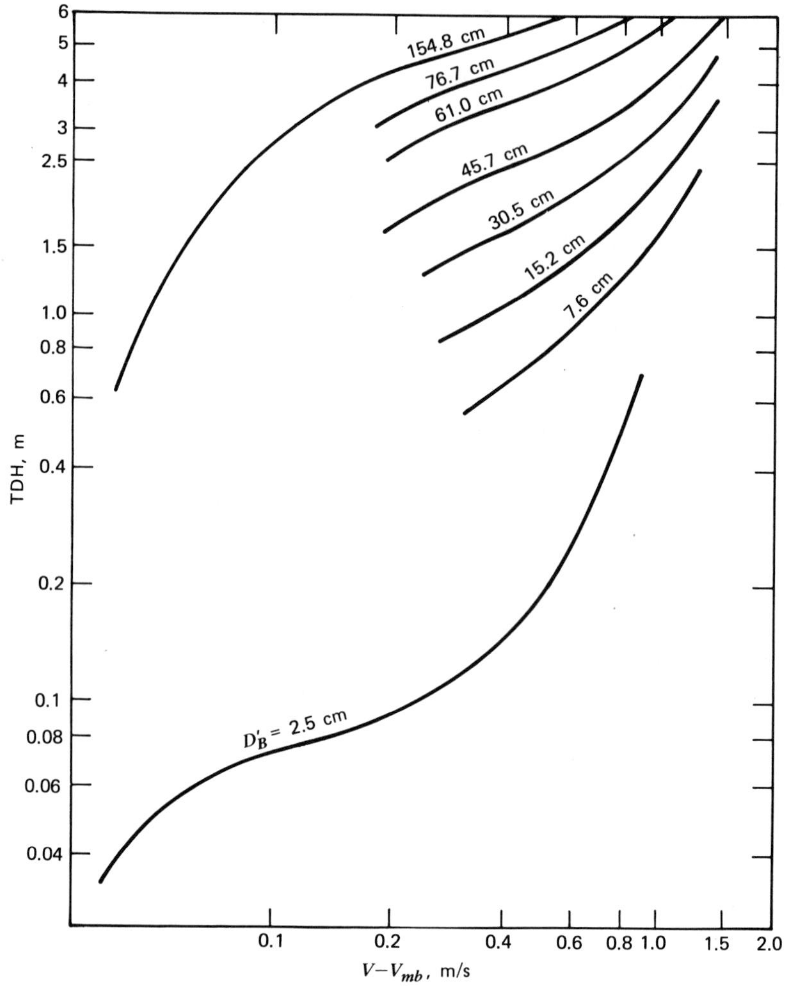

Figure 6. Transport disengaging height above bubbling fluidized beds.

sulting from a detached void is observed, within experimental error, to be about half the penetration depth. Since the grid is the source of the bubbles, its holes should be reasonably small (considering cost, plate strength, and possible pluggage) and of such total area that the pressure drop of the fluidizing gas passing through the holes is sufficient to assure gas distribution to all the holes (130–133). As in the case of a perforated plate distillation tray, bed weepage flow down through the holes is to be avoided. The pressure drop through a flat plate grid necessary to assure that all the holes are bubbling must be at least 30% of the pressure drop through the bed atop the grid. This criterion establishes the grid hole velocity and, in conjunction with the lowest anticipated fluidizing gas volume, determines the total hole area. In passing up through the bed the bubbles inevitably merge. Their fluid mechanics must be understood in order to relate to the gas–solids contact occurring within the bed, and to the size of the bubbles bursting on the bed surface.

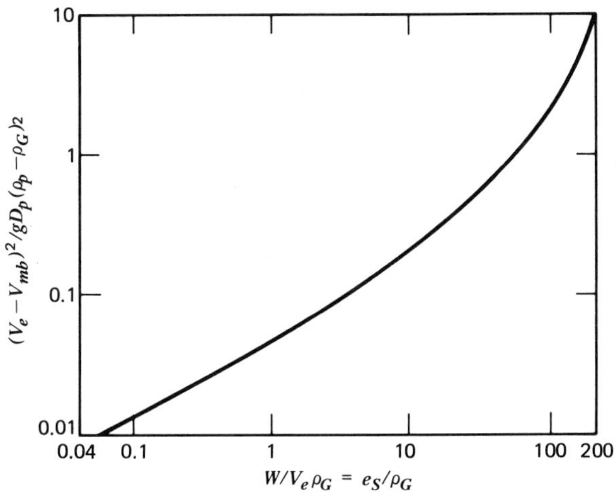

Figure 7. Maximum dilute phase entrainment in vertical gas–solids upflow. At values of the ordinate below 0.04: $(\text{ordinate}/0.0448)^{1.875}$ = abscissa.

Fluid Mechanics of Bubble Flow

Bubbles or gas voids rise in a fluidized bed as they are displaced with an inflow of solids from their perimeter (114,134). Since free flowing and/or incipiently fluidized bulk solids have shallow angles of repose, their walls cannot stand at 90°; the solids slide down the bubble's walls into its bottom where all the peripheral streams collide to form a wake as illustrated in Figure 1. Observations of this downflow of solids in a shell around the bubble have shown it to occupy an annular thickness of one fourth of the bubble diameter so that the overall diameter within which a bubble can rise freely as it would in a bed of infinite diameter can be defined as 1.5 D_B.

Since the peripheral surface of the bubble is simply a layer of particles, it is at first difficult to understand why the particles do not fall from its roof and annihilate the bubble. Simple bed support experiments are shown in Figure 10 (135), and experiments using the classical approach are described in ref. 136.

In Figure 10A the air rate is raised to the point of incipient fluidization and in Figure 10B–F this same gas rate is passed through the bed in the opposite direction. Note that in position D the solids do not slide to their angle of repose but instead are held at 90° and on reaching F the bed is held up without solids falling from what is now its lower side or conversely the upper surface of a bubble in a fluidized bed. When the surface of a bed is traversed by an incipiently fluidizing flow the particles cannot separate from each other. This not only explains the bubble's surface stability but also the integrity of the walls of a bed-penetrating jet as in Figure 8.

Classical drag measurements revealed drag on a downstream particle is reduced owing to the presence of an adjacent upstream particle (136). This means that a particle cannot fall from the roof of a bubble; if it fell, it would immediately be followed by the particle above it, and that by the particle still further above, ie, the entire mass or bed above the bubble would have to collapse as a unit. For this to occur, the excess gas could not be passed through the bed unless the bed were physically held down or restrained at its upper surface.

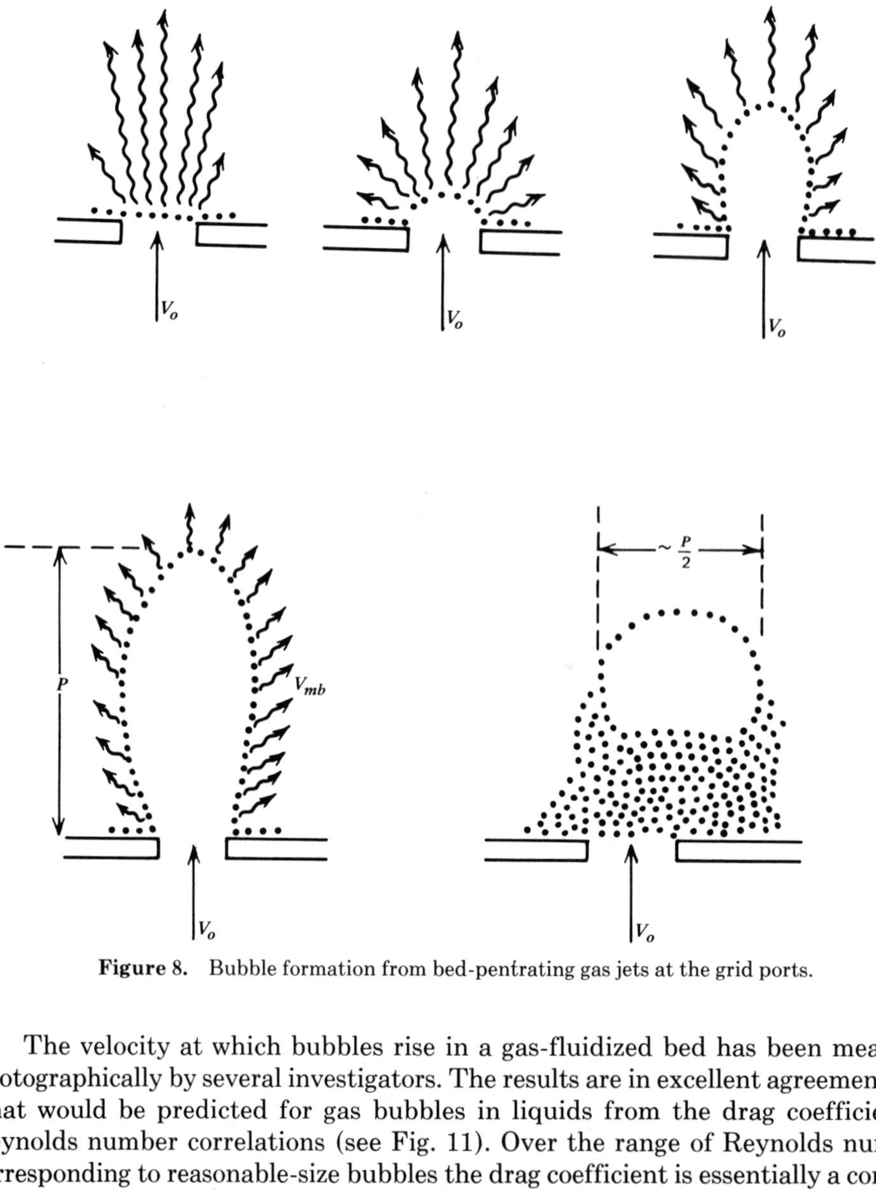

Figure 8. Bubble formation from bed-pentrating gas jets at the grid ports.

The velocity at which bubbles rise in a gas-fluidized bed has been measured photographically by several investigators. The results are in excellent agreement with what would be predicted for gas bubbles in liquids from the drag coefficient vs Reynolds number correlations (see Fig. 11). Over the range of Reynolds numbers corresponding to reasonable-size bubbles the drag coefficient is essentially a constant so that simple substitution shows, if gas density is small relative to the bed density:

$$C_D = \frac{4gD_B(\rho_B - \rho_G)}{3\rho_B V_B^2}$$

$$\therefore V_B = \sqrt{\frac{4gD_B(\rho_B - \rho_G)}{3C_D\rho_B}}$$

or

$$V_B = 4.01\sqrt{D_B}$$

This has been corroborated in experiments with freely bubbling beds.

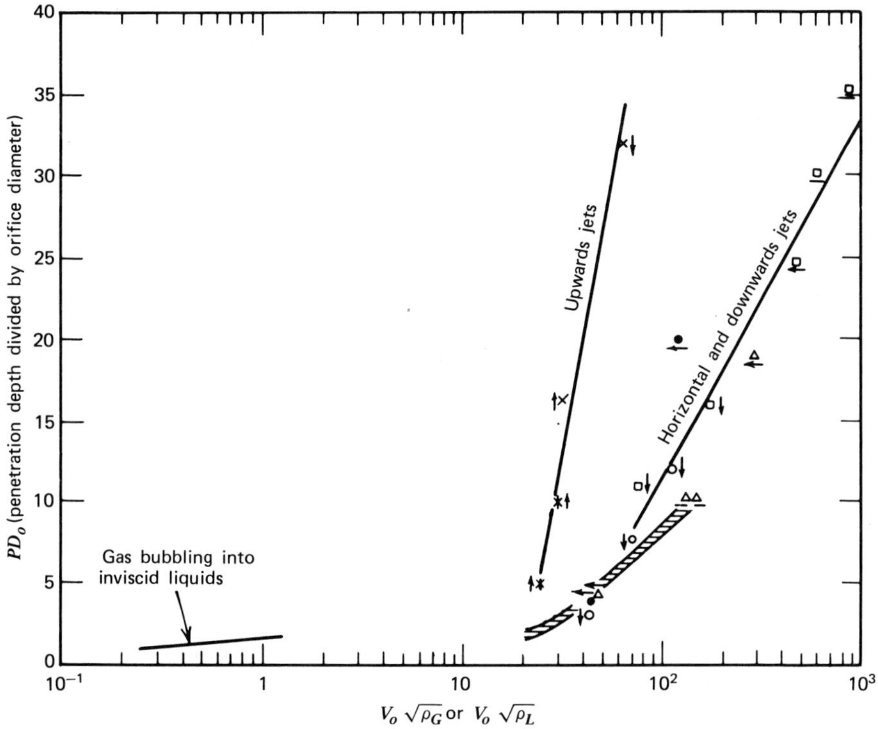

Figure 9. Jet penetration into fluid–particle media (129).

In slugging beds the full width of the downflowing solids shell (Fig. 1) is restricted and the velocity of bubble rise then approximately one half that in a freely bubbling bed (138). In most instances this becomes a matter for consideration only in the design or scale-up of small pilot- or bench-scale, fluidized-bed processes.

Rate of Bubble Growth by Merger

That bubbles must grow by merger as they rise through the bed is obvious from the large and less frequent surface eruptions relative to a much higher frequency of small voids initiated from an usual multitude of grid ports. Growth by simple gas expansion resulting from the pressure reduction between bottom and top of a fluidized bed is generally insignificant.

From the solids inflow model of Figure 1 it is obvious that a bed must be exceptionally homogeneous to expect the shell of downflowing solids around a bubble to be flowing at equal rate in every plane. Any bed nonuniformity can cause a shift in the bubble shape or position. Only the prior passage of another bubble can alter local densities or distributions so as to make bed solids in one area more readily flowable in a given direction than the bed solids in an adjacent area. The solids inflow model, therefore, obviates a simple mechanism of bubble merger. If two bubbles get close enough so their shells of downflowing solids begin to interact, the touching shells represent a local downflowing stream of solids faced with more than one path to the nearest void. The stream could be squeezed to the point of being insufficient to satisfy

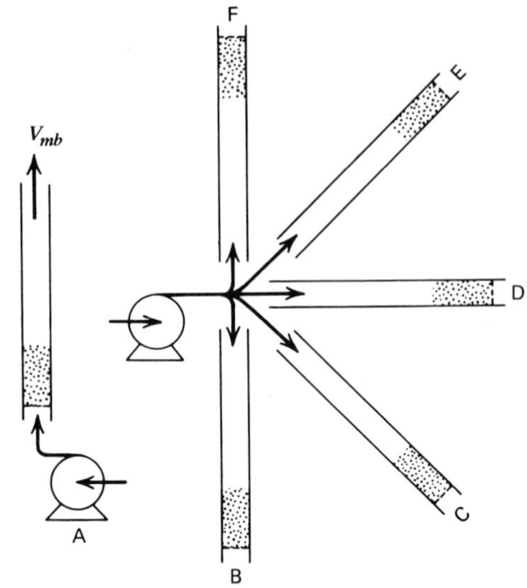

Figure 10. Bed support experiments (135). See text.

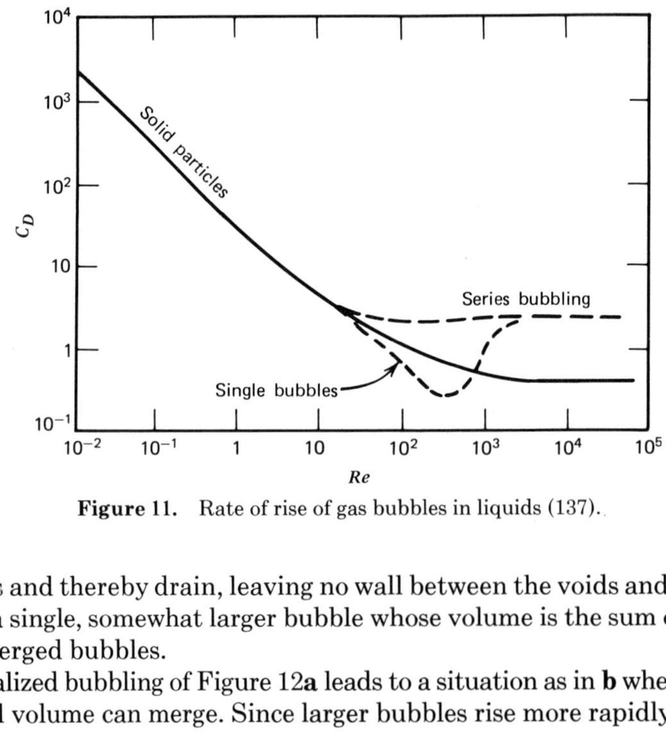

Figure 11. Rate of rise of gas bubbles in liquids (137).

both bubbles and thereby drain, leaving no wall between the voids and hence the appearance of a single, somewhat larger bubble whose volume is the sum of the volumes of the two merged bubbles.

The idealized bubbling of Figure 12**a** leads to a situation as in **b** where two bubbles of unit initial volume can merge. Since larger bubbles rise more rapidly, these double volume bubbles catch up and merge with other unit volume bubbles to yield bubbles of thrice the initial bubble volume. These newer bubbles will rise even more rapidly and catch up with bubbles of one or two times the volume of the initial bubble resulting in bubbles of at most five times the volume of the initial bubble. The bubble of fivefold volume can now catch up with bubbles of one, two, or three times the volume of the

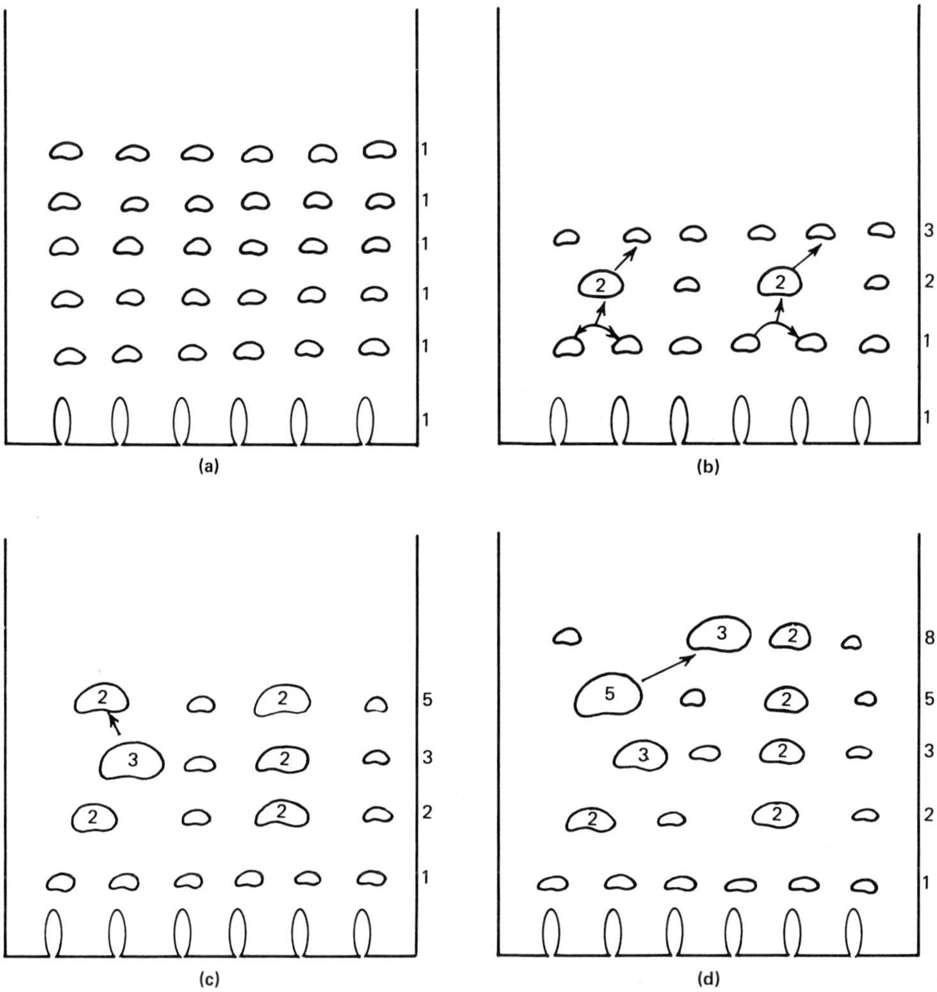

Figure 12. The catch-up mechanism of bubble growth (numbers represent relative sizes of bubbles). (**a**) Idealized bubbling; (**b**) bubbles merging; (**c**) threefold volume of the initial bubble; (**d**) fivefold volume of the initial bubble.

initial bubble resulting in bubbles of at most eight times the volume of the initial bubble as illustrated in Figures 12**c**–**d**. Carrying on this process of overtaking bubbles results in a sequence of maximum multiples of the initial bubble volume in which each multiple is the sum of the two previous bubbles. This sequence, illustrated in Figure 13, is the well known Fibonacci series (139).

The levels at which the maxima exist represent the summation of the diameters of their forebearers and their diameters are proportional to the cube root of their volumes, therefore, the ratio of merged bubble diameter to initial bubble diameter is equal to the cube root of the number of initial bubbles consumed in the merger, and also that the level at which the merged bubbles exists relative to the height (or diameter) of the initial bubble is equal to the summation of the cube root of the number of initial bubbles consumed in the merged bubble (125). For the case of the maximum

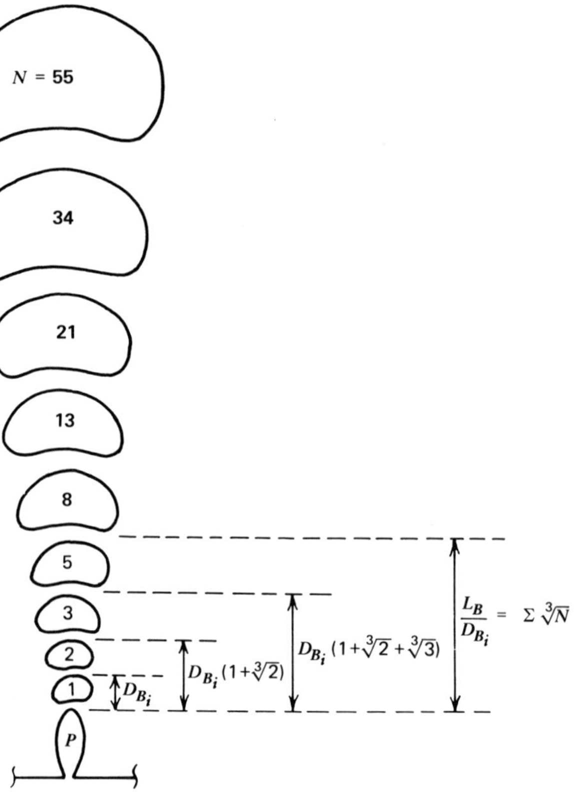

Figure 13. Maximum bubble growth by the catch-up mechanism resulting in a Fibonacci sequence (numbers represent relative sizes of bubbles). N = no. of bubbles of dia D_{B_i} merged. $D_B/D_{B_i} = \sqrt[3]{N}$.

size of merged bubble this is illustrated analytically in Figure 13 and shown graphically in Figure 14.

The mechanism of Figures 13 and 14 appears in good agreement with experimental observations as illustrated in Figure 15 where the empirical bubble growth

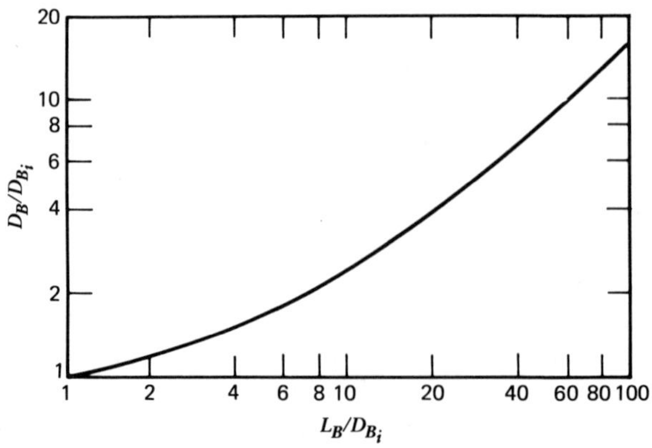

Figure 14. Bubble growth by merger represented by the Fibonacci sequence.

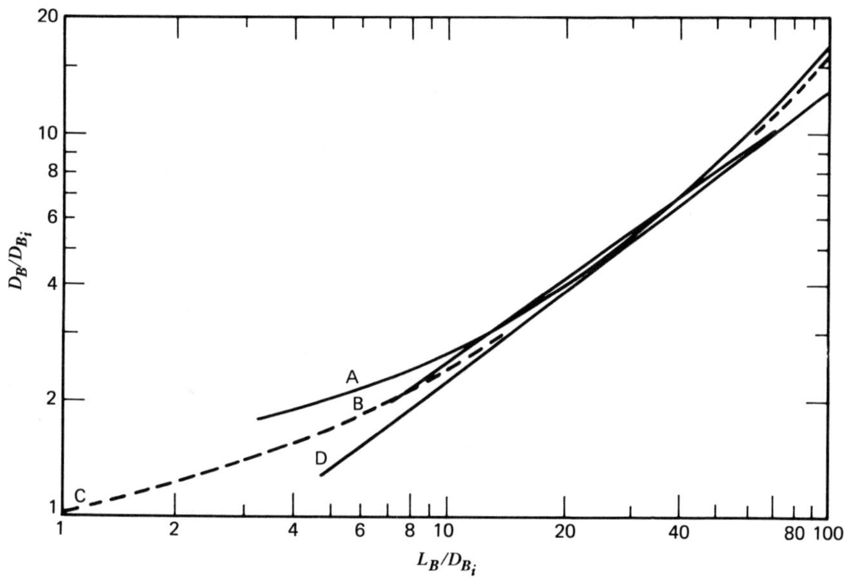

Figure 15. Comparison between empirical bubble growth correlations and the catch-up mechanism represented by the Fibonacci sequence. A, ref. 140; B, ref. 141; C, Fibonacci (to a first approximation: D_B/D_{B_i} = 0.15, L_B/D_{B_i} = 0.85); D, ref. 142.

relationships proposed by several workers (140–142) are superimposed on the curve representing the Fibonacci series. Use of Figure 15 to determine the maximum bubble diameter D_B at any bed level L_B above the grid requires determination of the initial bubble diameter D_{B_i}, which could exist at the grid level as a result of individual ($P/2$) or merged jets. Figure 15 cannot be extrapolated beyond the maximum attainable stable bubble size.

Maximum Stable Bubble Size

The bed support experiments shown in Figure 10 and those of ref. 136, based on particle drag force measurements, demonstrate that a bed interface (and hence a bubble) should be fundamentally stable against collapse as long as it is traversed by a superficial velocity equal to its incipient fluidization rate. Because the inflowing solid's shell volume usually greatly exceeds the incipient fluidization rate, there appears to be no limit to the attainable bubble size, or dome, apt to collapse. If the dome cannot collapse amid free-flowing bed solids, then as the bubble grows it is limited only by particles leaving the shell and being entrained into the bubble void. Such entrainment, or particle pickup, most likely occurs from the bubble walls as the result of the relative velocity between gas and surface particles at the interface (113). The relative flow of fluid past the particles at the bubble wall is twice the shell or bubble velocity, ie, against the downward velocity of bulk solids the bubble fluid (either gas or liquid) rises at approximately an equal velocity. This allows calculation of the minimum bubble size necessary to stir up the solids interface and thus thwart bubble appearance or growth.

Pickup velocity is approximately twice saltation (143–144), and bubble velocity is approximately equal to saltation velocity. This approximate equality has given results in reasonable agreement with a broad range of observations reported to date (125). For example, 80 μm particles of sand fluidized with air could sustain a maximum bubble diameter of the order of 61 cm, whereas when fluidized with water the maximum bubble size would be indiscernable. Sand particles of 600 μm dia when fluidized with water permit a maximum stable bubble size of only 0.64 cm, and 3 mm lead particles permit a water bubble of 18 cm.

Gas–Solids Contact

From grid design, operation of superficial velocity, and fluid particle properties, it is possible to calculate the initial bubble size at the grid, the maximum stable size, and the bed depth over which the bubbles may grow from their initial to their stable diameter. Once having reached their maximum stable diameter any further unlikely mergers would also lead to collapse, so that bubble diameter may be considered constant once having reached the stable size. The bubbles represent a flow superimposed on the superficial incipient bubbling velocity passing up through the bed, and they are purged continuously as they rise. Their local size, velocity, and residence times are calculable from grid to bed surface, and it is possible to calculate the degree to which they are purged before bursting at the surface, ie, to assure that no bubble gas by-passes contact with the bed solids. It may be assumed that the minimum bed depth required to avoid any feed–gas breakthrough (ie, 100% bubble purging) represents the minimum bed depth required for the desired reaction (145). This is represented graphically in Figure 16 for a freely bubbling bed.

In order to avoid entirely scale-up considerations in a pilot plant, superficial velocity and grid details must be identical to those anticipated in the commercial reactor. In addition, the diameter of the shell of downflowing bed solids surrounding the rising bubbles represents the minimum pilot-reactor diameter necessary to simulate free bubbling and avoid approach to slugging (sluggishness). Free bubbling must be simulated hydrodynamically, and the rates of gas permeation from bubble into surrounding bed and reverse should be equal. This becomes significant or controlling only with a coarse and easily permeated bed that has a high incipient fluidization velocity. The gas permeation or cloud diameter (146) is calculable from the depth of gas flow at incipient fluidization velocity over the time interval required for the bubble to rise a distance of one bubble diameter. Since the bubble rises at a velocity equal to four times the square root of its diameter it follows that:

$$\frac{\text{thickness of gas penetrated cloud}}{\text{thickness of downflowing solids shell}} = \frac{V_{mb}}{\sqrt{D_B}}$$

or

$$\text{shell OD} = 1.5\ D_B \text{ and cloud OD} = D_B + 0.5\sqrt{D_B V_{mb}}$$

Application of free shell or cloud criteria in scale-up or scale-down uses the relationship between bubble diameter and bed depth shown in Figure 15 with the limitation of the system's maximum stable bubble size.

A conservative approach to a minimal-risk, pilot-plant reactor free of scale-up considerations would suggest it have the larger of either cloud or shell diameter surrounding the system's maximum stable bubble.

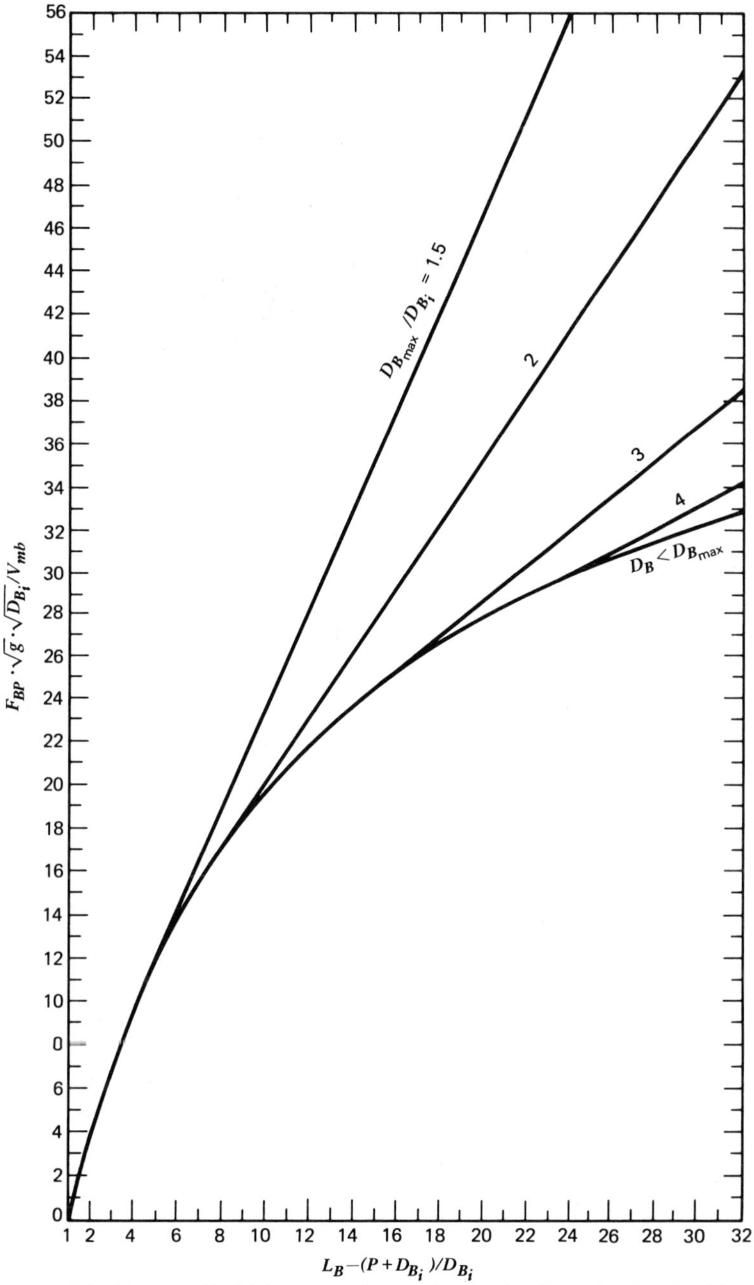

Figure 16. Degree of bubble gas purging during its rise through a fluidized bed.

Mixing and Heat Transfer of Solids

The superficial gas velocity minus the incipient bubbling velocity is approximately equal to the volumetric bulk solids movement across any unit bed cross section per unit time. This amounts to a relatively substantial mass movement, hence a fluidized

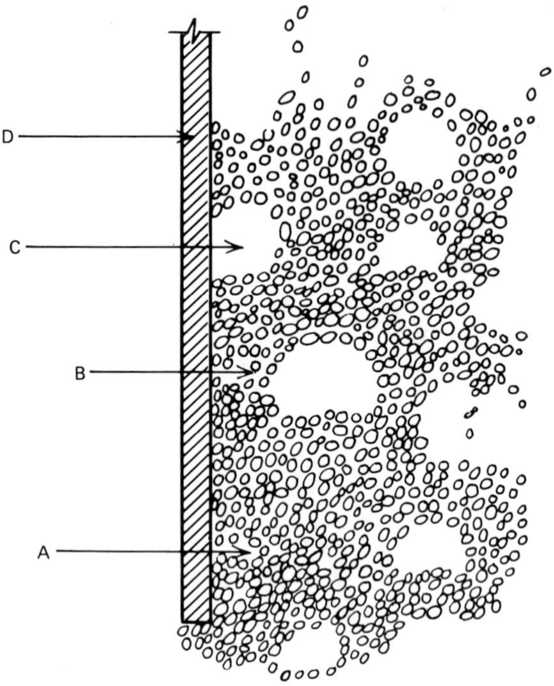

Figure 17. Regions of heat transfer encountered by surfaces immersed in fluidized beds of granular solids. A, stationary solids; B, moving solids; C, no solids; D, heat transfer surface.

bed exhibits reasonably uniform particle size distribution and bed temperature throughout its volume. Reasonable quantitative estimates of such local solids mixing rates (146–148) are of importance principally in determining allowable solids' feed rates. In order to avoid accumulation at any feed pipe location the bed mixing rate must be able to remove the fed material as rapidly as it enters. Such solids' mixing rates have been correlated reasonably well (149).

The substantial heat capacity of the bed solids relative to the gas inventory represents an enormous flywheel, which coupled with the high solids mixing rate leads to a rather uniform temperature throughout the bed. The heat transfer between gas and solids is nearly instantaneous owing primarily to the high particle surface area per unit of bed volume; however, the transfer of heat between the bulk bed and the vessel walls, or any other heat-transfer surface, represents a composite of mechanisms such that the average heat-transfer coefficient is 56.7–567 W/(m^2·K) [10–100 Btu/(h·ft^2·°F)] depending on particle size, bubble size, fluid properties, and superficial fluidizing velocity. The homogeneity and relative uniformity of bed temperature make a fluidized bed an attractive vehicle in which to conduct exothermic as well as endothermic reactions controlled by immersed boiler tubes, exchangers, platecoils, or other heat transfer surfaces when the bed walls do not offer sufficient area for cooling or heat via a fluid circulated through a surrounding jacket.

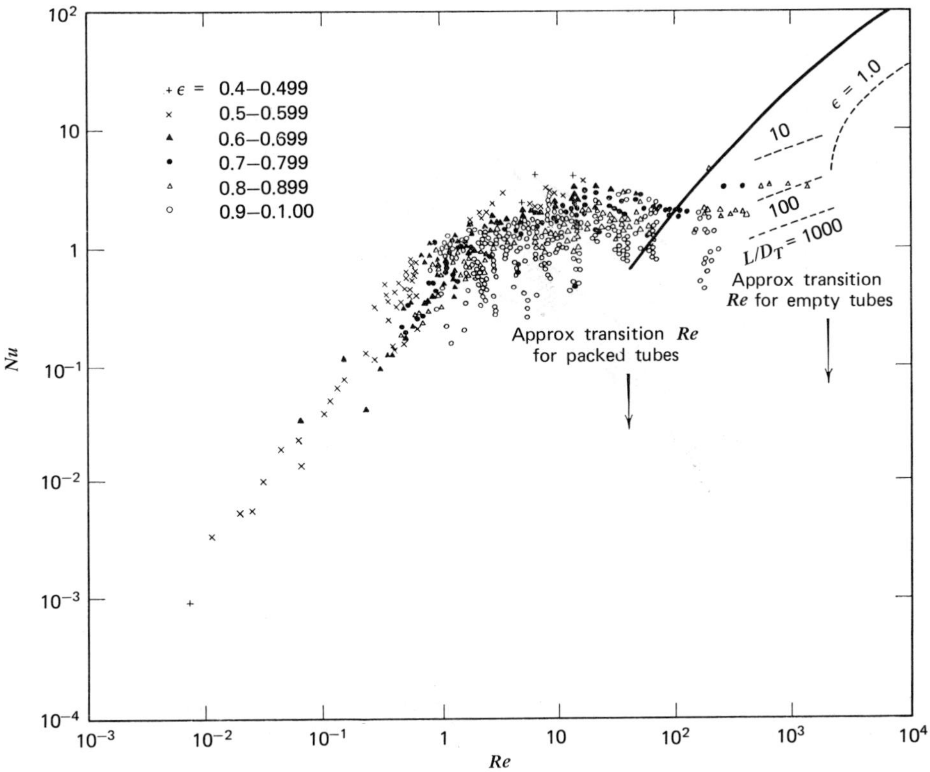

Figure 18. Heat transfer between fluid beds and tube walls. —, data for flow of fluids through fixed beds (151); - - -, correlation for flow of fluids through pipe (152). Sources of data are refs. 153–162.

Type of data	Value of *Re*	Value of *Nu*
fixed- and fluid-bed	$D_P v_\epsilon \rho_f / \mu_f$	hD_P/K_f
empty tube pipe flow	$D_T v_\epsilon \rho_f / \mu_f$	hD_T/K_f

The transfer of heat from the bed to an immersed or wall surface depends on whether the surface is bathed in stationary solids, in moving solids, or in a bubble void (150) as illustrated at points A, B, and C in Figure 17. Overall coefficients averaging these local mechanisms are summarized in Figure 18. Conductive transfer to or from stationary solids as at A in Figure 17, as well as transfer between bubble gas and metal surface as at C, are nearly negligible relative to the rate occurring when the surface is wiped by the shell of solids flowing down around a bubble passing within a distance of a quarter of its diameter from the transfer surface, as at B in Figure 17.

Although advantageously exploitable in industrial installations only under relatively rare circumstances, a number of experimental investigations (163–167) of the heat transfer coefficients under conditions such as at B in Figure 17 have also been correlated in terms of Nusselt and Reynolds numbers as shown in Figure 19. The cross hatched area in Figure 19 encompasses the overall heat transfer data of Figure 18.

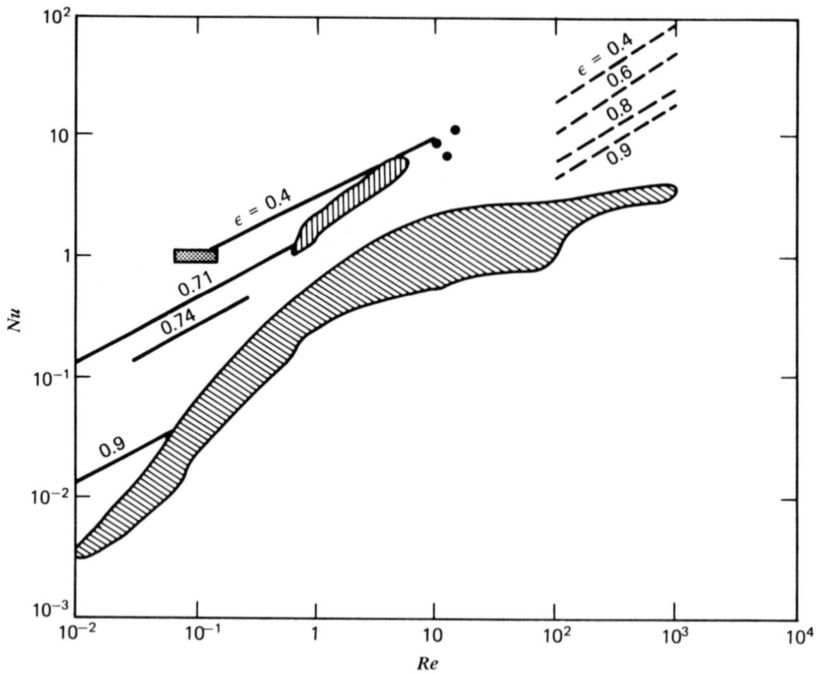

Figure 19. Comparison between overall bed coefficients and local flowing dense-phase coefficients. —, moving solids (163); - - -, liquid fluidized (164); ▥, moving solids (165); ▧, overall, gas fluidized, ϵ = 0.5 –0.95, see Fig. 18; ●, ref. 166; ■, ref. 167.

Nomenclature

C_D	= drag coefficient, dimensionless (see Figs. 3 and 4)
D_B	= bubble diameter
D_B'	= bubble diameter bursting at bed surface
D_{B_i}	= initial bubble diameter at grid level
D_o	= grid hole diameter
$D_{B_{max}}$	= maximum stable, or attainable, bubble diameter
D_p	= particle diameter
e_s	= maximum entrainable lb of solids/ft^3 of gas, or (16 kg/m^3)
F_{BP}	= volume fraction of bubble gas purged
f	= friction factor (see Fig. 3)
g	= gravitational constant
h	= heat transfer coefficient
K_f	= thermal conductivity of fluidizing medium
L_B	= bed depth
Nu	= Nusselt number, dimensionless (see Figs. 18–19)
P	= jet penetration depth
$\Delta P/L$	= pressure drop per unit length, psi/ft, or (22.6 kPa/m)
Re	= Reynolds number, dimensionless
TDH	= transport disengaging height
v	= superficial fluidizing medium velocity
v_ϵ	= superficial fluidizing medium velocity when bed voidage is ϵ
V_B	= bubble rise velocity
V_e	= effective superficial velocity governing rate of entrainment
V_{mb}	= superficial velocity at point of incipient bubbling
V_o	= fluidizing medium velocity through grid hole
V_t	= particle terminal or free fall velocity

W = weight rate of solids entrained, lb/(s·ft^2), or [4.88 kg/(s·m^2)]
ϵ = fractional void volume in bed
ρ_B = bed or bulk density
ρ_f = fluid or medium density
ρ_G = gas density
ρ_L = liquid density
ρ_p = apparent particle density
μ_f = viscosity of fluidizing medium

BIBLIOGRAPHY

"Fluidization" in *ECT* 1st ed., Suppl. 1, pp. 365–400, by F. A. Zenz, Consultant; "Fluidization" in *ECT* 2nd ed., Vol. 9, pp. 398–445, by F. A. Zenz, Consultant.

1. E. V. Murphree and co-workers, *Ind. Eng. Chem.* **35,** 768 (1943); E. V. Murphree and co-workers, *Trans. AIChE* **41,** 19 (1945).
2. P. C. Keith, *Oil Gas J.* **45**(6), 102 (1946).
3. *Chem. Eng.* **72,** 56 (Oct. 25, 1965).
4. L. Khodkevich and I. Panamanski, *Int. Chem. Eng.* **5,** 415 (1965).
5. A. K. Reed, L. Tewsbury, and G. R. Smithson, Jr., *Environ. Sci. Technol.* **4,** 432 (1970).
6. *Chem. Week* **94,** 96 (June 27, 1964).
7. Yu. G. Mamedaliev and co-workers, *Int. Chem. Eng.* **5,** 467 (1965).
8. L. Reh, *Chem. Eng. Prog.* **67**(2), 58 (1971).
9. *Br. Chem. Eng.* **10,** 710 (1965).
10. V. G. Shlyakhtov and V. S. Stretl'son, *Int. Chem. Eng.* **8,** 651 (1968); J. Cibrowski and S. Wronski, *Int. Chem. Eng.* **2,** 105 (1962).
11. J. Trojan and V. Vanacek, *Br. Chem. Eng.* **10,** 756 (1965); *Chem. Eng.* **68**(13), 74 (1961).
12. *Chem. Week* **85**(13), 68 (Sept. 26, 1959); U.S. Pat. 2,891,094 (June 16, 1959), O. C. Karkalitis, C. M. Vanderwaart, and F. H. Megson (to American Cyanamid Co.).
13. H. S. Murthy and co-workers, *Chem. Age India* **14,** 653 (1963).
14. *Ind. Eng. Chem. Proc. Des. Dev.* **2**(3), 214 (1963).
15. B. F. Ives, H. J. Cullinan, Jr., and J. Y. Yang, *Nucl. Technol.* **18,** 29 (1973).
16. S. K. Battacharyya and B. N. Avastni, *Ind. Eng. Chem. Proc. Des. Dev.* **2**(1), 45 (1963).
17. *Chem. Week* **93,** 69 (Oct. 19, 1963).
18. R. Remurez, *Chem. Eng.,* 97 (Sept. 12, 1977); L. A. Hernandez and P. Harriott, *Environ Sci. Technol.* **10,** 454 (1976).
19. *Chem. Eng.* **70,** 92 (Apr. 15, 1963); H. M. Rowson, *Br. Chem. Eng.* **8,** 180 (1963).
20. J. A. Thurley and O. Sitani, *Br. Chem. Eng. Proc. Technol.* **17,** 319 (1972).
21. *Chem. Week* **105,** 30 (Aug. 9, 1969).
22. *Chem. Eng.* **76,** 90 (Dec. 1, 1969).
23. S. B. Alpert, E. S. Johanson, and S. C. Schuman, *Chem. Eng. Prog.* **60**(6), 35, 58, 69 (1964); *Chem. Eng.* **72,** 78 (Dec. 6, 1965).
24. *Electr. Rev. (London)* **176,** 39 (1965).
25. *Chem. Eng.* **85,** 114 (Aug. 14, 1978).
26. R. E. Singiser, A. L. Heiser, and E. B. Drillig, *Chem. Eng. Prog.* **62**(6), 107 (1966).
27. *Chem. Week* **87**(10), 56 (Sept. 3, 1960); *ibid.* **89**(2), 48 (July 8, 1961); J. Gaynor, *Chem. Eng. Prog.* **56**(7), 75 (1960); *Chem. Eng.* **66**(28), 100 (1959); N. P. Chopey, *Chem. Eng.,* 36 (July 12, 1971); *Chem. Eng. News,* 37 (Jan. 25, 1971).
28. *Chem. Eng.* **71,** 22 (Aug. 31, 1964).
29. G. Volpicelli, L. Massimilla, and F. A. Zenz, *AIChE Symp. Ser.* **62**(67), 56 (1966).
30. *Chem. Eng.* **77,** 58 (Aug. 10, 1970).
31. *Hydrocarbon Process. Pet. Refiner.* **44**(11), 275 (1965).
32. *Br. Chem. Eng. Process Technol.* **17,** 383 (1972); *Chem. Eng. News* **43,** 37 (Dec. 6, 1965); *Chem. Eng.* **69,** 74 (Apr. 30, 1962).
33. *Chem. Eng. News* **43,** 37 (Dec. 6, 1965).
34. *Chem. Eng.* **85,** 72, 77 (Aug. 14, 1978).
35. P. T. Cichy, J. S. Ultman, and T. W. F. Russell, *Ind. Eng. Chem.* **61**(10), 8 (1969); S. M. Yoon and D. Kunii, *Ind. Eng. Chem. Prod. Des. Dev.* **9**(4), 563 (1970).

36. J. P. Forsman, *Hydrocarbon Process.* **51,** 130 (Nov. 1972); *Chem. Proc.* **41,** 18 (Feb. 1978); *Chem. Eng.* **80,** 72 (Nov. 26, 1973); J. C. Davis, *Chem. Eng.* **85,** 25 (Jan. 2, 1978); *ibid.,* 57 (Dec. 4, 1978).
37. *Br. Chem. Eng.* **11,** 799 (Aug. 1966).
38. *Br. Chem. Eng.* **8,** 800 (1963); **11,** 2 (1966).
39. *Chem. Week* **93,** 37 (Aug. 31, 1963).
40. *Chem. Eng. News* **56,** 26, 28 (Jan. 30, 1978); *Chem. Week* **122,** 35 (Jan. 25, 1978).
41. *Chem. Process.* **41,** 53 (Nov. 1978); *Chem. Eng.* **85,** 39 (Nov. 6, 1978); *Br. Chem. Eng.* **12,** 811 (1967).
42. *Chem. Eng. News* **51,** 33 (Sept. 3, 1973).
43. V. D. Chase, *Pop. Sci.* **210,** 91 (Jan. 1977).
44. *Chem. Eng.* **85,** 107 (Aug. 14, 1978); P. Marnell, *Chem. Eng.* **79,** 102 (Nov. 13, 1972).
45. *Chem. Eng.* **73,** 32 (Aug. 29, 1966).
46. W. Mounce and R. S. Rubin, *Chem. Eng. Prog.* **67**(8), 81 (1971); H. H. Stotler and W. Volk, *Br. Chem. Eng.* **16,** 1117 (Dec. 1971); C. R. Griswold and R. P. van Driesen, *Hydrocarbon Process.* **45**(5), 153 (1966).
47. H. Littman and R. G. Bacile, *AIChE Symp. Ser.* **62**(67), 42 (1966).
48. *Chem. Week* **95,** 63 (Nov. 28 1964); R. F. Schwab and W. N. Doyle, *Chem. Eng. Prog.* **66**(9), 49 (1970).
49. Brit. Pat. 1,187,372 (Apr. 8, 1970), G. J. Harris (to Fisons Fertilizers Ltd.); *Br. Chem. Eng.* **15,** 585 (1970).
50. *Chem. Eng.* **79,** 54 (May 15, 1972).
51. *Chem. Eng. News* **56,** 21 (Oct. 16, 1978); *Chem. Eng.* **78,** 39 (May 3, 1971).
52. R. F. Cheers and T. T. Furman, *Br. Chem. Eng.* **16,** 27 (Dec. 1971).
53. *Chem. Eng. News* **43,** 42 (Mar. 29, 1965).
54. *Chem. Eng. News* **51,** 24 (May 14, 1973).
55. *Chem. Eng.* **69,** 90 (Sept. 3, 1962).
56. K. A. Andrianov, *Br. Chem. Eng.* **11,** 927 (1966).
57. Yu. G. Mamedaliev, *Int. Chem. Eng.* **4,** 382 (1964).
58. *Br. Chem. Eng.* **8,** 414 (1963).
59. N. Y. Chen and co-workers, *Ind. Eng. Chem. Process Des. Dev.* **16**(3), 247 (1977); *Ind. Res.* **13,** 23 (May 1971); *Chem. Week* **100,** 55 (Apr. 15, 1967); *Chem. Eng. News* **49,** 85 (Feb. 15, 1971).
60. *Chem. Week* **111,** 68 (Nov. 22, 1972).
61. *Chem. Week* **112,** 27 (Apr. 11, 1973).
62. *Chem. Week* **88**(24), 74 (June 17, 1961); *Br. Chem. Eng.* **16,** 17 (1971).
63. *Br. Chem. Eng.* **14,** 1041 (1969).
64. *Chem. Eng.* **70,** 134 (Nov. 11, 1963).
65. *Chem. Eng. News* **56,** 30 (Jan. 16, 1978).
66. S. Tone and co-workers, *J. Chem. Eng. Jpn.* **7**(1), 44 (1974).
67. *Chem. Eng.* **85,** 44 (Dec. 18, 1978); *ibid.* **80,** 78 (June 11, 1973); *Br. Chem. Eng.* **15,** 1191 (Sept. 1970).
68. J. Ciborowski, *Int. Chem. Eng.* **2,** 105 (1962).
69. U.S. Pat. 2,642,393 (June 16, 1953), A. G. Gehm and L. I. Purcell (to Limestone Products Corporation of America).
70. *Chem. Week* **108,** 52 (Feb. 10, 1971); P. G. Thornhill, *Can. Mining Metall. Bull.* **54,** 601 (Aug. 1961).
71. *Chem. Eng. News* **37**(45), 50 (Nov. 9, 1959); *Chem. Eng.* **77,** 17 (Nov. 30, 1970).
72. H. K. Pargal, Ph.D. thesis, University of Colorado, Boulder, Colo., 1954.
73. *Ind. Eng. Chem.* **59**(3), 64 (1967).
74. G. G. Copeland and J. E. Janway, Jr., *Tappi* **47**(6), 175A (1964); *Chem. Week* **97,** 31 (July 31, 1965).
75. T. N. Sycheva and E. V. Donat, *Khim. Prom.* (6), 59 (1974); *Chem. Proc.* **41,** 60 (Nov. 1978).
76. *Hydrocarbon Process,* **46**(11), 210 (1967); U.S. Pat. 2,914,575 (Nov. 24, 1959), R. E. Feathers and R. H. Rogerson (to Columbia-Southern Chemical Corporation); U.S. Pat. 2,914,576 (Nov. 24, 1959), R. M. Vancamp and A. P. Muren, Jr. (to Columbia-Southern Chemical Corporation); U.S. Pat. 2,951,103 (Aug. 30, 1960), A. C. Ellsworth and R. M. Vancamp (to Columbia-Southern Chemical Corporation); U.S. Pat. 2,952,714 (Sept. 13, 1960), J. E. Milam, W. E. Makris, and R. I. McGreevy (to Columbia-Southern Chemical Corporation); U.S. Pat. 2,957,924 (Oct. 25, 1960), T. W. Heskell and F. C. Dehn (to Columbia-Southern Chemical Corporation).
77. *Chem. Eng.* **85,** 87 (Jan. 2, 1978); 60 (Aug. 14, 1978); P. A. Witt, Jr., *Chem. Eng.* **78,** 71 (Oct. 4, 1971).

78. L. C. Nardison and E. J. Dowd, *Chem. Eng. Prog.* **73,** 31 (Aug. 1977); *Chem. Process.* **39,** 13 (Aug. 1976).
79. *Chem. Eng.* **78,** 78 (May 17, 1971).
80. *Chem. Eng. News* **54,** 23 (Aug. 2, 1976); *Chem. Week* **111,** 37 (Aug. 2, 1972); *Chem. Eng.* **80,** 60 (Jan. 8, 1973).
81. *Chem. Eng.* **83,** 45 (Apr. 26, 1976).
82. E. Guccione, *Chem. Eng.* **73,** 122 (Feb. 14, 1966).
83. R. A. Labine, *Chem. Eng.* **67**(3), 96 (Feb. 8, 1960).
84. V. S. de Marchi, *AIChE Symp. Ser.* **62**(67), 15 (1966).
85. *Chem. Week* **113,** 32 (Sept. 5, 1973); **94,** 66 (Mar. 7, 1964); *Chem. Eng.* **86,** 49 (May 28, 1973).
86. *Chem. Week* **110,** 39 (May 3, 1972).
87. *Chem. Eng.* **85,** 95 (Oct. 23, 1978).
88. S. Ushio, *Chem. Eng.* **78,** 107 (Sept. 20, 1971).
89. *Br. Chem. Eng.* **10,** 710 (1965); T. N. Shakhtakhtinskii, K. I. Sadykhava, and Z. M. Farberg, *Int. Chem. Eng.* **6,** 674 (1966).
90. *Chem. Eng. News* **47,** 62 (Sept. 22, 1969); *Chem. Eng.* **77,** 101 (Oct. 19, 1970).
91. *Chem. Week* **98,** 87 (Mar. 19, 1966); F. Weinrohec, *Chem. Eng.* **72,** 180 (Oct. 11, 1965).
92. *Br. Chem. Eng.* **15,** 735 (June 1970).
93. *Chem. Week* **91,** 139 (Sept. 15, 1962).
94. J. H. Oxley and co-workers, *Nucl. Appl.* **1,** 567 (1965).
95. U.S. Pat. 3,020,148 (Apr. 5, 1960), W. A. Jenkins and H. W. Jacobson (to E. I. du Pont de Nemours & Co., Inc.).
96. *Bus. Week,* 76 (Nov. 16, 1974); *preprint 42D, 60th Annual AIChE Meeting, New York, Nov. 26–30, 1967.*
97. I. E. Knudson, H. E. Hootman, and N. M. Levitz, *Nucl. Sci. Eng.* **20,** 259 (1964); L. A. Anastasia and W. J. Mecham, *Ind. Eng. Chem. Process Des. Dev.* **4,** 338 (1965).
98. *Chem. Eng. News* **40,** 42 (Aug. 20, 1962).
99. S. G. Brandenberg, *Nucl. Technol.* **18,** 177 (May 1973).
100. *Chem. Eng. Prog. Symp. Ser. No. 67* **62**(67), 76 (1966).
101. M. G. Geiger, Ph.D. thesis, Purdue University, Lafayette, Ind., 1953–1954; *Chem. Eng. News* **48,** 39 (May 11, 1970).
102. *Chem. Week* **95,** 93 (Aug. 22, 1964); M. D. Rosenzweig, *Chem. Eng.* **78,** 105 (Oct. 18, 1971); E. F. Edwards and T. Weaver, *Chem. Eng. Prog.* **61**(1), 21 (1965).
103. *Chem. Process.* **41,** 47 (Nov. 1978); *Chem. Eng.* **85,** 113 (Apr. 10, 1978).
104. *Chem. Eng. News* **40,** 41 (Aug. 13, 1962); *Chem. Eng.* **69,** 60 (Aug. 20, 1962).
105. *Chem. Week* **119,** 40 (Sept. 1, 1976); *Chem. Eng.* **85,** 51 (Feb. 13, 1978).
106. I. H. Kim and L. Lewis, *Ind. Eng. Chem. Process Des. Dev.* **2**(2), 148 (1963).
107. U.S. Pat. 3,927,174 (Dec. 20, 1971), G. G. Copeland (to Copeland Systems, Inc.).
108. *Chem. Week* **121,** 37 (Dec. 21, 1977); C. W. Zielke, R. T. Struck, and E. Gorin, *Ind. Eng. Chem. Process Des. Dev.* **8**(4), 552 (1969).
109. G. Roberts and J. A. Phinney, *Oil Gas J.* **45,** 72, 139 (1947).
110. L. J. Jolley and J. E. Stanton, *J. Appl. Chem.* **2**(Suppl. 1), (1952); *Chemie Ingenieur Technik,* Vol. 24, 2nd ed., Suppl. Springer-Verlag, Berlin, Ger., Feb. 1952, pp. 57–128.
111. F. A. Zenz and D. F. Othmer, *Fluidization and Fluid–Particle Systems,* annotated version, Reinhold Publishing Corp., New York, 1966, p. 281.
112. R. H. Wilhelm and M. Kwauk, *Chem. Eng. Prog.* **44,** 201 (1948).
113. A. M. Squires, *paper presented at 54th Annual AI.Ch.E. Meeting, New York, Dec. 6, 1961; Chem. Eng. Prog. Symp. Ser.* **58**(38), 57 (1962).
114. F. A. Zenz, *Hydrocarbon Process.* **46**(4), 171 (1967).
115. A. O. Oman and K. M. Watson, *Natl. Pet. News* **36,** R795 (1944).
116. P. C. Carman, *Trans. Inst. Chem. Eng. (London)* **15,** 150 (1937); *J. Soc. Chem. Ind. (London)* **57,** 225 (1938).
117. B. A. Bakhmeteff and N. V. Fedoroff, *J. Appl. Mech.* **4,** A97 (1937).
118. S. Ergun, *Chem. Eng. Prog.* **48,** 89 (1952); *Ind. Eng. Chem.* **41,** 1179 (1949).
119. M. Leva, T. Shirai, and C. Y. Wen, *Genie Chim.* **75**(2), 33 (1956); J. F. Frantz, *Chem. Eng. Prog. Symp. Ser.* **62**(62), 21 (1966).
120. C. Y. Wen and Y. H. Yu, *Chem. Eng. Prog. Symp. Ser.* **62**(62), 100 (1966).
121. C. E. Lapple and C. B. Shepherd, *Ind. Eng. Chem.* **32,** 605 (1940).
122. F. A. Zenz, *Pet. Refiner* **36**(8), 147 (1957).

123. "Cyclone Separators," *Emissions Control Manual, Pub. No. 931,* American Petroleum Institute, Washington, D.C., May 1975, Ch. 11.
124. F. A. Zenz and N. A. Weil, *AIChE J.* **4,** 472 (1958); *Hydrocarbon Process.* **53,** 119 (1974).
125. F. A. Zenz, *Chem. Eng.,* 81 (Dec. 19, 1977).
126. J. Werther in J. F. Davidson and D. L. Keairns, eds., *Fluidization,* Cambridge University Press, New York, 1978, pp. 7–12.
127. W. C. Yang and D. L. Keairns in ref. 126, pp. 208–214.
128. F. A. Zenz, *Inst. Chem. Eng. (London) Symp. Ser.* (30), 136 (1968).
129. D. Harrison and L. S. Leung, *Trans. Inst. Chem. Eng. (London)* **39,** 409 (1961).
130. J. W. Higby, *Chemie. Ing. Technol.* **36,** 228 (1964).
131. *Proceedings of the Question and Answer Session on Refining Technology,* National Petroleum Refineries Association, Washington, D.C., 1970, p. 86.
132. P. N. Rowe and F. A. Zenz, *Sch. Sci. Rev.* **53**(182), 94 (1971).
133. F. A. Zenz and D. F. Othmer, *Fluidization and Fluid–Particle Systems,* Reinhold Pub. Co., New York, 1960, p. 171.
134. J. S. M. Botterill, J. S. George, and H. Besford, *Chem. Eng. Prog. Symp. Ser.* **62,** 7 (1966).
135. P. V. Danckwerts, *paper presented at Symposium on Fluidization held at AERE, Harwell, Eng., Oct. 5, 1959.*
136. P. N. Rowe and G. A. Henwood, *Trans. Inst. Chem. Eng. (London)* **37,** 43 (1961).
137. D. W. Van Krewelen and P. J. Hoftijzer, *Chem. Eng. Prog.* **44,** 529 (1948).
138. J. M. Matsen and B. L. Tarmy, *Chem. Eng. Prog. Symp. Ser.* **66**(101), 1 (1970); *Chem. Eng. Sci.* **24,** 1743 (1969).
139. F. A. Zenz, *Fibonacci Q.* **16**(2), 171 (1978).
140. J. Werther, *paper presented at the International Fluidization Conference, Asilomar, Calif., June 15–20, 1975;* D. L. Keairns, ed., *Fluidization Technology,* Vol. 1, Hemisphere Publishing Corporation, Washington, D.C., 1976.
141. C. Chavarie and J. R. Grace, *Ind. Eng. Chem. Fundam.* **14,** 75 (May 1975).
142. P. N. Rowe, *paper presented at the International Fluidization Conf., Asilomar, Calif., June 15–20, 1975.*
143. P. J. Jones and L. S. Leung, *Ind. Eng. Chem. Process Des. Dev.* **17**(4), 571 (1978).
144. F. A. Zenz, *Ind. Eng. Chem. Fundam.* **3**(1), 65 (1964).
145. Ref. 133, Chapt. 8.
146. P. N. Rowe, B. A. Partridge, and E. Lyall, *Chem. Eng. Sci.* **18,** 973 (1964); P. N. Rowe, *Chem. Eng. Prog.* **60**(3), 75 (1964).
147. S. Katz and F. A. Zenz, *Pet. Refiner* **33**(5), 203 (1954).
148. Ref. 133, Chapt. 9.
149. E. Talmor and R. F. Benenati, *AIChE J.* **9,** 536 (1963).
150. F. A. Zenz, *Proc. Third Int. Heat Transfer Conf.* **VI,** 311 (1966).
151. M. Leva, *Ind. Eng. Chem.* **39,** 857 (1947); D. A. Plautz and H. F. Johnstone, *AIChE J.* **1,** 193 (1955).
152. W. H. McAdams, *Heat Transmission,* 2nd ed., McGraw-Hill Book Co., Inc., New York, 1942.
153. R. N. Bartholomew, Ph.D. thesis, University of Micigan, Ann Arbor, Mich., 1950; *Chem. Eng. Prog. Symp. Ser.* **48**(4), 3 (1952).
154. L. H. Collins, M.S. thesis, Massachusetts Institute of Technology, Cambridge, Mass., 1946.
155. W. M. Dow, Ph.D. thesis, Illinois Institute of Technology, Chicago, Ill., 1949; *Chem. Eng. Prog.* **47,** 637 (1951).
156. H. Fischer and E. F. Dillon, B.S. thesis, Massachusetts Institute of Technology, Cambridge, Mass., 1947.
157. W. Lazor and S. A. Murray, M.S. thesis, Massachusetts Institute of Technology Cambridge, Mass., 1947.
158. M. Leva, M. Weintraub, and M. Grummer, *Chem. Eng. Prog.* **45,** 563 (1949).
159. H. S. Mickley and C. A. Trilling, *Ind. Eng. Chem.* **41,** 1135 (1949).
160. W. H. Millick and A. S. Humphrey, M.S. thesis, MIT, 1948.
161. R. V. Trense, Ph.D. thesis, Northwestern University, Evanston, Ill., 1954.
162. R. W. Urie, M.S. thesis, MIT, 1948.
163. N. K. Harakas and K. O. Beatty, *Chem. Eng. Prog. Symp. Ser.* **59**(41), 122 (1963).
164. R. Raju, Ph.D. Thesis, Andhra University, Waltair, India, 1959.

165. J. S. M. Botterill and J. R. Williams, *Trans. Inst. Chem. Eng. (London)* **41,** 217 (1963); J. S. M. Botterill, *Fluid Bed Heat Transfer,* Academic Press, Inc., New York, 1975.
166. V. D. Dunsky, S. S. Zabrodsky, and A. I. Tamarin, *Proc. Third Int. Heat Transfer Conf.* **IV,** 293 (1966).
167. A. A. H. Drinkenburg, N. J. J. Huige, and K. Rietema, *Proc. Third Int. Heat Transfer Conf.* **IV,** 271 (1966).

General References

J. M. Dallavalle, *Micromeritics,* Pitman Publishing Co., New York, 1948.

P. Reboux, *Phenomenes De Fluidisation,* Assoc. Francaise de Fluidisation, Paris, Fr., 1954.

D. F. Othmer, ed., *Fluidization,* Reinhold Publishing Co., New York, 1956.

M. Leva, *Fluidization,* McGraw-Hill, New York 1959.

F. A. Zenz and D. F. Othmer, *Fluidization and Fluid-Particle Systems,* annotated edition, Pemmco Publications, Garrison, New York, 1966.

J. F. Davidson and D. Harrison, *Fluidised Particles,* Cambridge University Press, New York, 1963.

Fluidisation, Joint Symposium of the Chemical Engineering Group and the Institution of Chemical Engineers, March 11–12, 1963, Society of Chemical Industry, London, Eng., 1964.

C. Orr, *Particulate Technology,* MacMillan Co., New York, 1966.

M. N. Rao, ed., *Fluidization and Related Processes,* Sree Saraswaty Press, Calcutta, India, 1966.

V. Vanecek, M. Markvart, and R. Drbohlav, *Fluidized Bed Drying,* Leonard Hill Publishing Co., London, Eng., 1966.

S. S. Zabrodsky, *Hydrodynamics and Heat Transfer in Fluidized Beds,* The M.I.T. Press, Cambridge, Mass., 1966.

A. A. H. Drinkenburg, ed., *Proceedings of the International Symposium on Fluidization,* Netherlands University Press, Amsterdam, The Netherlands, 1967.

J. M. Pirie, ed., *Fluidisation,* Institute of Chemical Engineers, London, Eng., 1968.

D. Kunii and O. Levenspiel, *Fluidization Engineering,* John Wiley & Sons, Inc., New York, 1969.

J. F. Davidson and D. Harrison, eds., *Fluidization,* Academic Press, Inc., New York, 1971.

D. Venkateswarlu and A. P. Rao, ed., *Particulate Technology,* Indian Institute of Technology Press, 1971.

M. Angelino and co-eds., *Fluidization and Its Applications,* Societe de Chimie Industrielle, Paris, Fr., 1973.

J. S. M. Botterill, *Fluid-Bed Heat Transfer,* Academic Press, Inc., New York, 1975.

D. L. Keairns, ed., *Fluidization Technology,* Vols. I and II, Hemisphere Publishing Corporation, Washington, D.C., 1975.

D. F. Davidson and D. L. Keairns, eds., *Fluidization,* Cambridge University Press, New York, 1978.

W. C. Patterson and R. Griffin, *Fluidized-Bed Energy Technology,* Inform Publishing Co., New York, 1978.

FREDERICK A. ZENZ
Consultant

FLUID MECHANICS

In fluid mechanics a strong contrast exists between the viewpoints of academicians and practicing chemical technologists. To the former, fluid mechanics is the art of obtaining very nearly exact solutions to well-posed mathematical problems that are related (perhaps remotely) to fluid-flow phenomena already fairly well understood. Emphasis is placed on the details of the flow and on geometries and situations sufficiently simple to be handled by the mathematical techniques currently available. To the technologist, fluid mechanics is the entire body of knowledge, theoretical and empirical, quantitative and qualitative, allowing analysis of the performance of complex plant equipment handling moving fluids. Calculation of the details of the flow is secondary to understanding the phenomena well enough to accomplish the process task. At times the technologist's needs are best satisfied by an empirical correlation; at other times the necessary skills consist largely of knowing how to fit the idealized solutions of the mathematician into a practical situation. Fluid mechanics is thus both a descriptive science of the phenomena that occur when fluids flow and a quantitative science showing how these phenomena may be described in mathematical terms.

This article attempts to strike a balance between these two aspects, with rather more emphasis than is customary on descriptions of phenomena and their significance. It is intended to provide a useful first understanding of flow phenomena and techniques and to provide an entry to more precise and detailed methods where they are called for. Although the main concern is the proper design and operation of plant equipment, the importance of preservation of the environment is recognized. Results from the formerly specialized fields of meteorology and oceanography are now occasionally needed by the technologist.

Fluid mechanics became an exact science with the application of the conservation principles. These include the law of the conservation of mass (including its detailed form: the law of conservation of individual species, electric charge, etc), the law of conservation of energy, and the law of conservation of force-momentum (Newton's second law). It differs from conventional mechanics in that it treats mainly bodies that are capable of unlimited deformation. Flow is regarded as primary and elasticity as secondary, in contrast to solid mechanics where elasticity is primary and flow is regarded as secondary. Instabilities in flow, eg, turbulence, play a large role in fluid mechanics whereas they are of minor concern in solid mechanics. There is, however, no essential difference between the two disciplines. In some instances the two converge, as in the treatment of extrusion of very viscous or plastic materials in which fluid-like and solid-like behavior are of equal importance.

The most useful mathematical formulation of a fluid flow problem is as a boundary value problem. This consists of two main parts: a set of differential equations to be satisfied within a region of interest and a set of boundary conditions to be satisfied on the surfaces of that region. Sometimes additional conditions are also of interest, eg, when one is investigating the stability of a flow.

The choice of the differential equation to be used within the region depends upon what approximations can legitimately be made. A central assumption to most situations is that the fluid can be treated as a continuum, ie, mass, momentum, and energy are strictly conserved for any subdivision of space, no matter how small, and the material properties of the fluid are not affected by the process of subdivision. This idealization allows one to define the velocity at a point V as the time rate of change of the center of mass of a material body. Similarly the rates of deformation or strain of the body constitute the components of a second-order tensor. Clearly if the subdivision were extended to molecular or mean-free-path dimensions, the velocity would become highly fluctuating and random. Thus, the results of fluid-mechanical calculations are limited to situations in which the changes of interest occur over length scales much larger than these dimensions. Where the continuum assumption breaks down, eg, in highly rarefied gases or in the behavior of submicron aerosol particles, recourse must be had to more general methods of calculation (see Dimensional analysis).

The boundary conditions for a situation are usually easy to state, at least where the boundaries are fixed in position. They usually correspond to the physical requirement that there is a specified flow through (or stress at) the boundary. Real fluids, which exhibit viscosity (a resistance to continuous deformation), also obey the principle of continuity of velocity, which asserts that the velocity of a flowing fluid does not suffer discontinuous changes at interfaces with other fluids or solids. As applied to flow past a stationary solid, it asserts that the fluid velocity decreases to zero as the wall is approached. In classical, inviscid hydrodynamics solid surfaces serve to confine or deflect the flow but offer no retardation to the velocity which is assumed to maintain its main-stream value vanishingly close to the surface. Although classical hydrodynamics is suitable for describing some features of flow, it fails completely in other key areas; its most notable failure is the incorrect prediction that a cylinder placed with its axis normal to the flow will experience no net force from the fluid (D'Alembert's Paradox). The assumption of a velocity decreasing continuously to zero at the surface and the introduction of shear stresses related to this gradient provide a more accurate prediction of flow phenomena consistent with experimental observations. This last serves to emphasize that fluid mechanics still depends significantly on experiment to verify that a given mathematical solution does describe reality. This is particularly true when, as is discussed later, multiple solutions are possible.

Equations of Motion

The starting point for obtaining quantitative descriptions of flow phenomena is Newton's second law. This states that the vector sum of forces acting on a body equals the rate of change of momentum of the body. This force balance can be made in many different ways. It may be applied over a body of finite size (see Integral Forms of Equations of Motion) or over each infinitesimal portion of the body. It may be utilized in a coordinate system moving with the body (the so-called Lagrangian viewpoint) or in a fixed coordinate system (the Eulerian viewpoint). Described below is derivation of the equations of motion from the Eulerian viewpoint using the Cartesian coordinate system. The equations in other coordinate systems are described in standard references (1–2).

General Equation of Motion. Neglecting relativistic effects, the rate of accumulation of mass within a Cartesian volume element $dx{\cdot}dy{\cdot}dz$ must equal the sum of the rates of inflow minus outflow. This is expressed by the equation of continuity:

$$\frac{\partial \rho}{\partial t} + \frac{\partial \rho u}{\partial x} + \frac{\partial \rho v}{\partial y} + \frac{\partial \rho w}{\partial z} = 0 \tag{1}$$

where ρ is the fluid density; u, v, and w, are the x-, y- and z-components of velocity; and t is the time.

Similarly, a force-momentum balance can be made on the same infinitesimal volume element. First it is necessary to have a firm idea of the nature of the forces that might be exerted on the element. There may be body forces which, in Newtonian mechanics, are pictured as acting at a distance and include gravitational, electrostatic, and magnetic forces and—if a noninertial frame of reference is used—those forces resulting from accelerations of the frame, eg, centrifugal force. In addition there are forces arising from the continuum stresses exerted on the faces or edges of the element. These stresses arise from intermolecular forces and motions of molecules which express themselves macroscopically as pressure, viscosity, yield stress, etc. These stresses behave mathematically as the components of a symmetrical, second-order tensor. At boundaries between phases additional forces must be included. These forces, which are generalizations of surface tension, are discussed in standard texts.

In Figure 1, the force balance in Cartesian coordinates for a body not intersected by phase boundaries is:

$$\begin{aligned}
\rho \frac{Du}{Dt} &= X + \frac{\partial \sigma_x}{\partial x} + \frac{\partial \tau_{yx}}{\partial y} + \frac{\partial \tau_{zx}}{\partial z} \\
\rho \frac{Dv}{Dt} &= Y + \frac{\partial \sigma_y}{\partial y} + \frac{\partial \tau_{zy}}{\partial z} + \frac{\partial \tau_{xy}}{\partial x} \\
\rho \frac{Dw}{Dt} &= Z + \frac{\partial \sigma_z}{\partial z} + \frac{\partial \tau_{xz}}{\partial x} + \frac{\partial \tau_{yz}}{\partial y}
\end{aligned} \tag{2}$$

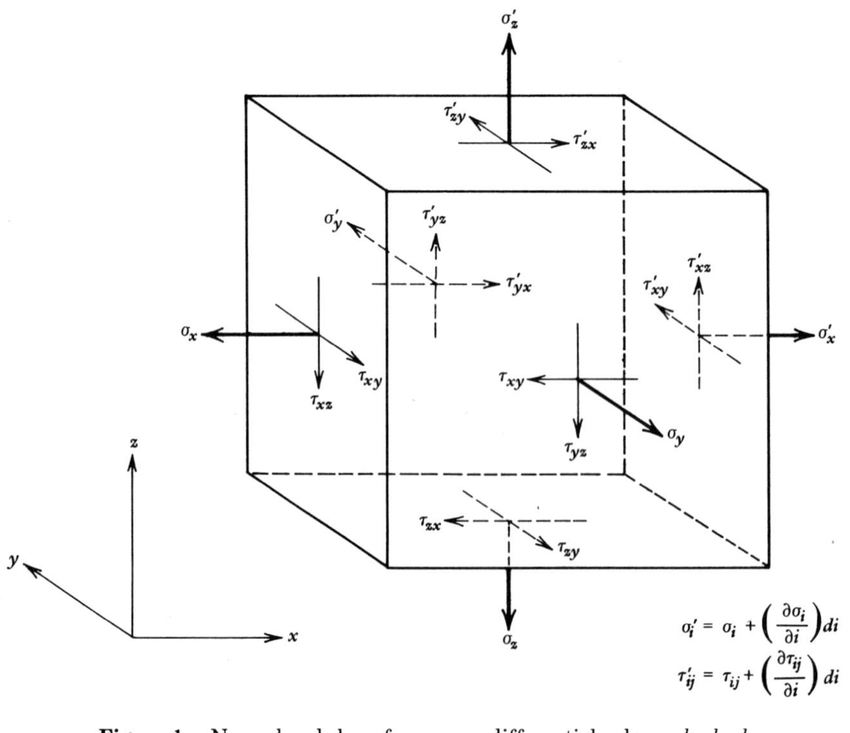

Figure 1. Normal and shear forces on a differential volume $dx{\cdot}dy{\cdot}dz$.

where Du/Dt, Dv/Dt and Dw/Dt are the components of acceleration; X, Y, Z are body forces per unit volume; σ_i is the normal stress on the i-face; and τ_{ij} is the shear stress on the i-face exerted in the j-direction. The operator D/Dt is called the material, total, or substantial derivative and can be expressed as:

$$\frac{D}{Dt} = \frac{\partial}{\partial t} + u\frac{\partial}{\partial x} + v\frac{\partial}{\partial y} + w\frac{\partial}{\partial z} \tag{3}$$

Equation 2 gives the relation between stresses and accelerations obtained from force/momentum balances. To proceed further requires use of the constitutive equations which codify the material properties through additional relations between the stresses (τ_{xy}, etc) and the rates of strain ($\partial u/\partial x$, etc). The constitutive equation for a given fluid is found empirically or theoretically by use of some theory of material properties. The simplest model is one in which the various stresses are expressed as linear combinations of the rates of strain. When the fluid is homogeneous and isotropic, this relation leads to the Navier-Stokes equations. Fluids that obey these equations are by definition Newtonian. The conditions of homogeneity and isotropy ensure that only two material constants, the shear viscosity μ and the dilational viscosity λ are needed to describe the fluid. By defining pressure as the negative of the average of the three normal stresses σ_i one finds that $\lambda = 2/3\ \mu$, hence only a single material viscosity μ is required. As defined, the pressure is usually identified with the thermodynamic pressure for purposes such as determining physical properties. Equations 4 and 5 show the constitutive equation so derived and equation 6 shows the form taken by the equations of motion when these are inserted into the force balance.

$$\begin{aligned}
\sigma_x &= -P + 2\mu\frac{\partial u}{\partial x} - \frac{2}{3}\mu\left(\frac{\partial u}{\partial x} + \frac{\partial v}{\partial y} + \frac{\partial w}{\partial z}\right)\\
\sigma_y &= -P + 2\mu\frac{\partial v}{\partial y} - \frac{2}{3}\mu\left(\frac{\partial u}{\partial x} + \frac{\partial v}{\partial y} + \frac{\partial w}{\partial z}\right)\\
\sigma_z &= -P + 2\mu\frac{\partial w}{\partial z} - \frac{2}{3}\mu\left(\frac{\partial u}{\partial x} + \frac{\partial v}{\partial y} + \frac{\partial w}{\partial z}\right)
\end{aligned} \tag{4}$$

$$\begin{aligned}
\tau_{xy} &= \tau_{yx} = \mu\left(\frac{\partial u}{\partial y} + \frac{\partial v}{\partial x}\right)\\
\tau_{yz} &= \tau_{zy} = \mu\left(\frac{\partial v}{\partial z} + \frac{\partial w}{\partial y}\right)\\
\tau_{zx} &= \tau_{xz} = \mu\left(\frac{\partial w}{\partial x} + \frac{\partial u}{\partial z}\right)
\end{aligned} \tag{5}$$

$$\rho\frac{Du}{Dt} = X - \frac{\partial P}{\partial x} + \frac{\partial}{\partial x}\left\{\mu\left[2\frac{\partial u}{\partial x} - \frac{2}{3}\left(\frac{\partial u}{\partial x} + \frac{\partial v}{\partial y} + \frac{\partial w}{\partial z}\right)\right]\right\} + \frac{\partial}{\partial y}\left[\mu\left(\frac{\partial u}{\partial y} + \frac{\partial v}{\partial x}\right)\right] + \frac{\partial}{\partial z}\left[\mu\left(\frac{\partial w}{\partial x} + \frac{\partial u}{\partial z}\right)\right] \tag{6}$$

The shear stresses are proportional to the viscosity, in accordance with experience and intuition. However, the normal stresses also have viscosity-dependent components, not an intuitively obvious result. For flow problems in which the viscosity is vanishingly small, the normal stress component is negligible, but for fluid of high viscosity, eg, polymer melts, it can be significant and even dominant.

For some materials the linear constitutive relation of Newtonian fluids is not accurate. Either stress depends on strain in a more complex way, or variables other than the instantaneous rate of strain must be taken into account. Such fluids are known collectively as non-Newtonian. Many different types of behavior have been observed, ranging from fluids for which the viscosity in the Navier-Stokes equation is a simple function of the shear rate to the so-called viscoelastic fluids, for which the constitutive equation is so different that the normal stresses can cause the fluid to flow in a manner opposite to that predicted for a Newtonian fluid.

Frictionless Flow. There is actually a constitutive equation that is even simpler than that for the Newtonian fluid. This arises when internal fluid friction is neglected. Although this is not precisely true of a real fluid (with the possible exception of liquid helium[3] at temperatures below 2 K), in many situations the flows calculated under this assumption are very close to those actually observed, at least in a major portion of the flow field. Examples are high speed flows around obstacles in regions well away from solid boundaries or from wake regions. The importance of frictionless flow theory lies in the wide variety of solutions available, and in the powerful techniques of calculation available. Notable is Bernoulli's theorem. This states (to take a useful special case) that for irrotational, time-independent, barotropic (ie, density is a function of pressure) flow in a gravitational field, the quantity in equation 7 is constant along any streamline:

$$\int \frac{dP}{\rho} + \frac{V^2}{2g_c} + \frac{gh}{g_c} = \text{constant} \tag{7}$$

Equation 7 should not be construed to mean that the constant is the same for all streamlines. Once the velocities are known, this immediately gives the pressure. For a constant density, irrotational flow, the velocity components themselves, are derivable from a velocity potential, ϕ:

$$u = \frac{\partial \phi}{\partial x} \quad v = \frac{\partial \phi}{\partial y} \quad w = \frac{\partial \phi}{\partial z} \tag{8}$$

where

$$\frac{\partial^2 \phi}{\partial x^2} + \frac{\partial^2 \phi}{\partial y^2} + \frac{\partial^2 \phi}{\partial z^2} = 0 \tag{9}$$

Equation 9 is Laplace's equation and occurs also in several other fields of mathematical physics. Where the flow problem is two-dimensional, the velocities are also derivable from a stream function, ψ.

$$u = \frac{\partial \psi}{\partial y} \quad v = -\frac{\partial \psi}{\partial x} \tag{10}$$

where

$$\frac{\partial^2 \psi}{\partial x^2} + \frac{\partial^2 \psi}{\partial y^2} = 0 \tag{11}$$

Both equation 11 and the two-dimensional counterpart of equation 9 can be solved by several standard mathematical techniques, one of the more useful being that of conformal mapping. A numerical solution is often more practical for complicated configurations.

Flow Phenomena

Flow Past Bodies. A fluid moving past a surface of a solid will exert a drag force on it. This force is usually manifested as a drop in pressure in the fluid. Locally, at the surface, the pressure loss stems from the stresses exerted by the fluid on the surface and the equal and opposite stresses exerted by the surface on the fluid. Both shear stresses and normal stresses can contribute; their relative importance depends upon the shape of the body and the relationship of fluid inertia to the viscous stresses, commonly expressed as a dimensionless number called the Reynolds number, $LV\rho/\mu$. The same phenomena which affect drag also affect heat and mass transfer to the surface (see Heat exchange technology; Mass transfer). Because of the importance of flows around bodies, a relatively detailed description of such flows and their associated pressure changes is given.

Flow Along Smooth Surfaces. When the flow is entirely parallel to a smooth surface, eg, in a pipe far from the entrance, only the shear stresses contribute to the drag, the normal stresses are directed perpendicular to the flow. The shear stress is usually expressed in terms of a dimensionless friction factor:

$$f = \frac{2\tau_w g_c}{\rho \overline{V}^2} \tag{12}$$

Two distinct regimes of flow are observed in a pipe. At low velocities the flow is laminar, so-called because the fluid flows in concentric cylindrical sheaths or *laminae* that do not mix with each other. A stream of dye introduced into the fluid proceeds down the pipe as a thread, spreading only slightly by molecular diffusion. The velocity profile is parabolic, with the velocity decreasing from twice the average velocity at the axis to zero at the wall. This is easily demonstrated by solving the Navier-Stokes equations, which take on a rather simple form in this configuration (see eqs. 30–31). The shear stress is found to be proportional to the velocity and to the viscosity of the fluid, ie, as fluid velocity rises, the friction factor drops in inverse proportion. At some high Reynolds number ($D\overline{V}\rho/\mu$), typically about 2100 in commercial pipe, an abrupt change occurs. Random instabilities, which decay at lower velocities, now grow, destroying the laminar flow pattern. At a Reynolds number of ca 10,000 the transformation is very nearly complete. The velocity over most of the pipe becomes fairly uniform, about equal to the average velocity. Only in a thin region near the wall are significant radial velocity gradients observed. A more quantitative description can be obtained by dividing the field conceptually into three regions: a turbulent core in which momentum, heat and mass are readily transported radially by strong eddying motions; a thin, very nearly laminar, sublayer hugging the wall; and a somewhat thicker transition or buffer layer bridging the two. Within the laminar sublayer, viscous stresses are important and the steady-state Navier-Stokes equations are valid. In the other regions they are inadequate. The fluctuating velocities in these regions can be described by the time-dependent Navier-Stokes equations, but the solution is enormously complex. (It has been estimated that modern computers would require a time equal to the age of the universe to adequately describe turbulent pipe flow at a Reynolds number of 10^7.) To avoid this complexity, the time-dependent equations can be averaged in the manner originally suggested by Reynolds. This yields equations that are similar to those for laminar flow, but which contain additional terms involving time-averaged products of the velocity fluctuations. Mathematically these terms can be treated as additional stresses and are termed the Reynolds stresses. Physically they can be thought of arising from the transport of momentum by the fluctuating eddies in a manner analogous to that by which viscous stresses arise from the transport of momentum by diffusing molecules. Much research in fluid mechanics has been devoted to developing physical models and mathematical descriptions for these stresses. Although some useful concepts have resulted, the solution of practical problems still rests heavily on empirical correlations developed for the configurations of interest.

As velocity continues to rise, the thicknesses of the laminar sublayer and buffer layers decrease, almost in inverse proportion to the velocity. The shear stress becomes almost proportional to the momentum flux (ρV^2) and is only a modest function of fluid viscosity. Heat and mass transfer to the wall, which formerly were limited by diffusion throughout the pipe, now are limited mostly by the thin layers at the wall. Both the heat and mass transfer rates are increased by the onset of turbulence and continue to rise almost in proportion to the velocity.

Figure 2 shows the friction factor both for smooth and roughened pipes. The higher frictional losses for rough pipes arise when the surface irregularities extend past the buffer zone and exert the higher drag characteristic of bluff bodies.

Flow Past Bluff Bodies. When a fluid passes around a bluff body, both the shear and normal stresses can contribute to the drag force exerted on the body. The force produced by the shear stresses is commonly referred to as skin drag to distinguish it from the form drag arising from an imbalance in normal stresses. Skin drag is dominant at low velocities, and form drag is dominant at high velocities. The combined drag force is usually expressed in terms of a dimensionless drag coefficient:

$$C_D = \frac{2Fg_c}{A_p \rho V_\infty{}^2} \tag{13}$$

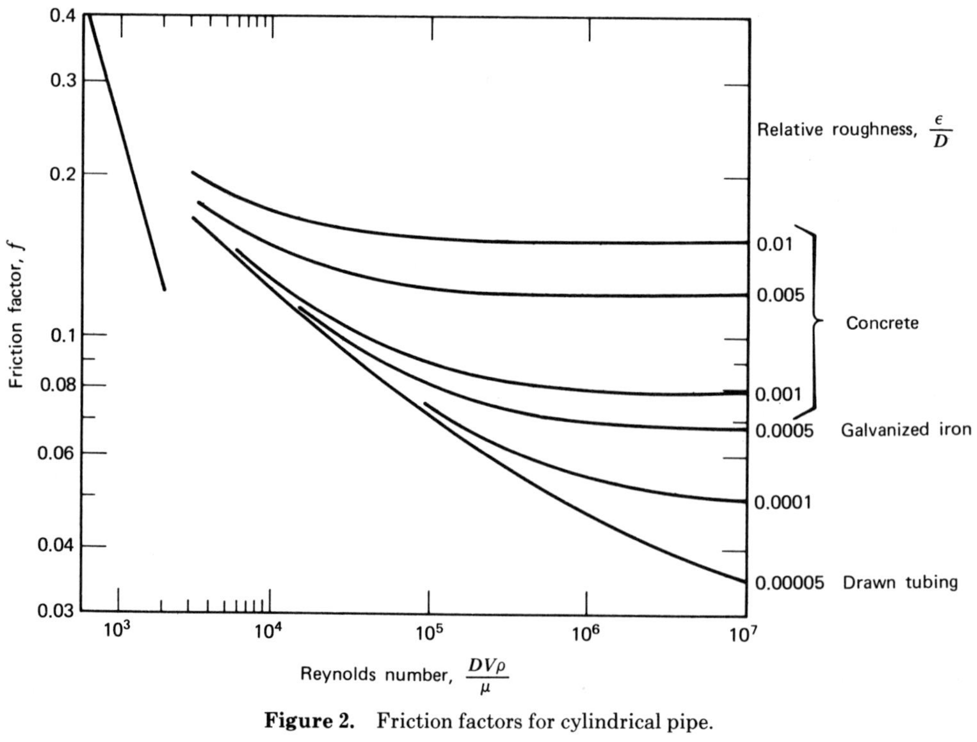

Figure 2. Friction factors for cylindrical pipe.

Flow around a cylinder with axis normal to the flow provides a good example of the many unusual features of such flows. These are illustrated in Figure 3.

At very low velocities the flow around the cylinder is symmetrical front-to-back as well as side-to-side. Fluid passing near the cylinder accelerates over the upstream face and decelerates over the downstream face. Normal stresses on these faces largely cancel each other and drag is due almost entirely to skin friction. Flow is laminar and the drag coefficient behaves like the friction factor in pipe flow, varying inversely with velocity. The drag force is directly proportional to the velocity and to fluid viscosity. As the velocity rises the flow becomes increasingly unsymmetrical. Under the influence of fluid inertia, the fluid streamlines crowd together on the upstream face of the cylinder and spread further apart on the downstream side. This imbalance is reflected in an imbalance in normal stresses and a growing importance of form drag. At a Reynolds number ($DV_\infty\rho/\mu$) of about 5, flow separation occurs. The fluid flowing next to the surface of the cylinder departs abruptly from the surface, the gap formed being filled by a pair of recirculating, laminar eddies in a bubble or wake behind the cylinder.

Separation is a result of the inability of the slowly moving fluid near the surface to move against an increasing pressure gradient. As the bulk of the fluid passes over the forward face of the cylinder and accelerates, it decreases in pressure. The fluid close to the surface, the boundary layer, also accelerates, although by a much smaller amount because of the retarding effects of viscous stresses. Over the rear face of the cylinder, the bulk fluid decelerates and pressure rises. This pressure gradient decelerates the already slowly moving fluid near the boundary, brings it to rest, and, as pressure continues to rise, reverses the flow at the surface and lifts the boundary layer (Fig. 4). The position at which the velocity becomes zero is the point of separation.

As the Reynolds number rises above about 40, the wake begins to display periodic instabilities, and the standing eddies themselves begin to oscillate laterally and to shed some rotating fluid every half cycle. These still laminar vortices are convected downstream as a vortex street. The frequency at which they are shed is normally expressed as a dimensionless Strouhal number which, for Reynolds numbers in excess of 300, is roughly constant:

$$Sr = \frac{n_s D}{V_\infty} \sim 0.2 \qquad (14)$$

The periodic shedding produces lateral forces of the same period on the cylinder. Should the cylinder be weakly supported and have a natural frequency close to the shedding frequency, it will oscillate strongly in concert with the vortex street. Such behavior is responsible for the singing of power lines, the oscillation of tall smokestacks and, most spectacularly, for the collapse in 1940 of the newly-built Tacoma Narrows suspension bridge under the influence of a steady 65 km/h wind.

Above a Reynolds number of about 100, the standing eddies are no longer apparent. The vortex street undergoes a laminar-to-turbulent transition, becoming fully turbulent at a Reynolds number of about 300. The drag coefficient has become nearly independent of Reynolds number. The pressure on the downstream side of the cylinder is almost constant, roughly equal to the low value at the point of separation. This results in a considerable imbalance of pressure between the upstream and downstream surfaces. Skin friction now provides only a minor part of the total drag. No further

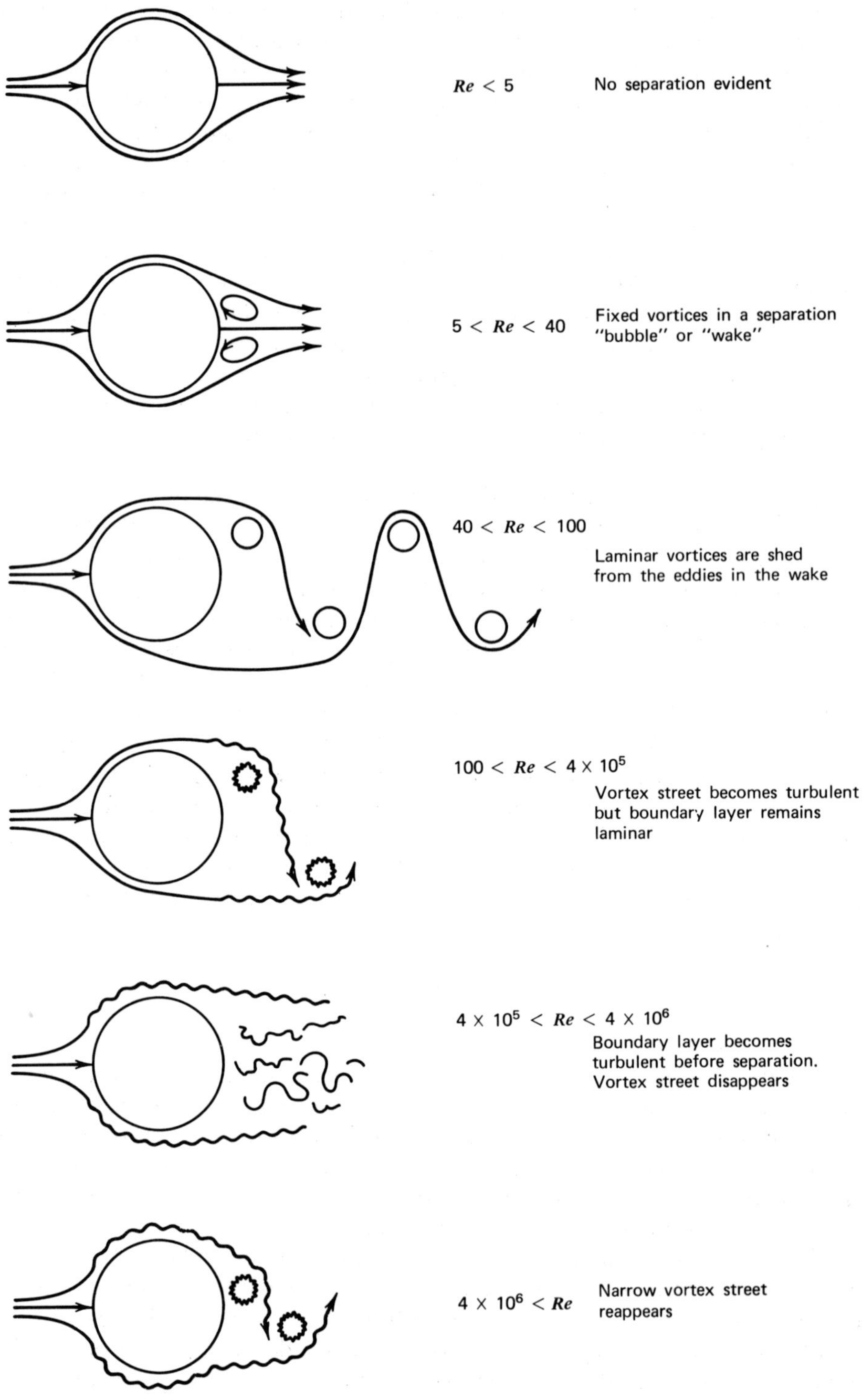

Figure 3. Flow past a circular cylinder (3).

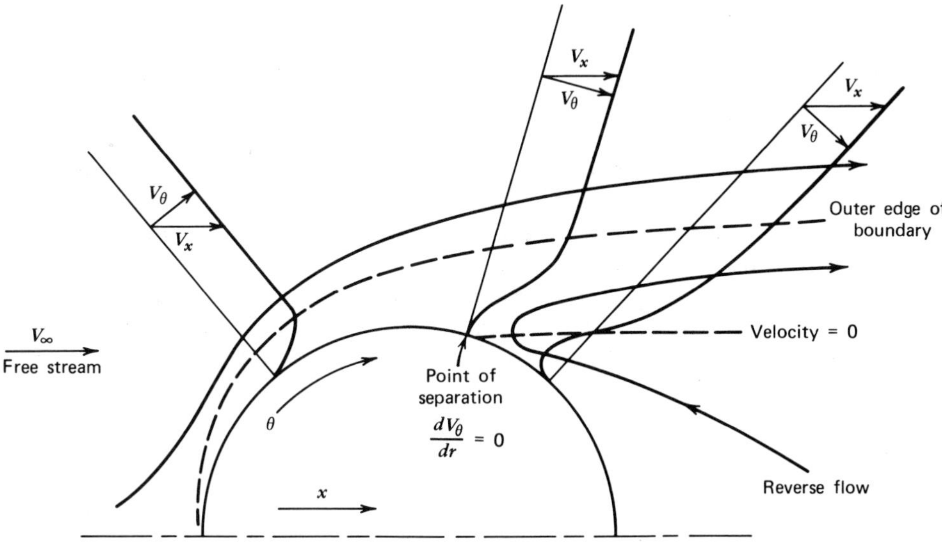

Figure 4. Boundary layer development around a circular cylinder.

changes occur until the Reynolds number reaches about 4×10^5, at which point the drag coefficient undergoes a sudden downward step change so large that the drag force actually decreases. This sudden change is caused by the development of turbulence within the boundary layer. Although the wake has been turbulent for some time, flow around the cylinder has remained laminar up to the point of separation. The onset of turbulence, by allowing a freer exchange of momentum between faster and slower moving regions at the surface, stabilizes the boundary layer against decelerations and postpones its separation. The separation point, which hitherto has been about 82° from the forward stagnation point of the cylinder, moves to about 110° and the partial recovery of pressure over the downstream surface reduces the net drag force. The onset of turbulence in the boundary layer is accompanied by the disappearance of the vortex street and its replacement by a narrower, more disorganized wake. The vortex street does not reappear until the Reynolds number exceeds 4×10^6. The laminar-to-turbulent transition can be initiated at somewhat lower Reynolds numbers by roughening the surface, eg, with sand or a trip wire. Sucking fluid through the surface can also postpone separation and reduce drag.

The behavior of the boundary layer also affects local heat transfer to the surface. At low Reynolds numbers, the thin boundary layer on the upstream face of the cylinder exhibits a higher heat transfer coefficient than does the thickened layer on the downstream face. Despite the onset of separated flow and the large energy losses in the wake, the upstream coefficient remains larger until the Reynolds number reaches 5×10^4. Above this point the washing of the surface by the shedding eddies makes the downstream coefficient larger. The point of maximum heat transfer occurs at the forward stagnation point (0°) for Reynolds numbers up to 5×10^4, at the rear stagnation point (180°) for Reynolds numbers between 5×10^4 and 2×10^5, and at about the 115° position for Reynolds numbers above 2×10^5. Minimum heat transfer always occurs near the point of separation.

Although the above description has concentrated on separation from a smooth surface, separation also occurs at sharp edges. Where separation is undesirable for a process reason, it can often be eliminated by redirecting the flow with turning vanes, ie, forcing it to hug the surface.

The Boundary Layer. At several points in the previous descriptions passing reference was made to the boundary layer—the fluid flowing close to the surface. The concept of the boundary layer was developed principally by Prandtl to facilitate the solution of the equations of motion in the presence of large velocity gradients perpendicular to the flow (4). It expresses the observation that the flow field often can be divided into two reasonably well-defined regions: a thin region close to the surface, the boundary layer, in which the gradient of tangential velocity is large and shear stresses are important; and a region outside the boundary in which velocity gradients are small and the potential flow solutions of inviscid flow theory, eg, Bernoulli's equation, are very nearly valid.

The transition between the two regions is gradual, not abrupt. The edge of the boundary layer is taken to be the point at which the velocity differs from that for inviscid flow by some small amount, say one percent. This approximation allows the main flow field and the normal stresses on the surface to be calculated by the many techniques applicable to Laplace's equation, with the boundary layer calculation then appended to yield the shear stresses at the surface. Extensive treatises have been devoted to analyzing flows in terms of boundary layer phenomena, particularly the growing boundary layers of entrance flows, in which a previously undisturbed fluid impinges on an obstruction (5). Although boundary layer analysis is widely applicable, it can fail badly in regions of strong deceleration where the boundary can separate under the influence of the unfavorable pressure gradient imposed by the main flow.

Entrance and Exit Effects. When a flowing fluid encounters an obstruction, the flow is disturbed and a new, steady-flow pattern is not reestablished until some distance downstream of the obstruction. The effects most commonly encountered are associated with the flow of fluids through ducts in which sudden changes in cross section occur. In Figure 5**a**, flow through a sudden contraction, except possibly at very low Reynolds numbers ($D\overline{V}\rho/\mu$), is normally accompanied by separation. The inertia of the converging flow causes it to contract to a smaller area than that of the duct. The area at the *vena contracta* (point of minimum area) depends upon the geometry, but for turbulent flows it is ca 60% of the duct's area. Flow down to the *vena contracta* is substantially frictionless—Bernoulli's equation can be used to describe the pressure change. Expansion beyond the *vena contracta* is not frictionless. The overall irreversible loss of energy depends on the geometry but is about $\overline{V}_2^2/4g_c$ for turbulent flow through a large, abrupt change in area. Where separation is undesirable it can be eliminated by rounding the inlet to the downstream duct.

Entrance flow is also accompanied by the growth of a boundary layer (Fig. 5**b**). As the boundary layer grows to fill the duct, the initially flat velocity profile is altered to yield the profile characteristic of steady-state flow in the downstream duct. For laminar flow in a tube, the distance required for the velocity at the center line to reach 99% of its asymptotic value is given by:

$$\frac{L}{D} = 0.057\left(\frac{D\overline{V}\rho}{\mu}\right) \tag{15}$$

For turbulent flow the entrance length is 50–100 pipe diameters. Throughout the

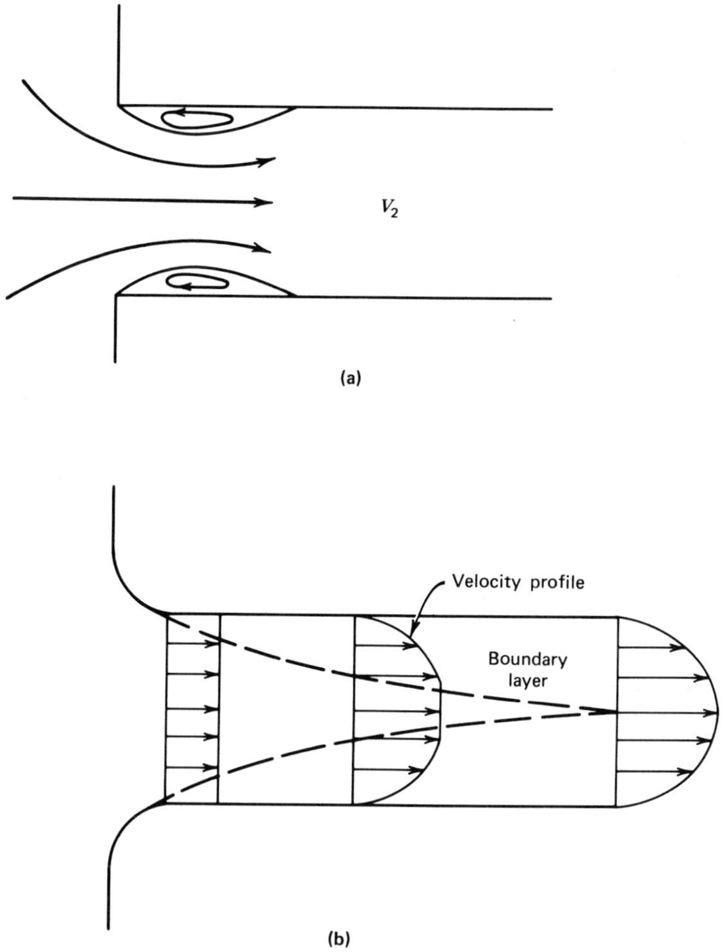

Figure 5. Entrance flows in a tube or duct. (**a**) Separation at sharp edge. (**b**) Growth of a boundary layer (illustrated for laminar flow).

entrance region the velocity gradient at the wall is larger than for the fully developed profile, producing a higher pressure drop and greater heat transfer per unit length than for fully developed flow. Although entrance phenomena are often ignored, they can be of significance, especially in capillary viscometers and compact (short tube) heat exchangers.

When a flow expands abruptly, separation will occur producing a jet of fluid flanked by recirculating eddies. For turbulent flows separation will occur in a diverging duct when the angle of divergence exceeds about 7°. When separation occurs, the ability of the fluid to recover pressure upon deceleration is seriously impaired since the kinetic energy is lost to friction. When the ratio of duct areas is large, all of the kinetic energy is lost and no pressure is recovered. Under these circumstances, the loss of energy per mass of fluid is $\overline{V}_1^2/g_c$ for laminar flow and $\overline{V}_1^2/2g_c$ for turbulent flow, where $\overline{V}_1$ is the average velocity in the upstream section of the duct.

When the entrance flow is bounded on only one side, the boundary layer can, in theory, grow indefinitely (albeit slowly) into the surrounding fluid. Flow past a flat plate illustrates the behavior. As the fluid passes the leading edge of the plate a boundary layer begins to grow. Initially this layer is laminar and its growth can be calculated easily by solving the Navier-Stokes equations. At a length defined by a Reynolds number ($LV\rho/\mu$) of about 4×10^5, the laminar boundary has grown so thick that it is unstable to fluctuations. Laminar flow disintegrates into vortices that break down into the random eddies of turbulence. This process is almost complete when the Reynolds number reaches ca 4×10^6. Turbulence allows a much freer transfer of momentum within the fluid; consequently, transition is accompanied by a considerable increase in the thickness of the boundary layer. The turbulent boundary layer is similar to that noted for turbulent flow in pipes, consisting of a laminar sublayer, a transition zone, and a much thicker turbulent zone. In both cases the velocity profiles and drag forces can be described by the so-called universal velocity distribution (5). Table 1 illustrates the boundary layer properties for a 100 m/s flow of air past a thin wing.

Flow Past Deformable Bodies. The flow of fluids past deformable surfaces is often important, eg, contact of liquids with gas bubbles or with drops of another liquid. Proper description of the flow must allow for both the deformation of these bodies from their shapes in the absence of flow and for the internal circulations that may be set up within the drops or bubbles in response to the external flow. Deformability is related to the interfacial tension and density difference between the phases, and internal circulation is related to the drop viscosity. A proper description of the flow involves not only the Reynolds number ($dV\rho/\mu$) but also other dimensionless groups, eg, the viscosity ratio (μ_d/μ_c), the Eötvos number ($g\Delta\rho d^2/\sigma$), and the Morton number ($g\mu_c^4\Delta\rho/\rho_c^2\sigma^3$) (6).

Where surface-active agents are present, the notion of surface tension and the description of the phenomena become more complex (see Surfactants and detersive systems). As fluid flows past a circulating drop (bubble), fresh surface is created continuously at the nose of the drop. This fresh surface can have a different concentration of agent (hence a different surface tension) than surface further downstream that was created earlier. Neither of these values need equal the surface tension developed in a static, equilibrium situation. A proper description of the flow under these circumstances involves additional dimensionless groups related to the concentrations and diffusivities of the surface active agents.

Because of their high buoyancy and frequently large size, gas bubbles rising in a liquid can deviate greatly from spherical shapes. Figure 6 illustrates the behavior observed. In pure liquids small bubbles rise faster than would be predicted from the

Table 1. Boundary Layer Properties for 100 m/s Flow of Air

Property	Value
distance to laminar turbulent transition, cm	6.0
thickness of laminar boundary layer at transition, cm	0.05
distance to complete transition, cm	60.0
thickness of turbulent boundary layer, cm	1.1
laminar sublayer, cm	0.002
buffer layer, cm	0.010
velocity at edge of laminar sublayer, m/s	19
buffer layer, m/s	53

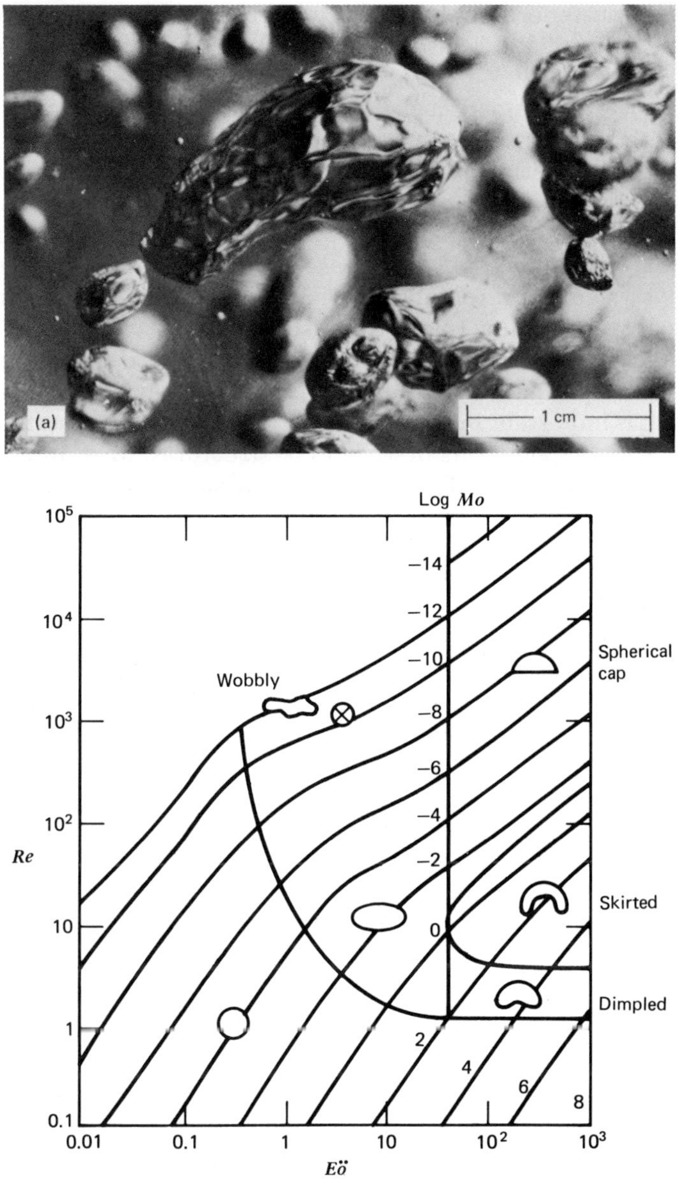

(b)

Figure 6. Shapes of drops and bubbles. (**a**) Bubble rising in sparged tower system: air-water. Courtesy of Shell Development Company. (**b**) Bubble and droplet regimes as described by Grace (6); ⊗ represents 0.5 cm gas bubble rising in water at 20°C.

drag correlations developed for solid spheres because internal circulation permits a higher fluid velocity at the surface and less drag than for the corresponding solid. Very large bubbles rise much more slowly than do undeformed spheres of the same volume. This is caused by deformation to a shape of large frontal area but small thickness. Very small bubbles obey Stokes law for solid (immobile) surfaces. This behavior is reflected also in the mass transfer coefficients for such bubbles, which are lower than would be expected from the correlations developed for larger, circulating bubbles (see Fluidization).

Free Flows—Jets, Wakes, and Plumes. When a smoothly flowing stream is disturbed by inserting a solid body or by injecting additional fluid, the disturbance is confined to a narrow region downstream of the initial point. When the disturbance is caused by the injection of a fluid at high velocity, the disturbed region is called a jet. When the velocity is lower than that in the main stream, the disturbance is termed a wake. The term plume is usually applied to a case in which an additional effect, such as buoyancy, plays a significant role and for which the turbulence characteristics of the main stream play the deciding role in its spread.

Jets are used industrially to perform many mixing operations ranging from gradual mixing of tanks to the rapid mixing needed in chemical reactors or flames. Mixing in turbulent wakes is important in the disposal of wastes at sea and in the dispersion of potential atmospheric pollutants downwind of structures. Most interest in plumes is centered around the behavior of gases emitted from smokestacks into the atmosphere (see Air pollution control methods).

The development of the flow field around a jet or wake may be regarded as a boundary layer phenomenon in which the jet grows by entraining fluid from its surroundings. The boundary in this case is the envelope over which the velocity is some small fraction (ca 1%) of the velocity of the jet along its center line. Jets may be either laminar or turbulent, although most important applications involve turbulent jets, which begin to form when the Reynolds number ($D\overline{V}\rho/\mu$) of the inlet stream exceeds a few hundred. The flow field around such a jet can be divided into three main regions (Fig. 7) (7). In the entry region the jet displays a diminishing core in which the inlet velocity profile is preserved. This is analogous to the entry region of a pipe but there is no pressure drop and no acceleration of flow. The entry region disappears about five nozzle diameters past the inlet. A transition region extends an additional two nozzle diameters, after which the jet enters the similarity region. In this region the velocity profile throughout the jet follows a similarity relationship:

$$\frac{u}{u_c} \cong e^{-4.6(y/\delta)2} \tag{16}$$

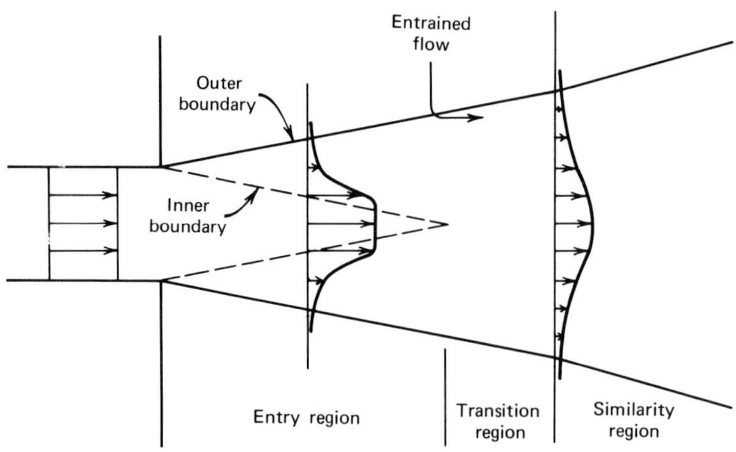

Figure 7. Development of a jet.

It has been experimentally verified that, initially, jets spread roughly linearly in the similarity region, ie, $\delta \alpha (x - x_0)$ where x_0 is the virtual point source of the jet. For both axially symmetric and plane parallel (slot) jets, the jet width parameter, δ, is given by:

$$\delta = 0.25x \tag{17}$$

In a free jet the absence of a pressure gradient makes the momentum flux at any cross section equal to the momentum flux at the inlet, ie, equations 16 and 17 define jet velocity at all points. For a cylindrical jet this leads to a center-line velocity that varies inversely with $(x - x_0)$, whereas for slot jets it varies inversely with the square root of $(x - x_0)$. As the jet proceeds still further downstream the turbulent entrainment initiated by the jet is gradually subordinated to the turbulence level in the surrounding stream and the jet, as such, disappears.

Mass and heat transport within the jet follow the same general pattern as does momentum transport. For gases it is found experimentally that the thermal or concentration jet spreads somewhat faster and decays more rapidly than does the momentum jet. All can be described by the turbulent equations of motion and useful results can be obtained by using eddy diffusivities, which are nearly constant across the jet but decay with time (distance downstream). All of the jets are gradually swallowed up in the surrounding fluid. The concentration jet, because of its association with a particular chemical species, remains identifiable longest, even though its behavior becomes completely dominated by the surrounding fluid. An important example of such a concentration jet is the plume, as might be released from a smokestack or vent. Except right at the source the behavior of the plume is dominated by its buoyancy and/or the turbulence of the fluid into which it is injected. Figure 8 describes several of the plume patterns that can be obtained as they are affected by the temperature gradient in the surrounding air. The quantitative description of the dispersion of such plumes, especially their ground level concentration, is still imperfectly developed and one must usually resort to empirical correlations. Where the effects of buoyancy are subordinate to the effects of wind velocity, useful results for dispersion over complex terrain or around structures can be obtained by wind tunnel experiments using scale models (see Modeling).

Most of the remarks above refer to unconfined or free flows. Many industrial applications involve the use of confined jets. It is customary to consider a jet confined when the ratio of the confinement radius to the source radius lies in the range 4–100. Below a ratio of 2, the jet does not develop its similarity profile before striking the wall whereas above a ratio of 100 the jet itself may usually be considered free. Under certain conditions, flow in confined jets is accompanied by the existence of a recirculation zone which significantly affects the jet behavior by returning material upstream (9). This recirculation can be particularly important in combustion processes.

Other Flow Phenomena. ***Compressible Flow.*** The flow of easily compressible fluids, ie, gases, exhibits features not evident in the flow of substantially incompressible fluid, ie, liquids. These differences arise because of the ease with which gas velocities can be brought to or beyond the speed of sound and the substantial reversible exchange possible between kinetic energy and internal energy. The Mach number, the ratio of the gas velocity to the local speed of sound, plays a central role in describing such flows.

Consider the converging-diverging nozzle described in Figure 9 and, for simplicity,

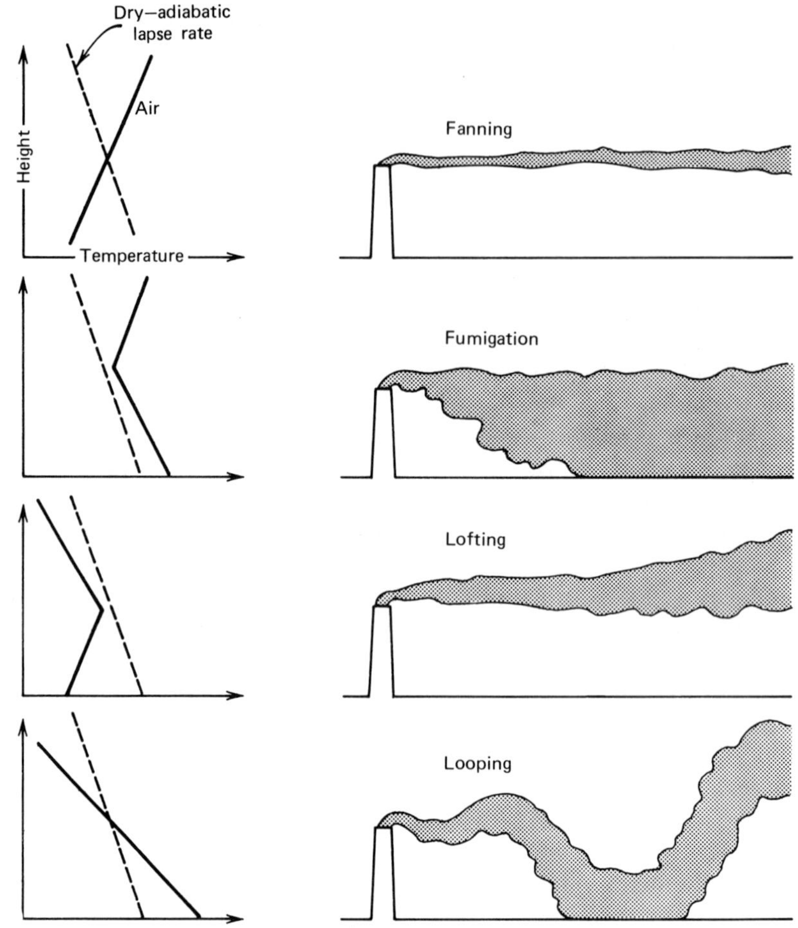

Figure 8. Characteristic plume patterns. Courtesy of *Scientific American* (8).

consider only frictionless, adiabatic, one-dimensional flow. When the pressure in the downstream reservoir is reduced, flow commences. In the converging section, velocity increases with distance and both pressure and temperature fall in accordance with the overall energy balance and the equation of state. In the diverging section, the fluid slows and some pressure is recovered B. As the downstream pressure is reduced further, the behavior remains qualitatively the same until a point is reached at which the velocity in the nozzle throat equals the local speed of sound. At this point it is found that there are two possible paths that the downstream flow can follow and remain thermodynamically reversible. One corresponds to recompression and deceleration to subsonic flow C, the other to further expansion and acceleration to supersonic flow G. Operation at an intermediate pressure in the reservoir D is possible only by sacrificing thermodynamic reversibility. Under these circumstances a standing, normal shock front is set up at some point in the diverging section. At this front, pressure, temperature, and entropy rise, and velocity drops abruptly over a distance of a few molecular mean-free paths. Flow changes from supersonic upstream of the shock to subsonic downstream. The relationship between the two conditions is established by conservation of energy and by conservation of momentum across the shock front.

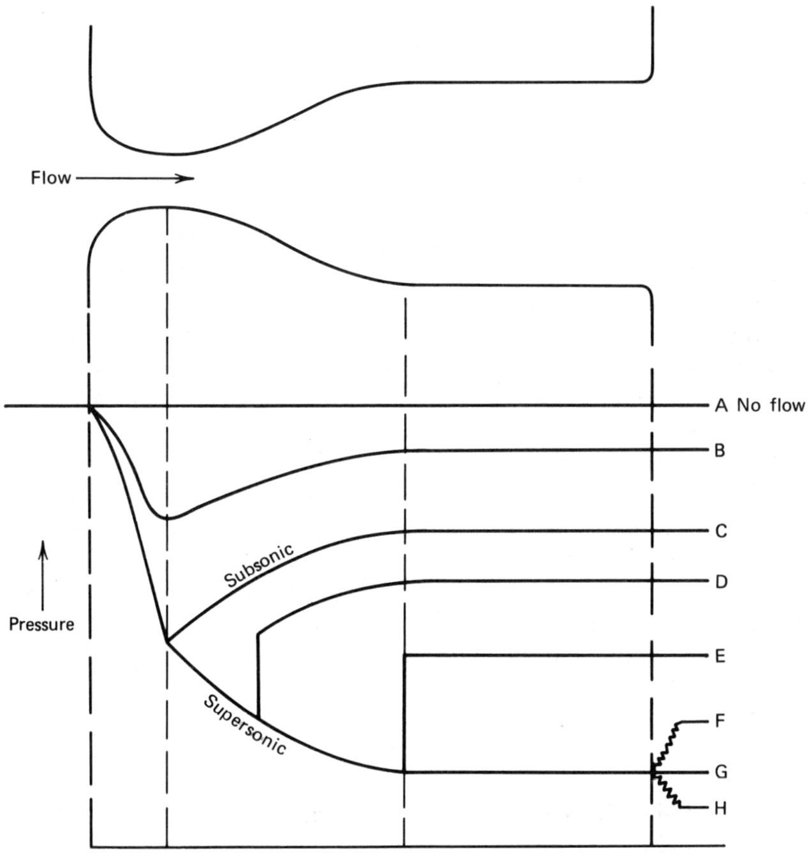

Figure 9. Compressible flow in a converging-diverging nozzle.

Further reductions in reservoir pressure move the shock front downstream until it reaches the outlet of the nozzle E. If the reservoir pressure is reduced further, the shock front is displaced to the end of the tube, and is replaced by an oblique shock F, no pressure change G, or an expansion fan H at the tube exit. Flow is now thermodynamically reversible all the way to the tube exit and is supersonic in the tube. In practice, frictional losses limit the length of the tube in which supersonic flow can be obtained to no more than 100 pipe diameters.

While the behavior downstream of the throat is displaying these complexities, nothing is changing upstream. Once the velocity at the throat becomes sonic, no downstream perturbation, eg, a sound wave, can be propagated upstream. Throughout the change from C to H the mass rate of flow remains constant at the value first established at C. An increase in flow can be accomplished only by raising the upstream pressure. In the absence of any temperature change in the upstream reservoir, the increase in flow results solely from a change in fluid density, not from a higher velocity in the throat. This isolation of flow rate from downstream effects can be employed to meter gases through orifices or nozzles at a constant rate, irrespective of downstream fluctuations. For ideal gases and where the upstream velocity is low, sonic flow through the throat is established whenever the ratio of upstream to downstream pressure across the constriction exceeds the quantity:

$$\left(\frac{\gamma + 1}{2}\right)^{\gamma/\gamma-1} \text{ where } \gamma = \frac{C_p}{C_v} \tag{18}$$

For simple gases this ratio is ca 2.

Phenomena analogous to shock waves in gases can occur in open-channel flow of liquids. The Froude number, $\sqrt{V^2/gh}$, is the ratio of fluid velocity to the velocity of a small surface wave and plays the same role in open-channel flow as the Mach number does in compressible flow. Thus liquid flowing under a sluice gate is often discharged at a shallow depth at high velocity, corresponding to a Froude number greater than unity. At a distance downstream the liquid is observed to undergo a hydraulic jump to a greater depth at a slower velocity, for which the Froude number is less than unity. The decreased liquid momentum appears as greater pressure (depth). In contrast to the normal shock, a great deal of turbulence is generated at the jump because of the sudden lateral expansion, and energy is dissipated through friction.

Flow in Porous Media. Flow of fluids through fixed beds of solids occurs in situations as diverse as oil field reservoirs, catalyst beds, sand filters, and absorption towers. The complex interconnected pore structure of such systems makes it necessary to use simplified models to make practical quantitative predictions. One of the more successful treatments of single-phase pressure drop through such systems employs the results for flow through tubes, using average velocities and tube diameters. The average velocity through the pores is related to the superficial velocity (the apparent velocity based on the entire cross section) via:

$$\overline{V} = \frac{V_s}{\epsilon} \tag{19}$$

For spherical particles the average diameter of the pores, defined as four times the pore volume divided by the surface area, can be shown to be:

$$\overline{D} = \frac{6\epsilon}{1 - \epsilon} d_p \tag{20}$$

The overall pressure drop is expressed as the sum of a laminar term proportional to $\overline{V}/\overline{D}^2$ and a turbulent term proportional to $\overline{V}^2/\overline{D}$ to yield the Ergun equation:

$$\frac{\Delta P}{L} = A \frac{\mu V_s}{d_p{}^2} \frac{(1 - \epsilon)^2}{\epsilon^3} + B \frac{\rho V_s^2}{d_p} \frac{1 - \epsilon}{\epsilon^3} \tag{21}$$

For randomly packed spherical particles, the constants A and B have been determined experimentally to be 150 and 1.75, respectively. For nonspherical particles, equivalent spherical diameters are employed and additional corrections for shape are introduced.

When two phases are present the situation is quite complex, especially in beds of fine solids where interfacial forces can be significant. In coarse beds, eg, packed towers, the effects are often correlated empirically in terms of pressure drops for the single phases taken individually.

Non-Newtonian Fluids: Die Swell and Melt Fracture. As mentioned earlier, for many fluids the Newtonian constitutive relation involving only a single, constant viscosity is not accurate. Either stress depends in a more complex way upon strain or variables other than the instantaneous rate of strain must be taken into account. Such fluids are known collectively as non-Newtonian. They are usually subdivided further on the basis of their behavior in simple shear flow, ie, flow between sliding planes or, to a good approximation, between two mutually rotating cylinders. Figure 10 illustrates the behavior of several of these. The types illustrated are examples of the generalized Newtonian fluid. This simple generalization uses the Navier-Stokes equations but takes viscosity to depend in some way on a single quantity—the rate of strain.

Pseudoplastic fluids are the most commonly encountered non-Newtonian fluids. Polymeric solutions, some polymer melts, and suspensions of paper pulp are examples. In simple shear flow, the constitutive relation for such fluids is:

$$\tau_{yx} = K\left(\frac{\partial u}{\partial y}\right)^n \tag{22}$$

The apparent viscosity, defined as $\tau_{yx}/(\partial u/\partial y)$, drops with increased rate of strain. Dilatant fluids follow a constitutive relation similar to that for pseudoplastics except that their viscosities increase with increased rate of strain, ie, $n > 1$ in equation 22. Dilatancy is observed in highly concentrated suspensions such as quicksand or beach sand. In the unsheared state the particles of such suspensions fit together closely. Movement can be accomplished only by separating the particles, causing the suspension to dilate. Bingham fluids display a linear stress-strain curve similar to Newtonian fluids, but with a nonzero intercept termed the yield stress, (eq. 23):

$$\eta\left(\frac{\partial u}{\partial y}\right) = \tau_{yx} - \tau_o, \text{ when } \tau_{yx} \geqslant \tau_o \tag{23}$$

$$\frac{\partial u}{\partial y} = 0, \text{ when } \tau_{yx} < \tau_o$$

The coefficient η is termed the modulus of rigidity. The viscosities of thixotropic fluids fall with time when subjected to a constant rate of strain, but recover upon standing. This behavior is associated with the reversible breakdown of structures within the fluid which are gradually reestablished upon cessation of shear. The smooth spreading of paint following the intense shear of a brush or spray is an example of thixotropic behavior. When viscosity rises with time at constant rate of strain, the fluid is termed

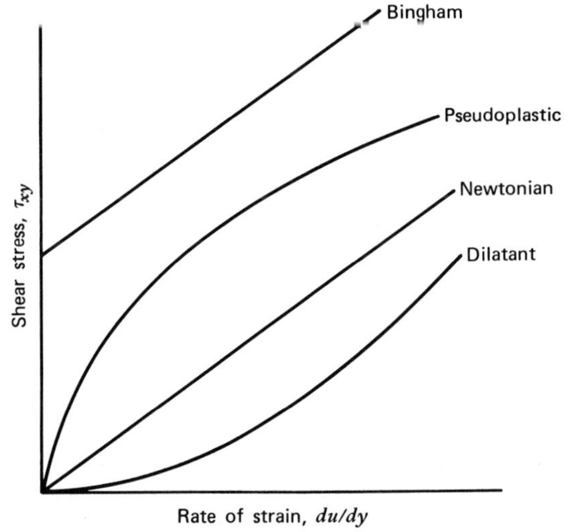

Figure 10. Fluid behavior in simple shear flow.

rheopectic. This behavior is much less common but is found in some clay suspensions, gypsum suspensions, and certain sols.

When these relatively simple fluids are conveyed in laminar flow in pipes, their behaviors can be deduced directly and quantitatively from their constitutive relations and the linear variation of shear stress with radius described in equation 51. Pseudoplastic fluids display lower viscosities in the high-shear region on the wall than at the pipe axis. Their velocity profiles are steeper at the wall and flatter at the axis than the parabolic shape developed by Newtonian fluids. Dilatant fluids show opposite behavior, tending to freeze at the wall and finger along the axis. Bingham fluids show no velocity gradient up to the radius at which the yield stress is reached, giving the impression of a solid core of material moving along the axis. Pipelines conveying thixotropic and rheopectic fluids are usually designed on a most viscous basis. For thixotropic fluids this corresponds to viscosity at time zero, but for rheopectic fluids the viscosity after extended shear is used.

The pressure drop accompanying pipe flow of such fluids can be described in terms of a generalized Reynolds number, which for pseudoplastic or dilatant fluids takes the form:

$$Re = \frac{D^n \overline{V}^{2-n} \rho}{8^{n-1} K \left(\dfrac{3n+1}{4} \right)^n} \tag{24}$$

The transition from laminar to turbulent flow occurs at Reynolds numbers varying from ca 2000 for $n \geq 1$ to ca 5000 for $n = 0.2$. In the laminar region the Fanning friction factor, Figure 2, is identical to that for Newtonian fluids. In the turbulent region the friction factor drops significantly with decreasing values of n, producing a family of curves.

In configurations more complex than pipes, eg, flow around bodies or through nozzles, additional shearing stresses and velocity gradients must be accounted for. More general equations for some simple fluids in laminar flow are described in reference 1.

Many industrially important fluids cannot be described in the simple terms presented above. Viscoelastic fluids are prominent offenders. These fluids exhibit memory. Such fluids flow when subjected to a stress, but recover part of their deformation when the stress is removed. Polymer melts and flour dough are typical examples. Both the shear stresses and the normal stresses depend on the history of the fluid. Even the simplest constitutive equations are complex, as exemplified by the Oldroyd expression for shear stress at low shear rates:

$$\tau + \lambda_1 \frac{d\tau}{dt} = \frac{\mu}{g_c} \left(\frac{d\dot{\gamma}}{dt} + \lambda_2 \frac{d^2\dot{\gamma}}{dt^2} \right) \tag{25}$$

The relaxation times, λ_1 and λ_2, describe, respectively, the times required to relieve stress on the cessation of strain and to relieve strain on the cessation of stress. The full Oldroyd tensor requires knowledge of eight material properties.

The development of significant normal stresses in such fluids can lead to bizarre behaviors when they are deformed. Stirred viscoelastic fluids can move radially inward and climb the stirring shaft (the Weissenberg effect) if the inward-directed normal stresses accompanying shear are greater than the centrifugal forces developed by the stirring. When viscoelastic polymer melts are extruded through dies, they swell, recovering in part the lateral dimensions they possessed upstream of the die. At some critical rate the polymer undergoes melt fracture. Stress is relieved in an unsymmetrical manner and the extruded rod takes on a screw-like form. As the flow increases, additional modes of instability appear and the rod takes on progressively more complex shapes.

Gas–Liquid Flow. When two or more fluids flow together, a much greater range of phenomena occurs as compared with flow of a single phase. In a conduit many of the technically significant phenomena have to do with the positions assumed by the phase boundaries, and these are governed by the flow conditions rather than by the walls of the conduit. In addition to the densities and viscosities, surface properties can be important. As might be expected, methods for quantitative calculation are poorly developed in comparison to those for single-phase flow. Consider, for example, pressure drop for fully developed, steady state, incompressible flow in smooth conduits. For single-phase flow, dimensional analysis shows that only two dimensionless numbers are involved. These are the Fanning friction factor $\tau_w/(\rho \overline{V}^2/2g_c)$ and the Reynolds number. A set of experiments in which only two measurable quantities are varied suffices to establish the desired empirical correlation. For fully developed, gas–liquid flow in horizontal pipes alone, there are at least six such groups and a six-dimensional space must be mapped. This has not yet been fully accomplished although thousands of experimental measurements have been made. The methods available run the gamut from empirical correlations only weakly based, if at all, on an understanding of the phenomena to correlations based on conceptual models that allow one to use simplifying relations connecting the variables of the multi-dimensional problem. Even with the best correlations, however, errors of a factor of two are not uncommon. Much of the data on which available correlations are based was obtained in small pipes (<10 cm in diameter) using air and water. There is still much uncertainty as to how or even whether such data can be extrapolated to large pipes conveying fluids having significantly different properties.

Despite these difficulties, many useful calculations can be made, particularly when one is interested in determining the effects of changes in an already existing operation. All of these calculations depend on an accurate picture of the flow regime, a phrase used to denote the significant configurational characteristics of the flow, eg, unidirectional, recirculating, steady, unsteady, liquid-dispersed, gas-dispersed, etc. For illustration, consider the simultaneous flow of air and water at various velocities in a 2.5-cm horizontal pipe. It is desirable to know such things as which phase is continuous, the degree of dispersion or atomization and the presence or absence of flow surges. At low velocities for both phases, both phases are continuous and flow smoothly; the flow is stratified as sketched in the lower left-hand corner of Figure 11. As either the liquid flow rate or gas flow rate is raised, the flow becomes unstable by a Kelvin-Helmholtz instability. This causes waves to rise at the phase interface, producing wave-stratified flow. With further increase in liquid flow rate, the waves grow in amplitude and at some flow rate, touch the upper wall of the pipe, destroying the continuity of the gas phase. This is plug flow or slug flow, the former term connoting the more gentle action where the gas bubble moves forward without intense agitation and the latter expressing the more violent action observed at higher gas flow rates. As liquid rate is raised still further, the gas plugs break up into rather small bubbles. The flow

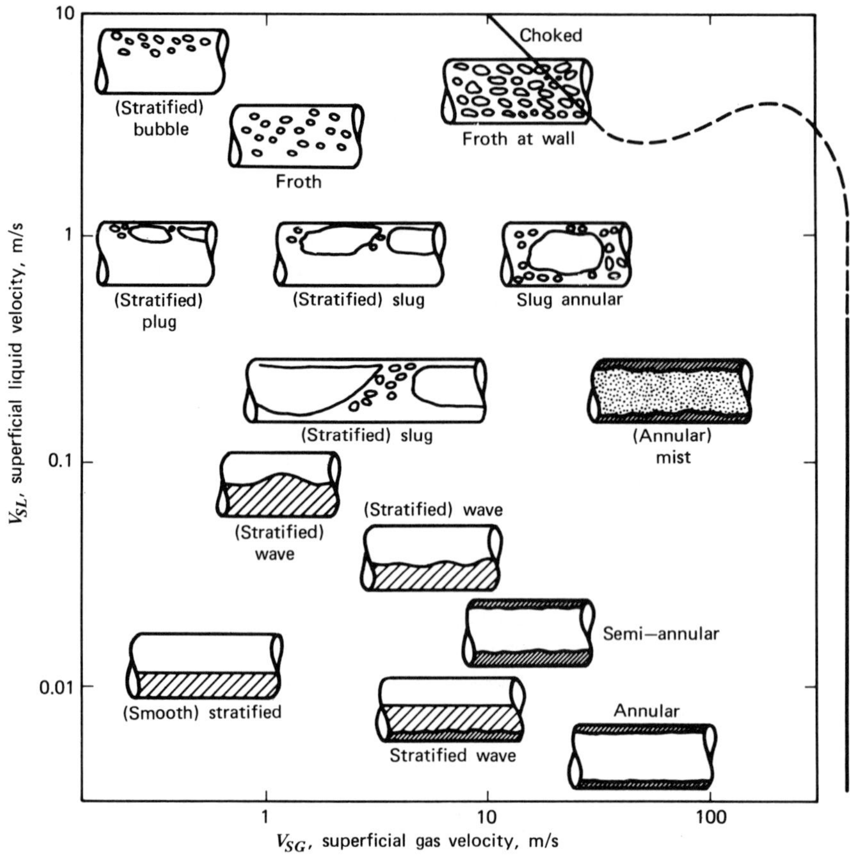

Figure 11. Flow regimes for air-water in a 2.5-cm horizontal pipe. Courtesy of Shell Development Company.

may then be described as bubble flow or froth flow according to whether the gas volume fraction is small or large, respectively (see Pipelines).

If the gas flow rate is increased, one eventually observes a phase transition for the above mentioned regimes. Coalescence of the gas bubbles becomes important and a regime with both continuous gas and liquid phases is reestablished, this time as a gas-filled core surrounded by a predominantly liquid annular film. Under these conditions there is usually some gas dispersed as bubbles in the liquid and some liquid dispersed as droplets in the gas. The flow is then annular. Various qualifying adjectives may be added to further characterize this regime. Thus, there are semiannular, pulsing annular, and annular mist regimes. Over a wide variety of flow rates, the annular liquid film covers the entire pipe wall. For very low liquid flow rates, however, there may be insufficient liquid to wet the entire surface, giving rise to rivulet flow.

At very high flow rates, phenomena occur that are equivalent to the sonic limited flow observed with gases. This is illustrated in Figure 11 by the envelope labeled choked flow. Without special effort, it is impossible to produce velocities in a pipe at rates higher than indicated by the envelope. It is especially noteworthy that the limiting gas velocity can be very low provided the liquid rate is high. Realization of this fact has often come as a surprise to a designer who has calculated higher velocities in some pipes.

In inclined or vertical pipes, the flow regimes are similar to those described for horizontal pipes when both gas and liquid flow rates are high. At lower flow rates, the effects of gravity are important and the regimes of flow are quite different. For liquid velocities near 30 cm/s and gas velocities near 150 cm/s, piston or vertical slug flow is observed (Fig. 12). Large gas bubbles, with approximately spherical top surfaces, bridge the pipe. The calculation of the rate of rise of such bubbles can be made fairly accurately by assuming frictionless flow near the bubble apex. The success of this rather simple calculation illustrates one of the advantages of having a firm knowledge of flow regimes.

The upward flow of gas and liquid in a pipe is subject to an interesting and potentially important instability. As gas flow increases, liquid holdup decreases and frictional losses rise. At low gas velocity the decrease in liquid holdup and gravity head more than compensates for the increase in frictional losses. Thus, an increase in gas velocity is accompanied by a decrease in pressure drop along the pipe, a potentially

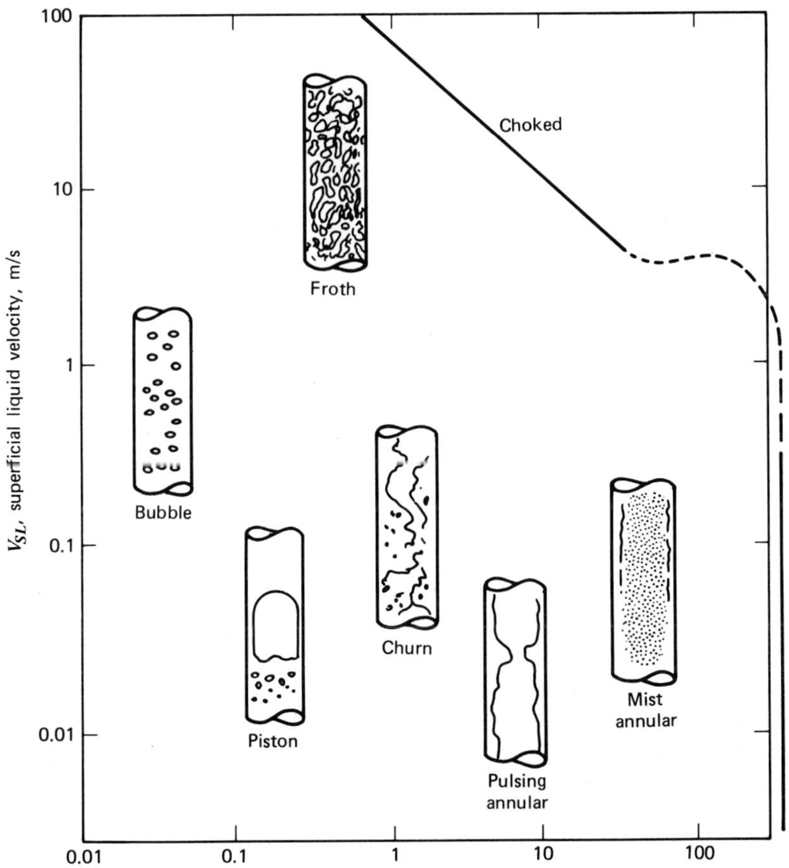

Figure 12. Flow regimes for air-water in a 2.5-cm vertical pipe. Courtesy of Shell Development Company.

unstable situation if the flows of gas and liquid are sensitive to the pressure drop in the pipe. Such a situation can arise in a thermosyphon reboiler which depends upon the difference in density between the liquid and a liquid–vapor mixture to produce circulation. The instability is manifested as cyclic surging of the liquid flow entering the boiler and of the vapor flow leaving it (see Piping systems).

Flow Instability. In many flow situations it is found that a mathematically valid solution to the Navier-Stokes equations is closely verified by experiment over some ranges of the variables, but when the variables are changed new flow patterns that are not in keeping with that solution are observed. The change is often rather sudden, eg, the laminar-turbulent transition. Such behavior occurs because solutions to the Navier-Stokes equations are generally not unique. When more than one solution exists, it is possible to observe one flow pattern under one set of circumstances and a different pattern under another. At the present stage of development of fluid mechanics, a decision as to whether a given solution applies can usually be made only by performing an experiment. In some cases, however, it is possible, by analyzing the equations of motion, to determine the criteria by which one flow pattern becomes unstable in favor of another. The mathematical technique used most often is linearized stability analysis, which starts from a known solution to the equations and then determines whether a small perturbation superimposed on this solution grows or decays as time passes.

Fluids in Motion. Many of the instabilities associated with fluids in motion are of the shear-flow type. In shear flow the velocity varies principally in a direction perpendicular to the flow direction, eg, pipe flow, boundary layer flow, jet flow, and wake flow. Such flows change from laminar to turbulent when the Reynolds number based on some characteristic length scale exceeds a critical value, which may be different for each flow configuration. In pipe flows and some types of boundary layer flows, the instability is usually referred to as Tollmien-Schlichting instability. This occurs in four steps: "First, small two-dimensional waves are formed and linearly amplified; second, the two-dimensional waves develop into finite three-dimensional waves and are amplified by nonlinear interactions; third, a turbulent spot forms at some localized point in the flow; and finally, the turbulent spot propagates until it fills the entire flow field with turbulence" (2). At low Reynolds numbers, viscosity damps out the instability. At high Reynolds numbers, however, viscosity provides the destabilizing mechanisms.

The phenomena are quite complex even for pipe flow. Efforts to predict the onset of instability have been made using linear stability theory. The analysis predicts that laminar flow in pipes is stable at all values of the Reynolds number. In practice, the laminar-turbulent transition is found to occur at a Reynolds number of about 2000, although by careful design of the pipe inlet it can be postponed to as high as 40,000. It appears that linear stability analysis is not applicable in this situation.

In jet flow and wake flow there occurs another type of instability called Kelvin-Helmholtz instability. This instability may be illustrated by considering the development of a small disturbance in the flow situation given in Figure 13. Suppose a small disturbance causes a slight waviness of the boundary between the two flows. The fluid on the convex sides of each flow moves slightly faster and that on the concave sides moves slightly slower. According to the Bernoulli equation, this disturbance decreases the pressure on the convex sides of each flow, and thus the initial disturbance is amplified. Kelvin-Helmholtz instability is also observed in horizonal, concurrent flows of stratified, immiscible fluids (Fig. 14). Raising of a wave at the interface forces the

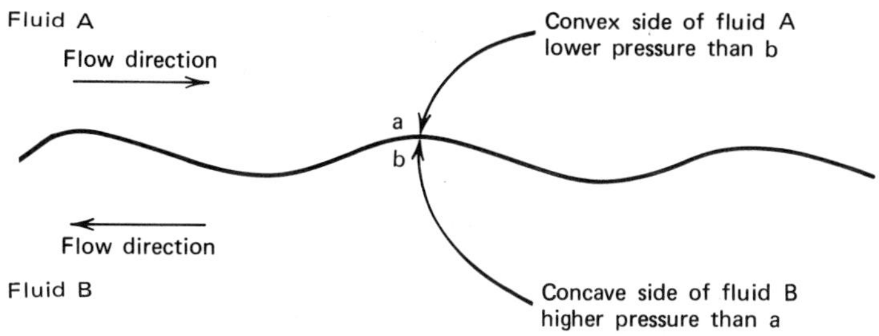

Figure 13. Sketch of Kelvin-Helmholtz instability.

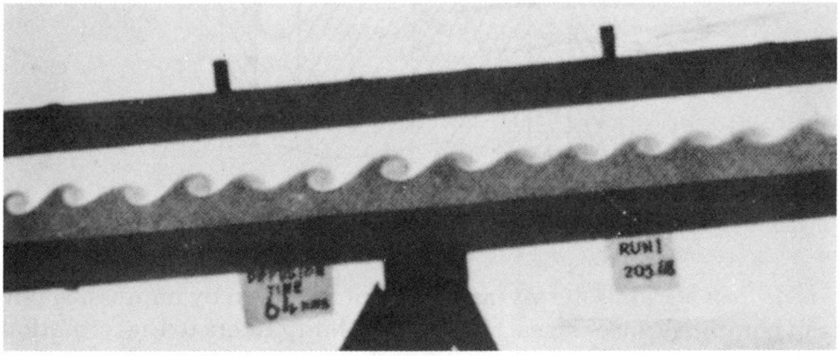

Figure 14. Shear instability in stably stratified fluid; lower-denser fluid is dyed. Courtesy of Cambridge University Press (10).

upper fluid to increase in velocity and drop in pressure as it passes the wave. If the pressure decrease is large enough, it overcomes the increase in potential associated with the raising of the wave and the disturbance grows.

An instability involving transition from one laminar flow pattern to another occurs in the flow between concentric, mutually rotating cylinders. When the inner cylinder is rotated at an angular velocity below a critical value, motion is purely circumferential (Couette flow). Above this value, however, centrifugal force destabilizes the flow and a series of laminar, cellular vortices known as Taylor cells is superimposed on the main flow (Fig. 15). The pattern of cells depends upon the geometry and whether the outer cylinder is stationary or also rotating. As the angular velocity is increased, the cells become wavy and then irregular. At still higher velocities the cells break up and the flow becomes turbulent. If only the outer cylinder is rotated, Taylor cells do not form, and this arrangement is customarily used in Couette-type viscometers so as to maximize their useful ranges of shear rates.

When a fluid of low viscosity is used to displace a fluid of higher viscosity from a porous medium or a pipe, the displacement can become unstable by a process known as fingering. Small fingers of the low viscosity fluid, once formed, become regions of lower pressure drop. The displacing fluid flows preferentially into these fingers which grow and ultimately reach the outlet, leaving a portion of the viscous fluid undisplaced. Such displacements are widely practiced to increase the productivities of oil fields, and much effort is devoted to minimizing fingering by proper adjustment of the physical and chemical properties of the displacing medium, as well as of the rate of displacement.

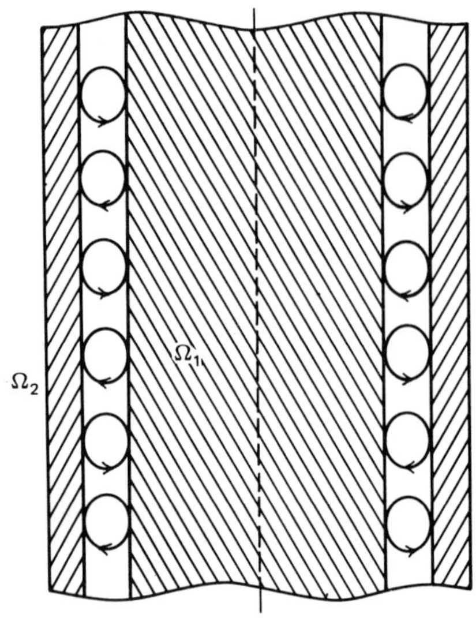

Figure 15. Flow pattern in rotating couette flow. Courtesy of Cambridge University Press (11).

Fluids at Rest. Fluids at rest may be set into motion by impressing upon them gradients in body or surface forces. Benard instability refers to the formation of convection cells within a fluid due to the action of a gravitational field on density differences induced by a temperature gradient in the fluid. Such behavior is observed, for example, when fluid is confined between two horizontal plates of which the lower is heated. The heated fluid on the bottom expands, resulting in an unstable density gradient and providing a driving force to redistribute the fluid. Redistribution is inhibited by viscosity, which slows the flow, and by thermal conductivity, which tends to even out the temperature difference. The onset of instability is described by a critical Rayleigh number:

$$Ra = \frac{g\left(\dfrac{\partial \rho}{\partial T}\right) \Delta T\ \ell_b^3}{\mu \alpha} \tag{26}$$

Hot fluid rises on one side and cold fluid falls on the other side of the cell. The rising fluid cools as it nears the top, and the falling fluid warms as it nears the bottom, thus maintaining a steady flow.

Cellular motions may also arise from gradients in surface tension caused by variations in temperature or concentration. This behavior is commonly referred to as Marangoni instability. To illustrate this, consider a solution from which a volatile solute is evaporating and suppose the surface tension of the solution decreases as the concentration of solute increases. Because the solute is evaporating, the surface has a lower concentration of solute than the bulk. If a small disturbance occurs and causes a spot of high solute concentration (low surface tension) to appear, this spot spreads. Spreading is a result of the attempt by the interface to minimize its free energy by expansion of regions of low surface tension and contraction of those of high surface tension. The local spreading is accompanied by movement of fluid from the bulk to fill the spreading spot. Since the new fluid has lower surface tension too, the original disturbance is amplified and convection cells are established. In some cases the driving force is so strong that the surface undergoes violent twitching with a large increase in the overall mass transfer rate. Such motion is commonly termed interfacial turbulence. If the direction of mass transfer is reversed, disturbances usually are not amplified and the surface remains quiescent. This effect is responsible for the occasionally observed large differences in mass transfer rates for absorption and desorption in otherwise the same system. The transfer of acetic acid between water and carbon tetrachloride is a classic example.

Circulation of fluid is promoted by surface tension gradients but inhibited by viscosity, which slows the flow, and by molecular diffusion, which tends to even out the concentration differences. The onset of instability is described by a critical Marangoni number, an analogue of the Rayleigh number:

$$Ma = \frac{\left(\frac{\partial \sigma}{\partial c}\right)\left(\frac{\partial c}{\partial x}\right) g_c \ell_m^2}{\mu \mathcal{D}} \tag{27}$$

Linear stability analysis has been successfully applied to derive the critical Marangoni number for several situations.

Surface tension is also responsible for the varicose or Rayleigh breakup of liquid strands into droplets. By virtue of surface tension the pressure within a strand is slightly higher than that in the ambient gas by the amount:

$$\Delta P = \frac{2\sigma}{D} \tag{28}$$

A small perturbation (reduction) in the radius results in a locally higher pressure than in the rest of the strand, causing liquid to flow away from that region, reducing the radius still further and making the thread unstable. Rayleigh analyzed this phenomenon for a low velocity, inviscid, cylindrical jet. He determined that one wave length of disturbance grows more rapidly than all others and thus, in the absence of forced oscillations, determines the drop diameter. For an inviscid fluid this is about twice the strand diameter. At higher flows a jet undergoes sinuous breakup in which the entire strand oscillates like a vibrating string. At still higher rates, as in commercial swirl atomizing nozzles, the liquid exits as a sheet that undergoes Kelvin-Helmholtz breakup into thread-like rings which in turn disintegrate further, via Rayleigh instability, into fine droplets. The drop size produced is a function primarily of nozzle diameter and fluid velocity. One widely used empirical correlation yields the relationship (12):

$$\overline{X} \propto D^{0.65} V^{-0.55} \rho_L^{-0.35} \sigma^{0.2} \mu_L^{0.15} \tag{29}$$

Atomization. A gas or liquid may be dispersed into another liquid by the action of shearing or turbulent impact forces that are present in the flow field. The steady-state drop size represents a balance between the fluid forces tending to disrupt the drop and the forces of interfacial tension tending to oppose distortion and breakup. When the flow field is laminar the ability to disperse is strongly affected by the ratio of viscosities of the two phases. Dispersion, in the sense of droplet formation, does not occur when the viscosity of the dispersed phase significantly exceeds that of the dispersing medium (13).

More commonly, atomization occurs under turbulent conditions. The mechanism of atomization and its quantitative description are still incompletely understood. It is possible that breakup occurs because of the impact forces exerted by the small, usually isotropic turbulent eddies in the manner proposed in ref. 14. The effects of these eddies are measured by the local rate of energy dissipation, but since this quantity is seldom known, correlations are often expressed in terms of average energy dissipation per unit mass or volume. In fact, in most important situations (eg, stirred tanks) breakup occurs in highly anisotropic regions, eg, the vortices shed from the turbine blades, where the local energy dissipation bears little relationship to the average and where nonturbulent mean shearing stresses may be large. In pipe flow it appears that, at least initially, breakup occurs quite close to the wall, quite likely as a result of mean flow shear. Drop size is related to velocity (gradient) and not to average energy dissipation. Under these circumstances the most practical approach is to employ empirical correlations derived for the geometries and residence times of interest.

Drop or bubble sizes are also strongly affected by the tendency of the drops or bubbles to coalesce once they enter regions of low turbulence or low mean shear. When coalescence is rapid, as in dispersion of gas in pure liquids, average bubble size seems to be determined by the average rate of energy dissipation and is insensitive to how the energy is applied. Correlations obtained for one geometry can be applied with reasonable success to other geometries. In many liquid–liquid systems where coalescence rates are lower, the drops made in the regions of intense shear can persist throughout. In a stirred vessel, for example, these drops are characteristic of conditions near the impeller and the same impeller operating at the same speed in two vessels of different sizes gives about the same size drops, despite the lower average energy dissipation in the larger vessel. In the extreme case of no coalescence, eg, emulsions stabilized by surface-active agents, drop sizes are determined by the highest shear rates available. In a stirred vessel this results in a sensitivity of drop size only to tip speed and not to impeller shape or energy input. In a practical situation the time needed to emulsify is as important as ultimate drop size, and impellers providing reasonable circulation times are required.

Solutions to Equations of Motion

Three basic approaches have been used to solve the equations of motion. For relatively simple configurations, direct solution is possible. For complex two-dimensional configurations, numerical methods can be employed. For many practical situations, particularly three-dimensional or one-of-a-kind configurations, scale modeling is employed and the results are interpreted in terms of dimensionless groups. This section outlines the procedures employed and the limitations of these approaches.

Exact Solutions to the Navier-Stokes Equations. As was true for the inviscid flow equations, exact solutions to the Navier-Stokes equations are limited to fairly simple configurations that allow for considerable simplification both in the equation and in the boundary conditions. For the important situation of steady, fully developed, laminar, Newtonian flow in a circular tube, for example, the Navier-Stokes equations reduce to:

$$\frac{dP}{dX} = \frac{\mu}{r}\frac{d}{dr}\left(r\frac{du}{dr}\right) \tag{30}$$

with boundary conditions:

$$\left(\frac{du}{dr}\right)_{r=0} = 0, \quad u(R) = 0 \tag{31}$$

The solution to this equation is the well-known Hagen-Poiseuille relationship (see eq. 54). Solutions for other simple flow situations are also available (1).

Numerical Solutions to the Equations of Motion. Usually an analytical solution to the equations of motion cannot be found; recourse must be had to numerical methods. The field of computational fluid dynamics (15) is now moderately mature. A vast repertory of techniques exist and solutions to many complex problems of real interest can be obtained without an exorbitant amount of effort. One still encounters, however, what one might term the dimensions barrier. Problems involving two space dimensions are usually tractable with existing computers and programs but the practical treatment of truly three-dimensional problems still appears to be far in the future. Many problems still remain in the treatment of turbulent systems although there are some useful methods even here. The numerical technique that is most suitable depends on what approximations can legitimately be made, on whether a time-dependent solution is required, on the nature of the boundary conditions, and especially on whether the boundaries are fixed in space or move with the flow. Both finite-difference and finite-element techniques have been usefully employed. There is extensive literature devoted to each (15). The finite element method has been used widely in fluid mechanics only recently although it has been used for more than two decades to solve problems in solid mechanics. Some of the methods used to solve the Navier-Stokes equation for viscous, incompressible flow are described below.

Finite-Difference Method. The most widely used numerical method for flow simulations is based on approximating the partial derivatives in the equations of motion by finite difference approximations. It is based on the assumption that the dependent variable can be expressed in an infinite Taylor series development everywhere in the flow field. The flow field or domain of interest is divided into a network of rectangular elements for two-dimensional flows (or volume elements for three-dimensional flows) as shown in Figure 16. If the values of the dependent variable are known at two neighboring grid points, the first derivative can be approximated from the Taylor series by discarding higher order terms. The first derivative, $\partial u/\partial x$, of the velocity u, can be replaced by the forward difference approximation:

$$\left(\frac{\partial u}{\partial x}\right)_{i,j} = \frac{u_{i+1,j} - u_{i,j}}{\Delta x} \tag{32}$$

Other finite-difference representations for the first derivative can also be derived. A second derivative requires the values of u to be known at three neighboring grid points. Thus, the centered difference approximation for $\partial^2 u/\partial x^2$ is:

$$\left(\frac{\partial^2 u}{\partial x^2}\right)_{i,j} = \frac{u_{i+1,j} + u_{i-1,j} - 2u_{i,j}}{(\Delta x)^2} \tag{33}$$

The accuracy of the approximations improves as the distances between the grid points are reduced. The finite difference method may be applied to the equations of motion in either the Eulerian or Lagrangian form. In the Eulerian formulation, a fixed grid is utilized whereas in the Lagrangian formulation, the coordinate system moves with the fluid. Eulerian formulations are simpler and hence more commonly used but may give poor representation of advection and discontinuities.

The equations of motion and continuity given in equations 1–6 may be written for two-dimensional flows as:

$$\frac{\partial u}{\partial t} + u\frac{\partial u}{\partial x} + v\frac{\partial u}{\partial y} + \frac{1}{\rho}\frac{\partial p}{\partial x} - \nu\left(\frac{\partial^2 u}{\partial x^2} + \frac{\partial^2 u}{\partial y^2}\right) = 0 \tag{34}$$

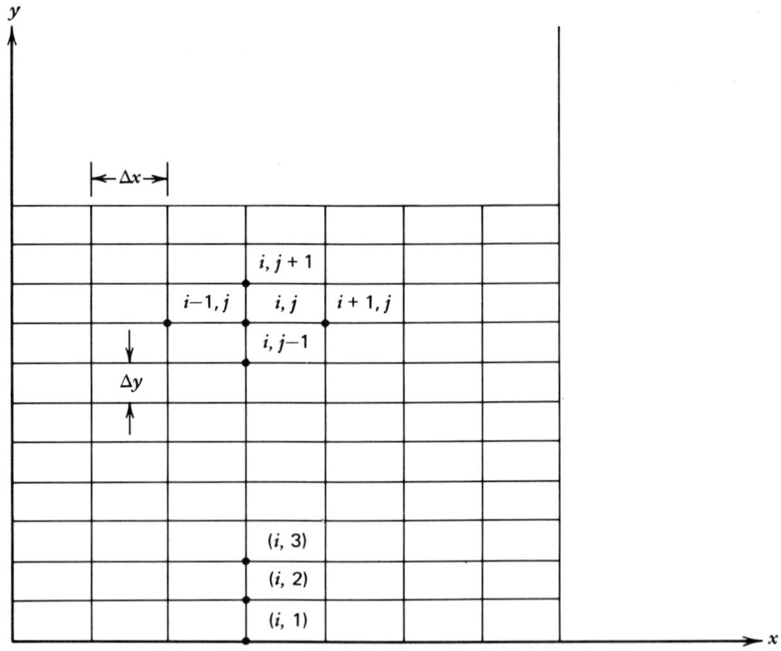

Figure 16. Grid structure for finite difference simulation.

$$\frac{\partial v}{\partial t} + u\frac{\partial v}{\partial x} + v\frac{\partial v}{\partial y} + \frac{1}{\rho}\frac{\partial p}{\partial x} - \nu\left(\frac{\partial^2 v}{\partial x^2} + \frac{\partial^2 v}{\partial y^2}\right) = 0 \tag{35}$$

$$\frac{\partial u}{\partial x} + \frac{\partial v}{\partial y} = 0 \tag{36}$$

These equations may be solved for the primitive variables u, v, P or may be rewritten in terms of the vorticity ζ and stream function ψ. The $\psi - \zeta$ approach has been used quite extensively, particularly in two dimensional flows. Thus, for two-dimensional flow, by eliminating the pressure terms from equations 34 and 35 and then introducing the stream function ψ defined by equation 13 and the scalar vorticity ζ defined by

$$\zeta = \frac{\partial u}{\partial y} - \frac{\partial v}{\partial x} \tag{37}$$

the following equations are obtained:

$$\frac{\partial \zeta}{\partial t} + \frac{\partial \psi}{\partial y}\frac{\partial \zeta}{\partial x} - \frac{\partial \psi}{\partial x}\frac{\partial \zeta}{\partial y} = \nu\left(\frac{\partial^2 \zeta}{\partial x^2} + \frac{\partial^2 \zeta}{\partial y^2}\right) \tag{38}$$

$$\frac{\partial^2 \psi}{\partial x^2} + \frac{\partial^2 \psi}{\partial y^2} = -\zeta \tag{39}$$

Equations 38 and 39 are simultaneous, second-order, partial differential equations of the elliptic type. It is possible to extend the $\psi - \zeta$ concept to three-dimensional flows. This requires the introduction of a vorticity vector $\vec{\zeta}$ and a vector potential, $\vec{\psi}$. The choice of whether to use the $\psi - \zeta$ (eqs. 38 and 39) system or the u-v-P (eqs. 34–36) system for two-dimensional flow problems or the $\vec{\psi} - \vec{\zeta}$ systems or the u-v-w-P system for three-dimensional flow problems depends on many factors. No general rules can be stated; one must draw on the experiences of others. The choice depends also on the detail of results required such as whether or not a solution for the pressure is required. For two-dimensional flow problems that do not involve free surfaces, the $\psi - \zeta$ system should be considered first. In the case of three-dimensional flow problems and flow involving free surfaces, the u-v-P or u-v-w-P systems offer advantages. Some discussion of the relative merits of the several systems is given in ref. 15. If the flow problem is two dimensional and the $\psi - \zeta$ system is chosen, there are many solution techniques to choose from. Two of the many methods that have been developed, are the Alternating Direction Implicit-Successive Over Relaxation (ADI-SOR) method and the Marker and Cell method (MAC) discussed below.

ADI-SOR. The ADI-SOR is described in detail in ref. 16. Both the steady and unsteady state solution can be obtained by solving equations 38 and 39. The steady-state solution is obtained as a limit of the unsteady flow. The finite-difference method allows one to calculate values of the vorticity and the stream function at every grid point of the flow-field grid structure (such as that shown in Fig. 16) at a discrete time t starting from the initial condition. Let n denote the number of time steps and Δt the size of the time step. The computational scheme for advancing the solution from

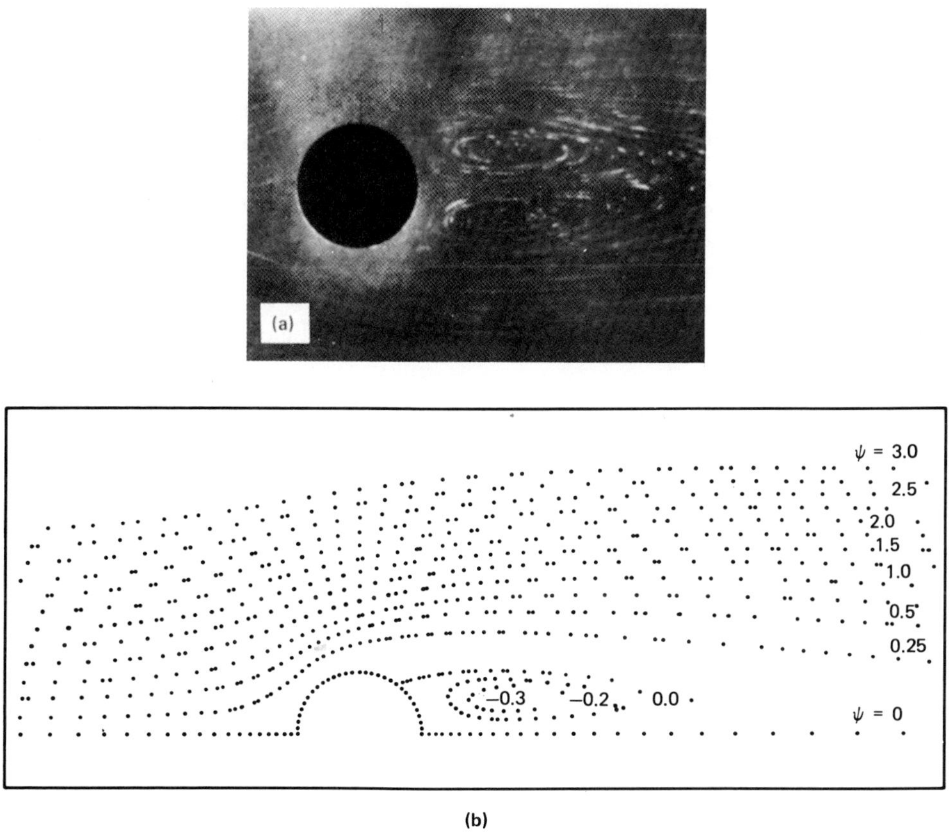

Figure 17. Comparison of (**a**) experimentally obtained flow pattern around a cylinder at $Re = 42$ (17) with (**b**) streamlines obtained by numerical simulation at $Re = 40$ (16). Courtesy of Nikon Butsuri Gakkai and Cambridge University Press.

$t = n\Delta t$ to time $t = (n + 1)\Delta t$ proceeds by calculating the new values of the vorticity $\zeta_{i,j}^{n+1}$ at all the grid points from the finite difference approximation to equation 38, the boundary conditions, and the values of the vorticity $\zeta_{i,j}^{n}$ and the stream function $\psi_{i,j}^{n}$ from the previous time step. Then, using the finite-difference approximation to equation 39, the boundary conditions, the new values of the vorticity ($\zeta_{i,j}^{n+1}$) just calculated, and the old values of the stream function ($\psi_{i,j}^{n}$), the new values of the stream function ($\psi_{i,j}^{n+1}$) are calculated at all the grid points.

The vorticity transport equation is solved by the ADI method. The time step, Δt, is divided into two half-steps and different finite-difference approximations are written for each half step. The stream function equation is solved by means of the SOR method. This method uses an over relaxation parameter to accelerate the rate of convergence to the numerical solution. An optimum value of this parameter can usually be determined experimentally. It is worthwhile for large problems (large number of grid points) since the stream function equation must be iterated at every iteration of the vorticity transport equation. In order to apply these procedures successfully on the grid structure shown in Figure 16, the ψ and ζ (Dirichlet-type boundary conditions) or their derivatives (Neumann-type boundary condition) must be prescribed at the boundaries.

Figure 17 compares streamlines calculated by ADI-SOR with those obtained experimentally for flow past a circular cylinder at a Reynolds number ($DV\rho/\mu$) of equation 40.

Marker and Cell (MAC). One of the more successful methods for solving the equations of motion in terms of the primitive variables is the Marker and Cell (MAC) method. This method has been used to study complex flow problems such as those involving free surfaces. The MAC grid layout is shown in Figure 18. The velocities are centered on the cell boundaries and the pressure is defined at the cell centers. The MAC technique is described in detail in ref. 18.

A Poisson equation for pressure is obtained by differentiating equations 34 and 35 with respect to x and y, respectively, and by adding the results. Starting from the boundary conditions and initial conditions that satisfy the continuity equation, the

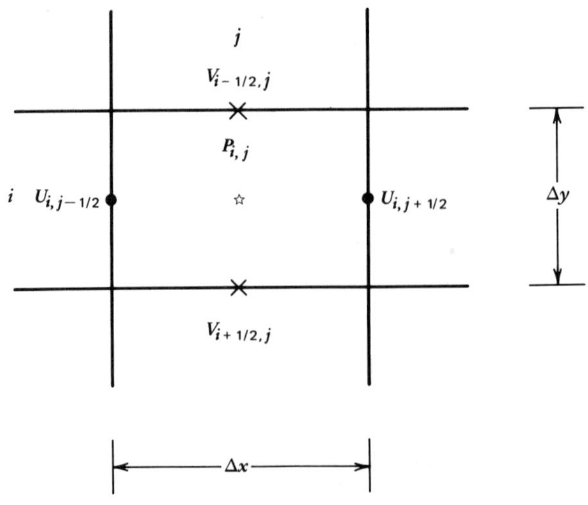

Figure 18. MAC grid structure.

steps used to advance through one time increment Δt begin with calculation of pressures by iterating the finite-difference equation corresponding to the Poisson equation for pressure. The velocities at the new time step are calculated from the finite difference equations corresponding to equations 34 and 35. If there is no free surface and/or one is not interested in obtaining fluid-element trajectories, this completes one Δt calculation. Otherwise, one uses the marker particles that move with the advection field. They do not participate in the calculation but they indicate fluid configuration for keeping track of free-surface position. Initially, marker particles are placed in the cells containing fluids. The new position of each particle is moved by $u_p \Delta t$ and $v_p \Delta t$ where the particle velocities u_p and v_p are calculated as weighted averages of velocities at the four nearest mesh points.

If the fluid has a free surface, its shape is not known *a priori*. The boundary conditions require zero tangential stress and a normal stress that balances any externally applied normal stress such as pressure. Fictitious velocities and pressures outside wall boundaries are used to handle wall-boundary conditions.

Figure 19 compares the configuration of the free surface at some time t computed by MAC with that obtained experimentally for the splash of a drop of liquid falling into another body of liquid.

Finite Element. The finite-difference method gives a pointwise approximation to the flow equations whose solution is improved as more grid points in the flow field are used. Its chief disadvantage is that it is difficult to use in flow regions with irregular

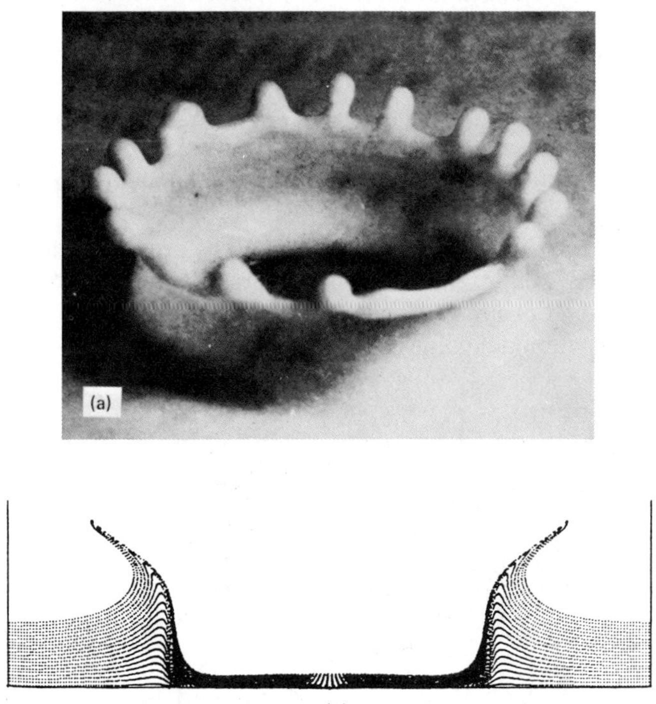

(b)

Figure 19. Comparison of (**a**) a photograph of the shallow pool splash of a milk drop (19) and (**b**) a calculation of the same splash by MAC method (20). Courtesy of Charles T. Brandford and the American Institute of Physics.

boundaries or unusual boundary conditions. The finite-element method, which does not have this disadvantage and possesses greater flexibility than the finite-difference method, is finding increased application in fluid flow problems. The finite-element method gives a piecewise approximation to the flow equations through division of the flow region into interconnected subregions or elements (see Fig. 20). The sizes and shapes of the elements can be varied to suit the boundaries and characteristics of the various regions (eg, smaller elements where gradients are known to be large).

Like the finite-difference method, the finite-element method reduces the equations of motion to a set of simultaneous algebraic equations. The procedure used to obtain this set of equations is more complex, however. The governing equations in the u-v-P system are equations 34–36. These are linearized by using the velocities from the previous iteration to approximate the nonlinear convective terms, eg, $u\partial u/\partial y$ is replaced by $u_n \partial u_{n+1}/\partial y$. Initially, where no previous iteration is available, a starting guess, eg, creeping flow, must be employed. The steps usually followed after the linearization are:

(*1*) Divide the flow region into elements (subregions) as illustrated in Figure 20. Elements can be of different shapes (triangles and quadrilaterals, eg, for two-dimensional flow fields, and tetrahedrons and hexahedrons, eg, for three-dimensional fields).

(*2*) Assign nodes to each element. Nodes are the points at which the field variables (eg, u,v,P) are evaluated and may lie at the vertices or on the sides of the various elements. The number of nodes dictates the degree of polynomials to be assumed.

(*3*) Assume an interpolation function to represent the field inside the element. Linear or quadratic interpolation functions are often used. The u velocity component, for example, may be represented by

$$u^{(e)} = \sum_{i=1}^{r} N_i(x,y)u_i \tag{40}$$

where u_i is the value of u at the nodes and N_i is the geometric function describing its variation in the region.

(*4*) Determine the matrix equations expressing the properties of the element using either the variational, weighted residuals, or energy balance approach.

(*5*) Develop the set of simultaneous algebraic equations from element properties and solve for unknown nodal values of the field variable.

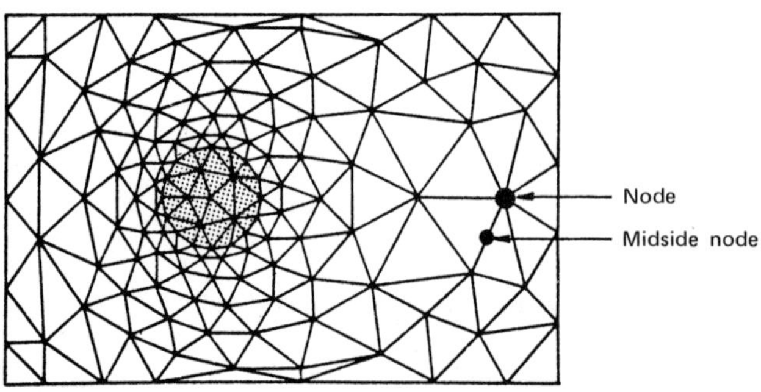

Figure 20. Finite element gridwork used in the analysis of the flow around a moving obstacle (20). Courtesy of Crane-Russak and Co., Inc.

The assembled equations, which are solved iteratively for the nodal values of u, v, and P can be written in compact form:

$$[\mathcal{K}]\begin{Bmatrix} u \\ \hdashline v \\ \hdashline P \end{Bmatrix} = [\mathcal{R}] \tag{41}$$

Integration of terms in equation 41 does not present any difficulty for Newtonian fluids. However, for non-Newtonian fluids and viscoelastic fluids in particular, the integration may be difficult. For power law fluids, finite-element solutions to a number of interesting problems have been developed.

Turbulent Flow Simulations. Most of the numerical simulations described in the literature are for laminar flow problems. An equally interesting but perhaps more difficult problem is the simulation of turbulent-flow motion. Only when computers are developed that are several orders of magnitude faster than those currently available can the time-dependent velocities be obtained by integrating the equations of motion directly. Since such an advance in computer technology is nowhere in sight, research in turbulent-flow simulations has depended much on Reynolds' suggestion of splitting up the velocity components into a time-average mean and a fluctuating part (21). This procedure results in the appearance of additional terms called Reynolds stresses which cannot be determined unless additional assumptions (often *ad hoc*) are made or semi-empirical relationships are used. This is called the closure problem of turbulence theory. A finite-difference method has been developed for solving the equations for turbulent motion in two dimensions using various types of Reynolds stress closure relations (22–23).

Modeling. The majority of technological flow problems are not solved by integrating the equations of motion. Instead, most are solved by carrying out laboratory experiments which are then correlated, ie, interpreted, so as to yield useful information about systems that may differ greatly in size and in fluid properties. For the behavior of the experimental model to duplicate that of a system of interest, two criteria must, in principle, be met: the experimental apparatus must be geometrically similar to the system of interest; and certain dimensionless groupings of variables must be duplicated on the two scales. There are two basic methods available for determining the dimensionless groups appropriate to a given situation: dimensional analysis, which can be applied when the equations governing the process are not known; and similarity analysis, which proceeds from the governing equations and offers physical insights into the meanings of the groups.

Dimensional analysis (qv) is a mathematical technique that proceeds from the general principle that physical laws must be independent of the units of measurement used to express them. If one quantity is related to a group of other quantities, the quantities comprising the group must be related in such a manner that the net units or dimensions of the group are the same as those of the dependent quantity. A dimensionless group can then be formed immediately by division. A useful tool, the Pi theorem, asserts that the number of dimensionless groups needed to describe a situation is equal to the total number of variables less the number of fundamental dimensions needed to express them. Fundamental dimensions are generally taken to be length, time, mass, temperature and heat content. The Pi theorem is a powerful tool because it limits the amount of experimental work needed to establish a general relationship. For example, consider the problem of determining the drag force on a smooth sphere around which a Newtonian fluid is flowing. The diameter of the sphere and the fluid viscosity, density, and velocity may vary. Each of the five variables can be expressed using various combinations of the three dimensions mass, length and time (force = mass × length/time^2). The Pi theorem leads immediately to the conclusion that only two dimensionless numbers are needed to describe the relationship. These may be taken to be a drag coefficient $2Fg_c/A_p\rho V_\infty^2$ and a Reynolds number $DV_\infty \rho/\mu$. An accurate set of measurements for one sphere in one fluid provides a universal relationship between these numbers that is applicable to all spheres in all Newtonian fluids.

The strength of dimensional analysis lies in its ability to limit the number of studies that need to be made and to handle situations in which the governing equations are not known. It can even handle lack of geometrical symmetry by using ratios of important dimensions as additional dimensionless groups. Its weakness lies in the need to know, *a priori,* which variables must be included, which can be ignored, and in its awkwardness when handling several variables that have the same dimensions, eg, densities in multiphase mixtures. This last difficulty reflects the fact that dimensional analysis is not capable of describing the functional relationship among the dimensionless groups, nor is it capable of describing how an empirical relationship might be extrapolated outside the range covered by the original data, nor is it even capable of selecting the best set of dimensionless groups, since products, sums, quotients, and differences of dimensionless groups are also dimensionless. To achieve the simplest, most meaningful relationship and to judge how it might best be extrapolated, it is necessary for the investigator to use previous experience, similarity analysis or physical insight.

Similarity analysis starts from the equation describing a system and proceeds by expressing all of the dimensional variables and boundary conditions in the equation in reduced or normalized form. Velocities, for example, are expressed in terms of some reference velocity in the system, eg, the average velocity. When the equation is rewritten in this manner certain dimensionless groupings of the reference variables appear as coefficients, and the dimensional variables are replaced by their normalized relatives. If another physical system can be described by the same equation with the same numerical values of the coefficients, then the solutions to the two equations (the normalized variables) are identical and either system is an accurate model of the other.

The principle can be illustrated by examining the Navier-Stokes equation for two-dimensional incompressible flow. The x-component of the equation is:

$$\rho \frac{Du}{Dt} = \rho g_x - \frac{\partial P}{\partial x} + \mu \left(2 \frac{\partial^2 u}{\partial x^2} + \frac{\partial^2 u}{\partial y^2} + \frac{\partial^2 v}{\partial x \partial y} \right) \tag{42}$$

All variables in the system can be expressed in reduced form. Velocity can be expressed as $u = V_o u'$ where V_o is a fixed reference velocity and u' is the dimensionless reduced velocity. Because time (T_o) is the quotient of length (L_o), and velocity (V_o), the equation can be manipulated to yield:

$$\rho' \frac{Du'}{Dt'} = \left[\frac{g_o L_o}{V_o^2} \right] \rho' g'_x - \left[\frac{P_o}{\rho_o V_o^2} \right] \frac{\partial P'}{\partial x'} + \left[\frac{\mu_o}{\rho_o L_o V_o} \right] \left(2\, \mu' \frac{\partial^2 u'}{\partial x'^2} + \dots \right) \tag{43}$$

Table 2. Some Dimensionless Numbers Used in Engineering

Dimensionless no.	Contributing effects	Area of use
Euler	inertia-pressure	compressible flow
Prandtl	conduction-convection	heat transfer
Nusselt	surface transfer-convection	heat transfer at boundary
Fanning	surface drag-inertia	drag at boundary
Weber	surface tension-inertia	multiphase contacting

The dimensionless quantities in brackets are, respectively, the reciprocal of the Froude number, the Euler number, and the reciprocal of the Reynolds number for the system.

Because the Navier-Stokes equation is basically a force balance stemming from Newton's law, the dimensionless groups that arise from it are often related conceptually to the forces acting upon a unit volume of fluid. It is correct to say that when all the dimensionless groups are matched between the two systems, then the ratio of viscous stresses to body forces or to fluid inertia in one system is the same as that ratio at the corresponding point of the second system; the flows are kinematically similar. It is often said, however, that the Reynolds number measures the ratio of fluid inertia to viscous forces. This is not correct, except perhaps in an average sense. The Reynolds number of a system is a single number, and the ratio of fluid inertia to viscous stresses may vary widely throughout the system. Perhaps the most useful general statement that can be made is that when a system's behavior is significantly affected by the interaction of fluid inertia and viscous drag, then the Reynolds number is an important parameter. When the system is affected by the interaction of gravity and inertia, eg, by a falling ball, the Froude number is important. It is the strength of similarity analysis that such groupings arise automatically from careful attention to the properties of the defining equation(s).

Dimensionless numbers are not the exclusive property of fluid mechanics but arise out of any situation describable by a mathematical equation. Some of the other important dimensionless groups used in engineering are listed in Table 2.

The principal difficulty faced by the experimenter in applying the results of dimensional or similarity analysis is that often there is insufficient freedom with respect to the physical properties of the modeling fluids to match all of the potentially important dimensionless groups. Construction and interpretation of the model rests largely upon the experience and judgment of the experimenter as to which groups can safely be ignored within the accuracy desired. In some cases useful results can be obtained even when none of the dimensionless groups can be duplicated. An example is the use of scale models in wind tunnels to study the turbulent dispersion of materials released to the atmosphere in the vicinity of complex structures such as refineries or orchards (24). It is not practical to maintain the Reynolds number by increasing wind velocity in inverse proportion to the reduction in scale. Instead, use is made of the observation that when the Reynolds number based on the characteristic dimension of the structure exceeds ca 10^4, one can reasonably expect flows to be similar on all scales, ie, turbulent fluctuations will be proportional to the mean velocity and the wakes of the structures will be geometrically similar (see also Fluidization). Although flow fields around small elements of the structure are not duplicated, the overall error from this approximation is less than would result from an attempt to calculate the dispersion by combining the correlations for individual structures.

Integral Forms of Equations of Motion

The solutions of some problems in fluid mechanics require the detailed integration of the differential equations of motion, whereas many others can be solved to a sufficient degree of accuracy by examining only the overall balances of mass, momentum, and energy. In applying these balances, an appropriate control volume is established first and the rates of accumulation of the quantities within the volume are balanced against their rates of generation within the volume and their rates of transport through the control surface. In most cases the volume of interest is a structure fixed in space, eg, a pipe or tank. In some cases, eg, shock waves, a moving coordinate system is more appropriate.

Conservation of Mass. The general equations for the conservation of mass are the scalar equations (see Fig. 21**a**):

$$\underbrace{\frac{\partial}{\partial t}\iiint c_i dB}_{\text{accumulation}} + \underbrace{\iint c_i V \cos\theta dS}_{\text{convection}} + \underbrace{\iint J_i dS}_{\text{diffusion}} - \underbrace{\iiint R_i dB}_{\text{chemical reaction}} = 0 \tag{44}$$

A separate equation is obtained for each component i. Consider the important case of steady-state flow of a single fluid through the pipe section of Figure 21**b**, where the flow is taken to be perpendicular to the cross sections S_1 and S_2. Making use of the concept of average velocity, one easily obtains:

$$S_1\overline{V}_1\rho_1 = S_2\overline{V}_2\rho_2 \tag{45}$$

where

$$\overline{V}_j = \frac{1}{\pi R_j^2}\int_0^{R_j} V_j 2\pi r dr \tag{46}$$

The conservation of mass gives comparatively little useful information until it is combined with the results of the momentum and energy balances.

Conservation of Momentum. The general equation for the conservation of momentum is:

$$\underbrace{\frac{\partial}{\partial t}\iiint \rho\vec{V} dB}_{\text{accumulation}} + \underbrace{\iint \rho\vec{V}V\cos\theta dS}_{\text{convection}}$$

$$- \underbrace{\iiint \vec{F}_B dB}_{\text{creation by body forces}} - \underbrace{\iint \vec{F}_s dS}_{\text{creation by surface forces}} = 0 \tag{47}$$

The creation terms embody the changes in momentum arising from external forces in accordance with Newton's second law: $F = ma$. The body forces arise from gravitational, electrostatic, and magnetic fields. The surface forces are the shear and normal forces acting on the fluid; diffusion of momentum, as manifested in viscosity, is included in these terms. In practice the vector equation is usually resolved into its

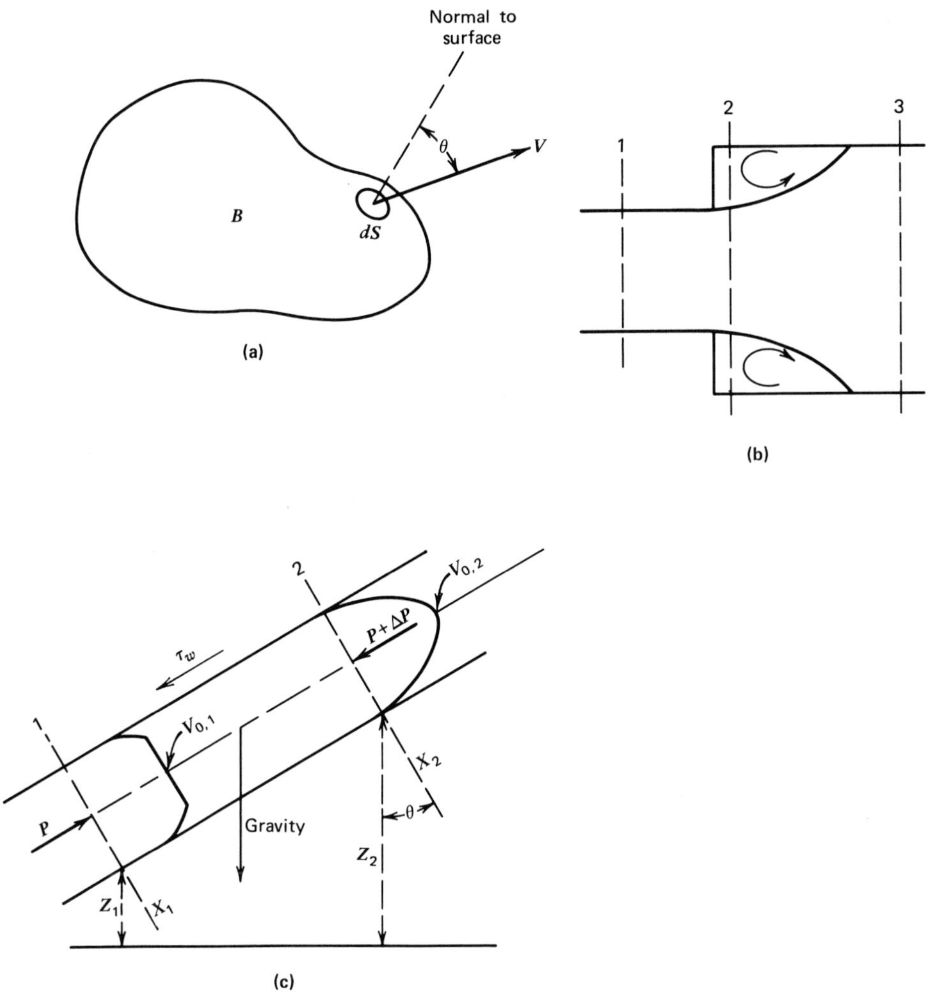

Figure 21. Control volumes for application of the integral equations of motion. (**a**) General. (**b**) Sudden expansion. (**c**) Pipe flow.

Cartesian components and the normal stresses are set equal to the pressures over those surfaces through which fluid is flowing.

To illustrate the use of the momentum balance, consider once again the situation shown in Figure 21**c** in which the control volume is bounded by the pipe wall and the cross sections 1 and 2. The forces acting on the fluid in the x-direction are the pressure forces acting on cross sections 1 and 2, the shear forces acting along the walls and the body force arising from gravity. The overall momentum balance is:

$$\left[-\Delta P + \frac{g}{g_c}\int_1^2 \rho dz\right]\pi R^2 - 2\pi R\int_1^2 \tau_w dx = \frac{1}{g_c}\int_0^R 2\pi r(\rho_2 V_2^2 - \rho_1 V_1^2)dr \quad (48)$$

The momentum terms can be rewritten using the mass balance to yield:

$$\frac{1}{g_c}\int_0^R 2\pi r(\rho_2 V_2^1 - \rho_1 V_1^2)dr = \frac{\pi R^2 W}{g_c}\left[\frac{\overline{V}_2^2}{\overline{V}_2} - \frac{\overline{V}_1^2}{\overline{V}_1}\right] = \frac{\pi R^2 W}{g_c}\left[\frac{\overline{V}_2}{\beta_2} - \frac{\overline{V}_1}{\beta_1}\right] \quad (49)$$

In general, $\overline{V^2} \neq (\overline{V})^2$. For laminar, Newtonian flow the radial velocity profile is parabolic and $\beta = 3/4$. For fully developed, turbulent flow the radial velocity profile is almost flat; $\beta = 0.95$–0.99, hence it can be taken as unity with little error.

When the fluid is incompressible and the velocity profiles are identical, $\rho_1 = \rho_2$, $\beta_1 = \beta_2$, the velocity terms disappear and τ_w is independent of x. Equation 48 reduces to:

$$\frac{1}{\Delta x}\left(-P - \rho \frac{g}{g_c} \Delta z\right) = -\frac{\Delta P^*}{\Delta x} = \frac{2\tau_w}{R} \tag{50}$$

A similar analysis can be carried out for any axisymmetric tube to yield:

$$-\frac{\Delta P^*}{\Delta x} = \frac{2\tau}{r} \text{ or } \frac{\tau}{\tau_w} = \frac{r}{R} \tag{51}$$

The shear stress is linear with radius. This result is quite general, applying to any axisymmetric, fully developed flow, laminar or turbulent. If the relationship between the shear stress and the velocity gradient is known, equation 46 can be used to obtain the relationship between velocity and pressure drop. Thus, for laminar flow of a Newtonian fluid, one obtains:

$$-\mu \frac{dV}{dr} = \tau = -\frac{r}{2}\left(\frac{\Delta P^*}{\Delta x}\right) \tag{52}$$

This integrates directly to yield a parabolic velocity profile:

$$V(o) - V(r) = \frac{r^2}{4\mu}\left(\frac{\Delta P^*}{\Delta x}\right) \tag{53}$$

Averaging the velocity using equation 41 yields the well-known Hagen-Poiseuille equation for laminar flow of Newtonian fluids in tubes:

$$\frac{\Delta P^*}{\Delta x} = \frac{32\mu \overline{V}}{g_c D^2} \tag{54}$$

The momentum balance can also be used to describe the pressure changes at a sudden expansion in turbulent flow (Fig. 21**b**). The control surface 2 is taken to be sufficiently far downstream that the flow is uniform but sufficiently close to surface 3 that wall shear is negligible. The additional important assumption is made that the pressure is uniform on surface 3. The conservation equations are then applied as follows:

$$\text{momentum: } (P_3 - P_2)S_2 = \frac{\rho}{g_c}(S_2 V_2^2 - S_1 V_3^2) \tag{55}$$

$$\text{mass: } V_1 S_1 = V_2 S_2 \tag{56}$$

At the inlet the velocity and pressure are presumed continuous, ie, $P_1 = P_3$, $V_1 = V_3$. Simple algebraic manipulations yield the result:

$$\frac{P_2 - P_1}{\rho} = \frac{V_1^2}{g_c}\left(\frac{S_1}{S_2}\right)\left[1 - \left(\frac{S_1}{S_2}\right)\right] \tag{57}$$

Conservation of Energy. The energy associated with a unit mass of the flowing fluid may be considered as the sum of its potential, kinetic, and internal energies:

$$E = \frac{g}{g_c} Z + \frac{V^2}{2g_c} + U \tag{58}$$

The equation describing the conservation of energy is the scalar equation:

$$\underset{\text{accumulation}}{\frac{\partial}{\partial t} \iiint \rho E dB} + \underset{\text{convection}}{\iint \rho E V \cos\theta dS} + \underset{\text{nonconvective flux}}{\iint J_E dS} = 0 \tag{59}$$

The nonconvective energy flux across the boundary is composed of two terms: a heat flux and a work term. The work term in turn is composed of two terms: useful work delivered outside the fluid, and work done by the fluid inside the control volume B on fluid outside the control volume B, the so-called flow work. The latter may be evaluated by imagining a differential surface moving with the fluid which at time zero coincides with a differential element of the surface S. During the time dt the differential surface sweeps out a volume $V\cos dSdt$ and does work on the fluid outside B at a rate of $PV\cos\theta dS$. The total flow work done on the fluid outside B by the fluid inside B is:

$$\iint PV\cos\theta dS \tag{60}$$

This is substituted into equation 59 to yield:

$$\frac{\partial}{\partial t} \iiint \rho E dB + \iint \left(E + \frac{P}{\rho}\right) \rho V\cos\theta dS + Q - W_E = 0 \tag{61}$$

where Q and W_E are the rates at which heat is added to the system and work, both mechanical and electrochemical, is done by the system on its surroundings. For the steady-state system of Figure 21**b**, the overall energy balance is:

$$Q - W_E + (U_1 - U_2) + \left(\frac{P_1}{\rho_1} - \frac{P_2}{\rho_2}\right) + \frac{g}{g_c}(Z_1 - Z_2) + \frac{1}{2g_c}\left(\frac{\overline{V_1^3}}{V_1} - \frac{\overline{V_2^3}}{V_2}\right) = 0 \tag{62}$$

In terms of enthalpy and average velocities, this becomes:

$$Q - W_E + (H_1 - H_2) + \frac{g}{g_c}(Z_1 - Z_2) + \frac{1}{2g_c}\left(\frac{\overline{V_1^2}}{\alpha_1} - \frac{\overline{V_2^2}}{\alpha_2}\right) = 0 \tag{63}$$

For laminar flow $\alpha = 0.5$. For fully developed turbulent flow $0.88 < \alpha < 0.98$, but can be taken to be unity with little error.

The total energy balance may be regarded as the first law of thermodynamics for flowing fluid. It is an essential tool in analyzing the performance of machines that interconvert heat, work, and internal energy. It contains, however, no explicit terms relating to such matters as dissipation of energy within the system by friction, and there must be recourse to more specific equations such as the mechanical energy "balance."

The Mechanical Energy Balance. The mechanical energy, often termed the engineering Bernoulli equation, is strictly speaking neither. Mechanical energy is not always a conserved quantity and the Bernoulli equation refers to a single streamline in an inviscid fluid. Instead, the mechanical energy equations can be viewed as a definition of a dissipation term ΔF which describes the irreversible conversion into heat of energy available for mechanical work.

Two approaches to this equation have been employed: (*1*) The scalar product is formed between the differential vector equation of motion and the vector velocity and the resulting equation is integrated (1). This is the most rigorous approach and for laminar flow yields an explicit equation for ΔF in terms of the velocity gradients within the system. (*2*) The overall energy balance is manipulated by asserting that the local irreversible dissipation of energy is measured by the difference:

$$
\begin{aligned}
-\delta(\Delta F) &= \delta Q - TdS \\
&= \delta Q - dU - Pd\left(\frac{1}{P}\right) - \sum \overline{F}_i dn_i
\end{aligned}
\tag{64}
$$

Both approaches yield the same general result, which is:

$$\Delta F = -W_M + \int_2^1 \frac{dP}{\rho} + \frac{g}{g_c}(Z_1 - Z_2) + \frac{1}{2g_c}\left(\frac{\overline{V_1^2}}{\alpha_1} - \frac{\overline{V_2^2}}{\alpha_2}\right) \tag{65}$$

The work term W_M is restricted to the mechanical work delivered to the outside via normal and shear forces acting on the boundary. Electrochemical work, ie, by electrolysis of the fluid, is excluded. Evaluation of the integral requires knowledge of the equation of state and the thermodynamic history of the fluid as it passes from S_1 to S_2, neither of which is needed in the overall energy balance.

In practice, the loss term ΔF is usually not determined by detailed examination of the flow field. Instead, the momentum and mass balances are employed to determine the pressure and velocity changes, these are substituted into the mechanical energy equation and ΔF is determined by difference. For the sudden expansion of a turbulent fluid depicted in Figure 21**b,** which delivers no work to the surroundings, application of equations 45, 57, and 65 yields:

$$\Delta F = \frac{(V_1 - V_2)^2}{2g_c} = \frac{V_1^2}{2g_c}\left(1 - \frac{S_1}{S_2}\right)^2 \geqslant 0 \tag{66}$$

Flow Measurement

There are dozens of flowmeters available for the measurement of fluid flow (25). The discussion in this section is limited to typical devices grouped according to the primary measurement used to determine flow. These primary measurements include: differential pressure, variable area, liquid level, electromagnetic effects, thermal effects, and light scattering. Most of the devices discussed are those used commonly in the process industries; a few, developed more recently for the measurement of turbulence, also are described (see also Rheological measurements).

Measurement by Differential Pressure. The most widely used devices for the flow of fluids are those that utilize a fixed constriction in the path of flow to produce a difference in pressure between the upstream and downstream measuring points. The devices discussed here are orifice meters, venturi meters, and target meters. The metering orifice (Fig. 22**a**) has been standardized in most countries as a centered, circular, square-edged opening in a thin plate. It is simple and convenient to use, but has the disadvantage that a large fraction of the pressure drop generated for the measurement of flow is dissipated as frictional energy.

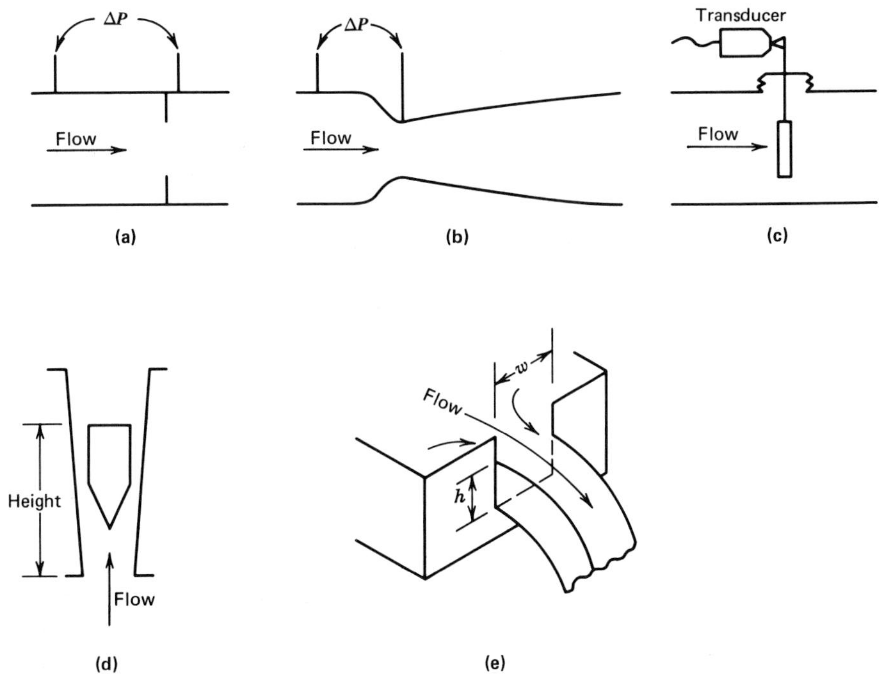

Figure 22. Common flowmeters. (**a**) Orifice. (**b**) Venturi. (**c**) Target. (**d**) Rotameter. (**e**) Weir.

The venturi meter (Fig. 22**b**) consists of a standardized conical nozzle followed by a gradually expanding downstream cone. The pressure change is measured between the upstream fluid and the narrow throat. The gradual expansion downstream of the throat allows the fluid to recover as much as 90% of the pressure difference that would otherwise be lost with a sudden expansion. The venturi meter is a rugged, precision meter but suffers from the very considerable disadvantages of high cost as well as large space requirements. In order that one installation may reproduce the results of another within a tolerable error, orifice and venturi installations have been rigorously standardized, largely as the result of work by the ASME Special Research Committee on Fluid Meters and the Joint AGA-ASME Committee on Orifice Coefficients. Their publications should be consulted before making an installation. To a reasonably good approximation, the pressure change measured is proportional to the quantity W^2/ρ_{fluid} for a turbulent fluid.

The target meter is a circular disk or target supported concentrically in the pipe thus giving an annular orifice (Fig. 22**c**). The force exerted on the target is proportional to the square of the flow rate. This force is transmitted to an external electrical or pneumatic transducer by the support rod which passes through a diaphragm seal. The open spaces at the top and bottom of the pipe makes the device particularly well suited to measuring fluids containing small amounts of gas or solids that tend to settle.

Measurement by Variable Area. Meters that operate on the principle of variable area incorporate an adjustable constriction in the path of the flow that may be varied so as to maintain a constant differential pressure. The most commonly used device of this type is the rotameter. This is a tapered vertical tube containing a free plummet or float (Fig. 22**d**). Fluid entering the lower, small end of the tube lifts the plummet until the velocity in the annulus between plummet and tube provides a dynamic force balance around the plummet. The level of the plummet indicates the rate of flow. Rotameters are supplied with glass and metal tubes to meet a variety of service conditions in metering clean liquids and gases. They can be used for manual or automatic control of flow, and can be adapted to indicating and recording instruments. For precise metering, a rotameter calibrated with the fluid that is to be metered is unsurpassed. With very large flow rates, especially for gases, the cost and bulk of the rotameter, as compared to an orifice meter, are important disadvantages. For a given position of the plummet, the mass rate of turbulent flow through the meter is proportional to the quantity: $[\rho_{fluid}\,(\rho_{plummet} - \rho_{fluid})]^{1/2}$. Corrections for nonturbulent flow are generally available from the meter manufacturer.

Measurement by Liquid Level. The flow rate of liquids flowing in open channels is often measured by the use of weirs. The most common type is the rectangular weir shown in Figure 22**e**. The flow rate across such a weir varies approximately with the quantity: $wh^{3/2}$. Other shapes of weirs are also employed. Standard civil engineering handbooks describe the precautions necessary for constructing and interpreting data from weirs.

Measurement by Electromagnetic Effects. The magnetic flow meter is a device which measures the potential developed when an electrically conductive flow moves through an imposed magnetic field. The voltage developed is proportional to the volumetric flow rate of the fluid and the magnetic field strength. The process fluid sees only an empty pipe so that the device has a very low pressure drop. The device is useful for the measurement of slurries and other fluid systems where an accumulation of another phase could interfere with flow measurement by other devices. The meter must be installed in a section of pipe that is much less conductive than the fluid. This limits its applicability in many industrial situations.

Measurement by Thermal Effects. When a fine wire heated electrically is exposed to a flowing gas, it is cooled and its resistance is changed. The hot wire anemometer makes use of this principle to measure both the average velocity and the turbulent fluctuations in the flowing stream. The fluid velocity V is related to the current I and the resistances R of the wire at wire (w) and gas (g) temperatures via:

$$\frac{I^2 R_w}{R_w - R_g} = A' + B'\sqrt{V} \tag{67}$$

The constants A' and B' can be calculated theoretically but, in practice, are usually determined experimentally.

Two modes of operation are available. In the simpler mode, a fixed current is applied to the wire. In the other mode, the sensor temperature (hence R_w) is maintained at a fixed value above that of the gas by means of a feedback amplifier. In both cases, velocity is measured by the voltage drop.

Because of its small size and portability, the hot-wire anemometer is ideally suited to measure gas velocities either continuously or on a troubleshooting basis in systems where excess pressure drop cannot be tolerated. Furnaces, smokestacks, electrostatic precipitators, and air ducts are typical areas of application. Its fast response to velocity or temperature fluctuations in the surrounding gas makes it particularly useful in studying the turbulence characteristics and rapidity of mixing in gas streams. The constant current mode of operation has a wide frequency response and relatively lower noise level, provided a sufficiently small wire can be used. Where a more rugged wire is required, the constant temperature mode is employed because of its insensitivity to sensor heat capacity. In liquids, hot-film sensors are employed instead of wires. The sensor consists of a thin metallic film mounted on the surface of a thermally and electrically insulated probe.

Laser Anemometry. When light is scattered from a particle moving at a velocity V its frequency is altered by the fraction: $(\cos \alpha' - \cos \beta')V/C$, where C is the velocity of light in the fluid and α' and β' are the angles between the particle track and the incident and scattered beams, respectively. This principle is employed in the laser anemometer, which uses a laser beam as the source of monochromatic light, particles in the fluid as scattering centers, and appropriate electronic circuitry to measure the frequency shift. This method has been successfully employed to study turbulent fluctuations and to measure local velocities in stirred vessels. It has an advantage over hot-wire/hot-film anemometers in that it does not disturb the flow, but the latter can measure over a smaller volume, with lower background noise and probably better frequency response. Complete hardware systems are commercially available. Because of its cost and the need for an optical path into the fluid, laser anemometry is largely confined to experimental work.

Nomenclature

A_p	= area of a body projected in a plane perpendicular to the flow
B	= control volume
C, C_i	= concentration, concentration of component i
C_D	= drag coefficient
C_p, C_v	= specific heat at constant pressure, constant volume
D	= diameter, eg, of pipe, nozzle, cylinder, strand
D	= average pore diameter in a porous bed
D_H	= diameter of helix
d	= diameter of drop or bubble
d_p	= diameter of particle in a porous bed
$\mathcal{D}$	= molecular diffusivity
E	= total energy of a unit mass of fluid
$E\ddot{o}$	= Eötvos number = $g\Delta\rho d^2/\sigma$
f	= Fanning friction factor
F	= force exerted on a body by a fluid
F_B, F_S	= body force, surface force acting on fluid in the control volume
F_i	= partial molal free energy of component i
ΔF	= irreversible energy dissipation per unit mass of fluid
g_c	= constant in Newton's second law = 9.81 m/s^2 = 32.17 ft/s^2
g_i	= component of gravitational acceleration in the i direction
h	= distance above datum plane; height of crest above weir; depth of liquid in channel
H	= enthalpy per unit mass of fluid
J_E	= nonconvective flux of energy through surface of control volume
J_i	= nonconvective flux of component i through surface of control volume
K	= constant in rheological equation for pseudoplastic or dilatant fluids
$\mathcal{K}$	= stiffness matrix
L	= characteristic length in a system; length to achieve a specified approach to fully developed flow; thickness of porous bed
l_b, l_m	= distance between plates (Benard instability); characteristic length (Marangoni instability)
Ma	= Marangoni number, $(\partial\sigma/\partial c)(\partial c/\partial x)g_c\ell_m^2/\mu\mathcal{D}$
Mo	= Morton number, $g\mu_c^4\Delta\rho/\rho_c^2\sigma^3$
n	= exponent or index in power-law rheological equation; number of time increments in finite difference or finite element calculation

n_i	= moles of component i per unit mass of fluid
n_s	= frequency of vortex shedding
P	= thermodynamic pressure
P^*	= $P + \rho gz/g_c$
Q	= heat added to a unit mass of fluid
r	= radial distance from axis
Ra	= Rayleigh number, $g(\partial\rho/\partial T)\Delta T \ell_b^3/\mu\alpha$
Re	= Reynolds number, $DV\rho/\mu$, $dV\rho/\mu$, etc
R, R_j	= radius of tube, radius at location j
R_i	= rate of generation of component i per unit volume
$\mathcal{R}$	= load matrix
S	= entropy; surface area
S_j	= cross-sectional area at location j
Sr	= Strouhal number = $n_s D/V_\infty$
t, Δt	= time; time increment
T, ΔT	= temperature, temperature difference
u, v, w	= components of velocity in the x, y, z directions
u_c	= velocity at the center line of a jet at distance x from the nozzle
$u_{i,j}$	= velocity component at point i,j
u_p, v_p	= velocities of marker points in the x,y directions
U	= internal energy
V	= velocity, characteristic system velocity
$\overline{V}$	= average velocity, eg, in a pipe
V_j	= velocity at location j; 1 = upstream, 2 = downstream
Vo, j	= velocity at centerline of pipe at location j
V_s	= superficial velocity in a porous bed
V_∞	= uniform velocity at a great distance from a bluff body
w	= width of weir
W_E	= work done by a fluid on its surroundings
W	= weight rate of flow
W_M	= mechanical work done by a fluid on its surroundings
$\overline{x}$	= Sauter mean drop size
x, y, z	= space coordinates used to describe flow. z also indicates height above datum plane
Δx, Δy, Δz	= widths of grid; differences between x, y, z coordinates
X, Y, Z	= body forces acting in x, y, z directions
α	= thermal diffusivity (Rayleigh number)
α_j	= factor expressing flux of kinetic energy in terms of average velocity = $(\overline{V_j^3}/\overline{V}_j^3)$
β_j	= factor expressing flux of momentum in terms of average velocity = $(\overline{V_j^2}/\overline{V}_j^2)$
$\dot{\gamma}$	= shear rate
δ	= characteristic width of a freely expanding jet at a distance x from the nozzle
ϵ	= porosity; size of roughness elements on pipe wall
$\zeta_{i,j}^n$	= vorticity at point i,j at time $n\Delta t$
$\vec{\zeta}$	= vector vorticity
θ	= angle from normal to surface or datum plane
η	= modulus of rigidity
μ, μ_c, μ_d	= shear viscosity; viscosity of continuous, dispersed phase
λ	= dilational viscosity; relaxation time
ν	= kinematic viscosity = μ/ρ
ρ, ρ_i	= density; density of fluid i
$\Delta\rho$	= density difference
ρ_c, ρ_d	= density of continuous phase, dispersed phase
σ	= interfacial tension
σ_i	= normal stress exerted in the i direction on a plane perpendicular to the i axis
τ, τ_w	= shear stress, shear stress at the wall
τ_{ij}	= shear stress exerted in the j direction on a plane perpendicular to the i axis
ϕ	= velocity potential function

ψ = stream function
$\vec{\psi}$ = vector potential
$\psi^n_{i,j}$ = stream function at point i,j at time $n\Delta t$
ω_1, ω_2 = density ratios $\rho_1/(\rho_1 + \rho_2)$, $\rho_2/(\rho_1 + \rho_2)$

BIBLIOGRAPHY

"Fluid Mechanics (Principles)" in *ECT* 1st ed., Vol. 6, pp. 614–640, by Thomas Baron, University of Illinois, and Mott Souders, Jr., Shell Development Company; "Fluid Mechanics (Flow Measurement)" in *ECT* 1st ed., Vol. 6, pp. 640–648, by Mott Souders, Jr., Shell Development Company; "Fluid Mechanics (Principles)" in *ECT* 2nd ed., Vol. 9, pp. 445–473, by Mott Souders, Chemical Engineer; "Fluid Mechanics (Flow Measurement)" in *ECT* 2nd ed., Vol. 9, pp. 473–483 by Mott Souders, Chemical Engineer.

1. R. B. Bird, W. E. Stewart, and E. H. Lightfoot, *Transport Phenomena,* John Wiley & Sons, Inc., New York, 1960.
2. R. S. Brodkey, *The Phenomena of Fluid Motions,* Addison-Wesley Publishing Co., Inc., Reading, Mass., 1967.
3. J. H. Lienhard, *Bulletin 300,* College of Engineering, Washington State University, Pullman, Wash., 1966.
4. L. Prandtl, *Uber Flussigkeitsbewegung bei sehr kleiner Reibung. Proc. Third Intern. Math. Congress, Heidelberg, 1904.*
5. H. Schlichting, *Boundary Layer Flow,* McGraw-Hill Book Company, New York, 1960.
6. J. R. Grace, T. Wairegi, and T. H. Nguyen, *Trans. Inst. Chem. Eng.* **54,** 167 (1976).
7. G. N. Abramovich, *The Theory of Turbulent Jets,* The M.I.T. Press, Cambridge, Mass., 1963.
8. J. Walker, *Sci. Am.* **238,** 168 (1978).
9. J. M. Beer and N. A. Chigier, *Combustion Aerodynamics,* John Wiley & Sons, Inc., New York, 1972.
10. S. A. Thorpe, *J. Fluid Mech.* **46,** 299 (1971).
11. D. Coles, *J. Fluid Mech.* **21,** 385 (1965).
12. R. A. Mugele and H. D. Evans, *Ind. Eng. Chem.* **43,** 1317 (1951).
13. H. J. Karam and J. C. Bellinger, *Ind. Eng. Chem. Fundam.* **7,** 576 (1967).
14. V. G. Levich, *Physicochemical Hydrodynamics,* Prentice-Hall, Inc., Englewood Cliffs, N.J., 1962.
15. P. J. Roache, *Computational Fluid Dynamics,* Hermosa Publishers, Albuquerque, N.M., 1976.
16. J. S. Son and T. J. Hanratty, *J. Fluid Mech.* **35,** 376 (1969).
17. S. Taneda, *J. Phys. Soc. Jpn.* **11,** 302 (1956).
18. F. H. Harlow and J. E. Welch, *The Physics of Fluids* **2,** 2182 (1965).
19. F. H. Harlow and J. P. Shannon, *J. Appl. Phys.* **38,** 3859 (1967).
20. C. A. Brebbia and J. J. Connor, ed., *Numerical Methods in Fluid Dynamics,* Crane, Russak and Co., Inc., New York 1974.
21. J. O. Hinzi, *Turbulence,* McGraw-Hill Book Co., Inc., New York, 1959.
22. A. D. Gosman and co-workers, *Heat and Mass Transfer in Recirculating Flows,* Academic Press, Inc., New York, 1969.
23. B. E. Launder and D. B. Spalding, *Mathematical Models of Turbulence,* Academic Press, Inc., New York, 1971.
24. J. E. Cermak, *J. Fluids Eng.* **97,** 9 (1975).
25. L. K. Spink, *Principles and Practice of Flow Meter Engineering,* The Foxboro Company, Foxboro, Mass., 1967.

A. M. Benson
G. Q. Martin
J. S. Son
C. V. Sternling
Shell Development Company

FLUORAPATITE, $Ca_{10}F_2(PO_4)_6$. See Fertilizers.

FLUORESCEIN. See Xanthene dyes.

FLUORESCENT PIGMENTS (DAYLIGHT). See Brighteners, fluorescent.

FLUORINE

Fluorine [*7782-41-4*], F_2, is a diatomic molecule that exists as a pale yellow gas at ordinary temperatures. It is formed by combination of the element F, atomic number 9, the first member of the Group VII halogens of the periodic table. The name is derived from the Latin word *fluere,* meaning to flow, alluding to the well-known fluxing power of the mineral fluorite, the most abundant naturally occurring compound of the element. The single naturally occurring isotope has an atomic weight of 18.9984. Other radioactive isotopes between atomic weight 17 and 22 have been artificially prepared and have half-lives between 4 s for ^{22}F, and 110 min for ^{18}F (1).

Fluorine has its place in the extreme upper right corner of the periodic system and is the most electronegative element and the most reactive nonmetal. The electron configuration of the atom is $1s^22s^22p^5$. The tendency to complete the outer shell with eight electrons explains the extreme chemical activity of fluorine.

The only feasible method for preparing elemental fluorine is by electrolysis. Fluorine was first isolated in 1886 by the French chemist Moissan (2) who applied a method originally suggested and unsuccessfully tried by Davey and Ampere in 1810–1812. Moissan's success was owing to the use of a dilute solution of potassium fluoride in completely anhydrous hydrogen fluoride, giving an electrolyte corresponding approximately to KF.12HF.

Fluorine was first produced commercially ca 50 years after its discovery. In the intervening time, fluorine chemistry was restricted to the development of various types of electrolytic cells on a laboratory scale. In World War II, the demand for uranium hexafluoride, UF_6, in the United States and England, and chlorine trifluoride, ClF_3, in Germany, led to the development of commercial fluorine-generating cells. Currently, the main use of fluorine is in the production of UF_6 for the nuclear electric power industry.

Fluorine does not occur free in nature, except for trace amounts in radioactive materials, but is widely found in combination with other elements and accounts for ca 0.065 wt % of the earth's crust (3). The most important natural source of fluorine for industrial purposes is the mineral fluorspar, CaF_2, contaning about 49% fluorine. The most significant minerals are listed in Table 1. Among the elements, fluorine is about thirteenth in abundance.

Table 1. Fluorine-Containing Minerals

Mineral	CAS Registry No.	Formula	Fluorine, %
fluorite (fluorspar)	*[14542-23-5]*	CaF_2	49
cryolite	*[15096-52-3]*	Na_3AlF_6	45
fluorapatite	*[1306-05-4]*	$Ca_5(PO_4)_3F$	3–4
topaz	*[1302-59-6]*	$Al_2SiO_4(OH, F)_2$	
sellaite	*[7783-40-6]*	MgF_2	61
villiaumite	*[7681-49-4]*	NaF	55
bastnaesite	*[1299-84-9,*	$(Ce, La)(CO_3)F$	9
fluorine hydrosilicates	*68909-12-6*, 68909-13-7*]*	n $Mg(OH, F)_2.4MgSiO_4$ + m $TiO_2.4Mg_2TiO_2$	

The largest reserves of fluorine are in phosphate rock deposits with low fluorine content. Reserves of fluorine in phosphate rock containing 3–4% fluorine are estimated at over 341×10^6 metric tons (see Table 2). Phosphate rock is found in many forms in practically all parts of the world. The largest known source of fluorine in the United States exists in phosphate rock deposits in Florida. The recovery of fluorine in the phosphate industry has not been economical; however, with the diminishing supply of high-grade fluorspar, it has been increasing in importance as a secondary source of fluorine. In 1977, ca 7% of domestic fluorine consumption was obtained from fluorosilicic acid recovered in the phosphate industry (4).

Current estimates are that the reserves of fluorspar will be depleted toward the end of the century. The U.S. Bureau of Mines estimates that 86×10^6 t of fluorine will be consumed worldwide by the year 2000. Since this is above the present known reserves of fluorspar, the world supply will have to be derived from new fluorspar deposits or phosphate rock and from more effective recovery and use by industry. It is anticipated that recovery from phosphate rock will become more prevalent, since these reserves are adequate to supply world demand well into the next century.

Total fluorine reserves in phosphate rock are estimated at 789×10^6 t (5). The U.S. Bureau of Mines (4) more conservatively estimated 341×10^6 t but this includes only that portion which is considered to be recoverable.

A complete survey of fluorine resources and locations of major reserves throughout the world can be found in ref. 3.

Physical Properties

Fluorine is a pale yellow gas that condenses to a yellowish-orange liquid at −188°C. It solidifies to a yellow solid at −220°C and turns white in a phase transition at −228°C. Fluorine has a strong odor which is easily detectable at concentrations as low as 20 ppb. The odor resembles that of the other halogens and is comparable to strong ozone.

Because of the extreme difficulty in handling fluorine, reported physical properties (see Table 3) show more than normal variations among investigators. A detailed summary and correlation of the physical, thermodynamic, transport, and electromagnetic properties of fluorine is given in ref. 20.

Table 2. Identified World Fluorine Resources in Fluorspar and Phosphate Rock in 1977[a]

Area	Fluorine reserves, t × 10^3	Other resources[b], t × 10^3	Total[b], t × 10^3
Fluorine in fluorspar[c]			
North America			
United States	2,450	12,700	15,150
Canada	820	450	1,270
Mexico	5,730	2,090	7,820
Total	*9,000*	*15,240*	*24,240*
South America			
Argentina	730	730	1,460
Brazil	820	820	1,640
Total	*1,550*	*1,550*	*3,100*
Europe			
France	1,270	450	1,720
GDR[d]	640	640	1,280
FRG[e]	540	540	1,080
Italy	1,000	3,270	4,270
Spain	1,540	910	2,450
UK	3,180	1,810	4,990
USSR	2,180	1,180	3,360
Total	*10,350*	*9,430*	*19,780*
Africa			
Republic of South Africa[f]	5,710	1,130	6,840
Southwest Africa	910	910	1,820
Kenya	1,180	540	1,720
Morocco	540	640	1,180
Zimbabwe Rhodesia	90	90	180
Tunisia	450	90	540
Total	*8,880*	*3,400*	*12,280*
Asia and Oceania			
Australia[g]	180	180	360
People's Republic of China	820	450	1,270
India	1,270	170	1,440
Democratic People's Republic of Korea	90	90	180
Republic of Korea	90	90	180
Mongolia	720	720	1,440
Thailand	1,540	3,990	5,530
Total	*4,710*	*5,690*	*10,400*
World total (rounded)	*34,000*	*35,000*	*69,800*
Fluorine in phosphate rock (fluorapatite)[h]			
United States	9,980	18,140	28,120
rest of world	61,680	251,200	312,880
World total (rounded)	*71,660*	*269,340*	*341,000*

[a] Ref. 4.
[b] Derived in collaboration with U.S. Geological Survey.
[c] Fluorine content obtained by multiplying tons of fluorspar by grade of ore and by 0.45.
[d] Includes Czechoslovakia and Romania.
[e] Includes Greenland and Sweden.
[f] South African proven reserves are estimated to be in excess of 12×10^6 t.
[g] Includes Turkey.
[h] It is assumed that two thirds of phosphate rock production will be wet-processed and that only 22% of contained fluorine will be recovered therefrom. Depending on locality, phosphate rock contains 3–4% fluorine.

Table 3. Physical Properties of Fluorine

Property	Value	Refs.
melting point, °C	−219.61[a]	6
	−217.9	7
	−223	8
boiling point, °C	−188.13[a]	6
	−187.7	7
	−187.0	8
	−188.22	9
	−188.03	10
solid transition temperature, °C	−227.60	6
critical temperature, °C	−129.2[a]	11
	−129.00	9
critical pressure, kPa[b]	5571	9, 11
heat of vaporization at −188.44°C and 98.4 kPa, J/mol[c]	6544	6
heat of fusion, J/mol[c]	510	6
heat of transition, J/mol[c]	727.6	
heat capacities, J/(mol·K)[c]		
solid at −223°C	49.338	6
at −238°C	31.074	6
	23.267	7
at −253°C	12.987	6
	9.372	7
liquid	57.312[a]	6
	45.35	7
gas, C_p	31.46[a]	12
	31.456	13
	31.380	14
	31.325	15
density of liquid at bp, kg/m³	1516[a]	16
	1514	17
density of solid, kg/m³	1900[d]	18
refractive index		
liquid at bp	1.2	19
gas at 0°C and 101.3 kPa[b]	1.000214	20
surface tension, liquid mN/m (= dyn/cm)		
at −193.26°C	14.81	21
at −192.16°C	14.60	16
viscosity, mPa·s (= cP)		
liquid at −187.96°C	0.257	16
at −203.96°C	0.414	16
gas at 0°C and 101.3 kPa[b]	0.218	24
	0.209	7
thermal conductivity, gas at 0°C and 101.3 kPa[b], W/(m·K)	24.77×10^{-7}	23
dielectric constant, E		
at −189.95°C	1.517	7
at −215.76	1.567	7
vapor pressure at 53.56 K, kPa[b]	0.22	6
at 63.49 K	2.79	
at 72.56 K	18.62	
at 83.06 K	80.52	
at 89.40 K	162.11	

[a] Generally accepted value.
[b] To convert kPa to mm Hg, multiply by 7.5.
[c] To convert J to cal, divide by 4.184.
[d] Mean estimated value.

Chemical Properties

Fluorine is the most reactive element, combining with most organic and inorganic materials at or below room temperature. Organic and hydrogen-containing compounds especially can burn or explode when exposed to fluorine. Fluorine forms compounds, in which it shows valence minus one, with all the elements except helium, neon, and argon. Elemental fluorine reacts directly with the noble gases xenon, radon, and krypton to form fluorides. Nitrogen and oxygen form fluorides but do not react directly with fluorine except in the presence of an electric discharge.

Fluorine is the most electronegative element with a normal potential of −2.85 V. Its extreme electronegativity is responsible for its ability to oxidize many elements to their highest oxidation state. The small size of the atom permits the arrangement of a large number of fluorine atoms around an atom of another element. These properties of high oxidation potential and small size allow the formation of many simple and complex fluorides in which elements show their highest oxidation states.

The reactivity of fluorine compounds varies from extremely stable compounds such as sulfur hexafluoride, nitrogen trifluoride, and the fluorocarbons to extremely reactive compounds such as the halogen fluorides. Another unique property of the nonionic metal fluorides is their volatility. Volatile compounds such as tungsten hexafluoride, WF_6, and molybdenum hexafluoride, MoF_6, are produced by the reaction of the metal with elemental fluorine.

Fluorine is the first member of the halogen family but in many ways its properties are not typical of the other halogens. Fluorine has only one valence, −1, whereas the other halogens form halogen fluorides with valences of 1, 3, 5, and 7. Fluorine has the lowest enthalpy of dissociation relative to the other halogens, which is in part responsible for its greater reactivity. Furthermore, the strength of the fluorine bond with other atoms is greater than that of the other halogens. Table 4 shows the bond energies for the four halogens, as well as the bond energies for their halides with hydrogen, carbon, boron, and aluminum.

From these data, it can be appreciated that when halogenation reactions are compared, the enthalpies of fluorination are much greater than those of other halogenations. Less energy is required to form fluorine atoms and much more energy is evolved in the formation of the fluorides. Therefore, fluorination reactions occur more readily with the liberation of intense heat and frequently occur in situations where other halogenations do not.

Table 4. Average Bond Energies, kJ[a]

	XX[b,c]	HX[d]	BX_3[d]	AlX_3[e]	CX_4[d]
F	157.8	569	645	582	456
Cl	243.7	431	444	427	327
Br	193.0	368	368	360	272
I	151.1	297	272	285	239

[a] To convert J to cal, divide by 4.184.
[b] X = halogen.
[c] Ref. 24.
[d] Ref. 25.
[e] Ref. 26.

Reactions. ***Metals.*** At ordinary temperatures, fluorine reacts vigorously with most metals to form fluorides. A number of metals, including aluminum, copper, iron, and nickel, form a protective surface film of the fluoride, so that the temperature must be increased to obtain further reaction. Susceptibility to attack depends, to a great extent, on the physical state of the metal. Powdered iron 0.84-mm size (20 mesh) is not attacked by liquid fluorine, whereas in the 0.14-mm size (100-mesh) it ignites and burns violently. There is no apparent reaction between liquid fluorine and powdered nickel as fine as 0.14 mm. Iron burns with fluorine at 672°C. Massive copper burns at 692°C, whereas copper wool ignites at a much lower temperature. Nickel burns in fluorine at 1147°C, and aluminum above its melting point (27).

Tin reacts completely with fluorine above 190°C to form tin tetrafluoride, SnF_4. Titanium reacts appreciably above 150°C at a rate depending upon the size of the particles. The conversion to titanium tetrafluoride, TiF_4, is complete above 200°C. Fluorine reacts with zirconium metal above 190°C. However, the formation of a coating of zirconium tetrafluoride, ZrF_4, prevents complete conversion. The reaction is 90% complete at 420°C (28). Tungsten powder reacts with fluorine at a temperature above 250°C to produce tungsten hexafluoride, WF_6.

Nonmetals. Sulfur reacts with fluorine to yield the remarkably stable sulfur hexafluoride, SF_6. A mixture of the lower fluorides such as disulfur difluoride, S_2F_2, disulfur decafluoride, S_2F_{10}, and sulfur tetrafluoride, SF_4, is also obtained as byproduct; operating conditions must be controlled to minimize their formation. In the range of 310–340°C, SF_4 is primarily obtained with essentially no SF_6 and only trace amounts of lower fluorides. Below 300°C and preferably at ca 275°C, SF_6 is the primary product. At 450–500°C, a mixture comprising ca 50% SF_4 and the lower sulfur fluorides is formed (see Fluorine compounds, inorganic—sulfur).

Silicon and boron burn in fluorine forming silicon tetrafluoride, SiF_4, and boron trifluoride, BF_3, respectively. Selenium and tellurium form hexafluorides, whereas phosphorus forms tri- or pentafluorides. Fluorine reacts with the other halogens to form eight interhalogen compounds (see Fluorine compounds, inorganic—halogens).

Water. Fluorine forms hydrofluoric acid, HF, and oxygen difluoride, OF_2, with oxygen. In dilute (<5%) caustic solutions, the reaction proceeds as follows:

$$2\,F_2 + 2\,NaOH \rightarrow OF_2 + 2\,NaF + H_2O$$

In the presence of excess caustic, the oxygen difluoride is gradually reduced to oxygen and fluoride:

$$OF_2 + 2\,NaOH \rightarrow 2\,NaF + O_2 + H_2O$$

The overall reaction under controlled conditions provides a method for the disposal of fluorine by conversion to a soluble salt:

$$2\,F_2 + 4\,NaOH \rightarrow 4\,NaF + O_2 + 2\,H_2O$$

Oxygen. Oxygen does not react directly with fluorine under ordinary conditions, although in addition to oxygen difluoride, three oxygen fluorides are known (29). Dioxygen difluoride, O_2F_2, trioxygen difluoride, O_3F_2, and tetraoxygen difluoride, O_4F_2, are produced in an electric discharge at cryogenic temperatures by controlling the ratio of fluorine to oxygen.

Nitrogen. Nitrogen does not react with fluorine under ordinary conditions and is often used as a diluent for moderate fluorinations. However, it can be made to produce nitrogen trifluoride, NF_3, by radiochemistry (30), glow discharge (31), or plasma (32) synthesis.

Noble Gases. Fluorine has the unique ability to react with the heavier noble gases (krypton, xenon, and radon) to form binary fluorides. With xenon at RT in uv radiation a difluoride, XeF_2, is formed (16–18) (21,33–34). A tetrafluoride, XeF_4, is obtained by mixing an excess of fluorine with xenon and heating to 400°C (33,35). Fluorine and xenon at 300°C under 6–6.8 MPa (60–67 atm) yield the hexafluoride, XeF_6 (36–37). Fluorine combines with radon at 400°C to yield a compound of low volatility, probably the difluoride (38). Krypton reacts with fluorine in an electric discharge at liquid air temperatures to yield the difluoride, KrF_2 (39).

Hydrogen. The reaction between fluorine and hydrogen is self-igniting and extremely energetic. It occurs spontaneously at ambient temperatures as evidenced by minor explosions which sometimes occur in fluorine-generating cells from the mixing of the H_2 and F_2 streams. The controlled high-temperature reaction of fluorine atoms (generated thermally or photolytically from fluorine gas) with hydrogen or deuterium is one of the most significant energy sources for the high power chemical laser. Over 50% of the energy released in the reaction of the hydrogen molecule with the fluorine atom appears preferentially in the vibrational energy mode of the newly formed diatomic molecules (40). On this basis, a substantial fraction of the exothermic energy of the overall hydrogen–fluorine reaction (perhaps 25%) could be released as laser radiation. Currently, the HF–DF laser is the most promising chemical laser system under development (41) (see Lasers; Hydrogen energy).

Ammonia. Ammonia (qv) reacts with excess fluorine in the vapor phase to produce N_2, NF_3, N_2F_2, HF, and NH_4F. This reaction is difficult to control in the vapor phase because of the intense heat of reaction, and in some cases only N_2 and HF are produced. Nitrogen trifluoride was obtained in 6% yields in a gas-phase reaction over copper (42). Yields of ca 60% are achieved by the reaction of fluorine and ammonia in a molten ammonium acid fluoride solution (43–44).

Organic Compounds. Generally, the reaction of fluorine with organic compounds is accompanied by ignition or violent explosion. Since the heats evolved are always high, heat removal is the main problem in direct fluorination. Reactions can be moderated by dilution with large amounts of an inert gas, such as nitrogen, by reducing reaction temperatures ($\leq$78°C), or by the presence of finely divided metal packing. The latter is used in the so-called catalytic fluorination of organic vapors, originally developed in the United States (45). It takes place over a metal reactor packing such as copper gauze, metal screen, wire, chips, ribbon, or shot, which control the reaction by reducing local heating.

Saturated hydrocarbons under controlled conditions react with elemental fluorine to give perfluorocarbons, usually accompanied by some fragmentation and polymerization (46–47). The fluorination of aromatic compounds gives degradation products, polymers, unstable unsaturated compounds, or highly fluorinated cyclohexane derivatives, but no aromatic compounds. Methanol and acetone give a variety of fluorinated carbonyl compounds (48–49) (see Fluorine compounds, organic-direct fluorination).

The direct fluorination of uracil (2,4-pyrimidinedione), $C_4N_2H_4O_2$, in aqueous solution to give 5-fluorouracil (5-fluoro-2,4-pyrimidinedione), $C_4N_2H_3FO_2$, has been recently reported (50) in yields of 8–78%, depending on reaction conditions. The higher yield was obtained by starting with an aqueous solution containing 1 wt % uracil. A diluted F_2 stream (33% F_2–67% N_2) was passed into the solution at 90°C until the total fluorine: uracil ratio was 1.67: 1.

Polymers. The dilution of fluorine with an inert gas such as nitrogen or helium significantly reduces its reactivity so that even reactions with hydrocarbon polymers at elevated temperatures can be controlled. High density polyethylene containers can be blow-molded with 1–10% fluorine in nitrogen mixtures to produce a fluorocarbon barrier layer on the inside of the container (51). The permeation rate of nonpolar solvent such as n-pentane can be reduced by a factor of almost 500 when containers are blow molded with a 1% fluorine-in-nitrogen mixture (52–53). Comparative permeation rate data for 24 common solvents in fluorinated and nonfluorinated high density polyethylene containers are presented in refs. 52–53. A mixture of 5–10% fluorine in nitrogen imparts a fluorocarbon barrier layer to polyester, polyamide, polyolefin, and polyacrylonitrite fibers, so that moisture transport, soil release, and soil redeposition are improved (54–55).

Carbon and Graphite. Fluorine reacts with amorphous forms of carbon, such as wood charcoal, to form carbon tetrafluoride, CF_4, along with small amounts of other fluorocarbons. The reaction initiates at ambient conditions but proceeds to elevated temperatures as the charcoal burns in fluorine. It is the basis for a simple, economical means of waste fluorine disposal (56–57).

Fluorine reacts with high-purity carbon or graphite at elevated temperatures under controlled conditions to produce a polycarbon monofluoride, $(CF_x)_n$. Compounds of various colors have been prepared with fluorine content ranging from $x = 0.1$ to $x = 1.3$ (58–62). The material was first obtained (58) in 1934, when graphite was heated to 420–460°C in a stream of fluorine to produce a gray product of composition $CF_{0.92}$. A white material of composition $CF_{1.12}$ was obtained (59) by accurate control of the reaction temperature to 627 ± 3°C. The polycarbon monofluorides can be utilized as a superior solid lubricant (63) and as a cathodic material in high-energy batteries (64).

Manufacture

Fluorine is produced by the electrolysis of anhydrous potassium bifluoride, KHF_2 or KF.HF, containing various concentrations of free HF. The fluoride ion is oxidized at the anode to liberate fluorine gas, and the hydrogen ion is reduced at the cathode to liberate hydrogen. Anhydrous HF cannot be used directly as the electrolyte because of its low conductivity.

Fluorine-generating cells are classified into three distinct types, depending on their operating temperature: low temperature cells (−80 to 20°C), medium temperature cells (60–110°C), and high temperature cells (220–300°C). Figure 1 is a melting point diagram for the KF–HF system and shows the three distinct areas of temperature operation and the corresponding HF concentrations in the electrolyte. Reference 65 presents a complete description of various laboratory and industrial fluorine cells worldwide.

Cells operating at low (2,66–67) and high (65,68) temperatures were developed first but were discontinued because of corrosion and other problems. The first medium temperature cell had an electrolyte composition corresponding to KF.3HF and op-

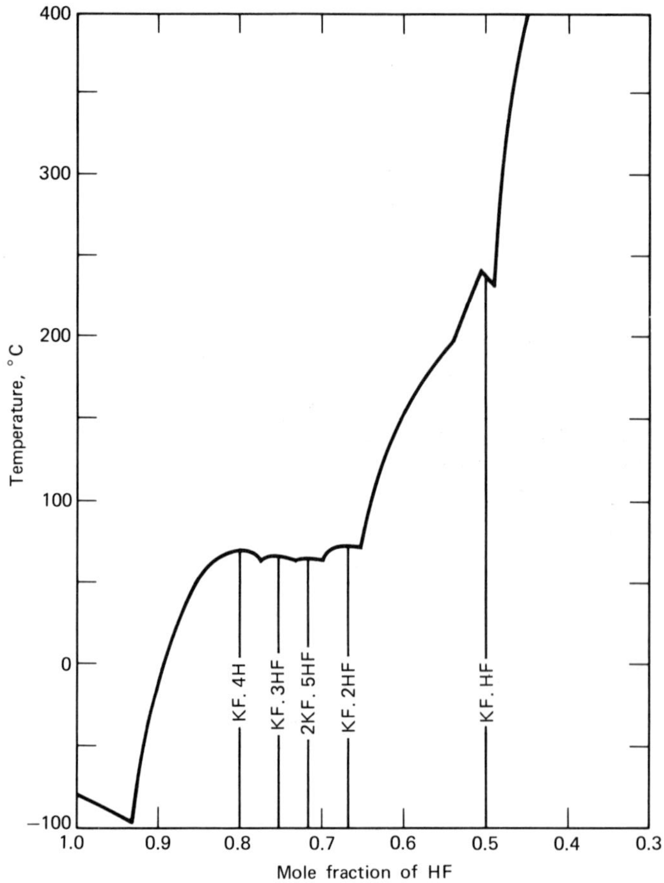

Figure 1. Phase diagram of the potassium fluoride–hydrogen fluoride system.

erated at 65–75°C with a copper cathode and nickel anodes. A later cell operated at 75°C and used KF.2.2HF or KF.2HF as electrolyte (69–70) and nickel and graphite as anode materials.

Commercial Cells. Today all commercial fluorine installations employ medium temperature cells with operating currents of ≥6000 A. Carbon anodes overcome the excessive corrosion encountered with metal. The medium temperature cell offers the following advantages over low and high temperature cells: (*1*) the vapor pressure of HF over the electrolyte is less; (*2*) the composition of the electrolyte can vary over a relatively wide range for only a small variation in the operation of the cell; (*3*) reduced corrosion; (*4*) tempered water can be used as cell coolant; and (*5*) the formation of a highly resistant film on the anode surface is considerably reduced compared to the high temperature cell.

The C and E type of the Atomic Energy Commission (AEC) (now Department of Energy) cell design (71–74) are used primarily in the United States and Canada. They were developed by the Union Carbide Nuclear Company under the auspices of the AEC as part of the overall manufacturing process for uranium hexafluoride, UF_6. Large fluorine-generating plants using this design were installed at the Paducah, Oak Ridge, and Portsmouth gaseous diffusion plants. The AEC designs have been made available to industry and are currently used by several commercial producers. The other cell type used in the United States is a proprietary design developed by Allied Chemical Corporation. It has a reported capacity of 5000 A and is used by Allied Chemical Corp. at its Metropolis, Illinois, plant.

AEC Cell. The main components are the cell tank, cell head, anode assembly, cathode assembly, screen diaphragm, and packing gland (72,74–75). The E-type and C-type, are similar in design and are both rated at 6000 A maximum capacity. The E-type incorporated design changes to improve heat removal efficiency and to prolong life of the anode-contact connection. The cell tank is constituted of 95-mm Monel plate surrounded by a thin-gauge Monel jacket to overcome corrosion problems encountered with a steel jacket. A recessed bolt carbon-plug fastens the anodes to a copper support bar. This design gives improved cell life over the C-type copper pressure-plate design using a steel anode support bar.

The cell head is fabricated from a 2.54-cm steel plate and has separate compartments for fluorine and hydrogen. The outlet-gas manifolds, hydrogen fluoride feed and purge lines, and electrical connections are on top of the head. The gas separation skirt (see below) is made of Monel. An insulating gasket maintains the seal between the tank and the head. The anode assembly consists of 32 carbon blades bolted onto a copper bar, each of which contains 3 copper conductor posts. The cathode assembly consists of three vertical, 0.6-cm parallel steel plates. The plates surround the anode assembly and are supported by three steel posts which also serve as conductors. Table 5 gives the typical operating characteristics of the AEC E-type cell.

Allied Cell. A cross section of the Allied Chemical Corp. cell is shown in Figure 2. The cell body is a rectangular steel box on wheels with an outside water-cooling jacket. A central partition divides the box lengthwise. The negative side of the d-c bus is connected directly to the cell container. The interior of the box and the central partition act as the cathodic area of the cell.

Cell Components. ***Anodes.*** The anode design and materials are key components in the design of fluorine cells (65). Initially, nickel was the most satisfactory metal for use up to 100°C but its high corrosion rate resulted in sludge formation and poor economics. Graphite is suitable for use around 260°C, and specially prepared hard carbon anodes (eg, baked carbon as opposed to graphite) can be used in the range of 70 105°C.

The anode connection is still the most serious cell design problem. The limiting factor in the operating life of a fluorine cell has always been the deterioration of the contact between the carbon anode and the copper anode bar. This contact is exposed

Table 5. AEC Cell Operating Characteristics

Characteristic	Value
current, A	6000
operating voltage, V	9–12
cell operating temperature, °C	90–105
hydrogen fluoride in electrolyte, %	40–42
effective anode area, m^2	3.9
anode current density, A/m^2	13
anodes	32
life, A·h	$40–80 \times 10^6$

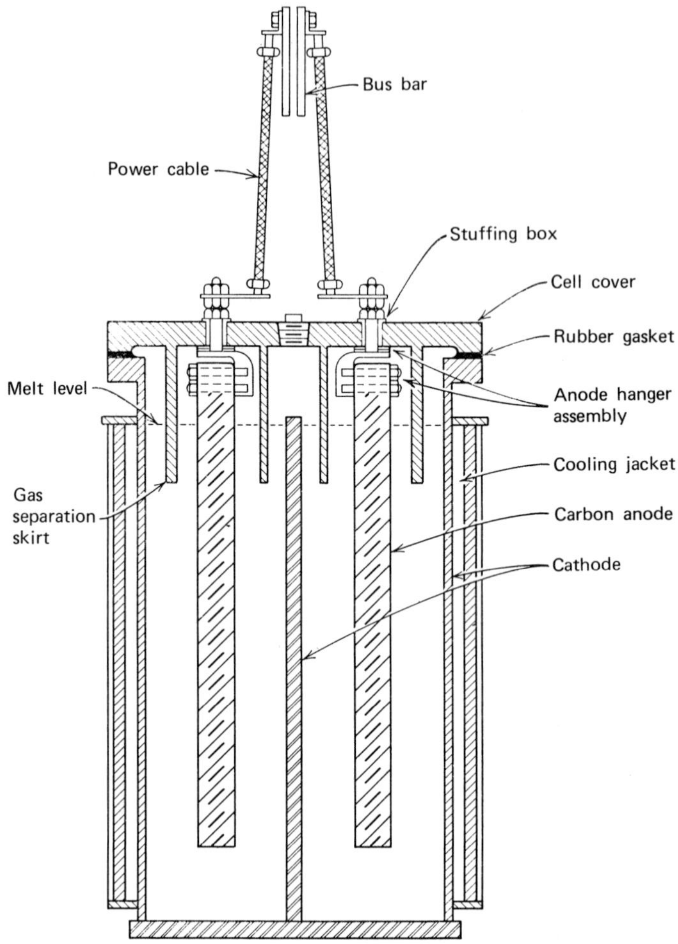

Figure 2. Fluorine generator. Courtesy of Allied Chemical Corp.

to fluorine and hydrogen fluoride and deteriorates by chemical and electrical action.

Since the start of the first large-scale fluorine cell operation, anode life has improved from a few thousand A·h to tens of millions of A·h service. Anode cell life of up to 85 MA·h is reported for AEC E-type cell design (76) which uses the so-called recessed bolt design.

Deterioration of the carbon anode itself also causes operating problems with the cells. The type of carbon used must provide low electrical resistance, high physical strength, resistance to corrosion attack, stability of the electrical contact with the bus bar, and resistance to polarization. Many types of carbon have been used with the Union Carbide YBD-grade carbon, a hard ungraphitized carbon made from petroleum coke, being the most satisfactory (see Carbon).

Cathodes. American fluorine manufacturers use Monel or steel cathodes. The early German investigators used magnesium cathodes without excessive corrosion in the high temperature cells.

Cell Body. American producers use welded steel or Monel construction for the cell body. The German cells used magnesium tanks consisting of an alloy of 98% Mg and 2% Mn. In the Allied cell the steel body is the cathode. Mild steel is satisfactory for the cell cover.

Skirt and Diaphragm. Skirts are used to separate the hydrogen and fluorine above the electrolyte. The solid metal skirt is welded to the cell cover plate and extends vertically downward 10–15 cm into the electrolyte. In the AEC cells, the skirt extends further into the electrolyte as a wire mesh. This extension is called the diaphragm and acts to direct the flow of gases as they are liberated. The AEC cells use Monel skirts and diaphragms. The Allied cell uses a magnesium alloy skirt and does not employ a diaphragm.

Insulation and Gasket. Polytetrafluoroethylene (PTFE) provides the most satisfactory electrical insulation. Concentric rings of PTFE and PTFE impregnated with calcium fluoride are used for the packing glands which support the anode and cathode posts. Rubber is used as the gasket material to form a seal between the cover and the cell body.

Jacket and Cooling Coils. Cells must be fitted with mild steel jackets and/or coils to remove heat during cell operation and to provide heat to maintain the electrolyte molten during shutdown. All commercial cell are totally jacketed. However, the accumulation of corrosive products can cause flow restrictions, resulting in decreased heat transfer capacity. This problem was overcome in the AEC E-type cell by using a water jacket constructed of thin-gauge Monel with vertical corrugations to provide strength. The AEC cells also use internal Monel tubes, manifolded to the external jacket, for additional heat transfer area. Since operating current in fluorine cells is limited by heat removal capability, improvements in jacket and tube design were a key factor in increasing the current-handling capacity of commercial cells.

Heat Transfer. Approximately 34.8 MJ (33,000 Btu) must be removed per kilogram of fluorine produced from any fluorine cell. This is accomplished by jacketing the cell and/or by using cooling tubes. The temperature of the cooling water should not drop below 58°C in order to avoid crystallization of bifluoride on the cell wall. The overall heat transfer coefficient, U (thermal conductance), depends on operating current and inlet cooling water temperature (72). At 3000 A and an inlet water temperature of 40°C, an average U value of 109.5 W/(m^2·K) [19.3 Btu/(h·ft^2·°F)] was calculated. For an operating current of 4000 A and a water temperature of 57°C, an average U value of 177.6 W/(m^2·K) [31.3 Btu/(h·ft^2·°F)] was calculated.

Polarization. In polarization, the potential required to operate the cell rises sharply, the current drops, and little or no fluorine is produced. This is a troublesome problem which is usually encountered during start-up and occasionally during the life of the cell; it can result in cell failure. Polarization is primarily associated with fluorine generation at the anode. The apparent cause is the formation on the anode surface of a gas film that acts as an electrical insulator. It is generally associated with localized high current densities at the anodes. In many cases, the loss of anode area by anode breakage or by failure of the electrical contact results in polarization.

Anode surface area may be lost by sludge deposits that partially cover the anodes, dispersed solids or impurities in the electrolyte which coat the anode and render portions of it ineffective, and electrolyte freezing on the anode. Polarization has also been attributed to rapid additions of HF and to the presence of impurities (particularly water) in the electrolyte, or to nonwetting of the anode surface by the electrolyte and to localized high current densities (66,77).

Polarization not attributed to physical reduction in anode area can usually be corrected by shutting the cell down and applying high voltage for 3–5 min. After a short cooling-off period, the cell, in most cases, behaves normally. Anode polarization can also be prevented by the use of highly permeable anodes (66), the addition of lithium fluoride (75), and the conditioning of new cells by starting up in 500 A increments for 2-h periods (75).

Raw Material. The principal raw material for fluorine production is high-purity anhydrous hydrofluoric acid. Each kilogram of fluorine generated requires ca 1.1 kg HF. The hydrofluoric acid consumed in fluorine production represents only ca 2% of the total United States hydrogen fluoride consumption. The commercial grade is acceptable for use as received, provided water content is less than 0.01%. Typical specifications for hydrofluoric acid are given in Table 6. Potassium bifluoride, KF.HF, is used as a raw material to charge the cells initially and for make-up when cells are rebuilt. A newly charged cell requires about 1400 kg KF.HF. Overall consumption of KF.HF/kg of fluorine generated is small. Commercial-grade flake potassium bifluoride is acceptable, and its specifications are given in Table 7.

Process. The generation of fluorine on an industrial scale is a complex operation (75,78). The basic raw material, anhydrous hydrogen fluoride, is stored in bulk and charged to a holding tank from which it is continuously fed to the cells. Electrolyte for the cells is prepared by mixing KF.HF with HF to form KF.2HF. The newly charged cells are depolarized by starting up at a low current, which is gradually increased at a conditioning station separate from the cell operating position until full current is obtained at normal voltages. The cells are series connected electrically, with dc supplied by a 6000 A low voltage rectifier. Rectifier capacity of ca 12 V is provided for each cell. Hydrogen fluoride content is maintained between 40 and 42% by continuous additions. The electrolyte level must be set and controlled at a definite level below the cell head in order to maintain a seal between the fluorine and hydrogen compartments. The cells are operated at 95–105°C and cooled with water at 75°C.

Table 6. Specifications for Hydrofluoric Acid

Assay	Wt %
HF, min	99.9
SO_2, max	0.01
H_2SiF_6, max	0.05
H_2O, max	0.05[a]
nonvolatile acid (as H_2SO_4), max	0.005

[a] Typical water content of commercial grade is below 0.001%.

Table 7. Specifications for Potassium Bifluoride

Assay	Wt %
KF.HF, min	99.3
KF, max	0.5
SO_4, max	0.01
Fe, max	0.03
Pb, max	0.005
weight loss at 105°C, max	0.05

Approximately 142.3 MJ/h (135,000 Btu/h) must be removed at a 6000 A operation. The hydrogen and fluorine gas leaving the cell contain ca 10 vol % HF. The individual gas streams from each cell are joined into separate hydrogen and fluorine headers for further processing in the plant. Demisters and filters are provided in the product gas streams to remove entrained electrolyte. The gas streams are then cooled to −110°C in Freon-cooled condensers to reduce the HF concentration to approximately 3 mol %. The condensed HF is recycled, and the hydrogen stream is scrubbed with a caustic solution and vented or burned. Several possibilities are available for utilizing the fluorine stream. If it is used in a process that can tolerate 3% HF, it is used as such. If lower HF levels are required, sodium fluoride towers or further cooling are employed to freeze out the HF and reduce the concentration to less than 0.2 mol %. Compressors or exhausters are normally required in both fluorine and hydrogen streams at some point in the system beyond the HF condensers.

Figure 3 presents the equipment flow sheet for a 10-t/d fluorine plant and Figure 4 gives the material balance data for this plant.

Equipment

Fluorine can be handled with a variety of materials (79–82). System cleanliness and passivation are critical to success. Materials and alloys, such as nickel, Monel, aluminum, magnesium, copper, brass, stainless steel, and carbon steel, are commonly used.

For liquid fluorine, low-temperature service, on the order of −190°C, all these materials except carbon steel are adequate. Reference 20 presents an extensive compilation of data on the corrosion rate of fluorine on aluminum, magnesium, brass, copper, steel, stainless steel, nickel, and Monel at up to 540°C and 1.83 MPa (265 psi).

Copper, brass, and steel are generally used for gaseous service at temperatures below 200°C. Monel and nickel are used at elevated temperatures. For critical applications and where there is any danger of temperature buildup, nickel or Monel should be used because of the stability of the nickel fluoride film. Highly fluorinated polymers, such as polytetrafluoroethylene, are the only polymers that are resistant to the gas under nominally static conditions. The PTFE must be free of any impurities, dirt, or foreign materials that could initiate ignition with fluorine. However, since fluoropolymers are more susceptible to ignition than metals, they should be used as little as possible in fluorine service, especially in contact with flowing streams at high velocities and pressures.

All equipment, lines, and fittings intended for fluorine service must be leak-tight, dry, and thoroughly cleansed of all foreign matter before use. The tightness of the system should be checked at least to its working pressure with helium. It should be flushed with a nonaqueous degreasing solvent, such as methylene chloride, thoroughly purged with a stream of dry nitrogen, and evacuated to make certain no volatiles are present. Any foreign matter not removed could react with fluorine and start a fire. After cleaning, the system should be filled with dry nitrogen.

The corrosion resistance of all materials used with fluorine depends upon the passivation of the system. This is a pickling operation intended to remove the last traces of foreign matter, and to form a passive fluoride film on the metal surface. The dry nitrogen in the system is slowly replaced with gaseous fluorine in small increments until the concentration and pressure approach working conditions.

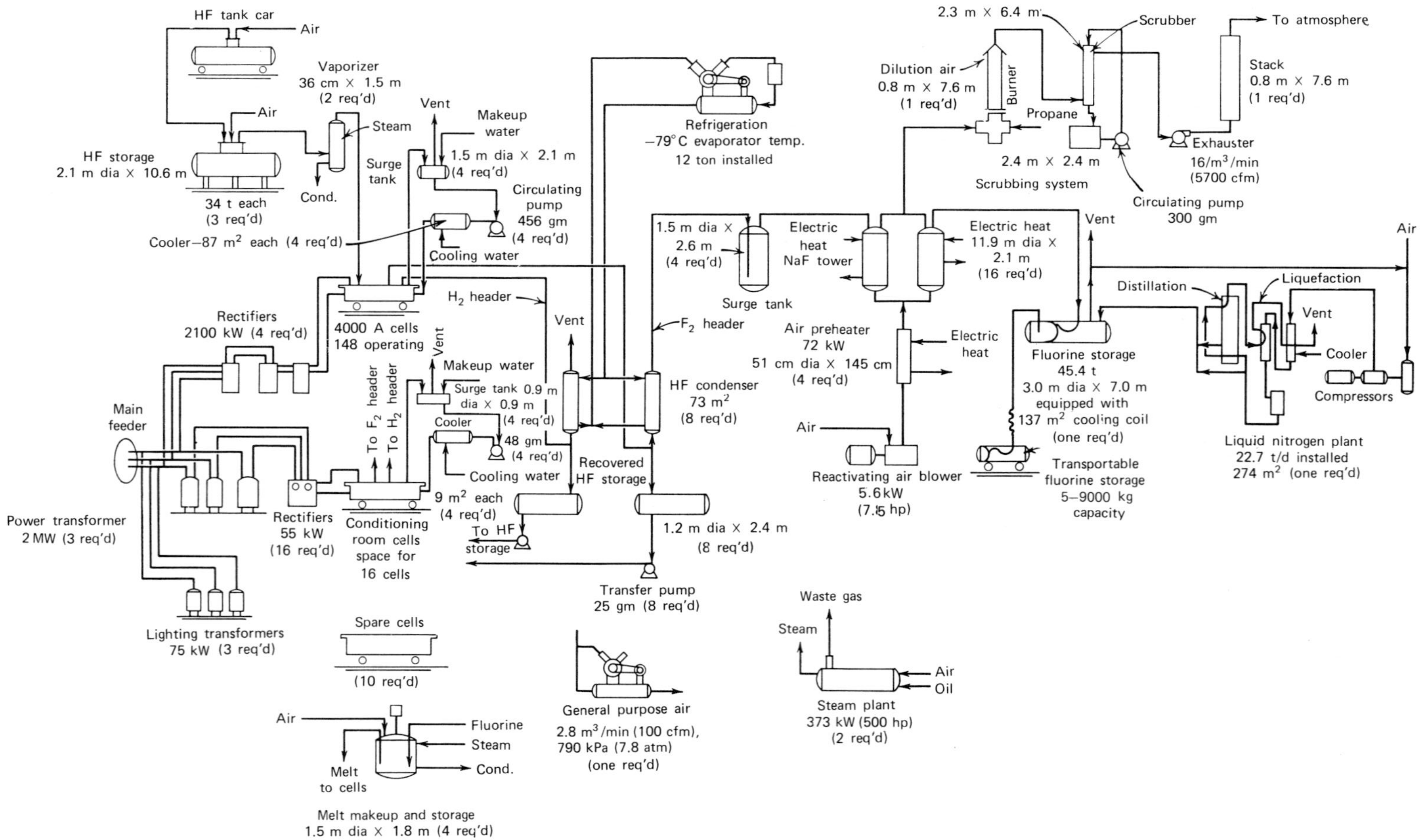

Figure 3. Equipment flow sheet of elemental fluorine production and liquefaction plant, 9 t/d capacity.

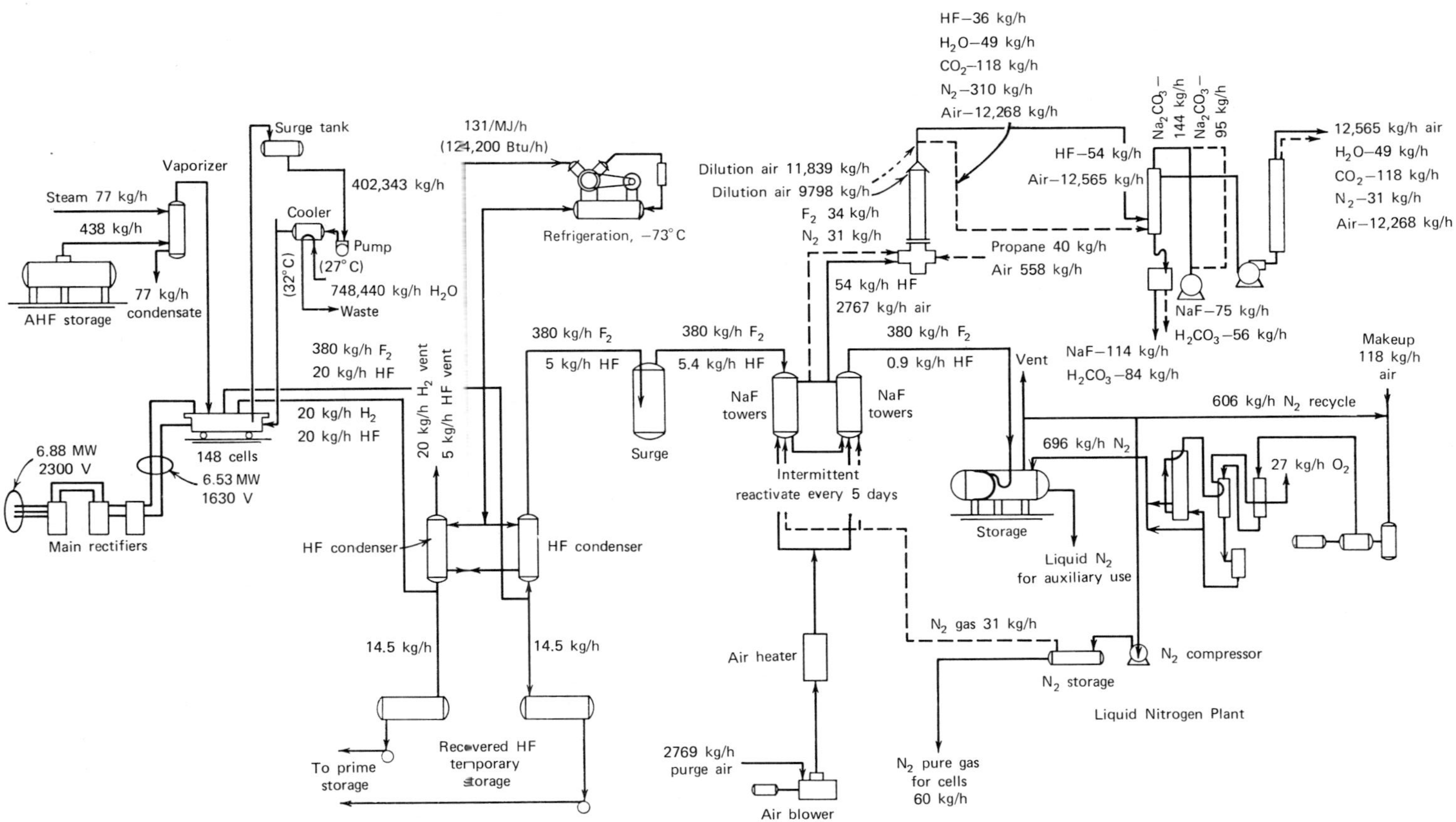

Figure 4. Material flow sheet of elemental fluorine production and liquefaction plant, 9 t/d capacity. Step 1: purging residual F_2 at rates indicated; all but a trace of residual F_2 is removed in 15 min; N_2 purge is maintained for 1 h to remove last traces. Step 2: HF removal at rates indicated; all but a trace of HF is removed in 10 h; air purge is maintained for 10 h.

Carbon steel is most commonly used for pipe and fittings (2.5 cm or greater) in gaseous fluorine service at ambient temperature conditions and pressures up to 2.86 MPa (415 psi).

Carbon steel or bronze-body gate valves are commonly used in gaseous fluorine service at low pressure. Plug valves, with Monel bodies and plugs, are recommended for moderate pressure service below 500 kPa (<5 atm). For valve-stem packing PTFE polymer is recommended and it must be maintained leak-tight. Bellows-type valves with Monel or stainless steel bellows are recommended for high pressure service, but not ball valves.

Compressors and blowers for gaseous fluorine service range in capacity to 4800 m^3/h (2800 ft^3 min) and 2.86 MPa (415 psi). These units vary in design from multistage centrifugal compressors to diaphragm and piston types. Standard commercial instrumentation and control devices are used in fluorine systems. Pressure is measured with Bourdon-type gauges or pressure transducers. Stainless steel or Monel construction is recommended for parts in contact with fluorine. Standard thermocouples are used for all fluorine temperature-measuring equipment, such as the stainless-steel shielded-type, inserted through a threaded compression fitting welded into the line. For high temperature service, nickel-shielded thermocouples should be used.

Shipping

Gas. Fluorine gas is packaged and shipped in steel cylinders conforming to ICC specification 3AA-1000 under a pressure of 2.86 MPa (415 psi). Cylinders containing 2.7, 2.2, 0.7, and 0.23 kg are available from Air Products and Chemicals, Inc. All cylinders are equipped with special fluorine valves, the outlets of which have a left-hand thread conforming to CGA (Compressed Gas Association) Specification 670. Current DOT regulations stipulate that cylinders must be shipped without safety relief devices and be equipped with valve protection caps. The quantity of fluorine is limited to 2.7 kg and the pressure must not exceed 2.86 MPa (415 psi) at 21°C.

Mixtures of 10% fluorine in nitrogen at 16.65 MPa (2415 psi) pressure are commercially available in cylinders and tube trailers from Air Products and Chemicals, Inc. A tube trailer of 10% fluorine can contain up to 250 kg of fluorine. This mixture is specifically intended for the blow-molding industry which uses a dilute fluorine mixture in nitrogen in the production of specialty high-density polyethylene containers. The shipment of fluorine mixtures at high pressure permits larger quantities of fluorine to be shipped per container under safe conditions.

Liquid. Fluorine was supplied by Allied Chemical Corporation as a liquid in tank trucks, which contained up to 2270 kg of fluorine. However, currently it is not available to the commercial market but is sold under contract to the United States government. The primary use of liquid fluorine was in the development of fluorine as a rocket fuel oxidizer by the United States government in the 1960s. With the discontinuance of this application, the market for liquid fluorine is very small.

Bulk shipment of liquid fluorine is authorized by the DOT under special permit No. 1479 which describes in detail the type of tank required and requires DOT approval of the carrier.

Economic Aspects

Total commercial production capacity of fluorine in the United States and Canada is estimated at over 5000 t/yr, of which 70–80% is devoted to uranium hexafluoride production. The primary raw material, HF, used in F_2 production is in ample supply and readily available.

Fluorine is sold as a gas in cylinders at 2.86 MPa (415 psi) pressure. The 1979 price in cylinders was $50.00/kg. Fluorine is also available from Air Products as a mixture diluted with nitrogen. Mixtures of 10% fluorine in nitrogen are available in cylinders and tube trailers at 16.65 MPa (2415 psi) pressure and contain 1.4 and 250 kg of fluorine, respectively.

The price in large volumes is determined by (*1*) the price of hydrofluoric acid, (*2*) power costs, ca 4.5 kW·h electricity is required per kg fluorine, (*3*) labor costs, (*4*) costs to maintain and rebuild cells, and (*5*) amortization of fixed capital: fluorine production capacity is highly capital intense. In addition, purification, compression, packaging, and distribution in cylinders increase the cost significantly.

Industrial Installations. At present, four commercial fluorine plants are operating in the United States and Canada with minimum capacities of 500 t/yr (see Table 8). A view of the cell room at the Air Products and Chemicals, Inc., Tamaqua, Pennsylvania, plant is shown in Figure 5. The DOE plants at Oak Ridge, Tennessee; Paducah, Kentucky; and Portsmouth, Ohio, are operating at reduced capacity and do not serve the commercial market. Commercial fluorine plants for captive use are operated by British Nuclear Fuels, Imperial Smelting Corporation in England, Ugine Kuhlmann in France, Montecatini-Edison in Italy, Kali-Chemie in Germany, and Asahi Glass Co., Ltd. in Japan.

Analytical and Test Procedures; Specifications

Direct analysis of gaseous fluorine is not possible by conventional methods because of its reactivity. Therefore fluorine is converted (83) quantitatively to chlorine by reaction with sodium chloride, followed by analysis of the effluent stream. First the sample is passed through a bed of sodium fluoride which retains the hydrogen fluoride. (This step can be eliminated if the fluorine is free of HF as received.) The sample is then passed over granular sodium chloride which reacts with fluorine to release an equivalent quantity of chlorine. The latter is reduced to chloride which is determined by the Volhard method. Impurities such as oxygen, nitrogen, carbon tetrafluoride, sulfur hexafluoride, and carbon dioxide are determined by conventional gas chromatography techniques. Hydrogen fluoride is determined by infrared analysis of a separate sample (see Analytical methods).

Table 8. United States and Canadian Fluorine Plants

Company	Location	Products
Air Products & Chemicals, Inc.	Tamaqua, Pa.	SF_6, SF_4, NF_3, BrF_3, ClF_3, IF_5, WF_6, ReF_6, gaseous F_2
Allied Chemical Corp.	Metropolis, Ill.	UF_6, SF_6, IF_5, liquid F_2
Eldorado Nuclear Limited	Port Hope, Ont., Can.	UF_6
Kerr-McGee Corp.	Sequoyah, Oklahoma	UF_6

Figure 5. Fluorine generator room. Courtesy of Air Products & Chemicals, Inc.

Fluorine in the atmosphere can be detected by chemical methods involving the displacement of halogens from halides as described previously. The oxidation of potassium iodide to produce free iodine is frequently used for detection of free fluorine. The Mast Oxidant Monitor is a continuous fluorine-monitoring instrument based on this reaction. The sensing of fluorine in the sample is accomplished by the oxidation–reduction of potassium iodide in the sensing solution. The reaction takes place at the cathode portion of the electrode support where H_2 is being generated. The reduction of H_2 by free I_2 causes a depolarization current to flow in an external circuit which is a measure of the free I_2 produced.

Specifications. Fluorine is sold in cylinders according to the specifications given in Table 9.

Health and Safety Factors

Fluorine, the most reactive element known, is a dangerous material that can be handled safely under proper precautions. For even short intervals of contact with low pressure fluorine, the operator should wear safety glasses, a neoprene coat and boots and clean neoprene gloves to afford overall body protection. This protection is effective against both fluorine and the hydrofluoric acid which may form by reaction with moisture in the air.

Face shields made of conventional materials or, preferably, transparent, highly fluorinated polymers, should be worn whenever operators approach equipment containing fluorine under pressure. A mask with a self-contained air supply should be available. Leaks in high pressure systems usually result in a flame from the reaction of fluorine with the metal. Shields should be provided for valves, pressure-reducing stations, and gauges. Valves are a particularly susceptible area for fluorine fires that can be initiated by foreign material accumulated at the valve seat. High pressure

Table 9. Specifications of Fluorine

Assay	Value
gas	
fluorine, mol %	98
oxygen, mol %	0.5
nitrogen, mol %	1.0
carbon tetrafluoride, mol %	0.1
sulfur hexafluoride, mol %	0.1
hydrogen fluoride, mol %	0.2
carbon dioxide, mol %	0.1
water, ppm	1.0
dew point, °C	40
liquid[a]	
fluorine, wt %, min	99.0
HF plus CF_4, wt % max	0.3
O_2 plus N_2, wt %, max	1.0

[a] Determined on a helium-free basis.

cylinders and valves should be remotely located with proper personnel protection and the latter should be operated with handle extensions.

Toxicity. Fluorine is extremely corrosive and irritating to the skin. Inhalation at even low concentrations irritates the respiratory tract; at high concentrations it results in severe lung congestion.

The American Conference of Governmental Industrial Hygienists (ACGIH) has established the TLV as 1 ppm or 2 mg fluorine per cubic meter of air. Fluorine has a sharp, penetrating odor detectable at levels well below the TLV. Therefore, it is unlikely that anyone would inadvertently inhale fluorine in harmful concentrations. Manifestations of overexposure to fluorine include irritation or burns of the eyes, skin, and respiratory tract. The following values for human exposure have been suggested (84): emergency exposure limits, EEL = 25.0 ppm for 5 min; emergency tolerance limits, ETL = 15.0 ppm for 10 min; threshold limit values, TLV = 1.0 ppm for 8 h.

Toxicity studies (85–87) established tolerance levels and degrees of irritations, indicating that the eye is the area most sensitive to fluorine. Fluorine is more toxic than hydrofluoric acid with a chronic exposure tolerance level of 1 ppm for fluorine and 7 ppm for hydrofluoric acid.

Comprehensive animal studies (88–90) determined a rat LC_{50} value of 3500 ppm–min for a single 5 min exposure and of 5850 ppm–min for a 15 min exposure. A no-effect concentration corresponded to a concentration–time value of about 15% of the LC_{50} levels.

Because of the corrosive effects and discomfort associated with inhalation of fluorine, chronic toxicity does not occur. Repeated exposures for a number of years to large dosages of fluoride result in fluorosis or fluoride poisoning. Daily absorption of 20–80 mg, or more, of fluorides into the body for 10–20 yr results in the eventual development of this severely crippling disease (91). It is caused by an increased fluorine content in the bones with degradation of the bone structure.

Burns. In the gaseous state, the skin burn of pure fluorine gas is identical with a second-degree thermal burn and differs considerably from that produced by hydrogen fluoride (88). Fluorine burns heal much more rapidly than a hydrofluoric acid burn.

Disposal. Fluorine can be disposed of by conversion to gaseous fluorocarbons or fluoride salts. The following methods have been successfully used: scrubbing with caustic solutions (92–93); reaction with charcoal (56–57,94); reaction with solid disposal agents such as alumina, limestone, lime, and soda lime (95–96); direct burning with fuels such as methane or propane (97); and reaction with superheated steam (98).

Burning with charcoal (41) and scrubbing with caustic solution (92–93) are practiced on an industrial scale. In a spray column, caustic scrubbing system, caustic potash, KOH, is preferred to caustic soda, NaOH, because of the higher solubility of the resulting potassium fluoride. Adequate solution contact and residence time must be provided in the scrub tower to ensure complete neutralization of the intermediate oxygen difluoride, OF_2. A minimum residence time of one minute is recommended to prevent OF_2 emission from the scrub tower.

Uses

Elemental fluorine is used captively by most manufacturers for the production of various inorganic fluorides (see Table 8). The market for gaseous fluorine is insignificant and the demand for liquid fluorine has virtually ceased with the discontinuance of its use as a rocket fuel oxidizer (see Explosives and propellants).

The main use of fluorine is in the manufacture of uranium hexafluoride, UF_6, by the following reaction:

$$UF_4 + F_2 \rightarrow UF_6$$

Uranium hexafluoride is used in the gaseous diffusion process for the separation and enrichment of uranium 235, which exists in low concentrations in natural uranium. It was first produced for the U.S. Atomic Energy Commission (now DOE) during World War II. The current market for UF_6 is in the production of enriched fuel for the nuclear electric power industry. Therefore, fluorine consumption is closely tied to the development of the nuclear power industry.

The other important use for elemental fluorine is in production of sulfur hexafluoride, SF_6, a gaseous dielectric for electrical and electronic equipment (see Fluorine compounds, inorganic—sulfur). Its high dielectric strength, inertness, thermal stability, and ease of handling have led to its increased use as an electrically insulating medium, permitting reductions in size, weight, and cost of a wide variety of electronic equipment. Elemental fluorine is also used to produce sulfur tetrafluoride, SF_4, by the reaction of sulfur and fluorine under controlled conditions. Sulfur tetrafluoride is a selective fluorinating agent used to produce fluorochemical intermediates in the pharmaceutical and herbicide industry.

Fluorine reacts with the halogens to produce several interhalogen compounds of commercial importance; namely, chlorine trifluoride, bromine trifluoride, and iodine pentafluoride. Chlorine trifluoride is used in the processing of UF_6 for the nuclear electric power industry. Bromine trifluoride is used in chemical cutting by the oil well industry. Iodine pentafluoride is used as a selective fluorinating agent to produce fluorochemical intermediates (see Fluorine compounds, inorganic—halogens).

Fluorination of tungsten and rhenium produces tungsten hexafluoride, WF_6, and rhenium hexafluoride, ReF_6, respectively. These volatile metal fluorides are used in the chemical vapor deposition industry to produce metal coatings and intricately shaped components (see Film deposition techniques).

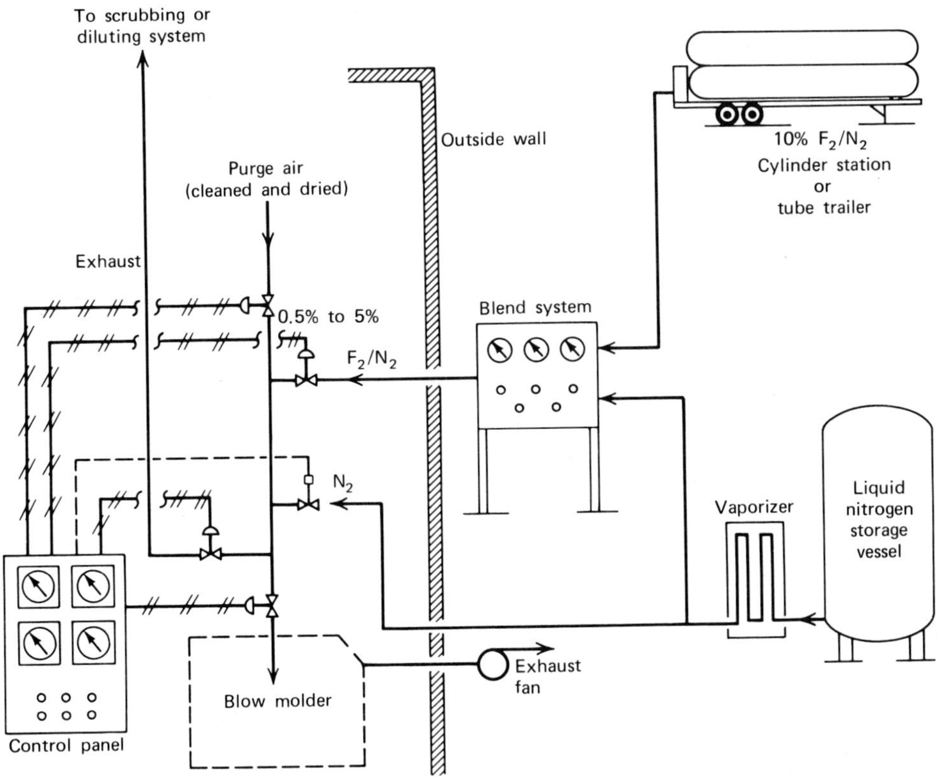

Figure 6. Airopak system.

Fluorine reacts with ammonia in the presence of ammonium acid fluoride to give nitrogen trifluoride, NF_3, the fluorine source in the high power HF/DF chemical lasers. Fluorine is also used in these laser systems but NF_3 is expected to be the predominant fluorine source because of its ease of handling as compared to fluorine.

A significant new use for fluorine has been developed for the blow-molding industry to produce solvent-resistant polyolefin containers (52–53). In this application, the air which is normally used to blow-mold containers is replaced by a low concentration of fluorine in a mixture with nitrogen. Containers, such as Airopak, produced by this process show outstanding resistance to nonpolar solvents. A major application for containers produced by this process is as vehicle fuel tanks. In this process, the fluorine is distributed as a 10% fluorine-in-nitrogen mixture and further diluted at the user's plant. The diluted supply simplifies the handling and distribution system at the user's plant and permits shipment of larger quantities of gaseous fluorine in tube trailer quantity. Figure 6 shows a typical blow-molding system to produce Airopak containers.

BIBLIOGRAPHY

"Fluorine" in *ECT* 1st ed., Vol. 6, pp. 656–667, by H. C. Miller and F. D. Lommis, Pennsylvania Salt Manufacturing Company; "Fluorine" in *ECT* 2nd ed., Vol. 9, pp. 506–525, by H. R. Neumark and J. M. Siegmund, Allied Chemical Corporation.

1. "Fluorine" in D. N. Lapedes, ed., *Encyclopedia of Science and Technology,* Vol. 5, McGraw-Hill, New York, 1977, pp. 389–393.
2. H. Moissan, *Comp. Rend.* **102,** 1534 (1886); **103,** 202, 256 (1886); *Gmelins Handbuch der anorganischen Chemie, System 5,* 8th ed., Deutsche Chemische Gesellschaft, Verlag Chemie, Berlin, Ger., 1926, pp. 4–16.
3. G. C. Fingers in M. Stacey, J. C. Tatlow, and A. G. Sharpe, eds., *Advances in Fluorine Chemistry,* Vol. 2, Butterworths, London, Eng., 1962.
4. *Fluorine, Mineral Commodity Profiles, MCP-20,* U.S. Bureau of Mines, Washington, D.C., Aug. 1978.
5. F. C. Noyes, *Min. Metall.* **25,** 495 (1944).
6. J. H. Hu, D. White, and H. L. Johnson, *J. Am. Chem. Soc.* **75,** 5642 (1953).
7. E. Kanda, *Bull. Chem. Soc. Jpn.* **12,** 473 (1937).
8. J. D. Collins, L. S. Stone, and P. A. Juyner, *Background Chemistry for Development of Liquid Rocket Oxidizers, AD-18-283,* Callery Chemical Company, Callery, Pa., 1953.
9. G. H. Cady and J. H. Hildenbrand, *J. Am. Chem. Soc.* **52,** 3829 (1930).
10. W. H. Claussen, *J. Am. Chem. Soc.* **56,** 614 (1934).
11. D. Horovitz, *A. Review of the Physical and Chemical Properties of Fluorine and Certain of Its Compounds, Report No. RMI-293-85,* Reaction Motors, Inc., Rockaway, N.J., 1950.
12. G. M. Murphy and J. E. Vance, *J. Chem. Phys.* **7,** 806 (1939).
13. V. N. Huff and S. Gordon, *Tables of Thermodynamics Functions for Analysis of Aircraft-Propulsion Systems, Tech. No. 2161,* National Advisory Committee for Aeronautics, Washington, D.C., Aug. 1950.
14. K. V. Butkov and R. B. Bozenbaum, *J. Phys. Chem. U.S.S.R.* **24,** 706 (1950).
15. L. G. Cole, M. Farber, and G. W. Eluerum, Jr., *J. Chem. Phys.* **20,** 586 (1952).
16. G. W. Elverum, Jr., and R. N. Doeschev, *J. Chem. Phys.* **20,** 1834 (1952).
17. R. L. Jarry and H. C. Miller, *J. Am. Chem. Soc.* **78,** 1553 (1956).
18. W. T. Ziegler and J. C. Mullins, *Calculation of the Vapor Pressure and Heats of Vaporization, and Sublimation of Liquids and Solids, Especially Below One Atmosphere, IV, Nitrogen and Fluorine, Technical Report No. 1,* Engineering Experiment Station, Georgia Institute of Technology, to Cryogenic Engineering Laboratory, Natl. Bur. Std., Boulder, Col., under NBS Contract CST-7404, Apr. 1963.
19. E. U. Franck, *Naturwissenschaften* **41,** 37 (1954).
20. J. F. Tompkins and co-workers, *The Properties and Handling of Fluorine, Technical Report No. ASD-TDR-62-273,* Air Products & Chemicals, Inc., Allentown, Pa., 1963.
21. J. L. Weeks, C. L. Chernick, and M. S. Matheson, *J. Am. Chem. Soc.* **84,** 4612 (1962).
22. E. U. Franck and W. Stober, *Z. Naturforsch.* **7a,** 822 (1952).
23. E. U. Franck and E. Wicke, *Z. Elektrochem.* **55,** 643 (1951).
24. I. G. Stamper and R. F. Barrow, *Trans. Faraday Soc.* **54,** 1592 (1958).
25. D. A. Johnson, *Some Thermodynamic Aspects of Inorganic Chemistry,* Cambridge University Press, New York, 1968, p. 158.
26. E. L. Muetterties and C. W. Tullock in W. L. Jolly, ed., *Preparative Inorganic Reactions,* Vol. 2, Interscience Publishers, New York, 1965, p. 243.
27. T. W. Godwin and C. F. Lorenzo, "Ignition of Several Metals in Fluorine," *Paper No. 740, American Rocket Society, 13th Annual Meeting ARS, New York, Nov. 17–21, 1958.*
28. H. M. Haendler and co-workers, *J. Am. Chem. Soc.* **76,** 2177 (1954).
29. A. G. Streng, *Chem. Rev.* **63,** 607 (1963).
30. V. A. Dmitrievskii, V. N. Cherednikov, and E. K. Illin, *Khim Vys. Energ.* **7**(3), 206 (1973).
31. I. V. Nikitin and V. Ya. Rosolovski, *Izv. Adad. Nauk USSR Ser. Khim* **7,** 1464 (1970).
32. U.S. Pat. 3,304,248 (Feb. 14, 1967), H. T. Fullan and H. V. Scklemian (to Stauffer Chemical Co.).
33. C. L. Chernick and co-workers, *Science* **138,** 136 (1962).
34. J. H. Holloway, *Chem. Commun.,* 22 (1966).
35. H. H. Claassen, H. Selig, and J. G. Malm, *J. Am. Chem. Soc.* **84,** 3593 (1962).
36. J. G. Malm, I. Sheft, and C. L. Chernick, *J. Am. Chem. Soc.* **85,** 110 (1963).
37. E. E. Weaver, B. Weinstock, and C. P. Knop, *J. Am. Chem. Soc.* **85,** 111 (1963).
38. P. R. Fields, L. Stein, and M. H. Zirin, *J. Am. Chem. Soc.* **84,** 4164 (1962).
39. A. V. Grosse and co-workers, *Chem. Eng. News* **41,** 47 (Jan. 7, 1963).
40. G. C. Pimentel and J. H. Packer, *J. Chem. Phys.* **51,** 91 (1961).
41. P. L. Klass, *Aviat. Week Space Technol.* **107,** 34 (Aug. 1977).

42. S. I. Morrow and co-workers, *J. Am. Chem. Soc.* **82,** 5301 (1960).
43. U.S. Pat. 4,091,081, (May 23, 1978), A. J. Woytek and J. T. Lileck (to Air Products and Chemicals, Inc.).
44. *Chem. Eng.* **84**(26), 116 (1977).
45. L. A. Bigelow, "The Action of Elementary Fluorine Chemistry, Vol. I, Academic Press, Inc., New York, 1950, Ch. 11, pp. 377–378.
46. E. H. Hadley and L. A. Bigelow, *J. Am. Chem. Soc.* **62,** 3302 (1940).
47. E. A. Tyczkowski and L. A. Bigelow, *J. Am. Chem. Soc.* **77,** 3007 (1955).
48. K. B. Kellogg and G. H. Cady, *J. Am. Chem. Soc.* **70,** 3968 (1948).
49. N. Fukuhara and L. A. Bigelow, *J. Am. Chem. Soc.* **63,** 788 (1941).
50. Can. Pat. 3,954,749 (Mar. 16, 1976), P. D. Schuman and co-workers (to PCR, Inc.).
51. U.S. Pat. 3,862,284 (Jan. 21, 1975), D. D. Dixson, D. G. Manly, and G. W. Recktenwald (to Air Products and Chemicals, Inc.).
52. A. J. Woytek and J. F. Gentilecore, "A New Blow Molding Process to Reduce Solvent Permeation of Polyolefin Containers," *Paper No. 13 presented at Advances in Blow Molding Conference, Rubber and Plastics Institute, London, Eng., Dec. 6, 1977.*
53. J. F. Gentilecore, M. A. Triolo, and A. J. Woytek, *Plast. Eng.* **34**(9), 40 (1978).
54. U.S. Pat. 4,020,223, (Apr. 26, 1977), D. D. Dixson and L. J. Hayes (to Air Products and Chemicals, Inc.).
55. U.S. Pat. 3,988,491 (Oct. 26, 1976), D. D. Dixson and L. J. Hayes (to Air Products and Chemicals, Inc.).
56. H. W. Schmidt, *Reaction of Fluorine with Carbon as a Means of Fluorine Disposal, NACA-RM-E57E02,* National Advisory Committee for Aeronautics, Washington, D.C., 1957.
57. H. W. Schmidt, *Design and Operating Criteria for Fluorine Disposal by Reaction with Charcoal, NASA Memo 1-27-59E,* National Aeronautics & Space Administration, Washington, D.C., 1959.
58. O. Ruff, D. Bretschneider, and F. Elert, *Z. Anorg. Chem.* **217,** 1 (1934).
59. U.S. Pat. 3,674,432 (July 4, 1972), R. J. Lagow and co-workers (to R. I. Patents, Inc.).
60. W. Rudorff, *Adv. Inorg. Chem. Radiochem.* **1,** 230 (1959).
61. G. R. Hennig, *Prog. Inorg. Chem.* **1,** 125 (1959).
62. N. Watanabe and K. Kumon, *Denki Kagaku* **35,** 19 (1967).
63. R. L. Fusaro and H. E. Sliney, *Graphite Fluoride: A New Solid Lubricant, NASA Technical Memorandum X-52624,* National Aeronautics & Space Administration, Washington, D.C., 1969.
64. H. F. Hunger and G. J. Keymack, *J. Electrochem. Soc.* **120,** 1161 (1973).
65. R. J. Ring and D. Royston, *A Review of Fluorine Cells and Fluorine Production Facilities, AUST. AEC, AAEC/E, AATCAZ, Issue 281,* 1973.
66. A. J. Rudge in A. Kuhn, ed., *Industrial Electrochemical Processes,* Elsevier Publishing Company, Amsterdam, The Netherlands, 1971, Chapt. 1.
67. A. J. Rudge, *Chem. Ind.* **22,** 504 (1956).
68. R. D. Fowler and co-workers, *Ind. Eng. Chem.* **39,** 3, 266 (1947).
69. G. H. Cady, H. S. Booth, eds., *Inorganic Synthesis,* Vol. 1, McGraw-Hill, New York, 1939, pp. 136–137.
70. C. H. Cady, D. A. Rogers, and C. A. Carlson, *Ind. Eng. Chem.* **34,** 4, 443 (1942).
71. S. H. Smiley and D. C. Brater, *USAEC Report TID-5295,* U.S. Atomic Energy Commission, Washington, D.C., L1956.
72. J. Dykstra and co-workers, *Ind. Eng. Chem.* **47,** 5, 883 1955.
73. B. W. Clark, *USAEC Report KY-326,* USAEC, Washington, D.C., 1960.
74. R. C. Kelley and W. E. Clark, eds., *USAEC Report TID4100, Suppl. 42, CAPE-55,* 1967, and *Suppl. 45, CAPE-486,* 1968, USAEC, Washington, D.C.
75. J. Dykstra, A. P. Huber, and B. H. Thompson, "Multi-Ton Production of Fluorine for Manufacture of Uranium Hexafluoride," *paper presented at Second United Nations International Conference on the Peaceful Use of Atomic Energy, A/CONF.15/P/524, June 1958.*
76. C. A. Powell, R. E. Cable, and W. K. Henderson, *Ind. Eng. Chem.* **52,** 7, 42A (1960).
77. N. Watanabe, M. Ishii, and S. Yaskizawa, *J. Electrochem. Soc. Jpn.* **29,** 3, 177 (1961).
78. J. Jacobson and co-workers, *Ind. Eng. Chem.* **47,** 5, 878 (1955).
79. J. R. McGuffey, R. Paluzelle and W. E. Muldrew, *Ind. Eng. Chem.* **54,** 5, 46 (1962).
80. J. M. Siegmund, *Chem. Eng. Prog.* **63,** 6, 88 (1967).

81. W. C. Robinson, ed., *Fluorine Systems Handbook,* Douglas Aircraft Co., Inc., Long Beach, Calif., 1967.
82. D. L. Endicott and L. H. Donahue, *Development and Demonstration of Criteria for Liquid Fluorine Feed System Components, Report AFRPL-TR-65-133,* McDonnell Douglas Astronautics Co., St. Louis, Mo., 1965.
83. N. S. Nikolaev and co-workers, English translation by J. Schmorak, *Analytical Chemistry of Fluorine,* Halsted Press, New York, 1972, Ch. VIII.
84. P. M. Ricca, *Am. Ind. Hyg. Assoc. J.* **31,** 22 (1970).
85. M. L. Keplinger and L. W. Suissa, *Am. Ind. Hyg. Assoc. J.* **29,** 10 (1968).
86. W. Machle and E. E. Evans, *J. Ind. Hyg. Toxicol.* **22,** 213 (1940).
87. J. S. Lyon, *J. Occup. Med.* **4,** 199 (1962).
88. H. E. Stockinger in C. Voegtlin and H. C. Hodge, eds., *The Pharmacology and Toxicology of Uranium,* McGraw-Hill Book Co., New York, 1949, Ch. 17.
89. N. Ericksen and co-workers, *A Study of the Toxicological Effects of the Inhalation of Gaseous Fluorine at Concentrations of Approximately 25, 8, 3 and 0.8 mg/m^3, United States Atomic Energy Reports 397, 407, 427, and 429,* University of Rochester, Rochester, N.Y., 1945.
90. N. Ericksen and co-workers, *A Study of the Lethal Effects of the Inhalation of Gaseous Fluorine at Concentrations from 100 ppm to 10,000 ppm, United States Atomic Energy Report 435,* University of Rochester, Rochester, N.Y., 1945.
91. R. Y. Eagers, *Toxic Properties of Inorganic Fluorine Compounds,* Elsevier Publishing Company, Ltd., Amsterdam, The Netherlands, 1969, p. 43.
92. R. Landau and R. Rosen, *Ind. Eng. Chem.,* **40,** 1239 (1948).
93. J. B. Ruch, *USAEC Report CF-60-4-38,* USAEC, Washington, D.C., 1960.
94. H. Pulley and R. L. Harris, *USAEC Report KY-638,* USAEC, Washington, D.C., 1972.
95. R. C. Liimatainer and W. L. Merchan, *Report ANL-5429,* Argonne National Laboratory, Northbrook, Ill., 1955.
96. J. D. Davratel, *USAEC Report RFP-1200,* USAEC, Washington, D.C., 1968.
97. S. G. Turnbull and co-workers, *Ind. Eng. Chem.* **39**(3), 286 (1947).
98. S. H. Smiley and C. R. Schmitt, *Ind. Eng. Chem.* **46,** 244 (1954).

General References

F. A. Cotton, ed., *Progress in Inorganic Chemistry,* Vol. II, Interscience Publishers, New York, 1960.

R. Y. Eagers, *Toxic Properties of Inorganic Fluorine Compounds,* Elsevier Publishing Company Ltd., London, Eng., 1969.

H. J. Emeleus, "Fluorine," *J. Chem. Soc.,* 441 (1942).

Gmelins Handbuch der Anorganischen Chemie, 8th ed., Suppl., Verlag Chemie, Weinheim, FRG, 1959, pp. 66–79.

M. Hudlicky, *Chemistry of Organic Fluorine Compounds,* Ellis Harwood Limited, Sussex, Eng., 1976.

A. T. Kuhn, ed., *Industrial Electrochemical Processes,* Elsevier Publishing Company Ltd., Amsterdam, The Netherlands, 1971.

Mellor's Comprehensive Treastise on Inorganic and Theoretical Chemistry, Vol. II, Suppl. I, Longmans, Green, New York, 1956, pp. 15–45.

N. S. Nikolaev and co-workers, English translation by J. Schmorak, *Analytical Chemistry of Fluorine,* Halsted Press, New York, 1972.

A. J. Rudge, *The Manufacture and Use of Fluorine and Its Compounds,* Oxford University Press, Inc., New York, 1962.

O. Ruff, "Fluorine and Its Compounds," *Chem. Ber.* **69A,** 181 (1936).

J. H. Simons, ed., *Fluorine Chemistry,* Vol. 1, Academic Press, Inc., New York, 1950.

C. Slesser and S. R. Schram, *Preparation, Properties, and Technology of Fluorine and Organic Fluoro Compounds, National Nuclear Energy Series, Div. VII,* Vol. I, McGraw-Hill Book Co., Inc., New York, 1951.

M. Stacey, J. C. Tatlow, and A. G. Sharpe, eds., *Advances in Fluorine Chemistry,* Vol. 2, Butterworth Inc., Washington, D.C., 1961.

ANDREW J. WOYTEK
Air Products & Chemicals, Inc.

FLUORINE COMPOUNDS, INORGANIC

INTRODUCTION

Fluorine is the most electronegative element and is much more reactive than the next most electronegative elements oxygen and chlorine. Fluorine reacts with virtually every element in the periodic table. Indeed, the Group VIII elements were commonly called inert gases until 1962 when xenon, radon, and krypton were shown to react with fluorine (see Helium-group gases). The unique properties of fluorine have caused the element to be called a superhalogen (1) and several of its compounds to be called superacids. (The term superacid (2) is used for systems with higher acidities than anhydrous sulfuric or fluorosulfuric acid. A number of fluorine species are superacids in HSO_3F or HSO_3F–SO_3 solutions; most attention is given to SbF_5–HSO_3F.)

The basic fluorine raw materials are fluorspar ore, CaF_2, and by-product fluorosilicic acid from phosphate production. Calcium fluoride reacts with sulfuric acid to produce hydrogen fluoride, the most common reagent for production of fluorine compounds. Elemental fluorine is produced by electrolysis of anhydrous potassium fluoride–hydrogen fluoride melts.

Caution: anhydrous or aqueous hydrogen fluoride or fluorides that are hydrolyzed to aqueous hydrogen fluoride can cause painful and severe burns, and even death. They should be handled with adequate safety equipment and extreme care by well-trained personnel (see Hydrogen fluoride).

The fluoride ion is a small anion, 0.136 nm (0.045 nm smaller than the chloride ion), and it is the least polarizable anion. The F^- and isoelectronic O^{2-} ions are the only anions of comparable size to many cations; they are about the same size as K^+ and Ba^{2+} and smaller than Rb^+ and Cs^+. The relatively small size of F^- allows higher coordination numbers and leads to different crystal forms, different solubilities, and higher bond energies than do the other halides. Bonds between fluorine and other elements are very strong whereas the fluorine–fluorine bond is much weaker than the chlorine–chlorine bond. (This bond weakness—relative to the second row elements—is also seen in O–O and N–N single bonds and is due to electronic repulsion.) A number of elements exhibit their highest oxidation state only as fluorides and oxidation states of 6 and 7 are not uncommon. Examples of volatile fluorides of high oxidation state include VF_5, CrF_5, TaF_5, WF_6, MoF_6, PtF_6, ReF_6, ReF_7, and IF_7. Many complex fluorides demonstrate coordination not found with other halides, eg, $NiF_6{}^{2-}$, $SiF_6{}^{2-}$, $TaF_7{}^{2-}$, $TaF_8{}^{3-}$, $PF_6{}^-$, $AsF_6{}^-$, $ZrF_7{}^{3-}$ and $PbF_7{}^{3-}$.

Fluorine's special properties lead to many applications. Its complexing properties account for its use as a flux in steelmaking and as an intermediate in aluminum manufacture. The reaction of fluorides with hydroxyapatite ($Ca_5(PO_4)_3OH$), which is found in tooth enamel, to form less soluble and/or more acid-resistant compounds, led to the incorporation of fluorides in drinking water and dentifrices (qv) to reduce dental caries. Because many fluorine compounds are very volatile, they are useful in applications such as the enrichment of uranium (UF_6) and the vapor deposition of metals (WF_6, MoF_6, ReF_6), and as unreactive dielectrics (SF_6). Because fluorine forms very stable bonds, its compounds can be both extremely reactive (F_2, HF, ionic MF_x) and extremely stable (CF, covalent MF_x). Fluorinated steroids, other fluorinated drugs and anesthetics have medical applications. Perfluorochemical emulsions are being used as short-term blood substitutes because of their ability to efficiently transport oxygen and carbon dioxide (see Fluorine compounds, organic; Emulsions). The stability and nontoxicity of some fluorine compounds are also demonstrated by studies reporting survival of animals in an atmosphere of 80% SF_6 and 20% oxygen. Fluorides including HF, BF_3, SbF_5, PF_5 and several complexes, eg, $BF_4{}^-$, $PF_6{}^-$, $SbF_6{}^-$, and $AsF_6{}^-$ are used in many applications in catalysis.

History

The names fluorine and fluorspar are derived from the Latin, *fluere,* meaning flow or flux. In 1529 the use of fluorspar as a flux was described. In 1670 the etching of glass by acid-treated fluorspar was reported. Elemental fluorine was isolated by Moissan in 1886 (3).

Fluorspar has been used as a flux in the steel industry since the introduction of the open-hearth process. Prior to 1930 hydrogen fluoride was used in limited quantities for glass etching, polishing, scale removal, and small volume production of fluorides.

In the 1930s the first significant commercial HF production was applied toward the production of aluminum and of chlorofluorohydrocarbons for refrigerants. World War II brought the rapid evolution of fluorine chemistry with the use of HF in alkylation catalysis to produce aviation gasoline and the use of fluorine to produce volatile UF_6 for isotopic enrichment. The use of UF_6 also stimulated development of fluorinated organic compounds for lubricants and seals that are resistant to UF_6.

Sources and Applications

The earth's crust consists of 0.09% fluorine. Among the elements it ranks about 13th in terrestial abundance.

The ores of most importance are fluorspar, CaF_2; fluorapatite, $Ca_5(PO_4)_3F$; and cryolite, Na_3AlF_6. Fluorspar is the primary commercial source of fluorine. Twenty-six percent of the world's high quality deposits of fluorspar are in the North American continent and most of that is in Mexico. United States production in 1972–1978 was 118,000–225,000 metric tons, most of which occurred in the Illinois–Kentucky area. Imported fluorspar represents about 80% of U.S. consumption and two-thirds of U.S. fluorspar imports are from Mexico. The majority of the fluorine in the earth's crust is in phosphate rock in the form of fluorapatite which has an average fluorine concentration of 3.5%. Recovery of these fluorine values as by-product fluorosilicic acid from phosphate production has grown steadily, partially because of environmental requirements. Recovery in 1978 contributed fluorine values equivalent to ca 100,000 t of fluorspar.

The steel industry continues to be the largest user of fluorspar and accounts for about 50% of total U.S. consumption. Most of the balance of the CaF_2 is treated with sulfuric acid to produce hydrogen fluoride. About 30% of all hydrogen fluoride is used to produce aluminum fluoride and synthetic cryolite for the Hall aluminum process. Chlorofluorocarbons that are used as refrigerants, solvents, aerosols, and in plastics utilize 40% of the hydrogen fluoride produced. Because of the concern that chlorofluorocarbons may be a cause of upper atmosphere ozone depletion, several federal agencies have banned chlorofluorocarbons in most aerosols. The balance of hydrogen fluoride is used in several applications including stainless steel pickling, inorganic fluorides, alkylation, uranium enrichment, and fluorine production. Hydrogen fluoride is used to convert uranium oxide to UF_4 which then reacts with elemental fluorine to produce volatile UF_6. The UF_6 is then isotopically enriched by gaseous diffusion or gas centrifuge processes for nuclear applications.

Synthesis

Most inorganic fluorides are prepared by the reaction of hydrofluoric acid with oxides, carbonates, hydroxides, chlorides, or metals. Routes starting with carbonate or hydroxide or oxide (if reactive) are the most common and the choice is determined by the most economical starting material. In many cases, the water produced by the reaction cannot be removed without at least partial hydrolysis of the metal fluoride. This hydrolysis frequently can be reduced by dehydrating in a stream of hydrogen fluoride. If hydrolysis is unavoidable, reaction of anhydrous HF with the metal or the metal chloride is required. The reaction of the metal with HF can be the most desirable if the metal is inexpensive relative to its salts and if it has an oxidation potential higher than hydrogen. If the metal is not finely divided, formation of a fluoride coating on the metal surface may occur which would slow the reaction. Another route to metal fluorides is by reaction of the metal or its salts with elemental fluorine or with interhalogen fluorides (see Halogens). These reactions occur rapidly and frequently are violent. Because hydrofluoric acid is much less expensive than fluorine, it normally is used whenever possible. However, many of the higher oxidation states can be achieved only by use of elemental fluorine.

Analysis and Characterization

The most popular device for fluoride analysis is the ion-selective electrode. Analysis using the electrode is rapid and is especially useful for dilute solutions and water analysis. Because the electrode responds only to free fluoride ion, care must be taken to convert complex fluoride ions to free fluoride to obtain the total fluoride value. The fluoride electrode also can be used as an end-point detector in titration of fluoride with lanthanum nitrate. Often volumetric analysis by titration with thorium nitrate or lanthanum nitrate is the method of choice. The fluoride is preferably steam distilled from perchloric or sulfuric acid to prevent interference. Fusion with a sodium carbonate–sodium hydroxide mixture or sodium may be required if the samples are covalent or insoluble.

Because fluorine has a nuclear spin of 1/2, nmr spectroscopy is an invaluable tool in characterization. The use of gas chromatography on the many volatile fluorine compounds also has greatly enhanced separation and identification. However, infrared spectroscopy remains a key technique in the study of fluorine compounds.

Nomenclature

IUPAC has recommended the prefix *fluoro* rather than the frequently used *fluo.* Unfortunately terms such as fluoborate, fluosilicate, and silicofluoride are frequently used rather than the preferred fluoroborate and fluorosilicate.

Safety, Toxicity, and Handling

Hazards associated with fluorides are severe. (*1*) Anhydrous or aqueous hydrogen fluoride is extremely corrosive to skin, eyes, mucous membranes and lungs; it can cause permanent damage and even death. Fluorides susceptible to hydrolysis can generate aqueous hydrogen fluoride. (*2*) Ingestion of excess fluorides may cause poisoning or damage to bones and/or teeth. (*3*) Fluorine-containing oxidizers can react with the body in addition to causing burns.

Hydrogen fluoride or compounds that can produce it and fluorine-containing oxidizers should be handled with adequate safety equipment and extreme care by well-trained personnel. Often the effect of skin exposure is not immediately evident, especially when dilute solutions are handled, and pain may develop hours later.

Fluorides in small quantities (1 ppm in water, 0.1% in dentifrices) have been shown to provide dramatic reductions in dental decay. Fluorides also show promise for bone treatment and in drugs—especially in steroids (qv). However, larger quantities of fluorides can lead to dental fluorosis, bone fracture, and even death. The oral LD_{50} for free fluoride ion in rats appears to be 50 to 100 mg F/kg body weight (based on LD_{50} values for several fluorides).

NIOSH recommends that occupational exposure be controlled so that no worker is exposed to a concentration of fluorides greater than 2.5 mg of fluoride (combined ionic fluorides, atomic weight 19) per cubic meter of air, determined as a time-weighted average exposure for up to a 10-h d or a 40-h wk (4).

BIBLIOGRAPHY

"Fluorine Compounds, Inorganic in *ECT* 1st ed., Vol. 6, pp. 667–668 by C. R. Hough, Polytechnic Institute of Brooklyn; "Fluorine Compounds, Inorganic" in *ECT* 2nd ed., Vol. 9, pp. 527–529, by J. B. Beal, Jr., Ozark-Mahoning Company.

1. L. Pauling, *The Nature of the Chemical Bond,* 3rd ed., Cornell University Press, Ithaca, N.Y., 1960, p. 82.
2. R. J. Gillespie, *Acc. Chem. Res.* **1,** 202 (1968).
3. H. Moissan, *Compt. Rend.* **102,** 1543 (1886).
4. *Criteria for a Recommended Standard—Occupational Exposure to Inorganic Fluorides, PB 246 692, NIOSH 76-103,* U.S. Department of Health, Education, and Welfare, 1975.

General References

The following reference works may be consulted for additional general information on fluorine chemistry and for characteristics of specific compounds.

P. Tarrant, ed., *Fluorine Chemistry Reviews,* Marcel Dekker Inc., New York, Vol. 1, 1967; Vol. 2, 1968; Vol. 3, 1969; Vol. 4, 1969; Vol. 5, 1971; Vol. 6, 1973; Vol. 7, 1974; Vol. 8, 1977.

J. H. Simons, ed., *Fluorine Chemistry,* Academic Press Inc., New York, Vol. 1, 1950; Vol. 2, 1954; Vol. 3, 1963; Vol. 4, 1965; Vol. 5, 1964.

M. Stacey and co-eds., *Advances in Fluorine Chemistry,* Butterworth Inc., Washington, D.C., Vol. 1, 1960; Vol. 2, 1961; Vol. 3, 1963; Vol. 4, 1965; Vol. 5, 1965; Vol. 6, 1970; Vol. 7, 1973.

G. Brauer, *Handbuch der Praparativen Anorganischen Chemie,* Ferdinand Enke, Stuttgart, Band 1, 1960; Band 2, 1962.

I. G. Ryss, *The Chemistry of Fluorine and Its Inorganic Compounds,* State Publishing House for Scientific, Technical, and Chemical Literature, Moscow, 1956; Eng. transl., AEC-tr-3927, Office of Technical Services, U.S. Department of Commerce, Washington, D.C., 1960.

F. A. Cotton, *Progress in Inorganic Chemistry,* Vol. 2, Interscience Publishers, Inc., New York, 1960.

Mellor's Comprehensive Treatise of Inorganic and Theoretical Chemistry, Vol. 2, Suppl. 1, Longmans Green & Co. Ltd., London, 1962.

Gmelin, *Handbuch der Anorganischen Chemie,* Fluor, System-Number 5, Verlag Chemie, GmbH, Weinheim, 1959.

M. F. A. Dove and A. F. Clifford, *Inorganic Chemistry in Liquid Hydrogen Fluoride,* Pergamon Press Ltd., Oxford, 1971.

E. Newbrun, ed., *Fluorides and Dental Caries,* Charles C Thomas Publishers, Springfield, Ill., 1975.

W. E. Jones and E. G. Skolnik, *Chem. Rev.* **76,** 563 (1976).

The following U.S. Government reports provide excellent data on production, consumption, uses, prices, and imports of both domestic and world fluorspar and other sources:

Minerals Yearbooks Chapter on Fluorspar (annual comprehensive reports but delayed approximately two years); Mineral Industry Surveys on Fluorspar, quarterly, and annual (less comprehensive but current), U.S. Department of the Interior, Bureau of Mines, Washington, D.C. 20241, Periodic.

CHARLES B. LINDAHL
DAYAL T. MESHRI
Ozark-Mahoning Company,
a subsidiary of the Pennwalt Corp.

ALUMINUM

Aluminum forms three fluorides, AlF, AlF_2, and AlF_3, and a series of fluoroaluminates of which cryolite, Na_3AlF_6, is the most important.

Aluminum Monofluoride and Aluminum Difluoride

Aluminum monofluoride [*13595-82-9*], AlF, is not known in the condensed state, since it disproportionates at lower temperatures to aluminum trifluoride and aluminum. A significant vapor pressure of aluminum monofluoride is observed when the trifluoride is heated with reducing agents such as aluminum or magnesium metal, or is in contact with the cathode in the electrolysis of fused salt mixtures. The heat of formation of aluminum monofluoride at 25°C is −264 kJ/mol (63.1 kcal/mol) and the free energy of formation is −290 kJ/mol (69.3 kcal/mol) (1). A process to make pure aluminum has been demonstrated in which aluminum monofluoride is formed at high temperature, transferred to a cooler zone, allowed to disproportionate (see Aluminum and aluminum alloys).

Aluminum difluoride [*13569-23-8*], AlF_2, has been detected in the high temperature equilibrium between aluminum and its fluorides (2). The heat of formation of AlF_2 is estimated to be −688 kJ/mol (164.4 kcal/mol) and the molecule is believed to be bent with a bond angle of 120° and an Al—F bond distance of 0.165 nm (1).

Aluminum Trifluoride

Aluminum trifluoride [*7784-18-1*], AlF_3, is a white crystalline solid used in large tonnages as a component of the electrolyte in which aluminum oxide is dissolved for the electrochemical manufacture and refining of aluminum.

Properties. The physical properties of anhydrous aluminum trifluoride are given in Table 1. The crystals are trigonal, usually forming rhombohedrons with vertex angle 88°49′. The space group is $R\ 32\text{-}D_3^7$ with 2 (AlF_3) per unit cell. The cell dimensions are a_0 = 0.504 nm and alpha 58°31′. The crystals are optically uniaxial positive and the mean refractive index is 1.376 (7). The Al atoms are coordinated to 6 F atoms in imperfectly octahedral symmetry with three Al—F distances at 0.170 nm and three at 0.189 nm. The F—F separation is 0.253 nm. The AlF_6 octahedra are connected to each other by their vertices, with Al—Al separation 0.351 nm, forming a giant molecule that accounts for the high melting point relative to the other aluminum halides. The open space in the lattice can be filled with H_2O molecules, resulting in high stability of the monohydrate, $AlF_3.H_2O$ [*13462-92-5*] (8).

Aluminum trifluoride has no liquid state under atmospheric pressure; it sublimes directly from the solid. The vapor is largely monomeric, but at the sublimation temperature of 455°C some dimer, Al_2F_6 [*17949-86-9*] is present at a mole fraction of 0.109. The AlF_3 molecule has the planar symmetry D_{3h} with the F—Al—F bond angle 120° and the Al—F bond distance 0.163 nm (1).

Aluminum trifluoride forms three well-defined hydrates: the nonahydrate [*15098-89-2*], $AlF_3.9H_2O$, crystallizes at low temperature and is not stable above 8°C. The trihydrate [*15098-87-0*], $AlF_3.3H_2O$ appears to exist in a soluble metastable (alpha) and a less soluble (beta) form (5). The alpha trihydrate can be made by the reaction of aluminum with aqueous hydrofluoric acid below 25°C. It slowly and irreversibly

Table 1. The Physical Properties of Anhydrous Aluminum Trifluoride

Property	Value	Reference
molecular weight	83.977	
composition		
fluorine, wt %	67.87	
aluminum, wt %	32.13	
sublimation temperature, °C	1278	1
transition temperature, °C	455	1
vapor pressure at 1000°C, Pa[a]	920	3
density, g/cm^3	3.10	4
solubility in water, g/L	4.1	5
dielectric constant	6	1
heat of transition at 455°C, kJ/mol[b]	0.667	1
heat of sublimation for crystals at 25°C, kJ/mol[b]	300	1
heat capacity, Cp, J/(mol·K)[b]		
25°C, alpha crystal	74.85	1
1000°C, beta crystal	100.5	1
heat of formation at 25°C, kJ/mol[b] alpha crystal	−1505	1
free energy of formation at 25°C, kJ/mol[b]	−1426	1
entropy at 25°C, J/(mol·K)[b]	66.23	1

[a] To convert Pa to mm Hg, multiply by 7.5×10^{-3}.
[b] To convert J to cal, divide by 4.184.

changes to the beta form when kept moist. The beta crystal (tetragonal, density 2.09 g/cm^3) may be obtained directly by crystallization from aqueous solution at 100°C. Aluminum trifluoride monohydrate, $AlF_3.H_2O$, is found in Cornwall, England (9), as the rare mineral fluellite [*12252-28-7, 15621-55-3*] (orthorhombic, Mohs hardness 3, density 2.17 g/cm^3, refractive index 1.47–1.51).

When aluminum trifluoride trihydrate is heated above 300°C, some pyrohydrolysis takes place:

$$2\ AlF_3 + 3\ H_2O \rightarrow 6\ HF + Al_2O_3$$

This can be avoided by reducing the water content to ca 10% at 200°C, and rapidly removing the remainder at about 700°C. By this procedure the water content can be brought below 3% and the Al_2O_3 content can be held below 8% (10). Aluminum oxide does not interfere when aluminum trifluoride is used in aluminum reduction cells, because it replaces part of the aluminum oxide of the electrolyte.

Aluminum trifluoride, unlike the other aluminum halides, is only sparingly soluble in water and organic solvents. Its solubility in aqueous hydrogen fluoride at various concentrations is given in Table 2.

Aluminum trifluoride dissolves slowly in strong sulfuric acid with the liberation of hydrogen fluoride and in strong aqueous alkalies with the formation of aluminate. Aluminum trifluoride is slowly attacked by fused alkalies, with formation of fluoride and aluminate.

A series of double salts with the fluorides of many metals and with the ammonium ion can be made by precipitation or by solid state reaction (see Fluoroaluminates).

Manufacture. In a typical wet process, alumina hydrate is added to aqueous 15% hydrofluoric acid in makeup tanks of steel lined with rubber and carbon brick using HF-resistant mortar. Solution of the alumina hydrate and formation of crystalline $AlF_3.3H_2O$ takes place in continuous-flow crystallizers with the slurry held for about

Table 2. Solubility of Aluminum Trifluoride in Aqueous Hydrogen Fluoride at 25°C[a]

HF concentration, %	g AlF_3 per 100 g solution
0	0.41
10	5.4
20	13.2
30	13.7
40	15.0
50	8.0
60	4.2
65	2.8

[a] Ref. 11.

3 hours at 93°C with continuous agitation. Continous dewatering on rubber-lined rotary suction filters gives a washed cake with ca 10% moisture. The product is calcined in rotating horizontal kilns (12). In a similar process, employing continuous reaction of hydrate and acid, hydrofluoric acid of higher concentration (13) is used.

In a dry process, partially dehydrated alumina hydrate reacts with HF gas at elevated temperature. The alumina may be packed in pellet form into vertical contact towers, but fluidized beds are more commonly used. Fluidization is maintained by the hydrogen fluoride gas and by the steam produced in the reaction. In one reported installation, the reactor is made of Inconel, divided into three superimposed zones by two horizontal sieve plates. Alumina trihydrate is fed into the top zone, and spent gases (water, hydrogen fluoride, silicon tetrafluoride, and dust) leave at this point. These gases may be used in the manufacture of cryolite (see under Fluoroaluminates). Hydrogen fluoride enters the bottom zone and receives heat from the departing aluminum trifluoride. The major reaction occurs in the middle zone to maintain a high temperature (14).

Aluminum trifluoride may also be made from fluorine values gained from waste gases from phosphate rock acidulation in fertilizer manufacture. The silicon tetrafluoride in the gas is first converted to fluorosilicic acid which then reacts with alumina hydrate (15):

$$H_2SiF_6 + 2\ Al(OH)_3 \rightarrow 2\ AlF_3.3H_2O + SiO_2$$

For economic operation loss of aluminum trifluoride to the filter cake must be low, and phosphorus contamination must be avoided. The conditions of operation are critical. The alumina is added slowly to fluorosilicic acid which should contain excess silica, with the F/Si ratio ca 5. With slow addition, a well-filterable silica is obtained, but the reaction time should not be long enough for conversion of alpha alumina hydrate, formed first, to the less soluble beta form (5,16). After removal of the precipitated silica, seed crystals are added and beta aluminum trifluoride crystallizes at an elevated temperature.

Economic Aspects. The volume of production can be estimated from the amount of hydrogen fluoride consumed by the aluminum industry, which was reported in 1978 (17). Assuming 60% of these fluorine values are converted to aluminum trifluoride with an overall yield of 90%, about 91,000 t appear to have been made in that year. This does not include recycled material or production from fluorosilicic acid. The demands for aluminum trifluoride can be expected to grow more slowly than the volume of aluminum production, because of greater efficiency in recycling fluorine values and reduction of fluoride losses. The demand for aluminum trifluoride will be greatly reduced if a recently tested chloride process for the production of aluminum becomes widely used (18).

The 1979 price for aluminum trifluoride, anhydrous technical grade, in bulk quantities, was 38.6¢/kg (19).

Aluminum trifluoride is shipped in 45-kg net weight paper bags and 90-t net weight hopper cars (20).

Specifications. Commercial technical aluminum trifluoride usually has the following composition (21):

assay as AlF_3, typical	90.0%
free alumina as Al_2O_3, typical	9.5
silica as SiO_2, max	0.1
iron as Fe_2O_3, max	0.1
sulfur as SO_2, max	0.32

The bulk density is 1.3 g/cm^3 loose and 1.6 g/cm^3 packed. The screen analysis may be 20% on 10 μm (140 mesh), 60% on 74 μm (200 mesh), and 90% on 44 μm (325 mesh) (22).

Analysis. A sample is fused with sodium or potassium carbonate in platinum and the aluminum determined gravimetrically with 8-hydroxyquinoline. Fluorine is determined by the method of Willard and Winter (23). The water content is determined by the loss in weight when 1 part of the sample is heated with 12 parts of PbO at 600°C for 2 h. Any water consumed by pyrohydrolysis of the fluoride is reconverted by reaction of HF with the lead oxide. The amount of aluminum oxide present is calculated from the aluminum found in excess over that required to combine with the fluoride present to form AlF_3 (10). Recently, procedures to determine fluorine directly in the carbonate fusion solution have been investigated using a lanthanum fluoride fluoride-specific electrode (24–25) (see Ion-selective electrodes).

Health and Safety Factors. Since aluminum trifluoride is only slightly soluble in water and body fluids, its toxicity is less than that of many inorganic fluorides. The lowest reported oral dose causing death in guinea pigs was 600 mg/kg (17). A TLV of 88 mg/m^3 (2.5 mg/ft^3) air has been set for maximum exposure in the workplace. Aluminum trifluoride is not combustible nor does it support combustion. There is some hazard of hydrogen fluoride being formed on exposure to high temperature in the presence of water, or on exposure to strong acids.

Uses. The principal use for aluminum trifluoride is as a makeup ingredient in the molten electrolyte of the aluminum reduction cell and in the electrolytic process for refining aluminum. This electrolyte consists mostly of cryolite, Na_3AlF_6 with 5–7% aluminum trifluoride, 5–7% calcium fluoride, and 0–7% lithium fluoride to which 2–6% aluminum oxide is added. Because the vapor above molten cryolite is mostly sodium tetrafluoroaluminate, $NaAlF_4$, the bath becomes depleted mainly in aluminum trifluoride. Typical electrolyte baths contain 23% aluminum trifluoride in the Pechiney process, and 48% in the Swiss AIAG Neuhausen process (see Aluminum and aluminum alloys).

Lower volume uses for aluminum trifluoride include incorporation into glass (optical, laser, and electrically conducting), low-melting glazes and enamels, and ceramics (electrical resistors), in fluxes (for casting, welding, brazing, and soldering), and catalyst compositions as inhibitor in fermentation processes, and in the manufacture of aluminum silicate fiber.

FLUOROALUMINATES

A great many compounds containing fluorine and aluminum combined in the negative ion are known. A common structural element is the hexafluoroaluminate octahedron, AlF_6^{3-}, in which the coordinate bonds can be described as being formed on the d^2sp^3 hybrid orbitals of aluminum. Some distortion of the octahedra may be present. When the fluorine to aluminum ratio is 6, as in cryolite, Na_3AlF_6, the AlF_6^{3-} ions are separate and are bound in position in the crystal lattice by balancing positive ions. Fluorine atoms may be shared between octahedrons. When opposite corners of each octahedron are shared with a corner of each neighboring octahedron, an infinite chain is formed as, for example, in Tl_2AlF_5 [*33897-68-6*]. In $KAlF_4$ [*14484-69-6*], all four equatorial fluorine atoms of each octahedron are shared and a layer lattice results. In AlF_3, all corners are shared and the crystal becomes a giant molecule of very high melting point (26). More complex relations exist; for example, in chiolite, $Na_5Al_3F_{14}$, one-third of the hexafluoroaluminate octahedra share four corners each and two-thirds share only two corners (27).

Cryolite

The most important fluoroaluminate is cryolite, Na_3AlF_6 (sodium hexafluoroaluminate), an important raw material for aluminum manufacturing. The natural mineral is accurately $3NaF.AlF_3$, but synthetic cryolite is often deficient in sodium fluoride.

Commercially exploitable deposits of cryolite have been found only in the south of Greenland at Ivigtut (28). Reports of cryolite deposits in the Tuva region of the USSR northwest of Mongolia (29) and in the Ukraine (30) do not indicate deposits of commercial significance. Other localities, not all authenticated, are Miask in the Ilmen Mountains, USSR; Sallent in the Pyrenees, Spain; and Pikes Peak, Colorado (31).

The deposit at Ivigtut was formed as a roughly egg-shaped mass estimated at over 2×10^6 t. It is coarse-grained aggregate carrying 10–30% of other minerals, including (in descending order of abundance) siderite, quartz, sphalerite, galena, chalcopyrite, and pyrite. Fluorspar [*14542-23-5*], ivigtite mica, and other rare fluoroaluminate minerals (see Table 3) are also frequently found. The deposit is probably of hydrothermal origin, translocated to its present position at a temperature below the transition point of cryolite (32). The mineral is named cryolite, meaning ice stone, because it resembles ice when immersed in water as a result of the closely matched refractive indexes.

Cryolite was mined by the open-pit method, beginning about 1850. Its first use was for the manufacture of soda, alumina hydrate, and calcium fluoride. The cryolite was ignited with limestone:

$$Na_2AlF_6 + 3\ CaCO_3 \rightarrow NaAlO_2 + Na_2CO_3 + 3\ CaF_2 + CO_2$$

The charge was lixiviated with water to form a solution of sodium aluminate and

Table 3. Naturally Occurring Fluoroaluminates[a]

Name	CAS Registry No.	Composition
cryolite	[*15096-52-2*]	Na_3AlF_6
chiolite	[*1302-84-7*]	$Na_5Al_3F_{14}$
cryolithionite	[*15491-07-3*]	$Na_3Li_3(AlF_6)_2$
thomsenolite, hagemannite	[*16970-11-9*]	$NaCaAlF_6.H_2O$
ralstonite	[*12199-10-9*]	$Na_{2-x}(Al_{2-x},Na_x)(F,OH)_6.yH_2O$
prosopite	[*12420-95-0*]	$CaAl_2(F,OH)_8$
jarlite, meta-jarlite	[*12004-61-4*]	$NaSr_3Al_3F_{16}$
weberite	[*12423-93-7*]	Na_2MgAlF_7
gearksutite	[*12415-96-2*]	$CaAl(F,OH)_5.H_2O$
pachnolite	[*15489-46-0*]	$NaCaAlF_6.H_2O$

[a] Refs. 28 and 32.

carbonate, which was then treated with carbon dioxide to decompose the aluminate:

$$2\ NaAlO_2 + CO_2 + 3\ H_2O \rightarrow 2\ Al(OH)_3 + Na_2CO_3$$

This process was operated by the Pennsylvania Salt Manufacturing Company (now Pennwalt Corporation) until about 1900, with an annual consumption of cryolite of ca 6000 t/yr (33).

Cryolite deposits were practically exhausted by 1962, and are today only a very minor contribution to the world's need for cryolite. Thus, in 1975 less than 2% of the U.S. demand for cryolite was provided as the natural mineral (34).

Properties. The physical properties of cryolite are given in Table 4. The crystals are monoclinic, departing only slightly from orthorhombic symmetry with an axial angle of 90°11′. The space group is $P2_1/m$. The [001] and [110] faces are usually dominant, giving the crystals a cubic appearance. Twinning is ubiquitous, and since the lamellae tend to be perpendicular, cleavage appears to be cubic. The fracture of individual crystals, however, is conchoidal or irregular. Because its refractive indexes are close to the refractive index of water, powdered cryolite becomes nearly invisible when immersed in water. However, since the optical dispersion is different for the two substances, the suspension shows Christiansen colors.

Cryolite mineral is usually white, but it may also be black, purple, or violet, and occasionally brownish or reddish. The luster is vitreous to greasy, sometimes pearly, and the streak is white. Identification is by optical characteristics, low hardness, and evolution of HF on treatment with strong sulfuric acid (31); x-ray diffraction is conclusive (42).

With increasing temperature, the crystallographic angles approach 90°, and at 565°C cryolite becomes cubic, with a small heat of transition. The unit cell of the cubic form has a length of 0.295 nm, with four formula units per cell. The space group is *Fm3m,* derived from the monoclinic crystal by a rotation and small translation of the AlF_6 octahedra into more symmetrical positions (37). A substantial density change takes place at the transition point, evidenced by a characteristic decrepitation (44). A second transformation of cryolite crystal is indicated by a change in slope of the heating curve, and a sharp rise in electrical conductivity at 880°C. The change in heat

Table 4. Physical Properties of Cryolite

Property	Value	Ref.
molecular weight	209.94	
composition, calculated, %		
fluorine	54.30	
aluminum	12.85	
sodium	32.85	
melting point, °C	1012	1
transition temperatures, °C		
monoclinic to cubic	565	1
second-order	880	1
dimensions of unit cell, nm		35
a	0.546	
b	0.561	
c	0.780	
vapor pressure of liquid, at 1012°C, Pa[a]	253	36
heat of fusion at 1012°C, kJ/mol[b]	107	1
heat of vaporization at 1012°C, kJ/mol[b]	225	36
heat of transition, kJ/mol[b]		
monoclinic to cubic, at 565°C	8.21	1
second-order, at 880°C	0.4	1
heat capacity, J/(mol·K)[b]		
monoclinic crystal at 25°C	215	1
cubic crystal at 560°C	281	1
liquid at 1012°C	395	1
entropy, monoclinic crystal, at 25°C, J/(mol·K)[b]	238	1
heat of formation, monoclinic crystal at 25°C, kJ/mol[b]	−3297	1
free energy of formation, monoclinic crystal at 25°C, kJ/mol[b]	−3133	1
density, g/cm^3		
monoclinic crystal, at 25°C	2.97	31
cubic crystal, from x-ray data, at 560°C	2.77	37
solid at 1012°C	2.62	38
liquid at 1012°C	2.087	39
hardness, Mohs	2.5	31
optical properties		
refractive index,		
alpha form	1.3385	
beta form	1.3389	
gamma form	1.3396	
biaxial character	positive	
axial plane	perpendicular to [010]	
angle of acute bisectrix to c-axis	43°54′	
electrical conductivity, (ohm^{-1} cm^{-1})		
liquid, at 1012°C	2.82	40
solid, at 400°C	4.0×10^{-6}	38
viscosity, liquid, at 1012°C, mPa·s(=cP)	6.7	41
surface tension, liquid in air, at 1012°C, mN/m (= dyn/cm)	125	42
solubility in water, g/L		36
at 25°C	0.042	
at 100°C	0.135	
activity product constant, in water, 25°C	1.46×10^{-34}	29

[a] To convert Pa to mm Hg, multiply by 7.5×10^{-3}.
[b] To convert J to cal, divide by 4.184.

content is very small, and the transition as temperature rises probably marks the onset of lattice disorder. This view is strengthened by the observation of the notable plastic character of the solid near the melting point (38).

Liquid cryolite is an equilibrium mixture of the products of the dissociation:

$$Na_3AlF_6 \rightleftharpoons 2\ NaF + NaAlF_4$$

The composition close to the melting point is estimated to be 65% Na_3AlF_6, 14% NaF, and 21% $NaAlF_4$ [*1382-15-3*]. Electric current is carried in liquid cryolite by Na^+ and by F^-. The mobility of AlF_4^- appears to be small. Evidence for this structure of cryolite includes electrical conductivity, density, thermodynamic data, cryoscopic behavior, and the presence of $NaAlF_4$ in the equilibrium vapor (45–46).

Molten cryolite dissolves many salts and oxides, forming solutions of lower melting point than the components. Figure 1 combines the melting point diagrams for cryolite–AlF_3 and for cryolite–NaF. Cryolite systems are of great importance in the Hall–Heroult electrolytic process for the manufacture of aluminum (see Aluminum and aluminum alloys).

Additional examples of the effectiveness of cryolite as a fused salt solvent for fluorides and oxides are given in Table 5.

The vapor from liquid cryolite is largely $NaAlF_4$, the vapor pressures of Na_3AlF_6, NaF, and $NaAlF_4$ near the melting point are about in the ratios 5:1:30. For this reason, the liquid tends to become depleted in AlF_3, and the composition of the aluminum cell electrolyte has to be regularly adjusted by the addition of AlF_3 (46,51).

In contact with moist air, liquid cryolite loses HF and is depleted in AlF_3.

$$2\ Na_3AlF_6 + 3\ H_2O \rightarrow 6\ NaF + 6\ HF + Al_2O_3$$

The pyrohydrolysis is notable but not complete, proceeding to ca 60% conversion with steam (52).

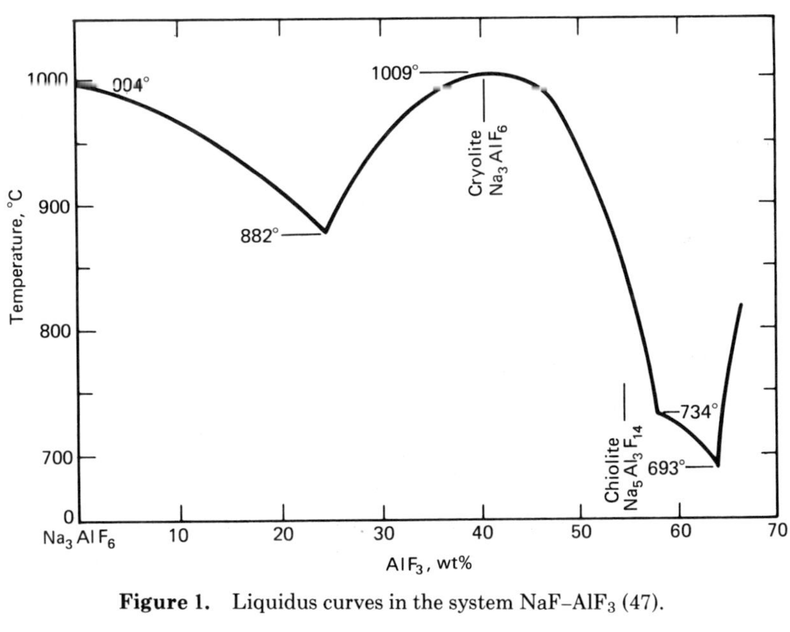

Figure 1. Liquidus curves in the system NaF–AlF_3 (47).

Table 5. Minimum-Melting Compositions Containing Cryolite[a]

Added component	Wt % of added component	Melting temp, °C
NaF	24.5	882
AlF_3	64	693
Al_2O_3	10.5	962
Li_3AlF_6	62	710
CaF_2	25.8	945
ZrO_2	14	969
MgO	7.5	902
CaO	11.3	896
ZnO	2.4	974
CdO	6.0	971
TiO_2	4.0	970
BaF_2	62.5	835
PbF_2	40	730
feldspar	70	830
NaF Al_2O_3	34.0 12.0	870
CaF_2 Al_2O_3	23.0 17.7	867
CaF_2 AlF_3	37.8 6.2	675
SiO_2 Al_2O_3	17 50	ca 800

[a] Refs. 46–50.

The more electropositive metals react with cryolite at high temperatures, liberating aluminum or aluminum monofluoride (53–54). The reduction of cryolite by sodium was used in the Netto aluminum process briefly in the 1890s (55). The reduction of cryolite by magnesium is a current method for the removal of magnesium in the refining of secondary aluminum.

The solubility of cryolite in water at 25°C is considerably less than that of aluminum trifluoride (0.41), and much less than that of sodium fluoride (4.03), demonstrating again a structure for cryolite other than that of a simple double salt and suggesting that the solution contains fluoroaluminate ions, probably AlF_6^{3-}. Further dissociation to ions of lower F:Al ratio may take place, probably with maintenance of an octahedral environment for Al by replacement of F^- with H_2O or OH^- (56). The solubility of cryolite in water is sharply decreased by the addition of sodium fluoride, strengthening the view that fluoride ion is a product of dissociation. The solubility is increased by acids, perhaps because of formation of HF_2^-, and by bases, perhaps by formation of AlO_2^-. Cryolite is very soluble in solutions of aluminum and iron salts (see Table 6); the explanation of this is not clear.

Cryolite is not combustible and does not support combustion. It may liberate hydrogen fluoride on exposure to strong acids, or in contact with water vapor at high temperatures.

Manufacture. Today, nearly all the cryolite consumed is synthetic. Its composition and properties are the same as those of natural cryolite, but the ratio $NaF{:}AlF_3$ is generally low, and it commonly contains oxygen or hydroxyl. For most uses, the defi-

Table 6. Solubility of Cryolite in Aqueous Solutions[a]

Solvent	Temp, °C	Solubility, g/100 g solvent
water	25	0.042
5 wt % $AlCl_3$	25	5.8
5 wt % $FeCl_3$	20	2.5
1.5 wt % HCl	20	0.38
NaOH, strong	hot	complete[b]

[a] Refs. 36, 57–58.
[b] Cryolite solubility limited by stoichiometry and by the solubility of NaF.

ciency in sodium fluoride is not a handicap. The presence of moisture, however, causes loss of fluorine as HF on heating. Losses of cryolite by dusting are higher for synthetic than for natural cryolite.

Many processes for the manufacture of cryolite have been described. The choice depends on the cost and quality of the available sources of soda, alumina, and fluorine. Starting materials include hydrogen fluoride kiln gases, aqueous hydrofluoric acid, fluorosilicic acid, and sodium fluorosilicate; aluminum oxide, sulfate and chloride, and alumina hydrate; and sodium carbonate, sodium chloride, and sodium aluminate.

The manufacture of cryolite is commonly integrated with the production of alumina hydrate and of aluminum trifluoride. The intermediate stream of sodium aluminate from the Bayer alumina hydrate process (see Aluminum and aluminum alloys) can be used along with aqueous hydrofluoric acid, or with hydrogen fluoride kiln gases, or with hydrogen fluoride-rich effluent from dry-process aluminum fluoride manufacture.

$$NaAlO_2 + Na_2CO_3 + 6\ HF \rightarrow Na_3AlF_6 + CO_2 + 3\ H_2O$$

The HF and Na_2CO_3 give a sodium fluoride solution. Bayer sodium aluminate solution is added in the stoichiometric ratio. Cryolite is precipitated at 30–70°C by passing in CO_2, until the pH reaches 8.5–10.0. Seed crystals are desirable. The slurry is thickened and filtered, or settled and decanted, or centrifuged. Cryolite is dried and calcined at 500–700°C. The weight ratio of fluorine to aluminum in the product should exceed 3.9 (theory, 4.2) (59).

A dry process for making cryolite involves passing gaseous HF over briquettes of alumina hydrate, sodium chloride, and sodium carbonate at 400–700°C, followed by sintering at 720°C (60).

Many ways have been described for making cryolite from low-value fluorine sources, such as the effluent gases from phosphorus or phosphate plants, or from low-grade fluorspar. In the manufacture of cryolite from fluorosilicic acid from superphosphate production, silica is separated by precipitation with ammonia, and the ammonium fluoride solution is added to a solution of sodium sulfate plus aluminum sulfate at 60–90°C to precipitate cryolite (61–62):

$$12\ NH_4F + 3\ Na_2SO_4 + Al_2(SO_4)_3 \rightarrow 2\ Na_3AlF_6 + 6\ (NH_4)_2SO_4$$

The ammonia values can be recycled or sold for fertilizer use.

In making cryolite from fluorine-containing by-products of phosphate manufacture, it is important to eliminate phosphorus from the product, since as little as 0.01% P_2O_5 in the electrolyte causes a 1–1.5% reduction in current efficiency for aluminum production (42).

A large amount of cryolite is recovered from waste material in the manufacture of aluminum. For example, the carbon lining of the electrolysis cells may contain 10–30% by weight of cryolite. The crushed liner may be extracted with sodium hydroxide or sodium carbonate solution and the cryolite precipitated with carbon dioxide (42). Gases from the operating cells, containing HF, CO_2, and fluorine-containing dusts, may be used for the carbonation (64).

Production, Economic Aspects. During production of the Greenland mine, ca 13,500 t was imported annually by the United States; world consumption was about 40,000 t (8). In 1975, 2624 t was imported from stockpiles in Denmark.

Since most synthetic cryolite is used for aluminum processes, the volume of production must be estimated. In 1978, 131,500 t of hydrogen fluoride was consumed by the aluminum industry (65). Assuming that 40% of these fluorine values was converted to cryolite with an overall yield of 90%, about 82,500 t appears to have been manufactured. This is ca 21.3 kg cryolite per ton of aluminum produced in that year (see Aluminum and aluminum alloys). Total production of cryolite should include recycled material, cryolite made from fluorosilicic acid, and the minor requirements for cryolite outside the aluminum industry.

The demand for cryolite will increase more slowly than the growth of the aluminum industry because of greater efficiency in the use of fluorine values. It will be greatly reduced if a recently tested chloride process for making aluminum becomes widely used (24).

The published price range for synthetic cryolite in April 1979 was $562–606/t, approximately twice the 1965 price of $287/t (66).

Specifications and Analysis. Natural cryolite has been specified for aluminum manufacture and for ceramics as 95% sodium aluminum fluorides calculated as Na_3AlF_6, and 4% other fluorides calculated as CaF_2. This is a flotation product passing 80% through a 44 μm sieve (325 mesh). A small amount of selected white lump, especially low in iron, has been sold for special ceramic purposes. Fines from the refining operation have been sold for insecticide preparation (67).

These specifications for cryolite as used intraplant by aluminum manufacturers vary with particular requirements and choice of starting materials. A typical analysis is given in Table 7. The bulk density is 1.4–1.5 g/cm^3 (90–95 lb/ft^3), the screen analysis is 65–75% <74 μm (-200 mesh) (68).

The analysis of natural cryolite for fluorine, aluminum, sodium, silica, calcium, and iron is performed by standard procedures, and from these values the assay of cryolite and the content of impurities can be calculated. In synthetic cryolite, water, sodium fluoride, alumina, and phosphorus are also determined (24,69). Fluoride can now be determined directly by means of a fluoride-selective lanthanum fluoride electrode after the cryolite has been fused with sodium carbonate and dissolved in water (24–25).

Health and Safety Factors. Although the relatively low solubility of cryolite in water suggests that the fluoride toxicity should be low, oral feeding tests on rats show the median lethal dose to be about the same as for more soluble fluorides. The reported LD_{50} values are for cryolite, 200 mg/kg, for sodium fluoride, 180 mg/kg, and for po-

Table 7. Typical Analysis of Commercial Cryolite

Component	Percent
cryolite as Na_3AlF_6	91
fluorine as F	48–52
sodium as Na	31–34
aluminum as Al	13–15
alumina as Al_2O_3	6.0
silica as SiO_2, max	0.70
calcium fluoride as CaF_2	0.04–0.06
iron as Fe_2O_3	0.10
moisture	0.05–0.15

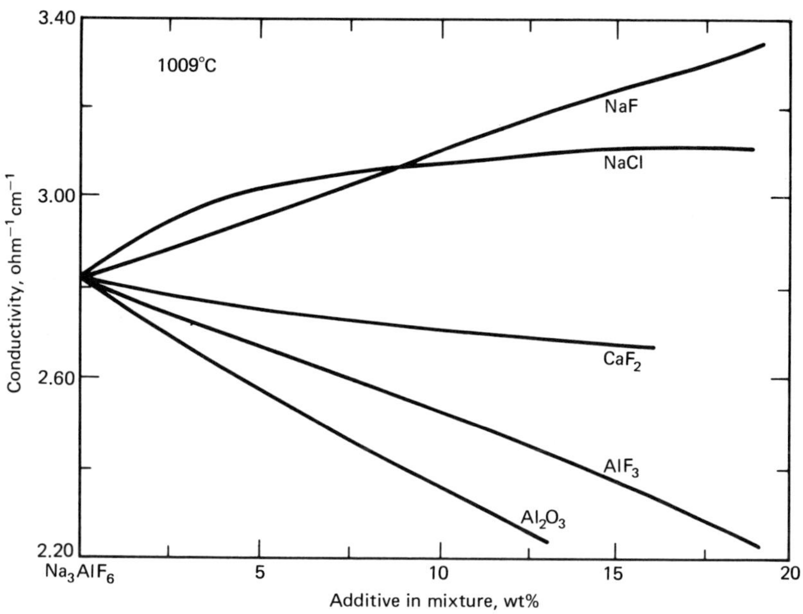

Figure 2. Effect of additives on the electrical conductivity of liquid cryolite (40,71–73).

tassium fluoride, 245 mg/kg (17). Apparently, stomach fluids are acid enough to bring the solubility of cryolite up to values comparable with other fluorides. Continued exposure may eventually produce symptoms of fluorosis. The toxicity to insects is in many cases high enough for control. Because of its variable composition, synthetic cryolite may show physiological activity greater than the natural mineral (70).

Uses. The need for cryolite in the manufacture of aluminum results from the effectiveness with which molten cryolite dissolves Al_2O_3 to provide a conducting bath for the electrolytic production of aluminum. Certain additives improve the physical and electrical properties of the electrolyte. For example, aluminum fluoride lowers the melting point (see Fig. 1). The amount that can be added, however, is limited because it also reduces the electrical conductivity (see Fig. 2). Calcium fluoride further depresses the melting point, with a less adverse effect on the conductivity (see Fig. 2). Too much calcium fluoride would raise the density of the melt too close to that of

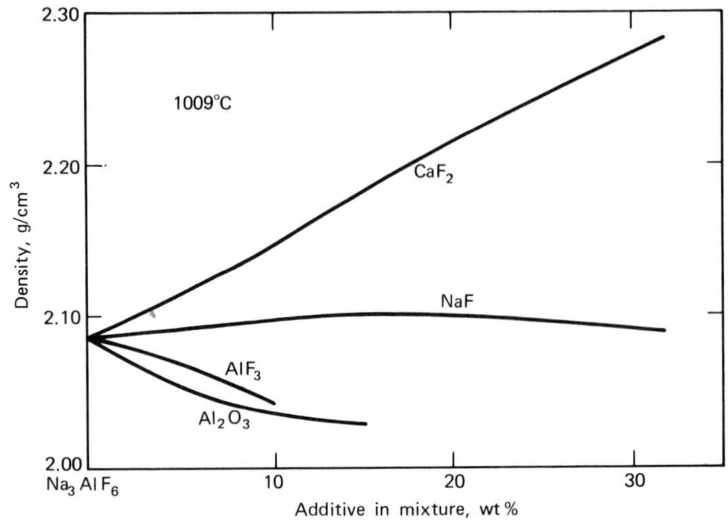

Figure 3. Effect of additives on the density of liquid cryolite.

liquid aluminum (ca 2.28 g/cm^3), inhibiting the separation of metal and electrolyte (see Fig. 3). Sodium fluoride improves density and conductivity, but melts high in sodium fluoride show a reduced current efficiency. Small amounts of lithium fluoride may also improve electrical conductivity. Compromises on all of these factors have led to electrolytes that consist of 80–85% cryolite, 5–7% aluminum fluoride, 5–7% calcium fluoride, 0–7% lithium fluoride, and 2–8% aluminum oxide.

A second significant use for cryolite in the aluminum industry is in the production of pure metal by electrolytic refining. A high-density electrolyte capable of floating liquid aluminum is needed, and compositions are used containing cryolite with barium fluoride to raise the density, and aluminum fluoride to raise the current efficiency (see Aluminum and aluminum alloys).

In the reworking of scrap aluminum, cryolite is added to the flux to remove magnesium by electrochemical displacement (see Recycling). Other metallurgical uses include flux in aluminizing steel, in the compounding of welding-rod coatings, and fluxes in processing a variety of metals.

In making glass (qv), cryolite functions as a powerful flux because of the excellent solvent power of the molten compound for the oxides of silicon, aluminum, and calcium, and because of the low-melting compositions it forms with many components. In the glass furnace, cryolite is the most stable fluorine carrier. It is used typically in amounts of 10–15% to make opal glass. In clear glass, cryolite speeds up batch melting, and, by reducing the melt viscosity, improves fining (50).

In enamels (qv), 2–15% cryolite lowers the surface tension of the melt and improves spreading (74). Cryolite also lowers the melting temperature and the viscosity, and has no adverse effect on acid resistance.

Cryolite is a filler for resin-bonded grinding wheels, varying from 1 to 2% of the weight of the wheel, and giving longer wheel life, reduced metal buildup on the wheel, and faster and cooler grinding action. These effects have not been fully explained. It has been proposed that cryolite is a reinforcing pigment for the resin-bond matrix, or that at the cutting surface cryolite fuses and has a lubricating action, or that melting of the cryolite allows controlled release of spent abrasive grains.

Cryolite, as a very finely ground powder, continues to be used in the control of chewing insects, especially on grapes. It has limited use on citrus fruits, beans, lettuce, and some other vegetables. Cryolite acts as a stomach poison to insects, but has low human and faunal toxicity, and does not concentrate in the food chain (75) (see Insect control technology).

Cryolite has been reported in use in the manufacture of artificial stone, as a component of roofing granules, in catalysts for olefin polymerization, in preparing antireflective coatings on glass, as a surface coating for laser mirrors, in pyrotechnic and propellant compositions, as a rimming agent for steel, and in self-lubricating bearings.

OTHER FLUOROALUMINATES

The fluoroaluminates of a large number of metals and of ammonium are known, in addition to ternary salts and hydrates; in some cases, partial replacement of fluorine by hydroxyl takes place. A number of naturally occurring fluoroaluminates have been found, mostly in the deposit at Ivigtut, Greenland (see Table 3).

The vapor from liquid cryolite is largely sodium tetrafluoroaluminate, $NaAlF_4$. An equivalent vapor, swept by a rapid passage of gas over a mixture of NaF and AlF_3 at 1000°C and then quenched, yields a sublimate of $NaAlF_4$. Above 470°C, this compound is transformed into a mixture of aluminum fluoride and chiolite, $Na_5Al_3F_{14}$ (76). As in other fluoroaluminates, in $NaAlF_4$ aluminum is in octahedral coordination with fluorine; a tetrahedral arrangement is calculated to be unstable (77). In $NaAlF_4$ and in the other fluoroaluminates with the same ratio of fluorine to aluminum, all four equatorial fluorine atoms of each octahedron are shared, and a layer lattice results. Crystals of similar composition may show lamellar cleavage, and certain fluoroaluminates have been proposed as mica replacements (78).

BIBLIOGRAPHY

"Aluminum Fluoride" in *ECT* 1st ed., under "Fluorine Compounds, Inorganic," Vol. 6, pp. 668–671 by R. G. Danehower, Pennsylvania Salt Manufacturing Co.; "Aluminum Fluorides" in *ECT* 2nd ed., under "Fluorine Compounds, Inorganic," Vol. 9, pp. 529–533, by John F. Gall, Pennsalt Chemicals Corporation; "Fluoroaluminates" in *ECT* 1st ed., under "Fluorine Compounds, Inorganic," Vol. 6. pp 671–675, by Isadore Mockrin, Pennsylvania Salt Manufacturing Co.; "Fluoroaluminates" in *ECT* 2nd ed., under "Fluorine Compounds, Inorganic," Vol. 9, pp. 534–548, by John F. Gall, Pennsalt Chemicals Corporation.

1. *JANAF Thermochemical Tables,* 2nd ed., Natl. Bur. Standards, NSRDS-NBS 37, 1971.
2. T. C. Ehlert and J. L. Margrave, *J. Am. Chem. Soc.* **86,** 3901 (1964).
3. W. P. Witt and R. F. Barrow, *Trans. Faraday Soc.* **55,** 730 (1959).
4. J. D. H. Donnay and W. Nowadic, *Crystal Data,* The Geological Society of America, New York, 1954, p. 416.
5. W. F. Ehret and F. J. Frere, *J. Am. Chem. Soc.* **67,** 64 (1945).
6. R. L. Thakur and R. Pepinsky, *Am. Mineral.* **37,** 695 (1952).
7. E. Staritsky and L. B. Asprey, *Anal. Chem.* **29,** 984 (1957).
8. F. Hanic and co-workers, *Acta Chim. Acad. Sci. Hung.* **32,** 309 (1962); *Chem. Abstr.* **58,** 3977f (1963).
9. J. D. Dana and co-workers, *The System of Minerology,* Vol. 2, John Wiley & Sons, Inc., New York, 1951, pp. 124–125.
10. J. K. Bradley, *Chem. Ind. London,* 1027 (1960).
11. I. V. Tananaev, *Zh. Obshch. Khim.* **8,** 1120 (1938).
12. J. R. Callaham, *Chem. Met. Eng.* **52**(3), 94 (1945).

13. U.S. Pat. 2,958,575 (Nov. 1, 1960), D. R. Allen (to The Dow Chemical Co.).
14. *Chem. Eng.* **72,** 92 (Jan. 18, 1965).
15. F. Weinratter, *Chem. Eng.* **71,** 132 (April 27, 1964).
16. U.S. Pat. 2,842,426 (July 8, 1958), E. M. Glocker (to W. R. Grace & Co.).
17. *The Toxic Substances List, 1974 ed.,* U.S. Dept. of Health, Education & Welfare, National Institute for Occupational Safety & Health, Rockville, Md., June 1974.
18. *Chem. Eng. News.* **57,** 23 (April 30, 1979).
19. *Chem. Market. Rep.,* 40 (April 20, 1979).
20. *Data Sheet, Aluminum Fluoride Technical Grade,* Stauffer Chemical Co., Industrial Chemical Division, Westport, Conn., 1975.
21. P. L. Brueckner, Private Communication, Allied Chemical, Industrial Chemicals Division, Morristown, N.J., Nov. 2, 1978.
22. *Data Sheet, Aluminum Fluoride, AlF_3,* Kaiser Aluminum & Chemical Sales, Inc., Hillside, N.J., 1965.
23. H. H. Willard and O. B. Winter, *Ind. Eng. Chem. Anal. Ed.* **5,** 7 (1933).
24. R. T. Oliver and A. G. Clayton, *Anal. Chim. Acta* **51,** 409 (1970).
25. *Analytical Methods Guide,* 6th ed., Orion Research Inc., Cambridge, Mass., 1973, p. 17.
26. P. J. Durrant and B. Durrant, *Introduction to Advanced Inorganic Chemistry,* John Wiley & Sons, Inc., New York, 1970, p. 570.
27. A. F. Wells, *Structural Inorganic Chemistry,* 4th ed., Clarendon Press, Oxford, 1975, pp. 388–390.
28. H. Pauly, *Met. Assoc. Acid. Magmat.* **I,** 393 (1974); *Chem. Abstr.* **84,** 47287 (1976).
29. V. S. Kudrin and M. A. Kudrina, *Tr. Mineral Muz. Akad. Nauk SSSR* (18), 80 (1968); *Chem. Abstr.* **70,** 21728 (1969).
30. E. P. Gurov and co-workers, *Dopov. Akad. Nauk Ukr. RSR, Ser. B,* **34**(7), 582 (1972); *Chem. Abstr.* **77,** 91099 (1972).
31. Ref. 9, pp. 110–113.
32. O. B. Bøggild, *The Mineralogy of Greenland,* C. A. Reitzels Forlag, Copenhagen, 1953, pp. 70–98, 439.
33. G. Lunge, *Sulfuric Acid and Alkali,* Vol. III, Gurney & Jackson, London, 1896, pp. 164–174.
34. *Minerals Yearbook 1975,* Vol. I, U.S. Bureau of Mines, U.S. Government Printing Office, Washington, 1977, p. 642; *Ibid.,* 1973, p. 540.
35. S. Naray-Szabo and K. Sasvari, *Z. Krist.* **99,** 27 (1938).
36. I. Tananaev and Sh. Talipov, *J. Gen. Chem. USSR* **9,** 1155 (1939).
37. E. G. Steward and H. P. Rooksby, *Acta Cryst.* **6,** 49 (1953).
38. G. J. Landon and A. R. Ubbelohde, *Trans. Faraday Soc.* **52,** 647 (1955).
39. J. O. Edwards and co-workers, *J. Electrochem. Soc.* **99,** 527 (1952).
40. E. W. Yim and M. Feinleib, *J. Electrochem. Soc.* **104,** 626 (1957).
41. A. Vayna, *Alluminio* **19,** 133 (1950).
42. E. Elchardus, *Compt. Rend.* **206,** 1460 (1938).
43. *X-Ray Diffraction Data Card 12-257,* American Society for Testing and Materials, Philadelphia, 1962.
44. P. P. Fedotiev and V. Ilyinskii, *Z. Anorg. Chem.* **80,** 113 (1913).
45. P. A. Foster, Jr. and W. E. Frank, *J. Electrochem. Soc.* **107,** 997 (1960).
46. L. M. Foster, *Ann. N.Y. Acad. Sci.* **79,** 919 (1960).
47. N. W. F. Phillips, R. H. Singleton, and E. A. Hollingshead, *J. Electrochem. Soc.* **102,** 648, 690 (1955).
48. A. Fenerty and E. A. Hollingshead, *J. Electrochem. Soc.* **107,** 993 (1960).
49. D. F. Weill and W. S. Fyle, *J. Electrochem. Soc.* **111,** 582 (1964).
50. I. Mockrin, *Ceram. Ind.* **52,** 65 (1949).
51. K. Grjotheim, H. Kuande, and K. Motzfeldt, *Light. Met.* **1,** 125 (1975).
52. G. Wendt, *Z. Erzbergbau Metallhuettenwes.* **8,** 207 (1955).
53. M. Feinleib and B. Porter, *J. Electrochem. Soc.* **103,** 231 (1956).
54. W. E. Haupin, *J. Electrochem. Soc.* **107,** 232 (1960).
55. J. D. Edwards, F. C. Frary, and Z. Jeffries, *The Aluminum Industry,* Vol. I, McGraw-Hill Book Co., Inc., New York, 1930, p. 10.
56. T. R. Scott, *J. Counc. Sci. Ind. Res. Australia* **19,** 442 (1946).
57. F. J. Frere, *J. Am. Chem. Soc.* **58,** 1695 (1936).
58. R. H. Carter, *J. Econ. Entomol.* **27,** 863 (1934).

59. U.S. Pat. 3,061,411 (Oct. 30, 1962), D. C. Gernes (to Kaiser Aluminum & Chemical Corp.).
60. U.S. Pat. 3,104,156 (Sept. 17, 1963), P. Saccardo and F. Gozzo (to Sicedison S.p.A.).
61. G. Tarbutton and co-workers, *Ind. Eng. Chem.* **50,** 1525 (1958).
62. U.S. Pat. 2,687,341 (Aug. 24, 1954), I. Mockrin (to Pennsylvania Salt Mfg. Co.).
63. U.S. Pat. 3,106,448 (Oct. 8, 1963), C. H. Whicher, A. G. Nickle, and A. K. Dolega-Kowaleski (to Aluminum Laboratories, Ltd.).
64. U.S. Pat. 3,065,051 (Nov. 20, 1962), H. Mader (to Vereinigte Metallwerke Ranshofen-Berndorf A.G.).
65. *Chem. Eng. News.* **59,** 9 (May 21, 1979).
66. *Chem. Mark. Rep.,* 45 (April 30, 1979).
67. *Cryolite,* Pennsylvania Salt Manufacturing Co. (now Pennwalt Corp.), 1950.
68. *Cryolite, Data Sheet,* Kaiser Aluminum and Chemical Sales, Inc., Oakland, Calif., 1965.
69. F. J. Frere, *Ind. Eng. Chem. Anal. Ed.* **6,** 124 (1934).
70. E. J. Largent, *J. Ind. Hyg. Toxicol.* **30,** 92 (1948).
71. K. Matiasovsky, M. Malinovsky, and S. Ordzovensky, *J. Electrochem. Soc.* **111,** 973 (1964).
72. W. B. Frank and L. M. Foster, *J. Phys. Chem.* **64,** 310 (1960).
73. J. D. Edwards and co-workers, *J. Electrochem. Soc.* **100,** 508 (1953).
74. R. Marker, *Glas Email Keramo Tech.* **4,** 117 (1957); R. Marker, *Glas Email Keramo Tech.* **5,** 178 (1957).
75. *Pesticide Dictionary—1979,* Meister Publishing Co., Willoughby, Ohio, p. D74.
76. E. H. Howard, *J. Am. Chem. Soc.* **76,** 2041 (1954).
77. J. L. Holm, *J. Chem. Ed.* **51,** 400 (1974).
78. U.S. Pat. 3,016,480 (Jan. 9, 1962), I. Mockrin (to Pennsalt Chemicals Corp.).

JOHN F. GALL
Philadelphia College of Textiles and Science

AMMONIUM

Two well-known salts of ammonia are the normal ammonium fluoride [*12125-01-8*], NH_4F, and ammonium bifluoride [*1341-49-7*], NH_4HF_2; the latter is sometimes named ammonium acid (or hydrogen) difluoride. Much of the commercial interest in the ammonium fluorides stems from their chemical reactivity as less hazardous substitutes for hydrofluoric acid.

Ammonium Fluoride

Ammonium fluoride is a white, deliquescent, crystalline salt. It tends to lose ammonia gas to revert to the more stable ammonium bifluoride. Its solubility in water is 45.3 g/100 g of H_2O at 25°C and its heat of formation is −466.9 kJ/mol (−116 kcal/mol). Ammonium fluoride is available principally as a laboratory reagent. If it is needed in large quantities, one mol of aqueous ammonia can be mixed with one mol of the more readily available ammonium bifluoride (1).

Ammonium Bifluoride

Properties. Ammonium bifluoride, NH_4HF_2, is a colorless, orthorhombic crystal (2). The compound is odorless; however, less than 1% excess HF can cause an acid odor. The salt has no tendency to form hydrates yet is hygroscopic if the ambient humidity is over 50%. A number of chemical and physical properties are listed in Table 1.

Corrosion. Ammonium bifluoride dissolves in aqueous solutions to yield the acidic bifluoride ion; the pH of a 5% solution is 3.5. In most cases, NH_4HF_2 solutions react readily with surface oxide coatings on metals, thus NH_4HF_2 is used in pickling solutions (see Metal surface treatments). Many plastics, such as polyethylene, polypropylene, unplasticized PVC, and carbon brick, are resistant to attack by ammonium bifluoride.

Manufacture. Anhydrous ammonium bifluoride containing 0.1% H_2O and 93% NH_4HF_2 can be made by dehydrating ammonium fluoride solutions and by thermally decomposing the dry crystals (7). Commercial ammonium bifluoride, which usually contains 1% NH_4F, is made by gas-phase reaction of one mole of anhydrous ammonia with two moles of anhydrous hydrogen fluoride (8); the melt that forms is flaked on a cooled drum. The cost of the material in 1978 was $1.01/kg.

Production of bifluoride from fluoride by-products from the phosphate industry (9) has had little if any commercial significance.

Precautions in Handling. Ammonium bifluoride, like all soluble fluorides, is toxic if taken internally. Hydrofluoric acid burns may occur if the material comes in contact with moist skin. Ammonium bifluoride solutions should be thoroughly washed from the skin with mildly alkaline soap as soon as possible; however, if contact has been prolonged, the affected areas should be soaked with 0.13% solution of Zephiran chlo-

Table 1. Physical and Chemical Properties of Ammonium Bifluoride, NH_4HF_2

		Reference
melting point, °C	126.1	3
boiling point, °C	239.5	3
index of refraction, n_D	1.390	3
solubility in water (25°C), wt %	41.5	3
solubility in 90% ethanol (25°C), wt %	1.73	3
specific gravity	1.50	3
standard heat of formation, kJ/mol[a]	−798.3	4
heat of fusion, kJ/mol[a]	19.1	4
heat of vaporization, kJ/mol[a]	65.3	4
heat of solution, kJ/mol[a]	20.3	4
heat of dissociation, kJ/mol[a] (to NH_3 and HF)	141.4	4
heat capacity, J/mol·K[a] C_p^o (25°C)	106.7	4
vapor pressure[b], $\log P_{Pa} = a - bT^{-1}$	$a = 11.72$, $b = 3370$ (153–207°C)	5
	$a = 9.38$, $b = 2245$ (207–245°C)	5
oral LD (guinea pig), mg/kg	150	6

[a] To convert kJ to kcal, divide by 4.184.
[b] To convert Pa to mm Hg, multiply by 7.5×10^{-3}.

ride, the treatment recommended for hydrofluoric acid burns (10). If the solution comes in contact with the eyes, they should be washed with water for at least 10 min and a physician should be consulted.

Applications. ***Uses Based on Reactions with Silicates.*** Ammonium bifluoride solubilizes silica and silicates by forming ammonium fluorosilicate [*16919-19-0*], $(NH_4)_2SiF_6$. Inhibited 15% hydrochloric acid containing about 2% ammonium bifluoride has been used to acidize oil wells in siliceous rocks to regenerate oil flow (11) (see Petroleum). In a new procedure, ammonium fluoride solution is made on site near the well bore from ammonium bifluoride and ammonia and mixed with methyl formate to prevent rapid consumption of most of the HF (12). The use of ammonium bifluoride is important in locations where dissolved silicates foul boiler tubes with scale that cannot be removed with usual cleaning aids. With the addition of 1–3% NH_4HF_2, the cleaning acids break down the scale so that it can be readily flushed out of the tubes, thus restoring the heat transfer capacity (13).

Rapid frosting of glass is accomplished in a concentrated solution of ammonium bifluoride and hydrofluoric acid with nucleating agents that assure uniform frosts (14). A single dip in an aqueous solution of NH_4HF_2, HF and sorbitol at <20°C for less than 60 s produces the low-specular-reflecting finish on television face plates and on glass for picture framing (15). Treating glass, eg, often badly weathered window panes, with 2–5% solutions of ammonium bifluoride results in a polishing effect. Glass ampoules for parenteral solutions (16) and optical lenses (17) are best cleaned of adhering particulate matter in dilute ammonium bifluoride solutions.

Textile Applications. Ammonium bifluoride is used as a sour or neutralizer for alkalies in commercial laundries and textile plants (see Drycleaning and laundering). Treatment also removes iron stain by forming colorless ammonium iron fluorides that are readily rinsed from the fabric, and it will not tender cloth even if excessive amounts are used (18).

Fluoride Salts and Metallurgical Applications. Ammonium fluorides react with many metal oxides or carbonates at elevated temperatures to form double fluorides; eg,

$$2\ NH_4HF_2 + BeO \xrightarrow{-H_2O} (NH_4)_2\,BeF_4 \xrightarrow{\Delta} BeF_2 + 2\ NH_3 + 2\ HF$$

The double fluorides decompose at even higher temperatures to form the metal fluoride, volatile NH_3 and HF. This reaction produces pure salts less likely to be contaminated with oxyfluorides. Beryllium fluoride, from which beryllium metal is made, has been produced this way for many years (19) (see Beryllium and beryllium alloys). In pickling of stainless steel and titanium, NH_4HF_2 is used with high concentrations of nitric acid to avoid hydrogen embrittlement. Ammonium bifluoride is used in acid dips for steel prior to phosphating and galvanizing, and for activation of metals before nickel plating (20–21). Ammonium bifluoride also is used in aluminum anodizing formulations. Ammonium bifluoride is used in treatments to provide corrosion resistance on magnesium and its alloys (22). Such treatment provides an excellent base for painting and good abrasion resistance, heat resistance, and protection from atmospheric corrosion. A minor use for ammonium bifluoride is in the preservation of wood (qv) (23).

BIBLIOGRAPHY

"Ammonium" in *ECT* 1st ed., under "Fluorine Compounds, Inorganic," Vol. 6, p. 676, by G. C. Whitaker, Harshaw Chemical Company; "Ammonium Fluoride" in *ECT* 2nd ed., under "Fluorine Compounds, Inorganic," pp. 548–549, by G. C. Whitaker, The Harshaw Chemical Company.

1. Fr. Pat. 1,546,234 (Nov. 15, 1968), (to Farbenfabriken Bayer A.-G.).
2. O. Hassel and H. Luzanski, *Z. Kristallogr.* **83,** 448 (1932).
3. R. C. Weast, ed., *Handbook of Chemistry and Physics,* 59th ed., The Chemical Rubber Co., Cleveland, 1978.
4. H. Schutza, M. Eucken, and W. Namesh, *Z. An. All. Chem.* **292,** 293 (1957).
5. L. N. Lazarev and B. V. Andronov, *J. Appl. Chem. USSR* **46,** 2087 (1973).
6. H. C. Hodge and F. A. Smith in J. H. Simon, ed., *Fluorine Chemistry,* Vol. 4 Academic Press, New York, 1965, p. 192.
7. U.S. Pat. 3,310,369 (Mar. 21, 1967), J. A. Peterson (to Hooker Chemical Corp.).
8. U.S. Pat. 2,156,273 (Apr. 28, 1939), A. R. Bozarth (to Harshaw Chemical Co.).
9. U.S. Pat. 3,501,268 (Mar. 17, 1970), R. J. Laran, A. P. Giraitix, and P. Kobetz (to Ethyl Corp.).
10. *Chemical Safety Data Sheet SD-25, Hydrofluoric Acid,* MCA, Washington, D.C., 1970.
11. H. K. van Poolen, *Oil Gas J.* **65,** 93 (Sept. 11, 1967).
12. U.S. Pat. 3,953,340 (Apr. 27, 1976), C. C. Templeton, E. H. Street, Jr., and E. A. Richardson (to Shell Oil Co.).
13. W. S. Midkiff and H. P. Foyt, *Mater. Perform.* **17**(2), 17 (1978).
14. *Glass Frosting and Polishing Technical Service Bulletin 667,* Harshaw Chemical Co.
15. U.S. Pat. 3,373,130 (Mar. 19, 1968), E. E. Junge and J. Chabal (to PPG Industries).
16. A. L. Hinson, *Bull. Parenter. Drug. Assoc.* **25,** 266 (1971).
17. R. L. Parkes and M. R. Browne. *Appl. Opt.* **17,** 1845 (1978).
18. *Control of Souring Operations, Special Report #7,* American Institute of Laundering, Joliet, Ill.
19. Brit. Pat. 833,808 (Apr. 27, 1960), A. R. S. Gough and E. W. Bennet (to the United Kingdom Atomic Energy Commission).
20. U.S. Pat. 3,767,582 (Oct. 23, 1973), G. A. Miller (to Texas Instruments, Inc.).
21. U.S. Pat. 3,296,141 (Jan. 3, 1967), W. A. Lieb and E. Billow (to R. O. Hull Co.).
22. L. F. Spencer, *Met. Finish.* **68**(10), 52 (1970); H. K. DeLong, *Met. Prog.* **97,** 105 (June 1970); *Met. Prog.* **98,** 43 (Mar. 1971); W. F. Higgins, *Light Met. Age* **17**(12), 8 (1959); A. E. Yaniv and H. Schick, *Plating* **55,** 1295 (1968).
23. E. Panck, *Am. Wood Preservers Assoc.* **59,** 189 (1963).

H. S. Halbedel
Thomas E. Nappier
The Harshaw Chemical Company

ANTIMONY

Antimony Trifluoride

Properties. Antimony trifluoride [*7783-56-4*], SbF_3, is a very hygroscopic, white, crystalline solid. It is very soluble in water, hydrofluoric acid, and polar organic solvents such as alcohols and ketones. Its solubility in water is 384.7 g/100 g at 0°C, 492.4 g/100 g at 25°C, and 563.6 g/100 g at 30°C. Solubility in alcohols at 25°C is 154 g SbF_3/100 mL CH_3OH, 33 g SbF_3/100 mL C_3H_7OH, and in acetone 55.3 g SbF_3/100 mL. It is practically insoluble in benzene, chlorobenzene, and heptane. The density of SbF_3 at 25°C is 4.385 g/cm^3. It does hydrolyze in water but the rate of hydrolysis is very slow and considerably less than $SbCl_3$.

It is a mild fluorinating reagent and does not react efficiently if used alone. But as used in the Swarts reactions (1), its effectiveness as a fluorinating reagent is drastically increased by addition of Cl_2, Br_2, or $SbCl_5$ to the reaction mixture (2). Antimony trifluoride can be used for the replacement of chlorine or bromine in halocarbons, hydrohalocarbons, and nonmetal and metal halides. Typical reactions can be summarized as follows:

$$RCCl_3 + SbF_3 \rightarrow RCCl_2F + RCClF_2 + RCF_3$$

In aliphatic compounds, the ease of fluorination is of the order of $CCl_3 > CCl_2F > CClF_2$. Other groups, eg, $C{=}CCl_2$ and $CHCl_2$, react but not readily.

$$PCl_3 + SbF_3 \xrightarrow{SbCl_5} PF_3 + SbCl_3.SbCl_5$$

$$SiHCl_3 + SbF_3 \xrightarrow{SbCl_5} SiHF_3 + SbCl_3.SbCl_5$$

In aqueous solutions SbF_3 reacts with many metal fluorides to form compounds such as $MSbF_4$ (M = Li [*72121-39-2*], Na [*34109-83-6*], K [*15273-81-1*], Cs [*36195-09-0*], and Tl [*54189-44-5*]) and M_2SbF_5 (M = K [*20645-41-4*], Cs [*40902-54-3*], and NH_4 [*32516-50-0*]). In addition, Na_3SbF_6 [*65176-04-7*] and MSb_4F_{13} (M = Tl [*60719-48-4*], Na [*56094-73-6*], K [*56094-72-5*], Rb [*12776-50-0*], Cs [*12775-92-7*], and NH_4 [*52015-24-4*]) have been reported (3).

Antimony trifluoride is not a suitable reagent for the replacement of hydrogen.

Preparation. Antimony trifluoride can be readily prepared by dissolving Sb_2O_3 in an excess of anhydrous hydrogen fluoride or in aqueous acid having 55% or higher strength hydrofluoric acid, followed by evaporation of the solution to dryness.

Uses. Early manufacturing processes for fluorocarbons (Isotrons (Pennwalt), Freons (DuPont), Genetrons (Allied), etc) used SbF_3 on a large scale but development of alternative routes to the Swarts reactions have greatly reduced its usage. Its main use is in the manufacture of antimony pentafluoride; the current market in the United States is less than 4.5 metric tons per year.

Antimony Pentafluoride

Properties. Antimony pentafluoride [*7783-70-2*], SbF_5, is a colorless, hygroscopic, very viscous liquid. Its viscosity at 20°C is 460 mPa·s (= cP) which is very close to the values for glycerol. The polymerization of high-purity SbF_5 at ambient temperature can be prevented by addition of 1% anhydrous hydrogen fluoride which can be removed by distillation prior to the use of SbF_5. The pure product melts at 7°C (4), boils at 142.7°C, and has a specific gravity (5) of 3.145 g/cm^3 at 15.5°C. The viscous, pure liquid can be handled in glass if moisture and air are carefully excluded. Commercial antimony pentafluoride is shipped in steel cylinders and polytetrafluoroethylene bottles (small quantities). Nickel is rapidly attacked by a mixture of SbF_5 and HF although there is little effect on mild steel or aluminum.

Preparation. Antimony pentafluoride can be prepared by direct fluorination of antimony powder or SbF_3. One can obtain SbF_5 by refluxing a mixture of $SbCl_5$ and anhydrous HF in a platinum flask (6):

$$SbCl_5 + 5\ HF \rightarrow SbF_5 + 5\ HCl$$

This reaction is not as simple as described in the above equation. Its solid forming mid-stage produces partially fluorinated compounds such as $SbCl_3F_2$ [*24626-20-8*], $SbCl_2F_3$ [*7791-16-4*], and $SbClF_4$ [*15588-48-4*] and also adducts such as $SbF_5.2SbF_3$ [*72121-45-0*] and/or $SbF_5.5SbF_3$ [*72121-46-1*]. These adducts are very stable and have higher melting points, 390 and 384°C, respectively. For the commercialization of this process the use of all-aluminum apparatus has been proposed.

Uses. Antimony pentafluoride is a moderate fluorinating reagent (7) and a powerful oxidizer. It spontaneously inflames phosphorus and sodium but it is practically inert towards arsenic. Powdered antimony reduces it to solid $SbF_5.2SbF_3$. It also reacts with water to form the solid dihydrate, $SbF_5.2H_2O$ [*65277-49-8*]. The latter reacts violently with an additional amount of water and forms a clear solution. The unusual stability of $SbF_5.2H_2O$ makes systematic study of hydrolysis very difficult. Antimony pentafluoride undergoes very slow hydrolysis in the presence of dilute NaOH solution and forms $Sb(OH)_6^-$. Sulfur dioxide and nitrogen dioxide react with SbF_5 to form the adducts $SbF_5.SO_2$ [*19344-14-0*] and $SbF_5.NO_2$ [*72121-47-2*], respectively. These adducts decompose in water.

Antimony pentafluoride is used to saturate double bonds in straight-chain olefins, cycloolefins, aromatic rings (8–10), and in the fluorination of halocarbons, CrO_2Cl_2, $MoCl_5$, WCl_6, PCl_3, P_4O_{10}, $SiCl_4$, $TiCl_4$, and SiO_2.

Antimony pentafluoride forms intercalation compounds with graphite (11–12) and fluorographite polymers (13), CF_x, where $x = 1.06$, which have much higher conductivity than graphite and fluorographite, respectively (14), and may have potential use as superconducting materials. When a mixture of O_2, F_2, and SbF_5 or NF_3, F_2, and SbF_5 is subjected to elevated temperature and pressure, it gives the dioxygenyl salt O_2SbF_6 [*51681-88-0*] (15) and perfluoroammonium salt NF_4SbF_6 [*16871-76-4*] (16), respectively. The dioxygenyl salt is used for the removal (17) of xenon, radon, and radon daughter elements from contaminated atmospheres.

Hexafluoroantimonates

Hexafluoroantimonic acid [*72121-43-8*], $HSbF_6.6H_2O$, is prepared by dissolving freshly prepared hydrous antimony pentoxide in hydrofluoric acid or adding the stoichiometric amount of 70% HF to SbF_5. Both of these reactions are exothermic and must be carried out carefully.

The superacid systems $HSO_3F.SbF_5$ [*33843-68-4*] and $HF.SbF_5$ [*16950-06-4*] (fluoroantimonic acid) are used frequently. Addition of SbF_5 drastically increases the acidities of HSO_3F and HF (18–19).

Anhydrous metal salts (20), $MSbF_6$, (M = Ag [*26042-64-8*], Tl [*41750-73-6*], Li [*18424-17-4*], K [*16893-92-8*], Rb [*57430-71-4*], and Cs [*16949-12-5*]) are prepared by dissolving metal fluorides in anhydrous HF and then adding stoichiometric amounts of SbF_5 to the metal fluoride solution. Hydrogen fluoride is removed by applying high vacuum to the system or bubbling dry N_2 through the solutions at elevated temperature. Hexafluoroantimonates are soluble in water and polar organic solvents.

BIBLIOGRAPHY

"Antimony Compounds" under "Fluorine Compounds, Inorganic" in *ECT* 1st ed., Vol. 6, pp. 676–677, by F. D. Loomis and C. E. Inman, Pennsylvania Salt Manufacturing Co.; "Antimony" under "Fluorine Compounds, Inorganic" in *ECT* 2nd ed., Vol. 9, pp. 549–551, by Wayne E. White, Ozark-Mahoning Company.

1. F. Swarts, *Bull. Acad. Roy. Belg.* **24,** 309 (1892).
2. H. H. Willard and O. B. Winter, *Ind. Eng. Chem. Anal. Ed.* **5,** 7 (1933).
3. I. G. Ryss, *The Chemistry of Fluorine and Its Inorganic Compounds,* State Publishing House for Scientific, Technical, and Chemical Literature, Moscow, USSR, 1956, Eng. transl., *AEC-tr-3927,* Office of Technical Services, U.S. Department of Commerce, Washington, D.C., 1960, pp. 283–295 (Part I).
4. O. Ruff and co-workers, *Ber.* **42,** 4021 (1909).
5. R. C. Shair and W. F. Shurig, *Ind. Eng. Chem.* **43,** 1624 (1951).
6. O. Ruff and W. Plato, *Ber.* **37,** 673 (1904).
7. D. T. Meshri and W. E. White, *George H. Cady Symposium, Milwaukee, Wisc., June 1970.*
8. E. T. McBee, P. A. Wisemen, and G. B. Bachman, *Ind. Eng. Chem.* **39,** 415 (1947).
9. U.S. Pat. 2,488,216 (Nov. 15, 1949), E. T. McBee, V. V. Lindgren, and W. B. Ligett (to Purdue Research Foundation).
10. U.S. Pat. 2,533,217 (May 15, 1951), F. B. Stilmar (to the United States of America as represented by Atomic Energy Commission).
11. J. M. Lalancette and J. Lafontaine, *J. Chem. Soc. Chem. Commun.,* 815 (1973).
12. A. A. Opalvskii, A. S. Nazarov, and A. A. Uminskii, *Zh. Neorg. Khimi* **19,** 1518 (1974), English ed., *Russ. J. Inorg. Chem.* **19,** 827 (1974).
13. L. B. Ebert, R. A. Muggins, and J. I. Brauman, *Mat. Res. Bull.* **2,** 615 (1976).
14. L. Vogel, *J. Mat. Sci.* **12,** 982 (1977).
15. J. B. Beal, Jr., C. Pupp, and W. E. White, *Inorg. Chem.* **8,** 828 (1969).
16. U.S. Pat. 3,708,570 (Jan. 2, 1973), W. E. Tolberg, R. S. Stringham, and R. T. Rewick (to Stanford Research Institute).
17. U.S. Pat. 3,829,551 (Aug. 13, 1974), L. Stein (to the United States Atomic Energy Commission).
18. R. J. Gillespie and T. E. Peel, *J. Am. Chem. Soc.* **95,** 5173 (1973).
19. R. J. Gillespie in V. Gold, ed., *Proton Transfer Reactions,* Chapman and Hall, London, Eng., 1975, p. 27.
20. D. T. Meshri and D. C. Contractor, *3rd Winter Fluorine Conference, St. Petersburg, Florida., Jan. 23, 1977.*

DAYAL T. MESHRI
CHARLES B. LINDAHL
Ozark-Mahoning Company,
a subsidiary of the Pennwalt Corp.

ARSENIC

Arsenic forms the binary compounds, arsenic trifluoride and arsenic pentafluoride (the only reported arsenic pentahalide), as well as a series of compounds and the acid of the very stable hexafluoroarsenate ion.

Great care should be exercised in the handling and use of all arsenic compounds because NIOSH has determined "inorganic arsenic" to be a carcinogen. Current and expected future handling requirements and environmental considerations may reduce or eliminate many present applications and restrict future uses.

Arsenic Trifluoride

Arsenic trifluoride [*7784-35-2*] is a colorless liquid with a melting point of −5.95 °C, a boiling point of 57.13°C (at 99 kPa or 742.5 mm Hg) (1), a specific gravity of 2.67, and a standard enthalpy of formation of −858.1 kJ/mol (−205.1 kcal/mol) (2). It can be prepared by fluorination of arsenic(III) oxide with sulfuric acid and calcium fluoride (3) or with hydrofluoric acid or fluorosulfuric acid. It is used as a fluorinating reagent, in ion implantation (qv), and in syntheses of arsenic pentafluoride and some hexafluoroarsenate compounds. It is produced by Ozark-Mahoning, with usage estimated to be less than 100 kg/yr.

Arsenic Pentafluoride

Arsenic pentafluoride [*7784-36-3*] melts at −79.8°C and boils at −52.8°C (4). The liquid has a density at the boiling point of 2.33 g/mL. The standard enthalpy of formation is −1237 kJ/mol (−295.6 kcal/mol), and the average bond strength is 387 kJ/mol (92.4 kcal/mol) compared to 484.1 kJ/mol (115.7 kcal/mol) for AsF_3 (5). Arsenic pentafluoride can be prepared by reaction of fluorine with arsenic trifluoride or arsenic. It is used in ion implantation, as a fluorinating reagent, and in syntheses of some hexafluoroarsenate compounds. Arsenic pentafluoride has been found to react with graphite to form AsF_5–graphite intercalation compounds (6) that have electrical conductivity as high as that of silver (7–8). Arsenic pentafluoride is produced by Ozark-Mahoning with usage estimated to be less than 100 kg/yr.

Hexafluoroarsenic Acid and the Hexafluoroarsenates

The MF_6^- ions of the group VA elements P, As, and Sb are similar in stability to the isoelectronic SF_6, SeF_6, and TeF_6, and SiF_6^{2-}, GeF_6^{2-}, and SnF_6^{2-} series.

The AsF_6^- ion is very stable toward hydrolysis in aqueous solution. It is not hydrolyzed by boiling a strongly basic solution almost to dryness (9), although it is hydrolyzed in sulfuric acid (10) or in boiling perchloric acid (9). Studies of the hydrolysis of AsF_6^- in concentrated sulfuric acid (10) and in base (11) at 193–222°C showed both to be first order in AsF_6^-. The hydrolysis of AsF_6^- in alkaline solution is slower than either PF_6^- or SbF_6^-.

Hexafluoroarsenic acid [*17068-85-8*] can be prepared by the reaction of arsenic acid with hydrofluoric acid or calcium fluorosulfate (12). The hexafluoroarsenates can be prepared directly from arsenates and hydrofluoric acid or by neutralization of $HAsF_6$. The reaction of 48% HF with potassium dihydrogen arsenate (V), KH_2AsO_4, gives potassium hydroxypentafluoroarsenate (V) [*17068-84-7*], $KAsF_5OH$, which hydrolyzes rapidly in water solution (9). Anhydrous HF reacts with KH_2AsO_4 or $KAsF_5OH$ to produce $KAsF_6$ [*17029-22-0*]

Because of the special properties of the hexafluoroarsenate ion there are a number of applications of hexafluoroarsenates in catalysis, electrolytes, and herbicides (qv). For example, onium hexafluoroarsenates (13) have been described as photoinitiators in the hardening of epoxy resins. Lithium hexafluoroarsenate [*29935-35-1*] has been used as an electrolyte in lithium organic electrolyte batteries (qv). Hexafluoroarsenates, especially alkali and alkaline earth metal salts or substituted ammonium salts, have been reported (14) to be effective as herbicides in killing weeds. Potassium hexafluoroarsenate has been reported (15) to be particularly effective against prickly pear.

Hexafluoroarsenic acid and several hexafluoroarsenates are produced by Ozark-Mahoning Company. Lithium hexafluoroarsenate is produced by Ozark-Mahoning Company and U.S. Steel Agrichemicals.

BIBLIOGRAPHY

"Arsenic" in *ECT* 1st ed., under "Arsenic Compounds", Vol. 2, pp. 119–123, by I. E. Campbell, Battelle Memorial Institute; "Arsenic" in *ECT* 2nd ed., under "Arsenic Compounds," Vol. 2, pp. 718–733, by G. O. Doak, Leon D. Freedman, and G. Gilbert Long, North Carolina State of the University of North Carolina at Raleigh.

1. H. Russell, Jr., R. E. Rundle, and D. M. Yost, *J. Am. Chem. Soc.* **63,** 2825 (1941).
2. A. A. Woolf, *J. Fluorine Chem.* **5,** 172 (1975).
3. O. Ruff, *Die Chemie des Fluors,* Springer-Verlag, Berlin, 1920, p. 27.
4. O. Ruff, A. Braida, O. Bretschneider, W. Menzel, and H. Plaut, *Z. Anorg. Allgem. Chem.* **206,** 59 (1932).
5. P. A. G. O'Hare and W. N. Hubbard, *J. Phys. Chem.* **69,** 4358 (1965).
6. L. Chun-Hsu, H. Selig, M. Rabinovitz, I. Agranat, and S. Sarig, *Inorg. Nucl. Chem. Lett.* **11,** 601 (1975).
7. E. R. Falardeau, G. M. T. Foley, C. Zeller, and F. L. Vogel, *Chem. Commun.,* 389 (1977).
8. G. M. T. Foley, C. Zeller, E. R. Falardeau, and F. L. Vogel, *Solid State Commun.* **24,** 371 (1977).
9. H. M. Dess and R. W. Parry, *J. Am. Chem. Soc.* **79,** 1589 (1957).
10. W. T. Lockhart, Jr., M. M. Jones, and D. O. Johnston, *J. Inorg. Nucl. Chem.* **31,** 407 (1969).
11. I. G. Ryss, V. B. Tul'chinskii and Y. A. Mazurov, *Izv. Sib. Otd. Akad. Nauk. SSSR Ser, Khim. Nauk,* 81 (1968).
12. U.S. Pat 3,875,292 (Apr. 1, 1975), R. A. Wiesboeck and J. D. Nickerson (to U.S. Steel Corp.).
13. Ger. Offen. 2,618,871 (Nov. 11, 1976), and 2,518,652 (May 2, 1974), J. V. Crivello (to General Electric Co.).
14. Belg. Pat. 659,342 (Aug. 5, 1965), T. N. Russell (to Pennsalt Chemicals Corp.).
15. P. E. Buckley, J. D. Dodd, and W. H. Culver, *Proc. West. Soc. Weed Sci.* **22,** 17 (1968).

CHARLES B. LINDAHL
Ozark-Mahoning Company,
a subsidiary of the Pennwalt Corp.

BARIUM

Barium Fluoride

Barium fluoride [*7787-32-8*], BaF_2, is a colorless solid. Reported melting points vary from 1290°C (1) to 1355°C (2), including values of 1301°C (3) and 1354°C (4). These differences may be due to impurities, reaction with the containers, and difficulty in accurate temperature measurements. The heat of fusion is 28 kJ/mol (6.8 kcal/mol) (5), the boiling point 2260°C (6), and the density 4.9 g/cm^3. The solubility in water is about 1.6 g/L at 25°C and 5.6 g/100 g (7) in anhydrous hydrogen fluoride. Barium fluoride can be prepared by reaction of the carbonate or the somewhat purer hydrate, $Ba(OH)_2$, with hydrofluoric acid. A typical analysis of the commercial product is 99% with a loss on ignition of 0.9%; sulfates as SO_4, 0.2%; hexafluorosilicate as SiF_4, 0.02%; heavy metals as lead, 0.02%; and iron, 0.005%.

Barium fluoride is used commercially in combination with other fluorides for arc welding (qv) electrode fluxes. However, this usage is limited because of the availability of less expensive naturally-occurring calcium fluoride.

The toxicity of barium fluoride has received only little attention. A value for oral LD_{Lo} of 350 mg/kg in guinea pigs has been reported (8).

OSHA has a time-weighted average (TWA) standard of 500 mg of Ba/m^3 for barium fluoride (8) in addition to a standard TWA of 2.5 mg of F/m^3 (9). NIOSH has issued a criteria document (9) on occupational exposure to inorganic fluorides.

BIBLIOGRAPHY

"Barium Fluoride" in *ECT* 1st ed., under Fluorine Compounds, Inorganic," Vol. 6, p. 677, by F. D. Loomis, Pennsylvania Salt Manufacturing Co; "Barium Fluoride" in *ECT* 2nd ed., under "Fluorine Compounds, Inorganic," Vol. 9, p. 551, by Wayne E. White, Ozark-Mahoning Company.

1. I. Barin and O. Knache, *Thermochemical Properties of Inorganic Substances,* Springer, Verlag, Berlin, 1973.
2. H. Kojima, S. G. Whiteway, and C. R. Masson, *Can. J. Chem.* **46,** 2698 (1968).
3. I. Jackson, *Phys. Earth Planet. Inter.* **14,** 143 (1977).
4. B. Porter and E. A. Brown, *J. Am. Ceram. Soc.* **45,** 49 (1962).
5. G. Petit and A. Cremieo, *C. R. Acad. Sci.* **243,** 360 (1956).
6. O. Ruff and L. LeBoucher, *Z. Anorg. Chem.* **219,** 376 (1934).
7. A. W. Jache and G. H. Cady, *J. Phys. Chem.* **56,** 1106 (1952).
8. *Registry of Toxic Effects of Chemical Substances,* NIOSH, Vol. II, 1977, p. 141.
9. *Criteria for a Recommended Standard-Occupational Exposure to Inorganic Fluorides,* PB 246 692, NIOSH 76-103, U.S. Department of Health, Education, and Welfare, 1975.

Charles B. Lindahl
Ozark-Mahoning Company,
a subsidiary of the Pennwalt Corp.

BORON

Boron Trifluoride

Boron trifluoride [*7637-07-2*], trifluoroborane, BF_3, was first reported in 1809 by Gay-Lussac and Thénard (1) formed by the reaction of boric acid and fluorspar at dull red heat. It is a colorless gas in a dry atmosphere, but fumes in the presence of moisture yielding a dense white smoke of irritating, pungent odor. It is widely used as an acid catalyst (2) for many types of organic reactions, particularly those of importance in the production of polymer and petroleum products. The gas was first produced commercially in 1936 by the Harshaw Chemical Company.

The boron atom in boron trifluoride is hybridized to the sp^2 planar configuration and consequently it is coordinatively unsaturated (a Lewis acid) and its chemistry is primarily concerned with satisfying this unsaturation by the formation of adducts that are nearly tetrahedral (sp^3) with Lewis bases. The electrophilic properties (acid strengths) of the trihaloboranes have been found to increase in the order $BF_3 < BCl_3 < BBr_3 < BI_3$ (3).

Physical Properties. The physical properties are listed in Table 1.

The molecule has a trigonal planar structure in which the F—B—F angle is 120° and the B—F bond distance is 0.1307 ± 0.0002 nm (12). Its symmetry point group is D_{3h}.

Nuclear magnetic resonance ^{11}B spectral studies of BF_3 have given a value of 9.4 ± 1.0 ppm for the chemical shift relative to $BF_3.O(C_2H_5)_2$ as the zero reference (13). Using methylcyclohexane as a solvent at 33.5°C with $BF_3.O(CH_2CH_3)_2$ as the internal standard, a value of 10.0 ± 0.1 ppm was obtained for the chemical shift (14). A value for the ^{19}F chemical shift of BF_3 in CCl_3F relative to CCl_3F is reported to be 127 ppm (15). The coupling constant $J_{^{11}B-^{19}F}$ is reported to be 15 ± 2 Hz for BF_3 (16). Additional constants may be found in refs. 3 and 17.

Solubilities are listed in Table 2.

Aqueous mineral acids react with BF_3 to yield the hydrates of BF_3 or the hydroxyfluoroboric acids, fluoroboric acid, or boric acid. Solution in aqueous alkali gives the soluble salts of the hydroxyfluoroboric acids, fluoroboric acids, or boric acid. Boron trifluoride is soluble in many organic solvents including saturated hydrocarbons, halogenated hydrocarbons, and aromatic compounds. In some organic liquids, boron trifluoride is soluble with reaction. It easily polymerizes unsaturated compounds such as butylene, styrene, or vinyl esters. Other molecules containing electron donating atoms (O, S, N, P, etc) (eg, alcohols, amines, phosphines, ethers) may dissolve BF_3 to produce soluble adducts.

Chemical Properties. In addition to the reactions listed in Table 3, boron trifluoride reacts with alkali or alkaline earth metal oxides, or their precursors (eg, KNO_3, Li_2CO_3), at 450°C to yield the trimer trifluoroboroxine $(BOF)_3$, MBF_4, and MF (29).

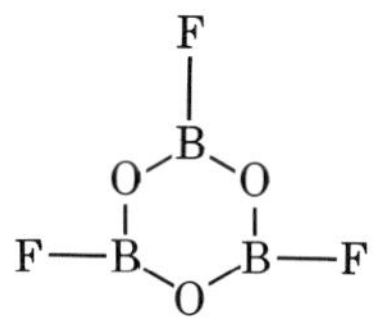

Table 1. Physical Properties of Boron Trifluoride

Property	Value	Ref.
molecular weight	67.8062	
melting point, °C	−128.37	4
boiling point, °C	−99.9	5
vapor pressure of liquid, kPa[a]		5
at 145 K	8.43	
at 170 K	80.19	
at 220 K	1156	
at 260 K	4842	6
triple point at 8.34 kPa[a], K	144.78 K	4
critical temperature, T_c, °C	−12.25 ± 0.03	6
critical pressure, P_c, kPa[a]	4984	6
density		
critical, d_c, g/cm³	ca 0.591	7
gas, STP, g/L	3.07666	8
limiting, L_N, g/L	3.02662	8
liquid, for 148.9 to 170.8 K, g/cm³	1.699–0.00445 (t + 125.0)	9
thermochemical values		
enthalpy of fusion, $\Delta H_{144.45}$, kJ/mol[b]	4.2417	10
enthalpy of vaporization, $\Delta H_{154.5}$, kJ/mol[b]	18.46	11
entropy, $S^\circ_{298.15}$ J/(mol·K)[b]	254.3	
Gibbs free energy of formation, $\Delta Gf^\circ_{298.15}$ kJ/mol[b]	−1119.0	12
enthalpy of formation, $\Delta Hf^\circ_{298.15}$ kJ/mol[b]	−1135.6	12
infrared absorption frequencies, cm⁻¹		
ν_1	888	
ν_2	696.7	
ν_3	1463.3	
ν_4	480.7	

[a] To convert kPa to mm Hg, multiply by 7.5.
[b] To convert J to cal, divide by 4.184.

Trifluoroboroxine is similarly obtained from alumina, titania, alkaline earths, silicates, glass, kaolin, feldspar, or oligoclase (29). The trimer is stable below −135°C but disproportionates to B_2O_3 and BF_3 at higher temperatures (30).

The reaction of metal hydrides with BF_3 depends upon the stoichiometry and the nature of the metal hydride. For example, LiH and $BF_3.O(C_2H_5)_2$ may form diborane (6) or lithium borohydride (31–32):

$$6\ LiH + 8\ BF_3.O(C_2H_5)_2 \rightarrow B_2H_6 + 6\ LiBF_4 + 8\ (C_2H_5)_2O$$

$$4\ LiH + 4\ BF_3.O(C_2H_5)_2 \rightarrow LiBH_4 + 3\ LiBF_4 + 4\ (C_2H_5)_2O$$

The first method is commonly used for preparing diborane.

Metal halides react with BF_3 (33) when heated to form BX_3 and the metal fluoride:

$$AlBr_3 + BF_3 \rightarrow BBr_3 + AlF_3$$

With the alkali halides, BX_3 and the respective alkali fluoroborates are obtained (34):

$$3\ KCl + 4\ BF_3 \rightarrow 3\ KBF_4 + BCl_3$$

Table 2. Solubilities of Boron Trifluoride

BF_3, g	Solvent, g	Temperature, °C	Product	CAS Registry No.	Ref.
369.4	water[a], 100[b]	6	$BF_3.H_2O$	[*15799-89-0*]	
(5.55 mol)	(5.45 mol)		$HBF_3(OH)$	[*16903-52-9*]	18
2.06	sulfuric acid, conc, 100	25			19
	nitric acid[a]	20	$HNO_3.2BF_3$	[*20660-63-3*]	20
	orthophosphoric acid[a]	25	$H_3PO_4.BF_3$	[*13699-76-6*]	21
2.18	hydrofluoric acid[c],	4.4			22
	hydrochloric acid, anhydrous (l)	24	miscible		24
(12.5 mol)	hydrobromic acid anhydrous (l), 87.5 mol		miscible		25

[a] Dissolves with reaction.
[b] A higher dilution results in a mixture of $H[BF_2(OH)_2]$, HBF_4, and H_3BO_3.
[c] Equations for the solubility of BF_3 in liquid HF at 24, 49, and 90°C and up to 6.8 kPa (51 mm Hg) are in ref. 23.

Table 3. Reactions of Boron Trifluoride

Reactant	Temperature, °C	Products	Formula	Ref.
sodium[a]		boron, amorphous, sodium fluoride	NaF	26
potassium[a]		boron, amorphous, potassium fluoride	KF	26
magnesium, molten alloys		no reaction		
calcium	1600	calcium hexaboride	CaB_6	
aluminum	1200	aluminum boride (1:12), tetragonal boron	AlB_{12}	
	further reaction at 1650	β-rhombohedral boron		
titanium	1600	titanium boride	TiB_2	27
zirconium	1600	zirconium boride	ZrB_2	27
copper, mercury, chromium, iron	RT or below	no reaction[b]		
hydrogen in tungsten arc		crystalline boron		
sodium nitrate, sodium nitrite	180	sodium fluoroborate, boric oxide	$NaBF_4$	28

[a] With incandescence.
[b] Even when subjected to pressure for a considerable length of time; also no reaction with red-hot iron.

Alkyl and arylboranes are obtained (35) from BF_3 with the appropriate Grignard reagent, alkylaluminum halide, or zinc alkyl (see also Organometallics):

$$BF_3 + 3\ RMgX \rightarrow BR_3 + 3\ MgXF$$

Alkylfluoroboranes result from the reaction of the appropriate alkylborane and BF_3 under suitable conditions (36):

$$BR_3 + 2\ (C_2H_5)_2O.BF_3 \rightarrow 3\ RBF_2 + 2\ (C_2H_5)_2O$$

Adducts of BF_3 with some organic compounds having labile hydrogen atoms in the vicinity of the atom bonding to the boron atom of BF_3 may form a derivative of BF_3 by splitting out HF. For example, β-diketones such as acetylacetone or benzoylacetone react with BF_3 in benzene (37):

$$BF_3 + CH_3COCH_2COCH_3 \rightarrow CH_3COCH{=}C(CH_3)OBF_2 + HF$$

As would be expected from electrophilic properties, no metal cations in Groups I, II, or III of the periodic table are reported to donate an electron pair to boron trifluoride. However, the hydride ion in sodium hydride is reported to bond with BF_3 to produce sodium hydridotrifluoroborate [*72264-77-8*], $Na(HBF_3)$ (38). A complex compound of indium(I) (Group III), cyclopentadienylindium(I), is reported to form a 1:1 adduct with BF_3 (39).

Carbon, in Group IV, serves as a Lewis base in a few of its compounds. In general, saturated aliphatic and aromatic hydrocarbons are stable in the presence of BF_3, whereas unsaturated aliphatic hydrocarbons, such as propylene or acetylene, are polymerized. However, some hydrocarbons and their derivatives have been reported to form adducts with BF_3. For example, several compounds having the perfluoromethyltrifluoroborate ion, $F_3CBF_3^-$, have been reported (40) to be derived from the reaction of $(CH_3)_3SnCF_3$ with BF_3 in CCl_4. Typical examples of adducts with unsaturated hydrocarbons are 1:1 adducts with tetracene and 3,4-benzopyrene (41), and 1:2 BF_3 adducts with α-carotene and lycopene (42).

In Group V, nitrogen compounds readily form molecular compounds with BF_3. Phosphorus compounds also form adducts with BF_3. Arsenic has been reported to serve as a donor to BF_3 in trimethylarsine (43).

Inorganic or organic compounds containing oxygen form many adducts with boron trifluoride, whereas sulfur and selenium have been reported to form only a few (44–46).

Boron trifluoride forms two hydrates, $BF_3.H_2O$ and $BF_3.2H_2O$ [*13319-75-0*] (also $BF_3.D_2O$ [*33598-66-2*] and $BF_3.2D_2O$ [*33598-66-2*]). According to nmr data reported (46), the latter compound is $H_3O^+F_3BOH^-$.

Most of the coordination compounds formed by trifluoroborane are with oxygen-containing organic compounds (see Table 4). Although the other boron halides frequently react with this type to split out hydrogen halide, boron trifluoride usually forms stable molecular compounds. The reason is that the back coordination of electrons from fluorine to boron forms a strong B—F bond which is 28% ionic (47).

It has been reported (48) that some adducts of alkyl ethers and/or alcohols are unstable and decompose at −80°C to yield BF_3, H_2O, and the polyalkene.

As expected, adducts of BF_3 have been reported with hydrogen sulfide, sulfur dioxide, thionyl fluoride, and the sulfur analogues of many of the kind of oxygen-containing organic molecules cited in Table 4. The carbonyl oxygen atom or the carbonyl sulfur atom is the donor to BF_3 in 1:1 adducts such as $CH_3COOCH_3.BF_3$ [*7611-14-5*], $CH_3COSCH_3.BF_3$ [*52913-04-9*], and $CH_3CSOCH_3.BF_3$ [*52912-98-8*] (49).

Dimethyl selenide forms $(CH_3)_2Se.BF_3$ [*420-18-8*] with BF_3 at −78°C (50).

Compounds containing fluorine and chlorine are also donors to BF_3. Aqueous fluoroboric acid and the tetrafluoroborates of metals, nonmetals, and organic radicals represent a large class of compounds in which fluoride ion is coordinating with trifluoroborane. Representative examples of these compounds are given in Table 5. Coordination compounds of boron trifluoride with the chlorides of sodium, aluminum,

Table 4. Boron Trifluoride Adducts with Oxygen-Containing Compounds

Donor	Adduct name	CAS Registry No.	Molecular formula
alcohols	ethanol trifluoroborane	*[353-41-3]*	$C_2H_5OH.BF_3$
	bis(ethanol) trifluoroborane	*[373-59-1]*	$2C_2H_5OH.BF_3$
	bis(2-chloroethanol) trifluoroborane	*[72985-81-0]*	$2ClCH_2CH_2OH.BF_3$
	benzyl alcohol trifluoroborane	*[456-31-5]*	$C_6H_5CH_2OH.BF_3$
acids	acetic acid trifluoroborane	*[753-53-7]*	$CH_3COOH.BF_3$
	bis(acetic acid) trifluoroborane	*[373-61-5]*	$2CH_3COOH.BF_3$
	stearic acid trifluoroborane	*[60274-92-2]*	$CH_3(CH_2)_{16}COOH.BF_3$
	bis(phenol) trifluoroborane	*[462-05-5]*	$2C_6H_5OH.BF_3$
ethers	diethyl ether trifluoroborane	*[109-63-7]*	$(C_2H_5)_2O.BF_3$
	tetrahydrofuran trifluoroborane	*[462-34-0]*	$(CH_2)_4O.BF_3$
	anisole trifluoroborane	*[456-31-5]*	$CH_3OC_6H_5.BF_3$
acid anhydride	acetic anhydride trifluoroborane	*[591-00-4]*	$(CH_3CO)_2O.BF_3$
esters	ethyl formate trifluoroborane	*[462-33-9]*	$HCOOC_2H_5.BF_3$
	phenyl acetate trifluoroborane	*[30884-81-6]*	$CH_3COOC_6H_5.BF_3$
ketones	acetone trifluoroborane	*[661-27-8]*	$(CH_3)_2CO.BF_3$
	benzophenone trifluoroborane	*[322-21-4]*	$(C_6H_5)_2CO.BF_3$
	acetophenone trifluoroborane	*[329-25-9]*	$C_6H_5COCH_3.BF_3$
aldehydes	acetaldehyde trifluoroborane	*[306-73-0]*	$CH_3CHO.BF_3$
	neopentanal trifluoroborane	*[306-78-5]*	$(CH_3)_3CCHO.BF_3$
	benzaldehyde trifluoroborane	*[456-30-4]*	$C_6H_5CHO.BF_3$

Table 5. Boron Trifluoride Adducts with Compounds Containing Chlorine and Fluorine

Name	CAS Registry No.	Molecular formula
potassium tetrafluoroborate	*[14075-53-7]*	KBF_4
hexaamminonickel tetrafluoroborate	*[13877-20-8]*	$[Ni(NH_3)_6](BF_4)_2$
nitrosyl tetrafluoroborate	*[14635-75-7]*	$NOBF_4$
acetylium tetrafluoroborate	*[2261-02-1]*	CH_3COBF_4
tetramethylammonium tetrafluoroborate	*[661-36-0]*	$(CH_3)_4NBF_4$
difluorobromine tetrafluoroborate	*[14282-83-8]*	BrF_2BF_4
anilinium tetrafluoroborate	*[15603-97-1]*	$C_6H_5NH_2HBF_4$

iron, copper, zinc, tin, and lead have been indicated (51); they are probably chlorotrifluoroborates.

Trifluoroborane also forms adducts with some of the transition elements. In Group VI-B molybdenum and tungsten have been reported (52) to form metal to boron bonds in cyclopentadienyl adducts such as $(C_5H_5)_2MH_2.BF_3$. Manganese and rhenium compounds (Group VII-B) are reported in refs. 52–53; cobalt, rhodium, iridium, and platinum compounds (Group VIII) are reported in refs. 53–55. A more detailed discussion of molecular compounds with trifluoroborane may be found in ref. 56.

Manufacture. Boron trifluoride is prepared by the reaction of a boron-containing material with a fluorine-containing substance in the presence of an acid. The traditional method uses borax, fluorspar, and sulfuric acid.

In another process fluorosulfonic acid is treated with boric acid:

$$3\ HSO_3F + H_3BO_3 \rightarrow BF_3 + 3\ H_2SO_4$$

For the preparation of small quantities, some of which is of high purity (57), numerous reactions are available.

Shipment and Handling. The gas is nonflammable and is shipped in steel cylinders under a pressure of 10.3–12.4 MPa (1500–1800 psi). An ICC Green Label is required. Cylinders containing 2.7 or 28 kg and tube trailers containing ca 5.4–5.9 metric tons are available. If boron trifluoride is compressed with oil as a compressor lubricant, it must not be used with oxygen under pressure nor with gauges, valves, or lines that are to be used with oxygen.

Inasmuch as the gas hydrolyzes readily, all equipment should be rinsed repeatedly with dry air before admitting boron trifluoride. Under anhydrous conditions, steel equipment is most satisfactory. At low pressures and temperatures, malleable iron, stainless steel, copper, brass, or glass may be used. For temporary low-pressure connections, polyethylene or polytetrafluoroethylene (PTFE) tubing may be employed. Permanent connections should have copper or stainless steel tubing.

In the presence of moisture, boron trifluoride may be handled in copper, Saran, PTFE, hard rubber, neoprene, paraffin wax, polyethylene, or Pyrex glass containers. Iron is only attacked slightly. Corrosion may be minimized by dilution with an inert gas.

At 600°C, stainless steel (304L) and Hastelloy N are attacked by BF_3; Hastelloy is more resistant (58).

Saran and PTFE serve as satisfactory gasket and packing materials, whereas rubber, fiber, polymerizable materials, or organic oxygen- and nitrogen-containing compounds must be avoided.

Because boron trifluoride is soluble in, and reacts with, many liquids, the gas must not be introduced into any liquid unless a vacuum break or similar safety device is employed.

Economic Aspects, Standards, and Analysis. The sole United States producer of boron trifluoride is the Allied Chemical Corporation. The 1978 price of boron trifluoride was $3.94–5.50/kg, depending upon its purity and the quantity purchased.

Commercial boron trifluoride is usually ca 99.0% pure. The common impurities are air, silicon tetrafluoride, and sulfur dioxide, with traces of hydrogen, sulfur trioxide, and hydrogen fluoride. Higher purity (99.5%) may be attained by fractional distillation.

An excellent procedure for sampling and making a complete analysis of gaseous boron trifluoride has been developed (59).

Health and Safety Factors. Boron trifluoride is primarily a pulmonary irritant. The toxicity of the gas to man has not been reported (60), but laboratory tests on animals gave results ranging from an increased pneumonitis to death. A TLV of 0.3 ppm has been suggested (61) with a ceiling value of 1 ppm (62). Prolonged inhalation produced dental fluorosis (63). High concentrations burn the skin similarly to hydrogen fluoride. No chronic effects have been observed in workers exposed to small quantities of the gas at frequent intervals over a period of years.

Uses. Boron trifluoride is an excellent Lewis acid catalyst for numerous types of organic reactions. Its advantages are its ease of handling as a gas and the absence of undesirable tarry by-products. As an electrophilic molecule, it is an excellent catalyst for Friedel-Crafts and many other types of reactions (64–66) (see Friedel-Crafts reactions).

In addition boron trifluoride and some of its adducts have widespread application as curing agents for epoxy resins, in the dyeing of polyesters, and in preparing alcohol-soluble phenolic resins (44).

Boron trifluoride catalyst is used under a great variety of conditions either alone in the gas phase or in the presence of many types of promoters. Many boron trifluoride coordination compounds are also used.

Boron trifluoride catalyst may be recovered by distillation, chemical reactions, or a combination of these methods. Ammonia or amines are frequently added to the spent catalyst to form stable coordination compounds that can be separated from the reaction products. Subsequent treatment with sulfuric acid releases boron trifluoride. An organic compound may be added that forms an adduct more stable than that formed by the desired product and boron trifluoride. This liberates the desired product. In another procedure, a fluoride is added to the reaction products to precipitate the boron trifluoride which is then released by heating. Selective solvents may also be employed in recovery procedures.

Boron trifluoride is also employed in nuclear technology by utilizing several nuclear characteristics of the boron atom. Of the two isotopes, ^{10}B and ^{11}B, only ^{10}B has a significant absorption cross section for thermal neutrons. It is used in $^{10}BF_3$ as a neutron-absorbing medium in proportional neutron counters and for controlling nuclear reactors (see Nuclear reactors).

Some of the complexes of trifluoroborane have been used for the separation of the boron isotopes and the enrichment of ^{10}B as $^{10}BF_3$ (67).

Boron trifluoride is used for the preparation of boranes (see Boron compounds). With alkali metal hydrides, diborane is obtained, organoboranes are obtained with a suitable Grignard reagent.

Boron trifluoride protects molten magnesium and its alloys from oxidation during casting, and is used as a flux for soldering magnesium (68) (see Solders). It has been used in mixtures to prepare boride surfaces on steel and other metals, and as a lubricant for casting steel. It is claimed that arc welding (qv) of titanium in the presence of BF_3 results in a joint having increased impact strength.

Trifluoroborane-*S,S,S*-tri-*n*-butyl trithiophosphate, $(C_4H_9S)_3PO.BF_3$, is reported to be a defoliant for cotton plants (69) (see Herbicides).

BIBLIOGRAPHY

"Boron Trifluoride" under "Fluorine Compounds, Inorganic" in *ECT* 1st ed., Vol. 6, pp. 678–684, by D. R. Martin, University of Illinois; "Boron Trifluoride" under "Boron" under "Fluorine Compounds, Inorganic" in *ECT* 2nd ed., Vol. 9, pp. 554–562, by Donald R. Martin, The Harshaw Chemical Company.

1. J. L. Gay-Lussac and J. L. Thénard, *Rech. Phys.* **2,** 38 (1811); *Ann. Chim. Phys.* **69,** 204 (1809).
2. J. A. Nieuwland, R. R. Vogt, and W. L. Foohey, *J. Am. Chem. Soc.* **52,** 1018 (1930).
3. H. C. Brown and R. R. Holmes, *J. Am. Chem. Soc.* **78,** 2173 (1956).
4. J. C. G. Calado and L. A. K. Staveley, *Trans. Faraday Soc.,* **67,** 1261 (1971).
5. E. Pohland and W. Harlos, *Z. Anorg. Allgem. Chem.* **207,** 242 (1932).
6. H. S. Booth and J. M. Carter, *J. Phys. Chem.* **36,** 1359 (1932).
7. R. F. Smith, *U.S.A.E.C.* **NAA-SR-5286,** (1960).
8. C. F. Rumold, Ph.D. thesis, Western Reserve University, Cleveland, Ohio, 1931.
9. E. Wiberg and W. Mäthing, *Ber.* **70B,** 690 (1937).
10. A. Eucken and E. Schröder, *Z. Physik. Chem.,* **341,** 307 (1938).
11. H. M. Spencer, *J. Chem. Phys.* **14,** 729 (1946).
12. D. R. Stull and H. Prophet, *Natl. Stand. Ref. Data Ser. Natl. Bur. Stand.* **37,** (1971).

13. T. P. Onak and co-workers, *J. Phys. Chem.* **63,** 1533 (1959).
14. M. F. Lappert and co-workers, *J. Chem. Soc. A.,* 2426 (1971).
15. T. D. Coyle, S. L. Stafford, and F. G. A. Stone, *J. Chem. Soc.,* 3103 (1961).
16. T. D. Coyle and F. G. A. Stone, *J. Chem. Phys.* **32,** 1892 (1960); I. S. Jaworiwsky and co-workers, *Inorg. Chem.* **18,** 56 (1979).
17. *Gmelins Handbuch der Anorganischen Chemie,* Bor., 8th ed., Vol. 13, Verlag Chemie, G.m.b.H., Weinheim/Bergstrasse, FRG, 1954, pp. 167–196.
18. S. Pawlenko, *Z. Anorg. Allgem. Chem.* **300,** 152 (1959).
19. N. N. Greenwood and A. Thompson, *J. Chem. Soc.,* 3643 (1959).
20. H. Gerding and co-workers, *Rec. Trav. Chim.* **71,** 501 (1952).
21. N. N. Greenwood and A. Thompson, *J. Chem. Soc.* 3493 (1959).
22. E. C. Hughes and S. M. Darling, *Ind. Eng. Chem.* **43,** 746 (1951).
23. R. J. Mikovsky, S. D. Levy, and A. L. Hensley, Jr., *J. Chem. Eng. Data* **6,** 603 (1961).
24. H. S. Booth and D. R. Martin, *J. Am. Chem. Soc.* **64,** 2198 (1942).
25. D. R. Martin and R. E. Dial, *J. Am. Chem. Soc.* **72,** 852 (1950).
26. K. I. Khachishvili and co-workers, *Zh. Neorg. Khim.* **6,** 1493 (1961).
27. P. Pichat, *C.R. Acad. Sci. Paris Ser. C* **265,** 385 (1967).
28. R. N. Scott and D. F. Shriver, *Inorg. Chem.* **5,** 158 (1966).
29. P. Baumgarten and W. Bruns, *Ber.* **B72,** 1753 (1939); *ibid.* **B74,** 1232 (1941).
30. H. D. Fisher, W. J. Lehmann, and I. Shapiro, *J. Phys. Chem.* **65,** 1166 (1961).
31. H. I. Schlesinger and co-workers, *J. Am. Chem. Soc.* **75,** 195 (1953).
32. H. I. Schlesinger and co-workers, *J. Am. Chem. Soc.* **75,** 199 (1953).
33. E. L. Gamble, *Inorg. Synth.* **3,** 27 (1950).
34. Br. Pat. 226,490 (Dec. 20, 1923), A. F. Meyerhofer.
35. E. Krause and R. Nitsche, *Ber.* **54B,** 2784 (1921).
36. B. M. Mikhaĭlov and T. A. Shchegoleva, *J. Gen. Chem. U.S.S.R.* **29,** 3404 (1959).
37. G. T. Morgan and R. B. Tunstall, *J. Chem. Soc.* **125,** 1963 (1924).
38. J. Goubeau and R. Bergmann, *Z. Anorg. Allgem. Chem.* **263,** 69 (1950).
39. J. G. Contreras and D. G. Tuck, *Inorg. Chem.* **12,** 2596 (1973).
40. R. D. Chambers, H. C. Clark, and C. J. Willis, *J. Am. Chem. Soc.* **82,** 5298 (1960).
41. W. I. Aalbersberg and co-workers, *J. Chem. Soc.,* 3055 (1959).
42. W. V. Bush and L. Zechmeister, *J. Am. Chem. Soc.* **80,** 2991 (1958).
43. D. C. Mente, J. L. Mills, and R. E. Mitchell, *Inorg. Chem.* **14,** 123 (1975).
44. H. S. Booth and D. R. Martin, *Boron Trifluoride and Its Derivatives,* John Wiley & Sons, Inc., New York, 1949.
45. P. Baumgarten and H. Hennig, *Chem. Ber.* **72B,** 1743 (1939).
46. C. Gascard and G. Mascherpa, *J. Chim. Phys. Phys. Chim. Biol.* **70,** 1040 (1973).
47. V. I. Durkov and S. S. Batsanov, *Zh. Strukt. Khim.* **2,** 456 (1961).
48. E. F. Mooney and M. A. Qaseem, *Chem. Commun.,* 230 (1967).
49. M. J. Bula, J. S. Hartman, and C. V. Raman, *J. Chem. Soc. Dalton Trans.,* 725 (1974).
50. W. A. G. Graham and F. G. A. Stone, *J. Inorg. Nuclear Chem.* **3,** 164 (1956).
51. Br. Pat. 486,887 (June 13, 1938), (to E. I. du Pont de Nemours & Co., Inc.).
52. M. P. Johnson and D. F. Shriver, *J. Am. Chem. Soc.* **88,** 301 (1966).
53. U.S. Pat. 3,330,629 (July 11, 1967), G. W. Parshall (to E. I. du Pont de Nemours & Co., Inc.).
54. R. N. Scott, D. F. Shriver, and D. D. Lehman, *Inorg. Chim. Acta* **4,** 73 (1970).
55. M. Fishwick and co-workers, *Inorg. Chem.* **15,** 490 (1976).
56. D. R. Martin and J. M. Canon, "Coordination Compounds of the Boron Halides" in G. A. Olah, ed., *Friedel-Crafts and Related Reactions,* Interscience Publishers, a division of John Wiley & Sons, Inc., New York, 1963, Vol. I, pp. 399–567.
57. H. S. Booth and K. S. Willson, *Inorg. Synth.* **1,** 21 (1939).
58. J. W. Koger, ORNL-TM-4172, 1972; *Nucl. Sci. Abstr.* **28,** 11211 (1973).
59. C. F. Swinehart, A. R. Bumblis, and H. F. Flisik, *Anal. Chem.* **19,** 28 (1947); *Ann. Proc. 35-0049,* Harshaw Chemical Company, Mar. 23, 1964.
60. K. H. Jacobson, R. A. Rhoden, and R. L. Roudabush, *DHEW (NIOSH) Publ.* 77 (1976).
61. T. R. Torkelson, S. E. Sadek, and V. K. Rowe, *Am. Ind. Hyg. Assoc. J.* **22,** 263 (1961).
62. *Handbook of Analytical Toxicology,* Chem. Rubber Co., Cleveland, Ohio, 1969, pp. 606–607.
63. C. J. Spiegl, *Natl. Nucl. Energy Ser. Div. VI* **1** (Book 4), 2291 (1953).

64. G. A. Olah, *Friedel-Crafts and Related Reactions,* Interscience Publishers, a division of John Wiley & Sons, Inc., New York, 1963, pp. 228–235.
65. Ref. 44, Chapt. 6.
66. A. V. Topchiev, S. V. Zavgorodnii, and Y. M. Paushkin, *Boron Fluoride and Its Compounds as Catalysts in Organic Chemistry,* Pergamon Press, New York, 1959.
67. A. A. Palko and J. S. Drury, *J. Chem. Phys.* **47,** 2561 (1967).
68. Belg. Pat. 450,730 (June, 1943), (to I. G. Farbenindustrie).
69. U.S. Pat. 3,103,433 (Sept. 10, 1963), A. D. F. Toy, R. B. Fearing, and C. L. Dewald (to Stauffer Chemical Co.).

DONALD R. MARTIN
University of Texas at Arlington

BORON—FLUOROBORIC ACID AND FLUOROBORATES

Fluoroboric Acid and the Fluoroborate Ion

Fluoroboric acid [*16872-11-0*], generally formulated as HBF_4, does not exist as a free, pure substance. The acid is stable only as a solvated ion pair, such as $H_3O^+BF_4^-$; the commercially available 48% HBF_4 solution approximates $H_3O^+BF_4^-.4H_2O$. Other names used infrequently are hydrofluoroboric acid, hydroborofluoric acid, and tetrafluoroboric acid. Salts of the acid are named as fluoroborates or occasionally borofluorides. Fluoroboric acid and its salts were investigated as early as 1809 by Berzelius, Gay-Lussac, and Thénard (1–2). The acid and many transition metal salts are used in the plating and metal finishing industries. Some of the alkali metal fluoroborates are used in fluxes.

Properties. Fluoroboric acid is stable in concentrated solutions, and hydrolyzes slowly in aqueous solution to hydroxyfluoroborates. For the stability of the fluoroborate species, see ref. 3. The equilibrium quotients (4–5) in 1 molal NaCl at 25°C show the strong affinity of boron for fluoride:

$$B(OH)_3 + F^- \rightleftharpoons BF(OH)_3^- \qquad \log Q = -0.36 \pm 0.19$$

$$B(OH)_3 + 2\,F^- + H^+ \rightleftharpoons BF_2(OH)_2^- + H_2O \qquad \log Q = 7.06 \pm 0.02$$

$$B(OH)_3 + 3\,F^- + 2\,H^+ \rightleftharpoons BF_3OH^- + 2\,H_2O \qquad \log Q = 13.689 \pm 0.003$$

$$B(OH)_3 + 4\,F^- + 3\,H^+ \rightleftharpoons BF_4^- + 3\,H_2O \qquad \log Q = 19.0 \pm 0.1$$

The hydrolysis of BF_4^- occurs stepwise to BF_3OH^-, $BF_2(OH)_2^-$, and $BF(OH)_3^-$. By conductivity measurements the reacton of boric acid and HF was found to form $H[BF_3(OH)]$ [*15433-40-6*] rapidly, but HBF_4 formed much more slowly from HBF_3OH. These studies demonstrate that BF_4^- is quite stable to hydrolysis and conversely is slow to form from BF_3OH^- and HF:

$$BF_4^- + H_2O \rightleftharpoons BF_3OH^- + HF$$

Kinetic results (5) and ^{19}F nmr experiments (6) illustrate clearly that the hydroxyfluoroborates are in rapid equilibrium and easily exchange fluoride.

Table 1 list some of the physical properties of fluoroboric acid. It is a strong acid in water, equal to most mineral acids in strength with pK_{H_2O} −4.9 as compared to −4.3 for nitric acid (9). The fluoroborate ion contains a nearly tetrahedral boron atom with almost equidistant B—F bonds in the solid state. Although lattice effects and hydrogen bonding distort the ion, the average B—F distance is 0.138 nm; the F—B—F angles are nearly the theoretical 109° (10–11). Raman spectra on molten $NaBF_4$ or $NaBF_4$(l) agree with the symmetrical tetrahedral structure (12).

The fluoroborate ion has traditionally been referred to as a noncoordinating anion. It has shown little tendency to form a coordinate–covalent bond with transition metals as do nitrates and sulfates. A few exceptional cases have been reported (13) in which a coordinated BF_4^- was detected by infrared or visible spectroscopy.

Hydroxyfluoroborates are products of the reaction of BF_3 with water; $BF_3.2H_2O$ [*13319-75-0*] is actually $H_3O^+BF_3OH^-$. Salts such as $NaBF_3OH$ [*13876-97-6*] are made by neutralizing the acid. The BF_3OH^- anions are distorted tetrahedra (14). In the HBO_2–HF system, $HBO_2.2HF$ was found to be $HBF_2(OH)_2$, dihydroxyfluoroboric acid [*17068-89-2*] (15).

Manufacture, Shipping, and Waste Treatment. Fluoroboric acid (48%) is made commercially by direct reaction of 70% hydrofluoric acid and boric acid, H_3BO_3. The reaction is exothermic and must be controlled by cooling.

The commercial product is usually a 48–50% solution which contains up to a few percent excess boric acid to eliminate any HF fumes and to avoid HF burns. Reagent-grade solutions are usually 40%. A 61% solution can be made from meta-boric acid, HBO_2, and 70% HF, and a lower grade by direct combination of fluorspar, CaF_2, sulfuric acid, and boric acid (16). The product contains a small amount of dissolved calcium sulfate. A silica-containing fluoroboric acid (0.11% SiO_2) is produced from inexpensive fluorosilicic acid (17). Boric acid is added to a 10% H_2SiF_6 solution and then concentrated in several steps to 45% HBF_4. Granular silicon dioxide must be filtered from the product.

Vessels and equipment must withstand the corrosive action of hydrofluoric acid. For a high-quality product the preferred materials for handling HBF_4 solutions are polyethylene, polypropylene, or a resistant rubber such as neoprene (see Elastomers,

Table 1. Physical Properties of Fluoroboric Acid

Property	Value	Ref.
heat of formation, kJ/mol[a]		
aqueous, 1 molal, at 25°C	−1527	
from boric oxide and aq HF	−123.34	2
BF_4^-, gas	−1765 ± 42	7
entropy of the BF_4^- ion, J/(mol·K)[a]	167	
specific gravity,		
48% soln	1.37	
42% soln	1.32	
30% soln	1.20	
surface tension, 48% soln at 25°C, mN/m (= dyn/cm)	65.3	
ir absorptions[b], cm^{-1}	ca 1100	8
	ca 530	

[a] To convert J to cal, divide by 4.184.
[b] Generally observed as strong absorptions.

synthetic). Where metal must be used, ferrous alloys with high nickel and chromium content show good resistance to corrosion. Impregnated carbon (Carbate) or Teflon can be used in heat exchangers. Teflon-lined pumps and auxilliary equipment are also good choices. Working in glass equipment is not recommended for fluoroboric acid or any fluoroborate.

Fluoroboric acid and some fluoroborate solutions are shipped as corrosive material, generally in polyethylene-lined steel pails and drums or in rigid nonreturnable polyethylene containers. Acid spills should be neutralized with lime or soda ash.

Waste treatment of fluoroborate solutions includes a pretreatment with aluminum sulfate to facilitate hydrolysis, and final precipitation of fluoride with lime (18). The aluminum sulfate treatment can be avoided by hydrolyzing the fluoroborates at pH 2 in the presence of calcium chloride; at this pH, hydrolysis is most rapid at elevated temperature (19).

Economic Aspects. In the United States fluoroboric acid is manufactured by Allied Chemical Corp., C. P. Chemical Co., Fidelity Chemical Products, The Harshaw Chemical Co., and Harstan Chemical Co. Research quantities of reagent grade are made by Alfa Inorganics, Inc., and Ozark-Mahoning Co. The price for 48% fluoroboric acid in truckload quantities in November, 1979 was \$1.23–1.32/kg. Although industry figures are not available, estimates for 1973 indicate that noncaptive production accounted for less than 90,000 kg of hydrofluoric acid (20).

Many specialty fluoroborates are available in research quantities for ca \$1/g (1978 price) from Alfa Inorganics, Inc., and Cationics, Inc.

Analysis. Fluoroboric acid solutions and fluoroborates are analyzed gravimetrically with the aid of nitron or by hydrolysis of the fluoroborate ion in a calcium chloride solution (2). A fluoroborate ion-selective electrode has been developed (21).

Toxicity. Fluoroborates are excreted mostly in the urine (22). Sodium fluoroborate is absorbed almost completely into the human bloodstream and over a 14-d experiment all of the $NaBF_4$ ingested was found in the urine. Although the fluoride ion is covalently bound to boron, the rate of absorption of the physiologically inert BF_4^- from the gastrointestinal tract of rats exceeds that of the physiologically active simple fluorides (23).

Uses. The main use of fluoroboric acid is as an intermediate in the manufacture of fluoroborate salts (see below). However, the Alcoa Alzak process for electropolishing aluminum requires substantial quantities of fluoroboric acid. A 2.5% HBF_4 solution is used to produce a highly reflective surface (24). The high solubility of many metal oxides in HBF_4 is a decided advantage in metal finishing operations (see Metal surface treatment). Before plating or other surface treatment, many metals are cleaned and pickled in fluoroboric acid solution; eg, continuous strip pickling of hot-rolled low-carbon steel is feasible in HBF_4 solutions (25). Nontempered rolled steel requires 80°C for 60 s at 130 g HBF_4/L whereas tempered rolled steel requires only 65°C for 60 s at 65 g HBF_4/L. The spent pickling solution was recovered by electrodialysis.

Fluoroboric acid is used as a stripping solution for the removal of solder and plated metals from less active substrates. It is also used in bright dipping solutions for Sn–Pb alloys in printed circuits and other electrical components (26). A number of fluoroborate plating baths (27) require pH adjustment with fluoroboric acid (see Electroplating).

A low-grade fluoroboric acid (16) is used in the manufacture of cryolite (28) for the electrolytic production of aluminum:

$$4\ Na_2SO_4.NaF + 5\ HBF_4 + 2\ Al_2O_3 + 9\ H_2O \rightarrow 4\ Na_3AlF_6 + 5\ H_3BO_3 + 4\ H_2SO_4$$

The boric and sulfuric acids are recycled to a HBF_4 solution by reaction with CaF_2. As a strong acid, fluoroboric acid is frequently used as an acid catalyst, eg, in synthesizing mixed polyol esters (29). This process provides an inexpensive route to confectioner's hard-butter compositions which are substitutes for cocoa butter in chocolate candies (see Chocolate). Epichlorohydrin is polymerized in the presence of HBF_4 for eventual conversion to polyglycidyl ethers (30) (see Chlorohydrins). A more concentrated solution, 61–71% HBF_4, catalyzes the addition of CO and water to olefins under pressure to form neo-acids (31) (see Carboxylic acids).

Alkali Metal, Ammonium, and Metal Fluoroborates

Properties. A summary of the chemical and physical properties of this group is given in Tables 2 and 3. Chemically these compounds are different from the transition metal fluoroborates usually separating in anhydrous form. This group is very soluble in water, except for the K, Rb, and Cs salts which are only slightly soluble. Many of the soluble salts crystallize as hydrates.

Lithium fluoroborate crystallizes from aqueous solutions as $LiBF_4.3H_2O$ [*39963-05-8*] and $LiBF_4.H_2O$ [*39963-03-6*]. The heat of dehydration of the monohydrate at 91°C is 70.9 kJ/mol (16.95 kcal/ mol); the melting point is 117°C (45). Magnesium, calcium, strontium, and barium fluoroborates crystallize as hydrates: $Mg(BF_4)_2.6H_2O$ [*19585-07-0*], $Ca(BF_4)_2.2H_2O$ [*27860-81-7*], $Sr(BF_4)_2.4H_2O$ [*27902-05-2*], and $Ba(BF_4)_2.2H_2O$ [*72259-09-7*]. These hydrated fluoroborates can be dehydrated completely to the anhydrous salts, which show decreasing stabilities: Ba > Sr > Ca > Mg.

The anhydrous magnesium salt, which is least stable thermally, forms MgF_2, which has the highest lattice energy. This has been confirmed by differential thermal analysis (dta) of the crystalline hydrates (46). Aluminum fluoroborate [*14403-54-4*], $Al(BF_4)_3.(H_2O)_n$, is soluble in strongly acid solutions and displays a tendency for fluoride exchange with BF_4^- to form aluminum fluorides. The aluminyl compound, $AlO^+BF_4^-$, is extremely hygroscopic and is prepared by the reaction of AlOCl, BF_3, and HF (47). Differential thermal analysis experiments show thermal decomposition beginning at 85°C, corresponding to removal of BF_3 and formation of AlOF.

Differential thermal analysis studies of ammonium fluoroborate showed the orthorhombic to cubic transition at 189 ± 5°C and BF_3 generation from 389 to 420°C (48). Sodium hydroxide reacts with NH_4BF_4 liberating ammonia and forming $NaBF_4$. When sodium fluoroborate was studied by infrared spectroscopy, sodium hydroxyfluoroborate, $NaBF_3OH$, was found to be present (49). Although pure sodium hydroxyfluoroborate is thermally unstable, decomposing to $Na_2B_2F_6O$ [*18953-03-2*] and H_2O, in a melt of $NaBF_4$–NaF no instability of the small amount of $NaBF_3OH$ present was detected. Fusion of $NaBF_4$ or KBF_4 with boric oxide generates BF_3 and complex borates such as KFB_4O_6 (50). Most fluoroborates decompose readily to give BF_3 when treated with sulfuric acid or when calcined (see Table 3 for dissociation pressure). Under strongly basic conditions the chemical equilibrium in shifted away from BF_4^- to borates and fluorides.

Manufacture. Fluoroborate salts are prepared commercially by several different combinations of boric acid and 70% hydrofluoric acid with oxides, hydroxides, carbonates, bicarbonates, fluorides, and bifluorides. Fluoroborate salts are substantially less corrosive than fluoroboric acid but the possible presence of HF or free fluorides cannot be overlooked. Glass vessels and equipment should not be used.

Table 2. General Properties of Metal Fluoroborates

Compound	CAS Registry No.	Molecular wt	Color	Physical form	mp, °C	Density[a], g/cm³	H_2O solubility, g/100 mL H_2O	Other solubility	Refs.
$LiBF_4$	[*14283-07-9*]	93.74	white				very soluble		2, 32
$NaBF_4$	[*13755-29-8*]	109.79	white	orthorhombic <240°C a = 0.68358, b = 0.62619, c = 0.67916 nm noncubic >240°C	406 dec	2.47	108 (26°C) 210 (100°C)	sl alcohol	13, 32–33
KBF_4	[*14075-53-7*]	125.92	colorless	rhombic <283°C a = 0.7032, b = 0.8674, c = 0.5496 nm cubic >283°C	530 dec	2.498	0.45 (20°C) 6.27 (100°C)	sl ethanol insol alkali	32–34
$RbBF_4$	[*18909-68-7*]	172.27		orthorhombic <245°C a = 0.7296, b = 0.9108, c = 0.5636 nm cubic >245°C	612 dec	2.820	0.6 (17°C) 10 (100°C)		32–34
$CsBF_4$	[*18909-69-8*]	219.71	white	orthorhombic <140°C a = 0.7647, b = 0.9675, c = 0.5885 nm cubic >140°C	555 dec	3.20	1.6 (17°C) 30 (100°C)		32–34
NH_4BF_4	[*13826-83-0*]	104.84	white	orthorhombic <205°C a = 0.7278, b = 0.9072, c = 0.5678 nm cubic >205°C	487 dec	1.871[b]	3.09 (−1.0°C) 5.26 (−1.5°C 10.85 (−2.7°C) 12.20 (0°C) 25 (16°C) 25.83 (25°C) 44.09 (50°C) 67.50 (75°C) 98.93 (100°C) 113.7 (108.5°C)	HF 19.89% (0°C)	32, 34–35 36
$NaBF_3OH$	[*13876-97-6*]			hexagonal a = 0.8084, c = 0.7958 nm		2.46			10

[a] Unless otherwise stated, at 20°C.
[b] At 15°C.

Table 3. Thermodynamic Data[a] for Metal Fluoroborates, kJ/mol[b]

Compound	ΔH_{diss}	$\log P_{Pa} = -aT^{-1} + b$			Lattice energy, $-U$	ΔH_{fus}	ΔH_f	Other	Refs.
		a	b^c	Temperature, °C					
$LiBF_4$	15.9	833	6.40	210–320	699		−1838.4	$\Delta H^d = -89.54$	37–39
$NaBF_4$	69.83	3650	8.75	400–700	657.3	13.6	−1843.5	$\Delta H^e = -134.1$	33, 38–40
KBF_4	121	6317	8.15	510–930	598	18.0	−1881.5	$\Delta H^f = -180.5$ $\Delta H_{sub} = 330$ S (25°) = 130 J/(mol·K) C_p (25°) = 112.1 J/(mol·K)	33, 38–39, 41–42
$RbBF_4$	112.8	5960	9.57	600–1000	577	19.6			33, 38,41
$CsBF_4$	112.5	5880	9.47	610–1040	556	19.2			33, 38,41
$NaBF_3OH$	77.0	4024	9.11	400–700			−1754		40
NH_4BF_4		2469	8.94		607[g]			$\Delta H_{sub} = 47.3$	43–44

[a] ΔH_{diss} = heat of dissociation, ΔH_{fus} = heat of fusion, ΔH_f = heat of formation (25°C), ΔH_{sub} = sublimation.
[b] Unless otherwise stated. To convert J to cal, divide by 4.184.
[c] To convert $\log P_{Pa}$ to $\log P_{mm\,Hg}$ subtract 2.12 from b.
[d] $LiF (s) + BF_3 (g) \rightarrow LiBF_4 (s)$, 25°C.
[e] $NaF (s) + BF_3 (g) \rightarrow NaBF_4 (s)$, 25°C.
[f] $KF (s) + BF_3 (g) \rightarrow KBF_4 (s)$, 25°C.
[g] At 260°C.

Sodium Fluoroborate. Sodium fluoroborate is prepared by the reaction of NaOH or Na_2CO_3 with fluoroboric acid (51), or by treatment of disodium hexafluorosilicate with boric acid.

Potassium Fluoroborate. Potassium fluoroborate is produced as a gelatinous precipitate by mixing fluoroboric acid with KOH or K_2CO_3. Alternatively, fluorosilicic acid is treated with H_3BO_3 in a 2:1 molar ratio to give HBF_3OH, which reacts with HF and KCl to yield 98% of KBF_4 in 98.5% purity (52). Commercial KBF_4 normally contains less than 1% KBF_3OH.

Ammonium and Lithium Fluoroborates. Ammonia reacts with fluoroboric acid to produce ammonium fluoroborate (53). An alternative method is the fusion of ammonium bifluoride and boric acid (54):

$$2\ NH_4HF_2 + H_3BO_3 \rightarrow NH_4BF_4 + 3\ H_2O + NH_3$$

The water and ammonia must be removed from the melt. Lithium hydroxide or carbonate react with HBF_4 to form $LiBF_4$.

Magnesium Fluoroborate. Treatment of magnesium metal, magnesium oxide, or magnesium carbonate with HBF_4 gives magnesium fluoroborate [*14708-13-5*]. The MgF_2 is filtered and the product is sold as a 30% solution.

Economic Aspects. Production figures are not available for individual salts but estimates of total inorganic fluorides can be made from *Chemical Marketing Reporter* profiles on hydrofluoric acid. In the United States the sodium, potassium, ammonium, and magnesium fluoroborates are sold by Allied Chemical Corp., Harshaw Chemical Co., and Kawecki-Berylco Chemical Co. Lithium fluoroborate is available from Foote Mineral Co. and Lithium Corporation of America. Small amounts of other fluoroborates are sold by Alfa Inorganics, Inc. and Ozark-Mahoning Co. November 1979 prices for truckload quantities are given below:

Fluoroborate	*Price, \$/kg*
$NaBF_4$	2.15–2.51
KBF_4	2.20
NH_4BF_4	2.61

Uses. Alkali metal and ammonium fluoroborates are used mainly for the high temperature fluxing action required by the metals processing industries (see Metal surface treatments; Welding). The tendency toward BF_3 dissociation at elevated temperatures inhibits oxidation in magnesium casting and aluminum alloy heat treatment.

The molten salts quickly dissolve the metal oxides at high temperatures to form a clean metal surface. Other uses are as catalysts and in fire-retardant formulations (see Flame retardants).

Potassium Fluoroborate. The addition of potassium fluoroborate to grinding wheel and disk formulations permits lower operating temperatures (5). Cooler action is desirable to reduce the burning of refractory materials such as titanium and stainless steels. Excellent results in grinding wheels are also obtained with $NaBF_4$ (56). A process for boriding steel surfaces with B_4C and KBF_4 as an activator improves the hardness of the base steel (57). Fluxes for aluminum bronze and silver soldering and brazing contain KBF_4 (58) (see Solders). Fire retardance is imparted to acrylonitrile polymers by precipitating KBF_4 within the filaments during coagulation (59). In polyurethanes KBF_4 and NH_4BF_4 reduce smoke and increase flame resistance (60). Both the potassium and ammonium salts improve insulating efficiency of intumescent coatings

(61). The endothermic characteristics of these fillers (release of BF_3) counteract the exothermic nature of the intumescent agents (nitroaromatic amines) in the coating. The sodium and potassium salts are claimed to have a synergistic effect with polyhalogenated aromatics that improve flame-retardant properties of polyesters (62). Elemental boron is prepared by the Cooper electrolysis of a KBF_4 melt with B_2O_3 and KCl (63). The boron may be up to 99.5% purity and, if KBF_4 containing the ^{10}B isotope is used, the product is ^{10}B which is used in the nuclear energy field as a neutron absorber (see Nuclear reactors).

Sodium Fluoroborate. Sodium fluoroborate can be used in the transfer of boron to aluminum alloys but the efficiency is lower than for KBF_4 (64). Sodium fluoroborate in an etching solution with sulfamic acid and H_2O_2 aids in removing exposed lead in printed circuit manufacture (65). During the annealing of galvanized iron (galvannealing), the surface becomes oxidized. The resulting oxide coating, which causes difficulty in soldering, can be removed by aqueous $NaBF_4$ or NH_4BF_4 (66). Work at Oak Ridge National Lab has shown that a $NaBF_4$—with 8 mol % NaF—salt mixture could be used as the coolant in the molten breeder reactor (67); in this molten salt at nearly 600°C the corrosion rate of Hastelloy N is about 8 μm/yr. Sodium fluoroborate acts as a catalyst for cross-linking cotton cellulose with formaldehyde (68); transesterification in the preparation of polycarbonates (69); and preparation of cyclic oligoethers from ethylene oxide (70). Sodium and lithium fluoroborates are effective flame retardants for cotton and rayon (71).

Ammonium Fluoroborate. Ammonium fluoroborate blends with antimony oxide give good results in flame-retarding polypropylene (72). The complete thermal vaporization makes ammonium fluoroborate an excellent gaseous flux for inert-atmosphere soldering (73). A soldering flux of zinc chloride and ammonium fluoroborate is used in joining dissimilar metals such as Al and Cu (74). Ammonium fluoroborate acts as a solid lubricant in cutting-oil emulsions for aluminum rolling and forming.

Miscellaneous. Flame-resistant cross-linked polyethylene can be made with a number of fluoroborates and antimony oxide. This self-extinguishing material may contain the fluoroborates of NH_4^+, Na^+, K^+, Ca^{2+}, Mg^{2+}, Sr^{2+}, or Ba^{2+} in amounts of 4–20% (75). Lithium fluoroborate is found in a number of newly developed batteries as an electrolyte, for example in the lithium–sulfur battery (76) (see Batteries). Magnesium fluoroborate catalyzes the epoxy treatment of cotton fabrics for permanent-press finishes (77) (see Textiles).

Transition Metal and Other Heavy Metal Fluoroborates

The physical and chemical properties are less well known for this series than for the alkali metal fluoroborates (see Table 4). Most transition metal fluoroborates are strongly hydrated coordination compounds and are difficult to dry without decomposition. Decomposition frequently occurs during the concentration of solutions for crystallization. The stability of the metal fluorides accentuates this problem. Loss of HF because of hydrolysis makes the reaction proceed even more rapidly. Even with low temperature vacuum drying to partially solve the decomposition, the dry salt readily absorbs water. The crystalline solids are generally soluble in water, alcohols, and ketones but only poorly soluble in hydrocarbons and halocarbons.

Differential thermal analysis in air on the crystalline hexahydrates of Zn, Cd, Fe, Co, and Ni fluoroborates show the loss of BF_3 and H_2O simultaneously at 195, 215,

Table 4. Properties of Metal Fluoroborates[a]

Compound	CAS Registry No.	Color	Specific gravity	Solubility	Miscellaneous
$Mn(BF_4)_2.6H_2O$	*[26044-57-5]*	pale pink	1.982	water, ethanol	
$Fe(BF_4)_2.6H_2O$	*[13877-16-2]*	pale green	2.038	water, ethanol	
$Co(BF_4)_2.6H_2O$	*[15684-35-2]*	red	2.081	water, alcohol	
$Ni(BF_4)_2.6H_2O$	*[14708-14-6]*	green	2.136	water, alcohol	
$Cu(BF_4)_2.6H_2O$	*[72259-10-0]*	blue	2.175	water, alcohol	
$AgBF_4.H_2O$	*[72259-11-1]*	colorless		water, less sol in alcohol, sol benzene, sol ether	dec 200°C, light sensitive
$Zn(BF_4)_2.6H_2O$	*[27860-83-9]*	white	2.120	water, alcohol	dehydrates at 60°C
$Cd(BF_4)_2.6H_2O$	*[27860-84-0]*	white	2.292	water alcohol	
$In(BF_4)_3.xH_2O$	*[27765-48-6]*	colorless		water	
$TlBF_4.H_2O$	*[72259-12-2]*	colorless		water	orthorhombic, $a = 0.947$, $b = 0.581$, $c = 0.740$ nm, light sensitive
$Sn(BF_4)_2.xH_2O$	*[72259-13-3]*	white		water	$Sn(BF_4)_2.SnF_2.$-$5H_2O$ crystallizes from soln
$Pb(BF_4)_2.H_2O$	*[26916-34-7]*	colorless			

[a] Crystalline solids; from refs. 2 and 78.

180, 185, and 205°C, respectively (46,79). The dta curves also indicate initial melting at 107, 117, and 150°C for Zn, Cd, and Fe fluoroborates, respectively. The anhydrous metal fluoride and/or oxide is usually isolated. The copper salt also decomposes with liberation of BF_3 and H_2O (80).

The water of hydration of these complexes can be replaced with other coordinating solvents. For example, the ethanol and methanol solvates were made by dissolving the hydrates in triethyl and trimethyl orthoformate, respectively (81–82). The acetic acid solvates are made by treating the hydrates with acetic anhydride (83). Conductivity and visible spectra (where applicable) of the Co, Ni, Zn, and Cu fluoroborates in *N,N*-dimethylacetamide (L) showed that all metal ions were present as the ML_6^{2+} cations (84). Solvated fluoroborate complexes of Cr^{3+}, Fe^{2+}, Co^{2+}, Ni^{2+}, Cu^{2+}, Cu^{+}, and Zn^{2+} in diethyl ether, nitromethane, and benzene solutions have been prepared. Solutions of $Ti(BF_4)_3$, $V(BF_4)_3$, and $Fe(BF_4)_3$ could not be prepared probably because of formation of BF_3 and the metal fluoride (85). Ammonia easily replaces the coordinated water; the products are usually tetrammine or hexammine complexes (2) (see Coordination compounds). The hexahydrate of $Ni(BF_4)_2$ was found to be stable from 25 to 100°C; solubility also was determined to 95°C (86). At 120°C the solid decomposed slowly to NiF_2 with loss of HF, H_3BO_3, and H_2O.

Manufacture. The transition and heavy metal fluoroborates can be made from the metal, metal oxide, hydroxide, or carbonate with fluoroboric acid. Because of the difficulty in isolating pure crystalline solids, these fluoroborates are usually available as 40–50% solutions, $M(BF_4)_x$. Most of the solutions contain about 1–2% excess fluoroboric acid to prevent precipitation of basic metal complexes. The solutions are usually sold in 19 and 57 L polyethylene containers.

In some cases, particularly with inactive metals, electrolytic cells are the primary method of manufacture of the fluoroborate solution. The manufacture of Sn, Pb, Cu, and Ni fluoroborates by electrolytic dissolution (87–88) is patented. A typical cell for continous production consists of a polyethylene-lined tank with tin anodes at the bottom and a mercury pool (in a porous basket) cathode near the top (88). Fluoroboric acid is added to the cell and electrolysis is begun. As tin fluoroborate is generated, differences in specific gravity cause the product to layer at the bottom of the cell. When the desired concentration is reached in this layer, the heavy solution is drawn from the bottom and fresh HBF_4 is added to the top of the cell continuously. The direct reaction of tin with HBF_4 is slow but can be accelerated by passing air or oxygen through the solution (89). The stannic fluoroborate is reduced by reaction with mossy tin under an inert atmosphere. In earlier procedures, HBF_4 reacted with hydrated stannous oxide.

Anhydrous silver fluoroborate [*1404-20-2*] is made by the addition of BF_3 gas to a suspension of AgF in ethylbenzene (90). An $AgBF_4.C_8H_{10}$ complex is precipitated with pentane and the complex is washed with pentane to give anhydrous $AgBF_4$.

Economic Aspects. Most fluoroborate solutions listed in Table 5 are manufactured by Allied Chemical Corp., Harstan Chemical Co., Harshaw Chemical Co., C. P. Chemical Co., and Fidelity Chemical Products. Research quantities are available from Alfa Inorganics, Inc., and Ozark-Mahoning Co. Prices are shown in Table 5; production figures are not available.

Uses. Metal fluoroborate solutions are used primarily as plating solutions and as catalysts. The Sn, Cu, Zn, Ni, Pb, and Ag fluoroborates cure a wide range of epoxy resins at elevated or ambient room temperature (91–92). In the textile industry zinc fluoroborate is used extensively as the curing agent in applying resins for crease-resistant finishes (93). Emulsions of epoxy resins (94), polyoxymethylene compounds (95), or aziridinyl compounds (96) with $Zn(BF_4)_2$ and other additives are applied to the cloth. After the excess is removed, the cloth is dried and later cured at a higher temperature. Similarly treated acrylic textiles using epoxy resins take on an antistatic finish (97), or the acrylic textiles can be coated with 20% $Zn(BF_4)_2$ which results in up to 5.5% added solids for a fire-resistant finish (98).

The use of silver fluoroborate as a catalyst or reagent often depends on the precipitation of a silver halide. Thus the silver ion abstracts a Cl^- from a rhodium chloride

Table 5. Commercial Metal Fluoroborate Solutions

Fluoroborate solution	CAS Registry No.	Formula	% metal	Specific gravity, g/cm^3	1979 price, \$/kg
antimony(II)	[*14486-20-5*]	$Sb(BF_4)_3$	12.8	1.42	
cadmium	[*14886-19-2*]	$Cd(BF_4)_2$	19.7	1.60	4.86
cobalt(II)	[*26490-63-1*]	$Co(BF_4)_2$	11.8	1.42	
copper(II)	[*38465-60-0*]	$Cu(BF_4)_2$	12.2	1.48	1.95–2.00
indium	[*27765-48-6*]	$In(BF_4)_3$	15.3	1.55	
iron(II)	[*13877-16-2*]	$Fe(BF_4)_2$	10.3	1.47	1.06
lead(II)	[*13814-96-5*]	$Pb(BF_4)_2$	28.9	1.75	1.38–1.47
nickel(II)	[*14708-14-6*]	$Ni(BF_4)_2$	11.2	1.47	3.16–3.25
tin(II)	[*13814-96-5*]	$Sn(BF_4)_2$	20.2	1.61	6.38–6.46
zinc	[*13826-88-5*]	$Zn(BF_4)_2$	11.0	1.39	1.25

complex, $[(C_6H_5)_3As]_2(CO)RhCl$, yielding the cationic rhodium fluoroborate [*30935-54-7*] hydrogenation catalyst (99). The complexing tendency of olefins for $AgBF_4$ has led to the development of chemisorption methods for ethylene separation (100–101). Copper(I) fluoroborate [*14708-11-3*] also forms complexes with olefins; hydrocarbon separations are effected by similar means (102).

The manufacture of linear polyester is catalyzed by Cd, Sn (103), Pb, Zn, or Mn (104) fluoroborates. The Beckmann rearrangement of cyclohexanone oxime to caprolactam is catalyzed by $Ba(BF_4)_2$ [*13862-62-9*] or $Zn(BF_4)_2$ [*13826-88-5*] (105). The caprolactam is polymerized to polyamide fibers with $Mn(BF_4)_2$ [*30744-82-2*] catalyst (106). Nickel and cobalt fluoroborates appear to be good catalysts for the polymerization of conjugated dienes to *cis*-1,4-polydienes; the cis configuration is formed in up to 96% yields (107–109).

Electroplating. Metal fluoroborate electroplating baths (27,110–111) are employed where speed and quality of deposition are important (see Electroplating). High current densities can be used for fast deposition and near 100% anode and cathode efficiencies can be expected. Since the salts are very soluble, highly concentrated solutions can be used without any crystallization. The high conductivity of these solutions reduces the power costs. The metal content of the bath is also easily maintained and the pH is adjusted with HBF_4 or aqueous ammonia. The disadvantages of using fluoroborate baths are treeing, lack of throwing power, and high initial cost. Treeing and throwing power can be controlled by additives; grain size of the deposits can also be changed. Metals currently being plated from fluoroborate baths are Cd, Co, Cu, Fe, In, Ni, Pb, Sb, and Zn. Recent studies on Fe (111,113), Ni (113), and Co (113) fluoroborate baths describe the compositions and conditions of operation as well as the properties of the coatings. Iron foils electrodeposited from fluoroborate baths and properly annealed have exceptionally high tensile strength (113).

The Fe, Co, and Ni deposits are extremely fine grained at high current density and pH. Electroless nickel, cobalt, and nickel–cobalt alloy plating from fluoroborate-containing baths yields a deposit of superior corrosion resistance, low stress, and excellent hardenability (114). Lead is plated alone or in combination with tin, indium, and antimony (115). Sound insulators are made as lead–plastic laminates by electrolytically coating Pb from a fluoroborate bath to 0.5 mm on a copper-coated nylon or polypropylene film (116) (see Insulation, acoustic). Steel plates can be simultaneously electrocoated with lead and PTFE (117). Solder is plated in solutions containing $Pb(BF_4)_2$ and $Sn(BF_4)_2$; thus the lustrous solder-plated object is coated with a Pb–Sn alloy (118).

BIBLIOGRAPHY

"Fluoroboric Acid" under "Fluorine Compounds, Inorganic" in *ECT* 1st ed., Vol. 6, pp. 684–688, by F. D. Loomis, Pennsylvania Salt Manufacturing Co.; "Fluoroboric Acid and Fluoborates" under "Fluorine Compounds, Inorganic" in *ECT* 2nd ed., Vol. 9, pp. 562–572, by H. S. Halbedel, The Harshaw Chemical Company.

1. J. W. Mellor, *Comprehensive Treatise on Inorganic and Theoretical Chemistry,* Vol. 5, Longman, Green and Co., New York, 1929, pp. 123–129.
2. H. S. Booth and D. R. Martin, *Boron Trifluoride and Its Derivatives,* John Wiley & Sons, Inc., New York, 1949, pp 87–165.
3. R. E. Mesmer, K. M. Palen, and C. F. Baes, *Inorg. Chem.* **12**(1), 89 (1973).

4. I. G. Ryss, *The Chemistry of Fluorine and Its Inorganic Compounds,* State Publishing House for Scientific, Technical, and Chemical Literature, Moscow, USSR, 1956; F. Haimson, English transl., *AEC-tr-3927,* U.S. Atomic Energy Commission, Washington, D.C., 1960, pp. 505–579.
5. C. H. Wamser, *J. Am. Chem. Soc.* **70,** 1209 (1948); **73,** 409 (1951).
6. R. E. Mesmer and A. C. Rutenberg, *Inorg. Chem.* **12**(3), 699 (1973).
7. R. D. Srinastava, M. O. Uy, and M. Faber, *J. Chem. Soc. Farad. Trans. 1* **70,** 1033 (1970).
8. H. Bonadeo and E. Silberman, *J. Mol. Spect.* **32,** 214 (1969).
9. J. Bessiere, *Anal. Chim. Acta* **52**(1), 55 (1970).
10. M. J. R. Clark, *Can. J. Chem.* **47,** 2579 (1969).
11. G. Brunton, *Acta Crystallogr. Sect. B* **24,** 1703 (1968).
12. A. S. Quist and co-workers, *J. Chem. Phys.* **54,** 4896 (1971); **55,** 2836 (1971).
13. M. R. Rosenthal, *J. Chem. Ed.* **50**(5), 331 (1973).
14. M. J. R. Clark and H. Linton, *Can. J. Chem.* **48,** 405 (1970).
15. I. Pawlenko, *Z. Anorg. Allgem. Chem.* **340**(3–4), 201 (1965).
16. H. W. Heiser, *Chem. Eng. Prog.* **45**(3), 169 (1949); U.S. Pats. 2,182,509–11 (Dec. 5, 1939), (to Alcoa).
17. U.S. Pat. 2,799,559 (July 16, 1957), T. J. Sullivan, C. H. Milligan, and J. A. Grady.
18. U.S. Pat. 3,959,132 (May 25, 1976), J. Singh (to Gilson Technical Services, Inc.).
19. U.S. Pat. 4,045,339 (Aug. 30, 1977), T. F. Korenowski, J. L. Penland, and C. J. Ritzert (to Dart Industries Inc.).
20. *Chemical Economics Handbook,* Stanford Research Institute, Menlo Park, Calif., 1975, p. 739.5030H.
21. D. C. Cornish and R. J. Simpson, *Meas. Contr.* **4**(11), 308 (1971).
22. E. J. Largent, "Metabolism of Inorganic Fluoride" in *Fluoridation as a Public Health Measure,* American Association for the Advancement of Science, Washington, D.C., 1954, pp. 49–78.
23. I. Zipkin and R. C. Likens, *Am. J. Physiol.* **191,** 549 (1957).
24. J. F. Jumer, *Met. Finish.* **56**(8), 44 (1958); **56**(9), 60 (1958).
25. R. M. Hudson, T. J. Butler, and C. J. Warning, *Met. Finish.* **74**(10), 37 (1976); U.S. Pat. 3,933,605 (Jan. 20, 1976), T. J. Butler, R. M. Hudson, and C. J. Warning (to U.S. Steel Corp.).
26. U.S. Pat. 3,888,778 (Mar. 13, 1973), M. Beckwith and G. F. Hau.
27. R. D. Mawiya and K. P. Joshi, *Indian Chem. J.* **6**(2), 19 (1971).
28. U.S. Pat. 2,925,325 (Feb. 16, 1960), J. Kamlet (to Reynolds Metals Co.).
29. U.S. Pat. 3,808,245 (Apr. 30, 1974), D. E. O'Connor and G. R. Wyness (to Proctor & Gamble Co.).
30. U.S. Pat. 3,305,565 (Feb. 21, 1967), A. C. Mueller (to Shell Oil Co.).
31. U.S. Pat. 3,349,107 (Oct. 24, 1967), S. Pawlenko (to Schering Akliengessellshaft).
32. R. C. Weast, ed., *Handbook of Chemistry and Physics,* Vol. 59, The Chemical Rubber Co., Cleveland, Ohio, 1978.
33. A. S. Dworkin and M. A. Bredig, *J. Chem. Eng. Data* **15,** 505 (1970).
34. M. J. R. Clark and H. Lynton, *Can. J. Chem.* **47,** 2579 (1969).
35. V. S. Yatlov and E. N. Pinaevskays, *Zh. Obshch. Khim.* **15,** 269 (1945); *Chem Abstr.* **40,** 3357 (1946).
36. H. Boch, *Z. Naturforsch.* **17b,** 426 (1962).
37. L. J. Klinkenberg, doctoral thesis, Leiden, FRG, 1937.
38. T. C. Waddington, *Adv. Inorg. Chem. Radiochem.* **1,** 158 (1959).
39. P. Gross, C. Hayman, and H. A. Joel, *Trans. Faraday Soc.* **64,** 317 (1968).
40. L. J. Klinkenberg, *Rec. Trav. Chim.* **56,** 36 (1937).
41. J. H. de Boer and J. A. H. Van Liempt, *Rec. Trav. Chim.* **46,** 24 (1927).
42. *JANAF Thermochemical Tables,* Clearinghouse for Federal Scientific and Technical Information, U.S. Dept. of Commerce, Springfield, Va., Dec. 1963.
43. A. W. Laubengayer and G. F. Condike, *J. Am. Chem. Soc.* **70,** 2274 (1948).
44. A. P. Altschuller, *J. Am. Chem. Soc.* **77,** 6515 (1955).
45. V. N. Plakhotnik, V. B. Tul'chinski, and V. K. Steba, *Russ. J. Inorg. Chem.* **22,** 1398 (1977).
46. T. V. Ostrovskaya and S. A. Amirova, *Russ. J. Inorg. Chem.* **15,** 338 (1970).
47. A. V. Pankratov and co-workers, *Russ. J. Inorg. Chem.* **17,** 47 (1972).
48. R. T. Marano and J. L. McAtee, *Thermochimica Acta* **4,** 421 (1972).
49. J. B. Bates and co-workers, *J. Inorg. Nucl. Chem.* **34,** 2721 (1972).
50. L. Maya, *J. Am. Ceram. Soc.* **60**(7–8), 323 (1977).
51. V. Pecak, *Chem. Prum.* **23**(2), 71 (1973).

52. Ger. Pat. 2,320,360 (Nov. 7, 1974), H. K. Hellberg, J. Massonne, and O. Gaertner (to Kali-Chemie Fluor G.m.b.H.).
53. U.S. Pat. 2,799,556 (Feb. 1, 1954), T. J. Sullivan and C. G. Milligan (to American Agriculture Chemical Co.).
54. H. S. Booth and S. Rhemar, *Inorganic Synthesis,* Vol. 2, McGraw-Hill Book Co., New York, 1946, p. 23.
55. U.S. Pat. 3,541,739 (Nov. 24, 1970), J. P. Bryon and A. G. Rolfe (to English Abrasives Limited).
56. U.S. Pat. 3,963,458 (June 15, 1976), M. T. Gladstone and S. J. Supkis (to Norton Co.).
57. G. von Matuschka, *Kunstofftechnik* **11**(11), 304 (1972).
58. USSR Pat. 495,178 (Dec. 15, 1975), V. Boiko.
59. U.S. Pat. 3,376,253 (Apr. 2, 1968), E. V. Burnthall and J. J. Hirshfeld (to Monsanto Co.).
60. Ger. Pat. 2,121,821 (Dec. 2, 1971), K. C. Frisch (to Owens Corning Fiberglass Co.).
61. P. M. Sawko and S. R. Riccitiello, *Tech. Brief ARC-11043,* NASA-Ames Research Center, Moffett Field, Calif., July 1977.
62. U.S. Pat. 3,909,489 (Sept. 30, 1975), D. D. Callander (to Goodyear Tire and Rubber Co.).
63. U.S. Pats. 2,572,248–9 (Oct. 23, 1951), H. S. Cooper (to Walter M. Weil).
64. J. D. Donaldson, C. P. Squire, and F. E. Stokes, *J. Mater. Sci.* **13,** 421 (1978).
65. U.S. Pat. 3,305,416 (Feb. 21, 1967), G. J. Kahan and J. L. Mees (to International Business Machines Corp.).
66. U.S. Pat. 3,540,943 (Nov. 17, 1970), E. M. Grogan (to U.S. Steel Corp.).
67. W. R. Huntley and P. A. Gnadt, *Report ORNL-TM-3863,* Oak Ridge Natl. Lab., Oak Ridge, Tenn., 1973.
68. L. Kravetz and G. R. Ferrante, *Text. Res. J.* **40,** 362 (1970).
69. Fr. Pat. 1,578,918 (Aug. 22, 1968), J. Borkowski.
70. J. Dale and K. Daasvet'n, *J. Chem. Soc. Chem. Commun.* (8), 295 (1976).
71. M. A. Kasem and H. R. Richard, *Ind. Eng. Chem. Prod. Res. Dev.* **11**(2), 114 (1972).
72. *Technical Bulletin FR175,* Harshaw Chemical Co., Cleveland, Ohio, 1975.
73. U.S. Pat. 2,561,565 (July 24, 1951), A. P. Edson and I. L. Newell (to United Aircraft Corp.).
74. Br. Pat. 1,181,753 (Feb. 18, 1970), (to Aluminum Co. of America).
75. U.S. Pat. 3,287,312 (Nov. 22, 1966), T. H. Ling (to Anaconda Wire and Cable Co.).
76. Ger. Pat. 2,334,660 (Jan. 23, 1975), H. Lauck.
77. T. Hongu, *S. Gakkaishi,* **26**(1), 38 (1970).
78. D. W. A. Sharp in M. Stacy, J. C. Tatlow, and A. G. Sharpe, eds., *Advances in Fluorine Chemistry,* Vol. 1, Academic Press, Inc., New York, 1960, pp. 68–128.
79. T. V. Ostrovskaya, S. A. Amirova, and N. V. Startieva, *Russ. J. Inorg. Chem.* **12,** 1228 (1967).
80. R. T. Marano and J. L. McAtee, *Therm. Anal. Proc. Int. Conf. 3rd, 1971* **2,** 335 (1972).
81. A. D. Van Ingen Schenau, W. L. Groenveld, and J. Reedijk, *Recl. Trav. Chim. Pays-Bas* **9,** 88 (1972).
82. P. W. N. M. Van Leeuwen, *Recl. Trav. Chim. Pays-Bas* **86,** 247 (1967).
83. U.S. Pat. 3,672,759 (July 4, 1972), T. Yamawaki and co-workers.
84. E. Kamienska and I. Uruska, *Bull. Akad. Pol. Sci. Ser. Sci. Chim.* **21,** 587 (1973).
85. D. W. A. Sharp and co-workers, *Proc. Int. Conf. Coord. Chem. 8th, Vienna,* 322 (1964).
86. V. N. Plakhotnik and V. V. Varekh, *Izv. Vyssh. Uchebn. Zaved. Khim. Khim. Tekhnol.* **16,** 1619 (1973).
87. U.S. Pat. 3,795,595 (Mar. 5, 1974), H. P. Wilson (to Vulcan Materials Co.).
88. U.S. Pat. 3,300,397 (Jan. 24, 1967), G. Baltakmens and J. P. Tourish (to Allied Chemical Corp.).
89. U.S. Pat. 3,432,256 (Mar. 11, 1969), H. P. Wilson (to Vulcan Materials Co.).
90. S. Buffagni and I. M. Vezzosi, *Gazz. Chim. Ital.* **97,** 1258 (1967).
91. U.S. Pat. 4,092,296 (May 30, 1978), R. A. Skiff.
92. U.S. Pat. 3,432,440 (Mar. 11, 1969), D. A. Shimp, W. F. McWhorter, and N. G. Wolfe (to Celanese Coatings Co.).
93. *Technical Bulletin ZBF873,* Harshaw Chemical Company.
94. A. Zemaitaitis and J. Zdanavicius, *Cellul. Chem. Technol.* **4,** 621 (1970).
95. U.S. Pat. 3,854,869 (Dec. 17, 1974), Y. Yanai (to Nisshin Spinning Co., Ltd.).
96. C. E. Morris and G. L. Drake, Jr., *Am. Dyestuff Rep.* **58**(4), 31 (1969).
97. Jpn. Pat. 71 11,080 (Mar. 20, 1971), S. Hiroaka and K. Mitsumura (to Mitsubishi Rayon Co., Ltd.).
98. U.S. Pat. 3,577,342 (May 4, 1971), L. I. Fidell (to American Cyanamid Co.).
99. U.S. Pat. 3,697,615 (Oct. 10, 1972), W. B. Hughes (to Phillips Petroleum Co.).

100. E. Rausz and S. Hulisz, *Chemik* **28**(7), 256 (1975).
101. H. W. Quinn and R. L. Van Gilder, *Can. J. Chem.* **48,** 2435 (1970).
102. U.S. Pat. 3,514,488 (May 26, 1970), C. E. Uebele, R. K. Grasselli, and W. C. Nixon (to Standard Oil Co. of Ohio).
103. Jpn. Pat. 714,030 (Dec. 3, 1971), Y. Fujita and T. Morimoto (to Mitsui Petrochemical Industries, Ltd.).
104. Jpn. Pat. 70 19,514 (July 3, 1970), I. Hiroi (to Toho Rayon Co., Ltd.).
105. Jpn. Pat. 76 04,163 (Jan. 14, 1976), J. Takeuchi, F. Iwata, and K. Kubo (to Ube Industries, Ltd.).
106. Jpn. Pat. 72 18,227 (May 26, 1972), S. Sugiura and co-workers (to Ube Industries, Ltd.).
107. Jpn. Pat. 73 06,185 (Feb. 23, 1973), T. Yamawaki, T. Suzuki, and S. Hino (to Mitsubishi Chemical Industries Co. Ltd.).
108. Fr. Pat 2,039,808 (Jan. 15, 1971), (to Mitsubishi Chemical Industries, Co., Ltd.).
109. Jpn. Pat. 72 06,411 (May 4, 1972), T. Yamawaki and co-workers (to Mitsubishi Chemical Industries Co., Ltd.).
110. *Plating Processes,* Harshaw Chemical Co., Mar. 1977.
111. Y. M. Faruq Marikan and K. I. Vasu, *Met. Finish.* **67**(8), 59 (1969).
112. F. Wild, *Electroplat. Met. Finish.* **13,** 331 (Sept. 1960).
113. E. M. Levy and G. J. Hutton, *Plating* **55**(2), 138 (1968).
114. U.S. Pat. 3,432,338 (Mar. 11, 1969), R. E. Sickles (to Diamond Shamrock Corp.).
115. N. J. Spiliotis, *Galvanotech. Oberflaechenschutz* **7**(8), 192 (1966).
116. Jpn. Pat. 76 02,633 (Jan. 10, 1976), J. Hara, R. Miyashata, and Y. Fukuoka (to Nippon Kayaku Co., Ltd.).
117. Ger. Pat. 2,146,908 (Mar. 23, 1972), K. Ishiguro and H. Shinohara (to Toyota Motor Co., Ltd.).
118. *Plating Processes, Tin–Lead Solder Alloy Fluoborate Plating Process for Printed Circuit Applications, HTPB5N 0272,* Harshaw Chemical Co.

H. S. HALBEDEL
THOMAS E. NAPPIER
The Harshaw Chemical Company

CALCIUM

Fluorine chemistry began with observations by Georgius Agricola as early as 1529 that fluorspar lowers the melting point of minerals and reduces the viscosity of slags. This property of fluxing (Latin *fluoere,* to flow) is the origin of the name fluorine. The term fluorspar correctly describes ores containing substantial amounts of the mineral fluorite [*14542-23-5*], CaF_2, but the word fluorspar is often used for the mineral itself (see also Calcium compounds).

Calcium Fluoride (Fluorspar)

Significant mining of fluorspar began in England about 1775 and in the United States after 1820. Substantial use of fluorspar began about 1880 in the basic open-hearth process for making steel. Large increases in demand came with the need for fluorides in the aluminum industry, starting about 1900. Progress in the understanding of fluorine chemistry led to a large fluorine chemicals industry based on hydrogen fluoride made from fluorspar for the production of refrigerants (1930), alkylation catalysis for gasoline (1942), feed materials for nuclear energy generation (ca 1942), aerosol propellants (ca 1942), fluoroplastics (ca 1942), and fluorocarbons for soil-repellant surface treatments (early 1950s). Fluorspar is used directly in the manufacture and finishing of glass, in ceramics and welding fluxes, and in the extraction and processing of nonferrous metals.

In the geochemistry of fluorine, the close match in ionic radii of fluoride ion (0.136 nm), hydroxide ion (0.140 nm), and oxide ion (0.140 nm) allows a sequential replacement of oxygen by fluorine in a wide variety of minerals. This accounts for the wide dissemination of the element in nature. The ready formation of volatile silicon tetrafluoride, the pyrohydrolysis of fluorides to hydrogen fluoride, and the low solubility of calcium fluoride and of calcium fluorophosphates, have provided a geochemical cycle in which fluorine may be stripped from solution by limestone and by apatite to form the deposits of fluorspar and of phosphate rock (fluoroapatite, approx. $CaF_2.3Ca_3(PO_4)_2$ [*1306-01-0*]) which are the world's main resources of fluorine (1).

On average, fluorine is about as abundant as chlorine in the accessible surface of the earth including oceans. The continental crust averages about 650 ppm fluorine. Igneous, metamorphic, and sedimentary rocks all show abundances in the range of 200 to 1000 ppm. Fluorspar is the principal source of fluorine for industry today, but enormous amounts of fluorine are extractable from phosphate rock, and future sources may include topaz (an aluminum fluorosilicate), bastnaesite (rare earth fluorocarbonate), and by-product selection from the mining of sulfide ores and barite.

Fluorspar deposits are commonly epigenetic, ie, the elements moved from elsewhere into the country rock. For this reason, fluorine mineral deposits are closely associated with extensive fault zones. In the United States, significant fluorspar deposits occur in the Appalachian Mountains and in the mountainous regions of the West, but the principal source is in the faulted carbonate rocks of Illinois and Kentucky where ca 80% of U.S. fluorspar is produced.

Worldwide, large exploitable deposits of fluorspar (see Table 1) are found in Mexico, Spain, Newfoundland, the United Kingdom, and South Africa. France, Germany, and probably the USSR have substantial reserves.

Table 1. World Fluorspar Reserves, 1976 (in 10^6 Metric Tons)[a]

Mexico	37.2
Republic of South Africa	29.0
other regions of Africa	19.1
United States	14.5
USSR	14.5
United Kingdom	12.7
Spain	10.9
Italy	7.2
Canada	5.4
France	5.4
People's Republic of China	5.4
Mongolia	4.5
other Eastern Bloc countries	4.5
unspecified	30.7
Total	*201*

[a] Ref. 2.

Properties. Some of the important physical properties of calcium fluoride are listed in Table 2. Pure calcium fluoride is without color. However, natural fluorite can vary from transparent and colorless to translucent and white, wine-yellow, green, greenish blue, violet-blue, and sometimes blue, deep purple, bluish-black, and brown. These color variations are produced by impurities and by radiation damage (color centers). A deep black fluorite has been observed which contains radiation halos and releases elemental fluorine when crushed. The color of fluorite is often lost upon heating, sometimes with luminescence. Mineral specimens are usually strongly fluorescent, and the mineral thus gives its name to this phenomenon. Specimens vary from well-formed crystals (optical grade) to massive or granular forms.

The crystal structure of fluorite gives name to the fluorite crystal type. The lattice is face-centered cubic, with each calcium ion surrounded by eight fluoride ions situated at the corners of a cube, and each fluoride ion lies within a tetrahedron defined by four calcium ions (25). The bonding is ionic. The unit cell (space group O_h^5) can be pictured as made up of eight small cubes, each containing a fluoride ion, and the eight forming a cube with a calcium ion on each corner and one in the center of each face (Fig. 1). The lattice constant is 0.54626 nm at 25°C (13). The habit is usually cubic, less frequently octahedral, rarely dodecahedral. Cleavage on the [111] planes is perfect. The crystals are brittle with flat-conchoidal or splintery fracture. Luster is vitreous, becoming dull in massive varieties.

Systems of metal oxides with calcium fluoride usually have a simple freezing point–composition diagram, commonly with a eutectic point and without abnormal lowering of the melting point (27). When silicates are present, the systems become more complicated, and a striking decrease in the viscosity of the glassy melts is observed. The viscosity most likely decreases because of depolymerization of chains or networks of SiO_4 tetrahedra by the replacement of oxide ion by the singly charged fluoride ion which is close in size and electronegativity to oxide ion (1). The benefits of calcium fluoride as a metallurgical flux result from both the freezing point depression and the decrease in slag viscosity.

Although stable at atmospheric temperature, calcium fluoride is slowly hydrolyzed by moist air above 1200°C, presumably to CaO and HF. Hopes to use this pyrohy-

Table 2. Physical Properties of Calcium Fluoride

Property	Value	Reference
molecular weight	78.08	
composition, % Ca	51.33%	
% F	48.67%	
melting point, °C	1402	3
boiling point, °C	2513	4
heat of fusion, kJ/mol[a]	23.0	5
heat of vaporization at bp, kJ/mol[a]	335	6
vapor pressure at 2100°C, Pa[b]	1013	6
heat capacity, C_p, kJ/(mol·K)[a]		
solid at 25°C	67.0	7
solid at mp	126	8
liquid at mp	100	8
entropy at 25°C, kJ/(mol·K)[a]	68.87	7
heat of formation, solid at 25°C, kJ/mol[a]	−1215	9–10
free energy of formation, solid at 25°C, kJ/mol[a]	−1164	11
thermal conductivity, crystal at 25°C, W/(m·K)	10.96	12
density, g/cm		
solid at 25°C	3.181	13
liquid at mp	2.52	14
thermal expansion, average, 25 to 300°C, per K	22.3×10^{-6}	15
compressibility, at 25°C and 101.3 kPa (= 1 atm)	1.22×10^{-8}	16
hardness		
Mohs scale	4	
Knoop, 500-g load	158	17
solubility in water, g/L at 25°C	0.146	18
refractive index at 24°C, 589.3 nm	1.43382	19
dielectric constant at 30°C	6.64	20
electrical conductivity of solid $(\Omega\cdot cm)^{-1}$		
at 20°C	1.3×10^{-18}	21
at 650°C	6×10^{-5}	22
at mp	3.45	23
optical transmission range, nm	150 to 8000	24

[a] To convert J to cal, divide by 4.184.
[b] To convert Pa to mm Hg, multiply by 7.5×10^{-3}.

drolysis to prepare hydrogen fluoride from steam and fluorspar have not been realized, presumably because of construction materials problems. The reaction is, however, significant in guarding against fluoride loss in ceramic and metallurgical procedures. Calcium fluoride is not attacked by alkalies or by reactive fluorine compounds, but is decomposed by hot, high-boiling acids, as in the reaction with concentrated sulfuric acid in the manufacture of hydrogen fluoride. Calcium fluoride is slightly soluble in cold dilute acids, and rather more soluble in solutions of aluminum halides.

Preparation. Very pure calcium fluoride for the manufacture of special glasses is made by the reaction of hydrofluoric acid with precipitated calcium carbonate.

The large amount of fluorine values released from phosphate rock in the manufacture of fertilizers (qv) gives a strong impetus to develop fluorine chemicals production from this source (see Phosphoric acid). Additional incentive comes from the need to control the emission of fluorine-containing gases. Most of the fluorine values

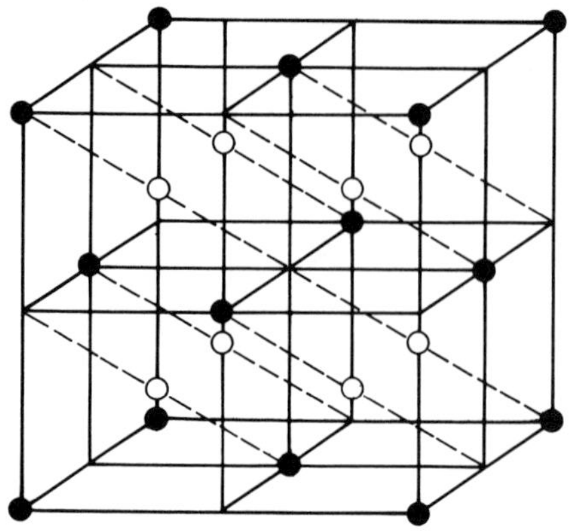

Figure 1. Structure of fluorite (26). Ca = ●, F = ○.

are scrubbed out as fluorosilicic acid, H_2SiF_6, which has limited usefulness. A procedure to convert fluorosilicic acid to calcium fluoride, developed by the U.S. Bureau of Mines, involves removing silica and impurities by precipitation with ammonia, followed by treating the ammonium fluoride filtrate with lime to form calcium fluoride. The ammonia evolved is recycled to the silica precipitation step (28).

Mining. In 1976 there were about 15 mines and 8 flotation and heavy-media plants operated by 11 companies, but only 4 of the beneficiating plants operated full time. In the United States, fluorspar extraction is almost always by shaft mining (29).

Fluorspar occurs in two distinct types of formation in the fluorspar district of southern Illinois and Kentucky: in vertical fissure veins and in horizontal bedded replacement deposits. A 61-m bed of sandstone and shale serves as a cap rock for ascending fluorine-containing solutions and gases. Mineralizing solutions come up the faults and form vein ore bodies where the larger faults are plugged by shale. Bedded deposits occur under the thick sandstone and shale roofs. Other elements of value associated with fluorspar ore bodies are zinc, lead, cadmium, silver, germanium, iron, and thorium. Ore has been mined as deep as 460 m in this district. Open-pit mining is used in Mexico and Kenya; and in the new workings at Pianciano, Italy.

Beneficiation. Preconcentration of crushed ore usually is accomplished by heavy media separation. Here, fluorite, barite, and valuable sulfide minerals are separated from waste in a cone containing a suspension of finely ground ferrosilicon (see Gravity concentration). The ratio of ferrosilicon to water is adjusted to give the suspension a density of 2.55–2.62 g/cm at the top of the cone and 2.85–3.1 g/cm at the bottom. Crude ore is introduced at the top of the cone. Heavy minerals such as fluorite, sulfides, and barite ($BaCO_3$) sink and are recovered at the bottom of the cone. Light minerals float and are carried away with the overflow. Ferrosilicon is recovered magnetically and returned to the process. The enriched material may have a composition of 40–50 wt % fluorite, 3–5 wt % zinc, and 20–30 wt % barite.

Froth flotation (qv) is the method of choice for the final purification of fluorite. In this procedure, air bubbles forced through a suspension of pulverized ore are induced to adhere to selected minerals and float them into a froth that is continuously skimmed off. From the flotation circuit, the fluorite is dewatered by filtration and dried in rotary kilns (29).

The fine-grained fluorspar produced at Pianciano, Italy, is concentrated only to metallurgical grade (80 wt % CaF_2). Preconcentration makes use of hydrocarbons and decanters. The concentrated pulp is converted to filter-press cake and then agglomerated into pellets before drying and calcining. Desulfurization is accomplished in the calciner by addition of a small amount of metallurgical coke and the maintenance of a high-oxygen atmosphere.

In steelmaking, the preferred form of fluorspar flux is a washed gravel, 0.6–5 cm in diameter, containing less than three percent water, and assaying 60 to 80% CaF_2 units (see below). Absence of fines is evidently necessary with oxygen blowing in the basic oxygen process. The higher CaF_2 ranges are hard to supply in large amounts from some sources of fluorspar, and utilization of fines remains a problem. For these reasons, fluorspar briquettes are produced on a large scale. In 1976, U.S. briquetting capacity was 510,000 metric tons per year. Fluorspar briquettes are formed between two counter-rotating rolls having pocketed or shaped mold faces. The briquettes are of uniform size, typically peach-stone shape, 2.5 cm long and 1.25 cm thick. Molasses is the usual binder, but tall oil, sodium silicate, or portland cement may be used. Additives such as MnO_2, Fe_2O_3, Al_2O_3, dolomite, hydrated lime, flue dust, feldspar, soda ash, olivine, ilmenite, and mill scale may be incorporated as specified by the steelmaker. The cost of briquetting may range from $5.50 to 13.23/metric ton (30) (see Pelleting and briquetting).

Economic Aspects. Pertinent statistics on the U.S. production, import, export, and consumption of fluorspar are given in Table 3. The United States for many years has relied on imports for more than 80% of its needs, most of them from Mexico, followed for many years by Spain and Italy. Since 1975, however, imports from the Republic of South Africa have increased so rapidly that it has become the second largest supplier. Imports from Mexico have dropped in recent years because Mexican export regulations favor domestic conversion of fluorspar to hydrogen fluoride for export to the United States.

Within the United States, the Illinois-Kentucky mining area is still the principal

Table 3. Fluorspar Statistics[a] (Thousands of Metric Tons)

	1951–1963[b]	1964–1968	1969–1973[b]	1974	1975	1977
United States production						
crude						
mine production	509	684	589	406	342	na
material milled or washed	489	599	619	371	364	na
beneficiated material recovered	179	220	206	188	120	na
finished[c]	185	228	222	182	127	138
imports[d] for consumption	498	788	1035	1212	953	885
exports[e]	0.81	7.6	6.6	5.3	1.2	
consumption[f]	601	936	1229	1382	1129	1024
World production	2023	2977	4418	4840	4620	4562

[a] Refs. 2, 31–33.
[b] Average.
[c] Shipments.
[d] Two-thirds acid spar (>97% CaF_2) and one-third metallurgical spar (97% CaF_2).
[e] Mainly to Canada.
[f] Reported.

source of fluorspar, producing 76% of finished fluorspar in 1964 and 71% in 1975 (31). Important discoveries may soon be exploited in Tennessee, Colorado, and the Lost River district of Alaska (2,32–33).

Table 4 gives a breakdown of world production of fluorspar. Table 5 gives U.S. prices between 1965 and 1978. Fluorspar is supplied in 22.5- and 45-kg bags and is shipped (34) in trucks (18–23 t), covered hopper cars (54–72 t), and barges (1080–1270 t, wet).

Grades, Quality Control. Fluorspar is marketed in four grades: metallurgical fluorspar (metspar) is sold as gravel, lump, or briquettes. The minimum acceptable assay is 60% effective calcium fluoride. The effective value is determined by subtracting from the contained calcium fluoride 2.5% for every percent of SiO_2 found in the complete analysis apparently based on the following stoichiometry (1):

$$2\,CaF_2 + SiO_2 \rightarrow 2\,CaO + SiF_4$$

Table 4. World Production of Finished Fluorspar, 1976 (Thousands of Metric Tons)[a]

Mexico	896
France	675
USSR	490
China	354
Mongolia	308
Republic of South Africa	288
Spain	255
United Kingdom	217
Thailand	195
United States	149
Czechoslovakia	90.0
German Democratic Republic	90.0
Canada	72.5
Federal Republic of Germany	66.1
Kenya	48.1
Tunisia	34.5
Romania	15.0
India	13.6
Republic of Korea	11.7
other countries[b]	12.9
Total	*4281*

[a] Ref. 2.
[b] In order of volume: Japan, Democratic Republic of Korea, Pakistan, Rhodesia, and Burma.

Table 5. U.S. Fluorspar Prices 1965–1978, $/Metric Ton

	1965[a]	1974	1976	1978
metallurgical, pellets, 70% effective CaF_2	41–43	77	92–101	101
ceramic				
88–90%		85–99	99–110	99–110
95–96%	35–37	93–105	105–117	105–120
97%		96–114	110–127	110–127
acid, dry basis 97% CaF_2	50	95–114	114–121	122–127

[a] 1965 analytical basis not exactly comparable with 1974–1978 basis.

The price standard is 70% effective calcium fluoride. As noted above, however, briquetted metallurgical fluorspar can have compositions tailored to the steelmaker's needs. In 1977 the average equivalent CaF_2 in briquettes was 75%.

Ceramic-grade fluorspar and acid-grade fluorspar have the typical analyses shown in Table 6. Both types are usually finely ground, the bulk of the powder passing a 0.23 mm (65 mesh) screen, and 22 to 81% held on a 44 μm (325 mesh) screen.

Optical-grade calcium fluoride, for special glasses and for growing single crystals, is supplied in purities up to 99.99% CaF_2, especially low in transition elements (36).

For process control and product specification, fluorspar is commonly analyzed for fluorine, calcium, silica, carbonate, sulfide, iron, barium, and where significant, for metal values.

Analytical Methods. Fluorite is readily identified by its crystal shape, usually simple cubes or interpenetrating twins, by its prominent octahedral cleavage, its relative softness, and the production of hydrogen fluoride when treated with sulfuric acid, evidenced by etching of glass. The presence of fluorite in ore specimens, or when associated with other fluorine-containing minerals, may be determined by x-ray diffraction.

Fluorine has for many years been determined by the Willard–Winters method in which finely ground ore, after removal of organic matter, is distilled with 72% perchloric acid in glass apparatus. The distillate, a dilute solution of fluorosilicic acid, is made alkaline to release fluoride ion, adjusted with monochloroacetic acid at pH 3.4, and titrated with thorium nitrate, using sodium alizarine sulfonate as indicator.

In recent years the direct determination of fluoride with ion-selective electrodes (qv) has allowed analysis of fluorspar without the tedious distillation step. The fluoride electrode uses a single crystal of lanthanum fluoride doped with europium in contact with the solution, and a high impedance detector similar to that used for glass electrode pH measurements. The finely ground fluorspar sample is dissolved in boiling aluminum chloride and hydrochloric acid solution, and the solution is buffered with ammonium sulfosalicylate. Ethylenediamine tetraacetic acid (EDTA) is added to displace fluoride ion from complexes with positive ions (37–39).

Fluorspar assay may be completed by fluoride determination alone, since the mineralogical grouping rarely includes fluorine minerals other than fluorite. Calcium

Table 6. Typical Analyses of Ceramic- and Acid-Grade Fluorspar, %[a]

Assay	Ceramic	Acid
CaF_2	90.0–95.5	96.5–97.5
SiO_2	1.2–3.0	1.0
$CaCO_3$	1.5–3.4	1.0–1.5
MgO		0.15
Pb		0.02
Zn		0.02
Fe_2O_3	0.10	0.10
P_2O_5		0.03
$BaSO_4$	1.5–3.2	0.2–1.3
R_2O_3	0.15–0.25	0.1–0.3

[a] Refs. 34–35.

can be determined as oxalate or by ion-selective electrodes (40). Silica can be determined in the residue from solution in perchloric acid–boric acid mixture by measuring the loss in weight on fuming off with hydrofluoric acid. Another method for determining silica in fluorspar is the ASTM Standard Test Method E 463-72.

Carbonate is measured by evolution of carbon dioxide on treating the sample with sulfuric acid. The gas train should include a silver acetate absorber to remove hydrogen sulfide, a magnesium perchlorate drying unit, and a CO_2-absorption bulb. Sulfide is determined by distilling hydrogen sulfide from an acidified slurry of the sample into an ammoniacal cadmium chloride solution, and titrating the precipitated cadmium sulfide iodimetrically.

Health and Safety Factors. The low solubility of calcium fluoride reduces the fluoride-related toxicity. Thus the saturation concentration of fluoride ion in water solution is 8.1 ppm, to be compared with the recommended water fluoridation level of one ppm fluoride ion. However, the solubility of calcium fluoride in stomach acid is higher, and continued oral ingestion of calcium fluoride could produce symptoms of fluorosis. NIOSH has recommended a maximum of 2.5 mg/m^3 for inorganic fluorides in air (time-weighted average). This is probably conservative for calcium fluoride (41).

A significant hazard results on contact of calcium fluoride with high concentrations of strong acids, causing evolution of toxic concentrations of hydrogen fluoride. A less recognized danger is the production of hydrogen fluoride by pyrohydrolysis when calcium fluoride is exposed to water vapor at high temperature, as in a direct-fired kiln.

Beneficiation facilities require customary air and water pollution control systems, with efficient control of dust emission, treatment of process water and proper disposal of tailings. In handling finished fluorspar, operators must avoid breathing fluorspar dust and contact of fluorspar with acids, observe proper disposal of spills, and use respirators and other protective equipment. Prolonged contact with fluorspar may irritate the skin and eyes. Fluorspar may be harmful when ingested (34).

Uses. The use patterns for fluorspar in the United States are shown in Table 7. The production of hydrogen fluoride accounts for about one-half of the fluorspar consumed in the United States. About one-half of the hydrogen fluoride is converted into chlorofluorocarbons to be used as refrigerant or pressurizing gases. The latter use was effectively discontinued in the United States in 1978 because the high chemical stability of these gases allowed their accumulation in the upper atmosphere with a presumed adverse effect on ozone concentration (2,42–43). Some fluorocarbons, such as vinyl fluoride, CH_2CHF, vinylidene fluoride, CH_2CF_2, and tetrafluoroethylene, CF_2CF_2, are converted to polymers. For 1 t of hydrogen fluoride, 2.2 t of acid-grade fluorspar are needed; 1 t of hydrogen fluoride produces ca 3 t of chlorofluorocarbon.

The aluminum industry uses fluorspar by way of hydrogen fluoride to make aluminum fluoride, AlF_3, and cryolite, Na_3AlF_6, for electrochemical reduction cells. The growth of this use has been slowed by increasing efficiency and recycling of fluorine values. In 1975 an average of ca 27.5 kg of fluorspar was used per metric ton of aluminum, but modern smelters have reduced this to 15.0 kg/t. A new chloride process now in the pilot plant stage may reduce fluorine requirements even further (2,44–45).

The demand for fluorspar to produce hydrogen fluoride for the manufacture of uranium hexafluoride remained steady in the years 1975 through 1977. The acid, and

Table 7. Uses of Fluorspar in the United States[a], 10^3 Metric Tons/Yr

Use	1975	1977
hydrofluoric acid	611	519
organic fluorine chemicals	305	254
aluminum industry	159	114
uranium hexafluoride	13	14
others[b]	134	136
steelmaking	471	504
basic oxygen furnaces	325	na
open-hearth furnaces	81	na
electric furnaces	65	na
iron and steel foundry	27	27
glass and fiberglass	8.4	9.07
enamel and pottery	1.8	1.81
primary aluminum and magnesium	0.66	0.90
welding-rod coatings	0.66	0.90
other uses or products	8.26	9.07
Total	*1129*	*1063*

[a] Refs. 2, 31–33.
[b] Includes steel pickling, petroleum alkylation, glass etching, and inorganic fluorides other than aluminum.

elemental fluorine derived from it, is used to convert U_3O_8 eventually to UF_6, which is the compound used in the gaseous diffusion process to separate the isotopes of uranium (see Nuclear reactors).

In steelmaking, the sharp rises in fluorspar demands observed in the 1960s because of the introduction of the basic oxygen process have tapered off because of more efficient use and the partial replacement of fluorspar by other fluxes. Thus the use of additional limestone can reduce the need for fluorspar, but the lower cost of the former may be offset by an increase in time between furnace tappings. In 1977, fluorspar demand was 5.05 kg/t in open-hearth furnaces, 4.50 kg/t in basic oxygen furnaces, and 2.30 kg/t in electric furnaces. The average for the steel industry was 4.20 kg/t crude steel (46).

The other demands for fluorspar have remained steady, as shown in Table 7, with acceptable substitutes for fluorspar not in evidence. Iron foundries use 6–8 kg fluorspar per ton metal. Glassmaking, largely in opaque glasses, consumes 22.6–226 kg fluorspar per 453 kg sand; and enamels up to 15 wt % (47).

A small but artistically significant use of fluorspar is in the production of vases, cups, and other ornamental objects, popularly known as Blue John, after the Blue John Mine, Derbyshire, England. Optical quality fluorite, sometimes from natural crystals, but more often artificially grown, is important in use as infrared transmission windows and lenses (24) and optical components of high-energy laser systems (48).

BIBLIOGRAPHY

"Calcium Fluoride" in *ECT* 1st ed., under "Fluorine Compounds, Inorganic," Vol. 6, pp. 689–692, by H. C. Miller, Pennsylvania Salt Manufacturing Co.; "Calcium Fluoride" in *ECT* 2nd ed., under "Fluorine Compounds, Inorganic," Vol. 9, pp. 573–582, by John F. Gall, Pennsalt Chemicals Corporation.

1. D. R. Shawe, ed., *Geology and Resources of Fluorine in the United States,* U.S. Geol Survey Prof. Paper 933, Washington, 1976, pp. 1–5, 18, 19, 82–87.

2. B. Coope, *Ind. Miner. London,* 39, 67 (Feb. 1978).
3. B. Porter and E. A. Brown, *J. Am. Ceram. Soc.* **45,** 49 (1962).
4. D. A. Schulz and A. W. Searcy, *J. Phys. Chem.* **67,** 103 (1963).
5. G. Petit and A. Cremieu, *Compt. Rend.* **243,** 360 (1956).
6. O. Ruff and L. Leboucher, *Z. Anorg. Allg. Chem.* **219,** 376 (1934).
7. S. S. Todd, *J. Am. Chem. Soc.* **71,** 4115 (1949).
8. B. F. Naylor, *J. Am. Chem. Soc.* **67,** 150 (1945).
9. F. D. Rossini and co-workers, *Natl. Bur. Stand. U.S. Circ.* **500** (1952).
10. G. T. Armstrong and L. A. Krieger, *Natl. Bur. Stand. U.S. Rept.* **7192,** 92 (July 1, 1961).
11. M. A. Mikhailov, *Izv. Sib. Otd. Akad. Nauk. SSSR* **1962**(9), 116; *Chem. Abstr.* **58,** 6258a (1963).
12. K. A. McCarthy and S. S. Ballard, *J. Appl. Phys.* **36,** 1410 (1960).
13. H. E. Swanson and E. Tatge, *Natl. Bur. Stand. U.S. Circ. 539* **1,** 69 (1953).
14. A. V. Grosse and C. S. Stokes, *U.S. Dep. Commer. Off. Tech. Serv. PB Rep. 161460,* (1960); *Chem. Abstr.* **56,** 9433a (1962).
15. O. J. Whittemore, Jr. and N. N. Ault, *J. Ceram. Soc.* **39,** 443 (1956).
16. E. W. Washburn, ed., *International Critical Tables,* Vol. 3, McGraw-Hill Book Co., Inc., New York, 1929, p. 50.
17. S. S. Ballard, L. S. Combes, and K. A. McCarthy, *J. Opt. Soc. Am.* **42,** 684 (1952).
18. D. W. Brown and C. E. Roberson, *J. Res. U.S. Geol. Surv.* **5,** 509 (1977).
19. *Natl. Bur. Stand. U.S., Tech. News Bull.* **47,** 91 (1963).
20. J. L. Pauley and H. Chessin, *J. Am. Chem. Soc.* **76,** 3888 (1954).
21. E. W. Washburn, ed., *International Critical Tables,* Vol. 6, McGraw-Hill Book Co., New York, 1929, p. 154.
22. R. W. Ure, Jr., *J. Chem. Phys.* **26,** 1365 (1957).
23. T. Baak, *J. Chem. Phys.* **29,** 1195 (1958).
24. *Data Sheet—IR Transmission Materials,* Barnes Engineering Co., Instrument Div., Stamford, Conn.
25. A. F. Wells, *Structural Inorganic Chemistry,* 3rd ed., Clarendon Press, Oxford, 1962, p. 77.
26. E. S. Dana and W. S. Ford, *A Textbook of Mineralogy,* 4th ed., John Wiley & Sons, Inc., New York, 1932, p. 41.
27. H. Krainer, *Radex Rundsch.,* 19 (1949); *Chem. Abstr.* **43,** 8630 (1949).
28. R. C. Kirby and A. S. Prokopovitsh, *Science* **191,** 717 (Feb. 1976).
29. *Minerva Mines,* Allied Chemical, Industrial Chemicals Div., Morristown, N.J., 1975.
30. H. B. Wood, *Eng. Min. J.* 81 (July 1978).
31. R. H. Singleton and J. E. Shelton, "Fluorspar" in *Minerals Yearbook 1975,* Vol. 1, U.S. Bureau of Mines, Washington, D.C., 1977, pp. 633–651.
32. H. B. Wood, *Eng. Min. J.* **178,** 132 (Mar. 1977).
33. G. Montgomery, *Eng. Min. J.* **179,** 92 (Mar. 1978).
34. P. L. Braekner, Private communication, Allied Chemical, Industrial Chemicals Div., Morristown, N.J., Nov. 1978.
35. *Data Sheet—Fluorspar,* Reynolds Chemicals, Richmond, Va., March 1978.
36. *Atomergic Chemetals Corp. Catalog,* Plainview, N.Y., Sept. 1978.
37. *Analytical Methods Guide,* 6th ed., Orion Research, Inc., Cambridge, Mass., Aug. 1973, p. 18.
38. R. T. Oliver and A. G. Clayton, *Anal. Chim. Acta,* **51,** 409 (1970).
39. R. B. Fisher, *J. Chem. Educ.,* **51,** 387 (1974). (Note: reference to lithium fluoride on p. 389 is in error. It should read lanthanum fluoride).
40. A. Hulanicki and M. Trojanowicz, *Anal. Chim. Acta* **68,** 155 (1974).
41. *Criteria for a Recommended Standard—Occupational Exposure to Inorganic Fluorides,* NTIS Document PB-246692, National Institute for Occupational Safety and Health, Rockville, Md., 1975.
42. *Drug Cosmet Ind., 44* (June 1978).
43. *Chem. Eng. News* **56**(43), 15 (Oct. 23, 1978).
44. W. A. Anderson and W. E. Haupin, "Aluminum and Aluminum Alloys" in M. Grayson and D. Eckroth, eds., *Kirk-Othmer Encylopedia of Chemical Technology,* 2nd ed., Vol. 2, John Wiley & Sons, Inc., New York, 1977, pp. 139, 157–158.
45. G. O. Bruno, *Min. Eng.* **30,** 1562 (Nov. 1978).
46. *Mineral Industries Survey—Fluorspar,* U.S. Bureau of Mines, June 13, 1978, p. 1.
47. *Industrial Minerals and Rocks,* 4th ed., Am. Inst. of Mining, Metallurgical and Petroleum Engineers, 1975, pp. 377–378.

48. C. B. Willingham and R. T. Newberg, *Exploratory Development of Fusion Cast Calcium Fluoride for 1.06 Micrometer Pulsed Laser Optics,* Progress Report, Sept. 25 to Dec. 25, 1976, NTIS Document COO-4029-2, Raytheon Co., for U.S. Energy Research and Development Administration, Jan. 1977.

JOHN F. GALL
Philadelphia College of Textiles and Science

COBALT

Cobalt Difluoride

Cobalt difluoride [*10026-17-2*], CoF_2, is prepared commercially by the reaction of cobalt carbonate with anhydrous hydrogen fluoride. The partially hydrated product is a lavender–pink, lumpy, solid mass that is dried and pulverized to obtain pink, fine powder. The di [*13455-27-1*], tri [*13762-15-7*], and tetra [*13817-37-3*] hydrates of cobalt difluoride have been obtained by the reactions of freshly prepared cobalt(II) hydroxide, cobalt(II) oxide, and cobalt(II) carbonate with aqueous hydrogen fluoride (1). Reaction of $CoCl_2$ with anhydrous hydrogen fluoride or heating hydrated CoF_2 under an atmosphere of anhydrous HF, or in the presence of P_2O_5 or H_2SO_4, produces contaminated CoF_2, may cause pollution problems, and seems to be economically unsound.

Cobalt difluoride is sparingly soluble in anhydrous HF, 0.036% at 14.2°C (2); moderately soluble in 13.5% aqueous HF, 9.6% at 20°C; and slightly soluble in water, 1.36% at 20°C. The melting point is about 1200°C, the boiling point is about 1400°C, and the specific gravity is 4.46 (2).

CoF_2 is used primarily for manufacture of CoF_3. Its price in 1979 was ca $80/500 g. Its shipping classification is NHDS (nonhazardous dry solid).

Cobalt Trifluoride

Cobalt trifluoride [*10026-18-3*], CoF_3, is a light brown powder. It is readily obtained by the action of elemental fluorine on $CoCl_2$ at 250°C or on CoF_2 at 150–180°C. Direct fluorination of CoF_2 produces 99.9% CoF_3 in quantitative yields.

The specific gravity of CoF_3 is 3.88 and the bulk density is 1.15. Melting and boiling points have not been reported owing to its dissociation at elevated temperature ($2\ CoF_3 \xrightarrow{600°C} 2\ CoF_2 + F_2$). It is a very hygroscopic material and reacts vigorously with water to liberate oxygen and a black precipitate containing cobalt. It reacts spontaneously with elements such as Na, Al, As, P, S, and I_2. Cobalt trifluoride is easily reduced to CoF_2 when heated at 400°C in an atmosphere of H_2.

As a "hard" fluorinating reagent (3), CoF_3 is used for the replacement of hydrogen by fluorine in halocarbons and hydrocarbons (4), the saturation of double bonds with fluorine (5–6), and the conversion of aromatic compounds to perfluorocyclic compounds. The principal by-products from the CoF_3 fluorination process are partially fluorinated hydrocarbons, and short-chain fragments. Cobalt trifluoride rarely causes polymerization of hydrocarbons.

Cobalt trifluoride converts metal oxides to higher valency metal fluorides. It is used internationally in nuclear technology for the estimation of uranium ore (mostly U_3O_8) (7):

$$U_3O_8 + 18\ CoF_3 \rightarrow 3\ UF_6 + 18\ CoF_2 + 4\ O_2$$

CoF_3 is a very powerful oxidizing agent and should be kept away from combustible materials and heat.

Commercial usage of cobalt trifluoride in the United States in 1979 was less than 225 kg.

The toxicity of CoF_2 and CoF_3 is similar to other cobalt and fluoride compounds. All cobalt salts are suspected carcinogens of connective tissue and lungs (8).

BIBLIOGRAPHY

"Cobalt Compounds," in *ECT* 1st ed., under "Fluorine Compounds, Inorganic," Vol. 6, p. 693, by F. D. Loomis, Pennsylvania Salt Manufacturing Co.; "Cobalt" in *ECT* 2nd ed., under "Fluorine Compounds, Inorganic," Vol. 9, pp. 582–583, by Wayne E. White, Ozark-Mahoning Co.

1. I. G. Ryss, *The Chemistry of Fluorine and Its Inorganic Compounds,* Part II, State Publishing House for Scientific, Technical and Chemical Literature, Moscow U.S.S.R., 1956; English Trans AEC-tr-3927, Office of Technical Services, U.S. Department of Commerce, Washington, D.C., 1960, pp. 659–665.
2. A. W. Jache and G. H. Cady, *J. Phys. Chem.* **56,** 1106 (1952).
3. D. T. Meshri and W. E. White, "Fluorinating Reagents in Inorganic and Organic Chemistry" in *Proceedings, George Cady Symposium, Milwaukee, Wis., June 1970.*
4. R. D. Fowler and co-workers, *Ind. Eng. Chem.* **39,** 292 (1947).
5. D. A. Rausch, R. A. Davis, and D. W. Osborne, *J. Org. Chem.* **28,** 494 (1963).
6. U.S. Pat. 2,670,387 (Feb. 23, 1954), H. B. Gottlieb and J. D. Park (to E. I. du Pont de Nemours & Co., Inc.).
7. Private communication, C. F. Trivisonno, Goodyear Atomic Corporation, Piketon, Ohio, Aug. 1978.
8. N. I. Sax, *Dangerous Properties of Industrial Materials,* 4th ed., Van Nostrand Reinhold Co., 1975.

DAYAL T. MESHRI
Ozark-Mahoning Company,
a subsidiary of the Pennwalt Corp.

COPPER

Copper(I) Fluoride

Several claims have been made regarding the synthesis (1–5), physical properties (6), and applications (7) of copper(I) fluoride [*13478-14-6*], CuF or Cu_2F_2. However its physical existence remains questionable. Mass spectral studies (8) of CuF_2 show no evidence of the presence of CuF. The only ions observed during this study were CuF^+ and CuF_2^+. The appearance potentials for these ions indicate that they were due to the ionization of $CuF_{2(g)}$ rather than that of $CuF_{(g)}$. Thermodynamic calculations (9) also show that copper(I) fluoride is thermodynamically unstable at RT. Therefore, it can be assumed that the earlier work describing copper(I) fluoride as a red crystalline powder referred to a mixture of copper(II) fluoride and metallic copper.

Copper(II) Fluoride

Anhydrous copper(II) fluoride [*7789-19-7*], CuF_2, can be prepared by permitting copper oxide or basic copper carbonate to react with anhydrous HF and then dehydrating it in the stream of anhydrous HF at the elevated temperatures. The other convenient method for the preparation of ultrapure (10), anhydrous CuF_2, is a direct fluorination of commercially available CuOHF [*13867-72-6*] or the action of a mixture of HF and BF_3 on $CuF_2.2H_2O$ [*13454-88-1*] (11).

Anhydrous copper(II) fluoride, mp 785 ± 10°C, is a white, very hygroscopic solid which forms blue dihydrate; $CuF_2.2H_2O$ when exposed to moist air. It is soluble in pyridine and ethyl acetate. The solubility of CuF_2 in water, 4.7 g/100 g H_2O at 20°C, increases with rise in temperature or in the presence of hydrofluoric acid (5). Although its solubility in 21.2% HF is 12.1% at 20°C, it is only sparingly soluble in anhydrous HF (0.01 g/100 g HF at 12.4°C) (12). It reacts with ammonia to form $CuF_2.5NH_3$ and its density is 4.85 ± 0.05 g/mL.

Copper(II) fluoride is used as a catalyst for the skeletal rearrangement of olefins (13), low temperature isomerization of pentane and hexane (14), as an electrode (15–18) in high-energy-density batteries (qv), as a fluorinating reagent (19, 20), catalyst for the synthesis of heterocyclic tetraaromatics (21), as a fungicide (qv) (22), as a selective herbicide (qv) (23), and as a termite repellent (24). The dihydrate may be used as a flux in casting gray iron.

BIBLIOGRAPHY

"Copper Compounds" under "Fluorine Compounds, Inorganic" in *ECT* 1st ed., Vol. 6, p. 693, by F. D. Loomis, Pennsylvania Salt Manufacturing Co.; "Copper" under "Fluorine Compounds, Inorganic" in *ECT* 2nd ed., Vol. 9, pp. 583–584, by Wayne E. White, Ozark-Mahoning Company.

1. J. J. Berzelius, *Pogg. Annal.* **I,** 8 (1824).
2. C. Poulenc, *Ann. Chim. Phys.* **2**(7), 1 (1894).
3. F. M. Jaeger, *Z. Anorg. Chem.* **27,** 26 (1901).
4. F. Ebert and H. Woitinek, *Z. Anorg. Chem.* **210,** 269 (1933).
5. H. Von Wartenberg, *Z. Anorg. Allgem. Chem.* **241,** 381 (1939).
6. H. M. Haendler and co-workers, *J. Am. Chem. Soc.* **76,** 2178 (1954).
7. U.S. Pat. 2,593,464 (Apr. 22, 1956), J. H. Krause (to Standard Oil of Indiana).

8. R. A. Kent, J. D. McDonald, and J. L. Margrave, *J. Phys. Chem.* **70,** 874 (1966).
9. I. H. Ryss, *Zh. Fiz. Khim.* **29,** 936 (1955).
10. J. R. Lundquist, *NASA CR-72571 Final Report,* Pacific Northwest Laboratories, Seattle, Washington, June 12, 1969.
11. U.S. Pat. 2,782,099 (Feb. 19, 1957), D. A. McCaulay (to Standard Oil of Indiana).
12. A. W. Jache and G. H. Cady, *J. Phys. Chem.* **56,** 1106 (1952).
13. U.S. Pat. 3,751,513 (Aug. 7, 1973), J. J. Tazuma (to Goodyear Tire & Rubber Co.).
14. Fr. Pat. 2,157,083 (July 6, 1973), T. Bernard (to Institut Francais du Petrole, des Carburants et Lubricants).
15. Ger. Offen. 2,211,416 (Jan. 18, 1973), K. D. Beccu, J. Mosetti, and P. Jonville (to Center National d'Etudes Spatiales).
16. Ger. Offen. 2,215,210 (Oct. 19, 1972), O. S. Savinow (to Honeywell Inc.).
17. U.S. Pat. 3,953,232 (Apr. 27, 1976), W. L. Roth and G. C. Farrington (to General Electric Co.).
18. J. H. Kennedy and J. C. Hunter, *J. Electrochem. Soc.* **123**(1), 10 (1976).
19. B. Leng and J. H. Moss, *J. Fluorine Chem.* **8,** 165 (1976).
20. J. H. Moss, R. Ottie, and J. B. Wilford, *J. Fluorine Chem.* **3,** 317 (1973).
21. T. Kauffmann and R. Otter, *Angew Chem.* **88,** 513 (1976).
22. H. Martin, R. L. Wain, and E. H. Wilkinson, *Ann. Appl. Biol.* **29,** 412 (1942).
23. I. G. Ryss, *The Chemistry of Fluorine and Its Compounds,* State Publishing House for Scientific, Technical and Chemical Literature, Moscow, USSR, 1956, Part II, AEC-tr-3927, Office of Technical Services, U.S. Department of Commerce, Washington, D.C., 1960, p. 643.
24. G. N. Wolcott, *P.R. Agric. Exp. Stn. Bull.* **73,** (1947).

DAYAL T. MESHRI
Ozark-Mahoning Company,
a subsidiary of the Pennwalt Corp.

GERMANIUM

Germanium forms both a difluoride and a tetrafluoride. It also forms a stable hexafluorogermanate complex ion, GeF_6^{2-}, that is present in the aqueous acid and a number of salts.

Germanium Difluoride

Germanium difluoride [*13940-63-1*] is a white solid, mp 110°C, and d_{23} = 3.7 g/cm^3. The difluoride can be vacuum distilled. In a mass spectrometer, ions corresponding to $(GeF_2)_n^+$, where n = 1–4, were observed at 361–403 K (1). At higher temperatures, GeF_2 disproportionates to GeF_4(g), Ge(s), and GeF(g). Presumably the GeF [*39717-71-0*] formed is unstable and subsequently condenses and disproportionates. The initially reported (2) orange–red solid became red–brown and finally the black of metallic Ge. The difluoride deliquesces (2) in moist air producing germanium(II) hydroxide. The fluoride is soluble in aqueous hydrofluoric acid giving a solution with the reducing properties expected of divalent germanium. The difluoride reacts with aqueous solutions of alkali metal fluorides to produce trifluorogermanites (3), eg, $CsGeF_3$ [*72121-41-6*] and $KGeF_3$ [*72121-42-7*]. The GeF_3^- ion is oxidized to GeF_6^{2-} either by oxygen in neutral solution or by its reduction of H^+ to hydrogen in hydrofluoric acid solution. Germanium difluoride is soluble in ethanol (2), forms a reversible complex with diethyl ether (2), and forms a $GeF_2.OS(CH_3)_2$ [*72121-40-5*] complex which is decomposed at 240°C (3).

Germanium difluoride can be prepared by reduction (2) of GeF_4 by metallic germanium or by reaction (1) of stoichiometric amounts of Ge and HF in a sealed vessel at 225°C.

Germanium Tetrafluoride

The tetrafluoride [*7783-58-6*] is a gas with a garlic-like odor and a reported (4) triple point of −50°C and 404.1 kPa (4.0 atm) and a vapor pressure near 100 kPa (ca 1 atm) at −36.5°C. Germanium tetrafluoride fumes strongly in air and is hydrolyzed in solution to form $GeFe_6^{2-}$ ions. Germanium tetrafluoride can be prepared (4–5) by thermal decomposition of $BaGeF_6$ [*60897-63-4*]. Direct fluorination of germanium has been reported to give GeF_4 of higher purity (2). Germanium tetrafluoride acts as a Lewis acid (6–7) to form complexes with many donor molecules. The tetrafluoride is commercially available in research quantities.

Fluorogermanates

Fluorogermanic acid [*16950-43-9*] solutions are prepared by reaction of germanium dioxide and hydrofluoric acid or by hydrolysis of germanium tetrafluoride. Addition of potassium fluoride, barium chloride (5), or other salts results in hexafluorogermanates such as K_2GeF_6 [*7783-73-5*] or $BaGeF_6$, both of which are stable at temperatures up to 500°C where $BaGeF_6$ starts to decompose to GeF_4 and BaF_2.

BIBLIOGRAPHY

"Germanium" in *ECT* 2nd ed., under "Fluorine Compounds, Inorganic," Vol. 9, pp. 584–585, by W. E. White, Ozark-Mahoning Company.

1. K. F. Zmbov, J. W. Hastie, R. Hauge, and J. L. Margrave, *Inorg. Chem.* **7,** 608 (1968).
2. N. Bartlett and K. C. Yu, *Can. J. Chem.* **39,** 80 (1961).
3. E. L. Muetterties, *Inorg. Chem.* **1,** 342 (1962).
4. L. M. Dennis and A. W. Laubengayer, *Z. Phys. Chem.* **130,** 520 (1927).
5. C. J. Hoffman and H. S. Gutowsky, *Inorg. Syn.* **4,** 147 (1953).
6. E. L. Muertterties, *J. Am. Chem. Soc.* **82,** 1082 (1960).
7. R. C. Aggarwal and M. Onyszchuk, *J. Inorg. Nucl. Chem.* **30,** 3351 (1968).

CHARLES B. LINDAHL
Ozark-Mahoning Company,
a subsidiary of the Pennwalt Corp.

HALOGENS

The halogen fluorides are binary compounds of bromine, chlorine, and iodine with fluorine. Of the eight known compounds listed in Table 1, only bromine trifluoride, chlorine trifluoride, and iodine pentafluoride have been of commercial importance. Properties and applications have been reviewed in refs.1–6, and reactions with organic compounds in ref. 7. A more recent review (8) covers the methods of preparation, properties, and analytical chemistry.

The halogen fluorides are best prepared by the reaction of fluorine with the corresponding halogen. They are very powerful oxidizing agents, with chlorine trifluoride approaching the reactivity of fluorine itself. They offer an advantage over fluorine in that they can be stored as liquids in steel containers and, unlike fluorine, high pressure is not required. Bromine trifluoride is used as an oxidizing agent in cutting tools used in deep oil-well drilling, whereas chlorine trifluoride is used to convert uranium to UF_6 in nuclear fuel processing.

Except for iodine pentafluoride, halogen fluorides have no commercial importance as fluorinating agents. Their extreme reactivity and the accompanying energy release of the reaction can be sufficient to disrupt C—C bonds and can result in explosive reactions or fires. In addition, both halogens are generally introduced into the compounds, giving rise to a complex mixture of products.

Physical Properties

The physical properties of the halogen fluorides are given in Table 2. Calculated thermodynamic properties can be found in the JANAF thermochemical tables (23).

Bromine Monofluoride. Bromine monofluoride is red to red–brown (4) and is very unstable, disproportionating rapidly into bromine and higher fluorides. Therefore, the measurement of its physical properties is difficult and the values reported in Table 2 are only approximate.

Bromine Trifluoride. Bromine trifluoride is a colorless liquid. The commercial grade is usually amber to red because of slight bromine contamination. The molecule has a distorted T structure (24). Infrared spectra data have been determined (25–27); vapor pressure data are given in ref. 28.

Table 1. The Halogen Fluorides Listed in Descending Order of Reactivity

Compound	CAS Registry No.	Formula
chlorine pentafluoride	*[13637-63-3]*	ClF_5
chlorine trifluoride	*[7790-91-2]*	ClF_3
bromine pentafluoride	*[7789-30-2]*	BrF_5
iodine heptafluoride	*[16921-96-3]*	IF_7
chlorine monofluoride	*[7790-91-2]*	ClF
bromine trifluoride	*[7787-71-5]*	BrF_3
iodine pentafluoride	*[7783-66-6]*	IF_5
bromine monofluoride	*[13863-59-7]*	BrF

Table 2. Physical Properties of the Halogen Fluorides[a]

Property	BrF	BrF_3	BrF_5	ClF	ClF_3	ClF_5	IF_5	IF_7
boiling point, °C	20	125.7	40.9	−100.1	11.75	−13.1	102	5.5
melting point, °C	−33	8.8	−60.6	−155.6	−76.3	−103	8.5	4.5
liquid density at 25°C, g/cm^3		2.803	2.463[b]	1.620[b]	1.825[b]	1.790[c]	3.252	2.669
critical temperature, °C					154.5	142.6		
$-\Delta H^{\circ}_{f(g)}$ at 25°C, kJ/mol[d]	58.5	255.4	443.9	56.4	164.5	254.6	839.3	961.0
$-\Delta G^{\circ}_{f(g)}$ at 25°C, kJ/mol[d]	73.6	229.1	351.5	57.7	124.4	163.0	771.6	841.4
heat of vaporization, kJ/mol[d]		42.8	30.6	20.1	27.50	22.21	35.92	24.7
heat of fusion, kJ/mol[d]		12.01	5.66		7.60		11.21	
specific heat, gas, J/(mol·K)[d]	32.9	66.5		32.0	65.2		99.1	136.3
specific heat, liquid, J/(mol·K)[d]		124.5						
specific conductivity of liquid at 25°C, Ω·cm		8.0×10^{-3}	9.1×10^{-8}	1.9×10^{-7}[e]	4.9×10^{-9}	1.25×10^{-9}[f]	5.4×10^{-6}	10^{-9}

[a] Compiled from refs. 9–22.
[b] At boiling point.
[c] At 20°C.
[d] To convert J to cal, divide by 4.184.
[e] At 145 K.
[f] At 256 K.

Bromine Pentafluoride. Bromine pentafluoride is a colorless liquid with the molecular structure of a tetragonal pyramid (5). The index of refraction n_D is 1.3529 (29). Infrared spectra have been determined (25,30); vapor pressure data are given in ref. 10.

Chlorine Monofluoride. Chlorine monofluoride is a colorless gas that condenses to a liquid with a slight yellow cast and freezes to a white solid. The infrared spectrum of gaseous chlorine monofluoride and the Raman spectrum of the liquid have been studied (31); vapor pressure data are given in ref. 10.

The following constant was developed for the equilibrium between chlorine monofluoride and fluorine gas (32):

$$K_P = \frac{(P_{ClF})(P_{F_2})}{(P_{ClF_3})}$$

The following values for K_P were determined:

Temperature, °C	*Pa (mm Hg)*
250	30 (0.22)
300	240 (1.8)
350	1450 (10.9)

Chlorine Trifluoride. Chlorine trifluoride is a pale yellow liquid or a colorless gas. It freezes to a white solid and undergoes a transition in the solid state at −82.66°C (10). The infrared and Raman spectra have been studied (26–27,33–34); vapor pressure data are given in ref. 35. The viscosity of the liquid is 0.448 mPa·s (= cP) at 290 K, and the surface tension is 26.6 mN/m (= dyn/cm) at 273 K (36). The density of the solid is 2.530 g/cm^3 at 153 K (37). The vapors of chlorine trifluoride are nonideal and this has been attributed to the following equilibrium:

$$Cl_2F_6 \rightleftharpoons 2\ ClF_3$$

The equilibrium constant at 24.2°C is 2.84 kPa (21.3 mm Hg) for the following (10):

$$K_P = \frac{P_{Cl_2F_6}}{(P_{ClF_3})^2}$$

Chlorine Pentafluoride. Chlorine pentafluoride is a colorless gas at RT. The ir and Raman spectra of the liquid and gas phase have been studied (30,33); vapor pressure data are given in ref. 17.

Iodine Pentafluoride. Iodine pentafluoride is a straw colored liquid; the ir and Raman spectra of the gas phase have been studied (18,39–40); vapor pressure data are given in refs. 13 and 41.

Iodine Heptafluoride. Iodine heptafluoride is a colorless liquid; the ir and Raman spectra of the gas have been studied (40 42); vapor pressure data are given in ref. 18.

Chemical Properties

Reactions. ***With Metals.*** All metals react to some extent with halogen fluorides, although several react only superficially to form an adherent fluoride film of low permeability that is a protection against further reaction. However, this protective capacity is lost at elevated temperatures; hence, each metal has a temperature above which it will continue to react. Mild steel reacts rapidly above 250°C. Copper and nickel lose their ability to resist reaction above 400 and 750°C, respectively.

Metals that form no protective fluoride film react readily with the halogen fluorides. Chlorine trifluoride reacts with finely divided metals at elevated temperatures to give HgF_2, AsF_5, CaF_2, TiF_3, CoF_3, PtF_5, and PbF_{3-4} (43). Titanium and molybdenum alloys and niobium metal react vigorously with ClF_3 (44). Molybdenum and tungsten react with BrF_3 to form the volatile MoF_6 and WF_6 (45–46). Chlorine trifluoride (47) converts Nb and Ta to their pentafluorides, and Mo and Re to their hexafluorides.

Uranium is converted by ClF_3, BrF_3, and BrF_5 to UF_6. The recovery of uranium from irradiated fuels has been the subject of numerous and extensive investigations sponsored by atomic energy agencies in a number of countries (48–56). The fluorides of the nuclear fission products are nonvolatile, hence the volatile UF_6 can be removed by distillation (see Nuclear reactors; Uranium).

The rapid reaction of ClF_3 and BrF_3 with metals is the basis of their commercial use in cutting pipe in deep oil wells (57–59). In this application, the pipe is cut by the high temperature reaction of the halogen fluoride and the metal (see under Uses).

With Nonmetals. Few elements withstand the action of interhalogen compounds at elevated temperatures and many react violently at or below ambient temperatures. The oxidation of the element proceeds to its highest valence state, whereas the halogen other than fluorine is reduced either to the element or a lower-valent interhalogen derivative. The oxidizing capacity of the interhalogens varies substantially from compound to compound. For example, chlorine trifluoride reacts vigorously with virtually every element at RT; on the other hand, iodine pentafluoride has a much milder oxidizing power. Thus chlorine trifluoride oxidizes xenon to xenon fluoride whereas iodine pentafluoride does not react (10). Furthermore, all stable halogen fluorides, except iodine pentafluoride, oxidize radon between −195 and 25°C (60) (see Helium-group gases, compounds).

In general, reactions of halogens with halogen fluorides yield mixtures (4). Bromine pentafluoride reacts with iodine at RT and with chlorine at 250–300°C, giving mixtures of interhalogen compounds. Bromine pentafluoride has been stored with bromine in steel cylinders at RT for extended periods of time without appreciable reaction. At elevated temperatures BrF_3 is formed. Bromine and chlorine trifluoride give bromine trifluoride (60–80% yields) and chlorine (61). Iodine and chlorine trifluoride produce IF_5 and ICl (61). Chlorine reacts with chlorine trifluoride to produce chlorine monofluoride. Bromine and iodine react with chlorine monofluoride to produce BrF_3 and IF_5, respectively. Bromine reacts with IF_5 on warming to give IBr and BrF_3 (4). Chlorine reacts with IF_7 to give ClF and interhalogens of iodine and chlorine (4).

Halogen fluorides react with sulfur, selenium, tellurium, phosphorus, silicon, and boron at RT to form the corresponding fluorides. Slight warming may be needed to initiate the reactions (4) but once started they proceed rapidly to completion accompanied by heat and light. The lack of protective film formation allows complete reaction.

With Inorganic Compounds. In an investigation of the reactions of BrF_3 with oxides (62–64), little or no reaction was found with the oxides of Be, Mg, Ce, Ca, Fe, Zn, Zr, Cd, Sn, Hg, Th, and the rare earths, whereas the oxides of Mo and Re formed stable oxyfluorides. Manganese dioxide reacted incompletely but $KMnO_4$ released oxygen quantitatively. Complete replacement of oxygen took place with oxides of B, Ti, V, Cr, Cu, Ge, As, Se, Nb, Sb, Te, I, Ta, W, Tl, Pb, Bi, and U at 75°C. An analytical method for the determinaton of oxygen in the latter group of oxides was developed based on these results.

Oxygen was partially replaced when P_2O_5, V_2O_5, and CrO_3 were dissolved in IF_5 to form POF_3, VOF_3, and CrO_2F_2. With WO_3 and MoO_3, $WO_3.IF_5$ and $2MoO_3.3IF_5$ complexes were formed (65). Reaction of excess IF_5 with $KMnO_4$ gives MnO_3F, IOF_3 [*19058-78-7*], and IO_2F [*28633-62-7*] (65).

Water reacts violently with all halogen fluorides. The hydrolysis process can be moderate by cooling or dilution. In addition to HF, the products may include oxygen, free halogens (except for fluorine), and oxyhalogen acids.

Fused silica and Pyrex glass are not significantly attacked by halogen fluorides up to 100°C if HF is absent.

Salts of halides other than fluorides react with halogen fluorides to produce the corresponding metal fluoride and release the free higher halogen. Filter paper moistened with KI solution darkens readily in the presence of ClF_3 and the bromine fluorides. This serves as a sensitive detector for leaks in equipment containing these halogen fluorides. If a metal exhibits more than one valence, reactions of halogen fluorides with halides (including fluorides) yields the fluoride in which the metal is usually at its highest valence. Chlorine trifluoride converts silver salts to AgF_2, cobalt compounds to CoF_3, and so on. Such reactions are useful in the preparation of fluorinated organic materials and also regenerate fixed-bed fluorinating agents without using fluorine itself (66). A comparison of the efficacy with which various halogen fluorides convert cobalt(II) chloride to cobalt(III) fluoride is indicated below (67).

fluorinating agent	ClF_3	BrF_5	BrF_3	IF_5
solid product	100% CoF_3	55% CoF_3	45% CoF_3	
		45% CoF_2	55% CoF_2	72% CoF_2

With Organic Compounds. Most organic compounds react vigorously with incandescence or even explosively with ClF_3 and BrF_3 (7). For this reason, only the less reactive iodine pentafluoride is used as fluorinating agent to any extent.

Inert diluents in which the halogen fluorides are soluble, such as carbon tetrachloride and methylene dichloride (68), have been used for control in liquid-phase reactions. Anhydrous hydrogen fluoride is a good diluent, since it does not react with halogen fluorides but is miscible in all proportions with them (69). Control of vapor-phase reactions may be improved by diluting the halogen fluoride with an inert gas such as nitrogen or argon. However, any reaction of the halogen fluorides with an organic compound in either the gas or liquid phase should be approached with extreme caution.

Bromine trifluoride in bromine solution reacts smoothly with bromofluoroethanes to give a clean, progressive substitution of the bromine by fluorine with no replacement of the hydrogen (70). The relative ease of replacement of bromine in various groups is $CBr_3 > CBr_2F > CHBr_2 > CF_2Br > CHBrF > CH_2Br$.

The reactions with IF_5 are more amenable to control to give good yields of identifiable products and lower losses owing to oxidative fragmentation. For example, the reaction of IF_5 with tetrafluoroethylene produces the telomer perfluoroethyl iodide [*354-64-3*]:

$$2\,I_2 + IF_5 + 5\,CF_2{=}CF_2 \rightarrow 5\,CF_3CF_2I$$

Yields exceed 98% based on $CF_2{=}CF_2$ using SbF_5 as a catalyst (71).

Aryl and alkyl isothiocyanates are converted in good yields by IF_5 in pyridine to thiobis(*N*-trifluoromethylamines) (72–73):

$$RN{=}C{=}S + IF_5 \rightarrow \underset{CF_3}{\underset{|}{RN}}\underset{CF_3}{\underset{|}{SNR}}$$

Fluorination of aromatic isothiocyanates occurs much more readily than of alkyl isothiocyanates. Alcohols treated with IF_5 in DMF give 30–70% yields of their respective formates.

Iodine pentafluoride fluorinates CCl_4 at RT to give $CClF_3$ and traces of CCl_2F_2 (74). It reacts with CHI_3 to yield CHF_3 and $CHIF_2$ (75) and with CI_4 to form C_2F_2 (75) and CIF_3 (76). With CBr_4 at 90°C, IF_5 forms some CBr_3F, 83% CBr_2F_2 and a little $CBrF_3$ (76).

Liquid Halogen Fluorides as Reaction Media. Bromine trifluoride and iodine pentafluoride are highly dimerized and behave as ionizing solvents:

$$(BrF_3)_2 \rightleftharpoons BrF_2^+ + BrF_4^-$$

$$(IF_5)_2 \rightleftharpoons IF_4^+ + IF_6^-$$

Antimony pentafluoride dissolves in both to form $BrF_2^+SbF_6^-$ and $IF_4^+SbF_6^-$ which act as acids. Potassium fluoride likewise forms $KBrF_4$ [*15705-87-0*] and KIF_6 [*20916-97-6*] which are both stable, white, crystalline solids (3,77–78). They dissociate at 200°C to KF and the corresponding halogen fluoride. Other salts are formed similarly (62,78–82).

Some of the acids and bases of these systems listed in Table 3.

The use of ClF_3 and BrF_3 as ionizing solvents has been studied (85–86). At 100°C and elevated pressures, significant yields of $KClF_4$ [*19195-69-8*], $CsClF_4$ [*15321-04-7*], $RbClF_4$ [*15321-10-5*], $KBrF_6$ [*32312-22-4*], $RbBrF_6$ [*32312-22-4*], and $CsBrF_6$ [*26222-92-4*] were obtained. Chlorine trifluoride showed no reaction with lithium fluoride or sodium fluoride.

Table 3. Some Acids and Bases Derived from Halogen Fluorides

Name	CAS Registry No.	Formula	Ref.
Acid			
difluorobromine hexafluoroantimonate	[*19379-47-6*]	BrF_2SbF_6	64
bis(difluorobromine) hexafluorostannate	[*72229-86-8*]	$(BrF_2)_2SnF_6$	64
difluorobromine hexafluoroniobate	[*72229-87-9*]	BrF_2NbF_6	83
difluorobromine hexafluorotantalate	[*35967-87-4*]	BrF_2TaF_6	83
difluorobromine hexafluorobismuthate	[*36608-81-8*]	BrF_2BiF_6	83
tetrafluoroiodine hexafluoroantimonate	[*41646-48-4*]	IF_4SbF_6	84
Base			
potassium hexafluorobromate	[*32312-22-4*]	$KBrF_6$	78
silver tetrafluorobromate	[*35967-89-6*]	$AgBrF_4$	78
barium tetrafluorobromate	[*35967-90-9*]	$Ba(BrF_4)_2$	78
potassium hexafluoroiodate	[*20916-97-6*]	KIF_6	77

Manufacture

Bromine Trifluoride. Bromine trifluoride is produced commercially by the reaction of fluorine with bromine in a continuous gas-phase process where the ratio of fluorine to bromine is maintained close to 3:1. It is also produced in a liquid-phase batch reaction where fluorine is added to liquid bromine at a temperature below the boiling point of bromine trifluoride.

Chlorine Trifluoride. Chlorine trifluoride is produced commercially by the continuous gas-phase reaction of fluorine and chlorine in a nickel reactor at ca 290°C. The ratio of fluorine to chlorine is maintained slightly in excess of 3:1 to promote conversion of the chlorine monofluoride to chlorine trifluoride. Sufficient time in the reactor must be provided to maintain high conversions to chlorine trifluoride. Temperature control is also critical since the equilibrium shift of chlorine trifluoride to chlorine monofluoride and fluorine is significant at elevated temperatures.

Iodine Pentafluoride. Iodine pentafluoride is produced by the reaction of iodine and fluorine. Since iodine has a high melting point, the reaction is either performed in a solvent or the reaction is maintained at a temperature where the iodine is liquid. In a continuous process using a solvent (87), ca 1% I_2 is dissolved in IF_5 and passed to a reactor where it is contacted with F_2 gas. The IF_5 is continuously discharged from the reactor where a small portion is taken off as product and the larger portion of the stream is recycled.

In another process (88), fluorine gas reacts under pressure with liquid I_2 held above its melting point (113°C) but below a temperature (150°C) that would result in the formation of significant amounts of IF_7. Fluorine is added continuously until all the iodine has been converted and yields of IF_5 in excess of 95% are reported. The reaction pressure is ca 300 kPa (3 atm) so that the IF_5 produced in the reaction is maintained as a liquid.

Economic Aspects

United States production of bromine trifluoride is several metric tons per year mostly used in oil-well cutting tools. Air Products and Chemicals, Inc. is the only United States producer. The 1979 price was ca $30/kg.

United States chlorine trifluoride production is several hundred metric tons per year with most of the product used in nuclear fuel processing. A large production plant for chlorine trifluoride was operated in Germany during World War II with a reported capacity of 5 t/d (89–90). Currently, Air Products and Chemicals, Inc. is the only United States producer. The 1979 price was ca $30/kg.

United States production of iodine pentafluoride is currently several hundred metric tons per year and could approach a thousand metric tons in the future. The two United States producers are Air Products and Chemicals, Inc. and Allied Chemical Corporation. The 1979 price was ca $25/kg.

Shipping, Specifications, and Analytical Methods

Bromine trifluoride is commercially available at a minimum purity of 98% (91). Free Br_2 is maintained at less than 2%. Other minor impurities are HF and BrF_5. Free Br_2 content estimates are based on color, with material containing less than 0.5% Br_2

having a straw color, and ca 2% Br_2 an amber-red color. Fluoride content can be obtained by controlled hydrolysis of a sample and standard analysis for fluorine content. Bromine trifluoride is too high boiling and reactive for gas chromatographic analysis. It is shipped as a liquid in steel cylinders in quantities of 91 kg or less. The cylinders are fitted with either a valve or plug to facilitate insertion of a dip tube. Bromine trifluoride is classified as an oxidizer and poison by DOT.

Chlorine trifluoride is commercially available at 99% min purity (91). Gas chromatography using a Teflon-packed stainless steel column detects impurities such as F_2, HF, ClF, and Cl_2 (92–93). A gas density detector eliminates the corrosion problems associated with thermal conductivity detectors. Chlorine trifluoride is shipped as a liquid under its own vapor pressure in steel cylinders in quantities of 82 kg per cylinder or less, and is available in cylinders fitted with dip tubes for liquid withdrawal. Chlorine trifluoride is classified as an oxidizer and poison by DOT.

Iodine pentafluoride is commercially available at a minimum purity of 98% (91). Iodine heptafluoride is the principal impurity and maintained at less than 2%. Free I_2 and HF are minor impurities. For analysis, gas chromatography through a Teflon-packed stainless steel column can be employed or a wet chemical analysis of the hydrolyzed IF_5 for free iodine, iodides, and fluorides can be used. Iodine pentafluoride is shipped as a liquid in steel cylinders and is available in various quantities up to 1350 kg cylinders. It is classified as an oxidizer and poison by DOT.

Handling

The halogen fluorides are highly reactive compounds and must be handled with extreme caution (94–98). The more reactive compounds, such as bromine trifluoride and chlorine trifluoride are hypergolic oxidizers and react violently and sometimes explosively with many organic and inorganic materials at RT. At elevated temperatures, they cause immediate ignition of most organic substances and many metals.

Materials of Construction. Nickel, Monel, copper, mild steel, 304 stainless steel, and aluminum have been found to be suitable metals of construction for handling halogen fluorides (44). Silver solder is acceptable; lead solder is not recommended. Nickel and Monel are more suitable for elevated temperatures. Steel is not dependable above 150°C. Gaskets may be made of soft copper or calcium fluoride-impregnated polytetrafluoroethylene. Packing and gasketing should have smooth surfaces and the surfaces should be free from organic greases and embedded impurities, which may ignite in the presence of halogen fluorides.

Equipment should be carefully and completely degreased and passivated with low concentrations of fluorine or the gaseous halogen fluoride before use. Special care should be taken that valves are completely disassembled and each part carefully cleaned.

Disposal. Moderate amounts of chlorine trifluoride or other halogen fluorides may be destroyed by burning with a fuel such as natural gas, hydrogen, or propane. The resulting fumes may be vented to tall stacks or to water or caustic scrubbers. Alternatively, they can be diluted with an inert gas and scrubbed in a caustic solution.

For further information on disposal of halogen fluorides see refs. 94–97.

Toxicity

The time-weighted average (TWA) concentrations for chlorine trifluoride and bromine pentafluoride have been established by the ACGIH in 1977 at 0.4 mg/m^3 and 0.7 mg/m^3, respectively, for daily exposure for 8 h/d.

NIOSH reports (99) the following toxicity levels for chlorine trifluoride: LC_{50} rats = 96 ppm/h; 480 ppm/40 min (100); LC_{50} mice = 176 ppm/h; and LC_{50} monkeys = 227 ppm/h.

No toxicity data have been reported on the other halogen fluorides, but they should all be regarded as highly toxic materials which are extremely irritating to all living tissue.

Uses

Chlorine trifluoride is utilized in the processing of nuclear fuels to convert uranium to gaseous uranium hexafluoride.

Bromine trifluoride and chlorine trifluoride are used in oil-well tubing cutters, (58,101). Chemical cutter tools are commercially available for use in wells at any depth. The cutter consists of three tubular chambers. The top chamber carries an explosive charge, the middle chamber contains the halogen fluoride, and the lower chamber contains a catalyst. At the extreme end is the cutter head which guides the halogen fluoride against the pipe to be cut. Arranged around the head is a row of evenly spaced orifices or nozzles. The cable used to lower the cutter into the hole also serves to carry an electric charge to set off the explosive. The force drives the chemical into the head where it jets out of the orifices under enormous pressure to impinge against the inner walls of the tube to be cut. The catalyst raises the temperature of the halogen fluoride to trigger a high-speed reaction so that the tube is cut in a fraction of a second. The cut is clean and unflared. Tension on the pipe at the top of the well aids in completing the separation (see Petroleum).

Iodine pentafluoride is an easily storable liquid source of fluorine with little of the hazards associated with other fluorine sources. It is used as a selective fluorinating agent for organic compounds. For example, it adds iodine and fluorine to tetrafluoroethylene to produce a useful telomer.

BIBLIOGRAPHY

"Halogen Fluorides" under "Fluorine Compounds, Inorganic" in *ECT* 1st ed., Vol. 6, pp. 694–695, by H. S. Booth Western Reserve University, and J. T. Pinkston, Harshaw Chemical Company; "Halogen Fluorides" under "Fluorine Compounds, Inorganic" in *ECT* 2nd ed., Vol. 9, pp. 585–598, by H. S. Halbedel, Harshaw Chemical Company.

1. H. S. Booth and J. T. Pinkston, *Chem. Rev.* **41,** 421 (1947).
2. H. J. Emeleus, "Halogen Fluorides—Recent Advances" in J. H. Simons, ed., *Fluorine Chemistry,* Vol. 2, Academic Press, Inc., New York, 1954, pp. 39–49.
3. A. G. Sharpe, *Q. Rev. Chem. Soc.* **4,** 115 (1950).
4. H. S. Booth and J. T. Pinkston, "The Halogen Fluorides" in J. H. Simons, ed., *Fluorine Chemistry,* Vol. 1, Academic Press, Inc., New York, 1950, pp. 189–224.
5. H. C. Clark, *Chem. Rev.* **58,** 869 (1958).
6. J. C. Bailer and co-workers, *Comprehensive Inorganic Chemistry,* Pergamon Press, Compendium Publishers, Elmsford, N.Y., 1973, Vol. 3, pp. 749–763.

7. W. K. R. Musgrave, "The Halogen Fluorides—Their Properties and Uses in Organic Chemistry" in M. Stacey, J. C. Tatlow, and A. G. Sharpe, eds., *Advances in Fluorine Chemistry,* Vol. 1, Academic Press, Inc., New York, 1960, pp. 1–28.
8. N. S. Nikolaev and co-workers, *Khimiyn Galoidnykh,* Soldinerii Ftorn, Moscow, USSR, 1968 (U.S. translation *AD-702-974*).
9. J. A. Dean, ed., *Lange's Handbook of Chemistry,* 11th ed., McGraw-Hill Book Co., New York, 1973, p. 4:139.
10. L. Stein in V. Gutmann, ed., *Halogen Chemistry,* Vol. 1, Academic Press, Inc., New York, 1967, p. 133.
11. D. Pilipovich and co-workers, *Inorg. Chem.* **6,** 1918 (1967).
12. Supplement to *Mellor's Comprehensive Treatise on Inorganic and Theoretical Chemistry,* Suppl. II, Part I, Longmans, Green and Co., London, Eng., 1956.
13. H. Selig, C. W. Williams, and G. J. Moody, *J. Phys. Chem.* **71,** 2739 (1967).
14. *National Bureau of Standards Technical Note 270-3,* U.S. Government Printing Office, Washington, D.C., 1968.
15. L. Stein, *J. Phys. Chem.* **66,** 288 (1962).
16. R. C. King and G. T. Armstrong, *J. Res. Natl. Bur. Stand.* **74A,** 769 (1970).
17. H. H. Rogers and co-workers, *J. Chem. Eng. Data* **13,** 307 (1968).
18. D. W. Osborne, F. Schreiner, and H. Selig, *J. Chem. Phys.* **54,** 3790 (1971).
19. O. Ruff and H. Krug, *Z Anorg. Alleg. Chem.* **190,** 270 (1930).
20. R. C. King and G. T. Armstrong, *J. Res. Nat. Bur. Stand.* **74A,** 769 (1970).
21. A. A. Banks, H. J. Emeleus, and A. A. Woolf, *J. Chem. Soc.,* 2861 (1949).
22. M. T. Rogers, J. L. Speirs, and M. B. Panish, *J. Am. Chem. Soc.* **78,** 3288 (1956).
23. *JANAF Thermochemical Tables, NSRDA-NBS37* 2nd ed., NBS, Washington, D.C., 1971.
24. D. W. Magnuson, *J. Chem. Phys.* **27**(1), 223 (1957).
25. H. M. Haendler and co-workers, *J. Chem. Phys.* **22,** 1939 (1954).
26. H. Selig, H. H. Claassen, and J. H. Holloway, *J. Chem. Phys.* **52,** 3517 (1970).
27. R. A. Frez, R. L. Redington, and A. L. K. Aljiburty, *J. Chem. Phys.* **54,** 344 (1971).
28. J. W. Grisard and G. D. Oliver, *The Vapor Pressure and Heat Vaporization of Bromine Trifluoride, K-25, Plant Report K-766,* U.C.C. Nuclear Co., June 8, 1951.
29. R. D. Long, J. J. Martin, and R. C. Vogel, *Ind Eng. Chem. Data Ser.* **3,** 28 (1958).
30. G. M. Begun, W. H. Fletcher, and D. F. Smith, *J. Chem. Phys.* **42,** 2236 (1965).
31. E. A. Jones, T. F. Parkinson, and T. G. Burke, *J. Chem. Phys.* **18,** 235 (1950).
32. H. Schmitz and H. J. Schumacker, *Z. Naturforsch Teil 2,* 363 (1947).
33. E. A. Jones, T. F. Parkinson, and R. B. Murray, *J. Chem. Phys.* **17,** 501 (1949).
34. K. Schaefer and E. Wicke, *Z. Elektrochem.* **52,** 205 (1948).
35. J. W. Grisard, H. A. Burnhardt, and G. D. Oliver, *J. Am. Chem. Soc.* **73,** 5725 (1951).
36. A. A. Banks, A. Davies, and A. J. Rudge, *J. Chem. Soc.,* 732 (1953).
37. R. D. Burbank and F. N. Bensey, *J. Chem. Phys.* **21,** 602 (1953).
38. K. O. Christie and co-workers, *Inorg. Chem.* **11,** 1679 (1972).
39. L. E. Alexander and I. R. Beattie, *J. Chem. Soc. A,* 3091 (1971).
40. H. H. Claassen, E. L. Gasner, and H. Selig, *J. Chem. Phys.* **49,** 1803 (1968).
41. C. J. Schack and co-workers, *J. Phys. Chem.* **72,** 4697 (1968).
42. R. K. Khanna, *J. Mol. Spectroscopy* **8,** 134 (1962).
43. W. Huckel, *Nachr. Akad. Wiss. Gottingen Math. Phys. Kl.* 36 (1946).
44. J. C. Grigger and H. C. Miller, *Met. Protect.,* 33 (Sept. 1964).
45. B. Cox, D. W. A. Sharp, and A. G. Sharpe, *Proc. Chem. Soc.* 1242 (1956).
46. N. S. Nikolaev and A. A. Opalovskii, *Zh. Neorgan. Khim.* **4,** 1174 (1959).
47. N. S. Nikolaev and E. G. Ippolitov, *Dokl. Akad. Nauk* **134,** 358 (1960).
48. H. A. Bernhardt, E. J. Barber, and R. A. Gustison, *Ind. Eng. Chem.* **51,** 179 (1959).
49. J. F. Ellis and L. H. Brooks, *U. K. Atomic Energy Authority* **8111-D,** (1959).
50. R. A. Gustison and co-workers, "The Chlorine Trifluoride Process" in F. R. Bruce and co-eds., *Progress in Nuclear Chemistry, Series 3: Process Chemistry,* McGraw-Hill Book Co., Inc., New York, 1956, pp. 281–285.
51. U.S. Pat. 3,012,849 (Dec. 12, 1961), F. L. Horn (to U.S. Atomic Energy Commission).
52. W. J. Mechan and co-workers, *Chem. Eng. Prog.* **53**(2), 72-F-77F (1957).
53. L. Stein and R. Vogel, *Ind. Eng. Chem.* **48,** 418 (1956).

54. G. Strickland, F. L. Horn, and R. Johnson, *U.S. At. Energy Comm. BNL-471 (T-107) 1957; BNL-457 (I-21)*, (1957).
55. U.S. Pat. 3,825,650 (July 23, 1974), R. A. Gustison and co-workers (to U.S. Atomic Energy Commission).
56. Gr. Pat. 2,209,628 (1971), P. Cousin and co-workers.
57. E. McGhee, *Oil Gas J.* **54**(14), 67 (1955).
58. Ger. Pat. 1,029,770 (May 14, 1958), W. G. Sweetman.
59. *Oil Forum*, 332 (Sept. 1955).
60. L. Stein, *J. Am. Chem. Soc.* **91,** 5396 (1969).
61. R. Burnett and R. E. Banks, "Fluorine Chemistry," *paper presented at British Symposium*, Nov. 30, 1949.
62. H. J. Emeleus and A. A. Woolf, *J. Chem. Soc.*, 164 (1950).
63. H. R. Hoekstra and J. J. Katz, *Anal. Chem.* **25,** 1608 (1953).
64. A. A. Woolf and H. J. Emeleus, *J. Chem. Soc.*, 2865 (1949).
65. E. E. Aynsley, R. Nichols, and P. L. Robinson, *J. Chem. Soc.*, 623 (1953).
66. E. G. Rochow and I. Kukin, *J. Am. Chem. Soc.* **74,** 1615 (1952).
67. J. F. Gall and co-workers, "Interhalogen Compounds of Fluorine," *papers presented at annual ACS Meeting, New York,* Sept. 1947.
68. U.S. Pat. 1,961,622 (June 5, 1934), H. S. Nutting and P. S. Petrie (to The Dow Chemical Co.).
69. U.S. Pat. 2,918,434 (Dec. 22, 1959), J. G. Gall and C. E. Inman (to Pennsalt Chemicals Corp.).
70. R. A. Davis and E. C. Larsen, *J. Org. Chem.* **32,** 3478 (1967).
71. U.S. Pat. 3,123,185 (May 5, 1964), R. E. Parsons (to E. I. du Pont de Nemours & Co., Inc.).
72. T. E. Stevens, *Tetrahedron Lett.* (17), 16 (1959).
73. T. E. Stevens, *J. Org. Chem.* **26,** 3451 (1961).
74. O. Ruff and R. Keim, *Z. Anorg. Allg. Chem.* **201,** 245 (1931).
75. J. H. Simons, R. L. Bond, and R. E. McArthur, *J. Am. Chem. Soc.* **62,** 3477 (1940).
76. A. A. Banks and co-workers, *J. Chem. Soc.*, 2188 (1948).
77. H. J. Emeleus and A. G. Sharpe, *J. Chem. Soc.*, 2206 (1949).
78. A. G. Sharpe and H. J. Emeleus, *J. Chem. Soc.*, 2135 (1948).
79. A. A. Woolf and H. J. Emeleus, *J. Chem. Soc.*, 1050 (1950).
80. A. A. Woolf, *J. Chem. Soc.*, 1053 (1950).
81. A. G. Sharpe, *J. Chem. Soc.*, 2901 (1949).
82. A. G. Sharpe and A. A. Woolf, *J. Chem. Soc.*, 798 (1951).
83. V. Gutman and H. J. Emeleus, *J. Chem. Soc.*, 1046 (1950).
84. A. A. Woolf, *J. Chem. Soc.*, 3678 (1950).
85. U.S. Pat. 3,143,391 (Aug. 4, 1964), T. L. Hurley, R. O. MacLaren, and E. D. Whitney (to Olin Matheson Chemical Corp.).
86. E. D. Whitney and co-workers, *J. Am. Chem. Soc.* **86,** 2583 (1964).
87. U.S. Pat. 3,367,745 (Feb. 6, 1968), H. G. Tepp (to Allied Chemical Corp.).
88. U.S. Pat. 4,108,966 (Aug. 22, 1978), J. T. Lileck (to Air Products and Chemicals, Inc.).
89. *Chem. Ind.* **57,** 1084 (1945).
90. H. R. Neumark, *Trans. Electrochem. Soc.* **91,** 367 (1947).
91. Bulletin, *Inorganic Fluorine Compounds*, Air Products and Chemicals, Inc., Allentown, Pa., 1974.
92. J. C. Million, C. W. Weber, and P. R. Kuehn, *Gas Chromatography of Some Corrosive Halogen-Containing Gases, Report No. K-1639,* Union Carbide Corporation, Nuclear Division, New York, 1966.
93. J. F. Ellis and G. Iveson, *The Application of Gas–Liquid Chromatography to the Analysis of Volatile Halogen and Interhalogen Compounds, Gas Chromatography,* Butterworths, London, pp. 300–309.
94. *Chlorine Trifluoride Handling Manual, AD266,121,* U.S. Department of Commerce, OTS, Washington, D.C., 1961.
95. R. L. Farrar, *Safe Handling of Chlorine Trifluoride, Report K-1416,* U.C.C. Nuclear Company, Oak Ridge, Tenn., 1960.
96. J. M. Siegmund and co-workers, *NASA Document N62-14523, AD 281-818,* NASA, Washington, D.C., 1962.
97. "Chemical Rocket/Propellant Hazards" in *Liquid Propellant Handling, Storage, and Transportation, CPLA Publication No. 194,* Vol. III, CPLA, May 1970, Chapt. 8.
98. *Handling Hazardous Materials, NASA SP-5032,* NASA, Washington, D.C., Chapt. 4.

99. *Registry of Toxic Effects of Chemical Substances,* 1976 ed., U.S. Dept. of Health, Education, and Welfare, Washington, D.C., 1976.
100. H. J. Horn and R. J. Weir, *A.M.A. Arch. Ind. Health* **12,** 515 (1955); **13,** 340 (1956).
101. *Chemical Cutter Bulletin,* McCullough Services, Baroid Division, NL Industries, Inc., Houston, Tex.

ANDREW J. WOYTEK
Air Products and Chemicals, Inc.

HYDROGEN

As the starting point for most fluorine products and in volume of production, hydrogen fluoride [*7664-39-3*] is the most important manufactured compound of fluorine. Both anhydrous hydrogen fluoride and its aqueous solutions are sold for use directly and as intermediates in making other fluorine-containing materials. The aqueous solution is usually called hydrofluoric acid; the anhydrous compound is preferably called hydrogen fluoride.

Anhydrous hydrogen fluoride is a colorless liquid or gas (bp 19.54°C), completely water soluble, and fuming strongly in contact with the atmosphere. The formula weight is 20.006, but hydrogen bonding between molecules produces extensive polymerization, and the liquid and gas show large departures from ideal behavior. The compound can cause human injury in any form of contact, and informed control of hazards in handling and use must be observed.

Hydrogen fluoride may have been made as early as 1670 when Schwankhard of Nürnberg etched glass with fluorspar and acid. Marggraf in 1764 showed that the etchant was a gas; Scheele showed that a new acid had been discovered; and Ampère in 1810 and Davy in 1813 and 1814 proved that the compound now known as hydrogen fluoride contained a new element (1).

The earlier quite limited uses for hydrofluoric acid, such as glass etching and polishing, foundry scale removal, and minor production of metal fluorides, predominated until the 1930s. In the period 1935 to 1940, the large increase in aluminum production brought an equivalent requirement for hydrogen fluoride, employed captively for aluminum refining. Anhydrous hydrogen fluoride was first made by Frémy in 1856 (2), but there was no significant commercial manufacture until 1931 when the first bulk shipment was made by Sterling Products Company for use in making chlorofluorohydrocarbons as refrigerating fluids. Under the stimulus of wartime need, new major uses for anhydrous hydrogen fluoride developed, including alkylation catalysis to manufacture aviation gasoline (Phillips Petroleum Company, 1942); manufacture of chlorofluorohydrocarbons for use in pressurized packaging (USDA, 1942); and the manufacture of uranium fluorides (Manhattan District Engineers, 1942). The growth of all of these industries, and of stainless steel pickling and of fluorine-containing plastics brought the volume of hydrogen fluoride production to the current high levels.

Physical Properties

The properties of anhydrous hydrogen fluoride are summarized in Tables 1–2. Additional data are available in the references cited. Enthalpy–temperature and entropy–temperature diagrams are shown in refs. 9 and 19; spectrographic data in ref.

Table 1. Properties of Anhydrous Hydrogen Fluoride

Property	Value	Reference
formula weight	20.006	
composition, wt %		
H	5.038	
F	94.96	
molecular weight		
saturated vapor at boiling point	78.24	3
saturated vapor at 100°C	49.08	3
boiling point at 101.3 kPa (1 atm), °C	19.54	4
melting point, °C	−83.55	5
cryoscopic constant K_f, mol/kg solvent, °C	1.52	5
ebullioscopic constant K_b, mol/kg solvent, °C	1.9	6
density		
liquid at 0°C, g/cm^3	1.002	
liquid at 25°C, g/cm^3	0.9576	7
saturated vapor at 25°C, mg/cm^3	3.553	3
saturated vapor at 34°C, mg/cm^3	3.979	3
vapor pressure at 25°C, MPa[b]	122.9	3
heat of vaporization at boiling point, 101.3 kPa (1 atm), J/mol[c]	7493	4
heat of fusion at melting point, J/mol[c]	3931	8
heat capacity, constant pressure		
liquid at 16°C, J/(mol·°C)[a,c]	50.6	8
vapor at 22°C, 101.3 kPa (1 atm), J/(mol·°C)[a,c]	456	4
heat of formation		
ideal gas at 25°C, kJ/mol[c]	−272.5	9–10
free energy of formation		
ideal gas at 25°C, kJ/mol	−274.6	9–10
entropy		
ideal gas at 25°C, J/(mol·°C)[a,c]	173.7	9–10
critical temperature, °C	188	11
critical pressure, MPa[b]	6.480	11
critical density, g/cm^3	0.29	11
viscosity, mPa·s (= cP) at 0°C	0.256	12
surface tension, mN/m (= dyn/cm) $\times 10^{-4}$ at 0°C	10.1	
refractive index,		
liq, 58930 nm, 25°C	1.1574	13
molar refractivity, cm^3	2.13	13
dielectric constant, at 0°C	83.6	14
dipole moment, HF molecule, C·m[d]	6.10×10^{-30}	15
bond length, H–F, nm	9.2	6
ionic character, %	43	16
electric conductivity, $(\mu\Omega\cdot cm)^{-1}$ at 0°C		6

[a] mol = 20.01 g.
[b] To convert MPa to atm, divide by 0.101.
[c] To convert J to cal, divide by 4.184.
[d] To convert C·m to debye, divide by 3.336×10^{-30}.

Table 2. Temperature Variation of Physical Properties of Anhydrous Hydrogen Fluoride

		Density, g/cm³			Heat capacity J/(mol·°C)[f]		
						Vapor	
Temperature, °C	Vapor pressure, kPa[a,b]	Saturated vapor[c]	Liquid[d]	Heat of vaporization, J/mol[e,f,g]	Liquid C_p[g]	C_v[h]	C_p[i]
−70	1.39		1.176		43.26		
−60	2.52		1.149		43.59		
−50	4.62		1.123		44.10		
−40	7.86		1.098		44.77		
−30	12.96		1.073		45.48		
−20	20.68		1.048		46.44		
−10	32.06		1.025		47.49		
0	48.26	0.00215	1.003	7042	48.61		
10	71.01	0.00252	0.98		49.96		
20	103.42	0.00317	0.968	7473	51.21	46.27	250
30	146.17	0.00398	0.945			47.69	805
40	203.40	0.00498	0.928			49.12	824
50	274.42	0.00619	0.908			50.54	350
60	370.95	0.00765	0.888			51.96	120
70	489.54	0.00939	0.867			53.38	50
80	641.23	0.01144	0.844			54.81	40
90	820.50	0.01385	0.820			56.23	30
100	1054.93		0.796			57.57	30

[a] To convert kPa to psi, multiply by 0.145.
[b] Refs. 3, 7, and 17.
[c] Ref. 3.
[d] Ref. 7.
[e] Calorimetric.
[f] To convert J to cal, divide by 4.184.
[g] Ref. 8.
[h] Ref. 7.
[i] Ref. 18.

9; and a critical review of thermodynamic properties of gaseous HF is given in ref. 4.

The properties of aqueous solutions of hydrogen fluoride are given in Tables 3–9, and Figures 1–2. An entropy–concentration diagram and adiabatic HF absorption diagrams are provided in refs. 29–30.

Table 3. Physical Properties of Aqueous 70% Hydrogen Fluoride

Property	Value	Reference
boiling point at 101.3 kPa (1 atm), °C	66.4	17
freezing point (solid phase is $HF.H_2O$), °C	−69	18
vapor pressure at 25°C	20 kPa (2.9 psia)	20
density at 25°C, g/cm³	1.22	

Table 4. Density of Aqueous Hydrogen Fluoride, g/cm^3 [a]

Hydrogen fluoride, wt %	Temperature, °C 0	20	40	60
0	1.000	0.998	0.992	0.983
20	1.080	1.070	1.055	1.035
40	1.159	1.135	1.12	1.10
60	1.235	1.21	1.18	1.15
65	1.248	1.225	1.19	1.16
70	1.258	1.23	1.195	1.16
75	1.262	1.23	1.195	1.155
80	1.259	1.22	1.18	1.135
85	1.240	1.18	1.135	1.11
90	1.178	1.105	1.06	1.04
100	1.003	0.968	0.928	0.888

[a] Refs. 7, 21–22.

Table 5. Partial Pressure of Hydrogen Fluoride over $HF–H_2O$ Solutions, Pa [a,b]

Hydrogen fluoride, wt %	Temperature, °C 0	20	40	60	80	100
0	0	0	0	0	0	0
10	4.0	18	68	216	500	1,490
20	12	55	201	633	1,760	4,360
30	40	169	595	1,790	4,760	11,400
50	488	1,653	4,770	12,200	27,900	58,700
70	5,490	15,370	39,300	88,200	108,600	343,000
100	48,500	103,041	202,000	370,300	640,010	1,050,000

[a] Ref. 20.
[b] To convert Pa to mm Hg, multiply by 7.5×10^{-3}.

Chemical Properties

The chemistry of hydrogen fluoride is dominated by five features: the great strength of the hydrogen–fluorine bond; the significant strength of secondary hydrogen bonds to fluorine; the absence of oxidation states of fluorine other than −1; strong polymeric association in the solid, liquid, and gas states; and a strong acid character in apparent contradiction with a low ionic dissociation in water solution.

The hydrogen fluoride molecule remains intact up to remarkably high temperatures. From the free energy of formation of hydrogen fluoride molecule and the heat capacities of hydrogen fluoride and its elements, it is calculated that only 40% is dissociated into hydrogen atoms and fluorine atoms at 4000°C. Next to N_2 and CO, HF is the most stable diatomic molecule. This fact, plus its high heat of formation, gives the combination of hydrogen and fluorine the highest specific impulse available to rocketry when stable chemical species are used (31) (see Explosives and propellants).

Above 200°C, the vapor of hydrogen fluoride is approximately monomolecular, but as the temperature is lowered, hydrogen bonding leads to chains of increasing

Table 6. Partial Pressure of Water over HF–H_2O Solutions, Pa[a,b]

Hydrogen fluoride, wt %	Temperature, °C					
	0	20	40	60	80	100
0	610	2,340	7,378	19,912	47,330	101,300
10	594	2,130	6,520	17,500	41,600	90,500
20	484	1,750	5,370	14,400	34,500	75,400
30	363	1,230	4,080	11,000	26,500	58,100
50	101	397	1,310	3,760	9,580	21,990
70	nil	13				0
100	0	0	0	0	0	0

[a] Ref. 20.
[b] To convert Pa to mm Hg, multiply by 7.5×10^{-3}.

Table 7. Heat of Dilution of Aqueous Hydrogen Fluoride Solution[a]

Hydrogen fluoride, wt %	Heat of dilution[b], J
0	970
10	937
20	912
30	874
40	820
50	736
60	623
70	502
80	380
90	217
100	0

[a] Ref. 23.
[b] Heat evolved when one gram of anhydrous HF is diluted with water to the concentration shown. To convert J to cal, divide by 4.184.

length, thus: —F—H—F—H—F—H—. Although these secondary hydrogen bonds are relatively weak (about 25 kJ/mol (6 kcal/mol) of H—F (32)), they have a marked effect on the properties of hydrogen fluoride. Thus the boiling point of hydrogen fluoride is much higher then those of the other hydrogen halides (HF, 19.5°C; HCl, −84°C; HBr, −67°C; HI, −35°C). The effects of hydrogen bonding persist into the vapor and cause large deviations from ideal gas behavior. Near its boiling point, hydrogen fluoride is one of the most imperfect gases known. In the solid state, hydrogen fluoride exists as zig-zag chains of molecules with spacing and angles as shown in Figure 3. In the liquid and gas states, rotation about bonds removes the conformation regularity. At the same time, the length of the hydrogen-bonded polymers decreases, and near the boiling point the number of units in the liquid and gas phases reaches an average of about 3.5 (12). The angle H—F—H, ca 120°, is close to the internal angle of a hexagon, and probably many of the molecular aggregates have the ring structure H_6F_6. Since fragments of different lengths are present, it is not meaningful to designate the formula as H_6F_6 or as H_2F_2, as is sometimes done. Because the mobile equilibria between the hydrogen fluoride polymers may have a significant time constant, measurements by dynamic and static methods may give different results.

Table 8. Specific Conductivity of Aqueous Hydrogen Fluoride Solutions, at 0°C, 1/(Ω·cm)

Hydrogen fluoride, wt %	Conductivity
70	7.9×10^{-1}
80	7.1×10^{-1}
85	6.3×10^{-1}
90	4.9×10^{-1}
92.5	3.8×10^{-1}
95	2.5×10^{-1}
96	1.95×10^{-1}
97	1.04×10^{-1}
98	9.4×10^{-2}
99	5.6×10^{-2}
99.5	3.4×10^{-2}
99.75	1.8×10^{-2}
99.9	5.7×10^{-3}
99.95	2.8×10^{-3}
100	$<1.6 \times 10^{-6}$

[a] Refs. 24–25.

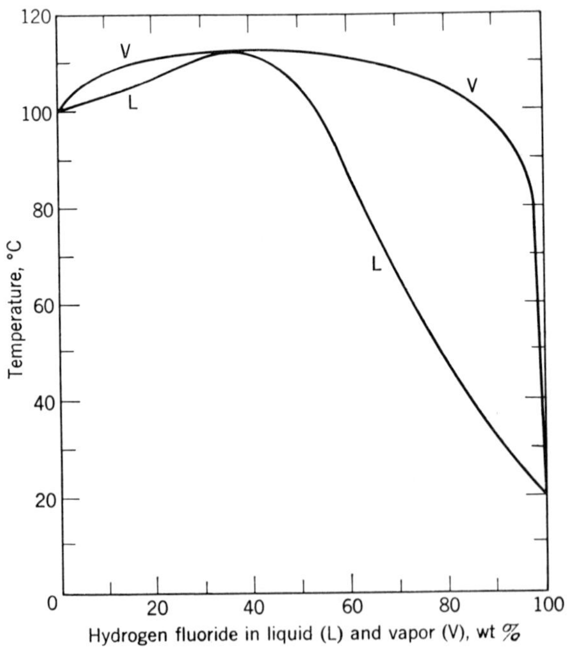

Figure 1. The boiling and liquid–vapor compositions of the system $HF.H_2O$ (26–27).

Although hydrogen fluoride in water solution is weakly dissociated (into H^+ and $(HF)_{x\text{-}1}F^-$), anhydrous hydrogen fluoride and high aqueous concentrations of HF are very strong acids in the proton donor sense (Brønsted-Lowry concept (33)). Thus the Hammett acidity function H_0 approaches −11. Since the Hammett scale is related to pH and is in a sense an extension of the pH scale, the high negative value of H_0 shows anhydrous hydrogen fluoride to be in the class of superacids. Addition of antimony

Table 9. Liquid and Vapor Compositions of Aqueous Solutions of Hydrogen Fluoride at the Boiling Points, 101.3 kPa (1 atm)[a,b]

wt % HF in liquid	0.0	10.0	20.0	30.0	40.0	50.0	60.0	70.0	80.0	90.0	100.0
wt % HF in vapor	0.0	2.0	6.77	19.3	43.5	74.3	95.5	98.8	99.2	99.5	100.0
boiling point, °C	100.0	102.8	106.8	110.3	112.1	105.1	88.0	66.1	47.6	32.0	19.57

[a] Ref. 26.

[b] There is a constant–boiling composition at 38.26% HF; boiling point 112°C at 98.6 kPa (740 mm Hg).

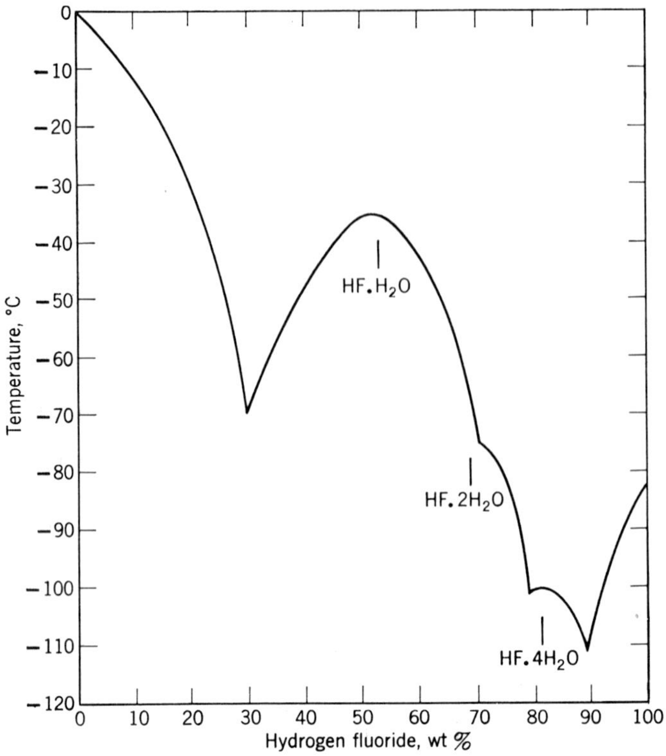

Figure 2. Freezing points of aqueous hydrogen fluoride solutions (28).

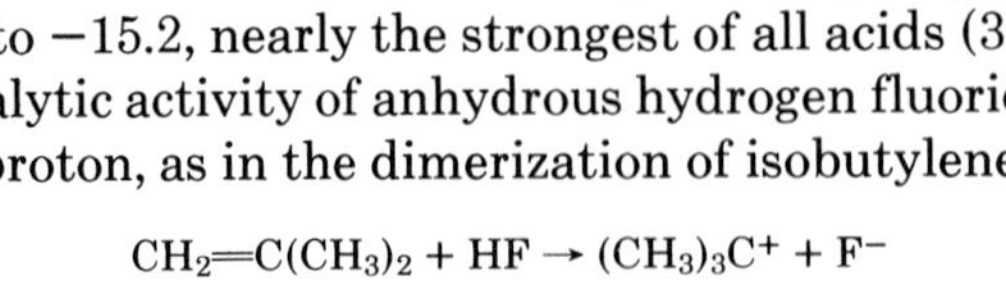

Figure 3. Conformation of hydrogen fluoride in the solid state (6).

pentafluoride to make a 3 *M* solution in anhydrous hydrogen fluoride raises the Hammett function to −15.2, nearly the strongest of all acids (34).

The strong catalytic activity of anhydrous hydrogen fluoride is attributed to its ability to donate a proton, as in the dimerization of isobutylene:

$$CH_2{=}C(CH_3)_2 + HF \rightarrow (CH_3)_3C^+ + F^-$$

$$(CH_3)_3C^+ + CH_2{=}C(CH_3)_2 \rightarrow (CH_3)_3C{-}CH_2C^+(CH_3)_2$$

$$\rightarrow (CH_3)_3\,CCH{=}C(CH_3)_2 + H^+$$

Because of its very high dielectric constant, its relatively high boiling point, long liquid range, and low viscosity, anhydrous hydrogen fluoride is an excellent solvent for ionic fluorides (Table 10). Generally, the soluble fluorides act as simple bases, fully ionized and increasing the concentration of HF_2^-:

$$HF + KF \rightarrow K^+ + HF_2^-$$

Table 10. Solubility of Metal Fluorides in Anhydrous Hydrogen Fluoride[a]

Fluoride	Temperature, °C	Solubility, g/100 g
LiF	12	10.3
NaF	11	30.1
KF	8	36.5
NH_4F	17	32.6
CaF_2	12	0.817
SrF_2	12	14.83
BaF_2	12	5.60
MgF_2	12	0.025
FeF_2	12	0.006
FeF_3	12	0.008
AlF_3	11	<0.002
SbF_5	25	miscible

[a] Ref. 12.

Many metal fluorides can then be isolated from hydrogen fluoride solution as salts of the HF_2^- anion.

The solubility of organic compounds in anhydrous hydrogen fluoride is often related to their proton-accepting capability. Most substances containing oxygen, nitrogen, or sulfur atoms that are not coordinatively saturated offer a lone electron pair capable of binding protons and are usually very soluble (6,35). Addition of a proton from the hydrogen fluoride may cause subsequent reactions including polymerization. On the other hand, the organic material can often be recovered unchanged from the hydrogen fluoride. In the simpler cases, ionic species are present and the solutions are electrically conducting:

$$C_2H_5OH + HF \rightarrow C_2H_5OH_2^+ + F^-$$

This behavior is the basis of electrochemical fluorination to produce a wide range of highly fluorinated products (36).

Anhydrous hydrogen fluoride is an active dehydrating agent. Wood and paper are charred on contact, and aldehydes undergo condensation by elimination of water. Anhydrous hydrogen fluoride is less powerful than sulfuric acid in this respect. For example, hydrogen fluoride does not rapidly dehydrate primary alcohols.

The small size of the fluoride ion permits it to participate in coordination structures of high rank. For example, tantalum and niobium form stable hexafluorotantalate and hexafluoroniobate ions and hydrogen fluoride will attack these usually acid-resistant metals. The attack of hydrofluoric acid on silicates is a special case. Ready formation of hexafluorosilicate ion is further promoted by decomposition to volatile silicon tetrafluoride.

As noted above, hydrogen fluoride in water solution is a weak acid. Two dissociation constants need to be considered:

$$K_1 = \frac{[H^+][F^-]}{[HF]} = 6.46 \times 10^{-4}\ M$$

$$K_2 = \frac{[HF_2^-]}{[HF][F^-]} = 5 \text{ to } 25\ M$$

The primary acid dissociation constant, K_1, shows hydrofluoric acid to be only about as strong as formic acid, whereas the other hydrohalic acids are almost completely dissociated. The apparent contradiction with hydrogen fluoride as a powerful proton donor has been discussed in a number of ways. The reduction in availability of protons from the hydrogen-bonded polymer should not be significant for the degree of polymerization that is likely, since the dissociation constants indicate that the hydrogen ion concentration in aqueous hydrofluoric acid is about 1/200th that in hydrochloric acid. A detailed thermodynamic comparison of the hydrohalic acids shows that the enthalpy changes for dissociation of the acids in water solution are not very different. However, there is a large negative entropy of hydration of fluoride ion in the case of hydrofluoric acid. This makes the Gibbs free energy change for the overall dissociation process actually positive, and the degree of dissociation in water solution is therefore small (37).

Manufacture

The preferred raw materials for the manufacture of hydrogen fluoride are acid-grade fluorspar and sulfuric acid:

$$CaF_2(s) + H_2SO_4 \rightarrow CaSO_4(s) + 2\ HF(g)$$

The rate and yield in commercial apparatus is influenced by the purity and fineness of the fluorspar, the concentration of the sulfuric acid, the ratio of sulfuric acid to fluorspar, the temperature of the reaction mixture, and the intimacy of mixing of the acid and spar. Since fluorine values dominate the raw material costs, the operating factors are optimized to give maximum yield of hydrogen fluoride based on fluorspar.

Acid-grade fluorspar is a finely ground flotation product having the following specifications:

Screen analysis: 2% on 149 μm (110 mesh), 13% on 74 μm (200 mesh), 55% through 44 μm (325 mesh) (38).

Chemical analysis:

CaF_2	minimum 96.5 to 98%
SiO_2	maximum 1.0%
S	maximum 0.05%
H_2O	maximum 0.05%
$CaCO_3$	principal remainder

Silica is a highly objectionable contaminant, since each 0.45 kg of silica consumes 1.18 kg fluorspar and 1.50 kg of sulfuric acid by the reaction:

$$SiO_2 + 2\ CaF_2 + 2\ H_2SO_4 \rightarrow SiF_4 + 2\ CaSO_4 + 2\ H_2O$$

When hydrogen fluoride containing SiF_4 is absorbed in water, a further loss of fluorine values occurs by the reaction:

$$SiF_4 + 2\ HF(aq) \rightarrow H_2SiF_6(aq)$$

Carbonates are troublesome in consuming sulfuric acid, in causing foaming in the generator, and in contributing carbon dioxide to the gas stream where it is a noncondensable diluent of the hydrogen fluoride. Sulfur-bearing contaminants of fluorspar, such as galena or pyrites, may generate hydrogen sulfide or sulfur dioxide, contaminating the hydrogen fluoride, and sometimes causing sulfur deposits in the gas handling equipment.

Other conditions being constant, the finer the fluorspar, the readier is the liberation of hydrogen fluoride from the spar and the better the yield. Fluorspar of the screen analysis given above has been used directly in the hydrogen fluoride kilns with good yields. Some manufacturers grind this material further. Such additional grinding is an expensive operation involving some handling loss of fluorspar, and the effect of this step must be analyzed on the basis of the improvement in yield it produces compared with other means of obtaining the same improvement.

Sulfuric acid of a concentration as low as 93% or as high as 99% has been used in hydrogen fluoride manufacture. The more dilute acid gives a more fluid reacting mass during initial mixing and a smoother liberation of hydrogen fluoride. Two factors offset this benefit: (*1*) the more dilute sulfuric acid is much more corrosive to the generator, and (*2*) the greater amount of water will leave with the hydrogen fluoride and increase problems in condensation.

The ratio of sulfuric acid to fluorspar charged to the generator is seldom the stoichiometric ratio. Depending on the relative cost of the two ingredients, one or the other is used in slight excess. In U.S. practice, economic operation demands the maximum release of hydrogen fluoride from the fluorspar and justifies the use of a 5–10% excess of sulfuric acid. A greater excess of sulfuric acid will result in a "wet" calcium sulfate residue, which may be strongly fuming unless the sulfuric acid is driven out of the generator with the hydrogen fluoride by high operating temperatures.

Because the reaction of calcium fluoride and sulfuric acid is endothermic, heat must be supplied. The theoretical heat absorbed is 1400 kJ/kg (603 Btu/lb). To effect a release of over 98% of the fluorine in the fluorspar, the reaction time is normally 30–60 minutes at 200–250°C, with the hydrogen fluoride leaving the generator at 100–150°C. The lowest operating temperatures at which good yields are obtained are desirable to reduce the energy requirement and to minimize corrosion of the generator.

The diffusion of sulfuric acid through the product layer of calcium sulfate to the unreacted core seems to be a significant barrier to the progress of the reaction (39). Therefore, constant and effective mixing is important. In present practice, hydrogen fluoride generators may be rotary kilns, horizontal and externally fired, with spar fed continuously at the forward end by a screw conveyor. Acid is also introduced at the forward end of the kiln, and calcium sulfate residue is removed through an air lock at the opposite end. Stationary reactors are also used. These consist of a long horizontal vessel of U-shaped cross section carrying a central shaft extending the full length and equipped with mixing and conveying paddles. As in the rotary kiln, spar and acid move concurrently down the length of the reactor. In recent constructions, a combination of the two types of reactors is used. A stationary prereactor receives spar and heated sulfuric acid and feeds the well-mixed but incompletely reacted components into a heated rotary kiln. This enables the kiln to be made of lower cost steel and is said to give improved yields (40).

Since hydrogen fluoride is readily absorbed by water, early manufacturers made an aqueous solution of the gases from the generator and prepared strong acid by fractionation, or when strengths above the atmospheric azeotrope (38%) were needed, by distillation with sulfuric acid. Current requirements are mainly for the anhydrous compound, and manufacturers have succeeded in controlling conditions in the generator to supply effluent gas low enough in water and noncondensables to permit direct condensation, with reboiling to give anhydrous hydrogen fluoride. Vent gases must still be scrubbed with water to recover additional hydrogen fluoride and to prevent air pollution.

Details of gas handling vary among manufacturers. In one published process (41), the hydrogen fluoride gases are freed from solids and sulfuric acid mist by cyclone separation and countercurrent scrubbing in the lower section of a precondenser. Scrubbing fluid is recirculated through the kiln after addition of 20% oleum to combine with water. The hydrogen fluoride vapors, now low in water and sulfuric acid, are condensed as 99% hydrogen fluoride in an amount representing 98% of the production from the kiln. This is redistilled to a purity above 99.9% hydrogen fluoride. The uncondensed gases are washed with fresh sulfuric acid to recover additional hydrogen fluoride. The final gaseous effluent is largely silicon tetrafluoride, which is absorbed in water and recovered as 30–35% fluorosilicic acid. The overall process is shown schematically in Figure 4.

Calcium sulfate from the kilns is commonly discarded, in view of the general availability of natural gypsum. Some, however, is neutralized with lime and sold for the manufacture of anhydrite cements.

The overall yield of hydrogen fluoride is 85–95% based on calcium fluoride; and 75–85% based on sulfuric acid. Production of one metric ton of hydrogen fluoride typically requires 1750 kg of steam at 1.03 MPa (150 psi) pressure, 220 kW·h of electricity, 1.5 GJ (3.6×10^5 kcal) of refrigeration at −10°C, 9.6 GJ (2.3×10^6 kcal) fuel, over 12 m^3 (ca 3000 gal) of process water, and 116 m^3 (ca 30,000 gal) of cooling water (42).

Recovery from Phosphate Rock. Recognizing the enormous supply of fluorine values existing in phosphate rock reserves, and the large amount of fluorine values that are processed daily through phosphate fertilizer plants, many studies have evaluated the feasibility of obtaining hydrogen fluoride from this source. Two economic obstacles exist: the fluorine values are released as silicon tetrafluoride, not easily convertible to hydrogen fluoride; and the collection of these fluorine values from many scattered fertilizer plants to provide feed for a hydrogen fluoride plant of economic capacity is expensive. Thus a hydrogen fluoride plant typically needs to produce at least 22,700 metric tons per year to be economically useful, but most fertilizer plants can supply no more than 9000 t of hydrogen fluoride equivalent annually.

A proposed process starts by the reaction of silicon tetrafluoride with recycled ammonium fluoride solution to form ammonium fluorosilicate:

$$SiF_4 + 2\ NH_4F \rightarrow (NH_4)_2SiF_6$$

When this solution is neutralized with ammonia, silica precipitates and more ammonium fluoride is formed:

$$(NH_4)_2SiF_6 + 2\ H_2O + 4\ NH_3 \rightarrow SiO_2 + 6\ NH_4F$$

The ammonium fluoride solution, except that which is recycled, is evaporated to a salt concentration of 94–95%. At this concentration, some conversion to ammonium bifluoride, NH_4HF_2, takes place by evaporation of ammonia. The evaporation is continued until a salt concentration of 98% is reached; the mixture solidifies at about 100°C. The solid mixture of ammonium fluoride and ammonium bifluoride, containing 60% fluorine is decomposed with 93–95% sulfuric acid at 180–190°C. The residue is mainly ammonium bisulfate, which can be converted to ammonium sulfate by neutralizing with ammonia. The economics of this process depend on a low-cost supply of silicon tetrafluoride, a premium market for precipitated silica, and a local market for ammonium sulfate. The process is handicapped by a large filtration and evaporation load.

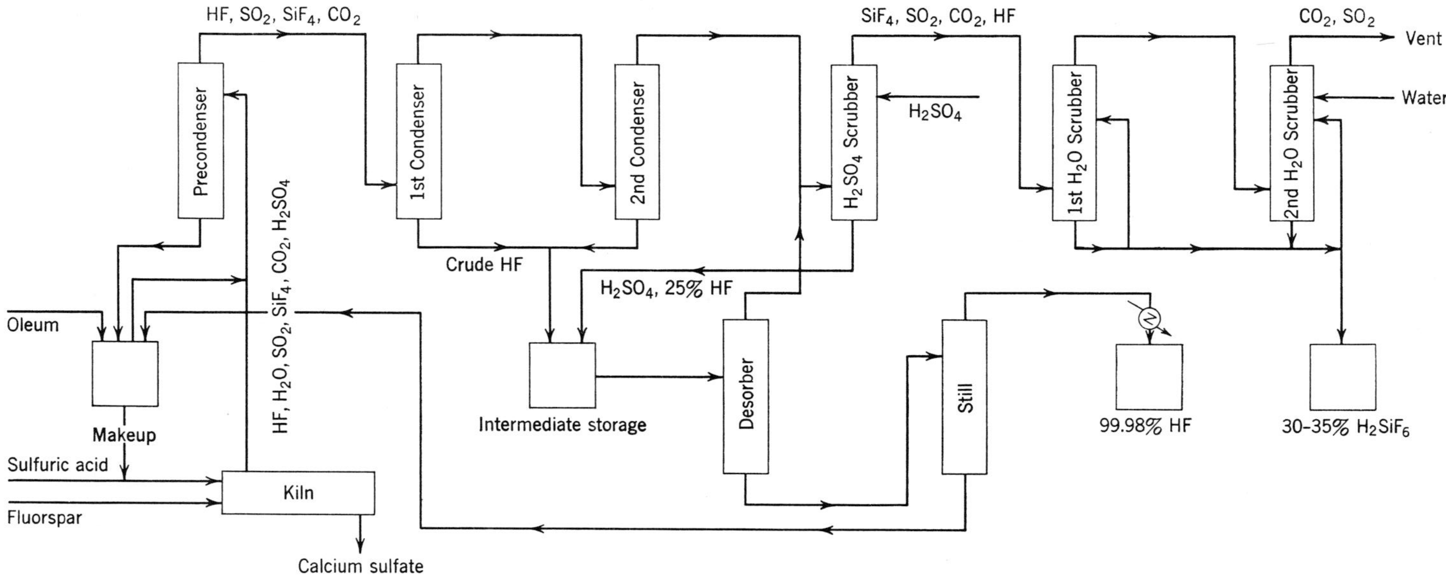

Figure 4. Schematic flow diagram for the manufacture of hydrogen fluoride.

A process developed by the U.S. Bureau of Mines starts from fluorosilicic acid:

$$3\ SiF_4 + 2\ H_2O \rightarrow 2\ H_2SiF_6 + SiO_2$$

Lime and silica are added:

$$3\ Ca(OH)_2 + H_2SiF_6 \rightarrow 3\ CaF_2 + SiO_2 + 4\ H_2O$$

The calcium fluoride and silica precipitate is filtered and pelletized. The pellets are fed to a 1050°C rotary kiln with steam. The reaction produces a gas stream of hydrogen fluoride, water vapor and a residue of calcium silicate:

$$CaF_2 + H_2O + SiO_2 \rightarrow CaSiO_3 + 2\ HF$$

Recovery of 80% of the fluorine in the fluorosilicic acid as hydrogen fluoride is claimed.

Buss, Ltd. has piloted a process in which fluorosilicic acid from the scrubber towers of fertilizer manufacture is concentrated, then vaporized as an $HF–SiF_4–H_2O$ mixture. The gases enter the middle of an absorption column. In the upper half, an organic solvent selectively absorbs hydrogen fluoride. Pure hydrogen fluoride vapors entering the bottom of the column strip small amounts of water and silicon tetrafluoride from the solvent. The hydrogen fluoride is desorbed from the solvent in a vacuum evaporator, liquefied, and purified by two-stage rectification. Silicon tetrafluoride in the overhead gas from the absorber is hydrolyzed to silica and fluorosilicic acid. The latter is recycled, and the silica can be sold for a credit to the processing costs.

Materials of Construction. Generalizations regarding materials of construction where hydrogen fluoride and its aqueous solutions are to be handled have sometimes proved misleading, and laboratory tests give at best only general guidance. Nevertheless, readily available materials have been found to meet each problem encountered. The presence of impurities such as sulfur compounds and fluorosilicic acid, and the aeration, velocity, and temperature of the hydrogen fluoride are important considerations in determining the appropriate material. In the manufacture of hydrogen fluoride, steel generators are successfully used. Pipes and other equipment handling the gases from the generator may be of steel as long as acid condensation from these gases is not less than 70% hydrogen fluoride. Weaker acids must be handled in nonferrous materials such as carbon, lead, bronze, or Monel. Lead is sensitive to the presence of fluorosilicic and cannot be used when more than 2–3% of that acid is present. Monel is adversely affected by aeration and the presence of sulfur compounds. Natural rubber and neoprene are useful for strengths below 60%, and particularly below 40%. Copper has been used for distillation of aqueous hydrogen fluoride solutions. For exacting laboratory work with aqueous solutions containing fluorosilicic acid, silver apparatus is excellent (26). Steel valves with Monel stems are generally satisfactory for anhydrous hydrogen fluoride, and bronze valves with Monel stems are suitable for aqueous hydrofluoric acid. A wide variety of materials have had limited use for gaskets, but polytetrafluoroethylene (PTFE) has proved to be the most satisfactory. Pump and valve packings of PTFE have given excellent performance. Lower strengths of aqueous hydrofluoric acid can be shipped and stored in polyethylene jugs or in polyethylene drums and steel overpacks (43). Glass and silicate ceramics are, of course, never acceptable.

Economic Aspects

In 1979, hydrogen fluoride was being manufactured in the United States by five companies in nine plants. Capacities of North American producers are shown in Table 11 (44). Plant capacity ranged from 5400–68,000 metric tons per year, with an average of 27,000 t/yr. One U.S. company operated a 68,000 t/yr hydrogen fluoride plant at Matamoros, Mexico; and another U.S. company has merchant capacity for 45,000 t in Canada (45).

Published production figures for hydrogen fluoride must be examined for the basis of presentation because of the large and changing proportion of captive uses to merchant production. A good index for hydrogen fluoride production is the consumption of acid-grade fluorspar, which is the source of nearly all hydrogen fluoride in the United States, and of which about 98% is used for this purpose (46). A conservative ratio of kg acid-grade fluorspar to 0.45 kg hydrogen fluoride (stoichiometric ratio for 100% CaF_2 is 1.95) gives the estimates of total U.S. production of hydrogen fluoride (100% basis) shown in Table 12 (45–46). Worldwide production is estimated in Table 13.

The growth rate of hydrogen fluoride production in the United States has dropped from about 10% annually in the 1940s and 1950s to 5% in the 1960s, to 3% in the early 1970s, and to the low current rate of about 0.5%. A sharp decrease in demand for hydrogen fluoride occurred in 1978 when the United States prohibited the use of chlorofluorocarbon gases in pressure packaging because of their supposed adverse effect on the concentration of ozone in the high altitude atmosphere (48–49) (see Air pollution; Ozone). This environmental question remains controversial. In the future the ban may be lifted, or it may be extended to prohibit the use of chlorofluorocarbons as refrigerants. Greater efficiency in the use of hydrogen fluoride in the aluminum industry, as described below, and production of aluminum fluoride and cryolite directly from fluorosilicic acid has further slowed the increase in demand for hydrogen fluoride. Other demands for hydrogen fluoride are about one-fourth of total use, and are likely to rise about in proportion to economic activity. Estimates of future growth in hydrogen

Table 11. North American Producers and Capacities of Hydrogen Fluoride, 1000 Metric Tons[a]

Producer	1979 capacity, 10^3 t
Allied	132
DuPont	68
Alcoa	50
Pennwalt	23
Harshaw	16
Essex	10
Alcan	45
Química Fluor	68
Industria Química de México	14
Quimobásicos	3
Total	*429*
Number of producers	*10*

[a] Ref. 44.

Table 12. U.S. Consumption of Hydrogen Fluoride Averaged, 1000 Metric Tons

1930–1939	6.3
1940–1949	37
1950–1959	97
1960–1964	165
1965–1969	253
1970	295
1971	303
1972	319
1973	332
1974	340
1975	386
1976	349
1977	293
1978	324

Table 13. World Production of Hydrogen Fluoride[a] 1973, 1000 Metric Tons × 10³

United States[b]	240
Soviet Bloc	100
France	73
Canada	72
FRG	81
Japan	65
Italy	59
United Kingdom	34
Mexico	54
Norway	18
Spain	15
People's Republic of China	7
India	5
Netherlands	6
Australia	5
Republic of South Africa	0.9
Total	*838*

[a] Ref. 47.
[b] U.S. figure differs from Table 10 mainly in the omission of imports.

fluoride consumption range from 1–4% annually. The lower rate is more probable, unless a major new use for fluorine chemicals arises. No such new use is now in sight.

Hydrogen fluoride prices have risen with general inflation (Table 14). Since no

Table 14. Price of Hydrogen Fluoride and Hydrofluoric Acid; Tank Cars, ¢/kg

	1950	1960	1970	1974	1975	1976	1977	1978	1979
anhydrous HF	36.4	39.7	54	81.6	90.4	93.7	98.1	98.1	98.1
hydrofluoric acid 70%	31.5	29.5	39.7	75.2	65.7	69.4	74.5	74.5	77.2

shortage of raw material is evident, and since manufacturing capacity is adequate, no extra-inflationary price increases are expected.

Health and Safety Factors

Hydrogen fluoride, anhydrous or aqueous, is a dangerous substance. However, years of experience in its manufacture and use have shown that it can be handled safely, provided the special hazards are recognized and the necessary precautions are taken. Excellent manuals describing equipment and procedures for the safe handling of hydrogen fluoride are available from the major manufacturers and from the Chemical Manufacturers' Association (9,43,50–53).

Anhydrous hydrogen fluoride, vaporized hydrogen fluoride, and even rather dilute solutions of the aqueous acid can be extremely damaging to skin, eyes, mucous membrane, and lungs. In certain concentration ranges of hydrogen fluoride, burns of the skin that are not immediately evident may be manifested later by deep-seated ulceration.

First-aid treatment after contact of hydrogen fluoride with the skin starts with immediate and thorough washing with water. To slow penetration of fluoride ion through the skin, the affected part should be immersed for at least one-half hour in ice-cold 7% alcohol, or in ice-cold saturated magnesium sulfate solution, or in ice-cold 0.13% benzalkonium chloride. Alternatively, the cold solutions may be applied by saturated compresses changed at least every two minutes. If a physician is not immediately available, a paste of powdered magnesium oxide and glycerol, freshly prepared, may then be applied. Medical treatment may include infiltration of calcium gluconate solution beneath the skin (53). Accidental inhalation of hydrogen fluoride vapors should be treated by immediately removing the exposed person from the contaminated atmosphere, and by administering 100% oxygen. The exposed person should avoid physical activity. After contact with eyes, first-aid treatment should be limited to thorough irrigation with clean water. Any significant hydrogen fluoride exposure is a medical emergency, and a physician should be called.

In work with hydrogen fluoride, the proper use of protective clothing, gloves, goggles, face shields, and respiratory devices is important, and a full knowledge of first-aid procedures is essential. Reference should be made to detailed instructions, such as those published by the Chemical Manufacturers' Association (53).

A danger not always recognized in working with hydrogen fluoride in the presence of metals is the chance of generating hydrogen, and the possibility of explosion, if the concentration of hydrogen fluoride drops below 60–65% through accidental or improper addition of water. Systems for dilution of hydrogen fluoride must be protected against return flow of diluted acid.

The recommended limit for occupational exposure to hydrogen fluoride in air is a time-weighted average of 1.5 mg HF/m^3 air for up to a 10-h workday, 40-h workweek; with a ceiling of 60 mg/m^3 averaged over 5 min (54).

Specifications, Packaging, and Analysis

Bulk shipments of hydrogen fluoride are limited mainly to anhydrous hydrogen fluoride and aqueous 70% hydrofluoric acid. The lower concentration limit is set by compatibility with steel; concentrations below ca 60% become corrosive to this material.

The primary suppliers of anhydrous fluoride ship in tank cars of 20–91 t capacity, specification 105A300 or 114A40; in tank trucks holding 20 t, specification MC311 or MC312. The product may be repackaged into steel cylinders holding 8–907 kg. The aqueous 70% solution is shipped in tank cars of 32–80 t capacity, in tank trucks of 20 t capacity, and in polyethylene-lined drums holding 114 or 208 L (30 or 55 gal). Both the anhydrous and aqueous acid must be labeled: Corrosive Liquid—White Label (43,50–52). Reagent grade hydrofluoric acid is supplied in polyethylene bottles or carboyls at 48%, and an especially pure electronic grade is made to 49%.

Anhydrous hydrogen fluoride is one of the purest chemicals in regular commercial distribution. A typical analysis is (55):

HF	99.99%
H_2SiF_6	0.001
SO_2	0.003
H_2SO_4	0.005
H_2O	0.0004

The analysis of hydrogen fluoride and its aqueous solutions includes the determination of hydrogen fluoride and water and of the commonly associated impurities: fluorosilicic acid, sulfuric and fluorosulfonic acids, sulfur dioxide, and metals (chiefly iron). For manufacturing control, fluorspar must be analyzed for moisture, calcium carbonate, silica, and calcium fluoride; and reactor residues, for calcium fluoride, hydrogen fluoride, and sulfuric acid.

Laboratory equipment for handling hydrogen fluoride during sampling and analysis is usually made of steel for high concentrations or of polyethylene for aqueous solutions. Samples of anhydrous hydrogen fluoride can be collected in evacuated steel cylinders. A cylinder should not be filled beyond 85% of its water capacity. The dilution of hydrogen fluoride for analysis requires care because of the heat evolved, the volatility of the hydrogen fluoride, and the hazards to the analyst. A typical procedure involves dilution with ice at the bottom of a cylindrical pipe in the upper section of which, supported on a grid, is additional ice to absorb any vapors. Frequently performed analyses include the following:

(*1*) Total acidity: by titration of the diluted sample with standard alkali to the phenolphthalein end point.

(*2*) Nonvolatile acidity (reported as H_2SO_4): by evaporation of the diluted sample in platinum on a steam bath and titration of the residue with standard alkali to the phenolphthalein end point.

(*3*) Sulfur dioxide: by iodimetric titration of the diluted sample.

(*4*) Iron: by evaporation of the diluted sample with excess sulfuric acid; fuming with sulfuric acid; reduction to ferrous iron; and colorimetric determination with *o*-phenanthroline.

(*5*) Silica (reported as H_2SiF_6): by evaporating the diluted sample in the presence of a small amount of sodium chloride; complexing fluoride with boric acid; formation of silicomolybdic acid; reduction to a blue heteropoly acid; and colorimetric estimation. An alternative procedure involves redissolving the residue, adding potassium chloride, and exact neutralization at 10°C, followed by titration with standard alkali at 60°C.

(*6*) Water: by conductance measurement on the anhydrous acid; or by Karl Fischer procedure.

(7) Total fluoride: by difference from the above measurements; or by the use of a fluoride electrode (56).

Details of some of these methods and of sampling procedures are given in ref. 52.

Uses

The traditional use for hydrofluoric acid (the etching and polishing of glass) has not dwindled, but has been superseded by the requirements of other industries. Again, because of the high proportion of hydrogen fluoride used captively, it is hard to fix use patterns exactly. The estimates of Table 15, however, show relative amounts and show trends in use. The relative importance of various uses has not changed much since the late 1950s, though the production of fluorinated organic compounds rose greatly in total amount until the mid-1970s.

The largest use for hydrogen fluoride is in the manufacture of aluminum fluoride and cryolite (sodium aluminum fluoride) as the fused salt solvent for aluminum oxide in the electrolytic manufacture of aluminum (see Aluminum compounds). About 0.35 t of hydrogen fluoride are needed to make 1 t of aluminum. This use has increased less rapidly than the expansion of aluminum production because of greater efficiency in recycling fluorine values, and because of the production of some of the fluoride salts from silicon tetrafluoride by-product of the manufacture of phosphate fertilizers.

The principal fluorine-containing organic compounds are the chlorofluoromethanes used as refrigerants and formerly as pressurizing gases; some chlorofluoroethanes used as foaming agents and cleaning fluids; fluorine-containing ethylene and propylene to be converted to polymers; and longer-chain fluorocarbons converted to soil-resistant coatings for textiles and leather. On average, ca 0.35 g of hydrogen fluoride are needed to produce one gram of fluorocarbon.

For uranium processing, hydrogen fluoride is used to make uranium tetrafluoride and to make elemental fluorine to convert the tetrafluoride to uranium hexafluoride, which is used in the diffusion process for the separation of uranium isotopes. The product of the uranium refinery, U_3O_8, is reduced by hydrogen to UO_2. Treatment with hydrogen fluoride gives UF_4, and reaction with fluorine produces UF_6. Stoichiometrically, 1 t of U_3O_8 would require 0.43 t of hydrogen fluoride for the overall

Table 15. Distribution of Uses for Hydrogen Fluoride in the U.S., %

	1957	1962	1967	1972	1977
aluminum production					
aluminum fluoride			32	26	
cryolite	40	40	12	12	40
chlorofluorocarbons, including pressurizing gases, refrigerants, and polymers	28	34	37	40	37
uranium processing	12	11	2	2	7
petroleum alkylation	4	5	3	4	5
stainless steel pickling	5	3	3	4	4
other uses, including glass etching, other fluorides, sand castings, rocket propellant stabilizers, electronic circuits, sours and stain removers, oil-well acidizing, and mineral analysis	11	7	11	12	7

conversion. Considerable hydrogen fluoride passes through the hydrofluorination step with the water of reaction. This has been returned to the market as aqueous hydrofluoric acid.

In the petroleum industry, anhydrous hydrogen fluoride is a liquid catalyst for alkylation of aromatic hydrocarbons to produce octane improvers for gasoline. This use continues to be important: in 1978, U.S. refiners scheduled eight new HF alkylation plants with a total capacity of 7950 m^3 (50,000 barrels) of alkylate per day for early completion (57). Total U.S. HF alkylation capacity in 1978 was 56,870 m^3 (357,700 barrels) per day, and 1.26–3.14 kg of hydrogen fluoride were used per cubic meter of alkylate. Anhydrous hydrogen fluoride is also a catalyst for disproportionation, isomerization, polymerization, and dehydration reactions.

The application of hydrofluoric acid to the removal of scale from stainless steels has increased somewhat more slowly than U.S. production of these steels, because of increasing imports. Glass etching, as for the inner surfaces of light bulbs, requires aqueous hydrofluoric acid. Hydrofluoric acid, to which ammonium fluoride is added, etches to a polished surface for the final treatment of cut glass. Other uses include preparation of fluorides other than aluminum; removal of sand from metal castings; reduction of the corrosiveness of nitric acid rocket propellant oxidizers; preparation of microelectronic circuits; laundry sours and stain removers; oil well acidizing; and mineral analysis.

BIBLIOGRAPHY

"Hydrogen Fluoride" in *ECT* 1st ed., under "Fluorine Compounds, Inorganic," Vol. 6, pp. 695–708, by A. S. Woodard, Pennsylvania Salt Manufacturing Co.; "Hydrogen Fluoride" in *ECT* 2nd ed., under "Fluorine Compounds, Inorganic," Vol. 9, pp. 610–625, by John F. Gall, Pennsalt Chemicals Corporation.

1. O. Kausch "Flussäure, Kieselflussäure und deren Metalsalze" in *Enke's Bibliothek für Chemie und Technik,* Vol. 24, Verlag von Ferdinand Enke, Stuttgart, 1936.
2. E. Frémy, *Ann. Chim. Phys.* **47**(3), 5 (1856).
3. R. L. Jarry and W. Davis, Jr., *J. Phys. Chem.* **57**, 600 (1953).
4. C. E. Vanderzee and W. W. Rosenberg, *J. Chem. Thermodyn.* **2,** 461 (1970).
5. R. J. Gillepsie and D. A. Humphreys, *J. Chem. Soc. A,* 2311, 1970.
6. T. A. O'Donnell "The Hydrogen Fluoride Solvent System" in *Comprehensive Inorganic Chemistry,* Vol. 2, Pergamon Press, 1973, pp. 1038–1054.
7. E. V. Frank and W. Spalthoff, *Z. Electrochem.* **61,** 348 (1957).
8. J.-H. Hu, D. White, and H. L. Johnston, *J. Chem. Thermodyn.* **2,** 461 (1970).
9. "Hydrogen Fluoride" in *The Matheson Unabridged Gas Data Book,* Matheson Gas Products, East Rutherford, N.J., 1974.
10. D. R. Stull and H. Prophet, *JANAF Thermochemical Tables,* 2nd ed., Natl. Bur. Standards Ref. Data Ser., NBS, NSRDS-NBS 37, U.S. Govt. Printing Office, Washington, D.C., 1971.
11. J. F. Mathews, *Chem. Rev.* **72**(1), 85, 97 (1972).
12. H. H. Hyman and J. J. Katz "Liquid Hydrogen Fluoride" in T. C. Waddington, ed., *Non-Aqueous Solvent Systems,* Academic Press, London, 1965, pp. 47–81.
13. A. J. Perkins, *J. Phys. Chem.* **68,** 654 (1964).
14. K. Fredenhagen and J. Dahmlos, *Z. Anorg. Allgem. Chem.* **178,** 272 (1929).
15. S. I. Chan, D. Ikenberry, and T. P. Das, *J. Chem. Phys.* **41,** 2107 (1964).
16. A. F. Wells, *Structural Inorganic Chemistry,* 3rd ed., Clarendon Press, Oxford, 1962, p. 33.
17. K. Fredenhagen, *Z. Anorg. Allgem. Chem.* **210,** 210 (1933).
18. E. U. Franck and F. Meyer, *Z. Electrochem.* **63,** 571 (1959).
19. R. M. Yabroff, J. C. Smith, and E. H. Lightcap, *J. Chem. Eng. Data* **9**(2), 178 (1964).
20. P. A. Munter, O. T. Aepli, and R. A. Kossatz, *Ind. Eng. Chem.* **41,** 1504 (1949).

21. E. W. Washburn, ed., *International Critical Tables,* National Research Council, Vol. III, McGraw-Hill, New York, 1928, p. 54.
22. *Hydrofluoric Acid,* Stauffer Chemical Co., Industrial Chemical Division, Westport, Conn., 1974.
23. *Hydrofluoric Acid—Handling and Engineering Data,* Pennsalt Chemicals Corp., Philadelphia, 1959.
24. K. Fredenhagen, *Z. Phys. Chem.* **128,** 1 (1927).
25. K. Fredenhagen and M. Wellman, *Z. Phys. Chem.* **A162,** 454 (1932).
26. P. A. Munter, O. T. Aepli, and R. A. Kossatz, *Ind. Eng. Chem.* **39,** 427 (1947).
27. W. S. Pappas, *A Continuous Boiling Point Analyzer and its Application to the Hydrogen Fluoride Water System,* A. E. C. Res. and Dev. Rept. K-1354 (TID-4500), Union Carbide Nuclear Co., Oak Ridge, Tenn., 1960.
28. G. H. Cady and J. H. Hildebrand, *J. Am. Chem. Soc.* **52,** 3843 (1930).
29. R. Vieweg, *Chem. Tech.* **15,** 12 (1963).
30. R. Vieweg, *Chem. Tech.* **15,** 734 (1963).
31. J. F. Gall, *ARS Am. Rocket Soc. J.* **29,** 95 (1959).
32. A. F. Clifford, H. C. Beachell, and W. M. Jack, *J. Inorg. Nucl. Chem.* **24,** 1549 (1962).
33. C. F. Bell and K. A. K. Lott, *Modern Approach to Inorganic Chemistry,* 3rd ed. Butterworth's, London 1972, pp. 143–145.
34. M. Kilpatrick and J. G. Jones in J. J. Lagowski, ed., *The Chemistry of Nonaqueous Solvents,* Vol. 2, Academic Press, New York, 1967, pp. 43–99.
35. J. H. Simons, *Fluorine Chemistry,* Vol. 1, Academic Press, New York, 1950, pp. 225–259.
36. J. Burdon and J. C. Tatlow, *Adv. Fluorine Chem.* **1,** 129 (1960).
37. R. T. Meyers, *J. Chem. Ed.* **53**(1), 17 (1976).
38. Private communication, P. L. Brueckner, Allied Chemical, Industrial Chemicals Div., Morristown, N.J., Nov. 1, 1978.
39. D. Candido and G. P. Mathur, *Ind. Eng. Chem. Process Des. Dev.* **13,** 20 (Jan. 1974).
40. *Chem. Week,* 47 (Jan. 6, 1971).
41. W. R. Rogers and A. Muller, *Chem. Eng. Prog.* **59,** 85 (1963).
42. W. L. Faith, D. B. Keyes and R. L. Clark, *Industrial Chemicals,* 4th ed., John Wiley & Sons, Inc., New York, 1975, pp. 462–467.
43. *Hydrofluoric Acid,* Allied Chemical, Industrial Chemicals Division, Morristown, N.J., 1968.
44. *Chem. Eng.* **62** (Dec. 17, 1979).
45. *Chem. Mark. Rep.,* 9 (Aug. 21, 1978).
46. R. H. Singleton and J. E. Shelton "Fluorspar" in *Minerals Yearbook,* Vol. 1, U.S. Bureau of Mines, Washington, D.C., 1975.
47. *Industrial Minerals and Rocks,* 4th ed., Am. Inst. of Mining Metallurgical Engineers, 1975, pp. 53–56.
48. *Chem. Eng. News,* (Oct. 23, 1978).
49. L. Dotto and H. Schiff, *The Ozone War,* Doubleday & Company, Inc., New York, 1978
50. *Hydrofluoric Acid, Anhydrous and Hydrofluoric Acid 70%,* E. I. du Pont de Nemours & Co., Inc., Wilmington, Del., 1970.
51. *Hydrofluoric Acid,* Pennwalt Corp., 1979.
52. *Hydrofluoric Acid,* Stauffer Chemical Co., Industrial Chemical Division, Westport, Conn., 1974.
53. *Hydrofluoric Acid, Chemical Data Sheet SD-25,* Manufacturing Chemists' Association, Washington, D.C., 1970.
54. *Criteria Document: Recommendations for an Occupational Exposure Standard for Hydrogen Fluoride,* Nat. Inst. for Occup. Safety and Health, U.S. Dept. of Health, Education and Welfare, Rockville, Md., 1976.
55. *Specification Sheet: Hydrofluoric Acid,* E. I. du Pont de Nemours & Co., Inc., Industrial Chemicals Dept., Wilmington, Del., Oct. 22, 1973.
56. *Analytical Methods Guide,* Orion Research, Inc., Cambridge, Mass., Aug. 1973.
57. A. Cantrell, *Oil Gas J.,* 113 (April 24, 1978).

JOHN F. GALL
Philadelphia College of Textiles and Science

IRON

Iron(II) Fluoride

Anhydrous iron(II) fluoride [*7789-28-8*], FeF_2, is an off-white to buff-colored powder. The latter appearance comes from partial oxidation of Fe^{2+} to Fe^{3+}. It does not decompose if heated in the presence of dry nitrogen and melts above 1000°C (1). It is sparingly soluble in water but the solubility can be increased by addition of hydrofluoric acid or any strong acid.

Anhydrous iron(II) fluoride is prepared by the action of anhydrous hydrogen fluoride on metallic iron powder or anhydrous iron(II) chloride. An excess of HF and by-product HCl can be removed by heating carefully in the absence of air. The other synthetic methods include heating a mixture of $FeCl_2$ with F_2 or a reduction of FeF_3 with hydrogen (2). It is quite likely that the latter processes give mixtures of Fe^{2+} and Fe^{3+} rather than a pure FeF_2.

Colorless crystals $FeF_2.4H_2O$ [*13940-89-1*] can be obtained by dissolving metallic iron or anhydrous FeF_2 into hydrofluoric acid. The crystals of $FeF_2.4H_2O$ are sparingly soluble in water and decompose to Fe_2O_3 if heated in the presence of air.

Anhydrous iron(II) fluoride is commercially available in 100 g to 5 kg lots. Because no major industrial application has been reported, it is used only for experimental purposes. It is shipped under chemical name in polyethylene containers.

Toxicity of iron(II) fluoride has not been determined but it should be treated as a weak source of fluoride ion.

Iron(III) Fluoride

Iron(III) fluoride [*7783-50-8*], FeF_3, the highest known fluoride of iron, is obtained in the form of light-greenish (lime green) rhombic crystals by the action of anhydrous hydrogen fluoride or elemental fluorine on iron(III) chloride. It is insoluble in alcohol, ether, and benzene, and sparingly soluble in anhydrous HF and water. The pH of the saturated solution of FeF_3 is between 3.0 and 3.5, depending upon the method of preparation. It sublimes at 1000°C.

Its major uses are as a fluorinating reagent (3), for pinhole prevention in cast iron (qv) (4), as a catalyst for xenon–fluorine compounds (5), burning-rate control catalyst, and a catalyst for aromatization, dealkylation, and polymerization, and in flame-retarding polymers (6) (see Flame retardants). Industrial demand for this important chemical varies from year to year. It is commercially available in 500 g to 450 kg lots. Current price of anhydrous FeF_3 is 500 g at $88 and 45 kg at $100/kg.

The hydrated form of iron(III) fluoride [*15469-38-2*], $FeF_3.3H_2O$, is easily prepared from Fe_2O_3 and hydrofluoric acid. Dehydration of $FeF_3.3H_2O$ gives a mixture of oxyfluorides.

In the presence of excess HF, complex ions such as FeF_4^- and FeF_6^{3-} are formed in solution. Upon neutralization with base, fluoroferrates are obtained; eg, neutralization with NaOH produces $NaFeF_4$ [*15274-99-4*] and Na_3FeF_6 [*20955-11-7*], respectively. Sodium hexafluoroferrate has been used as a fluorinating reagent (7).

Mixed-Valency Iron Fluoride Hydrate

A mixed-valency iron fluoride heptahydrate [*39312-70-4*], $FeF_5.7H_2O$, pale yellow crystals, is prepared by dissolving iron powder in 40% HF in the presence of air (8). There are no known applications of mixed-valent iron fluoride.

BIBLIOGRAPHY

"Iron Compounds" in *ECT* 1st ed., under "Fluorine Compounds, Inorganic," Vol. 6, p. 709, by F. D. Loomis, Pennsylvania Salt Manufacturing Co.; "Iron" in *ECT* 2nd ed., under "Fluorine Compounds, Inorganic," Vol. 9, pp. 625–626, by Wayne E. White, The Ozark-Mahoning Company.

1. W. Biltz and E. Rahlfs, *Z. Anorg. Chem.* **166,** 374 (1927).
2. I. G. Ryss, *The Chemistry of Fluorine and its Inorganic Compounds,* State Publishing House for Scientific, Technical and Chemical Literature, Moscow, 1956, English translation, AEC-tr-3927, Office Technical Services, U.S. Department of Commerce, Washington, D.C., 1960, p. 665.
3. S. Okazaki, *Nippon Kagaku Zasshi* **89,** 1054 (1968).
4. Jpn. Kokai 75 17,173 (June 19, 1975), T. Kusaka (to Hinoshita Rare Metal Institute).
5. B. Z. Slivnik, *Inorg. Nucl. Chem. Herbert H. Hyman Memorial* Volume, 173 (1976).
6. Ger. Offen. 2,531,816 (Feb. 12, 1976), E. Dorfman, R. R. Hindersinn, and W. T. Schwartz (to Hooker Chemical and Plastic Corp.).
7. B. Cornils, M. Rasch, and G. Schiemann, *Chem. Ztg. Chem. Appl.* **92**(5), 137 (1968).
8. K. J. Gallagher and M. T. Ottaway, *J. Chem. Soc. Dalton* 978 (1975).

DAYAL T. MESHRI
Ozark-Mahoning Company,
a subsidiary of the Pennwalt Corp.

LEAD

Lead(II) Fluoride

Lead difluoride [*7783-46-2*], PbF_2, is prepared by the action of hydrofluoric acid on lead carbonate or lead hydroxide. It can also be prepared by precipitation from lead nitrate or lead acetate solutions with ammonium fluoride or potassium fluoride. It is a white crystalline powder, mp 824°C and bp 1292°C. Its solubility in water at 25° is 0.066 g/100 g H_2O and in anhydrous hydrogen fluoride 2.62 g/100 g anhydrous HF (1). The pH of a saturated aqueous solution at 25°C is 5.3 (2), and the activity is 2.69×10^{-8} (3). It is a mild fluorinating reagent, and is used in the conversion of molybdenum and tungsten oxide to oxyfluorides, as a raw material for lead tetrafluoride, in the manufacture of special grades of glass (qv) and as an ingredient in underwater paints (see Coatings, marine).

A very high-purity material (99%) is commercially available (see Fine chemicals).

Lead(IV) Fluoride

Lead tetrafluoride [*7783-59-7*], PbF_4, is prepared by the actions of elemental fluorine (4–5) on lead difluoride, or sulfur tetrafluoride (6) with lead dioxide, or from lead dioxide and anhydrous hydrogen fluoride (7). The earlier procedure (8–9) described for the preparation of lead tetrafluoride *in situ* by reaction of lead tetraacetate with anhydrous hydrogen fluoride is now known to involve even more reactive species such as lead diacetate difluoride [*301-04-2*] (10).

Lead tetrafluoride is a white crystalline powder with a density of 6.7 g/cm^3 and melting point ca 600°C. This compound is highly moisture sensitive and turns brown in the presence of moist air owing to its hydrolysis to lead dioxide and hydrogen fluoride. It should be handled in a dry box or under an atmosphere of dry N_2.

Lead tetrafluoride produced by various routes, including *in situ* methods, is a very effective fluorinating reagent for olefins. It can be used for liquid-phase, low temperature selective fluorine addition to the double bond of both halogenated and hydrocarbon olefins (7,11). This compound has been utilized in the preparation of the biologically active fluorinated steroidal hormones (qv) (12–13). Lead fluorides should be handled with great care because they are highly toxic due to the presence of both lead and fluoride ions.

A typical liquid-phase fluorination is carried out by mixing the olefin, lead dioxide, and anhydrous hydrogen fluoride at −78°C and then allowing the mixture to warm slowly to room temperature. Precautions are necessary as the reaction may be a vigorous one.

BIBLIOGRAPHY

"Lead Compounds" under "Fluorine Compounds, Inorganic" in *ECT* 1st ed., Vol. 6, p. 709–710, by F. D. Loomis, Pennsylvania Salt Manufacturing Co.; "Lead" under "Fluorine Compounds, Inorganic" in *ECT* 2nd ed., Vol. 9, p. 626–627, by Wayne E. White, Ozark-Mahoning Company.

1. A. W. Jache and G. H. Cady, *J. Phys. Chem.* **56,** 1106 (1952).
2. R. H. Carter, *Ind. Eng. Chem.* **20,** 1195 (1928).

3. T. DeVaries and R. W. Ivelt, *J. Am. Chem. Soc.* **63,** 2821 (1941).
4. H. V. Wartenberg, *Z. Anorg. Allgem. Chem.* **244,** 337 (1940).
5. J. Bornstein and L. Skarlos, *J. Am. Chem. Soc.* **90,** 5046 (1968).
6. E. R. Bissell and D. B. Fields, *J. Org. Chem.* **29,** 1591 (1964).
7. A. L. Henne and T. P. Waalkes, *J. Am. Chem. Soc.* **67,** 1639 (1945).
8. O. Dimroth and W. Bockemüller, *Ber.* **64,** 516 (1931).
9. J. Bornstein and co-workers, *J. Am. Chem. Soc.* **85,** 1609 (1963).
10. D. D. Tanner and P. Van-Bostelen, *J. Am. Chem. Soc.* **94,** 3187 (1972).
11. A. L. Henne and T. H. Newby, *J. Am. Chem. Soc.* **70,** 130 (1948).
12. Ger. Pat. 1,167,828 (Apr. 16, 1964), K. Bruckner and H. J. Mannhardt (to E. Merck A.G.).
13. A. Bowers and co-workers, *J. Am. Chem. Soc.* **84,** 1050 (1962).

DAYAL T. MESHRI
Ozark-Mahoning Company,
a subsidiary of Pennwalt Corporation

LITHIUM

Lithium Fluoride

Properties. Lithium fluoride [*7789-24-4*], LiF, is a white nonhygroscopic crystalline material that does not form a hydrate. Although lithium fluoride is an alkali fluoride, its properties are more similar to the alkaline earth fluorides. The solubility in water is quite low and chemical reactivity is low, similar to that of calcium fluoride and magnesium fluoride. Several chemical and physical properties of lithium fluoride are listed in Table 1. At high temperatures, lithium fluoride hydrolyzes to hydrogen fluoride. A bifluoride [*12159-92-1*], LiF.HF, which forms on reaction of LiF with hydrofluoric acid, is unstable to loss of HF in the solid form.

Manufacture

Lithium fluoride is manufactured by the reaction of lithium carbonate or lithium hydroxide with dilute hydrofluoric acid. If the lithium carbonate is converted to the soluble bicarbonate, insolubles can be removed by filtration and a purer lithium fluoride can be made on addition of hydrofluoric acid (12). High purity material can also be made from other soluble lithium salts such as the chloride or nitrate with hydrofluoric acid or ammonium bifluoride (13).

Optical crystals of high purity lithium fluoride are grown by use of the Stockbarger process (10) in sizes to 25 cm dia × 25 cm high (14). Typical commercial material contains 99.2% LiF; typical impurities include Li_2CO_3 and Fe_2O_3 at <0.1% levels, and SO_4^{2-}, PO_4^{3-}, and heavy metals as Pb at <0.01% levels. The price as of July 1978 was $5.73/kg in truckload quantities. Although figures are not published, annual production is probably less than 100 metric tons. Lithium fluoride toxicity relative to use in

Table 1. Properties of Lithium Fluoride

Property	Value	Reference[a]
melting point, °C	848	1
boiling point, °C	1681	1
solubility, g/100 g solvent		
water, 25.4°C	0.133	2
water, 81.8°C	0.150	2
acetic acid 25°C	0.084	3
acetic acid 50°C	0.152	3
liquid HF, 12°C	10.3	4
tetrahydrofuran, 25°C	0.6	5
crystalline form	cubic (NaCl)	6
	a_0 = 0.401736 nm	
density, g/cm^3 (20°C)	2.635	6
index of refraction	1.3915	6
lattice energy, kJ/mol[b]	1020 ± 10	7
standard heat of formation, kJ/mol[b]	−613.0	1
standard entropy, J/(mol·K)[b]	35.9	1
heat capacity, C_p°, J/(mol·K)[b]	42.01	1
heat of fusion, kJ/mol[b]	27.09	1
heat of vaporization, kJ/mol[b]	213	8
heat of hydration, kJ/mol[b]	1.023	8
Debye temperature, °C	449	9

[a] Properties listed in this table can be supplemented by the comprehensive collections in refs. 10 and 11.

[b] To convert kJ to kcal, divide by 4.184.

thermoluminescent dosimetry is discussed in ref. 15 in which 10 mg/d is proposed as the maximum permissible daily intake for the average human body. Ingestion of 200 mg/kg of body weight is lethal to guinea pigs (16).

Uses. Lithium fluoride is used primarily in the ceramic industry to reduce firing temperatures and improve resistance to thermal shock, abrasion, and acid attack (see Ceramics). A major use of LiF is in flux compositions with other fluorides, chlorides, and borates for metal joining (17) (see Solders).

Lithium fluoride is an essential component of the fluorine cell electrolyte; 1% LiF in the KF.2HF electrolyte improves the wettability of the carbon anodes and lowers the tendency of the cells to depolarize (18). Thermoluminescent radiation dosimeters used in personnel and environmental monitoring and in radiation therapy contain lithium fluoride powder, extruded ribbons, or rods (19).

Molten lithium fluoride is used in salt mixtures for an electrolyte in high temperature batteries (FLINAK) (20), and as a carrier in breeder reactors (FLIBE) (21) (see Nuclear reactors).

Large high purity crystals are cut into windows and refracting components for use in x-ray monochromators (14), and in the vacuum uv, uv, visible, and ir ranges.

BIBLIOGRAPHY

"Lithium Fluoride" in *ECT* 1st ed., under "Fluorine Compounds, Inorganic," Vol. 6, pp. 709–710, by F. D. Loomis, Pennsylvania Salt Manufacturing Co.; "Lithium Fluoride" in *ECT* 2nd ed., under "Fluorine Compounds, Inorganic," Vol. 9, p. 627, by G. C. Whitaker, The Harshaw Chemical Company.

1. *JANAF Thermochemical Tables,* Clearinghouse for Federal Scientific and Technical Information, U.S. Dept. of Commerce, Springfield, Va., Dec. 1963.
2. C. B. Stubblefield and R. O. Bach, *J. Chem. Eng. Data* **17,** 491 (1972).
3. J. Emsley, *J. Chem. Soc. A,* 2511 (1971).
4. A. W. Jache and G. W. Cady, *J. Phys. Chem.* **56,** 1106 (1952).
5. Brit. Pat. 787,771 (Dec. 18, 1957), (to Metropolitan Vickers Electric Co.).
6. C. A. Hutchison and H. L. Johnson, *J. Am. Chem. Soc.* **62,** 3165 (1940).
7. D. F. C. Morris, *Acta Crystallogr.* **9,** 197 (1956).
8. M. C. Ball and A. A. Norbury, *Physical Data for Inorganic Chemists,* Longman, Inc., New York, 1974.
9. W. W. Scales, *Phys. Rev.* **112,** 49 (1958).
10. *Gmelins Handbuch der Anorganischen Chemie,* 8th ed., Deutsche Vol. 6, Verlag-Chemie, Weinheim/Bergstrasse, 1960, pp. 305–327.
11. J. W. Mellor, *Comprehensive Treatise on Inorganic and Theoretical Chemistry,* Vol. 2, Suppl. 2, Longman, Green and Co., New York, 1961, pp. 174–178.
12. D. C. Stockbarger, *Rev. Sci. Instrum.* **7,** 133 (1936).
13. U.S. Pat. 3,132,922 (May 12, 1954), R. D. Goodenough and T. G. Cook (to Dow Chemical Co.).
14. *Harshaw Optical Crystals,* Harshaw Chemical Co., Solon, Ohio, 1967, pp. 32–33.
15. N. C. Spoor, *Ann. Occup. Hyg.* **11**(1), 23 (1968).
16. H. C. Hodge and F. N. Smith in J. H. Simon, ed., *Fluorine Chemistry,* Vol. 4, Academic Press, New York, 1965, p. 199.
17. U.S. Pat. 3,958,979 (May 25, 1976), A. R. Valdo (to Ethyl Corp.).
18. J. T. Pinkston, *Ind. Eng. Chem.* **39,** 255 (1947).
19. F. M. Cox, *Proc. 2nd Int. Conf. on Luminescence Dosimetry,* Oak Ridge Nat. Lab., CONF 680920, Sept. 1968; F. M. Cox, A. C. Lucas, and B. M. Kaspar, *Health Phys.* **30,** 135 (1976); J. F. Valley, C. Pache, and P. Lerch, *Helv. Phys. Acta* **49**(2), 171 (1976).
20. G. L. Green, J. B. Hunt, and R. A. Sutula, *U.S. Nat. Tech. Inform. Serv., A.D. Rep. 1973, No. 758001.*
21. C. D. Scott and W. L. Carter, *AEC Accession No. 43422, Rept. No. ORNL-3791, Oak Ridge Nat. Lab.,* 1966.

H. S. HALBEDEL
THOMAS E. NAPPIER
The Harshaw Chemical Company

MAGNESIUM

Magnesium Fluoride

Properties. Magnesium fluoride [*7783-40-6*], MgF_2, is a fine white crystalline powder with low chemical reactivity. This relative inertness makes possible some of its uses, eg, stable permanent films to alter light transmission properties of optical and electronic materials. The reaction with sulfuric acid is so sluggish and incomplete that magnesium fluoride is not a suitable substitute for calcium fluoride in manufacturing hydrogen fluoride. Magnesium fluoride resists hydrolysis to hydrogen fluoride up to 750°C (1). Bimetallic fluorides (such as $KMgF_3$ [*28042-61-7*]) are formed on fusion of MgF_2 alkali metal and ammonium fluorides (2). Chemical and physical properties are listed in Table 1. MgF_2 is birefringent and only mildly affected by high energy radiation, making possible optics for the uv region.

Manufacture. Magnesium fluoride is manufactured by the reaction of hydrofluoric acid and magnesium oxide or carbonate.

$$MgO + 2\,HF \rightarrow MgF_2 + H_2O$$

$$MgCO_3 + 2\,HF \rightarrow MgF_2 + CO_2 + H_2O$$

Formation of a gelatinous precipitate that is difficult to filter can be avoided by addition of magnesium oxide to the acid solution. In order to increase particle size it is often necessary to keep the solution hot for several hours; however, this problem is avoided by heating an intimate mixture of ammonium bifluoride with magnesium

Table 1. Chemical and Physical Properties

Property	Value	Reference
melting point, °C	1263	3
boiling point, °C	2227	3
standard heat of formation, kJ/mol[a]	−112.4	4
standard entropy of formation, J/(mol·K)[a]	178	5
heat of fusion, kJ/mol[a]	58.2	3
heat of vaporization, kJ/mol[a]	264	5
lattice energy, kJ/mol[a]	2920	5
heat capacity, 25°C, J/(mol·K)[a]	61.59	3
free energy of solution, kJ/mol[a]	40.2	5
density, g/cm³	3.127	6
index of refraction[b]	n_o = 1.37770 (589 nm)[b]	7
	n_e = 1.38950[b]	7
crystalline form (sellaite)	tetragonal	6
	a = 0.4623 nm	
	c = 0.3052 nm	
solubility, g/100 g of solvent		
25°C, water	0.013	8
12°C, hydrogen fluoride	0.025	9
25°C, acetic acid	0.681	10

[a] To convert kJ to kcal, divide by 4.184.
[b] *o*, ordinary; *e*, extraordinary.

carbonate to 150–400°C (11). Particles of MgF_2 similar in size to those of the magnesium carbonate are obtained.

$$MgCO_3 + NH_4HF_2 \xrightarrow{150\text{–}400°C} MgF_2 + NH_3 + CO_2 + H_2O$$

The same results are obtained by adding magnesium carbonate to an aqueous solution of ammonium bifluoride and ammonium hydroxide and warming to 60°C (12). The resulting precipitate is ammonium magnesium fluoride [*35278-29-6*] which settles rapidly.

$$MgCO_3 + 3\ NH_4F \xrightarrow{aq} NH_4MgF_3 + (NH_4)_2CO_3$$

$$NH_4MgF_3 \xrightarrow[4\ h]{620°C} MgF_2 + NH_3 + HF$$

Magnesium fluoride is a by-product of the manufacture of metallic beryllium and uranium. The beryllium or uranium fluorides are intimately mixed with magnesium metal in magnesium fluoride-lined crucibles. On heating, a Thermite-type reaction takes place to yield the desired metal and MgF_2 (13). Part of the magnesium fluoride produced in this reaction is then used as a lining for the crucibles used in the process.

$$BeF_2 + Mg \rightarrow Be + MgF_2$$

A commercial grade of magnesium fluoride containing approximately 96–98% MgF_2 is manufactured by Kawecki-Berylco Industries in the United States. Imported technical grades suitable for fluxes containing 94–96% magnesium fluoride are available from American Fluoride Corporation, Atomergic Chemetals Company, and Fine Chemical Co.

Magnesium fluoride optical crystals are made by hot-pressing (14) high quality MgF_2 powder. The optical quality powder is made by the NH_4HF_2 method described above (11) or by reaction of magnesium bicarbonate and hydrofluoric acid (15). Lead fluoride can also be used in purification of MgF_2 for optical crystals (16). Such optical crystals are manufactured by the Harshaw Chemical Co.

Toxicity. The lethal dose of MgF_2 to guinea pigs by ingestion is 1000 mg/kg (17).

Uses. Established uses of magnesium fluoride are as fluxes in magnesium metallurgy and in the ceramics industry. A proposed use is the extraction of aluminum from arc-furnace alloys with Fe, Si, Ti, and C (18). The molten alloy in reacting with magnesium fluoride volatilizes the aluminum and magnesium which are later separated above the melting point of MgF_2. A welding (qv) flux for aluminum (19) as well as fluxes for steel (20) contain MgF_2.

Optical windows of highly purified magnesium fluoride which transmit light from the vacuum ultraviolet (140 nm) into the infrared (7) are recommended for use as ultraviolet optical components for use in space exploration.

BIBLIOGRAPHY

"Magnesium Fluoride" treated in *ECT* 1st ed., under Fluorine Compounds, Inorganic," Vol. 6, p. 709, by F. D. Loomis, Pennsylvania Salt and Manufacturing Co.; "Magnesium Fluoride" treated in *ECT* 2nd ed., under "Fluorine Compounds, Inorganic," Vol. 9, pp. 627–628 by G. C. Whitaker, The Harshaw Chemical Company.

1. D. R. Messier, *J. Am. Ceram. Soc.* **48,** 452, 459 (1965).
2. I. G. Ryss, *The Chemistry of Fluorine and Its Inorganic Compounds,* State Publishing House for Scientific and Technical Literature, Moscow, 1956; English translation by F. Haimson for the U.S. Atomic Energy Commission, *AEC-tr-3927,* 1960, p. 812.
3. *JANAF Thermochemical Tables,* Clearinghouse for Federal Scientific and Technical Information, U.S. Dept. of Commerce, Springfield, Va., 22151, 1966.
4. E. Ruelzitis, H. M. Fedar, and W. N. Hubbard, *J. Chem. Phys.* **68,** 2978 (1964).
5. M. C. Ball and A. A. Norbury, *Physical Data for Inorganic Chemicals,* Longman, Inc., New York, 1974.
6. C. Palache, H. Berman, and C. Frandel, *Danas System of Mineralogy,* 7th ed., Vol. 2, John Wiley & Sons, Inc., New York, 1951, p. 38.
7. The Harshaw Chemical Company, Crystal and Electronics Dept., Solon, Ohio.
8. D. D. Ikrami, A. S. Paramzin, and A. Kubr, *Russ. J. Inorg. Chem.* **16,** 425 (1971).
9. A. W. Jache and G. W. Cady, *J. Phys. Chem.* **56,** 1106 (1952).
10. C. J. Emsley, *J. Chem. Soc. (A),* 2511 (1971).
11. U.S. Pat. 3,357,788 (Dec. 12, 1967), J. F. Ross (to General Electric Co.).
12. U.S. Pat. 3,848,066 (Nov. 12, 1974), C. D. Vanderpool and M. B. MacInnis (to G.T.E. Sylvania).
13. H. E. Thayer, Proc. *U. M. Internat. Conference, Peaceful Uses of Atomic Energy,* 2nd, Geneva, **4,** 22 (1958); W. E. Dennis and E. Proudfoot, *U.K. At. Energy R&D* **B** (C) TN-88 (1954); *Chem. Abstr.* **53,** 14595 (1959).
14. U.S. Pat. 3,294,878 (Dec. 29, 1960), E. Carroll and co-workers (to Eastman Kodak Co.).
15. U.S. Pat. 3,920,802 (Nov. 18, 1975), R. H. Moss, C. F. Swinehart, and W. F. Spicuzza (to Kewanee Oil Co.).
16. U.S. Pat. 2,498,186 (Feb. 21, 1950), D. C. Stockbarger and A. A. Blanchard (to Research Corp.).
17. H. C. Hodges and F. A. Smith in J. H. Simons, ed., *Fluorine Chemistry,* Vol. 4, Academic Press, New York, 1965, p. 199.
18. G. S. Layne and co-workers, *Light Met. Age* **30**(3,4), 8 (1972).
19. U.S. Pat. 2,552,104 (May 8, 1951), M. A. Miller and W. E. Haupin (to Alcoa).
20. Jpn. Pat. 7,556,339 (Sept. 19, 1973), T. Tanigaki, T. Koshio, and T. Enomoto (to Nippon Steel Corp.).

THOMAS E. NAPPIER
H. S. HALBEDEL
The Harshaw Chemical Company

MERCURY

Mercury(I) Fluoride

Mercury(I) fluoride [*13967-25-4*], Hg_2F_2, also known as mercurous fluoride, is probably a white solid when pure; however, it is contaminated usually with small amount of oxides that give it a yellowish appearance. Several methods for its preparation are described in ref. 1: adding sodium fluoride to mercury(I) nitrate solution; heating a mixture of Hg_2Cl_2 and NaF; and dissolving Hg_2O in dilute hydrofluoric acid. The most practical method for obtaining reasonably pure Hg_2F_2 uses slow addition of freshly prepared mercury(I) carbonate to a stirred solution of 48 wt % hydrofluoric acid; the resulting mixture is evaporated to dryness in a platinum dish (2). This method gives 98% pure mercury(I) fluoride.

Mercury(I) fluoride is light sensitive and decomposes in water at 15°C. It can be dried in vacuum over H_2SO_4. It melts at 570°C, decomposes below 650°C, and its specific gravity at 15°C is 8.73. Though sparingly soluble in anhydrous HF, it is much more soluble in aqueous HF. It is classified as a soft fluorinating reagent (3)—similar in activity to AgF—but it does not form complex compounds and its total fluorine is available for exchange reactions. Fluorination by Hg_2F_2 is catalyzed by the presence of I_2 (2). Monobromides or monoiodides are easily converted by Hg_2F_2 into fluorides. All mercury compounds are highly toxic and should be handled carefully.

Mercury(I) fluoride has remained a laboratory curiosity reagent for a long time. No commercial applications have been developed, and it is rarely available from a commercial source.

Mercury(II) Fluoride

Mercury(II) fluoride [*967-25-4*], HgF_2, also known as mercuric fluoride, is a white hygroscopic solid that instantly turns yellow upon exposure to moisture or air. It should be handled in a dry box or under an atmosphere of nitrogen, and in the dark as it appears to be light sensitive. It melts at 645°C, has a specific gravity of 8.95, and is sparingly soluble in anhydrous HF.

Anhydrous mercury(II) fluoride can be prepared by passing elemental fluorine over $HgCl_2$ or HgO (4) at 270°C. The white solid dihydrate, $HgF_2.2H_2O$ [*26453-89-4*], can be obtained by treating HgO several times with 50% HF. Mercuric fluoride is produced *in situ* by condensing anhydrous HF over HgO (2) or $HgCl_2$ (5).

Mercuric fluoride is a moderate-to-hard fluorinating reagent, depending upon the reaction media and conditions. It reacts with metals, eg, Cu, Pb, Sn, Mg, Cr, and As, at an elevated temperature and converts them to fluorides. It is also used for fluorination of organic compounds (3,6–7), and for electrophilic addition to olefins (4).

Mercury(II) fluoride is commercially available in 100 g–45 kg lots. However, its use has been limited to selective applications owing to the toxicity of the reagent and its high cost and handling requirements. Current worldwide demand is less than 225 kg/yr.

BIBLIOGRAPHY

"Mercury Fluorides" under "Fluorine Compounds, Organic" in *ECT* 1st ed., Vol. 6, pp. 747–748, by E. T. McBee and O. R. Pierce, Purdue University; "Mercury" under "Fluorine Compounds, Inorganic" in *ECT* 2nd ed., Vol. 9, p. 628, by Wayne E. White, Ozark-Mahoning Company.

1. I. G. Ryss, *The Chemistry of Fluorine Compounds,* State Publishing House for Scientific, Technical, and Chemical Literature, Moscow, USSR, 1956, Engl. transl., *AEC-tr-3927,* Vol. II, Office of Technical Services, U.S. Department of Commerce, Washington, D.C., 1960, p. 634.
2. A. L. Henne and M. W. Renoll, *J. Am. Chem. Soc.* **60,** 1960 (1938).
3. D. T. Meshri and W. E. White, *G. H. Cady ACS Symposium, Milwaukee, Wisc., June 1970.*
4. F. Swarts, *Bull. Classe. Sci. Acad. Roy. Belg.* **7,** 438 (1921).
5. O. Ruff and co-workers, *Ber.* **51,** 1752 (1918).
6. F. Swarts, *Bull. Acad. Roy. Belg.* **22,** 105 (1936).
7. R. N. Haszeldine and B. R. Steele, *J. Chem. Soc.,* 1592 (1953).

DAYAL T. MESHRI
Ozark-Mahoning Company,
a subsidiary of the Pennwalt Corp.

MOLYBDENUM

Molybdenum has four binary fluorides: molybdenum trifluoride [*20193-58-2*], MoF_3; molybdenum tetrafluoride [*23412-45-5*], MoF_4; molybdenum pentafluoride [*13819-84-6*], MoF_5; and molybdenum hexafluoride [*7783-77-9*], MoF_6. Molybdenum also forms two oxyfluorides, MoO_2F_2 [*13824-57-2*] and $MoOF_4$ [*14459-59-7*]. None of these compounds has been of commercial significance and use is limited to research laboratories.

Molybdenum Hexafluoride

Molybdenum hexafluoride, the best known of the aforementioned compounds, is a volatile metal fluoride having properties very similar to tungsten hexafluoride. It is a solid and has a melting point of 17.4°C and a boiling point of 35°C (1). It exists as a body-centered cubic system near room temperature and undergoes a phase transition to an orthorhombic structure below −9.6°C (2–3). The calculated density of the solid is 2.88 g/cm^3 at 21°C and the density of the liquid is 2.544 g/cm^3 (2). The heat of fusion is 4.435 kJ/mol (1.06 kcal/mol) and the heat of vaporization is 27.7 kJ/mol (6.63 kcal/mol) (2). Thermodynamic formation data are shown in Table 1 (4). An ex-

Table 1. Thermodynamic Formation Data of MoF_6

	MoF_6 (l)	MoF_6 (g)
$-\Delta H_f$, kJ/mol, 25°C	1626	1558
(kcal/mol)	(388.6)	(372.4)
$-\Delta G_f$, kJ/mol, 25°C	1511	1468
(kcal/mol)	(361.2)	(350.8)
$-\Delta S°$, kJ/K		302
(kcal/K)		(72.1)

tensive compilation of the physical properties of molybdenum hexafluoride has been prepared (2).

Molybdenum hexafluoride is most commonly produced in the laboratory by the reaction of molybdenum metal with fluorine at elevated temperatures (5). It reacts vigorously with organic materials and rapidly hydrolyzes to produce HF. Its reduction by hydrogen at elevated temperatures to yield molybdenum metal is the basis of the use of MoF_6 in chemical vapor deposition operations to produce metallic molybdenum deposits. The chemical vapor deposition of molybdenum and molybdenum–tungsten alloys by chemical vapor deposition is described in refs. 6 and 7 (see Film deposition techniques). The lack of a significant market for MoF_6 has resulted in the discontinuance of its manufacture.

BIBLIOGRAPHY

1. C. L. Rollinson "Chromium, Molybdenum, and Tungsten" in J. C. Bailer and co-eds., *Comprehensive Inorganic Chemistry,* Vol. 3, Pergamon Press, Compendium Publishers, Elmsford, N.Y., 1973, p. 565.
2. M. Carles, "L'Hexafluorure de Molybdene, MoF_6," *Commissariat a L'Energie Atomique, Pierrelatte (France), Report CEA-BIB-124,* NASA Technical Translation TT-F-12, 702, Sept. 1968, pp. 1–25.
3. S. Siegel and D. A. Northrup, *Inorg. Chem.* **5,** 2187 (1966).
4. J. E. Fergusson, "Halide Chemistry of Chromium, Molybdenum, and Tungsten," in *Halogen Chemistry,* Vol. 3, Academic Press, New York, 1967, p. 276.
5. Private conversation with Dayal T. Meshri, The Ozark-Mahoning Company, Oct. 1979.
6. J. G. Donaldson and H. Kenworthy, *Electrodeposition Surf. Treat.* **2,** 435 (1973/74).
7. A. M. Shroff and G. Delval, *High Temp. High Pressure* **3,** 695 (1971).

ANDREW J. WOYTEK
Air Products & Chemicals, Inc.

NICKEL

Nickel Difluoride

Nickel difluoride [*10028-18-9*], NiF_2, a pale-yellow solid, is the only known stable binary compound of nickel and fluorine. It is prepared by the action of elemental fluorine or anhydrous hydrogen fluoride on anhydrous $NiCl_2$ at elevated temperatures. It is also prepared by heating nickel fluoride tetrahydrate [*13940-83-5*], $NiF_2.4H_2O$, in a stream of anhydrous hydrogen fluoride at 500°C for several hours. $NiF_2.4H_2O$ is prepared by the reaction of $NiCO_3$ with aqueous HF. Earlier reports that crystalline hydrates of CoF_2 and NiF_2 can be easily dehydrated are incorrect (1). The other methods for the preparation of anhydrous NiF_2 are the fluorination of metal salts with excess SF_4 (2) or ClF_3 (3) at elevated temperatures, or the reaction of $NiCO_3$ with anhydrous HF at 250°C (4).

Anhydrous nickel fluoride melts at ca 1100°C and slowly changes to oxide if heated in a humid atmosphere. It is slightly soluble in water (ca 2.5% at 25°C) and more soluble in aqueous hydrogen fluoride, but it is insoluble in anhydrous HF.

NiF_2 is used in the manufacture of varistors (5), as a catalyst for hydrofluorination (6) in the synthesis of XeF_6 (7), and in the preparation of high-purity elemental fluorine gas (8).

There is very little commercial demand for nickel fluoride. All work to date has been done on a laboratory scale and is the subject of patents. Current United States market demand for anhydrous NiF_2 is less than 45 kg/yr. It is available in 100 g–45 kg quantities. The current price for 99% pure nickel fluoride is \$45/100 g and \$200/500 g. Small quantities are stored and shipped in polyethylene bottles, whereas large amounts are shipped in fiber drums lined with polyethylene bags.

Nickel and its compounds should be treated as hazardous materials and may cause nasal and lung cancer (8). LD_{50} (mouse iv) for NiF_2 is 130 mg/kg (9).

Other Nickel Fluorides

There is some evidence for the existence of a higher valency binary fluoride of nickel, NiF_3 [*18642-20-1*], but it has not been fully identified or characterized. A brownish material which had a slight oxidizing power and which decomposed immediately after exposure to air was observed during the fluorination of anhydrous NiF_2 at 200°C (10). A similar brown material was obtained by electrolysis of NiF_2-HF solution (11). The brown material liberated iodine from KI solution and turned yellow. A black compound that was obtained by adding AsF_5 to K_2NiF_6 solution in anhydrous hydrogen fluoride was claimed to be an impure NiF_3 that decomposed to NiF_2 during purification process (12).

Nickel Fluoride Complexes

Complex salts of nickel, M_2NiF_6 (M = Na [*21958-95-2*], K [*17218-47-2*], Rb [*17218-48-3*], and Cs [*17218-49-4*]) and M_3NiF_6 (M = Na [*22707-99-9*], K [*14881-07-3*], Rb [*72151-96-3*], and Cs [*72138-72-8*]), have been prepared by the action of fluorinating reagents, eg, F_2, ClF_3, or XeF_2, on a mixture of nickel halides and alkali metal halides (13–14). A typical reaction is as follows:

$$NiX_2 + n\ MX \xrightarrow{F_2} M_nNiF_6$$

(M = alkali metal, X = Cl or F, n = 2 or 3)

These complex salts can be used as fluorinating reagents (12) in anhydrous HF, and as sources of elemental fluorine (8). Recently bis(tetrafluoroammonium) hexafluoronickelate [*63105-40-8*], $(NF_4)_2NiF_6$ (15) was prepared from Cs_2NiF_6 and NF_4SbF_6 by metathesis in anhydrous HF. It has a potential application as a high-energy solid propellant oxidizer (see Explosives and propellants).

BIBLIOGRAPHY

"Nickel Fluoride" under "Fluorine Compounds, Inorganic" in *ECT* 1st ed., Vol. 6, p. 710, by F. D. Loomis, Pennsylvania Salt Manufacturing Co.; "Nickel" under "Fluorine Compounds, Inorganic" in *ECT* 2nd ed., Vol. 9, pp. 628–629, by Wayne E. White, Ozark-Mahoning Company.

1. I. G. Ryss, *The Chemistry of Fluorine and Its Inorganic Compounds,* State Publishing House for Scientific, Technical, and Chemical Literature, Moscow, USSR, 1956, Eng. transl., *AEC-tr-3927,* Office of Technical Services, U.S. Department of Commerce, Washington, D.C., 1960, pp. 657–659 (Part II).
2. U.S.S.R. Pat. 495,279 (Dec. 15, 1975), A. P. Kostyuk and L. M. Yagupolskii (to Odessa Polytechnic Institute); *Chem. Abstr.* **84,** 1079486 (1976).
3. Y. I. Nikonorou and co-workers, *Izv. Sib. Otd. Akad. Nauk SSSR Ser. Khim. Nauk* (3), 88 (1976); *Chem. Abstr.* **85,** 865415 (1976).
4. U.S. Pat. 3,836,634 (Sept. 17, 1974), J. Saldick (to FMC Corp.).
5. Jpn. Kokai 75 139,395 (Nov. 7, 1975), M. Matsuura and co-workers (to Matsushita Electrical Industrial Co.).
6. U.S.S.R. Pat. 466,202 (Apr. 5, 1975), Kh. U. Usmanov and co-workers (to Tashkent State University); *Chem. Abstr.* **83,** 85068h (1975).
7. B. Zemva and J. Slivnik, *Vestn. Slov. Kem. Drus.* **19**(1–4), 43 (1972); *Chem. Abstr.* **79,** 38115K (1973).
8. L. B. Asprey, *J. Fluorine Chem.* **7**(1–3), 359 (1976).
9. *Metal Toxicity in Mammals,* Vol. 2, Plenum Publishing Corp., New York, 1978, p. 293.
10. L. Stein, Argonne National Laboratory, Argonne, Ill., private communication, July 1979.
11. L. Stein, J. M. Neil, and G. R. Alms, *Inorg. Chem.* **8,** 2472 (1969).
12. T. L. Court and M. F. A. Dove, *J. Chem. Soc. Chem. Commun.,* 726 (1971).
13. S. V. Zemskov and co-workers, *Izv. Sib. Otd. Akad. Nauk. SSSR Ser. Khim. Nauk* (3), 83 (1976).
14. W. Klemm and E. Huss, *Z. Anorg. Chem.* **258,** 221 (1949).
15. U.S. Pat. 4,108,965 (Aug. 22, 1978), K. O. Christe (to Rockwell International Corp.).

DAYAL T. MESHRI
Ozark-Mahoning Company,
a subsidiary of the Pennwalt Corp.

NITROGEN

Nitrogen has three binary fluorides: nitrogen trifluoride [*7783-54-2*], NF_3; tetrafluorohydrazine [*10036-47-2*], N_2F_4; and difluorodiazine [*10578-16-2*], N_2F_2. There are numerous other nitrogen fluorine compounds, the most significant of which are the perfluoroammonium salts based on the NF_4^+ cation. Of all the nitrogen fluorine compounds, only NF_3 has been of commercial importance. Nitrogen trifluoride is currently being used as the fluorine source in high-power chemical lasers.

Nitrogen Trifluoride

Physical Properties. Nitrogen trifluoride, NF_3, is a colorless gas, liquefying at 101.3 kPa (1 atm) and −129.0°C (1) and solidifying at −206.8°C (2). High purity NF_3 has little odor, but commercial material contaminated with traces of active fluorides has a pungent, musty odor. The critical-state constants for NF_3 are: T_c = 39.25°C (1), P_c = 4.53 MPa (44.7 atm) (1), and V_c = 123.8 cm³/mol (3). The heat of vaporization is 11.59 kJ/mol (2769 cal/mol) (2) and the heat of fusion is 398 J/mol (95.1 cal/mol) (2). Nitrogen trifluoride undergoes a low temperature solid transition at −216.5°C and the heat of transition is 1.513 kJ/mol (361.7 cal/mol) (2). The density of the liquid at its normal boiling point is 1.533 g/cm³ (1). Nitrogen trifluoride is very slightly soluble in water. The solubility is 1.4×10^{-5} mol NF_3 per mol H_2O at 101.3 kPa (1 atm) and 25°C (4–5). The standard heat of formation of NF_3 at 25°C is 131.5 J/mol (−31.43 cal/mol) (6). The heat of capacity of the gas is 53.39 J/(mol·K) [12.76 cal/(mol·K)] at 25°C (7). The vapor pressure of NF_3 in the temperature range of 89 to 233 K has the following relationship (1):

$$\log P_{\text{kPa}} = 5.90445 - \frac{501.913}{T - 15.37}\left(\log P_{\text{mm Hg}} = 6.77966 - \frac{501.913}{T - 15.37}\right)$$

The infrared spectrum of NF_3 has been investigated (8–11) and extensive tabulation and correlation of all the physical properties of NF_3 is available (12).

Chemical Properties. Nitrogen trifluoride is an oxidizer that is thermodynamically stable except at elevated temperature. At temperatures up to ca 350°C, its reactivity is comparable to oxygen. At higher temperature, its reactivity is similar to fluorine owing to appreciable dissociation into NF_2 and F. The thermal dissociation of NF_3 has been studied by a number of investigators (13–16) and was found to peak in the temperature range of 1100 to 1500 K.

Nitrogen trifluoride acts primarily upon the elements as a fluorinating agent, but not a very active one at lower temperatures. At elevated temperatures, NF_3 pyrolyzes with many of the elements to produce N_2F_4 and the corresponding fluoride. The pyrolysis of NF_3 over copper turnings produces N_2F_4 in a 62–71% yield at 375°C (17). Pyrolysis over carbon is more favorable (18–19).

Hydrogen reacts with NF_3 with the rapid liberation of large amounts of heat and is the basis for the use of NF_3 in high-energy chemical lasers. The flammability limit of NF_3/H_2 mixtures is 9.4–95 mol % NF_3. Nitrogen trifluoride reacts with organic compounds but generally an elevated temperature is required to initiate the reaction. Under these conditions, the reaction will often proceed explosively and great care must be exercised when exposing NF_3 to organic compounds. Therefore, NF_3 has found little use as a fluorinating agent for organic compounds. The reactions of NF_3 with the elements, and various inorganic and organic substances are summarized in ref. 12.

Manufacture and Economics. Nitrogen trifluoride can be formed from a wide variety of chemical reactions. However, only two processes have been technically and economically feasible for large-scale production. These processes are: (*1*) the electrolysis of molten ammonium acid fluoride; and (*2*) the direct fluorination of the ammonia in the presence of molten ammonium fluoride. In the electrolytic process, NF_3 is produced at the anode and H_2 is produced at the cathode. In a divided cell of 4 kA with nickel anodes, extensive dilution of the gas streams with N_2 was used to prevent explosive reactions between NF_3 and H_2 (19).

A recent process produced NF_3 directly by the reaction of NH_3 with F_2 in the presence of molten ammonium acid fluoride (20). The process uses a specially designed reactor wherein the gaseous F_2 is contacted with the molten salt which acts as a transport medium for supplying NH_3 to the reaction zone and carrying away the intense heat of reaction (21). Since H_2 is not generated in this process, the hazards associated with the reactions between NF_3 and H_2 are eliminated.

U.S. production is currently less than 25 metric tons per year. Production capacity is expected to increase significantly with the development of the high-energy chemical laser. Currently, Air Products & Chemicals, Inc., and the Rocketdyne Division of Rockwell International are the chief commercial producers in the United States. The 1979 price is approximately $100/kg at current production levels. Some reduction in price is anticipated as requirements increase.

Specifications and Analysis. Nitrogen trifluoride is shipped as a high pressure gas at 7 MPa (1000 psig) and is available in tube trailers and cylinders (22).

Nitrogen trifluoride is sold according to the following specifications (22):

nitrogen trifluoride	98.6 mol %
oxygen	0.6 mol %
nitrogen	0.1 mol %
carbon tetrafluoride	0.15 mol %
nitrous oxide	0.2 mol %
sulfur hexafluoride	0.05 mol %
reactive fluorides	<0.3 wt %

Analysis of NF_3 for the above impurities can be performed on the gas chromatograph (23). Active fluorides are determined by scrubbing with a basic solution and wet-chemical analysis for fluoride content (23).

Handling and Toxicity. Nitrogen trifluoride gas is noncorrosive to the common metals at temperatures below 70°C and can be used with steel, copper, nickel, aluminum, and stainless steel. The corrosion rate of NF_3 on these materials significantly increases with moisture or HF. Nitrogen trifluoride is compatible with the fluorinated polymers such as Teflon, Kel-F, and Viton at ambient conditions. Extensive data on the corrosion rates of gaseous and liquid NF_3 to a variety of metals and nonmetals are compiled in ref. 12.

Nitrogen trifluoride is a toxic substance and is most hazardous by inhalation. The threshold limit value is 29 mg/m^3 as established by the ACGIH (24). Since NF_3 has very little odor, it cannot be detected by its odor at concentrations within the TLV. Therefore, adequate personnel protection or monitoring must be provided when handling NF_3. Commercial halogen leak detectors can be used to detect NF_3 at approximately 1 ppm (25). However, they are nondiscriminating and will detect other halogenated gases as well. Long-path ir cells are sensitive to NF_3 in the 900 cm^{-1} region at concentrations below 1 ppm (25).

The inhalation toxicity of NF_3 on rats, mice, dogs, and monkeys using NF_3 concentrations from 100 to 40,000 ppm and exposure periods from 10 to 40,000 min have been studied extensively (26–35). These studies provide the basis of emergency exposure limits (EEL) that have been proposed for NF_3. The NAS-NRC Committee on Toxicology recommends that the EEL for NF_3 be the following: 10 min, 2250 ppm; 30 min, 750 ppm; 60 min, 375 ppm. Gaseous NF_3 is considered to be innocuous to the skin and a minor irritant to the eyes and mucous membranes.

Uses. The only significant current use of NF_3 is as a fluorine source for the hydrogen fluoride and deuterium fluoride (HF/DF) high-energy chemical lasers (qv). The HF/DF lasers are the most promising of the chemical lasers under development since a substantial fraction (ca 25%) of the energy of the reaction between H_2/D_2 and F_2 can be released as laser radiation (37–38). The use of NF_3 is preferred to F_2 because of its comparative ease of handling at ambient temperatures as compared to F_2. Therefore, its storage and handling do not require the precautions which are necessary for the large-scale use of fluorine.

Minor amounts of NF_3 are used as chemical intermediates in the production of N_2F_4 and a series of perfluoroammonium metal salts. Nitrogen trifluoride was used as an oxidizer in rocketry in the early 1960s, but this application was not commercialized.

Tetrafluorohydrazine

Tetrafluorohydrazine, N_2F_4, is a colorless gas of musty odor, condensing at −73°C at 101.3 kPa (1 atm), freezing at −163°C, with a critical temperature of 36°C, and a critical pressure of 7.8 MPa (77 atm). The gas slowly reacts with water. When heated from room temperature at 101.3 kPa to 300°C, dissociation to NF_2 radicals increases from 0.05 to 90%.

Tetrafluorohydrazine was produced in commercial quantities by Air Products and Chemicals, Inc. in the early 1960s for use in rocketry. The process involved the pyrolysis of NF_3 over a fluidized bed of carbon at ca 400–500°C (19). Currently, N_2F_4 is not available commercially and is produced in small quantities under government contract.

Table 1. Perfluoroammonium Salts

Compound	CAS Reg. No.	Reference
NF_4SbF_6	*[16871-76-4]*	39–42
NF_4AsF_6	*[16871-75-3]*	40, 43–44
$NF_4Sb_2F_{11}$	*[58702-89-9]*	40
$NF_4Sb_3F_{16}$	*[72203-31-7]*	40
$(NF_4)_2TiF_6$	*[61128-92-5]*	45
$(NF_4)_2SnF_6$	*[61587-66-4]*	46
NF_4SnF_5	*[61587-75-5]*	46
NF_4BiF_6	*[61587-71-1]*	41
NF_4BF_4	*[15640-93-4]*	36, 47
NF_4PF_6	*[58702-88-8]*	48
NF_4GeF_5	*[58702-86-6]*	48

Perfluoroammonium Salts. A new series of compounds based on the perfluoroammonium cation NF_4^+ have recently been synthesized. This cation was first reported in the reaction of NF_3, F_2, and SbF_5 in liquid HF at 200°C and up to 8.6 MPa (85 atm) to produce a mixture of NF_4SbF_6, and the antimonate and fluoride salts of metals from the reaction vessel. These salts are important since they are stable in dry atmospheres up to ca 200°C but rapidly decompose above 300°C to yield NF_3 and F_2. Therefore, the salts are convenient solid sources of F_2 and NF_3. Considerable work has been done to prepare and characterize perfluoroammonium salts shown in Table 1.

BIBLIOGRAPHY

"Nitrogen Trifluoride" in *ECT* 1st ed., under "Fluorine Compounds, Inorganic," Vol. 6, p. 710; "Nitrogen" in *ECT* 2nd ed., under "Fluorine Compounds, Inorganic," Vol. 9, pp. 629–630, by Wayne E. White, Ozark-Mahoning Company.

1. R. L. Jarry and H. C. Miller, *J. Phys. Chem.* **60,** 1412 (1956).
2. L. Pierce and E. L. Pace, *J. Chem. Phys.* **23,** 551 (1955).
3. D. N. Seshadri, D. S. Viswanath, and N. R. Kuloor, *Indian J. Technol.* **8**(5), 153 (1970).
4. J. T. Ashton and co-workers, *J. Chem. Soc. A* **8,** 1793 (1968).
5. C. R. S. Dean, A. Finch, and P. J. Gardner, *J. Chem. Soc., Dalton Trans.* (23), 2722 (1973).
6. A. N. Syverud, *Final Tech. Report, AFRPL-TR-70-20 (AD867952),* Contract FO4611-67-C-009, 1970.
7. *JANAF Thermochemical Tables,* Dow Chemical Co., Midland, Mich., June 1969.
8. P. N. Schatz and I. W. Levin, *J. Chem. Phys.* **29,** 475 (1958).
9. E. L. Pace and L. Pierce, *J. Chem. Phys.* **23,** 1248 (1955).
10. M. K. Wilson and S. R. Polo, *J. Chem. Phys.* **20,** 1716 (1952).
11. C. R. Bailey, J. B. Hale, and J. W. Thompson, *J. Chem. Phys.* **5,** 274 (1937).
12. R. E. Anderson, E. M. VanderWall and R. K. Schaplowsky, "Nitrogen Trifluoride," *USAF Propellant Handbook, AFRPL-TR-77-71,* Contract FO4611-76-C-0058, 1977.
13. K. O. McFadden and E. Tschuikow, *J. Phys. Chem.* **77,** 1475 (1973).
14. G. L. Schott, L. S. Blair, and J. D. Morgan, Jr., *J. Phys. Chem.* **77,** 2823 (1973).
15. E. A. Karko and co-workers, *J. Chem. Phys.* **63,** 3596 (1975).
16. P. J. Evans and E. Tschuikow-Roux, *J. Chem. Phys.* **65,** 4202 (1976).
17. C. B. Colburn and A. Kennedy, *J. Am. Chem. Soc.* **80,** 5004 (1958).
18. *Chem. Eng. News* **38,** 85 (Sept. 19, 1960).
19. U.S. Pat. 3,235,474 (Feb. 15, 1966), J. F. Tompkins, Jr., and E. S. J. Wang (to Air Products and Chemicals, Inc.).
20. U.S. Pat. 4,108,966 (May 22, 1968), A. J. Woytek and J. T. Lileck (to Air Products and Chemicals, Inc.).
21. *Chem. Eng.* **84**(26), 116 (1977).
22. *Specialty Gas Data Sheet, Nitrogen Trifluoride,* Air Products and Chemicals, Inc., Allentown, Pa., 1977.
23. L. A. Dee, "Analysis of Nitrogen Trifluoride," *AFRPL-TR-76-20 (AD-A022887),* Air Force Rocket Propulsion Laboratory, Edward AFB, Calif., April 1976.
24. "Threshold Limit Values for Chemical and Physical Agents in the Workroom Environment with Intended Changes for 1975," *Proceedings ACGIH, Cincinnati, Ohio, 1975.*
25. F. J. Pisacane and E. Baroody, "Nitrogen Trifluoride: Its Chemistry, Toxicity, and Safe Handling," *Report No. NSWC/WOL/MP 76-24,* Naval Surface Weapons Center, Nov. 1976.
26. E. H. Vernot and C. C. Haun, "Acute Toxicology and Proposed Emergency Exposure Limits of Nitrogen Trifluoride," *AMRL-TR-69-130-Paper No. 13 (AD 710062),* Syste-Med Corp., Dayton,Ohio, Contract F33615-70-V-1046, Dec. 1969. Reprinted from the *Proceedings of the 5th Annual Conference on Atmospheric Contamination in Confined Spaces, Sept. 16–18, 1969, AMRL-TR-69-130 (AD 709994),* Dayton, Ohio, Dec. 1969, pp. 165–171.
27. E. H. Vernot and co-workers, *Toxicol. Appl. Pharmacol.* **26,** 1 (1973).
28. T. R. Torkelson and co-workers, *Toxicol. Appl. Pharmacol.* **4,** 770 (1962).

29. D. J. Reed, F. N. Dost, and C. H. Want, "Inorganic Fluoride Propellant Oxidizers, Vol. 1: Their Effects Upon Seed Germination and Plant Growth," *Interim Report (May 15, 1964–May 15, 1966), AMRL-TR-66-187, Vol. 1 (AD 667-556),* Oregon State Univ., Corvallis Ore., Contract AF 33(615)-1767, Nov. 1967.
30. F. N. Dost and co-workers, "Metabolism and Pharmacology of Inorganic and Fluorine Containing Compounds," *Final Report (July 1, 1964–June 30, 1967), AMRL-TR-67-224 (AD 681-161),* Oregon State Univ., Corvallis, Oregon, Contract AF 33(615)-1799, Aug. 1968.
31. F. N. Dost, D. J. Reed and C. H. Wang, *Toxicol. Appl. Pharmacol.* **17,** 585 (1970).
32. G. M. Cianko and C. E. Schoettlin, "Toxicological Range-Finding on New Propellants," *Rept. No. R-3796 (AMRL-TDR-63-65), 15 Dec. 1961–17 Dec. 1962,* Rocketdyne, Contract AF 33(657)-7773, July 1963.
33. "Quarterly Progress Report on Chemical and Propellant Processing," *Rept. No. P-63-10 (Mar. 15–May 15, 1963),* Rohm and Haas Co., Huntsville, Ala., Contracts DA 01-021-ORD-, 11878, 11879, and 12341, Aug. 19, 1963.
34. G. L. Coppoc, and S. J. Leger, "Effect of Nitrogen Trifluoride on Plasma Concentrations of Lactate, Methemoglobin, and Selected Enzymes," *Rept. for April–June 1968, SAM-TR-70-42 (AD 711 044),* School of Aerospace Medicine, Brooks AFB, Texas, July 1970.
35. J. D. MacEwen and E. H. Vernot, "Toxic Hazards Research Unit Annual Technical Report: 1969," *AMRL-TR-69-84 (AD 699 594),* SysteMed Corp., Dayton, Ohio, Contract F33615-67-C-1025, Sept. 1969.
36. K. O. Christie, C. J. Schack, and R. D. Wilson, *J. Fluorine Chem.* **8,** 541 (1976).
37. E. R. Schulman, W. G. Burwell, and R. A. Meinzer, "Design and Operation of Medium Power CW HF/DF Chemical Lasers," AIAA Paper No. 74-546, *AIAA 7th Fluid and Plasma Dynamics Conf.* (June 1974).
38. P. J. Klass, *Aviat. Week Space Technol.* **107,** 34 (Aug. 1976).
39. W. E. Tolberg and co-workers, *Inorg. Nucl. Chem. Lett.* **2,** 79 (1966).
40. W. E. Tolberg and co-workers, *Inorg. Chem.* **6,** 1156 (1967).
41. K. O. Christie, R. D. Wilson, and C. J. Schack, *Inorg. Chem.* **16,** 937 (1977).
42. U.S. Pat. 3,708,570 (Jan. 1977), W. E. Tolberg, R. S. Stringham, and R. T. Rewick (to Stanford Research Institute).
43. K. O. Christie, J. P. Guertin, and A. E. Pavlath, *Inorg. Nucl. Chem. Lett.* **2,** 83 (1966).
44. J. P. Guertin, K. O. Christie, and A. E. Pavlath, *Inorg. Chem.* **5,** 1921 (1966).
45. K. O. Christie and C. J. Schack, *Inorg. Chem.* **16,** 353 (1977).
46. K. O. Christie, C. J. Schack, and R. D. Wilson, *Inorg. Chem.* **16,** 849 (1977).
47. S. P. Mishra and co-workers, *Inorg. Chem.* **14,** 1103 (1975).
48. K. O. Christie, C. J. Schack, and R. D. Wilson, *Inorg. Chem.* **15,** 1275 (1976).

General References

C. J. Hoffman and R. G. Neville, *Chem. Rev.* **62,** 1 (1962).
A. W. Jache and W. E. Shite "Nitrogen Trifluoride" in C. A. Hunpel, ed., *Encyclopedia of Electrochemistry,* Reinhold Publishing Corp., New York, 1964, pp. 856–859.
R. E. Anderson, E. M. VanderWall, and R. K. Schaplowsky, "Nitrogen Trifluoride," *USAF Propellant Handbook, AFRPL-TR-77-71,* Contract FO4611-76-C-0058, 1977.

ANDREW J. WOYTEK
Air Products and Chemicals, Inc.

OXYGEN

Oxygen Difluoride

Oxygen difluoride [*7783-41-7*] (fluorine monoxide), OF_2, is the most stable binary compound of oxygen and fluorine. Under ambient conditions, it is a colorless gas that condenses to a pale yellow liquid at −145°C (1) and freezes at −224°C (2). Oxygen difluoride is a powerful oxidizer that has attracted considerable attention as an ingredient of high-energy rocket propellant systems (see Explosives and propellants). Several comprehensive reviews of the physical and chemical properties of OF_2 (3–5) and its handling (6) are available.

Physical Properties. An extensive tabulation of the physical properties of OF_2 is available (4). Selected data are: mp −224°C (2); bp, −145°C (1); critical temperature −58°C (7); density of liquid, in g/mL from −145 to −153°C, d = 2.190–0.00523 × K (8); heat of formation 31.8 kJ/mol (7.6 kcal/mol) (9); and heat of vaporization 11.1 kJ/mol (2.65 kcal/mol) (10).

Spectroscopic investigations have shown that OF_2 is bent with equivalent O—F bonds. The O—F distance is 0.139–0.141 nm and the FOF angle is 103–104° (11–13). Measurements of the dipole moment have yielded values of 0.6–1.3 × 10^{-30} C·m (0.18–0.40 D) (12,14–15). The ir (16–18), uv (19), mass (20), and nmr (21) spectra of OF_2 have been reported.

Chemical Properties. The kinetics of decomposition of OF_2 by pyrolysis in a shock tube are different, as a result of surface effects, from those obtained by conventional decompositon studies. Dry OF_2 is stable up to 250°C (22).

Reactions with Metals. Many common metals react with OF_2, but the reaction stops after a passive metal fluoride coating is formed (3–4).

Reactions with Nonmetallic Elements and Inorganic Compounds. Mixtures of OF_2 with carbon, CO, CH_4, H_2, or H_2O vapor explode when ignited with an electrical shock. Elemental B, Si, P, As, Sb, S, Se, and Te react vigorously on slight warming to produce fluorides and oxyfluorides. Oxides such as CrO_3, WO_3, As_2O_3, and CaO react with OF_2 to form fluorides. The corresponding chlorides react with OF_2 to form the respective fluorides and liberate free chlorine in the process (3–4).

In aqueous solution, OF_2 oxidizes HCl, HBr, and HI (and their salts), liberating the free halogens. Oxygen difluoride reacts slowly with water and a dilute aqueous base to form oxygen and fluorine. The rate of this hydrolysis reaction has been determined (23).

Nitric oxide and OF_2 inflame on contact; emission and absorption spectra of the flame have been studied (24). Oxygen difluoride oxidizes SO_2 to SO_3, but under the influence of ultraviolet irradiation it forms sulfuryl fluoride, SO_2F_2 [*2699-79-8*], and pyrosulfuryl fluoride, $S_2O_5F_2$ [*37240-33-8*] (25). Photolysis of SO_3–OF_2 mixtures yields the peroxy compound FSO_2OOF [*13997-94-9*] (25–26).

Oxygen Difluoride as a Source of the OF Radical. The existence of the ·OF radical [*12061-70-0*] was first reported in ref. 27. This work was later refuted (28). The ·OF radical was produced by photolysis of OF_2 in a nitrogen or argon matrix at 4 K. The existence of the ·OF radical was deduced from a study of the kinetics of decomposition of OF_2 and the kinetics of the photochemical reaction (25–26):

$$OF_2 + SO_3 \xrightarrow[350\ \text{nm}]{h\nu} FSO_2OOF$$

The existence of the ·OF radical was further established by use of ^{17}O-labeled compounds and ^{17}O nmr studies to verify the mechanism of the above reaction. The ·OF radical was postulated as an intermediate (29):

$$OF_2 + h\nu \rightarrow F\cdot + \cdot OF$$

$$F\cdot + SO_3 \rightarrow FSO_3\cdot$$

$$FSO_3\cdot + \cdot OF \rightarrow FSO_2OOF$$

Carbonyl fluoride, COF_2, and oxygen difluoride react in the presence of cesium fluoride catalyst to give bis(trifluorylmethyl)trioxide [*1718-18-9*], CF_3OOOCF_3 (30). The following mechanism was proposed:

$$COF_2 + CsF \rightarrow Cs^+OCF_3^-$$

$$OF_2 + CF_3O^- \rightarrow CF_3OOF + F^-$$

$$CF_3OOF + CF_3O^- \rightarrow CF_3OOCF_3 + F^-$$

$$CF_3OOF + COF_2 \rightarrow CF_3OOOCF_3$$

although CF_3OOF [*34511-13-2*] was not detected (30). Additional evidence for the validity of this mechanism used ^{17}O in nmr studies; CF_3OOF has been isolated from the reaction in the presence of excess OF_2 (31).

Reactions with Organic Compounds. Tetrafluoroethylene and OF_2 react spontaneously to form C_2F_6 and COF_2. Ethylene and OF_2 may react explosively, but under controlled conditions monofluoroethane and 1,2-difluoroethane can be recovered (32). Benzene is oxidized to quinone and hydroquinone by OF_2. Methanol and ethanol are oxidized at room temperature (4). Organic amines are extensively degraded by OF_2 at room temperature, but primary aliphatic amines in a fluorocarbon solvent at −42°C are smoothly oxidized to the corresponding nitroso compounds (33).

The reaction of OF_2 with various unsaturated fluorocarbons has been examined (34) and it is claimed that OF_2 can be used to chain-extend fluoropolyenes, convert functional perfluorovinyl groups to acyl fluorides and/or epoxide groups, and act as a monomer for an addition-type copolymerization with diolefins.

Preparation. The synthesis of OF_2 was first achieved by the electrolysis of molten KHF_2 in the presence of water (35). Subsequently, the electrolysis of aqueous HF was also found to produce OF_2 (36–37).

The most satisfactory method of OF_2 generation is probably the fluorination of aqueous NaOH (3,22,38–42):

$$2\,F_2 + 2\,NaOH \rightarrow OF_2 + 2\,NaF + H_2O$$

This method can achieve yields of greater than 60% (43) and has been used for the commercial production of OF_2 (8). The NaOH concentration, however, must be kept low to avoid the loss of product by a secondary reaction:

$$OF_2 + 2\,OH^- \rightarrow O_2 + 2\,F^- + H_2O$$

An economic study of the preparation of OF_2 is available (44).

Analytical Procedures. Oxygen difluoride may be determined conveniently by quantitative application of ir, nmr, and mass spectroscopy. Purity may also be assessed by vapor pressure measurements. Wet-chemical analyses can be conducted either by digestion with excess NaOH, followed by measurement of the excess base (2) and the fluoride ion (45–46), or by reaction with acidified KI solution, followed by measurement of the liberated I_2 (4).

Handling and Safety Factors. Oxygen difluoride can be handled easily and safely in glass and in common metals (stainless steel, copper, aluminum, Monel, and nickel) from cryogenic temperatures to 200°C (4). At higher temperatures only nickel and Monel are recommended. The compatibility of OF_2 with process equipment depends largely upon the cleanliness of the equipment; contaminants such as dirt, moisture, oil, grease, scale slag, and pipe dope must be avoided. Equipment should be passivated with elemental fluorine before contact with OF_2.

Based on present knowledge, OF_2 must be regarded as a highly poisonous gas, somewhat more toxic than fluorine. It has a foul odor with a limit of detectability of 0.1–0.5 ppm. Repeated exposure of rats to 0.5 ppm OF_2 produced death; repeated exposure to 0.1 ppm, however, caused no discernible effects.

Dioxygen Difluoride

Dioxygen difluoride [*7783-44-0*], O_2F_2, was prepared by passing a 1:1 mixture of O_2 and F_2 through a high-voltage electric discharge tube cooled by liquid nitrogen. Dioxygen difluoride has also been prepared by ultraviolet irradiation of O_2 and F_2 (47–48) and by radiolysis of liquid mixtures of O_2 and F_2 at 77 K with 3 Mev bremsstrahlung (49).

Physical Properties. Since O_2F_2 is unstable, it is difficult to purify. Consequently, some of the physical properties presented here are open to question. Selected data are: density, in g/mL, from −87 to −156°C, d = 2.074–0.00291 T (47); heat of formation 19.8 kJ/mol (4.73 kcal/mol) (50); and heat of vaporization 19.2 kJ/mol (4.58 kcal/mol) at −57°C (50).

The structure of O_2F_2 is that of a nonlinear FOOF chain, with the following molecular constants (51): O—O distance, 0.122 nm; OOF angle, 109°30′; dihedral angle, 87°30′; dipole moment, 4.8×10^{-30} C·m (1.44 D). Additional physical and spectral data are summarized in ref 4.

Chemical Properties. The bond distance of O—O is relatively short (121.7 + 0.3 pm) and that of O—F is relatively long (157.5 + 0.3 pm) (51). The weakest bond in O_2F_2 is thus the O—F bond and the mechanisms of reaction of O_2F_2 can probably be explained by the formation of F· and ·OOF and not two ·OF radicals. The ·OOF radical [*15499-23-7*] is a feasible intermediate as it has been shown to exist at low temperatures (51–54). If O_2F_2 is allowed to react quickly with other compounds, simple fluorination usually results. The controlled reactions of O_2F_2, however, yield products that appear to be formed via an ·OOF intermediate.

Simple Fluorination Reactions (55). Some examples of O_2F_2 acting mainly as a fluorinating agent are given below:

$$N_2O \xrightarrow{O_2F_2} FNO + FNO_2 + FONO_2$$

[*7789-26-6*]

$$SO_2 \xrightarrow{O_2F_2} SO_2F_2 \text{ (main product)}$$

$$S \xrightarrow{O_2F_2} \cdot SOF + SO_2F_2$$

$$SO_3 \xrightarrow{O_2F_2} FSO_2\cdot + F_2S_2O_6$$

$$F_2SO \xrightarrow{O_2F_2} F_4SO$$

$$H_2SO_4 \xrightarrow{O_2F_2} FSO_2OF$$

[13536-85-1]

Reactions Involving an ·OOF Intermediate. In controlled reactions of O_2F_2 with various compounds, ^{17}O tracer studies and other techniques have shown that the first step in the reaction appears to be:

$$FOOF \rightarrow \cdot OOF + F\cdot$$

For example:

$$SO_2 + O_2F_2 \rightarrow FSO_2OOF$$

[13997-94-9]

where the proposed mechanism (56) is:

$$SO_2 + F\cdot \rightarrow FSO_2\cdot$$

$$FSO_2\cdot + \cdot OOF \rightarrow FSO_2OOF$$

Another example:

$$CF_3CF{=}CF_2 + O_2F_2 \rightarrow CF_3CF(OOF)CF_3 + CF_3CF_2CF_2OOF$$

in which the proposed mechanism (57) involves the transfer of an OOF group.

The formation of a new class of compounds (dioxygenyl, O_2^+) is also thought to take place via an ·OOF intermediate (58):

$$O_2F_2 \rightarrow \cdot OOF + F\cdot$$

$$\cdot O_2F + BF_3 \rightarrow O_2^+BF_4^-$$

A number of other fluorides have been shown to form O_2^+ compounds by reaction with O_2F_2.

Uses

Oxygen difluoride has never been a compound with major chemical uses in industry, it is mainly a laboratory chemical. It has been suggested that it may be used as an oxidizer for rocket applications and has been used for small tests in this area.

Higher Oxygen Fluorides

Several higher oxygen fluorides, O_3F_2 [*16829-28-0*] (47,59), O_4F_2 [*12020-93-8*] (60), O_5F_2 [*12191-79-6*] (61), and O_6F_2 [*12191-80-9*] (61), and radicals such as $\cdot O_3F$ (62–63) have been reported. Only ·OF, OF_2, O_2F_2, ·OOF, and O_4F_2, however, have been satisfactorily characterized. The structure, even the existence, of "O_3F_2" has been the subject of considerable concern. From cryogenic mass spectroscopy, it appears that O_3F_2 consists of loosely bonded $\cdot O_2F$ and ·OF radicals (64). The ^{19}F nmr spectrum of O_3F_2 suggests an O_3F_2 model consisting of O_2F_2 and interstitial oxygen (65). However, ^{19}F and ^{17}O nmr (7,66), and other studies have shown that O_3F_2, as reported in the literature, is actually a mixture of O_4F_2 and O_2F_2.

Little is known about O_4F_2. It has been reported to behave similarly to O_2F_2 in that it can act as a fluorinating agent or a source of the ·OOF radical. In fact, it appears to be a better source of the ·OOF radical than O_2F_2 in its reactions with SO_2 and BF_3.

BIBLIOGRAPHY

"Oxygen Compounds" in *ECT* 1st ed., under "Fluorine Compounds, Inorganic," Vol. 6, pp. 710–711; "Oxygen" in *ECT* 2nd ed., under "Fluorine Compounds, Inorganic," Vol. 9, pp. 631–635, W. B. Fox and R. B. Jackson, Allied Chemical Corporation.

1. J. Schnitzlstein and co-workers, *J. Phys. Chem.* **56,** 233 (1952).
2. O. Ruff and K. Clusius, *Z. Anorg. Allgem. Chem.* **190,** 267 (1930).
3. H. R. Leech "Compounds of Fluorine with Oxygen" in Mellor, ed., *Comprehensive Treatise on Inorganic and Theoretical Chemistry,* Suppl. II, Part I, Longmans, Green & Co., Inc., New York, 1956, pp. 186–193.
4. A. G. Streng, *Chem. Rev.* **63,** 607 (1963).
5. R. B. Jackson, *Oxygen Difluoride Handling Manual,* Report No. NASA-CR-72401, Allied Chemical Corp., Morristown, N.J., Dec. 1970.
6. R. F. Muraca, J. Neff, and J. S. Whittick, *Physical Properties of Liquid Oxygen Difluoride and Liquid Diborane—A Critical Review,* Report No. NASA-CR-88519, SRI-951581-4, Jet Propulsion Lab., Calif. Inst. of Tech., Pasadena, Stanford Research Inst., Menlo Park, Calif., July 1967.
7. R. Anderson and co-workers, *J. Phys. Chem.* **56,** 473 (1952).
8. *Oxygen Difluoride,* Product Data Sheet, General Chemical Division, Allied Chemical Corporation, Morristown, N.J.
9. W. Evans, T. Munson, and D. Wagman, *J. Res. Natl. Bur. Std.* **55,** 147 (1955).
10. O. Ruff and W. Menzel, *Z. Anorg. Chem.* **198,** 39 (1931).
11. A. Hilton and co-workers, *J. Chem. Phys.* **56,** 473 (1952).
12. L. Pierce, R. Jackson, and N. Dicianni, *J. Chem. Phys.* **35,** 2240 (1961).
13. J. Ibers and V. Schomaker, *J. Phys. Chem.* **57,** 699 (1953).
14. J. Bransford, A. Kunkel, and A. Jache, *J. Inorg. Nucl. Chem.* **14,** 159 (1960).
15. R. Dodd and R. Little, *Nature* **188,** 737 (1960).
16. H. Bernstein and J. Powling, *J. Chem. Phys.* **18,** 685 (1960).
17. E. Jones and co-workers, *J. Chem. Phys.* **19,** 337 (1951).
18. A. Nielsen, *J. Chem. Phys.* **19,** 379 (1951).
19. A. Glissman and H. Schumacher, *Z. Physik. Chem.* **324,** 328 (1934).
20. V. Dibeler, R. Reese, and J. Franklin, *J. Chem. Phys.* **27,** 1296 (1957).
21. H. Agahigian, A. Gray, and G. Vickers, *Can. J. Chem.* **40,** 157 (1962).
22. G. Brauer, *Handbuch der Preparativen Anorganischen Chemie,* Ferdinand Enke, Stuttgart, 1954.
23. S. N. Misra, and G. H. Cady, *Kinetics of Hydrolysis of Oxygen Difluoride,* Report No. TR-70, University of Washington Department of Chemistry, Seattle, Jan. 1972.
24. P. Goodfriend and H. Woods, *J. Chem. Phys.* **39,** 2379 (1963).
25. G. Franz and F. Neumayr, *Inorg. Chem.* **3,** 921 (1964).
26. R. Gath, E. Staricco, J. Sicre, and H. J. Schumacher, *Angew. Chem.* **75,** 137 (1963).
27. O. Ruff and W. Z. Menzel, *Z. Anorg. Allg. Chem.* **217,** 85 (1934).
28. P. Frisch and H. J. Schumacher, *Z. Anorg. Allg. Chem.* **229,** 423 (1936); (Leipzig) **B34,** 322 (1936); **B37,** 18 (1937).
29. I. J. Solomon, A. J. Kacmarek, and J. Raney, *J. Phys. Chem.* **72,** 2262 (1968).
30. L. R. Anderson and W. B. Fox, *J. Am. Chem. Soc.* **89,** 431B (1967).
31. I. J. Solomon and co-workers, *Inorg. Chem.* **11,** 195 (1972).
32. R. Rhein and G. Cady, *Inorg. Chem.* **3,** 1644 (1964).
33. R. Merritt and J. Ruff, *J. Am. Chem. Soc.* **86,** 1342 (1964).
34. M. S. Toy, *Utilization of Oxygen Difluoride for Syntheses of Fluoropolymers,* Report No. Patent-3,931,132, Pat. Appl.-45,549, NASA, Pasadena Office, Calif., Jan. 1976.
35. P. Lebeau and A. Damiens, *Compt. Rend.* **185,** 652 (1927).
36. A. Englebrecht and E. Nachbaur, *Monatsh. Chem.* **90,** 367 (1959).
37. J. A. Donohue, T. D. Nevitt, and A. Zletz, *Adv. Chem. Ser.* **54,** 192 (1966).
38. G. Rohrbach and G. H. Cady, *J. Am. Chem. Soc.* **69,** 677 (1947).
39. D. Yost, *Inorg. Synth.* **1,** 109 (1939).
40. W. Koblitz and H. Schumacher, *Z. Physik. Chem.* **B25,** 283 (1934).
41. P. Lebeau and A. Damiens, *Compt. Rend.* **188,** 1253 (1929).
42. A. Borning and K. E. Pullen, *Inorg. Chem.* **8,** 1791 (1969).

43. G. H. Cady, *J. Am. Chem. Soc.* **57,** 246 (1935).
44. F. L. Hyman and J. F. Tompkins, *An Economic Study of Oxygen Difluoride,* Final Report No. NASA-CR-117317, Air Products and Chemicals, Inc., Allentown, Pa., June 1970.
45. H. Willard and C. Horton, *Anal. Chem.* **22,** 1190 (1950).
46. H. Willard and C. Horton, *Anal. Chem.* **24,** 862 (1952).
47. S. Aoyama and S. Sakuraba, *J. Chem. Soc. Japan* **59,** 1321 (1938).
48. A. Kirshenbaum, A. Grosse, and J. Astor, *J. Am. Chem. Soc.* **81,** 6398 (1959).
49. C. D. Wagner and co-workers, *J. Am. Chem. Soc.* **91,** 4702 (1969).
50. A. Streng, *J. Am. Chem. Soc.* **85,** 1380 (1963).
51. R. Jackson, *J. Chem. Soc.,* 4585 (1962).
52. R. W. Fessenden and R. H. Schuler, *J. Chem. Phys.* **44,** 434 (1966).
53. A. Arkell, *J. Am. Chem. Soc.* **87,** 4057 (1965).
54. R. D. Sprately, J. J. Turner, and G. C. Pimentel, *J. Chem. Phys.* **44,** 2063 (1966).
55. P. H. Kassai and A. D. Kirshenbaum, *J. Am. Chem. Soc.* **87,** 3069 (1965).
56. I. J. Solomon, A. J. Kacmarek, and J. M. McDonough, *Chem. Eng. Data* **13,** 529 (1968).
57. I. J. Solomon, A. J. Kacmarek, and J. Raney, *Inorg. Chem.* **7,** 1221 (1968).
58. I. J. Solomon and co-workers, *J. Am. Chem. Soc.* **90,** 6557 (1968).
59. J. N. Keith and co-workers, *Inorg. Chem.* **7,** 320 (1968).
60. A. D. Kirshenbaum and A. V. Grosse, *J. Am. Chem. Soc.* **81,** 1277 (1959).
61. A. V. Grosse, A. G. Streng, and A. D. Kirshenbaum, *J. Am. Chem. Soc.* **83,** 1004 (1961).
62. A. G. Streng and A. V. Grosse, *J. Am. Chem. Soc.* **88,** 169 (1966).
63. A. D. Kirshenbaum and A. V. Grosse, *Production, Isolation, and Identification of the* $\cdot OF$, $\cdot O_2F$, *and* $\cdot O_3F$ *Radicals,* Research Institute, Temple University, Philadelphia, Pa., June 1964.
64. T. J. Malone and H. A. McGee, *J. Phys. Chem.* **71,** 3060 (1967).
65. J. W. Nebgen, F. I. Metz, and W. B. Rose, *J. Am. Chem. Soc.* **89,** 3118 (1967).
66. I. J. Solomon and co-workers, *J. Am. Chem. Soc.* **89,** 2015 (1967).

IRVINE J. SOLOMON
IIT Research Institute

PHOSPHORUS

The majority of the fluorine in the earth's crust is present in the form of the phosphorus fluoride fluoroapatite [*1306-05-4*], $Ca_5(PO_4)_3F$. Phosphate rock deposits contain an average concentration of 3.5 wt % fluorine. During phosphate processing these fluorine values are partially recovered as by-product fluorosilicic acid. The amount of fluorosilicic acid recovered has grown steadily, in part because of environmental requirements.

The compounds phosphorus trifluoride [*7783-55-3*] PF_3, phosphorus pentafluoride [*7647-19-0*] PF_5, phosphorus oxyfluoride [*13478-20-1*] POF_3, and phosphorus thiofluoride [*2404-52-6*] PSF_3 were prepared prior to 1900 by Thorpe and Moissan. However, diphosphorus tetrafluoride [*13537-32-1*], P_2F_4, was first reported in 1966 (1–3). The most widely studied of these are PF_5 and PF_3. The mixed chlorofluorides PCl_xF_y where $x + y = 3$ and 5 have also been studied. Some physical properties of PF_5, PF_3, and POF_3 are given in Table 1.

Phosphorus Pentafluoride

Phosphorus pentafluoride was first prepared in 1876 by Thorpe through fluorination of phosphorus pentachloride with arsenic trifluoride. Other routes to PF_5 have included fluorination of PCl_5 by HF, AgF, benzoyl fluoride, SbF_3, PbF_2, or CaF_2 (9). It can also be made by allowing PF_3 to react with fluorine, chlorine (10), chlorine in contact with calcium fluoride (11), or by thermal decomposition of diazonium hexafluorophosphates or sodium hexafluorophosphate [*21324-39-0*] (12).

Phosphorus pentafluoride has been commercially available (Ozark-Mahoning) in small quantities for many years. In the 1960s a group now a part of U.S. Steel Agrichemicals developed technology to produce PF_5, PF_3, POF_3, hexafluorophosphoric acid [*16940-81-1*] HPF_6, difluorophosphoric acid [*13779-41-4*] HPO_2F_2, and monofluorophosphoric acid [*13537-32-1*] H_2PO_3F (13–16). Based on this technology they established a pilot plant and attempted to develop commercial markets, especially for PF_5. They produced POF_3 and PF_3 by fluorination of phosphoric and phosphorous acids, respectively, with calcium fluorosulfate fluoride, $CaF(SO_3F)$, made by reaction

Table 1. Properties of Phosphorus Penta-, Oxy-, and Trifluorides

Property	PF_5	POF_3	PF_3	Refs. PF_5	Refs. POF_3	Refs. PF_3
melting point, °C	−91.6	−39.1	−151.5	4	5	5
boiling point, °C	−84.8	−39.7	−101.8	4	5	5
density, liquid, at bp, g/mL			1.6			
critical temperature, °C	>25	73.3	−2.05		5	5
critical pressure, MPa[a]		4.23	4.33		5	5
heat of fusion, kJ/mol[b]	12.1	14.9		4	5	
heat of vaporization, kJ/mol[b]	16.7	23.2	16.5	4	6	5
heat of formation, $-\Delta H$, kJ/mol[b]	1210		946	7	5	8

[a] To convert MPa to atm, divide by 0.101.
[b] To convert kJ to kcal, divide by 4.184.

of CaF_2 and SO_3. The POF_3 could then be treated with HF to form PF_5 and hexafluorophosphoric acid. This activity was terminated in 1972.

Phosphorus pentafluoride is a colorless gas which fumes in contact with moist air and reacts immediately with water to hydrolyze, first to POF_3 and then to the fluorophosphoric acids. Although PF_5 is probably stable in completely dry glass, glass is not recommended for storage as even a trace of moisture generates HF which regenerates moisture converting the glass to POF_3 and SiF_4.

Phosphorus pentafluoride behaves as a Lewis Acid showing electron-accepting properties, generally in a ratio of 1:1 with Lewis bases. It forms complexes with amines, ethers, nitriles, sulfoxides, and other bases. These complexes are frequently less stable than the similar BF_3 complexes, probably owing to steric factors. Because it is a strong acceptor, PF_5 is an excellent catalyst especially in ionic polymerizations. Phosphorus pentafluoride is also used as a source of phosphorus for ion implantation (qv) in semiconductors (17).

Phosphorus Trifluoride

Phosphorus trifluoride was prepared by Moissan in 1884 by reaction of copper phosphide with lead fluoride. It is usually prepared by fluorination of PCl_3 with CaF_2 (10), AsF_3, SbF_3, AgF, PbF_2, ZnF_2, or NaF. Laboratory-scale syntheses have been published (18–19), and PF_3 is commercially available (Ozark-Mahoning).

Phosphorus trifluoride is an almost odorless gas which does not fume in air and reacts slowly with water but rapidly with base. It may be very toxic, and great care should be taken in handling it. It reacts with ferrohemoglobin to form an unstable complex (20). Presumably, its action may be similar to carbon monoxide poisoning; and oxygen may be helpful in treatment. Phosphorus trifluoride forms many complexes by donating the lone electron pair on the phosphorus atom. It forms many complexes similar to CO complexes (see Carbonyls) such as $Ni(PF_3)_4$ [*13859-65-9*] (21), $Fe(PF_3)_5$ [*13815-34-4*] (22), $(PF_3)_2PtCl_2$ [*15977-33-0*] (23), $Mo(PF_3)_6$ [*15339-46-5*] (24) and $Cr(PF_3)_6$ [*26117-61-3*] (24). Although AsF_3 and SbF_3 have acceptor properties and function as Lewis acids, no evidence has been found for molecular complex formation by PF_3 as a Lewis acid nor for stable fluorophosphites (PF_4^-).

Phosphorus Oxyfluoride

Phosphorus oxyfluoride is a colorless gas which is susceptible to hydrolysis. It can be formed from PF_5 by reaction with water, and then undergoes further hydrolysis to form a mixture of fluorophosphoric acids. It reacts with HF to form PF_5. It can be prepared by fluorination of phosphorus oxytrichloride with HF, AsF_3, SbF_3.

FLUOROPHOSPHORIC ACIDS

The three primary fluorophosphoric acids, monofluorophosphoric acid (**1**), difluorophosphoric acid [*13779-41-4*] (**2**), and hexafluorophosphoric acid [*16940-81-1*] (**3**) were discovered in 1927 by Lange (25). In 1963 *sym*-difluorodiphosphoric acid [*44801-72-1*] (**4**), was reported (26). The three primary acids can be prepared by reaction of phosphoric acid or phosphoric anhydride (27) with HF or phosphorus oxyfluoride (14–17) with HF or water, or both. These acids are in reversible equilibrium with each other, HF, and phosphoric acid. The interrelationships of the acids are as follows:

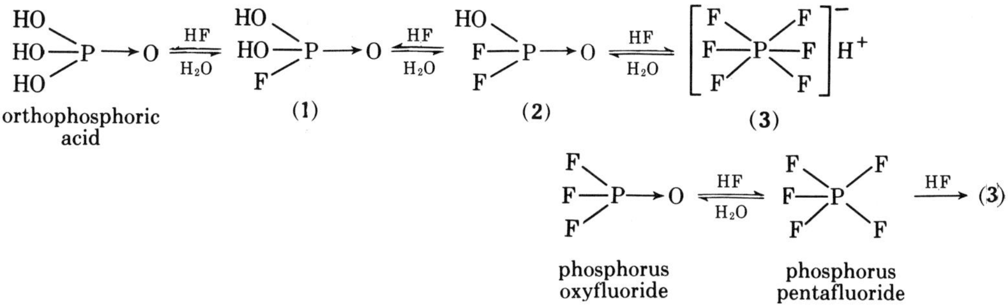

orthophosphoric acid (1) (2) (3)

phosphorus oxyfluoride phosphorus pentafluoride

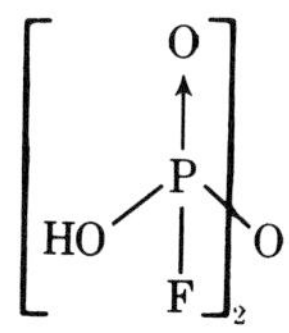

(4)

Numerous applications of the fluorophosphoric acids are referenced in the patent literature, and all three acids are commercially available (Ozark-Mahoning). The mono and difluoro acids can be made as anhydrous or hydrated liquids. The hexafluoro acid is stable only as an aqueous solution. Anhydrous hexafluorophosphoric acid may be prepared at reduced temperatures and pressures but it dissociates rapidly into PF_5 and HF at 25°C (28). The commercial hexafluorophosphoric acid is an aqueous solution. When diluted with water all these acids hydrolyze producing orthophosphoric acid. The hexafluoro acid is the most stable of the three fluorophosphoric acids.

A number of salts of the monofluoro and hexafluoro acids are known, whereas the difluoro acid salts are less stable toward hydrolysis. The most well-known of these is sodium monofluorophosphate which is the world's most widely used dentifrice additive for the reduction of tooth decay. In the United States both sodium monofluorophosphate and stannous fluoride are widely used for this application (see Dentifrices). The hexafluorophosphates can usually be prepared by neutralization of the appropriate base by the hexafluoro acid. The monofluorophosphates are usually prepared by other methods (29) because neutralization of the acid usually results in extensive hydrolysis.

Because HF is present in the aqueous acids and can be generated by hydrolysis of the anhydrous acids, glass should be avoided in handling or processing. Plastics can generally be used for laboratory work although there is evidence for migration of the acids through both thin plastic and the usual plastic bottle closures. The acids are generally shipped and stored in 2SL/6D (5 or 55 gal) heavy plastic drums with steel overpacks. Aluminum is also satisfactory for storage and use of concentrated solutions of the difluoro acid and hexafluoro acid.

Experimentation with test animals and laboratory and plant experience indicate that the fluorophosphoric acids are less toxic and dangerous than hydrogen fluoride (30). However, they contain, or can hydrolyze to, hydrofluoric acid and must be treated with the same care as hydrofluoric acid. Rubber gloves and face shields are essential for all work with these acids, and full rubber dress is necessary for handling larger quantities. The fumes from these acids contain HF.

Monofluorophosphoric Acid

Monofluorophosphoric acid (**1**) is a colorless, nonvolatile, viscous liquid with practically no odor. On cooling it does not crystallize but sets to a rigid glass at −78°C. It has a density of d_4^{25} = 1.818 g/mL. Little decomposition occurs up to 185°C under vacuum but it cannot be distilled. An aqueous solution shows the normal behavior of a dibasic acid with the first neutralization point in 0.05 *N* solution at pH 3.5 and the second at pH 8.5. Conductance measurements, however, indicate H_2PO_3F behaves as a monobasic acid in aqueous solution (31). The permanent end point shows the stability of the PO_3F^{2-} ion at this pH. Slow hydrolysis occurs at low pH to give orthophosphoric and hydrofluoric acids. The kinetics of the hydrolysis have been studied (32–33). Equilibrium concentrations of H_2PO_3F at varying concentrations have been determined by nmr studies (34). The anhydrous acid causes rapid swelling of cellulose, and acts as a polymerization catalyst. It has no oxidizing character.

Preparation. Monofluorophosphoric acid is one of the hydrolysis products of POF_3. It is the primary product of reaction between theoretical amounts of phosphoric anhydride and aqueous hydrogen fluoride. A reaction can be written to give the anhydrous acid with 69% aqueous hydrogen fluoride.

$$P_2O_5 + 2\,HF + H_2O \rightarrow 2\,H_2PO_3F$$

However, the product of this reaction usually contains up to 20% each of orthophosphoric and difluorophosphoric acids. More complete conversion to monofluorophosphoric acid results from reaction with 40% hydrogen fluoride. Although diacid is formed, it more readily hydrolyzes to give an aqueous solution of primarily monofluorophosphoric acid, with some orthophosphoric acid and HF.

Difluorophosphoric Acid

Difluorophosphoric acid (**2**) is a mobile, colorless liquid. It fumes on contact with air, probably due to HF formation. The anhydrous diacid has a reported mp of −96.5 ± 1°C (35) and −91.3 ± 1°C (36) and a density d_4^{25} of 1.583 g/mL. It partially decomposes on heating above 80–100°C. An extrapolated normal boiling point is 116°C (35) although it boils at 107–111°C (36–37) with decomposition. Vapor pressure data are available (35–36) although the data above ca 80°C are probably unreliable.

A freshly made solution behaves as a strong monobasic acid. Neutralized solutions slowly become acidic because of hydrolysis to monofluorophosphoric and hydrofluoric acids. The anhydrous acid undergoes slow decomposition on distillation at atmospheric pressure, reacts with alcohols to give monofluorophosphoric acid esters, and is an alkylation and a polymerization catalyst.

The commercially available difluorophosphoric acid has a formula approximating $HPO_2F_2.H_2O$ and contains some monofluorophosphoric acid (**1**) and some hexafluorophosphoric acid (**3**) as well as HF. It is primarily used in catalytic applications.

Hexafluorophosphoric Acid

Hexafluorophosphoric acid (**3**) is present under ambient conditions only as an aqueous acid as the anhydrous acid dissociates rapidly to HF and PF_5 at 25°C (28). The commercially available (Ozark-Mahoning) HPF_6 is ca 60% HPF_6 with HF, HPO_2F_2, HPO_3F, and H_3PO_4 in equilibrium equivalent to about 11% additional HPF_6. The acid is a colorless liquid which fumes considerably owing to HF. Frequently, reduced phosphates give it a dark honey color. This color can be removed by oxidation with a small amount of nitric acid. When the acid is diluted, it slowly hydrolyzes to the other fluorophosphoric acids and finally phosphoric acid. In concentrated solutions, the hexafluoro acid establishes equilibrium with its hydrolysis products in relatively low concentration. The hexafluoro acid also forms a hexahydrate, $HPF_6.6H_2O$ [*40209-76-5*] (38), mp ca 31.5°C. This has been isolated in good yield when a hexafluoro acid solution of this concentration is cooled rapidly to prevent renewed equilibration. The hexahydrate forms hard, coarse crystals with a cubic structure (39) which are very hygroscopic. However, when the crystals are melted, an equilibrium is again established. A liquid $HPF_6.(C_2H_5)_2O$ [*4590-57-2*] boiling at 114°C and melting at −35°C has been reported (40). Hexafluorophosphoric acid is used catalytically, as a fluorinating reagent, as a fungicide (qv), and extensively in preparation of numerous hexafluorophosphates.

The PF_6^- ion can be determined by precipitation with nitron or tetraphenylarsonium chloride (41).

FLUOROPHOSPHATES

The most studied and used fluorophosphates are sodium monofluorophosphate and a number of salts of the very stable hexafluorophosphate anion. The difluorophosphates are less stable hydrolytically.

Monofluorophosphates

Monofluorophosphates are probably the best characterized series of fluoroxy salts known at present. The PO_3F^{2-} ion is stable in neutral or slightly alkaline solution. The alkali metal and ammonium monofluorophosphates are soluble in water but the alkaline earth salts are only slightly soluble.

Monofluorophosphates of ammonium, lithium, sodium, potassium, silver, calcium, strontium, barium, mercury, lead, and benzidine are described in ref. 42, the nickel, cobalt, and zinc salts in ref. 43, and cadmium, manganese, chromium, and iron monofluorophosphates in ref. 44.

Many of the monofluorophosphates are very similar to the corresponding sulfates (45).

The monofluorophosphates can be prepared by neutralization of monofluorophosphoric acid (**1**). Sodium monofluorophosphate [*7631-97-2*] is prepared commercially (29) by fusion of sodium fluoride and sodium metaphosphate and the potassium salt [*14104-28-0*] can be prepared similarly. Insoluble monofluorophosphates can be readily prepared from reaction of nitrate or chloride solutions with sodium monofluorophosphate. Some of the salts described in the literature have been prepared by metathetical reactions between silver monofluorophosphate [*66904-72-1*] and metal chlorides.

Molten alkali metal monofluorophosphates are reactive and corrosive. They hydrolyze generating HF and react with many metals and ceramics. They readily dissolve metal oxides and are very effective metal surface cleaners and fluxes. They also have bactericidal and fungicidal properties (46). However, the main commercial application among monofluorophosphates is of sodium monofluorophosphate in dentifrices.

Sodium monofluorophosphate, mp 625°C, is soluble in water to the extent of 42 g/100 g solution. The pH of a 2% solution is between 6.5 and 8.0. Dilute solutions are stable indefinitely in the absence of acid or cations that form insoluble fluorides.

Sodium monofluorophosphate is used in dentifrices at a concentration of 0.76 wt % which produces the desired fluoride level of 1000 ppm. Although the mechanism of its efficacy in reducing dental decay is not completely understood (47), it almost certainly reacts with the apatite of the tooth converting it to fluoroapatite which is less soluble in mouth acids (see Dentifrices).

The National Formulary (48) specifications for sodium monofluorophosphate require a minimum of 12.1% fluoride as PO_3F^{2-} (theoretical 13.2%) and a maximum of 1.2% fluoride ion reflecting unreacted sodium fluoride. Analysis for PO_3F^{2-} is by difference between total fluoride in the product less fluoride ion as determined by a specific ion electrode.

The oral LD_{50} of sodium monofluorophosphate in rats is 888 mg/kg.

Sodium monofluorophosphate forms a decahydrate [*7727-73-3*], $Na_2PO_3F.10H_2O$, mp 9°C. Commercial sodium monofluorophosphate can be purified using this salt. The commercial salt is dissolved in about 10% more than the theoretical water and filtered. The filtrate is cooled to 0°C and allowed to stand. If crystals do not form, the solution may be seeded with $Na_2PO_3F.10H_2O$ or $Na_2SO_4.10H_2O$. The resulting crystals are filtered cold and washed with small quantities of ice water. Attempts to remove the water of hydration by heating lead to hydrolysis of the sodium monofluorophosphate. The water can be extracted by multiple extractions with ethyl alcohol (49).

Calcium, strontium, and barium monofluorophosphates can be precipitated by reaction of solutions of their chlorides or nitrates with sodium monofluorophosphate. The salts obtained are $CaPO_3F.2H_2O$ [*37809-19-1*], $SrPO_3F.H_2O$ [*72152-36-4*], and $BaPO_3F.xH_2O$ [*58882-62-5*] where $x < 1$. The solubility of $CaPO_3F.2H_2O$ is 0.417 g/100 mL solution at 27°C (50). It can be partially dehydrated to the hemihydrate [*72152-38-6*] but further attempts cause hydrolysis to CaF_2 and phosphate. Heating $SrPO_3F.H_2O$ to 450°C gives anhydrous $SrPO_3F$ [*66546-46-1*] but $BaPO_3F.xH_2O$ retains water even at 500°C (51).

Difluorophosphates

Difluorophosphates have limited applications largely because of the hydrolytic instability of the $PO_3F_2^-$ ion. The ammonium salt can be prepared from ammonium fluoride and phosphoric anhydride.

$$3\ NH_4F + P_2O_5 \rightarrow NH_4PO_2F_2 + (NH_4)_2PO_3F$$

The $NH_4PO_2F_2$ can be extracted from the solid reaction product with boiling methanol (52). Alkali metal difluorophosphates are prepared from the hexafluorophosphates by one of the following fusion reactions (53):

$$KPF_6 + 2\ KPO_3 \rightarrow 3\ KPO_2F_2$$

$$3\ KPF_6 + 2\ B_2O_3 \rightarrow 3\ KPO_2F_2 + 4\ BF_3$$

Even though the $PO_2F_2^-$ is considered to be hydrolytically unstable, hydrolysis is slow in a neutral solution. However, in a solution initially 0.1 *N* in NaOH, at 70°C, $NaPO_2F_2$ is quantitatively hydrolyzed to give the PO_3F^{2-} and F^- ions within 10 min (54).

A number of organic nitrogen-containing basic compounds give insoluble difluorophosphates. Among these is nitron which can be used for the gravimetric determination of $PO_2F_2^-$ (55).

Potassium and other metal difluorophosphates have been reported as stabilizing agents in chloroethylene polymers (56).

Hexafluorophosphates

There is a great deal of interest in the hexafluorophosphate anion [*16919-18-8*] as shown by ca 3500 references in the 1972–1976 periodic *Chemical Abstracts Index*. Most of these references are organic hexafluorophosphates with the primary applications in catalysis, especially photopolymerization. A number of the compounds are diazonium compounds (see Photoreactive polymers).

The hexafluorophosphates are among the most stable halogen complexes known. The highly symmetrical PF_6^- ion is stable to boiling aqueous alkali and is decomposed only slowly in acidic solutions at ambient temperatures. It does hydrolyze rapidly at elevated temperatures in acid (57). The stability of the PF_6^- ion can be compared to the isoelectronic SF_6. The hexafluorophosphates can be decomposed yielding PF_5 although in many cases only at temperatures where the PF_5 reacts with the metal containers. 4-Chlorobenzenediazonium hexafluorophosphate can be decomposed to PF_5, N_2, and fluorobenzene at 120°C (58) and is a convenient source for laboratory amounts of PF_5 as well as a frequently used catalyst.

Many of the organic and inorganic hexafluorophosphates can be prepared by reaction of hexafluorophosphoric acid with the appropriate base. Another method involves reaction of the appropriate chloride and PCl_5 with anhydrous HF (59).

Potassium hexafluorophosphate [*17084-13-8*], KPF_6, and ammonium hexafluorophosphate [*16941-11-0*], NH_4PF_6, are the most readily available PF_6^- salts. The KPF_6 salt melts with slow decomposition to PF_5 and KF at ca 575°C. The density of KPF_6 is 2.55 g/mL, and its solubility in water is 3.56 g/100 g solution at 0°C, 8.35 g/100 g solution at 25°C and 38.3 g/100 g solution at 100°C (60). A solution of KPF_6 is neutral and stable against hydrolysis unless the pH is reduced to about 3. The salt has a minimum intraperitoneal lethal dose of 1120 mg/kg for female albino mice (61). Potassium hexafluorophosphate is a soluble neutral salt which can be used in syntheses of other PF_6^- salts as a substitute for HPF_6 (**3**). Ammonium hexafluorophosphate has a water solubility of 60.4 g/100 g solution at 25°C (62).

Anhydrous silver hexafluorophosphate [*26042-63-7*], $AgPF_6$, as well as other silver fluoro salts, is unusual in that it is soluble in benzene, toluene, and *m*-xylene and forms 1:2 molecular crystalline complexes with these solvents (63). Olefins also form complexes with $AgPF_6$ and this characteristic has been used in the separation of olefins from paraffins (64). It also is used as a catalyst. Lithium hexafluorophosphate [*21324-40-3*], $LiPF_6$, as well as KPF_6, and other PF_6^- salts are used as electrolytes in lithium anode batteries (qv).

Substituted ammonium hexafluorophosphates having compositions represented by the formula $RR'R''R'''NPF_6$ (where R, R′, R″, R‴ may be alkyl, aryl, alkaryl groups, or H) are of decreasing water solubility corresponding to the number of hydrogens on the ammonium group replaced by the organic radicals. Unusual thermal stability is found in the quaternary compounds when the R groups are the low alkyl radicals. For example, tetramethylammonium hexafluorophosphate [*558-32-7*] is stable up to about 400°C.

FLUOROPHOSPHATE ESTERS

The esters of monofluorophosphoric acid are of great interest because of their cholinesterase inhibiting activity which causes them to be highly toxic nerve gases and also gives them medical activity (see Cholinesterase inhibitors). The most studied is the bis(1-methylethyl)ester of phosphorofluoridic acid [*55-91-4*] (**5**) also known as diisopropyl phosphorofluoridate, DFP, and as the ophthalmic ointment or solution, Isoflurophate USP. It is used as a parasympathomimetic agent, and as a miotic in glaucoma and convergent strabismus. It was developed during World War II as a nerve gas (65). It is prepared by reaction of PCl_3 and isopropanol, followed by chlorination to diisopropyl chlorophosphate and conversion to the desired product with NaF (66).

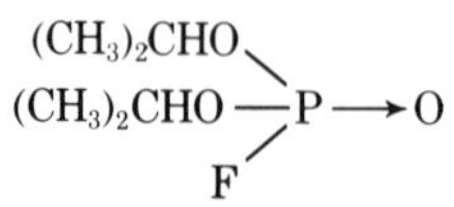

(**5**)

In the nerve gas known as Sarin [*107-44-8*] or GB (**6**), one of the isopropoxy groups of DFP is replaced by a methyl group. Sarin is more toxic than DFP. Sarin has an oral LD_{50} in rats of 550 μg/kg, and DFP has an oral LD_{50} in rats of 6 mg/kg (67) (see Chemicals in war).

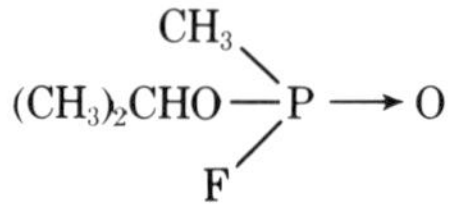

(**6**)

BIBLIOGRAPHY

"Phosphorus Compounds" under "Fluorine Compounds, Inorganic" in *ECT* 1st ed., Vol. 6, pp. 711–721, by Wayne E. White, Ozark-Mahoning Company; "Phosphorus" under "Fluorine Compounds, Inorganic" in *ECT* 2nd ed., Vol. 9, pp. 635–649, by W. E. White and C. Pupp, Ozark-Mahoning Company.

1. R. W. Rudolph, R. C. Taylor, and R. W. Parry, *J. Am. Chem. Soc.* **88,** 3729 (1966).
2. M. Lustig, J. K. Ruff, and C. B. Colburn, *J. Am. Chem. Soc.* **88,** 3875 (1966).
3. L. B. Centofanti and R. W. Rudolph, *Inorg. Syn.* **12,** 281 (1970).
4. S. Johnson, Ph.D. thesis, Purdue University, Lafayette, Ind., 1953.
5. T. D. Farr, *Phosphorus—Properties of the Element and Some of Its Compounds, Chem. Eng. Rept. No. 8 of Tennessee Valley Authority,* Part XI, U.S. Government Printing Office, Washington, D.C., 1950.
6. G. Tarbutton, E. P. Egan, Jr., and S. G. Frary, *J. Am. Chem. Soc.* **63,** 1782 (1941).

7. C. J. Hoffman, *Phosphorus–Fluorine Oxidizers, PB-150613-1,* Part 7, Propulsion Chemistry Part II, Lockheed Aircraft Corp., Missiles and Space Div., Burbank, Calif., 1959.
8. H. C. Duus and D. P. Mykytiuk, *J. Chem. Eng. Data* **9,** 585 (1964).
9. E. L. Muetterties and co-workers, *J. Inorg. Nucl. Chem.* **16,** 52 (1960).
10. Brit. Pat. 822,539 (Oct. 28, 1959), K. C. Brinker (to E. I. du Pont de Nemours & Co., Inc.).
11. U.S. Pat. 2,810,629 (Oct. 22, 1957), E. L. Muetterties (to E. I. du Pont de Nemours & Co., Inc.).
12. Ger. Pat. 814,139 (Sept. 20, 1951), H. Jonas (to Farbenfabriken Bayer A.G.).
13. U.S. Pat. 3,402,019 (Sept. 17, 1968), H. L. Bowkley and R. B. Thurman (to Armour Agricultural Chemical Company).
14. U.S. Pat. 3,428,422 (Feb. 18, 1969), R. A. Wiesboeck (to USS Agri-Chemicals, Inc.).
15. U.S. Pat. 3,634,034 (Jan. 11, 1972), J. D. Nickerson and R. A. Wiesboeck (to United States Steel Corporation).
16. U.S. Pat. 3,728,435 (Apr. 17, 1973), R. A. Wiesboeck (to United States Steel Corporation).
17. R. G. Wilson and D. M. Jamba, *Appl. Phys. Lett.* **22,** 176 (1973).
18. C. J. Hoffman, *Inorg. Syn.* **4,** 149 (1953).
19. A. A. Williams, *Inorg. Syn.* **5,** 95 (1957).
20. G. Wilkinson, *Nature (London)* **168,** 514 (1951).
21. G. Wilkinson, *J. Am. Chem. Soc.* **73,** 5501 (1951).
22. R. J. Clark, *Inorg. Chem.* **3,** 1395 (1964).
23. J. Chatt and A. A. Williams, *J. Chem. Soc.,* 3061 (1951).
24. Th. Kruck, *Z. Naturforsch.* **196,** 164 (1964).
25. W. Lange, *Chem. Ber.* **60,** 962 (1927).
26. K. B. Boerner, C. Stoelzer and A. Simon, *Ber.* **96,** 1328 (1963).
27. L. C. Mosier and W. E. White, *Ind. Eng. Chem.* **43,** 246 (1951).
28. U.S. Pat. 2,718,456 (Sept. 20, 1955), A. J. Mulder and W. C. B. Smithuysen (to Shell Development Co.).
29. U.S. Pat. 2,481,807 (Sept. 13, 1949), C. O. Anderson (to Ozark-Mahoning Co.).
30. J. M. Godwin and W. E. White, *Chem. Eng. News* **28,** 2721 (1950).
31. Y.-T. Chen, T.-C. Li, and K.-C. Yin, *Sci. Sinica (Peking)* **13,** 1719 (1952).
32. I. G. Ryss and V. B. Tul'chimskii, *Dokl. Akad. Nauk SSSR* **142,** 141 (1962).
33. L. N. Devonshire and H. H. Rowley, *Inorg. Chem.* **1,** 680 (1962).
34. D. P. Ames and co-workers, *J. Am. Chem. Soc.* **81,** 6350 (1959).
35. W. Lange and R. Livingston, *J. Am. Chem. Soc.* **72,** 1280 (1950).
36. A. S. Lenskii, A. D. Shaposhnikova, and A. S. Allilueva, *Zh. Prikl. Khim. (Leningrad)* **35,** 760 (1962).
37. G. Tarbutton, E. P. Egan, Jr., and S. G. Frary, *J. Am. Chem. Soc.* **63,** 1782 (1941).
38. U.S. Pat. 2,488,298 (Nov. 15, 1949), W. Lange and R. Livingston (to Ozark-Mahoning Co.).
39. H. Bode and G. Teufer, *Acta Cryst.* **8,** 611 (1955).
40. Ger. Pat. 812,247 (Aug. 27, 1951), H. Jonas (to Farbenfabriken Bayer A.G.).
41. H. E. Affsprung and U. S. Archer, *Anal. Chem.* **35,** 1912 (1963).
42. W. Lange, *Chem. Ber.* **62,** 793 (1929).
43. H. C. Goswami, *J. Indian Chem. Soc.* **14,** 660 (1937).
44. E. B. Singh and P. C. Sinha, *J. Indian Chem. Soc.* **41,** 407 (1964).
45. I. V. Mardirosova and co-workers, *Izv. Akad. Nauk SSSR Neorg. Mater.* **9,** 970 (1973).
46. J. M. Godwin, Masters thesis, University of Tulsa, Tulsa, Oklahoma, 1950.
47. E. J. Duff, *Caries Res.* **7,** 79 (1973).
48. *The United States Pharmacopeia XIX–National Formulary XIV,* 4th Suppl., United States Pharmacopeial Convention, Inc., Rockville, Md., Jan. 31, 1978; USP XX–NF XV, 1980.
49. G. F. Hill and L. F. Audrieth, *Inorg. Syn.* **3,** 108 (1950).
50. H. H. Rowley and J. E. Stuckey, *J. Am. Chem. Soc.* **78,** 4262 (1956).
51. Vu Quang Kinh, *Bull. Soc. Chim. Fr.,* 1466 (1962).
52. W. Lange in W. C. Fernelius, ed., *Inorganic Syntheses,* Vol. II, McGraw-Hill Book Co., Inc., New York, 1946, pp. 155–158.
53. Ger. Pat. 813,848 (Sept. 17, 1951), H. Jonas (to Farbenfabriken Bayer A.G.).
54. I. G. Ryss and V. B. Tul'chimskii, *Zh. Neorgna. Khim.* **7,** 1313 (1962); *ibid.,* **9,** 831 (1964); *Chem. Abstr.* **57,** 6852 (1962); *ibid.,* **61,** 1308 (1964).
55. W. Lange, *Chem. Ber.* **62,** 786 (1929).
56. U.S. Pat. 2,846,412 (Aug. 5, 1958), C. B. Havens (to The Dow Chemical Co.).

57. W. Lange and E. Mueller, *Chem. Ber.* **63,** 1058 (1930).
58. R. Schmutzler, "Fluorides of Phosphorus" in M. Stacey, J. D. Tatlow, and A. G. Sharpe, eds., *Advances in Fluorine Chemistry,* Vol. 5, Butterworth & Co., Inc., Washington, D.C., 1965, pp. 31–287.
59. M. M. Woyski in L. F. Audrieth, ed., *Inorganic Syntheses,* Vol. III, McGraw-Hill Book Co., Inc., New York, 1950, pp. 111–117.
60. J. N. Sarmousakis and M. J. D. Law, *J. Am. Chem. Soc.* **77,** 6518 (1955).
61. F. A. Smith and co-workers, *Toxicol. Appl. Pharmacol.* **2,** 54 (1960).
62. A. V. Nikolaev and co-workers, *Izv. Sib. Otd. Akad. Nauk SSSR Ser. Khim Nauk,* 48 (1976).
63. D. W. A. Sharp and A. G. Sharpe, *J. Chem. Soc.,* 1855 (1956).
64. U.S. Pat. 3,189,658 (June 15, 1965), H. W. Quinn (to The Dow Chemical Co.).
65. *Chemical Warfare Service TDMR 832*, Edgewood Arsenal, Apr. 1944.
66. U.S. Pat. 2,409,039 (Oct. 8, 1946), E. E. Hardy and G. M. Kosolapoff (to Monsanto Chemical Co.).
67. *Registry of Toxic Effects of Chemical Substances,* 1977 ed., Vol. II, NIOSH, U.S. Department of Health, Education, and Welfare, Washington, D.C., pp. 667, 686.

General References

R. Schmutzler, "Fluorides of Phosphorus" in M. Stacey, J. D. Tatlow, and A. G. Sharpe, eds., *Advances in Fluorine Chemistry,* Vol. 5, Butterworth & Co., Inc., Washington, D.C., 1965, pp. 31–287.

G. I. Drozd, "Phosphorus Fluorides," *Usp. Khim.* **39,** 3 (1970).

W. Lange, "The Chemistry of the Fluoro Acids of Fourth, Fifth, and Sixth Group Elements" in J. H. Simons, ed., *Fluorine Chemistry,* Vol. I, Academic Press, Inc., New York, 1950, pp. 125–188.

E. L. Muetterties and co-workers, "Chemistry of Phosphorus Fluorides," *J. Inorg. Nucl. Chem.* **16,** 52 (1960).

S. Johnson, "Some Chemical and Physical Properties of Phosphorus Pentafluoride," Ph.D. thesis, Purdue University, Lafayette, Ind., 1953.

CHARLES B. LINDAHL
Ozark-Mahoning Company,
a subsidiary of the Pennwalt Corp.

POTASSIUM

The two stable salts of potassium with fluorine of commercial significance are the normal fluoride [*7789-23-3*], KF, and potassium bifluoride [*7789-29-9*], KHF_2.

Potassium Fluoride

Properties. Anhydrous potassium fluoride is a white hygroscopic salt that forms two hydrates, $KF.2H_2O$ [*13455-21-5*] and $KF.4H_2O$ [*34341-58-7*]. The tetrahydrate exists at temperatures below 17.7°C. The dihydrate is stable at room temperature and starts to lose water above 40°C. Temperatures in the order of 250–300°C are required to remove the last few percent of water in a reasonable period of time. Potassium fluoride does not pyrohydrolyze at temperatures as high as 1000°C (1). Chemical and physical properties of KF are summarized in Table 1.

Halogen exchange with KF is not successful in acetic acid (9). Hydrogen bonding of the acid hydrogen with the fluoride ion was postulated to cause acetate substitution for the halide; however, the products of dissolved KF in acetic acid are potassium acetate and potassium bifluoride (10). Thus KF acts as a base rather than as a fluorinating agent in acetic acid.

Manufacture. Commercial KF is manufactured from potassium hydroxide and hydrofluoric acid followed by drying in a spray dryer or flaking from a heated drum. The KF assay is typically 97–99%; major impurities are $KF.2H_2O$ and either potassium carbonate or potassium bifluoride. The 1978 price of the anhydrous salt was $2.69/kg and that of the reagent grade dihydrate was $6.54/kg. Potassium fluoride can be purified by passing anhydrous HF through the melt (11).

Toxicology. By ingestion, the lethal dose of potassium fluoride in guinea pigs is 250 mg/kg body weight. The LD_{50} orally for rats is 245 mg/kg body weight (12). Ingestion of potassium fluoride may cause vomiting, abdominal pains, and diarrhea.

Table 1. Physical and Chemical Properties of Potassium Fluoride

		Reference
melting point, °C	857	2
boiling point, °C	1505	2
specific gravity (25°C)	2.48	2
solubility in H_2O, g/100 g (18°C)	92.3	3
solubility in liquid HF, g/100 g (8°C)	36.5	3
solubility in acetic acid, g/100 g (25°C)	28	4
solubility in methanol, g/100 g (25°C)	2.3	5
standard heat of formation (KF), kJ/mol[a]	−567.4	2
standard heat of formation ($KF.2H_2O$), kJ/mol[a]	−1159	6
heat of fusion, kJ/mol[a]	28.2	2
heat of vaporization, kJ/mol[a]	173	6
heat of dehydration ($KF.2H_2O$), kJ/mol[a]	58.2	7
entropy, J/(mol·K)[a] (25°C)	66.6	2
lattice energy, kJ/mol[a]	813.8	8
specific heat, J/(kg·K)[a] (0°C)	833	2
at (50°C)	854	2

[a] To convert kJ to kcal, divide by 4.184.

Uses. Potassium fluoride is used in the manufacture of silver solder fluxes and in fluxes for various metallurgical operations (see Solders and brazing alloys). In tin deposition from halogen plating baths, KF is used to complex tetravalent tin to form K_2SnF_6 [*16893-93-9*] which may be filtered from the solution (13). Highly purified KF is formed into single crystals by the Stockbarger process (14); the principal use of the crystals is in the studies of fundamental properties and defects in alkali halide crystals.

For many types of replacement of halogens by fluorine in organic compounds, KF is the most frequently used fluoride. As a fluorinating agent, KF must be as anhydrous as possible and very finely divided. The potassium chloride or bromide by-products from reactions with organic chlorides or bromides deposit on the potassium fluoride crystal surfaces, significantly retarding the reaction. Polar solvents such as dimethyl sulfoxide or formamide, and rapid stirring are useful in overcoming this drawback. Ball-milling the reaction mixture is also helpful in speeding up the reaction. This fluorination process is used commercially in the manufacture of sodium fluoroacetate, a very useful rat poison sold as "1080" (see Poisons, economic). An alkyl haloacetate is converted to an alkyl fluoroacetate. The latter ester is hydrolyzed with sodium hydroxide to the final product. Similarly, fluoroacetamide, a systemic insecticide, is made from chloroacetamide. Organic fluorides (15) that may be prepared from corresponding bromides or chlorides using potassium fluoride include: monofluoroalkanes; *alpha*-fluoro-esters, -ethers, and -alcohols; acyl fluorides; sulfonyl fluorides; and 1-fluoro-2,4-dinitrobenzene. Potassium fluoride behaves as a base when used as a catalyst for reactions such as dehydrohalogenation (16), Michael addition (17), and the Knoevenagel reaction (18). Polyurethane foams can be formed when KF is used as a catalyst for the reaction of adiponitrile carbonate and a polyester polyol (19) (see Urethane polymers).

Potassium Bifluoride

Properties. Other names for the compound are potassium hydrogen difluoride and potassium acid fluoride. This white crystalline salt is a soft, waxy solid. The crystal forms of potassium bifluoride are tetragonal and cubic (20). The bifluoride ion in KHF_2 averages 0.2292 nm between fluoride ions in the F—H—F group (21). At elevated temperatures, potassium bifluoride exhibits an appreciable vapor pressure of HF. At 440°C, KHF_2 is decomposed completely to KF and HF; this decomposition is a convenient means of obtaining very pure HF. Other chemical and physical properties are summarized in Table 2.

Manufacture. Potassium bifluoride is produced from potassium hydroxide or potassium carbonate and hydrofluoric acid. The concentrated solution is cooled and allowed to crystallize. The crystals are separated centrifugally and dried. The commercial product consists typically of 99.7% KHF_2 and 0.2 KF. Potassium bifluoride is available in the United States in 180-kg drums at \$1.93/kg (1978).

Toxicology and Handling. The lethal dose by ingestion in guinea pigs is 150 mg/kg body weight (12). The TLV for KHF_2 is 2.5 mg/m^3 (24). Potassium bifluoride crystals may break down readily to a fine white powder that is readily airborne. In this form, the salt is quite irritating to the nasal passages, eyes, and skin. Therefore, the hands and eyes should be protected and acid dust masks should be worn while handling. As an acid fluoride, KHF_2 can cause superficial hydrofluoric acid-type burns. Areas of

Table 2. Physical and Chemical Properties of Potassium Bifluoride, KHF_2

		Reference
melting point, °C	238.8	22
specific gravity	2.37	23
solubility in H_2O, g/100 g (20°C)	39.2	
crystal transition temperature, °C	196.7	22
standard heat of formation, kJ/mol[a]	−920.4	6
heat of fusion, kJ/mol[a]	6.6	22
heat of dissociation (to KF + HF, 226.8°C), kJ/mol[a]	77.5	23
entropy, J/(mol·K)[a] (25°C)	104.3	6
heat capacity, $C_p°$, J/(mol·K)[a] (25°C)	76.86	6
lattice energy, kJ/mol[a]	641.8	23

[a] To convert kJ to kcal, divide by 4.184.

skin that have been in contact with potassium bifluoride should be washed as soon as possible with mildly alkaline soaps or borax-containing hand cleaners. If there has been contact with the eyes, they should be washed well with water and a physician should be consulted.

Uses. A major use for potassium bifluoride is in the electrolyte for cells in fluorine manufacture (25). Sufficient hydrogen fluoride is dissolved with KHF_2 to bring the total HF content up to 40–42 wt %. This mixture approximates the formula, KF.2HF; it is molten at 90°C, the operating temperature of the cell. Fluxes for a wide variety of metal joining applications utilize KHF_2, usually in combination with potassium pentaborate, boric acid, and other fluorides and chlorides such as $ZnCl_2$ (26). Solutions of KHF_2 and citric acid are used to etch aluminum prior to coating with an acrylic copolymer emulsion; this opaque white coating resists chipping and is useful for automotive, architectural, and decorative applications (27). Tetrahydrofuran is polymerized to poly(tetramethylene glycol) with fuming sulfuric acid and potassium bifluoride (28).

BIBLIOGRAPHY

"Potassium Compounds" in *ECT* 1st ed., under "Fluorine Compounds, Inorganic," Vol. 6, pp. 721–722, by D. C. Whitaker, Harshaw Chemical Company; "Potassium" in *ECT* 2nd ed., under "Fluorine Compounds, Inorganic," Vol. 9, pp. 649–650, by D. C. Whitaker, The Harshaw Chemical Company.

1. D. L. Deadman, J. S. Machin, and A. W. Allen, *J. Am. Ceram. Soc.* **44**(3), 105 (1961).
2. *JANAF Thermochemical Tables,* Clearinghouse for Federal, Scientific, and Technical Information, U.S. Dept. of Commerce, Springfield, Va., 1964.
3. A. W. Jache and G. W. Cady, *J. Phys. Chem.* **56,** 1106 (1952).
4. J. Emsley, *J. Chem. Soc. A,* 2511 (1971).
5. R. E. Harner, J. B. Sydnor, and E. S. Gilbreath, *J. Chem. Eng. Data* **8,** 411 (1963).
6. F. D. Rossini and co-workers, *NBS, Circ. 500,* U.S. Government Printing House, Washington, D.C., 1952.
7. J. Bell, *J. Chem. Soc.,* 72 (1940).
8. H. Vaino and M. Kanko, *Ann. Univ. Turku.* **Ser. AI,** (40), 3 (1960); *Chem. Abstr.* **61,** 13963 (1964).
9. J. H. Clark and J. Emsley, *J. Chem. Soc. Dalton Trans.,* 2129 (1975).
10. V. Kazakov and V. G. Kharchuk, *Zh. Obshch. Khim.* **45,** 2744 (1975).
11. H. Kojima, S. G. Whiteway, and C. R. Masson, *Can. J. Chem.* **46,** 2968 (1968).
12. H. C. Hodge and F. A. Smith in J. H. Simons, ed., *Fluorine Chemistry,* Vol. 4, Academic Press, New York, 1965, p. 200.
13. I. Rajagonal and K. S. Rajams, *Met. Finish.* **76**(4), 43 (1978).

14. D. C. Stockbarger, *J. Opt. Soc. Am.* **14,** 448 (1927).
15. A. R. Basbour, L. F. Belf, and M. W. Bruxton in M. Stacy and co-eds., *Advances in Fluorine Chemistry,* Vol. 3, Butterworth, Washington, D.C., 1963, pp. 181–250.
16. F. Naso and L. Ronzini, *J. Chem. Soc. Perkin Trans. 1,* 340 (1974); J. H. Clark and J. M. Miller, *J. Am. Chem. Soc.* **99,** 498 (1977); J. H. Clark, J. Emsley, and O. P. A. Hoyta, *J. Chem. Soc. Perkin Trans. 1,* 1091 (1977).
17. I. Belski, *Chem. Commun.,* 237 (1977).
18. L. Rand, J. V. Swisher, and C. J. Cromin, *J. Org. Chem.* **27,** 3505 (1962); L. Rand, D. Haidukewych, and R. J. Dohinski, *J. Org. Chem.* **31,** 1272 (1966).
19. U.S. Pat. 3,766,147 (July 31, 1972), L. G. Walgemuth (to Atlantic Richfield Co.).
20. R. Kruh, K. Fuwa, and T. E. McEver, *J. Am. Ceram. Soc.* **78,** 4526 (1956).
21. H. L. Carrell and J. Donohue, *Isr. J. Chem.* **10**(2), 195 (1972).
22. M. L. Davis and E. F. Westrum, Jr., *J. Phys. Chem.* **65,** 338 (1961).
23. T. C. Waddington, *Trans. Faraday Soc.* **54,** 25 (1958).
24. *Proceedings, Am. Conf. Gov. Ind. Hyg.,* Cincinnati, 1977.
25. S. P. Vavalides and co-workers, *Ind. Eng. Chem.* **50**(2), 178 (1958).
26. U.S. Pat. 2,829,078 (Apr. 1, 1958), H. B. Aull and A. S. Cross; U.S. Pat. 3,958,979 (May 29, 1976), A. R. Valdo (to Ethyl Corp.); Jpn. Kokai 75 113,449 (Sept. 5, 1975), K. Motoyoshi, M. Kume, and Y. Amano (to Sumitomo Electric Industries).
27. U.S. Pat. 3,849,208 (Aug. 13, 1973), M. N. Marosi (to Convertex, Ltd.).
28. Jpn. Kokai 73 01,100 (Jan. 9, 1973), K. Matsuzawa, Y. Suzuki, and K. Ohya (to Mitsubishi Chemical Industries Co.).

H. S. Halbedel
Thomas E. Nappier
The Harshaw Chemical Company

RHENIUM

Rhenium has six binary fluorides: rhenium(II) bifluoride [*25604-91-5*], rhenium(III) trifluoride [*25605-13-4*], rhenium(IV) tetrafluoride [*15192-42-4*], rhenium(V) pentafluoride [*30937-52-1*], rhenium(VI) hexafluoride [*10049-17-9*], and rhenium(VII) heptafluoride [*17029-21-9*]. Rhenium also forms at least four oxyfluorides, $ReOF_4$ [*17026-29-8*], $ReOF_5$ [*23377-53-9*], ReO_2F_3 [*57246-89-6*], and ReO_3F [*25813-73-4*], and several complex fluorides. Of these compounds, only rhenium hexafluoride is of commercial importance; it is used in chemical vapor deposition to deposit metallic rhenium.

Rhenium Hexafluoride

Physical Properties. Rhenium hexafluoride, ReF_6, is a yellow solid that undergoes a phase transition from a low temperature orthorhombic form at −3.5°C (1) to a cubic form having a glassy appearance. Physical properties of rhenium hexafluoride are given in Table 1.

Table 1. Physical Properties of Rhenium Hexafluoride

Property	Value	Reference
melting point, °C	18.5	1–2
boiling point, °C	33.7	1–2
transition point, °C	−3.45	2
	−1.9	1
color, solid	yellow	
liquid	yellow	
vapor	colorless	
specific gravity	3.58	3
heat of fusion, kJ/mol[a]	4.635	1
heat of transition, kJ/mol[a]	8.487	1
heat of vaporization, kJ/mol[a]	28.75	1
entropy of fusion, J/(mol·K)[a]	60.46	1
entropy of transition, J/(mol·K)[a]	31.48	1
entropy of vaporization, J/(mol·K)[a]	93.4	2
heat of sublimation,		
above transition, kJ/mol[a]	32.657	2
below transition, kJ/mol[a]	41.407	2
vapor pressure at 20.3°C, kPa[b]	61	1
solubility in liquid HF, mol/kg	1.75	4

[a] To convert kJ to kcal, divide by 4.184.
[b] To convert kPa to mm Hg, multiply by 7.5.

Chemical Properties. Rhenium hexafluoride is very reactive and hydrolyzes to give hydrogen fluoride, rhenium dioxide, and perrhenic acid, $HReO_4$.

Rhenium hexafluoride is very stable when dry and can be handled in glass. It can be used with a variety of metals owing to the formation of a passive fluoride film that protects the metal from further corrosion.

Little has been done to study the chemistry of rhenium hexafluoride. It does not react with oxygen at 500°C (1). It reacts with nitric oxide to produce the complex salt $NOReF_6$ [*60447-76-9*] (5), with potassium fluoride to produce K_2ReF_8 [*57300-90-0*] (6), and it is reduced by alkali-metal iodides in liquid sulfur dioxide solution to produce complex salts M_2ReF_6 (M = Na [*12021-61-3*], K [*16962-12-2*], Rb [*16962-13-3*], and Cs [*16962-14-4*]) (7).

Rhenium hexafluoride is reduced by hydrogen at an elevated temperature to yield rhenium metal and hydrogen fluoride (8–10). Temperatures in the range of 550 to 750°C and pressures less than atmospheric are normally used. The reaction is the basis of the major use of rhenium hexafluoride in chemical vapor deposition (CVD) (see Film deposition techniques).

Manufacture and Economics. Rhenium hexafluoride is produced commercially by the reaction of rhenium powder with fluorine at approximately 100°C. The reaction is conducted in the presence of excess rhenium metal to convert any rhenium heptafluoride, which is produced as by-product, to the hexafluoride. Other impurities such as hydrogen fluoride or other metal fluorides are controlled by the quality of the fluorine and rhenium raw materials used in the process.

U.S. production is less than 100 kg per year with essentially all the production going to CVD. The 1979 price is ca $3000/kg and is dependent upon the price of rhenium metal. Because of the high value of the material, it is shipped in small quantities. Stainless steel or steel cylinders containing 0.45 kg are commercially available.

Handling and Safety Factors. Rhenium hexafluoride is irritating and corrosive to the upper and lower respiratory tract, eyes, and skin. Since it rapidly hydrolyzes to hydrogen fluoride, it is very corrosive to the skin and will produce burns typical of hydrofluoric acid. Therefore, extreme care should be used in the handling of the material and personnel should avoid breathing of the vapors or contact with either the liquid or vapor. No toxicity data are currently available on rhenium hexafluoride and ReF_6 should be handled with the same precautions used in handling anhydrous HF.

Uses

The CVD of rhenium by the thermochemical decomposition of rhenium hexafluoride to produce a metallic rhenium deposit is the major commercial use of ReF_6. Because of the high cost of rhenium, it is often used as a low percentage composite with tungsten. The addition of rhenium to tungsten improves the ductility and the high temperature properties of the deposit (11). Tungsten–rhenium alloys produced by chemical vapor deposition exhibit superconducting transition temperatures (T_c) substantially higher than those of arc-melted alloys (12) (see Superconducting materials).

Other Rhenium Compounds

The physical and chemical properties of the binary fluorides, ReF_4, ReF_5, and ReF_7, are presented in refs. 3 and 13. Data on the existence of ReF_2 and ReF_3 are reported in ref. 14 and properties of the oxyfluorides are presented in refs. 2, 15–18.

BIBLIOGRAPHY

1. J. G. Malm and H. Selig, *J. Inorg. Nucl. Chem.* **20,** 189 (1961).
2. G. H. Cady and C. B. Hargreaves, *J. Chem. Soc.,* 1563, 1568 (1961).
3. *Rhenium Chemicals—Their Properties and Applications,* Cleveland Refractory Metals, Cleveland, Ohio, 1969.
4. B. Frlec and H. H. Hyman, *Inorg. Chem.* **6,** 1596 (1967).
5. N. Bartlett, S. P. Beaton, and N. K. Jha, *Chem. Commun.,* 168 (1966).
6. E. G. Ippolitouo, *J. Inorg. Chem. USSR* **7,** 485 (1962).
7. R. D. Peacock, *J. Chem. Soc.,* 467 (1957).
8. U.S. Pat. 3,565,676 (Apr. 1, 1968), R. A. Holzl (to Fanstell Metallurgical Corp.).
9. F. W. Hoertel and J. G. Donaldson, *J. Electrodeposition Surf. Treat.* **2,** 343 (1974).
10. J. I. Federer and A. C. Schaffhauser, *J. IEEE Conf. Rec Thermion. Convers. Spec. Conf., Pap. Annu. Conf., 9th,* IEEE, New York, pp. 74–81.
11. W. L. Roberts, *High Temp. Mater., Pap. Plansee Semin., 6th, 1969,* pp. 880–884.
12. D. S. Easton and co-workers, *Philos. Mag.* **30,** 1117 (1974).
13. J. C. Bailar and co-workers, *Comprehensive Inorganic Chemistry,* Vol. 3, Pergamon Press, Elmsford, N.Y., 1973, pp. 966–978.
14. R. L. Johnson and B. Siegel, *Technical Rept. July–Oct. 68,* Aerospace Corp., Aug. 15, 1969.
15. O. Ruff and W. Kwansik, *Z. Anorg. Chem.* **219,** 65 (1934).
16. E. E. Aynsley and M. L. Hair, *J. Chem. Soc.,* 3747 (1958).
17. A. Englebrecht and A. V. Grosse, *J. Am. Chem. Soc.* **76,** 2042 (1954).
18. J. H. Simons, ed., *Fluorine Chemistry,* Academic Press, New York, 1964, pp. 75–78.

ANDREW J. WOYTEK
Air Products & Chemicals, Inc.

SILVER

Silver Fluoride

Anhydrous silver fluoride [*7775-41-9*], AgF, 99.8% pure, is a golden-yellow, hygroscopic solid. Attempts to prepare a colorless material have not been successful. The color is thought to be owing to the presence of silver oxide and free silver.

Silver fluoride is light sensitive and has a specific gravity of 5.852. It melts at 435°C into a black liquid which boils at ca 1150°C. Unlike the other silver halides, it is extremely soluble in water (182 g/100 g H_2O) and anhydrous hydrogen fluoride (83.2 g/100 g anhydrous HF at 11.9°C). It is slightly soluble in absolute methanol (1.5 g/100 mL methanol).

A hot concentrated solution of silver fluoride dissolves fine silver metal and produces silver subfluoride, Ag_2F. Silver fluoride forms explosive adducts with ammonia (1–2) and, therefore, all of the reactions involving liquid or gaseous ammonia should be carried out with extreme precautions. In the AgF–HF–H_2O system (3) several solid phases are formed: AgF.2H_2O [*72214-21-2*], AgF.4H_2O [*2242-42-6*], 3AgF.2HF [*72318-57-1*], AgF.2HF [*12444-84-7*], AgF.3HF [*12444-85-8*], AgF.5HF [*12444-86-9*], and 6AgF.7HF.2H_2O [*72318-56-0*]. In addition, AgF.3HF is formed in the absence of water at 0°C. When AgF.3HF is slowly warmed to 25°C under nitrogen atmosphere, it produces AgF.HF or $AgHF_2$ [*12249-52-4*].

Silver fluoride is used as a cathode material in batteries (4), antimicrobial agent (5), and fluorinating reagent (6–13).

In absence of solvent, AgF acts as a poor fluorinating reagent for halocarbons owing to the formation of double salt: AgF.AgX (X = Cl [*72785-04-7*], Br [*72785-05-8*], and I [*72785-06-9*]) (14).

Silver fluoride can be prepared (15) by dissolving silver carbonate or oxide in anhydrous hydrogen fluoride or aqueous hydrofluoric acid, evaporating to dryness and then treating it with methanol and ether.

Silver fluoride is commercially available in quantities from 100 g to 500 kg.

Silver Subfluoride

Silver subfluoride [*1302-01-8*], Ag_2F, yellowish-green solid is obtained by heating a concentrated solution of AgF with metallic silver powder, or the electrolysis of a solution containing AgF, aqueous HF, NH_4F at 50°C in a platinum dish with silver electrode.

It decomposes in water but is stable in alcohol and a saturated solution of AgF. It is not light sensitive. When heated above 100°C, it disproportionates into Ag and AgF.

No commercial application of silver subfluoride is known.

Silver Difluoride

Silver difluoride [*7783-95-1*], AgF_2, is a black crystalline powder. Although there are some reports (16–17) in the earlier literature describing silver difluoride as a dark-brown or yellow powder, the color discrepancy stems from the presence of AgF or Ag_2O in AgF_2. It melts at ca 690°C and its dissociation pressure is 10 kPa (0.1 atm) at 700°C.

Silver difluoride is a powerful oxidizing agent. It liberates ozone when heated with a dilute aqueous acid. It also liberates iodine from KI solution, oxidizes manganous salts to permanganate in either acid or alkaline solutions, and chromic salts to chromates. Ethyl alcohol is oxidized to acetaldehyde. The latter reaction should be carried out with extreme precautions. It is a hard (18) fluorinating reagent and is used for addition of F_2 to haloalkanes (19), for substitution of hydrogen by fluorine in halocarbons (20) and hydrohalocarbons (21–22), and for converting carbon monoxide into carbonyl fluoride.

It can be prepared by the action of elemental fluorine on AgF or AgCl at 200°C or on silver powder at 300°C.

Silver difluoride is commercially available in quantities of 100 g to 50 kg.

Hazards. Silver difluoride may react explosively with organic compounds, quite often after an induction period. Small-scale experiments should be tried first and proper steps should be taken so that heat of the reaction is dissipated.

Silver Fluoro Salts

Silver fluoro salts, silver hexafluoroantimonate [*26042-64-8*], $AgSbF_6$, silver tetrafluoroborate [*14104-20-2*], $AgBF_4$, and silver hexafluorophosphate [*26042-63-7*], $AgPF_6$, have gained potential industrial importance owing to their catalytic activities, and they are used in the separation of olefin–paraffin mixtures (23), and nitration of aromatic compounds (24). The anhydrous salts are slightly soluble in anhydrous HF but very soluble in water. They are also soluble in benzene, toluene, and *m*-xylene, and form adducts.

Silver fluoro salts can be prepared by the action of the Lewis acids SbF_5, PF_5, or BF_3, on AgF solution in anhydrous HF. These salts are commercially available from 100-g to 50-kg quantities.

Toxicity. All silver fluoro salts, silver difluoride, silver fluoride, silver subfluoride are irritant and toxic. Skin contact and inhalation should be avoided.

BIBLIOGRAPHY

"Silver Compounds" under "Fluorine Compounds, Inorganic" in *ECT* 1st ed., Vol. 6, pp. 730–731, by F. D. Loomis, Pennsylvania Salt Manufacturing Co.; "Silver" under "Fluorine Compounds, Inorganic" in *ECT* 2nd ed., Vol. 9, pp. 661–662, by Wayne E. White, Ozark-Mahoning Company.

1. W. Biltz and E. Rahlfs, *Z. Anorg. Chem.* **166,** 351 (1927).
2. L. J. Olmer and Dervin, *Compt. Rend.* **175,** 1058 (1922).
3. H. J. Thomas and A. W. Jache, *J. Inorg. Nucl. Chem.* **13,** 54 (1960).
4. Jpn. Kokai 75 131,034 (Oct. 16, 1975), T. MaSaaki (to Japan Storage Battery Co., Ltd.).
5. *Chem. Week,* 52 (Jan. 14, 1961).
6. P. Gielow and A. Haas, *Chem. Ztg.* **95,** 1010 (1971).
7. E. D. Bergmann and I. Shalhak, *J. Chem. Soc.,* 1418 (1959).
8. H. J. Emeleus and D. E. MacDuffie, *J. Chem. Soc.,* 2597 (1961).
9. J. S. Simons, D. F. Herman, and W. H. Pearlson, *J. Am. Chem. Soc.* **68,** 1672 (1946).
10. A. D. Britt and W. B. Moniz, *J. Am. Chem. Soc.* **91,** 6204 (1969).
11. F. J. Buckle, F. L. M. Pattison, and B. C. Saunders, *J. Chem. Soc.,* 1471 (1949).
12. D. T. Meshri and W. T. Miller, *Abst. #14, 158th ACS National Meeting, New York, Sept., 1969.*

13. W. T. Miller, R. A. Snider, and D. T. Meshri, *4th Winter Fluorine Conference, Daytona Beach, Florida, Feb., 1979.*
14. A. K. Barbour, L. J. Belf, and M. W. Buxton, "The Preparation of Organic Fluorine Compounds by Halogen Exchange" in M. Stacey, J. C. Tatlow, and A. G. Sharpe, eds., *Advances in Fluorine Chemistry,* Vol. 3, Butterworths, Inc., Washington, D.C., 1963, pp. 182–253.
15. F. A. Anderson, B. Bak, and A. Hillberet, *Acta Chem. Scand.* **7,** 236 (1953).
16. H. Jockusch, *Naturwissenschaften* **22,** 561 (1934).
17. O. Ruff and M. Giese, *Z. Anorg. Chem.* **219,** 143 (1934).
18. D. T. Meshri and W. E. White, *George H. Cady ACS Symposium, Milwaukee, Wisc., June, 1970.*
19. D. A. Rausch, R. A. Davis, and D. W. Osborne, *J. Org. Chem.* **28,** 494 (1963).
20. E. T. McBee and L. D. Bechtol, *Ind. Eng. Chem.* **39,** 380 (1947).
21. W. B. Burford, III and co-workers, *Ind. Eng. Chem.* **39,** 379 (1947).
22. E. T. McBee and co-workers, *Ind. Eng. Chem.* **39,** 310 (1947).
23. U.S. Pat. 3,189,658 (June 15, 1965), H. W. Quinn (to The Dow Chemical Co.).
24. G. Olah and S. Kahn, *J. Am. Chem. Soc.* **83,** 4564 (1961).

DAYAL T. MESHRI
Ozark-Mahoning Company,
a subsidiary of the Pennwalt Corp.

SODIUM

Sodium has two fluorides, sodium fluoride [*7722-88-5*] and sodium bifluoride [*1333-83-1*].

Sodium Fluoride

Sodium fluoride, NaF, is a white, free flowing crystalline powder, mp 992°C, bp 1704°C, with a solubility of 4.2 g/100 g water at 10°C, and 4.95 g/100 g water at 93.3°C. The purity of the commercial material is about 98%.

Sodium fluoride is normally manufactured by the reaction of hydrofluoric acid with soda ash (sodium carbonate), or caustic soda (sodium hydroxide), with control of pH essential and proper agitation necessary to obtain the desired crystal size. The crystals are centrifuged, dried, sized, and packaged. Reactors are usually constructed of carbon brick and lead-lined steel, with process lines of stainless, plastic or plastic-lined steel; diaphragm, plug cock, or butterfly valves are preferred.

The salt is packaged in 45-kg multi-wall bags or fiber drums of 45-, 170- or 181-kg. It is available in both powdered and granular forms with densities of 1.04 and 1.44 g/cm^3 (65 and 90 lb/ft^3), respectively. Only the powdered grade is authorized by and registered with the EPA for use in pesticide formulations, with the further proviso that it must be tinted blue or green, or otherwise discolored. The word poison appears on all labels together with first aid information.

Both sodium fluoride and sodium bifluoride are poisonous if taken internally.

Dust inhalation and skin or eye contact may cause irritation of the skin, eyes or respiratory tract, and should be avoided by the use of proper protective equipment (1).

Fluoridation of potable water supplies for the prevention of dental caries is one of the principal uses for sodium fluoride (see Water, municipal water treatment). Use rate for this application is of the order of 0.7 to 1.0 mg/L of water as fluoride or 1.5 to 2.2 mg/L as NaF (2). It is also applied topically to teeth as a 2% solution. Other uses are as a flux for deoxidizing (degassing) rimmed steel (qv), and in the resmelting of aluminum (qv). It is also used in the manufacture of vitreous enamels, in pickling stainless steel, in wood preservation compounds, casein glues, in the manufacture of coated papers, in heat-treating salts, and as a component of laundry sours.

Sodium Bifluoride

Sodium bifluoride (sodium acid fluoride, sodium hydrogen fluoride), $NaHF_2$ or NaF.HF, is a white, free-flowing fine granular material. Its solubility in water is 3.7 g/100 g solution at 20°C, and 16.4 g/100 g at 80°C. It decomposes at temperatures above 160°C to give sodium fluoride and hydrogen fluoride. Commercial material is ca 99% pure.

The same reactants are used for manufacture as for sodium fluoride, with an excess of acid required to crystallize the bifluoride. The crystals are dewatered, dried, sized and packaged. Cooling of the reaction is necessary to avoid overheating and decomposition. Reactors and auxiliary equipment are the same as for sodium fluoride.

The dried salt is shipped in 45-kg multi-wall bags and in 57-, 170- and 180-kg fiber drums. Densities range from ca 0.70 g/cm^3 (44 lb/ft^3) to 1.2 g/cm^3 (75 lb/ft^3) for crystalline material.

Sodium bifluoride, by itself, or in conjunction with other materials is a good laundry sour since, in the concentrations used, it will not create a pH below 4.0 and thus causes no damage to textile fibers, although it removes iron stains (see Drycleaning and laundering). Leather bleaching, cleaning of stone and brick building faces are other uses for this material (3).

BIBLIOGRAPHY

"Sodium Fluoride" treated in *ECT* 1st ed., under "Fluorine Compounds, Inorganic," Vol. 6, p. 731, by F. D. Loomis, Pennsylvania Salt Manufacturing Co.; "Sodium Bifluoride" treated in *ECT,* 1st ed., under "Fluorine Compounds, Inorganic," Vol. 6, pp. 731–732, by J. E. Dodgen, Pennsylvania Salt Manufacturing Co.; "Sodium Fluoride" and "Sodium Bifluoride," treated in *ECT* 2nd ed., under "Fluorine Compounds, Inorganic," Vol. 9, pp. 662–663, by J. Griswold, Allied Chemical Corp.

1. N. I. Sax, *Dangerous Properties of Industrial Materials,* 4th ed., Van Nostrand Reinhold, New York, 1975.
2. *Fluoridation Engineering Manual,* EPA, Office of Water Programs, 1972.
3. M. Windholz, ed., *Merck Index,* ninth ed., 1976.

KENNETH WACHTER
Olin Corporation

SULFUR—SULFUR FLUORIDES

The known binary compounds of sulfur and fluorine range in character from ephemeral to rocklike and provide excellent examples of the influence of electronic and structural factors on chemical reactivity. These marked differences are also reflected in their diversified technological utility.

Sulfur Hexafluoride

Sulfur hexafluoride [*2551-62-4*], SF_6, molecular weight 146.07, is a colorless, odorless, tasteless gas. It is not flammable and not very reactive. Its high chemical stability and good electrical characteristics have led to its widespread use in various electrical and electronic equipment such as circuit breakers, capacitors, transformers, microwave components, etc. Other properties of the gas have led to its limited usage in a variety of unique applications ranging from medical applications to space research.

Sulfur hexafluoride was first prepared in 1902 by Moissan and Lebeau (1) by burning sulfur in a fluorine atmosphere. In 1937 it was discovered that its dielectric strength is much higher than that of air (2), and soon thereafter it was introduced as an insulating material for cables, capacitors (3), and transformers (4). Sulfur hexafluoride has been commercially available from the Allied Chemical Corporation since 1948. It is also produced in France, Italy, the Federal Republic of Germany, Japan, and probably the USSR.

Physical Properties

Sulfur hexafluoride is a good dielectric because of a favorable combination of physical properties (see Table 1) which permits maintenance of high gas density and consequently high dielectric strength at low temperatures. The transformation from SF_6(I) to SF_6(II) takes place at −171.2°C, and the heat of transformation is 1.612 kJ/mol (385.3 cal/mol) (17).

The vapor pressure of the liquid is calculated as follows (10) for range −50 to 45.6°C, standard % deviation ± 0.18:

$$\log P_{\text{kPa}} = 0.87652594 - 816.48995\ T + 0.029287342\ T$$
$$-0.40107549 \times 10^{-4}\ T^2$$
$$+0.7142667 \frac{(319.802 - T)}{T} \log (319.802 - T)$$

T is in Kelvin. To convert kPa to mm Hg, multiply by 7.5.

Equations for the calculation of sublimation pressure can be found in refs. 5 and 24.

Heats of vaporization were calculated from the Clapeyron-Clausius equation, using consistent units (15). Several heats of vaporization are shown below.

Table 1. Physical Properties[a] of Sulfur Hexafluoride, SF_6(II)

Property	Value	Refs.
sublimation point, °C	−63.9	5
heat of sublimation, kJ/mol[b]	23.59	6
triple point, °C	−50.52	7
pressure at triple point, kPa[c]	225.31	7
critical temperature, °C	45.55	8–9
critical pressure, MPa[d]	3.759	9–10
critical density, g/cm³	0.737	10–12
density, g/cm³		
solid at −195.2°C	2.863	13
liquid	1.336	10, 14
gas	6.0886×10^{-3}	15
vapor pressure of saturated liquid, MPa[d]	2.3676	15
heat of formation, kJ/mol[b]	−1221.66	
free energy of formation, kJ/mol[b]	−1117.73	16
heat of vaporization, kJ/mol[b]	9.6419	15
entropy, kJ/mol·K)[b]	291.874	16
heat capacity, J/(mol·K)[b]		
liquid at −43°C	119.5	17
gas	97.234	15
surface tension at −20°C, mN/m (= dyn/cm)	8.02	18
viscosity, mPa·s (= cP)		
liquid	0.277	14
gas	0.01576	17
thermal conductivity, W/(m·K)		
liquid	0.0583	19
gas	0.01415	20
sound velocity, gas, m/s	136	14
refractive index, n_D	1.000783	21
dielectric constant		
liquid	1.81	22
gas	1.00204	23
loss tangent (liquid)	0.001	22

[a] All data refer to 25°C and 101.3 kPa (1 atm), unless otherwise stated.
[b] To convert J to cal, divide by 4.184.
[c] To convert kPa to mm Hg, multiply by 7.5.
[d] To convert MPa to atm, divide by 0.101.

Temperature, °C	ΔH_{vap}, *kJ/mol*
45.6	0
40	5.608
20	9.807
0	12.23
−20	13.58
−40	14.94

To convert kJ to kcal, divide by 4.184.

Chemical Properties. With few exceptions, SF_6 is chemically inert at RT and atmospheric pressure. Thermodynamically, SF_6 is unstable relative to reaction with many materials, including water, but these reactions are kinetically impeded by the fluorine shielding of the sulfur. Sulfur hexafluoride does not react with alkali hydroxides, ammonia, or strong acids.

At elevated temperatures SF_6 forms the respective fluorides and sulfides with many metals (25). In quartz, it starts to decompose at 500°C (1); in copper or stainless steel, it is less stable (26).

The thermal stability of SF_6 in the presence of materials of construction is of growing importance in view of its potential use in dry-type transformers operating at 200–250°C. The stability of SF_6 at 200 and 250°C in the presence of aluminum, copper, silicon steel, and mild steel is compared in Table 2 (14).

Careful exclusion of moisture from the system improves the stability of sulfur hexafluoride in the presence of most materials.

Sulfur hexafluoride is more stable in arcs (27) than fluorocarbons such as C_2F_6, or refrigerants like CCl_2F_2, but less stable than CF_4, BCl_3, or SiF_4. Carbon tetrafluoride exposed to 1000 C shows only a few percent decomposition, whereas SF_6 under the same conditions decomposes to SOF_2 and SF_4 to the extent of 10 mol %. Hexafluoroethane exposed to only 50 C is 50% decomposed. In spite of its decomposition, the dielectric strength of SF_6 remains the same.

The main products of SF_6 arc decomposition in the presence of air are SOF_2, SF_4, and SOF_4 plus metal fluorides and sulfides (28).

Electrical Properties. The electrical properties of SF_6 stem primarily from its effectiveness as an electron scavenger. To accomplish electrical breakdown in a dielectric gas, primary electrons must gain sufficient energy to generate appreciable secondary electrons on molecular impact. Sulfur hexafluoride interferes with this process by capturing the primary electrons resulting in the formation of SF_6^- or SF_5^- ions and F atoms:

$$SF_6 \rightarrow (SF_6^-)^* \text{ activated} \rightarrow SF_6^-$$

$$(SF_6^-)^* \rightarrow SF_5^- + F$$

The $(SF_6^-)^*$ activated complex is stable against autodissociation during tens of microseconds and thus can be stabilized by collisions.

Although production of SF_6^- is the primary process, formation of SF_5^- ions increases with temperature and at 200°C the SF_5^-:SF_6^- ratio is 1:25 (29). In addition to high dielectric strength, SF_6 can rapidly interrupt heavy currents at high voltages. At 550 or 765 kV, circuit breaker ratings can be 38,000 and 5,000 MVA (million volt amperes) respectively (30). Both SF_6 and C_2F_6 have a more rapid arc recovery than nitrogen (31); six S—F bonds, mean energy 3.4–3.8 eV, are available for fast energy absorption. Both SF_6 and SF_4 have high stability in arcs (27).

Table 2. Stability of Sulfur Hexafluoride in Various Materials of Construction

Material	Decomposition, %/yr	
	200°C	250°C
aluminum		0.006
copper	0.18	1.4
silicon steel	0.005	0.01[a]
mild steel	0.2	ca 2

[a] Estimated.

Paschen's Rule and Breakdown Voltage. As the pressure decreases to vacuum conditions, the breakdown voltage (BDV) increases, resulting in a minimum. Figure 1 summarizes data compiled from the available literature (32–33) for uniform electric fields, 60 Hz or dc at 25°C and pressure ≤200 kPa (ca 2 bar), gap ≤0.3 mm. Table 3 compares BDV data of SF_6 with those of other dielectrics. For optimum utility, a compromise is needed between low boiling point and high BDV.

From the data in Table 4, it appears that at 300–400 kPa (3–4 atm), SF_6 gas is equivalent to transformer oil under uniform field conditions. A BDV deviation from Paschen's rule appears as pressure increases. It is obeyed well only in the case of uniform fields at pressures up to ca 400 kPa (4 atm).

Actually, the description of SF_6 electrical properties needed for practical applications is more complex. Not only BDV but also corona-onset voltages must be considered, particularly for the more usual nonuniform fields. Figure 2 illustrates this for a point-to-plane electrode configuration where extensive prebreakdown corona occurs before sparkover (35).

The theory and application of SF_6 BDV and corona-onset voltages (COV) has been studied in uniform and in nonuniform electric fields (36). The ionization potentials of SF_6 and electron attachment coefficients are the basis for one set of correlation equations. A critical field exists at 89 kV/(cm·kPa) above which coronas can

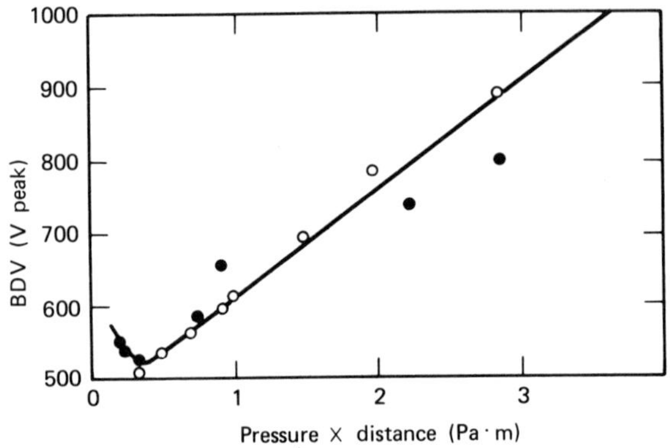

Figure 1. Paschen's rule plot for SF_6. Compiled from refs. 32 (○) and 33 (●). To convert Pa to μm Hg, multiply by 7.50.

Table 3. Sulfur Hexafluoride Compared with other Dielectrics[a,b]

Compound	Molecular wt	bp, °C	Relative BDV[c]
SF_6	146.05	−64 (subl)	2.7
CF_2Cl_2	120.9	−29	2.8
CF_3Cl	104.5	−81	1.44
CF_4	88.0	−128	1.14
N_2	28	−194	1.0

[a] Ref. 34.
[b] Conditions: 60 Hz, 0.5 cm gap, 5 cm spheres, P = 101.3 kPa (1 atm), t = ca 25°C.
[c] BDV = breakdown voltage.

Table 4. Effect of Pressure on Breakdown Voltage of SF_6 [a]

	Gap	
Pressure, kPa [b]	0.5 cm, kV	1.0 cm, kV
SF_6		
101.3	45.0	86.0
202.6	79.0	161.0
304.0	112.5	218.0
405.2	143.5	268.0
transformer oil		
101.3	110.0	218.0

[a] Ref. 34.
[b] To convert kPa to atm, divide by 101.3.

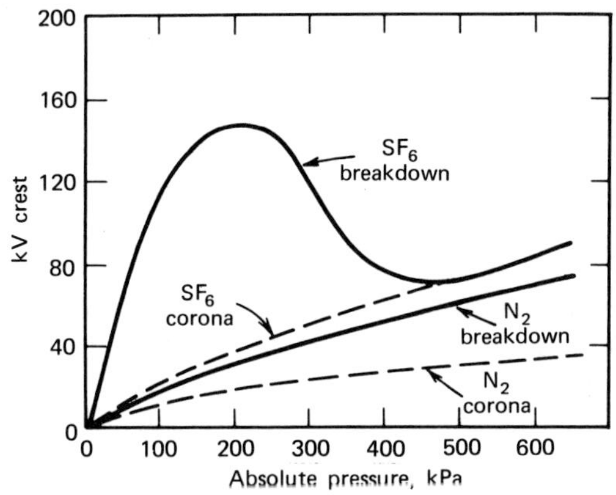

Figure 2. Sparkover and corona onset voltages for SF_6 and N_2. To convert kPa to atm, divide by 101.

appear. Relative field uniformity is characterized in terms of electrode radii of curvature. Peak voltages up to 100 kV can be sustained. A second BDV analysis (37) also uses electrode radii of curvature in rod–plane data at 60 Hz, and can be used to correlate results up to 150 kV. With d-c voltages (38), a similarity rule can be used to treat BDV in fields up to 500 kV/cm at pressures of 101–709 kPa (1–7 atm). It relates field strength, SF_6 pressure, and electrode radii to coaxial electrodes with 2.5-cm gaps. At elevated pressures and large electrode areas, a fall-off from this rule appears. The BDV properties of liquid SF_6 are described in two studies (39–40). In sphere gaps (39), the impulse voltage change with the appearance of a liquid phase was observed, and no discontinuity appeared. At a 0.254 cm gap, and ca 3 MPa (30 atm) a peak BDV value of 270 kV was found. USSR workers have completed an extensive liquid-phase study of BDV in compressed SF_6 liquids (40). Data are given for SF_6 densities of 1.3–6.3 g/cm^3 and are presented as isotherms at various field strengths. Bubble formation in the liquid caused SF_6 dissociation and consequent dielectric failure.

High Frequency Dielectric Strength. This property is important in microwave power uses such as radar (see Microwave technology). Since SF_6 has zero dipole moment, its dielectric strength is substantially constant as frequency increases. In early work (41) at 1.2 MHz, SF_6 had a dielectric strength of 2.3–2.5 relative to N_2. At 3 GHz, SF_6

has about ten times the power-carrying capacity of air (42) whereas at 9.375 GHz in a wave-guide assembly the power-carrying capacity was 7.5 times that of air (41).

Particle Contamination. In assembling large, high voltage equipment such as coaxial lines, contamination by metal particles may occur which may decrease the dielectric strength under various conditions by five to tenfold (43–44). Metal needles are the worst contaminants and electrostatic traps or adhesive areas have been designed to cope with them (45).

In some cases particles have been added to electrical systems to improve heat removal, for example with an SF_6-fluidized particulate bed to be used in transformers (46). This process appears feasible, using PTFE (polytetrafluoroethylene) particles of low dielectric constant. For a successful application, practical problems such as fluidizing narrow gaps must be solved.

Manufacture and Quality Control. Sulfur hexafluoride is manufactured by combining sulfur vapor with pure elemental fluorine (47). It is then given a preliminary scrubbing with caustic. Any disulfur decafluoride, S_2F_{10}, formed is decomposed by heating the product to 400°C to give SF_4 and SF_6. The SF_4 and any remaining lower fluorides of sulfur are removed by a second caustic scrubber. The gas is then intensively dried in a sulfuric acid tower, distilled, and packaged. A routine manufacturing quality-control test establishes the absence of toxic impurities. In this test, mice are exposed to an atmosphere of 80% sulfur hexafluoride and 20% oxygen for 16–20 h and must show no visible effects from the gas.

Economic Aspects; Shipping. Over the thirty years during which SF_6 has been commercially available, consumption has increased gradually as dielectric uses have broadened. The estimated current worldwide annual consumption is $2.5–4.5 \times 10^3$ metric tons. The 1979 United States price was ca \$6–8/kg.

Sulfur hexafluoride is packaged as a liquefied gas in DOT 3AA 2015 steel cylinders containing 52 kg. Larger quantities are available in tube trailers carrying 9070 kg in several DOT SP 5454 cylinders (21).

Specifications; Analytical Methods. Sulfur hexafluoride is made to rigid specifications; the only permissible impurities are traces of air, carbon tetrafluoride (0.05 wt % max), and water (9 ppm by wt max; dew point −45°C max). The product meets ASTM specification D 2472-72 (21).

Sulfur hexafluoride is analyzed chromatographically using a molecular sieve column. With an electron-capture detector, a sensitivity of 10^{-3} to 10^{-4} ppb is possible (48–50).

Health and Safety Factors. Sulfur hexafluoride is a nonflammable, relatively unreactive gas that has been described as physiologically inert (51–52). The current OSHA standard maximum allowable concentration for human exposure in air is 6000 mg/m^3 (TWA, time-weighted average) (53). The Underwriters Laboratories classification is Toxicity Group VI. It should be noted, however, that breakdown products of SF_6, as produced by electrical decomposition of the gas, are toxic. If SF_6 is exposed to electrical arcing, provision should be made to absorb the toxic components by passing the gas over activated alumina or molecular sieves (qv) (21,54).

Applications. For use as a gaseous dielectric, other specific properties are needed in addition to high breakdown strength, and a compromise must be made between electrical and mechanical requirements. For example, CCl_3F can have a relative dielectric strength 40% greater than SF_6 but its boiling point at 24°C excludes outdoor gaseous applications. It cannot maintain the high gas density needed for high dielectric strength under winter conditions. Other desirable properties include low toxicity, thermal stability toward materials of construction, good heat transfer, and absence of electrically conducting carbon. Sulfur hexafluoride has a good balance of properties (see Table 1), with good dew points and chemical attributes. Alone or with CO_2 it can suppress carbon formation in halocarbons under arcing conditions (55–56).

Circuit Breakers. Current interruption is essential in high voltage equipment when overloads or other emergencies occur. Circuit breakers consist of relays having contacts equipped with SF_6 jets and high voltage transformers holding fault-sensing coils that can activate a relay contact for each phase of current (57). At 60 Hz, a 765 kV (root-mean-square, rms) breaker can have a rating of 50,000 MVA (30). High pressure SF_6 jets (1.7 MPa or 17 atm) extinguish the arcs generated on opening the contacts. Sulfur hexafluoride can also act as the insulation for the sensing transformer and supply pressure to activate the relay contacts. At higher pressures and winter temperatures, heaters serve to prevent SF_6 condensation.

High Voltage Coaxial Lines. Sulfur hexafluoride is the main insulation for power transmission in high voltage coaxial lines which can move power above ground, underground, or under water (58–60). Voltage ratings range between 65 and 500 kV with 1200 kV lines being tested. Above ground, powers of 300 MVA have been transported. At present each phase needs one coaxial line but three-phase lines in one envelope are under development (58). Lines now handle a-c power, and d-c lines are being tested. To reduce costs, the use of SF_6–N_2 mixtures is projected (61). A long coaxial line with a dielectric constant very near unity is possible before reactive compensation is needed. Actually, owing to costs, lines of only a few hundred meters are used where land values are high or where passage under multilane highways or other power lines is necessary.

Mini Substations. Development of SF_6-insulated lines and circuit breakers made possible development of a compact electric substation that requires one-tenth the land area of conventional designs (58,62). Other advantages of the mini station are freedom from effects of weather and pollution, and less risk of vandalism, (21,63). A substation unit is now being tested in which SF_6 insulates 500 and 400 kV d-c equipment (64).

Transformers. Units insulated with SF_6 are used mainly in circuit breakers. Some SF_6 power transformers have been designed but the temperature limitation of ca 200°C restricts this use. The problem in high power units is heat transfer through the gas; this can be alleviated with a fluidized particulate bed (46). Although this approach appears feasible, problems in fluidizing narrow spaces must still be overcome.

Other Electrical Uses. With SF_6 insulation, wave guides can transport seven to tenfold more microwave power which results in doubling radar ranges (21,41–42). Voltage ratings in Van de Graaff generators and linear accelerators are also increased by replacing N_2 or air with SF_6 (21). An early application was in r-f power capacitors (65). Incorporation of SF_6 in polyethylene-insulated cables increases the starting voltage for tree formation breakdown processes in the plastic (66).

Nonelectrical Uses. Because of its inertness under normal conditions, SF_6 is used as an air-flow tracer (67). It has also been proposed as a refrigerant, either alone (68) or with $CHClF_2$ (69) or CHF_3 as an azeotrope (70). In pneumothorax treatments, it can replace argon with an increase in lung residence time (51). Owing to its low sound velocity, it can improve the performance of loudspeakers at lower pitch registers (71).

At elevated temperatures or under laser radiation, SF_6 becomes a source of fluorine atoms. In operation of a chemical laser, it abstracts hydrogen from suitable molecules (72–73). Mixed with air or CO_2 in the amount of a few tenths of a percent (0.22 vol %), SF_6 protects molten magnesium during its casting process, resulting in reduced slag and an improved metal surface (74–75).

Sulfur Tetrafluoride

Sulfur tetrafluoride [*7783-60-0*], SF_4, molecular weight 108.06, is a highly reactive colorless gas that fumes in moist air and has an irritating odor that resembles sulfur dioxide. Interest in this compound as a fluorinating agent was spurred by its unique ability to replace oxygen in compounds containing carbonyl groups. It was first reported in 1929 (76).

Physical Properties. Sulfur tetrafluoride has the structure of a distorted trigonal bipyramid with hybrid sp^3d orbitals and an unshared electron pair (77). The FSF bond angles have been found to be 101° and 187° with bond distances of 0.1646 and 0.1545 nm (78).

Selected physical properties are given in Table 5. The nmr data (81) and ir and Raman spectra (82) have been determined. The thermodynamic functions of the material have been calculated from spectral data (83).

Chemical Properties. Sulfur tetrafluoride reacts instantly with water to give hydrofluoric acid and thionyl fluoride:

$$SF_4 + H_2O \rightarrow SOF_2 + 2\ HF$$

With alcohols, mixtures of alkyl fluorides and alkyl ethers are obtained (84). Alcohols bearing electron-withdrawing groups can, however, be converted to the corresponding fluorides in high yield (85). Sulfur tetrafluoride replaces carbonyl oxygen with fluorine (84,86).

$$\rangle C{=}O + SF_4 \longrightarrow \rangle CF_2 + SOF_2$$

Table 5. Physical Properties of Sulfur Tetrafluoride

Property	Value	Ref.
molecular weight	108.055	
melting point, °C	−121.0	79
boiling point, °C	−38	80
critical temperature, °C	90.9	80
surface tension at −73°C, mN/m (= dyn/cm)	257	79
density, liquid, at −73°C, g/cm^3	1.9190	79
vapor pressure at 25°C, MPa[a]	2.0219	79
heat of vaporization, kJ/mol[b]	26.4	79
heat of formation at 25°C, kJ/mol[b]	−781.1	16
free energy of formation at 25°C, kJ/mol[b]	−740.4	16
entropy at 25°C, J/(mol·K)[b]	300.7	16
dipole moment, C·m[c]	2.11×10^{-30}	78

[a] To convert MPa to atm, divide by 0.101.
[b] To convert J to cal, divide by 4.184.
[c] To convert C·m to debye, divide by 3.3366×10^{-30}.

Sulfur tetrafluoride reacts with most inorganic oxides and sulfides to give the corresponding fluorides (87).

$$SF_4 + SnS_2 \rightarrow SnF_4 + 3\,S$$

$$5\,SF_4 + I_2O_5 \rightarrow 2\,IF_5 + 5\,SOF_2$$

$$3\,SF_4 + UO_3 \rightarrow UF_6 + 3\,SOF_2$$

With Lewis acids such as BF_3, SbF_5, PF_5, AsF_5, AsF_3, and GeF_4 (88) adducts are formed. Sulfur tetrafluoride adds a fluoride ion to form SF_5^- in the presence of an ionic fluoride of low crystal lattice energy (89–90).

An excellent review of SF_4 in organic fluorination is given in ref. 91.

Preparation, Economics. In the laboratory, sulfur tetrafluoride is made most conveniently by combining SCl_2 with NaF suspended in acetonitrile at ca 77°C (92). For commercial production, SF_4 is made by direct combination of sulfur with elemental fluorine (93).

Commercial applications of SF_4 are limited; it is available in quantity at $150–200/kg (1978 price) (94).

Toxicity. Sulfur tetrafluoride has an inhalation toxicity comparable to phosgene. The current OSHA standard maximum allowable concentration for human exposure in air is 0.4 mg/m^3 (TWA) (53). On exposure to moisture, eg, on the surface of skin, sulfur tetrafluoride liberates hydrofluoric acid and care must be taken to avoid burns.

Other Sulfur Fluorides

Although seven other binary sulfur fluorides have been synthesized and characterized, proof of the existence of several members of this group was dependent on modern instrumental methods of analysis because of their extreme instability. Disulfur decafluoride, however, is very stable, and perhaps best noted for its extreme toxicity.

As a group, these materials have no technological utility because of their instability, toxicity, and difficulty of preparation. An excellent review of these compounds (excluding S_2F_{10}) is given in ref. 95.

Disulfur Decafluoride. Disulfur decafluoride [5714-22-7], S_2F_{10}, is an extremely toxic, colorless, volatile liquid (96). Electron diffraction studies show the molecule to be composed of two octahedral (SF_5) groups joined by a sulfur—sulfur bond. The S—F bond distance is 0.156 nm, almost identical to that of SF_6, and the S—S distance is 0.221 nm (97). Table 6 summarizes the known physical properties.

Disulfur decafluoride does not react with water, mercury, copper, or platinum at ambient temperatures. It is decomposed by molten KOH to give a mixture of potassium compounds of sulfur and fluorine. The gas reacts vigorously with many other metals and silica at red heat (96). At ca 156°C it combines with Cl_2 or Br_2 to form SF_5Cl or SF_5Br (100–101). At ca 200°C, S_2F_{10} is almost completely thermally decomposed into the hexa- and tetrafluoride (102).

In the laboratory, S_2F_{10} is prepared by the photochemical reduction of SF_5Cl with hydrogen (103).

$$2\,SF_5Cl + H_2 \xrightarrow[h\gamma]{} S_2F_{10} + 2\,HCl$$

Table 6. Physical Properties of S_2F_{10}

Property	Value	Ref.
molecular weight	254.13	
melting point, °C	−55	98
boiling point[a], °C	28.7	96
critical temperature[b], °C	165	14
density, liquid, at 25°C, g/cm³	2.08	96
surface tension at 25°C, mN/m (= dyn/cm)	13.9	96
heat of vaporization, kJ/mol[c]	29.18	96
heat of formation[b,d] at 25°C, kJ/mol[c]	−2.08	99
entropy[b,d] at 25°C, J/(mol·K)[c]	444.93	99
Trouton constant	3.0	96
dielectric constant at 35°C	2.042	96

[a] Vapor pressure calculated.
[b] Estimated.
[c] To convert J to cal, divide by 4.184.
[d] Ideal gas.

Disulfur decafluoride is more toxic than phosgene. The current OSHA standard maximum allowable concentration for human exposure in air is 0.25 mg/m³ (TWA) (53).

No commercial uses for this compound have developed but it is available commercially in research quantities (104).

Thiothionyl Fluoride and Difluorodisulfane. Thiothionyl fluoride [*16860-99-4*], $S{=}SF_2$, and difluorodisulfane [*13709-35-8*], FSSF, are isomeric compounds which may be prepared as a mixture by the action of various metal fluorides on sulfur vapor or S_2Cl_2 vapor. Chemically, the two isomers are very similar and extremely reactive. However, in the absence of catalytic agents and other reactive species, FSSF is stable for days at ordinary temperatures and $S{=}SF_2$ may be heated to 250°C without significant decomposition (105). Physical properties of the two isomers are given in Table 7. All sulfur fluorides, other than SF_6, must be considered extremely toxic.

Table 7. Physical Properties of S_2F_2 Isomers[a]

Property	Value	
	FSSF	SSF_2
molecular weight	102.13	102.13
melting point, °C	−133	−164.6
boiling point, °C	15	−10.6
heat of vaporization, kJ/mol[b]	24.80	22.72
entropy of vaporization, J/(mol·K)[b]	86.67	78.08
heat of capacity at 25°C, J/(mol·K)[b]	64.0	41.50
entropy[c] at 25°C, J/(mol·K)[b]	299	290.77
heat content at 25°C, kJ/mol[b]	14.017	13.342
free energy, kJ/(mol·K)[b]	72.48	73.26

[a] Ref. 105.
[b] To convert J to cal, divide by 4.184.
[c] Ideal gas.

Difluoromonosulfane and Difluorodisulfane Difluoride. Difluoromonosulfane [*13814-25-0*] (sulfur difluoride), SF_2, and its dimer disulfane tetrafluoride [*27245-05-2*], SF_3SF, are both extremely unstable compounds which have only a fleeting existence except under rigorously controlled laboratory conditions. These compounds may be prepared by passing SCl_2 vapor over HgF_2 at 150°C (106).

Difluoro (Tri, Tetra) Sulfanes. Difluorotrisulfane [*31517-17-6*], FSSSF, and difluorotetrasulfane [*31517-18-7*], FSSSSF, have been identified as the constituents of the yellow oil obtained when sulfur vapor reacts with AgF. Their existence was demonstrated with the help of nmr and mass spectroscopy (107–108).

BIBLIOGRAPHY

"Sulfur Compounds" under "Fluorine Compounds, Inorganic" in *ECT* 1st ed., Vol. 6, pp. 732–734, by H. C. Miller, Pennsylvania Salt Manufacturing Co.; "Sulfur Hexafluoride" in *ECT* 1st ed., Suppl. 2, pp. 793–802, by Whitney H. Mears, Allied Chemical & Dye Corporation; "Sulfur" under "Fluorine Compounds, Inorganic" in *ECT* 2nd ed., Vol. 9, pp. 664–676, by James A. Brown, Allied Chemical Corporation.

1. H. Moissan and P. Lebeau, *Ann. Chim. Phys.* **26,** 145 (1902).
2. E. E. Charlton and F. S. Cooper, *Gen. Electr. Rev.* **40,** 438 (1937).
3. B. M. Hokhberg and co-workers, *J. Tech. Phys.* (*USSR*) **12,** 3 (1942).
4. Br. Pat. 532,670 (Jan. 29, 1941), (to The British Thompson-Houston Co., Ltd.).
5. W. E. Schumb and E. L. Gamble, *J. Am. Chem. Soc.* **52,** 4302 (1930).
6. D. M. Yost and H. Russell, Jr., *Systematic Organic Chemistry,* Prentice-Hall, Inc., Englewood Cliffs, N.J., 1944, pp. 297–309.
7. V. P. Borisoylekskii, L. I. Strokovskii, and I. S. Zhizuleva *Zh. Fiz. Khim.* **48,** 119 (1974).
8. L. A. Makarevich, E. S. Sokolova, and G. A. Sorena, *Zh. Fiz. Khim.* **42,** 22 (1968).
9. K. E. MacCormack and W. G. Schneider, *Can. J. Chem.* **29,** 699 (1951).
10. W. H. Mears, E. Rosenthal, and J. V. Sinka, *J. Phys. Chem.* **73,** 2254 (1969).
11. L. A. Makarevich and O. N. Sokolova, *Zh. Fiz. Khim.* **47,** 763 (1973).
12. D. Balzarini and P. Palffy, *Can. J. Phys.* **52,** 2007 (1974).
13. T. G. Pearson and R. L. Robinson, *J. Chem. Soc.,* 1427 (1933).
14. Unpublished data, Allied Chemical Corp., Morristown, N.J.
15. E. Rosenthal, *Sulfur Hexafluoride-Thermodynamic Properties,* unpublished study, Specialty Chemicals Div., Allied Chemical Corp., Morristown, N.J., 1969.
16. *JANAF Thermochemical Tables, NSRDS-NBS-37,* 2nd ed., NBS, Washington, D.C., June 1971.
17. A. Eucken and E. Schroder, *Z. Phys. Chem.* **B41,** 307 (1938).
18. J. Neudorffer, *Ann. Chim.* (*Paris*) **8,** 501 (1953).
19. P. Grassmann and W. Tauscher, *Allied Chemical Research Contract,* Institute of Heat and Engineering, EIDG Technical University, Zurich, Switz., 1967.
20. W. A. Tauscher, *Kaltetechnik-Klimatisierin* **24,** 67 (1972).
21. *Technical Bulletin SFBR-1* (*Sulfur Hexafluoride*), Specialty Chemicals Div., Allied Chemical Corp., Morristown, N.J., 1973.
22. D. Berg, *J. Chem. Phys.* **31,** 572 (1959).
23. H. E. Watson, G. C. Rao, and K. L. Ramaswamy, *Proc. R. Soc. London Ser. A* **132,** 569 (1931).
24. B. Genot, *J. Chim. Phys. Physiochem. Biol.* **68,** 111 (1971).
25. F. A. Cotton and G. Wilkinson, *Advanced Inorganic Chemistry,* 3rd ed., Interscience Publishers, a division of John Wiley & Sons, Inc., New York, 1962, p. 419.
26. D. K. Padma and A. R. Vasuderamurthy, *J. Fluor. Chem.* **5,** 181 (1975).
27. J. P. Manion, J. A. Philosophos, and M. B. Robinson, *IEEE Trans.,* **E1-2**(1), 1 (Apr. 1967).
28. C. Boudene and co-workers, *Rev. Gen. Electr.* (Special No.), 45 (June 1974).
29. F. C. Fehsenfeld, *J. Chem. Phys.* **53,** 2000 (1970).
30. R. N. Yerkley and C. F. Cromer, *IEEE Trans.* **PAS89**(8), 2065 (1970).
31. M. Hudis, *CP74-090-7, IEEE Power Engineering Society, Winter Meeting, New York, Jan. 27–Feb. 1, 1974.*
32. G. Luxa and co-workers, *1975 CIGRE International Conference on Large High Tension Electrical Systems, Item D, Paschen Curve for* SF_6.

33. S. Schreier, *IEEE Trans.* **PAS83,** 468 (1964).
34. P. R. Howard, *Proc. Inst. Elect. Eng.* **104**(Pt. A), 123 (1957).
35. C. N. Works and T. W. Dakin, *Trans. AIEE* **72**(Pt. 1), 682 (1953).
36. T. Nitta and Y. Shibuya, *IEEE Trans.* **PAS90,** 1965 (1971).
37. A. A. Azer and P. P. Comsa, *IEEE Trans.* **EI-8**(4), 136 (1973).
38. I. M. Bortnik and C. M. Cooke, *IEEE Trans.* **PAS91,** 2196 (1972).
39. C. N. Works, T. W. Dakin, and R. W. Rogers, *N.A.S.N.R.C. Publ.* (1080), 69 (1963).
40. Y. V. Torshin, *Conf. Paper 118, Third International Conference on Gas Discharges, IEE London, Eng., Sept. 9–12, 1974.*
41. T. Anderson and co-workers, *AIEE Conf. Paper 52-82,* 1957.
42. J. W. Sutherland, *Electron. Eng.,* 538 (1955).
43. C. M. Cook, R. E. Wootton, and A. H. Cookson, *IEEE Trans.* **PAS96**(3), 768 (1977).
44. A. H. Cookson, O. Farish, and G. M. Sommerman, *IEEE Trans.* **PAS9**(4), 1329 (1972).
45. J. G. Trump, *IEEE Trans.* **NS-14,** 113 (1962).
46. *Gas Insulated Fluidized Bed Transformer, Final Report EL-302, Project 479-1,* Electric Power Research Inst. (EPRI), Buffalo, N. Y., May 1977, Fig. 5-1.
47. U.S. Pat. 3,336,111 (Aug. 15, 1967), W. E. Watson, H. G. Tepp, and M. H. Cohen (to Allied Chemical Corp.).
48. J. E. Lovelock and S. B. Lipsky, *J. Am. Chem. Soc.* **82,** 860 (1960).
49. *Ibid.,* 431 (1960).
50. P. G. Simonds and co-workers, *Analyt. Chem.* **44,** 860 (1972).
51. H. Hodge, W. G. Swalbach, and H. Rahn, *J. Am. Pharm. Assoc.* **47,** 197 (1959).
52. D. Lester and L. A. Greenberg, *Arch. Ind. Hyg. Occup. Med.* **2,** 348 (1950).
53. *Threshold Limit Values for Chemical Substances and Physical Agents in the Workroom Environment* Am. Conf. of Gov. Ind. Hygienists, Cincinnati, Ohio, 1977.
54. W. C. Schumb, J. G. Trump, and G. L. Priest, *Ind. Eng. Chem.* **41,** 1348 (1949).
55. U.S. Pat. 4,052,559 (Oct. 4, 1977), W. H. Mears and S. R. Orfeo (to Allied Chemical Corp.).
56. U.S. Pat. 4,071,461 (Jan. 31, 1978), W. H. Mears and S. R. Orfeo (to Allied Chemical Corp.).
57. T. Ushio, I. Shimura, and G. Tominago, *IEEE Trans.* **PAS89,** 2615 (1970).
58. S. D. Barrett, *The A to Z of SF_6, Special Report Electric Light and Power,* TID ed., Dec. 1972.
59. *Electr. World Trans. Dist.* **54,** (Feb. 15, 1972).
60. B. O. Pedersen, H. C. Doepken, and D. C. Bolin, *IEEE Trans.* **PAS90,** 2631 (1971).
61. R. Nakata and co-workers, "An Underground High Voltage Direct Current Transmission Line," paper presented at *IEEE Underground Dist. and Trans Conf., Dallas, Tex., Apr. 1–5, 1974.*
62. *Electr. World* **61,** (Feb. 23, 1976).
63. *IEEE Trans.* **PAS94,** (1975).
64. E. E. Fischer, A. Glassanos, and N. G. Hingorani, *Electr. World Trans. Dist.* **40,** (Feb. 1, 1977).
65. B. M. Hochberg, *Elektrichestvo* **3,** 15 (1947).
66. T. Kojima and co-workers, *Showa Wire Cable Rev.* **221,** 11 (1972).
67. P. J. Drivas and F. H. Shair, *Atmos. Environ.* **8,** 1155 (1974).
68. R. Plank, *Kaltetechnik* **8,** 302 (1956).
69. U.S. Pat. 3,642,639 (Feb. 15, 1972), K. P. Murphy and R. F. Stahl (to Allied Chemical Corp.).
70. U.S. Pat. 3,719,603 (Mar. 6, 1973), R. F. Stahl (to Allied Chemical Corp.).
71. U.S. Pat. 2,797,766 (July 2, 1957), H. W. Sullivan (to D. Bogen and Co.).
72. D. N. Kaye, *New Sci.* **14,** 65 (Jan. 14, 1971).
73. D. J. Spenser and co-workers, *Int. J. Chem. Kinet.* **1,** 493 (1969).
74. J. W. Fruehling and J. D. Hanawalt, *Am. Foundry Soc. Trans.,* 159 (1969).
75. O. Schlem, *Giesserei* **19,** 558 (1971).
76. J. Fischer and W. Jaenckner, *Z. Angew. Chem.* **42,** 810 (1929).
77. F. H. Cotton, J. W. George, and J. S. Waugh, *J. Chem. Phys.* **28,** 994 (1958).
78. W. M. Tolles and W. D. Gwinn, *J. Chem. Phys.* **36,** 1119 (1962).
79. E. Brown and P. L. Robinson, *J. Chem. Soc.,* 3147 (1955).
80. *Sulfur Tetrafluoride, Tech. Bulletin 2B,* E. I. du Pont de Nemours & Co., Inc., Wilmington, Del.
81. J. Bacon and R. J. Gillespie, *Can. J. Chem.* **41,** 1016 (1963).
82. K. O. Christe and co-workers, *Spectrochim. Acta* **32A,** 1141 (1976).
83. M. Radhakrishnan, *Z. Naturforsch.* **18a,** 103 (1963).
84. W. R. Hasek, W. C. Smith, and V. A. Engelhardt, *J. Am. Chem. Soc.* **82,** 543 (1960).

85. U.S. Pat. 2,980,740 (Apr. 18, 1961), W. R. Hasek and A. C. Haven, Jr. (to E. I. du Pont de Nemours and Co., Inc.).
86. U.S. Pat. 2,859,245 (Nov. 4, 1958), W. C. Smith (to E. I. du Pont de Nemours and Co., Inc.).
87. A. L. Oppegard and co-workers, *J. Am. Chem. Soc.* **82,** 3835 (1960).
88. N. Bartlett and P. L. Robinson, *Proc. Chem. Soc. (London),* 230 (1957).
89. R. Tunder and B. Siegel, *J. Inorg. Nucl. Chem.* **25,** 1097 (1963).
90. C. W. Tullock, D. D. Coffman, and E. L. Muetterties, *J. Am. Chem. Soc.* **86,** 357 (1964).
91. G. A. Boswell, Jr., and co-workers in W. G. Dauben and co-eds., *Organic Reactions,* Vol. 21, John Wiley & Sons, Inc., New York, 1974, pp. 1–124.
92. F. S. Fawcett and C. W. Tullock in J. Kleinberg, ed., *Inorganic Syntheses,* Vol. 7, McGraw-Hill Book Co., Inc., New York, 1963, pp. 119–124.
93. U.S. Pat. 3,399,036 (Aug. 27, 1968), S. Kleinberg and J. F. Tompkins, Jr. (to Air Products and Chemicals, Inc.).
94. Unpublished information, Air Products and Chemicals, Inc., Tamaqua, Pa.
95. F. Seel in H. J. Emeleus and G. G. Sharpe, eds., *Advances in Inorganic and Radio Chemistry,* Vol. 16, Academic Press, Inc., New York, 1974, pp. 297–333.
96. K. B. Denbigh and R. W. Gray, *J. Chem. Soc.,* 1346 (1934).
97. R. B. Harvey and S. H. Bauer, *J. Am. Chem. Soc.* **75,** 2840 (1953).
98. N. R. S. Hollies and R. L. McIntosh, *Can. J. Chem.* **29,** 494 (1951).
99. G. Pass, *J. Appl. Chem.* **19,** 77 (1969).
100. B. Cohen and A. G. MacDiarmid, *Inorg. Chem.* **4,** 1782 (1965).
101. T. A. Kovacina, A. D. Berry, and W. B. Fox, *J. Fluor. Chem.* **7,** 430 (1976).
102. W. R. Trost and R. L. McIntosh, *Can. J. Chem.* **29,** 508 (1951).
103. H. L. Roberts, *J. Chem. Soc.,* 3183 (1962).
104. Unpublished information, PCR Research Chemicals, Inc., Gainesville, Florida.
105. F. Seel, *Chimia* **22,** 79 (1968).
106. F. Seel, H. Heinrich, and W. Gombler, *Chimia* **23,** 73 (1969).
107. F. Seel and co-workers, *Z. Anorg. Allgem. Chem.* **380,** 262 (1971).
108. F. Seel, R. Budenz, and D. Werner, *Ber.* **97,** 1369 (1964).

RICHARD E. EIBECK
WHITNEY MEARS
Allied Chemical Corporation

SULFUR—FLUOROSULFURIC ACID

Fluorosulfuric acid [*7789-21-1*], HSO_3F, is a mobile, almost colorless liquid that fumes strongly in moist air and has a sharp odor. It may be regarded as a mixed anhydride formed from sulfuric and hydrofluoric acids.

Fluorosulfuric acid was first identified and characterized in 1892 (1). It is a strong acid and is employed as a catalyst and chemical reagent in various chemical processes including alkylation (qv), acylation, polymerization, sulfonation, isomerization, and production of organic fluorosulfates (see also Friedel-Crafts reactions).

Properties

Physical properties of fluorosulfuric acid are shown in Table 1.

Fluorosulfuric acid is soluble in acetic acid, ethyl acetate, nitrobenzene, and diethyl ether and insoluble in carbon disulfide, carbon tetrachloride, chloroform, and tetrachloroethane. Many inorganic and organic substances dissolve in fluorosulfuric acid; the physical and chemical properties of such solutions have been extensively investigated (5–8). Of particular note is the solution of antimony pentafluoride in fluorosulfuric acid which is regarded as being a superacid possessing protonating power orders of magnitude greater than 100% sulfuric acid. The ir, Raman, and nmr spectra of fluorosulfuric acid have been reported (9).

Fluorosulfuric acid is extremely stable to heat but decomposes above 900 °C (10); vapor-phase dissociation into hydrogen fluoride and sulfur trioxide probably occurs. Several reviews of the chemistry and properties of fluorosulfuric acid have been published (11–13).

Reactions

The reaction of fluorosulfuric acid with water is violent and exothermic; it proceeds as follows:

$$\textit{fast hydrolysis: } HSO_3F + H_2O \rightleftharpoons H_2SO_4 + HF$$

$$\textit{ionization: } HSO_3F + H_2O \rightleftharpoons H_3O^+ + SO_3F^-$$

$$\textit{slow hydrolysis: } SO_3F^- + H_2O \rightleftharpoons HSO_4^- + HF$$

Table 1. Physical and Chemical Constants of Fluorosulfuric Acid

Property	Value[a]	Refs.
molecular wt	100.07	
boiling point, °C	162.7	1–2
freezing point, °C	−88.98	3
density, g/mL	1.726	2
viscosity, mPa·s (= cP)	1.56	2
dielectric constant	ca 120	2
specific conductance, 1/(Ω·m)	1.085×10^{-6}	2
heat of formation[b], ΔH_f, kJ/mol[c]	792.45	4

[a] All values at 25°C.
[b] From SO_3 and HF.
[c] To convert kJ to kcal, divide by 4.184.

The extent of the initial hydrolysis depends on how the water is added and at what temperature. Hydrolysis is reduced at slower addition rates and lower temperatures. The hydrolysis subsequent to the initial fast reaction is slow, presumably because part of the acid is converted to fluorosulfate ions (SO_3F^-) which hydrolyze slowly even at elevated temperatures. The hydrolysis in basic solution has also been studied (14).

Under controlled conditions, hydrates of HSO_3F containing one, two, and four molecules of water have been observed (15). The acid can be dehydrated to give $S_2O_5F_2$ [*13036-75-4*] (16).

The pure acid does not react in the cold with sulfur, selenium, tellurium, carbon, silver, copper, zinc, iron, chromium, or manganese, but slowly dissolves mercury and tin (17). At higher temperatures, lead, mercury, tin, and sulfur react rapidly; eg:

$$S + 2\ HSO_3F \rightarrow 3\ SO_2 + 2\ HF$$

Precipitated (hydrated) silica reacts vigorously with fluorosulfuric acid to give silicon tetrafluoride (18) but in the absence of moisture glass is not attacked (17). Alkali and alkaline earth metal chlorides are readily converted to fluorosulfates by treatment with fluorosulfuric acid (7,10,19).

Electrolysis of fluorosulfuric acid produces either $S_2O_6F_2$ [*13709-32-5*] (20) or SO_2F_2 [*13036-75-4*] plus OF_2 (21), depending on specific conditions. Various reactions of fluorosulfuric acid with inorganic compounds are shown in Table 2, with organic compounds in Table 3.

Table 2. Reactions of Fluorosulfuric Acid with Inorganic Compounds

Reactant	Product		Ref.
	Name	Formula	
nitrogen	nitrosyl fluorosulfate	FSO_3NO	22
oxides	nitryl fluorosulfate	FSO_3NO_2	23
H_3BO_3	boron trifluoride	BF_3	24
$KClO_4$	perchloryl fluoride	ClO_3F	25
$KMnO_4$	manganese fluoride oxide	MnO_3F	26
As_2O_3	arsenic trifluoride	AsF_3	27
As_2O_5	arsenic pentafluoride	AsF_5	27
CrO_3	difluorodioxo chromium	CrO_2F_2	28
P_4O_{10}	phosphoryl fluoride	POF_3	27
H_2S in ethanol	monothiosulfuric acid	$H_2S_2O_3$	29

Table 3. Reactions of Fluorosulfuric Acid With Organic Compounds

Reactant	Product		Refs.
	Name	Formula	
benzene[a]	benzenesulfonic acid	$C_6H_5SO_3H$	17, 30
	diphenyl sulfone	$(C_6H_5)_2SO_2$	30
	benzenesulfonyl fluoride	$C_6H_5SO_2F$	31
aliphatic amines	amidosulfuric acids	R_2NSO_3H or $RNHSO_3H$	32
aromatic compound	arylsulfonyl fluorides	$ArSO_2F$	
carboxylic acids	acid fluorides	RCOF	33
alcohol or alkene	alkyl fluorosulfates	$ROSO_2F$	17, 34–35
perhaloolefins or perhaloalkyl iodides	perhaloalkyl fluorosulfates	R_fOSO_2F	36–37

[a] Depending upon reaction conditions and proportions of reagents.

Fluorosulfuric acid may be used for the preparation of diazonium fluorosulfates, $ArN_2^+SO_3F^-$ (38), which decompose on heating to give aryl fluorosulfates (32,39). Aryl fluorosulfates are also obtained from arylsulfonyl chlorides and fluorosulfuric acid (31). Alkyl and other organofluorosulfates form during electrolysis of fluorosulfuric acid in the presence of organic species (40–41).

Preparation and Manufacture

Fluorosulfuric acid was first prepared by combining anhydrous HF with cooled, anhydrous SO_3 in a platinum container (1). It has also been prepared from ionic fluorides or fluorosulfates and sulfuric acid (17,42). The reaction of chlorosulfuric acid with ionic fluorides also gives fluorosulfuric acid (43).

Commercially, fluorosulfuric acid is made by processes utilizing the product as a solvent. Solutions of HF and SO_3 in fluorosulfuric acid are mixed in stoichiometric quantities or SO_3 and HF are separately introduced into a stream of fluorosulfuric acid to produce essentially pure HSO_3F. Some of the product is then recycled to serve again as the solvent for HF and SO_3 (44).

Specifications; Shipping

Commercial fluorosulfuric acid contains ca 98% HSO_3F with ca 1% H_2SO_4 and lesser amounts of sulfur trioxide and dioxide. Fluorosulfuric acid is available in tank transports (18 metric tons) and tank cars (27–45 t) only from the chemicals company of the Allied Chemical Corp. Price of fluorosulfuric acid is ca $730 per metric ton.

Analysis

Fluorosulfuric acid may contain some H_2SO_4 and either SO_3 or HF. The free sulfur trioxide can be titrated with water; the end point is determined conductimetrically. The sulfuric acid content is determined from the specific conductivity of the liquid at the point in the titration where no free SO_3 or excess water is present. If HF is present, a known amount of SO_3 is added to the acid and the excess SO_3 is determined as above.

Health and Safety Factors

Fluorosulfuric acid is a strong acid capable of causing severe burns similar to those experienced with sulfuric and hydrofluoric acids. In addition, the fumes of fluorosulfuric acid are extremely irritating, and breathing of the fumes is to be avoided. Precautions and first-aid measures generally observed in handling strong sulfuric acid and hydrofluoric acid are applicable to fluorosulfuric acid. Small containers of fluorosulfuric acid should be well cooled before opening and precautions taken to relieve any gas pressure that may have developed. In the laboratory, fluorosulfuric acid may be handled in glass if water is excluded; otherwise, containers of inert polymers or platinum should be used. For larger-scale equipment, iron or steel (not stainless steel) may be used. For additional precautions in handling large quantities of fluorosulfuric acid, see refs. 45 and 46.

Uses

Fluorosulfuric acid serves as catalyst in the alkylation (qv) of branched-chain paraffins (47–52) and aromatic compounds (53), and in the polymerization of monoolefins (54) and rosin (55). Addition of strong Lewis acids, such as SbF_5, TaF_5, and NbF_5, to fluorosulfuric acid markedly increases the system acidity and catalytic activity (56–63) (see Catalysis).

As a reagent, fluorosulfuric acid has been employed in the preparation of boron trifluoride (24), silicon tetrafluoride (64–65), alkyl fluorosulfates (17,34), arenesulfonyl fluorides (31), acyl fluorides (66), sulfamic acid (67), and diazonium fluorosulfates (38). Among its other uses are the removal of small amounts of organic fluorides from petroleum alkylate made by the hydrogen fluoride process (68), the removal of HF from process exhaust gases (69), the removal of HF from F_2 (70), and as a constituent of baths for electropolishing metals (71).

Derivatives

The nonmetallic derivatives of fluorosulfuric acid (other than organic) are generally made indirectly, although complex fluorosulfates of the Group V elements and of xenon can be made directly (72–73), as can the NO^+ and NO_2^+ salts (22–23).

Peroxydisulfuryl difluoride, FSO_2OOSO_2F, prepared from fluorine and SO_3 (74), is a ready source of fluorosulfate radicals ($FSO_2O\cdot$) (75) which react with many substances to form stable fluorosulfates (76–77). By using the route $X_2 + n\ S_2O_6F_2 \rightarrow 2\ X(OSO_2F)_n$ (n = 1 or 3), compounds of the type $FOSO_2F$ [*13536-85-1*] (78), $BrOSO_2F$ [*13997-93-8*], and $I(OSO_2F)_3$ [*13709-37-0*] (79) have been prepared.

Fluorosulfates of most metallic elements have been prepared either directly or indirectly but none have any technological significance.

The physical properties of some fluorosulfates are summarized in Table 4.

Ammonium fluorosulfate is produced from ammonium fluoride by reaction with sulfur trioxide, oleum, or potassium pyrosulfate, $K_2S_2O_7$ (42). Solutions of ammonium fluorosulfate show little evidence of hydrolysis and the salt may be recrystallized from hot water. Ammonium fluorosulfate absorbs anhydrous ammonia to form a series of liquid amines that contain 2.5–6 moles of ammonia per mole of salt (66).

Table 4. Physical Properties of Some Fluorosulfates

Salt	CAS Registry No.	Appearance	mp, °C	Soly in water[a]	Soly in other solvents[a]
NH_4SO_3F	[*13446-08-7*]	long colorless needles	245	s	sl s ethanol; v s methanol
$LiSO_3F$	[*13453-75-3*]	white powder	360	v s	v s ethanol, ether, acetone,
$LiSO_3F.3H_2O$		long shiny needles	60–61		amyl alcohol, ethyl acetate; i ligroin
$NaSO_3F$	[*14483-63-7*]	shiny leaflets, hygroscopic		s	s ethanol, acetone; i ether
KSO_3F	[*13455-22-6*]	short white prisms	311	6.9[b]	sl s methanol
$RbSO_3F$	[*15587-05-0*]	colorless needles	304	s	sl s methanol
$CsSO_3F$	[*13530-70-6*]	colorless rhombic	292	2.23[b]	

[a] i = insoluble; sl s = slightly soluble; s = soluble; v s = very soluble.
[b] In g/100 mL water.

Sodium fluorosulfate may be prepared by the action of fluorosulfuric acid on powdered, ignited sodium chloride (10) or of sulfur trioxide on sodium fluoride (42). At red heat, sodium fluorosulfate decomposes to sodium sulfate, sulfuryl fluoride, sodium fluoride, and sulfur trioxide.

In general, the alkali metal fluorosulfates may be prepared from the ammonium salt by evaporating a solution containing that salt and an alkali metal hydroxide (66). The solubilities of some Group I and II fluorosulfates in fluorosulfuric acid have been determined (80).

BIBLIOGRAPHY

"Fluosulfonic Acid" under "Fluorine Compounds, Inorganic" in *ECT* 1st ed., Vol. 6, pp. 734–738, by W. S. W. McCarter, Pennsylvania Salt Manufacturing Co.; "Fluorosulfuric Acid" under "Fluorine Compounds, Inorganic" in *ECT* 2nd ed., Vol. 9, pp. 676–681, by Richard E. Eibeck, Allied Chemical Corporation.

1. T. E. Thorpe and W. Kirman, *J. Chem. Soc.*, 921 (1892).
2. J. Barr, R. J. Gillespie, and R. C. Thompson, *Inorg. Chem.* **3,** 1149 (1964).
3. R. J. Gillespie, J. B. Milne, and R. C. Thompson, *Inorg. Chem.* **5,** 468 (1966).
4. G. W. Richards and A. A. Woolf, *J. Chem. Soc. A,* 1118 (1967).
5. A. A. Woolf, *J. Chem. Soc.,* 2840 (1954); *ibid.,* 433 (1955).
6. R. J. Gillespie and co-workers, *Can. J. Chem.* **40,** 675 (1962); **41,** 148, 2642 (1963); **42,** 502, 1433 (1964).
7. R. J. Gillespie and co-workers, *Inorg. Chem.* **3,** 1149 (1964).
8. R. J. Gillespie and co-workers, *Inorg Chem.* **4,** 1641 (1965); **8,** 63 (1969).
9. R. J. Gillespie and E. A. Robinson, *Can. J. Chem.* **40,** 644, 675 (1962); R. Savoie and P. A. Giguere, *Can. J. Chem.* **42,** 277 (1964).
10. O. Ruff, *Chem. Ber.* **47,** 646 (1914).
11. R. J. Gillespie, *Acc. Chem. Res.* **1**(7), 202 (1968).
12. R. C. Thompson in G. Nickless, ed., *Inorganic Sulphur Chemistry,* Elsevier Scientific Publishing Co., Amsterdam, The Netherlands, 1968, pp. 587–606.
13. A. W. Jache in H. J. Emeleus and A. G. Sharpe, eds., *Advances in Organic Chemistry and Radiochemistry,* Vol. 16, Academic Press, Inc., New York, 1974, pp. 177–200.
14. I. G. Ryss and A. Drabkina, *Kinet. Katal.* **7,** 319 (1966); *Chem. Abstr.* **65,** 3082b (1966).
15. R. C. Paul, K. K. Paul, and K. C. Malhotra, *Inorg. Nucl. Chem. Lett.* **5,** 689 (1969).
16. R. C. Paul, K. K. Paul, and K. C. Malhotra, *J. Inorg. Nucl. Chem.* **31,** 2614 (1969).
17. J. Meyer and G. Schramm, *Z. Anorg. Allgem. Chem.* **206,** 24 (1932).
18. L. J. Belf, *Chem. Ind. (London),* 1296 (1955).
19. P. Bernard, Y. Parent, and P. Vast, *C. R. Acad. Sci. Ser. C* **269,** 767 (1969).
20. J. M. Shreeve and G. H. Cady, *J. Am. Chem. Soc.* **83,** 4521 (1961).
21. H. Schmidt and H. D. Schmidt, *Z. Anorg. Allgem. Chem.* **279,** 289 (1955).
22. D. R. Goddard, E. D. Hughes, and C. K. Ingold, *J. Chem. Soc.,* 2559 (1950).
23. W. Lange, *Ber.* **60B,** 967 (1927).
24. U.S. Pat. 2,416,133 (Feb. 18, 1947), D. Young and J. Pearson (to Allied Chemical Corp.).
25. G. Barth-Wehrenalp, *J. Inorg. Nucl. Chem.* **2,** 266 (1956).
26. A. Engelbrecht and A. V. Grosse, *J. Am. Chem. Soc.* **76,** 2042 (1954).
27. E. Hayek, A. Aignesberger, and A. Engelbrecht, *Monatsh. Chem.* **86,** 470, 735 (1955).
28. A. Engelbrecht, *Angew. Chem. Int. Ed. Engl.* **4,** 641 (1965).
29. M. Schmidt and G. Talsky, *Z. Anorg. Allgem. Chem.* **303,** 210 (1960).
30. J. H. Simons, H. J. Passino, and S. Archer, *J. Am. Chem. Soc.* **63,** 608 (1941).
31. W. Steinkopf and co-workers, *J. Prakt. Chem.* **117,** 1 (1927).
32. Ger. Pat. 532,394 (Aug. 8, 1930), W. Lange.
33. W. Traube and A. Krahmer, *Chem. Ber.* **B52,** 1293 (1919).
34. Ger. Pats. 342,898 (Oct. 25, 1921), 346,245 (Dec. 27, 1921), W. Traube.
35. G. Olah, J. Nishimura, and Y. Mo, *Synthesis* (11), 661 (1973).
36. U.S. Pats 3,254,107 (May 31, 1966), 3,255,228, 3,255,229 (June 7, 1966), M. Hauptschein and M. Braid

(to Pennsalt Chemicals Corp.), 3,083,220 (Mar. 26, 1963), E. L. Edens (to E. I. du Pont de Nemours & Co., Inc.), 2,878,156 (Mar. 17, 1959), R. A. Davis (to The Dow Chemical Co.), 2,628,927 (Feb. 17, 1953), J. D. Calfee and P. A. Florio (to Allied Chemical Corp.).
37. M. Hauptschein and M. Braid, *J. Am. Chem. Soc.* **83,** 2502 (1961).
38. U.S. Pat. 1,847, 513 (Mar. 1, 1932), W. Hentrich, M. Hardtmann, and H. Ossenbeck (to General Aniline Works).
39. W. Lange and E. Müller, *Chem. Ber.* **B63,** 2653 (1930).
40. J. P. Coleman and D. Pletcher, *Tetrahedron Lett.* (2), 147 (1974).
41. D. Pletcher and C. Smith, *Chem. Ind. (London)* **8,** 371 (1976).
42. W. Traube, *Chem. Ber.* **46,** 2525 (1913).
43. U.S. Pat. 2,312,413 (Mar. 2, 1943), R. K. Iler (to E. I. du Pont de Nemours & Co., Inc.).
44. U.S. Pats. 2,430,963 (Nov. 18, 1947), R. Stephenson and W. Watson (to Allied Chem. Corp.); U.S. Pat. 3,957,959 (May 18, 1976), R. Wheatley, D. Treadway, and R. Toennies (to E. I. du Pont de Nemours & Co., Inc.).
45. *Fluosulfonic Acid, Tech. Serv. Bull. TA-30711,* Specialty Chemicals Div., Allied Chemical Corp., Morristown, N.J.
46. *Chemical Safety Data Sheets SD20 (1973) and SD25 (1970),* Manufacturing Chemist's Association, Washington, D.C.
47. Br. Pat. 537,589 (June 27, 1941), (to Standard Oil Development Co.).
48. U.S. Pat. 2,313,103 (Mar. 9, 1943), C. L. Thomas (to Universal Oil Products Co.).
49. U.S. Pat. 3,778,489 (Dec. 11, 1973), P. T. Parker and I. Mayer (to Esso Research and Engineering Co.).
50. U.S. Pat. 3,922,319 (Nov. 25, 1975), J. W. Brockington (to Texaco, Inc.).
51. U.S. Pat. 3,928,487 (Dec. 23, 1975), D. A. McCauley (to Standard Oil Co.).
52. U.S. Pat. 4,008,178 (Feb. 15, 1977), J. W. Brockington (to Texaco, Inc.).
53. U.S. Pat. 2,428,279 (Sept. 3, 1947), V. N. Ipatieff and C. B. Linn (to Universal Oil Products Co.).
54. U.S. Pat. 2,421,946 (June 10, 1947), V. N. Ipatieff and C. B. Linn (to Universal Oil Products Co.).
55. U.S. Pat. 2,419,185 (Apr. 15, 1947), C. A. Braidwood and A. G. Hovey (to Reichold Chemicals).
56. U.S. Pat. 3,594,445 (July 20, 1971), P. T. Parker (to Esso Research and Engineering Co.).
57. U.S. Pat. 3,636,129 (Jan. 18, 1972), P. T. Parker and C. N. Kimerlin, Jr. (to Esso Research and Engineering Co.).
58. U.S. Pat. 3,678,120 (July 18, 1972), H. S. Bloch (to Universal Oil Products Co.).
59. U.S. Pat. 3,708,553 (Jan. 2, 1973), G. A. Olah (to Esso Research and Engineering Co.).
60. D. T. Roberts, Jr. and L. E. Calihan, *J. Macromol. Sci. Chem.* **7,** 1629 (1973).
61. U.S. Pat. 3,819,743 (June 25, 1974), D. A. McCauley (to Standard Oil Co.).
62. U.S. Pat. 3,925,495 (Dec. 9, 1975), P. G. Rodewald (to Mobil Oil Corp.).
63. U.S. Pat. 3,984,352 (Oct. 5, 1976), P. G. Rodewald (to Mobil Oil Corp.).
64. Can. Pat. 448,662 (May 25, 1948), A. C. Hopkins, Jr., R. M. Stephenson, and W. E. Watson (to Allied Chem. Corp.).
65. Br. Pat. 755,692 (Aug. 22, 1956), A. J. Edwards (to National Smelting Co., Ltd.).
66. W. Traube, J. Horenz, and F. Wunderlich, *Chem. Ber.* **B52,** 1228 (1919).
67. W. Traube and E. Brehmer, *Chem. Ber.* **B52,** 1284 (1919).
68. U.S. Pat. 2,428,753 (Oct. 7, 1947), C. B. Linn (to Universal Oil Product Co.).
69. U.S. Pat. 2,434,040 (Jan. 6, 1948), B. F. Hartman (to Socony-Vacuum Oil Co.).
70. Br. Pat. 824,427 (Dec. 2, 1959), H. R. Leech and W. H. Wilson (to Imperial Chemical Industries, Ltd.).
71. C. B. F. Young and K. R. Hesse, *Met. Finish.* **45**(2), 63, 84 (1947); **45**(3), 64 (1947).
72. R. C. Paul and co-workers, *J. Inorg. Nucl. Chem.* **34,** 2535 (1972).
73. D. D. DesMarteau and M. Eisenberg, *Inorg. Chem.* **11,** 2641 (1972).
74. F. B. Dudley and G. H. Cady, *J. Am. Chem. Soc.* **79,** 513 (1957).
75. *Ibid.,* **85,** 3375 (1963).
76. F. Aubke and D. D. DesMarteau, *Fluorine Chem. Rev.* **8,** 74 (1977).
77. R. A. DeMarco and J. M. Shreeve, *Adv. Inorg. Chem. Radiochem.* **16,** 115 (1974).
78. J. E. Roberts and G. H. Cady, *J. Am. Chem. Soc.* **81,** 4166 (1959).
79. *Ibid.,* **82,** 352 (1960).
80. R. Seeley and A. W. Jache, *J. Fluorine Chem.* **2**(3), 225 (1973).

Richard E. Eibeck
Allied Chemical Corporation

TANTALUM

Tantalum forms a single binary fluoride, tantalum pentafluoride [*7783-71-3*].

Tantalum Pentafluoride

Tantalum pentafluoride, TaF_5, is a white solid having a mp of 97°C and a bp of 229°C (1). The vapor pressure of TaF_5 is given by $\log P_{kPa} = 7.649 - 2834/T$ ($\log P_{mm\ Hg} = 8.524 - 2834/T$) for temperatures of 80–230°C (1). Its heat of vaporization is 54.4 kJ/mol (13 kcal/mol) (1). Tantalum pentafluoride is produced by the direct fluorination of tantalum metal or by the reaction of hydrogen fluoride with tantalum pentachloride.

The chemistry of TaF_5 has been reviewed extensively (1–4). The production of tantalum metal by reduction of TaF_5 with hydrogen has been studied (5) and represents a potential application in the chemical vapor deposition (CVD) industry.

Tantalum pentafluoride is available from The Ozark-Mahoning Company. It is sold at approximately $600/kg (1978), and total U.S. production is less than 500 kg/yr. The purity of the commercial material is greater than 99% TaF_5 (6).

BIBLIOGRAPHY

"Tantalum" in *ECT* 2nd ed., under "Fluorine Compounds, Inorganic," Vol. 9, p. 681, by Wayne E. White, Ozark-Mahoning Company.

1. J. H. Canterford and R. Cotton, *Halides of the Second and Third Row Transition Metals,* John Wiley & Sons, London, 1968.
2. F. Fairbrother, "Halides of Niobium and Tantalum" in V. Gutmann, ed., *Halogen Chemistry,* Vol. 1, Academic Press, London, 1966, p. 123.
3. F. Fairbrother, *The Chemistry of Niobium and Tantalum,* Elsevier Scientific Publishing Co., London, 1967.
4. D. Brown, "Chemistry of Niobium and Tantalum" in J. C. Bailer and co-eds., *Comprehensive Inorganic Chemistry,* Vol. 3, Permagon Press, Compendium Publishers, Elmsford, N.Y., 1973, p. 565.
5. N. B. Pobedah, Yu. M. Korolov, and A. V. Rychagov, *Metall. i Metalloved Chist. Met.* **11,** 23 (1975); *Chem. Abstr.* **85,** 8049713 (1976).
6. Private conversation with Dayal T. Meshri, The Ozark-Mahoning Company, Oct. 1978.

ANDREW J. WOYTEK
Air Products and Chemicals, Inc.

TIN

The main binary tin fluorides are stannous fluoride and stannic fluoride. Because the stannous ion (+2) is readily oxidized to the stannic ion (+4), most reported tin and fluorine complexes are of stannic tin and fluorostannates.

Stannous Fluoride

Stannous fluoride [*7783-47-3*], SnF_2, is a white crystalline salt that has a mp of 215°C (1) and a bp of 850°C, and is readily soluble in water and hydrogen fluoride. At 20°C stannous fluoride dissolves in water to a concentration of 30–39% and in anhydrous hydrogen fluoride to 72–82% (2–4).

The pH of a freshly prepared 0.4% solution of stannous fluoride is between 2.8 and 3.5. Initially clear aqueous solutions become cloudy on standing owing to hydrolysis and oxidation. The insoluble residue is a mixture containing stannous and stannic species, fluoride, oxide, oxyfluorides, and hydrates.

Stannous fluoride probably was first prepared by Scheele in 1771 and was described by Gay Lussac and Thenard in 1809. Early preparation of SnF_2 was by reaction of aqueous hydrofluoric acid with stannous oxide. Commercial production of stannous fluoride is by reactions of stannous oxide with aqueous hydrofluoric acid, and metallic tin with anhydrous hydrogen fluoride (5). Stannous fluoride is used widely in dentifrices (qv) and other dental preparations because of its anticaries effect.

The first commercial fluoride dentifrice (6) contained SnF_2 and precipitated the acceptance of a number of toothpastes as effective decay-preventive dentifrices by the Council of Dental Therapeutics of the American Dental Association. The chemistry (7) involved in cavity prevention is thought to be reaction of stannous fluoride with the hydroxyapatite, $Ca_5(PO_4)_3OH$, of the tooth to form the more insoluble fluoroapatite, $Ca_5(PO_4)_3F$. More concentrated solutions of stannous fluoride react with hydroxyapatite to produce $Sn_3F_3PO_4$ [*12592-27-7*] (8) (see Dentifrices).

Fluorostannites and Fluorostannates

Complexes of the type $MSnF_3$ where M is NH_4 [*15660-29-4*], Na [*13782-22-4*], K [*13782-23-5*], and Cs [*13782-25-7*] have been crystallized from aqueous solutions (9–11). Solutions of these salts deposit tin(II) oxide crystals indicating hydrolysis but not oxidation. From molten mixtures of SnF_2 with NaF, RbF, and CsF, both the $MSnF_3$ (M = Na [*13782-22-4*], K [*13782-23-5*], Rb [*13782-24-6*], and Cs [*13782-25-7*]) and the fluorostannate salts, $MSnF_5$ (M = Na [*58179-42-3*], K [*58179-40-1*], Rb [*72264-75-6*], and Cs [*72264-76-7*]), were obtained (12).

Stannic Fluoride

Stannic fluoride [*7783-62-2*], SnF_4, is a white solid that sublimes at 705°C and hydrolyzes in water to form insoluble stannic acid. It can be prepared by reaction of fluorine and probably ClF_3 or BrF_3 with virtually any tin(II) or tin(IV) compound (eg, Sn, SnO, SnO_2, SnS, SnS_2 (13), and $SnCl_2$ (14)). Reaction of $SnCl_4$ with HF (15) forms $SnCl_4.SnF_4$. This can be decomposed by heating to 750°C where pure SnF_4 sublimes. Tin tetrafluoride forms numerous complexes as a Lewis acid.

Hexafluorostannates

The hexafluorostannate anion [*21340-04-5*], SnF_6^{2-}, is readily formed and is stable over a wide pH range. Numerous hexafluorostannates have been prepared by dissolving stannates in excess hydrofluoric acid, dissolving stannic acid in excess HF and neutralizing, or by reaction of salts with SnF_4. Although many of the hexafluorostannates were prepared as early as 1857, there has been relatively little recent work on these stable and generally water-soluble compounds other than spectral studies of the SnF_6^{2-} anion.

BIBLIOGRAPHY

"Tin Fluoride" treated in *ECT* 1st ed., under "Tin Compounds," Vol. 4, p. 160, by H. Richter, Metal & Thermit Corp.; "Tin" treated in *ECT* 2nd ed., under "Fluorine Compounds, Inorganic," Vol. 9, pp. 682–683, by Wayne E. White, Ozark-Mahoning Company

1. J. J. Dudash and A. W. Searcy, *High Temp. Sci.* **1,** 287 (1969).
2. W. H. Nebergall, J. C. Muhler and H. G. Day, *J. Am. Chem. Soc.* **74,** 1604 (1952).
3. J. B. Beal, Jr., Ph.D. thesis, Texas A and M University, College Station, Texas, 1963.
4. J. E. Gilliland, M.S. thesis, Oklahoma State University, Stillwater, Okla., 1960.
5. U.S. Pats. 2,924,508 (Feb. 9, 1960); 2,955,914 (Oct. 11, 1960); 3,097,063 (July 9, 1963) J. E. Gilliland, R. Ray, and W. E. White (to Ozark-Mahoning Company).
6. J. C. Muhler and co-workers, *J. Am. Dental Assoc.* **50,** 163 (1955).
7. E. J. Duff, *Caries Res.* **7,** 79 (1973).
8. S. H. Y. Wei, *J. Dent. Res.* **53,** 57 (1974).
9. E. L. Muetterties, *Inorg. Chem.* **1,** 342 (1962).
10. W. B. Schaap, J. A. Davis, and W. N. Nebergall, *J. Am. Chem. Soc.* **76,** 5226 (1954).
11. J. D. Donaldson and J. D. O'Donoghue, *J. Chem. Soc.,* 271 (1964).
12. J. D. Donaldson, J. D. O'Donoghue and R. Oteng, *J. Chem. Soc.,* 3876 (1965).
13. H. M. Haendler and co-workers, *J. Am. Chem. Soc.* **76,** 2179 (1954).
14. A. A. Woolf and H. J. Emeleus, *J. Chem. Soc.,* 2864 (1949).
15. H. J. Emeleus in J. H. Simons, ed., *Fluorine Chemistry,* Vol. 1, 1950, Academic Press, New York, pp. 1–76.

CHARLES B. LINDAHL
DAYAL T. MESHRI
Ozark-Mahoning Company,
a subsidiary of the Pennwalt Corp.

TITANIUM

Titanium has two known fluorides, TiF_3 and TiF_4. Attempts to prepare the difluoride have not been successful even though the other dihalides, $TiCl_2$, $TiBr_2$, and TiI_2, are well known.

Titanium Trifluoride

Titanium trifluoride [*13470-08-1*], TiF_3, can be obtained by dissolving titanium metal in hydrofluoric acid (1–2) and by passing anhydrous hydrogen fluoride over titanium trihydride (3) at 700°C or over heated titanium powder. Titanium trichloride reacts with liquid anhydrous hydrogen fluoride at room temperature and gives a crude product that can be purified by sublimation in a high vacuum at 930–950°C; at temperatures above 950°C, it slowly disproportionates into TiF_4 and titanium metal.

Titanium trifluoride is a sky-blue-colored crystalline solid, with a magnetic moment of 16.2×10^{-24} J/T (1.75 μ_β), and a specific gravity of 2.98 at 25°C. It is insoluble in water but is fairly soluble in dilute acids and alkalies. It is oxidized to TiO_2 on heating at 100°C in air. No commercial application of titanium trifluoride has been reported.

Titanium Tetrafluoride

Titanium tetrafluoride [*7783-63-3*], TiF_4, is a white solid which sublimes at 284°C and has a density of 2.798 at 20.5°C. It is extremely hygroscopic and is highly soluble in water. It may be prepared by the action of elemental fluorine (4) on the metal at 250°C, or on titanium dioxide at 350°C, or by reaction of anhydrous hydrogen fluoride on titanium tetrachloride. The evaporation of an aqueous solution gives $TiF_4.2H_2O$ [*60927-06-2*] crystals. An aqueous solution hydrolyzes slowly and produces $TiOF_2$ [*13537-16-1*] (5). It dissolves in aqueous hydrofluoric acid to form TiF_6^{2-} from which the sparingly soluble alkali salts can be easily obtained.

Titanium tetrafluoride is a mild fluorinating reagent and fluorinates silicochloroform to silicofluoroform at 100°C. It forms adducts (some quite stable) with ammonia, several amines, pyridine, and ethanol, and it has more ionic character than the tetrachloride. Use of topical applications of TiF_4 for the prevention of dental caries is being explored (6–8) (see Dentifrices).

There is no commercial demand for TiF_4. It is available in 100 g to 45.4 kg quantities, and total U. S. consumption is less than 450 kg/yr.

Fluorotitanates

Although titanium tetrafluoride forms the stable TiF_6^{2-} ion in aqueous hydrofluoric acid solutions, fluorotitanic acid [*17439-11-1*], H_2TiF_6 (similar to H_2SiF_6) has not been isolated. It probably exists in ionized form. Several fluorotitanates of the formula, M_2TiF_6 (where M = Li [*19193-50-1*], Na [*17116-13-1*], K [*16919-27-0*], Rb [*16962-41-7*], Cs [*16919-28-1*], NH_4 [*16962-40-6*], Tl [*26460-00-4*]) have been prepared, isolated, and found to be stable at room temperature (9). Fluorotitanates have been used in abrasive grinding wheels and for incorporating titanium into aluminum alloys (10).

Fluorotitanates of K, Rb, and Cs are slightly soluble in water, whereas Li, Na, NH_4 fluorotitanates are quite soluble in water.

Although titanium compounds are considered physiologically inert (11), fluorides are considered toxic; therefore, extreme care should be used in handling large amounts of titanium fluorides.

BIBLIOGRAPHY

"Titanium" in *ECT* 2nd ed., under "Fluorine Compounds, Inorganic," Vol. 9, pp. 683–684, by Wayne E. White, Ozark-Mahoning Company.

1. M. E. Straumanis and J. I. Ballass, *Z. Anorg. Chem.* **278,** 33 (1955).
2. P. H. Woods and L. D. Cockerell, *J. Am. Chem. Soc.* **80,** 1534 (1958).
3. P. Ehrlich and G. Pietzka, *Z. Anorg. Chem.* **275,** 121 (1954).
4. H. M. Haendler and co-workers, *J. Am. Chem. Soc.* **76,** 2177 (1954).
5. K. S. Vorres and F. B. Dutton, *J. Am. Chem. Soc.* **77,** 2019 (1955).
6. B. Regolati and co-workers, *Helv. Odontol. Acta* **18**(2), 92 (1974); *Chem. Abstr.* **82,** 80734 Y (1975).
7. A. S. Mundorff, M. F. Little, and B. G. Bibby, *J. Dent. Res.* **51,** 1567 (1972).
8. A. J. Reed and B. G. Bibby, *J. Dent. Res.* **55,** 357 (1976).
9. B. Cox and A. G. Sharpe, *J. Chem. Soc.*, 1783 (1953).
10. *Data Sheet DA-40723,* General Chemical Division, Allied Chemical Co.
11. N. I. Sax, *Dangerous Properties of Industrial Materials,* 4th ed., Van Nostrand Reinhold Co., New York, 1975.

DAYAL T. MESHRI
CHARLES B. LINDAHL
The Ozark-Mahoning Company,
a subsidiary of the Pennwalt Corp.

TUNGSTEN

Tungsten has two binary fluorides, tungsten(VI) hexafluoride [*7783-82-6*] and tungsten(IV) tetrafluoride [*13766-47-7*]. In addition, there are two tungsten(VI) oxyfluorides, WOF_4 [*13520-79-1*] and WO_2F_2 [*14118-73-1*], and a number of complex oxyfluorides are known (1–2). Of these compounds, tungsten hexafluoride is the best known.

Tungsten Hexafluoride

Physical Properties. Tungsten hexafluoride, WF_6, is a colorless gas at ordinary temperature, and is the heaviest of all known gases (ten times as heavy as air). At ca 100 kPa (1 atm) it condenses to a pale yellow liquid at 17.1°C, and below 2°C it forms a white solid. Tungsten hexafluoride exists as a symmetrical octahedral structure at near room temperature and in a phase of lower symmetry below −8.5°C (3). Various spectroscopic studies have been reported: Raman and uv spectra (4), Raman and ir spectra (5), and ir spectra (6). The physical properties of tungsten hexafluoride are given in Table 1.

Chemical Properties. Tungsten hexafluoride is readily hydrolyzed by water to give tungsten trioxide and hydrogen fluoride. It is a strong fluorinating agent and reacts with many metals at room temperature.

Tungsten hexafluoride reacts with ionic halides to form complex salts: KF, RbF,

Table 1. Physical Properties of Tungsten Hexafluoride

Property	Value	Reference
boiling point, °C	17.2	7
triple point, °C, 55.1 kPa[a]	2.0	8
liquid density at 15°C, g/cm^3	3.441	7
transition point, °C	−8.4	3
heat of vaporization, kJ/mol[b]	26.502	8
heat of fusion, kJ/mol[b]	1.758	8
heat of transition, kJ/mol[b]	5.861	8
heat of sublimation, kJ/mol[b]		
above transition	32.448	8
below transition	38.137	8
entropy of vaporization, J/(mol·K)[b]	91.3	8
entropy of fusion, J/(mol·K)[b]	6.07	8
entropy of transition, J/(mol·K)[b]	22.11	8
specific heat at 25°C, J/(mol·K)[b]	118.92	8
vapor pressure at 17°C (calc.), kPa[a]	100.49	8
vapor pressure equation		
liquid; T = 2.0 to 17.1°C	$\log P_{kPa} = 6.760 - 1380.5/T$ ($\log P_{mm\,Hg} = 7.635 - 1380.5/T$)	8
solid; T = −8.2 to 2.0°C	$\log P_{kPa} = 7.883 - 168.9/T$ ($\log P_{mm\,Hg} = 8.758 - 168.9/T$)	8
T = −60 to −8.2°C	$\log P_{kPa} = 9.076 - 2006.0/T$ ($\log P_{mm\,Hg} = 9.951 - 2006.0/T$)	8

[a] To convert kPa to mm Hg, multiply by 7.5.
[b] To convert kJ to kcal, divide by 4.184.

and CsF each react with WF_6 to form K_2WF_8 [*57300-87-5*], Rb_2WF_8 [*57300-88-6*], and Cs_2WF_8 [*57300-89-7*], respectively (9); the alkali halides and WF_6 in sulfur dioxide form $NaWF_6$ [*55822-76-9*], KWF_6 [*34629-85-1*], $RbWF_6$ [*53639-97-7*], and $CsWF_6$ [*19175-38-3*] (10); and KF and WF_6 in liquid IF_5 react to produce K_2WF_8 (11).

Tungsten hexafluoride is reduced by hydrogen at an elevated temperature to tungsten and hydrogen fluoride (12–15). Reduction generally requires temperatures of 550–750°C and pressures of <100 kPa (ca 1 atm). This reaction is the basis of the major use of tungsten hexafluoride in chemical vapor deposition (CVD) (see Film deposition techniques).

Manufacture and Economics. Tungsten hexafluoride is produced commercially by the reaction of tungsten powder with gaseous fluorine at a temperature in excess of 350°C. Tungsten hexafluoride is the major product of the reaction, and there are no by-products when high purity tungsten powder and fluorine are used.

U.S. production is several metric tons per year, with essentially all of the product used in CVD. Air Products and Chemicals, Inc. is the only U.S. producer. The 1979 price is approximately $60/kg, and the price is very sensitive to tungsten metal pricing.

Specifications. The use of tungsten hexafluoride in CVD requires a high purity product, essentially free of all metallic contaminants. High purity product is currently available with total metallic impurity levels of less than 50 ppm and overall tungsten hexafluoride content of greater than 99.8% (16). Hydrogen fluoride is present in concentrations less than 0.1%.

Tungsten hexafluoride is shipped as a liquid under its own vapor pressure in steel cylinders in quantities of 45 kilograms per cylinder or less. It is classified as a corrosive liquid by the DOT.

Handling and Toxicity. Tungsten hexafluoride is irritating and corrosive to the upper and lower airways, eyes, and skin. It is extremely corrosive to the skin, producing burns typical of hydrofluoric acid.

Monel and nickel are the preferred materials of construction; however, copper, brass, steel, and stainless steel can be used at room temperature with little effect. System components should be thoroughly clean and dried before use and passivated with a fluoride film to protect the metals from further corrosion.

Uses. The CVD of tungsten by the thermochemical decomposition of tungsten hexafluoride forms a dense, high purity coating of tungsten on both metallic and nonmetallic substrates. It is used also to fabricate solid tungsten pieces, such as tubing or crucibles. Tungsten coatings are used by the electronics industry for the metallization of silicon devices (17) and as a shield against x-ray or gamma radiation. Tungsten hexafluoride also is used to form hard tungsten carbide coatings on steel (18). Composite coatings of tungsten and rhenium are produced by the simultaneous chemical vapor deposition from their hexafluorides (15) and the addition of rhenium improves the ductility and high temperature properties of the deposit.

A recent patent (19) reports the potential use of chemical-vapor-deposited tungsten surfaces as a highly efficient solar energy (qv) adsorber. The tungsten crystal growth, which takes place as prominent dendrites ranging up to 60 μm in height, captures better than 96% of the light incident on it and effectively accumulates the heat. Its key advantage over the other solar absorbers is its ability to hold its heat at higher operating temperatures, ca 500°C. This application could offer a significant new use for tungsten hexafluoride (see Solar energy).

BIBLIOGRAPHY

1. J. C. Bailar and co-eds., *Comprehensive Inorganic Chemistry,* Vol. 3, Pergamon Press, Compendium Publishers, Elmsford, N.Y., 1973, pp. 749–763.
2. M. Stacey, J. C. Tatlow, and A. G. Sharpe, *Advances in Fluorine Chemistry,* Vol. 5, Butterworths and Co., Washington, D.C., 1965, pp. 46–69.
3. S. Siegal and D. A. Northrup, *Inorg. Chem.* **5,** 2187 (1966).
4. K. N. Tanner and A. B. F. Duncan, *J. Am. Chem. Soc.* **73,** 1164 (1951).
5. T. G. Burke, D. F. Smith, and A. H. Nielsen, *J. Chem. Phys.* **20,** 447 (1952).
6. J. Gaunt, *Trans. Faraday Soc.* **49,** 1122 (1953).
7. J. A. Dean, ed., *Lange's Handbook of Chemistry,* 12th ed., McGraw-Hill Book Co., 1979.
8. G. H. Cady and G. B. Hargreaves, *J. Chem. Soc.,* 1563, 1568 (1961).
9. B. Cox, D. W. Sharp, and A. G. Sharpe, *J. Chem. Soc.,* 1242 (1956).
10. G. B. Hargreaves and R. D. Peacock, *J. Chem. Soc.,* 4214 (1957); *J. Chem. Soc.,* 4390 (1958).
11. G. B. Hargreaves and R. D. Peacock, *J. Chem. Soc.,* 2170 (1958).
12. J. F. Berkeley, A. Brenner, and W. Reid, *Summary Technical Report,* National Bureau of Standards, Washington, D.C., March 1967.
13. *Final Report,* Utah Univ., Salt Lake City Div. of Materials Science and Engineering, Dec. 1971.
14. U.S. Pat. 3,565,676 (Feb. 23, 1971), R. A. Holzl (to Fansteel Metallurgical Corporation).
15. J. I. Federer and A. C. Schaffhauser, *Thermionic Conversion Specialist Conference, 9th, Miami Beach, Fl., 1970,* Conference Research Papers 74–81.
16. *Inorganic Fluorine Compounds,* Air Products and Chemicals Bulletin, Allentown, Pa., 1974.
17. J. M. Shaw and J. A. Amick, *Solid State Technology,* Dec. 1971, pp. 53–55.
18. N. J. Archer, *Proceedings Conference Chemical Vapor Deposition,* 5th International Conference, Electrochemical Society, Princeton, N.J., 1975.
19. U.S. Pat. 4,005,698 (Feb. 1, 1977), J. J. Cuomo, J. M. Woodall, and J. F. Ziegler (to IBM Corp.).

General References

J. H. Simons, ed., *Fluorine Chemistry,* Vol. 5, Academic Press, New York, 1964.

J. W. Mellor, *A Comprehensive Treatise on Inorganic and Theoretical Chemistry,* Vol. 11, John Wiley & Sons, Inc., New York, 1962.

ANDREW J. WOYTEK
Air Products and Chemicals, Inc.

ZINC

Zinc–fluorine compounds include zinc fluoride and its hydrates, and a number of fluorozincates.

Zinc Fluoride

Anhydrous zinc fluoride [*7783-49-5*], ZnF_2, melts at 872–910°C, and can be prepared by slowly drying the tetrahydrate [*13986-18-0*], $ZnF_2.4H_2O$, in a current of anhydrous hydrogen fluoride to minimize hydrolysis and formation of the oxide. There is x-ray evidence for dihydrate formation during dehydration of the tetrahydrate (1). Zinc fluoride tetrahydrate is prepared by reaction of ZnO with aqueous HF.

Zinc fluoride tetrahydrate has a water solubility of about 1.6 g/100 cm^3 solution at 25°C. Addition of HF increases the solubility to 11.8 g/100 cm^3 in a 29% HF solution.

Anhydrous ZnF_2 has a solubility of only 0.024 g/100 g anhydrous HF at 14.2°C (2). The tetrahydrate loses water at temperatures above 75°C.

Zinc fluoride of ca 96% purity was commercially produced for use as a flux in metallurgy. Current production is only on a small scale. Zinc fluoride has been used as a mild fluorinating reagent in replacement of chlorine in halogenated hydrocarbons (3–4). It is also used as a catalyst in several applications including cyclization processes (4).

The only reported (5) toxicity data on zinc fluoride is an LD_{Lo} of 280 mg/kg for subcutaneous administration in frogs. OSHA has a standard time-weighted average (TWA) of 2.5 mg (F)/m^3. NIOSH has issued a criteria document (6) on occupational exposure to inorganic fluorides.

Fluorozincates

Fluorozincates of the formula $MZnF_3$ (M = Na [*18251-84-8*], K [*13827-02-6*], Rb [*29987-38-0*], Cs [*29507-53-7*], and NH_4 [*14972-88-4*]) and M_2ZnF_4 (M = K [*37732-22-2*], Rb [*35944-46-8*], and Cs [*72161-48-9*]) have been reported (7). The $KZnF_3$ and $NaZnF_3$ salts are used as catalysts in alginate dental impression materials (see Dental materials) (8).

BIBLIOGRAPHY

"Zinc Fluoride" in *ECT* 1st ed., under "Fluorine Compounds, Inorganic," Vol. 6, p. 738, by F. D. Loomis, Pennsylvania Salt Manufacturing Co.; "Zinc" in *ECT* 2nd ed., under "Fluorine Compounds, Inorganic," Vol. 9, pp. 684–685, by W. E. White, Ozark-Mahoning Company.

1. E. A. Secco and R. R. Martin, *Can. J. Chem.* **43,** 175 (1965).
2. A. W. Jache and G. H. Cady, *J. Phys.Chem.* **56,** 1106 (1952).
3. A. Sekiya and N. Ishikawa, *Bull. Chem. Soc. Jpn.* **51,** 1267 (1978).
4. U.S. Pat. 3,728,405 (Sept. 14, 1970), J. Allan (to E. I. du Pont de Nemours & Co., Inc.).
5. *Registry of Toxic Effects of Chemical Substances,* Vol. II, NIOSH, 1977, p. 964.
6. *Criteria for a Recommended Standard-Occupational Exposure to Inorganic Fluorides,* PB 246 692, NIOSH 76-103, U.S. Department of Health, Education, and Welfare, 1975.
7. O. Schmitz-Dumont and A. Bornefeld, *Z. Anorg. Allg. Chem.* **287,** 120 (1956).
8. U.S. Pat. 2,769,717 (Nov. 6, 1956), J. Cresson (to L. D. Caulk Company).

CHARLES B. LINDAHL
The Ozark-Mahoning Company,
a subsidiary of the Pennwalt Corp.

ZIRCONIUM

Three fluorides of zirconium, ZrF_2 [*7783-49-5*], ZrF_3 [*13814-22-7*], and ZrF_4 [*7783-64-4*], and several fluorozirconates have been reported.

Zirconium Difluoride

Zirconium difluoride, ZrF_2, has been isolated in rare gas–solid matrices (1). It can be prepared by use of Knudsen cell techniques.

Zirconium Trifluoride

Zirconium trifluoride, ZrF_3, is prepared by reaction of zirconium hydride, ZrH_2, with a mixture of hydrogen and anhydrous hydrogen fluoride at 750°C (2). It can be prepared also by an electrolysis of Zr metal in KF–NaF melts (3). Zirconium trifluoride is stable at ambient temperature, but decomposes at 300°C. It is slightly soluble in hot water, and readily soluble in inorganic acids.

Zirconium Tetrafluoride

A few of the physical properties of zirconium tetrafluoride, ZrF_4, are shown on Table 1.

Anhydrous zirconium tetrafluoride is prepared by the action of anhydrous hydrogen fluoride on $ZrCl_4$ (4), by thermal decomposition of $(NH_4)_3ZrF_7$ [*17250-81-6*] (5), or by the action of elemental fluorine on ZrO_2 at 525°C (6). Zirconium tetrafluoride trihydrate [*14517-16-9*], $ZrF_4.3H_2O$, can be prepared by dissolving ZrF_4 in aqueous hydrofluoric acid and evaporating the solution to dryness. The monohydrate [*14956-11-3*], $ZrF_4.H_2O$, is prepared by removing two moles of water from $ZrF_4.3H_2O$ under vacuum.

The major industrial applications of zirconium tetrafluoride are in molten salt reactor experiments (MSRE), as a catalyst for the fluorination of chloroacetone to chlorofluoracetones (7–8), a catalyst for olefin polymerization (9), for the conversion of a mixture of formaldehyde, acetaldehyde, and ammonia in the ratio of 1:1.3:3 to pyridine (10), as an inhibitor for the combustion of NH_4ClO_4 (11), in dental applications (12), and in rechargeable electrochemical cells (13).

Anhydrous ZrF_4 of technical grade (97% ZrF_4 + 3% HfF_4) and very high-purity material containing less than 0.02% HfF_4 is commercially available in 500 g to 1000 kg quantities. Its toxicity is similar to other sparingly soluble fluorides.

Table 1. Properties of Zirconium Tetrafluoride

Property	Value
sublimation point, °C	640
boiling point, °C	905
specific gravity at 25°C	1.388
solubility, cold H_2O	very slight
aqueous HF	slight

Fluorozirconates

Zirconium tetrafluoride reacts with metal fluorides in aqueous media, or in the presence of dilute HF and forms several complexes containing fluorozirconate ions such as ZrF_5^-, ZrF_6^{2-}, ZrF_7^{3-}, $Zr_2F_9^-$ etc. Several fluorozirconates of alkali, ammonia, and transition metals are known (14–17). Pyrolysis of these complexes produces zirconium oxide, metal fluorides, and hydrogen fluoride (18).

Diammonium, calcium, dipotassium, and disodium fluorozirconates ($(NH_4)_2ZrF_6$ [*16919-31-6*], $CaZrF_6$ [*30868-51-0*], K_2ZrF_6 [*16923-95-8*], and Na_2ZrF_6 [*16925-26-1*]) are used in the manufacture of abrasive grinding wheels (see Abrasives). Other industrial uses of fluorozirconates are in aluminum metallurgy, ceramic industry, glass manufacturing, and in electrolytic cells.

BIBLIOGRAPHY

"Zirconium Fluorides" treated under "Zirconium Compounds" in *ECT* 1st ed., Vol. 15, pp. 297–298, by W. B. Blumenthal, National Lead Company; "Zirconium" under "Fluorine Compounds, Inorganic" in *ECT* 2nd ed., Vol. 9, pp. 685–686, by Wayne E. White, Ozark-Mahoning Company.

1. R. H. Hauge, J. L. Margrave, J. W. Hastie, *High Temp. Sci.* **5**(2), 89 (1973).
2. P. Ehrlich, F. Ploeger, and E. Koch, *Z. Anorg. Allg. Chem.* **333,** 209 (1964).
3. Y. V. Baimakov, *Freiberg. Forschungsh. B* **118,** 43 (1967).
4. U.S. Pat. 2,602,725 (July 8, 1952), H. A. Wilhelm and K. A. Walsh (to the United States of America, as represented by the Atomic Energy Commission).
5. H. M. Haendler, C. M. Wheeler, Jr., and D. W. Robinson, *J. Am. Chem. Soc.* **74,** 2352 (1952).
6. H. M. Haendler and co-workers, *J. Am. Chem. Soc.* **76,** 2177 (1954).
7. U.S. Pat. 2,807,646 (Sept. 24, 1957), C. B. Miller and C. Woolf (to Allied Chemicals & Dye Corp.).
8. U.S. Pat. 2,805,121 (Sept. 3, 1975), C. Woolf (to Allied Chemicals & Dye Corp.).
9. U.S. Pat. 3,165,504 (Jan. 12, 1965), J. P. Hogan (to Phillips Petroleum Co.).
10. Jpn. Kokai 76 63,176 (June 1, 1976), W. Yasuo and co-workers (to Koei Chemical Co. Ltd.).
11. A. P. Galazkova, *Dokl. Akad. Nauk. SSSR* **213,** 622 (1973).
12. B. M. Shrestha, S. A. Mundorff, and B. G. Bibby, *J. Dent. Res.* **51,** 1561 (1972).
13. U.S. Pat. 3,725,128 (Nov. 26, 1974), S. Senderoff (to Union Carbide).
14. N. M. Karasev, L. N. Sidorov; Y. M. Korenev, *Zh. Fiz. Khim. USSR* **50,** 1629 (1976).
15. C. Y. Champarna and co-workers, *Rev. Chim. Miner* **11,** 7351 (1974).
16. M. L. Kotsar and co-workers, *Tr. Mosk. Khim. Tekhnol. Inst.* **75,** 49 (1973).
17. R. L. Davidovich and co-workers, *Izv. Akad. Nauk SSSR Neorg. Mater.* **6,** 493 (1970).
18. Y. N. Sklyadnev and M. A. Mikhailov, *Izv. Sib. Otd. Akad. Nauk SSSR Ser. Khim. Nauk,* 128 (1971).

DAYAL T. MESHRI
Ozark-Mahoning Company,
a subsidiary of the Pennwalt Corp.

FLUORINE COMPOUNDS, INORGANIC—URANIUM. See Fluorine; Uranium and uranium compounds.

FLUORINE COMPOUNDS, ORGANIC

INTRODUCTION

The research of Swarts in Belgium in the early 1900s underlies organic fluorine chemistry and the industry that has grown around it. Swarts and co-workers studied the reaction of antimony trifluoride with chlorine-containing hydrocarbons in great detail as a means of replacing chlorine with fluorine. From this work, after a lapse of several decades, originated the manufacture of refrigerants and aerosols (qv) based on chlorofluoromethanes and ethanes (see Refrigeration).

A second impetus to the expansion of the field was the advent of World War II and the development of the atomic bomb. The desired radioactive isotope of uranium, ^{235}U, was prepared by a separation process, termed gaseous diffusion, of the mixed isotopes of UF_6 (see Diffusion separation methods; Uranium). Since these compounds were almost as reactive as fluorine, a need arose for inert organic materials to be used as liquid seals, greases, and plastics. Consequently, processes were developed to produce stable materials which later evolved into the modern production of Teflon (polytetrafluoroethylene [*9002-84-0*]) and its copolymers such as Viton (DuPont) and Fluorel (3M). Another material, poly(chlorotrifluoroethylene) [*9002-83-9*], although it played a prominent role for several years, has been displaced by the materials mentioned above. A detailed description of the fluorine research performed on the Manhattan Project may be found in ref. 1.

The organic fluorine chemical industry comprises a large and thriving business with production concentrated in the manufacture of refrigerants and aerosols. Recently (1974), considerable criticism has arisen over the use of chlorofluorohydrocarbons as aerosols in that their release into the atmosphere can eventually cause depletion of the ozone layer. Although the subject remains controversial, certain restrictions on chlorofluorocarbon uses have been imposed and there is little doubt that this will retard industrial applications (see Air pollution; Ozone).

Nomenclature

The three acceptable nomenclature systems that are designed to limit the cumbersome use of numerical locants for highly fluorinated materials are: (*1*) the system based on the rules of the International Union of Pure and Applied Chemistry (IUPAC), (*2*) the Chemical Abstracts System (CAS), and (*3*) the per system. Examples of these systems are given in Table 1. The IUPAC and CAS rules are essentially identical and can be used interchangeably. The per system is applied if all or nearly all of the hydrogen atoms in an organic compound are replaced by fluorine, thereby obviating citation of a locant for each halogen atom. The per prefix indicates replacement of all hydrogen atoms in a parent compound of a substituting group by fluorine unless the replacement of hydrogen alters the identity of a functional group as in perfluoro-1-propanol [*72301-80-5*] for $CF_3CF_2CF_2OH$.

Table 1. Comparison of Acceptable Nomenclature Systems for Fluorine Compounds

Compound	CAS Reg. No.	IUPAC	CAS	Per
$CF_3CF_2CF_3$	[*76-19-7*]	octafluoropropane	same as IUPAC	perfluoropropane
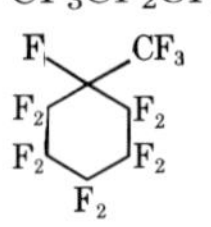	[*355-02-2*]	(trifluoromethyl) undecafluorocyclohexane	undecafluoro(trifluoromethyl)cyclohexane	perfluoromethylcyclohexane
$CF_3CH_2CF_3$	[*690-39-1*]	1,1,1,3,3,3-hexafluoropropane	same as IUPAC	
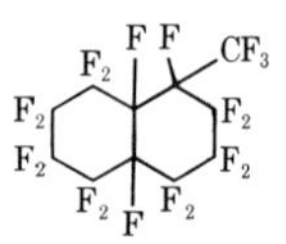	[*51294-16-7*]	1-(trifluoromethyl)-heptadecafluorodecahydronaphthalene	heptadecafluoro-1-(trifluoromethyl)decahydronaphthalene	perfluoro-1-methyldecahydronaphthalene
CF_3COOH	[*76-05-1*]	trifluoroethanoic acid	trifluoroacetic acid	perfluoroacetic acid
CF_3CH_2COOH	[*2516-99-6*]	3,3,3-trifluoropropanoic acid	same as IUPAC	
CF_3CH_2OH	[*75-89-8*]	2,2,2-trifluoroethanol	same as IUPAC	
$(CF_3CF_2CF_2)_2O$	[*356-62-7*]	heptafluoropropoxyheptafluoropropane	bis(heptafluoropropyl)-ether	perfluoropropyl ether
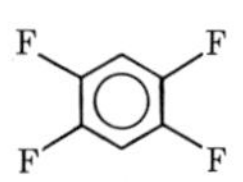	[*28016-01-5*]	1,2,4,5-tetrafluorobenzene	same as IUPAC	
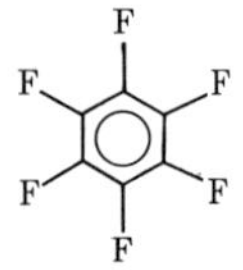	[*392-56-3*]	hexafluorobenzene	same as IUPAC	perfluorobenzene
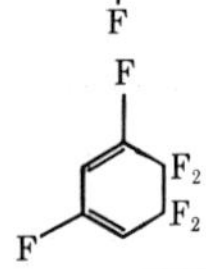	[*38136-60-6*]	1,3,5,5,6,6-hexafluorocyclohexa-1,3-diene	same as IUPAC	

The Nomenclature Committee of the Division of Fluorine Chemistry of the American Chemical Society proposes that the use of perfluoro be abandoned in favor of the letter prefix, F, which would have similar meaning to per. This proposal has not been officially accepted.

In the general literature, the term fluorocarbon is used rather indiscriminately to describe organic compounds containing elements other than fluorine and carbon. This is misleading and compounds such as CCl_3CF_3 [*26523-64-8*], $CHClF_2$ [*75-45-6*], or $CHBrF_2$ [*1511-62-2*] (commonly used chemicals) are correctly termed chlorofluoro- and bromofluorohydrocarbons, respectively.

Physical Properties

Substitution of fluorine for hydrogen in an organic molecule has a profound influence on the molecule's chemical and physical properties. Several factors that are characteristic of fluorine and that underlie these effects are: the large electronegativity of fluorine, its small size (slightly larger than hydrogen), the low degree of polarizability of the carbon–fluorine bond, and their weak intermolecular forces. These effects are illustrated in Tables 2 and 3.

Table 2. Boiling Points of Halomethanes

Chlorohydrocarbon	bp, °C	Fluoro-hydrocarbon	CAS Registry No.	bp, °C	Difference per F atom, °C
CH_3Cl	−24	CH_3F	[*593-53-3*]	−78	54
CH_2Cl_2	40	CH_2F_2	[*75-10-5*]	−52	46
$CHCl_3$	61	CHF_3	[*75-46-7*]	−83	48
CCl_4	77	CF_4	[*75-73-0*]	−128	51

Table 3. Boiling Points of Hydrocarbons and Fluorocarbons

Hydrocarbon	bp, °C	Fluorocarbon	CAS Registry No.	bp, °C
CH_4	−161	CF_4		−128
C_2H_6	−88	C_2F_6	[*76-16-4*]	−78
C_3H_8	−45	C_3F_8		−38
C_4H_{10}	0.6	C_4F_{10}	[*355-25-9*]	−5
C_7H_{16}	98	C_7F_{16}	[*355-57-9*]	82
C_6H_6	80	C_6F_6		82
C_6H_{12} (cyclic)	81	C_6F_{12} (cyclic)	[*355-68-0*]	52

The replacement of chlorine by fluorine results in a nearly constant boiling point drop of ca 50°C for every chlorine atom that is replaced (see Table 2). This effect is the result of low dispersion forces of internal pressure. In Table 3, a similar effect is apparent even though the molecular weight of the fluorocarbon is much higher than its corresponding hydrocarbon. One other significant property difference, attributed to weak intermolecular forces, can be found in the very low surface tensions of fluorocarbons as compared to hydrocarbons and water:

Compound	Surface tension at 20°C, mN/m (= dyn/cm)
perfluoroheptane	13.6
perfluoro-methylcyclohexane	15.4
perfluoro-1,4-dimethylcyclohexane [*374-77-6*]	16.3
octane	21.8
benzene	28.9
methyl bromide	41.5
water	72.8

This effect is utilized in the preparation of commercial oil and stain-repellent textile finishes (see Waterproofing).

In contrast to the above phenomena, the viscosities of fluorocarbons are considerably higher than those of the corresponding hydrocarbons. This can be explained by the greater stiffness of the fluorocarbon chain arising from the large repulsive forces between molecules, and from their greater density. Fluorocarbons have a high viscosity index, ie, the viscosity drops rapidly with increasing temperature, attributable to the large decrease in density that occurs simultaneously. The result is a rapid increase in the free space or holes in the liquid causing a rapid increase in fluidity. This is illustrated as follows with perfluoroheptane:

temperature, °C	0.4	27.5	32.1	38.2	45.0
viscosity, mPa·s (= cP)	1.367	0.861	0.805	0.783	0.664

Preparation

Various general methods are briefly described for the synthesis of fluorinated compounds; background information is contained in the general references.

Halogen Exchange. The exchange of a nonfluorine halogen atom in an organic compound for a fluorine atom is the most widely used method of fluorination. The relative ease of replacement follows the general order I > Br > Cl. The most commonly used fluorinating agents are the fluorides of alkali metals (especially KF), antimony, and mercury. Two other widely used exchange reagents are hydrogen fluoride and sulfur tetrafluoride; the latter also is effective in replacing carbonyl oxygen and hydroxyl groups with fluorine.

Antimony trifluoride, SbF_3, can be used in the following preparations (2):

$$RCOCl \xrightarrow{SbF_3} RCOF$$

$$RSO_2Cl \xrightarrow{SbF_3} RSO_2F$$

$$\rangle C{=}C(R)CX_3 \xrightarrow{SbF_3} \rangle C{=}C(R)CF_3$$

$$RCX_2R \xrightarrow{SbF_3} RCF_2R$$

$$CHX_2OCH_2X \xrightarrow{SbF_3} CHF_2OCH_2F$$

The limitations of this reagent are several. It cannot be used to replace a single halogen atom with the exception shown above or to replace a halogen attached to a carbon–carbon double bond. Fluorination of functional group compounds, eg, esters, sulfides, ketones, acids, and aldehydes, produces decomposition products caused by scission of the carbon chains.

The effectiveness of antimony trifluoride is increased if it is used in conjunction with chlorine or antimony pentachloride. The formation of either $SbCl_2F_3$ or $SbF_5.SbCl_5$ probably accounts for the increased reactivity (2).

Antimony pentafluoride, SbF_5, is a highly active fluorinating agent and is generally used to fluorinate only completely halogenated compounds, since those containing hydrogen, as well as halogen, undergo decomposition. In the case of halogenated olefins or aromatic compounds (3), addition as well as replacement occurs. Thus hexachlorobenzene forms a fluorochlorocyclohexene:

$$C_6Cl_6 \xrightarrow{SbF_5} \underset{[27458\text{-}17\text{-}9]}{C_6Cl_2F_8}$$

Potassium fluoride, KF, is the most frequently used of the alkali metal fluorides, although the reactivity of alkali fluorides is in the order CsF > RbF > KF > NaF > LiF (4). It can be used to convert alkyl halides and sulfonyl chlorides to fluorides, and to exchange halogen in various alcohols, acids, and esters (5).

$$RX \xrightarrow{KF} RF$$

$$RSO_2Cl \xrightarrow{KF} RSO_2F$$

$$CHCl_2CH_2OH \xrightarrow{KF} \underset{[359\text{-}13\text{-}7]}{CHF_2CH_2OH}$$

$$CHCl_2CO_2CH_3 \xrightarrow{KF} \underset{[433\text{-}53\text{-}4]}{CHF_2CO_2CH_3}$$

Another example is the rodenticide, sodium monofluoroacetate [*62-74-81*], which is prepared by reaction of KF with chloroacetic acid ester (see Poisons, economic).

In recent years, the preparation of fluoroaromatics by reaction of KF with perhaloaromatics, primarily perchlorobenzene, has received considerable attention. Two methods were developed and included either the use of a solvent, such as *N*-methylpyrrolidinone (6), or the absence of a solvent (7). These methods plus findings that various fluorobenzene derivatives are effective fungicides (8) prompted the Imperial Smelting Corporation of Great Britain to develop a commercial process for the production of polyfluorobenzenes (9) (see Fungicides). Although the details of the process are not available, it apparently uses a mixture of potassium and sodium fluorides in a solvent.

Mercuric fluoride, HgF_2, is used in the following conversions (2,10):

$$RX \xrightarrow{HgF_2} RF \ (X = Br \text{ or } I)$$

$$\text{polyhalides} \xrightarrow{HgF_2} \text{polyfluorides}$$

Because alkyl chlorides react very slowly, chloroform or methylene chloride can be used as solvents. Oxygen-containing compounds, eg, esters, ethers, and alcohols, appear to inhibit the fluorination.

Mercurous fluoride, Hg_2F_2, is less effective than HgF_2. The addition of chlorine or iodine to the reagent increases its reactivity owing to formation of the complex $HgF_2.HgX_2$ (2,10).

Hydrogen fluoride, HF, is a comparatively ineffective exchange agent and will replace only active halogens, eg, (11):

$$C_6H_5CCl_3 \xrightarrow{HF} C_6H_5CF_3$$

[98-08-8]

$$RCOCl \xrightarrow{HF} RCOF$$

When used with antimony pentachloride, the reactivity of HF is comparable to $SbCl_2F_3$; accordingly, a continuous fluorination process is possible:

$$SbCl_5 + 3\ HF \rightarrow SbCl_2F_3 + 3\ HCl$$

$$SbCl_2F_3 + CCl_4 \rightarrow CCl_2F_2 + SbCl_4F$$

[75-71-8]

$$SbCl_4F + 2\ HF \rightarrow SbCl_2F_3 + 2\ HCl$$

Since the antimony halides serve as fluorine carriers, the actual fluorination agent is hydrogen fluoride. This process is the principal one used in the production of the chlorofluorohydrocarbons for aerosols and refrigerants.

Another use of hydrogen fluoride, although not in halogen exchange, is the reaction with ethylenes or acetylenes to form the addition product, ie:

$$CH{\equiv}CH + HF \rightarrow CH_2{=}CHF$$

[75-02-5]

Reaction conditions must be carefully controlled since HF is an excellent polymerization catalyst.

Sulfur tetrafluoride, SF_4, replaces halogen in haloalkanes, haloalkenes, and aryl chlorides, but is only mildly effective (even at elevated temperatures) in the presence of a Lewis acid catalyst. The reagent is most often used in the replacement of carbonyl oxygen with fluorine (12–13). Aldehydes and ketones react readily, particularly if no α-hydrogen atoms are present, but acids, esters, acid chlorides, and anhydrides are very sluggish. However, these reactions can be catalyzed by Lewis acids, (HF, BF_3, etc).

$$C_6H_5CHO + SF_4 \rightarrow C_6H_5CHF_2$$

[455-31-2]

$$C_6H_5CO_2H + SF_4 \xrightarrow[\text{temp}]{\text{elevated}} C_6H_5CF_3$$

Halogen fluorides include compounds such as IF_3, IF_5, ClF, etc, of which a few, ClF, ClF_3, BrF_3, and IF_5, are used to some extent. They act both as halogen exchange agents and, in the case of the monofluorides, addition agents to unsaturated bonds (14).

Replacement of Hydrogen. Three methods of substitution of a hydrogen atom by fluorine are: (*1*) reaction of a C—H bond with elemental fluorine (direct fluorination); (*2*) reaction of a C—H bond with a metal fluoride, eg, AgF_2 or CoF_3; and (*3*) electrochemical fluorination in which the reaction occurs at the anode of a cell containing fluoride.

Direct Fluorination. The principal disadvantage of the use of elemental fluorine as a fluorination agent is the exothermicity of the reaction. A considerable degree of carbon–carbon bond scission can occur as well as polymer formation. In order to prevent these complications, the fluorine is diluted with nitrogen and the reaction zone is constructed so that good heat conductivity is possible (see Direct fluorination).

The fluorination reaction is best described as a radical-chain process:

$$F_2 \rightleftharpoons 2\,F\cdot$$

$$RH + F\cdot \rightarrow R\cdot + HF$$

$$R\cdot + F_2 \rightarrow RF + F\cdot$$

Initiation occurs readily at room temperature. If the molecule contains unsaturation, addition of fluorine takes place (14):

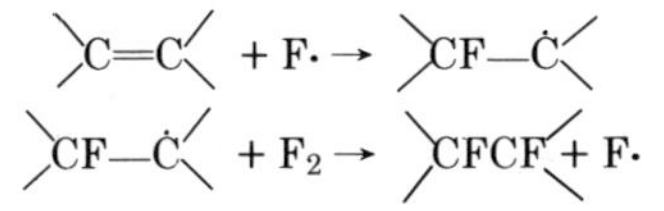

Reaction With a Metal Fluoride. A second technique for hydrogen substitution is the reaction of a higher metal fluoride with a carbon–hydrogen bond:

$$2\,CoF_2 + F_2 \rightarrow 2\,CoF_3$$

$$RH + 2\,CoF_3 \rightarrow RF + 2\,CoF_2 + HF$$

The principal advantage to this approach is that the heat evolved for each carbon–fluorine bond formed, 192.5 kJ/mol (46 kcal/mol), is much less than that obtained in direct fluorination, 435.3 kJ/mol (104 kcal/mol), and the reaction yields are much higher. Only two metal fluorides are of practical use, AgF_2 and CoF_3.

The reactivity of the metal fluoride appears to be associated with the oxidation potential of the metal ion (Ag^+ or Co^{2+}), the greatest reactivity with the highest potential. In order of decreasing effectiveness, they are: $AgF_2 > CoF_3 > CaF_2 > MnF_3 > PbF_4 > BiF_5 > CrF_3 > HgF_2 > FeF_3 > CuF_2 > SnF_4$.

The reaction is conducted by passing fluorine through a bed of AgCl or $CoCl_2$ at an elevated temperature to form the higher-valent fluorides. The organic reagent then is passed through the bed to effect fluorination. Thus a semicontinuous process can be realized. In general, the method is used for the preparation of fluorocarbons since any unsaturation or functionality in the starting material is removed. It can be used also to fluorinate polychlorohydrocarbons whereby replacement of chlorine as well as hydrogen is carried out. For a general discussion of this method see ref. 1.

Electrochemical Fluorination. The electrochemical synthesis of fluorinated organic compounds (15) involves the electrolysis of an organic substrate in liquid anhydrous hydrogen fluoride at a voltage below that for the liberation of fluorine. The reaction is limited by temperature (usually 0°C) and by the solubility of the substrate in HF. This method is used to fluorinate aryl halides, sulfonyl halides, ethers, carboxylic acids, and amines. The product is a fluorocarbon having no residual hydrogen; olefins and carbocyclics, as well as heterocyclic compounds, are saturated. Side reactions, resulting in reduced yields, include cleavage of carbon–carbon bonds and polymer formation (see Electrochemical processing).

Electrochemical fluorination is used successfully on a commercial scale to produce perfluoroacyl fluorides, perfluorosulfonyl fluorides, and perfluoroamines.

Chemical Properties and Applications

Substitution of fluorine in an organic molecule has two principal effects on the molecule's chemical properties: the chemical stability, eg, resistance to hydrolysis, is greatly increased; and the reactivity of functional groups is drastically altered resulting from the large inductive effect of fluorine. These effects become more pronounced as the degree of fluorine substitution is increased, and reach their maxima in the fluorocarbons and their derivatives. A brief description of the more common classes of fluorine compounds follows.

Fluorinated Alkanes. As the fluorine content increases, the chemical reactivity decreases until complete fluorination, after which they are inert to most chemical agents, including the highly reactive element, fluorine. Their lack of reactivity leads to their use in certain commercial applications.

Fluorinated Olefins. In electrophilic addition reactions, the reactivity of the unsaturated linkage is reduced by the inductive effect of fluorine. Nucleophilic additions are enhanced by this same effect; amines, phenols, alcohols, and many other nucleophiles add readily to the carbon–carbon double bond. Free-radical addition of halogen halides proceeds easily using peroxide or thermal initiation (see Initiators). Some halides, especially those derived from iodine, eg, ICl, react by an ionic mechanism.

Fluoroolefins undergo free radical polymerization producing a wide range of polymers derived from tetrafluoroethylene [*116-14-3*], vinylidene fluoride [*75-38-7*], vinyl fluoride [*75-02-5*], and hexafluoropropylene [*116-15-4*]. These polymers are used as coatings, gaskets, and seals which can withstand high temperatures and contact with chemical environments that quickly destroy other organic polymers.

Fluorinated Aromatic Hydrocarbons. Many aromatic fluorocarbon derivatives, eg, hexafluorobenzene, pentafluorobenzene [*363-72-4*], pentafluorotoluene [*771-56-2*], and perfluoronaphthalene [*313-72-4*], are now available (16). The fluorine atoms in these molecules readily undergo nucleophilic substitution as contrasted to those in perfluoroalkanes. The hydrogen atoms may be replaced by electrophiles, but this is usually quite difficult, requiring extreme reaction conditions.

Fluorinated Acids. This class of compounds is characterized by the strength of the fluorocarbon acids, eg, CF_3CO_2H approaches the strength of a mineral acid. This property results from the strong inductive effect of fluorine and is markedly less when the fluorocarbon group is moved away from the carboxyl group. Generally, their reactions are similar to organic acids and they find application, particularly trifluoroacetic acid and its anhydride [*407-25-0*], as promoters for the preparation of esters and ketones and in nitration reactions.

Fluorinated Drugs. Beginning in the 1950s with the use of highly fluorinated halocarbons and ethers as low flammability, fast acting inhalation anesthetics, the preparation of fluorinated compounds for use in medicine has increased rapidly (17–18). This great interest in such substances is based on the following considerations:

(*1*) fluorine most closely resembles bioactive hydrogen analogues with respect to steric requirements at receptor sites; (*2*) fluorine alters electronic effects, owing to its high electronegativity; (*3*) fluorine imparts improved oxidative and thermal stability to the parent molecule; and (*4*) fluorine improves lipid solubility in membranes, thereby increasing rates of absorption and transport, *in vivo*.

At this time, many classes of drugs have been developed and used successfully in the treatment of disease. Some of these are listed below along with some important examples of each class.

Inhalation Anesthetics. Examples of highly fluorinated halocarbons and ethers are enflurane [*13838-16-9*] (**1**), isoflurane [*26675-46-7*] (**2**), fluoroxene [*406-90-6*] (**3**), fluothane [*151-67-7*] (**4**), and methoxyflurane [*76-38-0*] (**5**) (see Anesthetics).

$ClCHFCF_2—O—CHF_2$ (**1**) $CF_3CHCl—O—CHF_2$ (**2**) $CF_3CH_2—O—CH{=}CH_2$ (**3**)

$CF_3CHBrCl$ (**4**) $Cl_2CHCF_2—O—CH_3$ (**5**)

Antineoplastic Agents. Drugs widely used in cancer chemotherapy are 5-fluorouracil [*51-21-8*] (**6**) and Halotestin [*76-43-7*] (**7**) (Fluoxymesterone) (see Chemotherapeutics, antimitotic).

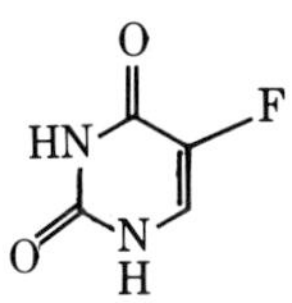
(**6**)

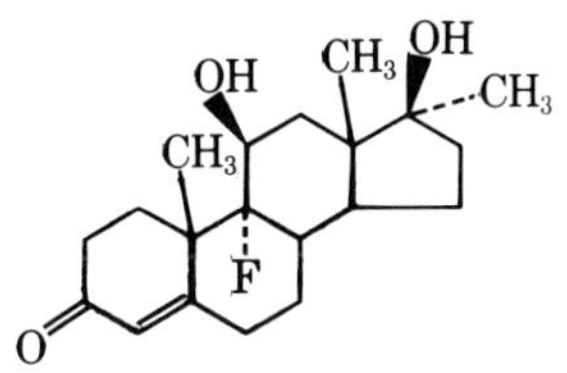
(**7**)

Steroids. More than 20 fluorinated steroids (qv) are currently available and are used as adrenocortical and anti-inflammatory agents, as progestational agents, and in androgenic hormone therapy, eg, dexamethasone [*50-02-2*] (**8**) and fluocinonide [*356-12-7*] (**9**) (see Immunotherapeutic agents).

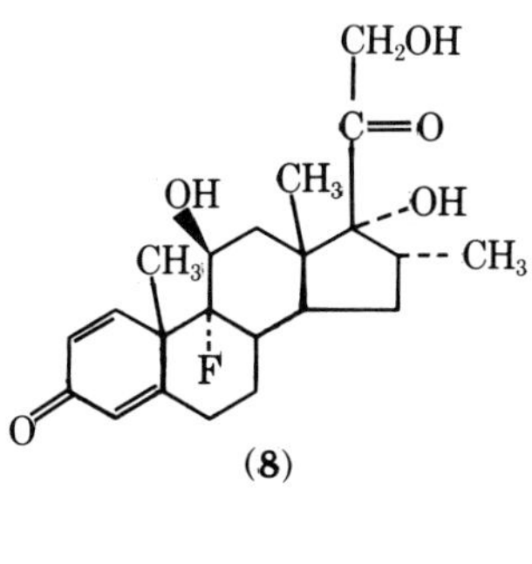
(**8**)

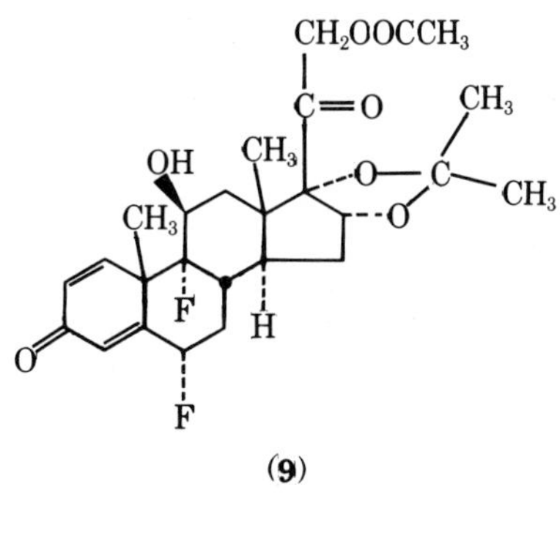
(**9**)

Analgesic Agents. Flufenisal [*22494-27-5*] (**10**), a direct analogue of aspirin, is 4–5 times more effective than aspirin; and Sulindac [*38194-50-2*] (**11**) is used for the treatment of rheumatoid arthritis (see Analgesics).

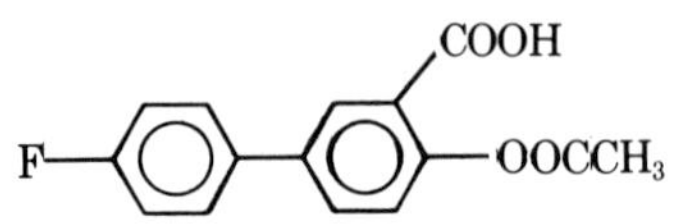

(**10**)

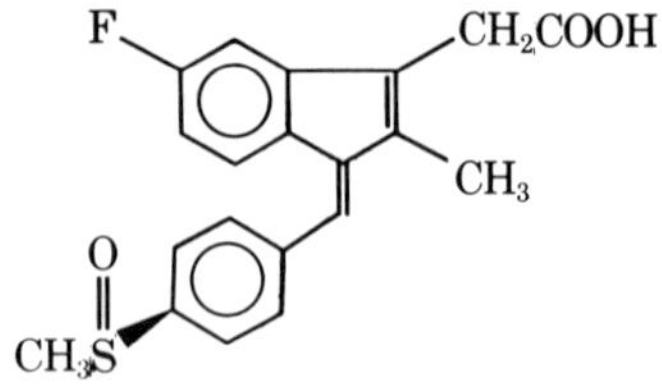

(**11**)

Antibiotics. Flucloxacillin [*5250-39-5*] (**12**) is a semisynthetic penicillin and a narrow spectrum antibiotic (see Antibiotics).

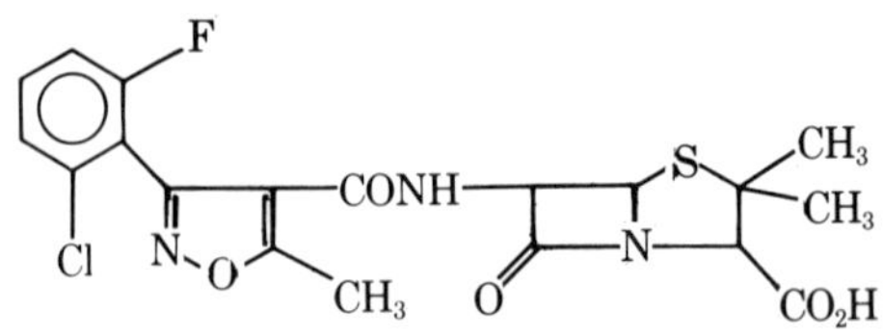

(**12**)

The application of fluorinated drugs is growing rapidly. Some of the newer uses being studied are antipsychotics, sedatives, hypnotics, muscle relaxants, and appetite depressants.

There are many other classes of functional fluorine compounds which encompass a broad area of chemistry (19–20).

BIBLIOGRAPHY

"Nomenclature, Physical Properties, Reactions and Methods of Fluorination," in *ECT* 1st ed., under "Fluorine Compounds, Organic," Vol. 6, pp. 735–751, by E. T. McBee, O. R. Pierce, and W. F. Edgell, Purdue University; "Introduction," in *ECT* 2nd ed., under "Fluorine Compounds, Organic," Vol. 9, pp. 686–704, by E. T. McBee, C. J. Norton, and T. Hodgins, Purdue University.

1. C. Slesser and S. R. Schram, *Preparation, Properties and Technology of Fluorine and Organic Fluorine Compounds,* McGraw-Hill Book Co., Inc., New York, 1951.
2. A. L. Henne and A. M. Whaley, "The Preparation of Aliphatic Fluorine Compounds" in R. J. Adams, ed., *Organic Reactions,* Vol. II, John Wiley & Sons, Inc., New York, 1964.
3. E. T. McBee, P. A. Wiseman, and G. B. Bachman, *Ind. Eng. Chem.* **39,** 415 (1947).
4. N. N. Vorozhtzov and G. G. Yakobson, *Khim. Nauk Prom.* **3,** 403 (1958).
5. E. Gryszkiewicz-Trochimowski, A. Sporzynski, and J. Wnuk, *Rec. Trav. Chim.* **66,** 413 (1947).
6. J. T. Maynard, *J. Org. Chem.* **28,** 112 (1963).
7. N. N. Vorozhtsov and co-workers, *Izvest. Akad. Nauk,* 1524 (1964).
8. G. C. Finger, F. H. Reed and L. R. Tehon, *Ill. State Geol. Surv. Circ.* **199** (1955).
9. W. Prescott, *Chem. Ind.* (2), 56 (1978).
10. W. Bockmuller, "Organic Fluorine Compounds" in *Newer Methods of Preparative Organic Chemistry,* Interscience Publishers, Inc., New York, 1948, pp. 229–248.
11. K. Wiechert, "Use of Hydrogen Fluoride in Organic Reactions" in ref. 10, pp. 315–368.
12. G. W. Parshall, *J. Org. Chem.* **27,** 4649 (1963).
13. W. C. Smith, *Angew Chem. Int. Ed. Engl.* **1,** 467 (1962).
14. W. K. R. Musgrave, *Adv. Fluorine Chem.* **1,** 1 (1960).

15. J. H. Simons, *J. Electrochem. Soc.* **95,** 47 (1949).
16. J. C. Tatlow, *Endeavour* **22,** 89 (1963).
17. R. Filler, "Fluorine-Containing Drugs" in R. E. Banks, ed., *Organofluorine Chemicals and Their Industrial Applications,* Ellis Horwood Ltd., Chichester, 1979.
18. *AMA Drug Evaluations,* 3rd ed., Publishing Sciences Group, Inc., Littleton, Mass., 1977.
19. W. A. Sheppard and C. M. Sharts, *Organic Fluorine Chemistry,* W. A. Benjamin, Inc., New York, 1964.
20. M. Hudlicky, *Chemistry of Organic Fluorine Compounds,* 2nd ed., Ellis Horwood, Ltd., Chichester, 1976.

General References

J. H. Fletcher, O. C. Denner, and R. B. Fox, *Nomenclature of Organic Compounds,* ACS Advances in Chemistry Series, Vol. 126, Washington, D.C., 1973.

J. H. Simons, ed., *Fluorine Chemistry,* Vols. 1–5, Academic Press, Inc., New York.

A. M. Lovelace, D. A. Rausch, and W. Postelnek, *Aliphatic Fluorine Compounds,* Reinhold Publishing Corp., New York, 1954.

P. Tarrant, ed., *Fluorine Chemistry Reviews,* Vols. 1–8, Marcel Dekker, Inc., New York.

J. C. Tatlow and co-eds., *Advances in Fluorine Chemistry,* Butterworths, London.

W. A. Sheppard and C. M. Sharts, *Organic Fluorine Chemistry,* W. A. Benjamin, Inc., New York, 1964.

R. E. Banks and M. G. Barbour, *Fluorocarbon and Related Chemistry,* Vol. 3, The Chemical Society, Burlington House, London, 1976.

M. Hudlicky, *Chemistry of Organic Fluorine Compounds,* 2nd ed., Ellis Horwood, Ltd., Chichester, 1976.

A. E. Pavlath and J. E. Leffler, *Aromatic Fluorine Chemistry,* Reinhold Publishing Corp., New York, 1962.

ODGEN R. PIERCE
Dow Corning Corporation

DIRECT FLUORINATION

Organic compounds containing fluorine are well known for their special properties, especially their inertness and very low boiling points and comparatively high melting points. Binary compounds of carbon and fluorine, commonly called fluorocarbons, can be traced back to Moissan who, in 1886, discovered and isolated fluorine (1). Initial attempts at reactions between fluorine and hydrocarbons were described as similar to combustion and the reaction products contained mostly carbon tetrafluoride and hydrogen fluoride:

$$C_xH_y + \left(\frac{4x + y}{2}\right) F_2 \rightarrow x\ CF_4 + y\ HF + \text{large amount of energy}$$

This reaction has often reached explosive proportions in the laboratory. In 1940–1965 several methods were devised for controlling it. Thus, for fluorination of hydrocarbons of low molecular weight at room temperature or below, yields as high as 80% of perfluorinated products have been reported together with partially fluorinated species (2–4). However, in a typical fluorination reaction involving elemental fluorine with complex hydrocarbons at elevated temperatures, appreciable cleavage of the carbon—carbon bonds occurs and the yields invariably are only a few percent.

The simplest solid fluorocarbon (5) is poly(fluoromethylidyne) $+CF+_n$, CFX (MarChem), with a layered structure (see Fig. 1). It can be synthesized by the reaction of fluorine with graphite at 300–600°C and fluorine pressures up to 101 kPa (1 atm) (6–7). Other solids (C_4F, C_2F, etc) have also been reported from the reaction of F_2 with graphite (8–9). This material is a grayish to white powder of variable stoichiometry and stable up to 600°C in air. It has a very low coefficient of friction, similar to Teflon, and is an excellent lubricant which can be used up to 600°C, surpassing Teflon which softens and decomposes above 350°C (see Polytetrafluoroethylene). When $+CF_x+_n$ decomposes above 650°C, it yields mainly carbon black (finely divided soot) and the inert gas, carbon tetrafluoride (10). Thus, it can be used safely at high temperatures without any risk of poisonous gases being produced on decomposition (5,11).

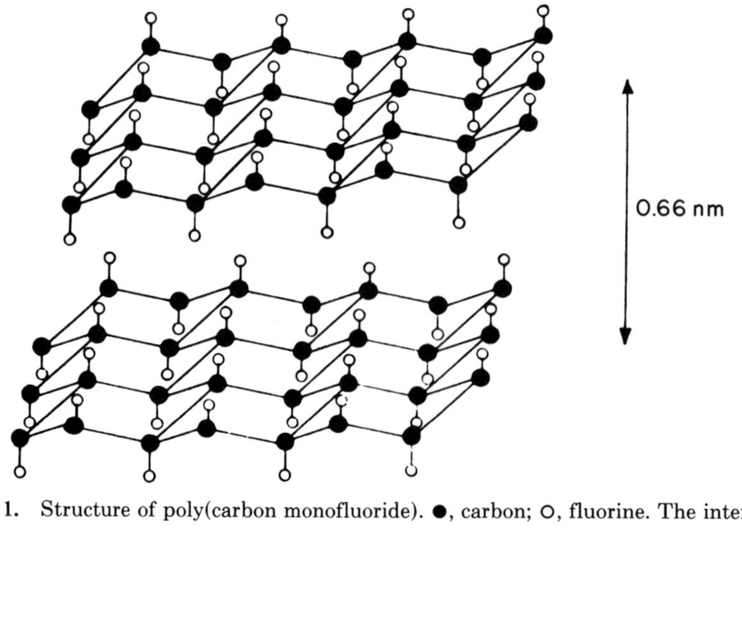

Figure 1. Structure of poly(carbon monofluoride). ●, carbon; ○, fluorine. The interstitial space is 0.66 nm.

Until 1970 the direct use of elemental fluorine was usually considered as the classical method of fluorination (12–13), whereas other approaches were regarded as modern methods. It was generally believed that, since the reaction of elemental fluorine with hydrocarbons is highly exothermic, the carbon—carbon bonds are first ruptured, creating free radicals and eventually leading to polymerization. Therefore, in most instances the expected product was a complex mixture of polymers and degradation products, including pure carbon and various fragments of the original hydrocarbon material.

During the last decade researchers have developed diverse fluorination methods which avoid these problems by using certain metal fluorides, inorganic fluorides, or halogen fluorides and electrochemical or indirect methods using certain fluorides as reagents. Hydrogen fluoride has been successfully used as a fluorinating agent and is, in fact, the most widely used industrial fluorinating agent at present.

Fluorination of organic compounds using high-valency metallic fluorides (14) can take two forms:

(*1*) Exchange of halogen with fluorine of the metal fluorides, MF_n:

$$\geq C{-}X + M^{+}F^{-} \longrightarrow \geq C{-}F + M^{+}X^{-}$$

where n = Cl, Br, or I and M = K, Sb, Ag, Hg_2, or Hg.

(*2*) Replacement of hydrogen with the fluorine of metal fluorides and addition to double bonds:

$$\geq C{-}H + 2\ MF_n \longrightarrow \geq C{-}F + HF + 2\ MF_{n-1}$$

$$>C{=}C< + 2\ MF_n \longrightarrow F{-}\overset{|}{\underset{|}{C}}{-}\overset{|}{\underset{|}{C}}{-}F + 2\ MF_{n-1}$$

High-valency metallic fluorides are very reactive compounds and most decompose in water. They are excellent reagents for fluorination of organic compounds and include CoF_3, AgF_2, MnF_3, CeF_4, PbF_4, and possibly BiF_5 and UF_6. At elevated temperatures, usually 100–400°C, they react with organic compounds producing the fluoro derivatives and the corresponding lower fluorides such as CoF_2 and AgF. Using cobalt trifluoride at up to 400°C, all hydrogen atoms of a hydrocarbon can be replaced by fluorine if the product is thermally stable. Similar addition of fluorine to unsaturated linkages and to aromatic nuclei takes place. These reactions are carried out with organic compounds either in the liquid or in the vapor state. When the organic compound is a liquid, the fluorinating agent (metal fluoride) is added to the heated sample with stirring. In some cases, the organic compound may be suspended or dissolved in an inert solvent, usually a high boiling fluorocarbon. In vapor-phase fluorination, a stream of the organic compound is swept over a bed of fluoride in a heated reactor. In general, fluorination has been more successful in the vapor phase than the liquid phase but yields tend to be low and extensive thermal degradation occurs at high temperatures.

Kinetic as well as thermodynamic problems are encountered in fluorination. The rate of reaction must be decelerated so that the energy liberated may be absorbed or carried away without degrading the molecular structure. A significant recent advance in direct fluorination is the La-Mar process (15–17).

Thermochemistry

Thermodynamic considerations are of utmost importance in fluorinations. Table 1 is based on JANAF data (18) for CH_4, which indicate an average carbon—hydrogen bond strength of 410.0 kJ/mol (98 kcal/mol) based on the atomization energy of CH_4. Table 2 is based on JANAF data for tetrafluoroethylene. This is, perhaps, a poor choice since the double bond in C_2F_4 is extremely weak. Reliable enthalpies and free energy functions for the average double bond are only available at 25°C.

The limiting parameter to be considered in attempting to develop a satisfactory method for controlling reactions of elemental fluorine is the weakest bond in the reactant compound. For hydrocarbons the average carbon—carbon single bond strength is 351.5–368.2 kJ/mol (84–88 kcal/mol). The overall reaction for the replacement of hydrogen by fluorine is exothermic enough [$\Delta G_{25} = -432.6$ kJ/mol (−103.4 kcal/mol)] for a fracture of carbon—carbon bonds if it were to occur via a concerted mechanism or on several adjacent carbon atoms simultaneously. This energy must be dissipated so as to avoid the fragmentation of the molecular skeleton. The comparison of 359.8 kJ/mol (86 kcal/mol) vs 430.9 kJ/mol (103 kcal/mol) has been cited in many previous discussions as an obvious basis to predict the failure of direct fluorination methods. For the rapid reaction rates, which were employed in most previous experiments, this is a valid argument.

It can be seen from Table 1 that there are no individual steps that are exothermic enough to break carbon—carbon bonds except the termination step 3a of −407.9 kJ/mol (−97.5 kcal/mol). Consequently, procedures or conditions are desirable that reduce the atomic fluorine concentration or decrease the mobility of hydrocarbon radical intermediates, and/or keep them in the solid state during reaction. It is necessary to reduce the reaction rate to the extent that these hydrocarbon radical intermediates have longer lifetimes permitting the advantages of fluorination in individual steps to be achieved experimentally. It has been demonstrated by epr (electron paramagnetic resonance) methods (19) that, with high fluorine dilution, various radicals do indeed have appreciable lifetimes.

The two possible initiation steps for the free-radical reaction are step 1b, and the combination of steps 1a and 2a from Table 1. The role of initiation step 1b in the reaction scheme is an important consideration in minimizing the concentration of atomic fluorine (20).

As indicated in Table 1, this process could be spontaneous at room temperature [$\Delta G_{25} = -24.4$ kJ/mol (−5.84 kcal/mol)] although the enthalpy is slightly positive.

Table 1. Thermodynamic Data[a] for Fluorination of CH_4, kJ/mol[b]

Step	Reaction	ΔH_{25}	ΔH_{325}	ΔG_{25}	ΔG_{325}
initiation 1a	$F_2 \rightarrow 2\,F\cdot$	157.7	161.0	123.6	87.4
1b	$F_2 + RH \rightarrow R\cdot + HF + F\cdot$	16.3	21.3	−24.4	−79.1
propagation 2a	$RH + F\cdot \rightarrow R\cdot + HF$	−141.4	−139.7	−151.2	−156.9
2b	$R\cdot + F_2 \rightarrow RF + F\cdot$	−25.5	−290.8	−284.9	−268.4
termination 3a	$R\cdot + F\cdot \rightarrow RF$	−446.8	451.8	−407.9	−356.0
3b	$R\cdot + R\cdot \rightarrow R-R$	−350.6	−347.5	−294.1	−240.6
overall reaction	$RH + F_2 \rightarrow RF + HF$	−430.5	−430.5	−432.6	−430.9

[a] Based on JANAF table data (18).
[b] To convert J to cal, divide by 4.184.

Table 2. Thermodynamic Data[a] for Fluorination of Double Bonds, kJ/mol[b]

Step	Reaction	$\Delta H_{25}{}^a$	$\Delta H_{325}{}^a$	$\Delta G_{25}{}^a$	$\Delta G_{325}{}^a$	$\Delta H_{25}{}^c$	$\Delta H_{25}{}^d$	$\Delta H_{325}{}^d$	$\Delta G_{25}{}^d$	$\Delta G_{325}{}^d$	$\Delta H_{25}{}^e$
initiation 4a	$F_2 \rightarrow 2\,F\cdot$	157.7	161.0	123.5	87.4	−157.7					
4b	$R_2C{=}CR_2 + F_2 \rightarrow \cdot R_2C{-}FCR_2 + F\cdot$	−156.9	−150.2	−151.0	−145.6	35.5	−78.5	−75.1	−75.3	−72.8	17.8
initiation and termination 5	$R_2C{=}CR_2 + F_2 \rightarrow R_2CF{-}FCR_2$	−674.8	−674.5	−624.2	−572.3	−515.9	−337.4	−336.8	−312.1	−286.1	−258.0
propagation 6a	$R_2C{=}CR_2 + F\cdot \rightarrow \cdot R_2C{-}FCR_2$	−314.6	−311.3	−274.5	−235.6	−194.6	−157.3	−155.6	−135.1	−118.0	−97.3
6b	$\cdot CR_2{-}CFR_2 + F_2 \rightarrow R_2CF{-}FCR_2 + F\cdot$	−359.8	−362.7	−349.4	−336.4	−290.0	−179.9	−181.6	−174.9	−168.2	−145.0
termination 7a	$F\cdot + \cdot CR_2{-}FCR_2 \rightarrow R_2CF{-}FCR_2$	−684.9	−524.2	−473.2	−424.2	−447.7	−259.0	−261.9	−236.4	−212.1	−223.8
7b	$F\cdot + F\cdot \rightarrow F_2$	−157.7	−161.0	−123.5	−87.4	−157.7					
7c	$\cdot CR_2{-}FCR_2 + \cdot CR_2FCR_2 \rightarrow FCR_2CR_2{-}CR_2FCR_2$					−435.1					−217.5
overall reaction	$R_2C{=}CR_2 + F_2 \rightarrow R_2CF{-}FCR_2$	−674.8	−674.5	−624.2	−572.4	−515.9	−337.4	−336.8	−312.12	−286.1	−258.0

[a] Based on JANAF table data (18).
[b] To convert J to cal, divide by 4.184.
[c] Based on bond strengths for average carbon—carbon double bond.
[d] Energy per carbon atom based on same data as [a].
[e] Energy per carbon atom based on same data as [c].

The validity of this step has not yet been conclusively established by spectroscopic methods which makes it an unsolved problem of prime importance. Furthermore, the fact that fluorine reacts at a significant rate with some hydrocarbons in the dark at temperatures below −78°C indicates that step 1b is important and may have very little or no activation energy at RT. At extremely low temperatures (ca 10 K) there is no reaction between gaseous fluorine and CH_4 or C_2H_6 (21).

A simple equilibrium calculation reveals that, at 25°C and atmospheric pressure, fluorine is less than 1% dissociated, whereas at 325°C an estimated 4.6% dissociation of molecular fluorine is calculated. Obviously, less than 1% of the collisions occurring at RT would result in reaction if step 1a were the only important initiation step. At 325°C the fluorine atom initiation step should become more important. From the viewpoint of energy control, as seen in Table 1, it would be advantageous to have step 1b predominate over step 2a and promote attack by molecular rather than atomic fluorine. Ambient or lower temperatures keep the atomic fluorine concentration low.

In the addition of fluorine to double bonds (see Table 2), the energetic situation is less severe, ie, the addition of fluorine to double bonds is only 251.4–292.9 kJ/mol (60–70 kcal/mol) exothermic per carbon—carbon bond. This energy is not sufficient to fracture the carbon skeleton if care is taken to keep addition from occurring on several adjacent carbon atoms simultaneously. Here, as in the case of hydrogen removal, the individual steps are less exothermic than the overall reaction. It has been established experimentally that less fragmentation occurs and, correspondingly, a higher yield is obtained with most conventional fluorination processes when an unsaturated rather than a saturated hydrocarbon is the starting material. This is owing to the greater exothermicity of the reaction with hydrogen, ie, 434.7 kJ/mol (103.9 kcal/mol) per saturated carbon atom as compared with 207–289.8 kJ/mol (50–70 kcal/mol) per unsaturated carbon atom. In the case of addition of fluorine to double bonds (see Table 2), the corresponding initiation step 4b, is probably exothermic by 20.7–190.4 kJ/mol (5–46 kcal/mol) and thus plays an important role. A second important possibility is the concerted mechanism of step 5 which is exothermic by 207.0–283.2 kJ/mol (50–58.4 kcal/mol) per carbon atom.

Steric Factors

Initially, most of the collisions of fluorine molecules with saturated or aromatic hydrocarbons occur at a hydrogen site or at a π-bond (unsaturated) site. When collision occurs at the π-bond, the double bond disappears but the single bond remains because the energy released in step 4b (Table 2) is insufficient to fracture the carbon—carbon single bond. Once carbon—fluorine bonds have begun to form on the carbon skeleton of either an unsaturated or alkane system, the carbon skeleton is somewhat sterically protected by the sheath of fluorine atoms. Figure 2, which shows the crowded helical arrangement of fluorine around the carbon backbone of polytetrafluoroethylene (PTFE), is an example of an extreme case of steric protection of carbon—carbon bonds (22).

The nonbonding electron clouds of the attached fluorine atoms tend to repel the oncoming fluorine molecules as they approach the carbon skeleton. This reduces the number of effective collisions, making it possible to increase the total number of collisions and still not accelerate the reaction rate as the reaction proceeds toward com-

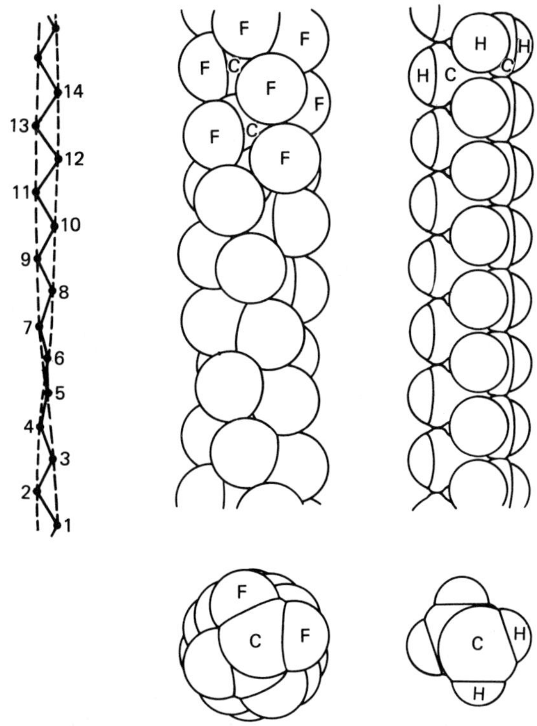

Figure 2. Models of polytetrafluoroethylene and polyethylene chains showing the protective effect of fluorine on the carbon—carbon bond.

pletion. This protective sheath of fluorine atoms provides the inertness of Teflon and other fluorocarbons. It also explains the fact that greater success in direct fluorination processes has been reported when the hydrocarbon to be fluorinated had already been partially fluorinated by some other process or was prechlorinated, ie, the protective sheath of halogens reduced the number of reactive collisions and allowed reactions to occur without excessive cleavage of carbon—carbon bonds or runaway exothermic processes.

Kinetic Control of the Rate of Elemental Fluorine Reaction

In direct fluorination processes, concentration, time, and temperature can be controlled. In most previous work, the fluorine was diluted with an inert gas such as nitrogen, helium, or even carbon dioxide. However, the concentration of fluorine in the reactor was kept at a constant level, usually 10% or greater, by premixing the inert gas with fluorine in the desired proportion and then introducing this mixture into the reactor. The rate of reaction between a hydrocarbon compound and a 10% fluorine mixture is relatively high and this very exothermic process can lead to fragmentation (see Table 3) and, in some cases, to combustion. The initial stages of reaction are most critical and nearly all the fragmentation occurs at this time. An initial concentration of 10% fluorine or more is, for most compounds, much too high for nondestructive fluorination.

Molecular relaxation processes such as vibrational or rotational relaxations or

Table 3. Thermodynamic Data[a] for Fluorination of C_2H_6 in kJ[b] per Carbon Atom Fragment

Step	Reaction	ΔH_{25}	ΔH_{325}	ΔG_{25}	ΔG_{325}
1	$H_3C—CH_3 + F_2 \rightarrow 2\ H_3CF$	−266.1	−268.2	−264.4	−261.5
2	$H_3C—CH_3 + 2\ F\cdot \rightarrow 2\ H_3CF$	−344.8	−348.5	−329.5	−305.4

[a] Based on JANAF table data (18).
[b] To convert J to cal, divide by 4.184.

thermal conduction make it possible to dissipate the energy released during fluorination. Such relaxation processes are capable of minimizing the chances that the energy required to break the weakest bond is appropriately localized if the reaction sites are widely distributed over the system. Therefore, in the initial stages of fluorination it is necessary to reduce the probability of adjacent reaction sites simultaneously occurring in the same molecule, or in adjacent molecules in a crystal, by diluting the reactants and relying on relaxation processes to distribute the energy over the entire system and thus avoid fragmentation.

A very low initial fluorine concentration is used in the La-Mar fluorination process (17). Initially, the reactor is passivated with helium or nitrogen and fluorine is then fed slowly into the system. If pure fluorine at low pressure is the incoming gas, the fluorine pressure may approach one atmosphere over an extended time period. The very low concentrations of fluorine in the system greatly decrease the probability of simultaneous fluorine collisions on the same molecule or on adjacent reaction sites. As previously discussed, reactant molecules are able to withstand more fluorine collisions, as they become more highly fluorinated, without decomposition because some sites are sterically protected, ie, collisions at carbon—fluorine sites are obviously nonreactive. The fluorine concentration may therefore be increased as the reaction proceeds to obtain a practical reaction rate. Actual dilution schemes to achieve successful fluorination must be individually tailored for specific reaction systems and may, in some cases, include a stepwise procedure.

Thus, for a successful fluorination process involving elemental fluorine, the number of collisions must be drastically reduced in the initial stages; the rate of fluorination must be slow enough to allow relaxation processes to occur and a heat sink must be provided to remove the reaction heat.

Most direct fluorination reactions with organic compounds are performed at or near room temperature unless: (*1*) reaction rates are so slow that even a fluorine pressure of 101.3 kPa (1 atm) does not lead to appreciable reaction; (*2*) a vapor-phase reaction is desired and it is necessary to heat the sample being treated to get enough vapor; or (*3*) reaction rates are so fast that excessive fragmentation, charring or decomposition occurs and a much lower temperature is desirable.

Examples of all three types are known. Thus in the production of partially fluorinated lubricating oils (23) fluorine gas is bubbled at 101 kPa (1 atm) through medium weight oils at 50, 75, and 100°C. For production of fluorocarbon coatings on natural or synthetic rubber or polypropylene, temperatures of 25–50°C are optimum. The fluorination of hot hydrocarbon vapors like gaseous naphthalene or gaseous anthracene (24) is not too satisfactory because fragmentation and charring lead to low yields.

Low temperature fluorination techniques (−78°C) are promising for the preparation of complex fluorinated molecules, especially where functional groups are present (25); eg, fluorination of hexamethylethane to perfluorohexamethylethane, of norbornane to perfluoro- and 1-hydroundecafluoronorbornane, and of adamantane to 1-hydropentadecafluoroadamantane.

Extremely low temperatures have been used to moderate the reactivity of fluorine still further. Hydrocarbons such as methane, ethane, ethylene, allene, benzene, or acetylene, can be co-condensed into a solid fluorine matrix at 20 K without reaction (26–27). This has permitted the study of the reactivity of molecular fluorine without monatomic free-radical chain reactions. Atom reactions can be initiated by uv photolysis.

Infrared absorption spectroscopy provides a convenient method of monitoring these reactions. Ethylene and propylene are particularly interesting since they can be kept indefinitely in solid fluorine in the dark but when exposed to near ir radiation, 1–2 μm, fluorine is added across the double bond. Ethylene gives the less stable trans conformer of 1,2-difluoroethane. Under identical conditions propylene reacts more rapidly which suggests that the presence of a methyl group has an accelerating effect. Further fluorination studies of ethylene show that, when the fluorine is diluted with argon, reaction with a single fluorine molecule leads to fluoroethylene and hydrogen fluoride; thus the double bond is not attacked. As the concentration is increased the production of *trans*-1,2-difluoroethane increases (28). The acceleration by methyl substitution was demonstrated by the reaction of propylene immediately upon condensation with fluorine. This suggests that little or no activation energy barrier exists for reaction of these molecules with molecular fluorine.

Experimental Techniques

In early reaction systems (2–3,29–30) the vaporized hydrocarbon was combined with nitrogen in a reactor and mixed with a nitrogen–fluorine mixture from a preheated source. The jet reactor (4) for low molecular weight fluorocarbons was an important improvement. The process takes place at around 200–300°C, and fluorination is carried out in the vapor state.

At 200–300°C many compounds, both organic and inorganic, are marginally stable, and certainly not in their lowest vibrational states. They may even undergo some pyrolytic decomposition. Thus the addition of extra energy produced by the interaction of fluorine with these compounds is likely to produce substantial fragmentation. Vaporization of high molecular weight hydrocarbons is rather difficult, and their fluorination was not attempted in the earlier pioneering studies.

The typical fluorination apparatus used for these reactions is simple in design (see Fig. 3) (31). It is essential that the materials of construction are resistant to fluorine (see Table 4) (32). The presence of even traces of oxygen or moisture can have a deleterious effect and, therefore, extreme precautions must be taken to eliminate these contaminants.

The connections are conveniently made of 0.635 cm OD copper tubing. When the fluorocarbon produced in the reaction is volatile, a cold finger-type trap can be placed between the reaction chamber and the trap to catch volatile products. The temperature of the trap must be high enough to pass unreacted or excess fluorine and nitrogen but cold enough to condense the reaction products. Before the reaction is started, the whole system is purged with helium or nitrogen for ca 30 min, and then 0.5–2 mL of fluorine and 50–100 mL of nitrogen per minute are passed through the system. Solid material has to be ground to a very fine powder to achieve complete fluorination.

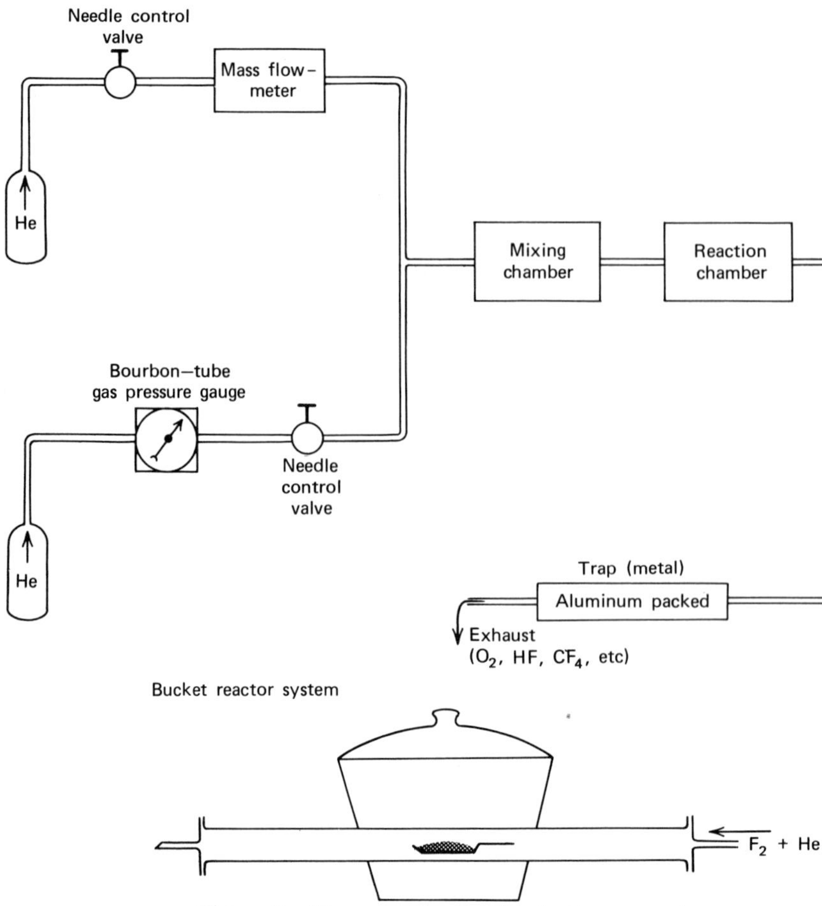

Figure 3. Diagram of typical fluorination apparatus.

A special cryogenic reactor (33) in which the reactions of fluorine with liquid and gaseous samples can be controlled at very low temperatures is shown in Figure 4. Reactants are volatilized into the reaction zone of the cryogenic reactor from the heated oil evaporator prior to initiation of the reaction. The main reaction chamber is a nickel tube, 2.54 cm in diameter, packed with copper turnings. The compartments (10.1 × 10.1 × 20.2 cm) are constructed of stainless steel and insulated with urethane foam and act as heat sinks. All connections are made of 0.635 cm copper or aluminum tubing. A sodium fluoride trap is used to remove the hydrogen fluoride from the reaction products. By cooling or warming the compartments, they can be used to create a temperature gradient along the reaction tube. As the products are highly fluorinated, they are usually volatile and tend to move through the reactor tube rapidly, depending on the temperature gradient. This provides a continually renewed surface of reactant at the optimum temperature for fluorination. Fluorinated copper turnings effectively increase the surface area of the compound exposed to fluorine. The individual zones of the reactor may be cooled with various solvent–solid carbon dioxide or with solvent–liquid nitrogen slushes. Preferably, the temperature is precisely regulated with an automatic liquid nitrogen temperature controller. In addition to the four-zone reactor shown in Figure 4, a multizone reactor can also be used; an eight-zone reactor has been found to be particularly efficient (34). Internal Freon cooling is very effective for controlling the temperatures of the various compartments (35).

Table 4. Materials Resistant to Dry[a] Fluorine

Temperature, °C	Fluorine-resistant materials[b]	
25–150	carbon (graphite, diamond)	polyethylene
	magnesium and its alloys	polypropylene
	iron and its alloys	rubber
	nickel and its alloys	polystyrene
	copper and its alloys	poly(vinyl chloride)
	metal oxides	Teflon
	CaF_2, LaF_3, LaB_6	Kel-F
up to 500	copper	
	nickel	metal oxides
	Monel	CaF_2
	carbon (graphite, diamond)	LaF_3
	nickel–aluminum alloys	LaB_6
up to 1000–1500	carbon (diamond)	CaF_2, LaF_3
	Al_2O_3, UO_2, ThO_2	LaB_6

At 2000°C carbon and other refractory metals react slowly to form gaseous mono- or difluorides and other species; refractory oxides form volatile fluorides + O_2; refractory carbides form metal fluorides + CF_4; etc.

[a] Moisture results in HF development and increased corrosion of metals.
[b] Protection of fluoride or fluorocarbon film.

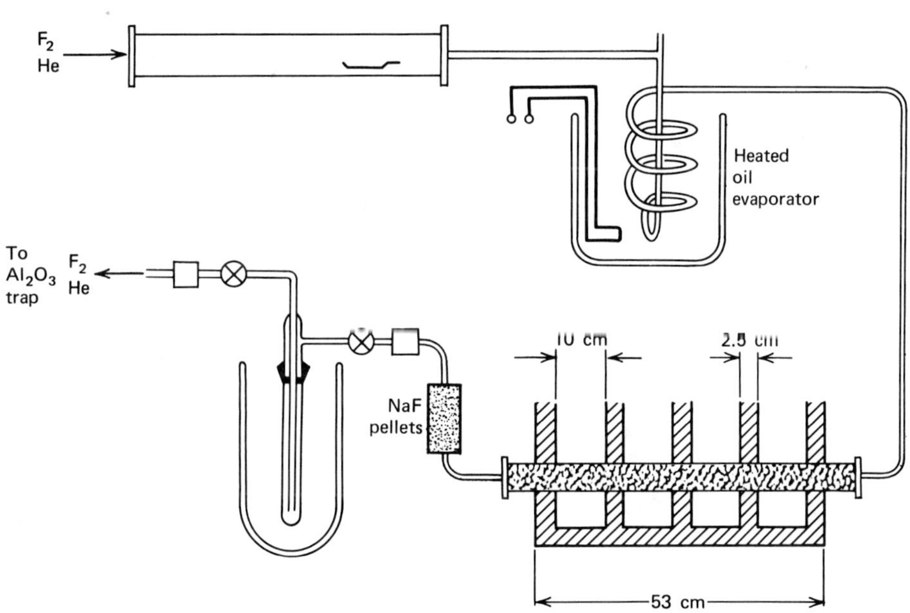

Figure 4. Cryogenic reactor system.

Oxygen or moisture has to be excluded because the presence of oxygen leads to cross-linking, presumably with epoxy bridges, to carbonyl groups which give acid fluorides, and to peroxides (36). Cross-linking obviously decreases the yield of pure perfluorocarbon. It can be detected by ir activity in the 1600–2000 cm^{-1} region and by noting the polymeric nature of the products, ie, high melting points, low vapor pressures, etc.

Applications

Direct fluorination, though it sometimes leads to degradation, cross-linking, and combustion can be controlled and useful products derived (37). Even though direct fluorination sometimes produces a very low yield of desired products, it may offer the only possible method of preparation. Oxyfluorination is a new technique (38) offering unique possibilities for the functionalization of fluorocarbons and for the preparation of fluoromembranes.

In 1954 the surface fluorination of polyethylene sheets by using a water-cooled heat sink was patented (39). Later patents covered the fluorination of PVC (40) and of polyethylene bottles (41). Studies of surface fluorination have been reported (42). The fluorination of polyethylene powder was described (24) as a fiery intense reaction, which was finally controlled by dilution with an inert gas at reduced pressures. Direct fluorination of polymers was achieved by Lagow and Margrave in 1970 (11,43). More recently, surface fluorinations of poly(vinyl fluoride), polycarbonates, polystyrene, and poly(methyl methacrylate), and the surface fluorination of containers have been described (44–45). Partially fluorinated poly(ethylene terephthalate) and polyamides like nylon have good soil release properties as well as high wettability (46–47).

The following companies manufacture organic fluorine compounds by direct fluorination techniques: Air Products and Chemicals, Inc., Tamaqua, Pa.; MarChem, Inc., Houston, Tex.; Ozark-Mahoning, Inc., Tulsa, Okla.; and PCR, Inc., Gainesville, Fla.

Simple and Complex Organic Molecules. A great variety of new perfluoro and partially fluorinated compounds has been synthesized (48–54). The direct fluorination of CS_2 (49) has produced several derivatives including SF_5CF_3, $(SF_3)_2CF_2$, $SF_5CF_2SF_3$, SF_3CF_3 and $F_3SCF_2SF_3$; and the treatment of $Si(CH_3)_4$ or $Ge(CH_3)_4$ with F_2 (50) can produce most of the Si–C–H–F or Ge–C–H–F species possible. The fluorination of $Hg(CH_3)_2$ yields $Hg(CF_3)_2$. In all these cases, the identities of the products have been confirmed by elemental analyses, and ir, mass, and nmr spectra. Low temperature fluorination methods (33,48–50) lead to high yields of acid fluorides (55), perfluoroethers (56), and other functionalized organic species (57).

Direct electrophilic fluorination (52) of adamantane gives fluoroadamantane. Elemental F_2 was used for the selective functionalization of carbons-14 and 17 of the steroid nuclei as shown below (52):

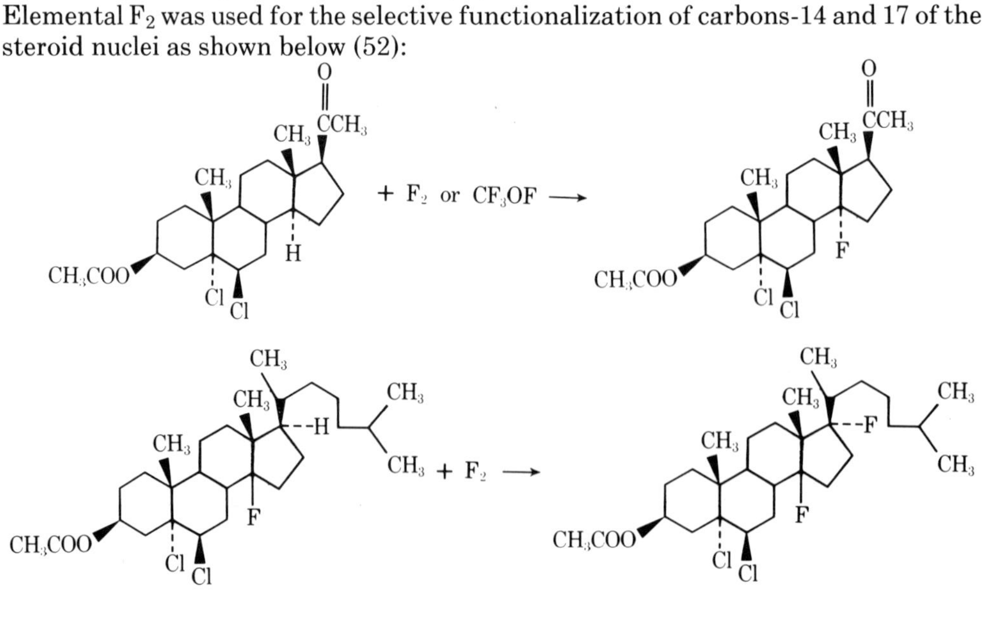

Furthermore, 5-fluorouracil can be prepared directly by low temperature fluorination of a Freon–uracil slurry (53).

Hydrocarbon Polymers. It is difficult to produce perfluorocarbon polymers by the usual methods. Many monomers, such as hexafluoropropylene, polymerize only slowly because of the steric hindrance of fluorine. Furthermore, some monomers are not very stable and are difficult to synthesize. Direct fluorination can be used for the direct synthesis of fluorocarbon polymers (58) and for producing fluorocarbon coatings on the surfaces of hydrocarbon polymers (11,31,39–43,58–59). Some hydrocarbon polymers and their hypothetical fluorocarbon perfluoro-analogues are shown below:

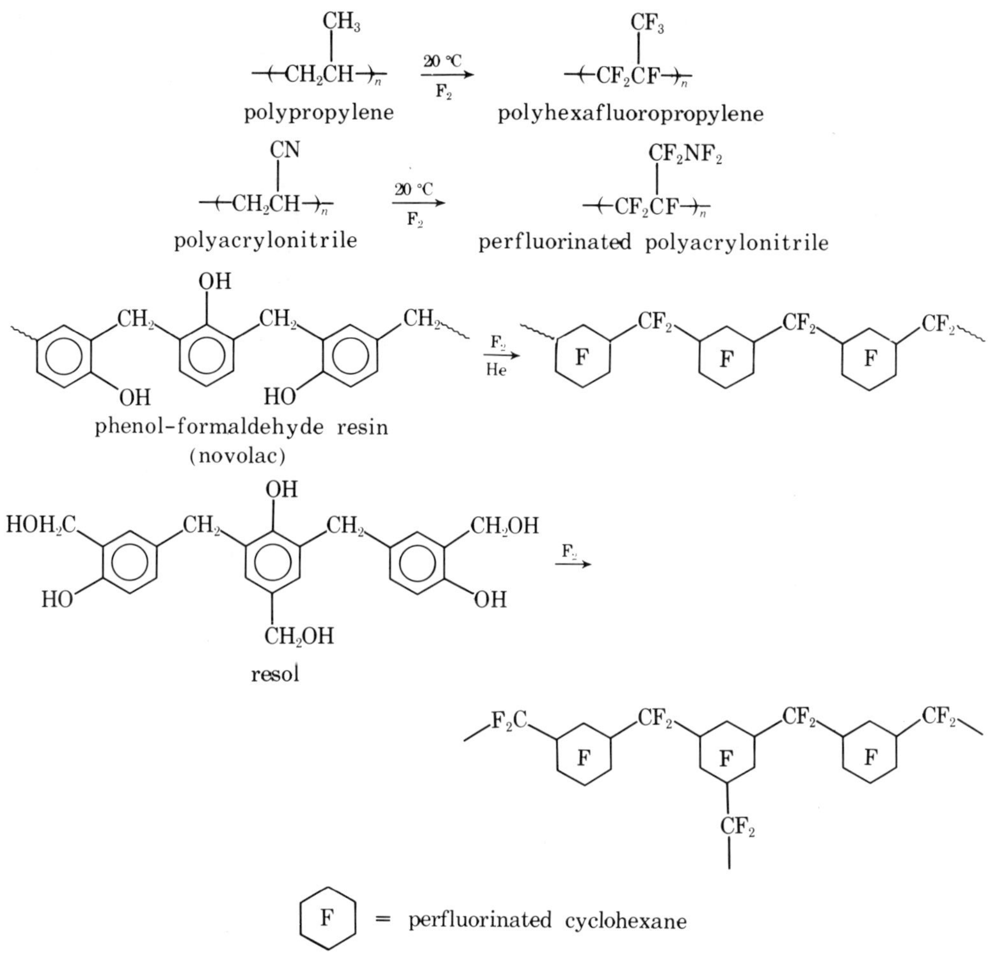

Thus fluorocarbon polymers can be produced with chemical compositions very similar to polytetrafluoroethylene by the direct reaction of fluorine with polyethylene and the perfluoro analogues of polypropylene and polystyrene can be prepared. These fluorocarbon polymers differ from the more familiar linear structures because carbon–carbon cross-linking occurs to a significant extent during fluorination. Most of these fluoropolymers are white solids with high thermal stability; some are stable in air as much as 200°C above the ignition temperatures of their corresponding hydrocarbon precursors. Hydrocarbon polymers such as polyethylene and the new surfaces formed by direct fluorination have been studied by esca which shows that the surface is truly converted to a fluorocarbon polymer (60) (see Analytical methods). Most of these fluorocarbon surfaces are inert and many of them have good lubricant properties. The fluorination of PVC has been followed with esca and $\leftarrow CF_x \rightarrow_n$ polymer was also identified (61).

If solid polymers or polymer particles larger than ca 100 μm are fluorinated, only surface fluorination results. Penetration of fluorine and the conversion of hydrocarbon to fluorocarbon can occur to depths of at least 0.1 mm. Fluorocarbon-coated objects should have many practical applications because the chemically adherent surface provides: increased thermal stability; resistance to oxidation and corrosive chemicals and solvents; decreased coefficient of friction, and thus decreased wear; and decreased permeability to gas flow. Unusual surface effects can be obtained by fluorinating the polymer surfaces only partially (62).

Natural and Synthetic Rubber. Fluorination of natural or synthetic rubber creates a fluorocarbon coating (31,63–64) which is very smooth and water-repellent (see Waterproofing). Rubber articles such as surgical gloves, O-rings, gaskets, and windshield wiper blades, can be fluorinated on the surface while the interior retains the elastic, flexible properties of the natural rubber. Fluorinated O-rings can be used without extra lubricant in corrosive atmospheres since the fluorocarbon is unreactive. In food-processing equipment, grease or lubricants are eliminated and do not contaminate the food products. Fluorinated O-rings have smooth surfaces, very low

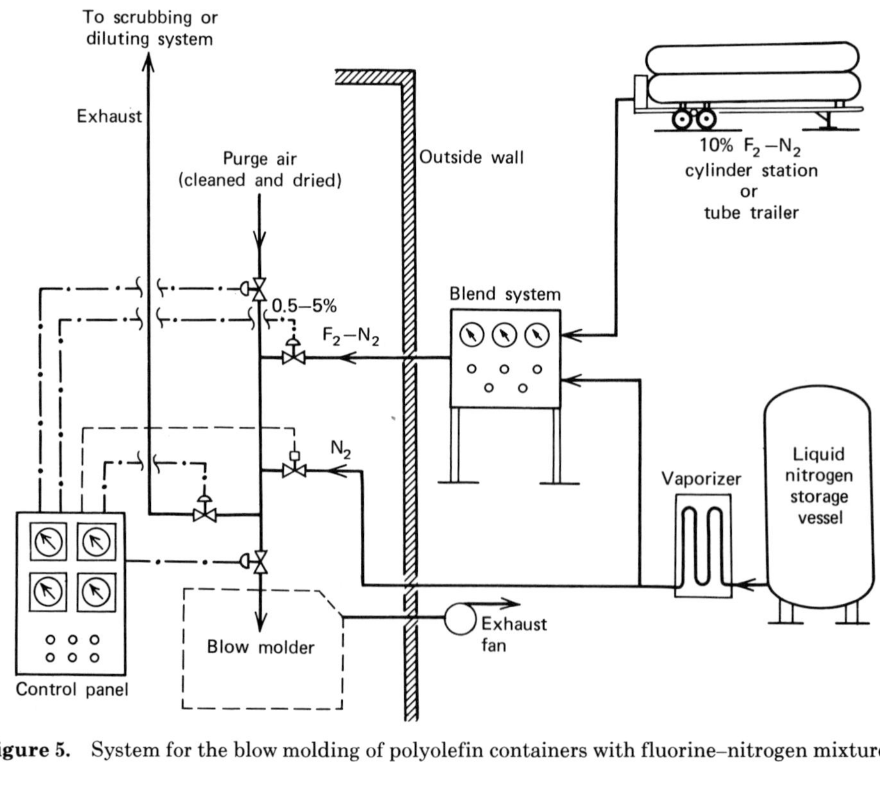

Figure 5. System for the blow molding of polyolefin containers with fluorine–nitrogen mixtures.

frictional coefficients, and enhanced thermal stabilities. Fluorinated windshield wiper blades have a very low coefficient of friction, run smoother with less squeak, their surface is more resistant to the sun's uv radiation and attack by ozone, and they require less electrical energy for operation.

Many applications of this technique are apparent in medicine, such as surgical rubber gloves, rubber sheets, drain tubes, catheters, etc. Since talcum or other lubricating powder often used with surgical gloves can cause allergic reactions, thin powderless gloves are desirable. Teflon or silicone films have proved unsuccessful for surgical gloves. However, an excellent direct fluorination process has been developed (64) whereby the inside surfaces are fluorinated under expanded conditions at elevated temperatures. A very smooth surface is obtained and powder is not required. At the same time, the outside surface remains rough so that surgical instruments can be held firmly without slippage. The tactile sensitivity of the tips of the fingers seems to be increased, and the problem of powder forming lumps or a mudlike slush inside the glove is eliminated.

Blow-Molded Containers. A surface-fluorination process has been developed for the blow-molding industry to produce solvent-resistant polyolefin containers (see Fig. 5). In this application, the air which is normally used to blow mold containers is replaced by a low concentration of fluorine in nitrogen. Airopak containers produced by this process show outstanding resistance to nonpolar solvents (44–45,65–66) and are being evaluated for use as vehicle fuel tanks.

BIBLIOGRAPHY

1. H. Moissan, *Le Fluor et ses Composes,* Steinbeil, Paris, Fr., 1900.
2. G. H. Cady and co-workers, *Ind. Eng. Chem.* **39,** 290 (1947).
3. R. N. Haszeldine and F. Smith, *J. Chem. Soc.,* 2689, 2787 (1950).
4. L. A. Bigelow in J. H. Simmons, ed., *Fluorine Chemistry,* Vol. 1, Academic Press, Inc., New York, 1970, p. 373; E. A. Tyczkowski and L. A. Bigelow, *J. Am. Chem. Soc.* **77,** 3007 (1955).
5. P. Kamarchik and J. L. Margrave, *Acc. Chem. Res.* **11,** 296 (1978); U.S. Pat. 3,519,657 (July 7, 1970), G. A. Olah (to The Dow Chemical Co.); L. B. Ebert, J. I. Brauman, and R. A. Huggins, *J. Am. Chem. Soc.* **96,** 7841 (1974).
6. O. Ruff and O. Bretschneider, *Z. Anorg. Allgem. Chem.* **217,** 1 (1934); W. Rudorff and G. Rudorff, *Z. Anorg. Allgem. Chem.* **253,** 281 (1947); W. Rudorff and G. Rudorff, *Chem. Ber.* **80,** 413 (1947); W. Rudorff and K. Brodersen, *Z. Naturforsch* **12b,** 575 (1957); W. Rudorff, *Adv. Inorg. Chem. Radiochem.* **1,** 230 (1959).
7. A. K. Kuriakose and J. L. Margrave, *J. Phys. Chem.* **69,** 2772 (1965); U.S. Pat. 3,674,432 (July 4, 1972), R. B. Badachhape and co-workers (to R. I. Patents, Inc.); R. B. Badachhape and co-workers, *J. Am. Chem. Soc.* **96,** 1268, 2628 (1974); R. B. Badachhape, V. K. Mahajan, and J. L. Margrave, *Inorg. Nucl. Chem. Lett.* **10,** 1103 (1974).
8. N. Watanabe and M. Ishii, *J. Electrochem. Soc. Jpn.* **29,** 364 (1961); N. Watanabe, Y. Koyama, and S. Yoshizawa, *J. Electrochem. Soc. Jpn.* **31,** 756 (1963); N. Watanabe and K. Kumon, *J. Electrochem. Soc. Jpn.* **35,** 19 (1967); H. Imoto and N. Watanabe, *Bull. Chem. Soc. Jpn.* **49,** 1736 (1976); M. Takashima and N. Watanabe, *Nippon Kagaku Kaishi,* 1222 (1976); N. Watanabe, Y. Kita, and T. Kawaguchi, *Nippon Kagaku Kaishi,* 191 (1977).
9. Y. Kita, N. Watanabe, and Y. Fujii, *J. Am. Chem. Soc.* **101,** 3823 (1979); R. J. Lagow and co-workers, *Inorg. Metalorg. Chem.* **2,** 145 (1972).
10. P. Kamarchik and J. L. Margrave, *J. Therm. Anal.* **11,** 259 (1977).
11. J. L. Margrave and co-workers, *IR-100 Award for CFX, a Lubricant Powder,* 1970; *Ind. Res.* **12,** 47 (1970).
12. C. M. Sharts, *J. Chem. Ed.* **45,** 3 (1968).
13. R. E. Banks, *Fluorocarbons and Their Derivatives,* Oldbourne Press, London, Eng., 1964, p. 87.

14. W. A. Shepherd and C. M. Sharts, *Organic Fluorine Chemistry,* W. A. Benjamin, New York, 1969; M. Hudlicky, *Chemistry of Organic Fluorine Compounds,* 2nd ed., Halsted Press, a division of John Wiley & Sons, Inc., New York, 1976; M. Stacey and J. C. Tatlow, "Exhaustive Fluorinations of Organic Compounds with High-Valency Metallic Fluorides" in *Advances in Fluorine Chemistry,* Vol. 1, Butterworths Scientific Publications, London, Eng., 1960, pp. 166–198.
15. R. J. Lagow and J. L. Margrave, *Proc. Natl. Acad. Sci.* **67,** 4, 8A (1970).
16. R. J. Lagow, Ph.D. thesis, Rice University, Houston, Tex., 1970.
17. B. Fegley, *MIT Technol. Eng. News,* 13 (Apr. 1973).
18. D. R. Stull and co-eds., *JANAF Thermochemical Tables,* 2nd ed., *NSRDS-NBS37,* U.S. Government Printing Office, Washington, D.C., June 1971, and subsequent revisions.
19. R. E. Florin and L. A. Wall, *Electron Spin Resonance Studies on Fluorination of Polymers, Abstract No. 8,* Fluorine Chemistry Div., *165th Amer. Chem. Soc. Meeting, Dallas, Tex., Apr. 10, 1973.*
20. W. T. Miller, S. D. Koch, and F. W. McLafferty, *J. Am. Chem. Soc.* **78,** 4992 (1956).
21. R. H. Hauge, J. Wang, and J. L. Margrave, *paper presented at First Winter Fluorine Conference, St. Petersburg, Fl., Jan. 1972.*
22. C. W. Bunn and E. R. Howell, *Nature (London)* **174,** 549 (1954).
23. R. B. Badachhape and J. L. Margrave, unpublished work, and work currently in progress, Rice University, Houston, Tex.
24. M. Okade and K. Makuuchi, *Ind. Eng. Chem. Prod. Res. Dev.* **8,** 334 (1969).
25. N. J. Maraschin and R. J. Lagow, *J. Am. Chem. Soc.* **94,** 8601 (1972).
26. R. H. Hauge and co-workers, *Proceedings of the Berlin Symposium on Matrix Isolation Spectroscopy, West Berlin, Ger., June 21–24, 1977.*
27. R. H. Hauge and J. L. Margrave, *Proceedings of the 10th Materials Research Symposium on Characterizations of High Temperature Vapors and Gases, Sept. 18–22, 1978,* U.S. Natl. Bur. Standards, Washington, D.C.
28. R. H. Hauge and co-workers, *J. Am. Chem. Soc.* **101,** 6950 (1979).
29. O. Ruff, *Die Chemie des Fluors,* Springer, Berlin, Ger., 1920.
30. W. Bockemuller, *Organische Fluorverbindungen,* F. Enke, Stuttgart, Ger., 1936.
31. U.S. Pat. 3,758,450 (Sept. 11, 1973), R. J. Lagow and J. L. Margrave (to R.I. Patents, Inc.); U.S. Pat. 3,775,489 (Nov. 27, 1973), R. J. Lagow and J. L. Margrave (to R.I. Patents, Inc.).
32. H. F. Priest and A. V. Grosse, *Ind. Eng. Chem.* **39,** 279 (1947); R. Landau and R. Rosen, *Ind. Eng. Chem.* **39,** 281 (1947).
33. N. J. Maraschin and co-workers, *J. Am. Chem. Soc.* **97,** 513 (1975).
34. U.S. Pat. 3,904,501 (Sept. 9, 1975), R. J. Lagow and co-workers (to Massachusetts Institute of Technology).
35. U.S. Pat. Appl. 699,391 (1973), R. J. Lagow and co-workers.
36. U.S. Pat. 3,480,667 (Nov. 25, 1969), W. R. Siegart and W. D. Blackley.
37. R. J. Lagow and J. L. Margrave, *Inorg. Chim. Acta* **10,** 9 (1974).
38. S. Inoue, J. L. Adcock, and R. J. Lagow, *J. Am. Chem. Soc.* **100,** 1948 (1978).
39. Br. Pat. 710,523 (June 16, 1954), A. J. Rudge.
40. U.S. Pat. 2,497,046 (Feb. 7, 1950), E. L. Kopra (to American Cyanamid Co.).
41. U.S. Pat. 2,811,468 (Oct. 29, 1957), S. P. Joffre (to Shulton, Inc.).
42. H. Schonhorn and R. Hansen, *J. Appl. Polym. Sci.* **12,** 1231 (1968).
43. *Chem. Eng. News* **48,** 40 (Jan. 12, 1970); *Chemistry* **43**(4), 30 (1970).
44. C. W. Rechtenwald, D. D. Dixson, and D. J. Manley, *IR-100 Award for Producing Containers with a Fluorinated Surface,* 1978.
45. *Ind. Res. Dev.* **12,** 102 (1978).
46. U.S. Pat. 3,988,491 (Oct. 26, 1976), and 4,020,223 (Apr. 26, 1977), D. D. Dixson and L. J. Hayes (to Air Products and Chemicals, Inc.).
47. L. J. Hayes, *J. Fluorine Chem.* **8,** 69 (1976).
48. N. J. Maraschin and R. J. Lagow, *Inorg. Chem.* **12,** 1459 (1973).
49. J. L. Adcock and R. J. Lagow, *J. Org. Chem.* **38,** 3617 (1973).
50. J. L. Adcock, R. A. Beh, and R. J. Lagow, *J. Org. Chem.* **40,** 3271 (1975); L. A. Shimp and R. J. Lagow, *J. Org. Chem.* **42,** 3437 (1976); E. K. S. Liu and R. J. Lagow, *J. Organomet. Chem.* **145,** 167 (1978); *J. Am. Chem. Soc.* **98,** 8270 (1976).
51. R. J. Lagow and J. L. Margrave, *J. Inorg. Nucl. Chem.* **35,** 2084 (1973); J. L. Margrave, A. P. Conroy, and R. B. Badachhape, *USAF Technical Report No. AFML-TR-71-271,* U.S. Government Printing Office, Washington, D.C., Feb. 1972.

52. D. H. R. Barton and co-workers, *J. Am. Chem. Soc.* **98,** 3035 (1976); D. H. R. Barton and co-workers, *J. Am. Chem. Soc.* **98,** 3036 (1976).
53. P. Tarrant and co-workers, *Chem. Eng. News* **49,** 40 (Sept. 27, 1971).
54. J. Kolonitsch, L. Barash, and G. A. Dolduras, *Chem. Eng. News* **49,** 40 (Sept. 27, 1971).
55. J. L. Adcock and R. J. Lagow, *J. Am. Chem. Soc.* **96,** 7588 (1974).
56. G. G. Gerhardt and R. J. Lagow, *J. Chem. Soc. Chem. Comm.,* 259 (1976).
57. J. L. Adcock and co-workers, *J. Fluorine Chem.* **7,** 197 (1976); L. Shimp and R. J. Lagow, *Inorg. Chem.* **16,** 2974 (1977).
58. J. L. Margrave and R. J. Lagow, *J. Polym. Sci. Polym. Lett. Ed.* **12,** 177 (1974); A. J. Otsuka and R. J. Lagow, *J. Fluorine Chem.* **4,** 371 (1974).
59. U.S. Pat. 3,647,613 (Mar. 7, 1972), J. L. Scotland (to British Resin Products Limited).
60. D. T. Clark and co-workers, *J. Polym. Sci. Polym. Chem. Ed.* **13,** 857 (1975); D. T. Clark and co-workers in L. H. Lee, ed., *Advances in Friction Wear,* Vol. 5A, Plenum Press, 1975, p. 373.
61. G. Parks, Ph.D. thesis, Rice University, Houston, Tex., 1976.
62. J. Pederson, M.A. thesis, Rice University, Houston, Tex., 1979.
63. Can. Pat. 1,002,689 (Dec. 28, 1976), R. J. Lagow and J. L. Margrave (to DAMW Associates, a partnership, U.S.A.); Br. Pat, 1,440,605 (Oct. 20, 1976), R. J. Lagow and J. L. Margrave (to DAMW Associates, U.S.A.).
64. U.S. Pat. 3,992,221 (Nov. 16, 1976), R. B. Badachhape, C. Homsy, and J. L. Margrave (to Vitek Inc., and MarChem, Inc.).
65. A. J. Woytek and J. F. Gentilecore, "A New Blow Molding Process to Resist Solvent Permeation of Polyolefin Containers," *Paper No. 13, presented at Advances in Blow Molding Conference, Rubber and Plastics Institute London, Eng., Dec. 6, 1977.*
66. J. F. Gentilecore, M. A. Triolo, and A. J. Woytek, *Plast. Eng.* **34,** 23 (1978).

JOHN L. MARGRAVE
R. H. HAUGE
R. B. BADACHHAPE
Rice University

R. J. LAGOW
University of Texas

FLUORINATED ALIPHATIC COMPOUNDS

Fluorinated derivatives of methane and ethane are commonly identified by their trademark followed by a series of numbers related to the formula of the compound. In this numbering system, the first digit on the right is the number of fluorine atoms in the compound, and the second digit from the right is one more than the number of hydrogen atoms. The third digit from the right is one less than the number of carbon atoms. When the digit is zero, it is omitted. The remaining available positions in the compound can be taken by chlorine atoms. For example, CCl_3F, CCl_2F_2, $CHClF_2$, $CCl_2F—CClF_2$, and $CClF_2—CClF_2$ are numbered 11, 12, 22, 113, and 114, respectively (see Aerosols). When bromine is present, the same rules apply except that the letter B is used, followed by a number that indicates the number of chlorine atoms replaced by bromine. For example, CF_2Br_2 and CF_3Br are coded 12B2 and 13B1, respectively. This numbering system has been extended to olefins and to compounds with more than two carbon atoms but the method for indicating various isomers becomes quite cumbersome.

Fluorocarbons and Fluorohydrocarbons

Properties. The physical properties of fluorocarbons appear abnormal when compared with those of their hydrocarbon counterparts (1–4). Fluorocarbons are much more volatile than hydrocarbons of similar molecular weight. For example, tetrafluoromethane, mol wt 88, boils at −128°C, whereas *n*-hexane, mol wt 86, boils at +69°C. Fluorocarbons containing up to four carbon atoms boil somewhat higher than their corresponding hydrocarbons; the reverse is true of fluorocarbons with more carbon atoms. Liquid fluorocarbons are two to three times as dense as hydrocarbons with the same carbon skeleton. The compressibility of fluorocarbons is greater than that of hydrocarbons. Their surface tension is low and their dielectric properties are good.

The absolute viscosities of fluorocarbons are higher than those of hydrocarbons but the kinematic viscosities are lower owing to higher densities. The physical properties of some fluorocarbons and fluorohydrocarbons are listed in Tables 1 and 2.

The low forces of attraction between fluorocarbon molecules account for the differences in hydrocarbon and fluorocarbon physical properties. Fluorocarbons boil only slightly higher than the noble gases of similar molecular weight, and they behave as nearly ideal liquids. It has been argued that the physical properties of fluorocarbons are normal but those of hydrocarbons are unusual (4).

The physical properties of fluorohydrocarbons reflect the importance of intermolecular hydrogen bonding. Fluorohydrocarbons boil higher than either their fluorocarbon or hydrocarbon counterparts. Within the series of fluorinated methanes and ethanes, the boiling point reaches a maximum for the structure containing approximately an equal number of hydrogens and fluorines where the number of possible hydrogen bonds is maximized. Fluorohydrocarbons have higher indexes of refraction, higher dielectric constants, and lower densities than their fluorocarbon counterparts.

The chemical and thermal stability of fluorocarbons is considerably higher in general than that of the corresponding hydrocarbons (1–2,9). Fluorocarbons are significantly less reactive than hydrocarbons toward all chemical reagents except alkali

Table 1. Some Physical Properties of Aliphatic Fluorocarbons and Fluorohydrocarbons

Formula	CAS Registry No.	Molecular weight	Boiling point, °C	Melting point, °C	Liquid density, g/mL at °C	Liquid refractive index, n_D at °C	Crit temp, °C	Crit pressure, MPa[a]	Crit vol, L/mol	Refs.
CF_4	[*75-73-0*]	88.00	−128.06	−183.6	1.613_{-130}	1.151_{-73}	−45.6	3.74	0.140	5–6
CHF_3	[*75-46-7*]	70.01	−82.16	−155.2	1.442_{-80}	1.215_{-73}	25.7	4.83	0.133	5–6
CH_2F_2	[*75-10-5*]	52.02	−51.6	−136	1.200_{-50}	1.190_{20}				6–7
CH_3F	[*593-53-3*]	34.03	−78.33	−141.8	0.884_{-80}	1.1727_{20}	44.6	5.86	0.113	6–7
CF_3CF_3	[*76-16-4*]	138.01	−78.2	−100.6	1.600_{-80}	1.206_{-73}	19.7	2.99	0.224	5–6
CHF_2CF_3	[*354-33-6*]	120.02	−48.5	−103		1.5012_{19}				8
CH_2FCF_3	[*811-97-2*]	102.03	−26.5	−101						8
CH_3CF_3	[*420-46-2*]	84.04	−47.4	−111.3	1.176_{-50}		73.1	3.76	0.194	6
CH_3CHF_2	[*75-37-6*]	66.05	−25.8	−117	1.023_{-30}	1.3011_{-72}	113.5	4.50	0.181	6, 8
CH_3CH_2F	[*353-36-6*]	48.06	−37.4	−143.2	0.818_{-37}	1.3033_{-37}	102.2	4.72		6–7
$CF_3CF_2CF_3$	[*76-19-7*]	188.02	−36.7	−183	1.350_{20}		71.9	2.68	0.299	6
$CF_3CF_2CF_2CF_3$	[*355-25-9*]	238.02	−2.2	−128	1.543_{20}		113.2	2.32	0.378	6
cyclo-C_4F_8	[*115-25-3*]	200.03	−5.85	−41.4	1.500_{25}	1.217_{25}	115.2	2.78	0.325	5–6
$CF_3(CF_2)_3CF_3$	[*678-26-2*]	288.03	29.2	−126	1.620_{20}		149	2.04	0.512[b]	6
$CF_3(CF_2)_4CF_3$	[*355-42-0*]	338.04	58	−86	1.680_{25}	1.2514_{22}	174.5	1.90	0.602[b]	6
$CF_3(CF_2)_5CF_3$	[*335-57-9*]	388.05	82.5	−51	1.733_{20}	1.262_{20}	201.6	1.62	0.664	6

[a] To convert MPa to atm, divide by 0.101.

[b] Estimated.

Table 2. Additional Physical Properties of Some Aliphatic Fluorocarbons and Fluorohydrocarbons[a]

Property	CF_4	CF_3—CF_3	CHF_3	C_4F_8 (cyclic)
heat capacity, J/(kg·K)[b]				
liquid at 25°C	1.23_{-80}	0.971_{-73}	6.50	1.11
vapor at constant pressure, 101.3 kPa[c], 25°C	0.707	0.761	0.736	0.795
latent heat of vaporization at bp, kJ/kg[a]	135.9	117.0	239.4	116.2
viscosity, at 25°C, mPa·s (= cP)				
liquid	0.09_{-73}	0.30	0.034	0.373
vapor, 101.3 kPa[c]	0.0172	0.0148	0.0146	0.0117
relative dielectric strength (nitrogen = 1)	1.06	2.02	1.04	2.86
dielectric constant				
liquid		2.34_{25}		
vapor, 50.65 kPa[c]	$1.0012_{24.5}$	1.0021_{23}	1.0073_{25}	1.0032_{25}
surface tension at −73°C, mN/m (= dyn/cm)	4	16	15	7_{25}
solubility in water at 101.3 kPa[c] and 25°C, wt %	0.0015		0.10	

[a] Refs. 5–6.
[b] To convert J to cal, divide by 4.184.
[c] To convert kPa to mm Hg, multiply by 7.5.

metals. Molten alkali metals or alkali-metal hydrocarbon complexes degrade most fluorocarbons and this reaction is used for their chemical analysis (10). Fluorocarbons are not affected by acids or oxidizing agents and are not hydrolyzed below 500°C. Carbon tetrafluoride decomposes only slowly at carbon arc temperatures, and it does not react with Cu, Ni, W, or Mo at 900°C. Fluorocarbons of higher molecular weight are less thermally stable but appreciable decomposition occurs only above 300°C. Partially fluorinated hydrocarbons are less stable and more reactive, especially when only one fluorine atom is present. Hydrogen fluoride can be eliminated by chemical or thermal action from fluorohydrocarbons containing two or more carbons.

Manufacture. The direct fluorination (qv) of hydrocarbons with elemental fluorine is extremely exothermic and difficult to control. Special methods including metal packing techniques, jet reactors, and high dilution have been developed to control the reaction but currently they have little industrial importance (1–2,11–12). The disadvantages of direct fluorination have been overcome by the use of fluorine carriers, in particular, high-valence metal fluorides such as cobalt trifluoride, CoF_3. This reagent replaces hydrogen and halogen atoms by fluorine and adds fluorine to double bonds and aromatic systems (1–2,13).

$$—CH_2— + 4\ CoF_3 \rightarrow —CF_2— + 4\ CoF_2 + 2\ HF$$

Cobalt trifluoride is generated *in situ* by passing fluorine over cobalt difluoride which is contained in a horizontal, mechanically agitated steel reactor. The compound to be fluorinated is passed through the reactor at 150–300°C as a vapor in a stream of nitrogen. After the reaction is completed, the CoF_3 is regenerated by adding fluorine, and the process is essentially continuous. Principally cyclic and higher molecular-weight acyclic fluorocarbons are prepared by this method. Potassium tetrafluorocobaltate, $KCoF_4$, has been introduced recently as a milder fluorinating agent (14).

Fluorocarbons are made commercially also by the electrolysis of hydrocarbons in anhydrous hydrogen fluoride (15–16). Nickel anodes and nickel or steel cathodes are used. Special porous anodes improve the yields. This method is limited to starting materials that are appreciably soluble in hydrogen fluoride. The most useful application has been in the preparation of fluoroalkyl carboxylic and sulfonic acids.

Fluorohydrocarbons are also prepared from acetylene or olefins and hydrogen fluoride (1–2), or from chlorocarbons and anhydrous hydrogen fluoride in the presence of an antimony fluoride catalyst (1–2,17). A commercial synthesis of 1,1-difluoroethane, an intermediate to fluoroolefins, is conducted in the vapor phase over an aluminum fluoride catalyst.

Perfluorocyclobutane is prepared by the thermal cyclodimerization of tetrafluoroethylene [*116-14-3*].

Health and Safety Factors. Completely fluorinated alkanes are essentially nontoxic (18). Rats exposed for four hours to 80% perfluorocyclobutane and 20% oxygen showed only slight effects on respiration, but no pathological changes in organs. However, some fluorochemicals, especially functionalized derivatives and fluoroolefins, can be lethal. Monofluoroacetic acid and perfluoroisobutylene [*382-21-8*] are notoriously toxic (18).

Uses. The chemical inertness, thermal stability, low toxicity, and nonflammability of fluorocarbons coupled with their unusual physical properties suggest many useful applications. However, the high cost of raw materials and manufacture has limited commercial production to a few, small-volume products. Carbon tetrafluoride is used as a degreasing agent for the plasma etching of semiconductor devices (see Ion implantation); it also has been used in low temperature refrigeration (qv) and in wind-tunnel studies. Hexafluoroethane and octafluoropropane have some applications as dielectric gases. Azeotropic combinations of fluorohydrocarbons and chlorofluorocarbons are used as refrigerants. Perfluorocyclobutane was used as an aerosol propellant for foods but has been supplanted by CF_3CClF_2. It is, however, used in minor amounts as a dielectric fluid and as a refrigerant when high stability is required. An interesting application still in the experimental stage is the use of liquid fluorocarbons for oxygen and carbon dioxide transport in artificial blood (19). Perfluorodecalin has received the most attention in this field (20).

The fluorocarbons of greatest commercial interest and volume are the high molecular weight fluoroplastics and elastomers derived from tetrafluoroethylene, hexafluoropropylene, vinylidene fluoride, and vinyl fluoride (21). Poly(tetrafluoroethylene) [*9002-84-0*], Teflon (trademark of E. I. du Pont de Nemours & Co.), and its copolymer [*25067-11-2*] with hexafluoropropylene, Teflon FEP, are fluorinated plastics, particularly notable for their outstanding chemical and thermal stability, electrical inertness, and nonflammability. Teflon has service temperatures in the range of −196 to 260°C. The copolymers [*9011-17-0*] of hexafluoropropylene and vinylidene fluoride, and the terpolymers [*25190-89-0*] of hexafluoropropylene, tetrafluoroethylene, and vinylidene fluoride (Viton fluoroelastomers (trademark of E. I. du Pont de Nemours & Co.)) are rubbers with excellent thermal, chemical, and oxidative stability. They remain useful elastomers for indefinite periods of continuous exposure in air up to about 230°C. A new perfluoroelastomer, Kalrez (trademark of E. I. du Pont de Nemours & Co.), is based on the copolymer [*39613-22-4*] of tetrafluoroethylene and perfluoromethyl vinyl ether (22). Its chemical resistance is similar to that of Teflon, and it is suitable for long-term service at 260–280°C and intermittant service to 316°C.

Several commercially important surfactants (qv) are based on perfluoroalkane carboxylic and sulfonic acids. The ammonium and potassium salts of these acids are powerful anionic surfactants which are used as corrosion inhibitors and wetting, leveling, emulsifying, and dispersing agents. The ammonium fluoroalkane carboxylates are particularly effective emulsifiers for fluorocarbon systems. They are widely used in the aqueous emulsion polymerization of fluorinated olefins(see also Emulsions). In general, the fluorocarbon surfactants are employed in strong oxidizing, concentrated alkaline, or high temperature environments where hydrocarbon surfactants would decompose. The fluorosurfactants are sold under the Zonyl or Fluorad trademarks (23–24).

Chlorofluorocarbons

Properties. The physical properties of aliphatic fluorine compounds containing chlorine are similar to those of the fluorocarbons (2,5,25). They usually have high densities and low boiling points, viscosities, and surface tensions. The irregularity in the boiling points of the fluorinated methanes and ethanes, however, does not appear in the chlorofluorocarbons. Their boiling points consistently increase with the number of chlorines present. The properties of some chlorofluorocarbons are shown in Tables 3 and 4.

Although the chlorofluorocarbons are not as stable as the fluorocarbons, they

Table 3. Some Physical Properties of Aliphatic Chlorofluorohydrocarbons

Formula	CAS Registry No.	Molecular weight	Boiling point, °C	Melting point, °C	Liquid density, g/mL at °C	Liquid refractive index, n_D at °C	Crit temp, °C	Crit pressure, MPa[a]	Crit vol, L/mol	Refs.
CCl_3F	[*75-69-4*]	137.36	23.82	−111	1.476_{25}	1.374_{25}	198.0	4.41	0.247	5
CCl_2F_2	[*75-71-8*]	120.91	−29.79	−158	1.311_{25}	1.287_{25}	112.0	4.11	0.217	5
$CClF_3$	[*75-72-9*]	104.46	−81.4	−181	1.298_{-30}	1.199_{-73}	28.9	3.87	0.181	5
$CHCl_2F$	[*75-43-4*]	102.92	8.92	−135	1.366_{25}	1.354_{25}	178.5	5.17	0.197	5
$CHClF_2$	[*75-45-6*]	86.47	−40.75	−160	1.194_{25}	1.256_{25}	96.0	4.97	0.165	5
CH_2ClF	[*593-70-4*]	68.48	−9.1	−133	1.271_{20}					8
$CCl_3—CCl_2F$	[*354-56-3*]	220.29	137	100	1.740_{25}					8
$CCl_2F—CCl_2F$	[*76-12-0*]	203.82	92.8	26	1.634_{30}	1.413_{25}	278	3.44	0.370	5
$CCl_3—CClF_2$	[*76-11-9*]	203.82	91.5	40.6	1.649_{20}					8
$CCl_2F—CClF_2$	[*76-13-1*]	187.38	47.57	−35	1.565_{25}	1.354_{25}	214.1	3.41	0.325	5
$CCl_3—CF_3$	[*354-58-5*]	187.38	45.8	14.2	1.579_{20}	1.361_{20}				7
$CClF_2—CClF_2$	[*76-14-2*]	170.92	3.77	−94	1.456_{25}	1.288_{25}	145.7	3.26	0.293	5
$CCl_2F—CF_3$	[*374-07-02*]	170.92	3.6	−94	1.455_{25}	1.309_{0}	145.6	3.29		7, 26
$CClF_2—CF_3$	[*76-15-3*]	154.47	−39.1	−106	1.291_{25}	1.214_{25}	80.0	3.12	0.259	5
$CHClF—CClF_2$	[*354-23-4*]	152.93	28.2		1.498_{10}	1.327_{20}				8
$CHF_2—CClF_2$	[*354-25-6*]	136.48	−10.2	−117	1.379_{20}					8
$CH_2Cl—CClF_2$	[*1649-08-7*]	134.94	46.8	−101	1.416_{20}	1.362_{20}				8
$CH_2Cl—CF_3$	[*75-88-7*]	118.49	6.1	−101	1.389_{0}	1.309_{0}				8
CH_3CClF_2	[*75-68-3*]	100.50	−9.8	−131	1.113_{25}		137.1	4.12		5

[a] To convert MPa to atm, divide by 0.101.

Table 4. Additional Physical Properties of Some Chlorofluorohydrocarbons[a]

Property	CCl_3F	CCl_2F_2	$CClF_3$	$CHCl_2F$	$CHClF_2$	$CCl_2F—CCl_2F$	$CCl_2F—CClF_2$	$CClF_2—CClF_2$	$CClF_2—CF_3$
heat capacity, J/(kg·K)[b]									
liquid at 25°C	0.870	0.971	1.033_{-30}	1.071	1.255		0.912	1.016	1.192
vapor at constant pressure, 101.3 kPa[c], 25°C	0.565	0.607	0.661	0.586	0.657		0.674_{60}	0.711	0.686
latent heat of vaporization at bp, kJ/kg[b]	180.3	165.1	148.4	242.1	233.5	155[d]	146.7	136.0	126.0
viscosity at 25°C, mPa·s (= cP)									
liquid	0.430	0.214	0.195_{-45}	0.313	0.198	121	0.68	0.36	0.193
vapor, 101.3 kPa[c]	0.0107	0.023	0.0145	0.0114	0.0127		$0.010_{10\ kPa}$[c]	0.012	0.0125
relative dielectric strength (nitrogen = 1)	3.71	2.46	1.65	1.85	1.27	5[d]	$3.9_{44\ kPa}$[c]	3.34	2.54
dielectric constant									
liquid	2.28_{29}	2.13_{29}		5.34_{28}	6.11_{24}	2.54_{25}	2.41_{25}	2.26_{25}	
vapor, 50.65 kPa[c]	1.0036_{24}	1.0032_{26}	1.0024_{29}	1.0070_{30}	1.0071_{25}			$1.0043_{26.8}$	1.0035_{27}
surface tension at 25°C, mN/m (= dyn/cm)	18	9	14_{-73}	18	8	23_{30}	17.3	12	5
solubility in water at 101.3 kPa[c] and 25°C, wt %	0.11	0.028	0.009	0.95	0.30	0.012[e]	0.017[e]	0.013	0.006

[a] From ref. 5.
[b] To convert J to cal, divide by 4.184.
[c] To convert kPa to mm Hg, multiply by 7.5.
[d] Estimated.
[e] At the vapor pressure.

are unusually stable compounds. Dichlorodifluoromethane, CCl_2F_2, is stable at 500°C in quartz; CCl_3F and $CHClF_2$ begin to decompose at 450 and 290°C, respectively (5). The pyrolysis of $CHClF_2$ at 650–700°C in metal tubes is the basis of a commercial synthesis of tetrafluoroethylene:

$$2\,CHClF_2 + heat \rightarrow CF_2{=}CF_2 + 2\,HCl$$

The chlorofluorocarbons react with molten alkali metals and CCl_2F_2 reacts vigorously with molten aluminum, but with most metals they do not react below 200°C. An exception is the dechlorination of chlorofluorocarbons with two or more carbon atoms in the presence of Zn, Mg, or Al in polar solvents. A commercial synthesis of chlorotrifluoroethylene [*79-38-9*] employs this reaction:

$$CClF_2{-}CCl_2F + Zn \xrightarrow{\text{alcohol}} CF_2{=}CClF$$

Most chlorofluorocarbons are hydrolytically stable; CCl_2F_2 being considerably more stable than either CCl_3F or $CHCl_2F$. Chlorofluoromethanes and ethanes disproportionate in the presence of aluminum chloride. For example, CCl_3F and CCl_2F_2 give $CClF_3$ and CCl_4; $CHClF_2$ disproportionates to CHF_3 and $CHCl_3$.

The carbon—chlorine bond in most chlorofluorocarbons can be homolytically cleaved under photolytic conditions (185–225 nm) to give chlorine radicals. This photochemical decomposition is the basis of the prediction that chlorofluorocarbons that reach the upper atmosphere deplete the earth's ozone shield (see below).

Manufacture. The most important commercial method for manufacturing chlorofluorocarbons is the successive replacement of chlorine by fluorine using hydrogen fluoride (1–2,17). The traditional, liquid-phase process uses antimony pentafluoride or a mixture of antimony trifluoride and chlorine as catalysts. This process is being replaced by a continuous vapor-phase process that employs gaseous hydrogen fluoride in the presence of chromium oxide or halide, ferric chloride, or thorium tetrafluoride catalysts. Carbon tetrachloride, chloroform, and hexachloroethane (or tetrachloroethylene plus chlorine) are commonly used starting materials for one- and two-carbon chlorofluorocarbons. The extent of chlorine exchange can be controlled by varying the hydrogen fluoride concentration, the contact time, or the reaction temperature.

$$CHCl_3 \xrightarrow{HF} CHCl_2F + CHClF_2 + CHF_3$$

$$CCl_4 \xrightarrow{HF} CCl_3F + CCl_2F_2 + CClF_3$$

$$CCl_3CCl_3 \xrightarrow{HF} CCl_2FCCl_3 + CCl_2FCCl_2F + CClF_2CCl_2F + CClF_2CClF_2$$

$$CCl_3CHCl_2 \xrightarrow{HF} CCl_2FCHCl_2 + CClF_2CHCl_2$$

The direct chlorination of fluorohydrocarbons and fluoroolefins has also been used commercially, eg, in the preparations of CH_3CClF_2 from CH_3CHF_2 and $CClF_2CClF_2$ from tetrafluoroethylene.

Economic Aspects. The estimated world and United States production of major industrial fluorochemicals is shown in Table 5. Trichlorofluoromethane, dichlorodifluoromethane, and chlorodifluoromethane account for at least 90% of the total production. Between 1964 and 1974 the production of $CHClF_2$ and CCl_2F_2 has increased at approximately 8 and 9% per annum, respectively. The production of CCl_3F has

Table 5. Estimated United States and World Production of Important Chlorofluorocarbons[a], Thousands of Metric Tons

Year	$CFCl_3$ (U.S.)	$CFCl_3$ (World)	CF_2Cl_2 (U.S.)	CF_2Cl_2 (World)	CF_2HCl (U.S.)	CF_2HCl (World)
1973	148	354	221	448	62	77
1974	158[b]	378	231[b]	473	64[b]	88
1975	121	323	178	420	60	74
1976	116	350	174	450		93
1977	94	331	155	424		97
1978	88[c]	308	148[c]	401	93[c]	

[a] Refs. 27–29.
[b] In addition to 29×10^3 t CCl_2FCClF_2, 10.5×10^3 t $CClF_2CClF_2$, and 6.4×10^3 t miscellaneous chlorofluorocarbons.
[c] Current commercial price ($/kg): $CFCl_3$ (0.97), CF_2Cl_2 (1.00), CF_2HCl (1.80).

increased about 4% whereas $CCl_2F—CClF_2$ has experienced a rapid growth of over 18% per annum. Since 1974 the production of CCl_3F and CCl_2F_2 has not increased, whereas, $CHClF_2$ has increased about 25%. 1,2-Dichlorotetrafluoroethane production has remained essentially constant since 1965 (27). An estimate of the distribution in 1973 according to use is shown in Table 6. The world market pattern has not changed significantly since 1973; however, it has changed dramatically in the United States since the ozone-depletion hypothesis was introduced in 1974 (see below). The chlorofluorocarbon sales to the aerosol industry in 1977 dropped to about 30% of the 1973 volume (29). United States and foreign trademarks and manufacturers are listed in Table 7.

Health and Safety Factors; Environmental Problems. The toxicity of aliphatic halofluorocarbons generally decreases as the number of fluorine atoms increases (18), as shown for the methane series in Table 8.

The Chlorofluorocarbon–Ozone Problem. The possibility that certain chlorofluorocarbons accumulated in the upper atmosphere can deplete the earth's ozone layer has had a major impact on the fluorochemicals industry (27–29). The chlorofluorocarbons of concern are CCl_2F_2 and CCl_3F which have been released in quantity and have the stability to reach the stratosphere. Perfluorocarbons, chlorofluoroethanes, and chlorofluorohydrocarbons, eg, $CHClF_2$, are not expected to contribute significantly to ozone depletion. Since aerosols account for about 75% of the release of CCl_2F_2 and CCl_3F (refrigerants and foams account for ca 14 and 12%, respectively), the propellant

Table 6. Estimated Consumption of Fluorocarbons by Use in 1973, % of Total Production[a]

Application	U.S.	World
aerosols and propellants	49	55
refrigerants	28	29
plastics and resins	4	7
solvents	5	3
foam blowing agents	7	4
exports and other uses	7	2

[a] Ref. 27.

Table 7. Trademarks and Manufacturers of Chlorofluorocarbons

Country and trademark	Manufacturer
France	
Flugene	Pechiney-Saint Gobain S.A.
Forane	Ugine
Federal Republic of Germany	
Frigedohn	Fluorwerke Dohna
Frigen	Farbwerke Hoechst
Fluogen	Chemische Fabrik von Heyden
German Democratic Republic	
Fridohna	VEB Chemiewerk Nunchritz
Italy	
Algofren	Montecatini
Japan	
Asahiflon	Asahi Glass
Daiflon	Daikin Kogyo
Flon	Mitsui Fluorochemicals
United Kingdom	
Arcton	Imperial Chemical Industries
Isecon	Imperial Smelting
United States	
Freon	E. I. du Pont de Nemours & Co., Inc.
Genetron	Allied Chemical Corporation
Isotron	Pennwalt Corporation
Ucon	Union Carbide Corporation
USSR	
Eskimon	
Khladon	

use has been the most severely criticized (see Aerosols). The EPA has ruled that aerosol products containing CCl_2F_2 and CCl_3F propellants cannot be manufactured in the United States after December 15, 1978 (see Air pollution; Ozone).

Uses. The largest commercial application of the chlorofluorocarbons has been in the aerosol field as a propellant for a wide variety of products. Personal products such as antiperspirants and hair sprays account for ca 85% of this use. The characteristic stability, nonflammability, low toxicity, and lack of odor provide a unique combination of properties desirable for propellants. The most common propellants are CCl_2F_2, CCl_3F, and $CClF_2CClF_2$, used either by themselves or in combination. The principal propellant for nonfood aerosols is CCl_2F_2. A combination of CCl_3F and CCl_2F_2 is used to reduce the propellant pressure and to increase the compatability of other compounds in the aerosol formulation. Approximately 60% of the total manufactured CCl_2F_2 and CCl_3F is used in aerosols. 1,2-Dichlorotetrafluoroethane is used primarily in the cosmetic field, often in combination with CCl_2F_2.

Another important use of chlorofluorocarbons is as refrigerants (see Table 9). The air-conditioning industry owes much of its tremendous growth to the chlorofluorocarbons. Air conditioning (qv) and refrigeration consume over 90% of the $CHClF_2$ manufactured and ca 32% of the CCl_2F_2 and 8% of the CCl_3F manufactured (see Refrigeration).

Approximately 30% of the trichlorofluoromethane is produced as a blowing agent for rigid and flexible polyurethane foams. The chlorofluorocarbon trapped in the rigid

Table 8. Toxicity of Halogenated Methanes[a]

Compound	TLV[b], ppm	Underwriters' Laboratory Classification[c]
CCl_4	10	3
$CHCl_3$	10[d]	3
CH_2BrCl	200	3
CBr_2F_2	100	4
CH_2Cl_2	100	4–5
CCl_3F	1000	5a
$CHClF_2$	1000	5a
$CHCl_2F$	10	
$CBrClF_2$		5a
CCl_2F_2	1000	6
$CClF_3$	1000	6
$CBrF_3$	1000	6

[a] Refs. 5 and 30.
[b] Except for CO_2, no compound has a higher TLV than 1000 ppm.
[c] The classification for CO_2 is 5a.
[d] Suspected carcinogen in man.

Table 9. Uses of Chlorofluorocarbons as Refrigerants

Refrigerant number	Compound(s)	Use
11	CCl_3F	in large industrial and commercial systems
12[a]	CCl_2F_2	home refrigeration, commercial freezers and display cases, dehumidifiers, automotive units
13	$CClF_3$	for temperatures of −70 to −100°C
22	$CHClF_2$	home air-conditioning units
500[b]	$(CH_3CF_2H + CCl_2F_2)$	refrigerant
502[b]	$(CCl_2F_2 + CF_3CClF_2)$	commercial freezers and display cases
503	$(CF_3H + CClF_3)$[b]	low temperature refrigerant

[a] Accounts for ca 50% of the total refrigerants sold today; introduced in 1931.
[b] Azeotropic combination.

foam significantly increases its insulation properties. Minor quantities of rigid polyurethane foam are manufactured by a frothing process using CCl_2FCClF_2 and $CClF_2CClF_2$. Extruded polystyrene foams use CCl_2F_2 as the principal blowing agent. Both $CClF_2CClF_2$ and CCl_2F_2 are used in low density polyethylene and polystyrene foams (see Foamed plastics).

The key solvent applications of chlorofluorocarbons are in the aerospace and electronics industry, where CCl_2FCClF_2 and CCl_3F are widely used as solvents and degreasing agents. Oil and greases are dissolved without affecting most plastic, elastomeric, or metal components. 1,1,2-Trichlorotrifluoroethane also is used in the dry cleaning (qv) of specialty items such as furs and leathers (see Solvents, industrial).

The use of $CHClF_2$, CCl_2FCClF_2, and CH_3CClF_2 in the synthesis of commercially important fluorovinyl monomers has been described (21). Other minor applications of chlorofluorocarbons include their use as dielectric fluids and in wind tunnels and bubble chambers.

Fluorocarbons Containing Other Halogens

Properties. The physical and chemical properties of bromo- and iodofluorocarbons are similar to those of the chlorofluorocarbons except for higher densities and generally decreased stability. The stability of these compounds decreases as the ratio of bromine or iodine to fluorine increases. The reactivity of carbon-halogen bonds toward exchange by fluorine or homolytic cleavage increases in the order C—Cl, C—Br, C—I. Iodofluorocarbons and most bromofluorocarbons readily lose iodine or bromine radicals under photolytic, thermal, or radical initiation to give the corresponding carbon-centered radical. The physical properties of several examples are shown in Tables 10 and 11.

Manufacture. Brominated fluoromethanes are prepared industrially by the halogen exchange of tetrabromomethane or by the bromination of CH_2F_2 or CHF_3 at elevated temperatures (1–2). Other bromo- or iodofluorocarbons can be prepared by halogenating suitable fluorocarbons, including fluoroolefins:

$$CF_3CH_2Cl + Br_2 \xrightarrow{500°C} CF_3CHBrCl$$

[151-67-7]

$$CI_4 + IF_5 \rightarrow CF_3I$$

[2314-97-8]

$$CF_2{=}CF_2 + Br_2 \rightarrow CBrF_2CBrF_2$$

[25947-30-7]

$$CF_2{=}CF_2 + I_2 + IF_5 \rightarrow CF_3CF_2I$$

[354-64-3]

$$CF_3(CF_2)_6CF_2I + Br_2 \xrightarrow{300°C} CF_3(CF_2)_6CF_2Br$$

[423-55-2]

The higher molecular weight perfluoroalkyl iodides are prepared by the telomerization of tetrafluoroethylene with lower molecular weight perfluoroalkyl iodides (33).

$$CF_3CF_2I + nCF_2{=}CF_2 \rightarrow CF_3(CF_2CF_2)_nCF_2I$$

Health and Safety Factors. Fluorocarbons containing bromine or iodine are more toxic than the corresponding chloro compounds. When the ratio of the fluorine to other halogens is high, the toxicity can be quite low, especially for bromofluorocarbons. Perfluoro-1-bromooctane *[423-55-2]* has an LD_{50} of greater than 64 mL/kg when administered into the gastrointestinal tract and has little effect when instilled into the lungs (34). Other examples are included in Table 8.

Uses. The most important industrial products of this class are $CBrClF_2$ and $CBrF_3$; both are used as fire-extinguishing agents. The latter is considerably more effective than CO_2, and is nontoxic as well as its decomposition products (35–36). It is used in commercial aircraft for inflight engine fires and in portable fire extinguishers for both military and civilian markets. The most effective fire extinguisher of this class is $BrCF_2CF_2Br$ but it is too expensive for this commercial use (see Plant safety).

Halothane, $CF_3CHClBr$, is a widely used anesthetic which has the advantages of nonflammability, high anesthetizing power, and general lack of postnarcotic effects.

Table 10. Physical Properties of Aliphatic Bromo- and Iodofluorohydrocarbons

Formula	CAS Registry No.	Molecular weight	Boiling point, °C	Melting point, °C	Liquid density, g/mL at °C	Liquid refractive index, n_D at °C	Refs.
CBr_3F	*[353-54-8]*	270.74	106	−75	2.765_{20}	1.5256_{20}	7, 8
CBr_2ClF	*[353-55-9]*	226.28	80				8
CBr_2F_2	*[75-61-6]*	209.83	24.5	−110	2.306_{15}		31
$CBrF_3$	*[75-63-8]*	148.92	−57.8	−168	1.538_{25}	1.238_{25}	5
$CBrCl_2F$	*[353-58-2]*	181.82	52				8
$CBrClF_2$	*[353-59-3]*	165.37	−4	−161	1.850_{15}		31
$CHBr_2F$	*[1868-53-7]*	191.84	64.9	26.5	2.421_{20}	1.4685_{20}	7
$CHBrF_2$	*[1511-62-2]*	130.92	−15		1.551_{16}		7
$CHBrClF$	*[593-98-6]*	147.38	−45	36.1	1.977_{0}	1.4144_{25}	7
$CBrCl_2—CF_3$	*[354-50-7]*	231.83	69.2		1.950_{20}	1.3977_{20}	8
$CBrF_2—CF_3$	*[354-55-2]*	198.92	−21		1.810_{0}	$1.2966_{29.8}$	8
$CHBrCl—CF_3$	*[151-67-7]*	197.39	50.2		1.860_{20}	1.3700_{20}	32
$CBrF_2—CBrF_2$	*[124-73-2]*	259.85	47.3	−110	2.163_{25}	1.367_{25}	5
$CHClF—CBrF_2$	*[354-06-03]*	197.39	52.5		1.864_{25}	1.3685_{25}	8
$CH_2Br—CClF_2$	*[421-01-2]*	179.40	68.4	−76	1.830_{20}	1.4018_{20}	8
$CH_2Br—CHF_2$	*[359-07-9]*	144.95	57.3	−75	$1.824_{18.5}$	$1.3940_{10.5}$	8
CF_3I	*[2314-97-8]*	195.91	−22.5		2.361_{-32}	1.379_{-42}	8
CCl_2FI	*[420-48-4]*	228.82	90.0	−107	2.313_{20}	1.510_{20}	8
CHF_2I	*[1493-03-4]*	177.92	−122	21.6	3.238_{-19}		7
$CHFI_2$	*[1493-01-2]*	285.83	100.3	−34.5	3.197_{22}		7–8
CH_2FI	*[373-53-5]*	159.93	53.4		2.366_{20}	1.491_{20}	7
$CClFI—CClF_2$	*[354-61-0]*	278.83	99		2.196_{25}	1.447_{25}	8
$CF_2I—CF_2I$	*[354-65-4]*	353.82	112		2.629_{25}	1.4895_{25}	8
$CF_2I—CF_3$	*[354-64-3]*	245.92	13		2.072_{28}	$1.3378_{0.5}$	8
$CH_2I—CF_3$	*[353-83-3]*	209.94	55		2.142_{25}	1.3981_{25}	8

Table 11. Additional Physical Properties of Aliphatic Bromofluorocarbons[a]

Property	$CBrF_3$	$CBrF_2$—$CBrF_2$
heat capacity, J/(kg·K)[b]		
liquid at 25°C	0.870	0.695
vapor at constant pressure, 101.3 kPa[c], 25°C	0.469	
latent heat of vaporization at bp, kJ/kg[b]	118.7	105[d]
viscosity at 25°C, mPa·s (= cP)		
liquid	0.157	0.72
vapor, 101.3 kPa[c]	0.0154	
relative dielectric strength (nitrogen = 1)	1.83	$4.02_{44\ kPa}$
dielectric constant, liquid		2.34_{25}
solubility in water at 101.3 kPa[c] and 25°C, wt %	0.03	
critical temperature, °C	67.0	214.5
critical pressures, MPa[e]	3.96	3.48
critical volume, L/mol	0.200	0.324
surface tension at 25°C, mN/m (= dyn/cm)	4	18

[a] From ref. 5.
[b] To convert J to cal, divide by 4.184.
[c] To convert kPa to mm Hg, multiply by 7.5.
[d] Estimated.
[e] To convert MPa to atm, divide by 0.101.

However, it can have a toxic effect if not properly administered (32) (see Anesthetics). Perfluorooctyl bromide has shown considerable promise as a radiopaque agent that is less irritating and safer to the lungs than currently available agents (34) (see Radiopaques).

Iodofluorocarbons have no direct commercial applications, although several telomer iodides, such as $CF_3(CF_2)_7I$ [*507-63-1*] and $CF_3(CF_2)_7CH_2CH_2I$ [*2043-53-0*], are intermediates in the manufacture of oil and water repellents, surfactants, and fire-extinguishing foams (36–37). The alcohols derived from perfluoroalkane carboxylic acids, eg, $CF_3(CF_2)_nCH_2OH$, or from the telomer iodides, eg, $CF_3(CF_2)_nCH_2CH_2OH$, are used to prepare fluorinated acrylate esters. Aqueous dispersions of the fluoroacrylate polymers, such as Scotchgard or Zepel, are used as soil, water, and oil repellents for fabrics (38–39) (see Waterproofing).

BIBLIOGRAPHY

"Aliphatic Fluorinated Hydrocarbons" under "Fluorine Compounds, Organic" in *ECT* 1st ed., Vol. 6, pp. 752–757, by A. F. Benning, E. I. du Pont de Nemours & Co., Inc., and L. J. Hals and W. H. Pearlson, Minnesota Mining & Manufacturing Co.; "Fluorinated Hydrocarbons" under "Fluorine Compounds, Organic" in *ECT* 2nd ed., Vol. 9, pp. 739–751, by R. C. Downing, E. I. du Pont de Nemours & Co., Inc.

1. W. A. Sheppard and C. M. Sharts, *Organic Fluorine Chemistry,* W. A. Benjamin, Inc., New York, 1969.
2. M. Hudlicky, *Chemistry of Organic Fluorine Compounds,* 2nd ed., Ellis Horwood Ltd., Chichester, Eng., 1976.
3. H. G. Bryce, "Industrial and Utilitarian Aspects of Fluorine Chemistry" in J. H. Simons, ed., *Fluorine Chemistry,* Vol. V, Academic Press, Inc., New York, 1965, pp. 297–492.
4. T. M. Reed, "Physical Chemistry of Fluorocarbons" in ref. 3, pp. 133–236.
5. *Freon Fluorocarbons, B-2,* E. I. du Pont de Nemours & Co., Inc., Wilmington, Del., 1969.
6. *Selected Values of Properties of Chemical Compounds,* Thermodynamics Research Center Data Project, Texas Engineering Experimental Sta., Texas A&M University, College Station, Tex., 1977.
7. R. Weast, ed., *Handbook of Chemistry and Physics,* CRC Press, Inc., West Palm Beach, Florida, 1978.

8. J. H. Simons, ed., *Fluorine Chemistry,* Vol. I, Academic Press, Inc., New York, 1950, pp. 463–548.
9. J. C. Tatlow, *Chem. Ind. (London),* 522 (July 1978).
10. P. Johncock, W. K. R. Musgrave, and A. Wiper, *Analyst (London)* **84,** 245 (1959).
11. J. M. Tedder, *Adv. Fluorine Chem.* **2,** 104 (1961).
12. R. E. Banks and M. G. Barlow, eds., *Fluorocarbon and Related Chemistry,* Vols. I–III, The Chemical Society, London, Eng., 1971, 1974, 1976.
13. M. Stacey and J. C. Tatlow, *Adv. Fluorine Chem.* **1,** 166 (1960).
14. A. J. Edwards and co-workers, *Chem. Commun.,* 1028 (1972).
15. S. Nagase, *Fluorine Chem. Rev.* **1,** 77 (1967).
16. A. J. Rudge, "Electrochemical Fluorination" in A. T. Kuhn, ed., *Industrial Electrochemical Processes,* Elsevier, Amsterdam, The Netherlands, 1971.
17. A. K. Barbour, L. J. Belf, and M. W. Buxton, *Adv. Fluorine Chem.* **3,** 181 (1963).
18. J. W. Clayton, Jr., *Fluorine Chem. Rev.* **1,** 197 (1967).
19. R. G. Reiss and M. LeBlanc, *Angew. Chem. Int. Ed. Engl.* **17,** 621 (1978).
20. L. C. Clark, Jr. and co-workers, "Intravenous Infusion of Cis–Trans Perfluorodecalin Emulsions in the Rhesus Monkey" in R. Filler, ed., *Biochemistry Involving Carbon—Fluorine Bonds,* American Chemical Society, Washington, D.C., 1976.
21. L. A. Wall, ed., *Fluoropolymers,* John Wiley & Sons, Inc., New York, 1972.
22. *Du Pont Kalrez® Perfluoroelastomer Parts,* E. I. du Pont de Nemours & Co., Inc., Wilmington, Del., 1978.
23. *Zonyl® Fluorosurfactants,* E. I. du Pont de Nemours & Co., Inc., Wilmington, Del., 1978.
24. *An Introduction to Fluorad® Brand Fluorochemical Surfactants,* 3M Co., St. Paul, Minn., 1975.
25. R. C. Downing, "Simple Organic Compounds Containing Fluorine" in J. S. Sconce, ed., *Chlorine, ACS Monograph No. 154,* Reinhold Publishing Co., New York, 1962; A. M. Lovelace and co-workers, "Aliphatic Fluorine Compounds," *ACS Monograph No. 138,* Reinhold, New York, 1958.
26. *Genetron Refrigerants,* Specialty Chemical Division, Allied Chemical Corp., Morristown, N.J.
27. *Preliminary Economic Impact Assessment of Possible Regulatory Action to Control Atmospheric Emission of Selected Hydrocarbons, EPA-450/3-75-073,* A. D. Little & Co., Cambridge, Mass., 1975.
28. *Halocarbons: Effects on Stratospheric Ozone,* National Academy of Sciences, Washington, D.C., 1976.
29. *Stratospheric Ozone Depletion by Halocarbons, Chemistry and Transport,* National Academy of Sciences, Washington, D.C., 1979.
30. *Threshold Limit Values for Chemical Substances in Work Room Air Adopted by ACGIH for 1978,* American Conference of Governmental Industrial Hygienists, Cincinnati, Ohio, 1978.
30. *Halocarbons: Environmental Effects of Chlorofluoromethane Release,* National Academy of Sciences, Washington, D.C., 1976.
31. J. M. Hamilton, Jr., *Adv. Fluorine Chem.* **3,** 117 (1963).
32. E. R. Larsen, *Fluorine Chem. Rev.* **3,** 1 (1969).
33. R. Chambers, *Fluorine in Organic Chemistry,* John Wiley & Sons, Inc., New York, 1973, pp. 176–178.
34. D. M. Long and co-workers, "Radiopaque Applications of Brominated Fluorocarbon Compounds in Experimental Animals and Human Subjects" in ref. 20.
35. *Du Pont Freon FE 1301 Fire Extinguishing Agent,* E. I. du Pont de Nemours & Co., Inc., Wilmington, Del., 1969.
36. J. M. Hamonack, "Fire Extinguishing Agents" in A. Standen, ed., *Kirk-Othmer Encyclopedia of Chemical Technology,* 2nd ed., Supplement Volume, John Wiley & Sons, Inc., New York, 1971, pp. 372–374.
37. H. C. Fielding, "Organofluorine Surfactants and Textile Chemicals" in R. E. Banks, ed., *Organofluorine Chemicals and their Industrial Applications,* Ellis Horwood Ltd., Chichester, Eng., 1979.
38. J. D. Reid and W. J. Connick, Jr., "Water Proofing and Water Repellency" in A. Standen, ed., *Kirk-Othmer Encyclopedia of Chemical Technology,* 2nd ed., Vol. 22, John Wiley & Sons, Inc., New York, 1970, pp. 145–146.
39. S. S. Smith, "Soil Release Finishes" in ref. 36, pp. 964–973.

BRUCE E. SMART
E. I. du Pont de Nemours & Co., Inc.

FLUOROETHANOLS

2-Fluoroethanol

2-Fluoroethanol [*371-62-0*] (ethylene fluorohydrin, β-fluoroethyl alcohol), FCH_2CH_2OH, is a colorless liquid with an alcohol-like odor; mp, −26.45°C; bp, 103.55°C; d_4, 1.1297; n_D^{18}, 1.13647; heat of combustion, −1214.0 kJ/mol (−290.16 kcal/mol) (1–2). Like the other fluoroethanols, it is miscible with water, stable to distillation, and low in flammability. It is the least acidic of the fluoroethanols, although more acidic than ordinary alcohols. Its most notable difference from the other fluoroethanols is its extreme toxicity (3–11). In mice, an LD_{50} of 10 mg/kg has been measured (9–11). The toxicity is due to its facile oxidation in animals to derivatives of fluoroacetic acid, a known inhibitor of the tricarboxylic acid cycle of respiration. No effective antidote to this poisoning is known, although ethanol appears to diminish the lethal effects of 2-fluoroethanol in rats and monkeys (8).

In its chemical reactions, 2-fluoroethanol behaves like a typical alcohol. Oxidation (11) yields fluoroacetaldehyde or fluoroacetic acid; reaction with phosphorus tribromide (11) gives 1-bromo-2-fluoroethane; addition to olefins results in ethers (12); and additions to isocyanates give carbamates (13). The alcohol can be prepared in 50% yield by the reaction of potassium fluoride with 2-chloroethanol at 175°C in high-boiling glycol solvents (14). Alternatively, the addition of hydrogen fluoride to ethylene oxide (15) or the fluorination of 2-bromoethyl or 2-chloroethyl acetate with silver, potassium, or mercuric fluoride followed by hydrolysis will give the alcohol (1–3). 2-Fluoroethanol is not currently produced in commercial quantities, although it is available in research quantities for ca $2/g (1979).

Because of its high toxicity, special procedures should be followed by users of 2-fluoroethanol. Suggested precautions include working with it in sealed reactors at subatmospheric pressure and careful monitoring to ensure that contamination of the surroundings is minimized (16). Another potential hazard is the formation of the alcohol as a minor by-product in reactions such as those involving boron trifluoride and ethylene oxide (17). Despite these problems, several potential uses for the alcohol and its derivatives have been reported. The alcohol has been used to control rodent populations (18) (see Poisons, economic) or, when labeled with ^{18}F as a radio-diagnostic agent (19) (see Radioactive drugs and tracers). Various derivatives show promise as herbicides (qv) or as agents to control mites and other plant pests (13,20–23) (see Insect control technology).

2,2-Difluoroethanol

2,2-Difluoroethanol [*359-13-7*], F_2CHCH_2OH, is a colorless liquid with an alcohol-like odor; mp, 28.2°C, bp, 96°C; d_1^{17}, 1.3084; n_D^{17}, 1.3320; heat of combustion, −1026 kJ/mol (−245.3 kcal/mol). It is stable to distillation and miscible with water and many organic solvents. As expected, its acidity lies between that of 2-fluoroethanol and 2,2,2-trifluoroethanol both in the gas phase (24) and in 50% aqueous ethanol solution (25), where its K_a of 1.0×10^{-12} is about 4.8 times smaller than that of trifluoroethanol.

2,2-Difluoroethanol is prepared by the mercuric oxide catalyzed hydrolysis of

2-bromo-1,1-difluoroethane (26–27). Its chemical reactions are similar to those of most alcohols. It can be oxidized to difluoroacetic acid (28); it forms alkoxides with alkali and alkaline earth metals (26); with alkoxides of other alcohols it forms mixed ethers such as the methyl ether, bp 47°C, or the ethyl ether, bp 66°C (26). The difluoromethyl ether (1,1,4,4-tetrafluoro-2-oxabutane), made from the alcohol and chlorodifluoromethane in aqueous base, has been investigated as an inhalation anesthetic (29–30) as have several ethers made by addition of the alcohol to various fluoroalkenes (31–32) (see Anesthetics). At present, neither the alcohol nor these ethers are commercially significant.

2,2,2-Trifluoroethanol

2,2,2-Trifluoroethanol [*75-89-8*], CF_3CH_2OH, is a colorless liquid with an ethanol-like odor; mp, −45°C; bp, 73.6°C; d_4^{25}, 1.3823; n_D^{20}, 1.2907; flash point (open cup), 41°C; flash point (closed cup), 33°C; no fire point (33); heat of combustion, −886.6 kJ/mol (−211.9 kcal/mol) (34); and dielectric constant (25°C), 26.14 (35). Many other physical and thermodynamic properties of the alcohol and its solutions have been published (33,36–39). It is the most acidic fluoroethanol with an ionization constant of 4.3×10^{-13} (40). It is stable to distillation and miscible with water and many organic solvents. It has the unusual property of dissolving most nylons (41) (see Polyamides) and polypeptides (42) (qv) at room temperature. Because of its excellent combination of physical and thermodynamic properties, 2,2,2-trifluoroethanol–water mixtures (also known as Fluorinols) have application as working fluids in Rankine-cycle engines being developed for recovering energy from waste heat sources (33,43–44) (see Power generation). Its high ionizing power and low specific conductance make the alcohol useful as a solvent for ionic reactions and conductometric titrations (45), and basic research into solvolysis mechanisms (46–48).

Chemically, 2,2,2-trifluoroethanol behaves as a typical alcohol. It can be converted to trifluoroacetaldehyde or trifluoroacetic acid by various oxidizing agents such as aqueous chlorine solutions (49) or oxygen in the presence of a vanadium pentoxide catalyst (50). Under basic conditions, it adds to tetrafluoroethylene and acetylene to give, respectively, 1,1,2,2-tetrafluoroethyl 2′,2′,2′-trifluoroethyl ether (51) and 2,2,2-trifluoroethyl vinyl ether which was used as the inhalation anesthetic Fluroxene. Its alkoxides react with bromoethane to give the ethyl ether (bp, 50.3°C). Similarly prepared is 2,2,2,2′,2′,2′-hexafluorodiethyl ether used as the convulsant drug flurothyl as a substitute for electric shock therapy. 2,2,2-Trifluoroethanesulfonic acid, useful in solvolysis studies, can be prepared from trifluoroethanol by the reaction of its methanesulfonate ester with sodium thiocyanate followed by oxidation with an aqueous chlorine solution (52).

Trifluoroethanol was first prepared by Swarts by the catalytic reduction of trifluoroacetic anhydride (53). Other methods include the catalytic hydrogenation of trifluoroacetamide (54), the lithium aluminum hydride reduction of trifluoroacetyl chloride (55) or of trifluoroacetic acid or its esters (56–57), and the acetolysis of 2-chloro-1,1,1-trifluoroethane followed by hydrolysis (55). More recently, the hydrogenation of 2,2,2-trifluoroethyl trifluoroacetate over a copper(II) oxide catalyst has been reported to give the alcohol in 95% yield (58).

Trifluoroethanol is manufactured and sold by the Halocarbon Products Corporation for ca $22.00/kg (1978). The commercial product is of very high purity with the main impurity ca 0.1% water.

Preliminary toxicity studies on trifluoroethanol show acute oral LD_{50}, 240 mg/kg; acute dermal LD_{50}, 1680 mg/kg; and acute inhalation L $(ct)_{50}$, 4600 ppm-h. Long-term subchronic inhalation exposure to 50 to 150 ppm of the alcohol has caused testicular depression in male rats, but no effects were noted at the 10 ppm level (33). Although the significance of the latter observations for human safety is unknown, it is recommended that continuous exposure to greater than 5 ppm or skin contact with it be avoided.

BIBLIOGRAPHY

"Fluoroethanols" in *ECT* 1st ed., under "Fluorine Compounds, Organic," Vol. 6, pp. 760–762, by John F. Nobis, Xavier University; "Fluoroethanols" in *ECT* 2nd ed., under "Fluorine Compounds, Organic," Vol. 9, pp. 751–752, by L. L. Ferstandig, Halocarbon Products Corporation.

1. F. Swarts, *Rec. Trav. Chim.* **33,** 252 (1914).
2. F. Swarts, *J. Chem. Soc.* **106,** 475 (1914).
3. E. Gryszkiewicz-Trochimowski, *Rec. Trav. Chim.* **66,** 427 (1947).
4. H. McCombie and B. C. Saunders, *Nature* **158,** 382 (1946).
5. E. V. Avdeeva and N. M. Dukel'skaya, *Vestn. Mosk. Univ. Ser. VI* **21**(4), 49 (1966); *Chem. Abstr.* **66,** 18287w (1967).
6. F. R. Johannsen and C. O. Knowles, *Comp. Gen. Pharmacol.* **5**(1), 101 (1974).
7. E. O. Dillingham and co-workers, *J. Pharm. Sci.* **62,** 22 (1973).
8. D. I. Peterson, J. E. Peterson and M. G. Hardinge, *J. Pharm. Pharmacol.* **20,** 465 (1968).
9. F. L. M. Pattison, *Toxic Aliphatic Fluorine Compounds,* Elsevier Publishing Co., New York, 1959, p. 65.
10. F. L. M. Pattison and co-workers, *J. Org. Chem.* **21,** 739 (1956).
11. B. C. Saunders, G. J. Stacey, and I. G. E. Wilding, *J. Chem. Soc.,* 773 (1949).
12. R. J. Koshar, T. C. Simmons, and F. W. Hoffmann, *J. Am. Chem. Soc.* **79,** 1741 (1957).
13. East Ger. Pat. 111,149 (Feb. 5, 1975), H. G. Werchan and co-workers.
14. F. W. Hoffmann, *J. Am. Chem. Soc.* **70,** 2596 (1948).
15. I. L. Knunyants, O. V. Kil'disheva, and I. P. Petrov, *J. Gen. Chem. USSR, Eng. Transl.* **19,** 95 (1949).
16. G. DiDrusco and F. Smai, *Quad. Ing. Chim. Ital.* **9**(11), 156 (1973).
17. C. T. Bedford, D. Blair, and D. E. Stevenson, *Nature* **267,** 335 (1977).
18. A. I. Kryl'tsov and co-workers, *Tr. Kaz. Nauch. Issled. Inst. Zashch. Rast.* **11,** 171 (1972); *Chem. Abstr.* **78,** 93627h (1973).
19. G. D. Robinson, Jr., *Radiopharm. Label Compounds, Proc. Symp.* **1,** 423 (1973).
20. U.S. Pat. 4,022,609 (May 10, 1977), D. E. Hardies and J. K. Rinehart (to PPG Industries, Inc.).
21. Jpn. Kokai 77 28,934 (Mar. 4, 1977), T. Yoshimoto and co-workers (to Mitsui Toatsu Chemicals, Inc.).
22. U.S. Pat. 3,852,464 (Dec. 3, 1974), D. E. Hardies and J. K. Rinehart (to PPG Industries, Inc.).
23. Fr. Pat. 1,604,978 (July 30, 1971), G. Rossi, G. Michieli, and P. Paolucci (Montecatini Edison S.P.A.).
24. J. H. J. Dawson and K. R. Jenning, *Int. J. Mass Spectrom. Ion Phys.* **25**(1), 47 (1977).
25. R. N. Haszeldine, *J. Chem. Soc.,* 1757 (1953).
26. F. Swarts, *Bull. Soc. Chim. Belg.* **11,** 731 (1902).
27. A. L. Henne and T. Midgley, *J. Am. Chem. Soc.* **58,** 882 (1936).
28. F. Swarts, *Chem. Zentr.* **II,** 709 (1903).
29. U.S. Pat. 3,769,433 (Oct. 30, 1973), R. C. Terrell (to Airco, Inc.).
30. U.S. Pat. 3,896,178 (July 22, 1975), R. C. Terrell (to Airco, Inc.).
31. U.S. Pat. 3,746,769 (July 17, 1973), R. C. Terrell (to Airco, Inc.).
32. U.S. Pat. 3,862,240 (Jan. 21, 1975), R. C. Terrell (to Airco, Inc.).
33. *Trifluoroethanol Brochure,* Halocarbon Products Corp., Hackensack, N.J., 1979.
34. V. P. Kolesov, I. D. Zenkov, and S. M. Skuratov, *Zh. Fiz. Khim.* **39,** 2474 (1965); *Chem. Abstr.* **64,** 1422a (1965).
35. J. M.Mukherjee and E. Grunwald, *J. Phys. Chem.* **62,** 1311 (1958).

36. J. Murto and E. Heino, *Suom. Kemistil. B* **39,** 263 (1966).
37. C. H. Rochester and J. R. Symonds, *J. Fluorine Chem.* **4,** 141 (1974).
38. A. Kivenen, J. Murto, and M. Lehtonen, *Suom. Kemistil. B* **41,** 359 (1968).
39. C. H. Rochester and J. R. Symonds, *J. Chem. Soc. Faraday Trans. I* **69,** 1274 (1973).
40. P. Ballinger and F. A. Long, *J. Am. Chem. Soc.* **81,** 1050 (1959).
41. Ger. Pat. 1,017,782 (Oct. 17, 1957), P. Schlack (to Farbwerke Hoechst A.G.).
42. M. Goodman, I. G. Rosen, and M. Safdy, *Biopolymers* **2,** 503, 519, 537 (1964).
43. U.S. Pat. 3,722,211 (Mar. 27, 1973), R. C. Conner and L. L. Ferstandig (to Halocarbon Products Corp.).
44. D. G. Shepherd, *Hydrocarbon Process.*, 141 (Dec. 1977).
45. N. Paetzold, *J. Polymer Sci. Part B* **1,** 269 (1963).
46. V. J. Shiner, Jr. and co-workers, *J. Am. Chem. Soc.* **91,** 4838 (1969).
47. D. A. daRoza, L. J. Andrews, and R. M. Keefer, *J. Am. Chem. Soc.* **95,** 7003 (1973).
48. D. S. Noyce, R. L. Castenson, and D. A. Meyers, *J. Org. Chem.* **37,** 4222 (1972).
49. U.S. Pat. 3,088,896 (May 7, 1963), M. Braid (to Pennsalt Chemicals Corp.).
50. U.S. Pat. 3,038,936 (June 12, 1962), M. Braid (to Pennsalt Chemicals Corp.).
51. A. L. Henne and M. A. Smook, *J. Am. Chem. Soc.* **72,** 4378 (1950).
52. R. K. Crossland, W. E. Wells, and V. J. Shiner, Jr., *J. Am. Chem. Soc.* **93,** 4217 (1971).
53. F. Swarts, *Compt. Rend.* **197,** 1201 (1933).
54. H. Gilman and R. G. Jones, *J. Am. Chem. Soc.* **70,** 1281 (1948).
55. A. L. Henne, R. M. Alm, and M. Smook, *J. Am. Chem. Soc.* **70,** 1968 (1948).
56. K. N. Campbell, J. O. Knobloch, and B. K. Campbell, *J. Am. Chem. Soc.* **72,** 4380 (1950).
57. D. R. Husted and A. H. Ahlbrecht, *J. Am. Chem. Soc.* **74,** 5422 (1952).
58. U.S. Pat. 4,072,726 (Feb. 7, 1978), H. R. Nychka and co-workers (to Allied Chemical Corp.).

GARY ASTROLOGES
Halocarbon Products Corporation

FLUORO ETHERS AND AMINES

The perfluoroaliphatic ethers and perfluoroalkyl tertiary amines, together with the perfluoroalkanes, comprise a class of materials known as perfluorinated inert fluids. They are colorless, odorless, dense, virtually nontoxic, and nonflammable compounds with an extremely nonpolar character. They are used primarily as dielectric heat-transfer and test-bath fluids in the electronics industry and as hydraulic fluids and lubricants for specialty applications.

Physical Properties

Numerous perfluoro ethers and tertiary amines, such as the examples shown in Table 1, have been prepared by electrochemical fluorination (1–4). Certain perfluoroalkyl polyethers have also been prepared by polymerization of hexafluoropropylene oxide (5) and photooxidation of hexafluoropropylene at low temperatures (6).

The perfluorinated inert liquids boil lower than their hydrocarbon analogues even though they have much higher molecular weights. The low boiling points of the per-

Table 1. Physical Properties of Some Perfluorinated Inert Liquids[a]

Name	CAS Registry No.	Molecular formula	bp, °C	d_4^{25}	n_D^{24}	Pour point, °C (ASTM D 97)
perfluoro-4-methylmorpholine	*[382-28-5]*	$CF_3N(CF_2)_2O(CF_2)_2$	51	1.70	1.267	−80
perfluorotriethylamine	*[359-70-6]*	$(C_2F_5)_3N$	71	1.73	1.262	−110
perfluoro-2-ethyltetrahydrofuran	*[356-48-9]*	$C_6F_{12}O$	56	1.69	1.263	−136
perfluorohexane	*[355-42-0]*	C_6F_{14}	58	1.68	1.252	−74
perfluoro-4-isopropylmorpholine	*[1600-71-1]*	$(CF_3)_2CFN(CF_2)_2O(CF_2)_2$	95	1.79	1.283	−112
perfluorobutyl ether	*[308-48-5]*	$(C_4F_9)_2O$	102	1.71	1.261	−100
FC-75	*[11072-16-5]*	$C_8F_{16}O$[b] (cyclic)	103	1.76	1.276	−93
perfluorooctane	*[307-34-6]*	C_8F_{18}	103	1.77	1.272	−42
perfluorotripropylamine	*[338-83-0]*	$(C_3F_7)_3N$	130	1.82	1.279	−52
perfluorononane	*[375-96-2]*	C_9F_{20}	123	1.80	1.276	−16
perfluorotributylamine	*[311-89-7]*	$(C_4F_9)_3N$	178	1.86	1.291	−50
perfluorohexyl ether	*[424-20-4]*	$(C_6F_{13})_2O$	181	1.81	1.278	−90
perfluorododecane	*[307-59-5]*	$C_{12}F_{26}$	175			42

[a] Many of these compounds are described in the General References.
[b] $C_nF_{2n}O$ represents a mixture of isomers of cyclic perfluoroaliphatic ethers.

fluoroaliphatic ethers and perfluoroalkyl tertiary amines are especially remarkable. As shown in Table 1, they boil very close to the perfluoroalkanes having the same number of carbon atoms. The hetero atoms apparently contribute little polarity to the molecules. The hetero atoms do have a marked effect on the freezing points, however. This is illustrated by the C_{12}-perfluoro ether and C_{12}-perfluoro tertiary amine in Table 1. These compounds freeze at −90 and −50°C, respectively, in contrast with the perfluoroalkane, $C_{12}F_{26}$, which is a solid at room temperature. The oxygen and nitrogen atoms provide increased internal flexibility in the molecules, thus lowering the freezing points. This characteristic is especially evident in the perfluoro polyethers; eg, an ether having the structure $CF_3O(C_3F_6O)_m(CF_2O)_nCF_3$ with an average molecular weight of 2600 has a pour point of −42°C (7) (see also Polyethers).

Many of the unusual properties of the perfluorinated inert fluids are the result of their low intermolecular forces. An example is their low surface tensions, which are 9–18 mN/m (= dyn/cm) at 25°C, enabling these liquids to wet any surface, including polytetrafluoroethylene. Their refractive indexes are lower than any other organic liquids, as are their acoustic velocities. They have isothermal compressibilities almost twice as high as water (8).

The absolute viscosities of the perfluorinated inert fluids are higher than the analogous hydrocarbons but the kinematic viscosities are lower because of the high densities of the fluorinated compounds. The viscosity index, ie, the change in viscosity with temperature, is much higher for the perfluorinated liquids than for hydrocarbons.

Thermal Stabilities. The perfluoroaliphatic ethers have thermal stabilities equivalent to the perfluoroalkanes when heated in stainless steel tubes. They undergo no quantitatively significant decomposition until they are heated above ca 400°C. The perfluoroalkyl tertiary amines are slightly less stable thermally and begin to decompose at ca 250°C (9).

Electrical Properties. The low polarizability of these inert liquids gives them excellent dielectric properties. Their dielectric strengths are about 40 kV (ASTM D 877); their dissipation factors are about 0.0001 at 1 MHz; their dielectric constants are ca 1.8; and their volume resistivities are ca 1×10^{15} Ω·cm (ASTM D 257) (8).

Chemical Properties

The inert character of the perfluoro ethers and tertiary amines is demonstrated by their lack of basicity or reactivity normally associated with hydrocarbon ethers or amines. They do not form salts with acids or complexes with boron trifluoride. They are not attacked by most oxidizing or reducing agents or strong acids or bases. Like the perfluoroalkanes, the perfluoro ethers and tertiary amines may react violently with fused alkali metals. Unlike the alkanes, however, they also react with aluminum chloride at ca 150–200°C with some substitution of fluorine by chlorine (10–11).

Solvent Properties. The solvent properties of the perfluorinated inert liquids are also unusual because of their nonpolar nature and low intermolecular forces. Water and alcohols are almost completely insoluble. Most hydrocarbons and nonfluorine-containing aliphatic halogen compounds are only slightly soluble. On the other hand, lower aliphatic hydrocarbons, such as pentanes and hexanes, and fluorine-containing liquids, such as benzotrifluoride and 1,1,2-trichloro-1,2,2-trifluoroethane, may be completely miscible with the perfluoro compounds at RT. Some typical liquid solubilities are shown in Table 2.

Table 2. Solubility Relationships For $C_8F_{16}O$ (Cyclic) and $(C_4F_9)_3N$ [a]

	Solubility at 20°C, g/100 mL			
Solvent	$C_8F_{16}O$ (cyclic) in solvent	Solvent in $C_8F_{16}O$ (cyclic)	$(C_4F_9)_3N$ in solvent	Solvent in $(C_4F_9)_3N$
acetone	4.8	1.2	0.9	0.6
benzene	2.6	3.8	0.3	0.2
benzotrifluoride	miscible	miscible	miscible	miscible
benzyl alcohol	0.2	0.4	insol	insol
carbon tetrachloride	20.2	36.5	2.4	15.0
chlorobenzene	1.8	3.2		0.4
chloroform	7.7	4.5	1.2	5.4
cyclohexane	7.7	8.4	1.8	2.2
dioxane	0.9	2.4	0.1	insol
ethyl ether	miscible	miscible	4.9	5.0
ethyl acetate	7.5	6.5	2.2	2.4
heptane	25.5	11.6	6.4	3.4
isopropyl alcohol	4.1	1.3	insol	insol
methanol	1.0	0.1	insol	insol
petroleum ether	miscible	miscible	33.2	7.0
refrigerant 113	miscible	miscible	miscible	miscible
stoddard solvent	5.9	3.0	1.4	
toluene	2.9	4.1	0.4	
turpentine	5.3	1.0	0.9	insol
xylene	3.0	3.0		1.0
water	insol	insol	insol	insol

[a] $C_8F_{16}O$ denotes a mixture of isomers of cyclic perfluoroaliphatic ethers.

The solubilities of gases such as oxygen, nitrogen, and carbon dioxide are extremely high as shown in Table 3. The solubility of oxygen in related compounds has led to possible uses as blood replacements (see Uses). These solubility relationships are consistent with predictions based on Hildebrand's solubility parameters (12). The solubility parameters of perfluorinated inert fluids are ca 10–12 $J^{1/2}/cm^{3/2}$ (5–6 $cal^{1/2}/cm^{3/2}$), the lowest known values for liquids.

Methods of Preparation

In the electrochemical fluorination process, organic reactants are dissolved in anhydrous hydrogen fluoride to form conductive solutions. The solutions are electrolyzed with direct current at 4–8 V in a single-compartment cell, usually with nickel

Table 3. Typical Solubility of Gases in Perfluorinated Inert Liquids at 25°C and 101 kPa (1 atm)

	Solubility	
Gas	mL gas/100 mL $C_8F_{16}O$ (cyclic)	mL gas/100 mL $(C_4F_9)_3N$
oxygen	48.8	38.9
nitrogen	33.4	28.4
carbon dioxide	192.0	152.0

electrodes. Essentially complete replacement of organic hydrogen atoms by fluorine atoms occurs at the anode and hydrogen gas is released at the cathode. A low temperature condenser strips HF and liquid products from the hydrogen stream. Liquid products, higher in density and insoluble in HF, are recovered by phase separation:

$$(C_4H_9)_3N + 27\ HF \xrightarrow{e} (C_4F_9)_3N + 27\ H_2$$

The products of electrochemical fluorination generally contain appreciable amounts of the perfluorinated analogues of the starting materials. In addition, because of extensive molecular rearrangement, many isomers and perfluorinated cleavage products are also present. The products are usually purified by treatment with base and distillation.

The polymerization of hexafluoropropylene oxide is carried out using fluoride ion catalyst (eg, cesium fluoride) and an aprotic solvent such as diglyme. An acyl fluoride end group terminates the polymer:

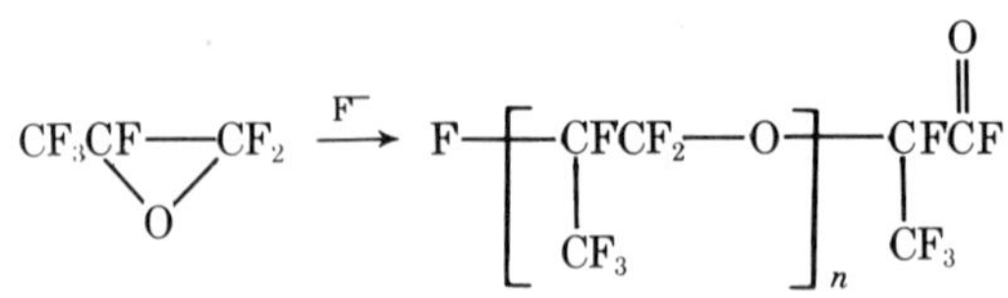

The reactive acyl fluoride group can be stabilized in several ways:

(*1*) Elimination of carbon monoxide and addition of a fluorine atom with antimony pentafluoride or aluminum fluoride catalyst:

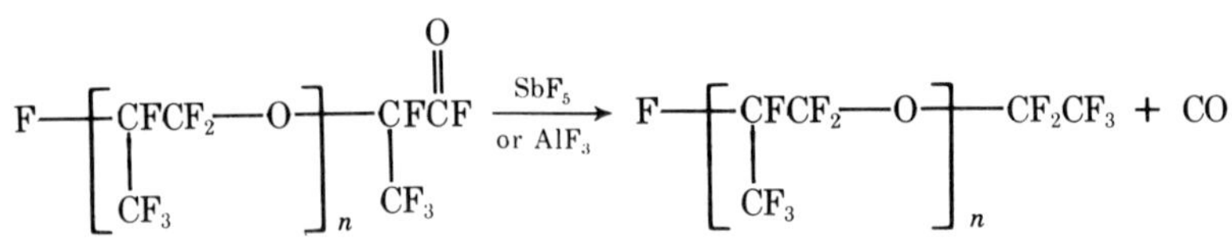

(*2*) Treatment with aqueous base and subsequent decarboxylation:

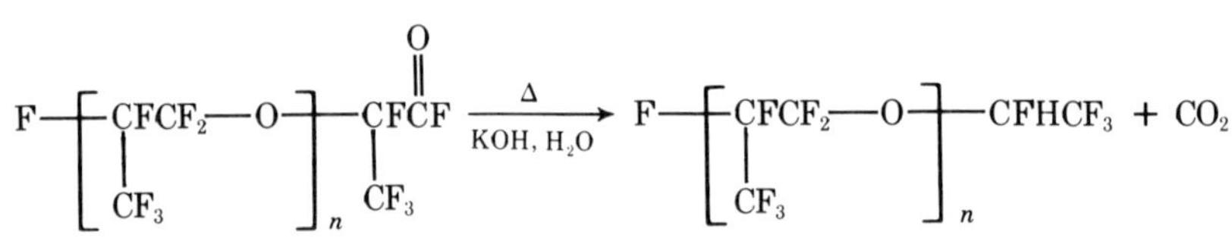

(*3*) Photolytic irradiation to eliminate carbonyl fluoride and carbon monoxide and approximately double the molecular weight:

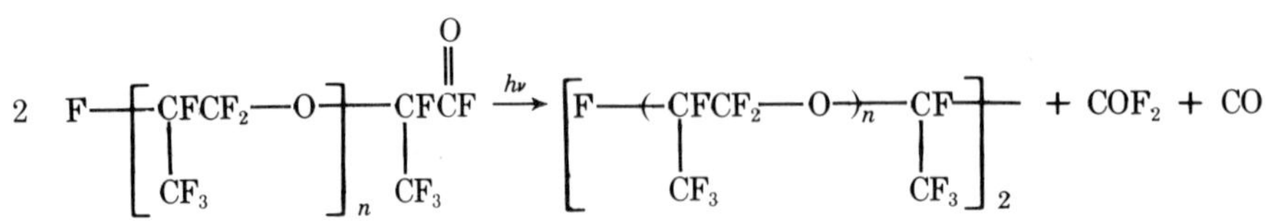

Photooxidation of hexafluoropropylene at low temperatures yields peroxides of molecular weights up to 7000 (6):

$$CF_3CF{=}CF_2 + O_2 \xrightarrow{h\nu} \text{—}O\text{—}(C_3F_6O)_n\text{—}(C_3F_6\text{—}O\text{—}O)_m$$

The end groups include (13):

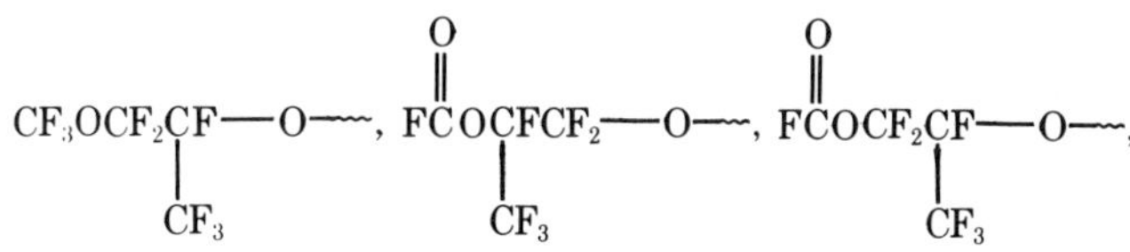

The peroxides are destroyed by heat or radiation and the polymers are end-capped by direct fluorination:

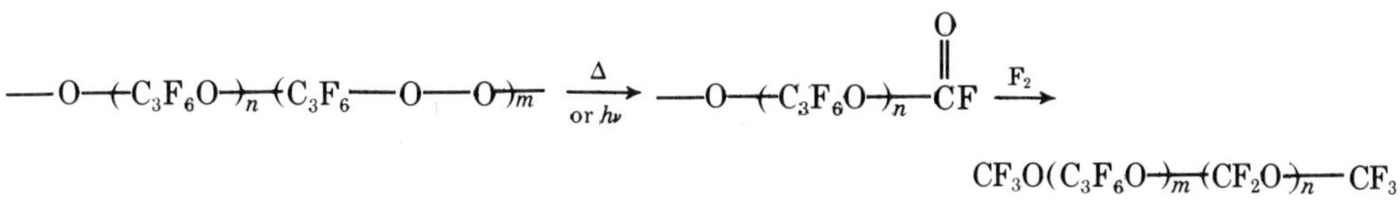

Characteristically, the polymeric perfluoro ethers are obtained in a wide distribution of molecular weights from either process. They are usually separated into various fractions by distillation.

Economic Aspects

Although information on production levels of perfluoro ethers and tertiary amines is not disclosed, the products are commercially available. They are main constituents of some members of the family of Fluorinert electronic liquids marketed by 3M Co. (8). The Fluorinert liquids have bps of 30–215°C and molecular weights of 290–820. Prices are \$22–66/kg. There are no other known sources of perfluoroalkyl tertiary amines but perfluoroalkyl polyethers are available as Krytox fluorinated oils from DuPont (14) and Fomblin fluorinated fluids from Montedison (15). The Krytox products have the general formula $F(CF(CF_3)CF_2O)_nC_2F_5$ with molecular weights of 2000–7000. The Fomblin fluids have the general formula $CF_3(OCF(CF_3)CF_2)_n(OCF_2)_mOCF_3$ and molecular weights of 1200–7000. The perfluoropolyether products have a price range of ca \$110–220/kg.

Health and Safety Factors

Animal studies have shown the perfluorinated inert fluids to be nonirritating to the eyes and skin and practically nontoxic by oral ingestion, inhalation, or intraperitoneal injection (8,16). Thermal degradation, however, can produce toxic decomposition products, including perfluoroisobutylene which has a reported LC_{50} of 0.5 ppm (6-h exposure in rats) (17). Such decomposition generally begins only at temperatures above 200°C as discussed above.

Uses

The unique combination of properties of the perfluorinated inert fluids makes them useful for applications involving their direct contact with sensitive materials. They are used extensively in the electronics industry as test baths because of their compatibility with semiconductor devices and their ability to evaporate completely without leaving residues (8). They are used as direct-immersion dielectric coolants for electrical devices and as heating media for mass production reflow soldering (18) (see Solders and brazing alloys). Their low solvent strengths and low refractive indexes have made them useful as mobile phases for high performance liquid chromatography, especially in the high precision determination of olefins in gasolines (19). They are also potentially useful as inert reaction media, especially for dissolved gas reactions. The higher molecular weight perfluoroalkyl polyethers have applications as nonflammable, inert hydraulic fluids (qv) and lubricants which are especially useful in the aerospace industry (14–15). They also show promise as working fluids for vapor-diffusion vacuum pumps.

The high solubility of oxygen in perfluorinated inert fluids, such as perfluorotripropylamine and perfluorobutyltetrahydrofuran [*40464-54-8*], has allowed their use as blood substitutes (20). One such product is Fluosol (Alpha Therapeutic Corp.) (21).

BIBLIOGRAPHY

"Fluoro Ethers and Amines" under "Fluorine Compounds, Organic" in *ECT* 1st ed., Vol. 6, pp. 762–763, by L. J. Hals and W. H. Pearlson, Minnesota Mining & Manufacturing Co.; "Fluoro Ethers and Amines" under "Fluorine Compounds, Organic" in *ECT* 2nd ed., Vol. 9, pp. 753–754, by L. J. Hals and W. H. Pearlson, Minnesota Mining & Manufacturing Co.

1. J. H. Simons and co-workers, *J. Electrochem. Soc.* **95,** 47 (1949).
2. J. H. Simons, "Fluorocarbons and Their Production" in J. H. Simons, ed., *Fluorine Chemistry,* Vol. I, Academic Press, Inc., New York, 1950, pp. 414–420.
3. J. Burdon and J. C. Tatlow, "The Electrochemical Process For the Synthesis of Fluoro Organic Compounds" in M. Stacey, J. C. Tatlow, and A. G. Sharpe, eds., *Advances in Fluorine Chemistry,* Vol. 1, Butterworths, London, Eng., 1960, pp. 129–165.
4. S. Nagase, "Electrochemical Fluorination" in P. Tarrant, ed., *Fluorine Chemistry Reviews,* Vol. 1, Marcel Dekker, New York, 1967, pp. 77–106.
5. H. S. Eleuterio, *J. Macromol. Sci. Chem.* **A6**(6), 1027 (1972).
6. D. Sianesi and co-workers, *Chim. Ind. (Milan)* **55**(2), 208 (1973).
7. L. Laurenson, *Res. Dev.* **28**(11), 61 (1977).
8. *Fluorinert Liquids,* technical notebook, 3M Co., St. Paul, Minn., 1978.
9. H. G. Bryce, "Industrial and Utilitarian Aspects of Fluorine Chemistry" in J. H. Simons, ed., *Fluorine Chemistry,* Vol. V, Academic Press, Inc., New York, 1964.
10. S. V. Sokolov and S. A. Mazalov, *Dokl. Acad. Sci. USSR* **162**(5), 1071 (1965).
11. G. V. D. Tiers, *J. Am. Chem. Soc.* **77,** 4837, 6703, 6704 (1955).
12. J. H. Hildebrand, J. M. Prausnitz, and R. L. Scott, *Regular and Related Solutions,* Van Nostrand Reinhold, New York, 1970, p. 207.
13. D. Sianesi and co-workers, *Wear* **18,** 85 (1971).
14. *Krytox, Technical Bulletins No. L5 and G5,* E. I. du Pont de Nemours & Co., Inc., Wilmington, Del.
15. *Fomblin y Fluorinated Fluids,* Montedison S.p.A., Milan, Italy, 1971.
16. J. W. Sargent and R. J. Seffl, *Fed. Proc.* **29,** 1699 (Sept.–Oct. 1970).
17. J. W. Clayton, Jr., "Fluorocarbon Toxicity and Biological Action," in ref. 4, pp. 225–232.
18. R. C. Pfahl, Jr., J. C. Mollendorf, and T. Y. Chu, *Weld. J.* **54**(1), 22 (1975).
19. J. C. Suatoni, H. R. Garber, and B. E. Davis, *J. Chrom. Sci.* **13,** 367 (1975).
20. *Science* **206,** 205 (1979).
21. *Chem. Eng. News,* 12 (Nov. 26, 1979).

General References

J. H. Simons, ed., *Fluorine Chemistry,* Vol. I, Academic Press, Inc., New York, 1950, pp. 486–491; Vol. II, 1954, pp. 242, 391, 396; Vol. V, 1964, pp. 311–314, 325, 332.

A. M. Lovelace and co-workers, *Aliphatic Fluorine Compounds,* Reinhold Publishing Corp., New York, 1958, pp. 155, 285.

M. Stacey, J. D. Tatlow, and A. G. Sharpe, eds., *Advances in Fluorine Chemistry,* Vol. 1, Butterworths Scientific Publications, London, Eng., 1960, pp. 139–145; Vol. 2, p. 134; Vol. 3, p. 32.

M. Hudlicky, *Chemistry of Organic Fluorine Compounds,* Halsted Press, a division of John Wiley & Sons, Inc., New York, 1976, p. 73.

R. E. Banks and M. G. Barlow, *Fluorocarbon and Related Chemistry,* Vol. 1, The Chemical Society, London, Eng., 1971, pp. 98, 139; Vol. 2, 1974, pp. 204, 252; Vol. 3, 1976, pp. 207, 266.

RICHARD D. DANIELSON
Minnesota Mining & Manufacturing Co.

HEXAFLUOROACETONE AND DERIVATIVES

Hexafluoroacetone

Hexafluoroacetone [*684-16-2*], CF_3COCF_3, is a reactive, nonflammable, gaseous perfluoro ketone used as a chemical intermediate. It was first prepared in 1941 in low yield by the direct fluorination of acetone (1) and remained a laboratory curiosity until the 1960s, when it was produced commercially in pilot-plant quantities by the Allied Chemical Corporation and DuPont. Today, it is widely used in the synthesis of polymers, medicinals, agriculture chemicals, and other compounds of practical and scientific interest. It has also found some commercial use as a component of high-performance polymers and in the manufacture of the solvent 1,1,1,3,3,3-hexafluoro-2-propanol.

Physical Properties. Physical constants for hexafluoroacetone are listed in Table 1. The boiling point of hexafluoroacetone, which is 84°C lower than the boiling point of acetone, indicates very little association in the liquid state. The dielectric constant and other physical properties support this conclusion. This lack of association is owing to the complete absence of hydrogen bonding and the low polarity of the carbonyl group caused by the opposing inductive effect of the fluorine atoms.

A melting point curve as a function of pressure has been recorded (3). Several equations have been proposed to calculate the vapor pressure (2,4,7) and liquid density (2,7) as a function of temperature. The most accurate appear to be the following equations:

$$\log_{10}P_{kPa} = 27.48332 - (1963.631/T) - 7.316256 \log_{10}T$$

$$d(\text{g/mL}) = (2.386 - 0.003491\ T) \pm 0.001$$

To convert kPa to mm Hg, multiply by 7.5.

Table 1. Physical Properties of Hexafluoroacetone

Property	Value	Refs.
mol wt	166.03	
boiling point at 101.3 kPa[a], °C	−27.28	2
melting point, °C	−125.45, −129	2–3
critical pressure, MPa[b]	2.834	4
liquid density, g/mL		
at 23.3°C	1.323	5
at 25°C	1.318	3
at 44.4°C	1.149	5
heat capacity of liquid at 0°C, kJ/(kg·K)[c]	1.610	2
heat capacity of vapor, kJ/(kg·K)[c]	1.730	2
molar heat of fusion at −125.45°C, kJ/mol[c]	8.38	2
molar heat of vaporization at −27.28°C, kJ/mol[c]	21.61	2
dielectric constant, 10 kHz		
solid at −128°C	1.910	6
liquid at 35°C	1.958	6
dipole moment, C·m[d]		
gaseous	2.16×10^{-30}	6
liquid	2.10×10^{-3}	6
entropy at −27.28°C, J/K[c]	375	2
heat of formation, kJ/mol[c]	1250[e]	7–9
bond length, nm		
F—C	0.1335	
C—C	0.1549	
C=O	0.1246	

[a] 1 atm.
[b] To convert MPa to psi, multiply by 145.
[c] To convert J to cal, divide by 4.184.
[d] To convert C·m to debye, divide by 3.336×10^{-30}.
[e] Calculated.

The structure of gaseous hexafluoroacetone has been determined by electron diffraction methods (8–9). It has C_s symmetry, with the two CF_3 groups being staggered with respect to one another. The vacuum uv spectrum in the region 60–215 nm indicates continuous absorption from 105 nm to shorter wavelengths (10). Absorption also occurs in the 220–400 nm region with a broad band at 245–355 nm, λ_{max} 302 nm (ϵ, 16.9) (11–12). The most noteworthy infrared absorption is the strong carbonyl stretching band at the unusually high frequency of 1807 cm^{-1} (13). The far ir spectrum exhibits two broad torsional bands at 72.5 and 39.0 cm^{-1} (14). Comparison of the Raman spectrum with the ir has led to a complete vibrational assignment (15).

The ^{19}F nmr spectrum of hexafluoroacetone in carbon tetrachloride shows a single resonance at −76.05 ppm δ relative to $CFCl_3$ used as an internal standard (negative sign indicates a shift to higher field) (16). The photoelectron spectrum was measured and used to determine that the ionization potential of hexafluoroacetone is 11.68 eV (17).

Chemical Properties. The chemical properties of hexafluoroacetone differ markedly from those of halogen-free ketones such as acetone. The inductive effect of the electronegative fluorine substituents causes the carbonyl group to be exceptionally electron deficient. As a consequence, hexafluoroacetone is very resistant to attack by electrophilic reagents. It is not protonated by strong acids (18) and can even be distilled from concentrated sulfuric acid. As would be expected, however, the electron-deficient carbonyl group is very susceptible to nucleophilic reagents, and this is the basis for the exceptional reactivity of hexafluoroacetone.

Hexafluoroacetone is thermally stable. Even prolonged heating at 300°C causes only negligible decomposition. At 550°C, a first-order decomposition occurs to give hexafluoroethane and carbon monoxide. Above 600°C, trifluoroacetyl fluoride and octafluoroisobutylene are also formed (19).

Reactions with Nucleophiles. Hexafluoroacetone reacts exothermically with water to form an acidic (pK_a = 6.58) solid (mp 49°) *gem*-diol, $(CF_3)_2C(OH)_2$ [*677-71-4*]. With more water, a liquid sesquihydrate [*13098-39-0*] which is a useful polymer solvent is formed (5). These hydrates are potent donors for hydrogen bonding and form stable complexes with acceptors such as ethers and amines. This ability is responsible for the high solvent power of hexafluoroacetone sesquihydrate for polyamides (20), polyesters (20), polyacetals (21), and polyols (22). Although the hydrate is quite acidic, it causes little degradation. For example, silk fibroin can be dissolved to form a stable solution in which the protein exists as a hydrogen-bonded random coil, whereas the same solute in trifluoroacetic acid is unstable (23). The deuterate [*72301-81-6*] of hexafluoroacetone is a useful nmr solvent for the study of polymers.

Aliphatic alcohols form stable hemiketals (24). Hexafluoroacetone ketals are obtained by treating a ketone with an alcohol and acid catalyst, or by alkylating the hemiketal in the presence of a base.

Compounds that contain hydroperoxide groups add across the carbonyl group of hexafluoroacetone to give peroxides that are oxidizing agents (25) and polymerization catalysts (26). Thiol compounds add similarly (27).

$$(CF_3)_2CO + ROH \rightarrow (CF_3)_2C(OH)OR \xrightarrow[NaOH]{R_2SO_4} (CF_3)_2C(OR)_2$$

$$(CF_3)_2CO + ROOH \rightarrow (CF_3)_2C(OH)OOR \quad R = H, \textit{tert}\text{-Bu}$$

$$(CF_3)_2CO + RSH \rightarrow (CF_3)_3C(OH)SR \quad R = H, \text{alkyl}$$

Ammonia and most primary (28) and secondary amines give stable *gem*-aminoalcohols. Hexafluoroacetone is dehydrated to the imine by pyridine and $POCl_3$ (28); thermal dissociation to the imine does not occur (29).

$$(CF_3)_2CO + NH_3 \rightarrow \underset{[31253\text{-}34\text{-}6]}{(CF_3)_2C(OH)NH_2} \xrightarrow[\text{pyridine}]{POCl_3} \underset{[1645\text{-}76\text{-}6]}{(CF_3)_2C{=}NH}$$

Carboxylic acid amides and ureas also add to hexafluoroacetone. Dehydration of the amide adducts gives *N*-acylimines (30–31). Certain urea adducts are reported to have fungicidal and herbicidal properties (32). Other nitrogen nucleophiles such as hydrazoic acid (34) and cyanic acid (35) also add readily to hexafluoroacetone.

Hexafluoroacetone does not form oximes or hydrazones when treated in the usual manner with hydroxylamines or hydrazines. Instead, adducts that resist dehydration are formed. Hexafluoroacetone oxime [*1645-76-7*] or hydrazone are prepared via the imine (29,35–36). Hexafluoroacetone oxime is a colorless liquid, bp 75°C, and is an excellent solvent for polyamide and polyester polymers (37). Hexafluoroacetone hydrazone [*1588-41-6*] is a colorless liquid, bp 96°C. It is a starting material for bis(trifluoromethyl)diazomethane, bp 12–13°C (35).

Active methylene compounds such as malononitrile, cyanoacetic esters, acetoacetic esters, and malonic esters add to the carbonyl group of hexafluoroacetone without loss of water to give alcohol adducts. These alcohols can be dehydrated with P_2O_5 (38).

Hexafluoroacetone reacts with Grignard reagents, organolithium compounds, and other organometallics to give tertiary alcohols. An example is the preparation of the perfluoroalcohol, $CF_2{=}CFC(CF_3)_2OH$ [*15052-92-3*], by reaction with trifluoroethenyllithium, $CF_2{=}CFLi$ (39).

Hydrogen fluoride forms an unstable perfluoroalcohol, $(CF_3)_2CFOH$ [*24427-67-6*]. Metal fluorides give the relatively stable heptafluoroisopropoxide anion, $(CF_3)_2CFO^-$ (40), and this anion has been used to prepare a wide variety of products (41–45) containing the heptafluoroisopropoxy group.

Friedel-Crafts Reactions. Hexafluoroacetone reacts with aromatic compounds under Friedel-Crafts conditions to give tertiary alcohols (46–48). Bisphenol derivatives are prepared by condensing two equivalents of a phenol with hexafluoroacetone in the presence of boron trifluoride or hydrogen fluoride (49–50). Heterocyclic aromatic compounds also react readily with hexafluoroacetone, sometimes without catalyst (51) (see Friedel-Crafts reactions).

Reactions with Olefins and Acetylenes. Hexafluoroacetone is a potent enophile. Its reacts with olefins that contain at least one allylic hydrogen atom with an allylic shift of the double bond to give unsaturated alcohols (52). Terminally substituted olefins are the most reactive but internal olefins also react. With isobutylene, mono, di, and tri-adducts are obtained, depending on the reaction conditions (53). Vinyl ethers form oxetanes which isomerize on heating to unsaturated alcohols (54).

Acetylenes containing α-hydrogen atoms form allenic carbinols (55) and alkoxy acetylenes form oxetes that rearrange to unsaturated esters (56).

$$(CF_3)_2CO + CH_3C{\equiv}CH \longrightarrow CH_2{=}C{=}CHC(CF_3)_2OH$$

[*20227-70-7*]

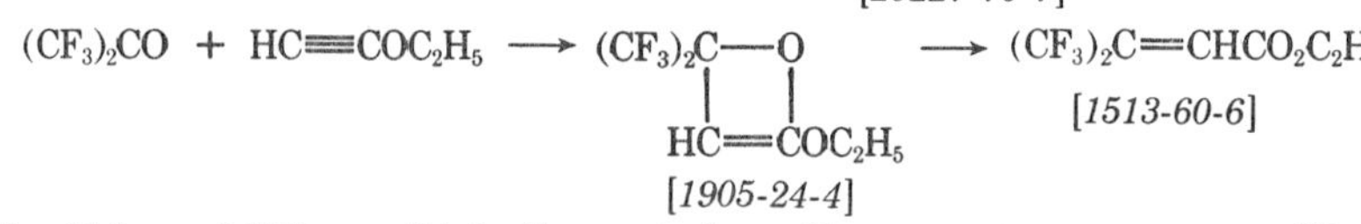

[*1905-24-4*] [*1513-60-6*]

Diels-Alder additions of 1,3-dienes to hexafluoroacetone occur readily at 100°C or lower (57). 2,3-Dimethylbutadiene gives a stable adduct. Cyclopentadiene (qv), on the other hand, gives a thermally unstable adduct that can be stabilized as the dibromide or rearranged to a carbinol (58).

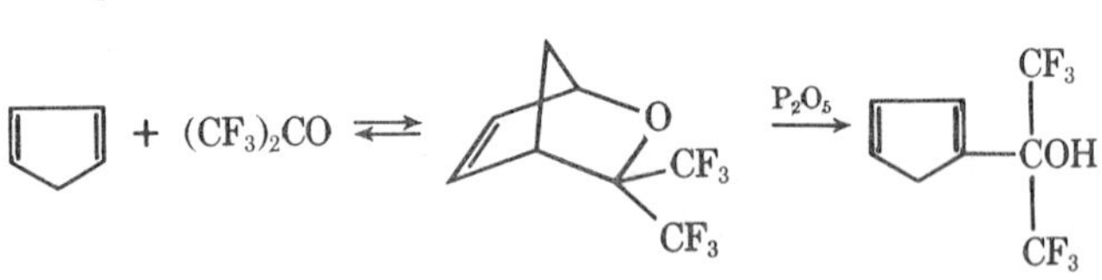

[*57859-03-7*] [*58751-23-8*]

Ketene reacts with hexafluoroacetone at −80°C without catalyst to form a β-lactone (59). Diphenylketene (60) and ketenimines (61) also undergo similar cycloadditions to oxetanes.

Photochemical Reactions. Hexafluoroacetone decomposes under uv irradiation to trifluoromethyl radicals and carbon monoxide. Small amounts of $CF_3CO\cdot$ have also been detected in low temperature flash photolysis (62). Photolysis of hexafluoroacetone offers a convenient source of trifluoromethyl radicals for chemical reactions. A precursor to these radicals is a photoexcited hexafluoroacetone. Irradiation causes the $n \rightarrow \pi^*$ promotion of an electron which results in a species with radical activity at oxygen (12). For example, photoexcited hexafluoroacetone forms cyclic adducts with fluoroolefins (63) and hydrocarbon olefins (64) in a reaction which has been proposed to involve diradical intermediates:

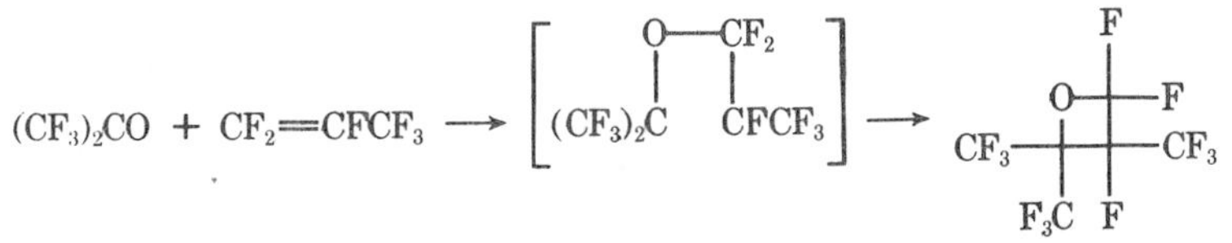

[*730-56-3*]

Radical Additions. Molar amounts of methyl radicals generated by decomposition of di-*tert*-butyl peroxide add exclusively to the carbonyl carbon of hexafluoroacetone to form a tertiary alcohol. This addition has been extended to other alkyl radicals, and can be carried out as a free-radical chain reaction. For example, cyclohexane adds efficiently to hexafluoroacetone with either peroxide catalyst or photochemical initiation. The radical adds to both the carbon and oxygen of the carbonyl group to give an alcohol and an ether but the alcohol product predominates:

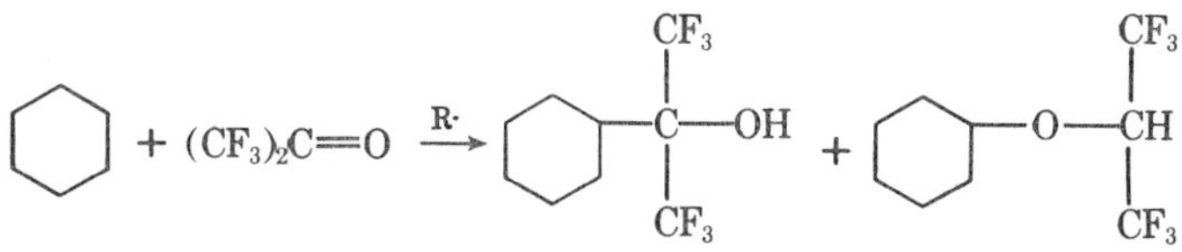

[*15232-99-2*] [*15233-00-8*]

Synthesis of Heterocyclic Compounds. Hexafluoroacetone is a useful starting material for oxygenated heterocyclic compounds containing *gem*-trifluoromethyl groups. Carbenes (65–66) and diazomethanes (67) give oxiranes. Oxazolidine derivatives have been prepared by reaction with sodium cyanide (68–69) and with amino acids (70), and 1,3-dioxolane derivatives have been prepared by reaction with isocyanates (71), orthoformates (72), and oxalyl fluoride (73).

In addition to Diels-Alder adducts, a number of other 6-membered heterocyclic derivatives have been prepared from hexafluoroacetone, including 1,3,5-oxadiazines and 1,3,5-dioxazines (74–75), and 1,3,5-oxathiazines (76).

Metal Complexes. Hexafluoroacetone forms complexes with several metals, including platinum (77–80), palladium (77), nickel (81), indium (82), osmium (83), rhodium (84), and iron (85). Thus hexafluoroacetone replaces one of the ligands of zero-valent palladium or nickel phosphine complexes. Heterocyclic compounds containing a metal atom in the ring have also been prepared (86).

Polymers from Hexafluoroacetone. Hexafluoroacetone does not form a homopolymer. However, it does copolymerize with other monomers, and it can also be used for the modification of polymeric substrates and the synthesis of fluorinated monomers.

Copolymers. Vinyl monomers can be copolymerized with hexafluoroacetone under free-radical conditions (87–88). Hydrogen-free monomers, such as tetrafluoroethylene, form linear copolymers, whereas hydrogen-containing monomers, such as ethylene, form copolymers containing numerous branches terminated with hexafluoroisopropyl alcohol groups. Other monomers containing less hydrogen, including vinyl fluoride and vinylidine fluoride form copolymers with proportionally fewer branches. Hexafluoroacetone forms copolymers with epoxides (89–90) and formaldehyde (91) under conditions of anionic initiation but these polymers have not attained any commercial significance.

Modified Polymeric Substrates. The radical-catalyzed addition of alkanes to hexafluoroacetone proceeds with remarkable efficiency when applied to polymeric substrates (92). Conventional peroxide catalysts serve to introduce hexafluoro-2-propanol groups in polymers, such as polyethylene or polypropylene, resulting in products that tend to be more soluble, lower melting, dyeable, and capable of hydrogen bonding to suitable acceptors. Unsaturated polymers such as poly(butadiene) have also been substituted with hexafluoroisopropyl alcohol groups by thermal ene reaction with hexafluoroacetone (93).

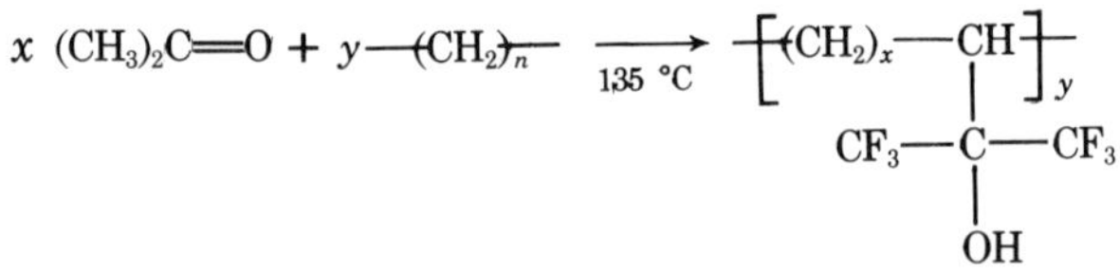

Monomers. Monomers prepared from hexafluoroacetone can be polymerized. This approach allows for considerable variety of structure and polymerization techniques. Vinyl monomers, epoxides, and bifunctional compounds for condensation polymers have been prepared. For example, reactions of the heptafluoroisopropyoxy anion prepared from hexafluoroacetone and potassium fluoride with acrylyl chloride (94) and glycidyl bromide (95) have led to the preparation of polymerizable acrylates and epoxides. A styrene monomer containing a pendant hexafluoroisopropyl alcohol group has also been prepared from hexafluoroacetone. The polymer from this monomer possesses a pronounced ability to form hydrogen bonds (96).

Polycarbonates, polyesters, and polyethers can be prepared from hexafluoroacetone bisphenol derivatives (50). A dianhydride, (**1**) prepared by the oxidation of the condensation product of hexafluoroacetone with *o*-xylene is used to prepare high-performance polyimides (see below).

Preparation, Manufacture, and Application. A number of laboratory syntheses for hexafluoroacetone have been developed, including oxidation of perfluoroisobutylene with potassium permanganate (7), isomerization of hexafluoropropylene oxide with Friedel-Crafts catalysts (98), oxidation of hexafluorothioacetone with nitrous oxide (99), and direct oxidation of hexafluoropropene with oxygen in the presence of a fluorinated Al_2O_3 (100).

Hexafluoroacetone is manufactured by the reaction of hexachloroacetone and hydrogen fluoride in the presence of a chromium catalyst (101). Currently (1979) it is produced by DuPont Co. in low volume for internal use only, mainly for the manufacture of the solvent 1,1,1,3,3,3-hexafluoro-2-propanol (102) and high performance polymers.

Health and Safety Factors. Hexafluoroacetone is moderately toxic. Its TLV has been established as 0.1 ppm (103). The inhalation toxicity for rats ranges from LC_{50} = 900 ppm for 30 min to 275 ppm for 3 h (104–106). The undiluted hydrate is a severe eye and skin irritant but aqueous concentrations below 50% produce only mild irritation (106). The oral toxicity of the hydrate is LD_{50} = 190 mg/kg for rats. Moderate central nervous system depression has been noted following administration (104).

Hexafluoroacetone hydrate was not mutagenic when tested according to the protocols of the Ames test but it did show an appreciable teratogenetic potential (106).

Derivatives and Related Compounds

Hexafluoroisopropyl Alcohol. Hexafluoroisopropyl alcohol [*920-66-1*], HFIP, C_3HF_6O, mol wt 167.04, bp 58.2°C, mp −3.4°C, $pK_a = 9.30$, is a volatile, stable liquid with high solvent power for polyamides, polyesters, and polyacetal resins (102). Its toxicity (rat oral LD_{50} = 1040 mg/kg, rat 1 h inhalation LC_{50} = 4600 ppm) is lower than that of hexafluoroacetone hydrate (106), and it is, therefore, a safer solvent to use. It is manufactured in limited quantities by the DuPont Co. by the catalytic hydrogenation of hexafluoroacetone, and is supplied in 5-kg glass bottles. Current uses include laboratory control in the manufacture of polyester resins and solvent welding for machined or molded plastic parts made of polyamides or polyesters.

Bisphenol AF. Bisphenol AF [*1478-61-1*] (4,4′-[2,2,2-trifluoro-1-(trifluoromethyl)ethylidene]bisphenol), $C_{15}H_{10}F_6O_2$, mol wt 336.24, is a white crystalline compound, mp 162°C, prepared by the reaction of hexafluoroacetone with phenol in the presence of hydrogen fluoride. It is used as a cross-linking component in the manufacture of fluoroelastomers.

4,4′-[Trifluoro-1-(trifluoromethyl)ethylidene]diphthalic Anhydride. The dianhydride [*1107-00-2*], (**1**), $C_{19}H_6F_6O_6$, mol wt 444.26, mp 252–253°C is obtained by the oxidation of the condensation product of hexafluoroacetone and *o*-xylene, and is used to prepare high-performance polyimide polymers (97).

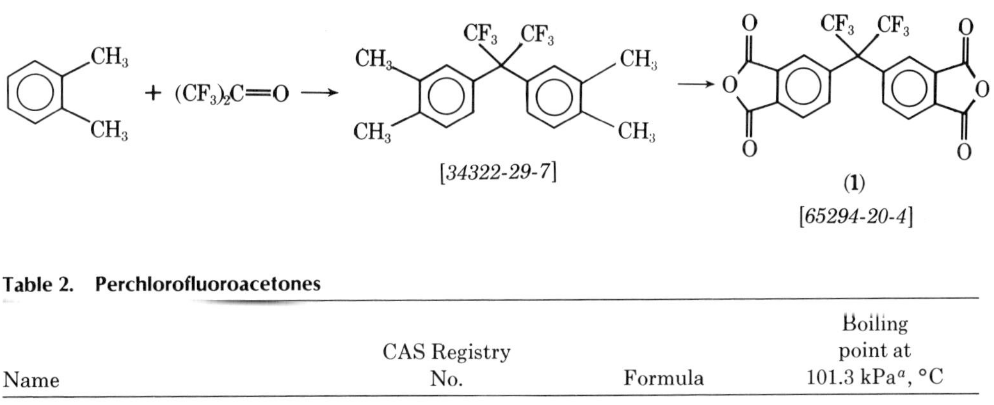

Table 2. Perchlorofluoroacetones

Name	CAS Registry No.	Formula	Boiling point at 101.3 kPa[a], °C
1,1,1,3,3,3-hexachloro-2-propanone	[*116-16-5*]	CCl_3COCCl_3	203.6
1,1,1,3,3-pentachloro-3-fluoro-2-propanone	[*2378-08-7*]	CCl_3COCl_2F	163.7
1,1,3,3-tetrachloro-1,3-difluoro-2-propanone	[*79-51-6*]	$CCl_2FCOCCl_2F$	123.9
1,1,3-trichloro-1,3,3-trifluoro-2-propanone	[*79-52-7*]	$CCl_2FCOCClF_2$	84.5
1,3-dichloro-1,1,3,3-tetrafluoro-2-propanone	[*127-21-9*]	$CClF_2COCClF_2$	45.2
1-chloro-1,1,3,3,3-pentafluoro-2-propanone	[*79-53-8*]	$CClF_2COCF_3$	7.8
1,1,1,3,3,3-hexafluoro-2-propanone	[*684-16-2*]	CF_3COCF_3	−27.3

[a] 1 atm.

Midaflur. Midaflur [*23757-42-8*] (2,5-dihydro-2,2,5,5-tetrakis(trifluoromethyl)-1*H*-imidazol-4-amine) (**2**) is the generic name of an aminoimidazoline that has been clinically tested as a tranquilizer and a muscle relaxant (107). It is a white crystalline solid, mp 159.7–160.4°C, bp 194°C$_{101.3\ kPa\ (1\ atm)}$, that is prepared by the reaction of hexafluoroacetone imine with sodium cyanide (108).

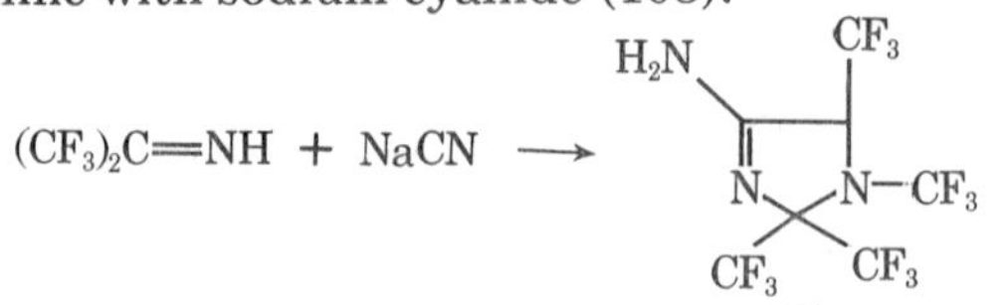

(2)

Perchlorofluoroacetones. Perchlorofluoroacetones are obtained by the incomplete replacement of chlorine atoms with fluorine when hexachloroacetone reacts with HF in the presence of a catalyst. Several of these compounds have been isolated and characterized and are listed in Table 2. These ketones, and especially those that contain at least two fluorine atoms, possess properties similar to hexafluoroacetone and undergo similar reactions. Some of these chlorofluoroacetones were manufactured in pilot-plant quantities by the Allied Chemical Corporation during the 1960s but their production has been discontinued.

BIBLIOGRAPHY

"Perchlorofluoroacetones" under "Fluorine Compounds, Organic" in *ECT* 2nd ed., Vol. 9, pp. 754–767, by Cyril Woolf, Allied Chemical Corporation.

1. N. Fukuhara and L. A. Bigelow, *J. Am. Chem. Soc.* **63,** 788 (1941).
2. A. C. Plaush and E. L. Pace, *J. Chem. Phys.* **47,** 44 (1967).
3. S. E. Babb, *J. Chem. Phys.* **50,** 5271 (1969).
4. K. P. Murphy, *J. Chem. Eng. Data* **9,** 259 (1964).
5. *Tech. Rept. DP-1, Hexafluoroacetone and Hexafluoroacetone Sesquihydrate,* E. I. du Pont de Nemours & Co., Inc., Wilmington, Del., 1965.
6. R. K. Chan, *Can. J. Chem.* **47,** 2253 (1969).
7. A. T. Morse, P. B. Ayscough, and L. C. Leitch, *Can. J. Chem.* **33,** 453 (1955).
8. G. A. Boulet, *Diss. Abstr.* **25,** 3283 (1964).
9. R. L. Hilderbrandt, A. L. Anderson, S. P. Bauer, *J. Phys. Chem.* **74,** 1586 (1970).
10. S. R. La Paglia and A. B. F. Duncan, *J. Chem. Phys.* **34,** 350 (1961).
11. E. L. Varetti and P. J. Aymonino, *An. Aso. Quim. Argent.* **59,** 435 (1971) (in Engl.); *Chem. Abstr.* **76,** 133802d (1972).
12. P. B. Ayscough and E. W. R. Steacie, *Proc. R. Soc. London Ser. A* **234,** 476 (1956).
13. E. L. Pace, A. C. Plaush, and H. V. Samuelson, *Spectrochem. Acta* **22,** 993 (1966).
14. K. D. Moeller and co-workers, *J. Chem. Phys.,* **47,** 2609 (1967).
15. M. Perttila, *Acta Chem. Scand. Ser A.* **28,** 933 (1974).
16. L. H. Sutcliffe and B. Taylor, *Spectrochem. Acta Pt. A* **28,** 619 (1972).
17. M. J. S. Dewar and S. D. Worley, *J. Chem. Phys.* **50,** 654 (1969).
18. G. A. Olah and C. V. Pittman, *J. Am. Chem. Soc.* **88,** 3310 (1966).
19. W. Batey and A. B. Trenwith, *J. Chem. Soc.,* 1388 (1961).
20. U.S. Pat. 3,245,944 (Apr. 12, 1966), W. J. Middleton (to E. I. du Pont de Nemours & Co., Inc.).
21. U.S. Pat. 3,227,674 (Jan. 4, 1966), W. J. Middleton (to E. I. du Pont de Nemours & Co., Inc.).
22. U.S. Pat. 3,153,004 (Oct. 13, 1964), W. J. Middleton (to E. I. du Pont de Nemours & Co., Inc.).
23. R. Longworth, *Nature (London)* **203,** 295 (1964).
24. I. L. Knunyants, T-Y. Chen, and N. B. Gambaryan, *Izv. Akad. Nauk SSSR Otd. Khim. Nauk,* 686 (1960).
25. R. D. Chambers and M. Clark, *Tetrahedron Lett.,* 2741 (1970).
26. U.S. Pat. 3,859,361 (Jan. 7, 1975), L. R. Anderson and co-workers (to Allied Chemical Corp.).
27. J. F. Harris, Jr., *J. Org. Chem.* **30,** 2190 (1965).

28. W. J. Middleton and H. D. Carlson in R. Breslow, ed., *Organic Syntheses,* Vol. 50, John Wiley & Sons, Inc., New York, 1970, p. 81.
29. W. J. Middleton and C. G. Krespan, *J. Org. Chem.* **30,** 1398 (1965).
30. L. N. Kryukov and co-workers, *Zh. Vses. Khim. Ova.* **21,** 232 (1976).
31. W. Steglich and co-workers, *Chem. Ber.* **107,** 1488 (1974).
32. U.S. Pat. 3,594,362 (July 20, 1971), K. Szabo (to Stauffer Chemical Co.).
33. I. L. Knunyants and co-workers, *Bull. Acad. Sci. USSR Div. Chem. Sci.,* 633 (1962).
34. F. W. Hoover, H. B. Stevenson, and H. S. Rothrock, *J. Org. Chem.* **28,** 1825 (1963).
35. W. J. Middleton in R. Breslow, ed., *Organic Syntheses,* Vol. 50, John Wiley & Sons, Inc., New York, 1970, p. 6.
36. Yu. V. Zeifman, N. P. Gambaryan, and I. L. Knunyants, *Bull. Acad. Sci. USSR Div. Chem. Sci.,* 435 (1965).
37. O. A. Mochalova, Ya. G. Urman, and I. Ya. Slonim, *Vysokomol. Soedin Ser. B* **9,** 798 (1967).
38. W. J. Middleton, *J. Org. Chem.* **30,** 1402 (1965).
39. P. Tarrant, R. H. Summerville, and R. W. Whitfield, *J. Org. Chem.* **35,** 2742 (1970).
40. F. W. Evans and co-workers, *J. Org. Chem.* **33,** 1837 (1968).
41. N. Ishikawa and co-workers, *Nippon Kagaku Kaishi,* 141 (1977).
42. J. M. Shreeve, D. A. Couch, and R. A. DeMarco, *J. Chem. Soc. D,* 91 (1971).
43. U.S. Pat. 3,761,501 (Sept. 25, 1973), C. Woolf, R. E. Dear, and D. E. Young (to Allied Chemical Corp.).
44. J. K. Ruff, A. R. Pitochelli, and M. Lustig, *J. Am. Chem. Soc.* **88,** 4513 (1966).
45. S. Andreades, *J. Org. Chem.* **27,** 4157 (1962).
46. B. S. Farah, E. E. Gilbert, and J. P. Sibilia, *J. Org. Chem.* **30,** 998 (1965).
47. W. A. Sheppard, *J. Am. Chem. Soc.* **87,** 2410 (1965).
48. B. S. Farah and co-workers, *J. Org. Chem.* **30,** 1003 (1965).
49. I. L. Knunyants and co-workers, *Zh. Vses. Khim. Ova.* **5,** 114 (1960).
50. U.S. Pat. 3,739,035 (June 12, 1973), J. L. Webb and W. L. Hall (to General Electric Co.).
51. U.S. Pat. 3,197,480 (July 27, 1965), D. C. England (to E. I. du Pont de Nemours & Co., Inc.).
52. R. L. Adelman, *J. Org. Chem.* **33,** 1400 (1968).
53. W. H. Urry, J. H. Y. Niu, and L. G. Lundsted, *J. Org. Chem.* **33,** 2302 (1968).
54. U.S. Pat. 3,164,610 (Jan. 5, 1965), H. R. Davis (to Minnesota Mining and Manufacturing).
55. U.S. Pat. 3,424,732 (Jan. 28, 1969), W. J. Middleton (to E. I. du Pont de Nemours & Co., Inc.).
56. W. J. Middleton, *J. Org. Chem.* **30,** 1307 (1965).
57. W. J. Linn, *J. Org. Chem.* **29,** 3111 (1964).
58. B. I. Martynov, L. T. Lantseva, and B. L. Dyatkin, *Zh. Org. Khim.* **11,** 2282 (1975).
59. I. L. Knunyants and Yu. A. Cheburkov, *Bull Acad. Sci. USSR Div. Chem. Sci.,* 640 (1960).
60. Z. Zubovics and N. Ishikawa, *J. Fluorine Chem.* **8,** 43 (1976).
61. A. M. Weidler-Kubanek and M. H. Litt, *J. Org. Chem.* **33,** 1844 (1968).
62. J. S. E. McIntosh and G. B. Porter, *Trans. Faraday Soc.* **64,** 119 (1968).
63. J. F. Harris and D. D. Coffman, *J. Am. Chem. Soc.* **84,** 1553 (1962).
64. E. W. Cook and B. F. Landrum, *J. Heterocyclic Chem.* **2,** 327 (1965).
65. R. B. Minasyan and co-workers, *Bull. Acad. Sci. USSR Div. Chem. Sci.,* 746 (1965).
66. D. M. Gale, W. J. Middleton, and C. G. Krespan, *J. Am. Chem. Soc.* **87,** 657 (1965).
67. J. R. Griffith and J. G. O'Rear, *Ind. Eng. Chem. Prod. Res. Dev.* **13,** 148 (1974).
68. U.S. Pat. 3,712,902 (Jan. 23, 1973), W. J. Middleton (to E. I. du Pont de Nemours & Co., Inc.).
69. W. J. Middleton and C. G. Krespan, *J. Org. Chem.* **32,** 951 (1967).
70. F. Weygand, K. Buger, and K. Engelhardt, *Ber.* **99,** 1461 (1966).
71. W. J. Middleton, D. C. England, and C. G. Krespan, *J. Org. Chem.* **32,** 948 (1967).
72. R. A. Braun, *J. Am. Chem. Soc.* **87,** 5516 (1965); *J. Org. Chem.* **31,** 1147 (1966).
73. T. S. Croft and J. J. McBrady, *J. Fluorine Chem.* **6,** 213 (1975).
74. M. E. Hermes and R. A. Braun, *J. Org. Chem.* **31,** 2568 (1966).
75. U. Utebaev, E. M. Rokhlin, and I. L. Knunyants, *Izv. Akad. Nauk SSSR Ser. Khim.,* 2260 (1974).
76. K. Burger, J. Albanbauer, and M. Eggersdorfer, *Angew. Chem.* **87,** 816 (1975).
77. B. Clarke and co-workers, *J. Chem. Soc. H,* 67 (1968).
78. J. Browning and co-workers, *J. Chem. Soc. Dalton Trans.,* 381 (1973).
79. M. Green and co-workers, *J. Chem. Soc. Chem. Commun.,* 451 (1975).
80. M. Green and co-workers, *J. Chem. Soc. Dalton Trans.,* 278 (1977).
81. R. Countryman and B. R. Penfold, *J. Cryst. Mol. Struct.* **2,** 281 (1972).

82. J. Clemens, M. Green, and F. G. A. Stone, *J. Chem. Soc. Dalton Trans.*, 375 (1973).
83. M. Green, M. Cook, and T. A. Kuc, *J. Chem. Soc. A,* 1200 (1971).
84. F. G. A. Stone and co-workers, *J. Chem. Soc. A,* 3166 (1970).
85. M. Green and A. Bond, *J. Chem. Soc. D,* 12 (1971).
86. H. D. Empsall and co-workers, *J. Chem. Soc. A,* 3472 (1971).
87. E. G. Howard and P. B. Sargent, *J. Macromol. Sci. Chem.* **1,** 1011 (1967).
88. U.S. Pat. 3,322,721 (May 30, 1967), E. G. Howard (to E. I. du Pont de Nemours & Co., Inc.).
89. U.S. Pat. 3,502,619 (Mar. 24, 1970), H. A. Smith (to The Dow Chemical Co.).
90. U.S. Pat. 3,518,216 (Apr. 25, 1967), E. G. Howard (to E. I. du Pont de Nemours & Co., Inc.).
91. U.S. Pat. 3,518,229 (June 30, 1970), V. A. Engelhardt (to E. I. du Pont de Nemours & Co., Inc.).
92. E. G. Howard, P. B. Sargent, and C. G. Krespan, *J. Am. Chem. Soc.* **89,** 1422 (1967).
93. U.S. Pat. 3,332,892 (July 25, 1967), W. J. Middleton (to E. I. du Pont de Nemours & Co., Inc.).
94. A. G. Pittman and D. L. Sharp, *Text. Res. J.* **35,** 190 (1965).
95. A. G. Pittman and D. L. Sharp, *J. Org. Chem.* **31,** 2316 (1966).
96. U.S. Pat 3,179,640 (Apr. 20, 1965), W. J. Middleton (to E. I. du Pont de Nemours & Co., Inc.).
97. U.S. Pat. 3,356,648 (Dec. 5, 1967), F. E. Rogers (to E. I. du Pont de Nemours & Co., Inc.).
98. U.S. Pat. 3,213,134 (Oct. 19, 1965), D. E. Morin (to Minnesota Mining and Manufacturing).
99. W. J. Middleton and W. H. Sharkey, *J. Org. Chem.* **30,** 1384 (1965).
100. Ger. Offen. 2,624,349 (Jan. 13, 1977), T. Tozuka and Y. Ohsaka (to Daikin Kogyo Co., Ltd.).
101. U.S. Pat. 3,164,637 (Jan. 5, 1965), H. R. Nychka and C. Woolf (to Allied Chemical Corp.); Fr. Pat. 1,372,549 (Sept. 18, 1964), F. W. Swamer (to E. I. du Pont de Nemours & Co., Inc.).
102. U.S. Pat. 3,418,337 (Dec. 24, 1968), W. J. Middleton (to E. I. du Pont de Nemours & Co., Inc.).
103. American Conference of Governmental Industrial Hygienists, *Documentation of Threshold Limit Values for Substances in Workroom Air,* 3rd ed., U.S. Government Printing Office, Washington, D.C., 1974, p. 324.
104. J. F. Brozelleca and D. Lester, *Toxicol. Appl. Pharmacol.* **7,** 592 (1965).
105. Unpublished data, Allied Chemical Corp., Morristown, N.J.
106. Unpublished data, Haskell Laboratory, Wilmington, Del.
107. R. Clark and co-workers, *Toxicol. Appl. Pharmacol.* **18,** 917 (1971); *Pharmacologist* **10,** 197 (1968).
108. W. J. Middleton and C. G. Krespan, *J. Org. Chem.* **35,** 1480 (1970).

General Reference

C. G. Krespan and W. J. Middleton in P. Tarrant, ed., *Fluorine Chemistry Reviews,* Vol. 1, Marcel Dekker, Inc., New York, 1967, p. 145.

WILLIAM J. MIDDLETON
E. I. du Pont de Nemours & Co., Inc.

FLUORINATED ACETIC ACIDS

Fluoroacetic Acid

Fluoroacetic acid [*144-49-0*], FCH_2COOH, is noted for its high toxicity to animals, including humans. It is sold in the form of its sodium salt as a rodenticide and general mammalian pest control agent (see Poisons, economic). The acid has mp, 33°C; bp, 165°C; heat of combustion, −715.8 kJ/mol (−171.08 kcal/mol) (1); and enthalpy of vaporization, 83.89 kJ/mol (20.05 kcal/mol) (2). Some thermodynamic and transport properties of its aqueous solutions have been published (3), as has the molecular structure of the acid as determined by microwave spectroscopy (4). Although first prepared by Swarts in 1896 (5), its unusual toxicity was not published until fifty years later (6). Independently, the acid was found to be the toxic constituent of a South African plant *Dichapetalum cymosum,* better known as gifblaar (7). Since then, at least two dozen other poisonous plant species have been found to contain it (8).

Chemically, fluoroacetic acid behaves like a typical carboxylic acid, although its acidity is higher than the average ($K_a = 2.2 \times 10^{-3}$) (9). It can be prepared from the commercially available sodium salt by distillation from sulfuric acid (10).

Sodium Fluoroacetate. Sodium fluoroacetate [*62-74-8*], FCH_2COONa, known commercially as Compound 1080 or as Fratol, is a hygroscopic white solid, mp, 200–202°C, which decomposes when heated above the melting point. Its solubility at 25°C in g/100 g solvent is: water, 111; methanol, 5; ethanol, 1.4; acetone, 0.04; and carbon tetrachloride, 0.004. Because its carbon–fluorine bond is unreactive under most conditions, this salt can be converted by standard procedures to typical carboxylic acid derivatives such as fluoroacetyl esters (11–12), fluoroacetyl chloride [*359-06-8*] (13), fluoroacetamide (see below) (14), or fluoroacetonitrile [*503-20-8*] (14).

Sodium fluoroacetate is usually made by displacing the halogen from an ester of bromo- or chloroacetic acid with potassium fluoride or, in one instance, antimony fluoride, followed by hydrolysis with aqueous sodium hydroxide (15–17). A commercial process for its manufacture from ethyl chloroacetate and potassium fluoride has been described (18). The ester, purified by distillation to remove traces of acid and water, is treated with oven-dried, finely powdered potassium fluoride in a well-stirred autoclave at 200°C for eleven hours. The resulting ethyl fluoroacetate [*459-72-3*] is then distilled into an agitated tank containing sodium hydroxide dissolved in methanol. The solid product is isolated by centrifugation, followed by vacuum drying. Throughout this process, all liquids are handled in a closed system of pipes and vessels, carefully inspected for leaks before each run. This is important since the intermediate fluoroacetate is highly toxic, and the starting chloroacetate is a lacrimator.

$$ClCH_2CO_2C_2H_5 + KF \rightarrow FCH_2CO_2C_2H_5 + KCl$$

$$FCH_2CO_2C_2H_5 + NaOH \rightarrow FCH_2CO_2Na + C_2H_5OH$$

In the United States, the Tull Chemical Company of Oxford, Alabama, is the only current commercial producer of sodium fluoroacetate. It is sold in 11.3-kg quantities packed in 227-g cans for about \$45/kg (1979). It is sometimes colored with the black dye, nigrosine.

Toxicity. Sodium fluoroacetate is one of the most effective all-purpose rodenticides known (18–19) (see Poisons, economic). It is highly toxic to all species of rats tested and can be used either in water solution or in bait preparations. Its absence of objectionable taste and odor and its delayed effects lead to its excellent acceptance by rodents. It is nonvolatile, chemically stable, and not toxic or irritating to the unbroken skin of workers. Rats do not appear to develop any significant tolerance to this compound from nonlethal doses. However, it is extremely dangerous to man, and to common household pets and farm animals, and should only be used by experienced personnel. The rodent carcasses should be collected and destroyed since they remain poisonous for a long period of time to any animal that eats them.

The unusually high toxicity of fluoroacetic acid and of other monofluorinated organic compounds that can be metabolized to fluoroacetate has stimulated much research into the mechanism of this toxicity (8,20–23). Fluoroacetate mimics acetate by being incorporated into the tricarboxylic acid cycle of cellular respiration where it becomes converted into fluorocitric acid. This acid inhibits the enzyme, aconitate hydratase, which normally catalyzes the dehydration of citric acid (qv). As a result, citric acid accumulates in the organism, and the energy-producing cycle is interrupted. Because of the time it takes for the fluorocitrate to form and accumulate, there is usually a latent time of at least an hour before the appearance of symptoms of fluoroacetate poisoning, eg, ventricular fibrillation or convulsions. This is advantageous in its use as a pesticide. One characteristic of fluoroacetate toxicity is the wide range in lethal doses for different species ranging from (LD_{50}, mg/kg) 0.06 in dogs, 0.2 in cats, 0.4 in sheep or rabbits, 2–10 in humans, 5 in rats, 7 in mice, to about 400 in toads (20,24). The only suggested antidotes for the poisoning are 1,2,3-propanetriol monoacetate (20,23), acetamide (20), and other acetate donors, but these only have an effect if administered before significant amounts of fluoroacetate have been converted to fluorocitrate. To determine if fluoroacetate poisoning has occurred, it is often desirable to detect the presence of small amounts of the poison in animal tissue. Although difficult, this can be done by spectrochemical methods (25), processes involving ion-selective fluoride electrodes (21,26) (see Ion-selective electrodes), or gas chromatography often combined with mass spectrometry (27–28) (see Analytical methods). Detection sensitivities down to 0.1 mg have been reported.

Fluoroacetamide. Fluoroacetamide [*640-19-7*], FCH_2CONH_2, is a white water-soluble solid, mp 108°C (14). It has been used as a rodenticide and has been reported to have a better acceptability to rats than sodium fluoroacetate (29). However, like the latter compound, its misuse has caused deaths to farm animals and pets (20).

Difluoroacetic Acid

Difluoroacetic acid [*381-73-7*], $F_2CHCOOH$, is a colorless liquid with a sharp odor; mp, 35°; bp, 134°; d_4^{10}, 1.539 g/mL; n_D^{20}, 1.3428 (30,31); flash point, 78°C (95% aqueous solution); enthalpy of vaporization, 67.82 kJ/mol (16.21 kcal/mol); and enthalpy of solution in water of the undissociated acid, −7.03 kJ/mol (−1.68 kcal/mol) (2). It is a moderately strong acid; recent determinations of its acid dissociation constant are 4.6×10^{-2} (32) and 3.5×10^{-2} (33). Its molecular structure in the gas phase has been determined by electron diffraction studies (34). Details of the acid's ir (35) and nmr (36) spectra also have been reported.

Difluoroacetic acid undergoes reactions typical of a carboxylic acid such as forming an ester when heated with an alcohol and sulfuric acid. Typical esters are methyl difluoroacetate [*433-53-4*], bp, 85.2°C, and ethyl difluoroacetate [*454-31-9*], bp, 99.2°C; d^{17} 1.18 g/mL. It can also be photochemically chlorinated to chlorodifluoroacetic acid [*76-04-0*] (see Photochemical technology), or brominated in the presence of iron to bromodifluoroacetic acid [*667-27-6*] (37–38).

The acid can be synthesized in several different ways. The reaction of tetrafluoroethylene with ammonia to give 2,4,6-tris(difluoromethyl)-*s*-triazine, followed by its alkaline hydrolysis has been reported to give the acid in 80% overall yield (31). The addition of diethylamine to tetrafluoroethylene gives, after partial hydrolysis, a 49% yield of the amide, $F_2CHCON(C_2H_5)_2$ [*56425-08-2*] which can be hydrolyzed in excellent yield to the acid (39). The same amide can be prepared in 60% yield by the addition of diethylamine to chlorotrifluoroethylene followed by hydrolysis and fluorination with KF in diethylene glycol. Another method that gives the acid in 86% yield is the permanganate oxidation of $CHF_2CH{=}CCl_2$ (40).

Difluoroacetic acid is much less toxic than fluoroacetic acid (41). It is available in research quantities for about $4.00/g.

Trifluoroacetic Acid

Physical Properties. Trifluoroacetic acid [*76-05-1*], CF_3COOH, is a colorless liquid with a sharp odor resembling that of acetic acid. Its physical properties are shown in Table 1. It is a very strong carboxylic acid with an acid dissociation constant at 25°C of 0.588 (9) or 0.32 (32). It is miscible with water, fluorocarbons, and most common organic solvents including methanol, benzene, carbon tetrachloride, acetone, ether, and hexane. Compounds with limited solubility in the acid include alkanes with more than six carbon atoms and carbon disulfide. It is a good solvent for proteins (qv) (50) and polyesters (qv). The viscosities, densities, and conductivities of solutions of the acid in acetic acid, water, and several other liquids have been studied (46).

Chemical Properties. Trifluoroacetic acid undergoes reactions typical of a carboxylic acid. The trifluoromethyl group is inert to most common reagents including lithium aluminum hydride, which gives trifluoroacetaldehyde and 2,2,2-trifluoroethanol (51–52). Common oxidizing agents do not attack the acid at room temperature except for potassium permanganate, which slowly oxidizes the anhydrous acid to carbon dioxide and other products (53). The acid is also slowly attacked by boiling

Table 1. Physical Properties of Trifluoroacetic Acid

Property	Value	Reference
freezing point, °C	−15.36	42
boiling point, °C	71.8	43
water azeotrope (20.6% H_2O) bp, °C	105.5	44
density at 25°C, g/mL	1.4844	45
heat of vaporization, kJ/mol	33.26	43
viscosity at 25°C, mPa·s (= cP)	0.813	46
dielectric constant at 25° (ϵ)	42.1	47
conductivity at 25°C, 1/(Ω·cm)	2600	46
surface tension at 25° mN/m (= dyn/cm)	13.44	48
heat of formation, liquid, kJ/mol[a]	−1060 ± 2	49

[a] To convert kJ to kcal, divide by 4.184.

25% aqueous sodium hydroxide to yield oxalate and fluoride ions (44). Although the acid is stable to temperatures above 250°C, its sodium salt will decompose above 205°C to give sodium fluoride, trifluoroacetyl fluoride [*354-34-7*], carbon monoxide, carbon dioxide, and other products (44). In ethylene glycol solution at 180°C the sodium salt can be made to decompose quantitatively to trifluoromethane [*75-46-7*] and carbon dioxide if a boric acid buffer is present (54). Except for a few instances like these, the reactions of trifluoroacetic acid closely parallel those of other carboxylic acids, but there are important differences; eg, its amides and esters are more easily hydrolyzed than is typical for carboxylic acids. This has led to the use of the acid and its anhydride [*407-25-0*] (55) in making derivatives of carbohydrates (56), amino acids (57) and peptides (57) from which the trifluoroacetyl protective group can be removed with relative ease. Its peracid [*359-48-8*], formed from the reaction of trifluoroacetic anhydride and hydrogen peroxide, is a stronger oxidizing agent than other peroxycarboxylic acids and gives better yields of epoxides from alkenes (58), esters from ketones (59), and nitrobenzenes from anilines (60) (see Peroxides and peroxy compounds). Trifluoroacetic acid and its anhydride are also useful as catalysts for reactions involving other carboxylic acids such as esterifications of alcohols or acylations of aromatic or other unsaturated compounds (45). The acid has been reported to be superior to sulfuric acid as a catalyst for the Beckmann rearrangement of oximes to amides (61). Owing to its low nucleophilicity, the acid has been used as a solvent for basic research into solvolysis mechanisms (62).

Preparation. Because of its stability to further oxidation, trifluoroacetic acid can be prepared by the oxidation of compounds containing a trifluoromethyl group bonded to carbon. Although first prepared by Swarts in 1922 by the oxidation of trifluoromethylcyclohexane or 3-aminobenzotrifluoride, later workers received better results from the alkaline permanganate oxidation of olefins such as 1,1,2-trichloro-3,3,3-trifluoropropene (40), or, more economically, 2,3-dichlorohexafluoro-2-butene which gives an 87% yield of the acid (63):

$$CF_3CCl{=}CClCF_3 \xrightarrow[2.\ H_2SO_4]{1.\ KMnO_4} 2\ CF_3COOH$$

The acid has been prepared by photochemical oxidation of ethanes such as 2-chloro-1,1,1-trifluoroethane or 2,2-dichloro-1,1,1-trifluoroethane with oxygen to give high yields of trifluoroacetyl chloride [*354-32-5*] which easily hydrolyzes to the acid (64–65). Another process involves the trimerization of trichloroacetonitrile to a triazine which can be fluorinated with a mixture of SbF_3 and $SbCl_5$ and then hydrolyzed to the acid (66). The reaction of trichloroacetyl chloride with hydrogen fluoride at 320°C over a chromium and nickel oxide catalyst has been reported to give trifluoroacetic acid in 92% yield (67).

Trifluoroacetic acid was produced commercially by the Minnesota Mining and Manufacturing Company by the electrolysis of mixtures containing acetyl fluoride, hydrogen fluoride, and sodium fluoride to give trifluoroacetyl fluoride, which upon hydrolysis gave the acid (68). Although a 71% yield is claimed, isolation of the low-boiling acid fluoride product from by-product hydrogen is costly. Improvements in this process have recently been patented (69–70) as well as processes involving the electrochemical fluorination of 2-chloroethanol (71) or chloroacetyl fluoride (72).

Economic Aspects. Trifluoroacetic acid is sold by the Halocarbon Products Corporation for ca \$15/kg (1978). The commercial product is of very high purity with the main impurity of ca 0.2% water.

Health and Safety Factors. Unlike fluoroacetic acid, trifluoroacetic acid presents no unusual toxicity problems. However, owing to its strong acidity, its vapors can be irritating to tissue, and the liquid acid can cause deep burns if allowed to contact the skin. The acid can be safely stored in containers made of glass or common corrosion-resistant alloys and metals such as stainless steel or aluminum.

BIBLIOGRAPHY

"Monofluoroacetic Acid" in *ECT* 1st ed., under "Fluorine Compounds, Organic," Vol. 6, pp. 764–766, by E. E. Hardy and J. H. Saunders, Monsanto Chemical Company; "Difluoroacetic Acid" in *ECT* 1st ed., under "Fluorine Compounds, Organic," Vol. 6, pp. 766–767, by M. G. Gergel and M. Revelise, Columbia Organic Chemicals Company; "Trifluoroacetic Acid" in *ECT* 1st ed., under "Fluorine Compounds; Organic," Vol. 6, pp. 767–768, by M. G. Gergel and M. Revelise, Columbia Organic Chemicals Company; "Monofluoroacetic Acid" in *ECT* 2nd ed., under "Fluorine Compounds, Organic," Vol. 9, pp. 767–770, by Edgar Hardy, Monsanto Research Corporation, J. H. Saunders, Mobay Chemical Company, and John B. Hynes, Hynes Chemical Research Corporation; "Difluoroacetic Acid" in *ECT* 2nd ed., under "Fluorine Compounds, Organic," Vol. 9, pp. 770–771, by John B. Hynes, Chemical Research Corporation; "Trifluoroacetic Acid" in *ECT* 2nd ed., under "Fluorine Compounds, Organic," Vol. 9, pp. 771–772, by Cyril Woolf, Allied Chemical Corporation.

1. F. Swarts, *Bull. Acad. R. Belg.* **35,** 849 (1898).
2. P. Haberfield and A. K. Rakshit, *J. Am. Chem. Soc.* **98,** 4393 (1976).
3. M. V. Kaulgud and G. H. Pandya, *Indian J. Chem. Sect. A* **14A**(2), 91 (1976).
4. B. P. Van Eijck, P. Brandts, and J. P. M. Maas, *J. Mol. Struct.* **44,** 1 (1978).
5. F. Swarts, *Bull. Acad. R. Belg.* **31,** 675 (1896).
6. F. L. M. Pattison, *Toxic Aliphatic Fluorine Compounds,* Elsevier Publishing Co., New York, 1959, p. 16.
7. J. S. C. Marais, *Onderstepoort, J. Vet. Sci. Animal Ind.* **20,** 67 (1944).
8. G. W. Miller, M. H. Yu, and M. Psenak, *Fluoride* **6**(3), 203 (1973).
9. A. L. Henne and C. J. Fox, *J. Am. Chem. Soc.* **73,** 2323 (1953).
10. F. L. M. Pattison, J. B. Stothers, and R. G. Woolford, *J. Am. Chem. Soc.* **78,** 2255 (1956).
11. F. L. M. Pattison, S. B. D. Hunt, and J. B. Stothers, *J. Org. Chem.* **21,** 883 (1956).
12. C. C. Price and W. G. Jackson, *J. Am. Chem. Soc.* **69,** 1065 (1947).
13. F. L. M. Pattison and co-workers, *Can. J. Technol.* **34,** 21 (1956).
14. F. J. Buckle, R. Heap, and B. C. Saunders, *J. Chem. Soc.,* 912 (1949).
15. B. C. Saunders and G. J. Stacey, *J. Chem. Soc.,* 1773 (1948).
16. E. D. Bergmann and I. Blank, *J. Chem. Soc.,* 3786 (1953).
17. Ref. 6, pp. 21–22.
18. R. L. Jenkins and H. C. Koehler, *Chem. Ind.* **62,** 232 (1948).
19. Ref. 6, p. 167.
20. R. A. Peters, *Fluoride* **6**(3), 189 (1973).
21. M. N. Egyed, *Fluoride* **6**(3), 215 (1973).
22. P. Buffa, V. Guarriero-Bobyleva, and R. Costa-Tiozzo, *Fluoride* **6,** 224 (1973).
23. Ref. 6, pp. 27–56, 208–210.
24. Ref. 6, pp. 3–4.
25. I. Schoenfeld and M. Lidji, *J. Forsensic Sci.* **13,** 267 (1968).
26. J. A. Peters and K. J. Baxter, *Bull. Environ. Contam. Toxicol.* **11**(2), 177 (1974).
27. H. M. Stahr, W. B. Buck, and P. F. Ross, *J. Assn. Off. Anal. Chem.* **57,** 405 (1974).
28. J. E. Peterson, *Bull. Environ. Contam. Toxicol.* **13,** 751 (1975).
29. T. Kusano, *J. Fac. Agric. Tottori Univ.* **10,** 15 (1975); *Chem. Abstr.* **84,** 160580b (1976).
30. F. Swarts, *Bull. Soc. Chim. Fr.,* 597 (1903).
31. A. L. Henne and R. L. Pelley, *J. Am. Chem. Soc.* **74,** 1426 (1952).
32. J. L. Kurz and J. M. Farrar, *J. Am. Chem. Soc.* **91,** 6057 (1969).

33. M. M. Kreevoy and co-workers, *J. Am. Chem. Soc.* **89**, 1201 (1967).
34. J. M. Bijen and J. L. Derissen, *J. Mol. Struct.* **27**, 233 (1975).
35. J. R. Barceló and C. Otero, *Spectrochim. Acta* **18**, 1231 (1962).
36. V. Barboiu, *Rev. Roum. Chim.* **19**, 363 (1974).
37. F. Swarts, *Chem. Zentr.* **II**, 709 (1903).
38. F. Swarts, *Chem. Zentr.* **I**, 1237 (1906).
39. N. N. Yarovenko and co-workers, *Obschei Khim.* **27**, 2246 (1957).
40. A. L. Henne, T. Alderson, and M. S. Newman, *J. Am. Chem. Soc.* **67**, 918 (1945).
41. Ref. 6, pp. 62–63.
42. H. H. Cady and G. E. Cady, *J. Am. Chem. Soc.* **76**, 915 (1954).
43. M. D. Taylor and M. B. Templeman, *J. Am. Chem. Soc.* **78**, 2950 (1956).
44. F. Swarts, *Bull. Acad. R. Belg. Classe Sci.* **8**, 343 (1922).
45. *Trifluoroacetic Acid Brochure,* Halocarbon Products Corp., Hackensack, N.J., 1967.
46. Y. Y. Fialkov and V. S. Zhikarev, *Zh. Obshch. Khim.* **33**, 3466, 3471, 3790 (1963).
47. J. H. Simons and K. E. Lorentzen, *J. Am. Chem. Soc.* **72**, 1426 (1950).
48. J. J. Jasper and H. L. Wedlick, *J. Chem. Eng. Data* **9**, 446 (1964).
49. V. P. Kolesov, G. M. Slavutskaya, and T. S. Papina, *Zh. Fiz. Khim.* **46**, 815 (1972).
50. J. J. Katz, *Nature* **174**, 509 (1954).
51. M. Braid, H. Iserson, and F. E. Lawlor, *J. Am. Chem. Soc.* **76**, 4027 (1954).
52. O. R. Pierce and T. G. Kane, *J. Am. Chem. Soc.* **76**, 300 (1954).
53. G. S. Fujioka and G. H. Cady, *J. Am. Chem. Soc.* **79**, 2451 (1957).
54. I. Auerbach, F. H. Verhoek, and A. L. Henne, *J. Am. Chem. Soc.* **72**, 299 (1950).
55. J. M. Tedder, *Chem. Rev.* **55**, 787 (1955).
56. E. J. Bourne and co-workers, *J. Chem. Soc.*, 2976 (1949).
57. F. Weygand, *Bull. Soc. Chim. Biol.* **43**, 1269 (1961).
58. W. D. Emmons and A. S. Pagano, *J. Am. Chem. Soc.* **77**, 89 (1955).
59. W. D. Emmons and G. B. Lucas, *J. Am. Chem. Soc.* **77**, 2287 (1955).
60. W. D. Emmons, *J. Am. Chem. Soc.* **76**, 3470 (1954).
61. U.S. Pat. 2,721,199 (Oct. 18, 1955), M. L. Huber (to E. I. duPont de Nemours & Co., Inc.).
62. P. E. Peterson and co-workers, *J. Am. Chem. Soc.* **87**, 5169 (1965).
63. A. L. Henne and P. Trott, *J. Am. Chem. Soc.* **69**, 1820 (1947).
64. R. N. Haszeldine and F. Nyman, *J. Chem. Soc.*, 387 (1959).
65. U.S. Pat. 3,883,407 (May 13, 1975), A. L. Dittman (to Halocarbon Products Corp.).
66. T. R. Norton, *J. Am. Chem. Soc.* **72**, 3527 (1950).
67. Ger. Offen 2,221,849 (Nov. 16, 1972), Ramanadin (to Rhone-Progil).
68. U.S. Pat. 2,717,871 (Sept. 13, 1965), H. M. Scholberg and H. G. Bryce (to Minnesota Mining and Manufacturing Co.).
69. U.S. Pat. 4,022,824 (May 10, 1977), W. V. Childs (to Phillips Petroleum Co.).
70. Jpn. Kokai 75 30,827 (Mar. 27, 1975), T. Suzuki and S. Yahara (to Mitsubishi Gas Chemical Co., Inc.).
71. U.S.S.R. Pat. 329,165 (Feb. 9, 1972), N. M. Arakelyan and S. E. Isabekyan.
72. Czech. Pat. 119,682 (Sept. 15, 1966), D. Frantisek.

GARY ASTROLOGES
Halocarbon Products Corporation

FLUORINATED HIGHER CARBOXYLIC ACIDS

Perfluoro Carboxylic Acids

Properties. The boiling points and densities of some straight-chain acids are listed in Table 1 (1).

The boiling points are about 50°C lower than those of the corresponding non-fluorinated acids. The density range of these acids is 1.5–1.8 g/cm^3. The acids with fewer than eight carbon atoms are liquids at room temperature. Because of the extremely electronegative character of fluorine, these compounds are strong acids, completely ionized in water solution. The acids exhibit the usual stability associated with fully fluorinated materials. They can be treated with strong alkali without hydrolysis of fluorine, and are not attacked by aqueous solutions of oxidizing or reducing agents.

The acids and their salts, especially the higher members, are very surface-active (2) and have a variety of surfactant uses (3). The physicochemical properties of aqueous solutions of fluorinated surfactants are described in ref. 4 (see Surfactants).

The reactions of the perfluoro acids are, in general, similar to those of hydrocarbon acids (see Carboxylic acids; Dicarboxylic acids). Salts are formed with the ease expected of strong acids. The heavy-metal salts are very hygroscopic. The metal salts are all water soluble and much more soluble in organic solvents than the salts of the corresponding hydrocarbon acids. Esterification takes place readily with primary and secondary alcohols. Acid anhydrides can be prepared by distillation of the acids from phosphorus pentoxide, and halides can be prepared by standard procedures. The amides are readily prepared by ammonolysis of the acid halides, anhydrides, or esters, and can be dehydrated to the corresponding nitriles (5).

Preparation. Many of the perfluoro carboxylic acids have been prepared by the electrochemical fluorination of the corresponding carboxylic acid halides (6) which involves the electrolysis of a solution of the acid halide in anhydrous hydrogen fluoride (see Electrochemical processing). The perfluoro acid fluoride is removed from the electrolytic cell as a gas or an insoluble liquid. Hydrolysis of the acid fluoride, followed by removal of fluoride ion and distillation, gives the acid. The acids may also be prepared from a perfluoro iodide, R_fCF_2I (R_f = perfluoroalkyl), by reaction with SO_3 (7) or $ClSO_3H$ (8) followed by hydrolysis. Dichromate oxidation of the olefin,

Table 1. Properties of Perfluoro Carboxylic Acids

Acid	CAS Registry No.	Formula	bp, °C	d_4^{20}, g/mL
perfluoroacetic	*[76-05-1]*	CF_3COOH	72.4	1.489
perfluoropropionic	*[422-64-0]*	C_2F_5COOH	96	1.561
perfluorobutyric	*[375-22-4]*	C_3F_7COOH	120	1.651
perfluorovaleric	*[2706-90-3]*	C_4F_9COOH	130	
perfluorocaproic	*[307-24-4]*	$C_5F_{11}COOH$	157	1.762
perfluoroheptanoic	*[375-85-9]*	$C_6F_{13}COOH$	175	1.792
perfluorocaprylic	*[335-67-1]*	$C_7F_{15}COOH$	189	
perfluorocapric	*[335-76-2]*	$C_9F_{19}COOH$	218	

$R_fCF_2CH{=}CH_2$, gives the acid R_fCF_2COOH (9). The polyfluoroalkoxyl acids of the type

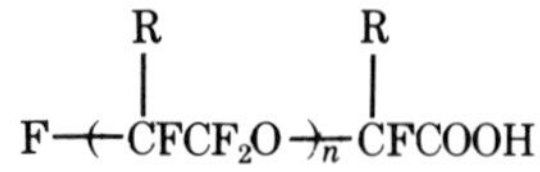

R = F [*694-17-7*]
R = CF_3 [*1187-93-5*]

are obtained from the polymerization of tetrafluoroethylene oxide or hexafluoropropylene oxide in the presence of alkali metal fluorides, with perfluoro acyl fluorides, followed by hydrolysis (10).

Amines of the formula $C_nF_{2n+1}CH_2NH_2$ can be prepared by the lithium aluminum hydride reduction of the corresponding amides or by hydrogenation of the nitriles. The analogous alcohols, $C_nF_{2n+1}CH_2OH$, can be prepared by the high pressure hydrogenation of the appropriate methyl or ethyl esters over a barium oxide-stabilized chromite catalyst or by lithium aluminum hydride reduction of the acids or acid chlorides. Perfluoro aldehydes can be formed by modification of the above reduction (5). The derivatives of perfluorobutyric acid have been the most thoroughly studied (1,11).

Uses. The patent literature reveals a number of diverse applications for the acids, eg, an improved method of engraving metal surfaces (12) (see Metal surface treatments); chemical treatment of carrier particles (13) and lubricants (14) in electrophotography (qv); corrosion (qv) inhibition by hydrophobic compounds (15) (see Waterproofing); removal of nitrogen oxides from a waste gas stream (16); purification of polyolefins (17) (see Olefin polymers); and codeposition of a metal and fluorocarbon resin in aqueous plating baths in the presence of fluorocarbon surfactants (18) (see Electroplating).

The major commercial use of the acids, especially ammonium perfluorocaprylate, is in the production of polytetrafluoroethylene colloidal aqueous dispersions (19). Conventional emulsifiers cause polymerization inhibition or produce polymers of unacceptable properties by chain transfer (see Emulsions).

Perfluoro Dicarboxylic Acids

The lower members of the series of perfluoro dicarboxylic acids are stable compounds. They have been synthesized by oxidation of the appropriate chlorofluoroolefins as well as by electrochemical fluorination. Perfluoroglutaric acid and perfluoroadipic acid can be prepared by exchange fluorination and subsequent oxidation of hexacyclopentadiene and perchlorobenzene, respectively. The acids are recovered from the oxidation mixture by acidification with sulfuric acid and prolonged extraction with ethyl ether.

Table 2 lists typical properties of some perfluoro dicarboxylic acids.

These acids are of potential interest in the preparation of polyamides (qv) and other fluorinated polymers. A detailed description of the chemistry of the perfluoro carboxylic acids and their derivatives appear in ref. 24.

Table 2. Properties of Perfluoro Dicarboxylic Acids

Acid	CAS Registry No.	Formula	mp, °C	bp[a], °C	bp of diester[a], °C	Reference
perfluoromalonic	[*1514-85-5*]	$CF_2(COOH)_2$			$58–59_{1.2\ kPa}$ (methyl)	20
perfluorosuccinic	[*377-35-8*]	$(CF_2)_2(COOH)_2$	87	$150_{2\ kPa}$	$89_{2\ kPa}$ (ethyl)	21
perfluoroglutaric	[*376-73-8*]	$(CF_2)_3(COOH)_2$	78–88	$134–138_{0.4\ kPa}$	$100_{4.5\ kPa}$ (methyl)	22
perfluoroadipic	[*336-08-3*]	$(CF_2)_4(COOH)_2$	70–100[b]		$70–71_{0.3\ kPa}$ (ethyl)	23
perfluorosuberic	[*678-45-5*]	$(CF_2)_6(COOH)_2$	154–158		$156–159_{3.6\ kPa}$ (ethyl)	

[a] To convert kPa to mm Hg, multiply by 7.5.
[b] Dihydrate.

Safety and Handling. All normal precautions for handling strong acids should be used with perfluoro acids. Inhalation of vapors or mists should be avoided. Ingestion or contact with the eyes should be prevented in order to avoid severe damage to the tissues. Heating perfluoro acids to ca 250°C may liberate toxic gases. The acute toxicity of perfluorooctanoic acid [*335-67-1*] in female albino mice has been reported (25). The compound was given a toxicity rating of "slightly toxic."

BIBLIOGRAPHY

"Heptafluorobutyric Acid" in *ECT* 1st ed., under "Fluorine Compounds, Organic," Vol. 6, p. 769, by M. G. Gergel and Max Revelise, Columbia Organic Chemicals Company, Inc.; "Other Perfluorocarboxylic Acids" in *ECT* 1st ed., under "Fluorine Compounds, Organic," Vol. 6, pp. 769–771, by L. J. Hals and W. H. Pearlson, Minnesota Mining & Manufacturing Co.; "Heptafluorobutyric Acid" in *ECT* 2nd ed., under "Fluorine Compounds, Organic," Vol. 9, p. 773, by M. G. Gergel, Columbia Organic Chemicals Company; "Other Perfluorocarboxylic Acids" in *ECT* 2nd ed., under "Fluorine Compounds, Organic," Vol. 9, pp. 773–775, by L. J. Hals and W. H. Pearlson, Minnesota Mining & Manufacturing Company.

1. E. A. Kauck and A. R. Diesslin, *Ind. Eng. Chem.* **43,** 2332 (1952).
2. H. M. Scholberg, R. A. Guenthner, and R. I. Coon, *J. Phys. Chem.* **57,** 923 (1953).
3. R. A. Guenthner and M. L. Vietor, *Ind. Eng. Chem. Prod. Res. Dev.* **1,** 165 (1962).
4. H. Shinoda, M. Hato, and T. Hryoshi, *J. Phys. Chem.* **76,** 909 (1972).
5. D. R. Husted and A. H. Ahlbrecht, *J. Am. Chem. Soc.* **75,** 1605 (1953).
6. J. H. Simmons and co-workers, *J. Electrochem. Soc.* **95,** 47 (1949).
7. Fr. Pat. 1,343,661 (Nov. 22, 1963), M. Hauptschein (to Pennwalt Corp.).
8. U.S. Pat. 3,238,240 (Mar. 1, 1966), M. Hauptschein (to Pennwalt Corp.).
9. U.S. Pat. 3,525,758 (Aug. 25, 1970), A. Kalsushimo (to Daikin Koygo).
10. U.S. Pat. 3,088,958 (May 7, 1963), R. Feltman (to E. I. du Pont de Nemours & Co., Inc.); U.S. Pat. 3,798,265 (Mar. 19, 1974) P. Bartlett (to E. I. du Pont de Nemours & Co., Inc.).
11. *3M Brand Perfluorocarboxylic Acids,* Minnesota Mining and Mfg. Co., St. Paul, Minn.
12. Jpn. Kokai 77 50,239 (Apr. 22, 1977), T. Ueno and T. Oguchi (to Tokyo Shiburn Electric Co., Ltd.).
13. U.S. Pat. 4,073,980 (Sept. 9, 1976), V. Wetsdale and J. Novotny (to Addressograph-Multigraph Corp.).
14. Neth. Pat. 75 02354 (Mar. 25, 1974), (to Xerox Corp.).
15. E. McCafferty, *Corros/74, Int. Corros. Forum 1974,* 72/1-72/6; *Chem. Abstr.* **84,** 171353f (1976).
16. Jpn. Kokai 76 112,784 (Oct. 5, 1976), K. Konno, K. Sato, and H. Watanabe (to Kureka Chemical Industry Co.).
17. U.S. Pat. 3,709,850 (Jan. 9, 1973), R. Harrison and D. Crum (to Phillips Petroleum).
18. U.S. Pat. 3,677,907 (July 18, 1972), H. Brown (to Udylite Corp.).
19. U.S. Pat. 2,559,752 (July 10, 1952), K. Berry (to E. I. du Pont de Nemours & Co., Inc.).
20. A. L. Henne and E. G. DeWitt, *J. Am. Chem. Soc.* **70,** 1548 (1948).
21. A. L. Henne and W. J. Zimmerschied, *J. Am. Chem. Soc.* **69,** 281 (1947).
22. *Ibid.,* **67,** 1235 (1945).
23. E. T. McBee, P. A. Wiseman, and G. B. Bachman, *Ind. Eng. Chem.* **39,** 415 (1947).
24. A. M. Lovelace, W. Postelnek, and D. A. Rauch, *Aliphatic Fluorine Compounds,* ACS Monograph 138, Reinhold Publishing Co., 1958.
25. F. A. Smith and co-workers, *Toxicol. Appl. Pharmocol.* **2,** 54 (1960).

RICHARD A. GUENTHNER
3M Company

FLUORINATED AROMATIC COMPOUNDS

This article covers the preparation, properties, reactions, and applications of three types of fluorinated aromatic compounds: ring-fluorinated aromatics, eg, fluorobenzenes, -biphenyls, -naphthalenes, and fused-ring systems; side-chain fluorinated aromatics, eg, benzotrifluorides and other poly(fluoroalkyl) benzenes; fluorinated nitrogen heterocyclics, eg, fluoropyridines, -quinolines, -pyrimidines, -triazines, and miscellaneous nitrogen systems.

Although known for over eighty years, aromatic fluorine compounds have had few practical applications until recently. Fluoroaromatics have found utility in drugs, crop protection chemicals, and numerous other uses. The special properties conferred by fluorine justify the higher costs required to produce fluoroaromatics. The unusual physiochemical and biological properties that fluorine imparts to aromatics result from the small size of fluorine (it is bio-isosteric with both the hydrogen atom and the hydroxyl group) and from its striking electronic properties. The latter includes its high electronegativity and its ability to alter polarity of adjacent groups, as well as to donate electrons by resonance. Other significant properties imparted to aromatics by fluorine include: enhanced stability of the C—F bond (absence of activating groups); high lability of fluorine (presence of nitro and other activating groups); high lipid solubility; hydrogen-bonding potential (acceptor role); and enzyme inhibition. The carbon–fluorine link in fluoroaromatics has been of considerable value as a label for metabolic, mechanistic, and structural studies.

Depending on which substituents are present, fluoroaromatic intermediates can be converted into fluorinated or fluorine-free products. Fluorine substitution can affect the biological spectrum of the parent aromatic or heterocyclic compound by enhancement of desired properties or by suppression of undesired properties. Fluorine-containing aromatics have been incorporated into drugs (hypnotics, tranquilizers, anti-inflammatory agents, analgesics, antibacterials, etc) and into crop protection chemicals (herbicides, insecticides, fungicides). Oxygen-regenerative chelate systems, liquid crystals, etc, represent new areas where fluoroaromatics have gained attention.

For fluorine-free products, the lability of fluorine in fluoronitrobenzenes and other activated molecules permits it to serve as a handle in hair-dye manufacturing operations. Other examples of labile fluorine include characterization of amino acids and peptides, protein solubilizing agents, antibody tagging agents, protein cross-linking agents, and diuretics. The discovery of new fluorination techniques, several of which have been adapted to industrial operations, are associated with these commercial developments (see Fluorine compounds, organic—Direct fluorination).

Preparative Methods

Ring-Fluorinated Aromatics and Heterocyclics. In contrast with other molecular halogens (Cl_2, Br_2), direct aromatic substitution with fluorine (F_2) gave violent reactions involving ring scission, addition, coupling, and polymerization. Consequently, indirect fluorination techniques (based on diazotization of anilines) initially were developed to prepare partially fluorinated aromatics. Recent advances in synthetic methods include discoveries of new fluorinating agents and modifications of known methods.

Diazotization Routes. Conventional Sandmeyer reaction conditions are not suitable to make fluoroaromatics. Phenols primarily result from the high solvation of fluoride ion in aqueous media.

Fluoroaromatics are produced on an industrial scale by diazotization of substituted anilines with sodium nitrite in anhydrous hydrogen fluoride, followed by *in situ* decomposition of the aryldiazonium fluoride (1–2). The decomposition temperature depends on the stability of the diazonium fluoride (3). The addition of pyridine (4) and other tertiary amines (5), as well as ammonium bifluoride (6), permits higher decomposition temperatures (>50°C) under atmospheric pressure with minimum hydrogen fluoride losses. High temperature *in situ* decomposition of highly stable *o*-aryldiazonium fluorides in hydrogen fluoride have been described (7).

The Balz-Schiemann reaction is a useful laboratory method for the preparation of fluoroaromatics. The water-insoluble diazonium fluoborate is filtered, dried, and thermally decomposed (dediazoniation) to give the aryl fluoride, nitrogen, and boron trifluoride (8). Aryldiazonium hexafluorophosphates, ArN_2PF_6, frequently give higher

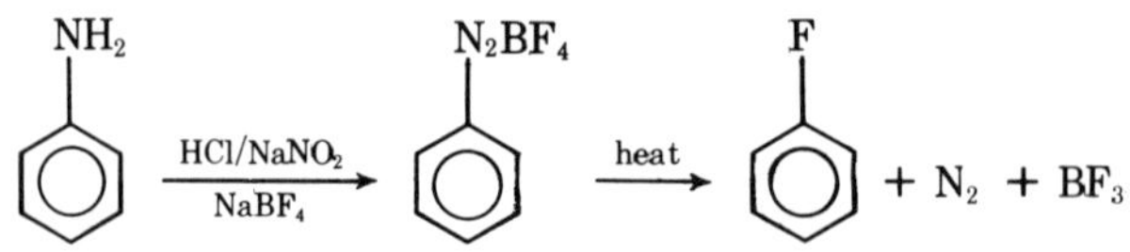

overall yields of the fluoroaromatic (9). A single fluorine atom can be introduced sequentially by the Balz-Schiemann reaction (via successive nitration, reduction, and diazotization) for a total of up to four fluorine atoms, eg, 1,2,4,5-tetrafluorobenzene.

Exchange Fluorination. Reactions of activated halo-aromatics and -heterocyclics with potassium fluoride represent a major development in industrial aromatic fluorine chemistry. G. C. Finger and co-workers initially prepared *o*- and *p*-fluoroaromatics that were activated by nitro or cyano groups by exchange fluorination in dipolar aprotic solvents, eg, dimethyl sulfoxide (DMSO), *N,N*-dimethylformamide (DMF), and dimethyl sulfone ($DMSO_2$) (10–11). Other useful solvents include tetramethylene sulfone (Sulfolane), *N*-methylpyrrolidinone (NMP), and *N,N*-dimethylacetamide (DMAC).

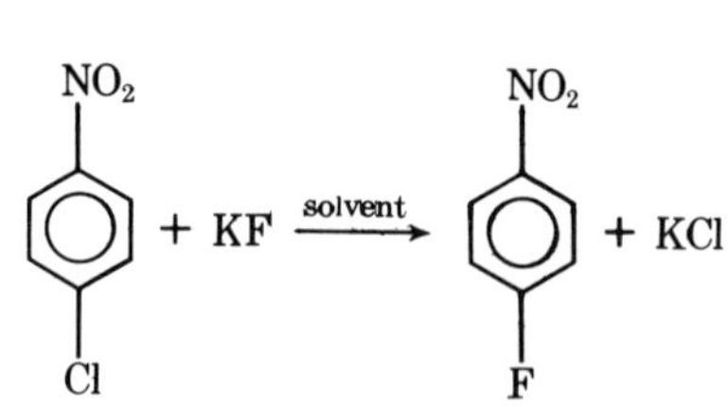

Features of the Finger technique include good fluorine utilization (1:1 stoichiometry), facile product separation, and potential recycling of potassium chloride (as KF) by treatment with hydrogen fluoride or fluorine. The role of aprotic solvents permits less solvation of fluoride ion (as compared with protic solvents), a kinetically significant amount of fluoride ion in solution, and greater insolubility of potassium chloride which, in turn, provides a further reaction driving force.

Of the alkali metal fluorides, potassium fluoride offers the best compromise between cost and effectiveness. Although cesium fluoride generally gives higher yields, its higher cost may be a potential drawback as an industrial fluorination tool. In contrast, inexpensive sodium fluoride consistently gives lower yields than potassium fluoride.

The degree of fluorination can be limited by the thermal stability of the solvent or by its reaction with basic potassium fluoride through proton abstraction. Such solvent-derived by-products can subsequently react with the starting material and/or main product.

Perchlorinated aromatics devoid of nitro groups, eg, hexachlorobenzene, also can be fluorinated in the presence of polar solvents (12) or under solvent-free conditions (450–540°C in an autoclave) (13).

$$C_6Cl_6 + KF \xrightarrow[\substack{230\text{–}240°C \\ 18\ h}]{\text{sulfolane}} C_6ClF_5 + C_6Cl_4F_2 + 1{,}3{,}5\text{-}C_6Cl_3F_3$$

Saturation–Rearomatization. The first commercial route to perfluorinated aromatics was based on a multistage saturation–rearomatization process (14). In the first stage, benzene is fluorinated by a high-valency oxidative metal fluoride (cobalt trifluoride) to give a mixture of polyfluorocyclohexanes. The latter is subjected to a combination of dehydrofluorination (with alkali) and/or defluorination (with heated iron, iron oxide, or nickel packing) to give hexa-, penta-, and tetrafluorobenzenes. Recent modifications of the first stage include the use of complex metal fluorides, eg, potassium tetrafluorocobaltate(III), that have been found to be milder and more selective fluorinating agents than cobalt trifluoride (15). An electrochemical reduction process for the rearomatization step also has been described (16). A related oxidative fluorination process features successive treatment of hexachlorobenzene with chlorine trifluoride (17) or fluorine (18), followed by dehalogenation with iron powder at 300°C.

Substitutive Aromatic Fluorination. New techniques effect substitutive aromatic fluorination with *tamed* fluorine, eg, elemental fluorine (19–20), trifluoromethyl hypofluorite (21–22), and xenon fluorides, primarily as XeF_2 (23). Some success has been achieved with chlorine trifluoride (24) and pentafluoride (25) and with silver difluoride (26).

Fluoronucleic bases and fluoronucleosides can be made by direct substitutive fluorination. For example, 5-fluorouracil [*4476-06-1*], a drug used in cancer chemotherapy, is prepared in high yield from the reaction of uracil with fluorine or trifluoromethyl hypofluorite (27–30) (see Chemotherapeutics, antimitotic).

Fluoroaliphatic Thermolytic Routes. A potential commercial process is the reaction of difluorocarbene (generated from $CHClF_2$ at 600°C) with cyclopentadiene to give fluorobenzene in 70% yield (31). Pyrolysis of tribromofluoromethane (CBr_3F) at 630–640°C gives hexafluorobenzene in 55% yield (32–33). The principal disadvantage of this process is the low weight yield of product; 90% of the costly CBr_3F that is charged is lost as bromine.

Miscellaneous Methods. Decarbonylation of benzoyl fluorides with *tris*(triphenylphosphine)rhodium chloride, $Rh[(C_6H_5)_3P]_3Cl$, in refluxing xylene (34), or thermal decarboxylation of aryl fluoroformates and thiolfluoroformates (35) represent novel complementary routes to fluoroaromatics. Neither method has been commercialized.

Single-step routes to *p*-fluoroaniline [*371-40-4*] (61% yield) based on fluoro-deoxygenation of nitrobenzene have been discovered. Techniques include hydrogenation (PtO_2 catalyst) in anhydrous HF at 50°C (36–37) or use of triphenylphosphine reagent in HF at 150°C (38).

Fluoroaromatics also can be synthesized by controlled potential electrolysis in the absence of hydrogen fluoride under the following conditions: platinum anode, +2.4 V; acetonitrile solvent; tetraethylammonium fluoride electrolyte (39). For example, anodic fluorination of benzene gave fluorobenzene in 36% yield and *p*-difluorobenzene in 5% yield. Yield enhancement and other improvements could have commercial implications.

Side-Chain Fluorinated Aromatics and Heterocyclics. ***Fluorination of Benzotrichlorides.*** Benzotrifluorides generally are prepared from trichloromethylaromatics with metal fluorides or hydrogen fluoride. Industrial processes feature reaction with hydrogen fluoride under high pressure (40), or at atmospheric pressure at low temperature (41), or under vapor-phase conditions (42).

Sulfur Tetrafluoride and Aromatic Carboxylic Acids. Benzotrifluorides also are prepared from aromatic carboxylic acids (and their derivatives) with sulfur tetrafluoride (SF_4) (43). Two equivalents of sulfur tetrafluoride are required:

$$ArCO_2H + SF_4 \rightarrow ArCOF + HF + SOF_2$$

$$ArCOF + SF_4 \rightarrow ArCF_3 + SOF_2$$

Typical temperature and operating time conditions are 100–150°C and 6 h; hydrogen fluoride is frequently used as a catalyst. The high cost of SF_4 and the incomplete use of fluorine justify its use only for inaccessible benzotrifluorides.

Perfluoroalkylation. A useful route to side-chain fluorinated aromatics features copper-assisted coupling of perfluoroalkyl iodides with aromatic halides (44–45). A complementary approach was aluminum chloride-catalyzed electrophilic alkylation by hexafluoroacetone; eg, *p*-fluoro-α,α-bis(trifluoromethyl)benzyl alcohol [*2402-74-6*] was obtained in 66% yield from fluorobenzene (46).

Ring-Fluorinated Benzenes

Fluorobenzene. Fluorobenzene [*462-06-6*] (monofluorobenzene), C_6H_5F, has a mol wt of 96.1, and is a colorless mobile liquid with a pleasant aromatic odor. Its thermal stability is of a high order; fluorobenzene undergoes no detectable decomposition when kept at 350°C for 24 h at pressures of up to 40.5 MPa (400 atm). Toxicity: oral (rat), $LD_{50} > 4$ g/kg; inhalation (mouse), LD_{50} 45 g/m^3 (2 h) (47).

Physical properties of fluorobenzene are listed in Table 1.

Reactions. *Electrophilic substitution.* Fluorobenzene electrophilic substitution reactions are more para directing than are the same chlorobenzene reactions, as is shown in Table 2 (49).

Nitration of fluorobenzene with concentrated nitric and sulfuric acid gives a 92:8 mixture of *p*- [*350-46-9*] and *o*-fluoronitrobenzene [*1493-27-2*]. The most attractive route to *o*- and *p*-fluoronitrobenzene is based on the exchange-fluorination of commercially available *o*- and *p*-chloronitrobenzene in numerous polar solvents (50–52). Solvent-free exchange fluorination of chloronitrobenzenes with potassium fluoride also has been described (53).

Application of the Friedel-Crafts ketone synthesis to fluorobenzene has commercial importance. For example, the reaction of 4-chlorobutyryl chloride with fluorobenzene gives 4-chloro-4′-fluorobutyrophenone [*3874-54-2*], $4\text{-}FC_6H_4\underset{\underset{O}{\|}}{C}(CH_2)_3Cl$

Table 1. Physical Properties of Fluorobenzene

Property	Value
melting point, °C	−42.22
boiling point, °C	84.73
density, 25°C, g/mL	1.0183
coefficient of expansion	0.00116
refractive index, n_D^{25}	1.4629
viscosity, mPa·s (= cP)	
9.3°C	0.653
19.9°C	0.585
80.9°C	0.325
surface tension, mN/m (= dyn/cm)	
9.3°C	28.49
20.0°C	27.71
34.5°C	25.15
latent heat of fusion, J/mol[a]	11,305.2
latent heat of vaporization, 25°C, J/mol[a]	34,576.6
specific heat, 25°C, J/mol[a]	146.3
critical temperature, °C	286.94
critical pressure, kPa[b]	4550.9
critical density, g/mL	0.269
dielectric constant, 30°C	5.42
dipole moment, C·m[c]	4.90×10^{-30}
heat of combustion, J/g[a]	−32,273.3
heat of formation, kJ/mol[a]	
vapor	−110.5
liquid	−145.2
solubility in water, 30°C, g/100 g	0.154
solubility of water in fluorobenzene, 25°C, g/100 g	0.322
boiling point of binary azeotrope, °C	
with 31 wt % *tert*-butyl alcohol	76.0
with 32 wt % methanol	59.7
with 30 wt % isopropyl alcohol	74.5
flash point (tag open cup), °C[d]	−13
vapor pressure, in °C and kPa[b]	Antoine equation: $\text{Log}_{10} P = 6.07687 - \frac{1248.083}{(t + 221.827)}$

[a] To convert J to cal, divide by 4.184.
[b] To convert kPa to mm Hg, multiply by 7.5.
[c] To convert C·m to debye (D), divide by 3.336×10^{-30}
[d] Ref. 48.

(54), which is the key intermediate in the synthesis of fluorobutyrophenone tranquilizers, eg, Haloperidol (see Monofluoroaromatic Drugs). Acetylation of fluorobenzene with acetyl chloride or acetic anhydride gives 4-fluoroacetophenone [*403-42-9*],

$$4\text{-}FC_6H_4\underset{\underset{O}{\|}}{C}CH_3$$

Table 2. Isomer Distribution From Electrophilic Substitution

	Fluorobenzene		Chlorobenzene	
	Para	*Ortho*	*Para*	*Ortho*
bromination	98.2	1.8	88.8	11.2
nitration	92	8	70	30
chlorination	84	16	61	39

(55), which is a principal intermediate in the synthesis of the anti-inflammatory agent, Flazalone (see Fluoroaromatic Drugs). Another reaction with pronounced para orientation is the formation of the insecticide, 1,1-bis(4-fluorophenyl)-2,2,2-trichloroethane [*475-26-23*], GIX (DFDT), from chloral and fluorobenzene. This DDT analogue was produced in Germany during World War II.

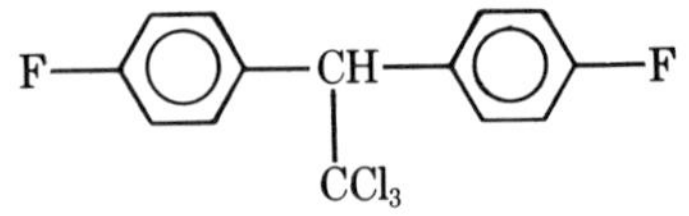

GIX (DFDT)

Nucleophilic displacement reactions. It is difficult to replace the fluorine atom in fluorobenzene by nucleophilic attack. Only a very slow reaction occurs with potassium amide at 80°C or with aqueous ammonia (with a cuprous chloride catalyst) at 250°C. The presence of activating groups, eg, nitro groups, makes aromatic fluorine very reactive in nucleophilic displacement reactions. This is illustrated in Table 3 by the relative fluorine/chlorine displacement ratios from the reaction of halonitrobenzenes with sodium methoxide in methanol (56).

Numerous applications have been based on the lability of fluoronitroaromatics: 4-fluoro-3-nitroaniline [*364-76-1*], a commercial hair-dye intermediate (57–58); 2,4-dinitrofluorobenzene [*70-34-8*] (Sanger's reagent), for amino acid characterization (59); 4-fluoro-3-nitrophenyltrimethylammonium iodide [*39508-27-5*], a protein solubilizing reagent (60); 4-fluoro-3-nitrophenylazide [*28166-06-5*], an antibody tagging reagent (61) and for industrial immobilization of enzymes (62). Other examples of biochemical applications (amino acid or peptide characterization; protein cross-linking reagent) include: 2,4-dinitro-5-fluoroaniline [*361-81-7*] (Bergmann's reagent) (63); 4-fluoro-3-nitrobenzoates (64–65); 4-fluoro-3-nitrobenzenesulfonic acid [*349-05-3*] (66); 4-fluoro-3-nitrophenyl sulfone [*51451-34-4*] (67); 1,5-difluoro-2,4-dinitrobenzene [*327-92-4*] (68); and 3,5-dinitro-2-fluoroaniline [*18646-02-1*] (69).

Table 3. Fluorine–Chlorine Displacement Ratios: Reaction of Halonitrobenzenes with Methoxide Ion

Aryl halide	Temperature, °C	F/Cl Ratio
o-fluoronitrobenzene	100	3.2×10^2
o-chloronitrobenzene	100	
p-fluoronitrobenzene	82	2.2×10^2
p-chloronitrobenzene	82	

Examples of reactive fluoroaromatics are not restricted to fluoronitrobenzenes. The fluorine-free diuretic, Furosemide, is prepared in 85% yield from 2-fluoro-4 chloro-5-sulfamoylbenzoic acid and furfurylamine at 95°C for 2 h (70):

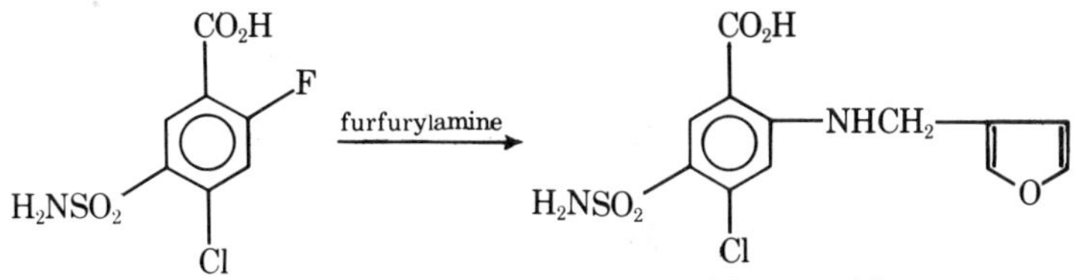

Furosemide

The design of activated aromatics containing labile fluorine has been recommended for use as a chemical marker to elucidate reaction mechanisms (71). This would complement the widely used method of labeling aromatic substrates with fluorine using ^{19}F nmr as a probe for product identification (72). Another important development is the incorporation of the ^{18}F isotope, which has a $t^{1/2}$ of 109.27 ± 0.06 min, into biologically active aromatics for radiopharmaceutical research (73) (see Radioactive drugs and tracers).

Less activated substrates such as fluorohalobenzenes also undergo nucleophilic displacement and thereby permit entry to other useful compounds. Bromine is preferentially displaced in *p*-bromofluorobenzene [*460-00-4*] by hydroxyl ion under the following conditions: calcium hydroxide, water, cuprous oxide catalyst, 250°C, 3.46 MPa (500 psi), to give *p*-fluorophenol [*371-41-5*] in 79% yield (74–75). *p*-Fluorophenol and its nuclear-substituted derivatives can be converted to the corresponding fluorophenoxyacetic acids, which exhibit promising plant growth regulator properties (76–78) (see Plant growth substances). Similar preferential bromine replacement by alkaline hydrolysis occurs with *o*-bromofluorobenzene [*1075-82-1*] to give *o*-fluorophenol [*367-12-4*]. *o*-Fluorophenol is a precursor to 3-fluorosalicylaldehyde [*394-50-3*], which is converted to cobalt bis(3-fluorosalicylaldehyde)ethyleneimine [*62207-65-5*] (Fluomine), a promising oxygen-regenerative system for aircraft (79) (see Chelating agents).

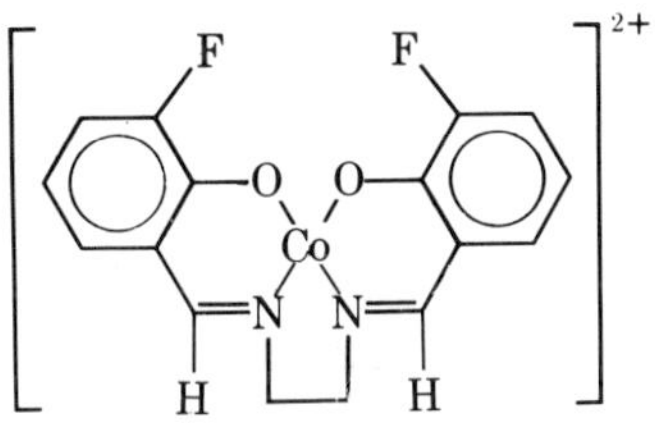

Fluomine

Fluoroaryl organometallics. Fluorobenzene does not form a Grignard reagent with magnesium. However, a recently discovered form of activated magnesium, which was prepared from magnesium chloride–potassium iodide–potassium, converted *p*-fluorotoluene [*352-32-9*] to *p*-tolylmagnesium fluoride (80). 4-Bromofluorobenzene [*460-00-4*] can be selectively converted to 4-fluorophenylmagnesium bromide [*352-13-6*].

Phenyllithium cannot be formed from fluorobenzene. Instead, the electronegativity of fluorine makes the ortho hydrogen sufficiently acidic to permit reaction with *n*-butyllithium in tetrahydrofuran at −50°C to give 2-fluorophenyllithium [*348-53-8*]. *Caution:* An isomer, 4-fluorophenyllithium [*1493-23-8*], was reported to be explosive in the solid state (81).

Manufacture. Fluorobenzene is produced by diazotization of aniline in anhydrous hydrogen fluoride at 0°C, followed by *in situ* decomposition of benzenediazonium fluoride at 20°C (1). According to German experience during World War II, the yield for 750 kg batches was 75–77% (2).

The spent hydrogen fluoride layer (which contains water and sodium bifluoride) from this process is treated with sulfur trioxide, and hydrogen fluoride is distilled for recycle to the next batch (82).

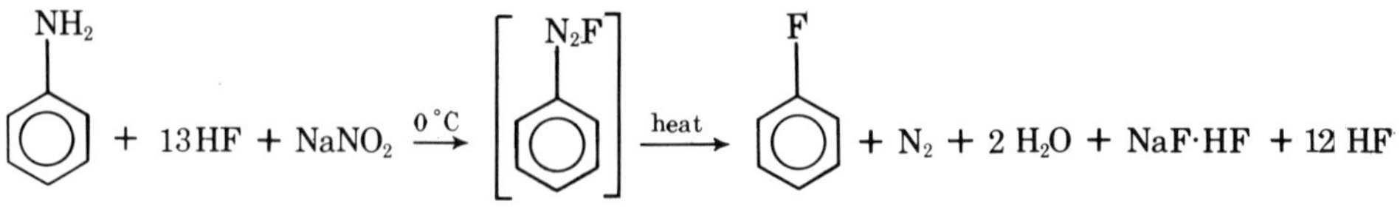

$$12\ HF + NaF{\cdot}HF + 2\ H_2O + 2\ SO_3 \longrightarrow 14\ HF + NaHSO_4 + H_2SO_4$$

Either nitrosyl chloride (83), nitrosyl fluoride–hydrogen fluoride (NOF.3HF; NOF.6HF) (84), or nitrous acid–hydrogen fluoride solutions (85–86) have been used in place of sodium nitrite in the diazotization step. Exchange fluorination of chlorobenzene is not a practical method for fluorobenzene synthesis. Chlorobenzene is inert to potassium fluoride upon refluxing in polar solvents.

Applications. *Crop-protection chemicals.* Low manufacturing costs of chloroaromatic crop-protection chemicals represents a major advantage over fluorinated analogues. However, chloroaromatics pose ecological problems such as pesticide persistency, toxicity, etc. Since ring-fluorinated aromatics offer potential agronomic advantages, eg, dosage, selectivity, and crop safety, a renewed interest in these compounds as crop-protection chemicals has developed (see Herbicides; Fungicides). Table 4 lists representative monofluoroaromatic examples. (For other examples, see Difluorobenzenes).

Drugs. The toxicology of fluoroaromatics differs from that of some fluoroaliphatics, eg, monofluoroacetic acid, and their derivatives; one review (87) states: "In general, monofluorinated aromatic compounds are not highly toxic. They are managed by the body as are the nonfluoroanalogues and with the C—F link intact."

Thus ring-fluorinated aromatics have found broad pharmaceutical applications, eg, in tranquilizers, hypnotics, sedatives, antibacterial agents, etc. Representative monofluorinated drugs are listed in Table 5 (see Fluorinated Biphenyls, Fluoronaphthalenes, and Fused-Ring Bicyclics for other examples).

Other biological applications. Salts of fluorohexidine, 1,6-di-(4′-fluorophenyldiguanido)hexane [*53415-55-0*], and related *p*-fluoroaniline-based anticaries agents represent a potential complementary approach to inorganic fluorides for the prevention of dental decay (see Dentifrices) (88). Dinitrofluorobenzenes exhibited the highest fungicidal potency from an evaluation of constitution–potency relationships in 12 fluoroaromatic classes (89–90).

Table 4. Monofluoroaromatic Crop-Protection Chemicals

Common name	CAS Registry No.	Structure	Application
flampropisopropyl	*[52756-22-6]*	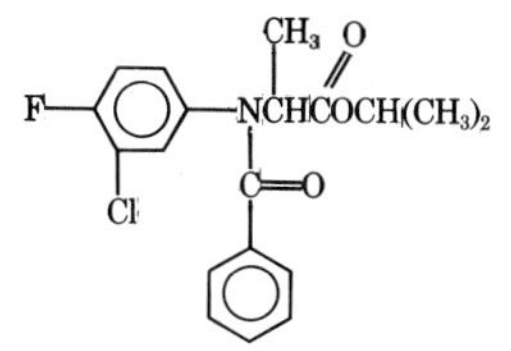	post-emergent herbicide
fluoronitrofen	*[13738-63-1]*	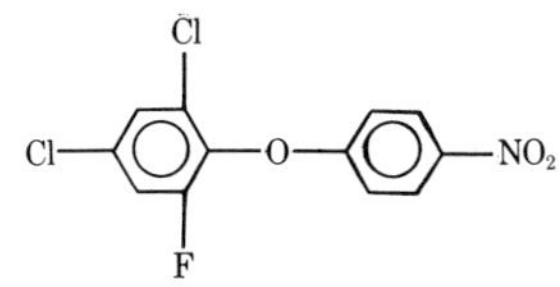	pre-emergent herbicide
fluoroimide	*[41205-21-4]*	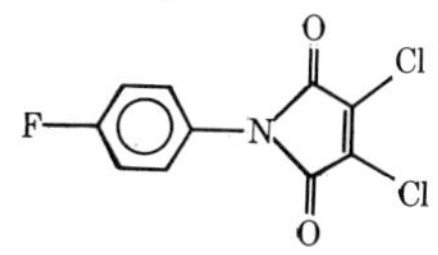	fungicide

Dyes. In contrast to benzotrifluorides, few commercial dye applications have been developed for ring-fluorinated aromatics. Fluorophenylhydrazines have been converted to (fluorophenyl)pyrazolones, which are disperse dyes for cellulose acetate and nylon (91) (see Dyes, application). The α-crystalline form of 2,9-difluoroquinacridone *[746-87-2]* pigment (prepared from *p*-fluoroaniline) has been incorporated with titanium dioxide in an automobile paint finish (92) (see Coatings, industrial).

Miscellaneous. Liquid crystals (qv) represent a potential use for fluoroaromatics. Two types have been evaluated: Schiff bases derived from *p*-fluoroaniline (93–94) or from *p*-fluorobenzaldehyde *[459-57-4]* (93), and esters based on *p*-fluorobenzoic acid *[456-22-4]* (95) or on 3-fluoro-4-octyloxybenzoic acid *[326-78-3]* (96).

DIFLUOROBENZENES

Several practical applications for difluorobenzenes have emerged. The analgesic and anti-inflammatory agent, Diflunisal *[22494-42-4]* (see Fluorobiphenyls) and a promising insecticide, Diflubenzuron *[35367-38-5]*, exemplify the growing interest in difluoroaromatics.

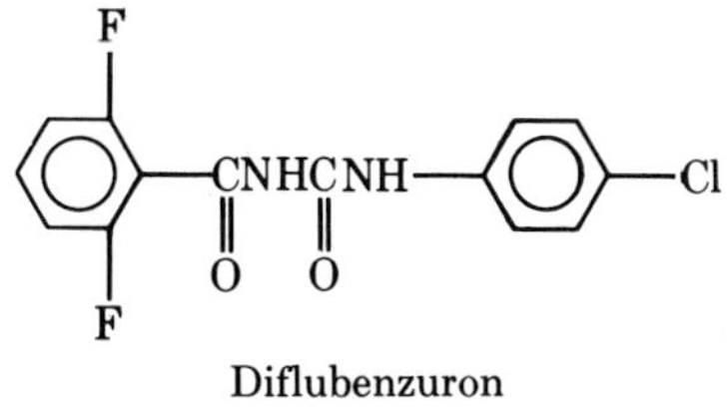

Diflubenzuron

Table 6 lists some of the physical properties of the difluorobenzenes.

Table 5. Mono-Fluorinated Aromatic Drugs

Common name	CAS Registry No.	Structure	Application
haloperidol	[*52-86-8*]	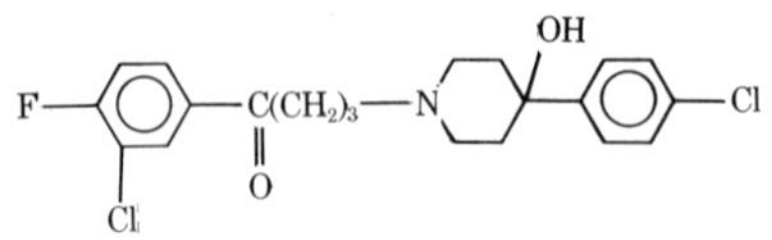	tranquilizer
flurazepam hydrochloride	[*1172-18-5*]	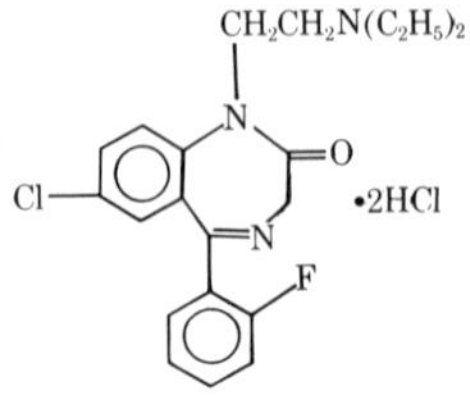	hypnotic; sedative
floxacillin	[*5250-39-5*]	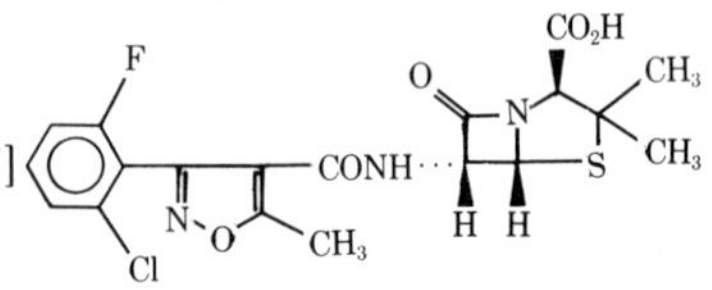	antibacterial
flazalone	[*21221-18-1*]	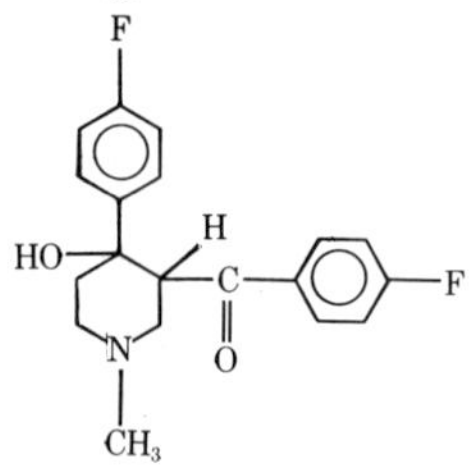	anti-inflammatory
fluspirilene	[*1841-19-6*]	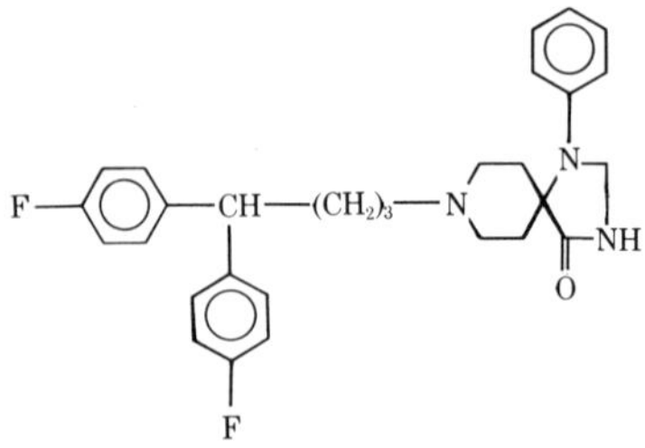	tranquilizer

1,2-Difluorobenzene. 1,2-Difluorobenzene (*o*-difluorobenzene) is prepared in 42% yield from *o*-fluoroaniline [*348-54-9*] by the Balz-Schiemann reaction (97). Nitration with concentrated nitric and sulfuric acid gives 3,4-difluoronitrobenzene [*369-34-6*]. Since fluorine atoms para to a nitro group are labile, amination with 28% ammonium hydroxide at 150–160°C for 6 h gives a 90% yield of 2-fluoro-4-nitroaniline [*369-35-7*].

1,3-Difluorobenzene. 1,3-Difluorobenzene (*m*-difluorobenzene) can be prepared from *m*-phenylenediamine in 31% yield by the Balz-Schiemann reaction (99). Diazotization of *m*-fluoroaniline [*372-19-0*] in the presence of ammonium bifluoride, tertiary amines, or dimethyl sulfoxide gives 46–73% yields of 1,3-difluorobenzene (5–6). Nitration of 1,3-difluorobenzene at 0°C forms 2,4-difluoronitrobenzene [*446-35-5*] in 92% yield. The latter also is prepared in 40–70% yield from 2,4-dichloronitrobenzene

Table 6. Properties of Fluorinated Aromatic and Heterocyclic Compounds

Compounds	CAS Registry No.	mol wt	Color	mp, °C	bp, °C/kPa[a]	Ionization potent	Refractive index, n_D^t	Specific gravity, d_4^t	Dipole moment, (C·m) 10^{-30}[b]	Surface tension, 20°C, mN/m (= dyn/cm)	Heat of formation, 25°C, kJ/mol[c]	Heat of combustion, 25°C, J/g[c]	Heat of vaporization, J/mol[c]	Flash point, °C	Toxicity, LD_{50}
Difluorobenzenes, $C_6H_4F_2$															
1,2-difluorobenzene	[367-11-3]	114.09	colorless	−34	91–92		1.4452^{20}	1.1496^{25}			−283.05	−25,939.1		7.2[e]	55[d]
1,3-difluorobenzene	[372-18-9]	114.09	colorless	−59.3	82–83		1.4410^{20}	1.1572^{20}	5.27	25.93	−298.53	−25,816.5		−11.1[e]	55[d]
1,4-difluorobenzene	[540-36-3]	114.09	colorless	−13	88–89		1.4421^{20}	1.1716^{20}		27.05	−295.4	−25,831.6		−11.7[e]	55[d]
Trifluorobenzenes, $C_6H_3F_3$															
1,2,4-trifluorobenzene	[367-23-7]	132.08	colorless		88	9.37				26.2				−5[e]	
1,3,5-trifluorobenzene	[372-38-3]	132.08	colorless	−5.5	75.5		1.4140^{20}	1.277^{20}		27.16					
Tetrafluorobenzenes, $C_6H_2F_4$															
1,2,3,4-tetrafluorobenzene	[551-62-2]	150.08	colorless	−42	95	9.01	1.4069^{20}	1.422^{25}							
1,2,3,5-tetrafluorobenzene	[2367-82-0]	150.08	colorless	−48	83		1.4011^{25}	1.393^{20}		23.99					
1,2,4,5-tetrafluorobenzene	[327-54-8]	150.08	colorless	4	90	9.39	1.4045^{20}	1.424^{25}		24.9				4[e]	
Pentafluorobenzene, C_6HF_5[f]	[363-72-4]	168.07	colorless	−48	85	9.37	1.3881^{25}	1.531^{20}					$36,672.8^{25}$ $32,070.4^{85}$		710[g]
Pentafluorophenyls															
bromopentafluorobenzene, C_6BrF_5	[344-07-7]	246.98	colorless	−29.9	135.6		1.4493^{25}	1.947^{25}							959[y]
pentafluoroaniline, $C_6F_5NH_2$	[771-60-8]	183.08	colorless	34	153–154		1.4441^{36}	1.744^{25}							390[h]
pentafluorophenol, C_6F_5OH[j]	[771-61-9]	184.07	colorless	34–36	143 $72–73_{6.4}$		1.4263^{26}	1.757^{25}							384[i] 283[k]
pentafluorotoluene, $C_6F_5CH_3$	[771-56-2]	182.09	colorless	−30	117–118		1.4016^{25}	1.440^{25}							322[l]
pentafluorobenzaldehyde, C_6F_5CHO	[653-37-2]	196.08	colorless	20	164–166 $67–70_2$		1.4483^{25}	1.619^{25}							320[y]
pentafluorobenzoic acid, $C_6F_5CO_2H$	[602-94-8]	212.08	white	103	220			1.944^{25}			−1,211.94	−2,553.20 kJ/mol			2500[y] 178[z] 100[m]

Table 6 (*continued*)

Compounds	CAS Registry No.	mol wt	Color	mp, °C	bp, °C/kPa[a]	Ion-ization potent	Refrac-tive index, n_D^t	Spe-cific gravity, d_4^t	Dipole mo-ment, (C·m) 10^{-30b}	Surface ten-sion, 20°C, mN/m (= dyn/cm)	Heat of forma-tion, 25°C, kJ/mol[c]	Heat of combus-tion, 25°C, J/g[c]	Heat of vapori-zation, J/mol[c]	Flash point, °C	Tox-icity, LD_{50}
Fluoronaphthalenes															
1-fluoronaphthalene, $C_{10}H_7F^n$	[*321-38-0*]	146.16		−8.5	215 $48_{0.008}$		$1.5939^{19.5}$	1.1256^{30}	4.74	39.7					380[o,bb]
2-fluoronaphthalene	[*323-09-1*]	146.16		60	212.5				4.97						
1,4-difluoronaphthalene, $C_{10}H_6F_2$	[*315-52-6*]	164.15		31.5	$95–99_{4.5}$										
octafluoronaphthalene, $C_{10}F_8$	[*313-72-4*]	264.1	white	87–88											
Benzotrifluorides															
4-chlorobenzotrifluoride, $ClC_6H_4CF_3^p$	[*98-15-7*]	180.56	colorless		139 $29.5_{1.33}$ $79_{13.33}$		1.4444^{25}	$1.3380^{25}_{15.5}$		25			207.1 J/g[139]	54[q]	5000[aa]
3-aminobenzotrifluoride, $H_2NC_6H_4CF_3^r$	[*98-16-83*]	116.13	colorless to yellow	5–6	187–188 $86_{2.67}$ $125_{16.0}$ $175_{69.3}$		1.4788^{20}	1.305^{25}		25			246.9 J/g[187–188]	85[s]	440[t] 690[d,cc] 220[g]
Fluoropyridines, C_5H_4FN															
2-fluoropyridine	[*372-48-5*]	97.09	colorless		$126_{100.4}$		1.4678^{20}	1.1281^{20}						24[s]	
3-fluoropyridine	[*372-47-4*]	97.09			105–$107_{100.3}$		1.4700^{20}	1.125^{25}						13[s]	
4-fluoropyridine	[*694-52-0*]	97.09	colorless	100[u]	$108_{100.0}$		1.4730^{20}								
$C_5H_2F_3N$															
2,4-difluoropyridine	[*34941-90-7*]	115.08		134–135[u]	104–105										
2,6-difluoropyridine	[*1513-65-1*]	115.08	colorless		$124.5_{99.1}$		1.4349^{25}	1.265^{25}						32[s]	
$C_5H_2F_3N$															
2,4,6-trifluoropyridine	[*3512-17-2*]	133.07			94–95										
C_5HF_4N															
2,3,5,6-tetrafluoropyridine	[*2875-18-5*]	151.06			102										
$C_5F_5N^v$															
pentafluoropyridine	[*70-16-3*]	169.05		−41.5	83.3		1.3856^{20}								

Perfluoroalkylpyridines								
2-trifluoromethylpyridine, $C_6H_4F_3N$	[*368-48-9*]	147.10			$143_{99.4}$	1.4144^{25}		
3-trifluoromethylpyridine	[*3796-23-4*]	147.10			113–115	1.4150^{25}		
4-trifluoromethylpyridine	[*3796-24-5*]	147.10			108–110	1.4144^{25}		
Fluoropyrimidines								
trifluoropyrimidine, 2,4,6-$C_4HF_3N_2$	[*696-82-2*]	134.06	colorless		98 60_{24}	1.4015^{25}		
tetrafluoropyrimidine, 2,4,5,6-$C_4F_4N_2$	[*767-79-3*]	152.06	colorless		89	1.3875^{25}		
Fluorotriazines								
2,4,6-trifluoro-1,3,5-triazine, $C_3F_3N_3$	[*675-14-9*]	135.05	colorless	−38	$72.4_{101.67}$	1.3844^{24}	1.60 ± 0.05^{25}	3.1^{w} 160^{x}
2,4,6-*tris*-(trifluoromethyl)-1,3,5-triazine, C_6F_9N	[*368-66-1*]	285.07	colorless	−24.8	95–96	1.3161^{25}	1.593^{25}	

[a] To convert kPa to mm Hg, multiply by 7.5
[b] To convert C·m to debye (D), divide by 3.336×10^{-30}.
[c] To convert J to cal, divide by 4.184.
[d] Inhalation (mouse), g/m^3 for 2 h, ref. 47.
[e] Open-cup.
[f] Critical temperature, 256 ± 1°C; critical pressure, 3 52 MPa (34.7 atm).
[g] Oral (mouse) mg/kg, ref. 47.
[h] ip rat, mg/kg, ref. 47.
[i] ip mouse, mg/kg, ref. 47.
[j] Ionization constant, K_a, 3.0×10^{-6} at 25°C.
[k] Subcutaneous mouse, mg/kg, ref. 47.
[l] Subcutaneous rat, mg/kg, ref. 47.
[m] Myotonic activity, mice, ED_{50}, mg/kg; ref. 48.
[n] Thermal coefficient in the range 30–40°C, $\alpha \times 10^5 = 79$.
[o] Oral, rats, mg/kg, ref. 47.
[p] Viscosity = 0.67 mPa·s (= cP) at 37.8°C and 0.39 mPa·s at 98.9°C; solubility in H_2O = 1.6 g/100 g; fire point (Cleveland open-cup) 68°C, ref. 98.
[q] Cleveland open-cup.
[r] Solubility in H_2O = 1.8 g/100 g.
[s] Tag closed-cup.
[t] Inhalation rat, LC_{50} mg/m^3 for 4 h, ref. 47.
[u] mp of HCl salt.
[v] Trouton's constant = 24.9; latent heat of vaporization = 36,338 J/mol.
[w] On inhalation (rat), ppm for 4 h, ref. 47.
[x] Skin (rabbit), mg/kg, ref. 47.
[y] Oral, mouse, mg/kg, ref. 48.
[z] ip, rat, mg/kg, ref. 48.
[aa] Oral, rats, mg/kg, ref. 98.
[bb] LD_{LO}.
[cc] LC_{50}.

and potassium fluoride in polar solvents (10–11). Reduction of 2,4-difluoronitrobenzene gives 2,4-difluoroaniline [*367-25-9*], a precursor to the analgesic/anti-inflammatory agent, Diflunisal.

1,4-Difluorobenzene. 1,4-Difluorobenzene (*p*-difluorobenzene) is prepared from *p*-phenylenediamine or *p*-fluoroaniline [*371-40-4*] by the Balz-Schiemann reaction in 27–40% yield (99).

TRIFLUOROBENZENES

Few applications for trifluorobenzenes have been developed. Some of their physical properties are listed in Table 6.

1,2,4-Trifluorobenzene. 1,2,4-Trifluorobenzene has been prepared in 55% yield from 2,4-difluoroaniline [*367-25-9*] by the Balz-Schiemann route. Nitration with a mixture of concentrated nitric acid and sulfuric acid at 20°C gives 2,4,5-trifluoronitrobenzene [*2105-61-5*] in 87% yield.

1,3,5-Trifluorobenzene. 1,3,5-Trifluorobenzene (*s*-trifluorobenzene) has been prepared in 60% yield by the Balz-Schiemann reaction with 3,5-difluoroaniline [*372-39-4*]. Nitration with a mixture of concentrated nitric and sulfuric acids at 70°C gave an 85% yield of 2,4,6-trifluoronitrobenzene [*315-14-0*].

TETRAFLUOROBENZENES

Few applications for tetrafluorobenzenes have been developed. Isomeric dichlorotetrafluorobenzenes have been studied as power fluids for Rankine-cycle external combustion engines (100). Some of the physical properties of the tetrafluorobenzenes are listed in Table 6.

1,2,3,4-Tetrafluorobenzene. 1,2,3,4-Tetrafluorobenzene has been prepared from fluorination of benzene with cobalt trifluoride and subsequent combination of the dehydrofluorination and defluorination steps (see Hexafluorobenzene for general process description). Nitration with mixed concentrated nitric and sulfuric acid gives 2,3,4,5-tetrafluoronitrobenzene [*5580-79-0*] in 75% yield.

1,2,3,5-Tetrafluorobenzene. 1,2,3,5-Tetrafluorobenzene has been prepared from 2,3,5-trifluoroaniline [*363-80-4*] in 43% yield by the Balz-Schiemann reaction. Nitration is accomplished with mixed concentrated nitric and sulfuric acid to give 2,3,4,6-tetrafluoronitrobenzene [*314-41-0*].

1,2,4,5-Tetrafluorobenzene. 1,2,4,5-Tetrafluorobenzene has been prepared from 2,4,5-trifluoroaniline [*57491-45-9*] by the Balz-Schiemann reaction in 38–46% yield or from pentafluorophenylhydrazine [*828-39-9*] with aqueous sodium hydroxide in 90–95% yield (101). Nitration of 1,2,4,5-tetrafluorobenzene with fuming nitric acid in 30% oleum results in hydrolytic displacement to give 2,5-difluorobenzoquinone [*327-55-9*]. However, fuming nitric acid saturated with boron trifluoride in sulfolane gave 2,3,5,6-tetrafluoronitrobenzene [*6257-03-0*] in 18% yield (102).

PENTA- AND HEXA-FLUOROBENZENES

Pentafluorobenzene. Pentafluorobenzene has been prepared by several routes: fluorination of benzene with cobalt trifluoride, etc (see Hexafluorobenzene); reductive dechlorination of chloropentafluorobenzene with 10% palladium-on-carbon in 82%

yield (103); and oxidation of pentafluorophenylhydrazine in aqueous copper sulfate at 80°C in 77% yield (104). Nucleophiles react with pentafluorobenzene to give para-substituted (relative to the hydrogen atom) tetrafluorophenyl products, p-XC_6F_4H (X = H; NH_2; $NHNH_2$; SH; OCH_3; SC_6H_5; OH). Nitration of pentafluorobenzene with concentrated nitric acid and boron trifluoride in sulfolane gave pentafluoronitrobenzene [*880-78-4*] in 82% yield (105). The vapor pressure of pentafluorobenzene has been measured over the range 20–85°C (106).

Hexafluorobenzene. The development of commercial routes to hexafluorobenzene included an intensive study of the chemistry of its derivatives. Particularly noteworthy was the development of high-temperature lubricants, heat transfer fluids, and radiation-resistant polymers. The scope of these studies was summarized in a recent monograph on fluoropolymers (107).

Hexafluorobenzene [*392-56-3*] (perfluorobenzene), C_6F_6, mol wt 186.06, is a colorless liquid with a sweet odor. It is miscible with many organic solvents and is immiscible with water. Hexafluorobenzene forms a 1:1 molecular complex with benzene, with a mp of 23.7°C. Hexafluorobenzene has good thermal stability; very slight decomposition occurs at 500°C in Nimonic 75 (alloy containing 85% nickel and 20% chromium) after 3 weeks. Toxicity: inhalation (mouse), LD_{50} 95 g/m^3 (2 h) (47).

Physical properties of hexafluorobenzene are given in Table 7.

Reactions. Hexafluorobenzene is very susceptible to attack by nucleophilic agents to given pentafluorophenyl compounds of the general formula, C_6F_5X, where X is OCH_3, NH_2, OH, SH, $NHNH_2$, $NHCH_3$, $N(CH_3)_2$, H, C_6H_5, CH_3, $CH_3CH{=}CH$, n-C_4H_9, C_6H_5S, etc. Radicals attack hexafluorobenzene and cause displacement of a fluorine atom. For example, phenyl radicals, generated by the decomposition of benzoyl

Table 7. Physical Properties of Hexafluorobenzene

Property	Value
melting point, °C	5.10
boiling point, °C	80.261
density, 25°C, g/mL	1.60682
refractive index, n_D^{25}	1.3761
latent heat of fusion, kJ/mol[a]	11.59
latent heat of vaporization	
at 25°C, kJ/mol[a]	35.69 ± 0.084
at bp, kJ/mol[a]	32.69
specific heat, 23°C, kJ/(mol·°C)[a]	0.221
critical temperature, °C	243.57 ± 0.03
critical pressure, MPa[b]	3.304 ± 0.005
coefficient of cubical expansion at 25°C	0.001412
heat of combustion, kJ/mol[a]	−2444.0 ± 1.2
heat of formation at 25°C, kJ/mol[a]	
liquid	−958.30
gas	−922.15
vapor pressure in °C and kPa[c]	Antoine equation: $\log_{10} P = 6.1422$ $1219.410/(t + 214.525)$

[a] To convert J to cal, divide by 4.184.
[b] To convert MPa to atm, divide by 0.101.
[c] To convert kPa to mm Hg, multiply by 7.5.

peroxide, react with hexafluorobenzene and give 2,3,4,5,6-pentafluorobiphenyl [*784-15-15*].

Manufacture. The commercial process features a three-stage saturation–rearomatization technique using benzene and fluorine gas as raw materials.

(*1*) Benzene is fluorinated with cobalt trifluoride at about 150°C, giving a mixture of fluorinated cyclohexanes containing 1 to 4 hydrogen atoms. During this process the cobalt trifluoride is converted into cobalt difluoride and the trifluoride has to be regenerated by passing fluorine gas over the difluoride before recycling.

(*2*) The mixture of polyfluorocyclohexanes is heated with an alkali metal hydroxide. During this process dehydrofluorination occurs and a mixture of polyfluorocyclohexenes and -hexadienes results.

(*3*) The third stage is the defluorination of the product produced in stage (*2*). This is accomplished by passing the product from stage (*2*), in vapor form, through a reactor packed with a defluorinating agent (eg, iron or nickel; magnetite–Fe_3O_4; or various metal fluorides) at 400–600°C. The product from stage (*3*) consists of hexafluorobenzene, pentafluorobenzene and the isomeric tetrafluorobenzenes. These products are purified by fractional distillation and chemical methods.

This technique has been successfully employed to prepare other perfluoroaromatics, eg, octafluorotoluene [*434-64-0*], perfluoroxylenes, fused-ring polycyclics, etc. However, principal problems with this method are: the complex nature of the process; its dependence on fluorine gas which is costly to produce; and the poor overall utilization of fluorine (since about one-half of the input fluorine is removed during the process).

Applications. Hexafluorobenzene is being considered for use in veterinary anesthesia (108). Also, a study of working fluids for automotive Rankine-cycle power units demonstrated that the hexafluorobenzene–pentafluoropyridine combination had the requisite stability at 382°C (109).

Derivatives

Numerous derivatives of pentafluorobenzene and hexafluorobenzene have found utility in analytical and other applications, and some of their physical properties are listed in Table 6.

Brompentafluorobenzene. Bromopentafluorobenzene has been prepared by aluminum bromide-catalyzed bromination of pentafluorobenzene in 20% oleum. Bromopentafluorobenzene is readily converted to pentafluorophenylmagnesium bromide [*879-05-0*]; the latter undergoes conventional Grignard reactions. Pentafluorophenyllithium [*1076-44-4*] can be synthesized from bromopentafluorobenzene and *n*-butyllithium or lithium amalgam in ether at 0°C. The preferred route is metallation of pentafluorobenzene with *n*-butyllithium at −65°C (110). *Caution:* A serious explosion occurred during hydrolysis (D_2O) of pentafluorophenyllithium (111).

Pentafluorophenylmagnesium bromide or lithium can be converted to other pentafluorophenyl organometallics by reaction with the corresponding metal chloride (112). Bis(pentafluorophenyl)phenylphosphine [*5074-71-5*], $(C_6F_5)_2C_6H_5P$, is offered commercially as a marker for mass spectral standardization (113). Pentafluorophenyldimethylsilanes, $C_6F_5Si(CH_3)_2X$ (X = Cl; NH_2; $N(C_2H_5)_2$), are available commercially for derivatization of sterols prior to chromatographic analysis (114) (see Analytical methods).

Pentafluoroaniline. Pentafluoroaniline has been prepared from amination of hexafluorobenzene with sodium amide in liquid ammonia or with ammonium hydroxide in ethanol (or water) at 167–180°C for 12–18 h. It is weakly basic ($pK_a = 0.28$) and dissolves only in concentrated acids. Liquid crystals have been prepared from Schiff bases derived from pentafluoroaniline (115).

Pentafluorophenol. Pentafluorophenol has been prepared from the reaction of hexafluorobenzene with potassium hydroxide in *t*-butyl alcohol. Pentafluorophenyl esters prepared from pentafluorophenol illustrate the key features of a rapid stepwise peptide synthesis technique (116). Commercial high-performance elastomers based on copolymerization of tetrafluoroethylene, perfluoro(methyl vinyl ether), and a third monomer incorporating a pentafluorophenoxy group as a cure site, give vulcanizates with good chemical and fluid resistance and high temperature oxidative resistance (117) (see Elastomers, synthetic).

Pentafluorotoluene. Pentafluorotoluene has been prepared from reaction of methyllithium with hexafluorobenzene or from pentafluorophenylmagnesium bromide with dimethyl sulfate. Photochlorination of boiling pentafluorotoluene gives 2,3,4,5,6-pentafluorobenzyl chloride [*653-35-0*]. Products based on pentafluorotoluene are used as derivatizing agents for gas chromatographic analysis of biologically active compounds via electron capture detection. Examples include: *O*-(2,3,4,5,6-pentafluorobenzyl)hydroxylamine hydrochloride [*57981-02-9*], $C_6F_5CH_2ONH_2.HCl$, (Florox reagent) for assay of ketosteroids (118–119) and 2,3,4,5,6-pentafluorobenzyl chloroformate [*53526-74-2*], for assay of physiologically-active tertiary amines (120).

$$C_6F_5CH_2O\underset{\underset{\displaystyle O}{\|}}{C}Cl$$

Pentafluorobenzaldehyde. Pentafluorobenzaldehyde can be prepared by reaction of *N*-methylformanilide with pentafluorophenylmagnesium bromide or pentafluorophenyllithium. Pentafluorobenzaldehyde is used as a reagent for gas chromatographic assay of biological amines such as catecholamines by conversion to the pentafluorobenzylimine-trimethylsilyl derivatives (121).

Pentafluorobenzoic Acid. The acid has been prepared in 80% yield by successive *n*-butyllithium exchange with chloropentafluorobenzene in diethyl ether, followed by carbonation (122). Derivatives of pentafluorobenzoic acid are being considered as potential analytical tools for gas chromatography. Applications include *N*-pentafluorobenzoyl derivatives of amphetamines and ephedrines (123) and ethylenebis-(dithiocarbamate) fungicide metabolites (see Fungicides) (124).

Fluorobiphenyls

4,4′-Difluorobiphenyl [*398-23-2*] was used in Germany for bronchial disorders (Antitussin) and in mixtures with 2,4,5-trimethylfluorobenzene [*400-01-1*] for skin disorders (Epiderm-Fluor) (125). Considerable interest has again developed in pharmaceutical applications for fluorobiphenyls, eg, analgesics and anti-inflammatory agents. Two common analgesics and anti-inflammatory drugs are listed in Table 8.

Mono- and difluorobiphenyls can be prepared by the Balz-Schiemann reaction (or modification in HF); eg, 4,4′-difluorobiphenyl was formed in 80% yield from 4,4′-

Table 8. Mono- and Difluorobiphenyl Drugs

Common name	CAS Registry No.	Structure	Application
Diflunisal	[*22494-42-4*]	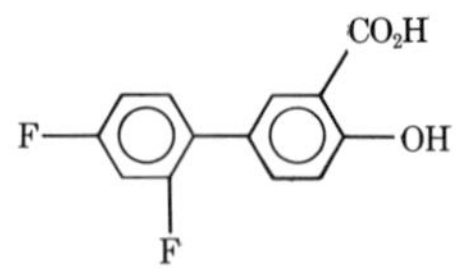	analgesic; anti-inflammatory
Flurbiprofen	[*5104-49-4*]	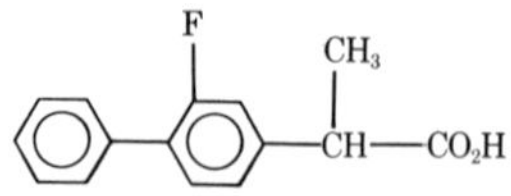	analgesic; anti-inflammatory

diaminobiphenyl by the Balz-Schiemann reaction (126). 2,4-Difluorobiphenyl [*2285-28-1*], a key precursor to the analgesic/anti-inflammatory agent, Diflunisal, is prepared by diazotization of 2,4-difluoroaniline [*367-25-9*] and subsequent coupling with benzene (127).

Decafluorobiphenyl [*434-90-2*], $C_6F_5C_6F_5$; mol wt, 334.1; mp, 68°C; bp, 206°C; can be prepared by Ullmann coupling of bromo- [*344-04-7*], chloro- [*344-07-0*] or iodopentafluorobenzene [*827-15-6*] with copper. This product shows good thermal stability; decafluorobiphenyl was recovered unchanged after 1 h below 575°C (128). Thermally stable fluids based on decafluorobiphenyl and derivatives exhibited greater oxidative stability than similar hydrocarbon compounds (129).

Fluoronaphthalenes and Fused-Ring Bicyclics

No commercial outlets for fluoronaphthalenes have materialized. However, fused-ring bicyclics can be used as anti-inflammatory agents, eg, Sulindac [*38194-50-2*], a monofluorinated indene-3-acetic acid. Physical properties of fluoronaphthalenes and fused-ring bicyclics are listed in Table 6.

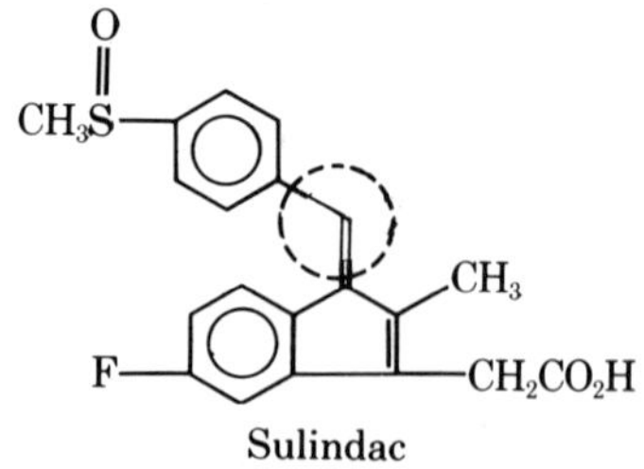

Sulindac

1-Fluoronaphthalene. 1-Fluoronaphthalene is prepared from 1-naphthylamine by the Balz-Schiemann reaction in 52% yield or by diazotization in anhydrous hydrogen fluoride in 82% yield. Electrophilic substitution occurs at the 4-position, eg, nitration with fuming nitric acid in acetic acid gave an 88% yield of 1-fluoro-4-nitronaphthalene [*341-92-4*].

2-Fluoronaphthalene. 2-Fluoronaphthalene is prepared in 54–67% yield from 2-naphthylamine by the Balz-Schiemann reaction or in 51% yield by pyrolysis of indene and chlorofluoromethane at 600°C (31).

1,4-Difluoronaphthalene. 1,4-Difluoronaphthalene is prepared from 4-fluoro-1-naphthylamine by a Balz-Schiemann reaction. 1,4-Difluoronaphthalene is used in chemical carcinogenesis studies as a synthon for highly-condensed difluoro-polycyclic aromatic hydrocarbons (130).

Octafluoronaphthalene. Octafluoronaphthalene, a white crystalline solid, is prepared in 53% yield by defluorination of perfluorodecahydronaphthalene [*306-94-5*] over iron or nickel at 500°C. Exchange fluorination of octachloronaphthalene with KF in sulfolane (235°C) gave 60% yield of octafluoronaphthalene. This product exhibits good stability to ionizing radiation (131).

Fused-Ring Polycyclic Fluoroaromatics

Partially-fluorinated polycyclic aromatic hydrocarbons are useful compounds for structure–activity relationships (SAR) of chemical carcinogens (130). These compounds can be made from the corresponding amino fused-ring polycyclic or from preformed fluoroaromatics, eg, 4-fluorophenylacetonitrile [*459-22-3*] (132).

Side-Chain Fluorinated Aromatics

Trifluoromethyl aromatics are used widely in the production of drugs, crop protection chemicals, germicides, dyes, etc. Some of the key intermediates to these products are benzotrifluoride, 4-chlorobenzotrifluoride, and 3-aminobenzotrifluoride which are reviewed following the general discussion of the properties of this class of compounds.

General Properties. ***Stability of Benzotrifluorides.*** The trifluoromethyl group is stable under different reaction conditions, eg, the multi-step classical transformation of benzotrifluoride to trifluoroacetic acid.

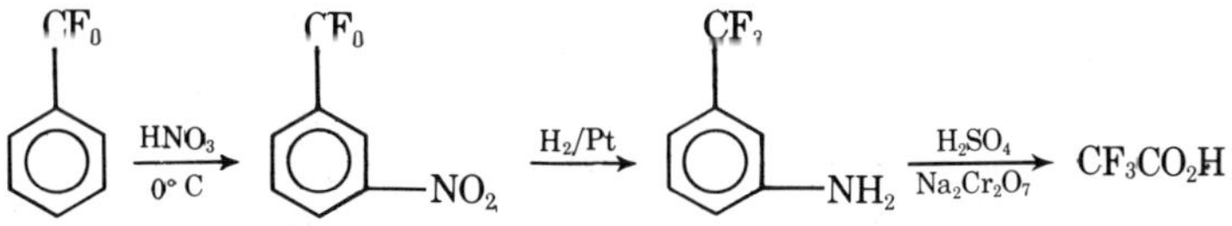

However, aluminum chloride effects chlorinolysis of benzotrifluoride to benzotrichloride (133). The stability of benzotrifluoride and its derivatives to thermal, hydrolytic, oxidative, and reducing conditions is discussed below.

Trifluoromethylphenyl organometallics. Care must be exercised in handling trifluoromethylphenyl organometallics. *o*-Trifluoromethylphenyllithium [*49571-35-5*] (from *o*-bromobenzotrifluoride [*392-83-6*]) has exploded during reflux in diethyl ether under nitrogen (134). Both *m*- [*368-49-0*] and *p*-trifluoromethylphenyllithium [*2786-01-8*] are explosive in the solid state (135). Explosions have been reported in the preparation of *o*- [*395-47-1*], *m*- [*402-26-6*] and *p*-trifluoromethylphenylmagnesium bromide [*402-51-7*] (134,136–137). *p*-Trifluoromethylphenylmagnesium chloride [*2923-41-3*], prepared from *p*-chlorobenzotrifluoride in tetrahydrofuran solvent on a 50-mol scale, may be more stable (138).

Thermal stability. Benzotrifluoride is stable at 350°C in the presence of iron or copper. Working fluids for external combustion engines (Rankine cycle) must exhibit thermal stability with engine materials at high temperatures. Some of the promising working fluids include 1,3-bis(trifluoromethyl)benzene [*402-31-3*] (139) or a mixture of perfluorotoluene [*434-64-0*], $CF_3C_6F_5$, and hexafluorobenzene [*392-56-3*] (140). The stability of the isomeric CF_3-substituted anilines has been established by differential thermal analysis (dta) (141): *m*-$CF_3C_6H_4NH_2$ [*98-15-7*], 223°C > *o*-$CF_3C_6H_4NH_2$ [*88-17-5*], 187°C > *p*-$CF_3C_6H_4NH_2$ [*455-14-1*], 155°C.

Hydrolytic stability. The trifluoromethyl group is sensitive to hydrolysis in acidic media. Benzotrifluoride is hydrolyzed to benzoic acid by heating with hydrobromic, hydrofluoric or >80% sulfuric acid. Reaction conditions and structural features for this reaction have been summarized (142). Benzotrifluorides are generally stable to base. Benzotrifluoride was recovered unchanged after heating (120–130°C) with sodium hydroxide. Although *m*-hydroxybenzotrifluoride [*98-17-9*] is stable to refluxing 50% sodium hydroxide, cold dilute alkali polymerizes the para isomer [*402-45-9*] to give (143).

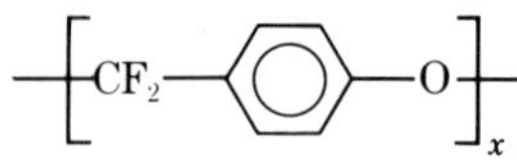

(Similar polymers are formed from 2,3,5,6-tetrafluoro-4-trifluoromethylphenol [*2787-79-3*] (144)). Photohydrolysis of hydroxy- and aminobenzotrifluorides in dilute acid and alkali, respectively, gives the corresponding hydroxy- and aminobenzoic acid in high yield (145). Benzotrifluoride is inert under these photohydrolytic conditions.

Oxidative stability. Benzotrifluoride resists ring oxidation. In contrast, chromic acid readily oxidizes 3-aminobenzotrifluoride to trifluoroacetic acid in 95% yield (146).

Stability to reducing agents. The trifluoromethyl group is inert to numerous reducing agents. Catalytic hydrogenation (platinum black) of benzotrifluoride gives trifluoromethylcyclohexane [*401-75-2*] (147). Benzotrifluoride was not reduced by lithium aluminum hydride (148). However, *o*-trifluoromethylbenzoic acid [*433-97-6*] and *m*-trifluoromethylbenzoic acid [*454-92-2*] are catalytically reduced (Raney nickel or cobalt alloys) to *o*- and *m*-toluic acid, respectively (149).

Reactions. Benzotrifluoride undergoes electrophilic substitution reactions, eg, halogenation, nitration, typical of an aromatic containing a strong electron-withdrawing group. The trifluoromethyl group (sometimes referred to as a pseudohalogen) is meta directing.

Halogenation. Liquid-phase monochlorination of benzotrifluoride gives pronounced meta orientation (150); in contrast, vapor-phase halogenation favors para substitution (151):

			$CF_3C_6H_4Cl$, %		
Conditions	*Catalyst*	*Temp, °C*	*Ortho* [*88-16-4*]	*Meta* [*98-15-7*]	*Para* [*98-56-6*]
liquid	$FeCl_3$	65	7	83	10
vapor	none	400	14	33	53

Sealed tube, photochemical, or dark chlorination (radical initiator) forms hexachloro(trifluoromethyl)cyclohexane; thermal dehydrochlorination (550°C) gives 2,4,6-

trichlorobenzotrifluoride [*567-59-9*] (152). Liquid-phase bromination (Br_2) provides 3-bromobenzotrifluoride [*401-78-5*] in 60% yield. Catalyst performance decreases in the order: $FeCl_3 > Fe > SbCl_5 > I_2$ (153). Silica gel serves as an hydrogen fluoride scavenger to suppress corrosion of glass reactors during halogenations (154). Benzotrifluoride bromination can also be accomplished with bromine chloride in the presence of a halogen carrier, eg, antimony pentachloride. This technique permits complete utilization of bromine (155):

$$ArH + BrCl \rightarrow Ar{-}Br + HCl$$

Hydrogen peroxide/hydrochloric acid reagent converts 2-aminobenzotrifluoride to 2-amino-5-chlorobenzotrifluoride [*121-50-6*], a dye intermediate (C. I. Azoic Diazo Component 17), without contamination by the 3-chloro isomer such as is observed with molecular chlorine (Cl_2) (156).

Nitration. Nitration of benzotrifluorides is an important industrial reaction. The nitro group can be reduced to aminobenzotrifluorides or converted to nitro-containing products. Mononitration of benzotrifluoride gives pronounced meta-orientation: 91% meta, 6% ortho, and 3% para (151). (Further nitration to 3,5-dinitrobenzotrifluoride can be effected under forcing conditions at 100°C.) Other commercially significant nitrations include conversion of 2-chlorobenzotrifluoride to give 2-chloro-5-nitrobenzotrifluoride [*777-37-7*] (dye and germicide precursor) and conversion of 4-chlorobenzotrifluoride to form 4-chloro-3-nitrobenzotrifluoride [*121-17-5*] and 4-chloro-3,5-dinitrobenzotrifluoride [*393-75-9*] (for crop-protection chemicals).

Nucleophilic displacement reactions. The strong electron-withdrawing effect of a trifluoromethyl group activates ortho and para halogen toward nucleophilic attack. For example, methoxydechlorination of 4-chloro-3-nitrobenzotrifluoride proceeds

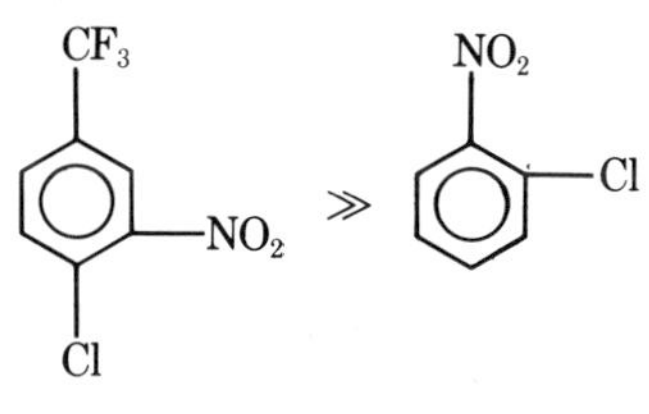

at a rate 364 times faster than the same reaction with 2-chloronitrobenzene (157). Such chlorine lability is utilized in the manufacture of crop-control chemicals containing trifluoromethyl and nitro groups; some are listed in Table 10.

Benzotrifluoride. Benzotrifluoride [*98-08-8*], (α,α,α-trifluorotoluene), $C_6H_5CF_3$, mol wt, 146.11; is a colorless liquid. Toxicity: oral (rat), LD_{50}, 1500 mg/kg; oral (mouse), LD_{LO}, 10,000 mg/kg; subcutaneous (frog), LD_{LO}, 870 mg/kg; intraperitoneal (mouse), LD_{LO}, 100 mg/kg (47).

A summary of the physical properties of benzotrifluoride is given in Table 9. Also see Table 6. Other physical properties of this compound and its derivatives, such as dipole moment, infrared and ultraviolet spectra, fluorine magnetic resonance, steric effect of the trifluoromethyl group and thermodynamic properties have been recently reviewed (159).

Benzotrifluoride was first synthesized by Swarts (1898) via the reaction of benzotrichloride and antimony trifluoride (160). Benzotrifluoride is produced by the high

Table 9. Physical Properties of Benzotrifluoride

Property	Value
mol wt	146.11
color	colorless
melting point, °C	−29.02
boiling point, °C	102.05
density, 25°C, g/mL	1.1814
coefficient of expansion (30–40°C)	0.00121
refractive index, n_D^{25}	1.4114
viscosity, mPa·s (= cP)	
38°C	0.488
99°C	0.282
surface tension, mN/m (= dyn/cm)	
20°C	23.39
latent heat of fusion, J/mol[a]	13,782.1
latent heat of vaporization, 102.05°C	
J/mol[a]	32,635.2
specific conductivity at 25°C, S/cm	1×10^{-7}
dielectric constant, 30°C	9.18
dipole moment, C·m[b]	8.54×10^{-30}
heat of combustion, J/g[a]	23,064.3
heat of formation, kJ/mol[a]	
vapor	−580.7
liquid	−618.4
boiling point of binary azeotrope,	
°C with 96.7 mol % bromine	58.1
solubility in water, g/100 g at	
room temperature	0.045
flash point (Cleveland open cup), °C[d]	15.6
fire point (Cleveland open cup), °C[d]	15.6
vapor pressure, in °C and kPa[c]	Antoine equation: $\log_{10} P = 6.0939 - \frac{1305.509}{(t + 217.280)}$

[a] To convert J to cal, divide by 4.184.
[b] To convert C·m to debye (D), divide by 3.336×10^{-30}.
[c] To convert kPa to mm Hg, multiply by 7.5.
[d] Ref.158.

pressure reaction of benzotrichloride with anhydrous hydrogen fluoride (AHF). Typical conditions include a 4:1 AHF/benzotrichloride mole ratio at 80–110°C and 1.52–1.55 MPa (220–225 psi) for 2–3 h to give 70–75% yields of benzotrifluoride (40,161). Atmospheric pressure at low temperature (41,162), and high temperature at vapor-phase (42) conditions also have been developed.

4-Chlorobenzotrifluoride. 4-Chlorobenzotrifluoride, a key feedstock to crop protection chemicals, is produced from the reaction of 4-chlorobenzotrichloride and anhydrous hydrogen fluoride under atmospheric or high pressure conditions (see discussion on benzotrifluoride manufacturing for key references). Nuclear chlorine in 4-chlorobenzotrichloride is inert under these conditions. (For physical properties and toxicity, see Table 6.)

3-Aminobenzotrifluoride. The standard manufacturing route to 3-aminobenzotrifluoride involves nitration of benzotrifluoride to 3-nitrobenzotrifluoride [*98-46-4*], followed by hydrogenation. A recent review of 3-aminobenzotrifluoride analytical methods describes techniques for assay of ortho- and para-isomer contaminants (163). The same handling precautions as for aniline should be exercised. (For physical properties and toxicity, see Table 6.)

The amine group of 3-aminobenzotrifluoride can be replaced by Cl, Br, I, F, CN, or OH groups by standard diazotization reactions. Phosgenation gives 3-trifluoromethylphenylisocyanate [*329-01-1*].

Applications. *Crop-protection chemicals.* Benzotrifluoride derivatives have gained wide acceptance as herbicides (Table 10).

Drugs. Until recently, trifluoromethyl-based pharmaceuticals were limited to phenothiazine tranquilizers and benzothiadiazine 1,1-dioxide diuretics (qv). However, new drugs have been developed (Table 11).

Table 10. Benzotrifluoride Crop-Protection Chemicals

Class	Common name	CAS Registry No.	Structure	Application
dinitroaniline	Trifluralin	[*1582-09-8*]	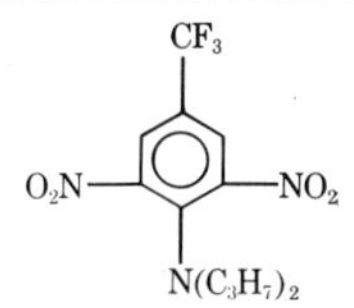	pre-emergent herbicide
	Dinitramine	[*29091-05-2*]	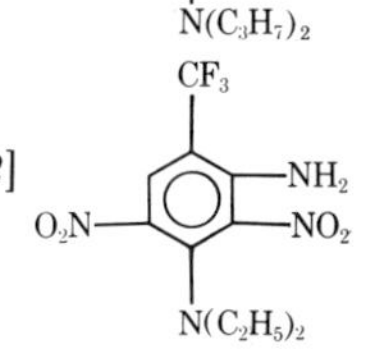	pre-emergent herbicide
substituted urea	Fluometuron	[*2164-17-2*]	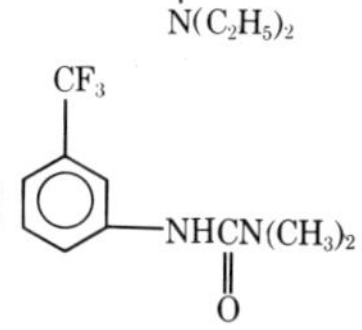	selective herbicide
diphenyl ether	Fluorodifen	[*15457-05-3*]	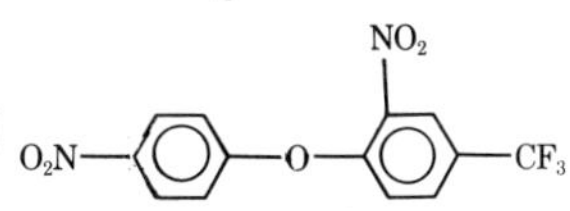	selective herbicide
	Acifluorfen	[*50594-66-6*]	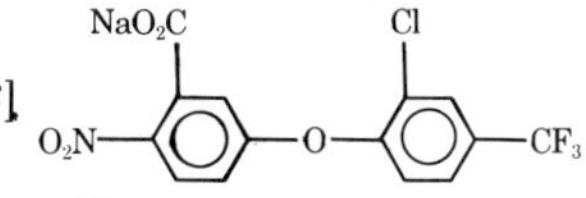	selective herbicide
substituted pyridazinone	Norflurazon	[*27314-13-2*]	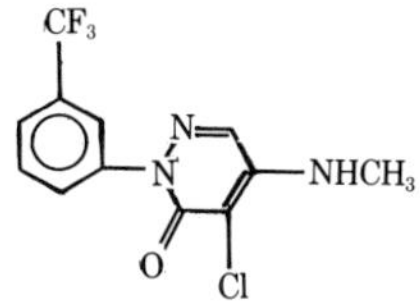	selective herbicide

Table 11. Benzotrifluoride Drugs

Common name	CAS Registry No.	Structure	Application
Triflupromazine	[*146-54-3*]	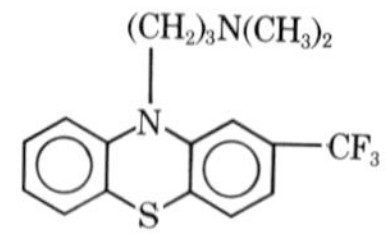	tranquilizer
Flumethiazide	[*148-561-1*]	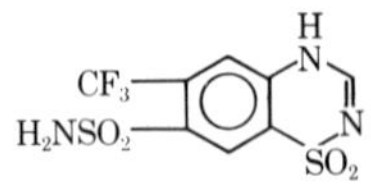	diuretic
Fenfluramine	[*404-82-0*]	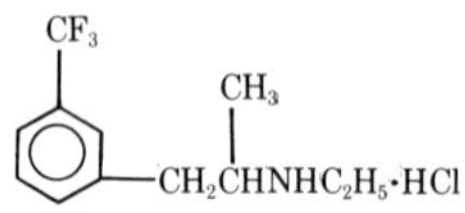	anorexigen
Flufenamic Acid	[*53-78-9*]	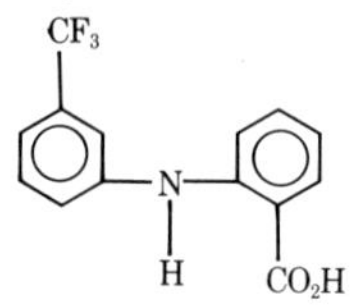	analgesic
Trifluperidol	[*749-13-3*].	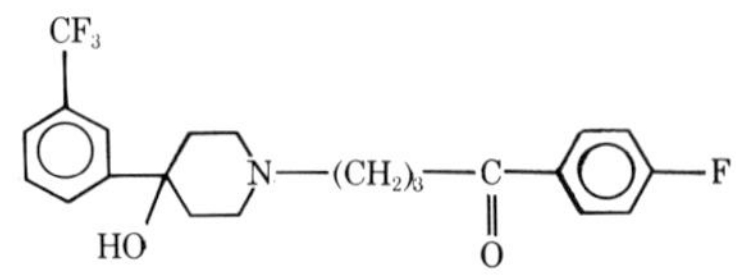	tranquilizer
Flumetramide	[*7125-73-7*]	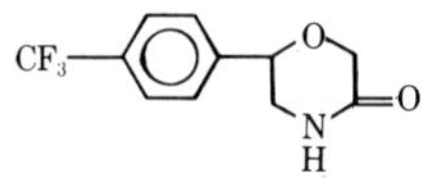	muscle relaxant

Germicides. Benzotrifluoride derivatives have also found wide use as antimicrobial agents in soaps (see Table 12) (see Disinfectants).

Miscellaneous biological applications. 4-Nitro-3-(trifluoromethyl)phenol [*88-30-2*] (TFM) is employed by the Canadian Bureau of Fisheries and the U.S. Fish and Wildlife Service as a lampricide for the control of parasitic sea lamprey in the Great Lakes.

Table 12. Benzotrifluoride Germicides

Common name	CAS Registry No.	Structure
Fluorosalan	[*4776-06-1*]	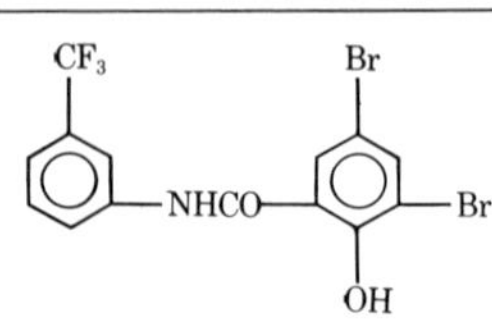
Cloflucarban	[*369-77-7*]	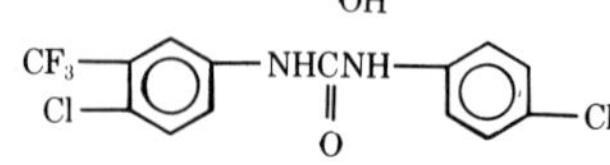

Dyes. Trifluoromethylphenyl groups are incroporated into azo, anthraquinone, and triphenylmethane dyes. The trifluoromethyl group is claimed to improve the brightness, tinctorial values, and lightfastness of dyes. The electron-withdrawing effects of this group also tend to modify the absorption of light by a dye in the visible and ultraviolet region. In azo dyes, aminobenzotrifluorides (Fast Base) are diazotized for subsequent coupling: 3,5-bis(trifluoromethyl)aniline [*328-74-5*], C. I. Azoic Diazo Component 16; 2-amino-5-chlorobenzotrifluoride [*445-03-4*], C. I. Azoic Diazo Component 17; 3-amino-4-ethylsulfonylbenzotrifluoride [*382-85-4*], C. I. Azoic Diazo Component 19; and 3-amino-4-chlorobenzotrifluoride [*121-50-6*], C. I. Azoic Diazo Component 49. 3-Trifluoromethylbenzoyl halides are incorporated into anthraquinone vat dyes, eg, Indanthrene Blue CLB [*6942-78-0*].

FLUORINATED NITROGEN HETEROCYCLICS

Commercial applications, eg, drugs, reactive dyestuffs, etc, have developed for fluorinated nitrogen heterocyclics. Activity in this area has focused on fluorinated pyridines, -quinolines, -pyrimidines, and -triazines. Some physical properties of fluorinated nitrogen heterocyclics are listed in Table 6.

MONOFLUOROPYRIDINES

Substitutive nuclear fluorination of pyridine with fluorine is not a practical route to fluoropyridines because of handling hazards and low yields. Some success has been achieved by the use of xenon difluoride to make mono- and difluoropyridines (164); however, the high cost of xenon difluoride limits its application.

2-Fluoropyridine. Diazotization of 2-aminopyridine in anhydrous hydrogen fluoride forms 2-fluoropyridine in 60% yield (82). Exchange-fluorination of 2-chloropyridine with potassium fluoride in polar solvents is sluggish (210°C for 21 d; 50–58% yield) (165). A novel solvent-free exchange-fluorination technique with potassium bifluoride proceeds in a smoother fashion (315°C/4 h; 74% yield (166)). Fluorine at the 2 position significantly reduces pyridine basicity more than at the 3-position (167):

	pK_a
pyridine	(+) 5.17
2-fluoropyridine	(−) 0.44
3-fluoropyridine	(+) 2.97

2-Fluoropyridine is readily hydrolyzed to 2-pyridone in 60% yield by reflux in 6 *N* hydrochloric acid (168). It is quite reactive with nucleophiles. For example, the halogen mobility ratio from the comparative methoxydehalogenation of 2-fluoropyridine and 2-chloropyridine was 85.5/1 at 99.5°C (169). This lability of fluorine has been utilized to prepare fluorine-free *O*-2-pyridyl oximes of 3-oxo steroids from 2-fluoropyridine for possible use as anti-fertility agents (170). Labile 2-fluoropyridine 1-oxide [*7613-14-1*] (171), 2-fluoro-3-nitropyridine [*1480-87-1*] (172) and 2-fluoro-5-nitropyridine [*456-24-6*] (172–173) have been used to characterize amino acids.

2-Fluoropyridine may become a useful reagent for synthetic applications. It reacts with methyl *p*-toluenesulfonate to give 1-methyl-2-fluoropyridinium *p*-toluenesulfonate [*58086-67-2*]. Such onium salts are used in the oxidation of alcohols, cross-coupling of Grignard reagents, Beckmann rearrangements of ketoximes, and nonenzymatic biogenetic-like synthesis of terpenes (174).

3-Fluoropyridine. 3-Fluoropyridine is prepared from 3-aminopyridine (Balz-Schiemann technique) in 50% yield. *Caution:* Earlier warnings concerning the thermal instability of 3-pyridyldiazonium fluoborate (175) were confirmed by a recent detonation involving this salt (176). Derivatives such as 3-fluoro-4-nitropyridine [*13505-01-6*] (177) or the 1-oxide [*769-54-0*] (178) have been used to characterize amino acids and peptides. 5-Fluoro-3-pyridinemethanol [*22620-32-2*] is an antilipolytic agent (179).

4-Fluoropyridine. 4-Fluoropyridine can be prepared in 54% yield by diazotization of 4-aminopyridine in anhydrous hydrogen fluoride and stabilization–storage as the hydrochloride salt (180). Free 4-fluoropyridine readily undergoes self-quaternization to give pyridyl pyridinium salts (181).

DIFLUOROPYRIDINES

2,4-Difluoropyridine. 2,4-Difluoropyridine can be prepared in 26% yield from 2,4-dichloropyridine and potassium fluoride in sulfolane and ethylene glycol initiator (182). The 4-fluorine is preferentially replaced by nucleophiles to give 2-fluoro-4-hydroxypyridine [*22282-69-5*], which is a precursor to 2-fluoro-3,5-dihalo-4-hydroxypyridine herbicides (183).

2,6-Difluoropyridine. 2,6-Difluoropyridine was first prepared in 1962 by a low-yield multistep route based on methyl 6-fluoropicolinate [*455-71-0*] (184). Two simplified routes have since been developed. The diazotization–fluorination of 2,6-diaminopyridine in anhydrous hydrogen fluoride gave a 62% yield of 2,6-difluoropyridine (6,185). This product is also prepared in 52% yield (200°C, 100 h) from 2,6-dichloropyridine and potassium fluoride in dimethyl sulfone or sulfolane (186) or under solvent-free conditions (400°C/16 h) in 80% yield (187). The latter observation attests to its high thermal stability. Nitration with mixed nitric and sulfuric acids provides a 79% yield of 3-nitro-2,6-difluoropyridine [*5860-02-1*], bp 218–20°C (188). Alkaline hydrolysis forms 2-fluoro-6-hydroxypyridine [*55758-32-2*], a precursor to Dowco 275, an insecticide-nematocide (189).

TRI-, TETRA- AND PENTAFLUOROPYRIDINES

2,4,6-Trifluoropyridine. 2,4,6-Trifluoropyridine can be prepared in 75% yield by catalytic hydrogenolysis (palladium-on-carbon, 280°C) of 3,5-dichloro-2,4,6-trifluoropyridine [*1737-93-5*] (190).

2,3,5,6-Tetrafluoropyridine. 2,3,5,6-Tetrafluoropyridine can be prepared in 75% yield from the hydrogenation of pentafluoropyridine under free radical (catalytic) or nucleophilic (lithium aluminum hydride) conditions (190–191).

Pentafluoropyridine. Pentafluoropyridine was first synthesized in 1960 in 27% yield by the defluorination of undecafluoropiperidine (prepared in low yield by the electrochemical fluorination of pyridine) over a nickel or iron surface at 560–610°C (192–193). The preferred route is the solvent-free exchange-fluorination (potassium fluoride) of pentachloropyridine at 480–500°C to give 69–83% yields of pentafluoropyridine (193–194). Pentafluoropyridine is a weak base, does not form a hydrochloride salt, and is more volatile (bp 83.3°C) than pyridine (bp 115°C). Pentafluoropyridine readily undergoes reaction with nucleophilic agents to give 4-substituted-2,3,5,6-tetrafluoropyridines as is illustrated in Figure 1.

The nucleophilic equivalent of the Friedel-Crafts reaction permits perfluoroalkylation of pentafluoropyridine with tetrafluoroethylene/cesium fluoride in sulfolane at 170°C to give perfluoro(pentaethylpyridine) [*20017-53-7*] (194). The latter undergoes valence isomerization by ultraviolet irradiation (>200 nm) to give a prismane, pentakis(pentafluoroethyl)-1-azatetracyclo[2.2.0.0.^{2,6}0^{3,5}]hexane (197). This photochemical capture technique is claimed to be a method for storage of solar energy (qv). The stored energy is released by heat and pyridine aromaticity is restored (198).

Pentafluoropyridine–hexafluorobenzene working fluids show the requisite stability at 382°C for automotive Rankine-cycle power units (109). Derivatives of 4-hydroxytetrafluoropyridine [*2693-66-5*], 4-hydroxy-3,5-dichloro-2,6-difluoropyridine [*2693-61-0*], and other 4-hydroxy-2,3,5,6-fluorohalopyridines exhibit promising herbicidal properties (199–200).

PERFLUOROALKYLPYRIDINES

Numerous patents have been issued for trifluoromethylpyridine derivatives as crop-protection chemicals (201–204). The standard synthesis method involves side-chain chlorination of a methylpyridine (picoline), followed by fluorination with pentavalent antimony chlorofluoride or anhydrous hydrogen fluoride (205). Fluorination of pyridinecarboxylic acids with sulfur tetrafluoride also has been successfully employed.

2-Trifluoromethylpyridine. 2-Trifluoromethylpyridine can be prepared in 54% yield from picolinic acid and sulfur tetrafluoride-hydrogen fluoride (206). 2-Trifluoromethylpyridine is a weak base; no hydrochloride salt is formed. However, 2-trifluoromethylpyridine 1-oxide [*22253-71-0*], bp 132–133°C/2.7 kPa (20 mm Hg), is prepared in 81% yield using 30% hydrogen peroxide–acetic acid (207).

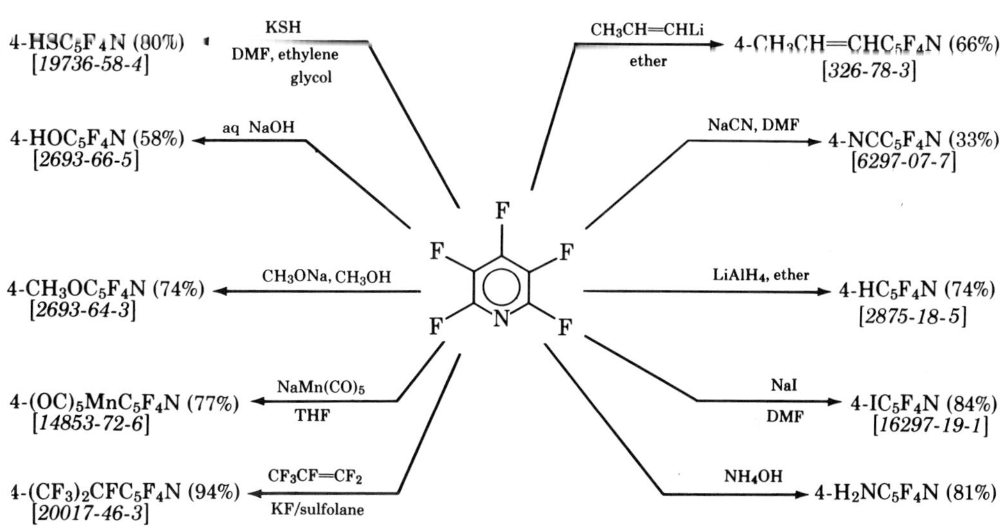

Figure 1. Reactions of pentafluoropyridine.

3-Trifluoromethylpyridine. 3-Trifluoromethylpyridine can be prepared in 25–65% yield from nicotinic acid and sulfur tetrafluoride (206,208). 3-Trifluoromethylpyridine 1-oxide [*22253-72-1*] has a mp at 75–77°C; a bp of 100–110°C at 0.2 kPa (1.5 mm Hg).

4-Trifluoromethylpyridine. 4-Trifluoromethylpyridine is prepared in 57% yield from isonicotinic acid and sulfur tetrafluoride. 4-Trifluoromethylpyridine 1-oxide [*22253-59-4*] melts at 175–177°C.

Fluoroquinolines

Some ring- and trifluoromethylquinolines exhibit biological properties, eg, Mefloquine [*53230-10-7*] is an antimalarial.

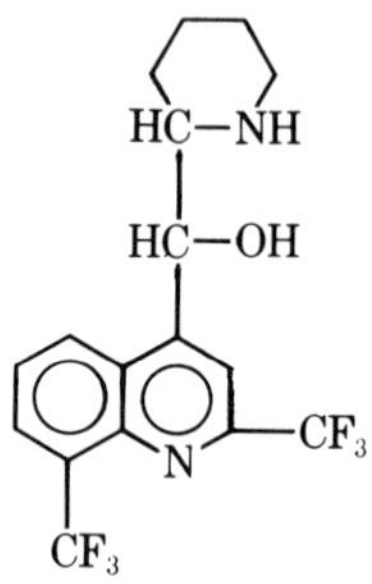

Mefloquine

The standard routes to monofluoroquinolines have been the Balz-Schiemann reaction from the corresponding aminoquinoline or the Skraup reaction from glycerol and a fluoroaniline. Exchange-fluorination also has been used. 2-Chloroquinoline and potassium fluoride in dimethylsulfone gave a 60% yield of 2-fluoroquinoline [*580-21-2*], C_9H_5FN; bp 133°C at 4 kPa (30 mm Hg); 75°C at 0.3 kPa (2 mm Hg); n_D^{25} 1.5827 (186). Likewise, heptachloroquinoline and potassium fluoride at 470°C for 17 h gave a 71% yield of heptafluoroquinoline [*13180-38-6*], C_9F_7N; mp 95–95.5°C; bp 205°C (209).

Fluoropyrimidines

Fluoropyrimidines find diverse use in cancer chemotherapy and other drug applications, as well as in fiber-reactive dyes.

MONOFLUOROPYRIMIDINES

5-Fluoropyrimidine derivatives are of tremendous importance in cancer chemotherapy, eg, 5-fluorouracil [*51-21-8*] (5-FU). The original 5-fluorouracil process featured a multistep low-yield route based on ethyl fluoroacetate (210). Direct fluorination (fluorine or trifluoromethyl hypofluorite) has now been achieved to give 80–92% yields of 5-FU (27–30).

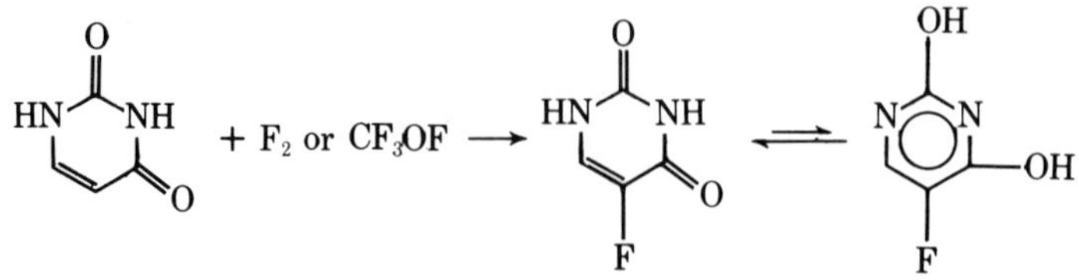

5-Fluorouracil

Other monofluoropyrimidines of biological interest are: 5-fluorocytosine [*2022-85-7*], an antifungal agent, and 2′-deoxy-5-fluorouridine [*50-91-9*] (5-FUDR), an antiviral and antineoplastic agent (see Chemotherapeutics, antibacterial and antimycotic; Chemotherapeutics, antimitotic; Chemotherapeutics, antiviral).

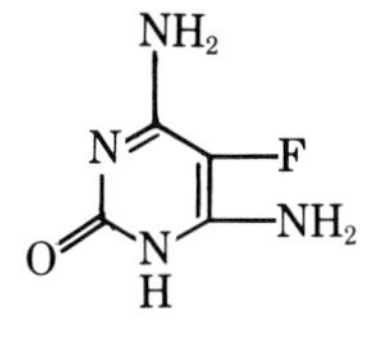

5-fluorocytosine

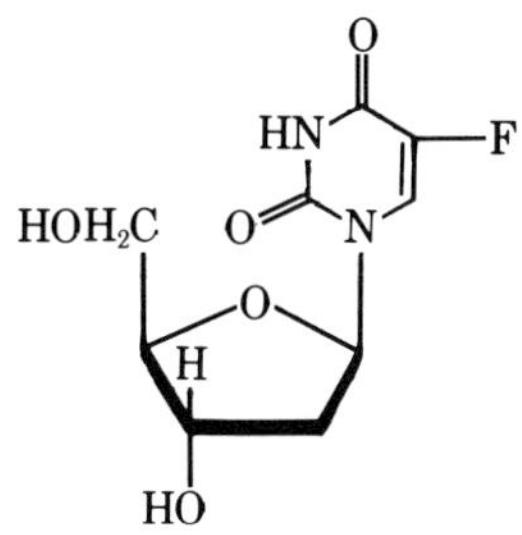

2′-deoxy-5-fluorouridine (5-FUDR)

TRI- AND TETRAFLUOROPYRIMIDINES

Table 6 lists some of the physical properties of polyfluoropyrimidines.

2,4,6-Trifluoropyrimidine. 2,4,6-Trifluoropyrimidine was first prepared in 1960 from 2,4,6-trichloropyrimidine and silver monofluoride (26,211). Potassium fluoride (in sulfolane or solvent-free conditions) is used instead of expensive silver fluoride since the former process is less expensive and more efficient (85% yield) (212–213).

2,4,5,6-Tetrafluoropyrimidine. 2,4,5,6-Tetrafluoropyrimidine [*767-79-3*], $C_4F_4N_2$ was first prepared by direct fluorination of 2,4,6-trifluoropyrimidine with silver difluoride in perfluorobutylamine solvent (26,211). A more direct route (85% yield) is the reaction of tetrachloropyrimidine and potassium fluoride in an autoclave at 480°C for 12 h (214).

Partial fluorination of 2,3,5,6-tetrachloropyrimidine with anhydrous hydrogen fluoride (autoclave or vapor phase) or sodium fluoride (autoclave, 300°C) gives high yields of 5-chloro-2,4,6-trifluoropyrimidine [*697-83-6*] (bp 114.5°C at 100 kPa (750 mm Hg)), a valuable commercial dye intermediate (215). The latter provides the 5-chloro-2,4-difluoropyrimidyl group for fiber-reactive dyes for cottons and cellulosics, eg, Levafix EA/PA (Bayer) and Drimarene R (216–217) (see Dyes, reactive). A typical structure of a fiber-reactive system is:

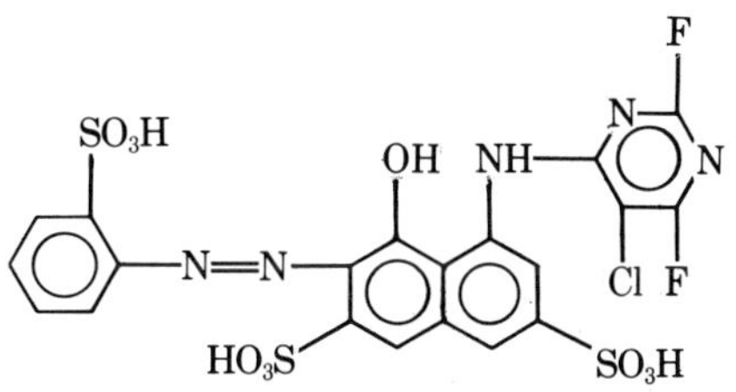

Fluorotriazines

Ring-fluorinated triazines are used in fiber-reactive dyes, and perfluoro-alkyl triazines are incorporated into elastomers and are offered commercially as mass spectral markers. Their physical properties are listed in Table 6.

2,4,6-Trifluoro-1,3,5-triazine (Cyanuric Fluoride). Cyanuric fluoride can be prepared from 2,4,6-trichloro-1,3,5-triazine (cyanuric chloride) with numerous fluorinating agents: pentavalent antimony chlorofluoride, silver monofluoride, potassium fluoride, sodium fluoride, potassium fluorosulfinate, and sulfur tetrafluoride (see Cyanuric and isocyanuric acids).

Fiber-reactive dyes based on cyanuric fluorides represent an emerging major application (218–219). Cyanuric fluoride has been offered commercially for biochemical applications, eg, as a specific reagent for tyrosine (220).

2,4,6-tris-(Trifluoromethyl)-1,3,5-Triazine. 2,4,6-*tris*-(Trifluoromethyl)-1,3,5-triazine is a colorless liquid which produces a burning sensation in the respiratory tract when inhaled. It can be prepared by trimerization of trifluoroacetonitrile (221) or fluorination of 2,4,6-*tris*-(trichloromethyl)-1,3,5-triazine with hydrogen fluoride/antimony pentachloride or antimony trifluoride/antimony pentafluoride (222). *tris*-(Perfluoroalkyl)-1,3,5-triazines are commercially offered as mass spectrometry internal reference standards for a wide mass range, 285–1485 (223). Numerous laboratories have investigated the synthesis and properties of perfluoroalkylenetriazine elastomers (224):

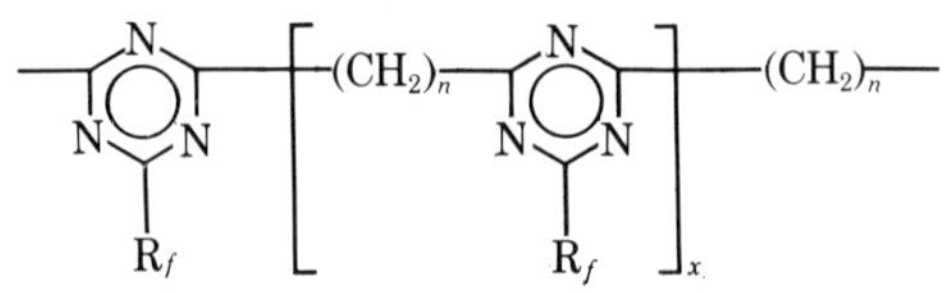

Miscellaneous Fluorinated Nitrogen Heterocyclics

The biological properties of fluorinated pyridazines and imidazoles have also attracted attention. Derivatives of 2,6-difluoropyridazine [*33873-09-5*] (225) and 3,4,5,6-tetrafluoropyridazine (226) exhibit herbicidal activity. The lability of fluorine in 2-fluoroimidazole [*57212-34-7*] has suggested that other 2-fluoro analogues, eg, 2-fluoro-L-histidine [*42310-02-1*] and 2-fluorohistamine [*50581-18-5*], might serve as affinity labels and inactivators of receptor sites and enzymes (227). 2-Fluorohistidine [*57372-70-0*] produces a mild soporific and anesthetic effect (228).

BIBLIOGRAPHY

"Fluorinated Aromatic Compounds" in *ECT* 2nd ed., under "Fluorine Compounds, Organic," Vol. 9, pp. 775–802, by A. K. Barbour, M. W. Buxton, and G. Fuller, Imperial Smelting Corporation, Ltd.

1. Ger. Pat. 600,706 (July 30, 1934), P. Osswald and O. Scherer (to I.G. Farbenind.).
2. F. O. Robitschek and B. H. Wilcoxon, *FIAT Final Report No. 998,* March 31, 1947.

3. R. L. Ferm and C. A. Vander Werf, *J. Am. Chem. Soc.* **72,** 4809 (1950).
4. G. A. Olah and J. Welch, *J. Am. Chem. Soc.* **97,** 208 (1975).
5. U.S. Pat. 4,096,196 (June 20, 1978), M. M. Boudakian (to Olin).
6. U.S. Pat. 4,075,252 (Feb. 21, 1978), M. M. Boudakian (to Olin).
7. U.S. Pat. 3,950,444 (April 13, 1976), W. Gay (to Olin).
8. D. T. Flood, *Org. Synth., Coll.* **II,** 295 (1943).
9. K. G. Rutherford, W. Redmond, and J. Rigamonti, *J. Org. Chem.* **26,** 5149 (1961).
10. G. C. Finger and C. W. Kruse, *J. Am. Chem. Soc.* **78,** 6034 (1956).
11. L. D. Starr and G. C. Finger, *Chem. Ind. London,* 1328 (1962).
12. G. Fuller, *J. Chem. Soc.,* 6264 (1965).
13. N. N. Vorozhtsov, Jr., V. E. Platonov, and G. G. Yakobson, *Isv. Akad. Nauk SSSR, Ser. Khim.,* (8), 1524 (1963).
14. M. Stacey and J. C. Tatlow in M. Stacey, J. C. Tatlow, and A. G. Sharpe, eds., *Advances in Fluorine Chemistry,* Vol. 1, Butterworths, London, 1960, p. 166.
15. P. A. Coe, R. G. Plevey, and J. C. Tatlow, *J. Chem. Soc. C,* 1060 (1969).
16. A. M. Doyle and A. E. Pedlar, *J. Chem. Soc. C,* 282 (1971).
17. R. D. Chambers, J. Heyes, and W. K. R. Musgrave, *Tetrahedron,* **19,** 891 (1963).
18. G. M. Brooke and co-workers, *J. Chem. Soc.,* 729 (1964).
19. V. Grakauskas, *J. Org. Chem.* **35,** 723 (1970); *J. Org. Chem.* **34,** 2835 (1969).
20. N. B. Kaz'mina and co-workers, *Dokl. Akad. Nauk SSSR* **194,** 1329 (1970).
21. D. H. R. Barton and co-workers, *Chem. Commun.,* 804 (1968).
22. J. Kollonitsch, L. Barash, and G. A. Doldouras, *J. Am. Chem. Soc.* **92,** 7494 (1970).
23. R. Filler, *Isr. J. Chem.* **17,** 71 (1978).
24. J. F. Ellis and W. K. R. Musgrave, *J. Chem. Soc.* 3608 (1950).
25. M. M. Boudakian, G. A. Hyde, and E. H. Kober, "Liquid Phase Aromatic Fluorination with Chlorine Pentafluoride," *162nd ACS Meeting, Washington, D.C. Division of Fluorine Chemistry, Sept. 1971,* Paper No. 6.
26. H. Schroeder, *J. Am. Chem. Soc.* **82,** 4115 (1960).
27. M. J. Robins and S. R. Naik, *J. Am. Chem. Soc.* **93,** 5277 (1971).
28. Ger. Offen. 2,014,216 (Oct. 14, 1971), I. L. Knunyants, L. S. German, and N. B. Kaz'mina (to Acad. Sci., USSR).
29. D. H. R. Barton and co-workers, *J. Org. Chem.* **37,** 329 (1972).
30. U.S. Pats. 3,954,758 (May 4, 1976), and 4,113,949 (Sept. 12, 1978), P. D. Schuman and co-workers (to PCR).
31. U.S. Pat. 3,499,942 (March 10, 1970), O. M. Nevedov and A. A. Ivashenko.
32. Y. Dĕsirant, *Bull. Sci. Acad. Roy. Belg.* **41,** 759 (1955).
33. J. M. Birchall and R. N. Haszeldine, *J. Chem. Soc.,* 13 (1959).
34. G. A. Olah and P. Kreienbühl, *J. Org. Chem.* **32,** 1614 (1967).
35. K. O. Christe and A. E. Pavlath, *J. Org. Chem.* **31,** 559 (1966).
36. U.S. Pat. 2,884,458 (April 28, 1959), D. A. Fidler (to Olin).
37. D. A. Fidler, J. S. Logan, and M. M. Boudakian, *J. Org. Chem.* **26,** 4014 (1961).
38. P. H. Scott and co-workers, *Tetrahedron Lett.* 1153 (1970); U.S. Pat. 3,558,707 (Jan. 26, 1971), J. W. Churchill, E. H. Kober, and P. H. Scott (to Olin).
39. I. N. Rozhkov, A. V. Bukhtiarov, and I. L. Knunyants, *Isv. Akad. Nauk SSSR Ser. Khim.,* 1130 (1972).
40. J. H. Brown, C. W. Suckling, and W. B. Whalley, *J. Chem. Soc. Suppl. Issue 1,* **S** 95 (1949).
41. U.S. Pat. 3,136,822 (June 9, 1964), L. J. Frainier, Jr. (to Olin).
42. U.S. Pat. 3,859,372 (Jan. 7, 1975), S. Robota (to Hooker Chemical).
43. G. A. Boswell, Jr. and co-workers, *Org. React.* **21,** 1 (1974).
44. Y. Kobayashi and I. Kumadaki, *Tetrahedron Lett.,* 4095 (1969); *Chem. Pharm. Bull.* **18,** 2334 (1970).
45. V. C. R. McLaughlin and J. Thrower, *Tetrahedron* **25,** 5921 (1969).
46. W. A. Sheppard, *J. Am. Chem. Soc.* **87,** 2410 (1965).
47. E. J. Fairchild, ed., *Registry of Toxic Effects of Chemical Substances—1977 Edition,* Vols. I and II, DHEW Pub. No. NIOSH 78-104-A, Cincinnati, 1977.
48. *Handbook of Aromatic Fluorine Compounds,* Olin Corp., Stamford, Conn., 1976.
49. G. Olah, A. Pavlath, and G. Varsanyi, *J. Chem. Soc.,* 1823 (1957).
50. N. Ishikawa, *Kogyo Kagaku Zasshi* **69,** 1484 (1966).

51. Brit. Pat. 1,469,700 (April 6, 1977), G. Fuller (to ISC Chemicals).
52. Brit. Pat. 1,360,327 (July 17, 1974), S. Misaki, M. Okamoto, and M. Hirata (to Daikin Kogyo).
53. U.S. Pat. 3,280,124 (Oct. 18, 1966), M. M. Boudakian and E. R. Shipkowski (to Olin).
54. P. A. Janssen and co-workers, *J. Med. Pharm. Chem.* **1,** 281 (1959).
55. H. C. Brown and G. Marino, *J. Am. Chem. Soc.* **84,** 1658 (1962).
56. R. E. Parker in M. Stacey, J. C. Tatlow, and A. G. Sharpe, eds., *Advances in Fluorine Chemistry,* Vol. 3, Butterworths, Washington, 1963, p. 63.
57. M. Bil, *J. Appl. Chem. Biotechnol.* **22,** 853 (1972).
58. U.S. Pat. 3,959,377 (May 25, 1976), M. Bil (to Clairol).
59. F. Sanger, *Biochem. J.* **39,** 507 (1945).
60. S. E. Drews, H. E. M. Magojo, and D. A. Sutton, *J. Chem. Soc., Perkin Trans. I,* 1283 (1975).
61. G. W. J. Fleet and R. Porter, *Nature,* 511 (1970).
62. K. J. Skinner, *Chem. Eng. News,* 23 (Aug. 18, 1975).
63. E. D. Bergmann and M. Bentov, *J. Org. Chem.* **26,** 1480 (1961).
64. R. W. Holley and A. D. Holley, *J. Am. Chem. Soc.* **74,** 1110 (1952).
65. F. F. Micheel, K. Weichbrodt, and J. Plenikowski, *Ann.* **581,** 238 (1953).
66. H. Zahn and K. H. Lebkücher, *Biochem. Z.* **334,** 134 (1961).
67. H. Zahn and H. Zuber, *Ber.* **86,** 172 (1953).
68. H. Zahn and J. Meinhoffer, *Makromol. Chem.* **26,** 126 (1958).
69. K. L. Kirk and L. A. Cohen, *J. Org. Chem.* **34,** 395 (1969).
70. K. Sturm and co-workers, *Chem. Ber.* **99,** 328 (1966).
71. H. Suschitzky, *Angew. Chem. Int. Edit.* **6,** 596 (1967).
72. W. J. Weigert and W. A. Sheppard, *J. Org. Chem.* **25,** 4006 (1976).
73. J. P. DeKleijn, *J. Fluorine Chem.* **10,** 341 (1977).
74. M. M. Boudakian and co-workers, *J. Org. Chem.* **26,** 4641 (1961).
75. G. G. Yakobson, T. D. Rubina, and N. N. Vorozhtsov, Jr., *Dokl. Akad. Nauk SSSR* **141,** 1395 (1961).
76. R. L. Weintraub, J. W. Brown, and J. A. Throne, *Agr. Food Chem.* **2,** 996 (1954).
77. G. C. Finger and co-workers, *J. Am. Chem. Soc.* **81,** 94 (1959).
78. U.S. Pat. 3,314,985 (April 18, 1967), F. F. Frulla and M. M. Boudakian (to Olin).
79. A. J. Adduci, *Chemtech,* 575 (1976).
80. R. D. Rieke and S. E. Bales, *J. Am. Chem. Soc.* **96,** 1775 (1974).
81. L. Bretherick, *Chem. Ind. London,* 1017 (1971).
82. U.S. Pat. 2,939,766 (June 7, 1960), J. W. Churchill (to Olin).
83. U.S. Pat. 2,563,796 (Aug. 7, 1951), W. H. Shenk, Jr. and G. R. Mellon (to Harshaw Chemical Company).
84. U.S. Pat. 3,160,623 (Dec. 8, 1964), L. G. Anello and C. Woolf (to Allied Chemical).
85. F. Seel, *Angew. Chem. Int. Ed.* **4,** 635 (1965).
86. Jpn. Kokai 74 81,330 (Aug. 6, 1974), S. Misaki and M. Okamoto (to Daikin Kogyo).
87. J. W. Clayton, Jr. "Pharmacology of Fluorides" in F. Smith, ed., *Handbook of Experimental Pharmacology,* Vol. XX, Springer-Verlag, New York, Chapt. 9, pt. 1, p. 492, 1966.
88. U.S. Pat. 4,001,393 (Jan. 4, 1977), R. L'Orange (to Blendax-Werke R. Schneider).
89. G. C. Finger, F. H. Reed, and L. R. Tehon, *Aromatic Fluorine Compounds as Fungicides,* Illinois State Geological Survey, Circular No. 199, Urbana, 1955.
90. R. S. Shiley and co-workers, *J. Fluorine Chem.* **5,** 371 (1975).
91. H. Hopff, *Chimia* **15,** 193 (1961).
92. Brit. Pat. 1,415,895 (Dec. 3, 1975), (to DuPont).
93. U.S. Pat. 3,979,321 (Sept. 7, 1976), A. Couttet, J. C. DuBois, and A. Zaum (to Thomson-CSF).
94. U.S. Pat. 4,018,507 (April 19, 1977), N. V. V. Raghavan (to American Micro-Systems).
95. J. P. Van Meter, *Eastman Org. Chem. Bull.* **45,** 1 (1973).
96. Ger. Offen. 2,618,609 (Dec. 2, 1976), M. Abert-Mellah, A. Zanne, and J. C. DuBois (to Thomson-CSF).
97. J. T. Minor and C. A. Vander Werf, *J. Org. Chem.* **17,** 1425 (1952).
98. *p-Chlorobenzotrifluoride,* Hooker Chemical Corp. Data Sheet No. 784-B, 1977.
99. G. Schiemann and R. Pillarsky, *Ber.* **62,** 3035 (1929).
100. S. K. Ray and G. Moss, *Adv. Energy Convers.* **6,** 89 (1966).
101. D. G. Holland, G. J. Moore, and C. Tamborski, *J. Org. Chem.* **29,** 3042 (1964).
102. P. L. Coe and A. E. Jukes, *Tetrahedron* **24,** 5913 (1968).

103. Brit. Pat. 996,498 (June 30, 1965), G. Fuller (to Imperial Smelting).
104. J. M. Birchall, R. N. Haszeldine, and A. R. Parkinson, *J. Chem. Soc.*, 4966 (1962).
105. P. L. Coe, A. E. Jukes, and J. C. Tatlow, *J. Chem. Soc. C*, 2323 (1966).
106. C. R. Patrick and G. S. Prosser, *Trans. Faraday Soc.* **60,** 700 (1964).
107. L. A. Wall, ed., *Fluoropolymers*, Vol. XXV, Wiley-Interscience, New York, 1972.
108. L. W. Hall, S. R. K. Jackson, and G. M. Massey, *Excerpta Med. Int. Congr. Ser.*, 347 (1974).
109. D. R. Miller and co-workers, *Intersoc. Energy Convers. Eng. Conf. Conf. Proc. 7th*, 315 (1972).
110. R. J. Harper, E. J. Soloski, and C. Tamborski, *J. Org. Chem.* **29,** 3042 (1964).
111. E. Kinsella and A. G. Massey, *Chem. Ind. London*, 1017 (1971).
112. S. C. Cohen and A. G. Massey in J. C. Tatlow, R. D. Peacock, and H. H. Hyman, eds., *Advances in Fluorine Chemistry*, Vol. 6, CRC Press, Cleveland, 1970, p. 83.
113. J. W. Eichelberger, L. E. Harris, and W. L. Budde, *Anal. Chem.* **47,** 995 (1975).
114. E. D. Morgan and C. F. Poole, *J. Chromatogr.* **104,** 351 (1975).
115. M. M. Murza and co-workers, *Zh. Org. Khim.* **13,** 1046 (1977).
116. I. Kisfaludy and co-workers, *Tetrahedron Lett.*, 1785 (1974).
117. G. H. Kalb, R. W. Quarles, and R. S. Graff, *Appl. Polym. Symposium*, (22), 127 (1973).
118. K. T. Koshy, D. G. Kaiser, and H. L. Van der Silk, *J. Chromatogr. Sci.* **13,** 97 (1975).
119. T. Nambara and co-workers, *J. Chromatogr.* **114,** 81 (1975).
120. U.S. Pat. 3,946,063 (May 23, 1976), J. Vessman, C. M. Svahn, and P. Hartvig (to Pierce Chemical).
121. J. C. Lhuguenot and B. F. Maume, *J. Chromatogr. Sci.* **12,** 411 (1974).
122. D. E. Pearson and co-workers, *Synthesis*, 127 (1978).
123. G. R. Wilkinson, *Anal. Lett.* **3,** 289 (1970).
124. W. H. Newsome, *J. Agric. Food Chem.* **26,** 1325 (1978).
125. G. Schiemann and B. Cornils, *Chemie and Technologie Cyclischer Fluorverbindungen*, Ferdinand Enke Verlag, Stuttgart, 1969, pp. 340–1.
126. G. Schiemann and E. Bolsted, *Ber.* **61,** 1403 (1928).
127. J. Hannah and co-workers, *J. Med. Chem.* **21,** 1093 (1978).
128. W. J. Pummer and L. A. Wall, *J. Chem. Eng. Data* **6,** 76 (1961).
129. M. W. Buxton, R. H. Mobbs, and D. E. Wotton, *J. Fluorine Chem.* **2,** 387 (1972/73).
130. M. S. Newman in P. O. P. Ts'o and J. A. DiPaolo, eds., *Chemical Carcinogenesis*, Marcel Dekker, N.Y., 1974, p. 177.
131. D. R. MacKenzie, F. W. Bloch, and P. H. Wiswall, *J. Phys. Chem.* **69,** 2526 (1965).
132. E. Boger, R. F. O'Malley, and D. J. Sardella, *J. Fluorine Chem.* **8,** 513 (1976).
133. A. L. Henne and M. S. Newman, *J. Am. Chem. Soc.* **60,** 1697 (1938).
134. *Product List No. 4,* Marshallton Research Labs, West Chester, Pa., Jan. 1973, p. 27.
135. L. Bretherick, *Chem. Ind. London*, 1017 (1971).
136. I. C. Appleby, *Chem. Ind. London*, 120 (1971).
137. L. Bretherick, *Handbook of Reactive Chemical Hazards*, CRC Press, Cleveland, 1975, p. 520.
138. J. Novotny, C. H. Collins, and F. W. Starks, *J. Pharm. Sci.* **62,** 910 (1973).
139. U.S. Pat. 3,707,040 (Jan. 2, 1973), R. C. Conner and L. L. Ferstandig (to Halocarbon Products).
140. U.S. Pat. 3,753,345 (Aug. 21, 1973), F. H. Cassidy and R. Garcia (to Aerojet-General).
141. E. Kühle and E. Klauke, *Angew. Chem. Int. Ed.* **16,** 735 (1977).
142. R. Filler and H. Novar, *Chem. Ind. London*, 1273 (1960).
143. R. G. Jones, *J. Am. Chem. Soc.* **69,** 2346 (1947).
144. V. C. R. McLoughlin and J. Thrower, *Chem. Ind. London*, 1557 (1964).
145. R. Grinter and co-workers, *Tetrahedron Lett.*, 3845 (1968).
146. R. Wächter, *Angew. Chem.* **67,** 305 (1955).
147. F. Swarts, *Bull. Sci. Acad. Roy. Belg.*, 399 (1920).
148. Y. Kobayashi and I. Kumadaki, *Acc. Chem. Res.* **11,** 197 (1978).
149. N. P. Buu-Hoï and co-workers, *Compt. Rend.* **257**(21), 3182 (1963).
150. U.S. Pat. 3,234,292 (Feb. 8, 1966), S. Robota and E. A. Belmore (to Hooker Chemical).
151. R. J. Albers and E. C. Kooyman, *Rec. Trav. Chim.* **83,** 930 (1964).
152. A. A. Ushakov and co-workers, *Zh. Org. Khim.* **12,** 158 (1963).
153. K. Inukai and T. Ueda, *Kogyo Kagaku Zasshi* **64,** 2156 (1961).
154. Jpn. Kokai 75 76,029, S. Misaki, Y. Furutaka, and T. Shimoike (to Daikin Kogyo) (June 21, 1975).
155. E. T. McBee, R. A. Sanford, and P. J. Graham, *J. Am. Chem. Soc.* **72,** 1651 (1950).
156. U.S. Pat. 4,008,278 (Feb. 15, 1977), M. M. Boudakian (to Olin).
157. D. T. Downing, R. L. Heppolette, and J. Miller, *Chem. Ind. London*, 1260 (1953).

158. *Benzotrifluoride,* Hooker Chemical Corp., Data Sheet No. 778-D, 1969.
159. R. Filler in J. C. Tatlow, R. D. Peacock, and H. H. Hyman, eds., *Advances in Fluorine Chemistry,* Vol. 6, CRC Press, Cleveland, Ohio, 1970, p. 1.
160. F. Swarts, *Bull. Acad. Roy. Belg. Cl. Sci.* **35,** 375 (1898)..
161. A. K. Barbour, L. J. Belf, and M. W. Buxton in M. Stacey, J. C. Tatlow, and A. G. Sharpe, eds., *Advances in Fluorine Chemistry,* Vol. 3, Butterworth & Co., Ltd., London, 1963, p. 181.
162. U.S. Pat. 4,045,502 (Aug. 30, 1977), S. K. Bhutani and W. A. Nichols (to Olin).
163. B. W. Lawrence, D. O. Tampsett, and D. J. A. Webb, *Manuf. Chem. Aerosol News* **41,** 37 (Jan. 1970).
164. S. P. Anand and R. Filler, *J. Fluorine Chem.* **7,** 179 (1976).
165. G. C. Finger and co-workers, *J. Org. Chem.* **28,** 1666 (1963).
166. M. M. Boudakian, *J. Heterocyclic Chem.* **4,** 381 (1967); U.S. Pat. 3,296,269 (Jan. 3, 1967), (to Olin).
167. H. C. Brown and D. McDaniel, *J. Am. Chem. Soc.* **77,** 3752 (1955).
168. L. Bradlow and C. Van der Werf, *J. Org. Chem.* **14,** 509 (1949).
169. G. Bressan and co-workers, *J. Chem. Soc. B,* 225 (1971).
170. U.S. Pats. 3,873,701 (March 25, 1975), and 3,816,406 (June 11, 1974), A. F. Hirsch (to Ortho Pharmaceuticals).
171. D. Sarantakis and co-workers, *Chem. Comm.,* 105 (1966).
172. A. Signor and co-workers, *Eur. J. Biochem.* **7,** 328 (1969).
173. W. Gruber, *Can. J. Chem.* **31,** 1020 (1953).
174. T. Mukaiyama, *Anqu. Chem. Int. Ed.* **18,** 707 (1979).
175. A. Roe and C. Hawkins, *J. Am. Chem. Soc.* **69,** 2443 (1947).
176. *Chem. Eng. News,* 44 (Oct. 16, 1967); *Chem. Eng. News,* 8 (Dec. 8, 1967).
177. T. Talik and Z. Talik, *Rocz. Chem.* **40,** 1187 (1966).
178. T. Talik and Z. Talik, *Rocz. Chem.* **38,** 785 (1964).
179. U.S. Pat. 3,637,714 (Jan. 25, 1972), L. A. F. Carlsson and co-workers (to Aktiebolaget Astra).
180. U.S. Pat. 3,703,521 (Nov. 21, 1972), M. M. Boudakian (to Olin).
181. P. B. Desai, *J. Chem. Soc. Perkin Trans. I,* 1866 (1973).
182. Ger. Offen 2,128,540 (Dec. 16, 1971), A. Nicolson (to Imperial Chemical Industries).
183. Brit. Pat. 1,479,721 (July 13, 1977), D. W. R. Headford, J. W. Slater, and R. L. Sunley (to Imperial Chemical Industries).
184. G. C. Finger and co-workers, *J. Org. Chem.* **27,** 3965 (1962).
185. U.S. Pat. 3,798,228 (March 19, 1974), M. M. Boudakian (to Olin).
186. J. Hamer and co-workers, *Rec. Trav. Chim.* **81,** 1059 (1962).
187. M. M. Boudakian, *J. Heterocycl. Chem.* **5,** 683 (1968).
188. F. Mutterer and C. D. Weis, *Helv. Chim. Acta* **59,** 229 (1976).
189. Ger. Pat. 2,263,429 (March 23, 1975), and U.S. Pat. 3,810,902 (May 14, 1974), R. H. Rigterink (to Dow).
190. R. D. Chambers, F. G. Drakesmith, and W. K. R. Musgrave, *J. Chem. Soc.,* 5045 (1965).
191. R. E. Banks and co-workers, *J. Chem. Soc.,* 575 (1965).
192. R. E. Banks, A. E. Ginsberg, and R. N. Haszeldine, *J. Chem. Soc.,* 1740 (1961).
193. J. A. Burdon and co-workers, *Nature* **196,** 231 (1960).
194. R. D. Chambers, J. Hutchinson, and W. K. R. Musgrave, *J. Chem. Soc.,* 3573 (1964).
195. R. E. Banks and co-workers, *J. Chem. Soc.,* 594 (1965).
196. R. D. Chambers and co-workers, *J. Chem. Soc. C,* 2221 (1968).
197. M. G. Barlow, J. G. Dingwall, and R. N. Haszeldine, *J. Chem. Soc. Chem. Commun.,* 1580 (1970).
198. Jpn. Pat. 78 64,833 (June 9, 1978), (to Battelle Development Corp.).
199. U.S. Pat. 3,850,943 (Nov. 26, 1974), R. D. Bowden, J. W. Slater, and B. G. White (to Imperial Chemical Industries).
200. U.S. Pat. 4,063,926 (Dec. 20, 1977), C. D. S. Tomlin, J. W. Slater, and D. Hartley (to Imperial Chemical Industries).
201. U.S. Pat. 3,787,420 (Jan. 22, 1974), F. E. Torba (to Dow).
202. Belg. Pat. 865,136 (Sept. 21, 1978), R. Nishiyama and co-workers (to Ishihara Sangyo).
203. Belg. Pat. 862,325 (June 27, 1978), (Ciba-Geigy).
204. U.S. Pat. 3,818,019 (June 18, 1974), R. H. Rigterink (to Dow).
205. E. T. McBee, H. B. Hass, and E. M. Hodnett, *Ind. Eng. Chem.* **39,** 389 (1947).
206. M. S. Raasch, *J. Org. Chem.* **27,** 1406 (1962).

207. Y. Kobayashi and I. Kumadaki, *Chem. Pharm. Bull.* **17,** 510 (1969).
208. Y. Kobayashi and E. Chinen, *Chem. Pharm. Bull.* **17,** 510 (1967).
209. R. D. Chambers and co-workers, *J. Chem. Soc. C,* 2328 (1966).
210. R. Duschinsky, E. Pleven, and C. Heidelberger, *J. Am. Chem. Soc.* **79,** 4559 (1957).
211. H. Schroeder and co-workers, *J. Org. Chem.* **27,** 2580 (1962).
212. U.S. Pat. 3,314,955 (April 18, 1967), M. M. Boudakian and C. W. Kaufman (to Olin).
213. U.S. Pat. 3,280,124 (Oct. 18, 1966), M. M. Boudakian, E. H. Kober, and E. R. Shipkowski (to Olin).
214. R. E. Banks, D. S. Field and R. N. Haszeldine, *J. Chem. Soc. C,* 1822 (1967).
215. Can. Pat. 844,625 (June 16, 1970), E. Klauke and H. S. Bien (to Farbenfabriken Bayer).
216. U.S. Pat. 3,669,951 (June 13, 1972), H. S. Bien and E. Klauke (to Farbenfabriken Bayer).
217. D. R. Hildebrand, *Chemtech,* 224 (1978).
218. Ger. Pat. 2,611,550 (Sept. 30, 1976), A. Seiler and G. Hegar (to Ciba-Geigy).
219. Brit. Pat. 1,461,125 (Jan. 13, 1977), I. K. Barben and C. V. Stead (to Imperial Chemical Industries).
220. K. Kurihara and co-workers, *Biochim. Biophys. Acta* **384,** 127 (1975).
221. W. L. Reilly and H. C. Brown, *J. Org. Chem.* **22,** 698 (1967).
222. E. T. McBee, O. R. Pierce, and R. O. Bolt, *Ind. Eng. Chem.* **39,** 391 (1947).
223. T. Aczel, *Anal. Chem.* **40,** 1917 (1968).
224. J. A. Young "Perfluoroalkylenetriazine Elastomers and Related Heterocyclic Polymers" in L. A. Wall, ed., *Fluoropolymers,* Wiley-Interscience, New York, 1972, Chapt. 9.
225. N. Ishikawa, K. Kuroda, and N. Onodera, *Kogyo Kagaku Zasshi* **74,** 1490 (1971).
226. U.S. Pat. 4,083,713 (April 11, 1978), R. Y. Yih, W. O. Johnson, and G. E. Kollman (to Rohm & Haas).
227. K. L. Kirk and L. A. Cohen, *J. Am. Chem. Soc.* **92,** 3060 (1971); **95,** 8389 (1973).
228. K. L. Kirk and L. A. Cohen, *J. Am. Chem. Soc.* **95,** 4619 (1973).

General References

A. E. Pavlath and A. L. Leffler, *Aromatic Fluorine Compounds,* ACS Monograph No. 155, Reinhold, N.Y., 1962.

W. A. Sheppard and C. M. Sharts, *Organic Fluorine Chemistry,* W.A. Benjamin, N.Y., 1969.

M. Hudlicky, *Chemistry of Organic Fluorine Compounds,* 2nd ed., E. Horwood, Chichester, 1976.

G. Schiemann and B. Cornils, *Chemie and Technologie Cyclischer Fluorverbindungen,* F. Enke Verlag, Stuttgart, 1969.

R. E. Banks, *Fluorocarbons and Their Derivatives,* 2nd ed., Elsevier, New York, 1970.

Handbook of Aromatic Fluorine Compounds, Olin Corporation, Stamford, Conn., 1976.

R. D. Chambers, *Fluorine in Organic Chemistry,* John Wiley & Sons, Inc., New York, 1973.

E. Forche "Houben-Weyl, Methoden der Organischen Chemie" in E. Mueller, ed., *Halogen-Verbindungen,* Georg Thieme Verlag, Stuttgart, 1962, pp. 1–502.

R. Filler in J. C. Tatlow, R. D. Peacock, and H. H. Hyman, eds., *Advances in Fluorine Chemistry,* Vol. 6, CRC Press, Cleveland, 1970, p. 1.

G. Fuller, *Manuf. Chem. Aerosol News,* **35,** 43 (May 1964); *ibid.* **35,** 45 (June 1964).

Y. Kobayashi and I. Kumadaki, *Acc. Chem. Res.* **11,** 197 (1978).

A. K. Barbour, L. J. Belf, and M. W. Buxton in M. Stacey, J. C. Tatlow, and A. G. Sharpe, eds., *Advances in Fluorine Chemistry,* Vol. 3, Butterworths, Washington, 1963, p. 181.

N. Ishikawa, *Yuki Gosei Kagaku Kyokai Shi* **25,** 808 (1967).

R. H. Shiley, D. R. Dickerson, and G. C. Finger, *Aromatic Fluorine Chemistry at the Illinois State Geological Survey,* Circular 501, Urbana, Ill., 1978.

A. Roe in R. Adams, ed., *Organic Reactions,* Vol. V, John Wiley & Sons, New York, 1949, Chapt. 4, p. 193.

H. Suschitzky in M. Stacey, J. C. Tatlow, and A. G. Sharpe, eds., *Advances in Fluorine Chemistry,* Vol. 4, Butterworths, Washington, D.C., 1964, p. 1.

M. M. Boudakian, "Halopyridines" in R. A. Abramovitch, ed., *The Chemistry of Heterocyclic Compounds,* Supplement, Vol. 14, Part 2, *Pyridine and Its Derivatives,* Wiley-Interscience, New York, 1974, Chapt. VI, p. 407.

G. G. Yakobson, T. D. Petrova, and L. S. Kobrina, *Fluorine Chem. Rev.* **7,** 115 (1974).

P. Bouchet, C. Coquelot, and J. Elguero, *Bull. Soc. Chim. France,* 171 (1977).

R. Filler, ed., *Biochemistry Involving Carbon–Fluorine Bonds,* ACS Symposium Series No. 28, American Chemical Society, Washington, D.C. 1976.
Ciba Foundation Symposium, *Carbon–Fluorine Compounds: Chemistry, Biochemistry and Biological Activities,* Elsevier, New York, 1972.
F. A. Smith, "Biological Properties of Compounds Containing the C—F Bond," *Chemtech,* 422 (1973); R. Filler, "Fluorinated Compounds of Medicinal Interest," *Chemtech,* 752 (1974).
"Pharmacology of Fluorides" in F. A. Smith, ed., *Handbook of Experimental Pharmacology,* Vol. XX, Springer-Verlag, N.Y., Part 1 (1966); Part 2 (1970).
R. Filler, *Fluorine Chem. Rev.* **8,** 1 (1977).

MAX M. BOUDAKIAN
Olin Chemicals

FLUOROALKOXYPHOSPHAZENES

The phosphazenes are compounds containing the —P=N— moiety as the basic unit in either cyclic or linear structures. In fluoroalkoxyphosphazenes, fluoroalkoxy groups are attached to the phosphorus atom of the basic unit. Commercial importance was established with a poly(fluoroalkoxyphosphazene) elastomer introduced by the Firestone Tire and Rubber Co. under the trademark PNF.

Preparation

The phosphazene system is obtained by the reaction of phosphorus pentachloride with ammonium chloride (1–2):

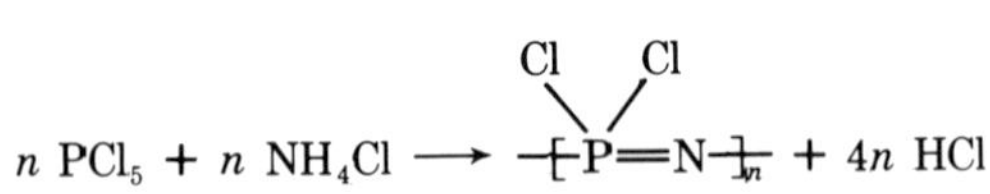

A mixture of oligomers is obtained whose composition can be varied widely by control of reaction conditions. The most useful oligomers are represented by the cyclic structures (1) and (2):

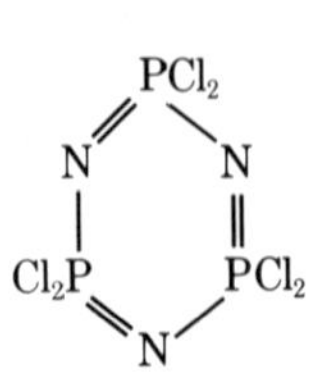

(1)
hexachlorocyclotriphosphazene
(trimer)

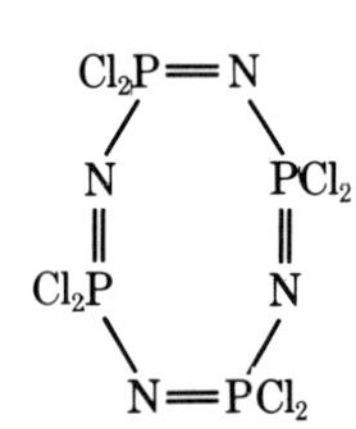

(2)
octachlorocyclotetraphosphazene
(tetramer)

Current technology permits very high conversion to the cyclic trimer and tetramer (3–4). Depending on reaction conditions, higher cyclic homologues, $(NPCl_2)_{5-9}$, as well as low molecular weight, open-chain poly(dichlorophosphazenes) (3–5) are obtained.

Polymers. High molecular weight, linear phosphazene polymers form at 250°C from the cyclic trimer:

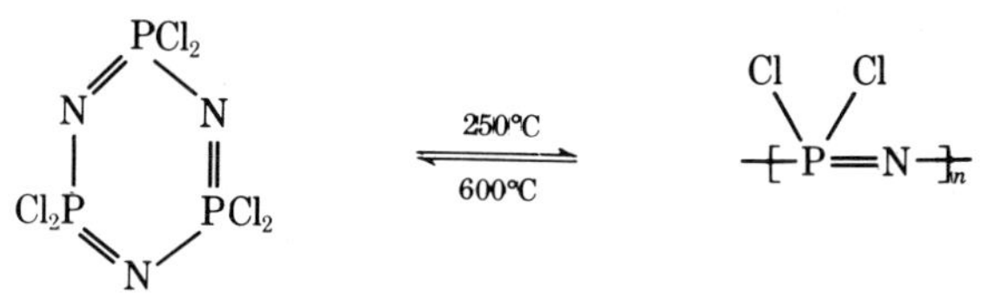

(1)

hexachlorocyclotriphosphazene poly(dichlorophosphazene)

This ring-opening polymerization has a classical ring–chain type of equilibrium. Poly(dichlorophosphazene) is a rubbery material which was termed inorganic rubber by early workers (6). However, the elastomeric qualities of poly(dichlorophosphazene) are short lived on exposure to air owing to a slow hydrolysis of the P—Cl bonds. During hydrolysis the polymer passes through stages of cross-linked elastomer, highly cross-linked resin, and finally, total hydrolytic degradation to ammonium chloride and ammonium phosphate. Because of this instability, much effort has been expended on the preparation of hydrolytically stable derivatives by displacing the chlorines with alkoxy or amino groups.

Alkoxyphosphazenes. These compounds are generally derived from the reaction of poly(dichlorophosphazene) with sodium alkoxides:

$$\left[\mathrm{P(Cl)_2{=}N}\right]_n + 2\ \mathrm{NaOR} \longrightarrow \left[\mathrm{P(OR)_2{=}N}\right]_n + 2\ \mathrm{NaCl} \qquad (2)$$

The alkoxyphosphazenes are hydrolytically stable materials, provided complete substitution is attained. The degree of stability is also dependent on the alkyl groups. A multitude of different polymers can be prepared by varying the alkoxide:

$$\left[\mathrm{P(Cl)_2{=}N}\right]_n + 2\ \mathrm{NaOCH_2CF_3} \longrightarrow \left[\mathrm{P(OCH_2CF_3)_2{=}N}\right]_n + 2\ \mathrm{NaCl} \qquad (3)$$

poly[bis(trifluoroethoxy)phosphazene] [*28212-50-2*]
(a thermoplastic, $T_g = -66°C$)

$$\left[\mathrm{P(Cl)_2{=}N}\right]_n + \mathrm{NaOCH_2CF_3} + \mathrm{NaOCH_2CF_2CF_2CF_3} \longrightarrow \left[\mathrm{P(OCH_2CF_3)(OCH_2CF_2CF_2CF_3){=}N}\right]_n + 2\ \mathrm{NaCl} \qquad (4)$$

poly(trifluoroethoxyheptafluorobutoxyphosphazene) [*37002-15-6*]
(an elastomer, $T_g = -77°C$)

Poly[bis(trifluoroethoxy)phosphazene] is termed a homopolymer in this context because it utilizes a single alkoxide species, whereas the poly(trifluoroethoxyheptafluorobutoxyphosphazene) is called a copolymer here because of its two different pendant groups. Multifunctionality is obtained simply by use of a mixture of differing alcoholates during the reaction with poly(dichlorophosphazene).

Reactions 3 and 4 represent a breakthrough in the development of poly(fluoroalkoxyphosphazenes), PFAP. The synthesis of poly[bis(trifluoroethoxy)phosphazene] in 1966 was the first successful preparation of a fully substituted, high molecular weight, hydrolytically stable poly(alkoxyphosphazene) (7). Earlier attempts to prepare alkoxyphosphazene polymers via nucleophilic displacement of chlorine resulted in incompletely substituted, degraded, or cross-linked polymers. Most of these difficulties can be ascribed to the utilization of insoluble, cross-linked poly(dichlorophosphazene). Thus, the key to the successful synthesis of poly[bis(trifluoroethoxy)phosphazene] was the use of soluble, noncross-linked poly(dichlorophosphazene).

The preparation of poly(trifluoroethoxyheptafluorobutoxyphosphazene) copolymer in 1968 (8) showed that elastomeric as well as plastic fluoroalkoxyphosphazenes were accessible merely by proper choice of pendant groups. Thus, an elastomeric polyphosphazene was obtained from poly(dichlorophosphazene) and a mixture of sodium trifluoroethoxide and heptafluorobutoxide (see eq. 4). This elastomeric copolymer is solvent resistant, flexible at low temperature, and stable at high temperature (see Elastomers, synthetic).

Inert Fluids

Fluid fluoroalkoxyphosphazenes are readily obtained from reaction of sodium fluoroalkoxides with either the cyclic trimeric or tetrameric dichlorophosphazene. The oily products (9) possess remarkable chemical and thermal stability. Cyclic derivatives such as structure (**3**) and related multiphosphazene ring compounds are stable above 300°C, highly resistant to oxidation, and can be refluxed with concentrated nitric acid or aqueous base without measurable degradation (9).

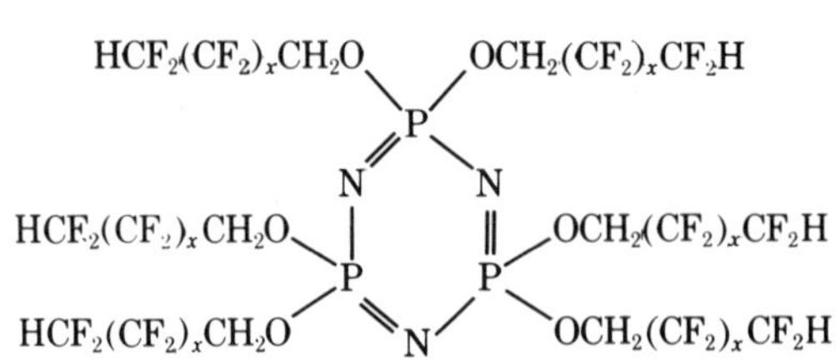

(**3**) (x = 1, 3, 5, 7, 9)

Thus far, no commercial applications have been established for these inert fluids. A study has been made of the utility of these liquids as fire-resistant hydraulic fluids (qv) and lubricants (10). Although many of the requirements for this application were met, further refinements are necessary before these materials can be used as hydraulic fluids. Other potential uses for these liquids are in the textiles area as protective and flame-retardant additives. However, the lower cost, nonfluorinated alkoxycyclophosphazenes appear to have a greater potential in these applications (see Flame-retardant textiles; Flame retardants, phosphorus compounds).

Homopolymers

Fluoroalkoxyphosphazene homopolymers are obtained by substitution of poly(dichlorophosphazene) with a single sodium fluoroalkoxide. They are generally thermoplastics and form flexible films which are resistant to solvents, chemicals, and water.

Poly[bis(trifluoroethoxy)phosphazene] is the most attractive of the fluorinated homopolymers because of its low cost and availability. It is soluble in acetone, tetrahydrofuran, ethyl acetate, ethylene glycol, dimethyl ether, or methyl ethyl ketone, and forms highly flexible films by slow evaporation of solvents from viscous solutions. These films are unaffected by moisture, glacial acetic acid, alcohols, pyridine, concentrated caustic solution, and high-intensity uv light (11). The thermal stability of this fluorinated polymer is considerably greater than that of a nonfluorinated poly(alkoxyphosphazene) such as poly[bis(ethoxy)phosphazene]. Homopolymers with longer-chain fluoroalkoxy pendant groups have also been investigated and show even greater solvent and chemical resistance (12). The combination of properties exhibited by these poly(fluoroalkoxyphosphazenes) has initiated interest in coatings applications. At present, no commercial applications have been established.

Copolymers

Reaction of two fluoroalkoxides of different chain lengths with poly(dichlorophosphazene) inhibits crystallization and, therefore, produces an elastomer. Several elastomers of this type have been prepared and show very good resistance to hydrocarbon solvents, good thermal stability, and excellent low temperature flexibility.

Poly(trifluoroethoxyheptafluorobutoxyphosphazene) has good high and low temperature properties. In addition, prolonged contact with boiling water, common organic solvents, concentrated sulfuric acid, or concentrated potassium hydroxide has no apparent effect (8). These properties were the stimulus for industrial (13–14) and government-sponsored (15–22) research on polymers of this type.

Manufacture

The basic preparative scheme follows the synthetic pathways outlined above. The polymeric $-\overset{|}{P}{=}N-$ backbone is attained by thermal polymerization of hexachlorocyclotriphosphazene. Subsequently, the chlorines are replaced nearly quantitatively by refluxing poly(dichlorophosphazene) in benzene with the sodium fluoroalkoxides in THF (see eq. 4). The longer-chain fluoroalkoxides are a mixture of telomeric alcohols derived from telomerization of tetrafluoroethylene with methanol.

Compounding. The unusual feature of poly(fluoroalkoxyphosphazene) elastomer is its ability to function well over a wide temperature range in harsh environments. A basic formulation consists of 100 parts by weight gum rubber, 30 parts reinforcing filler, 6 parts magnesium oxide, 2 parts stabilizer, and a curing agent.

The choice of reinforcing filler is dictated by the requirements of a particular application (see Fillers). For best thermal stability, neutral or basic fillers are better than untreated or acidic fillers. Inorganic fillers such as Quso WR-82, Tullanox 500, or Nulok 321-L yield compounds having the best resistance to aging at high temperature. Carbon black-filled compounds do not age as well but offer improved mechanical properties, flex life, and abrasion resistance. Carbon black, FEF (N-550) and Shawinigan acetylene blacks, gave the best results.

Selection of filler is also important for optimum low temperature flexibility and compression-set resistance. In general, the more highly reinforcing fillers have an adverse effect on low temperature flexibility and compression set. Carbon black FEF probably affords the best balance between low temperature and mechanical properties. For good low and high temperature performance, a combination of Quso WR-82 and FEF black appears suitable.

Thermal stability is enhanced by magnesium oxide (MgO) or a stabilizer such as zinc-8-quinolinolate (1–2 pt). The MgO serves as an acid acceptor and promotes a higher state of cure which in turn improves thermal stability (see Heat stabilizers).

For most applications a balance of several important qualities must be achieved, and judicious compounding is of utmost importance. A PNF formulation that affords a good balance of normal physical properties, toughness, processibility, compression set, fuel resistance, low temperature flexibility, and high temperature stability is shown in Table 1.

Processibility. Poly(fluoroalkoxyphosphazene), if compounded properly, can be handled on standard types of rubber processing equipment including calenders, extruders, mills, and internal mixers. Processing is most conveniently achieved by mixing the polymer and reinforcing agents in an internal mixer and adding the curatives on a rubber mill. To facilitate mill mixing and mill release, small amounts of FEF black, silicone gum rubber and Teflon-8A are often added to a Quso WR82-filled compound.

Vulcanization and Stress–Strain Properties. Postpolymerization substitution of the poly(phosphazene) system offers a variety of cure sites on the polymer chain. Sulfur cures yield excellent normal stress–strain properties but poor thermal stability. Peroxide or radiation cures provide the best thermal stability (see Radiation curing). A variety of commercial peroxides, such as Varox, Vulcup, or Dicup, have been employed.

Table 2 shows properties attainable with the different cure systems (23).

The normal stress–strain properties of a vulcanizate are primarily controlled by peroxide concentration or cure state and the reinforcing agent. These effects are

Table 1. Formulation of PNF Compound 231[a]

Composition	Parts by wt
PNF	100
Quso WR82	23
FEF black	12
Silane A-174	1
MgO	6
zinc 8-quinolinolate	2
Teflon 8A	3
Vulcup 40KE[b]	0.9

[a] Cure, 5 min at 188°C; postcure, 4 h at 188°C.
[b] Peroxide curing agent.

Table 2. Properties of Poly(fluoroalkoxyphosphazenes) with Different Cures

	A[a]	B[a]	C[b]	D	E
formulation, parts by wt					
polymer (PNF)	100	100	100	100	100
Tullanox 500	30	30	30	30	30
MgO	6	6		6	6
ZnO			5		
Dicup 40C	1	3			
sulfur			0.3		
DTDM[c]			1		
methyl zimate[d]			2		
butyl zimate[d]			2		
radiation dose, kGy[e]				10	30
stress–strain properties					
50% modulus of elasticity, MPa[f]	0.86	2.00	3.88	1.21	2.00
100% modulus of elasticity, MPa[f]	3.10	7.59	9.14	3.45	6.76
tensile strength, MPa[f]	13.88	11.10	16.21	11.72	10.55
elongation at break, %	170	117	310	175	125
Shore A hardness	60	70	76	65	70

[a] Cured 30 min at 160°C, postcured 24 h at 100°C.
[b] Cured 10 min at 160°C.
[c] 4,4′-Dithiodimorpholine.
[d] Zinc salts of respective dialkyldithiocarbamates.
[e] To convert Gy to rad, multiply by 10^2.
[f] To convert MPa to psi, multiply by 145.

demonstrated by the data in Table 3 (25). The changes in properties shown are consistent with a change in cross-link density.

The effect on physical properties of varied filler concentration is illustrated in Table 4 for the organosilylated silica Tullanox 500 (23). The profound influence of reinforcing agents is clearly evident by a comparison of properties of the unfilled (A) and filled (B–E) stocks. Optimum tensile strength was achieved at 40 parts of Tullanox 500.

Table 3. Effect of Peroxide Concentration on Poly(fluoroalkoxyphosphazene) Properties

	A	B	C	D
formulation[a], parts by wt				
PNF	100	100	100	100
Tullanox 500	30	30	30	30
MgO	6	6	6	6
Dicup 40C	1	2	3	6
stress–strain properties				
50% modulus of elasticity, MPa[b]	0.86	1.55	2.00	2.76
100% modulus of elasticity, MPa[b]	3.10	5.94	7.59	9.83
tensile strength, MPa[b]	13.88	10.34	11.10	10.69
elongation at break, %	170	190	117	109
Shore A hardness	60	67	70	72
compression set[c], %	56	38	34	33

[a] ASTM D 412: cure, 30 min at 160°C; postcure, 24 h at 100°C.
[b] To convert MPa to psi, multiply by 145.
[c] ASTM D 395, 70 h at 135°C.

Table 4. Effect of Filler Concentration on Poly(fluoroalkoxyphosphazene) Properties

	A	B	C	D	E
formulation[a], parts by wt					
PNF	100	100	100	100	100
Tullanox 500		10	20	40	50
MgO	6	6	6	6	6
Dicup 40C	1	1	1	1	1
stress–strain properties					
50% modulus of elasticity, MPa[b]	0.14	0.34	0.52	1.38	4.13
100% modulus of elasticity, MPa[b]	0.41	1.21	1.72	4.34	10.43
tensile strength, MPa[b]	1.69	4.55	11.03	16.41	13.52
elongation at break, %	190	150	200	175	130
Shore A hardness	21	30	44	76	88
compression set[c], %	20	36	43	80	89

[a] ASTM D 412: cure, 30 min at 160°C; postcure, 24 h at 100°C.
[b] To convert MPa to psi, multiply by 145.
[c] ASTM D 395, 70 h at 135°C.

For most applications, retention of mechanical properties at elevated temperatures is probably the most critical requirement. Although good thermal stability is inherent in poly(fluoroalkoxyphosphazene) elastomers, compounding additives greatly influence vulcanizate aging at high temperatures. Selection of filler is of utmost importance. Acidic reinforcing agents adversely affect thermal stability. The best-aging compounds contain basic or neutral fillers, magnesium oxide, and a small amount of stabilizer, as shown below for a formulation cured 30 min at 171°C and postcured 4 h at 117°C.

Formulation	*Parts by wt*
PFAP	100
Quso WR82	30
zinc 8-quinolinolate	2
MgO	6
Vulcup 40KE	0.8

An example of the aging properties attainable with such a formulation is given in Table 5. The test temperature of 200°C represents a practical limit for utilization of PFAP compounds for short periods. Lower service temperatures (150–175°C) permit serviceability over much longer time periods.

Table 5. Heat Aging of a Poly(fluoroalkoxyphosphazene)[a]

Stress–strain properties	Unaged	120 h at 200°C
50% modulus of elasticity, MPa[b]	3.18	2.19
100% modulus of elasticity, MPa[b]	9.97	6.46
tensile strength, MPa[b]	10.75	7.01
elongation at break, %	108	110

[a] A polymer with trifluoroethoxy and mixed telomeric fluoroalkoxy groups such as [$—OCH_2(CF_2)_xCF_2H$].
[b] To convert MPa to psi, multiply by 145.

The excellent low temperature flexibility of poly(fluoroalkoxyphosphazene) elastomer is best depicted by the Gehman curves in Figure 1 (23). The Gehman torsional-wire apparatus (ASTM D 1053) measures flexibility as a function of temperature by observation of the recovery of twist angle of a sample of rubber during a 10-s period after a 180° twist. It also measures the temperature at which relative modulus values occur, ie, T_{10} is the temperature at which the modulus is ten times the RT modulus. It is clear that the PNF remains flexible down to very low temperatures and can thus be compared to fluorosilicone. This is in contrast to the other thermally stable and solvent-resistant elastomers which become embrittled and useless at much higher temperatures.

Table 6 shows the stress–strain properties of PNF compound 231 (see Table 1) unaged and after 240 h at 149°C.

In addition to the properties reviewed, PNF also shows good resistance to ozone, N_2O_4, radiation, and fire, and is an excellent damping material.

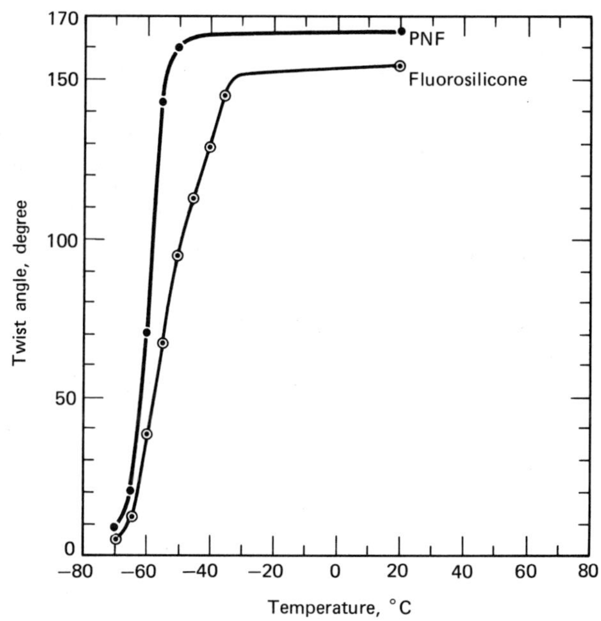

Figure 1. Gehman Flexure of PNF as a function of temperature, ASTM D 1053. To convert MPa to psi, multiply by 145.

	PNF	*Fluorosilicone*
T_2, °C	−54	−39
T_5, °C	−56	−48
T_{10}, °C	−58	−54
T_{100}, °C	−65	−66
freezing point, °C	−66	−67
100% modulus of elasticity, MPa	4.67	4.27
tensile strength, MPa	9.94	8.88
elongation at break, %	185	185
Shore A hardness	50	64

Table 6. Stress–Strain Properties of PNF Compound 231, Unaged and Aged

Stress–strain properties	Unaged	After 240 h at 149°C
50% modulus of elasticity, MPa[a]	3.66	3.43
100% of modulus of elasticity, MPa[a]	9.17	8.79
tensile strength, MPa[a]	11.93	11.93
elongation at break, %	130	137
Shore A hardness	75	
compression set, %	20	
NBS abrasive index		57
tear strength at 23°C, N/cm (ppi)		26.3 (150)
low temperature flexibility, Gehman T_{10}, °C		−56.7
fuel resistance, % vol change		
ASTM fuel A		6
ASTM fuel B		14

[a] To convert MPa to psi, multiply by 145.

Characterization

The solution characterizations of poly(fluoroalkoxyphosphazenes) (24–25) have been studied extensively. Extremely high molecular weights (ca 10^6) and very broad molecular weight distribution are distinctive features of these elastomers. A typical gel permeation chromatography (GPC) curve using DMF as a solvent and Styragel columns, is shown in Figure 2. The GPC trace clearly shows the broad molecular weight distribution (high $\overline{M}_w/\overline{M}_n$) and indicates that this distribution is due to a small amount of a low molecular weight tail. Actually, 86% of the polymer is a high molecular weight fraction with a polydispersity of only 3.0. This work (25) was performed on a poly(trifluoroethoxyoctafluoropentoxyphosphazene) and is in agreement with work performed on the other poly(fluoroalkoxyphosphazenes) (24).

Although fractionations of the phosphonitrilic fluoroelastomers by molecular weight have been achieved, these fractions were still broad in molecular weight distribution. Table 7 shows molecular weights as obtained by light scattering ($\overline{M}_w$) on various fractions of a poly(trifluoroethoxyoctafluoropentoxyphosphazene) (25). These

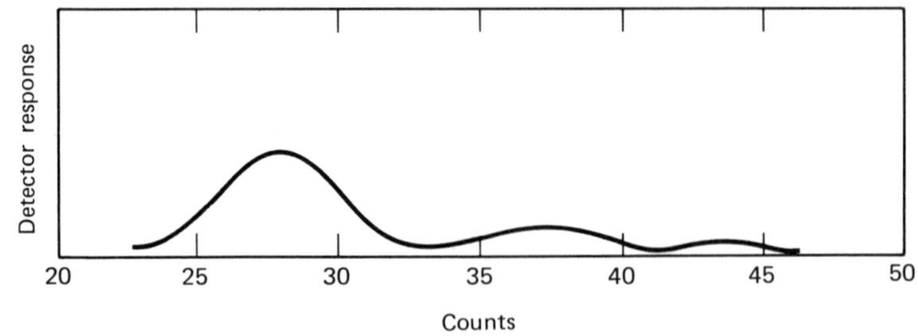

Figure 2. Typical GPC curve using DMF solvent and Styragel columns. $\overline{M}_n$, number-average molecular weight; $\overline{M}_w$, weight-average molecular weight.

	$\overline{M}_n$	$\overline{M}_w$	$\overline{M}_w/\overline{M}_n$
total polymer	123,700	3,514,500	28.4
high mol wt fraction	1,370,600	4,085,800	3.0

Table 7. Molecular Weights of Fractionated Poly(fluoroalkoxyphosphazene)

Fraction	$\overline{M}_w$	$\overline{M}_n$	$\overline{M}_w/\overline{M}_n$
total polymer	8,160,000	267,000	30.6
A	12,610,000	552,000	22.8
B	7,710,000	671,000	11.5
C	7,380,000	646,000	11.4
D	6,610,000	594,000	11.1
E	7,300,000	495,000	14.8
F	5,550,000	390,000	14.2
G	2,270,000	315,000	7.2

data are in general agreement with the GPC results, ie, the polymer shows high molecular weight and broad molecular weight distribution. However, the high polydispersity of the individual fractions make it difficult to formulate further macrostructural conclusions based on solution property measurements. A tentative conclusion is that the poly(fluoroalkoxyphosphazenes) are relatively linear with only low degrees of chain branching.

Elucidation of the microstructure showed that a poly(trifluoroethoxyoctafluoropentoxyphosphazene) could consist of one or more of the following units:

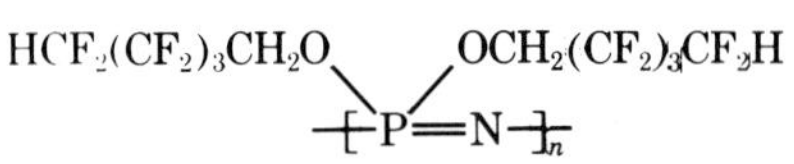

(4) [52956-85-1]

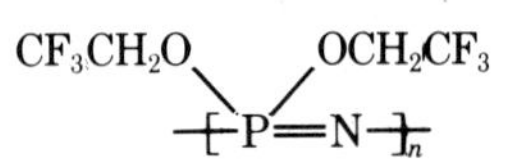

(5)

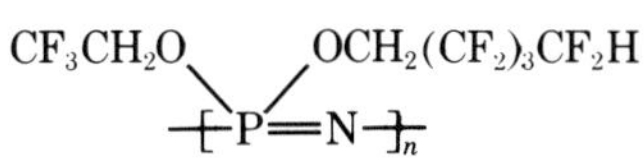

(6) [57579-81-4]

Proton nmr is used routinely to determine the overall composition of the pendant groups but does not reveal the relative amounts of the structures (**4**), (**5**), and (**6**). Phosphorus nmr, however, could possibly distinguish these units. A ^{31}P nmr on a 50:50 trifluoroethoxyoctafluoropentoxy polymer showed a single peak which would indicate the presence of very high concentration of structure (**6**). However, a poly[bis(trifluoroethoxy)phosphazene] also gave a single peak at practically the identical position of the copolymer. It must be concluded that the ^{31}P nmr chemical shifts for structures (**4–6**) are not sufficiently different to permit any conclusions on microstructure of the poly(trifluoroethoxyoctafluoropentoxyphosphazene) (see Analytical methods).

The poly(fluoroalkoxyphosphazene) raw gum polymer has a relatively high specific gravity (1.75) and a low Mooney viscosity (ML-4 at 100°C = 15). Although the bulk viscosity is low, the polymer is very tough and exhibits high elastic recovery and little tendency to cold flow. The polymer is soluble in ketones, tetrahydrofuran, methanol, and acetonitrile. A glass transition of −68°C for this polymer indicates good low temperature flexibility.

Table 8. Fluid Resistance of PNF

Fluid	Resistance[a]
acids (aq)	P/G
alcohols	P/G
aliphatic hydrocarbons	E
alkaline solutions	F/E
ammonia (aq)	E
animal and vegetable oils	F
aromatic hydrocarbons	E
brake fluids (ester base)	U
brake fluids (petroleum and silicone base)	E
chlorinated hydrocarbons	F
diester synthetic lubricants (Anderol 774, ASTM 101, etc)	G/E
ethylene–propylene gear lubricants	G/E
ethylene–propylene (liquefied)	E
gasoline	
leaded	E
nonleaded	E
hydraulic fluids	
petroleum	E
aryl phosphate esters (Fyrquel 220)	E
alkyl phosphate esters (Skydrol 500)	U
silicate ester	E
water–glycol	E
hydrazine	G/E
liquid oxygen (impact test)	G/E
oxygenated solvents	U
water	F/G

[a] E, excellent; G, good; F, fair; P, poor; U, unsatisfactory.

Applications

PNF elastomers are serviceable at a broad range of temperatures and have excellent solvent resistance and good mechanical properties. At present, fluorosilicone is the only other oil-resistant elastomer that functions below −40°C. However, its flex and tear resistance are inferior. PNF is tougher than fluorosilicone and complements the currently available solvent-resistant elastomers (see Table 8). Applications are in the aerospace, aircraft, industrial, automotive, and oil exploration industries. Products utilizing PNF include oil seals, gaskets, diaphragms, hoses, shock mounts, foams, and coatings.

Elastomeric fluoroalkoxyphosphazene copolymers (PNF) are sold by The Firestone Tire and Rubber Company at $88/kg (1979) as compounds or masterbatches. The polymers are utilized in specialty applications where premium performance is critical and the unique combination of properties of the elastomer is required.

BIBLIOGRAPHY

1. J. Liebig, *Ann. Chem.* **11,** 139 (1834).
2. H. Rose, *Ann. Chem.* **11,** 131 (1834).
3. M. Becke-Goehring and E. Fluck, *Angew. Chem. Int. Ed.* **1,** 281 (1962).
4. J. Emsley and P. B. Udy, *J. Chem. Soc. A,* 768 (1971).

5. L. G. Lund and co-workers, *J. Chem. Soc.*, 2542 (1960).
6. H. N. Stokes, *Am. Chem. J.* **19,** 782 (1897).
7. H. R. Allcock, R. L. Kugel, and K. J. Valan, *Inorg. Chem.* **5,** 1709 (1966).
8. S. H. Rose, *J. Polym. Sci. Part B* **6,** 837 (1968).
9. R. Rätz and co-workers, *J. Am. Chem. Soc.* **84,** 551 (1962).
10. H. Lederle, E. Kober, and G. Ottmann, *J. Chem. Eng. Data* **11,** 221 (1966).
11. H. R. Allcock, *Phosphorus–Nitrogen Compounds,* Academic Press, Inc., New York, 1972, p. 356.
12. U.S. Pat. 3,838,073 (Sept. 24, 1974), S. H. Rose and K. A. Reynard (to Horizons, Inc.).
13. D. P. Tate, *J. Polym. Sci. Symp.* (48), 33 (1974).
14. J. C. Vicic and K. A. Reynard, *J. Appl. Polym. Sci.* **21,** 3185 (1977).
15. S. H. Rose, K. A. Reynard, and J. R. Cable, *Synthesis of New Low Temperature Petroleum Resistant Elastomers, AMMRC CR70-1,* Horizons, Inc., Cleveland, Ohio, Jan. 1970, AD704332.
16. K. A. Reynard, R. W. Sicka, and S. H. Rose, *Synthesis of New Low Temperature Petroleum Resistant Elastomers, AMMRC 73-18,* Horizons, Inc., Cleveland, Ohio, June 1973, AD766745.
17. K. A. Reynard and co-workers, *Elastomers for Service with Engine Lubricants and Hydraulic Fluids, Naval Air Systems Command Contract N00019-73-C-0406,* Horizons, Inc., Cleveland, Ohio, Mar. 1974, AD781785.
18. G. S. Kyker and J. A. Beckman, *Phosphazene Fluoroelastomers Applications, Army Materials and Mechanics Research Center Contract DAAG-46-74-C-0066,* The Firestone Tire & Rubber Company, Akron, Ohio, Mar. 1975.
19. T. A. Antkowiak, *Phosphonitrilic Fluoroelastomer Fuel Hose-Utilization of Extruded Tubes, U.S. Army Mobility Equipment Research and Development Center Contract DAAK-70-76-C-0239,* The Firestone Tire & Rubber Company, Akron, Ohio, Oct. 1977.
20. K. A. Reynard, R. W. Sicka, and J. C. Vicic, *Phosphonitrilic Fluoroelastomers for Lip Seal Applications, Army Materials and Mechanics Research Center Contract DAAG-46-74-C-0056,* Horizons, Inc., Cleveland, Ohio, Mar. 1975.
21. P. Touchet and P. E. Gatza, *J. Elast. Plast.* **9,** 8 (1977).
22. A. Wilson, *J. Coated Fabrics* **1,** 233 (1978).
23. G. S. Kyker and T. A. Antkowiak, *Rubber Chem. Technol.* **47,** 32 (1974).
24. G. L. Hagnauer and N. S. Schneider, *J. Polym. Sci. A-2,* **10,** 699 (1972).
25. D. W. Carlson and co-workers, *J. Polym. Sci. Polym. Chem. Ed.,* 1379 (1976).

David P. Tate
T. A. Antkowiak
Firestone Tire and Rubber Co.

PERFLUOROALKYLENE TRIAZINES

Perfluoroalkylene triazines consist of *sym*-triazine rings with perfluorinated substituents in the 2, 4, and 6 positions. The first member of the homologous series is *tris*-trifluoromethyl-*sym*-triazine [*368-66-1*] (**1**).

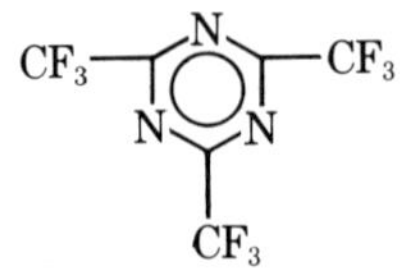

(**1**)

Interest in these triazine compounds was generated during research sponsored by the U.S. Air Force in the mid 1950s (1–3). A rapid, quantitative synthesis of *sym*-triazines resulted from these investigations of new fluorine-containing chemicals which were made by the electrochemical fluorination of hydrocarbons. Fluorocarbon amidines were included (4). These were prepared and observed to form triazines when heated under anhydrous conditions:

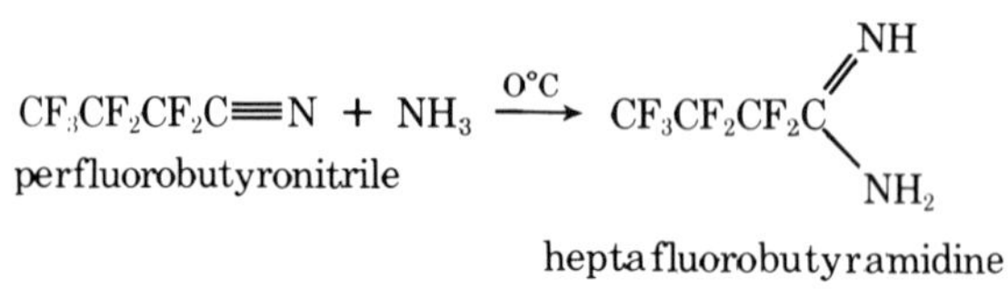

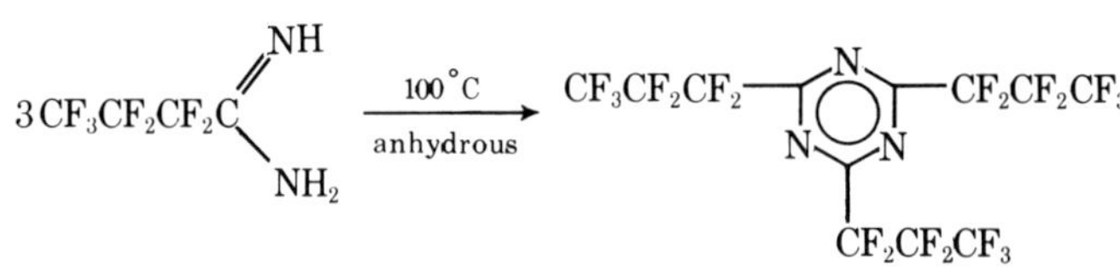

[*21674-38-4*]

Perfluoroalkyl *sym*-triazines are very resistant to high temperature oxidation and have become the basis of considerable research on fluids and elastomeric polymers (5–6).

Fluids

Physical Properties. Some physical properties important for aircraft and missile systems are listed in Table 1.

Structure (**2**) [*52844-28-7*] represents a typical nonflammable, high temperature hydraulic fluid (7–9). Boiling points of these compounds range 387–407°C, and pour points from −46 to −19°C. Oxygen ether linkages in the perfluoro radicals located between the rings or in the pendent positions broaden the fluid range by lowering the melting point without impairing high temperature stability.

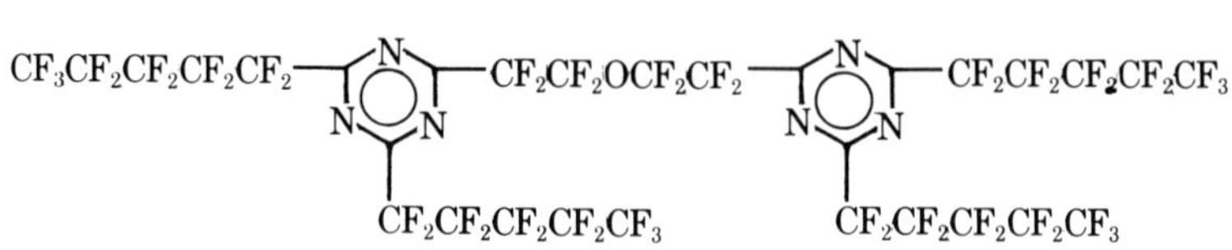

(**2**)

Table 1. Physical Properties of Fluorinated *sym*-Triazines

Property	Value
density, g/mL	1.5–1.8
melting point, °C	−100 to 30
boiling point, °C	98–407 and above
viscosity, $m^2/s \times 10^6$	0.3–1000 and above
dielectric constant, 60–10^6 Hz	1.98–2.03
volume resistivity, Ω·cm	2.7×10^{13}
flash point	nonflammable
spontaneous ignition temperature	nonflammable

Chemical Properties. The perfluoroalkylene triazine fluids are inert toward oxidizing agents and strong acids such as H_2SO_4, P_2O_5, and HCl, but are attacked by HNO_3, and amines and other strong alkalis, especially in the presence of water. Compounds of silver, tin, antimony, arsenic, and copper cause rearrangements of the substituents on the triazine ring at high temperatures. The triazine fluids are not easily degraded at use temperatures, but can cause slight corrosion of metals and glass. The fluoroalkyl phosphines developed as oxidation–corrosion inhibitors for poly(hexafluoropropylene epoxide) fluids (9–12) also are of benefit for perfluorocarbon ether triazine fluids.

Elastomers

Elastomeric polymers with excellent rubberlike properties have been prepared and formulated for potential use as electrical wire insulation and seals for hydraulic, lubricating, and fuel systems of aircraft (13–16) (see Elastomers, synthetic). The large number of the perfluorocarbon and perfluorocarbonether nitriles allows a wide variety of chemical composition in the elastomer series (see structure (**3**)).

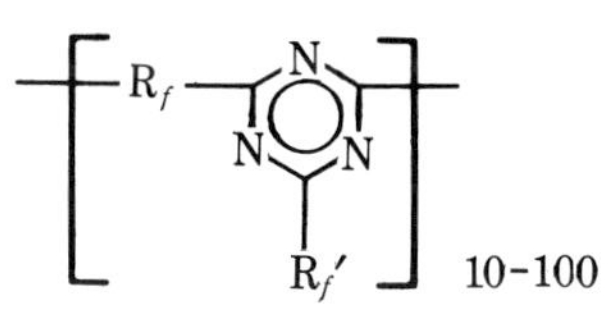

(**3**)

where: R_f = perfluoroalkylene or perfluoroalkyleneoxy and R_f' = perfluoroalkyl or perfluoroalkyloxy.

Physical Properties. High polymers in the range of 150,000–200,000 $\overline{M}_n$ (number-average molecular weight) are water-clear tough gums that can be processed on conventional equipment associated with the rubber industry. The low temperature flexibility varies with the polymer chemistry. For example, the —CF_2CF_2— backbone with the —CF_3 pendent group has a glass transition temperature (T_g) of +40°C, whereas the —$(CF_2)_8$— backbone with —C_3F_7 pendent group ranges from −15 to −22°C. Linear perfluorocarbon ether segments in the backbone lower the T_g to the −50 to −100°C range. Elastomeric gums with pendent perfluoroalkylene nitrile cross-linking sites have also been prepared. The nitrile can be trimerized with tetraphenyl tin (17) or arsenic or antimony catalysts (18). Elastomers reinforced with fumed silica, carbon black, and asbestos powder display properties that are typical of other fluorinated elastomers (10–14 MPa or 1450–2030 psi tensile strength, 200–500% elongation, and 30–70 Shore A Durometer hardness points) (15). Useful elastomeric properties remain after 1000 h at 260°C and shorter times in the 350–400°C range.

Chemical Properties. Inert behavior in oxidizing environments is evident from the nonflammable properties of the fluids and elastomers. The elastomers have survived 420°C in air for short periods of time and are resistant to the swelling action of many lubricating oils, fuels, and hydraulic fluids. The polymers and vulcanizates are chemically attacked by alkalis such as ammonia, amines, pyridines, and caustic solutions. Compounds of tin, silver, cadmium, barium, and antimony cause ring opening and rearrangements of these systems but can also be used in catalytic quantities for trimerization of pendent and terminal nitriles. Perfluoroalkyloxy nitriles with alpha trifluoromethyl substitutions form triazines with greatly enhanced resistance to hydrolysis in high humidity and strongly alkaline conditions.

Manufacturing Prospects

The chemistry required to prepare perfluoroalkylene triazines is well established (7–8,13). Starting materials are the perfluorinated carboxylic acids or acid fluorides available from electrochemical fluorinations or oligomerizations of perfluorinated monomers (19–20). The corresponding nitriles are obtained in high yields by dehydration of the amides with P_2O_5 or α,α,α-trichlorotoluene. Heating the nitriles with ammonia or other catalysts give the *sym*-triazines.

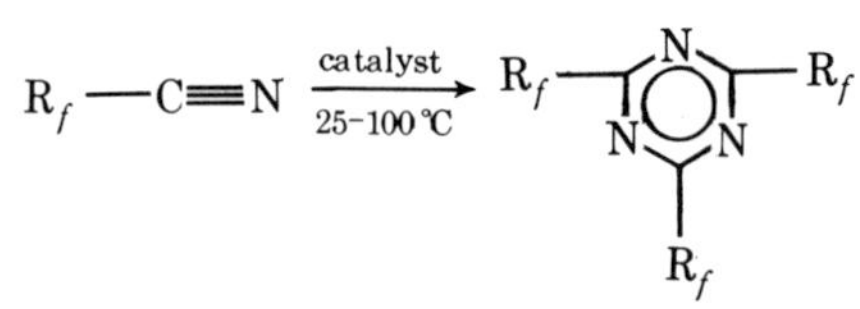

For specific 2,4,6 substitutions on the triazine ring see Figure 1, which illustrates a three-step process for trimerization. First, the R_f-nitrile adds ammonia to become the R_f-amidine. Then the R_f-amidine and R_f'-nitrile combine to form the R_f, R_f'-imidoylamidine. Finally, R_f''-carboxylic acid anhydride acylates the imidoylamidine and removes the elements of water to close the ring to a specific R_f, R_f', R_f''-*sym*-triazine. Bisimidoyl amidines and dinitriles form high polymers by the same sequence of reactions. Polymerization requires days at 25°C owing to the slow rate of nitrile addition to amidine. Accelerating the reaction rate by raising the temperature causes side reactions that lead to gel and linear —C=N— structures. Coordination of the

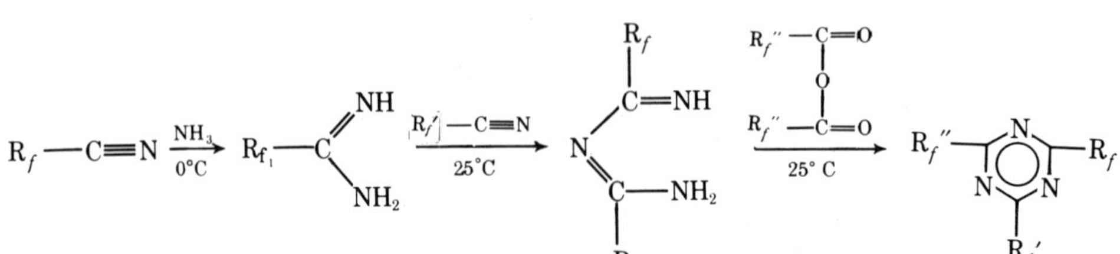

(4)

Figure 1. Synthesis of symmetric triazines (8).

imidoyl amidine structure (4) by metal cations such as Ag^+ has allowed the preparation of a soluble high polymer at 40–80°C in a few hours (21).

Health and Safety Factors

Perfluoroalkylene triazine fluids and elastomers are inert and no health or safety problems have been reported. However, the high vapor pressure is reason for cautious handling and the thermal degradation fragments formed above 400°C are significant hazards associated with the finished fluids and elastomers. Exposure of white rats to the fumes given off from the $—(CF_2)_3—$, $—C_3F_7$ elastomers at 420°C caused death within 24 h from lung swelling and bleeding.

Uses

The *tris*-trifluoromethyl-*sym*-triazine as well as the *tris*-perfluoroethyl [*858-46-8*]-, propyl-, heptyl [*21674-38-4*]-, and nonyl [*57101-59-4*]-*sym*-triazines are utilized as mass spectroscopic standards. The perfluoroalkylene(perfluoroalkyl)-*sym*-triazines and perfluoroalkyleneoxy(perfluoroalkyloxy)-*sym*-triazines are suitable for nonflammable hydraulic fluids in the −25 to +300°C temperature range. They may also find uses as specialty heat-transfer fluids and lubricating oils.

The polymeric forms of the fluids have excellent properties for use as O-ring seals, gaskets, and hose for aircraft hydraulic and lubricating systems. Initial applications will be in components where specific properties are required for high temperature performance such as with seals for hydraulic systems of aircraft wheel brakes. These triazine elastomers also have potential use as fuel-tank sealants with long life at 300°C.

BIBLIOGRAPHY

1. W. L. Reilly and H. C. Brown, *J. Am. Chem. Soc.* **78,** 6032 (1956).
2. W. L. Reilly and H. C. Brown, *J. Org. Chem.* **22,** 698 (1956).
3. U.S. Pat. 3,086,946 (Apr. 23, 1963), H. C. Brown (to Research Corporation of New York).
4. U.S. Pat. 2,676,985 (Apr. 27, 1954), D. R. Husted (to 3M Company).
5. C. E. Snyder, *ASLE Trans.* **14**(3), 237 (1971).
6. U.S. Pat. 3,369,002 (Feb. 13, 1968), W. R. Griffin (to U.S. Air Force).
7. U.S. Pat. 3,888,854 (June 10, 1975), P. D. Schuman and E. C. Stump, Jr. (to PCR, Inc.).
8. Ger. Offen. 2,145,174 (Mar. 23, 1972), J. L. Zollinger (to 3M Company).
9. U.S. Pat. 4,011,267 (Mar. 8, 1977), C. Tamborski and C. E. Snyder, Jr. (to U.S. Air Force).
10. U.S. Pat. 4,043,926 (Aug. 23, 1977), C. Tamborski and C. E. Snyder, Jr. (to U.S. Air Force).
11. U.S. Pat. 4,097,388 (June 27, 1978), C. E. Snyder, Jr. and C. Tamborski (to U.S. Air Force).
12. C. E. Snyder, Jr. and C. A. Svisco, *Lub Engr.* **35,** 451 (1979).
13. E. Dofman and co-workers, *Rubber Chem. Technol.* **39** (Pt. 2), 1175 (1966).
14. W. R. Griffin, *Rubber Chem. Technol.* **39**(Pt. 2), 1178 (1966).
15. T. L. Graham, *Rubber Age* **108,** 43 (Aug. 1969).
16. USSR Pat. 145,748 (May 25, 1961), M. P. Krasuskaya and I. L. Knunyants.
17. U.S. Pat. 3,728,344 (Apr. 17, 1973), W. E. Emerson and E. Dorfman (to Hooker Chemical Corp.).
18. U.S. Pat. 3,470,176 (Sept. 30, 1969), J. L. Zollinger (to 3M Company).
19. U.S. Pat. 3,317,484 (May 2, 1967), C. G. Fritz and J. L. Warnell (to E. I. du Pont de Nemours & Co., Inc.).

20. U.S. Pat. 3,250,807 (May 10, 1966), C. G. Fritz and E. P. Moore (to E. I. du Pont de Nemours & Co., Inc.).
21. U.S. Pat. 4,102,872 (July 25, 1978), W. R. Griffin (to U.S. Air Force).

WARREN R. GRIFFIN
Air Force Materials Laboratory

PERFLUOROALKANE SULFONIC ACIDS

The perfluoroalkane sulfonic acids were first reported in 1954. Trifluoromethanesulfonic acid was obtained by oxidation of bis(trifluoromethylthio)mercury with aqueous hydrogen peroxide (1) (see also Sulfonic acids). The preparation of a series of perfluoroalkane sulfonic acids by electrochemical fluorination (ECF) of alkane sulfonyl halides was also disclosed in the same year (2).

$$RSO_2F \xrightarrow[\text{ECF}]{\text{HF}} R_fSO_2F \xrightarrow[\text{aq KOH}]{} R_fSO_3K \xrightarrow{H_2SO_4} R_fSO_3H$$

CF_3SO_3H may also be prepared from trifluoromethanesulfenyl chloride (3).

$$CF_3SSCF_3 \xrightarrow{Cl_2} CF_3SCl \xrightarrow[H_2O]{Cl_2} \underset{[421\text{-}83\text{-}0]}{CF_3SO_2Cl} \xrightarrow[\text{aq NaOH}]{} \underset{[2926\text{-}30\text{-}9]}{CF_3SO_3Na}$$

or by the oxidation of methyl trifluoromethyl sulfide under a variety of conditions (4–5).

The boiling points of a series of perfluoroalkane sulfonic acids are listed in Table 1 (2).

Trifluoromethanesulfonic Acid

The first member of the series, CF_3SO_3H, has been studied extensively. Two excellent review articles (6–7) cover the chemistry of the acid and its derivatives in great detail.

Trifluoromethanesulfonic acid is available as Fluorochemical Acid FC-24 (3M Co.). The technical brochure (8) is an excellent source for applications, safety, and handling data of the acid.

Properties. Trifluoromethanesulfonic acid is a stable, hygroscopic liquid which fumes in air. An equimolar amount of water with the acid results in a stable, distillable monohydrate, mp 34°C, bp 96°C at 0.13 kPa (1 mm Hg) (9). Measurement of conductivity of strong acids in acetic acid has shown the acid to be one of the strongest simple protic acids known, similar to fluorosulfonic and perchloric acid (10). Trifluoromethanesulfonic acid is miscible in all proportions in water, and is soluble in many polar organic solvents such as dimethylformamide, dimethyl sulfoxide, and acetonitrile. In addition, it is soluble in alcohols, ketones, ether, and esters, but these generally are not suitably inert solvents. The acid with diethyl ether gives a colorless, liquid oxonium complex, which with further heating, gives the ethyl ester and ethylene. Reaction with ethanol gives the ester, but in addition, dehydration and ether formation occurs.

Table 1. Boiling Points of a Series of Perfluoro Alkane Sulfonic Acids.

Compound	CAS Reg. No.	Boiling point[a] °C/kPa	°C/101 kPa
CF_3SO_3H	*[1493-13-6]*	60/0.4	166
$C_2F_5SO_3H$	*[354-88-1]*	81/2.9	175[b]
n-$C_4F_9SO_3H$	*[59933-66-3]*	76-84/0.13	200[b]
n-$C_5F_{11}SO_3H$	*[3872-25-1]*	110/0.67[c]	212[b,d]
n-$C_6F_{13}SO_3H$	*[355-46-4]*	95/0.47	225[b]
n-$C_8F_{17}SO_3H$	*[1763-23-1]*	133/0.8	249
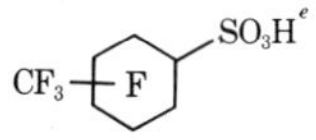	*[72441-89-5]*	120/0.4	241
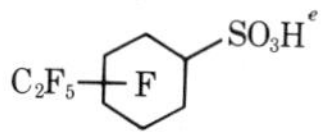	*[41242-13-1]*		254

[a] To convert kPa to mm Hg, multiply by 7.5.
[b] Estimated boiling point.
[c] n-$C_5F_{11}SO_3H.H_2O$.
[d] n-$C_5F_{11}SO_3H$.
[e]  = *cyclo*-C_6F_{12}.

The strongly electron-attracting $CF_3SO_3^-$ group makes the esters excellent alkylating agents. Methyl trifluoromethanesulfonate *[333-27-7]* is more than 10^4 times as reactive as methyl toluenesulfonate in acetolysis (11). Trifluoromethyl trifluoromethane sulfonate *[3582-05-6]* was obtained by the reaction of trifluoromethanesulfonic acid with fluorosulfuric acid (12). The trifluoromethylating ability of the triflate (see below) is limited, benzene at 100°C giving only small amounts of benzotrifluoride.

Derivatives. Alkyl esters of trifluoromethanesulfonic acid, commonly called triflates have been prepared primarily from the silver salt *[51098-28-3]* and an alkyl iodide, or by reaction of the anhydride with alcohol (9,13).

The metallic salts of trifluoromethanesulfonic acid, also called triflates, are prepared by reaction of the acid with the corresponding hydroxide or carbonate. The salts are hygroscopic but can be dehydrated at 100°C under vacuum. The sodium salt has a melting point of 248°C and decomposes at 425°C. The rare earth salts are also obtained from the acid and rare earth metal oxides (14). Organic trifluoromethanesulfonate salts are quantitatively prepared from the acid and amines (9). Triflate salts show promise as electrolytes for nonaqueous electrochemical cells (15), as catalysts for epoxy resins (16), and latent catalysts for cationically polymerizable monomers (17–18).

Trifluoromethanesulfonic acid anhydride *[358-23-6]*, bp 84°C, is prepared by refluxing the acid over an excess of phosphorus pentoxide (19). It reacts instantaneously with ammonia, aqueous ammonium hydroxide, or amines, to form trifluoromethanesulfonamides. The anhydride is generally a more effective esterification promotor than trifluoroacetic acid anhydride. A number of methods have appeared that allow the introduction of the $CF_3SO_2^-$ group on carbon; the synthetic utility of these derivatives in organic chemistry has been reviewed (20).

Higher Perfluoroalkane Sulfonic Acids

The longer chain perfluoroalkane sulfonic acids are very hygroscopic oily liquids. Distillation of the acid from a mixture of its salt and H_2SO_4 gives hydrated mixtures with melting points above 100°C. The acids show the same polar solvent solubilities as trifluoromethanesulfonic acid but are quite insoluble in benzene, heptane, carbon tetrachloride, and the perfluorinated liquids. Many of the higher perfluoroalkane sulfonic acids have been prepared by electrochemical fluorination as previously described (13). α,ω-Perfluoroalkane disulfonic acids can be prepared by aqueous alkali permanganate oxidation of the bis-sulfone, $R_fSO_2(CF_2CF_2)_nSO_2R$ (21).

The longer chain acids, particularly $C_8F_{17}SO_3H$ and higher, are surface-active agents in aqueous media. Derivatives of the acids have unique surface-active properties and have formed the basis for a number of commercial fluorochemical surfactants (22). The chemical and thermal stability of the perfluorosulfonate salts as well as their surface activity has led to their use in strong acid and base environments. Very low surface tensions in aqueous media, less than 20 mN/m (= dyn/cm), are obtained with most fluorochemical surfactants derived from the perfluoro sulfonic acids. This has led to their use as wetting and leveling agents (see also Surfactants).

Safety and Storage

Extreme caution should be used when handling the perfluoroalkane sulfonic acids. Eye irritation studies with albino rabbits suggest extreme irritation and permanent eye damage could occur following eye contact, even if the eyes are immediately flushed with water. Acute inhalation toxicity studies (albino rats) indicate that high vapor of mist concentration can cause significant respiratory irritation. All contact of the acids and alkyl esters with the skin should be avoided. Normal procedures for treatment of strong acid burns should be used.

Contact of the perfluoroalkane sulfonic acids with cork, rubber, cellulosics, and plasticized materials will cause discoloration and deterioration of those materials. Samples are best stored in glass ampuls or glass bottles with Kel-F or Teflon plastic screw-cap linings.

Trifluoromethanesulfonic acid in contact with stainless steel, iron or aluminum typically causes a weight loss of about 20 mg/($m^2 \cdot d$). Use of glass equipment where possible is recommended.

BIBLIOGRAPHY

1. R. N. Hazeldine and J. M. Kidd, *J. Chem. Soc.*, 4228 (1954).
2. P. W. Trott and co-workers, *126th National Meeting of the American Chemical Society, New York, 1954,* Abstracts p. 42-M.
3. R. N. Hazeldine and J. M. Kidd, *J. Chem. Soc.*, 2901 (1955).
4. R. N. Hazeldine and co-workers, *Chem. Commun.*, 249 (1972).
5. R. B. Ward, *J. Org. Chem.* **30,** 3009 (1965).
6. A. Senning, *Chem. Rev.* **65,** 385 (1965).
7. R. D. Howells and J. D. McCown, *Chem. Rev.* **77,** 69 (1977).
8. *FLUORAD® Brand Fluorochemical Acid FC-24,* 3M Company, St. Paul, Minn.
9. T. Gramstad and R. N. Hazeldine, *J. Chem. Soc.*, 469 (1957).
10. T. Gramstad, *Tidsskr. Kremi Bergves. Metall.* **19,** 62 (1959); *Chem. Abstr.* **54,** 14102 (1960).
11. R. L. Hansen, *J. Org. Chem.* **30,** 4322 (1965).

12. G. A. Olah and T. Ohyama, *Synthesis* (5), 319 (1976).
13. U.S. Pat. 2,732,398 (Jan. 24, 1956), T. J. Brice and P. W. Trott (to 3M Co.).
14. U.S. Pat. 3,796,738 (Mar. 12, 1974), K. F. Thom (to 3M Co.).
15. W. H. Tredman and D. N. Bennion, *J. Chem. Engr. Data* **16,** 368 (1971).
16. U.S. Pat. 3,842,019 (Oct. 15, 1974), J. E. Kropp (to 3M Co.).
17. U.S. Pat. 3,998,763 (Dec. 21, 1976), B. Bohnel (to 3M Co.).
18. U.S. Pat. 3,907,706 (Sept. 3, 1975), J. Robbins (to 3M Co.).
19. J. Burden and co-workers, *J. Chem. Soc.*, 2574 (1957).
20. J. B. Hendrickson, D. D. Sternbach, and K. W. Bair, *Acc. Chem. Res.* **10,** 306 (1977).
21. U.S. Pat. 3,346,606 (Oct. 10, 1967), R. B. Ward (to E. I. duPont de Nemours & Co., Inc.).
22. R. A. Guenthner and M. L. Vietor, *Ind. Eng. Chem., Product Research and Development* Vol. 1, pp. 165–169, 1962.

RICHARD A. GUENTHNER
3M Company

PERFLUOROALKYLSULFUR FLUORIDES

Compounds containing fluorine bound to both carbon and sulfur with no other elements present are relatively rare and little known substances. Less than fifty examples of these materials are known and none of them have any current commercial utility. In general, they are difficult and costly to prepare. Their properties range from thermally stable and chemically inert to extremely unstable.

The only member of this group worthy of mention is the parent compound trifluoromethylsulfur pentafluoride [*373-80-8*], CF_3SF_5, which is a good gaseous dielectric material (1), and this use has been mentioned in two patents (2–3). A bibliography on CF_3SF_5 is available in ref. 4.

BIBLIOGRAPHY

"Perfluoroalkylsulfur Fluorides" under "Fluorine Compounds, Organic" in *ECT* 2nd ed., Vol. 9, pp. 802–805 by Richard E. Eibeck, Allied Chemical Corporation.

1. R. Geballe and F. S. Linn, *J. Appl. Phys.* **21,** 592 (1950).
2. U.S. Pat. 2,989,577 (June 20, 1961), D. Berg (to Westinghouse Electric Corp.).
3. Ger. Pat. 1,108,779 (June 15, 1961), D. Berg and T. E. Brown (to Westinghouse Electric Corp.).
4. L. S. Cove and R. J. Seffl, *Proceedings of the 6th Electrical Insulation Conference, New York, Sept. 13–16, 1965.*

RICHARD E. EIBECK
Allied Chemical Corporation

PERFLUOROEPOXIDES

Perfluoroepoxides were first prepared approximately twenty years ago by the DuPont Company. Subsequent work on these compounds has taken place throughout the world and is the subject of a number of reviews (1–3). The main use of these epoxides is as intermediates in the preparation of other fluorinated monomers. Although the polymerization of the epoxides has been described (4–6), the resulting homopolymers are of relatively minor commercial significance (see Epoxidation). Almost all work on perfluoroepoxides has been with only three compounds, tetrafluoroethylene oxide, (TFEO); hexafluoropropylene oxide (HFPO); and perfluoroisobutylene oxide (PIBO). Most of this work has dealt with HFPO which is the most versatile and valuable of this class of materials.

Physical Properties

In general, the perfluoroepoxides have boiling points that are quite similar to those of the corresponding fluoroalkenes. They can be distinguished easily from the olefins by ir spectroscopy, specifically by the lack of olefinic absorption and the presence of a characteristic epoxide band in the 1550 cm^{-1} region. Little physical property data concerning these compounds have been published (see Table 1). The solubility and heats of solution of HFPO in some organic solvents have been measured (20).

Chemical Properties

There are three general reactions of perfluoroepoxides. These are pyrolyses (thermal reactions), electrophilic reactions, and by far the most important, reactions with nucleophiles and bases.

Thermal Reactions. Those perfluoroepoxides that contain a CF_2 group in the epoxide ring undergo a smooth decomposition at relatively mild, neutral conditions (140–220°C) to give a perfluorocarbonyl compound and difluorocarbene (13,21) (eq.

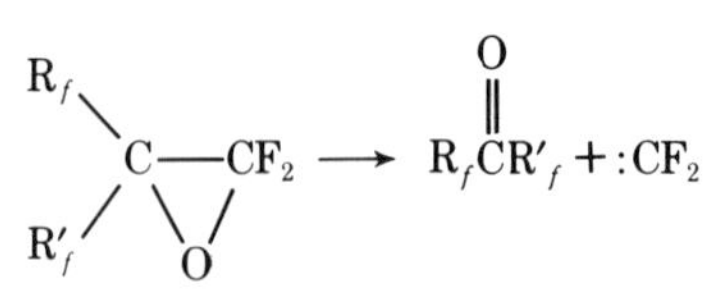

(1)

R'_f = Perfluoroalkyl

1). The difluorocarbene produced in this way reacts with a variety of compounds (22). The ease of decomposition is a function of the R_f and R'_f groups of the epoxide. Those epoxides that do not contain a CF_2 grouping in the epoxide ring have much greater thermal stability and do not undergo thermal decomposition even at 300°C (13).

Electrophilic Reactions. Perfluoroepoxides are quite resistant to electrophilic attack. However, they react readily with Lewis acids to give ring-opened carbonyl compounds (13,23) (eq. 2). The ring opening of HFPO by sulfur trioxide at 150°C has

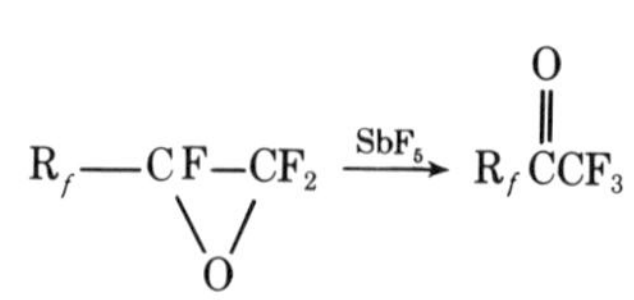

(2)

$R_f = CF_3, C_5F_{11}$

been interpreted as an example of an electrophilic reaction (24) (eq. 3).

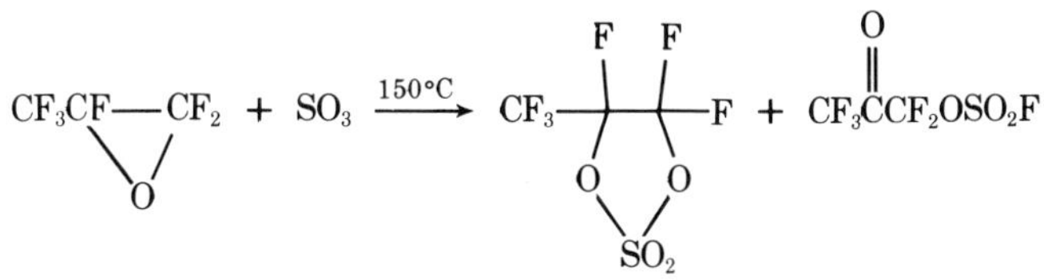

(3)

Nucleophilic Reactions. The strong electronegativity of fluorine permits the facile reaction of perfluoroepoxides with nucleophiles. These reactions comprise the majority of the reported reactions of this class of compounds. Nucleophilic attack on the epoxide ring takes place at the more highly substituted carbon atom to give ring-opened products. Fluorinated alkoxides are intermediates in these reactions and are in equilibrium with fluoride ion and a perfluorocarbonyl compound. This process is illustrated by the reaction of methanol and HFPO (eq. 4).

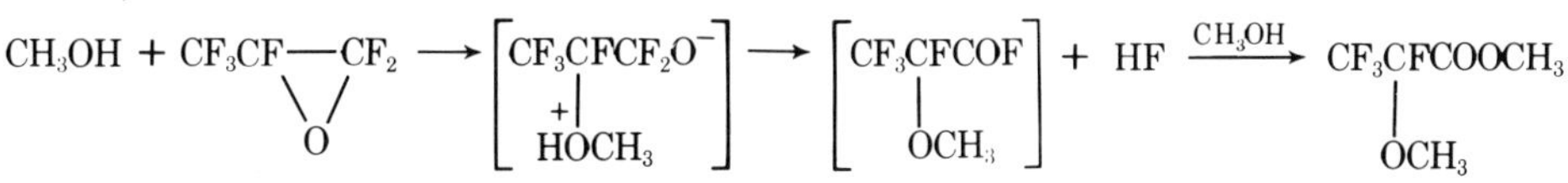

(4)

methyl 2,3,3,3-tetrafluoro-2-methoxypropanoate

Although TFEO is by far the most reactive epoxide of the series, all the reported perfluoroepoxides undergo similar ring-opening reactions. The most important reactions of these epoxides are those with fluoride ion and with the weakly nucleophilic perfluoroalkoxides prepared by reaction of fluoride ion either with perfluorocarbonyl compounds or epoxides. The reactions shown, using PIBO as an example, illustrate the general scheme which permits the oligomerization of these epoxides (eq. 5).

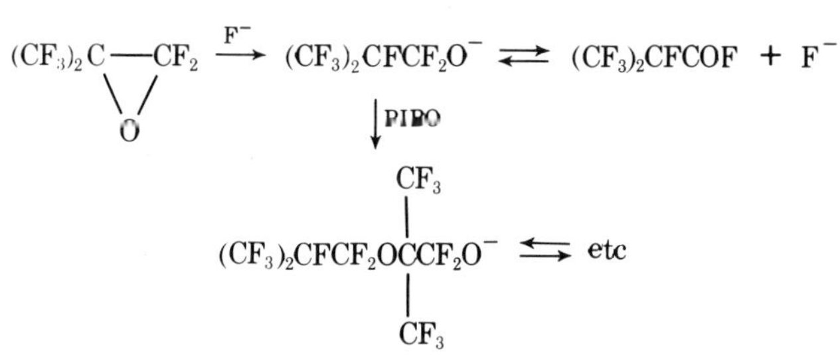

(5)

Although many oligomeric products have been prepared by this procedure and variations of it, no truly high polymers have been reported (4,6).

Preparation

A large number of methods are used to prepare perfluoroepoxides. All of these methods must contend with the great chemical reactivity of the epoxide product, especially in subsequent ionic and thermal reactions which result in the loss of the desired epoxide.

Table 1. Some Physical Properties of Perfluorocarbon Epoxides

	CAS Registry Number	Molecular structure	bp, °C	Infrared absorption, μm	Other properties	Reference
tetrafluoroethylene oxide	[*694-17-7*]	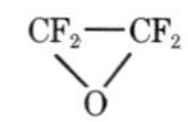	−63.5	6.21	mp, −118°C	7–8
hexafluoropropylene oxide	[*428-59-1*]	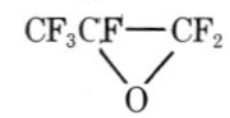	−27.4	6.43		9
trifluoroglycidyl fluoride	[*24419-82-7*]	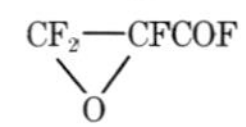	16 (extrap.)	6.61 5.35 (COF)	heat of vap., 28.9 kJ/mol[a]	10
perfluoro-1,2-epoxybutane	[*3709-90-8*]	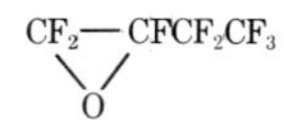		6.48		11
perfluoro-2,3-epoxybutane	[*773-29-5*]	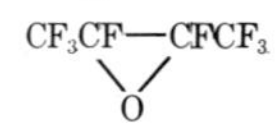		6.65		9
perfluoroisobutylene oxide	[*707-13-1*]	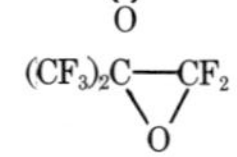	3	6.66	mp, −122°C	9
perfluoro-3,4-epoxybutyryl fluoride	[*5068-88-2*]	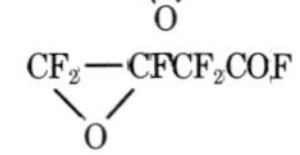		6.65		12
perfluoro-1,2-epoxycyclobutane	[*13324-28-2*]	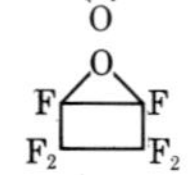		6.3		13
perfluoro-4,5-epoxy-1-pentene	[*15453-08-4*]	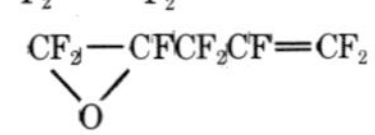	37	6.45 5.65 (C=C)		14
perfluoro-1-oxaspiro[2.3]hexane	[*53389-66-5*]	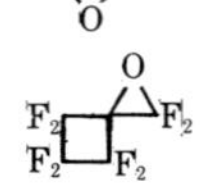	18–21	6.55		15
perfluoro-1,2-epoxycyclopentane	[*710-70-3*]	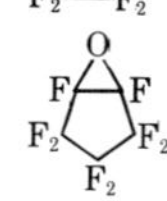	26.5	6.55		16

perfluoro-2-methyl-2,3-epoxypentane	[*788-67-0*]	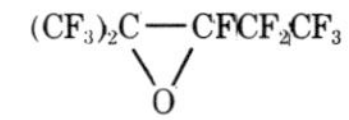	57	6.85		9
perfluoro-2,3-epoxy-4 methylpentane	[*788-50-1*]	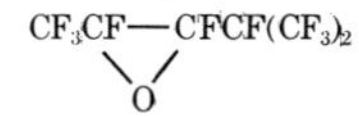	53	6.61		9
2,3-bis(trifluoromethyl)perfluoro-2,3-epoxybutane	[*1708-78-7*]	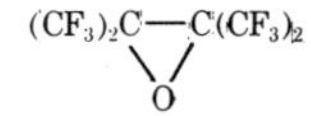	53–54			17
perfluoro-5,6-epoxy-1-hexene	[*15453-10-8*]	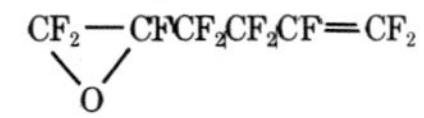	59			14
perfluoro-1,2-epoxycyclohexane	[*5927-67-3*]	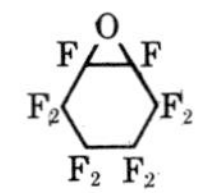	53–55	6.7		16
perfluoro-1,2-epoxyheptane	[*847-76-7*]	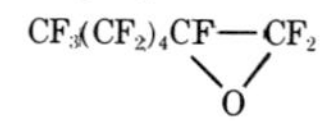	80–88			11
perfluoro-7,8-epoxy-1-octene	[*72264-78-9*]	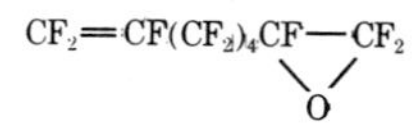	105			18
perfluoro-1,2:7,8-diepoxyoctane	[*13714-88-0*]	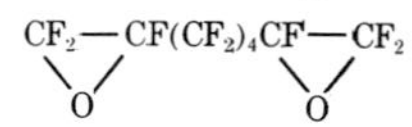	104	6.45	n_D^{20} 1.2900 d_4^{20} 1.7220	18–19
perfluoro-2,3-epoxy-2,3,5-trimethylhexane	[*748-45-8*]	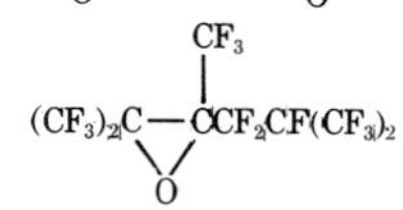		6.95		9
perfluoro-3,4-epoxy-2,3,5-trimethylhexane	[*2355-27-3*]	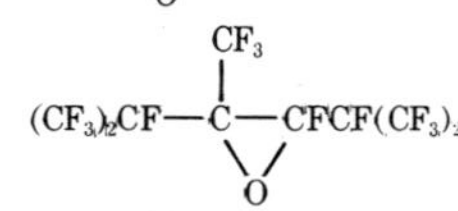	36–39 (0.13 kPa)[b]	7.21		9
perfluoro-1,2:9,10-diepoxydecane	[*13714-90-4*]	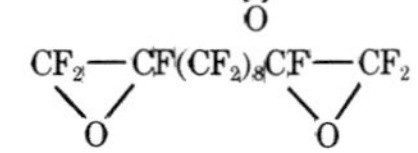	88 (5.3 kPa)[b]		n_D^{20} 1.3030 d_4^{20} 1.8260	18

[a] To convert kJ to kcal, divide by 4.184.
[b] To convert kPa to mm Hg, multiply by 7.5.

The reaction of perfluoroalkenes with alkaline hydrogen peroxide is a good general method for the preparation of perfluoroepoxides with the exception of the most reactive of the series, TFEO (eq. 6). Usually the perfluoroalkene is allowed to react at

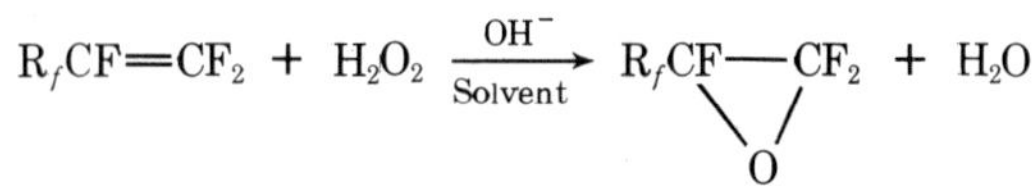
(6)

low temperatures with a mixture of aqueous hydrogen peroxide, base and a cosolvent to give a low conversion of the alkene (25). These conditions permit reaction of the water-insoluble alkene and minimize the subsequent ionic reactions of the epoxide product. A variation of this scheme using a peroxycarbimic acid has been reported (26) (see Epoxidation).

The direct oxidation of fluoroalkenes is also an excellent general synthesis procedure for the preparation of perfluoroepoxides (eq. 7). This method makes use of the low reactivity of the epoxide products to both organic and inorganic free radicals. The

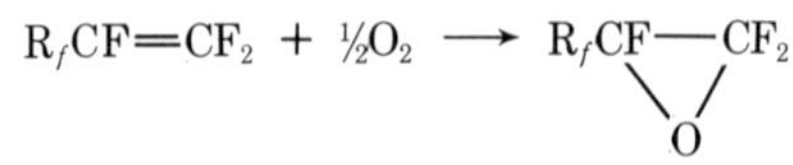
(7)

oxidation may be carried out with an inert solvent thermally (27), with a sensitizer such as bromine (7), with ultraviolet radiation (28), or over a suitable catalyst (29). Principal by-products of all these oxidation processes are the acyl fluoride products derived from oxidative cleavage of the perfluoroalkene (eq. 8). Perfluoroepoxides have also been prepared by anodic oxidation of fluoroalkenes (30), the low-temperature oxidation of fluoroalkenes with potassium permanganate (31), by addition of difluorocarbene to perfluoroacetyl fluoride (32) or hexafluoroacetone (33), epoxidation of fluoroalkenes with oxygen difluoride (34) or peracids (35), and the photolysis of substituted 1,3-dioxolan-4-ones (36).

$$R_fCF{=}CF_2 + O_2 \rightarrow R_fCOF + COF_2 \qquad (8)$$

Tetrafluoroethylene Oxide (TFEO)

TFEO has been prepared only by a process employing oxygen because of its extreme reactivity to ionic reagents. This reactivity may best be illustrated by its low-temperature reaction with the weak nucleophile, dimethyl ether, to give either of two products (37) (eq. 9). Reaction of TFEO with acid fluorides and fluoride ion yields

(9)

oligomers with the structure, $R_fCF_2O(CF_2CF_2O)_nCF_2COF$ (37–38). The epoxide yields a waxy solid polymer when exposed to high-energy radiation (37,39) or when treated with amines at low temperature (37,40). The extreme chemical reactivity and facile rearrangement to perfluoroacetyl fluoride have been deterrents to the large scale development of TFEO.

Hexafluoropropylene Oxide (HFPO)

HFPO is the most important of the perfluoroepoxides and has been synthesized by almost all of the methods noted above. Many attempts have been made to polymerize HFPO (4,6). The most successful has been the reaction of HFPO with fluoride ion at low temperature to give a series of oligomeric acid fluorides which have been end-capped to yield stable fluids (eq. 10). Materials of this type have been sold by the

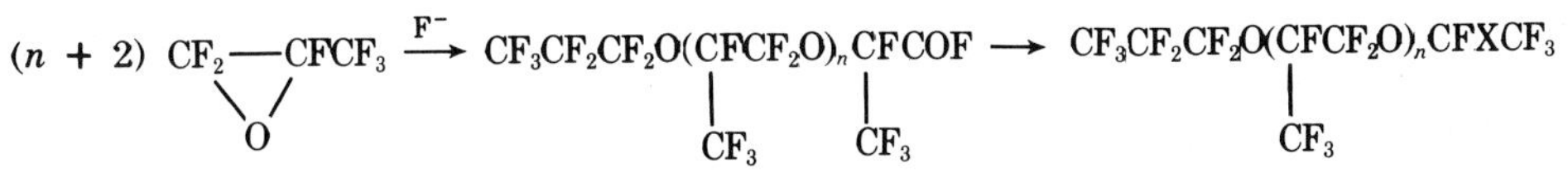
(10)

X = H,F

DuPont Company under the Freon E and Krytox trademarks. Perfluorinated materials structurally similar to the above have been prepared by Montedison by the low-temperature irradiation of either hexafluoropropylene or tetrafluoroethylene with oxygen followed by heating and have been sold as Fomblin liquids (41).

HFPO reacts with a large number of acyl fluorides in a general reaction to give 2-alkoxytetrafluoropropionyl fluorides which in turn may be converted to trifluorovinyl ethers (eq. 11). These ethers readily copolymerize with tetrafluoroethylene and other

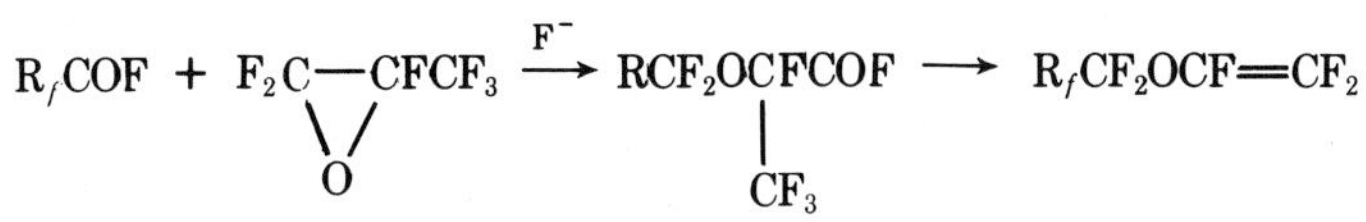
(11)

fluoroalkenes to give commercially significant plastics, elastomers, and ion-exchange resins, eg, Teflon PFA, Kalrez, and Nafion.

Recent publications have described the use of HFPO to prepare acyl fluorides (42), fluoroketones (43), and fluorinated heterocycles (44), as well as serving as a source of difluorocarbene for the synthesis of numerous cyclic and acyclic compounds (45). The nmr spectrum of HFPO is given in ref. 46.

Perfluoroisobutylene Oxide (PIBO)

PIBO has been prepared primarily by the addition of difluorocarbene to hexafluoroacetone or by reaction of alkaline hydrogen peroxide with perfluoroisobutylene. The small amount of published work on PIBO deals with its oligomerization (47), isomerization to perfluoroisobutyryl fluoride (16), conversion to perfluoro-*t*-butyl alcohol (48), and reaction with nucleophiles (49). PIBO has been reported to be as toxic as perfluoroisobutylene (50). The nmr spectrum of PIBO is reported in ref. 51.

BIBLIOGRAPHY

1. D. Sianesi and co-workers, *Gazz. Chem. Ital.* **98,** 265, 277, 290 (1968).
2. P. Tarrant and co-workers, *Fluorine Chem. Rev.* **5,** 77 (1971).
3. N. Ishikawa, *Yuki Gosei Kagaku Kyokai Shi* **35,** 131 (1977).
4. H. S. Eleuterio, *J. Macromol. Sci. Chem.* **A6,** 1027 (1972).
5. J. T. Hill and J. P. Erdman, *Am. Chem. Soc. Polym. Prepr.* **18,** 100 (1977).
6. J. T. Hill, *J. Macromol. Sci.* **A8,** 499 (1974).
7. Brit. Pat. 931,857 (July 17, 1963), (to E. I. du Pont de Nemours & Co., Inc.).
8. V. Caglioti and co-workers, *J. Chem. Soc.,* 5430 (1964).

9. Brit. Pat. 904,877 (Sept. 5, 1962), (to E. I. du Pont de Nemours & Co., Inc.).
10. W. Stuckey, J. Heicklen, and V. Knight, *Can. J. Chem.* **47,** 2329 (1969).
11. U.S. Pat. 3,213,134 (Oct. 19, 1965), D. E. Morin (to Minnesota Mining and Manufacturing Co.).
12. Brit. Pat. 1,034,492 (June 29, 1966), (to E. I. du Pont de Nemours & Co., Inc.).
13. D. P. Carlson and A. S. Milian, *4th International Fluorine Symposium, Estes Park, Colo., July 1967.*
14. U.S. Pat. 3,366,610 (Jan. 30, 1968), B. C. Anderson (to E. I. du Pont de Nemours & Co., Inc.).
15. R. J. DePasquale, K. B. Baucom, and J. R. Patton, *Tetrahedron Lett.*, 1111 (1974).
16. U.S. Pat. 3,321,515 (May 23, 1967), E. P. Moore and A. S. Milian (to E. I. du Pont de Nemours & Co., Inc.).
17. E. M. Rokhlin and co-workers, *Dokl. Akad. Nauk SSSR* **161,** 1356 (1965).
18. I. L. Knunyants and co-workers, *Zh. Obsch. Khim.* **36,** 1981 (1966).
19. U.S.S.R. Pat. 545,645 (Mar. 25, 1977), D. S. Rondarev and co-workers.
20. V. A. Gubanov and co-workers, *Zh. Fiz. Khim.* **48,** 2084 (1974).
21. R. C. Kennedy and J. B. Levy, *J. Fluorine Chem.* **7,** 101 (1976).
22. P. B. Sargeant, *J. Am. Chem. Soc.* **91,** 3061 (1969); V. V. Karpov, V. E. Platonov, and G. G. Yakobson, *Izv. Akad. Nauk SSSR, Ser. Khim.*, 981 (1975).
23. Jpn. Kokai 78 25,512 (Mar. 9, 1978), Y. Osaka and H. Takashi (to Daiken Kogyo).
24. I. L. Knunyants, V. V. Shokina, and E. I. Mysov, *Izv. Akad. Nauk SSSR Ser. Khim.*, 2725 (1973).
25. U.S. Pat. 3,358,003 (Dec. 12, 1967), H. S. Eleuterio and R. W. Meschke (to E. I. du Pont de Nemours & Co., Inc.).
26. Ger. Offen. 2,557,655 (June 30, 1977), R. A. Sulzbach and F. Heller (to Hoescht).
27. S. V. Kartsov and co-workers, *Izv. Akad. Nauk SSSR Ser. Khim.*, 2230 (1975).
28. V. Caglioti, M. Lenzi, and A. Mele, *Nature* **201,** 610 (1964); D. Sianesi, A. Pasetti, and C. Corti, *Makromol. Chem.* **86,** 308 (1965).
29. U.S. Pat. 3,775,438 (Nov. 27, 1973), R. J. Cavanaugh (to E. I. du Pont de Nemours & Co., Inc.); U.S. Pat. 3,775,439 (Nov. 27, 1973), G. M. Atkins, Jr., (to E. I. du Pont de Nemours & Co., Inc.); U.S. Pat. 3,775,440 (Nov. 27, 1973), R. J. Cavanaugh and G. M. Atkins, Jr. (to E. I. du Pont de Nemours & Co., Inc.).
30. U.S. Pat. 4,014,762 (Mar. 29, 1977), H. Millower (to Hoescht).
31. I. L. Knunyants and co-workers, *Izv. Akad. Nauk SSSR Ser. Khim.*, 2780 (1967).
32. W. Mahler and P. R. Resnick, *J. Fluorine Chem.* **3,** 451 (1973/74).
33. U.S. Pat. 3,338,978 (Aug. 29, 1967), E. P. Moore (to E. I. du Pont de Nemours & Co., Inc.).
34. U.S. Pat. 3,622,601 (Nov. 23, 1971), J. W. Dale (to Monsanto); U.S. Pat. 3,639,429 (Feb. 1, 1972), V. Weinmayr (to E. I. du Pont de Nemours & Co., Inc.).
35. Jpn. Kokai 77 108,914 (Sept. 12, 1977), Y. Oda, K. Uchida, and S. Morikawa (to Asahi Glass).
36. T. S. Croft, *J. Fluorine Chem.* **7,** 438 (1976).
37. J. L. Warnell, *4th International Fluorine Symposium, Estes Park, Colorado, July 1967.*
38. U.S. Pat. 3,250,806 (May 10, 1966), J. L. Warnell (to E. I. du Pont de Nemours & Co., Inc.).
39. P. Barnaba and co-workers, *Chim. Ind.* (*Milan*) **47,** 1060 (1965).
40. Can. Pat. 778,490 (Feb. 13, 1968), J. L. Warnell (to E. I. du Pont de Nemours & Co., Inc.).
41. D. Sianesi, *Am. Chem. Soc. Polym. Prepr.* **12,** 411 (1971); D. Sianesi and co-workers, *Chim. Ind.* (*Milan*) **55,** 208 (1973).
42. N. Ishikawa and S. Sasaki, *Chem. Lett.* 483, 1407 (1976).
43. T. Martini, *Tetrahedron Lett.*, 1857, 1861 (1976).
44. N. Ishikawa and S. Sasaki, *Bull. Chem. Soc. Jpn.* **50,** 2164 (1977).
45. V. M. Karpov, V. E. Platonov and G. G. Yacobsen, *Izv. Akad. Nauk SSSR Ser. Khim.*, 2295 (1976).
46. J. K. Ruff and R. F. Merritt, *J. Org. Chem.* **30,** 3968 (1965).
47. J. T. Hill, *8th International Fluorine Symposium, Kyoto, Japan, Aug. 1977.*
48. U.S. Pat. 3,385,904 (May 28, 1968), F. J. Pavlik (to Minnesota Mining and Manufacturing Co.).
49. I. L. Knunyants and co-workers, *Izv. Akad. Nauk SSSR Ser. Khim.*, 1133 (1972).
50. I. L. Knunyants, V. V. Shokina, and I. V. Salakhov, *Khim. Geterotsikl. Soedin.* **2,** 873 (1966).
51. F. J. Pavlik and P. E. Toren, *J. Org. Chem.* **35,** 2054 (1970).

PAUL R. RESNICK
E. I. du Pont de Nemours & Co., Inc.